ADVANCED DAIRY CHEMISTRY—1: PROTEINS

ADVANCED DAIRY CHEMISTRY—1: PROTEINS

Edited by

P.F. FOX

Department of Food Chemistry
University College,
Cork, Ireland

ELSEVIER APPLIED SCIENCE
LONDON and NEW YORK

ELSEVIER SCIENCE PUBLISHERS LTD
Crown House, Linton Road, Barking, Essex IG11 8JU, England

WITH 97 TABLES AND 107 ILLUSTRATIONS

British Library Cataloguing in Publication Data
Advanced Dairy Chemistry.—Vol. 1: Proteins
I. Fox, P.F.
637

ISBN 1-85166-761-x

Library of Congress Cataloging-in-Publication Data
Advanced dairy chemistry/edited by P.F. Fox.
p. cm.
Includes bibliographical references and index.
Contents: 1. Proteins
ISBN 1-85166-761-X (v. 1)
1. Milk—Composition. 2. Dairy products—Composition. I. Fox, P.F.
SF251.A38 1992
637'.1—dc20 92-4335
CIP

Typeset by Alden Multimedia, Northampton.
Printed in Great Britain at the University Press, Cambridge.

CONTENTS

PREFACE TO THE FIRST EDITION

Because of its commercial and nutritional significance and the ease with which its principal constituents, proteins, lipids and lactose, can be purified free of each other, milk and dairy products have been the subject of chemical investigation for more than a century. Consequently, milk is the best-described in chemical terms, of the principal food groups. Scientific interest in milk is further stimulated by the great diversity of milks—there are about 4000 mammalian species, each of which secretes milk with specific characteristics. The relative ease with which the intact mammary gland can be isolated in an active state from the body makes milk a very attractive subject for biosynthetic studies. More than any other food commodity, milk is a very versatile raw material and a very wide range of food products are produced from the whole or fractionated system.

This text on Proteins is the first volume in an advanced series on selected topics in Dairy Chemistry. Each chapter is extensively referenced and, it is hoped, should prove a useful reference source for senior students, lecturers and research personnel. The selection of topics for 'Proteins' has been influenced by a wish to treat the subject in a comprehensive and balanced fashion. Thus, Chapters 1 and 2 are devoted to an in-depth review of the molecular and colloidal chemistry of the proteins of bovine milk. Although less exhaustively studied than those of bovine milk, considerable knowledge is available on the lactoproteins of a few other species and an inter-species comparison is made in Chapter 3. The biosynthesis of the principal lactoproteins is reviewed in Chapter 4. Chapters 5 to 8 are devoted to alterations in the colloidal state of milk proteins arising from chemical, physical or enzymatic modification during processing or storage, viz. enzymatic coagulation, heat-induced coagulation, age gelation of sterilized milks and chemical and enzymatic changes in cold-stored raw milk. Milk and dairy products provide 20–30% of protein in 'western' diets and are important world-wide in infant nutrition: lactoproteins in particular, are considered in Chapter 9. The increasing significance of 'fabricated' foods has created a demand for 'functional' proteins: Chapters 10 to 12 are devoted to the technology, functional properties and food applications of the caseinates and various whey protein products.

Because of space constraints, it was necessary to exclude coverage of the more traditional protein-rich dairy products: milk powders and cheese. It is hoped to devote sections of a future volume to these products.

I wish to thank sincerely the 13 other authors who have contributed to this text and whose cooperation made my task as editor a pleasure.

P.F. Fox

PREFACE TO THE SECOND EDITION

Considerable progress has been made on various aspects of milk proteins since *Developments in Dairy Chemistry 1—Proteins* was published in 1982. *Advanced Dairy Chemistry* can be regarded as the second edition of *Development in Dairy Chemistry* which has been updated and considerably expanded. Many of the original chapters have been revised and updated, e.g. 'Association of Caseins and Casein Micelle Structure', 'Biosynthesis of Milk Proteins', 'Enzymatic Coagulation of Milk', 'Heat Stability of Milk', 'Age Gelation of Sterilized Milks' and 'Nutritional Aspects of Milk Proteins'. Chapter 1 in *Developments*, i.e. 'Chemistry of Milk Proteins', has been subdivided and extended to 4 chapters: chemistry and physico-chemical properties of the caseins, β-lactoglobulin, α-lactalbumin and immunoglobulins. New chapters have been added, including 'Analytical Methods for Milk Proteins', 'Biologically Active Proteins and Peptides', 'Indigenous Enzymes in Milk', 'Genetic Polymorphism of Milk Proteins', 'Genetic Engineering of Milk Proteins', 'Ethanol Stability of Milk' and 'Significance of Proteins in Milk Powders'. A few subjects have been deleted or abbreviated; the three chapters on functional milk proteins in *Developments* have been abbreviated to one in view of the recently published 4th volume of *Developments in Dairy Chemistry—4—Functional Milk Proteins*.

Like its predecessor, the book is intended for lecturers, senior students and research personnel and each chapter is extensively referenced.

I would like to thank all the authors who contributed to the book and whose cooperation made my task a pleasure.

P.F. Fox

LIST OF CONTRIBUTORS

A.T. Andrews
Biotechnology Group, Department of Food Science and Technology, University of Reading, Whiteknights, PO Box 226, Reading RG6 2AP, UK

G. Bengtsson-Olivecrona
Department of Medical Biochemistry and Biophysics, University of Umea, S-90187, Umea, Sweden

L. Bjorck
Department of Food Science, Swedish University of Agricultural Sciences, Box 7051, S-750 07 Uppsala, Sweden

K. Brew
Department of Biochemistry and Molecular Biology, University of Miami School of Medicine, Miami, Florida 33101, USA

E.M. Brown
Eastern Regional Research Center, United States Department of Agriculture, Philadelphia, Pennsylvania 19118, USA

L.K. Creamer
New Zealand Dairy Research Institute, Palmerston North, New Zealand

D.G. Dalgleish
Department of Food Science, University of Guelph, Guelph, Ontario N1G 2W1 Canada

N.Y. Farkye
California Polytechnic State University, San Luis Obispo, California 93407, USA

H.M. Farrell Jr.
Eastern Regional Research Center, United States Department of Agriculture, Philadelphia, Pennsylvania 19118, USA

A. Flynn
Department of Nutrition, University College, Cork, Republic of Ireland

P.F. Fox
Department of Food Chemistry, University College, Cork, Republic of Ireland

P. Gaye
Laboratoire de Biologie Cellulaire et Moléculaire, Institut National de la Recherche Agronomique, INRA-CRJ, 78350 Jouy-en-Josas, France

R. Grappin
Institut National de la Recherche Agronomique (INRA), Station de Recherches en Technologie et Analyses Laitières, BP 89, 39 800 Poligny, France

J.A. Grobler
Department of Biochemistry and Molecular Biology, University of Miami School of Medicine, Miami, Florida 33101, USA

F. Grosclaude

Laboratoire de Génétique Biochimique, Institut National de la Recherche Agronomique, 78350 Jouy-en-Josas, France

S.G. Hambling

Department of Biochemistry, University of Edinburgh Medical School, Hugh Robson Building, George Square, Edinburgh EH8 9XD, UK

L. Hambraeus

Department of Nutrition, University of Uppsala, Box 551, S-751 22 Uppsala, Sweden

V.R. Harwalkar

Food Research Centre, Agriculture Canada, Ottawa, Canada K1A 0C6

D.S. Horne

The Hannah Research Institute, Ayr KA6 5HL, UK

T.F. Kumosinski

Eastern Regional Research Center, United States Department of Agriculture, Philadelphia, Pennsylvania 19118, USA

R. Jimenez-Flores

Department of Food Science and Technology, University of California, Davis, California 95616, USA

B.L. Larson

Department of Animal Sciences, University of Illinois, Urbana, Illinois 61801, USA

P. Martin

Laboratoire Génétique Biochimique, Institut National de la Recherche Agronomique, INRA-CRJ, 78350 Jouy-en-Josas, France

A.S. McAlpine

Department of Biochemistry, University of Edinburgh Medical School, Hugh Robson Building, George Square, Edinburgh EH8 9XD, UK

T.B. Mepham

Department of Physiology and Environmental Science, University of Nottingham, Sutton Bonington, Loughborough, Leicestershire LE12 5AD, UK

J.C. Mercier

Laboratoire Génétique Biochimique, Institut National de la Recherche Agronomique, INRA-CRJ, 78350 Jouy-en-Josas, France

D.M. Mulvihill

Department of Food Chemistry, University College, Cork, Republic of Ireland

D.F. Newstead

New Zealand Dairy Research Institute, Palmerston North, New Zealand

K.F. Ng-Kwai-Hang

Department of Animal Science, Macdonald College of McGill University, 21 111 Lakeshore Road, Ste Anne de Bellevue, P.Q., Canada H9X 1C0

S. Oh
Department of Food Science and Technology, University of California, Davis, California 95616, USA

T. Olivecrona
Department of Medical Biochemistry and Biophysics, University of Umea, S-90187 Umea, Sweden

B. Ribadeau-Dumas
Institut National de la Recherche Agronomique (INRA), Station de Recherches Laitieres, C.R.J., 78352 Jouy-en-Josas, Cedex, France

T. Richardson
Department of Food Science and Technology, University of California, Davis, California 95616, USA

H.S. Rollema
Netherlands Institute for Dairy Research, P.O. Box 20, 6710, BA, Ede, The Netherlands

L. Sawyer
Department of Biochemistry, University of Edinburgh Medical School, Hugh Robson Building, George Square, Edinburgh EH8 9XD, UK

H. Singh
Department of Food Technology, Massey University, Palmerston North, New Zealand

H.E. Swaisgood
Southeast Dairy Foods Research Center, North Carolina State University, Raleigh, North Carolina 26796-7624 USA

1

ANALYTICAL METHODS FOR MILK PROTEINS

RÉMY GRAPPIN

Institut National de la Recherche Agronomique (INRA), Station de Recherches en Technologie et Analyses Laitières, BP 89, 39 800 Poligny, France

&

BRUNO RIBADEAU-DUMAS

Institut National de la Recherche Agronomique (INRA), Station de Recherches Laitières, Centre de Recherches de Jouy en Josas, 78352 Jouy-en-Josas Cedex, France

ABBREVIATIONS

ABTS	Ammonium 2,2′-azino-di (3-ethylbenzothiazoline)-6-sulfonate
A & M	Aschaffenburg & Mullen
AOAC	Association of Official Analytical Chemists
AP	Alkaline phosphatase
AUFS	Absorption Units Full Scale
BSA	Bovine serum albumin
CMP	Caseino macropeptide
Cn	Casein
CV	Coefficient of variation
DBC	Dye binding capacity
ELISA	Enzyme-linked immunosorbent assay
FMOC	9-Fluorenyl methyl chloroformate
FPLC™	Fast protein liquid chromatography
FT-IR	Fourier transform infrared
Gu.HCl	Guanidinium chloride
HA	Hydroxyapatite
HI	Hydrophobic interaction
HPLC	High performance liquid chromatography
HTST	High temperature short time
IDF	International Dairy Federation
IEF	Isoelectric focusing
Ig	Immunoglobulin
IR	Infrared

ISO	International Standardization Organization
LF	Lactoferrin (lactotransferrin)
LP	Lactoperoxidase
MESA	Mercaptoethane sulfonic acid
MFGM	Milk fat globule membrane
MIR	Mid-infrared
Mr	Relative molecular weight
NFDM	Non-fat dry milk
NIR	Near-infrared
NPN	Non-protein nitrogen, or non-protein nitrogenous substances
OPA	Orthophtaldialdehyde
PAGE	Polyacrylamide gel electrophoresis
Pg	Plasminogen
PITC	Phenylisothiocyanate
PTC	Phenyl thio-carbamyl
r	Repeatability, or coefficient of correlation
R	Reproducibility, or $1/R$ = reflectance
RP-HPLC	Reversed phase HPLC
RSD_r	Repeatability relative standard deviation
RSD_R	Reproducibility relative standard deviations
s_r	Repeatability standard deviation
s_R	Reproducibility standard deviation
$s_{y,x}$	Accuracy or residual standard deviation
SD or S	Standard deviation
SDS	Sodium dodecylsulfate
SE	Standard error
TCA	Trichloroacetic acid
TNBS	Trinitrobenzene sulfonic acid
UHT	Ultra high temperature
WP	Whey proteins
WPC	Whey protein concentrate
WPI	Whey protein index
α-La	α-Lactalbumin
β-Lg	β-Lactoglobulin
λ	Wavelength

1. INTRODUCTION

This review is an up-dated version of the paper recently published by Ribadeau-Dumas and Grappin (1989) on the same subject. It concerns all types of analyses dealing with proteins in milk, dairy and a few non-dairy products. Only the methods that are in current use or those which have been recently published and might become of general use will be considered. For all of them, the reader will be supplied with basic principles, schematic description, references and critical evaluation.

Compared with other food products, milk is a fairly simple fluid. It has been thoroughly studied since the beginning of the 19th century. Its composition and the main characteristics of its various constituents are now well known. In particular, the amino acid sequences of its 7 main protein components have been elucidated. There is today no other food product whose proteins are so well characterized. This makes their analysis straightforward in raw milk. However, as soon as technological treatments have been applied, any quantitative measurement, except nitrogen determination, becomes far more difficult. In particular, protein denaturation, which is not a one-step phenomenon, leads, for a given protein, to products which may differ according to the treatment, often with an ultimate transformation into insoluble aggregates. Furthermore, a number of chemical reactions may occur during the processing of milk, dairy products and non-dairy products which lead to covalent modifications of proteins. In a number of food products, milk proteins have been intentionally fragmented into peptides and amino acids by proteinases and peptidases. Of course it is not possible to know from which protein free amino acids originated. Theoretically, the origin of any peptide with more than 5 amino acids residues, provided it can be isolated, can usually be established if it is derived of any milk protein. However, even the rough characterization of a milk protein hydrolysate is a difficult and long task which can only, for the moment, be performed in few research laboratories. The procedures allowing the determination of milk proteins as a whole and individually in milk and dairy products will be reviewed as well as those used to detect and determine them in non-dairy products. Then, some applications of milk protein analysis in dairy and non-dairy products will be given.

2. Determination of Total Protein in Milk and Dairy Products

As in many other fields, the nature and number of analyses performed to assess the composition of milk and dairy products have changed dramatically during the past 20 years. Fat, which was considered to be the most valuable component in milk since the very beginning of the dairy industry, is now being replaced by proteins which have higher nutrional and economic values. In most countries with a highly developed dairy industry, protein content is now included in milk quality payment schemes and breeding programmes.

Progress in milk protein research, as well as the applications of protein testing to dairy husbandry and quality control in the dairy industry, were for a long time hampered because no rapid and accurate method of analysis was available. A major breakthrough occurred with the introduction of the dye-binding methods in the 1960s, followed by the development of the infrared (IR) techniques which had the advantage of measuring directly all the major milk compounds: fat, protein and lactose.

This chapter will first deal with new developments concerning the determination of nitrogen, mainly by the Kjeldahl method, then the two most important indirect methods which have been used in dairy laboratories, dye-binding and IR methods, will be described. In central testing laboratories, IR methods have now replaced the dye-binding method. However, because this latter technique is simple, does not require expensive equipment and is still very valuable for quality control in small dairy plants, it will be fully described. Because most of the studies on these techniques have been devoted to milk protein testing, relatively little information is available on their application to dairy products. Extensive reviews on the numerous methods for measurement of the protein content of milk have been published by Bosset *et al.* (1976) and Guillou *et al.* (1987).

2.1. Direct Methods (Nitrogen Determination)

In milk, as well as in other foodstuffs, nitrogen is the element that essentially characterizes proteins. As a consequence, nitrogen determination has always been used as a standard method for the estimation of the protein content of foods.

2.1.1. The Kjeldahl Method

This method is now internationally recognized as the reference method for measuring the protein content of milk products, and is listed as such in the Codex Alimentarius.

2.1.1.1. Principle. In 1883, Kjeldahl discovered that, by heating organic compounds in concentrated sulfuric acid, nitrogen is converted quantitatively into ammonium sulfate and can subsequently be estimated as ammonia by distillation and titration after addition of sodium hydroxide. During digestion, carbon is transformed into CO_2 and hydrogen into H_2O. Detailed information concerning the various reactions involved during digestion can be found in the review of McKenzie and Murphy (1970).

2.1.1.2. Development. Since its discovery, this method has been studied extensively and the procedure revised periodically to improve both the digestion rate and the accuracy. The objective is to convert, as quickly as possible, the totality of the organic nitrogen, even the most refractory compounds, into NH_4^+ without loss of nitrogen by pyrolytic decomposition of ammonia. It is now well established that a high digestion rate and good nitrogen recovery (over 99%) can be achieved only with a suitable ratio of K_2SO_4/H_2SO_4 to reach a high boiling temperature, and with the addition of a catalyst. Several concentrations of K_2SO_4, types of catalyst (selenium, mercury, copper) and oxidizing agent (H_2O_2) have been tested. If, with hard-to-digest compounds, mercury is certainly the most efficient catalyst, it no longer tends to be used because of current environmental concerns. Copper sulfate, which was first proposed by Rowland (1938) for milk analysis, seems to be a good alternative.

Following a study by Rexroad and Cathey (1976), recent studies on foodstuffs (Kane, 1984) and on milk (IDF, 1986) have shown that HgO and $CuSO_4$ give identical results. In the latest version of IDF Standard 20 A/1986 (1986) for the determination of nitrogen in milk by the macro-Kjeldahl method (5 ml of milk), the following ratios are prescribed: K_2SO_4/H_2SO_4 = 15 g/25 ml; $CuSO_4$, 5 H_2O/H_2SO_4 = 0·05 g/25 ml. Compared to the original $CuSO_4$ concentration recommended by Rowland (1938) (i.e. 0·2 g/5 ml), a 12-fold lower concentration was adopted to prevent the formation of ammoniacal complexes with copper, which may lead to a relative underestimation by about 1% of the nitrogen content.

It is important to bear in mind that a catalyst cannot be exchanged without reconsidering the whole procedure, and especially the K_2SO_4/H_2SO_4 ratio, and the total mineralization time. It is also recommended to consider the total mineralization time given in a Standard Method as a minimum. The heating time should not be reduced if the clearing time (when the digest becomes clear) is short, as for instance with low-fat milk samples. On the other hand, for samples with a high fat or protein content, the amount of H_2SO_4 has to be increased because organic material consumes H_2SO_4 and the total mineralization time should be extended if the clearing time is longer than that given in the Standard Method.

For routine analysis, the classical Kjeldahl flask and gas burner, or electric heater with separate distillation apparatus, are progressively being replaced by block digestion and steam distillation apparatus. These types of equipment give results similar to those obtained with the standard method. However, Barbano *et al.* (1990*a*) found that to achieve good reproducibility, different total mineralization times may be necessary to compensate for differences in line voltage between laboratories.

Preliminary trials (Jeunet, R., 1989, pers. comm.) with a new device, the Maxidigest (Prolabo, Paris) using microwaves as a means for heating individual flasks gave the same results as the conventional procedure. The Kjel-Fast (CEM Corp., Mattews, NC) which also uses microwaves and mineralizes the sample contained in a quartz vessel in 6 min was tested successfully by Barbano and Clark (1990). Both systems use H_2O_2 to facilitate digestion. The traditional titration using an indicator solution can be carried out by automatic pH titration or specific electrode (Pailler, 1982). The Flow Injection Analysis (FIA) technique which is an automated colorimetric measurement of NH_4^+, has been successfully used by Courroye *et al.* (1989) for the analysis of various nitrogen fractions from cheese. Fully automatic instruments, like the Kjelfoss (Foss-Electric, Hillerod, Denmark), or the Kjel-Auto (MRK Inc., Tokyo, Japan) which combine in a single instrument all the steps of the procedure, i.e. mineralization, distillation and titration, allow the determination of nitrogen with good precision and accuracy at a rate of 20 samples/hour (Grappin & Jeunet, 1976*a*).

2.1.1.3. Analytical performance. When performed correctly, the Kjeldahl method is assumed to give the true nitrogen content of milk. However, many reports of collaborative studies have shown that frequently large discrepancies between laboratories occur. To improve the reproducibility, the new IDF standard requires that two accuracy tests be performed regularly. First, a test based upon the analysis of tryptophan or a similar refractory compound, like phenacetin, to check the mineralization efficiency, with a percentage of N recovery better than 98%.

Lysine-HCl is an alternative, but difficulties have been reported by Kane (1984); nicotinamide, which is also a difficult-to-digest material, does not give an excellent recovery with copper sulfate. Second, a test based on the analysis of an ammonium salt (sulphate or oxalate) is performed to check the distillation and the titration steps, with a percentage of N recovery between 99 and 100%.

To check their procedure, laboratories may take advantage of using reference materials like an $NH_4H_2PO_4$ solution or a milk powder with a certified N content supplied by specialized organizations like the American National Institute of Standards and Technology, or the European Bureau Communautaire de Référence (Brussels).

A collaborative study on the IDF Standard 20 A/1986 (Grappin & Horwitz, 1988), carried out with the participation of 24 reporting laboratories representing 12 countries, found a repeatability relative standard deviation (RSD_r) of 0·5% and a reproducibility relative standard deviation (RSD_R) of 1%; within the United States, Barbano and Clark (1989) found, respectively, RSD_r = 0·31% and RSD_R = 0·44% with 11 participating laboratories.

2.1.2. The Dumas and Related Methods

In these methods, organic and inorganic nitrogen are converted after combustion at high temperature in a furnace into nitrogen gas which is determined by gas chromatography, volumetrically or by thermal conductivity or even chemiluminescence. Automatic apparatus, able to perform fast and reliable analyses (e.g. the Carlo Erba NA 1 500 machine determines nitrogen in 3 min) are commercially available. One machine was tested by Lunder (1974) on casein and whole dried milk and gave reliable data.

Recently, two instrument manufacturers (LECO Corp., St Joseph, USA and Foss-Electric, Hillerod, Denmark) have marketed automated instruments (FP-428 LECO and HERAUS N Analyzer) for the measurement in 3–7 min of nitrogen gas by thermal conductivity via the complete combustion of the food products. A collaborative study on the determination of crude proteins in feeds by the LECO instrument (Sweeney, 1989) has shown good precision (RSD_r = 0·59% and RSD_R = 1·10%). No results are now available concerning the precision of the analysis of liquid products containing relatively low N concentrations.

2.1.3. Conversion Factor—Terminology

To estimate the amount of protein in milk and milk products, it is necessary to convert nitrogen into protein, by multiplying the nitrogen content by a factor, called the Kjeldahl conversion factor. A value of 6·38 for this factor, originally proposed a century ago by Hammarsten and Sebelien on the basis of the nitrogen content (15·67%) of purified acid-precipitated casein, is generally accepted and was confirmed in the latest IDF Standard.

However, this method of determining the protein content raises two important questions. First, the terminology 'protein content' is not fully correct, since the proportion of non-protein nitrogen (NPN) within and between dairy products, varies from 3 to 8% in milk and up to 25–30% in whey. To avoid confusion, the

TABLE 1
Protein Content and Kjeldahl Factor of Milk (Karman & Van Boekel, 1986)

Protein	*Concentration (g/liter)*	*Without carbohydrate*		*With carbohydrate*	
		N%	*Kjeldahl factors*	*N%*	*Kjeldahl factors*
α_{s1}-Cn	10·0	15·77	6·34		
α_{s2}-Cn	2·6	15·83	6·30		
β-Cn	9·3	15·76	6·34		
κ-Cn	3·3	16·26	6·15	15·67	6·38
γ-Cn	0·8	15·87	6·30		
β-Lg	3·2	15·68	6·38		
α-La	1·2	16·29	6·14		
β-SA	0·4	16·46	6·07		
Ig	0·8	16·66	6·00	16·14	6·20
PP,8F,8S	0·5	15·30	6·54		
PP3	0·3	16·97	5·89	15·27	6·55
Lactoferrin	0·1	17·48	5·72	16·29	6·14
Transferrin	0·1	17·00	5·88	16·10	6·21
MFGM	0·4	15·15	6·60	14·13	7·08
Milk	33·0	15·87	6·30	15·76	6·34

term 'crude protein' has to be used to express the nitrogenous matter in milk. Its quantitative expression is represented by the amount of total nitrogen multiplied by 6·38, and is expressed in g per 100 g (or kg or liter) of milk or milk product.

Second, the conversion factor is not constant, but is highly dependent on the amino acid composition of the protein fraction. Recently, using the primary structure of milk proteins, Karman and van Boekel (1986) have shown that for cows' milk, the conversion factor should be 6·34 instead of 6·38, and different factors should be used for casein (6·34), paracasein (6·29), rennet whey proteins (6·45), acid whey proteins (6·30), and NPN (3·60). Recently, these figures were corrected slightly by Van Boekel and Ribadeau-Dumas (1987). For individual protein fractions, the variability of the factor is even greater (Table 1). In their study, they demonstrated that experimental determination of the Kjeldahl factor on (pure) protein fractions leads to substantial discrepancies from the theoretical values obtained from amino acid sequences (Table 2), mainly because it is difficult to obtain pure fractions and to accurately measure the ash content.

Even though the term 'crude protein' and the conversion factor of 6·38 are widely used and accepted for milk and milk products, a more accurate definition and methodology for measuring the 'true' protein content should be promoted. Several authors (Szijarto *et al.*, 1973; Verdi *et al.*, 1987; Robertson and van der Westhuizen, 1990) have pointed out that the NPN content and the NPN/Total N ratio vary extensively according to several biological factors (species, feeding, etc.) and the major proportion of the NPN fraction (urea) has no nutritional nor economic value.

TABLE 2
Experimental Kjeldahl Factors for Isolated Milk Protein (Karman & Van Boekel, 1986)

Protein	*% Non-protein ash*[a]	*Corrected %N*[b]	*Experimental Kjeldahl factor*	*Theoretical Kjeldahl factor*
α_s-Casein	2·16	15·55	6·43	6·33
β-Casein	3·78	16·41	6·10	6·34
κ-Casein	2·11	14·84	6·74	6·38
Total casein	1·40	15·62	6·40	6·34
β-Lactoglobulin	12·60	14·97	6·68	6·38

[a]The difference between the ash content and the sum of PO_4 and SO_4 contents taken as non-protein ash (for β-lactoglobulin only the SO_4 content was taken).
[b]Corrected for fat and water content and non-protein ash.

2.2. Indirect Methods

2.2.1. Definition and Characteristics

2.2.1.1. Definition. According to Grappin (1984), a method of analysis is designated 'indirect' when it does not measure directly the component it is intended to quantify, but instead measures one or more quantities or properties which are functionally linked to that component. For instance, the Amido Black method is an indirect method because it does not directly determine the protein content, but measures the amount of dye that is bound to proteins. The same holds for instrumental methods such as IR techniques. As a consequence, indirect methods require standardization against a reference method to convert the instrumental signal into component concentration. We may point out that, following this definition, the Kjeldahl method could also be considered as an indirect method since it measures nitrogen and not the proteins as such, even though it is accepted as a reference for the determination of the protein content in milk products.

2.2.1.2. Analytical attributes and calibration. Because of the influence of sample matrix, the quantities or properties which are measured by indirect methods are not usually exactly proportional to the component concentration, introducing therefore systematic errors in the final test result.

Figure 1 illustrates the various criteria involved in the overall accuracy of an indirect method (IDF, 1987). When the results of both reference and indirect methods are expressed in the same units, if we except random errors (i.e. instrument repeatability), systematic errors can be divided into two components:

- The exactness of calibration, which relates to the closeness of agreement, at each concentration level, between the indirect method value and the estimated mean of the true values of all the individual samples at the corresponding level.
- The accuracy of estimate, or simply accuracy, which relates to the closeness of agreement between individual test results obtained with the instrument

exactly calibrated and with the reference method. The ellipse in Fig. 1, which represents approximately the 95% confidence interval within which the true value is expected to lie for any result of the indirect method, illustrates the accuracy of the indirect method. The mathematical expression of the accuracy is given by the residual standard deviation $s_{y,x}$. This important attribute gives the best estimate of the predicting value of the indirect method.

To counteract the influence of known or unknown factors, which are common to a whole population of samples, the results of the indirect method will have to be checked regularly against the reference method. By definition (IDF, 1987), the calibration of an instrument concerns the adjustment of the instrumental signal so that, at each level of the component concentration, the mean of individual test results given by the instrument is closely approximate to the true value of the component concentration. Practically, the calibration of an indirect method can be

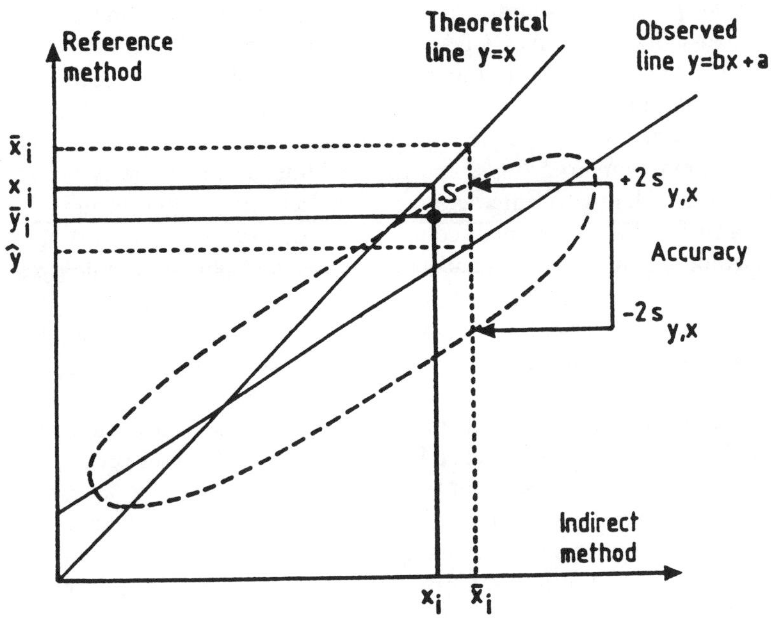

Fig. 1. Breakdown of the overall accuracy of an indirect method (IDF, 1987). x_i = Single instrumental value for sample *S*; $\bar{x}_i$ = arithmetic mean of several determinations of sample *S* with the instrument; $\hat{y}$ = estimated mean reference value for all the samples with an instrument level $\bar{x}_i$; $\bar{y}_i$ = 'true value' of the component to be measured for sample *S*; $s_{y,x}$ = standard deviation from the regression. The mathematical model of the components of the total error on *x* is:

		↗ $(\bar{x}_i - \bar{y}_i)$		
$(x_i - \bar{y}_i)$	= $(x_i - \bar{x}_i)$	+ Accuracy of the mean		
Overall accuracy	Repeatability	↘ $\bar{x}_i - \hat{y}$	+	$(\hat{y} - \bar{y}_i)$
		Exactness of calibration		Accuracy

achieved by calculating the relationship between the component concentration obtained by the reference method on a set of samples or by reference materials, and the instrument signal.

2.2.2. Dye-Binding Methods

Following the fundamental work of Fraenkel-Conrat and Cooper (1944) on the use of dyes for the determination of acid and basic groups in proteins, quantitative determination of milk protein by dye-binding was introduced by Schober and Hetzel (1956) and by Udy (1956). This method was the first routine method available for breeding programmes and payment schemes, and is still used for control in the dairy industry. A brief review of the method was published by Van Reusel and Klijn (1987) in the IDF monograph on rapid indirect methods for measurement of the major components of milk.

2.2.2.1. Principle. The dye-binding procedures used for milk protein testing employ different azo-sulfonic acid dyes: CI Acid Black 1 (or Amido Black), CI Acid Orange 12, and CI Acid Orange 10 (or Orange G). The mechanism of the reaction between the dye and the proteins is now relatively well, it not fully, understood. In acid buffer solution below the isoelectric point of the proteins, the positive charges of both the terminal amino groups, and the histidine, arginine and lysine residues, combine in a stoichiometric interaction with the negative sulfonic group of the dye, to form an insoluble dye–protein complex. In addition, Lakin (1974) has shown that hydrophobic interactions occur between free and bound anionic dyes.

The differences between dyes, with respect to their sensitivity (change of absorbancy per unit change of protein weight), are partly associated with their abilities to form hydrophobic interactions. Amido Black has the highest sensitivity followed by Acid Orange 12 and Acid Orange 10.

The quantity of dye bound per unit weight of protein, called the dye-binding capacity (DBC) of proteins, is dependent on the ratio of dye to proteins. Several workers (Fraenkel-Conrat & Cooper, 1944; Dolby, 1961) have shown that an excess of dye must be present in solution to get both a complete reaction between dye and proteins and a proportionality between milk protein content and the amount of dye which precipitates. Similarly, if the dye : protein ratio is too high, the DBC increases (Alais *et al.*, 1961; Ashworth & Chaudry, 1962). The various milk proteins or nitrogen fractions do not react in a similar way with the dye. Several authors have found similar results concerning the DBC of milk protein fractions (Alais *et al.*, 1961; Ashworth & Chaudry, 1962; Tarassuk *et al.*, 1967; McGann, 1978). In Table 3 are reported the findings of Tarassuk *et al.* (1967) and Ashworth and Chaudry (1962). It is important to point out that whey proteins have a higher DBC (ca. 27% more) than the caseins and that NPN does not bind dye, or rather, it remains in the supernatant (Alais *et al.*, 1961; Tarassuk *et al.*, 1967).

2.2.2.2. Analysis. The measurement of protein content is based upon the determination of the amount of unbound dye after removal of the dye–protein precipitate.

TABLE 3
Dye-Binding Capacity of Milk Proteins, as mg Amido-black (AB) or Orange G (OG) per g Protein[a]

	AB[b]		*OG*[c]		*AB*[c]	
Milk	350	(1·08)	179	(0·95)	354	(1·04)
Casein	325		189		339	
Whey proteins	410	(1·26)	247	(1·31)	447	(1·32)
Proteose-peptone					350	(1·03)
Casein						
Alpha	345	(1·06)	199	(1·05)	373	(1·10)
Beta	290	(0·89)	170	(0·90)	318	(0·94)
Kappa	310	(0·95)	193	(1·02)	359	(1·06)
Whey proteins						
α-Lactalbumin	440	(1·35)				
β-Lactoglobulin	425	(1·31)	252	(1·33)	456	(1·35)
Serum albumin	430	(1·32)				
Immunoglobulin	320	(0·98)				

[a]Numbers in parenthesis refer to the ratio to casein.
[b]Tarassuk *et al.* (1967).
[c]Ashworth & Chaudry (1962).

Provided that the ratio of dye to protein remains within certain limits, a linear relationship exists between the protein content and the quantity of unbound dye. However, McGann and Murphy (quoted in McGann, 1978) have demonstrated that, with Acid Orange 12, the relationship between absorbance and protein content is best given using a second power quadratic equation. A similar observation was made recently in the authors' laboratory (Leray, O., 1989, pers. comm.) for the Amido Black method, with individual milk, reconstituted or ultrafiltered milk samples.

To improve the reproducibility of results between laboratories, a centralized system of calibration using reference milk samples has been used in France (Grappin & Jeunet, 1976*b*), Poland (Michalak & Oczkowicz, 1975) and Australia (Clarke, 1979, 1988). International standards for dye-binding methods are available from the AOAC and IDF.

2.2.2.3. Instrumentation. To perform a protein test by dye-binding, no specific instrument is necessary except a centrifuge and a spectrophotometer. For routine determination in a central testing laboratory, fully automated equipment using Amido Black was first developed in Holland in 1960 and then in several other European countries, particularly in France. Specific manual instruments have been marketed since 1960: the Udy system (Udy Analyser Co., Boulder, CO, USA) using Acid Orange 12, and the Pro-Milk MKII (Foss Electric, Hillerod, Denmark) and Pro-o-Mat (Funke-Gerber, Berlin, Germany), using Amido Black.

2.2.2.4. Factors affecting the accuracy of protein determination in milk. Factors

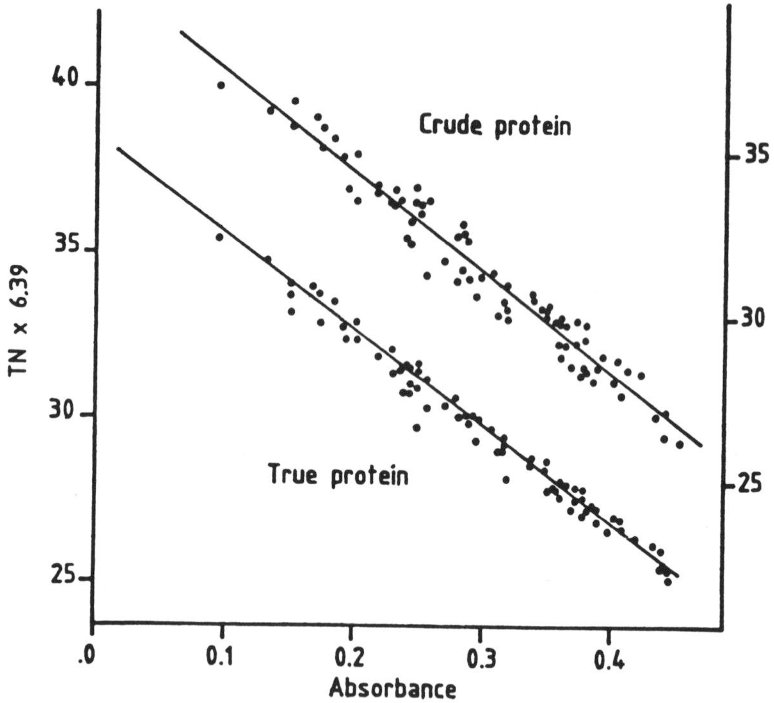

Fig. 2. Determination of true protein content and crude protein of individual milk samples by the amido-black dye-binding method (IDF, 1987). True protein: $r = -0{\cdot}994$, $s_{y,x} = 0{\cdot}31$ g/kg; crude protein: $r = -0{\cdot}985$, $s_{y,x} = 0{\cdot}51$ g/kg.

influencing the selection of the dye to be used, the methodology and the reaction conditions were reviewed by Tarassuk *et al.* (1967) and Sherbon (1978). Great care should be taken to ensure the purity of the dye, especially Amido Black which sometimes contains a large proportion of sodium chloride (Lakin, 1970) and traces of other dyes. Amido Black 10 B for electrophoresis is generally recommended. Acid Orange 12 can be obtained pure.

Experience has shown that, despite the fact that the various milk protein fractions have slightly different DBCs (Table 3), the main source of discrepancy between dye-binding and Kjeldahl methods is the variation in NPN content. In a population of individual milk samples, the accuracy SD ($s_{y,x}$) is lowered by 60% when true protein is used for reference instead of crude protein (Fig. 2). Variations in the ratio NPN/Total N in milk explain almost entirely the differences in calibration observed throughout the year and also between goat and cow milks (Lakin, 1970; Grappin & Jeunet, 1976*b*; Grappin *et al.*, 1979). However, for individual goat milks, Grappin and Jeunet (1979) have shown that 67% of the differences between Kjeldahl true protein contents and Amido Black readings are explained by the variation of the ratio whey protein : true protein in the milks (Fig. 3). As expected, a slight but significant influence of mastitis on accuracy was found by Grappin *et*

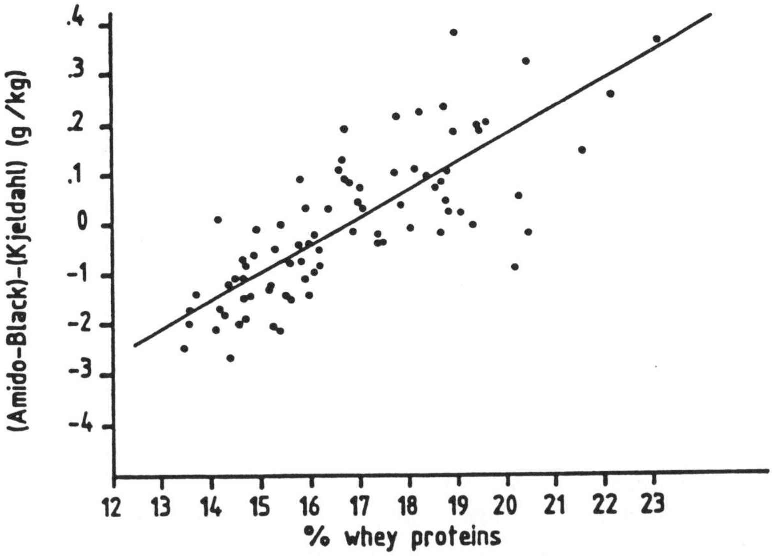

Fig. 3. Relationship between the differences Amido Black readings—true protein Kjeldahl values (*y* axis) and the percentage of whey protein in true proteins (*x* axis) of 81 individual goat milk samples (Grappin & Jeunet, 1979). Coefficient of correlation: $r = 0{\cdot}82$.

al. (1970) and by Waite and Smith (1972). For instance, on quarter milk samples, an increase of 1 log somatic cell count increases the Amido Black result by a relative 0·47% (Grappin *et al.*, 1970).

2.2.2.5. Analytical performances.

- Precision: several collaborative studies have been undertaken to evaluate the precision parameters of the dye-binding methods for milk protein testing (Sherbon, 1975; Grappin *et al.*, 1980). A repeatability value of 0·03 g/100 g can easily be achieved by experienced laboratories. Concerning the reproducibility, the relatively poor result of $R = 0{\cdot}215$ g/100 g, found by Grappin *et al.* (1980), can be dramatically improved ($R = 0{\cdot}068$) when standard materials are used for instrument calibration (Grappin, 1984).
- Accuracy: according to the definition given in Section 2.2.1, the milk protein content can be estimated with a $s_{y,x}$ varying from 0·03 g/100 g to 0·06 g/100 g, according to the origin of the milk (individual or herd milk) and the method used (manual or automatic).
- Application to dairy products: the dye-binding methods are also applicable to a large variety of dairy products, like cheeses (Steinsholt, 1976), chocolate drinks, cultured buttermilk and half and half (Sherbon & Luke, 1968), dry milk and ice-cream (Sherbon & Luke, 1969) and wheys (Roeper & Dolby, 1971). It is clear that, for each type of product, the procedure has to be adapted and a specific calibration is required to reflect both the modification in the procedure (volume ratio of test sample/dye solution) and the possible

alteration of the DBC of the protein mixture following heat treatment or proteolysis.

2.2.3. *Infrared Methods*

2.2.3.1. *Basic principle of IR measurements.*

—The IR spectrum: three regions characterize the IR spectrum according to the wavelength or wavenumber of the radiation: the near IR region (NIR) from 0·7 to 2·5 μm (from 14 285 to 4000 cm^{-1}), the mid IR region (MIR), from 2·5 to 25 μm (4000 to 400 cm^{-1}) and the far IR region, from 25 up to 100 μm (from 400 to 100 cm^{-1}). The energy of electromagnetic radiation is proportional to the frequency of the radiation, i.e. inversely proportional to the wavelength. When a molecule is submitted to IR radiation, energy will be absorbed only if the frequency of the radiation corresponds to the frequency of one of the fundamental vibrations of the molecule (stretching vibrations at high frequencies and bending deformations at low frequencies).

The vibration energy which characterizes a chemical group (e.g. C—H, O—H, C=O) is dependent upon both the bond strength and the masses of the two atoms of the group. Fundamental vibrations of the molecules take place mainly in the MIR region and absorption in the NIR occurs at wavelengths corresponding to either harmonic frequencies, or combination frequencies of the fundamental vibrations. These NIR absorption bands are generally quite broad, allowing use of rather large spectral pass bands for measurements. Their intensities are weak, compared to the signals obtained in the MIR region from fundamental vibrations of the molecules, but they are sufficiently important to allow quantitative analysis. Because NIR bands are less specific for a given molecule than wavelengths in MIR, one can anticipate that quantitative analysis of a specific component will require more wavelengths to obtain accurate measurements.

Both techniques have been used for the analysis of milk and milk products; MIR essentially for the analysis of milk by transmission and NIR for the analysis of dairy products by diffuse reflexion. Comparing these two techniques, Rudzik (1985) concluded that they are complementary for the analysis of dairy products.

—Quantitative analysis: the measurement of the IR energy absorbed by a sample can be made either in the transmission mode, if the product is in solution or is sufficiently transparent to IR radiation, or in the reflection mode for opaque or solid samples. Most of the MIR instruments designed for liquid samples measure the transmitted light directly, while NIR instruments are usually built to measure the IR light which is either diffusely reflected from the product surface for solid samples or from the surface of the sample holder for liquids, or transmitted through a cuvette for liquids.

With a single component solution, the amount of IR energy that is absorbed by the sample (absorbance) is exponentially proportional to the component concentration and follows the Beer-Lambert law. If I_0 is the intensity of the incident beam and I is the intensity of the transmitted or reflected beam, the absorbance is $A_\lambda = \log I_o/I = E_\lambda \times C \times l$, where the ratio I/I_o represents either the transmittance or the reflectance, E the extinction coefficient of the component, C the concentration, λ the measurement wavelength and l the path length of the cell, for transmittance measurement.

The absorbance of n absorbing components is then expressed by the equation:

$$A_\lambda = (E_{\lambda 1} \times C_1 + E_{\lambda 2} \times C_2 + \ldots + E_{\lambda n} \times C_n) \times l$$

where $C_1, \ldots, C_n$ are the concentrations of the n components, and $E_{\lambda 1}$, $E_{\lambda 2}, \ldots, E_{\lambda n}$ are the extinction coefficients of the n components at the wavelengths $\lambda_1, \lambda_2, \ldots, \lambda_n$.

In conventional MIR instruments, to correct any variation of the system response (source brightness, temperature of sample, soil on the cuvette, etc.) and to diminish, as much as possible, the influence of interfering components like water, the measurements are made by reference to the amount of IR energy absorbed, either by water at the same wavelength as the measuring wavelength, or by the sample at a close wavelength at which there is only a slight absorption by the component being measured. For NIR instruments using reflectance, the reference is usually obtained by the intensity of the incident beam reflected by a ceramic disk.

2.2.3.2. Analysis of proteins by MIR transmission. Since the work of Goulden (1964), who first studied the analysis of milk by MIR absorption, a large number of papers assessing new instruments, or dealing with more fundamental works, have been published. In the IDF monograph on indirect methods for milk analysis, Biggs *et al.* (1987) reviewed the current status of knowledge concerning this technique. Much of the information given in this chapter is drawn from this monograph. A comprehensive work on this method was published by Sjaunja (1982) and a review of the pre-calibration checks, calibration setting and maintenance was published by Barbano and Clark (1989).

- Principle: in MIR, there is a strong absorption band, called 'amide II', at $\sim 6{\cdot}46\,\mu$m (1550 cm^{-1}) by a peptide bond. This absorption orginates from C—N stretching vibration (40%), and from N—H bending deformation (60%) (Fig. 4). The peptide bond also shows other absorption bands at near 1650 cm^{-1} (6·1 m, 'amide I', due mainly to C=O stretching vibration) and at 3300 cm^{-1} (3·0 μm, N—H stretching vibration). As illustrated by Fig. 5, the determination of protein concentration in milk is based upon the 'specific' absorption of the peptide linkages at 6·46 μm. Although proteins are the major absorbing compounds at this wavelength, the absorbance is influenced to a greater or lesser extent by the other major milk compounds (fat and lactose) and by minor soluble elements. As most of the instruments are designed to make simultaneous determinations of fat, protein and lactose (and sometimes water), the best accuracy for protein determination will be obtained if the primary protein signal is corrected for the absorption by fat and lactose at the protein wavelength. With modern instruments, automatic correction is

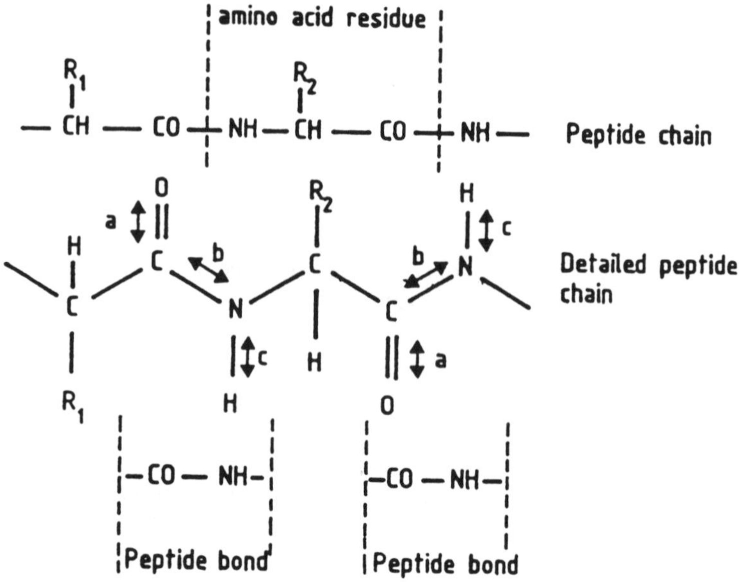

Fig. 4. Mid-IR absorption bands of the peptide bond: a, amide I(C=O stretch); b, amide II (C–N stretch); c, amide II (N–H bending); ↔, direction of the light-induced dipole; R_1, R_2, amino acid side chains.

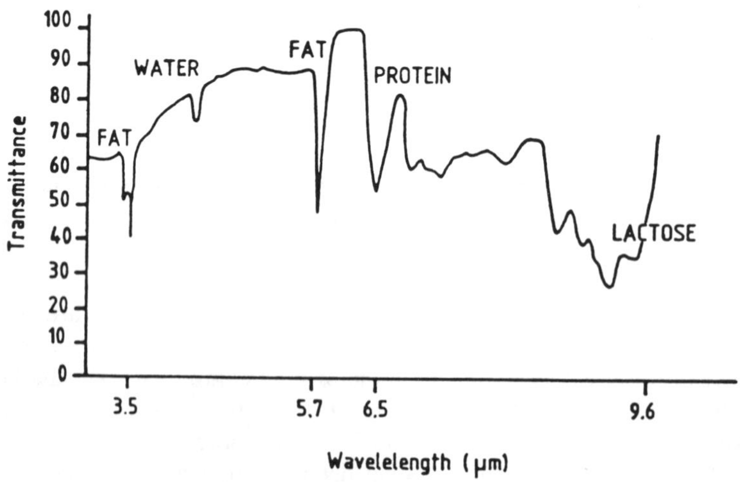

Fig. 5. The infrared transmission spectrum of milk versus water (courtesy of Multispec Ltd, IDF, 1987).

achieved by setting internal intercorrection factors for each component at each wavelength. Intercorrection factors can be determined either with specifically prepared samples to characterize each of the intercorrection factors or by multiple regression of uncorrected instrument signal versus chemical data on a large number of milk samples (Barbano & Clark, 1989). To minimize the light scattering influence of fat globules, the sample has to be homogenized prior to analysis to reduce the globule size below 1 μm.

- Analysis: a warmed sample (40°C), thoroughly mixed, and if necessary blended and/or diluted, is pumped through a 1, 2 or 3 stage valve homogenizer or an ultrasonic homogenizer. The specific wavelengths are focused through the cell containing the sample and the infrared radiation is chopped by a half-section mirror to produce alternating sample and reference signals. The difference between the reference and the sample signals is then collected by an IR detector and the signal output is processed by a log/lin converter and corrected for interference. Component concentration is finally displayed on the instrument. Before analysis, the instrument must be calibrated by comparison with a reference method. Detailed calibration procedures are given by instrument manufacturers and can be found in the paper of Barbano and Clark (1989). A provisional international standard (IDF, 1990) has been produced by an IDF/ISO/AOAC group of experts.
- Instrumentation: following the pioneering work of Goulden, the first commercially available IR instrument allowing the direct measurement of fat, protein and lactose in milk, was a conventional double-beam spectrophotometer with two cells (a reference cell with water and a sample cell), with a servosystem and a diffracting grating. This instrument, named the Infra-Red Milk Analyzer (IRMA), was manufactured by Grubb Parsson (UK). With the outstanding progress made during the last 20 years in electronics, interference narrow band filters and microprocessors, simpler, more reliable, automatic and accurate instruments are now available. All modern instruments are now equipped with a single cell and optical filters. They may utilize either a single or a double beam optical system, and may use either electronic ratioing or a servo system to estimate the I/I_0 ratio. Figure 6 illustrates different optical systems used in IR instruments and Table 4 gives the main characteristics of commercial instruments currently available for the analysis of milk and milk products.
- Factors affecting the accuracy of protein determination in milk:

  —Physico-chemical factors: if we exclude the influence of instrumental factors like temperature, linearity, water vapor in the optical console, homogenization efficiency, etc., and if we assume that the instrument is correctly calibrated (including the correction for fat and lactose contents), the accuracy of milk protein testing will be mainly influenced by the variation in the proportion of NPN and by the presence of carboxylic acids. The relationship between infrared absorption at 6·46 μm and true protein concentration measured by Kjeldahl is relatively independent of the amino acid composition of the protein, since the ratio N content/number of peptide

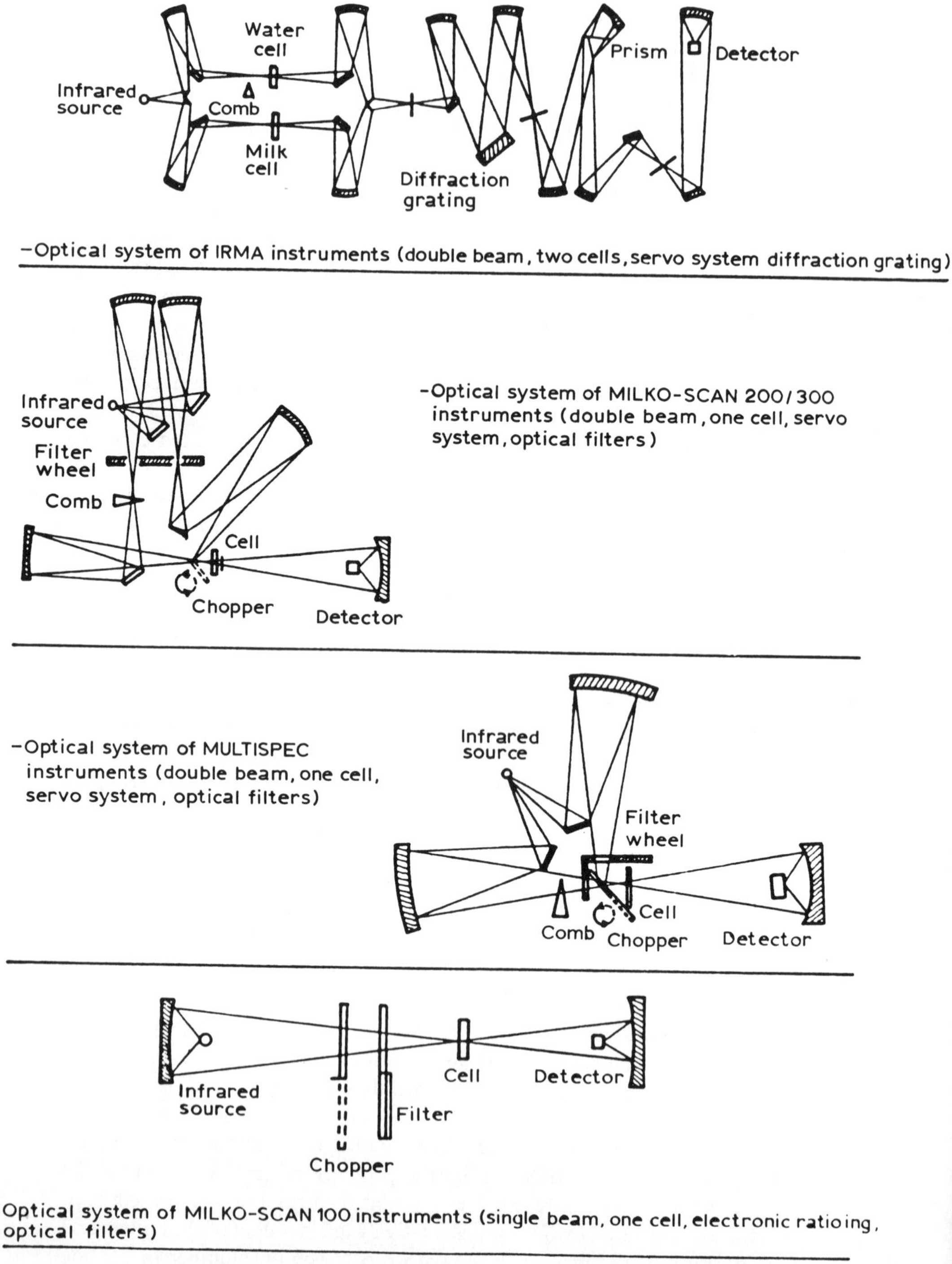

Fig. 6. Optical systems of different mid-IR instruments (IDF, 1987).

linkages is relatively constant. On the other hand, because the NPN fraction is not measured by infrared, any variation in the proportion of NPN will influence the accuracy of an instrument calibrated in crude protein (total N × 6·38).

The presence of ionized carboxyl group, COO^-, gives rise to an absorption band at the protein wavelength. The main indigenous source of such groups is citrate. Gaudillere and Grappin (1982) and Sjaunja and Anderson (1985) have shown that natural variations in the citrate content of individual milks explain between 40 and 60% of the difference between IR and Kjeldahl true protein results, and an increase of 0·01 g/100 g in the concentration of citric acid changes the protein reading by +0·075 g/100 g.

The formation of carboxylic acids by fermentation of lactose may also cause interference absorption at the protein wavelength (Goulden, 1964). Similarly, it has been demonstrated that lipolysis causes an increase of the protein signal by 0·01 g/100 g per millimole of free fatty acid (Sjaunja, 1982). Grappin and Jeunet (1979) have clearly demonstrated that in fact most of the interfering compounds are present in the soluble phase of milk (Fig. 7).

—Biological factors: any biological factor (e.g. stage of lactation, mastitis, breed, species, feeding, season, etc.), that is known to influence one of the physico-chemical characteristics mentioned above, will in its turn cause systematic errors in protein measurement by IR methods. According to the review of Biggs *et al.* (1987), only the influence of few factors has been demonstrated clearly. Goat milk, which has a lower concentration of citrate, requires a different calibration than cows' milk for true protein analysis (Grappin *et al.*, 1979). Season/feeding, as well as species (goat versus cow) or breed (Jersey versus others), which influence the proportion of NPN in milk, will have a significant influence on the accuracy of the method when the apparatus is calibrated for crude protein (Grappin & Jeunet, 1979; Sjaunja, 1982). Whenever possible, adjustment of the instrument calibration will therefore be necessary.

- Analytical performance: since the first instrument tested by Goulden (1964), real progress in the performance of instruments has been achieved (Biggs, 1979*a*).

  —Repeatability: with modern instruments, an excellent repeatability SD of 0·02 g/100 g, for milk fat, protein and lactose analysis is now commonly obtained.

  —Accuracy: following the definition given in Section 2.2.1, milk protein content can be estimated with a $s_{y,x}$ varying from 0·08 g/100 g for individual milks to 0·04 g/100 g for herd milks. One can assume that with silo milk, accuracy will be even better and limited mainly by the repeatability of both the Kjeldahl and the IR methods.

- Application to dairy products: to analyse viscous or semi-solid products with high protein content, dilution and vigorous blending of a weighted sample is necessary before testing with the commercial instruments. Because of the

TABLE 4
Major Characteristics of Mid-infrared Instruments Marketed Since 1965 (1DF, 1987)

Manufacturer	*Instrument*	*Year*	*No. of cell*	*Beam system*	*Measurement system*	*Monochrometer system*	*Homogenization stages*	*Components*	*Notes*
Grubb Parsson, UK	IRMA	1965	2	Double	Servo	Grating	–	Fat, protein, lactose	Obsolete
–	Mini IRMA	1978	2	–	Ratio	Optical filter	–	Fat, protein, lactose, water	–
Foss Electric, DK	Milko-Scan 300	1974	1	–	Servo	– –	1	Fat, protein	
–	Milko-Scan 203	1975	1	–	–	– –	1	Fat, protein lactose	
–	Milko-Scan 102/101 104	1979	1	Single	Ratio	– –	2	Fat, protein, lactose, water	
–	Milko-Scan 203B	1979	1	Double	Servo	– –	2	Fat A, protein, lactose	
–	Milko-Scan 103B	1980	1	Single	Ratio	– –	2	Fat B, protein, lactose	
–	Milko-Scan 104 A/B	1961	1	–	–	– –	2	Fat A, fat B, protein, lactose	
–	Milko-Scan 104 Auto	1983	1	–	–	– –	3	See Milko-Scan 104	
–	Milko-Scan 104 Speed	1983	1	–	–	– –	3	See Milko-Scan 104	
–	Milko-Scan 605	1984	1	–	–	– –	3	Fat A, fat B, protein, lactose	
–	Milko-Scan 133/133A/133B	1985	1	–	–	– –	1	Fat A or/and B,	Recirculating homogenization
REIL[a]	Milko-Scan 104 A/B	1986	1	Single	Ratio	– –	4	Fat A, Fat B, protein, lactose	

Multispec, UK	Multispec	1977	1	Double	Servo	–	–	2	Fat A, fat B, protein, lactose
–	Multispec Auto	1978	1	–	–	–	–	2	Fat A, fat B, protein, lactose
–	Micro-null	1984	1	–	–	–	–	3	Fat A, fat B, protein, lactose
–	Multispec II	1984	1	–	–	–	–	2	Fat A, fat B, protein, lactose
–	Dairy Lab	1986	1	Single	Ratio	–	–	2	Fat A, fat B, protein, lactose and others

[a]Under technical assistance from Foss Electric.

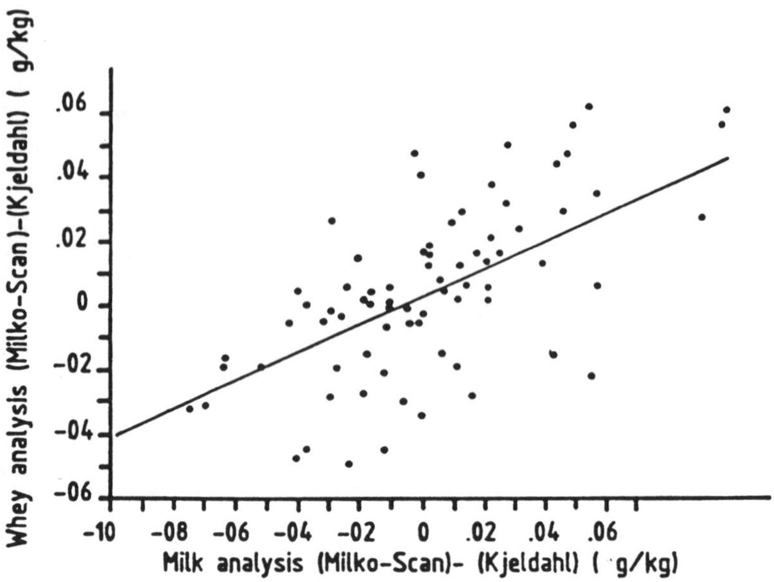

Fig. 7. Relationship between Milkoscan protein readings – true protein Kjeldahl values, for 81 individual goat milk samples (x) and the corresponding whey samples (y) (Grappin & Jeunet, 1979). Coefficient of correlation, $r = 0{\cdot}65$.

interference of other compounds and the influence of the technological process, specific calibration of the instrument is necessary for each kind of product. Moreover, better results will be obtained before any hydrolysis of the proteins occurs. Although Milko-Scan or Multispec milk analysers are commonly used by the dairy industry for quality control of their products, little information is available on the analytical performance of the method. On cottage cheese, Biggs (1979*b*) found a SD of difference between reference and instrumental protein results of 0·17 g/100 g. In a survey on a large variety of dairy products, Karasch (1986) reported good repeatability and accuracy with a Milko-Scan 104 except for whipping cream, dried whey and yoghurt.

2.2.3.3. Analysis of proteins by NIR reflection. Rapid measurement of the chemical composition of food products by NIR reflection spectroscopy has been studied and used mainly in the field of cereals and plant analysis. Despite the fact that the application of this technique to the analysis of milk products was investigated by Goulden (1957), it has received limited attention from dairy scientists. However, with the development of reliable commercial instruments, the dairy industry is now increasingly using NIR methods to monitor the quality of dairy products, e.g. moisture content of milk powder; moisture, protein and fat contents of fresh cheeses and yogurt (Egli & Meyhack, 1984). For detailed information concerning theory, instrumentation, calibration and application of NIR spectroscopy to food analysis, the textbook of Osborne and Fearn (1988) can be usefully consulted.

- Principle: milk has an NIR spectrum, which is very similar to that of water, and the determination of wavebands that are 'specific' to milk proteins can be achieved only by mathematical treatment of a collection of milk spectra. From the reflectance values recorded by the instrument at the different wavelengths available (for instance up to 19 with the Technicon Infraalyzer 450) and the corresponding component concentration measured by a reference method, the most common way of selecting wavelengths is to calculate a step-by-step multiple regression using the equation:

$$C = K_1 \log(1/R_1) + K_2 \log(1/R_2) + \ldots + K_n \log(1/R_n) + K_0$$

where C is the component concentration, K is a constant equivalent to a coefficient of reflectance at the corresponding wavelength, K_0 is the bias, $1/R$ is the reflectance at the corresponding wavelength, and n is the number of selected wavelengths.

Since many wavelengths give redundant information, the optimum number of wavelengths is obtained when the addition of a new filter does not improve the residual standard deviation of the multiple regression.

A different approach was followed by Robert *et al.* (1987), who applied multivariate analysis (principal component analysis and correspondence factorial analysis). On a population of 38 milk samples from cow, goat, ewe, colostrum and mastitic animals, they found that within the limits 2030–2350 nm, where there is no intense absorption band for water, two wavelengths, 2050 and 2180 nm, could be assigned to milk proteins.

In a one-year study, Launay *et al.* (1986) found that, amongst the 13 filters that were selected during the study by multiple regression analysis, four wavelengths were constantly used: 2180, 2100, 1820 and 1450 nm.

These bands were also selected by Jeunet and Grappin (1985) in a study on individual milks. Two wavelengths, 2100 and 1820 nm, which have negative coefficients, are not specific to proteins.

Table 5 gives an example of the wavelengths that have been assigned by different authors for NIR protein analysis.

- Analysis and instrumentation: most of the research and practical work done on milk and milk product analysis has been carried out with the Infraalyzer instrument and in some cases with Pacific Scientific/Neotec, or Dickey-John (Auburn, IL, USA) instruments.

Sample preparation requires different approaches according to the nature of the product. For milk analysis, after homogenization to limit the light scattering effect of fat globules, the sample is placed into a temperature-controlled (40 ± 0·1°C) holder with a quartz window and a ceramic diffuse reflector. Powder samples are simply placed in a sample holder and pressed against a quartz window. Solid and pasty products, like cheese, should have a uniform surface, and are placed in an open holder.

Wavelengths can be selected either by a holographic grating for research work (e.g. Technicon Infraalyzer 500), or by narrow band interference filters in routine analysis. An intermediate solution is offered by tilting filters with a

TABLE 5
Wavebands Assignment for NIR Protein Analysis of Dairy Products

Dried milk (Goulden, 1957)	*Liquid milk* (Jeunet & Grappin, 1985; Launay *et al.*, 1986)	*Dried milk* (Baer *et al.*, 1983)	*Casein and dried cheese* (Frank & Birth, 1982)	*Liquid milk* (Robert *et al.*, 1987)
1 180			1 170	
			1 290	
1 450–1 600	1 450		1 500	
1 730		1 730	1 700	
1 820	1 820	1 820		
1 930		1 980		
	2 100		2 050	2 050
3 050	2 180	2 190	2 175	2 180
2 280				
2 320		2 310		

variable angle of incidence which provides some flexibility in the wavelength setting.

Reflected light originating from the irradiated sample is measured either by two lead sulfide detectors placed in an integrating sphere (Infraalyzer, Technicon) to collect as much as possible of the reflected energy, or by four detectors placed at an angle of 45°C above the sample (Pacific Scientific Instrument).

The incident or reference beam intensity is measured automatically by the detectors after rotation of the mirror for each sample measurement. A ceramic disk is used as an external reference and the reflectance values obtained at the different selected wavelengths are stored and processed to give the component concentration.

- Analytical attributes and factors affecting the accuracy of protein testing in milk: compared to MIR techniques, very little work has been done to assess the performances of the NIR instruments for the analysis of milk and to thoroughly evaluate the physico-chemical and biological factors that may influence the response of NIR analysis. A value of 0·023 g/100 g was given by Jeunet and Grappin (1985) for the repeatability SD of the Infraalyzer Dairy 400 and an accuracy SD ($s_{y,x}$) of 0·021 g/100 g for individual milk samples. Using a Neotec Model 6350, Sato *et al.* (1987) obtained a SD of prediction of 0·098 g/100 g on 60 individual milk samples versus 0·075 when the samples are analyzed by a Milko-Scan 203 B. On individual samples of cow and goat milks, Jeunet and Grappin (1985) found that lipolysis does not interfere and that only species significantly influences the protein results. Conversely to MIR, the accuracy SD is slightly lower (0·021 versus 0·025 g/100 g) when the instrument is calibrated for crude instead of true protein. Similarly, a better estimate was obtained by Baer *et al.* (1983) on non-fat dry milk when the Kjeldahl method was used as a reference, rather than the dye-binding method.
- Application to dairy products: NIR techniques have been evaluated for the analysis of moisture, fat and protein mainly for various milk powders (Baer *et al.*, 1983; De Vilder & Bossuyt, 1983; Frankhuizen & van der Veen, 1985) and cheese (Frankhuisen & van der Veen, 1985; Frank & Birth, 1982). An example of the findings of Frankhuizen and van der Veen (1985), concerning protein determination on milk powders and processed cheese, is given in Table 6. To obtain reliable results, sample preparation is extremely important. For protein measurement, De Vilder and Bossuyt (1983) pointed out that the granular structure of the milk powder affects the results.

2.3. Amino Acid Composition of Milk and Dairy Products

Direct amino acid analysis after acid hydrolysis can give a good estimation of the protein content of any food product if the content of peptides and free amino acids is negligible. This condition can be considered as fulfilled for dairy products such as industrial caseins and caseinates, whey protein concentrates (WPC), lactalbumins, coprecipitates, and partially fulfilled for good quality milks, wheys, milk

TABLE 6
Near Infrared Reflectance Data for Protein in Some Dairy Products (Frankhuizen & Van der Veen, 1985)

Product	*Calibration*				*Prediction*			
	n_c	R_c^2	SE_c	$Range_c$	n_p	R_p^2	SE_p	$Range_p$
Skim milk powder	159	0·97	0·27	32·9–41.2	19	0·95	0·20	34·0–36·8
Butter milk powder	84	0·97	0·21	28·9–34·6	20	0·97	0·21	29·2–34.6
Skim milk powder and buttermilk powder	254	0·98	0·34	38·9–41·2	20	0·97	0·46	30·4–40·6
Denatured milk powder	92	0·98	0·34	27·1–40·1	20	0·95	0·42	28·5–39·7
Milk powder with non-milk fat	61	1·00	0·18	17·6–26·0	10	0·96	0·25	18·4–25·4
Processed cheese 40	50	0·90	0·31	20·0–24·0	10	0·89	0·35	19·9–23·7

n: Number of samples.
R^2: Coefficient of determination.
SE: Standard deviation of estimate.

TABLE 7
Recovery of Nitrogen from Amino Acid Analysis of Casein Carried Out by Seven Laboratories

	Between laboratories		*Within laboratories*
Mean %	*SE*	*CV %*	*CV%*
99·4	2·9	2·9	0·8

The protein content ($N \times 6{\cdot}38$) of the sample was 96·7 ± 0·8 (mean ± SEM of the 7 laboratories). SEM, standard error of the mean; SE, standard error; CV, coefficient of variation (from Sarwar *et al.*, 1983).

and whey powders. However, it is certainly not the best method as far as rapidity (at least 1 h for one sample) and accuracy are concerned. An indication of the accuracy of protein determination by this technique can be deduced from the results of inter- and intra-laboratory analyses performed in 1983 on a single casein sample (Sarwar *et al.*, 1983). Each laboratory analyzed the sample for dry matter and nitrogen by AOAC (1975) procedures. Three hydrolyses (6N HCl, performic acid + 6N HCl, and 4·2N NaOH) were carried out in duplicate for determination of stable amino acids, sulfur amino acids and tryptophan, respectively. Table 7 shows a good correlation between the nitrogen content (Kjeldahl) and the recovery of nitrogen calculated from quantitative analysis of 18 amino acids plus ammonia.

Amino acid composition of a food product is most useful for nutritionists. It can sometimes be used for the detection of adulteration, e.g. detection of added WPC in non-fat dry milk (Greenberg & Dower, 1986). Therefore, some information will be given below on the amino acid analysis of milk and dairy products, although the methodology does not differ significantly from that used for other food products and has not changed much for many years.

2.3.1. Sample Preparation

Sampling is easy for most dairy products (liquids and powders). For cheese it is convenient to freeze-dry and grind a fairly large sample in order to get an homogeneous powder.

Many amino acid analyzers separate the amino acids from the hydrolyzed sample by ion exchange. This technique does not necessitate very clean products. In particular, carbohydrates, even in large proportion, and salts, do not cause serious problems. However, the lipid content must be low. Centrifugal skimming is quite sufficient for liquid dairy products unless milk has been homogenized. In other cases, the products must be dried and ether-extracted. The levels of salts and lipids in the samples for analysis by HPLC after pre-column derivatization must be negligible.

Although amino acid analysis after direct acid hydrolysis can give fairly accurate determination of cysteine + cystine, most people prefer to transform these amino acids to more stable derivatives. For food products, this is usually achieved by performic acid oxidation. Part of the sample is treated with a mixture of formic acid and hydrogen peroxide at low temperature and the excess of reagents is removed by evaporation. This converts cystine and cysteine into cysteic acid

(2 moles/mole cystine) and methionine into methionine sulfone (MacDonald *et al.*, 1985). As this procedure totally destroys tryptophan and partially destroys tyrosine and histidine, complete analysis of a sample requires three different determinations to be carried out on three hydrolysates: acid hydrolysates of the native and oxidized sample and alkaline hydrolysate for tryptophan determination.

2.3.2. Hydrolysis

Food products are usually hydrolyzed at 110°C for 22 or 24 h in the presence of 6N HCl (for stable and sulfur amino acids) at a ratio, mg protein/ml acid, varying between 0·4 and 4·0 (Sarwar *et al.*, 1983). The lower ratios are used for carbohydrate-rich products. Hydrolysis is performed in tubes or flasks, either sealed, or evacuated and sealed, or evacuated, nitrogen flushed and sealed, or under reflux. After hydrolysis, the acid is evaporated and the residue taken with pH 2·2 citrate buffer for conventional amino acid analyzers. Filtration through 0·45 μm filter is performed. Some analysts add phenol (50 μl to 1 ml/100 ml) to the 6N HCl to reduce tyrosine losses; the authors do not think it is really useful. Comparison has been made between conventional acid hydrolysis, as above, and hydrolysis at 145°C for 4 h for foods and feeds (Lucas & Sotelo, 1982; Gehrke *et al.*, 1985). For casein, results were quite similar, with slightly higher values for Ile and lower values for Ser and Thr at the higher temperature. A slightly higher recovery of total amino acids was obtained at 110°C.

In order to obtain more precise data for sensitive or more difficult-to-release amino acids (Thr, Ser, Met, Tyr; Val and Ile when bonds involving these 2 residues occur), 3 hydrolysis times are performed (e.g. 22, 44, 88 h at 110°C; 4, 8 and 12 h at 140°C). The values for the amino acids of the former group are extrapolated to time zero, while the longest time only is taken into account for Val and Ile.

For high sensitivities, gaseous 5·7 N HCl is frequently used for protein hydrolysis because of the lower background obtained and the lower amount of acid needed (Bidlingmeyer *et al.*, 1984).

Hydrolysis for determination of Trp is performed at alkaline pH since this amino acid is destroyed during acid hydrolysis. Two methods for the determination of Trp in foods and feeds, including casein, have been published recently and the conditions for hydrolysis have been evaluated (Sato *et al.*, 1984; Nielsen & Hurrell, 1985); the results were similar. About 50 mg protein were hydrolyzed in a pyrex tube or a polypropylene liner contained in the pyrex tube (the two methods gave similar results) with 4 ml of 4·2N NaOH. The addition of partially-hydrolyzed starch (0·17 g) improved Trp recovery. One drop of l-octanol (which could also be used for acid hydrolysis) prevented foaming during evacuation of the air, which could be performed with a tap-water vacuum pump. The sealed tube was placed at 110°C for 20 h. After cooling, the hydrolysate was adjusted to pH 4·0 with precooled 6N HCl and diluted to 20 ml with a pH 4·2–4·3 citrate buffer. This solution could be conserved for at least 2 weeks at 5°C without further Trp loss. Some destruction of Trp occurred during hydrolysis but this could be corrected on the basis of the recovery of an internal standard. In the inter- and intra-laboratory study mentioned above (Sarwar *et al.*, 1983), it was shown that inter-laboratory variation of Trp (CV up to 24%) was greater than that for all other amino acids (CV up to 10%). The values for Trp obtained by one laboratory using the colorimetric methods of Spies and Chambers (1948) were comparable to those using a method close to that described above.

It seems that destruction of Trp during hydrolysis with HCl is due to the presence of chlorine in HCl. Hydrolysis of proteins with sulfonic acids in a reducing medium (e.g. with mercaptoethane sulfonic acid, MESA) is considered to protect Trp from destruction. However, this procedure is little used: even under these conditions, the presence of carbohydrates affects Trp recovery. Furthermore, MESA is not volatile.

2.3.3. Amino Acid Analysis

Several new methods for amino acid determination after acid hydrolysis have appeared during the last 10 years. They present two main advantages over the conventional ion exchange method followed by the ninhydrin reaction: somewhat faster analyses and higher sensitivity. They include, in particular, ion exchange with new post-column derivatization reagents and reversed-phase high performance liquid chromatography (RP-HPLC) with pre-column derivatization.

2.3.3.1. Conventional techniques. The conventional ion-exchange/ninhydrin procedure is by far the most widely employed for amino acid analysis on foods. Improvements include: automatic samplers of various types, narrow bore shorter columns (made of glass or metal) with resin particles of small diameter (for increased sensitivity and reduced buffer consumption, with concomitant higher pressure), more general use of a single column with stepwise elution (3 or 4 different buffers), shorter reaction times with ninhydrin at a higher temperature (e.g. 120°C), sophisticated colorimeters and computerization of data collection.

2.3.3.2. Ion exchange with new reagents for post-column derivatization. Fluorescamine, and O-pthaldialdehyde (OPA) + 2-mercaptoethanol react with primary amino acids only, giving fluorescent derivatives which permit high sensitivity. The latter is now the most popular reagent. It provides a 100-fold increase in sensitivity over ninhydrin. However, the detection of secondary amino acids (e.g. proline) requires an additional pump and a post-column reactor to convert the amino acids to primary amino acids with chloramine T or sodium hypochlorite (Böhlen & Schroeder, 1982).

2.3.3.3. Reversed-phase HPLC (RP-HPLC). RP-HPLC with pre-column derivatization with OPA (Cooper *et al.*, 1984), dansyl chloride (Wilkinson, 1978), dabsyl chloride (Vendrel & Aviles, 1986) or phenylisothiocyanate (PITC) are the most popular at present in protein chemistry since they allow high sensitivity (picomolar level) and fast analyses. OPA, of course, presents the same drawback as mentioned above; determination of secondary amines requires an oxidation step prior to reaction with OPA. Furthermore, OPA derivatives of Gly, Lys and hydroxyLys decompose easily. The reaction time must be short and accurately controlled.

The other reagents mentioned above, especially PITC which converts all amino acids, including proline, into fairly stable phenylthiocarbamyl (PTC) derivatives, are being used increasingly. For pure protein samples, separation of PTC-amino acid on C18 reversed-phase columns, 3·9 mm × 15 cm, with particles of 3 mm and gradient elution, requires ~12 min. The total duration of the analysis, including column washing and equilibration, requires only ~20 min (Bidlingmeyer *et al.*, 1984). The procedure was developed commercially by Waters (Milford, MA) as the Pico-Tag method which was recently assessed for the determination of amino acids in individual food items, such as casein, and 41 recipe-prepared foods (Hagen *et al.*, 1989). Performic oxidation enabled quantification of most amino acids besides Cys and Met, except for His, Tyr, Phe and Trp. While Cys and Trp were totally destroyed in samples that were not oxidized, recovery of Met was often complete. For casein, coefficients of variation between duplicate analyses were less than 3·5%, except for His (5·6%). The amino acid contents as determined by the Pico-Tag method agreed well with published values when normalized for differences in protein content. An entirely automatic instrument which performs hydrolysis, PITC derivatization, HPLC separation, data collection and analysis, is now available from Applied Biosystems (Foster City, CA).

Recently, the Varian company (Hicksville, NY) has developed a new automatic system for amino acid analysis of protein hydrolyzates by RP-HPLC after precolumn derivatization with 9-fluorenyl methyl chloroformate (FMOC). The amino acid derivatives thus obtained are quite stable, unlike PTC-amino acids, and display a high fluorescence which gives extremely high sensitivity. Furthermore, it seems that derivatization is not influenced by the presence of salts in the hydrolyzate.

RP-HPLC with direct fluorimetric detection can be used to determine tryptophan in food products after alkaline hydrolysis (Nielsen & Hurrell, 1985).

2.3.3.4. Accuracy of data from amino acid analysis. As an example, further details are given of an inter-laboratory assay, already mentioned above, carried out in 1983 by seven North American laboratories (Sarwar *et al.*, 1983) on seven food products (casein, egg white, beef, soy isolate, rapeseed concentrate, pea flour and wheat flour) using conventional techniques. The results for casein are shown in Table 8. The inter-laboratory variation (estimated as CV) for all amino acids except cystine (17·6%), tryptophan (14·3%) and glutamic acid (10·3%) was less than 10% (3·6–9·1%), whereas the intra-laboratory variation for all amino acids was lower than 3% (0·6–2·6%). Adjustment of amino acid data to constant nitrogen recovery was reported to improve the comparability between laboratories. Application of a similar correction to the data in Table 8 (CV1 versus CV between laboratories) reduced inter-laboratory variability for all amino acids, but not for ammonia.

TABLE 8
Accuracy of Casein Amino Acid Analysis Performed by Seven Laboratories

Amino acid	*Mean g/16g · N*	*Between laboratories*			*Within laboratories*
		SE	*CV1%*	*CV2%*	*CV %*
Arg	3·71	0·34	9·1	8·9	1·8
His	2·97	0·18	6·0	5·2	1·6
Ile	5·36	0·26	4·8	3·8	1·3
Leu	10·16	0·44	4·3	2·5	0·7
Lys	8·44	0·42	5·0	4·3	0·6
Met	3·02	0·13	4·2	3·1	0·8
Cys	0·47	0·08	17·6	17·0	2·6
Phe	5·47	0·39	7·1	5·8	1·5
Tyr	6·04	0·42	6·9	5·6	1·2
Thr	4·64	0·32	7·0	5·4	1·5
Trp	1·31	0·19	14·3	14·2	1·1
Val	6·85	0·30	4·4	2·6	1·4
Ala	3·30	0·19	5·8	4·3	1·8
Asx	7·71	0·31	4·0	1·9	0·9
Glx	24·00	2·53	10·5	10·1	1·3
Gly	2·00	0·13	6·4	5·5	1·3
Pro	11·72	0·78	6·6	5·0	1·9
Ser	6·10	0·22	3·6	2·1	1·6
NH_3	1·98	0·19	9·6	10·7	5·9

SE, standard error; CV, coefficient of variation; CV1, between laboratories; CV2, CVs for adjusted amino acid data (corrected to a total nitrogen recovery of 100% by each laboratory (from Sarwar *et al.*, 1983).

3. SEPARATE DETERMINATION OF WHOLE CASEIN, WHEY PROTEINS, β-LACTOGLOBULIN AND NPN IN MILK AND DAIRY PRODUCTS

3.1. The Aschaffenburg-Rowland Procedure

As early as 1938, Rowland (1938) published a method for quantitative determination of different protein fractions, whole casein, albumins, globulins and proteose-peptones, in cows' milk. Aschaffenburg and Drewry (1959) improved this method on the basis of paper electrophoresis used as an analytical technique and proposed a procedure to quantitate β-Lg (an albumin), residual albumins (mainly α-La), globulins (mainly immunoglobulins) and proteose-peptones. It must be recalled that proteose-peptones constitute a heterogeneous group of peptides or proteins which are not made insoluble by heating milk or whey at pH 4·6 but are precipitated by 12% TCA. Both methods use several precipitating agents. Nitrogen is determined in milk and supernatants by the Kjeldahl technique, the protein contents being obtained by using the 6·38 coefficient (see above). The Aschaffenburg and Drewry procedure was examined (Farah, 1979) on the basis of polyacrylamide gel electrophoresis of the fractions obtained in the different steps, as far as the non-casein-proteins of milk were concerned. This study indicated that casein precipitation with Na acetate and acetic acid, as performed by Aschaffenburg and Drewry, gives, indeed, a supernatant devoid of caseins. On the other hand, it is well known that whole casein so prepared, provided it is washed at pH 4·7, is only slightly contaminated by whey proteins (below 5%). However, this is not true for human casein (Brignon *et al.*, 1984). This study also showed that the Aschaffenburg and Drewry fraction 'β-Lg plus NPN', which remains soluble at pH 2·0 in the presence of 20 g Na_2SO_4/ml milk, indeed contains β-Lg as the only protein. Since it has been known for a long time that the NPN fraction contains most if not all the low molecular weight nitrogenous molecules present in milk (amino acids, small peptides, nucleotides, vitamins, urea, creatinine, ammonia, etc.), the following fractions can be safely determined according to Aschaffenburg and Drewry (1959) with the new conversion factors recently determined (Karman & Van Boekel, 1986; Van Boekel & Ribadeau-Dumas, 1987):

—crude protein content of milk: (nitrogen content) × 6·35
—content in milk of non-protein nitrogenous compounds: (nitrogen content of the 12% TCA supernatant) × 3·60
—whole casein content of milk: (nitrogen of milk minus that of the pH 4·7 supernatant) × 6·36
—whey protein content of milk: (nitrogen content of the above supernatant minus that of the 12% TCA supernatant of milk) × 6·28
—β-Lg content of milk: (nitrogen content of the Na_2SO_4 supernatant, pH 2, of milk minus that of the 12% TCA supernatant) × 6·29

According to the data given by Farah (1979), the other protein fractions which could be determined by the Aschaffenburg–Drewry method appear to be meaningless.

The detailed procedure has been clearly described by Aschaffenburg and Drewry (1959), taking into account, for the calculations, the dilutions, the volume of the precipitates and the change of volume caused by salt addition. Changing the volumes of test samples and reagents, even if ratios remain the same, may change the results slightly (Barbano *et al.*, 1990*b*).

Note that the Kjeldahl factors mentioned above give protein contents which include bound phosphate and carbohydrates. Therefore, the figures obtained should be slightly higher than those deduced from amino acid analysis (except for β-Lg which contains neither phosphate nor carbohydrate). Also as mentioned above, according to international agreements, a coefficient of 6·38 is currently used for crude and true protein analysis.

Since the measurement of true protein is becoming more and more a routine method, Barbano *et al.* (1990*b*) improved the classical procedure by developing a direct method. Instead of performing the precipitation in an Erlenmeyer, the TCA precipitation is made directly in a Kjeldahl flask. The precipitated proteins are removed by filtration and measured directly by Kjeldahl. This procedure, which requires one N measurement instead of two, gave better precision values than the indirect method: $RSD_r = 0{\cdot}285\%$, versus $= 0{\cdot}0483\%$ and $RSD_R = 0{\cdot}702\%$, versus $= 1{\cdot}051\%$.

3.2. Determination of Casein and Whey Proteins by Indirect Methods

Because in milk, both whole casein and serum proteins behave similarly in the presence of dyes or when submitted to IR radiation, it is usually necessary to separate caseins either by acid precipitation or by rennet coagulation and then to perform at least two measurements to estimate the concentration of casein and/or serum proteins in milk. However, with the new developments in IR spectroscopy it might be possible to measure directly casein in milk, providing 'specific' wavelengths are found.

3.2.1. Dye Binding Methods

McGann *et al.* (1972) developed a procedure for measuring casein in milk after acid precipitation (acetic acid-sodium acetate) and separation of the non-casein fraction by filtration. Determination involves three measurements: milk, milk + 1 ml of water, and milk + 1 ml of the non-casein filtrate. A correction factor of 0·04 has to be used to get an accurate casein reading when using a Pro-milk instrument calibrated for crude milk protein analysis. A standard deviation of 0·022 g/100 g was obtained with bulk milk samples by comparison with the Kjeldahl method.

A different procedure using rennet clotting was developed by several authors (Kristoffersen *et al.*, 1974; Mickelsen & Shukri, 1975). This method involves initial separation of the rennet-coagulable casein fraction by filtration or centrifugation after a small amount of rennet has been added to the milk, followed by analysis on milk and whey. Whey protein content can be measured directly using the same procedure and the same calibration as for milk protein analysis. The rennet casein content is then obtained by subtracting the whey protein content from the milk protein content; a correction factor might be necessary to get the exact rennet casein value (Kristoffersen *et al.*, 1974). In their experimental work on goat milk, Jeunet and Grappin 1985, and other unpublished experiments), used a slightly different procedure. The whey protein content was assayed by using a test sample volume equivalent to three times the volume used for milk analysis in order to get a protein concentration in the reaction mixture that was similar to that of milk.

For cows' milk, the whey true protein content (WP) is obtained using the formula: $WP = 0{\cdot}36 \times (AB) - 2{\cdot}10$, where (AB) is the Amido Black reading (in g/kg) given by the instrument calibrated for the measurement of true protein in milk. To obtain the correct whey protein content in milk, the test result on whey has to be multiplied by a factor close to 0·95 (Grappin & Jeunet, 1979), which takes into account the volume of the precipitate obtained during clotting. The rennet casein content of milk is then obtained by subtracting the corrected true protein content of whey from the true protein content of milk.

3.2.2. Infrared Methods

With the exception of Sato *et al.* (1987), who tried to measure directly milk casein with a NIR 'transflectance' instrument calibrated in per cent casein, most of the

TABLE 9
Accuracy Standard Deviation ($s_{y,x}$ in g/100 g) for the Determination of Milk True Protein, Whey True Protein and Rennet Casein in Milk using the Amido-Black, Milko-Scan and Infraalyzer Techniques

Source	*Method*	*Milk (true protein)*	*Whey (true protein)*	*Difference milk − whey (rennet casein)*
81 Individual goat milks[a]	Amido-Black	0·040	0·018	0·040
	Milk-Scan 300	0·035	0·026	0·025
40 Individual cow milks[b]	Amido-Black	0·035	0·024	0·028
	Milko-Scan 104	0·044	0·032	0·018
20 individual cow milks[c]	Amido-Black	0·043	0·015	0·059
	Milko Scan 104	0·035	0·026	0·024
	Infraalyzer 400	0·025	0·023	0·018

[a]Grappin and Jeunet (1979).
[b]Michalak and Oczkowicz (1975).
[c]Jeunet and Grappin (1985).

studies have been carried out using indirect measurements of the casein content. Because in the classical acid precipitation technique, the presence of dissociated carboxyl groups (COO^-) from acid reagents interfere at the protein wavelength(s), Jeunet and Grappin (1985) and Karman *et al.* (1987) preferred the procedure based on rennet precipitation. This method also has the advantages of a negligible sample dilution and a measurement of the casein fraction which is closely related to the cheesemaking process.

Despite the interference of the acid reagent, Sjaunja and Schaar (1984) developed a MIR procedure, using a single calibration setting and a multiple regression equation, to determine milk casein concentration from the analysis of milk and acid filtrate. Barbano and Dellavelle (1987) significantly improved the method using aqueous phosphoric acid, which does not give ionizable carboxyl group, at a concentration such that there is no dilution nor concentration.

To measure the protein content of whey, the IR instrument has to be calibrated by reference to the Kjeldahl method. If the 6·46 μm wavelength in MIR is suitable for both milk and whey analysis, a specific calibration involving different wavelengths and different coefficients may be necessary with the NIR technique (Jeunet & Grappin, 1985).

In each case, the rennet casein content is obtained by subtracting the whey protein content from the milk protein content.

3.2.3. Analytical Performances

For the Amido Black method and IR techniques, the repeatability standard deviation for the measurement of protein in whey is close to the values found for milk analysis. Because of the lower protein concentration in whey, one might expect a higher coefficient of variation. Table 9 shows examples of the accuracy of the measurement of whey proteins and rennet casein by Amido Black and IR instruments, using the Kjeldahl method as a reference. On herd milk, Barbano and

Delavelle (1987) obtained a similar accuracy with their method to measure acid casein ($s_{y,x}$ = 0·020 g/100 g) whereas Sjaunja and Schaar (1984) reported higher SD (0·048 g/100 g) on individual milks using a different procedure.

The excellent accuracy of the IR methods for measuring rennet casein will be appreciated, when compared to the other methods. One reason might be that most of the non-protein components which interfere at the protein wavelength(s) for both milk and whey analysis are in the soluble phase (see Fig. 7 for MIR) and their influence is eliminated by subtracting the whey protein test result from the milk protein test result.

4. Separate Determination and Characterization of Individual Proteins in Milk and Dairy Products

Since the technological and nutritional behavior of a protein is unique, more and more people would like to have quantitative and qualitative information (e.g. the content in the various proteins, genetic variants) on some or each main protein species in food products. Such information can now be obtained, at least in some laboratories, on raw milk and sometimes on many simple dairy products such as pasteurized and sterilized milks, caseins, caseinates, whey and WPC. The situation is more complex for fermented dairy products (cheese, yoghurt, etc.) in which partial degradation of proteins has occurred. However, information on the occurrence and amount of a milk protein can give useful information on technological treatments and fraudulent adulterations (e.g. mixtures of milks from different species).

The main techniques that have been employed during last 10 years to characterize and determine the main proteins in milk and some dairy products are electrophoresis, liquid column chromatography and immunochemical methods. For the milk enzymes of interest, determinations are usually made by activity measurements.

4.1. Electrophoresis

4.1.1. Electrophoresis Without SDS

This is the most widely used procedure for phenotyping individual cows and the most rapid one for determining the main proteins in raw milk.

4.1.1.1. Dairy species phenotyping. Electrophoresis has been used to phenotype milk proteins from cow, water buffalo, goat and ewe. We will consider here only what has been done on bovine milk.

Electrophoresis is performed in either starch, polyacrylamide or polyacrylamide-agarose. The presence of urea and a reducing agent (2-mercaptoethanol or dithiothreitol) is required for satisfactory separation of caseins. The former reagent allows dissociation of intermolecular aggregates while the latter cleaves intermolecular disulfide bridges in α_{s2}- and κ-Cns.

Genetic variants of a protein arise from heritable modifications involving the gene coding for it. As these are rare events, they lead only to a few modifications within a species. The most common modifications are point mutations; a single base in the gene DNA is replaced by another. This may lead to the substitution of one amino acid residue by another in the polypeptide chain. This substitution can be detected by electrophoresis if it involves at least one amino acid carrying an electric charge, that is, in milk proteins, Asp, Glu, P-Ser, P-Thr, Arg, Lys or His. However, Asp and Glu have a negative charge only above the pK of their lateral carboxyl groups, i.e. above pH 3 to 4, while the positive charge of His appears below the pK of the imidazole ring NH, i.e. below pH 6. P-Ser and P-Thr are negatively charged and Arg and Lys are positively charged throughout the entire pH range which can be used in electrophoresis. In other words, substitutions of P-Ser, P-Thr, Arg and Lys by a neutral amino acid can be detected both at alkaline and acid pH values (usually pH 8·6 and 3) on electrophoresis, while an acid pH must be used when His is involved. Unfortunately, only about one-third of the mutations involve charged amino acids and there is no simple method for detecting the other two-thirds. Studies on genetic variants of milk proteins have received renewed interest during the last 10 years; indeed correlations have been found between the occurrence of certain variants and some interesting technological features of milk (see Chapter 10).

Among the common variants of milk proteins, (α-La B, β-Lg A and B, α_{s1}-Cn B and C, α_{s2}-Cn A, β-Cn A^1, A^2 and B), only β-Cns A^1 and A^2, which differ by a histidine residue, have to be differentiated by electrophoresis at acid pH.

Methods have been published for the simultaneous characterization of the variants of the main whey proteins (α-La, β-Lg) and caseins (α_{s1}, α_{s2}, β and κ) by electrophoresis at alkaline pH. Unfortunately, this does not allow differentiation of β-Cn A^1 and A^2. Furthermore, the electrophoretic patterns so obtained are not easily interpretable. Nowadays, most people perform 3 different electrophoretic separations for each sample to be tested: at alkaline and acid pH values on milk or pH 4·7-precipitated casein in the presence of urea and reducing agent, for typing caseins and at alkaline pH on milk or acid whey, without the last two reagents, for typing whey proteins.

Electrophoresis is usually run for 4 h (or overnight) at alkaline pH and overnight at acid pH. A large number of samples can be analyzed simultaneously. Staining is performed using Coomassie Blue (R or G) or Amido Black.

Information concerning the chromatographic procedures that can be used for phenotyping or quantifying bovine milk proteins, the frequencies of the various alleles of each main milk protein in different breeds and the association of genetic variants with some milk characteristics can be found in the publications by McKenzie (1971), Swaisgood (1982), Grosclaude (1979, 1988), Ng-Kwai-Hang *et al.* (1984, 1987), Ng-Kwai-Hang and Kroeker (1984), McLean *et al.* (1982, 1984) and Gripon *et al.* (1975) (see Chapter 10).

4.1.1.2. Quantitation of milk proteins. This has mainly been performed on individual fresh bovine milks for genetic purposes. The techniques are identical with those mentioned above. However, as densitometric scanning is required, acrylamide or acrylamide-agarose gels are preferred to starch gels which are not transparent. Separate electrophoresis at alkaline pH values are carried out on pH 4·7-precipitated casein (with urea and reducing agent) and whey. When vertical slab gels are used, para-κ-Cn, which often occurs in milk in small amounts, cannot be determined since it migrates upwards at alkaline pH. Amido Black 10 B or Coomassie Blue G 250 are used to stain proteins. For determination of the absolute amount of proteins following scanning densitometry, which gives areas proportional to the amounts, it is necessary to know their dye binding capacities. These are: 345, 290, 310, 425, 440, 430 and 320 mg dye/g protein with Amido Black for αs-, β-, κ-Cns, β-Lg, α-La, BSA and Igs, respectively (see Table 3, Tarassuk *et al.*, 1967), αs-Cn being the sum $\alpha_{s1} + \alpha_{s2}$. With Coomassie Blue G 250, the following relative figures have been determined: α_{s1}B = 1·00; α_{s1}C, 0·94; α_{s2}, 0·98; βA^1, 0·69; βA_2, 0·66; βB, 0·71; para-κ, 0·76 (McLean *et al.*, 1982, 1984). In the last case, whey proteins were determined by radial immunodiffusion. Furthermore, as horizontal gels were used, the concentrations of α_{s1}-, α_{s2}- and β-Cns were determined in samples not treated with chymosin (CV = 5% expressed as a percentage of total casein) and those of γ- and κ-Cns in chymosin-treated samples (CV = 14%).

On electrophoresis, α_{s1}-, α_{s2}- and κ-Cns give 2, 4 and 1 major and several minor bands, respectively. The several κ-Cn bands are transformed by chymosin into one (or two) para-κ-Cn bands. When κ-Cn is determined directly on vertical gels (without chymosin treatment), only the major band is taken into consideration.

4.1.2. Electrophoresis in the Presence of SDS

SDS strongly binds to proteins, mainly through hydrophobic interactions. The amount that is fixed is approximately proportional to the weight of protein: ~1·4 g SDS/g protein (Reynolds & Tanford, 1970). Thus, any protein molecule will bind a large number of SDS molecules, each of which carries a negative charge. The indigenous net charge of the protein at any pH is thus made negligible. Any protein should migrate at the same velocity towards the anode in free-flow electrophoresis in the presence of SDS. However, in zone electrophoresis, in particular in acrylamide gels, the larger a protein, the lower its electrophoretic mobility because of the sieving action of the gel.

This technique is widely used to determine Mr of proteins since the logarithm of Mr is proportional to the distance of migration (Shapiro *et al.*, 1967). Mr is easily obtained by running an electrophoregram of the protein(s) of interest and of a mixture of proteins of known molecular weights, and by plotting log Mr versus distance of migration.

By using polyacrylamide gels with an acrylamide concentration increasing linearly from cathode to anode, it is possible to cover a wider range of Mr values than with homogeneous gels. In this case, a linear relationship exists between log Mr and log (distance of migration) (Rodbard *et al.*, 1971).

Contrary to what would be expected from their molecular weights, the 4 bovine caseins can be separated by SDS-PAGE in the presence of a reducing agent (Green & Pastewka, 1976; Trieu-Cuot, 1981; Miranda, 1983; Creamer & Richardson, 1984), giving four distinct bands corresponding, in the order of increasing mobility, to: α_{s1}-, α_{s2}-, β- and κ-Cns. Furthermore, the following other proteins, all of which have mobilities higher than those of the caseins, can also be distinguished by SDS-PAGE of milk: γ1-Cn, β-Lg, α-La + para-κ-Cn, γ2- + γ3-Cns in the order of increasing mobilities (Miranda, 1983). To the authors' knowledge, this method, which is widely used in the authors' laboratory for other purposes, has never been employed for quantitative determination of milk proteins. However, it could give, by scanning densitometry of a single gel, the absolute amounts of the main proteins present in individual or bulk milks.

4.1.3. Isoelectric Focusing (IEF)

Although IEF was applied to casein fractions in 1969, the first clear identification of the various protein components which can be separated from whole casein by this technique appeared in a paper by Trieu-Cuot and Gripon (1981). IEF was performed in 1-mm thick polyacrylamide gels containing ampholytes, 7M urea and 0·1% 2-mercaptoethanol. The following components were identified, in order of decreasing isoelectric pHs: γ-Cns (the 3 known components), κ-Cn (2 components), α_{s2}-Cn (the 4 known components), β-Cn and α_{s1}-Cn (several components, including α_{s0}). IEF was used for phenotyping in a single run all milk proteins in ultrathin-layer polyacrylamide gels (Seibert *et al.*, 1985). The following variants could be detected: α_{s1}-caseins A, B, C; α_{s2}-CnA; β-Cns A^1, A^2, A^3, B, C; κ-Cn A, B, α-La B; β-Lgs A, B, C. Table 10 gives the isoelectric pHs that have been measured (Trieu-Cuot & Gripon, 1981; Seibert *et al.*, 1985).

The method described by Seibert *et al.* (1985) appears quite promising. It is particularly short: 15 min prefocusing and 30 min focusing. A number of samples can be treated simultaneously. Prior to focusing, defatted milk was diluted 1 : 11 with a solution of 8M urea in distilled water containing 3% (v/v) 2-mercapto-ethanol.

Rapid identification of the genetic variants of milk proteins using the PhastSystem (Pharmacia) was recently reported by two groups. Vegarud *et al.* (1989) analyzed separately the caseins (10% Servalytes, pH 4·0–6·0, 4·5–5·0, 5·0–7·0, 1/1/1, v/v/v; 8M urea, 2·5% Triton X-100, and 2-mercaptoethanol + urea in the samples) and whey proteins (Phastgel IEF, pH 3–9 without urea) from 350 individual milk samples. All the variants mentioned above were detected. Bovenhuis and Verstege (1989) separated all the milk proteins on a single gel (16% ampholytes, pH 4·2–4·9, 4·5–5·4, 3·5–5·0, 1/1/1, v/v/v; 8M urea; 0·8% Triton X-100, and 2-mercaptoethanol + urea in the samples of whole milk). The genetic variants mentioned above formed sharp distinct bands except for α-La B and β-Lg A which focused together. Thousands of milk samples were analyzed successfully in this way and one person could analyze 240 milk samples a day.

4.1.4. Two-Dimensional Electrophoresis

This technique is especially useful for analyzing qualitatively complex mixtures of proteins, by taking advantage of two different criteria simultaneously, generally isoelectric pH (or electrophoretic mobility) and Mr. It was applied to caseins by Trieu-Cuot and Gripon (1981). They used isoelectric focusing (pH 4 to 9) in the

TABLE 10
pHi Values of Milk Proteins Determined by Isoelectric focusing

Protein	*Interval at probability level of 5% after correction*[a]	
	Seibert et al. *(1985)*	*Trieu-Cuot and Gripon (1981)*
α_{s1}-Cn A	4·16–4·40	—
B	4·23–4·47	4·44–4·76
C	4·27–4·49	—
α_{s2}-Cn A	4·83–5·13	—
β-Cn A^1	4·68–4·96	—
A^2	4·60–4·84	4·83–5·07
A^3	4·50–4·74	—
B	4·78–5·10	—
C	4·97–5·29	—
κ- Cn A	5·43–5·81	5·45–5·77
B	5·54–6·12	—
α- La B	4·66–4·89	—
β- Lg A	4·64–4·90	—
B	4·72–4·98	—
C	4·77–5·13	—
γ-Cn1	—	5·55–5·87
γ-Cn2	—	6·38–6·72
γ-Cn3	—	6·01–6·29

[a]Correction for pH shift due to urea was made by subtracting the pH difference between an ampholyte solution and the same solution plus urea, ampholytes and urea being at the same concentration as in the gel.

first dimension and PAGE (from 1% to 28% acrylamide) in the presence of 0·1% SDS and 4·9M urea. Good resolution was obtained for α_{s1}-Cn (2 fractions with same Mr), α_{s2}-Cn (several fractions with same Mr), β-Cn, κ-Cn (3 fractions with same Mr), γ2 and γ3 fractions (same Mr) and γ1.

Miranda (1983) showed that two-dimensional electrophoresis of milk proteins (fraction insoluble in 12% TCA) separated the caseins mentioned above as well as para-κ-Cn, β-Lg and α-La. UHT milk gave a similar pattern. However κ-Cn, α-La and β-Lg produced fainter spots. With sterilized milk, only α_{s1}- and β-Cns remained visible. A high Mr component, which could not penetrate the gel, appeared. It may correspond to a polymer produced by Maillard reaction between milk proteins and lactose (Andrews & Cheeseman, 1971).

The main interest of two-dimensional electrophoresis is its high resolution which can be exploited to study complex hydrolysates of milk proteins. Thus, Trieu-Cuot and Gripon (1981) and Miranda (1983) used it to follow cheese ripening, and proteolytic breakdown of milk proteins during in-vivo digestion, respectively.

4.2. Column Chromatography

A review dealing with the chromatographic separation of milk proteins was published by Yaguchi and Rose (1971). At that time, only gel filtration and ion

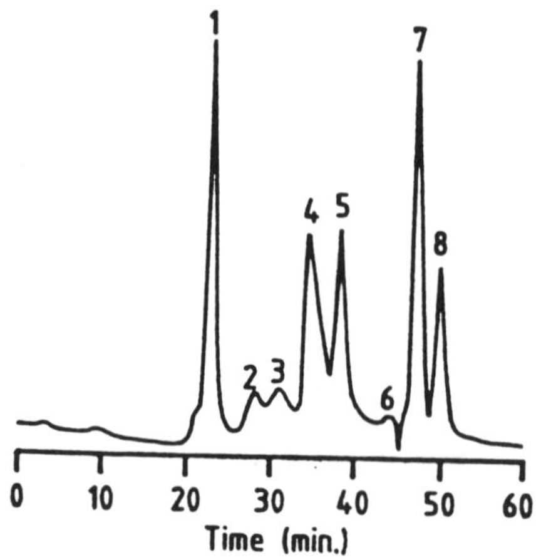

Fig. 8. FPLC Gel filtration of skim milk on a tandem of 7·5 × 300 mm Toyo-Soda 3000 SW plus 2000 SW columns (Dimenna & Segall, 1981). Flow rate: 0·5 ml/min; 10 μl injected; eluent: 0·05M phosphate buffer, pH 6·80, containing 0·1M Na_2SO_4; detection at 280 nm. 1, Cns; 2, IgGs; 3, BSA; 4, β-Lgs; 5, α-La; no protein in 6–8.

exchange in open columns were used. Nowadays, the tremendous advances in the technology of chromatographic stationary phases and equipment have led to an almost complete superseding of open columns by HPLC columns in which stationary phases made of spherical, small-diameter particles are used at high pressure. Fast analyses at high sensitivity can thus be obtained. This is especially interesting in the analytical field. Furthermore, new procedures, such as reversed-phase chromatography and chromatofocusing, have appeared. An extensive review article has been recently published by Gonzales-Liano *et al.* (1990) which describes the analysis of most dairy products by HPLC and FPLC.

4.2.1. Gel Filtration

Direct analysis of skim-milk in phosphate buffer, pH 7·0 at 4°C on Sephadex G 200 separated caseins into 3 fractions ($\kappa + \alpha_s + \beta$, $\alpha_s > \beta$, $\beta < \alpha_s$). β-Lg and α-La followed caseins in this order. Resolution was poor (Yaguchi & Tarassuk, 1967). Several attempts at separating skim milk proteins on high performance gel filtration columns have been made (Dimenna & Segall, 1981; Shimazaki & Sukegawa, 1982; Gupta, 1983). The authors believe that the best results for analytical purposes were obtained by Dimenna and Segal (1981) using a tandem of 2 TSK columns from Toyo-Soda (Biorad, Richmond, CA) (Fig. 8). Gupta (1983) studied the fractionation of skim milk proteins on a TSK 3000 SW column, either in their native state, or after denaturation by SDS and 2-mercaptoethanol. According to Gupta, the first procedure allowed quantitation of the individual whey proteins within 10%. A method of calculation was given for determining, by using the second procedure, the concentration of the individual proteins in a mixture of casein and whey proteins.

Indeed gel permeation is not suitable for analyzing skim milk and whole casein.

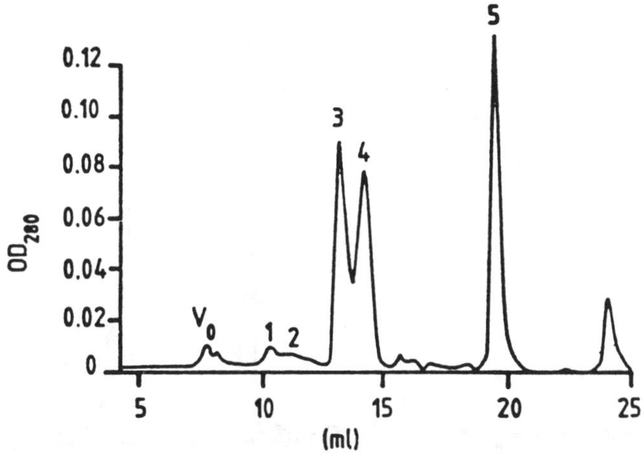

Fig. 9. FPLC separation of whey proteins by gel filtration on a Superose 12 column (Andrews *et al.*, 1985). Fifty microliters of fresh acid whey injected; flow rate: 0·5 ml/min; eluent: 0·1M Tris-HCl buffer pH 7·0 containing 0·5M NaCl and 10 mM NaN_3. 1, Igs; 2, BSA; 3, β-Lgs; 4, α-La; 5, orotic acid.

In milk, caseins occur as micelles of various sizes, together with some soluble casein (aggregates). Removal of calcium dissociates the micelles into large polydisperse aggregates containing various proportions of the different caseins. Finally, in the presence of urea and reducing agent, the monomeric caseins are obtained. Their Mr values are too close to each other and too close to those of some of the whey proteins (β-Lg, Igs light chains) for satisfactory separation in gel permeation.

However, whey proteins are quite well separated by such a method which permits their quantitation (Dimenna & Segall, 1981; Shimazaki & Sukegawa, 1982; Gupta, 1983; Humphrey, 1984; Andrews *et al.*, 1985). In a study by Andrews *et al.* (1985), using a Superose 12 column (Pharmacia, Uppsala, Sweden), satisfactory separation of the main whey proteins and orotic acid was obtained (Fig. 9). Urea measurements led to concentration values of Igs = 0·51, BSA = 0·56, β-Lg (B + A) = 3·25, α-La = 1·25 mg/ml, in good agreement with previously published values.

4.2.2. Ion Exchange

No satisfactory result has been obtained for fractionating skim-milk proteins directly by ion exchange. However, both whole casein and whey proteins can be separated under conditions suitable for quantitative analyses.

With open columns, caseins were separated on anion or cation exchangers (e.g. Mercier *et al.*, 1968; Annan & Manson, 1969) in the presence of urea and 2-mercapto-ethanol. Anion exchangers are now by far the most frequently employed. Using DEAE-cellulose, a procedure was developed by Davies and Law (1977*a*) for quantitative analysis of the components of reduced and alkylated (iodoacetamide) whole casein. Satisfactory separation of γ2-, γ1- + γ3-, κ-, β-, α_{s2}- and α_{s1}-Cns, eluted in that order by a NaCl gradient in TRIS buffer, pH 8·6, containing 6M urea,

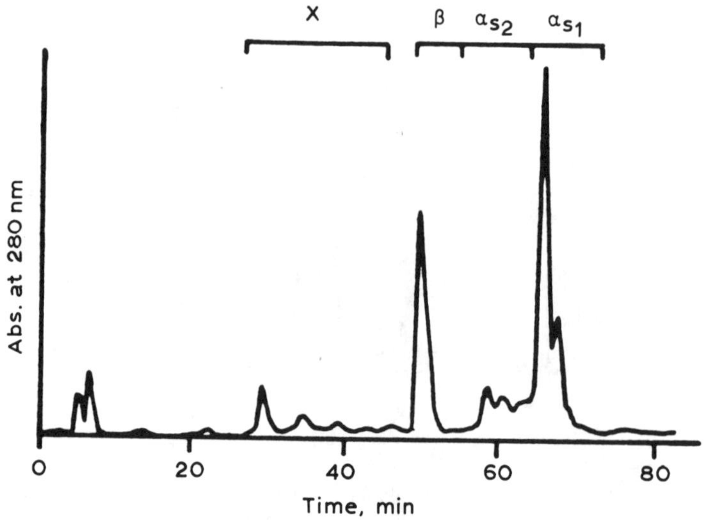

Fig. 10. FPLC separation of whole casein by anion exchange on a DEAE-TSK-5PW column (Aoki *et al.*, 1987). Conditions are given in the text. Column size: 7·5 × 75 mm.

was obtained. The protein content of each casein fraction was determined by a micro-biuret method, from the specific extinction coefficients of the corresponding copper–protein complexes. This procedure was used to determine the proteins mentioned above in whole casein from a number of individual, herd bulk and creamery milks (Davies & Law, 1977*a*, *b*, 1980). The agreement between duplicate analyses and the average recovery (95·66 ± 1·3%) were both satisfactory, but long runs (~ 20 h) were necessary. Quite similar fractionations were recently obtained from reduced non-alkylated whole casein by high performance anion exchange on DEAE-TSK-5PW (Toyo-Soda, Japan) or Mono Q HR 5/5 (Pharmacia, Uppsala, Sweden) columns under conditions similar to those mentioned above but with the addition of a reducing agent to buffers, with elution times of ~ 1 h (Humphrey & Newsome, 1984; Visser *et al.*, 1986; Guillou *et al.*, 1987). Limited quantitative data were given by Humphrey and Newsome (1984): in the sample of casein analyzed, α_{s1}-casein constituted 39% (w/w) and β-casein 36% (w/w) of the total proteins. Protein recovery determined for these two caseins was greater than 96%. An excellent separation of whole casein was obtained by Aoki *et al.* (1987) using a DEAE-TSK-5PW column, like Humphrey and Newsome (1984), but under different conditions (Fig. 10). The buffer was 0·02M imidazole, 3·3M urea, 0·08M NaCl, pH 8·0, and elution was achieved by a linear NaCl gradient from 0·08 to 0·28M over 70 min at 0·5 ml/min, room temperature. Whole casein (1%) was dissolved in the starting buffer containing 20mM 2-mercaptoethanol and left overnight at 5°C; 50 l were injected. Quantitative determination of the different caseins was made by peak area measurements using the following extinction coefficients:

$E^{1\%}_{1\text{cm}}$: α_{s1}-Cn, 10·5; α_{s2}-Cn, 11·0; β Cn, 4·6; κ-Cn, 9·5 (Swaisgood, 1982)

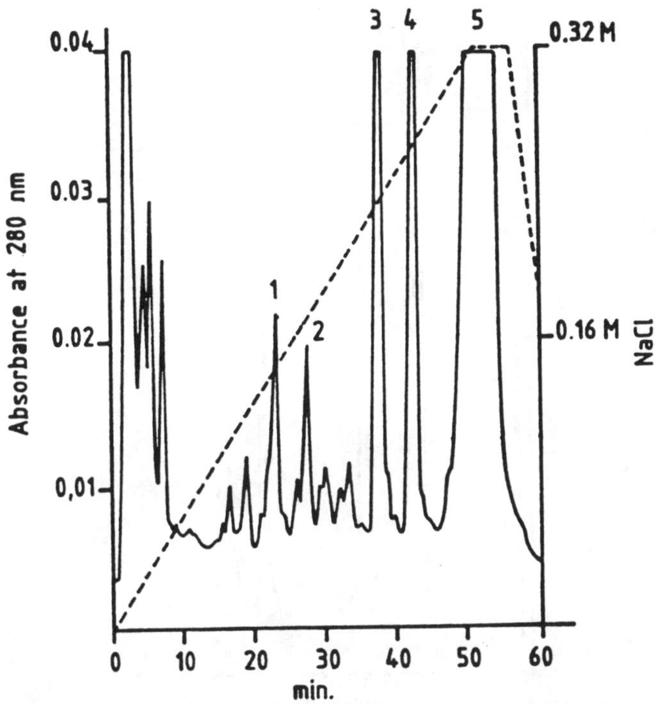

Fig. 11. FPLC separation of individual whole casein (κ A/B, β C/A[1], α_{s2} A, α_{s1} B) by anion exchange on a Mono Q column (Guillou *et al.*, 1987). Casein sample in 5×10^{-3}M Tris-HCl, 4·5M urea buffer pH 8·0, 8×10^{-4}M dithiothreitol; flow rate: 1 ml/min; 40°C; elution with the same solution as above with a 0–0·32M NaCl gradient. 1, κ_0-Cn B; 2, κ_0-Cn A; 3, β-CN C; 4, β-CN A[1]; 5, α_{s2}-, α_{s1}-, α_{s0}-Cns. Overloading allows clear visualization of the κ-Cn fractions, but decreases resolution between α_{s2}-(left part of the main peak), α_{s1}-(main peak), α_{s0}-(right part of main peak) Cns.

The following figures (means of two determinations), expressed as per cent of whole casein, were obtained: α_{s1} -Cn: 38·2; α_{s2} -Cn: 11·0; β-Cn: 39·5; κ-Cn: 11·3. Figure 11 shows the separation of an individual casein sample on a mono Q column in the presence of urea and reducing agent. In this case, some genetic variants were separated (Guillou *et al.*, 1987).

Although it has been known for a long time that satisfactory separation of whey proteins can be obtained on DEAE-cellulose, no quantitative data are available on the protein composition of whey deduced from chromatography on open anion exchange columns. By anion exchange HPLC, with Toyo-Soda DEAE-5PW or Pharmacia Mono Q HR 5/5 columns good separations were obtained for the main whey proteins (Humphrey, 1984; Humphrey & Newsome, 1984; Andrews *et al.*, 1985; Manji *et al.*, 1985) (Fig. 12). Concentrations determined from chromatography of a sample of acid whey on mono Q column were 2·1, 1·3 and 0·8 mg/ml for β-Lg A, β-Lg B and α-La, respectively (Humphrey & Newsome, 1984).

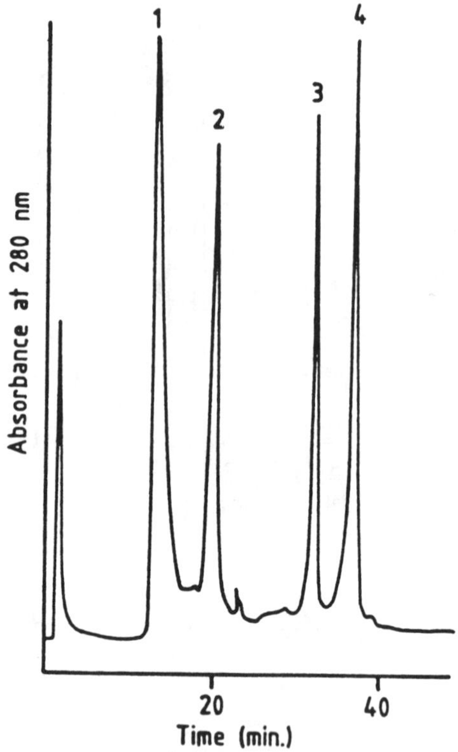

Fig. 12. FPLC separation of acid whey by anion exchange on a Mono Q column (Humphrey & Newsome, 1984). Acid whey sample; flow rate: 0·5 ml/min; elution in 0·02M piperazine buffer, pH 6·0, with 0–0·4M NaCl gradient. 1, Orotic acid; 2, α-lactalbumin; 3, β-lactoglobulin B; 4, β-lactoglobulin A.

4.2.3. Chromatography on Hydroxyapatite (HA)

This stationary phase is especially interesting for fractionating caseins, since it separates them according to their phosphate content. A number of casein samples from herd milks were analyzed on open columns yielding the following fractions: γ + para-κ-Cns + unidentified chymosin-resistant components, unidentified minor protein, β-Cn, α_{s1}- + α_{s2}-Cns. The quantitative contribution of chymosin-resistant components to the κ-Cn fraction was deduced from chromatography of chymosin-treated whole casein. The analysis time was ~10 h. HA-HPLC on a column of Bio-Gel HPHT (Biorad, Richmond, CA) gave quite similar results with better resolution in ~45 min. Partial separation of κ-A and κ-B variants was observed. α_{s2}-Cn was poorly separated from α_{s1}-Cn (tailing part of the α_{s1}-casein peak) but a different elution gradient could produce better separation (Kawasaki *et al.*, 1986; Visser *et al.*, 1986). The main problem appears to be the life-time of HA-HPLC columns.

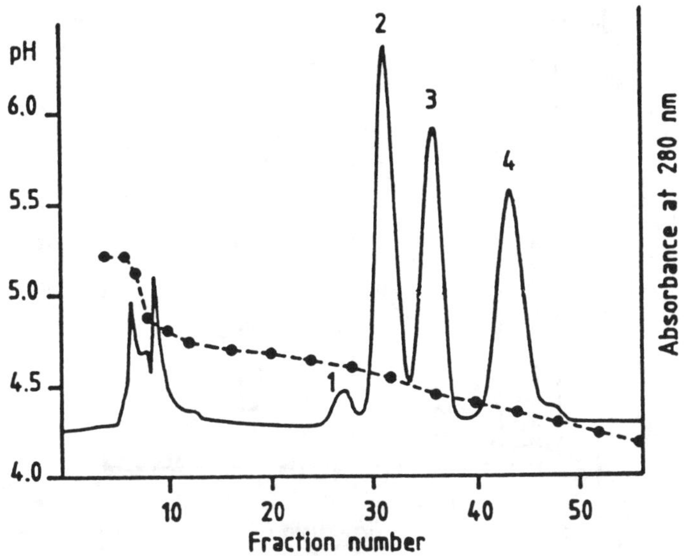

Fig. 13. Chromatofocusing separation of acid whey (Pearce & Shanley, 1981). Sample of milk serum concentrated and dialyzed against starting buffer; columns of Polybuffer exchanger 94 equilibrated in 0·025M histidine/HCl buffer pH 5·2; 5 ml of Polybuffer 74 (diluted 1 : 10 in water and adjusted to pH 4·2) were run on the column before the sample was loaded. Elution with Polybuffer/HCl. Flow rate: 0·32 ml/min. 1, BSA; 2, β-Lg B; 3, β-Lg A; 4, α-La.

4.2.4. Cromatofocusing

To the authors' knowledge, this powerful technique has been employed only for analytical and preparative separations of whey proteins on an open column of Polybuffer Exchanger 94 (Pharmacia, Uppsala, Sweden). With a pH gradient from 5·2 to 4·2, excellent separation of the main whey proteins was obtained from acid whey (Pearce & Shanley, 1981, Fig. 13). In the authors' opinion, fast and resolutive separation could be obtained with the Mono P column supplied by Pharmacia.

4.2.5. Hydrophobic Interaction (HI) and Reversed-Phase (RP) HPLC

Although both techniques rely upon hydrophobic interactions between a stationary phase and the solutes to be fractionated, their applications are quite different. With HI-HPLC, fixation of the solutes on the phase is carried out in aqueous solution at high ionic strength and elution is achieved by lowering the ionic strength of the mobile phase. In RP-HPLC, fixation occurs in aqueous solution of low ionic strength and elution is obtained by increasing the hydrophobicity of the mobile phase (increasing the proportion of a low-polarity solvent). In HI-chromatography, the stationary phase is made of a polar material to which non-polar groups (such as phenyl) are covalently attached, while the phases used in RP-HPLC are made of silica with non-polar group (such as C8 or C18 alkyl chains) similarly attached. Unsubstituted residual silianol groups are subsequently 'end-capped' by groups such as trimethylsilyl: in theory, RP-HPLC stationary phases can therefore interact with solutes only by hydrophobic interactions.

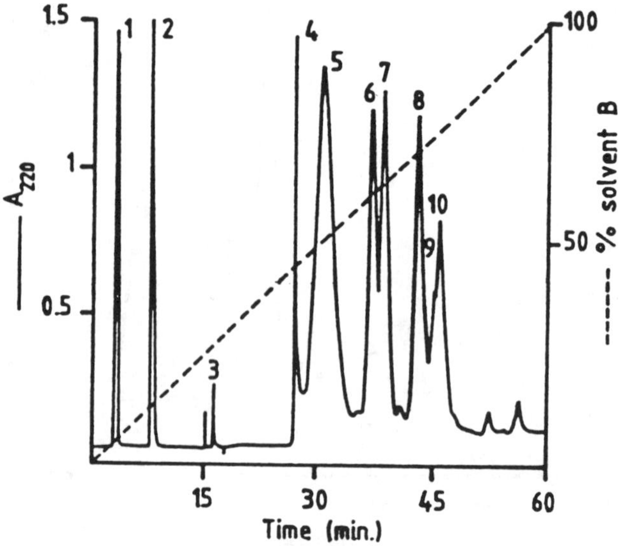

Fig. 14. RP-HPLC separation of an individual whole casein on a μ-Bondapak C 18 column (Carles, 1986). Casein sample: 1·25 mg in 50 μl of 10 mM Na phosphate buffer, pH 7·2, 1% 2-mercaptoethanol, 0·1% NaN_3; flow rate: 1·5 ml/min; solvent A: 10 mM Na phosphate, 10 mM SDS, pH 7·2; solvent B: 2-propanol/solvent A (2/1; v/v), 10 mM SDS, pH 7·2. 1, NaN_3 and phosphate; 2, mercaptoethanol; 3,4, no protein; 5, α_{s1}-casein; 6,7, β-casein variants (substitution Pro in 6 → Leu in 7); 8, α_{s2}-casein; 9,10, κ-casein.

Whole casein and whey proteins were each separated by Chaplin (1986) on a phenyl-Superose HR 5/5 HI column from Pharmacia. With whole casein, elution was performed with a 0·8–0·5 M, pH 6·0, sodium phosphate gradient containing 3·75 M urea. For whey proteins, a 1·5–0·05 M gradient of ammonium sulfate in 0·05 M sodium phosphate, pH 7·0, was used. Run times were ~ 50 and 30 min, respectively. In the authors' opinion, the poor resolution thus obtained prevents analytical applications.

A few studies have been made on the behavior of whole casein (Carles, 1986; Visser *et al.*, 1986) and whey proteins (Pearce, 1983; Humphrey, 1984) in RP-HPLC. For caseins, Visser *et al.* (1986) compared 2 columns from Biorad (Richmond, CA): a 25-cm RP-318 (C18-alkylated, silica-based, 33 nm pores, 5 mm particle size) and a 7·5-cm TSK-Phenyl RP (phenyl hydroxylated-polyether, 100 nm pore size, 10 mm particle size). A sample of whole casein in a pH 7 buffer solution containing urea and 2-mercaptoethanol was left standing for ~ 1 h and then injected. Solvent A was 0·1% TFA in 10% CH_3CN; solvent B was 0·1% TFA in 90% CH_3CN. Elution was achieved by a linear gradient from A to B, with recording at 220 nm. The best resolution was obtained with the RP-318 column, the caseins being eluted in the following order: κ (3 peaks), α_{s2}, α_{s1}, $\beta + \gamma$ (several peaks), in ~ 30 min. Starting with reduced whole casein, Carles (1986) obtained good resolution on a 30-cm μ-Bondapak C18 column at 40°C with SDS as a counter-ion (Fig. 14). The caseins were eluted in the following order: α_{s1}, β, α_{s2}, κ.

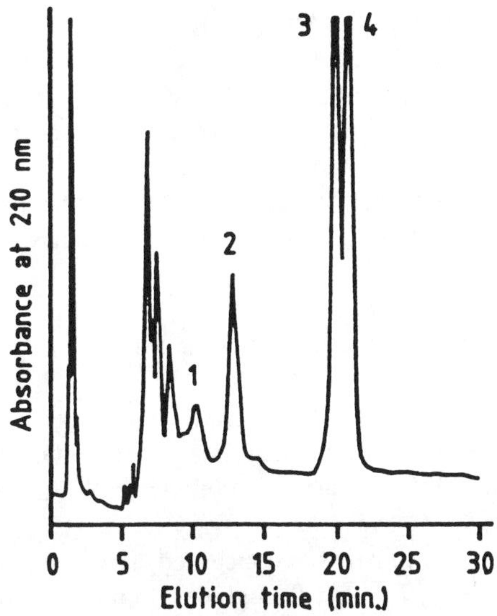

Fig. 15. RP-HPLC separation of defatted cheese whey on a Spherisorb C6 column (Pearce, 1983). Sample of whole, defatted cheese whey adjusted to pH 2·1; flow rate: 1 ml/min; solvent A: 0·15 M NaCl/HCl pH 2·1; solvent B: acetonitrile. Elution by multistage linear gradient from 0 to 48% B. 1, BSA; 2, α-La; 3, β-Lg B; 4, β-Lg A.

Using this procedure, two genetic variants of β-casein, differing by a substitution Pro → Leu, were separated and characterized from an individual casein sample.

Figure 15 shows the fractionation of cheese whey on a C6 column (Pearce, 1983).

4.3. Immunological Techniques

Whey proteins behave like other monomeric or oligomeric globular proteins: they are highly antigenic and their determination by usual techniques such as Mancini's double diffusion (Mancini *et al.*, 1963), rocket-immunoelectrophoresis (Laurell, 1966), enzyme-linked immunosorbent assay (ELISA) (Ruitenberg *et al.*, 1976), radioimmunoassay (Bennett & Mohla, 1976) is straightforward. Although quite feasible, the determination of the different caseins by immunological techniques is more difficult because these proteins are poor antigens and they occur as aggregates (self aggregation or aggregation with each other).

A few examples of the use of immunological techniques are given below. McLean *et al.* (1984) used radial immunodiffusion for the determination of α-La and β-Lg in a study of the effect of milk protein genetic variants on milk yield and composition. The same technique was used extensively for assaying Igs (antisera against bovine IgG1 and IgG2 are now commercially available) (e.g. Norman & Hohenboken, 1981). Ig determination is useful for detecting colostrum in milk. LF was determined in milk by radial immunodiffusion, rocket immunoelectrophoresis (e.g. Welty *et al.*, 1976), and by ELISA in human serum (Hetherington *et al.*, 1983). A radioimmunoassay was developed for α-La (Beck & Tucker, 1977). β-Cn was quantified in a number of individual human milk samples by rocket immunoelectrophoresis (Chtourou *et al.*, 1985). Using the same technique and specific antisera against the four goat caseins (α_{s1}, α_{s2}, β, κ), Grosclaude *et al.* (1987), quite recently, found a Mendelian polymorphism underlying quantitative variations of goat α_{s1}-Cn.

Immunological techniques are by far the most specific tools for determining proteins, provided they are used carefully. Their use in dairy technology will likely increase in the near future.

4.4. Determination of Some Indigenous Milk Enzymes

For many years, and still now, the determination of milk enzymes has been used to assess milk quality (detection of mastitis, heat treatment evaluation). The reader could refer to Kitchen (1985) for a review on the milk indigenous enzymes (see Chapter 8).

As far as the detection of mastitis is concerned, Kitchen (1985) found that this disease induced an elevated N-acetyl-β-D-glucosaminidase activity in milk. Rapid and simple diagnostic tests have been developed for routine use in mastitis monitoring programmes. As indicated earlier, the level of catalase in milk varies with the somatic cell count, and thus a measure of its activity has been also used to detect mastitis (Kitchen, 1985). This is often performed using a 'Catalasemeter', a device which measures the time taken for a filter paper disc soaked in the sample and immersed in H_2O_2 solution to evolve enough oxygen to bring the disc to the surface. The catalase activity is inversely proportional to this time (Dubois *et al.*, 1982). However, the indications given by this assay, although correlated to some extent to the hygienic quality of milk (microorganisms too have catalase activities), are often difficult to interpret: indigenous catalase activity is highly dependent upon individuals, breed, feeding and lactation period. Its level is high in colostrum and in late lactation milk.

The efficiency of pasteurization is monitored routinely by the alkaline phosphatase (AP) test which was discovered in 1933. All milks contain bovine AP. It appears that this enzyme is more heat-resistant than non-spore forming pathogens which can be present in milk. A number of methods for measuring AP activity in milk and dairy products have been published (Richardson, 1983; Seng Kwee, 1985). Most are based on the release by the enzyme of phenol from disodium phenyl phosphate or para-nitrophenyl phosphate, or phenolphtalein from phenolphtalein monophosphate. Phenol is measured colorimetrically after reaction with 2,6-dichloroquinone chloroimide, which gives indophenol blue. Phenolphtalein is detected by NaOH addition.

These tests must be performed carefully since they can easily give false responses: occurrence of phenol in chemicals and glassware, interfering coloured materials, 'reactivation'. Milks can give a negative phosphatase test immediately after heating and cooling, or when stored at 4°C or below. However, the same milks can give a positive test when inadequately stored, e.g. at or above 10°C for extended periods. Such 'reactivated' AP can be differentiated from residual AP: storage at 34°C in the presence of Mg salts gives 4 to 10 times more AP activity from reactivation than the same product stored without Mg salts (Richardson, 1983). The National Health and Medical Research Council Food Standards Code (1987) prescribes the use of the method of Aschaffenburg and Mullen (1949) (A & M) for the analysis of AP to test for adequate pasteurization. This method is based on the colorimetric determination of p-nitrophenol released from a phosphoester. The code mentioned above states that pasteurized cows' milk shall not liberate more than 10 mg p-nitrophenol/ml of milk. A modification of the A & M procedure, which is adapted to goat milk, was recently published (Williams & Nottingham, 1990). Indeed, raw goat milk contains substantially lower amounts of AP than cow milk.

From the work of Griffiths (1986) it appears that lactoperoxidase (LP) determination offers the most promising method for detecting heat treatments of the order 76°C for 15 s. This enzyme was formerly determined by colorimetric measurement of the reaction products of guiacol or o-dianisidine. The best substrate of LP now appears to be ABTS (Boehringer, Germany) which can be used as indicated by Shindler and Bardsley (1975). It was reported that, depending on the temperature of inactivation, some LP activity may be restored after storage. The more severe the heat treatment, the smaller the degree of subsequent regeneration. However, storage at 4°C for 24 h yielded no restoration of activity following heating at 65 to 80°C with a holding time of 15 s. LP activity does not vary greatly among raw milk samples (Griffiths, 1986).

Plasmin activity of milk appears to be beneficial in some dairy products and detrimental in others. This means that a number of dairy manufacturers should be interested in a simple test for assessing plasmin activity in milk. A rapid, sensitive assay for plasmin in dairy products is available (Richardson, 1983). However, both the substrate (N-succinyl-L Ala-L Phe-L Lys-7-amino-4-methyl coumarin) and the required equipment (fluorimeter) are expensive. Furthermore, milk contains plasmin, plasminogen (Pg), Pg activator, plasmin inhibitors and, likely, Pg-activator inhibitor. Thus, the plasmin content is of little significance, since Pg, always present in milk in higher amount, can give plasmin during milk processing. Pg determination is quite feasible: plasmin assays are performed before and after Pg activation by urokinase. However, the fate of Pg and plasmin during processing is unknown: Pg could be converted into plasmin, but Pg activator and plasmin may also be inactivated.

5. Some Applications of Milk Protein Analysis

A number of such applications have been given throughout this review. These applications will be classified here and some that have not been mentioned previously will be added.

5.1. Milk Protein Analysis in Dairy Products

5.1.1. Raw Milk

Milk is analyzed routinely in many countries for breeding purpose and when it is paid for according to its composition and hygienic quality. In this case, protein determination is especially important. In addition, the presence of colostrum, which may cause problems in cheese manufacture, is often sought (detection of Igs by immunological methods). Information on the levels of some enzymes, indicating the occurrence of mastitic milk (N-acetylglucosaminidase, catalase) or potential proteolytic activity (Pg, plasmin) is often useful. The action of plasmin on α_{s2}-Cn was studied by Le Bars and Gripon (1989). The determination of specific peptides from α_{s2}-Cn by RP-HPLC was recently used to determine the activity of plasmin (Le Bars, D., 1991, pers. comm.). Quite recently, Igarashi (1989) used γ-casein determination to investigate proteolysis in milk. The soluble fraction obtained from 1·5 ml skim milk in a solvent system (50% ethanol, v/v; 0·4 M Na thiocyanate: 0·15 M $CaCl_2$) was chromatographed on a small DEAE-cellulose column and eluted with 0·02 M Tris-HCl buffer, pH 8·0, containing 0·03 M NaCl and 6M urea. The absorbance of the recovered fraction (5 ml), which corresponded to γ-casein only, was recorded at 280 min. This method is applicable also to heated skim milk.

A number of individual milks have been, and still are, analyzed to find relationships between genetic variants and technological properties.

5.1.2. Processed Fluid Milk

Heat treatments affect both milk microbial flora and whey proteins; caseins are very resistant to such treatments. As mentioned above, the extent of inactivation of some milk enzymes can indicate the extent of heat treatments. The effects of UHT and HTST pasteurization of milk on the properties of its proteins have been studied. The solubility of caseins is reduced by UHT treatment. Analysis of whey protein nitrogen shows significant protein denaturation. No significant losses in nutritive value were found, and differences in viscosity and emulsification capacity were small (Douglas *et al.*, 1981). It has been suggested that residual plasmin activity (this enzyme is fairly heat-resistant) in UHT milk may be one of the factors involved in gelation during storage (see Chapter 8).

5.1.3. Dry Milk Products

Protein analyses are extensively used for the classification and evaluation of products such as milk powders, non-fat dry milks (NFDM), caseins, caseinates and whey protein concentrates (WPC). In particular, the extent of heat denaturation of whey proteins is often measured. Most methods determine the proportion of soluble whey proteins under various conditions. Whey proteins are considered to be in their native state when they remain soluble after heat treatments. Harland and Ashworth (1947) developed a turbidimetric method for routine estimation of undenatured whey proteins in heat-treated milk and NFDM. It was standardized by Kuramoto *et al.* (1959) and modified by Leighton (1962) who introduced the whey protein index (WPI) for NFDM. WPI is defined as milligrams of whey protein nitrogen soluble in saturated NaCl solution per gram of NFDM. Quantitative analysis by SDS-PAGE was recently compared with these procedures. The described method appears to be effective for obtaining the ratio of casein to whey protein in dairy products and detecting adulteration of NFDM (or fluid milk) with WPC (Basch *et al.*, 1985). This type of adulteration can also be detected from amino acid analysis (Greenberg and Dower, 1986). Kneifel and Ulberth (1985) compared methods for the evaluation of heat treatment and the residual undenatured whey protein nitrogen content of NFDM using different procedures to precipitate casein and denatured whey proteins. They found a coefficient of correlation of 0·99 ($P < 0{\cdot}01$) between HPLC (gel filtration) and nitrogen analysis of the soluble fraction. Greiner *et al.* (1985) developed a rapid immunoturbidimetric method for the screening of whey proteins in NFDM and buttermilk using anti-whole whey serum. The coefficient of variation was ~4% and the minimum detectable level was 3% whey protein. A higher sensitivity (detection level: less than 1% whey protein added) was obtained by Olieman and van den Beden (1983) by HPLC-gel permeation determination of the caseino-macropeptide (CMP); however, this is more time-consuming. The presence of rennet whey in NFDM and buttermilk powder can also be detected by RP-HPLC. The method (Olieman & van Riel, 1989) is based, as above, on the detection of the CMP at 210 nm. The detection limit is 0·2% of rennet whey solids.

5.1.4. Cheese and Milk Protein Hydrolyzates

As indicated earlier, electrophoresis and electrofocusing are convenient methods for following the extent of proteolysis during cheese manufacture. Furthermore, these methods often allow the identification of the proteinases which have been involved during ripening. In a study of Camembert cheese ripening, Trieu-Cuot and Gripon (1982) showed, from the identification of several degradation products, that rennet, aspartyl- and metallo-proteinases from *P. caseicolum*, and plasmin were involved. A number of milk protein hydrolysates (from cheese or other sources) were studied in the authors' laboratory. Good information was usually obtained by treating the sample in 2% TCA and analyzing the insoluble fraction by electrophoresis or ion-exchange FPLC on one hand and the soluble fraction by RP-HPLC on the other.

In any case, the degree of proteolysis can be estimated by using TNBS (trinitrobenzene sulfonic acid), a reagent specific for free NH_2 groups, and colorimetry (Adler-Nissen, 1979).

5.1.5. Detection of Bovine Milk in Milk (or Cheese) from Other Species

Methods are necessary for detecting fraudulent adulteration of human, goat, sheep and water buffalo milks with bovine milk, which is cheaper. Such methods do exist and are based either on immunological procedures or, more often, on electrophoresis (e.g. Addeo, 1984; Sanchez Ramon & Olano, 1984). Cattaneo *et al.* (1989) detected goat milk in sheep milk by electrophoresis on acrylamide-agarose and densitometry of β-lactoglobulin. Cattaneo *et al.* claim that their method is less ambiguous than the immunoradial-agar diffusion method (Levieux, 1978) for goat milk addition over 5%. An improved procedure for the detection of bovine milk in Pecorino cheese was published by Addeo *et al.* (1990). By isoelectric focusing in the pH range 7–10 they were able, from densitometric measurement of para-κ-casein after silver staining, to detect 0·5% bovine milk in the cheese. The same authors (Moio *et al.*, 1989) developed isoelectric focusing on PhastSystem (Pharmacia) for detecting bovine milk in ovine cheese.

Rodriguez *et al.* (1990) published an ELISA procedure to detect 1–50% cow milk in sheep milk and cheese. Antibodies were raised against bovine whole casein, recovered by immunoadsorption on a whole casein column, biotinylated and rendered cow milk-specific by mixing them with ovine and caprine whole caseins.

5.1.6. Detection of Non-Milk Proteins in Dairy Products

This is usually possible only when the occurrence of a known product is suspected. Detection of soybean proteins in milk products was studied by Hähnel (1984) using SDS-PAGE, IEF and immunodiffusion with antisera to native and heat-denatured soya proteins. The highest sensitivity was obtained with the latter.

5.2. Milk Protein Analysis in Non-Dairy Products

Some food-stuffs (e.g. milk chocolate) must contain milk proteins in quantities usually above a certain limit while some foods must contain none (e.g. some meat products). Although electrophoresis can be used in some instances, immunological techniques are most interesting in both cases. Their potential applications and limitations have been studied by Willner (1984). It must be realized that the antisera used must be able to recognize denatured (or both native and denatured) milk proteins.

Using isoelectric focusing and laser densitometry (Braun *et al.*, 1990) reported sensitive determination of skim milk powder, whole casein, whey proteins and total milk proteins in feeding stuffs (milk replacers).

6. Conclusion

As far as milk protein analysis is concerned, the present status and future trends can be described briefly:

—The crude protein content of milk and of many dairy products is today widely determined by IR techniques, sometimes with local reference to a dye-binding procedure, and national or international reference to the Kjeldahl method. With the new developments in instrumentation, the authors believe that the Dumas method could in the future supersede the Kjeldahl method, at least for routine determination of nitrogen in milk and milk products.

—The content in some protein fractions (e.g. casein in cheese milk, whey protein in WPC) is demanded by many manufacturers. As these fractions cannot be easily distinguished by their physical characteristics from the other milk proteins, it is likely that their separation will always be required prior to determination by IR or dye-binding methods. Furthermore, NPN assays necessitate nitrogen determination.

—In the field of rapid indirect methods, new techniques like NMR were not listed among the methods because few or no results are available and their potential use by dairy laboratories is still uncertain. On the other hand, another type of IR spectrometer, the Fourier transform infrared spectrometer (FT-IR), which uses a Mickelson interferometer instead of dispersing devices for wavelength discrimination, has recently gained interest among scientists for food analysis (Wilson, 1990). Compared to other instruments, FT-IR has the advantage of providing a high signal/noise ratio, high energy, accurate frequency measurement and rapid scan, making spectral ratioing a useful feature and allowing the use of a full spectrum instead of a few wavelengths (Sanders, 1984).

In addition, FT-IR spectrometers are well adapted for analyses which result in low-energy throughput to the detector, like diffuse or internal reflectance. Internal reflectance, also called attenuated total reflectance (ATR) or multiple internal reflectance, uses a crystal (Ge or ZnSe) which is in close contact with the liquid or the semi-solid sample, for the propagation of the IR beam. For analyses of milk and milk products, FT-IR spectrometry is currently being evaluated by several research laboratories and the first results should be available during 1992.

Among the recent innovations in IR spectroscopy which could be applicable to food analysis and particularly to dairy products, we may list the fixation and evaporation of liquid samples on a sulfur porous solid support which results in a 100-fold increase of the analyte concentration (Meurens & Paulus, 1988). Many manufacturers of NIR instruments now offer the possibility to connect the spectrometer to optical fibers for remote sample analysis. New trends in NIR instrumentation have been summarized by McClure & Hoy (1989).

—Among the various methods available for the characterization and determination of individual milk proteins, electrophoresis or isoelectric focusing are, without doubt, the most suitable methods for routine analysis of a number of samples. However, more care should be taken when they are used for quantitative measurements. Immunological procedures will certainly be employed more frequently in the near future, when good and cheap polyclonal or monoclonal antibody preparations are commercially available.

—Column chromatography does not appear to be suitable yet for analysis of individual milk proteins on a large scale because of their slowness ($\sim$30 min per sample). However, automatic equipments capable of working days and nights are available. RP-HPLC is the most powerful technique for characterizing peptides and protein hydrolyzates. Recent studies show that separation of peptides and proteins by RP- or IE-HPLC can be performed in seconds or minutes using pellicular, non-porous stationary phases (Kalghatgi & Horvath, 1987; Kato *et al.*, 1987).

REFERENCES

Addeo, F. (1984). Detection and determination of bovine milk in Pecorino cheese. *Latte*, **9**, 37–44.

Addeo, F., Moio, L., Chianese, I., Stingo, G. & Di Luccia, A. (1990). Improved procedure for detecting bovine and ovine milk mixtures in cheese by isoelectric focusing of para-kappa-casein. *Milchwissenschaft*, **45**, 221–4.

Adler-Nissen, J. (1979). Determination of the degree of hydrolysis of food protein hydrolyzates by trinitrobenzene sulfonic acid. *J. Agric. Food Chem.*, **37**, 1256–62.

Alais, C., Ribadeau-Dumas, B. & Saint-Lèbe, L. (1961). Etude de la réaction des protéines du lait avec le Noir-Amido. Application au dosage rapide des matières azotées du lait. *Lait*, **41**, 241–73.

Andrews, A.T. & Cheeseman, G.C. (1971). Properties of aseptically packed UHT milk: casein modification during storage and studies with model systems. *J. Dairy Res.*, **38**, 193–207.

Andrews, A.T., Taylor, M.D. & Owen, A.J. (1985). Rapid analysis of bovine milk proteins by HPLC. *J. Chromatogr.*, **348**, 177–85.

Annan, W.D. & Manson, W. (1969). A fractionation of the α_s-casein complex of bovine milk. *J. Dairy Res.*, **36**, 259–68.

AOAC (1975). *Official Methods of Analysis*, 12th edn, Association of Official Analytical Chemists, Washington, DC.

Aoki, T., Yamada, N., Tomita, I., Kako, Y. & Imamura, T. (1987). Caseins are cross-linked through their ester phosphate groups by colloidal calcium phosphate. *Biochim. Biophys. Acta*, **911**, 238–43.

Aschaffenburg, R. & Drewry, J. (1959). New procedure for the routine determination of non-casein proteins of milk. *Proc. XV Intern. Dairy Congr., London*, **3**, 1631–7.

Aschaffenburg, R. & Mullen, J.E.C. (1949). A rapid and simple phosphatase test for milk. *J. Dairy Res.* **16**, 58–67.

Ashworth, U.S. & Chaudry, M.A. (1962). Dye-binding capacity of milk proteins for Amido-Black 10 B and Orange G. *J. Dairy Sci.*, **45**, 952–7.

Baer, R.J., Frank, J.F. & Loewenstein, M. (1983). Compositional analysis of non-fat dry milk by using near infrared diffuse reflectance spectroscopy. *J. Assoc. Off. Anal. Chem.*, **66**, 858–63.

Barbano, D.M. & Clark, J.L. (1989). Infra-red milk analysis. Challenge for the future. *J. Dairy Sci.*, **72**, 1627–36.

Barbano, D.M. & Clark, J.L. (1990). Total nitrogen content of milk: a rapid microwave Kjeldahl digestion. *J. Dairy Sci.*, **73** (suppl. 1), 83.

Barbano, D.M. & Delavelle, M.E. (1987). Rapid method for determination of milk casein content by infra-red analysis. *J. Dairy Sci.*, **70**, 1524–8.

Barbano, D.M., Clark, J.L., Dunham, C.E. & Fleming, J.R. (1990*a*). Kjeldahl method for determination of total nitrogen content of milk: Collaborative study. *J. Assoc. Off. Anal. Chem.*, **73**, 849–59.

Barbano, D.M., Lynch, J.M. & Fleming, J.R. (1990*b*). Direct and indirect determination of true protein content of milk by Kjeldahl analysis: Collaborative study. *J. Assoc. Off. Anal. Chem.*, **74**, 281–8.

Barry, J.G. & Donnelly, W.J. (1980). Casein compositional studies. 1. The composition of casein from Friesian herd milks. *J. Dairy Sci.*, **47**, 71–82.

Basch, J.J., Douglas, F.W., Procino, L.G., Holsinger, V.H. & Farrell, H.M. (1985). Quantification of casein and whey proteins of processed milks and whey protein concentrates; application of gel electrophoresis and comparison with Harland-Ashworth procedure. *J. Dairy Sci.*, **68**, 23–31.

Beck, N.F.G. & Tucker, H.A. (1977). Relationships between radioimmunoassays of alpha lactalbumin and prolactin in bovine skim milk. *J. Dairy Sci.*, **60**, 542–5.

Bennet, R.M. & Mohla, C. (1976). A solid-phase radioimmunoassay for the measurement of lactoferrin in human plasma: variations with sex, age, and disease. *J. Lab. Clin. Med.*, **88**, 156–66.

Bidlingmeyer, B.A., Cohen, S.A. & Tarvin, T.L. (1984). Rapid analysis of amino acids using pre-column derivatization. *J. Chromatogr.*, **336**, 93–104.

Biggs, D.A. (1979*a*). Performance signification for infrared milk analysis. *J. Assoc. Off. Anal. Chem.*, **62**, 1211–4.

Biggs, D.A. (1979*b*). Applications of infra-red instrumentation in cheese analysis. Estimation of fat, protein and total solids in cottage cheese with Multispec M. In *Proceedings from the First Biennal Marschall International Cheese Conference*, ed. Miles Laboratories Inc., Madison, WI, USA, pp. 409–14.

Biggs, D.A., Johnsson, G. & Sjaunja, L.O. (1987). Analysis of fat, protein, lactose and total solids by infra-red absorption. *IDF Bull.*, **208**, 21–30.

Böhlen, P. & Schroeder, R. (1982). High-sensitivity amino acid analysis: methodology for the determination of amino acid composition with less than 100 picomoles of peptides. *Anal. Biochem.*, **126**, 144–52.

Bosset, J.O., Blanc, B. & Plattner, E. (1976). Le dosage des protéines du lait et de ses principaux dérivés: méthodes et appareillages tirés de la littérature parue entre 1964 et 1974. *Trav. Chim. Alim. Hyg.*, **67**, 226–61.

Bovenhuis, H. & Verstege, A.J.M. (1989). Improved method for phenotyping milk protein variants by isoelectric focusing using PhastSystem. *Neth. Milk Dairy J.*, **43**, 447–51.

Braun, F., Krause, I. & Klostermeyer, H. (1990). Efficient determination of skim milk powder, casein, whey protein and total milk protein in compound feeding stuffs by isoelectric focusing and laser densitometry. *Milchwissenschaft*, **45**, 3–9.

Brignon, G., Chtourou, A. & Ribadeau-Dumas, B. (1984). Contamination of human whole casein by whey proteins. Quantitative aspects. *Lait*, **64**, 537–43.

Carles, C. (1986). Fractionation of bovine caseins by reversed-phase high-performance chromatography: identification of a genetic variant. *J. Dairy Res.*, **53**, 35–41.

Cattaneo, T.M.P., Saita, A.S. & Toppino, P.M. (1989). Riconoscimento di latti di specie diverse in miscela: metodi elettroforetici ed immunologici. *Sci. Tech. Latt. Casear.*, **40**, 173–85.

Chaplin, L.C. (1986). Hydrophobic interaction fast protein liquid chromatography of milk proteins. *J. Chromatogr.*, **363**, 329–35.

Chtourou, A., Brignon, G. & Ribadeau-Dumas, B. (1985). Quantification of β-casein in human milk. *J. Dairy Res.*, **52**, 239–47.

Clarke, T. (1979). Standard reconstituted milks for Pro-Milk and Kjeldahl tests. *Aust. J. Dairy Technol.*, **34**, 121–3.

Clarke, T. (1988). A reference milk system for instrumental measurement of milk fat and protein. *J. Dairy Res.*, **55**, 355–60.

Cooper, J.D.H., Lewis, M.T. & Turnell, D.C. (1984). Pre-column o-phtalaldehyde derivatization of amino acids and their separation using reversed-phase high-performance liquid chromatography. *J. Chromatogr.*, **285**, 490–4.

Courroye, M., Berdague, J.L. & Leray, O. (1989). Application de la technique FIA (Flow Injection Analysis) au dosage des fractions azotées du fromage. *Le Lait*, **69**, 233–41.

Creamer, L.K. & Richardson, T. (1984). Anomalous behaviour of bovine αs1- and β-caseins on gel electrophoresis in sodium dodecyl sulfate buffer. *Arch. Biochem. Biophys.*, **234**, 476–86.

Davies, D.T. & Law, A.J.R. (1977*a*). An improved method for the quantitative fractionation of casein mixtures using ion-exchange chromatography. *J. Dairy Res.*, **44**, 213–21.

Davies, D.T. & Law, A.J.R. (1977*b*). The composition of whole casein from the milk of Ayreshire cows. *J. Dairy Res.*, **44**, 447–54.

Davies, D.T. & Law, A.R.J. (1980). The content and composition of protein in creamery milks in south-west Scotland. *J. Dairy Res.*, **47**, 83–90.

De Vilder & Bossuyt (1983). Practical experiences with an Infra-Alyzer 400 in determining the water, protein and fat content of milk powder. *Milchwissenschaft*, **37**, 65–9.
Dimenna, G.P. & Segall, H.J. (1981). High-performance gel-permeation of bovine skim milk proteins. *J. Liq. Chromatogr.*, **4**, 639–49.
Dolby, R.M. (1961). Dye binding methods for estimation of protein in milk. *J. Dairy Res.*, **28**, 43–55.
Douglas, F.W., Greenberg, R., Farrell, H.M. & Edmonson, L.F. (1981). Effect of ultra-high temperature pasteurization on milk proteins. *J. Agric. Food Chem.*, **29**, 11–15.
Dubois, G., Charbonneau, R. & Gagnon, M. (1982). Dosage automatisé de la catalase dans le lait. *Sci. Aliments*, **2**, 245–9.
Egli, H.R. & Meyhack, U. (1984). Measurements of the principal constituents of solid and liquid milk products by means of near infra red analyses. In *Challenges to Contemporary Dairy Analytical Techniques*, The Royal Society of Chemistry, London.
Farah, Z. (1979). Examination of Aschaffenburg and Drewry procedure for determination of non-casein proteins of milk by discontinuous polyacrylamide electrophoresis. *Z. Lebensm. Unters. Forsch.*, **168**, 394–6.
Fraenkel-Conrat, H. & Cooper, M. (1944). The use of dyes for the determination of acid and basic groups in proteins. *J. Biol. Chem.*, **154**, 239–46.
Frank, J.F. & Birth, G.S. (1982). Application of near infrared reflectance spectroscopy to cheese analysis. *J. Dairy Sci.*, **65**, 1110–16.
Frankhuizen, R. & van Der Veen, N.G. (1985). Determination of major and minor constituents in milk powder and cheese by near-infrared reflectance spectroscopy. *Neth. Milk Dairy J.*, **39**, 191–207.
Gaudillere, T. & Grappin, R. (1982). Influence de l'acide citrique sur le dosage des protéines du lait par la méthode infra-rouge. Mémoire de fin d'études, SPILAB, Douai.
Gehrke, C.W., Wall, L.L., Absheer, J.S., Kaiser, F.E. & Zumwalt, R.W. (1985). Sample preparation for chromatography of amino acids: acid hydrolysis of proteins. *J. Assoc. Off. Anal. Chem.*, **68**, 811–21.
Gonzales-Liano, D., Polo, C. & Ramos, M. (1990). Update on HPLC and FPLC analysis of nitrogen compounds in dairy products. *Lait*, **70**, 255–77.
Goulden, J.D.S. (1957). Diffuse reflexion spectra of dairy products in the near infra-red region. *J. Dairy Res.*, **24**, 242–51.
Goulden, J.D.S. (1964). Analysis of milk by infra-red absorption. *J. Dairy Res.*, **31**, 273–84.
Grappin, R. (1984). Assessment and optimization of indirect instrumental methods for testing major constituents in milk and dairy products. In *Challenges to Contemporary Dairy Analytical Techniques*, The Royal Society of Chemistry, London.
Grappin, R. & Horwitz, W. (1988). Determination of nitrogen content in milk by the Kjeldahl method using copper sulfate: Interlaboratory study. *J. Assoc. Off. Anal. Chem.*, **71**, 893–7.
Grappin, R. & Jeunet, R. (1976*a*). Essai de l'appareil Kjel-Foss utilisé pour le dosage de l'azote dans le lait. *Rev. Lait. Fr.*, **346**, 591–7.
Grappin, R. & Jeunet, R. (1976*b*). Calibrage des méthodes de dosage de routine des protéines du lait. *Rev. Lait. Fr.*, **347**, 663–7.
Grappin, R. & Jeunet, R. (1979). Méthode de routine pour le dosage de la matière grasse et des protéines du lait de chèvre. *Lait*, **59**, 345–60.
Grappin, R., Jeunet, R. & Roguinsky, M. (1970). Influence de la mammite sur la composition de la matière azotée des laits de quartiers et sur les dosages de matière axotée par la méthode Noir Amido et l'appareil Infra Red Milk Analyser. *Lait*, **50**, 491–510.
Grappin, R., Jeunet, R. & Le Dore, A. (1979). Determination of the protein content of cows' and goats' milk by dye-binding and infra-red methods. *J. Dairy Sci.*, **62** (Suppl. 1), 38.
Grappin, R., Packard, V.S., Ginn, R.E. & Mellema, J. (1980). Precision of the Pro-Milk method in routine determination of protein in dairy testing laboratories. *J. Food Prot.*, **43**, 52–3.
Green, M.R. & Pastewka, J.V. (1976). Molecular weights of three mouse milk caseins by

sodium dodecyl sulfate-polyacrylamide gel electrophoresis and κ-like characteristics of a fourth casein. *J. Dairy Sci.*, **59**, 1738–45.

Greenberg, R. & Dower, H.J. (1986). Detection of added whey protein concentrate in non-fat dry milk by amino acid analysis. *J. Agric. Food Chem.*, **34**, 30–2.

Greiner, S.P., Kellen, G.J. & Carpenter, D.E. (1985). A rapid immunoturbidimetric method for whey proteins in non-fat dry milk and buttermilk. *J. Food Sci.*, **50**, 1106–9.

Griffiths, M.S. (1986). Use of milk enzymes as indices of heat treatment. *J. Food. Prot.*, **49**, 696–9.

Gripon, J.C., Desmazeaud, M.J., Le Bars, D. & Bergere, J.L. (1975). Etude du rôle des microorganismes et des enzymes au cours de la maturation des fromages. 2. Influence de la présure commerciale. *Lait*, **548**, 502–12.

Grosclaude, F. (1979). Polymorphism of milk proteins: some biochemical and genetical aspects. In *Proceedings 18th International Conference of Animal Blood Grps*, vol. 1, pp. 54–92.

Grosclaude, F. (1988). Le prolymorphisme génétique des principales lactoprotéines bovines. *INRA Prod. Anim.*, **1**, 5–17.

Grosclaude, F., Mahé, M.F., Brignon, G., Di Stasio, L. & Jeunet, R. (1987). A mendelian polymorphism underlying quantitative variations of goat αs1-caseins. *Génét. Sél. Evol.*, **19**, 399–412.

Guillou, H., Pélissier, J.P. & Grappin, R. (1986). Méthodes de dosage des protéines du lait de vache. *Lait*, **66**, 143–75.

Guillou, H., Miranda, G. & Pélissier, J.P. (1987). Analyse quantitative des caséines dans le lait de vache par chromatographie liquide rapide d'échange d'ions (FPLC). *Lait*, **67**, 135–48.

Gupta, B.B. (1983). Determination of native and denatured milk proteins by high-performance size-exclusion chromatography. *J. Chromatogr.*, **282**, 463–75.

Hagen, S.R., Frost, B. & Augustin, J. (1989). Precolumn phenylisothiocyanate derivatization and liquid chromatography of amino acids in food. *J. Assoc. Off. Anal. Chem.*, **72**, 912–16.

Hähnel, W.C. (1984). The detection of soyabean proteins in milk products. Thesis, Ludwig-Maximilian-Universität, Munchen, Germany.

Harland, H.A. & Ashworth, U.S. (1947). A rapid method for estimation of whey proteins as an indication of baking quality of nonfat dry-milk solids. *Food Res.*, **12**, 247–51.

Hetherington, S.V., Spitznagel, J.K. & Quie, P.G. (1983). An ELISA for measurement of lactoferrin. *J. Immunol. Methods*, **65**, 183–90.

Humphrey, R. (1984). Whey proteins. In *Handbook of HPLC for the Separation of Amino Acids, Peptides, and Proteins*, vol. 2, ed. W.S. Hancock. CRC Press, Boca Raton, pp. 471–8.

Humphrey, R.S. & Newsome, L.J. (1984). High performance ion-exchange chromatography of the major bovine milk proteins. *N.Z. J. Dairy Sci. Technol.*, **19**, 197–204.

IDF (1986). Provisional standard 20 A. Milk. Determination of nitrogen content (Kjeldahl method) and calculation of crude protein content. International Dairy Federation, Brussels.

IDF (1987). Monograph on rapid indirect methods for measurement of the major components of milk. Bulletin No. 208. International Dairy Federation, Brussels.

IDF (1990). International provisional standard 141A. Whole milk. Determination of milk fat, protein and lactose content. Guide for the operation of mid-infra-red instruments. International Dairy Federation, Brussels.

Igarashi, Y. (1989). A method for determination of γ-casein and its use for investigating proteolysis in bovine milk. *J. Dairy Res.*, **56**, 619–29.

Jeunet, R. & Grappin, R. (1985). Evaluation de l'Infralyser Dairy pour le dosage des principaux constituants du lait. *Tech. Lait.*, **1003**, 53–8.

Kalghatgi, K. & Horvath, C. (1987). Rapid analysis of proteins and peptides by reversed-phase chromatography. *J. Chromatogr.*, **398**, 335–9.

Kane, P.F. (1984). Comparison of HgO and $CuSO_4$ as digestion catalysts in manual Kjeldahl determination of crude protein in animal feeds: collaborative study. *J. Assoc. Off. Anal. Chem.*, **67**, 869–77.
Karasch, U. (1986). Determination of constituents of milk products with an infra-red laboratory instrument. Thesis, Justus-Liebig-Universität, Giessen, DE. *Dairy Sci. Abst.*, **48**, 881.
Karman, A.H. & Van Boekel, M.A.J.S. (1986). Evaluation of the Kjeldahl factor for conversion of the nitrogen content of milk and milk products to protein content. *Neth. Milk Dairy J.*, **40**, 315–36.
Karman, A.H., Van Boekel, M.A.J.S. & Arentsen-Stasse, A.P. (1987). *Neth. Milk Dairy J.*, **41**, 175–87.
Kato, Y., Kitamra, T., Mitsui, A. & Hashimoto, T. (1987). High performance ion exchange chromatography of proteins on non-porous ion exchangers. *J. Chromatogr.*, **398**, 327–34.
Kawasaki, T., Ikeda, K., Takahashi, S. & Kuboki, Y. (1986). Further study of hydroxyapatite high-performance liquid chromatography using both proteins and nucleic acids, and a new technique to increase chromatographic efficiency. *Eur. J. Biochem.*, **155**, 249–57.
Kitchen, B.J. (1985). Indigenous milk enzymes. In *Developments in Dairy Chemistry-3-Lactose and Minor Constituents*, ed. P.F. Fox. Applied Science Publishers, London, pp. 239–80.
Kjeldahl, J. (1883). En ny Methode til Kvaelstofvestemmelse i organiske Stoffer. *Z. Anal. Chem.*, **22**, 366–82.
Kneifel, W. & Ulberth, F. (1985). Zur Methodik der Hitzklassifizierung von Magemilchpulver unter Berücksichtigung der HPLC-Analyse. *Milchwissenschaft*, **40**, 265–9.
Kristoffersen, T., Koo, K.H. & Slatter, W.L. (1974). Determination of casein by the dye method for estimation of cottage cheese curd yield. *Cult. Dairy Prod. J.*, **9**, 12–14.
Kuramoto, S., Jenness, R., Coulter, S.T. & Choi, P. (1959). Standardization of the Harland-Ashworth test for whey protein nitrogen. *J. Dairy Sci.*, **42**, 28–38.
Lakin, A.L. (1970). Methods, apparatus: new product research, process development and design. *Chem. Ind.*, **38**, 1227–9.
Lakin, A.L. (1974). A rapid method for estimating total protein in milk. *Public. Short Communication IDF Congr., New Delhi*, **1F**, 303–04.
Launay, D., Devaux, M.F., Bertrand, D. & Robert, P. (1986). Etude méthodologique de l'analyse du lait par spectroscopie infrarouge et proche infrarouge. Rapport Adrian, Nantes.
Laurell, C.B. (1966). Quantitative estimation of proteins by electrophoresis in agarose gels containing antibodies. *Anal. Biochem.*, **15**, 45–52.
Le Bars, D. & Gripon, J.-C. (1989). Specificity of plasmin towards bovine αs2-casein. *J. Dairy Res.*, **56**, 817–21.
Leighton, F.R. (1962). Determination of whey protein index of skim milk powder. *Aust. J. Dairy Technol.*, **17**, 186–8.
Levieux, D. (1978). Détection immunologique des mélanges de laits de diverses espèces. Proc. XX Int. Dairy Cong., Conferences 15 ST, Paris.
Lucas, B. & Sotelo, A. (1982). Amino acid determination in pure proteins, foods, and feeds using two different acid hydrolysis methods. *Anal. Biochem.*, **123**, 349–56.
Lunder, T.L. (1974). Determination of total nitrogen in foodstuffs according to Dumas, by means of the 'Micro Rapid N' automatic analyzer. *Lab. Pract.*, **23**, 170–2.
MacDonald, J.L., Krueger, M.W. & Keller, J.H. (1985). Oxidation and hydrolysis determination of sulfur amino acids in food and feed ingredients: collaborative study. *J. Assoc. Off. Anal. Chem.*, **68**, 826–9.
Mancini, G., Vaerman, J.P., Carbonera, A.O. & Heremans, J.F. (1963). A single radial-immunodiffusion method for the immunological quantitation of proteins. *Protides Biol. Fluids*, AA, 370–84.
Manji, B., Hill, A., Kakuda, Y. & Irvine, D.M. (1985). Rapid separation of milk whey proteins by anion exchange chromatography. *J. Dairy Sci.*, **68**, 3176–9.

McClure, W.F. & Hoy, R.M. (1989). Near infra-red instrumentation: status, trends and futuristic concepts. In *Proceedings of the 2nd International Near Infra-red Spectroscopy Conference*, ed. M. Iwanmoto & S. Kawano. Korin Publishing, Tokyo, pp. 65–79.

McGann, T.C.A. (1978). Automated physico-chemical methods for the analysis of milk. A review of major advances (1960–1978). *Ir. J. Food Sci. Technol.*, **2**, 141–56.

McGann, T.C.A., Mathiassen, A. & O'Connell, J.A. (1972). Applications of the Pro-Milk MK II. Part III. Rapid estimation of casein in milk and protein in whey. *Lab. Pract.*, **21**, 628–50.

McKenzie, H.A. (1971). Methods for zone electrophoresis of milk proteins. In *Milk Proteins*, vol. II, ed. H.A. McKenzie. Academic Press, London, pp. 483–87.

McKenzie, H.A. & Murphy, W.H. (1970). General methods and elemental analysis. Determination of total nitrogen. In *Milk Proteins*, vol. I, ed. H.A. McKenzie. Academic Press, New York, pp. 154–61.

McLean, D.M., Graham, E.R.B. & McKenzie, H.A. (1982). Estimation of casein components by gel electrophoresis. *Proc. 21st Intern. Dairy Congr., Moscow*, **1** (2), 221.

McLean, D.M., Graham, E.R.B. & Ponzoni, R.W. (1984). Effect of milk protein genetic variants on milk yield and composition. *J. Dairy Res.*, **51**, 531–46.

Mercier, J.C., Maubois, J.L., Poznanski, S. & Ribadeau-Dumas, B. (1968). Fractionnement préparatif des caséines de vache et de brebis par chromatographie sur DEAE- cellulose en milieu urée et 2-mercaptoéthanol. *Bull. Soc. Chim. Biol.*, **50**, 521–30.

Meurens, M. & Paulus, F. (1988). Near infra-red reflectance spectroscopy of dry extract on solid support. *Analysis*, **16**, 53–4.

Michalak, W. & Oczkowicz, H. (1975). The application of a central system of controlling apparatuses for determining the main milk components and its effects on the comparability of results obtained. *Pr. Mater. Zootech.*, **8**, 103–12.

Mickelsen, R. & Shukri, N.A. (1975). Measuring casein by dye-binding. *J. Dairy Sci.*, **58**, 311–12.

Miranda, G. (1983). Etude cinétique de la protéolyse in vivo des lactoprotéines bovines dans l'estomac du rat: effet du traitement thermique. Thesis, University of Paris VII.

Moio, L., Di Luccia, A. & Addeo, F. (1989). Fast isoelectric focusing of milk proteins on small ultrathin polyacrylamide gels containing urea. *Electrophoresis*, **10**, 533–5.

National Health and Medical Research Council (1987). Food standards, Code 241.

Ng-Kwai-Hang, K.F. & Kroeker, E.M. (1984). Rapid separation and quantification of major caseins and whey proteins of bovine milk by polyacrylamide gel electrophoresis. *J. Dairy Sci.*, **67**, 3052–6.

Ng-Kwai-Hang, K.F., Hayes, J.F., Moxley, J.E. & Monardes, H.G. (1984). Association of genetic variants of casein and milk serum proteins with milk fat, and protein production by dairy cattle. *J. Dairy Sci.*, **67**, 835–40.

Ng-Kwai-Hang, K.F., Hayes, J.F., Moxley, J.E. & Monardes, H.G. (1987). Variation in milk protein concentration associated with genetic polymorphism and environmental factors. *J. Dairy Sci.*, **70**, 563–70.

Nielsen, H.K. & Hurrell, R.F. (1985). Tryptophan determination of food proteins by HPLC after alkaline hydrolysis. *J. Sci. Food Agric.*, **36**, 893–907.

Norman, L.M. & Hohenboken, W.D. (1981). Genetic differences in concentration of immunoglobulins G1 and M in serum and colostrum of cows and in serum of neonatal calves. *J. Anim. Sci.*, **53**, 1465–72.

Olieman, C. & van den Beden, J.W. (1983). A sensitive HPLC method of detecting and estimating rennet whey total solids in skim milk powder. *Neth. Milk Dairy J.*, **37**, 27–36.

Olieman, C. & van Riel, J.A.M. (1989). Detection of rennet whey solids in skim milk powder and buttermilk powder with reversed-phase HPLC. *Neth. Milk Dairy J.*, **43**, 171–84.

Osborne, B.G. & Fearn, T. (1988). *Near Infra-Red Spectroscopy in Food Analysis*. Longman Scientific & Technical, Harlow and John Wiley & Sons, Inc., New York.

Pailler, F. (1982). Dosage après minéralisation des protéines par électrode spécifique de l'ion ammonium. Comparaison avec la méthode Kjeldahl. Application aux denrées alimentaires. *Ann. Falsif. Expert. Chim. Toxicol.*, **75**, 431–9.

Pearce, R.J. (1983). Analysis of whey proteins by high performance liquid chromatography. *Aust. J. Dairy Technol.*, **38**, 114–17.

Pearce, R.J. & Shanley, R.M. (1981). Analytical and preparative separation of whey proteins by chromatofocusing. *Aust. J. Dairy Technol.*, **36**, 110–14.

Rexroad, P.R. & Cathey, R.D. (1976). Pollution-reduced Kjeldahl method for crude protein. *J. Assoc. Off. Anal. Chem.*, **59**, 1213–17.

Reynolds, J.A. & Tanford, C. (1970). The gross conformation of protein-sodium dodecyl sulfate complexes. *J. Biol. Chem.*, **245**, 5161–5.

Ribadeau-Dumas, B. & Grappin, R. (1989). Milk protein analysis. *Lait*, **69**, 357–416.

Richardson, B.C. (1983). Variation of the concentration of plasmin and plasminogen in bovine milk. *N.Z. J. Dairy Sci. Technol.*, **18**, 247–52.

Robert, P., Bertrand, D., Devaux, M.F. & Grappin, R. (1987). Multivariate analysis applied to near-infrared spectra of milk. *Anal. Chem.*, **59**, 2187–191.

Robertson, N.H. & van der Westhuizen, E. (1990). Non-protein nitrogen content of milk as influenced by various factors. *S. Afr. J. Dairy Sci.*, **22**, 1–8.

Rodbard, D., Kapadia, G. & Chrombach, A. (1971). Pore gradient electrophoresis. *Anal. Biochem.* **40**, 135–57.

Rodriguez, E., Martin, A., Garcia, T., Hernandez, P.E. & Sanz, B. (1990). Detection of cows' milk in ewes' milk and cheese by an indirect enzyme-linked immunosorbent assay. *J. Dairy Res.*, **57**, 197–205.

Roeper, J. & Dolby, R.M. (1971). Estimation of the protein content of wheys by the amido black method. *N.Z. J. Dairy Sci. Technol.*, **6**, 65–8.

Rowland, S.J. (1938). The determination of the nitrogen distribution in milk. *J. Dairy Res.*, **9**, 42–6.

Rudzik, L. (1985). MIR-NIR ein Methodenvergleich. *Deutsche. Milchwitschaft.*, **8**, 225–8.

Ruitenberg, E.J., Steerenberg, P.A., Brosi, B.J.M. & Buys, J. (1976). Reliability of the enzyme linked immunosorbent assay (ELISA) for the serodiagnosis of *Trichinella spiralis* infections in conventionally raised pigs. *J. Immunol. Methods*, **10**, 67–83.

Sanchez Ramon, M. & Olano, A. (1984). Quantification of cow's milk in goat's milk by polyacrylamide-gel electrophoresis using αs1-casein as internal standard. *J. Dairy Sci.*, **67** (suppl. 1), 72.

Sanders, R.A. (1984). Application of Fourier transform infra-red spectroscopy in the field of foods and beverages. In *Analysis of Foods and Beverages*, Academic Press, New York, pp. 554–83.

Sarwar, G., Christensen, D.A., Finlayson, A.J., Friedman, M., Hackler, L.R., MacKenzie, S.L., Pellett, P.L. & Tkachuk, R. (1983). Inter- and intra-laboratory variation in amino acid analysis of food proteins. *J. Food Sci.*, **48**, 526–31.

Sato, H., Seino, T., Kobayashi, T., Murai, A. & Yugari, Y. (1984). Determination of the tryptophan content of feed and feedstuffs by ion exchange liquid chromatography. *Agric. Biol. Chem.*, **48**, 2961–9.

Sato, T., Yoshino, M., Furukawa, S., Somey, Y., Yano, N., Uozumi, J. & Iwamoto, W. (1987). Analysis of milk constituents by the near infra-red spectrophotometric method. *Jpn J. Zootech. Sci.*, **58**, 698–706.

Schober, R. & Hetzel, H.F. (1956). Uber ein einfache, kolorimetrische Bestimmung von Milchproteinen. *Milchwissenschaft*, **11**, 123–6.

Seibert, B., Erhardt, G. & Senft, B. (1985). Procedure for simultaneous phenotyping of genetic variants in cow's milk by using isoelectric focusing. *Anim. Blood Groups Biochem. Genet.*, **16**, 183–91.

Seng Kwee, W. (1985). An improved Aschaffenburg and Mullen alkaline phosphatase text for coloured milk products. *Aust. J. Dairy Technol.*, **40**, 27–30.

Shapiro, A.L., Vinuela, E. & Maizel, J.V. (1967). Molecular weight estimation of polypep-

tide chains by electrophoresis in SDS-polyacrylamide gels. *Biochem. Biophys. Res. Commun.*, **28**, 815–20.
Sherbon, J.W. (1975). Collaborative study of the Pro-Milk method for the determination of protein in milk. *J. Assoc. Off. Anal. Chem.*, **58**, 770–6.
Sherbon, J.W. (1978). Recent developments in determining protein content of dairy products by dye-binding. *J. Dairy Sci.*, **61**, 1274–8.
Sherbon, J.W. & Luke, H.A. (1968). Collaborative study of the dye-binding method applied to chocolate milk drinks, cultured buttermilk and half-and-half. *J. Assoc. Off. Anal. Chem.*, **51**, 811–16.
Sherbon, J.W. & Luke, H.A. (1969). Comparison of the dye-binding and Kjeldahl methods for protein analysis of non-fat dry milk and ice cream. *J. Assoc. Off. Anal. Chem.*, **52**, 138–42.
Shimazaki, K. & Sukegawa, K. (1982). Chromatographic profiles of bovine milk whey components by gel filtration. *J. Dairy Sci.*, **65**, 2055–62.
Shindler, J.S. & Bardsley, W.G. (1975). Steady-state kinetics of lactoperoxidase with ABTS as chromogen. *Biochem. Biophys. Res. Commun.*, **67**, 1307–12.
Sjaunja, L.O. (1982). Studies on milk analyses of individual cow milk samples. Report 56, Swedish University of Agricultural Sciences, Uppsala.
Sjaunja, L.O. & Schaar, J. (1984). Determination of casein in milk by infra-red spectrophotometry. *Milchwissenschaft*, **39**, 288–90.
Sjaunja, L.O. & Andersson, I. (1985). Laboratory experiments with a new IR milk analyzer, the Milko-Scan. *Acta Agric. Scand.*, **35**, 345–52.
Spies, J.R. & Chambers, D.C. (1948). Chemical determination of tryptophan. *Anal. Chem.*, **20**, 30–9.
Steinsholt, A.K. (1976). Dairy Research Institute, Agricultural University of Norway, Report 193.
Swaisgood, H.E. (1982). Chemistry of milk protein. In *Developments in Dairy Chemistry: 1 —Proteins*, ed. P.F. Fox. Applied Science Publishers, London, pp. 1–61.
Sweeney, R.A. (1989). Generic combustion method for determination of crude protein in feeds: Collaborative study. *J. Assoc. Off. Anal. Chem.*, **72**, 770–4.
Szijarto, R.J., Biggs, D.A. & Irvine, D.M. (1973). Variability of casein, serum protein and nonprotein nitrogen in plant milk supplies in Ontario. *J. Dairy Sci.*, **56**, 45–51.
Tarassuk, N.P., Abe, N. & Moats, W.A. (1967). The dye-binding of milk proteins. US Dept. of Agriculture, Washington, DC, Tech. Bull., 1369.
Trieu-Cuot, P. (1981). Etude des caséines et de leurs produits de dégradation par méthodes électrophorétiques. Application à l'étude de l'affinage du camembert. Thesis, University of Paris VII.
Trieu-Cuot, P. & Gripon, J.-C. (1981). Electrofocusing and two-dimensional electrophoresis of bovine caseins. *J. Dairy Res.*, **48**, 303–10.
Trieu-Cuot, P. & Gripon, J.-C. (1982). A study of proteolysis during Camembert cheese ripening using isoelectric focusing and two-dimensional electrophoresis. *J. Dairy Res.*, **49**, 501–10.
Udy, D.C. (1956). A rapid method for estimating total protein in milk. *Nature*, **178**, 314–15.
Van Boekel, M.A.J.S. & Ribadeau-Dumas, B. (1987). Addendum to the evaluation of the Kjeldahl factor for conversion of the nitrogen content of milk and milk products to protein content. *Neth. Milk Dairy J.*, **41**, 281–4.
Van Reusel, A. & Klijn, C.J. (1987). Automated methods for routine analysis of raw milk. The dye-binding method for determination of the protein content of milk. *IDF Bull.*, **208**, 17–20.
Vegarud, G.E., Molland, T.S., Brovold, M.J., Devold, T.G., Alestrom, P., Steine, T., Rogne, S. & Langsrud, T. (1989). Rapid separation of genetic variants of caseins and whey proteins using urea-modified gels and fast electrophoresis. *Milchwissenschaft*, **44**, 689–91.

Vendrel, J. & Aviles, F. (1986). Complete amino acid analysis of proteins by dabsyl derivatization and reversed-phase liquid chromatography. *J. Chromatogr.*, **358**, 401–13.

Verdi, R.J., Barbano, D.M., Della Valle, M.E. & Senyk, G.F. (1987). Variability in true protein, casein, non-protein nitrogen and proteolysis in high and low somatic cell milks. *J. Dairy Sci.*, **70**, 230–42.

Visser, S., Slangen, K.J. & Rollema, H.S. (1986). High performance liquid chromatography of bovine caseins with the application of various stationary phases. *Milchwissenschaft*, **41**, 559–62.

Waite, R. & Smith, G.M. (1972). Measurement of the protein content of milk from mastitic quarters by the Amido Black method. *J. Dairy Res.*, **39**, 195–201.

Welty, F.K., Smith, K.L. & Schanbacher, F.L. (1976). Lactoferrin concentration during involution of the bovine mammary gland. *J. Dairy Sci.*, **59**, 224–31.

Wilkinson, J.M. (1978). The separation of dansyl amino acids by reversed-phase high performance liquid chromatography. *J. Chromatogr. Sci.*, **16**, 547–52.

Williams, D.J. & Nottingham, S.M. (1990). Suitability of a modification to the Aschaffenburg and Mullen alkaline phosphatase test for goat's milk: collaborative study. *Aust. J. Dairy Technol.*, **45**, 21–3.

Willner, I. (1984). Potential applications and limitations of immunological determination of milk proteins in foods and animal feeds. Thesis, Technishe Universität, München, Germany.

Wilson, R.H. (1990). Fourrier Transform mid-infra-red spectroscopy for food analysis. *Trends in Anal. Chem.*, **9**, 127–31.

Yaguchi, M. & Rose, D. (1971). Chromatographic separation of milk proteins: A review. *J. Dairy Sci.*, **54**, 1725–43.

Yaguchi, M. & Tarassuk, N.P. (1967). Gel filtration of acid casein and skim milk on Sephadex. *J. Dairy Sci.*, **50**, 1985–8.

2

CHEMISTRY OF THE CASEINS

HAROLD E. SWAISGOOD

Southeast Diary Foods Research Centre, North Carolina State University, Raleigh, North Carolina, 27695-7624, USA

1. INTRODUCTION

Milk protein is widely consumed as a human food, particularly in developed countries. The long history of its inclusion in the human diet and the relative ease with which the major proteins can be purified from raw milk have contributed to its early and extensive characterization by biochemists. Consequently, milk proteins are probably the best characterized of all food proteins; for example, the primary structures of all bovine milk proteins are known and even the three-dimensional structures of the major whey proteins are known. Nevertheless, the molecular characteristics, particularly the tertiary structure, of the caseins remain as a challenge since these proteins are not typical and apparently cannot be crystallized. In the past, they have at times been inaccurately classified as denatured or random coil proteins. Although the caseins are apparently more flexible and less stable than typical globular or rod-shaped proteins, they appear to have some secondary structure and a definite unordered tertiary structure.

The discussion in this chapter is principally a revision of a previous chapter by the same author (Swaisgood, 1982a). However, the discussion will be limited to the caseins. The focus will be on the composition of bovine milk, methods for casein isolation and identification, their amino acid compositions and primary structures, and various physicochemical characteristics, including secondary structures and various interactions. Numerous reviews of casein chemistry have appeared previously (Jenness *et al.*, 1956; Brunner *et al.*, 1960; Lindqvist, 1963*a*,*b*; Thompson *et al.*, 1965; McKenzie, 1967, 1970; Rose *et al.*, 1970; McKenzie, 1971; Swaisgood, 1973; Gordon & Kalan, 1974; Whitney *et al.*, 1976; Whitney, 1977; Brunner, 1981; Swaisgood, 1982*a*; Eigel *et al.*, 1984); however, the author will attempt to summarize the most pertinent aspects and update the progress since the last review (Swaisgood, 1982*a*).

2. Protein Composition of Milk

2.1. Overall Composition

Normal bovine milk contains 30–35 g protein/litre. About 80% of these proteins are present in casein micelles which are large spherical complexes containing 92% protein and 8% inorganic salts, principally calcium phosphate (Schmidt, 1980; Swaisgood, 1985; Whitney, 1988). Although this discussion focuses on the caseins, their properties in milk are not independent of their interactions with milk salts, particularly calcium and calcium salts. Thus, of the total 30 mM calcium in milk, roughly 65% exists in the casein micelle and only 10% is present as free calcium ions (White & Davies, 1958; Holt, 1985). Essentially all of the caseins are in the micelles which can be physically isolated by centrifugation which yields a micelle pellet or by size exclusion chromatography. Other classical methods of fractionating milk proteins are isoelectric precipitation of the caseins at pH 4·6 and 20°C or by limited proteolysis of micellar κ-casein with chymosin, resulting in coagulation of the altered casein micelles and separation of a whey fraction. Since milk proteins are readily separated into casein and whey protein fractions, these fractions have been commonly used as the starting materials for further purification of individual proteins. Accordingly, the schematic diagram (Fig. 1) depicting the origin and concentrations of various individual milk proteins of skim milk is based on these two major fractions.

Caseins, which account for 76–86% of the total protein, represent four gene products: α_{s1}-casein, α_{s2}-casein, β-casein and κ-casein. Some additional heterogeneity arises from post-translational processing such as phosphorylation, glycosylation and limited proteolysis. Most notable are the varying levels of phosphorylation of α_{s2}-casein and glycosylation of κ-casein and the proteolytic action of plasmin on β-casein, yielding γ-caseins and proteose-peptones. For simplicity and following a historical perspective, the term γ-caseins will be used here for mixtures while reserving the β-casein nomenclature for specific peptides. The revised nomenclature for various caseins is listed in Table 1.

2.2. Occurrence of Genetic Variants

The heterogeneity of the caseins is further complicated by the fact that they are products of co-dominant allelic autosomal genes. Genetic variants of β-lactoglobulin were the first to be recognized (Aschaffenburg & Drewry, 1957) and discovery of genetic variants of caseins followed shortly thereafter. Currently, there are five known variants of α_{s1}-casein (Thompson *et al.*, 1962; Grosclaude *et al.*, 1966; Eigel *et al.*, 1984), four of α_{s2}-casein (Grosclaude *et al.*, 1976*a*, 1979), seven of β-casein (Aschaffenburg, 1961; Peterson & Koppler, 1966; Aschaffenburg *et al.*, 1968) and two of κ-casein (Neelin, 1964; Schmidt, 1964; Woychik, 1964). The pattern of inheritance of the various casein polymorphs supports the concept of a close linkage of the four gene loci for these proteins (Grosclaude *et al.*, 1964; King *et al.*, 1965; Larsen & Thymann, 1966). Thus, the four genetic variants are not

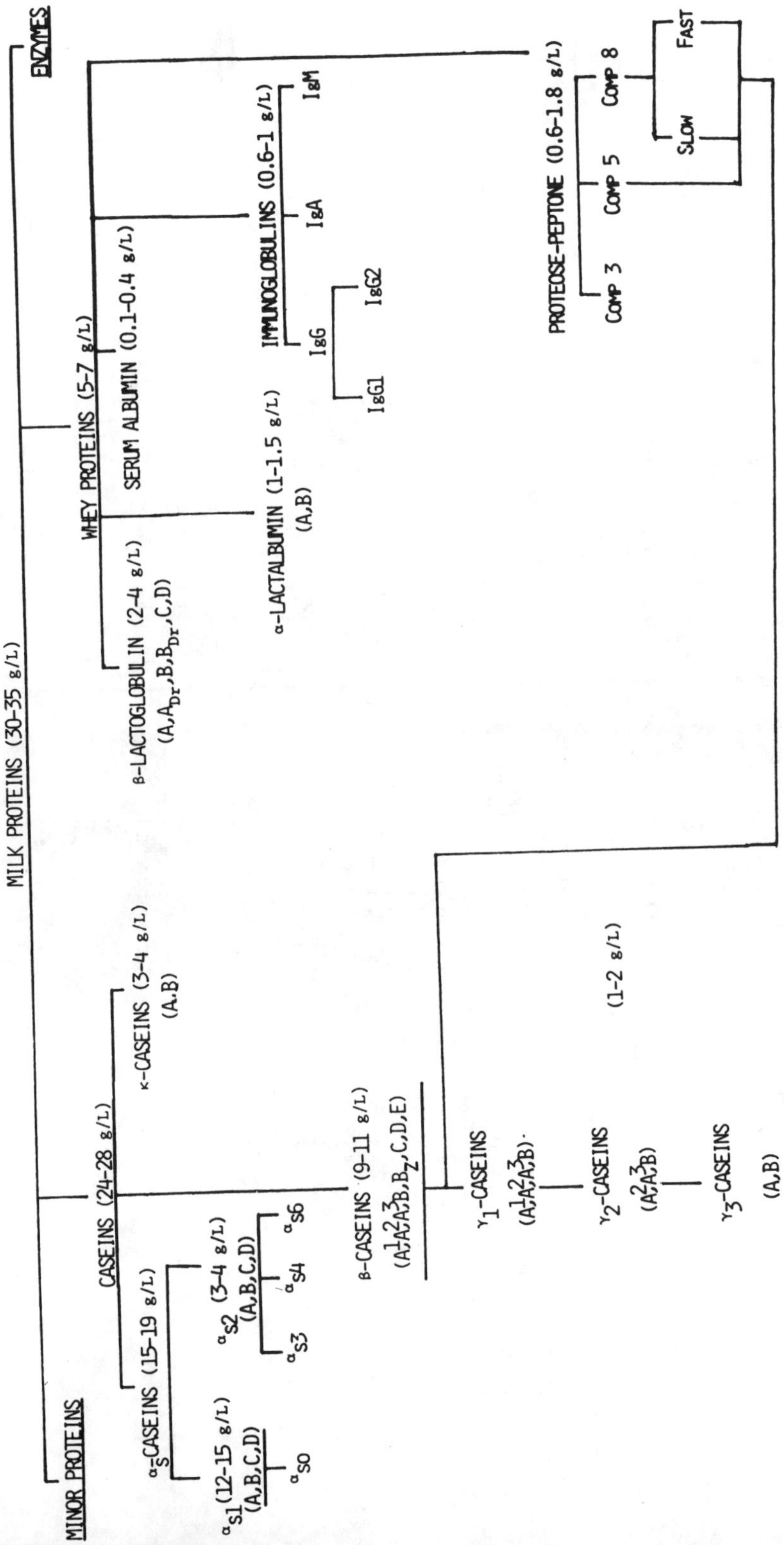

Fig. 1. Distribution of fractions and proteins in bovine milk.

TABLE 1
Comparison of Former and Revised Casein Nomenclature[a]

α_{s1} Caseins		α_{s2}-Caseins		β-Caseins		β-Casein peptides		κ-Caseins	
Former	*Revised*	*Former*	*Revised*	*Former*	*Revised*	*Former*	*Revised*	*Former*	*Revised*
α_{s0}-CN	α_{s1}-CN B-9P	α_{s2}-CN A	α_{s2}-CN A-13P	β-CN A^1	β-CN A^1-5P	γ_1-CN A^1	β-CN A^1-1P(f29–209)	κ-CN A	κ-CN A-1P
α_{s1}-CN A	α_{s1}-CN A-8P	α_{s3}-CN	α_{s2}-CN A-12P	β-CN A^2	β-CN A^2-5P	γ_1-CN A^2	β-CN A^2-1P(f29–209)	κ-CN B	κ-CN B-1P
α_{s1}-CN B	α_{s1}-CN B-8P	α_{s4}-CN	α_{s2}-CN A-11P	β-CN A^3	β-CN A^3-5P	γ^1-CN A^3	β-CN A^3-1P(f29–209)		
α_{s1}-CN C	α_{s1}-CN C-8P	α_{s6}-CN	α_{s2}-CN A-10P	β-CN B	β-CN B-5P	γ_1-CN B	β-CN B-1P(f29–209)		
α_{s1}-CN D	α_{s1}-CN D-9P			β-CN C	β-CN C-4P	γ_2-CN A^2	β-CN A^2(f106–209)		
α_{s1}-CN E	α_{s1}-CN E-8P			β-CN D	β-CN D-4P	γ_2-CN A^3	β-CN B(f106–209)		
				β-CN E	β-CN E-5P	γ_2-CN B	β-CN B(f106–209)		
						γ_3-CN A	β-CN A(f108–209)		
						γ_3-CN B	β-CN B(f108–209)		
						Component 5	β-CN-5P(f1–105)		
						Component 5	β-CN-5P(f1–107)		
						Component 8-slow	β-CN-1P(f29–105)		
						Component 8-slow	β-CN-1P(f29–107)		
						Component 8-fast	β-CN-4P(f1–28)		

[a]The revised nomenclature is that recommended by the Nomenclature Committee of the American Dairy Science Association (Eigel *et al.*, 1984).

observed with random or equal frequencies. In western breeds, as shown in Table 2, α_{s1}-casein B, α_{s2}-casein A (data not shown), and β-casein A^1 and A^2 occur with a much higher frequency than other genetic variants. κ-Caseins A and B occur with more equal frequency. Identification of genetic variants has assumed new importance with the discovery that composition of milk and the protein's functionality are different for different variants (McLean *et al.*, 1984; Schaar, 1984; Schaar *et al.*, 1985; Van Eenennaam & Medrano, 1991).

3. Isolation and Identification of the Caseins

This discussion will not review all of the methods developed for isolation of the caseins; however, a brief description of some of the classical methods as well as some recently developed chromatographic methods will be presented. The reader should consult the reviews cited for discussion of other methods presented. In general, the classical methods are based on solubility differences in various solvent systems and yield substantially purified individual proteins. The use of an agent to disrupt protein–protein interactions is essential for isolation of highly purified individual caseins.

3.1. Classical Methods of Isolation

Historically, the classical methods of isolation also served as the operational definition of the casein fraction obtained as designated by the Milk Protein Nomenclature and Methodology Committee of the American Dairy Science Association. Some characteristics of these various protein fractions are given in Table 3. Fractions obtained by differential solubility exhibit varying degrees of heterogeneity; thus, highly purified individual caseins are usually obtained by some chromatographic procedure.

Whole casein, which is often used as the starting material for purification of individual caseins, can be isolated by several methods as noted above. Frequently, it is obtained by isoelectric precipitation by adjustment of skim milk (usually diluted with distilled water) to pH 4·6 at 20°C with HCl. Whole casein can also be prepared by isolating the micelles by sedimentation or size exclusion chromatography. For example, centrifugation at 45 000 g for 90 min at 37°C following adjustment of skim milk to 0·07 M $CaCl_2$ at pH 6·6–6·8 yields a micelle pellet which was designated as 'first cycle casein' (Waugh *et al.*, 1962). Variations of these methods are discussed and compared by McKenzie (1971). More recently, casein micelles have been obtained by size exclusion chromatography using controlled-pore glass as a support (McNeill & Donnelly, 1987).

A useful fractionation of whole casein by differential solubility is obtained by separating that which is soluble in 0·25 M Ca^{2+} at pH 7 and 37°C from that which is not (Waugh & Von Hippel, 1956). This fractionation is particularly meaningful since the calcium-insensitive fraction forms protein–calcium complexes ('micelles') with, and thereby stabilizes, the calcium-sensitive fraction at Ca^{2+} concentrations

TABLE 2

Gene Frequencies of the Polymorphs of the Major Milk Proteins for Common Breeds[c]

Polymorph	*Holstein-Friesian*				*Guernsey*			*Jersey*			*Ayrshire*			*Brown Swiss*	
α_{s1}-Caseins															
Location	US[a]	US[b]	US[b]	Holland	US[a]	US[b]	US, UK[b]	US[a]	UK, US[b]	Germany[b]	US[a]	US[b]	UK[a]	US[a]	US[b]
Number tested	541	542	754	693	296	400	170	287	351	148	257	98	52	282	250
Gene frequency															
A	0·003	0·05	< 0·01	0	0	0	0	0	0	0	0	0	0	0	0
B	0·94	0·87	0·91	0·99	0·79	0·70	0·77	0·74	0·72	0·92	1·00	1·00	1·00	0·98	0·94
C	0·06	0·08	0·08	0·01	0·21	0·30	0·23	0·26	0·28	0·08	0	0	0	0·02	0·006
D	0	0	0	0	0	0	0	0	0	0	0	0	0	0	0
β-Casteins															
Location	US[a]	US[b]	UK[b]		US[a]	US[b]		US[a]	US[b]	UK[b]	US[a]	US[b]	UK[b]	US[a]	US[b]
Number tested	526	632	85		262	196		298	37	47	202	45	29	235	23
Gene frequency															
A^1	0·49	0·31	0·66		0·06	0·01		0·09	0·22	0·09	0·67	0·72	0·60	0·15	0·14
A^2	0·49	0·62	0·24		0·88	0·98		0·54	0·49	0·63	0·32	0·28	0·40	0·72	0·66
A^3	0·01	0·05	0·04		0	0		0	0	0	0·01	0	0	0	0
A	0·99	0·98	0·94		0·94	0·99		0·63	0·71	0·72	1·00	1·00	1·00	0·87	0·80
B	0·01	0·02	0·06		0·01	0·01		0·37	0·29	0·28	0	0	0·10	0·18	
C	0	0·001	0		0·05	< 0·01		0·003	0	0	0	0	0	0·03	0·02
D	0	0	0		0	0		0	0	0	0	0	0	0	0
κ-Caseins															
Location	US[a]	US[b]	UK[b]		US[a]	US[b]		US[a]	US[b]	UK[b]	US[a]	UK[b]	US[a]	South Africa	
Number tested	543	138	189		297	47		297	48	70	211	52	290	30	
Gene frequency															
A	0·75	0·85	0·83		0·59	0·74		0·12	0·10	0·24	0·70	0·80	0·41	0·70	
B	0·25	0·15	0·17		0·41	0·26		0·88	0·90	0·76	0·30	0·20	0·59	0·30	
β-Lactoglobulins															
Location	US[a]	US[b]		Australia[b]	US[a]	US[b]	Australia[b]	US[a]		Australia[b]	US[a]		Australia[b]	US[a]	
Number tested	494	406		108	278	200	68	270		448	191		104	259	
Gene frequency															
A	0·50	0·46		0·51	0·38	0·30	0·44	0·36		0·26	0·17		0·38	0·33	
B	0·50	0·54		0·49	0·62	0·70	0·56	0·64		0·64	0·83		0·62	0·67	
C	—	—		—	—	—	—	—		0·10	—		—	—	

[a] Data of Li and Gaunt (1972).
[b] Selected from the summary given by Aschaffenburg (1968).
[c] This table is adapted from Swaisgood (1982*b*) and appeared in Chapter 1 of *Developments in Dairy Chemistry–1*, 1982, Applied Science Publishers, London.

comparable with those of milk. Of course, both fractions are very heterogeneous. The calcium-insensitive fraction (Fraction S) contains mostly κ-casein, which is known to be the 'protective colloid' proposed many years ago by Linderstrom-Lang (1929) to stabilize the micelle, as well as some β-casein and λ-caseins. (The λ-casein fraction is most likely a mixture of peptides derived from the caseins by proteolysis.) The calcium-sensitive fraction is composed of α_{s1}-, α_{s2}- and β-caseins.

Due to the tenacity of casein interactions, an agent capable of dissociating various casein complexes must be used to purify individual caseins from crude fractions. Historically, urea is most commonly used for this purpose. For example, fractionation of whole casein from 6·6 M urea solution (Hipp *et al.*, 1952) can be achieved by dilution to 4·63 M urea yielding a precipitate of whole α-casein (a mixture of α_{s1}-, κ- and λ-caseins). β-Casein is obtained from the supernatant by further dilution to 1·7 M urea at pH 4·7. The γ-caseins are obtained from the final supernatant by addition of ammonium sulphate.

Whole casein or the α-casein fraction can be fractionated by differential solubility in urea solutions upon addition of strong acids. Thus, substantially purified κ-casein has been prepared by precipitation of other caseins from a concentrated urea solution upon adjustment to 12% trichloroacetic acid (Swaisgood & Brunner, 1962) or to pH 1·3–1·5 with H_2SO_4 (Zittle & Custer, 1963). Other caseins can also be obtained in this fractionation; for example, using whole α-casein as the starting material yields substantially purified α_s-caseins as the precipitate and γ-caseins are obtained by a modification of the TCA procedure (Swaisgood & Brunner, 1962). The α_{s1}-caseins can be obtained in relatively pure form by precipitation from 6·6 M urea upon adjustment to pH 4·5 and dilution to 3·3 M urea if the calcium-sensitive fraction is used as starting material (Fox & Guiney, 1972).

Differential solubility in ethanol has also been used to fractionate the caseins (Hipp *et al.*, 1952). For example, a substantially purified κ-casein fraction is obtained by precipitation from 50% (v/v) ethanol at pH 7·2 with ammonium acetate (McKenzie & Wake, 1961).

3.2. Identification by Gel Electrophoresis

Historically, the identification and nomenclature of milk proteins were defined by their free electrophoretic mobility. With the advent of zone electrophoresis in gels, much greater resolution became possible. The mobility in gels is directly related to the fractional volume accessible to the molecule; thus, resolution occurs due to both electrophoresis and molecular sieving (Swaisgood, 1975). Inclusion of urea in the gels to dissociate the casein complexes provided a powerful analytical tool for identification of individual caseins (Wake & Baldwin, 1961). This development provided the impetus for identification, isolation and characterization of the various caseins and the discovery of genetic variants.

Alkaline urea-gel electrophoresis in either starch (SGE) or polyacrylamide (PAGE) gels are the most commonly used methods (Wake & Baldwin, 1961; Thompson, 1970; Swaisgood, 1975). Following the original recommendation of Wake and Baldwin (1961), a distinct sharp zone common to the milk of most

TABLE 3
Characteristics and Nomenclature of Casein Fractions and Components

Protein fraction	*Characteristics*	*Composition*	*Number of zones in alkaline urea–gel electrophoresis*
Whole casein	Precipitated from skim milk at pH 4·6 and 20°C.[a] Sediments as micelles at 3°C or 37°C in 0·07 M $CaCl_2$ at pH 6·6–6·8	Contains all the caseins	18–20[b]
Whole α-casein	Soluble in 6·6 M urea, but insoluble in 4·6 M urea (pH 4·6–4·8)	Contains α_s-, κ- and λ-caseins	8–10
α_s-Caseins	Fraction of whole α-casein precipitating in 0·4 M $CaCl_2$ at pH 7 and 0–4°C.[a] Stabilized against precipitation by 0·02 M calcium by κ-casein. Also precipated from 6·6 M urea solutions of whole α-casein by addition of 12% TCA or H_2SO_4 to pH 1·3–1·5 and 2·2 M urea	Contains α_{s1}- and α_{s2}-caseins	5–7
α_{s1}-Casein	Major component of α_s-casein fraction	Homogeneous as isolated by DEAE-cellulose chromatography in 3·3 M urea	One zone for each polymorph or post-translational variant
α_{s2}-Casein	Minor component of α_s-casein fraction	Homogeneous as isolated by DEAE-cellulose chromatography in 3·3 M urea	One zone for each polymorph or post-translational variant

β-Casein	Soluble in 6·6 M urea but insoluble in 1·7 M urea, at pH 4·6.[a] Temperature-dependent solubility at low calcium concentration. (0·01–0·03 M: soluble at 4°C)	Homogeneous as isolated by DEAE-cellulose chromatography in 3·3 M urea	Alkaline urea gel gives one band for A^1, A^2 and A^3. Acid urea gel resolves A^1, A^2, A^3, C and **B**, but not D and B. B_Z is not resolved from **B** under any conditions
Fraction S	Fraction of whole casein soluble in 0·25 M $CaCl_2$ at pH 7 and 37°C	Contain κ-, β- and λ-caseins	6–10
κ-Casein	Major constituent of Fraction S. Fraction of whole α-casein soluble in 0·4 M $CaCl_2$ at pH 7 and 0–4°C.[a] The only milk protein capable of stabilizing α_s-caseins against precipitation by 0·02 M calcium and affected by chymosin in the primary phase of proteolysis	Homogeneous as isolated by DEAE-cellulose chromatography in 3·3 M urea	One zone for each polymorph when disulphide bonds are reduced
Whole γ-casein	Fraction of whole casein soluble in 3·3 M urea but insoluble in 1·7 M urea, at pH 4·7 on addition of ammonium sulphate[c]	Contains γ_1-, γ_2- and γ_3-caseins	3–6 zones by alkaline urea-disc gel
β-Casein (f29–209) (γ_1-Casein)	The fraction of whole γ-casein eluting from DEAE-cellulose with 0·02 M phosphate buffer at pH 8·3	Homogeneous	Alkaline urea-disc gel gives one band for A^1, A^2 and A^3. These can be resolved by acid urea–disc gels
β-Casein (f106–209) and β-casein (f108–209) (γ_2- and γ_3-caseins)	Fractions of whole γ-casein separable from γ_1-casein by ion-exchange chromatography	Homogeneous	One zone for each polymorph. γ_2-A and γ_3-**B** are not separated

TABLE 3—*contd.*

Protein fraction	*Characteristics*	*Composition*	*Number of zones in alkaline urea–gel electrophoresis*
λ-Caseins	An ill-defined fraction of whole α-casein soluble in 0·4 M $CaCl_2$ at pH 7 and 0–4°C.[a] Not capable of stabilizing α_s-caseins	Heterogeneous (possibly contains component 8)	?

[a]Definition of the fraction assigned by the 1965 Nomenclature Committe report (Brunner *et al.*, 1960).
[b]The number of zones observed depends upon, among other things, the milk sources, e.g. individual milk or pooled milks.
[c]Definition of the fraction assigned by the 1970 Nomenclature Committe report (Thompson *et al.*, 1965).

TABLE 4
Relative Mobilities of Various Caseins as Determined by Alkaline Urea-Starch or Polyacrylamide Gel Electrophoresis

Protein	R_m^a	
	Alkaline SGE	*Alkaline PAGE*
α_{s1}-Caseins		
B-9P	1·18 [c]	1·28 [b,d]
A-8P	1·18 [e]; 1·15 [f]	1·22 [e]
D-8P	?	1·15 [e]
B-8P	1·10 [e]; 1·10 [f]	1·13 [e]
C-8P	1·07 [e]; 1·07 [f]	1·10 [e]
α_{s2}-Caseins A		
A-13P	1·11 [c]; 1·07 [b,g]	1·13 [b,d]
A-12P	1·06 [c]; 1·04 [b,g]	1·04 [b,h]; 1·07 [b,d]
A-11P	1·00	1·00
A-10P	0·97 [b,g]	0·92 [b,d]
β-Caseins		
A[1], A[2] & A[3]-5P	0·80 [e]; 0·80 [f]; 0·80 [i]	0·65 [e]; 0·69 [b,h]
B-5P	0·76 [e]; 0·73 [f]; 0·77 [i]	0·61 [e]; 0·65 [b,h]
D-4P	?	0·57 [e]
C-4P	0·70 [e]	0·54 [e]
κ-Caseins		
A-1P	0·60 [f]; 0·58 [b,g]; 0·53 [i]	0·45 [b,j]; 0·40 [b,d]
B-1P	0·52 [f]; 0·52 [b,g]; 0·46 [i]	0·40 [b,j]; 0·36 [b,d]
γ-Caseins		
γ_1-A β-CN A[1], A[2] & A[3]-1P(f29–209)	?	0·40 [b,k]
β-CN B-1P (f29–209)	?	0·36 [b,k]
β-CN A[2] & A[3] (f106–209)	?	0·28 [b,k]
β-CN B(f106–209)	?	0·16 [b,k]
β-CN A(f108–209)	?	0·40 [b,k]
β-CN B(f108–209)	?	0·28 [b,k]

[a]Values are relative to the migration of α_{s2}-CN A-11P.
[b]Calculated by the author from figures appearing in the reference.
The other superscripts refer to the following references: [c]Annan & Manson (1969); [d]Davis & Law (1977); [e]Thompson (1970); [f]Schmidt (1964); [g]Snoeren *et al.* (1980); [h]Kiddy (1975); [i]El-Negoumy (1966); [j]Schmidt *et al.* (1966); [k]Groves (1969).

individual cows is usually chosen as a reference for calculation of relative mobilities. The zone originally chosen has now been identified as α_{s2}-CN A-11P (α_{s4}-casein) (Whitney *et al.*, 1976). Representative values for the electrophoretic mobilities of various caseins are listed in Table 4. These values exhibit some variation between laboratories since the degree of resolution is dependent on many experimental variables; however, when a specific set of conditions is chosen, the values can be very reproducible (El-Negoumy, 1966). Also, the value would change if a different genetic variant of α_{s2}-casein was present. Fortunately, α_{s2}-CN A-11P is quite prevalent in western breeds. Some forms of α_{s2}-casein are not resolved from certain species of α_{s2}-casein and some forms of κ-casein are not resolved from β-caseins(f29–209) (formerly known as γ_1-caseins). Alkaline-urea PAGE of κ-caseins yields multiple zones due to post-translational incorporation of varying amounts of carbohydrate (Schmidt, 1964; Mackinlay *et al.*, 1966; Schmidt *et al.*, 1966). Perhaps the most conclusive method of identifying κ-casein is to observe the effect of chymosin treatment on the electrophoretic pattern (Kim *et al.*, 1969; Van Eenennaam & Medrano, 1991) since the enzyme specifically cleaves the Phe_{105}—Met_{106} peptide bond. Alkaline-urea PAGE does not resolve β-CN A^1, A^2 or A^3 variants or the corresponding γ-caseins; hence, these variants were first discovered, and are now identified, by acid-urea PAGE (Peterson & Koppler, 1966).

3.3. Fractionation and Identification by Column Chromatography

3.3.1. Fractionation of Caseins

Due to differences in their isoelectric points and, consequently, net charge at other pH values, the caseins can be fractionated by ion-exchange chromatography. As with zone electrophoresis, the casein complexes must be dissociated for effective resolution of individual proteins. For example, DEAE-cellulose anion exchange chromatopraphy in urea was one of the first methods used to fractionate whole casein (Ribadeau-Dumas, 1961). The sample was applied in pH 7, 0·01 M imidazole–4·5 M urea buffer and eluted with a 0–0·6 M NaCl gradient in pH 7, 0·02 M imidazole–3·3 M urea buffer (Ribadeau-Dumas, 1961; Ribadeau-Dumas *et al.*, 1964). However, this method does not completely resolve whole casein, so purified α_{s1}-and β-casein were obtained by chromatography of crude α_{s1}- or β-casein fractions under similar conditions (Thompson & Kiddy, 1964; Thompson & Pepper, 1964) or by an additional ion-exchange chromatographic purification under different conditions (Garnier *et al.*, 1964).

β-Casein appears to be tenaciously associated with κ-casein (Swaisgood & Brunner, 1962), so the complete elimination of κ-casein during chromatographic purification of β-casein is very difficult (Garnier *et al.*, 1964; Thompson, 1966). Reduction of the disuphide bonds in κ-casein apparently weakens this interaction, allowing improved chromatographic resolution. Thus, DEAE-cellulose chromatography of whole casein using a 0–0·3 M NaCl gradient in pH 7, 0·01 M imidazole–3·3 M urea buffer containing 2-mercaptoethanol gives the best resolution of individual proteins (Thompson, 1966). In fact, κ-casein components cannot be purified by ion-exchange chromatrography without first reducing the disulphide bonds. As would be predicted from their isoelectric points, the order of elution from DEAE-cellulose is γ-caseins, κ-caseins, β-caseins, α_{s2}-caseins and finally the α_{s1}-caseins.

Recently, anion-exchange chromatographic isolation of individual caseins similar to the DEAE-cellulose procedure has been scaled-up by using a QAE Zeta Prep 250 cartridge (Ng-Kwai-Hang & Pélissier, 1989). Using a stepwise elution with NaCl from 0 to 0·3 M, rather than a gradient, this method resolved roughly 13 g of whole casein into six fractions containing in order of their elution, γ-caseins,

κ-caseins, β-caseins, β-caseins, α_{s1}-caseins and α_{s2}-caseins, and α_{s1}-casein, respectively. These relatively homogeneous fractions should be easily purified by classical ion-exchange methods (see above) or by newer HPLC methodologies (see below).

Chromatographic resolution of individual molecular components of κ-casein is particularly challenging since, in addition to the two genetic variants, post-translational incorporation of both phosphate and oligosaccharide can occur. Moreover, artifacts can be introduced by long exposure to cencentrated urea solutions due to carbamylation of lysyl residues (Kim *et al.*, 1969) or by hydrolysis of N-acetylneuraminic acid (AcNeu) residues during exposure to strong acid in the urea–H_2SO_4 method of preparation (Doi *et al.*, 1979*a*). Hence, urea which has been freshly purified by an ion-exchange treatment should be used in all isolation procedures. Also, the urea–H_2SO_4 method has been modified by reducing the amount of acid used (Doi *et al.*, 1979*a*). It appears that variable amounts of both phosphate (Vreeman *et al.*, 1977, 1986) and AcNeu are post-translationally incorporated. Since many forms are present in only small amounts, isolation of the various molecular forms of κ-casein requires the use of crude preparations as the starting materials (Schmidt *et al.*, 1966; Woychik *et al.*, 1966; Vreeman *et al.*, 1977; Doi *et al.*, 1979*a*, *b*; Takeuchi *et al.*, 1985; Vreeman *et al.*, 1986). The components elute from DEAE-cellulose in order of their net charge, with molecules containing one phosphate group and lacking carbohydrate eluting first. Most recently, nine components were isolated by DEAE-cellulose chromatography in imidazole-urea-2-mercaptoethanol with a NaCl gradient (Takeuchi *et al.*, 1985) and ten components were obtained by a two-step chromatographic procedure using DEAE-Sepharose with a NaCl gradient in Tris-urea-2-mercaptoethanol followed by a pH gradient elution from DEAE-cellulose (Vreeman *et al.*, 1986).

Agents other than urea have been used to dissociate casein complexes during DEAE anion-exchange chromatography. For example, incorporation of 20% (v/v) dimethylformamide in buffers yields chromatography results similar to those obtained with urea (Mackinlay & Wake, 1965).

Chromatographic methods other than ion-exchange can be used to isolate κ-casein or κ-casein-rich fractions. A number of years ago, the author observed that, without disulphide reduction, κ-casein is not completely dissociated in concentrated protein denaturants (Swaisgood *et al.*, 1964). Accordingly, substantially purified κ-casein has been isolated by gel chromatography on Sephadex G-150 in pH 8·6 tris-citrate-6M urea buffer from which it eluted in the void volume (Yaguchi *et al.*, 1966). Of course, this procedure cannot separate the components of κ-casein but this preparation can be used as starting material for further fractionation with ion-exchange media (Doi *et al.*, 1979*a*). Chromatographic methods that are specific for certain residues or surface features unique to specific proteins have also been exploited for the isolation of specific caseins. Since only κ-caseins and α_{s2}-caseins contain sulphydryl groups when whole casein is reduced, these proteins can be isolated from other caseins by covalent chromatography at pH 7 in urea on thiol matrices such as Thiol-Sepharose (Nijhuis & Klostermeyer, 1975). Further ion-exchange chromatography of this fraction should separate the κ- and α_{s2}-caseins. Unique binding of polyanions to a positively charged cluster on the surface of κ-caseins has also been observed which results in selective binding to matrices such as κ-carrageenan or cation-exchange resins such as Amberlite (Snoeren *et al.*, 1977).

The γ-casein fraction can be resolved by DEAE-cellulose chromatography in the absence of urea (Groves *et al.*, 1962; Groves & Gordon, 1969). Thus, whole casein is absorbed and the column eluted with 5 mM sodium phosphate, pH 8·3, which removes both β-CN A^2, A^3 and B(f106–209) and β-CN A and B(f108–209). When the buffer is changed to 20 mM phosphate, pH 8·3, β-CN A^1, A^2, A^3 and B(f29–209) are eluted (Groves & Gordon, 1969); an operation that was used originally to define this γ-casein fraction (Whitney *et al.*, 1976).

3.3.2. Quantification of Proteins in the Casein Fraction

Because of the difficulty of quantifying protein-stained zones following gel electrophoresis, the excellent resolution of more recent chromatographic methods make the latter the preferred techniques for determination of the amounts of various milk proteins in a sample. Initial studies of the quantities of various caseins in whole casein employed DEAE-cellulose chromatography in imidazole-urea buffers without disulphide reduction (Ribadeau-Dumas *et al.*, 1964). Resolution was improved by reducing the disulphide and subsequent alkylation of the protein (Rose *et al.*, 1969; Davies & Law, 1977). For example, nearly complete resolution of γ-, κ-, β-, α_{s2}- and α_{s1}-casein (in order of elution) was achieved by DEAE-cellulose chromatography at 4°C in pH 8·6, 5 mM Tris, 6 M urea using a 0·03–0·22 M NaCl gradient (Davies & Law, 1977). Quantification of κ-casein remained questionable due to incomplete resolution. With the advent of HPLC-type matrices, similar resolution could be achieved in a much shorter time—usually about 20 min (Humphrey & Newsome, 1984; Barrefors *et al.*, 1985). Thus, whole casein has been analysed using a Mono Q FPLC column with the protein in pH 7·0, 10 mM imidazole–3·3 M urea buffer containing 10 mM 2-mercaptoethanol (Barrefors *et al.*, 1985) or with a TSK DEAE-5PW column using a pH 8·0, 20 mM Tris–4·5 M urea buffer (Humphrey & Newsome, 1984). More recently, rather good resolution of various κ-casein species was achieved using Mono Q FPLC (Dalgleish, 1986; Guillou *et al.*, 1987).

Varying degrees of resolution of genetic and post-translationally modified variants are obtained by the above chromatographic methods. Generally, the genetic variants are not resolved except for those variants differing by multiple charges. Usually, α_{s1}-CN B-9P appear as a shoulder on the α_{s1}-casein peak although in some cases it is resolved. As noted previously, the post-translationally modified forms of κ-casein are the most difficult to separate.

Whole casein can also be resolved by chromatography on hydroxyapatite (Donnelly, 1977; McGann *et al.*, 1979). Caseins can be applied and eluted in buffers containing a disulphide reducing agent or the protein can be reduced and alkylated prior to chromatography similar to the methodology for ion-exchange chromatography. For example, protein is loaded onto the column in pH 6·8, 5 mM phosphate–6 M urea and eluted first with 100 mM phosphate–6 M urea followed by a linear gradient of 100 to 450 mM phosphate–6 M urea at 20°C. The proteins elute according to their phosphate content, viz. κ-caseins, followed in order by β-caseins, α_{s1}-caseins, and α_{s2}-casein. Resolution does not appear to be as good as that obtained with the latest ion-exchange methods.

4. Primary Structure and Chemical Composition of the Caseins

4.1. Chemical Composition

Since the previous review (Swaisgood, 1982*a*), the primary structures of the four caseins, including most of the common genetic variants, have been refined and substantiated by analysis of DNA sequences (see below). Hence, the compositions deduced from these primary structures can be reported with great accuracy. The compositions of specific genetic variants which are commonly observed in milks of western cattle are listed in Table 5 for each of the four caseins. Also shown are the compositions for the γ-caseins which occur naturally in milk as a result of limited proteolysis of β-caseins by plasmin.

Compositions of other genetic variants are given in Table 6 by listing the number of residues for those residues that are changed from the variant given in Table 5. The amino acid compositions are not unlike those of typical globular proteins. Even the average hydrophobicities are only slightly higher than the range of values (4·18 to 4·60 kJ/residue) obtained for a majority of globular proteins; fibrous proteins typically have much lower average hydrophobicities. The caseins also exhibit a relationship between average hydrophobicity and molecular weight similar to that for other proteins that undergo self-association. The α_{s2}-caseins are the least hydrophobic and possess the highest charge frequency (e.g. 0·36 for α_{s2}-CN A-11P) while β-caseins are significantly more hydrophobic than other caseins (Table 5).

Caseins are unique as a result of their post-translational modifications. All caseins are phosphorylated to varying extents at seryl and occasionally threonyl residues. Genetic variants of κ-, α_{s1}-, and β-caseins usually contain a characteristic number of phosphoseryl residues, *viz.* 1, 8 or 9 and 4 or 5, respectively (Tables 1, 5, and 6) (Eigel *et al.*, 1984; Whitney, 1988). A minor form of α_{s1}-casein occurs due to the occasional phosphorylation of Ser 41 (Manson *et al.*, 1977) due to its occurrence in the less frequently phosphorylated sequence, SXD. α_{s1}-Casein D contains a phosphothreonyl residue (Grosclaude *et al.*, 1972; Ribadreau-Dumas *et al.*, 1973); the minor component of this variant would have 10 phosphate groups and is named α_{s1}-CN D-10P (Eigel *et al.*, 1984). The α_{s2}-casein family of proteins exhibits the greatest variability in the extent of phosphorylation; for example, α_{s2}-CN A has 10–13 phosphaseryl residues (Brignon *et al.*, 1976, 1977; Eigel *et al.*, 1984; Whitney, 1988). At present it is not known whether this variability results from different extents of phosphorylation by caseiñ kinases or from varying degrees of dephosphorylation by phosphatases. The major components of κ-casein

TABLE 5
Chemical Composition of the Commonly Occurring Caseins[a]

Acid	*α_{s1}-CN B-8P*	*α_{s2}-CN A-11P*	*κ-CN B-1P*	*β-CN A^2-5P*	*β-CN A^2-1P(f29–209)*	*β-CN A^2(f106–209)*	*β-CN A(f108–209)*
Asp	7	4	3	4	4	2	2
Asn	8	14	8	5	3	1	1
Thr	5	15	14	9	8	4	4
Ser	8	6	12	11	10	7	7
SerP	8	11	1	5	1	0	0
Glu	25	24	12	19	12	5	5
Gln	14	16	14	20	20	10	10
Pro	17	10	20	35	34	21	21
Gly	9	2	2	5	4	2	2
Ala	9	8	15	5	5	2	2
$\frac{1}{2}$ Cys	0	2	2	0	0	0	0
Val	11	14	11	19	17	10	10
Met	5	4	2	6	6	4	4
Ile	11	11	13	10	7	3	3
Leu	17	13	8	22	19	14	14
Tyr	10	12	9	4	4	3	3
Phe	8	6	4	9	9	5	5
Trp	2	2	1	1	1	1	1
Lys	14	24	9	11	10	4	3
His	5	3	3	5	5	4	3
Arg	6	6	5	4	2	2	2
Pyr or Glu	0	0	1	0	0	0	0
Total residues	199	207	169	209	181	104	102
Molecular weight	23 623	25 238	19 006	23 988	20 523	11 824	11 559
H_{ave} (kJ/residue)[b]	4·89	4·64	5·12	5·58	5·85	6·23	6·29

[a]Based on their primary structures.
[b]Values for individual residues taken from Bigelow (1967).

TABLE 6
Composition of Casein Variants as They Differ from the Composition of the Variants Listed in Table 5[a]

Variant	*Composition*
Differing from α_{s1}-CN B	
α_{s1}-CN A	Asn_6, Glu_{23}, Ala_8, Val_9, Leu_{14}, Phe_6, Arg_5
α_{s1}-CN C	Glu_{24}, Gly_{10}
α_{s1}-CN D	Ala_8, ThrP
α_{s1}-CN E	Gln_{13}, Glu_{24}, Gly_{10}, Lys_{15}
Differing from κ-CN B	
κ-CN A	Asp_4, Ala_{14}, Thr_{15}, Ile_{12}
Differing from β-CN A^2	
β-CN A^1	Pro_{34}, His_6
β-CN A^3	Gln_{21}, His_4
β-CN B	Pro_{34}, His_6, Ser_{10}, Arg_5
β-CN C	Glu_{18}, $SerP_4$, Ser_{12}, Lys_{12}, Pro_{34}, His_6
β-CN D	$SerP_4$, Lys_{12}
β-CN E	Glu_{18}, Lys_{12}

[a]See the reviews in Eigel *et al.* (1984) and Whitney (1988) and the primary structures given in Figs 2, 5 and 6.

contain only one phosphoseryl residue but some evidence indicates a minor amount of doubly phosphorylated protein (Vreeman *et al.*, 1977).

In addition to post-translational phosphorylation, the κ-caseins are also glycosylated. The result is extensive heterogeneity of this family of proteins, as indicated earlier.

This heterogeneity is manifested by the numerous zones observed by alkaline-urea PAGE (Swaisgood, 1975, 1982*a*) and the 8–10 fractions obtained by ion-exchange chromatography (Doi *et al.*, 1979*b*; Takeuchi *et al.*, 1985; Vreeman *et al.*, 1986). Nevertheless, the non-glycosylated form is the major component and the sum of κ-CN B-1P and κ-CN B-1P(3AcNeu) accounts for about 80% of whole κ-CN B (Vreeman *et al.*, 1986). In normal milk, the carbohydrate moiety is composed of three monosaccharides: AcNeu, Gal and GalNAc (Tran & Baker, 1970; Fiat *et al.*, 1972; Doi *et al.*, 1979*b*). Structural work in a number of laboratories (Tran & Baker, 1970; Fiat *et al.*, 1972; Fournet *et al.*, 1979; Jollés & Fiat, 1979; Saito *et al.*, 1980) has established the presence of either trisaccharide

$$\text{AcNeu}\xrightarrow{\alpha\text{-}2,3}\text{Gal}\xrightarrow{\beta\text{-}1,3}\text{GalNAc}\xrightarrow{\beta\text{-}1}\text{Thr}$$

or tetrasaccharide

$$\begin{array}{c}\text{AcNeu}\xrightarrow{\alpha\text{-}2,3}\text{Gal}\xrightarrow{\beta\text{-}1,3}\text{GalNAc}\xrightarrow{\beta\text{-}1}\text{Thr}\\ \qquad\qquad\qquad\quad \big|\ \alpha\text{-}2,6\\ \qquad\qquad\qquad\quad \text{AcNeu}\end{array}$$

carbohydrate moieties attached to threonyl residues of κ-casein. Further studies

(Jollés *et al.*, 1972*a*, 1973; Kanamori *et al.*, 1980; Doi *et al.*, 1980; Zevaco & Ribadeau-Dumas, 1984) identified the number and sites of glycosylation as threonyl residues 131,133,135,136 and 142 in κ-CN A-1P and 131,133,135 and 142 in κ-CN B-1P. Serine 141 may also be glycosylated (Zevaco & Ribadeau-Dumas, 1984). Thus, the minor components of κ-casein can contain variable numbers of tri- and tetrasaccharide moieties attached to at least 1 to 4 threonyl residues (Doi *et al.*, 1979*b*; Takeuchi *et al.*, 1985; Vreeman *et al.*, 1986). The carbohydrate moieties in κ-casein from colostrum are more complex and contain *N*-acetylglucosamine (GlcNAc) in addition to the three usual monosaccharide constituents (Fournet *et al.*, 1979; Doi *et al.*, 1980).

Two other features of casein composition are important to their structure and their chemical interactions. One is the cystinyl and cysteinyl residues in κ- and α_{s2}-caseins which allow these proteins to participate in sulphydryl–disulphide interchange. Another unique characteristic is the relatively large amount of prolyl residues, especially in β-casein, which should significantly affect protein structure since these residues interrupt α-helical and β-sheet structures and usually occur in β-turns.

Certain physical–chemical parameters can be calculated on the basis of accurate composition data. Values for several of these parameters computed from the composition of each casein as determined by its primary structure are listed in Table 7. The predicted values agree rather well with those available from experimental observations. The net charge was calculated for each protein at pH 6·6 since this is roughly the pH of milk. The net charge for κ-casein represents only that for the protein, exclusive of the carbohydrate moiety. For example, κ-CN β-1P(3AcNeu) would have a net charge of roughly −5. The pKs used in the calculation of these values were obtained from the fit to titration data (Ho & Waugh, 1965; Nozaki & Tanford, 1967; Creamer, 1972). Agreement betwen the composition and the number of various ionisable residues is relatively good. Furthermore, independent determination of the net charge at pH 6·6 for α_{s1}-casein B by electrophoresis (Schmidt, 1969) and proton release (Waugh *et al.*, 1971) gave values of −22·8 and −22, respectively. Similar data from proton release for β-casein A indicated a net charge at pH 6·6 of −13 (Waugh *et al.*, 1971). Binding of ions, particularly calcium, will alter the net charge on these molecules. Such binding modulates the balance of electrostatic forces and hydrophobic interactions which is manifested in the self-association and/or inter-association of the caseins.

Experimental values for the *isoionic points* are: 5·16 (Ho & Waugh, 1965) and 5·05 for α_{s1}-casein, 5·30 (Creamer, 1972) and 5·35 (Ho & Waugh, 1965) for β-casein and 5·37 (Ho & Waugh, 1965) for κ-casein. Relative zonal electrophoretic mobilities (Table 4) can also be explained on the basis of net charge and size. For example, the net charge ranking of α_{s1}-caseins is D > B > A ~ C > E; however, α_{s1}-CN A-8P has a deletion of 13 residues resulting in a higher charge density and lower hydrodynamic radius. Hence, the expected relative mobilities would be A > D > B > C > E (Whitney, 1988). In summary, all of the experimental data and calculations indicate dissociation constants for all prototropic groups in the

TABLE 7
Physical-Chemical Characteristics of Casein Calculated From Composition

Protein	*Charge at pH 6·6*[a]	*Isoionic pH*[b]	*Partial specific volume*[c]	*Absorptivity*[d] *(cm²/mg) 280 nm*	*Molecular weight*
α_{s1}-Casein					
A-8P	−21·0	4·94	0·722	1·13	22 077
B-8P	−21·9	4·94	0·725	1·05	23 623
C-8P	−20·9	4·97	0·725	1·06	23 551
D-9P	−23·5	4·88	0·723	1·05	23 734
α_{s2}-Casein					
A-10P	−12·2	5·45	0·722	1·11	25 157
A-11P	−13·8	5·37	0·721	1·10	25 238
A-12P	−15·5	5·30	0·720	1·10	25 319
A-13P	−17·1	5·23	0·718	1·10	25 400
β-Casein					
A^3-5P	−13·8	5·07	0·742	0·46	23 979
A^2-5P	−13·3	5·14	0·742	0·46	23 988
A^1-5P	−12·8	5·22	0·742	0·46	24 028
B-5P	−11·8	5·29	0·742	0·46	24 097
C-4P	−9·2	5·46	0·742	0·46	23 946
κ-Casein					
A-1P	−3·0	5·61	0·734	0·95	19 038
B-1P	−2·0	5·90	0·734	0·95	19 006

[a] Calculated using the relation $Z_H = (Z_H)_{max} - r$, where $r = \Sigma_i(n_i k_i/a_H)/(1 + k_i/a_H)$. The apparent pK_i values used where: α-carboxyl, 3·6 (Nozaki & Tanford, 1967); phosphoseryl, $pK_1 = 1{\cdot}5$, $pK_2 = 6{\cdot}4$ (Creamer, 1972); β, γ-carboxyl, 4·9 (Ho & Waugh, 1965); histidyl, 6·6 (Nozaki & Tanford, 1967); and α-amino, 7·4 (Nozaki & Tanford, 1967) except for κ-casein where the more common value for β,γ-carboxyl of 4·6 was used (Nozaki & Tanford, 1967). If a pk_{app} of 4·9 is used the isoionic point is 5·84 for κ-A and 6·06 for κ-B.
[b] Calculated as above noting that $Z_H = 0$ at the isoionic pH.
[c] Calculated according to McMeekin *et al.* (1949).
[d] Calculated according to Wetlaufer (1962).

caseins that are in the range of values for model compounds, i.e. there are no anomolous dissociation constants required to reconcile electrochemical and compositional data. Consequently, casein structure must allow for ready accessibility of these groups to the solvent.

Calculated partial specific volumes and specific absorptivities are also in accord with available experimental values. Measured values for specific volumes and absorptivities, respectively, are: α_{s1}-caseins, 0·728 ml/g (McKenzie & Wake, 1959) and 1·01 cm²/mg (Thompson & Kiddy, 1964; Thompson *et al.*, 1969); β-caseins, 0·741 ml/g (McMeekin *et al.*, 1949; Noelken & Reibstein, 1968) and 0·46 cm²/mg (Thompson & Pepper, 1964) and κ-caseins 0·739 (Vreeman *et al.*, 1981) and 0·96 cm²/mg (Talbot & Waugh, 1970).

Finally, it should be noted that the molecular weights listed in Table 7 correspond to κ-CN A-1P and B-1P. Each trisaccharide chain of the structure given would add 657 daltons and each tetrasaccharide of the listed structure would add 948 daltons to the molecular weight. For example, κ-CN B-1P(3 AcNeu) would have a mass of 20 977 daltons if it contained three trisaccharides.

4.2. Primary Structures

Knowledge of the sequence of residues in a protein chain provides much more insight into the molecular structure than does composition alone. Since the native structure is an equilibrium thermodynamic state representing the lowest free energy for inter-residue and residue–solvent interactions for a particular sequence, in principle it should be possible to compute the three-dimensional structure from the primary structure. In practice, the extrapolation from primary to tertiary structure is not yet possible, although considerable progress has been made in recent years. Some predictions of the three-dimensional structure of caseins based on such computations are presented in Chaper 13 of this volume.

Also, some success has been achieved in the prediction of secondary structures based on the primary structure. Some years ago, Chou and Fasman (1974*a,b*, 1977) developed an empirical method for prediction of secondary structure by correlation of the X-ray crystallographic structures for a large number of proteins with their respective primary structures. The empirical methods for prediction of secondary structures have since been refined (Lim, 1974; Garnier *et al.*, 1978); however, their accuracy is still far from perfect (Garnier, 1990).

Enzymatic post-translational modification of a protein usually requires a recognition sequence and accessibility of this sequence to the enzyme. Residues at or near β-turns are especially accessible. Consequently, sequence information and structural predictions based on the sequence are extremely useful for locating the most likely sites of post-translational modification.

Elucidation of the gene structure for the caseins has also permitted some speculation on their evolution (Bonsing & Mackinlay, 1987). Comparison of the genomic sequences from a number of species indicates that the calcium-sensitive caseins (α_{s1}-, α_{s2}-and β-caseins) are members of a gene family that arose from a common ancestral gene (Bonsing & Mackinlay, 1987) whereas the κ-casein gene has similarities with that of γ-fibrinogen (Thompson *et al.*, 1985). These comparisons also allow identification of homology and conserved structure which suggest important features for the function of the protein.

In the discussion to follow, the reader should remember the distinction between random coil and unordered structure. All globular proteins contain some degree of unordered structure; nevertheless, the conformation in these regions is fixed with respect to the ψ and ϕ angles of the peptide bonds *vis-a-vis* the rapid fluctuation of these bond angles for a random coil polypeptide.

4.2.1. α_{s1}-Caseins

The primary structures of the known genetic variants are shown in Fig. 2. These sequences were obtained chemically for each of the genetic variants (Mercier *et al.*, 1971; Grosclaude *et al.*, 1972, 1973, 1976*b*). The primary structure of α_{s1}-CN B has

1
H.Arg-Pro-Lys-His-Pro-Ile-Lys-His-Gln-Gly-Leu-Pro-Gln- Glu-Val-Leu-Asn-Glu-Asn-Leu-

21
Leu-Arg-Phe-Phe-Val-Ala(Variant B,C,D,E,)
-Pro-Phe-Pro-Glu-Val-Phe-Gly-Lys-Glu-Lys-Val-Asn-Glu-Leu
(Variant A)

41
Ala(Variants A,B,C,E) Gln(Variants A,B,C,D)
Ser-Lys-Asp-Ile-Gly-*SerP*-Glu-*SerP*-Thr-Glu-Asp-Gln- -Met-Glu-Asp-Ile-Lys- -Met-
ThrP(Variant D) Glu(Variant E)

61
Glu-Ala-Glu-*SerP*-Ile-*SerP*-*SerP*-*SerP*-Glu-Glu-Ile-Val-Pro-Asn-*SerP*-Val-Glu-Gln-Lys-His-

81
Ile-Gln-Lys-Glu-Asp-Val-Pro-Ser-Glu-Arg-Tyr-Leu-Gly-Tyr-Leu-Glu-Gln-Leu-Leu-Arg-

101
Leu-Lys-Lys-Tyr-Lys-Val-Pro-Gln-Leu-Glu-Ile-Val-Pro-Asn-*SerP*-Ala-Glu-Glu-Arg-Leu-

121
His-Ser-Met-Lys-Glu-Gly-Ile-His-Ala-Gln-Gln-Lys-Glu-Pro-Met-Ile-Gly-Val-Asn-Gln-

141
Glu-Leu-Ala-Tyr-Phe-Tyr-Pro-Glu-Leu-Phe-Arg-Gln-Phe-Tyr-Gln-Leu-Asp-Ala-Tyr-Pro-

161
Ser-Gly-Ala-Trp-Tyr-Tyr-Val-Pro-Leu-Gly-Thr-Gln-Tyr-Thr-Asp-Ala-Pro-Ser-Phe-Ser-

181 199
Glu(Variant A,B,D)
Asp-Ile-Pro-Asn-Pro-Ile-Gly-Ser-Glu-Asn-Ser- -Lys-Thr-Thr-Met-Pro-Leu-Trp.OH
Gly(Variant C,E)

Fig. 2. Primary structures of the known genetic variants of α_{s1}-casein (Mercier *et al.*, 1971; Grosclaude *et al.*, 1972, 1973, 1976*b*). Sites of post-translational phosphorylation are in italics.

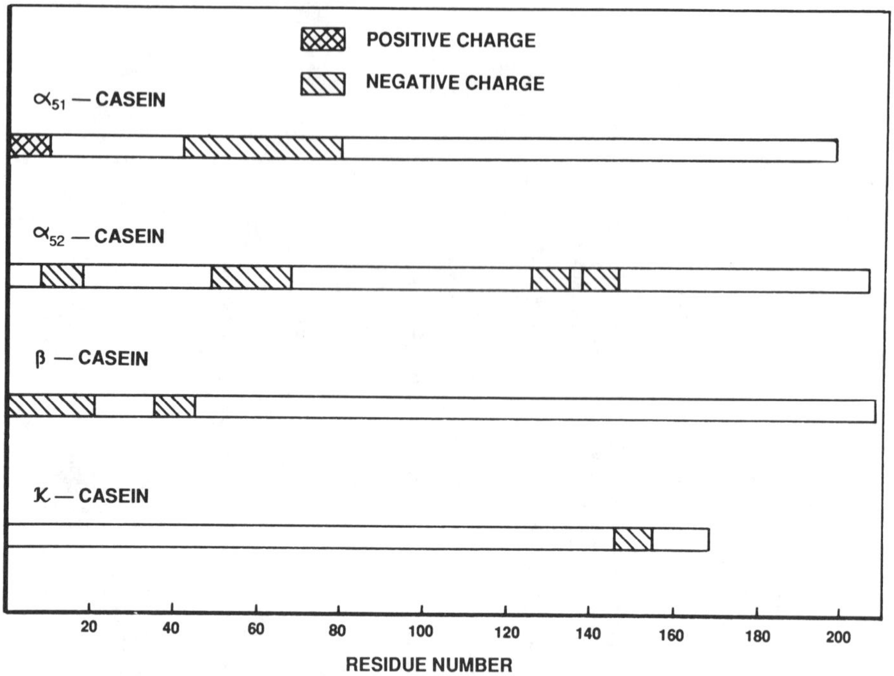

Fig. 3. Ten residue sequences of bovine milk proteins with a charge density of 0·5 or greater at the pH of milk.

since been substantiated from the sequence of the cDNA for this variant as obtained by two independent studies (Nagao *et al.*, 1984; Stewart *et al.*, 1984). The cDNA sequences obtained are identical except for the base in the 'wobble' position corresponding to A163 and P185. On the basis of the cDNA sequence, the primary structure given in the previous review (Swaisgood, 1982*a*) has been corrected by the change Q30E, with the appropriate corrections to Tables 5 and 7.

In common with other caseins, hydrophobic and charged residues are not uniformly distributed in the sequence. The unique clustering of these residues becomes more apparent by comparison of their distribution in casein sequences with that for the compact globular whey prtoeins β-lactoglobulin and α-lactalbumin (see Fig. 7). Three hydrophobic regions are discernible, roughly including residues 1–44, 90–113 and 132–199. However, clustering of the phosphoseryl residues is perhaps the most unique feature of the primary structure of the calcium-sensitive caseins. The cluster sequence EpSXpSpSpSEE, where X is I or L and pS is SerP, occurs in all known variants of α_{s1}- and β-caseins. Similar sequences, pSpSpSEE and pSpSpSEEpSAE, occur in α_{s2}-caseins. Sequences such as these, which are present in the polar domains of the calcium-sensitive casein, apparently are encoded by a number of short exons, thus allowing for their duplication and the rapid evolution of these proteins (Bonsing & Mackinlay, 1987).

The region encompassing residues 41–80 represents a very polar domain within

```
1
H.Lys-Asn-Thr-Met-Glu-His-Val-SerP-SerP-SerP-Glu-Glu-Ser-Ile-Ile-SerP-Gln-Glu-Thr-Tyr-

21
Lys-Gln-Glu-Lys-Asn-Met-Ala-Ile-Asn-Pro-Ser-Lys-Glu-Asn-Leu-Cys-Ser-Thr-Phe-Cys-

41
Lys-Glu-Val-Val-Arg-Asn-Ala-Asn-Glu-Glu-Glu-Tyr-Ser-Ile-Gly-SerP-SerP-SerP-Glu-Glu-

61
SerP-Ala-Glu-Val-Ala-Thr-Glu-Glu-Val-Lys-Ile-Thr-Val-Asp-Asp-Lys-His-Tyr-Gln-Lys-

81
Ala-Leu-Asn-Glu-Ile-Asn-Gln-Phe-Tyr-Gln-Lys-Phe-Pro-Gln-Tyr-Leu-Gln-Tyr-Leu-Tyr-

101
Gln-Gly-Pro-Ile-Val-Leu-Asn-Pro-Trp-Asp-Gln-Val-Lys-Arg-Asn-Ala-Val-Pro-Ile-Thr-

121
Pro-Thr-Leu-Asn-Arg-Glu-Gln-Leu-SerP-Thr-SerP-Glu-Glu-Asn-Ser-Lys-Lys-Thr-Val-Asp-

141
Met-Glu-SerP-Thr-Glu-Val-Phe-Thr-Lys-Lys-Thr-Lys-Leu-Thr-Glu-Glu-Glu-Lys-Asn-Arg-

161
Leu-Asn-Phe-Leu-Lys-Lys-Ile-Ser-Gln-Arg-Tyr-Gln-Lys-Phe-Ala-Leu-Pro-Gln-Tyr-Leu-

181
Lys-Thr-Val-Tyr-Gln-His-Gln-Lys-Ala-Met-Lys-Pro-Trp-Ile-Gln-Pro-Lys-Thr-Lys-Val-
                                                (Leu)

201                 207
Ile-Pro-Tyr-Val-Arg-Tyr-Leu.OH
```

Fig. 4. Primary structure of α_{s2}-casein A.11P (Bigelow, 1967; Stewart *et al.*, 1987). Sites of post-translational phosphorylation identified in the species containing 11 orthophosphate groups are in italics.

an approximate net charge of −20·6 at pH 6·6 while the remainder of the molecule would have very little net charge. This characteristic segregation of like charges in polar domains of casein molecules is illustrated in Fig. 3 which indicates those regions having a charge density > 0·5. In addition to being essentially neutral, the remainder of the protein is rather hydrophobic, suggesting a structure composed of a more globular hydrophobic domain with an attached highly solvated polar domain as proposed in the previous review (Swaisgood, 1982*a*). Such domain structure is supported by recent hypothetical structures obtained by energy minimization calculations (Kumosinski *et al.*, 1991).

4.2.2. α_{s2}-Caseins

The primary structure of α_{s2}-casein A-11P is given in Fig. 4. The sequence determined chemically (Brignon *et al.*, 1977) and shown in the previous review (Swaisgood, 1982*a*) has been corrected by making the substitution E87Q as indicated by cDNA sequencing (Stewart *et al.*, 1987). Also, the cDNA sequence indicates that residue 193 should be Leu rather than Trp; however, the author chooses not to make this substitution until there is further confirmation of the structure. The latter substitution would change the specific absorptivity of this protein. Interestingly, the primary structure (Brignon *et al.*, 1977), as well as the cDNA sequence (Stewart *et al.*, 1987), suggests that this protein resulted from a large gene duplication and supports the concept of rapid evolution of these proteins.

This protein is the most hydrophilic of all the caseins (Table 5) with three clusters of anionic phosphoseryl and glutamyl residues in its structure (Figs 3 and 4), viz.

```
1
H.Arg-Glu-Leu-Glu-Glu-Leu-Asn-Val-Pro-Gly-Glu-Ile-Val-Glu-SerP-Leu-SerP-SerP-SerP-Glu-

                                    |→ γ1-caseins                 (Variant C)
21                                  |                             Ser      Lys
Glu-Ser-Ile-Thr-Arg-Ile-Asn-Lys-Lys-Ile-Glu-Lys-Phe-Gln-     -Glu-    -Gln-Gln-Gln-
                                                              SerP     Glu
                                                              (Variants A,B)
41
Thr-Glu-Asp-Glu-Leu-Gln-Asp-Lys-Ile-His-Pro-Phe-Ala-Gln-Thr-Gln-Ser-Leu-Val-Tyr-

61                Pro(Variants A2, A3)
Pro-Phe-Pro-Gly-Pro-Ile-     -Asn-Ser-Leu-Pro-Gln-Asn-Ile-Pro-Pro-Leu-Thr-Gln-Thr-

                  His(Variants C,A1, and B)

81
Pro-Val-Val-Val-Pro-Pro-Phe-Leu-Gln-Pro-Glu-Val-Met-Gly-Val-Ser-Lys-Val-Lys-Glu-

                                   |———————→ γ3-caseins
                                   |
101               His(Variants A1,A2, B,C)
Ala-Met-Ala-Pro-Lys-     -Lys-Glu-Met-Pro-Phe-Pro-Lys-Tyr-Pro-Val-Glu-Pro-Phe-Thr-
                     |Gln(Variant A3)
                     |——————————————————→ γ2-caseins
121 Ser(Variants A,C)
Glu-   -Gln-Ser-Leu-Thr-Leu-Thr-Asp-Val-Glu-Asn-Leu-His-Leu-Pro-Leu-Pro-Leu-Leu-
    Arg(Variant B)

141
Gln-Ser-Trp-Met-His-Gln-Pro-His-Gln-Pro-Leu-Pro-Pro-Thr-Val-Met-Phe-Pro-Pro-Gln-

161
Ser-Val-Leu-Ser-Leu-Ser-Gln-Ser-Lys-Val-Leu-Pro-Val-Pro-Gln-Lys-Ala-Val-Pro-Tyr-

181
Pro-Gln-Arg-Asp-Met-Pro-Ile-Gln-Ala-Phe-Leu-Leu-Tyr-Gln-Glu-Pro-Val-Leu-Gly-Pro-

201                         209
Val-Arg-Gly-Pro-Phe-Pro-Ile-Ile-Val.OH
```

Fig. 5. Primary structures of β-caseins A^1, A^2, A^3, B and C (Brignon *et al.*, 1970; Grosclaude *et al.*, 1972; Ribadeau-Dumas *et al.*, 1972; Grosclaude *et al.*, 1973, 1974*a*,*b*; Jimenez-Flores *et al.*, 1987; Stewart *et al.*, 1987; Carles *et al.*, 1988). Sites of post-translational phosphorylation are in italics. Sites of post-translational hydrolysis by plasmin yielding the resulting γ-caseins and components 5 [(β-CN-5P(f1–105 and 107)] and 8[β-CN-1P(f29–105 and 107) and β-CN-4P(f1–28)] are also indicated.

residues 8–12, 56–63 and 129–133. Only two areas of structure are relatively hydrophobic, including roughly residues 160–207 of the C-terminal sequence and residues 90–120 of the central structure (Fig. 7). However, the C-terminal sequence, although relatively hydrophobic, also possesses a large net positive charge. Thus, the C-terminal domain has a net charge of +9·5 at pH 6·6, while the more hydrophilic N-terminal 68 residues has a net charge of −21. Hence, the α_{s2}-casein structure is characterized by distinct domains of high net charge including, in particular, the three anionic phosphoseryl clusters; consequently, the structure and resulting physical properties of this protein are extremely sensitive to ionic strength and concentrations of cations such as protons and calcium.

Like the α_{s1}-caseins, one of the genetic variants, α_{s2}-CN D, results from a deletion in the sequence (Grosclaude *et al.*, 1979); however, in this case one of the anionic phosphoseryl clusters is deleted. The nine-residue sequence deleted is believed to be residues 50–58, 51–59 or 52–60. The resulting protein should have different structural and physical properties due to the absence of this phosphoseryl cluster; for example, it should be less sensitive to calcium and more hydrophobic than the other α_{s2} variants.

1
PyroGlu-Glu-Gln-Asn-Gln-Glu-Gln-Pro-Ile-Arg-Cys-Glu-Lys-Asp-Glu-Arg-Phe-Phe-Ser-Asp-

21
Lys-Ile-Ala-Lys-Tyr-Ile-Pro-Ile-Gln-Tyr-Val-Leu-Ser-Arg-Tyr-Pro-Ser-Tyr-Gly-Leu-

41
Asn-Tyr-Tyr-Gln-Gln-Lys-Pro-Val-Ala-Leu-Ile-Asn-Asn-Gln-Phe-Leu-Pro-Tyr-Pro-Tyr-

61
Tyr-Ala-Lys-Pro-Ala-Ala-Val-Arg-Ser-Pro-Ala-Gln-Ile-Leu-Gln-Trp-Gln-Val-Leu-Ser-

81
Asn-Thr-Val-Pro-Ala-Lys-Ser-Cys-Gln-Ala-Gln-Pro-Thr-Thr-Met-Ala-Arg-His-Pro-His-

101 105 106
Pro-His-Leu-Ser-Phe-Met-Ala-Ile-Pro-Pro-Lys-Lys-Asn-Gln-Asp-Lys-Thr-Glu-Ile-Pro-

121 Ile(Variant B)
Thr-Ile-Asn-Thr-Ile-Ala-Ser-Gly-Glu-Pro-*Thr*-Ser-*Thr*-Pro-*Thr*- -Glu-Ala-Val-Glu
Thr(Variant A)

141 Ala(Variant B)
Ser-*Thr*-Val-Ala-Thr-Leu-Glu- -*SerP*-Pro-Glu-Val-Ile-Glu-Ser-Pro-Pro-Glu-Ile-Asn-
Asp(Variant A)

161 169
Thr-Val-Gln-Val-Thr-Ser-Thr-Ala-Val.OH

Fig. 6. Primary structures of known genetic variants of κ-casein (Jollès *et al.*, 1970; Brignon *et al.*, 1972; Jollès *et al.*, 1972*b*; Mercier *et al.*, 1972*a*, 1973; Stewart *et al.*, 1984; Alexander *et al.*, 1988). Sites of post-translational phosphorylation and glycosylation are in italics.

4.2.3. β-Caseins

Primary structures for five of the most common genetic variants of this protein are shown in Fig. 5. The structures shown, corrected from that in Swaisgood (1982*a*), are based on both chemical sequencing of the protein (Bringnon *et al.*, 1970; Grosclaude *et al.*, 1972; Ribadeau-Dumas *et al.*, 1972; Grosclaude *et al.*, 1973, 1974*a,b*; Carles *et al.*, 1988) and, for A^2, the cDNA sequence (Jimenez-Flores *et al.*, 1987; Stewart *et al.*, 1987). Results of recent chemical sequencing (Carles *et al.*, 1988) are in complete agreement with the cDNA sequence; however, the variant was reported to be β-CN A^2-5P but it must have been β-CN A^1-5P since residue 67 was His rather than Pro. Agreement between the two reported cDNA sequences (a corrected sequence was provided by R. Jimenez-Flores) is excellent with only silent base changes in the codons being observed, 6 out of the 11 being in the wobble position, except for the codon for residue 93 (CTG (Jimenez-Flores *et al.*, 1987) versus ATG (Stewart *et al.*, 1984)). The author has chosen to place Met at position 67, in agreement with the latter codon, since this residue was found by chemical sequencing.

Of all the caseins, β-casein is the most hydrophobic. The highly charged, N-terminal domain containing the anionic phosphoserine cluster is clearly distinct from a very hydrophobic C-terminal domain (Figs 3, 4 and 7). At pH 6·6, the N-terminal 21- residue polar domain will have a net charge of about −11·5, while the large hydrophobic C-terminal domain will essentially have no net charge. Consequently, this molecule is very amphipathic with a polar domain comprising one-tenth of the chain but carrying one-third of the total charge and a hydrophobic domain consisting of the C-terminal three-fourths of the molecule. Comparison of cDNA sequences for β-caseins from different species suggests conservation of the polar N-terminal and hydrophobic C-terminal domains are important for biological function (Bonsing & Mackinlay, 1987). The actual sequence in the C-terminal domain appears relatively unimportant as long as the hydrophobicity is conserved. The molecule also has a high frequency of prolyl residues (0·17) which may have a major effect on the structure.

4.2.4. κ-Caseins

The primary structures for both known genetic variants of κ-casein are given in Fig. 6. These structures, corrected from Swaisgood (1982*a*), are based on chemical sequencing of the protein (Jollès *et al.*, 1970, 1972*b*; Brignon *et al.*, 1972; Mercier *et al.*, 1972*a*, 1973) and both cDNA (Stewart *et al.*, 1984) and genomic DNA (Alexander *et al.*, 1988) sequences for κ-CN A. Both the primary structure and the DNA sequences suggest that this protein is not part of the same family as the calcium-sensitive caseins, but rather it may have a common ancestor with γ-fibrinogen (Jollès *et al.*, 1986; Bonsing & Mackinlay, 1987). Comparison of κ-casein with the calcium-sensitive caseins indicates two unique features of this protein, viz. the absence of phosphoseryl residue clusters and the glycosylation of threonyl residues. Consequently, κ-casein does not bind calcium to the same extent as the other caseins; hence, its physical properties, such as solubility, are not sensitive to the presence of this ion. Like other caseins, particularly β-casein, however, its structure is extremely amphipathic (Hill & Wake, 1969). This protein comprises a very hydrophobic N-terminal domain encompassing roughly residues 1–105 (para-κ-casein) and a C-terminal polar domain. For example, there are no cationic residues in the 53-residue C-terminal sequence and its net charge at pH 6·6 is roughly -10 or -11 depending on the variant. In addition, each AcNeu would add a negative charge, e.g. κ-CN A-1P(3 AcNeu)(f117–169)—which constitutes a large portion of the minor κ-casein components (Vreeman *et al.*, 1986)—would have a net charge of -14.

Functional restraints on the structure of κ-casein, including: (1) the required interaction with the calcium-sensitive caseins, (2) the amphipathic structure with an inert polar domain that is insensitive to calcium, and (3) a recognition sequence specific for limited proteolysis by chymosin, are reflected by comparison of cDNA sequences from various species indicating conservation of this general structure (Bonsing & Mackinlay, 1987). Distinct separation of hydrophobic and polar domains has been appreciated for some time as a consequence of the specific chymosin-catalysed hydrolysis of the Phe105—Met106 peptide bond which occurs during cheesemaking. Release of the polar glycomacropeptide domain from κ-casein, which occurs on the surface of casein micelles (Schmidt, 1980; Swaisgood, 1985) eliminates the polar electrostatic and steric stabilization of the micelle surface, with an increase in surface hydrophobicity allowing micelles to associate, leading to clot formation.

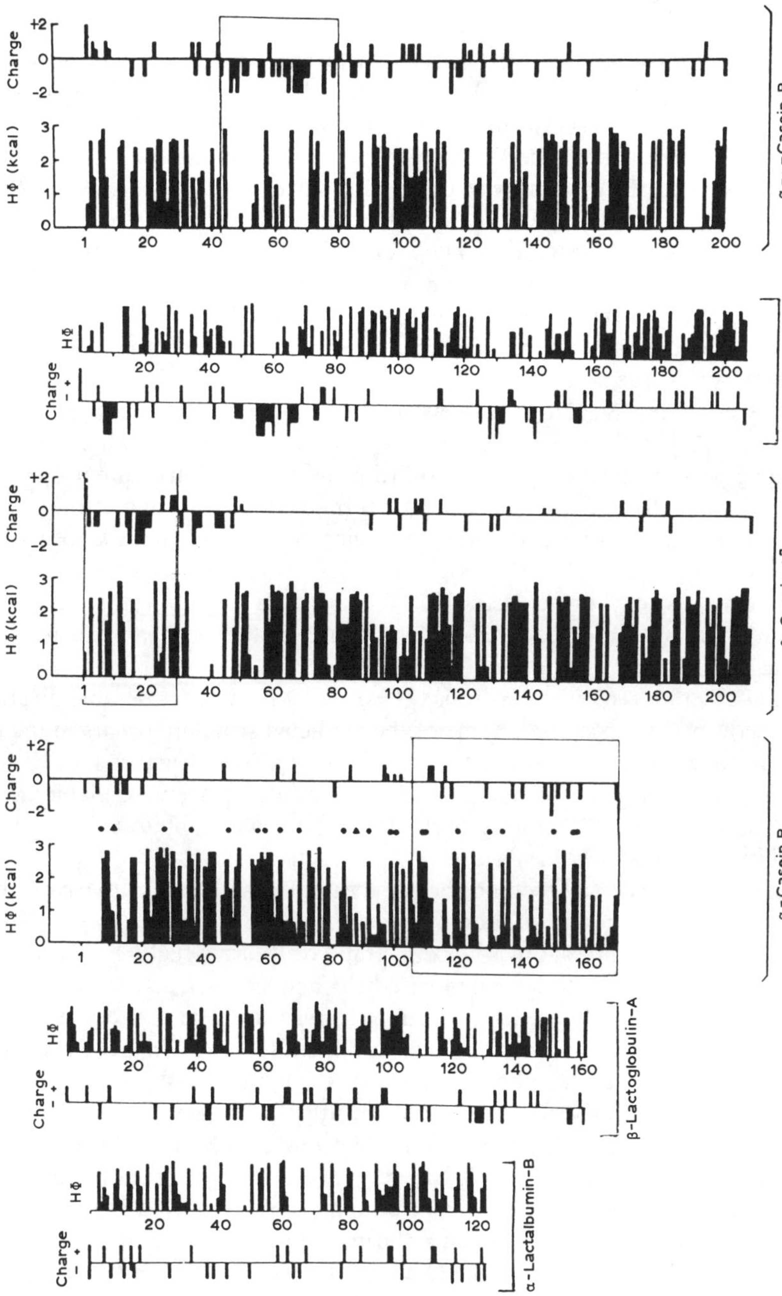

Fig. 7. Schematic illustration of the linear chain distribution of charged and hydrophobic amino-acid residues in the major milk proteins.

4.3. Secondary Structures

Lacking the possibility of direct measurement of secondary structures by X-ray crystallography, the amounts and location of such ordered structures in the caseins have been estimated by: (1) using various spectral methods for determination of the amount of this structure and (2) using algorithms that predict secondary structures from knowledge of the primary structure. Results obtained from CD and Raman spectroscopy measurements and various prediction methods (Chou & Fasman, 1974*a*,*b*, 1977; Lim, 1974; Garnier *et al.*, 1978; Garnier, 1990) based on the primary structure are summarized in Table 8. Obviously, an accurate assessment of the amount of ordered structure in these proteins is difficult at present. Nevertheless, it is also apparent that the statement that caseins exhibit random coil or unordered structure is not true.

These estimates suggest that κ-casein may contain 10–20% α-helix, 20–30% β-structure and 15–25% turns. Several structural motifs have also been suggested, including possible antiparallel and parallel β-sheets or βαβ structure in the hydrophobic domain (Raap *et al.*, 1983) and a β-turn–β-strand–β-turn motif centered on the chymosin sensitive Phe105-Met106 region. The latter motif appears to be conserved in κ-caseins from various species as would be expected for specific sensitivity to aspartyl proteinases (Holt & Sawyer, 1988). Using the more restrictive criterion of conserved predicted structure in the protein sequences from various species, Holt and Sawyer (1988) noted several regions of conserved α-helix (3%), β-structure (16%) and β-turns (11%) even though considerable divergence of the sequence has occurred. Some of the predicted structure occurs in the polar macropeptide domain; however, the stability of ordered structure in a region of such high net charge and apparent hydration would seem questionable. In fact, NMR studies suggest that this part of the molecule exhibits a great deal of flexibility (Rollema *et al.*, 1984, 1988).

Prediction and spectral methods both indicate the calcium-sensitive caseins also exhibit significant amounts of ordered structure (Table 8). In fact, roughly 50% of the residues may be present in α-helix, β-strand or β-turn structures. The calcium-sensitive caseins appear to have a more α-helix and less β-structure than κ-casein, particularly α_{s1}- and α_{s2}-caseins. A distinguishing feature of conserved predicted structure in the calcium-sensitive caseins as opposed to κ-casein is the helix–loop–helix motif centered on the phosphoseryl cluster with the cluster being on or near a β-turn (Holt & Sawyer, 1988). Consequently, the conserved predicted helix contents of α_{s1}-, α_{s2}- and β-caseins are quite high (24%, 17% and 13%, respectively) as compared to κ-casein (Holt & Sawyer, 1988). The hydrophobic domain of β-casein has conserved predicted α-helix and β-structure allowing for possible formation of structural motifs that may be important to the overall architecture of this domain. Although smaller, the C-terminal hydrophobic domain of α_{s1}-casein appears to have a conserved β-turn–β-strand–β-turn motif. Such structural motifs in the hydrophobic domains of the calcium-sensitive and κ-caseins may form key architectural units that direct the interaction of these proteins.

TABLE 8
Secondary Structure of the Caseins

Protein	*Sequence prediction*			*CD spectra*		*Raman spectra*		
	α-Helix	*β-Structure*	*β-Turn*	*α-Helix*	*β-Structure*	*α-Helix*	*β-Structure*	*β-Turn*
κ-Casein	23[b]	31[b]	24[b]	14[b]	31[b]			
	16[c]	17[c]	14[c]					
α_{s1}-Casein	22[d]	7·5[d]		13–20[e]	17[e]	13[e]	20[e]	34[e]
β-Casein	12[d]	11[d]		12–20[d]	0[d]	7–13[e]	19–22[e]	23–35[e]
	10[f]	13[f]		1–11[f,g]	13–16[f,g]			
	9–16[h]	20–26[h]	15–31[h]	7–10[h]	17–33[h]			

[a]Per cent residues in structures.
[b]Obtained by analysis of ORD data.
The other superscripts refer to the following references: [c]Loucheux-Lefebvre *et al.* (1978); [d]Raap *et al.* (1983); [e]Creamer *et al.* (1981); [f]Byler *et al.* (1988); [g]Andrews *et al.* (1979); [h]Graham *et al.* (1984).

4.4. Post-translational Modification as Related to Primary Structure

Phosphorylation of the folded protein chain, catalysed by casein kinases (Bingham, 1979), occurs in the Golgi apparatus (Turkington & Topper, 1966). The recognition sequence for phosphorylation is Ser/Thr-X-Glu/SerP, with an apparent preference for seryl residues (Mercier *et al.*, 1972*b*; Mercier, 1981). For example, the sequence TXE occurs four times in α_{s2}-CN A-11P. Nevertheless, when Ala 53 is replaced by Thr in the structure of α_{s1}-CN D, the latter residue is phosphorylated. Also, Ser 41 is sometimes phosphorylated in α_{s1}-caseins where it occurs in the sequences SKD (Manson *et al.*, 1977). However, the sequence TXD must not be a recognition site since this sequence has never been observed to be phosphorylated although it occurs in α_{s1}-caseins (Thr 49), α_{s2}-caseins (Thr 72, Thr 138) and β-caseins (Thr 41).

Other considerations are important for post-translational phosphorylation, including secondary and/or tertiary structure and other post-translational modifications. As noted above, there appears to be a preference for helix–loop–helix motifs with the phosphorylation site near or in β-turns (Holt & Sawyer, 1988). These factors are probably responsible for the lack of phosphorylation of some of the recognition sequences in α_{s2}- and κ-caseins and most likely in typical globular proteins such as β-lactoglobulin and α-lactalbumin (Swaisgood, 1982*a*). κ-Casein is particularly interesting since usually only Ser149 in the sequence SPE is phosphorylated, while Ser127 occurs in the sequence SGE and Thr135 in the sequence TXE (Fig. 6). Post-translational glycosylation, when it occurs first, may preclude phosphorylation by sterically hindering access to the recognition site by casein kinases. It should be noted that a minor species of κ-casein with two SerP residues and no carbohydrate has been identified (Vreeman *et al.*, 1977).

Glycosylation of κ-casein most likely also occurs in the Golgi apparatus. Secondary and/or tertiary structures appear to be especially important factors determining the site of glycosylation since most of the threonyl residues in κ-casein and none in the calcium-sensitive caseins are modified. The sites glycosylated (Thr 131,133,135,136,142, Ser 141) are in the polar domain and in the region of predicted turns (Loucheux-Lefebvre *et al.*, 1978; Holt & Sawyer, 1988). Moreover, comparison of the predicted structures of human and bovine macropeptides indicates that the human domain, which contains roughly three times more carbohydrate, has twice as many residues in β-turn regions (Fiat *et al.*, 1980).

4.5. Manifestations of Limited Proteolysis of the Caseins

Casein structures exhibit more chain flexibility than typical globular preoteins; consequently, they are more susceptible to proteolysis (Swaisgood, 1989). Susceptibility of a particular peptide bond depends upon both the enzyme specificity and chain flexibility in that region (Swaisgood, 1989; Swaisgood & Catignani, 1987); therefore, specific regions of a protein's structure, particularly sequences between domains, are most rapidly hydrolysed. As a result, limited proteolysis will produce specific large peptides that are characteristic of both the protein structure and the

TABLE 9
Some Peptides or Domains, Derived from Caseins by Limited Proteolysis, that have Functional or Biological Activities

Peptide or domain	*Enzyme(s)*	*Function or activity*
β-CN X(f29–209)	Plasmin	hydrophobic
(f106–209)		hydrophobic
(f108–209)		hydrophobic
(f1–105 or 107)		heat stable, very soluble
(f29–105 or 107)		heat stable, very soluble
(f1–28)		heat stable, very soluble
	Chymosin	
κ-CN X(f1–105)	very specific	hydrophobic, low solubility
κ-CN X(f106–169)	very specific	very polar, very soluble
α_{s1}-CN B, C(f24–199)	slow release	increased hydrophilicity
α_{s1}-CN B, C, D, E(f1–23)	slow release	emulsification
β-CN X(f1–52)	Chymotrypsin	may form an amphipathic helix
β-CN X(f196–209)		bitter
X(f202–209)	Trypsin or a mixture	extremely bitter
X(f203–209)	of proteinases: more	bitter
X(f202–207)	extensive proteolysis	bitter
α_{s1}-CN B, C, D, E (f23–84)		bitter
α_{s2}-CN A(f174–181)		bitter
	Pepsin and/or intestinal enzymes	
β-CN A^2A^3(f60–70)	(β-casomorphin-11)	weak opioid
β-CN X(f60–66)	(β-casomorphin-7)	opioid
β-CN X(f60–65)	(β-casomorphin-6)	weak opioid
β-CN X(f60–64)	(β-casomorphin-5)	strong opioid
β-CN X(f60–63 amide)	(morphiceptin)	very strong opioid
α_{s1}-CN X(f90–96)	(α-exorphin)	opioid

proteinase present. A partial list of peptides or domains that result from varying degrees of proteolysis of the caseins is presented in Table 9. Such peptide domains are inherently present in milk due to the natural occurrence of plasmin (Kaminogawa *et al.*, 1972; Eigel *et al.*, 1979; Eigel, 1981; Andrews, 1983) which is particularly active on the calcium-sensitive caseins, especially β-caseins (Andrews, 1983). Consequently, various sections of the C-terminal hydrophobic domain of β-casein, viz. β-CN X(f29–209), β-CN X(f106–209) and β-CN X(f108–209), as well as the N-terminal polar domain, viz. β-CN(f1–105 or 107), β-CN(f29–105 or 107) and β-CN(f1–28) are naturally present in milk. As would be expected, the polar and hydrophobic domains are quite different in their physical–chemical properties; thus, the polar domains are very heat stable (proteose-peptones) and the hydrophobic domain can be extracted in organic solvents (Reimerdes & Herlitz, 1979). Certain peptide domains are also released from α_{s1}-caseins (El-Negoumy, 1973) and α_{s2}-caseins (Visser *et al.*, 1989) by plasmin and are present in the protose-peptone and λ-casein fractions.

As noted earlier, the action of chymosin on κ-casein is very specific, hydrolysing only one bond and separating the hydrophobic and polar domains of that protein. Over a longer time period, chymosin also hydrolyses some specific bonds in the calcium-sensitive caseins. Chymosin action on α_{s1}-casein, which is the most well characterized, has been shown to occur during the cheese ripening process (Kaminogawa *et al.*, 1980; Creamer *et al.*, 1981). Cleavage occurs at residues 23 or 24 and removes the hydrophobic N-terminal domain, resulting in the more hydrophilic α-CN B or C(f25–199), known as α_s-I, which contributes to a change in the rheological properties of the cheese. The N-terminal domain, α_{s1}-CN(f1–23), is also released and exhibits emulsifying activity in the pH range 3–5 (Shimizu *et al.*, 1986).

Chymotrypsin slowly releases the N-terminal polar domain from β-caseins, thus separating the hydrophobic and polar domains (Chapin *et al.*, 1988). The polar domain, β-CN X(f1–52), is capable of forming an amphipathic α-helix in its C-terminal half which may have important functionality characteristics (Chapin *et al.*, 1988).

More extensive hydrolysis of the calcium-sensitive caseins, particularly by neutral or alkaline proteinases, is frequently accompanied by the production of bitter peptides (Matoba *et al.*, 1970; Minamiura *et al.*, 1972; Hashimoto *et al.*, 1980; Fukui *et al.*, 1983; Miyake *et al.*, 1983; Otagiri *et al.*, 1983; Kanehisa *et al.*, 1984; Shinoda *et al.*, 1985). Such bitterness is frequently observed in certain types of cheeses. Bitter peptides have been identified in tryptic hydrolysates of caseins corresponding to α_{s1}-CN B,C,D,E(f23–84) and α_{s2}-CN A(f174–181) (Matoba *et al.*, 1970). However, the most common source of bitter peptides is the C-terminal 14-residue sequence of β-caseins. Various bitter-tasting peptides have been isolated from proteolytic digests of caseins with reported sequences closely corresponding to the C-terminal of β-casein, thus, in fact, probably representing C-terminal peptides of this protein (Matoba *et al.*, 1970; Minamiura *et al.*, 1972). More detailed studies of bitterness have been performed with numerous synthetic analogs of the C-terminal sequence of β-casein (Hashimoto *et al.*,1980; Fukui *et al.*, 1983; Miyake *et al.*, 1983; Otagiri *et al.*, 1983; Kanehisa *et al.*, 1984; Shinoda *et al.*, 1985). The C-terminal peptide, RGPFPIIV, corresponding to β-CN X(f202–209), is extremely bitter (250 times more bitter than caffeine) with a threshold of 0·004 mM (Kanehisa *et al.*, 1984). Other peptides such as β-CN X (f196–209), β-CN X (f203–209) or β-CN X(f202–207) are less bitter, with thresholds ranging from 0·13 to 0·34 mM (Hashimoto *et al.*, 1980; Kanehisa *et al.*, 1984; Shinoda *et al.*, 1985) indicating the importance of sequence, an N-terminal positive charge and C-terminal hydrophobic residues for peptide bitterness.

Recent investigations have suggested that extensive in-vivo digestions of food proteins may release bioactive peptides. In particular, such digestion of α_{s1}- and β-caseins has been shown to release peptides with strong affinity for opiate receptors (Table 9) (Brantl *et al.*, 1979; Chang *et al.*, 1979; Henschen *et al.*, 1979; Zioudrou *et al.*, 1979; Loukas *et al.*, 1983; Chang *et al.*, 1985; Meisel, 1986; Yoshikawa *et al.*, 1986). Initially, a peptide with opioid activity in guinea pig ileum was isolated from casein digests and identified as β-CN X(f60–66) and named β-casomorphin-7 since it was derived from β-casein (Brantl *et al.*, 1979; Henschen *et al.*, 1979). Peptides of various lengths from this region of β-casein ranging from β-CN X(f60–63) to β-CN X(f60–66), named β-casomorphin-4, -5, -6 and -7, have been sythesized and examined for various opioid activities (Chang *et al.*, 1979; Yoshikawa *et al.*, 1986). In addition, the amide of β-casomorphin-4 [(β-CN X(f60–63 amide)], called morphiceptin, was synthesized and shown to have μ receptor binding and ileum assay activities as potent as that of morphine (Chang *et al.*, 1979). Moreover, these peptides have all been identified in enzymatic digests of casein, even including morphiceptin (Brantl *et al.*, 1979; Henschen *et al*, 1979; Chang *et al.*, 1985). Furthermore, these peptides and particularly a precursor, β-CN A^2, A^3(f60–70), (β-casomorphin-11), were identified in the duodenal chyme of minipigs on a diet of casein (Meisel, 1986). The relative opioid activity of these peptides, as measured by μ receptor binding and the ileum assay, is morphiceptin > β-casomorphin-5 > β-casomorphin-7 ≈ β-casomorphin-4 > β-casomorphin-6 > β-casomorphin-11 (Chang *et al.*, 1979; Meisel, 1986, Yoshikawa *et al.*, 1986). Liberation of these peptides most likely requires intestinal mucosal enzymes in addition to the gastric and pancreatic enzymes. Also, enzymes that amidate peptides are likely to occur in the pancreas and the intestine since amidated peptides are widely distributed in the gastrointestinal tract (Chang *et al.*, 1985). Given that μ receptors are present in the brain and predominate in the ileum and that such small hydrophobic peptides would be expected to cross the blood-brain barrier, release of these peptides may have considerable physiological significance (Brantl *et al.*, 1979; Chang *et al.*, 1979; Meisel, 1986; Yoshikawa *et al.*, 1986).

Examination of casein digests for opioid activity has also resulted in the identification of a bioactive peptide derived from α_{s1}-casein, viz. α_{s1}-CN X(f90–96) (Zioudrou *et al.*, 1979). Synthesis of this peptide and several analogs indicated that the N-terminal Arg residue is necessary for its activity (Loukas *et al.*, 1983; see Chapter 7).

5. Physicochemical Characteristics of the Caseins

Unlike many typical globular proteins, such as the whey proteins, caseins are not present in milk as individual molecular structures, but, rather, as large protein complexes that also incorporate milk salts, particularly calcium salts. Thus, the 'native' structure is actually a protein complex resulting from interaction of all the individual caseins. Nevertheless, an understanding of the physicochemical behaviour of the individual caseins provides for an understanding of their individual structures as well as for insight into how these structures may interact to form the larger structure known as the casein micelle. Hence, this discussion will focus on the properties of the indiviual proteins. Due to their propensity for interaction or self-association, it is difficult to find conditions under which native monomer forms can be observed. Consequently, the actual characteristics of monomers as they would exist in a physiological buffer at 37°C remains questionable; however, attempts have been made to obtain the characteristics of native monomers. Some of these results, together with characteristics obtained under strongly dissociating conditions, are listed in Table 10.

5.1. α_{s1}- and α_{s2}-Caseins

Characteristics of the self-association of the α_{s1}-caseins are quite different from those of β- and κ-caseins. This protein exhibits progressive consecutive self-association to dimers, tetramers, hexamers, etc., with the degree of association being strongly dependent upon the pH and ionic strength of the solution (Ho & Waugh, 1965; Payens & Schmidt, 1965, 1966; Schmidt & van Markwijk, 1968; Swaisgood & Timasheff, 1968; Schmidt, 1970*a*,*b*, 1985). At pH 6·6 and ionic strengths greater than 0·003, the monomers exists in a rapidly equilibrating equilibrium with oligomers, with increasing ionic strength resulting in increasing association constants (decreasing free energy) and the appearance of larger n-mers in the equilibrium (Ho & Waugh, 1965; Schmidt & van Markwijk, 1968; Schmidt, 1970*b*). The free energies for formation of the various oligomers are similar so that all species exist in appreciable concentrations, but there are significant differences. For example, at 0·2 ionic strength, dimers and tetramers are favoured while the formation of larger oligomers becomes progressively less favourable; hence, there is a definite break in the apparent molecular weight versus concentration curve at roughly 1 mg/ml. As the ionic strength decreases, the repulsive electrostatic free energy contribution becomes predominant such that at 0·01 ionic strength only a monomer–dimer equilibrium is observed and at 0·003 ionic strength only monomers are present (Ho & Waugh, 1965; Schmidt, 1970*b*). Likewise, as the pH increases, the electrostatic repulsive free energy increases, resulting in smaller association constants (Swaisgood & Timasheff, 1968) (smaller negative free energies) yielding a lowered degree of association at a particular concentration as compared to pH 6·6. The larger association constants, and resulting much stronger association, of α_{s1}-CN C as compared to α_{s1}-CN B can be explained by the change in electrostatic free energy (Schmidt, 1970*a*) due to its smaller net charge (Table 7, Fig. 2); however, α_{s1}-CN D behaves identically to α_{s1}-CN B although its net charge is greater than that of α_{s1}-CN B. It should be noted that the α_{s1}-CN B to α_{s1}-CN D substitution (A53pT) occurs in the polar domain, whereas the α_{s1}-CN B to α_{s1}-CN C substitution (E 192 G) occurs in the hydrophobic domain which is most likely to be in the association contact surface.

Hydrodynamic properties of native α_{s1}-casein monomers are not like those of typical globular proteins, but also definitely not those of a random coil (Table 10). Assuming a typical proteins hydration, Simha's shape factor derived from intrinsic viscosity indicates a prolate ellipsoid with $a/b = 7$ (Tanford, 1961), which seems unreasonably asymmetric on the basis of other observations, including electron microscopy (Desako *et al.*, 1980). However, if the monomer is assumed to be spherical, the calculated hydration, ~ 3 g H_2O/g, is much larger than that for globular proteins. As expected, when the protein has an extremely high net charge, pH 12, or is dissolved in 6 M guanidinium chloride (Gdn Cl), it exhibits random coil

TABLE 10
Hydrodynamic Properties of Proteins in the Casein Fraction

Protein	*Experimental values*					*Calculated values*	
	$s_{20,w} \times 10^{13}$ *(s^{-1})*	[η] *(ml/g)*	R_e[a] *(nm)*	δ_{max}[a] *(g H_2O/g)*	*Simha's shape factor*[m]	*Globular*[n] $s_{20,w} \times 10^{13}$ *(s^{-1})*	*Random coil*[b] R_e *(nm)*
α_{s1}-*Casein*							
Native monomer	2·4[c]	10[d]	$3·4_{[\eta]}$ $2·4_s$ 3·2[e]	$3·3_{[\eta]}$	8	2·53	4·3
Monomer: pH 12	1·35[c,f]	19·5[f]	$4·2_{[\eta]}$ $4·3_s$		—	—	4·3
Monomer: 6 M Gdn.Cl	?	19·2[d]	4·2			—	4·3
α_{s2}-*Casein*							
Native monomer	?	11·4[g]	3·6	3·9	9	2·70	4·4
β-*Casein*							
Native monomer	1·5[h,i]	23[h,j]	$4·4_{[\eta]}$ $3·7_s$	$8·4_{[\eta]}$ $6·1_s$	18	2·38	4·5
Monomer: 6 M Gdn.Cl	?	22·2[j]	4·4		—	—	4·5
κ-*Casein*							
Native polymer	15·6[k]	9·5[k]	$9·8_s$ $9·9_{[\eta]}$	3·1 2·9	8		
Native polymer (SH form)	12·0[l]	15·6[l]	$11·1_s$(117) $11·4_{[\eta]}$	$7·2_s$ 5·5			
Monomer: pH 12	1·4[k]	15·1[k]	$3·6_{[\eta]}$ $3·4_s$	— —	— —	2·11	4·1
5 M Gdn.Cl	1·88[k]	?	2·5	—	—	2·11	4·1
67% Acetic Acid, 0·15 M NaCl	1·26[k]	?	3·7	—	—	2·11	4·1

[a] These values were calculated by the author, except where a reference is given.
[b] Calculated from the relationship $R_e = 0·789(\alpha^2 \beta^2 n/6)^{0·5}$, where n is the number of residues in the chain, $\alpha = 0·85\, n^{0·055}$, and $\beta = 8·4\, A$. This relation specifically applies to 6 M Gdn.Cl (e.g. see Leach *et al.*[207]).

[c] Swaisgood & Timasheff (1968).
[d] Ho & Chen (1967).
[e] Ho & Waugh (1965).
[f] Schmidt *et al.* (1967).
[g] Snoeren *et al.* (1980).
[h] Payens & van Markwijk (1963).
[i] Evans *et al.* (1971).
[j] Noelken & Reibstein (1968).
[k] Swaisgood *et al.* (1964).
[l] Vreeman *et al.* (1981).

[m] Simha's shape factor was calculated assuming the hydration of an average globular protein (Squire and Himmel, 1979).
[n] Calculated from the relationship $s_{20,w} = 0·01\, M^{0·667}\, (1 - vp)/v^{-0·33}$ (Teller *et al.*, 1979; Leach *et al.*, 1980).

behaviour, as indicated by the agreement between observed and calculated values (Table 10).

Primary structure, predicted secondary structure and physicochemical properties are all consistent with an amphipathic structure composed of a hydrophobic and a polar domain. Both domains must exhibit more flexibility than is typical of globular proteins, especially the polar domain which is highly charged. The anionic clusters in the polar domain are probably responsible for the sensitivity of its physicochemical properties to pH and ionic strength. Interactions of the hydrophobic domain are probably responsible for the associative characteristics of this protein; however, the interactions must not be exclusively hydrophobic since dissociation does not occur at low temperatures. It is possible that some H-bonding or even extension of the secondary structure occurs at hydrophobic surface contact sites which may give greater specificity and directionality to self-association or interactions with κ- and β-caseins.

In comparison to α_{s1}-casein, α_{s2}-caseins are more hydrophilic, containing three anionic clusters, may contain inter- or intramolecular disulphide bonds and have 40% fewer prolyl residues. The primary structure suggests an N-terminal negatively charged polar domain containing two anionic clusters, an internal hydrophobic domain, followed by another polar domain with an anionic cluster, and finally a C-terminal positively charged hydrophobic domain. With such an amphipathic highly charged structure, it is not surprising that the associative properties are very dependent on the ionic strength (Snoeren *et al.*, 1980). It also associates less extensively than α_{s1}-casein, probably as a result of stronger electrostatic repulsive forces and lower hydrophobicity. Nevertheless, it does exhibit consecutive, and apparently isodesmic, self-association with the extent of association reaching a maximum at 0·2–0·3 ionic strength and then decreasing (Snoeren *et al.*, 1980). The unexpected decrease in association at higher ionic strengths may be due to ionic suppression of an electrostatic interaction between the N-terminal and the C-terminal domains (Snoeren *et al.*, 1980).

5.2. β-Caseins

Like κ-caseins, the presence of polar and hydrophobic domains is clearly distinct; however, β-caseins are the most hydrophobic and contain more prolyl residues than any other casein. Furthermore, the polar domain has only one anionic cluster. Hence, a molecular structure dominated by hydrophobic interactions of its surface and less sensitive to ionic strength than the α_{s1}- and $P\alpha_{s2}$-caseins would be expected. These characteristics appear to be manifested in the physicochemical properties of this protein.

Unlike α_{s1}-casein, the self-association of β-caseins is extremely temperature dependent; consequently, only monomers are observed at 0–4°C (Payens & van Markwijk, 1963). Under these conditions, the hydrodynamic behaviour is very different from other caseins and, in fact, approaches that of a random coil (Table 10). The Stokes radius, R_e = 3·7 nm, determined by gel chromatography (Schmidt & Payens, 1972), agrees well with those obtained by sedimentation and viscosity, and is also consistent with the sizes of spherical particles observed by electron microscopy (4–5 nm) (Andrews *et al.*, 1979). Such hydrodynamic properties for a molecule with the molecular weight of β-casein can only be explained by unusual hydration or large asymmetry (Tanford, 1961). For example, small angle X-ray scattering yields a radius of gyration, R_G, of 4·6 nm (Schmidt & Payens, 1972; Andrews *et al.*, 1979) (equivalent to $R_e = 3{\cdot}1$ nm if $\xi = 0{\cdot}67$) which corresponds to a rod nearly 16 nm in length if asymmetry is assumed (Tanford, 1961) and likewise the frictional ratio derived from the sedimentation coefficient of Simha's shape factor from viscosity corresponds to a prolate ellipsoid with $a/b \approx 12$ (Tanford, 1961). This large asymmetry is not consistent with the ratio of the coefficient of sedimentation velocity concentration dependence, K_s, to the intrinsic viscosity, $[\eta]$, which is 1·35 (Schmidt & Payens, 1972) or with the electron microscopic observations (Andrews *et al.*, 1979). Rather, a more flexible structure with high solvation, 6–8 g H_2O/g (Table 10), is indicated. On the other hand, the molecule does not behave like a random coil. Calculations of R_G from viscosity or sedimentation data give values of 5·1 nm and 5·5 nm (Schmidt & Payens, 1972) which are significantly larger than that observed. Also, the Stokes radius calculated for a random coil (Table 10) is larger than the experimental value (3·7 nm).

As the temperature increases above 4–5°C, β-caseins undergo a highly cooperative, reversible, rapidly equilibrating discrete self-association yielding large polymers with a narrow size distribution, similar to the formation of detergent micelles (Schmidt & Payens, 1972; Niki *et al.*, 1977; Andrews *et al.*, 1979; Arima *et al.*, 1979; Buchheim & Schmidt, 1979; Evans *et al.*, 1979; Takase *et al.*, 1980; Schmidt, 1985; Thurn *et al.*, 1987; Kajiwara *et al.*, 1988). The transport properties for this monomer–polymer micelle equilibria have been treated theoretically using a shell model for the micelle with a continuous distribution of intermediates between the monomer and largest polymer micelle (Tai & Kegeles, 1984). Like detergent micelle formation, there appears to be a critical concentration above which micelles are formed, ranging from less than 0·5 mg/ml to about 2 mg/ml (Schmidt & Payens, 1972; Niki *et al.*, 1977; Evans *et al.*, 1979). The 'CMC' depends on the temperature, ionic strength and pH; most likely, the amount of electrostatic repulsion as reflected by the surface zeta potential. The size of the polymer micelle has been characterized by the number of monomers in the polymer (n = 15–60 (Schmidt & Payens, 1972; Buchheim & Schmidt, 1979; Takase *et al.*, 1980; Thurn *et al.*, 1987; Kajiwara *et al.*, 1988)), the radius of gyration (R_G = 7·3–13·5 nm (Andrews *et al.*, 1979; Thurn *et al.*, 1987; Kajiwara *et al.*, 1988)), the Stokes radius (R_e = 15 nm (Niki *et al.*, 1977; Thurn *et al.*, 1987)) and radius observed by electron microscopy (R = 8–17 nm (Arima *et al.*, 1979; Buchheim & Schmidt, 1979)). Increasing ionic strength shifts the equilibrium towards the polymer micelle but affects n only slightly (Schmidt & Payens, 1972; Takase *et al.*, 1980), whereas increasing the temperature shifts the equilibrium position and increases n (Takase *et al.*, 1980). At room temperature, n appears to be roughly 40–50. Obviously, there is a difference between the values for R_G determined by small angle X-ray scattering (Andrews *et al.*, 1979; Kajiwara *et al.*, 1988) and small angle neutron scattering (Thurn *et al.*, 1987) and values for R_e determined by dynamic light scattering

(Thurn *et al.*, 1987) and gel chromatography (Niki *et al.*, 1977). The theoretical ratio, R_G/R_e, for a hard sphere is 0·775 (Thurn *et al.*, 1987; Kajiwara *et al.*, 1988), while that observed for polymer micelle is less than 0·6, suggesting the immobilization of water in a soft outer layer surrounding a more dense core (Kajiwara *et al.*, 1988).

Several observations suggest that the monomer in equilibrium with the polymer micelle may have undergone a small conformational change and may be more compact than that at low temperature. Evans *et al.* (1979) confirmed Garnier's (1966) earlier observation of a conformational change producing more α-helix as the temperature increases, but the data of Evans *et al.* were obtained with carboxyacyl derivatives that do not polymerize, suggesting that the conformational change must be a preliminary step to polymerization. Viscosity data exhibit downward curvature at low concentrations at 30°C, extrapolating to give $[\eta] = 4{\cdot}6$ ml/g, and Re for the monomer as measured by gel chromatography at low concentrations decreases with increasing temperature (Niki *et al.*, 1977; Arima *et al.*, 1979). If the value 4·6 ml/g is accepted for the intrinsic viscosity of the monomer, δ_{smax} at this temperature would be 1·1 ml/g; thus, monomer β-casein remains more hydrated and expanded in the polymer micelle as suggested on the basis of other studies (Andrews *et al.*, 1979).

From the structure and characteristics described above one would predict that the hydrophobic domain associates to form the core of the polymer micelle. The interaction must be predominately hydrophobic as indicated by the temperature dependence, particularly the negative heat capacity change (Pearce, 1975; Takase *et al.*, 1980). This conclusion is supported by the observation of a temperature and concentration dependent flourescene emission maximum shift indicating transfer of Trp 143, which is in the hydrophobic domain, to an increasingly apolar environment (Pearce, 1975) and loss of chain flexibility in the hydrophobic domain upon polymer micelle formation as indicated by NMR data (Andrews *et al.*, 1979). Moreover, removal of the C-terminal three hydrophobic residues, IIV, greatly reduces the association (Thompson *et al.*, 1967; Evans *et al.*, 1979) and removal of 20 C-terminal residues eliminated binding of the hydrophobic surface probe, ANS (Barry & Creamer, 1975).

Except for the anionic phosphate cluster in the polar domain of β-casein, the similarity of the amphipathic structure with that of κ-casein is quite interesting. *Vis-a-vis* α_{s1}-caseins that exhibit consecutive association equilibria, these two proteins are characterized by monomer–polymer micelle association equilibria. Furthermore, dephosphorylated β-caseins, like κ-caseins, will stabilize α_{s1}-caseins in the presence of calcium (Yoshikawa *et al.*, 1975).

5.3. κ-Caseins

A physiological role played by this protein is the stabilization of calcium-sensitive caseins in the presence of calcium salts in milk. Its structure is clearly amphipathic, but without the anionic phosphate cluster in its polar domain, κ-casein remains soluble in the presence of calcium at all temperatures. Also, the hydrophobic domain of κ-casein as compared to β-casein is less hydrophobic, has a lower frequency of prolyl residues and probably contains more secondary structure. Hence, although this protein forms polymer micelles like β-casein, the equilibria are not very sensitive to temperature or ionic strength (Swaisgood *et al.*, 1964; Vreeman, 1979; Vreeman *et al.*, 1981). Furthermore, interactions of the hydrophobic domain, either with itself or hydrophobic domains of other caseins, are probably less dominated by hydrophobic interactions, with participation of more H-bonding and ion-pair interactions and possibly extension of secondary structure. The combination of high hydrophobicity and charge frequency (Talbot & Waugh, 1970) (but no net charge) and the conserved predicted α-helix and β-structure (Holt & Sawyer, 1988) in the N-terminal 34 residues is particularly interesting in this regard.

Both cysteinyl residues are in the hydrophobic domain and occur in predicted loops or β-turns and thus should be readily accessible for oxidation or sulphydryl–disulphide interchange. Some years ago, the author showed that κ-casein obtained without reduction was apparently randomly crosslinked by intermolecular disulphide bonds to give oligomers (M_W = 125 000) with the smallest detectable oligomer having a weight of 60 000 daltons (Swaisgood *et al.*, 1964), corresponding to a trimer. The existence of disulphide-crosslinked oligomers has since been substantiated (Talbot & Waugh, 1970; Pepper & Farrell, 1988; Farrell, 1988) with the further suggestion that during biosynthesis reduced monomers first interact with the calcium-sensitive caseins to form micelles, followed by random crosslinking by oxidation (Pepper & Farrell, 1982). Both disulphide-crosslinked oligomers and reduced κ-caseins are capable of forming polymer micelles and stabilizing calcium-sensitive caseins (Talbot & Waugh, 1970; Vreeman, 1979).

Very little is known of the physical properties of κ-casein monomers under physiological conditions although, at least for reduced κ-casein, only monomers should be present below the CMC. Cleavage of the disulphide bonds by reduction or β-elimination at pH 12 permits characterization of the monomer in strong dissociating agents (Swaisgood & Brunner, 1963; Swaisgood *et al.*, 1964). As expected, the hydrodynamic properties in such solvents approach that for random coil behaviour (Table 10).

The monomer–polymer micelle equilibrium, at least for reduced κ-casein, is characterized by a CMC which varies from 0·53 to 0·24 mg/ml with ionic strengths from 0·1 to 1·0 (Vreeman, 1979; Vreeman *et al.*, 1981). This variation can be completely accounted for by the change in electrostatic free energy. Hydrodynamic studies of the polymer micelle indicate a spherical shape; for example, the $K_s/[\eta]$ ratio is that of a sphere (Vreeman *et al.*, 1981). A rather narrow size distribution in NaCl solutions or phosphate buffers at pH 7·0 is observed for micelles formed with either disulphide-crosslinked (Swaisgood *et al.*, 1964) or reduced κ-casein (Vreeman, 1979; Vreeman *et al.*, 1981). There appear to be small differences in the average polymer size depending on the preparation and environmental conditions. For example, data from sedimentation and viscosity measurements in Table 10 give molecular weights of 570 000 and 650 000 by application of the Scheraga-Mandelkern equation for reduced and native polymer, respectively. (The former value is identical to that obtained by sedimentation equilibrium (Scheraga & Mandelkern, 1953).) The reason for these small differences is not known; however, it should be noted that the reduced polymer studies were with the carbohydrate-free κ-casein in NaCl, while the native polymer included all species and the data were obtained in sodium phosphate buffer. Furthermore, similar analysis of data for the reduced whole κ-casein polymer in an imidazole-KCl buffer yielded a weight of 650 000 daltons, whereas isolated species appeared to have slightly smaller polymer sizes (Doi *et al.*, 1979*a*). In any case, the polymer micelle under these conditions appears to be composed of 30–35 monomer units. Lateral interactions of the hydrophobic domains produce a dense apolar core surrounded by flexible, less densely packed polar domains, thus accounting for the large hydration values (Table 10). Flexibility of the polar domains is also indicated by NMR studies of κ-casein and micelles (Rollema *et al.*, 1984, 1988).

Polymer micelle formation appears to assume another dimension in the presence of milk salts. Thus, the limiting polymer as measured by light scattering is much larger, approaching a value of $2{\cdot}5 \times 10^6$ daltons in simulated milk ultrafiltrate (SMUF) (Thurn *et al.*, 1987). Moreover, analysis of small angle neutron scattering data suggests a model composed of star-shaped submicelles, each containing 9 monomers, interacting at the star's points to give a total extent of association of about 140 (Thurn *et al.*, 1987). Although smaller, the submicelles appear similar to the polymers observed in phosphate or NaCl buffers. It is possible that the polar domain chains, which form a flexible, loose surface of such polymer micelles, could interact, possibly by calcium bridging, to yield larger supermicelles in SMUF.

5.4. Interactions of Caseins with Calcium

No discussion of the chemistry of caseins is complete without at least a brief description of their interaction with calcium. Spectroscopic studies have shown that calcium binds primarily to the phosphoseryl residues (Ho & Waugh, 1965; Ono *et al.*, 1976, 1980). This result is expected on the basis of their calcium sensitivities in relationship to composition and primary structures of the four caseins. Thus, α_{s2}-casein is the most calcium sensitive, precipitating at Ca^{2+} concentrations less than 2 mM (Toma & Nakai, 1973; Aoki *et al.*, 1985), while α_{s1}-casein precipitates in the range of 3–8 mM Ca^{2+} (Schmidt, 1969; Bingham *et al.*, 1972; Toma & Nakai, 1973; Dalgleish & Parker, 1980; Parker & Dalgleish, 1981; Aoki *et al.*, 1985; Farrell *et al.*, 1988) and β-casein precipitates in the range of 8–15 mM Ca^{2+} at 37°C (Schmidt, 1969; Parker & Dalgleish, 1981; Farrell *et al.*, 1988) but remains in solution at concentrations up to 400 mM at 1°C (Farrell *et al.*, 1988). Of course, κ-casein remains soluble under all these conditions. The calcium concentration that initiates precipitation decreases with increasing pH and increasing temperature but increases with increasing ionic strength and protein concentration (Schmidt, 1969; Bingham *et al.*, 1972; Dalgleish & Parker, 1980; Parker & Dalgleish, 1981). The Ca^{2+} concentration required to initiate precipitation of α_{s1}-casein at pH 6·6–7·0, 0·08–0·1 ionic strength and 37–40°C is roughly 3–5 mM. Independent of ionic strength and temperature, each protein initiates precipitation when the number of site-bound calcium ions is roughly equivalent to the number of phosphoseryl residues (Schmidt, 1969; Dalgleish & Parker, 1980; Parker & Dalgleish, 1981). Recently, the principle of thermodynamic linkage between solubility and salt binding free energies was used to evaluate calcium salting-out and salting-in of α_{s1}- and β-caseins (Farrell *et al.*, 1988). This analysis yielded salt binding constants (186 M^{-1} for α_{s1}-CN B, 156 M^{-1} for β-CN C) and moles of salt bound per mole of protein (8 and 4, respectively) that were similar to equilibrium binding constants and stoichiometries obtained by other methods.

Nevertheless, there is substantial evidence that calcium also binds to other sites, carboxylate residues in particular. Thus, at the point of minimum solubility the number of calcium ions bound exceeds the number of phosphoseryl residues; for example, approximately 13 moles of site-bound Ca^{2+} per mole of α_{s1}-CN B are required for complete precipitation (Creamer & Waugh, 1966). Direct evidence for calcium binding to carboxylate residues comes from changes in the infrared spectra (Ono *et al.*, 1976, 1980; Byler & Farrell, 1989). Location of these residues with respect to the phosphoseryl clusters is unknown, but it is tempting to suggest the cooperative participation of both phosphoseryl and carboxylate residues in anionic clusters capable of binding multiple calcium ions.

Calcium binding to these proteins is affected by pH, temperature and ionic strength in a manner consistent with the effects on the Ca^{2+} concentration required to initiate precipitation. Thus, equilibrium binding constants increase with increasing pH and temperature, but decrease with increasing ionic strength (Dalgleish & Parker, 1980; Parker & Dalgleish, 1981; Baumy & Brule, 1988; Jang & Swaisgood, 1990). These constants appear to be similar for α_{s1}- and β-caseins, varying from 0·4 to 1·9 litres/mmol at pH 6·6–7·0 and 0·1 ionic strength for the average binding constant per site (Dickson & Perkins, 1971; Parker & Dalgleish, 1981; Baumy & Brule, 1988).

Detailed analyses of the effects of calcium binding on α_{s1}-caseins have indicated several steps in the equilibria. Between 0 and 1 mM Ca^{2+}, calorimetry indicates an exothermic process (Holt *et al.*, 1975), infrared spectroscopy shows binding only to phosphoseryl residues (Ono *et al.*, 1976) and difference spectroscopy indicates transfer of both tyrosyl and tryptophanyl residues from an aqueous to an apolar environment (Ono *et al.*, 1976). As the Ca^{2+} concentration is increased from 1 to 3 mM, calorimetry suggests a second exothermic phase followed by an increasingly endothermic reaction (Holt *et al.*, 1975), infrared spectroscopy shows binding to carboxylate residues as well as phosphoseryl residues (Ono *et al.*, 1976), turbidity increases slightly to a plateau level (Holt *et al.*, 1975), and electron microscopy reveals increasing numbers of bent-chain polymers (Dosaka *et al.*, 1980). Difference spectroscopy shows that burying of the aromatic chromophores is abated (Ono *et al.*, 1976). Finally, between 3 and 5 mM Ca^{2+}, the reaction becomes very endothermic (Holt *et al.*, 1975), binding primarily to carboxylate residues continues (Ono *et al.*, 1976), the turbidity increases dramatically (Holt *et al.*, 1975), and eventually precipitation occurs. The initial exothermic conformational change may reflect H-bond formation, while the endothermic aggregation results primarily from hydrophobic interactions (Holt *et al.*, 1975). These results suggest that binding of Ca^{2+} to high affinity phosphoseryl clusters in the polar domain alters its interaction with the hydrophobic domain bringing about a conformational change in that domain which allows some association to occur. Further binding to carboxyl residues throughout the structure reduces the electrostatic repulsion and interaction of the hydrophobic domains leads to formation of large aggregates.

Recently, the interaction of Ca^{2+} and calcium salts with 'submicelles' prepared immediately from casein micelles that had never been cooled was characterized by analytical affinity chromatography (Jang & Swaisgood, 1990). The equilibrium binding constants exhibited pH and temperature dependencies similar to those found for α_{s1}- and β-caseins. However, slightly larger binding constants were obtained; for example, roughly 10 litres/mmol at pH 6·7 and 20°C in the absence of NaCl. Correction for the ionic strength effect does not appear to account for the total difference. Higher affinities exhibited by submicelles as measured by this technique could result from selective measurement of binding to highest affinity sites or to structural differences in the anionic clusters of submicellar sites. Furthermore, intriguing differences were found between binding characteristics of calcium in Ca^{2+} buffer solutions, and in milk salt solutions, suggesting higher affinities in the latter.

6. Projection of Future Work

During the past 10 years the primary structures of the casein genetic variants have been refined, cDNAs for the caseins have been prepared, more insight on the secondary structures has been provided and a better understanding of the interactions of these proteins has been realized. There are several areas that the author believes hold particular promise for major advances in our knowledge of casein chemistry in the next decade. The advancing technologies for protein structure determination by NMR analyses and by molecular modelling may allow actual three-dimensional models of these proteins to be produced for the first time. In addition, with the recombinant DNA technology available and with the cDNA for caseins in hand, it is now possible to prepare specific site mutants to address specific questions about the function of a particular residue. Combination of these techniques should provide valuable tools for exploring protein structures and interactions.

References

Alexander, L.J., Stewart, A.F., Mackinlay, A.G., Kapelinskaya, T.V., Tkach, T.M. & Gorodetsky, S.I. (1988). *Eur. J. Biochem.*, **178**, 395.
Andrews, A.T. (1983). *J. Dairy Res.*, **50**, 45.
Andrews, A.L., Atkinson, D., Evans, M.T.A., Finer, E.G., Green, J.P., Phillips, M.C. & Robertson, R.N. (1979). *Biopolymers*, **18**, 1105.
Annan, W.D. & Manson, W. (1969). *J. Dairy Res.*, **36**, 259
Aoki, T., Toyooka, K. & Kako, Y. (1985). *J. Dairy Sci.*, **68**, 1624.
Arima, S., Niki, R. & Takase, K. (1979). *J. Dairy Res.*, **46**, 281.

Aschaffenburg, R. (1961). *Nature*, **192**, 431.
Aschaffenburg, R. (1968). *J. Dairy Res.*, **35**, 447.
Aschaffenburg, R. & Drewry, J. (1957). *Nature*, **180**, 376.
Aschaffenburg, R., Sen, A. & Thompson, M.P. (1968). *Comp. Biochem. Physiol.*, **25**, 177.
Barrefors, P., Ekstrand, B., Fägerstam, L., Larsson-Raznikiewicz, M., Schaar, J. & Steffner, P. (1985). *Milchwissenschaft*, **40**, 257.
Berry, G.P. & Creamer, L.K., (1975). *Biochemistry*, **14**, 3542.
Baumy, J.J. & Brule, G. (1988). *Lait*, **68**, 409.
Bigelow, C.C. (1967). *J. Theoret. Biol.*, **16**, 187.
Bingham, E. (1979). *J. Dairy Res.*, **46**, 181.
Bingham, E.W., Farrell, H.M., Jr. & Carroll, R.J. (1972). *Biochemistry*, **11**, 2450.
Bonsing, J. & Mackinlay, A.G. (1987). *J. Dairy Res.*, **54**, 447.
Brantl, V., Teschemacher, H., Henschen, A. & Lottspeich, F. (1979). *Hoppe-Seyler's Z. Physiol Chem.*, **360**, 1211.
Brignon, G., Ribadeau-Dumas, B., Grosclaude, F. & Mercier, J.-C. (1970). *Eur. J. Biochem.*, **22**, 179.
Brignon, G., Mercier, J.-C.,. Ribadeau-Dumas, B. & Das, B.C. (1972). *FEBS Lett.*, **27**, 301.
Brignon, G., Ribadeau-Dumas, B. & Mercier, J.-C. (1976). *FEBS Lett.*, **71**, 111.
Brignon, G., Ribadeau-Dumas, B., Mercier, J.-C., Pélissier, J.P. & Das, B.C. (1977). *FEBS Lett.*, **76**, 274.
Brunner, J.R. (1981). *J. Dairy Sci.*, **64**, 1038.
Brunner, J.R., Ernstrom, C.A., Hollis, R.A., Larson, B.L., Whitney, R. McL. & Zittle, C.A. (1960). *J. Dairy Sci.*, **43**, 901.
Buchheim, W. & Schmidt, D. G. (1979). *J. Dairy Res.*, **46**, 277.
Byler, D.M. & Farrell, H.M., Jr. (1989). *J. Dairy Sci.*, **72**, 1719.
Byler, D.M., Farrell, H.M., Jr. & Susi, H. (1988). *J. Dairy Sci.*, **71**, 2622.
Carles, C., Huet, J.-C. & Ribadeau-Dumas, B. (1988). *FEBS Lett.*, **229**, 265.
Chang, K.-J., Killian, A., Hazum, E., Cuatrecasas, P. & Chang, J.-K. (1979). *Science*, **212**, 75.
Chang, K.J., Su, Y.F., Brent, D.A. & Chang, J.K. (1985). *J. Biol. Chem.*, **260**, 9706.
Chapin, L.C., Clark, D.C. & Smith, L.J. (1988). *Biochim. Biophys. Acta*, **956**, 162.
Chou, P.Y. & Fasman, G.D. (1974*a*). *Biochemistry*, **13**, 211.
Chou, P.Y. & Fasman, G.D. (1974*b*). *Biochemistry*, **13**, 222.
Chou, P.Y. & Fasman, G.D. (1977). *J. Mol. Biol.*, **115**, 135.
Creamer, L.K. (1972). *Biochim. Biophys. Acta*, **271**, 252.
Creamer, L.K. & Waugh, D.F. (1966). *J. Dairy Sci.*, **49**, 706 (Absract).
Creamer, L.K., Richardson, T. & Parry, D.A.D. (1981) *Arch. Biochem. Biophys.*, **211**, 689.
Creamer, L.K., Zoerb, H.F., Olson, N.F. & Richardson, T. (1981). *J. Dairy Sci.*, **65**, 902.
Dalgleish, D.G. (1986). *J. Dairy Res.*, **53**, 43.
Dalgleish, D.G. & Parker, T.G. (1980). *J. Dairy Res.*, **47**, 113.
Davies, D. R. & Law, A.J.R. (1977). *J. Dairy, Res.*, **44**, 213.
Desako, S., Kimura, T., Taneya, S., Sone, T., Kaminogawa, S. & Yamauchi, K. (1980). *Agric. Biol. Chem.*, **44**, 2443.
Dickson, I.R. & Perkins, D.J. (1971). *Biochem. J.*, **124**, 235.
Doi, H., Ibuki, F. & Kanamori, M. (1979*a*). *J. Dairy Sci.*, **62**, 195.
Doi, H., Kawaguchi, N., Ibuki, F. & Kanamori, M. (1979*b*). *J. Nutr. Sci. Vitaminol.*, **25**, 95.
Doi, H., Kobatake, H., Ibuki, F. & Kanamori, M. (1980). *Agric. Biol. Chem.*, **44**, 2605.
Donnelly, W.J. (1977). *J. Dairy Res.*, **44**, 621.
Dosako, S., Kimura, T., Taneya, S., Sone, T., Kaminogawa, S. & Yamauchi, K. (1980). *Agric. Biol. Chem.*, **44**, 2443.
Eigel, W.N. (1981). *Int. J. Biochem.*, **13**, 1081.
Eigel, W.N., Hofmann, C.J., Chibber, B.A.K., Tomich, J.M., Keenen, T.W. & Mertz, E.T. (1979). *Proc. Natl. Acad. Sci. USA*, **76**, 2244.

Eigel, W.N., Butler, J.E., Ernstrom, C.A., Farrell, H.M., Hawalkar, V.R., Jenness, R. & Whitney, R. McL. (1984). *J. Dairy Sci.*, **67**, 1599.
El-Negoumy, A.M. (1966). *Anal. Biochem.*, **15**, 437.
El-Negoumy, A.M. (1973). *J. Dairy Sci.*, **56**, 1486.
Evans, M.T.A., Irons, L. & Jones, M. (1971) *Biochim. Biophys. Acta*, **229**, 411.
Evans, M.T.A., Phillips, M.C. & Jones, M.N. (1979). *Biopolymers*, **18**, 1123.
Farrell, H.M., Jr. (1988). In *Fundamentals of Dairy Chemistry*, 3rd Edn, ed. N.P. Wong. Van Nostrand Reinhold, New York, pp. 461–510.
Farrell, H.M. Jr., Kumosinski, T. F., Pulaski, P. & Thompson, M.P. (1988). *Arch. Biochem. Biophys.*, **265**, 146.
Fiat, A.-M., Alais, C. & Jollés, P. (1972). *Eur. J. Biochem.*, **27**, 408.
Fiat, A.-M., Jolles, J., Aubert, J.-P., Loucheux-Lefebvre, M.-H. & Jolles, P. (1980). *Eur. J. Biochem.*, **111**, 333.
Fournet, B., Fiat, A.-M., Alais, C. & Jollés, P. (1979). *Biochim. Biophys. Acta* **576**, 339.
Fox, P.F. & Guiney, J. (1972). *J. Dairy Res.*, **39**, 49.
Fukui, H., Kanehisa, H., Ishibashi, N., Miyake, I. & Okai, H. (1983). *Bull. Chem. Soc. Jpn.*, **56**, 766.
Garnier, J. (1966). *J. Mol. BIol.*, **19**, 586.
Garnier, J. (1990). *Biochimie*, **72**, 513.
Garnier, J., Ribadeau-Dumas, B. & Mocquot, G. (1964). *J. Dairy Res.*, **31**, 131.
Garnier, J., Osguthorpe, D.J. & Robson, B. (1978). *J. Mol. Biol.*, **120**, 97.
Gordon, W.G. & Kalan, E.B. (1974). In *Fundamentals of Dairy Chemistry*, ed. B.H. Webb, A.H. Johnson & J.A. Alford. AVI Publishing Co., Westport, CT, pp. 87–124.
Graham, E.R.B., Malcolm, G.N. & McKenzie, H.A. (1984). *Int. J. Biol. Macromol.*, **6**, 155.
Grosclaude, F., Garnier, J., Ribadeau-Dumas, B. & Jeunet, R. (1964). *C.R. Acad. Sci.*, **259**, 1569.
Grosclaude, F., Pujolle, J., Garnier, J. & Ribadeau-Dumas, B. (1966). *Ann. Biol. Anim. Biochim. Biophys.*, **6**, 215.
Grosclaude, F., Mahé, M.-F., Mercier, J.-C. & Ribadeau-Dumas, B. (1972). *Eur. J. Biochem.*, **26**, 328.
Grosclaude, F., Mahé, M.-F. & Ribadeau-Dumas, B. (1973) *Eur. J. Biochem.*, **40**, 323.
Grosclaude, F., Mahe, M.-F. & Mercier, J.-C. (1974*a*). *Ann. Génét. Sel. Anim.*, **6**, 305.
Grosclaude, F., Mahé, M.-F. & Voglino, G.-F. (1974*b*). *FEBS Lett.*, **45**, 3.
Grosclaude, F., Mahé, M.-F., Mercier, J.-C., Bonnemaire, J. & Tessier, J.H. (1976*a*). *Ann. Genet. Sel. Anim.*, **8**, 481.
Grosclaude, F., Mahé, M.-F., Mercier, J.-C., Bonnemaire, J. & Tessier, J.H. (1976*b*). *Ann. Genet. Sel. Anim.*, **8**, 461.
Grosclaude, F., Jourdier, P. & Mahé, M.-F. (1979). *J. Dairy Res.*, **46**, 211.
Groves, M.L. (1969) *J. Dairy Sci*, **52**, 1155.
Groves, M.L. & Gordon, W.G. (1969). *Biochim. Biophys. Acta*, **194**, 421.
Groves, M.L. McMeekin, T.L., Hipp, N.J. & Gordon, W.G. (1962). *Biochim. Biophys. Acta*, **57**, 197.
Guillou, H., Miranda, G. & Pelissier, J.-P. (1987). *Lait*, **67**, 135.
Hashimoto, A., Aoyagi, H. & Izumiya, N. (1980). *Bull. Chem. Soc. Jpn.*, **53**, 2926.
Henschen, A., Lottspeich, F., Brantl, V. & Teschemacher, H. (1979). *Hoppe-Seyler's Z. Physiol. Chem.*, **360**, 1217.
Hill, R.J. & Wake, R.G. (1969). *Nature*, **221**, 635.
Hipp, N.J., Groves, M.L., Custer, J. H. & McMeekin, T.L. (1952). *J. Dairy Sci*, **35**, 272.
Ho, C. & Chen, A.H. (1967). *J. Biol. Chem.*, **242**, 551.
Ho, C. & Waugh, D.F. (1965). *J. Am. Chem. Soc.*, **87**, 110.
Holt, C. (1985). In *Developments in Dairy Chemistry*—3, ed. P.F. Fox. Elsevier Applied Science, London, pp. 143–81.
Holt, C. & Sawyer, L. (1988). *Protein Engr.*, **2**, 251.
Holt, C., Parker, T.G. & Dalgleish, D.G. (1975). *Biochim. Biophys. Acta*, **379**, 638.

Humphrey, R.S. & Newsome, L.J. (1984). *New Zealand J. Dairy Sci. Technol.*, **19**, 197.
Jang, H.D. & Swaisgood, H.E. (1990). *Arch. Biochem. Biophys.*, **283**, 318.
Jenness, R., Larson, B.L., McMeekin, T.L., Swanson, A.M., Whitnah, C.H. & Whitney, R. McL. (1956). *J. Dairy Sci.*, **39**, 536.
Jimenez-Flores, R., Kang, Y.C. & Richardson, T. (1987). *Biochem. Biophys. Res. Commun.*, **142**, 617.
Jollés, P. & Fiat, A.-M. (1979). *J. Dairy Res.*, **46**, 187.
Jollés, J., Alais, C. & Jollés, P. (1970). *Helv. Chim. Acta*, **53**, 1918.
Jollés, J., Schoentgen, F., Alais, C., Fiat, A.-M. & Jollés, P. (1972*a*). *Helv. Chim. Acta*, **55**, 2872.
Jollés, J., Schoentgen, F., Alais, C. & Jollés, P. (1972*b*).*Chimia*, **26**, 645.
Jollés, J., Fiat, A.-M., Alais, C. & Jollés, P. (1973). *FEBS Lett.*, **30**, 173.
Jollés, P., Levy-Toledano, S., Fiat, A.-M., Soria, C., Gillessen, D., Thomaidis, A., Dunn, F.W. & Caen, J.P. (1986). *Eur. J. Biochem.*, **158**, 379.
Kajiwara, K., Niki, R., Urakawa, H., Hiragi, Y., Donkai, N. & Nagura, M. (1988). *Biochim. Biophys. Acta*, **955**, 128.
Kaminogawa, S., Mizobuchi, H. & Yamauchi, K. (1972). *Agric. Biol. Chem.*, **36**, 2163.
Kaminogawa, S., Yamauchi, K. & Yoon, C.-H. (1980). *J. Dairy Sci.*, **63**, 223.
Kanamori, M., Kawaguchi, N., Ibuki, F. & Doi, H. (1980). *Agric. Biol. Chem.*, **44**, 1855.
Kanehisa, H., Miyake, I., Okai, H., Aoyagi, H. & Izumiya, N. (1984). *Bull. Chem. Soc. Jpn.*, **57**, 819.
Kiddy, C.A. (1975). In *Methods of Gel Electrophoresis of Milk Proteins*, ed. H.E. Swaisgood. American Dairy Science Association, Champaign, IL, pp. 16–17.
Kim, Y.K., Yaguchi, M. & Rose, D. (1969). *J. Dairy Sci.*, **52**, 316.
King, J.W.B., Aschaffenburg, R., Kiddy, C.A. & Thompson, M.P. (1965). *Nature*, **206**, 324.
Kumosinski, T.F., Brown, E.M. & Farrell, H.M. Jr. (1991). *J. Dairy Sci.*, **74**, 2899.
Larsen, B. & Thymann, M. (1966). *Acta Vet. Scand.*, **7**, 189.
Leach, B.S., Collawn, J.F., Jr. & Fish, W.W. (1980). *Biochemistry*, **19**, 5741.
Li, F.H.F. & Gaunt, S.N. (1972). *Biochem. Genet.*, **6**, 9.
Lim, V.I. (1974). *J. Mol. Biol.*, **88**, 873.
Linderstrøm-Lang, K. (1929). *C.R. Trav. Lab. Carlsburg, Ser. Chim.*, **17**, 1.
Lindqvist, B. (1963*a*). *Dairy Sci. Abstr.*, **25**, 257.
Lindqvist, B. (1963*b*). *Dairy Sci. Abstr.*, **25**, 299.
Loucheux-Lefebvre, M.-H., Aubert, J.-P. & Jollés, P. (1978). *Biophys. J.*, **23**, 323.
Loukas, S., Varoucha, D., Zioudrou, C., Streaty, R.A. & Klee, W.A. (1983). *Biochemistry*, **22**, 4567.
Mackinlay, A.G. & Wake, R.G. (1965). *Biochim. Biophys. Acta*, **104**, 167.
Mackinlay, A.G., Hill, R.J. & Wake, R.G. (1966). *Biochim. Biophys. Acta*, **115**, 103.
Manson, W., Carolan, T. & Annan, W.D. (1977). *Eur. J. Biochem.*, **78**, 411.
Matoba, T., Hayashi, R. & Hata, T. (1970). *Agric. Biol. Chem.*, **34**, 1235.
McGann, T.C.A., Kearney, R.D. & Donnelly, W.J. (1979). *J. Dairy Res.*, **46**, 307.
McKenzie, H.A. (1967). *Adv. Protein Chem.*, **22**, 55.
McKenzie, H.A. (1970). *Milk Proteins*, Vol. 1. Academic Press, New York.
McKenzie, H.A. (1971). *Milk Proteins*, Vol. 2. Academic Press, New York.
McKenzie, H.A. & Wake, R.G. (1959). *Aust. J. Chem.*, **12**, 734.
McKenzie, H.A. & Wake, R.G. (1961). *Biochim. Biophys. Acta*, **47**, 240.
McLean, D.M., Graham, E.R.B., Ponzoni, R.W. & McKenzie, H.A. (1984). *J. Dairy Res.*, **51**, 531.
McMeekin, T.L., Groves, M.L. & Hipp, N.J. (1949). *J. Am. Chem. Soc.*, **71**, 3298.
McNeill, G.P. & Donnelly, W.J. (1987). *J. Dairy Res.*, **54**, 19.
Meisel, H. (1986). *FEBS Lett.*, **196**, 223.
Mercier, J.-C. (1981). *Biochimie*, **63**, 1.
Mercier, J.-C., Grosclaude, F. & Ribadeau-Dumas, B. (1971). *Eur. J. Biochem.*, **23**, 41.

Mercier, J.-C., Uro, J., Ribadeau-Dumas, B. & Grosclaude, F. (1972*a*). *Eur. J. Biochem.*, **27**, 535.
Mercier, J.-C., Grosclaude, F. & Ribadeau-Dumas, B. (1972*b*). *Milchwissenschaft*, **27**, 402.
Mercier, J.-C., Brignon, G. & Ribadeau-Dumas, B. (1973). *Eur. J. Biochem.*, **35**, 222.
Minamiura, N., Matsumura, Y., Fukumoto, J. & Yamamoto, T. (1972). *Agric. Biol. Chem.*, **36**, 588.
Miyake, I., Kouge, K., Kanehisa, H. & Okai, H. (1983). *Bull. Chem. Soc. Jpn.*, **56**, 1678.
Nagao, M., Maki, M., Sasaki, R. & Chiba, H. (1984). *Agric. Biol. Chem.*, **48**, 1663.
Neelin, J.M. (1964). *J. Dairy Sci.*, **47**, 506.
Ng-Kwai-Hang, K.F. & Pélissier, J.P. (1989). *J. Dairy Res.*, **56**, 391.
Nijhuis, H. & Klostermeyer, H. (1975). *Milchwissenschaft*, **30**, 530.
Niki, V.R., Takase, K. & Arima, S. (1977). *Milchwissenschaft*, **32**, 577.
Noelken, M. & Reibstein, M. (1968). *Arch. Biochem. Biophys.*, **123**, 397.
Nozaki, Y. & Tanford, C. (1967). *Meth. Enzymol.*, **11**, 715.
Ono, T., Kaminogawa, S., Odagiri, S. & Yamauchi, K. (1976). *Agric. Biol. Chem.*, **40**, 1717.
Ono, T., Yahagi, M. & Odagiri, S. (1980). *Agric. Biol. Chem.*, **44**, 1499.
Otagiri, K., Miyake, I., Ishibashi, N., Fukui, H., Kanehisa, H. & Okai, H. (1983). *Bull. Chem. Soc. Jpn.*, **56**, 1116.
Parker, T.G. & Dalgleish, D.G. (1981). *J. Dairy Res.*, **48**, 71.
Payens, T.A.J. & Schmidt, D.G. (1965). *Biochim. Biophys. Acta* **109**, 214.
Payens, T.A.J. & Schmidt, D.G. (1966). *Arch. Biochem. Biophys.*, **115**, 136.
Payens, T.A.J. & van Markwijk, B.W. (1963). *Biochim. Biophys. Acta*, **71**, 517.
Pearce, K.N. (1975). *Eur. J. Biochem.*, **58**, 23.
Pepper, L. & Farrell, H.M., Jr. (1982). *J. Dairy Sci.*, **65**, 2259.
Peterson, R.F. & Koppler, F.C. (1966). *Biochem. Biophys. Res. Commun.*, **22**, 388.
Raap, J. Kerling, K.E.T., Vreeman, H.J. & Visser, S. (1983). *Arch. Biochem. Biophys.*, **221**, 117.
Reimerdes, E.H. & Herlitz, E. (1979). *J. Dairy Res.*, **46**, 219.
Ribadeau Dumas, B. (1961). *Biochim. Biophys. Acta*, **54**, 400.
Ribadeau Dumas, B., Maubois, J.L., Mocquot, G. & Garnier, J. (1964). *Biochim. Biophys. Acta*, **82**, 494.
Ribadeau Dumas, B., Brignon, G., Grosclaude, F. & Mercier, J.-C. (1972). *Eur. J. Biochem.*, **25**, 505.
Ribadeau Dumas, B., Mercier, J.-C. & Grosclaude, F. (1973). *Neth. Milk Dairy J.*, **27**, 304.
Rollema, H.S., Vreeman, H.J. & Brinkhaus, J.A. (1984). In *XXII Congress Ampere Committee*, ed. K.A. Muller, R. Kind & J. Roos. Ampere Committee, Zurich, p. 494.
Rollema, H. S., Brinkhuis, J.A. & Vreeman, H.J. (1988). *Neth. Milk Dairy J.*, **42**, 233.
Rose, D., Davies, D.T. & Yaguchi, M. (1969). *J. Dairy Sci.*, **52**, 8.
Rose, D., Brunner, J.R., Kalan, E.B., Larson, B.L., Melnychyn, P., Swaisgood, H.E. & Waugh, D.F. (1970). *J. Dairy Sci.*, **53**, 1.
Saito, T., Itoh, T. & Adachi, S. (1980). *Agric. Biol. Chem.*, **44**, 1023.
Schaar, J. (1984). *J. Dairy Res.*, **51**, 397.
Schaar, J., Hansson, B & Pettersson, H.-E. (1985). *J. Dairy Res.*, **52**, 429.
Scheraga, H.A. & Mandelkern, L. (1953). *J. Am. Chem. Soc.*, **75**, 179.
Schmidt, D.G. (1964). *Biochim. Biophys. Acta*, **90**, 411.
Schmidt, D.G. (1969). On the association of α_{s1}-casein. Thesis, Utrecht.
Schmidt, D.G. (1970*a*). *Biochim. Biophys. Acta*, **221**, 140.
Schmidt, D.G. (1970*b*). *Biochim. Biophys. Acta*, **207**, 130.
Schmidt, D.G. (1980). *Neth. Milk Dairy J.*, **34**, 42.
Schmidt, D.G. (1985). In *Developments in Dairy Chemistry—1*, ed. P.F. Fox. Elsevier Applied Science, London, p.61.
Schmidt, D.G. & Payens, T.A.J. (1972). *J. Coll. Interface Sci.*, **39**, 655.
Schmidt, D.G. & van Markwijk, B.W. (1968). *Biochim. Biophys. Acta*, **154**, 613.
Schmidt, D.G., Both, P. & de Koning, P.G. (1966). *J. Dairy Sci.*, **49**, 776.

Schmidt, D.G., Payens, T.A.J., van Markwijk, B.W. & Brinkhuis, J.A. (1967). *Biochem. Biophys. Res. Commun*, **27**, 448.
Shimizu, M., Lee, S.W., Kaminogawa, S. & Yamauchi, K. (1986). *J. Food Sci.*, **51**, 1248.
Shinoda, I., Fushimi, A., Kato, H., Odai, H. & Fukui, S. (1985). *Agric. Biol. Chem.*, **49**, 2587.
Snoeren, T.H.M., Van der Spek, C.A. & Paynes, T.A.J. (1977). *Biochim. Biophys. Acta*, **490**, 255.
Snoeren, T.H.M., Van Markwijk, B. & Van Montfort, R. (1980). *Biochim. Biophys. Acta*, **622**, 268.
Squire, P.G. & Himmel, M.E. (1979). *Arch. Biochem. Biophys.*, **196**, 165.
Stewart, A.F., Willis, I.M. & Mackinlay, A.G. (1984). *Nucl. Acids Res.*, **12**, 3895.
Stewart, A.F., Bonsing, J., Beattie, C.W., Shah, F., Willis, I.M. & Mackinlay, A.G. (1987). *Mol. Biol. Evol.*, **4**, 231.
Swaisgood, H.E. (1973). *CRC Critical Reviews in Food Technology*, **3**, 375.
Swaisgood, H.E. (ed.), (1975). *Methods of Gel Electrophoresis of Milk Proteins*. American Dairy Science Association, Champaign, IL.
Swaisgood, H.E. (1982*a*). In *Developments in Dairy Chemistry—1*, ed. P.F. Fox. Applied Science Publishers, London, pp. 1–59.
Swaisgood, H.E. (1982*b*). In *Handbook of Processing and Utilization in Agriculture*, ed. I.A. Wolff. CRC Press, Boca Raton, FL, pp. 553–69.
Swaisgood, H.E. (1985). In *Food Chemistry*, 2nd ed. O.R. Fennema. Marcel Dekker, New York, pp. 791–827.
Swaisgood, H.E. (1989). In *Milk Proteins*, ed. C.A. Barth & E. Schlimme. Steinkopff Verlag Darmstadt (Springer-Verlag, New York), pp. 192–210.
Swaisgood, H.E. & Brunner, J.R. (1962). *J. Dairy Sci.*, **45**, 1.
Swaisgood, H.E. & Brunner, J.R. (1963). *Biochem. Biophys. Res. Commun*, **12**, 148.
Swaisgood, H.E. & Catignani, G.L. (1987). *Meth., Enzymol.*, **135**, 596.
Swaisgood, H.E. & Timasheff, S.N. (1968). *Arch. Biochem. Biophys.*, **125**, 344.
Swaisgood, H.E., Brunner, J.R. & Lillevik, H.A. (1964). *Biochemistry*, **3**, 1616.
Tai, M.-S. & Kegeles, G. (1984). *Biophys. Chem.*, **20**, 81.
Takase, K., Niki, R. & Arima, S. (1980). *Biochim. Biophys. Acta*, **622**, 1.
Takeuchi, M., Tsuda, E., Yoshikawa, M., Sasaki, R. & Chiba, H. (1985). *Agric. Biol. Chem.*, **49**, 2269.
Talbot, B. & Waugh, D.F. (1970). *Biochemistry*, **9**, 2807.
Tanford, C. (1961) *Physical Chemistry of Macromolecules*. John Wiley & Sons, New York.
Teller, D.C., Swanson, E. & de Haen, C. (1979). *Meth. Enzymol.*, **61**, 103.
Thompson, M.P. (1966). *J. Dairy Sci.*, **49**, 792.
Thompson, M.P. (1970). *J. Dairy Sci.*, **53**, 1341.
Thompson, M.P. & Kiddy, C.A. (1964). *J. Dairy Sci.*, **47**, 626.
Thompson, M.P. & Pepper, L. (1964). *J. Dairy Sci.*, **47**, 633.
Thompson, M.P., Kiddy, C.A., Pepper, L. & Zittle, C.A. (1962). *Nature*, **195**, 1001.
Thompson, M.P., Tarassuk, N.P., Jenness, R., Lillevik, H.A., Ashworth, U.S. & Rose, D. (1965). *J. Dairy Sci.*, **48**, 159.
Thompson, M.P., Kalan, E.B. & Greenberg, R. (1967). *J. Dairy Sci.*, **50**, 767.
Thompson, M.P., Gordon, W.G., Boswell, R.T. & Farrell, H.M., Jr. (1969). *J. Dairy Sci.*, **52**, 1166.
Thompson, M.D., Dave, J.R. & Nakhasi, H.L. (1985). *DNA*, **4**, 263.
Thurn, A., Burchard, W. & Niki, R. (1987). *Coll. Polymer Sci.*, **265**, 653.
Toma, S.J. & Nakai, S. (1973). *J. Dairy Sci.*, **56**, 1559.
Tran, V.D. & Baker, B.E. (1970). *J. Dairy Sci.*, **53**, 1009.
Turkington, R.W. & Topper, Y.J. (1966). *Biochim. Biophys. Acta*, **127**, 366.
Van Eenennaam, A.L. & Medrano, J.F. (19). *J. Dairy Sci.*, **74**, 1491.
Visser, S., Slangen, K.J., Alting, A.C. & Vreeman, H.J. (1989). *Milchwissenschaft*, **44**, 333.
Vreeman, H.J. (1979). *J. Dairy Res.*, **46**, 271.

Vreeman, H.J., Both, P., Brinkhauis, J.A. & Van der Spek, C.A. (1977). *Biochim. Biophys. Acta*, **491**, 93.
Vreeman, H.J., Brinkhuis, J.A. & Van der Spek, C.A. (1981). *Biophys. Chem.*, **14**, 185.
Vreeman, H.J., Visser, S., Slangen, C.J. & Van Riel, J.A.M. (1986). *Biochem. J.*, **240**, 87.
Wake, R.G. & Baldwin, R.L. (1961). *Biochim. Biophys. Acta*, **47**, 225.
Waugh, D.F. & Von Hippel, P.H. (1956). *J. Am. Chem. Soc.*, **78**, 4576.
Waugh, D.F., Ludwig, M.L., Gillespie, J.M., Melton, B., Foley, M. & Kleiner, E.S. (1962). *J. Am. Chem. Soc.*, **84**, 4929.
Waugh, D.F., Slattery, C.W. & Creamer, L.K. (1971). *Biochemistry*, **10**, 817.
Wetlaufer, D.B. (1962). *Adv., Protein Chem.*, **17**, 303.
White, J.C.D. & Davies, D.T. (1958). *J. Dairy Res.*, **25**, 236.
Whitney, R. McL. (1977). In *Food Colloids*, ed. H.D. Graham. AVI Publishing Company, Westport, CT, pp. 66–151.
Whitney, R. McL. (1988). In *Fundamentals of Dairy Chemistry*, 3rd edn, ed. N.P. Wong. AVI book, Van Nostrand Reinhold, New York, pp. 81–169.
Whitney, R. McL., Brunner, J.R., Ebner, K.E., Farrell, H.M., Jr., Josephson, R. V., Morr, C.V. & Swaisgood, H.E. (1976). *J. Dairy Sci.*, **59**, 785.
Woychik, J.H. (1964). *Biochem. Biophys. Res. Commun.*, **16**, 267.
Woychik, J.H., Kalan, E.B. & Noelken, M.E. (1966). *Biochemistry*, **5**, 2276.
Yaguchi, M., Davies, D.T. & Kim, Y.K. (1966). *J. Dairy Sci.*, **51**, 473.
Yoshikawa, M., Sugimoto, E. & Chiba, H. (1975). *Agric. Biol. Chem.*, **39**, 1843.
Yoshikawa, M., Tani, F., Yashimura, T. & Chiba, H. (1986). *Agric. Biol. Chem.*, **50**, 2419.
Zevaco, C. & Ribadeau-Dumas, B. (1984). *Milchwissenschaft*, **39**, 206.
Zioudrou, C., Streaty, R.A. & Klee, W.A. (1979). *J. Biol. Chem.*, **254**, 2446.
Zittle, C.A. & Custer, J.H. (1963). *J. Dairy Sci.*, **46**, 1183.

3

CASEIN ASSOCIATION AND MICELLE FORMATION

H.S. Rollema

Netherlands Institute for Dairy Research, Ede, The Netherlands

1. Introduction

Milk is a complex biological fluid containing relatively high amounts of proteins and minerals. Its primary biological function is to act as a supply of nutrients for the newborn. To serve this function the milk proteins are designed in such a way that they form complexes with a relatively large amount of calcium phosphate, which readily coagulate in the stomach of the newborn. The major part of the milk proteins, together with calcium phosphate, occur in the form of large colloidal particles, the casein micelles. Therefore, milk can be considered as a two-phase system consisting of a micellar and a serum phase. The distribution of the main constituents between the two phases is given in Table 1.

The unique arrangement of the caseins and calcium phosphate in the casein micelles is a consequence of the peculiar properties of the different casein components. These properties are reflected in the self-association behaviour of the individual caseins.

Because of their importance for the functional behaviour of dairy products, caseins and casein micelles have been studied for a long time. In the last decade the properties of caseins and casein micelles have been reviewed several times (Payens & Vreeman, 1982; Schmidt, 1982; McMahon & Brown, 1984; Ruettiman & Ladish, 1987; Farrell, 1988; Whitney, 1988).

This chapter deals with self-association of the individual caseins, metal ion binding to the caseins, properties of casein micelles, different models for the casein micelle and finally with non-bovine caseins and casein micelles. The emphasis in this chapter is on the more recent literature; for a review of older literature the reader is referred to reviews, listed in the previous paragraph.

TABLE 1
The Distribution of the Major Casein Micelle Constituents Between the Serum and Micellar Phases of Bovine Milk at pH 6·7 and 20°C (data from Davies & Law, 1983; Holt, 1985)

Component	*Micellar phase (g/litre)*	*Serum phase (g/litre)*
α_{S1}-Casein	10·9	0·7
α_{S2}-Casein	3	0·1
β-Casein	9	1·3
κ-Casein	2·9	0·5
Calcium	0·8	0·4
Phosphate	0·9	1·1
Citrate	0·1	1·8

2. SELF-ASSOCIATION OF BOVINE CASEINS

Caseins are prone to association due to their high hydrophobicity and peculiar charge distribution (see Chapter 2). The major interactions involved in these processes are of electrostatic and hydrophobic nature. A delicate balance between these forces, depending on experimental conditions such as pH, temperature and ionic strength, determines the type and extent of association.

2.1. α_{s1}-Casein

This casein component has the highest charge at milk pH (i.e. −20 at pH 6·6) and it is, with the exception of α_{s2}-casein, the least hydrophobic amongst the caseins. The hydrophobic residues are distributed more or less uniformly along the polypeptide chain. The same applies to the charge, except for the region between residues 40 and 80 where a cluster of ester phosphates is situated.

At low ionic strength (0·003–0·01 M) and neutral or alkaline pH the molecule occurs as a monomer (Swaisgood & Timasheff, 1968; Schmidt, 1970*a*). Self-association is initiated by increasing ionic strength. The association depends on pH in a way that an increase in pH reduces association. Temperature has little effect on the association.

From sedimentation studies it was concluded (Payens & Schmidt, 1966) that at neutral pH the association of the genetic variant B of α_{s1}-casein proceeds by a series of consecutive steps:

$$(\alpha_{S1})_i + \alpha_{S1} \rightleftharpoons (\alpha_{S1})_{i+1}$$

For the pH range 8·0–9·0, a mechanism was proposed in which the dimer is the elementary associating unit (Swaisgood & Timasheff, 1968).

From dynamic and static light scattering studies of α_{s1}-casein association at 35°C, a value for the structure factor of α_{s1}-casein polymers, characteristic for rod-like chains, was obtained (Thurn *et al.*, 1987*a*). This result is opposed to the earlier suggestion of spherical polymers (Schmidt, 1970*a*).

The association behaviour of the genetic variants of α_{s1}-casein C(192, Glu→Gly) and D(53, Ala→SerP) was compared with that of variant B (Schmidt, 1970*b*). The C variant showed stronger association, possibly due to the reduction in charge. The D variant, however, did not differ from the B variant in spite of the extra negative charge.

2.2. α_{s2}-Casein

This minor component is the least hydrophobic of the caseins and is highly phosphorylated. The ester phosphates are spread over three clusters in the polypeptide chain. At neutral pH the charges on the protein (net charge of −13 to −18 depending on the phosphoserine content) are distributed in such a way that the C-terminal part carries a positive charge whereas the N-terminal part is strongly negatively charged.

Based on light scattering studies, Snoeren *et al.* (1980) described the polymerization of α_{s2}-casein by a series of consecutive steps:

$$(a_{S2})_i + \alpha_{S2} \rightleftharpoons (\alpha_{S2})_{i+1}$$

The association behaviour could be described by an isodesmic mechanism, i.e. the equilibrium constant has the same value for all association steps. The degree of association strongly depends on ionic strength. With increasing ionic strength the degree of association first increases, but above an ionic strength of 0·2 M, the degree of association decreases with a further increase in ionic strength. This peculiar ionic strength dependence has been attributed to the non-uniform distribution of positive and negative charges along the polypeptide chain, resulting in a positively charged cluster on a particle with net negative charge (Snoeren *et al.*, 1980). Apart from hydrophobic bonds, the interactions between the particles in the polymer are asssumed to be composed of an attractive force due to the binding of the positive cluster of one casein molecule to the negative part of an adjacent one and of a repulsion due to the overall negative charge of the casein molecules. At low ionic strength the repulsive force dominates. On increasing the ionic strength the repulsive force is reduced and the attractive force starts to dominate. Ultimately, at high ionic strength the latter force is reduced and the degree of association decreases.

2.3. β-Casein

This protein is the most hydrophobic casein and has a relatively low charge at neutral pH (net charge of −12 at pH 6·6). The molecule has a negatively charged N-terminal part and a strongly hydrophobic C-terminal region. This distribution of hydrophobic and charged residues, characteristic of an amphiphile, is in fact reflected in its micellar-like association behaviour.

Ultracentrifugation and light scattering studies have shown that the association of β-casein can be described by a monomer–polymer equilibrium:

$$n\beta \rightleftharpoons \beta_n$$

with a narrow size distribution of the polymers (Payens & van Markwijk, 1963;

TABLE 2
The Degree of Association, *n*, of Bovine β-Casein at Different Temperatures and Ionic Strengths

Temperature (°C)	*Ionic strength*	*pH*	n	*Reference*
5	0·4	6·7	17·5	Kajiwara *et al.* (1988)
8·5	0·2	7·5	22	Payens & van Markwijk (1963)
10	0·4	6·7	12	Takase *et al.* (1980)
15	0·4	6·7	38	Kajiwara *et al.* (1988)
15	0·4	6·7	40	Thurn *et al.* (1987*b*)
15	0·4	6·7	22	Takase *et al.* (1980)
20	0·1	7·0	14	Evans *et al.* (1979)
21	0·05	7·0	13	Payens & Vreeman (1982)
21	0·2	7·0	23	Payens & Vreeman (1982)
22	0·2	6·7	58	Buchheim & Schmidt (1979)
25	0·4	6·7	59	Kajiwara *et al.* (1988)

Payens *et al.*, 1969). In the ultracentrifuge the sedimentation pattern consists of two peaks and in electron microscopy a bimodal size distribution was found for β-casein (Buchheim & Schmidt, 1979). Studies of the association of β-casein at low protein concentration have shown that the association is also characterized by a critical micelle concentration ranging from 0·3 to 0·7 mg/ml depending on temperature and ionic strength (Schmidt & Payens, 1972; Takase *et al.*, 1980).

The association of β-casein is strongly influenced by temperature and ionic strength. At low temperature and ionic strength only monomers occur; an increase in temperature or ionic strength leads to an increased association (Schmidt & Payens, 1972).

The number of monomers, *n*, in the β-casein micelle depends on temperature, ionic strength and probably pH. Values for *n* reported by different investigators are given in Table 2; it is apparent that there is still some controversy concerning the size of β-casein micelles.

For the shape of the polymer, a thread-like structure (Payens & van Markwijk, 1963), a spherical particle (Andrews *et al.*, 1979; Buchheim & Schmidt, 1979; Thurn *et al.*, 1987*b*) and an oblate ellipsoid approaching spherical shape at high degrees of association (Kajiwara *et al.*, 1988) have been proposed.

The polymerization of β-casein represents a typical example of the self-association of an amphiphile. The driving forces in micelle formation are the hydrophobic interactions between the C-terminal segments. These forces are opposed by repulsions between the charged head-groups (N-terminal segments), which limit the micellar size. The hydrophobic interactions are responsible for the temperature dependence, the repulsions for the ionic strength effects. This interpretation of the association mechanism of β-casein is supported by observations that removal of the last three C-terminal residues reduces association (Evans *et al.*, 1979) and that removal of the last 20 C-terminal residues of the protein completely abolishes its ability to associate (Berry & Creamer, 1975).

2.4. κ-Casein

Within the casein group, κ-casein takes second place as far as hydrophobicity is concerned and it possesses the lowest net charge at neutral pH (i.e. −3·9 at pH 6·6). The C-terminal, or macropeptide, part of the molecule carries a negative net charge and is less hydrophobic than the positively charged N-terminal domain.

The character of the protein is therefore clearly amphiphilic, but not to the same extent as in the case of β-casein. Due to its low ester phosphate content, κ-casein is soluble in calcium-containing media. This property and its unique structure are responsible for its important role in the stabilization of casein micelles.

It has been long recognized that both oxidized (S-S-containing) and reduced (SH-containing) forms of κ-casein polymerize (Swaisgood *et al.*, 1964; Talbot & Waugh, 1970; Slattery & Evard, 1973), the reduced form showing less association than the oxidized form.

Sedimentation studies of the association of reduced carbohydrate-free κ-casein showed that this protein associates according to a monomer–polymer mechanism similar to that of β-casein (Vreeman *et al.*, 1977; Vreeman, 1979; Vreeman *et al.*, 1981):

$$n\kappa \rightleftharpoons \kappa_n$$

At pH 7 and 20°C, the degree of polymerization, n, is approximately 30 and it is virtually independent of temperature between 4 and 20°C and of ionic strength between 0·1 and 1·0 M (Vreeman *et al.*, 1981). At low protein concentrations, a critical micelle concentration (typically 0·2–0·5 mg/ml) is observed which decreases with increasing ionic strength. From sedimentation and hydrodynamic measurements it was found that the polymer is a spherical particle with a diameter of 23 nm (Vreeman *et al.*, 1981).

From small-angle neutron scattering and light scattering studies of κ-casein association at 15°C in simulated milk ultrafiltrate, a degree of association of 140 was reported and it was suggested that the polymer is composed of star-like submicelles, each containing 9 casein molecules (Thurn *et al.*, 1987*b*). However, in this study the experiments were performed in the absence of reducing agent(s) and the κ-casein sample was not depolymerized, which appears to be essential to obtain reliable results (Vreeman *et al.*, 1981).

^{1}H-NMR studies of carbohydrate-free and carbohydrate-containing κ-casein fractions demonstrated that the polymer consists of a rigid core surrounded by a layer of flexible polypeptide regions (Rollema *et al.*, 1988). The N-terminal part of the molecule consititutes the core whereas the C-terminal part is highly mobile. The hinge between these two regions is located between the residues 86 and 96. This structure was observed both in phosphate buffer and in simulated milk ultrafiltrate.

The association of κ-casein is substantially different from that of β-casein. In the case of κ-casein, polymers of a fixed size are formed, irrespective of temperature and ionic strength. This behaviour is partly due to the relatively low charge of the C-terminal region. In addition, it has been suggested that entropic repulsion between the negatively charged C-terminal domains and spacial requirements could play a role in the limitation of micellar size (Vreeman, 1979). The association of κ-casein has been compared to the assembly of virus coat proteins and it has been argued that the polymerization of κ-casein can be described by theories developed for virus assembly (Vreeman *et al.*, 1981).

3. Metal Ion Binding to Caseins

All casein components are able to bind different bivalent metal ions (Thompson *et al.*, 1969; Dickson & Perkins, 1971; Baumy & Brule, 1988). Because of its relevance to the structure and stability of casein micelles, the binding of calcium to the different casein components has been studied extensively (Waugh *et al.*, 1971; Dalgleish & Parker, 1980; Ono *et al.*, 1980; Parker & Dalgleish, 1981).

The phosphate groups of the phosphoserine residues constitute the main calcium binding sites in the caseins. Apart from binding to these sites, calcium ions also interact with other groups, such as the carboxylate groups of aspartate and glutamate residues (Dickson & Perkins, 1971; Byler & Farrell, 1989). The calcium binding capacity of the casein components decreases in the order $\alpha_{s2} > \alpha_{s1} > \beta > \kappa$-casein, corresponding to their decreasing phosphoserine content.

Binding of calcium to the caseins at neutral pH reduces their negative charge, which diminishes the electrostatic repulsion between the molecules. In this way, calcium binding promotes hydrophobic interactions which can lead to increasing association and ultimately to precipitation (Thompson *et al.*, 1969; Toma & Nakai, 1973; Aoki *et al.*, 1985). For κ-casein, however, such effects are not observed because of its low phosphoserine content. The solubility of α_{s2}-, α_{s1}-and β-caseins is strongly affected by the binding of calcium ions. Upon addition of calcium these proteins remain soluble until a critical amount of calcium is bound, after which an additional increase in calcium ion concentration leads to a sharp decrease in their solubility. An exception to this behaviour is observed for β-casein and α_{s1}-casein variant A at low temperature. The β-casein–calcium complex does not precipitate at all under these conditions. The α_{s1}-casein variant A, in which a relatively hydrophobic region (residues 14–26) is deleted, shows with increasing calcium concentration an initial decrease in solubility followed by full restoration of solubility at high calcium concentrations. Removal of the first 23 N-terminal residues from α_{s1}-casein abolishes its calcium sensitivity completely (Kaminogawa *et al.*, 1980). This observation and the temperature dependent behaviour of α_{s1}-casein variant A and β-casein illustrate the importance of hydrophobic interactions in the formation of a precipitate.

κ-Casein is able to protect the other caseins from calcium-induced precipitation (Thompson *et al.*, 1969; Green, 1971; Toma & Nakai, 1973). This effect is responsible for the solubility of the caseins in milk, which contains a concentration of calcium sufficient to precipitate α_{s1}-, α_{s2}- and β-casein. Apparently, in the presence of calcium, casein complexes are formed with properties differing from those of the polymers of the individual caseins. Moreover, it has been reported that these complexes have higher affinities for calcium than the individual components

(Jang & Swaisgood, 1990). These specific interactions between the different casein components in the presence of calcium form the basis for the formation of casein micelles.

4. Bovine Casein Micelles

Casein micelles are relatively large aggregates of the four casein components, calcium phosphate and small amounts of magnesium, sodium, potassium and citrate. In electron microscopy the micelles appear as more or less spherical particles with a relatively wide size distribution. The majority of the micelles have diameters of approximately 15 nm, but these constitute only a low mass fraction. The size distribution continues up to diameters of approximately 600 nm. Relatively high values are found for the voluminosity and hydration, indicating that the micelle is a rather porous particle (McMahon & Brown, 1984; Farrell, 1988).

The presence of calcium phosphate is essential for the integrity of the micelles. The presence of κ-casein plays an important role for the stability of casein micelle suspensions; specific cleavage of this protein by chymosin induces a coagulation reaction. Attempts to explain the stability of casein micelle suspensions by colloid-chemical theories have not been successful (Payens, 1979). The current view is that electrostatic and steric repulsions opposing the van der Waals forces are responsible for the stability of the micelles (Hill & Wake, 1969; Holt, 1975; Walstra, 1979, 1990).

Since the specific arrangement of proteins and minerals in the micelle is responsible for its properties, much research has been devoted to the structure and the functional properties of the casein micelle. This section will deal with a number of important topics in this field and with the different models proposed for the structure of the casein micelle.

4.1. Location of Casein Components in the Casein Micelle

To study the location of the different caseins in the micelle, two different approaches have been used: electron microscopy of specifically-labelled casein micelles and determination of the casein composition of size-fractionated micelles.

The results from electron microscopic studies are rather confusing. From an investigation in which an acid-silver methenamine staining method for carbohydrates was used, it was concluded that glycosylated κ-casein as a whole is distributed throughout the casein micelles, but that the more highly glycosylated components are located in the surface layer (Kudo *et al.*, 1979).

Lectin-labelled gold markers were used in two studies. In one case, hardly any intramicellar location of the labelled glycosylated κ-casein was observed; it was found in a bridging network interconnecting the micelles (Horisberger & Vonlanthen, 1980). In the other case, uniform distribution of glycosylated κ-casein throughout micelles of all sizes was observed (Horisberger & Rouvet-Vauthey, 1984).

From an electron microscopic study of artificial micelles containing gold-labelled α_{s1}-, β- or κ-casein it was concluded that κ-casein is located at the micellar surface whereas α_{s1}- and β-caseins are uniformly distributed throughout the casein micelle (Schmidt & Both, 1982).

The use of immunological techniques combined with electron microscopy led to an opinion that in large micelles κ-casein is located at the periphery while in small micelles there is uniform distribution of κ-casein (Carroll & Farrell, 1983). However, other investigators, using essentially the same combination of techniques, came to the conclusion that there is no peripheral location of κ-casein and that the protein is uniformly distributed in the micelles, its concentration being lower in large micelles than in small ones (Horisberger & Vauthey, 1984).

The results obtained by electron microscopy appear to depend on the method of labelling and/or specimen preparation techniques. Therefore, no unequivocal conclusions concerning the location of the different casein components in the casein micelle can be drawn from these studies.

More consistent results have been obtained from studies of the casein micelle composition as a function of micellar size. Compositional studies of casein micelles fractionated by permeation chromatography on controlled-pore glass showed that the fractional content of α_s- and β-caseins, decreases and that of κ-casein increases with decreasing micellar size (McGann *et al.*, 1980; Donnelly *et al.*, 1984). A more or less linear dependence of the fractional content of the three proteins on the surface to volume ratio of the micelles was observed. Davies and Law (1983) fractionated casein micelles by differential centrifugation at 4 and 20°C and determined the casein composition of the different fractions. At both temperatures, the α_{s1}-casein content was virtually constant, α_{s2}-casein content decreased slightly, β-casein showed an appreciable decrease and the κ-casein content increased strongly with decreasing micellar size. These results were confirmed and amplified by Dalgleish *et al.* (1989) who fractionated casein micelles by ultracentrifugation, determined the size of the fractions by photon correlation spectroscopy and analysed the micelles with respect to their protein and mineral composition. From these data, the fractional content of each casein component in the surface layer and in the interior of the micelle was calculated. According to this study the outer layer of the micelle is composed of nearly equimolar amounts of $\alpha_s$casein (sum of α_{s1}- and α_{s2}-casein) and κ-casein and a small amount of β-casein while the interior of the micelle consists of equimolar amounts of α_s- and β-caseins with possibly a very small amount of κ-casein. The dependence of the fractional content of the different caseins on micellar size is shown in Fig. 1. The differences between the two data sets in Fig. 1 illustrate the effect of the two different techniques used for the size determination of the micelles. While these differences will have no consequences for the qualitative conclusions concerning the location of the caseins in the micelle, they do, however, affect the values for the fractional contents of the caseins in the surface layer and in the interior of the micelle.

The results described above, i.e. that all casein components are present at the surface of the micelle, are in accordance with the result of Heth and Swaisgood (1982) who covalently coupled casein micelles to glass beads, removed unbound

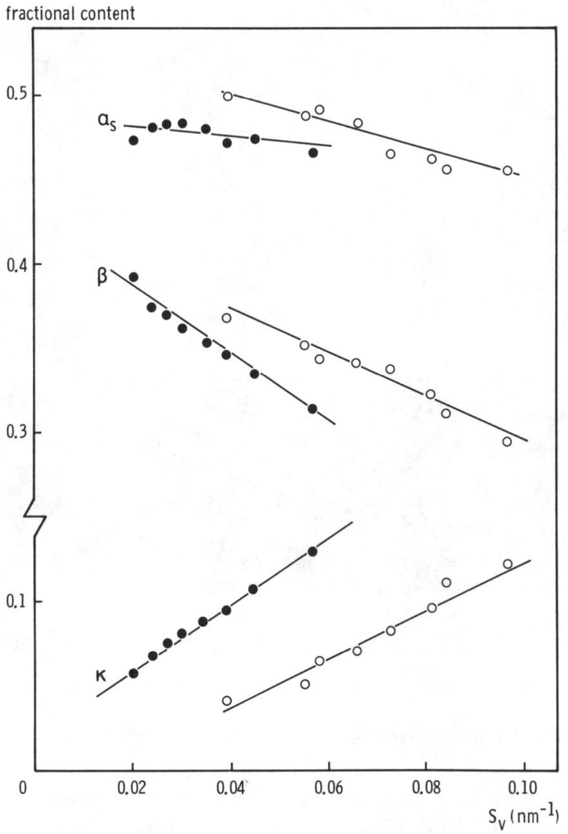

Fig. 1. The fractional casein contents of size-fractionated casein micelles as a function of the micellar surface to volume ratio, S_V. Casein components are indicated in the figure; α_S refers to the sum of α_{S1}- and α_{S2}casein. (●): Data set from Dalgleish *et al.* (1989), size determination by photon correlation Spectroscopy; (○): data set from Donnelly *et al.* (1984), size determination by electron microscopy.

proteins, analysed the bound fraction and found that all casein components were able to react with the glass surface.

Concerning the distribution of glycosylated versus unglycosylated κ-casein, less uniformity exists in the literature. In two cases, no difference was observed in the distribution of the two κ-casein fractions (Donnelly *et al.*, 1984; Dalgleish, 1985). In other cases either a higher proportion of glycosylated κ-casein in larger micelles was claimed (Creamer *et al.*, 1973) or it was found that the amount of glycosylated protein was inversely related to micellar size, except for the largest micelles, where a relatively high proportion of glycosylated κ-casein was found (Slattery, 1978).

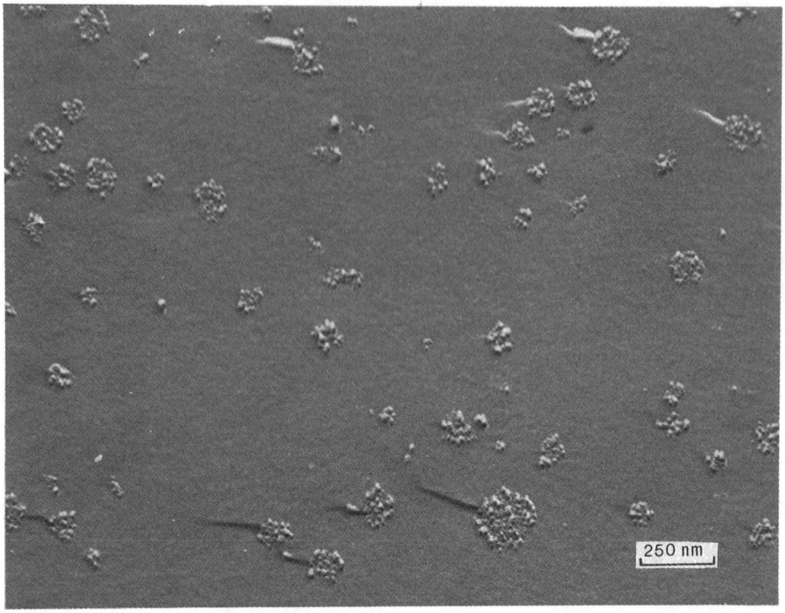

Fig. 2. Electron micrograph of casein micelles prepared by the freeze-fracturing technique.

4.2. Structure of the Casein Micelle

In electron micrographs of casein micelle specimens prepared by different techniques the micelles show an inhomogeneous structure built up from particles of approximately 10 nm diameter (Shimmin & Hill, 1964; Calapaj, 1968; Schmidt & Buchheim, 1970; Knoop *et al.*, 1973; Kalab *et al.*, 1982). Two typical examples are shown in Figs 2 and 3. These observations have been the basis for the concept of submicelles and for models in which it is assumed that casein micelles are composed of a number of discrete subunits. However, electron microscopy has been subject to criticism because with this technique values for the voluminosity of casein micelles are obtained which are considerably lower than those determined by methods like light scattering, sediment volume measurement, diffusion coefficient determination and viscometry (Walstra, 1979). This discrepancy could be due to shrinkage of the micelles during the specimen preparation procedure. An alternative explanation could be that in electron micrographs only the more dense protein domains are visible. In the latter case the raspberry-like appearance of casein micelles could also be interpreted as a structure consisting of compact protein domains connected to each other by less dense protein material.

Supplementary information on the internal structure of casein micelles has been obtained from small angle neutron scattering (SANS) and small angle X-ray scattering (SAXS) experiments. The SANS experiments indicated the existence of a substructure in the casein micelle (Stothart & Cebula, 1982; Stothart, 1989). The

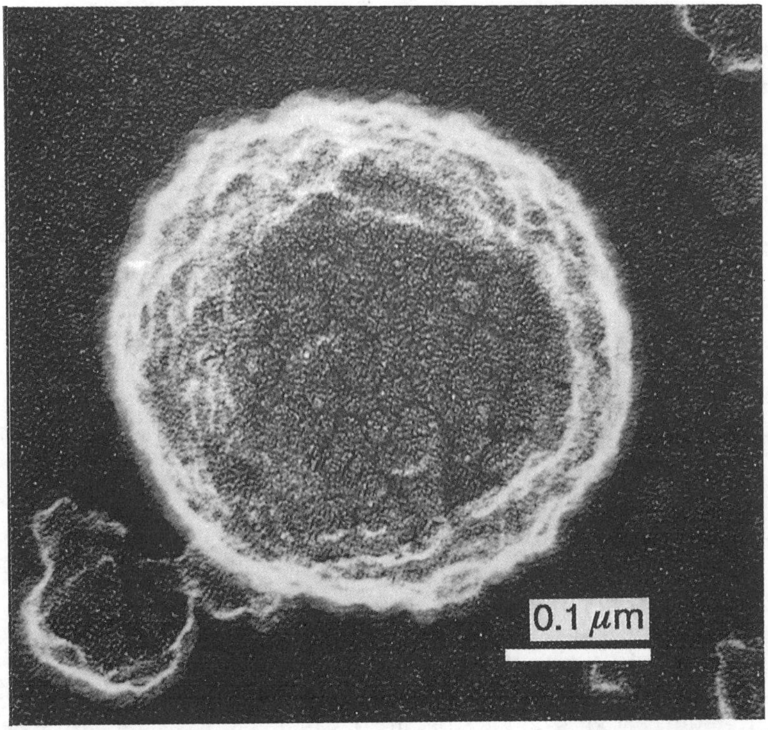

Fig. 3. Electron micrograph of a casein micelle prepared by the rotatory shadowing technique. By courtesy of Dr M. Kalab.

scattering data could be fitted by a model assuming a hexagonal close packing of spheres with dimensions similar to those of caseinate particles.

SAXS investigations of caseinate in the presence and absence of calcium provided similar results (Kumosinski *et al.*, 1988). For caseinate particles in the absence of calcium, indications were found for an inhomogeneous distribution of electron density. A structure for the submicelle was proposed in which a core of high electron density is surrounded by a concentric shell of low electron density. The experimental data in the presence of calcium were interpreted in terms of a model in which the calcium caseinate micelle is built up from submicelles (the particles observed in the absence of calcium) with a possible overlap of the low electron density regions.

The dynamic behaviour of native casein micelles and of calcium caseinate (i.e. whole casein in the presence of calcium) have been studied by nuclear magnetic resonance (NMR) spectroscopy (Griffin & Roberts, 1985; Rollema *et al.*, 1988; Kakalis *et al.*, 1990). The ^{1}H-NMR spectrum of casein micelles was found to be composed of a component with a large line width and a component with a relatively small line width. The presence of the latter component provides evidence that parts of the polypeptide chains in the micelle have a relatively high flexibility and mobility. The major part of the flexibile regions was assigned to the C-terminal part of κ-casein (Griffin & Roberts, 1985). Additionally, it was shown that only approximately half of the κ-casein has mobile segments and that there is a small contribution from other casein components to the flexible regions of the micelle (Rollema *et al.*, 1988). These results provide experimental evidence for the hypothesis that the C-terminal part of κ-casein protrudes into solution from the surface of the micelle (Hill & Wake, 1969; Holt, 1975; Walstra, 1979, 1990). These flexible polypeptide chains, forming a water-rich proteinaceous layer around the micelle, are responsible for a steric stabilization of the casein micelle.

Calcium depletion of the micelles resulted in a dramatic increase in intensity of the mobile component in the NMR spectrum (Rollema & Brinkhuis, 1989). This observation shows that the free caseinate particles or submicelles exhibit a high degree of conformational flexibility, which is lost upon incorporation in the casein micelle. This could be due to calcium phosphate induced protein–protein interactions between the submicelles in the casein micelle.

From ^{1}H-, ^{13}C- and ^{31}P-NMR studies of whole casein in the presence and the absence of calcium it was concluded that the particles formed in these systems have a rather loose structure, allowing a considerable segmental motion of the constituting polypeptide chains (Kakalis *et al.*, 1990).

4.3. Dissociation of Casein Micelles

Apart from direct studies on the structure of casein micelles, many investigations were carried out using more indirect approaches. In these experiments, casein micelles were dissociated by varying the solvent conditions and the changes in the composition of the micelles were studied.

Gradual removal of calcium from micelles at constant pH initially results in a release of caseins without major effects on the size of the micelles, but ultimately leads to complete micellar disaggregation (Lin *et al.*, 1972; Griffin *et al.*, 1988).

The caseinate obtained on removing calcium from casein micelles consists of particles with different composition. By high resolution gel permeation chromatography two fractions were obtained, one containing α_s- and κ-caseins and the other α_s- and β-caseins (Ono *et al.*, 1983*a*), with molecular weights of 500 kDa and 100–200 kDa, respectively (Ono & Takagi, 1986). The content of the α_s-/κ-casein fraction increased with decreasing micellar size.

Disaggregation of casein micelles by urea, followed by gel permeation chromatography, resulted in a calcium phosphate-containing protein fraction and a fraction with a very low calcium phosphate content (Aoki, *et al.*, 1986). The calcium phosphate-rich fraction contained α_s- and β-caseins; the other fraction consisted of α_{s1}-, β- and κ-caseins. In experiments where calcium phosphate was added to different casein mixtures it was observed that the amount of the calcium phosphate-rich fraction was positively related to the phosphoserine content of the constituent casein components (Aoki *et al.*, 1987).

Removal of micellar calcium or phosphate by dialysis against phosphate-free or calcium phosphate buffers containing different amounts of calcium induced dissociation of micellar casein (Holt *et al.*, 1986). In these experiments, performed at pH 6·7 and 20°C, it was observed that up to 30% of the micellar phosphate could be removed without noticeable changes in serum casein levels. In the experiments with calcium phosphate buffers, dissociation of micellar casein occurred at calcium concentrations below 1 mM. The degree of retention of the individual caseins in the micelle was in the order $\alpha_{s2} > \alpha_{s1} > \beta > \kappa$, corresponding with the decreasing ester phosphate content of the different caseins.

An alternative way to disturb the structure of the casein micelles is acidification, which results in a gradual dissolution of micelles calcium phosphate. Such a depletion of the micellar calcium phospate is accompanied by a concomitant release of casein components (Rose, 1968). In some studies it was reported that the released casein consisted predominantly of β-casein (Snoeren *et al.*, 1984; van Hooydonk *et al.*, 1986). However, in these studies the composition of the serum phase, including whey proteins, was studied by gel electrophoresis, which could have obscured other serum casein components. In fact, other studies in which the serum phase was analysed by column chromatography demonstrated that all casein components dissociate from casein micelles upon acidification (Roefs *et al.*, 1985; Dalgleish & Law, 1988). Dalgleish and Law (1988) who studied the acidification-induced casein dissociation at 4, 20 and 30°C, showed that the amount of serum casein depends strongly on temperature. At 4 and 20°C, the protein content and the composition of the serum phase varied appreciably with pH. However, at 30°C the serum casein composition was virtually constant at pH values above 5·4 and the amount of dissociated casein was very low. The pH-induced release of colloidal calcium phosphate is independent of temperature between 4 and 30°C (Dalgleish & Law, 1989). Apparently the major part of the micellar calcium phosphate can be removed from casein micelles without a massive dissociation of the constituent caseins; actually at 30°C the amount of liberated casein is negligible.

Although acidification affects micellar parameters such as voluminosity, hydrodynamic diameter and molecular weight, the integrity of the micellar particle as such is little influenced (Snoeren *et al.*, 1984; Roefs *et al.*, 1985; van Hooydonk *et al.*, 1986; Visser *et al.*, 1986; Vreeman *et al.*, 1989). Moreover, the dynamic behaviour of casein micelles as judged by NMR was found to be independent of pH in the range 5·8–7·5 (Rollema & Brinkhuis, 1989), showing that acidification does not result in an increased mobility of parts of the proteins consitituting the micelle. All these observations indicate that protein–protein interactions between submicelles play an important role with respect to the integrity of the micelle. Obviously, depending on the solvent conditions, there is a delicate balance between protein–protein interactions and calcium phosphate-induced binding. The decrease in protein charge induced by acidification favours hydrophobic protein–protein interactions, which then compensate for the loss of calcium phosphate. If, additionally, the temperature is raised, the hydrophobic interactions become so strong that the calcium phosphate can be removed without liberation of caseins. At milk pH and room temperature, the calcium phosphate interactions dominate and are essential for the integrity of the micelle.

4.4. Artificial casein micelles

To study the relationship between the properties of casein micelles and their content of proteins and inorganic constituents, numerous studies dealing with the preparation and the properties of artificial micelles have been reported.

It was shown that addition of calcium to mixtures of α_{s1}- and κ-caseins or α_{s1}-, β- and κ-caseins at 37°C produces aggregates which resemble native casein micelles as judged by electron microscopy (Schmidt *et al.*, 1974). Calcium-induced micelle formation in mixtures of α_s-, β- and κ-caseins was confirmed by sedimentation studies (Slattery, 1979). However, these micelles appeared to be temperature-sensitive and dissociated at low temperatures. Addition of phosphate to the system abolished this temperature sensitivity. In addition, a submicelle–micelle equilibrium was observed at certain concentrations of calcium and phosphate.

Reversibility of the micelle formation process was demonstrated in experiments where artificial micelles were prepared by dialysis against simulated milk ultrafiltrate or milk. When caseinate obtained from large or small casein micelles was used as starting material, the reconstituted micelles showed a sedimentation behaviour and a size distribution similar to the original one (Choate *et al.*, 1959; Ono *et al.*, 1983*b*).

Extensive studies on the properties of artificial casein micelles as a function of their composition and degree of phosphorylation were performed by Schmidt and co-workers (Schmidt & Koops, 1977; Schmidt *et al.*, 1977, 1979; Schmidt & Poll, 1989). From these studies it became evident that addition of calcium phosphate to mixtures of α_s-, β- and κ-caseins over a wide range of protein compositions produces micellar particles that are similar to the casein micelles in milk as judged by electron microscopy. The micellar size decreased with increasing content of κ-casein. At constant κ-casein content, the ratio of β- to α_{s1}-casein had little influence on the micellar size. Micelle formation was also observed in the case of mixtures of κ-casein with either α_{s1}- or β-casein. Artificial micelles prepared from whole casein with varying amounts of calcium, phosphate and citrate showed an increasing size with increasing amount of micellar calcium phosphate. With increasing degree of dephosphorylation of whole casein, a decreasing micellar size, concomitant with a drastically diminishing percentage of micellar casein, was observed. At 70% dephosphorylation, only 15% of the casein was incorporated into the micelles. This last effect was due to the dephosphorylation of α_{s1}- and β-caseins; dephosphorylation of κ-casein had no effect. Phosphorylation of the caseins led to an increased amount of micellar calcium phosphate, but surprisingly also to a lower percentage of micellar casein.

Aoki (1989) prepared artificial casein micelles by the addition of varying amounts of calcium and phosphate to whole casein and determined the amount and composition of the casein fraction cross-linked by calcium phosphate obtained by high-performance gel permeation chromatography of urea-dissociated micelles. At a calcium and phosphate concentration above 10 and 12 mM, respectively, casein cross-linked by calcium phosphate became apparent. The incorporation of the individual casein components in this fraction decreased in the order $\alpha_{s2} > \alpha_{s1} > \beta$, which parallels their decreasing degree of phosphorylation.

Ono and Obata (1989) prepared artificial micelles from the two casein fractions obtained by gel chromatography of calcium-depleted micellar casein. By varying the ratio of the two fractions, one of which consisted of α_s- and κ-caseins and the other of α_s- and β-caseins, they were able to produce micelles of different sizes; the size of the micelles decreased with increasing content of the κ-casein-containing fraction.

4.5. Micellar Calcium Phosphate

Milk contains appreciable amounts of calcium, magnesium, inorganic phosphate and citrate, which are distributed to different extents between the micellar and serum phases (Table 1). The cations and anions in the micellar phase constitute the so-called colloidal or micellar calcium phosphate, which, by its interaction with the caseins, is essential for the integrity of the casein micelle.

In electron micrographs, calcium phosphate obtained from ultrafiltrates of acidified milks was very similar to that of enzymatically digested casein micelles (McGann *et al.*, 1983). In both materials, particles of approximately 2·5 nm diameter were observed. Electron micrographs and diffraction patterns of micellar calcium phosphate showed that in this material there is no order beyond 15 Å (Lyster *et al.*, 1984).

The composition and properties of the mineral phase of casein micelles have been studied in experiments in which calcium phosphate was withdrawn from the micelles by the addition of EDTA (Holt, 1982; Griffin *et al.*, 1988), dialysis (Holt *et al.*, 1986) and by acidification (Chaplin, 1984; van Hooydonk *et al.*, 1986; Dalgleish & Law, 1989). In all cases, plots of micellar calcium versus micellar phosphate are to a good approximation linear and show a positive intercept, which is generally interpreted to be due to calcium ions bound to casein ester phosphate groups. Different values have been reported for the slope of these plots. In the studies where EDTA was used for calcium sequestration, slope values ranging from 1·61 to 1·98 were reported. In the studies using pH-induced calcium depletion, similar slope values, in the range of 1·61–1·85, were found. Calcium depletion by dialysis yielded a slope value of 1·98. From the pH dependence of the calcium and phosphate concentrations in the milk serum phase, an invariable solubility product was calculated for a brushite-like calcium phosphate (Holt, 1982; Chaplin, 1984).

As to the definition of the micellar calcium phosphate, two fundamentally different views have been put forward which lead to differing opinions concerning the structure of the micellar calcium phosphate. If the ester phosphate groups of the caseins together with their bound calcium are assumed not to be part of the calcium phosphate matrix, the calcium phosphate would be of a tricalcium phosphate type. This is based on the calcium to phosphate molar ratio of approximately 1·5 (Holt, 1982; Schmidt, 1982). On the other hand, if one assumes that the casein ester phosphates are an integral part of the calcium phosphate matrix, resulting in a calcium to phosphate ratio close to 1·0, the micellar calcium phosphate would be of a brushite type (Holt *et al.*, 1982, 1989).

By X-ray absorption and infrared spectroscopy it was observed that the micellar calcium phosphate closely resembles brushite (Holt *et al.*, 1982, 1989). The spectroscopic data appear to have solved the problem concerning the structure of the micellar calcium phosphate. However, it should be realized that the micellar system contains ester phosphates, magnesium and citrate, all interacting with and/or forming part of the calcium phosphate lattice. In fact, studies on the formation of calcium phosphates showed that caseins at relatively low concentrations (10^{-8}–10^{-7} M) strongly influence the kinetics of the formation of different calcium phosphate phases and that in certain cases the presence of casein leads to a change in the type of precipitating phase (van Kemenade & de Bruyn, 1989*a,b*). These considerations, together with the fact that micellar calcium phosphate lacks any long range order, render all parallels drawn between the structure of micellar calcium phosphate and that of pure calcium phosphates questionable. The micellar calcium phosphate could be of unique type.

5. Models for the Structure of Bovine Casein Micelles

The casein micelle occupies a unique position among biological systems because of the numerous different models that have been proposed for its structure. This situation has probably developed due to the complexity and the relatively large size of the casein micelles, which prohibit a direct and unequivocal determination of the structure. Moreover, the majority of the models were based on experimental data covering a limited range of micellar properties.

Some of the models describe the micelle as an aggregate of caseins with an outer layer differing in composition from the interior. In these coat-core models the structure of the inner part is not explicitly specified. Models which specify the mode of aggregation of the different caseins belong to the class of the internal structure models. A third type of model is the submicelle model, in which the micelle is assumed to be composed of discrete subunits. Some typical examples of the three types of models are shown in Figs 4–7.

5.1. Coat-Core Models

Waugh and Noble (1965) presented the first coat-core model of casein micelles. In this model, the core of the micelle was considered to consist of spherical particles formed by α_s- and β-caseins, whereas the coat consisted of κ-casein.

Based on experimental data on the association of caseins, Payens (1966) proposed a model (Fig. 4) in which the core of the micelle consists of compactly folded α_{s1}-casein molecules attached to a more loose network of β-casein. In this model, the surface of the micelle is covered with κ-casein and calcium phosphate is located both on the surface and in the interior of the micelle.

Parry and Carroll (1969), who were not able to localize κ-casein at the surface

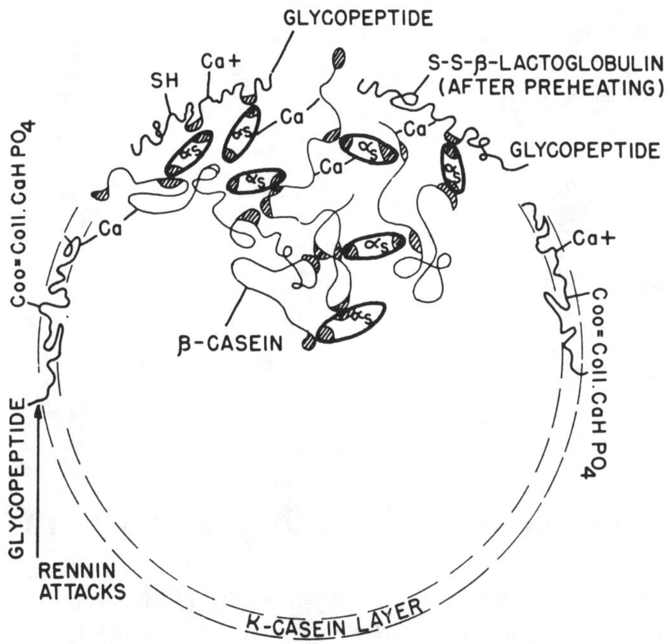

Fig. 4. Coat-core model of the casein micelle proposed by Payens. (From Payens (1966), Reproduced by permission of the Journal of Dairy Science.)

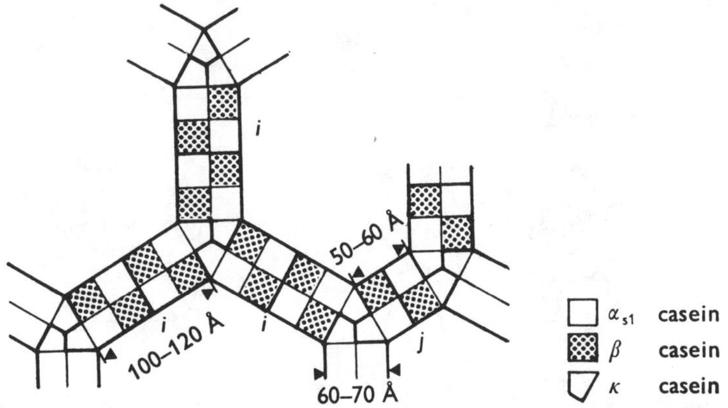

Fig. 5. Part of the protein network in the casein micelle according to the internal structure model proposed by Garnier and Ribadeau-Dumas. (From Garnier & Ribadeau-Dumas (1970), Reproduced by permission of the Cambridge University Press.)

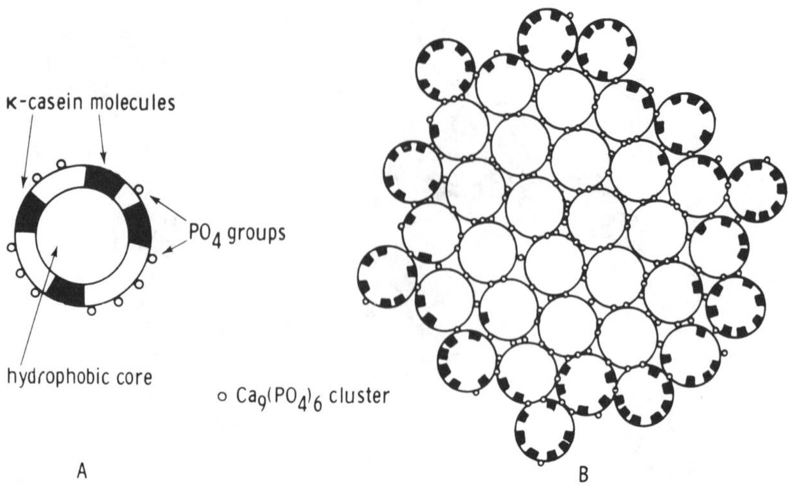

Fig. 6. Submicelle model proposed by Schmidt (1982): (a) submicelle; (b) casein micelle.

of casein micelles by electron microscopy using ferritin-labelled κ-casein antibodies, assumed that κ-casein is located in the interior of the micelle, forming a nucleation centre for α_s-, β-caseins and calcium phosphate. In this model the surface of the micelle consists of α_s- and β-caseins supplemented by some calcium phosphate.

In 1987, Paquin and co-workers proposed a model which was based on the results of experiments using monolayer techniques for the characterization of two protein fractions obtained by gel chromatography of EDTA-dissociated casein micelles (Britten *et al.*, 1986; Paquin *et al.*, 1987). Their model envisages the core of the micelle as a framework of α_s-casein and calcium phosphate to which β-casein

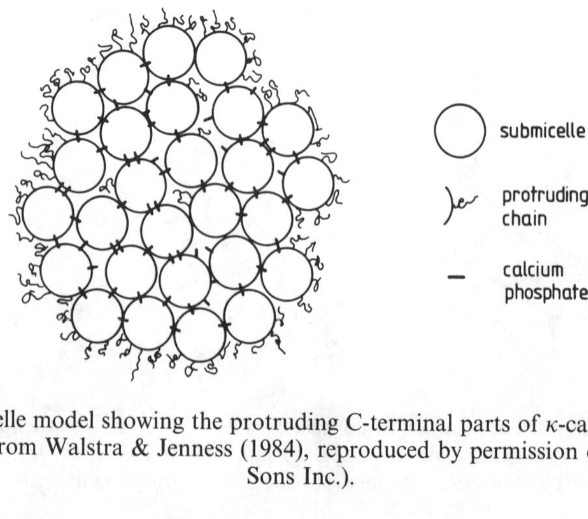

Fig. 7. Submicelle model showing the protruding C-terminal parts of κ-casein as proposed by Walstra. (From Walstra & Jenness (1984), reproduced by permission of John Wiley & Sons Inc.).

is bound by hydrophobic interactions. The core is surrounded by particles containing α_s-, β- and a relatively high proportion of κ-casein.

5.2. Internal Structure Models

The first internal structure model was presented by Rose (1969). Rose assumed that an end-to-end association of β-casein first occurs in micelle formation. Subsequently, α_s-casein molecules are bound to the β-casein polymers while κ-casein binds to the α_s-casein, forming aggregates of limited size. Upon cross-linking by calcium phosphate, these casein aggregates form the micelles in which the particles are oriented in such a way that κ-casein is directed to the periphery of the micelle.

Assuming a different mode of association of the caseins, Garnier and Ribadeau-Dumas (1970) proposed a model (Fig. 5) which depicts the casein micelle as a three-dimensional porous network of protein aggregates. In this network, three κ-casein molecules act as nodes and the branches consist of α_s- and β-caseins. In this model, no essential role was assigned to calcium phosphate; only a possible binding to the casein network was suggested.

5.3. Subunit Models

The first subunit model was advanced by Morr (1967). This model was based on the results of an investigation on the influence of urea and oxalate treatment on the sedimentation behaviour of casein micelles. In this model, the casein micelle is assembled from casein particles held together by calcium phosphate. Each of these subunits has a core of α_s- and κ-casein and a surface covered by α_s- and κ-casein. All subunits in the micelle have identical protein compositions.

A model in which the casein micelle is composed of subunits of variable composition was proposed by Slattery and Evard (1973). This model was based on observations on the influence of calcium on the sedimentation behaviour of the particles formed in mixtures of caseins. Some of the subunits in this model consist of α_s- and β-caseins, while the rest also contain κ-casein. The presence of α_s- and β-caseins gives rise to hydrophobic regions on the surface of the subunits. If present, κ-casein is accumulated in a hydrophilic region of the surface. The hydrophobic regions are responsible for the mutual binding of the subunits in the micelle, whereas the hydrophilic κ-casein areas are exposed to the solvent. The growth of a micellar particle will come to an end when the surface of the micelle is covered by the hydrophilic κ-casein regions. In this way, the amount of κ-casein available will govern the size distribution of the micelles. Although the authors were aware of the importance of calcium and phosphate ions with respect to the formation of casein micelles, calcium phosphate was not considered to be a separate phase in the micelle (Slattery, 1979). Calcium binding to the caseins was considered to modulate the hydrophobic interactions between the submicelles by the reduction of their negative charge. The presence of inorganic phosphate would contribute to inter-subunit binding by electrostatic interactions through calcium phosphate salt bridges between the ester phosphates of adjacent casein molecules.

Schmidt and Payens (1976) presented a model in which the subunits are assumed to be linked together exclusively by calcium phosphate. In this model the subunits have a hydrophobic core surrounded by polar regions of the constituent proteins. The protein composition of the subunits was not defined, except for the requirement that part of the κ-casein should be located at the surface of the micelle. Later, Schmidt (1980) adopted the concept of variable subunit composition from the model of Slattery and Evard, but maintained the view that only calcium phosphate is responsible for the binding of the submicelles. Subsequently this model was extended by a more detailed description of the inorganic phase of the micelles (Schmidt, 1982). It was assumed that a calcium ion is bound to virtually all of the ester phosphate groups and that some of these groups are linked in pairs by $Ca_9(PO_4)_6$ cluster (Fig. 6).

An extension of the subunit model described above was proposed by Walstra, in which the concept of steric stabilization of the micelle by κ-casein was incorporated (Walstra & Jenness, 1984; Walstra, 1990). In this model the C-terminal parts of the κ-casein molecules are sticking out from the micellar surface into the surrounding solvent (Fig. 7). This concept is developed further in Chapter 16.

From studies of artificial micelles prepared from protein fractions obtained by gel chromatography of calcium-depleted micelles, Ono and Obata (1989) proposed a micelle model in which these protein fractions consititute the building blocks of the micelle. In their model, the micellar core is composed of aggregates of α_s- and β-caseins while the coat consists of particles containing equimolar amounts of α_s- and κ-caseins. The two types of subunits are linked by calcium phosphate. In the coat particles, κ-casein is directed outwards whereas α_s-casein is responsible for the binding to the core subunits.

5.4. Discussion

The preceding historical overview of the different casein micelle models illustrates the diversity of opinions concerning the structure of the micelle. The merits of the different models have been discussed repeatedly, but each time with a different outcome (Garnier, 1973; Slattery, 1976; Schmidt, 1982). In the following, the different models will be discussed in the light of some well established structural and functional properties of casein micelles, such as the location of the different casein components in the micelle, the role of calcium phosphate in micellar integrity and the dissociation behaviour of the micelles induced by acidification and calcium sequestration.

The generally accepted surface location of κ-casein disqualifies the models of Parry and Carroll (1969), Rose (1969), Morr (1967) and Garnier and Ribadeau-Dumas (1970). In the models of Waugh and Noble (1965), Payens (1966), Paquin *et al.*, (1987) and Ono and Obata (1989), the micellar surface is more or less completely covered with κ-casein. The experimental evidence that considerable amounts of α_s-casein are located at the surface of the micelle does not support these models.

Consequently, we are left with two types of models: the model of Slattery and

Evard (1973) and the models of Schmidt (1982) and Walstra (1990). Both types of models are actually very similar in that they both assume subunits with variable protein composition as the building blocks of the micelle. The main difference between the two models is the mechanism of inter-subunit interaction. In the model of Slattery and Evard (1973), the subunits interact by hydrophobic binding initiated and modulated by calcium binding to the caseins. A secondary interaction is of electrostatic nature mediated by inorganic phosphate. Schmidt (1982), on the other hand, took the line that there are no inter-subunit protein–protein interactions and that the submicelles are linked together only by calcium phosphate. The disruption of casein micelles upon calcium sequestration at constant pH is accounted for by both types of models. When calcium depletion is achieved by acidification, the overall size of the micelles is hardly affected in spite of significant dissociation of caseins and calcium phosphate. The solubilization of calcium phosphate is independent of temperature between 4 and 30°C whereas the amount of liberated casein decreases with increasing temperature. Actually, at 30°C hardly any casein is dissociated at pH values where the major part of calcium phosphate is removed from the micelle. These observations cannot be described with a model in which the submicelles interact only through calcium phosphate. As discussed in Section 4.3, it is necessary to incorporate, apart from calcium phosphate binding, additional interactions between the submicelles, most probably hydrophobic interactions. The balance between these two forces will govern the stability of the micelle.

The presence of protein–protein interactions between submicelles implies that a proteinaceous phase exists between the subunits. Strictly speaking, this would mean a shift of the pure submicellar model towards a coat-core model in which the core consists of regions with different protein density. However, such a reasoning runs the risk of getting stuck in semantics. Apart from this, there are experimental indications for a subunit-like structure of the casein micelle.

In conclusion, it can be stated that the experimental data known to date are best described by a combination of the models of Slattery and Evard (1973), Schmidt (1982) and Walstra (1990). This combination implies that a calcium phosphate phase and hydrophobic interactions between submicelles are responsible for the integrity of the casein micelles. The model of Walstra enhances the concepts of Slattery and Evard and Schmidt by taking into account the stabilizing role of κ-casein, in accordance with the results of NMR studies of casein micelles (see Section 4.3).

6. Non-Bovine Caseins and Casein Micelles

Although there is much literature on the protein composition of milks of different species (Jenness, 1982), relatively few studies have appeared on the association behaviour of the individual caseins and on the properties of the casein micelles. Therefore, a comprehensive inter-species comparison concerning the functional properties of caseins and casein micelles can not be given yet. Nevertheless, it is worthwhile to compare the information available with that of bovine caseins and casein micelles.

With respect to casein association behaviour the best studied system up to now is that of the human caseins. Human milk contains little or no α_s-casein. The β-casein fraction consists of different components containing up to 5 ester phosphate groups, denoted by 0-P through 5-P (Groves & Gordon, 1970). Human κ-casein has an appreciably higher carbohydrate content than the bovine protein.

Sedimentation studies of human κ-casein have shown that the protein exhibits little tendency to polymerize. At pH 7, between 6°C and 37°C and in the presence of 0–37 mM calcium, no changes in sedimentation behaviour were observed (Azuma *et al.*, 1984). However, in the presence of human β-casein, a β–κ complex is formed with a sedimentation coefficient of 12S.

The self-association properties of human β-casein are reminiscent of those of bovine β-casein as was shown in studies of the 0-P, 3-P and 5-P components (Sood *et al.*, 1985, 1988, 1990). In the ultracentrifuge a monomer peak together with a polymer peak were observed, indicating a monomer–polymer equilibrium. Self-association increases with increasing temperature. The association of human β-caseins starts at a significantly higher temperature than that of bovine β-casein. At low ionic strength in the absence of calcium, the onset of association is at about 20°C for the 0-P component and 24°C for the 3-P and 5-P components. At constant temperature the association is positively correlated to ionic strength; the 0-P component shows a stronger ionic strength dependence than the 3-P and 5-P components, which show a similar dependence. For the 0-P, 3-P and 5-P components at high ionic strength and 37°C, the polymer was reported to contain 75, 48 and 95–105 monomers, respectively. In a recent study, Slattery *et al.* (1989) reported that in the absence of calcium, at 25°C and low ionic strength, the 0-P and 5-P human β-casein components initially associate to submicelle-like particles consisting of 20–30 monomers. A higher aggregation stage, corresponding to 12–15 submicelles, was observed at higher temperatures.

Azuma *et al.* (1989) observed that the calcium binding ability of the different human β-casein components was correlated to the number of ester phosphate groups. In addition, they demonstrated that the components with low phosphorylation levels are able to protect the highly phosphorylated ones against calcium-induced precipitation. However, their observations that the nonphosphorylated component did not exhibit this protective function and that it had a high calcium binding ability are rather unexpected.

Calcium-induced formation of micelles in systems containing human κ- and β-caseins was demonstrated by Azuma *et al.* (1985). When combined with κ-casein in the presence of calcium, the 3-P and 5-P β-casein components were able to form micelles but the 0-P component was not able to do so. Size fractionation of these micelles showed that small micelles were relatively rich in κ-casein whereas the distribution of the different β-casein components showed little dependence on micellar size. Small micelles were relatively rich in highly glycosylated κ-casein components.

With goat casein micelles, which, like human micelles, are poor in α_s-caseins, gel filtration experiments indicated that the κ-casein content is inversely related to micellar size and that low temperature-induced dissociation of β-casein occurs (Ono & Creamer, 1986).

The protein aggregates in sheep, goat and horse caseinate, obtained by EDTA-induced dissolution of casein micelles, were studied by high performance gel chromatography (Ono *et al.*, 1989) and the elution patterns were compared with that of bovine casein. All caseins, except the equine casein, showed three peaks with retention times similar to those observed for bovine casein. For bovine casein the peaks, with increasing retention time, were in an earlier investigation assigned to α_{s1}–κ, α_{s1}–β complexes and α_{s1}-casein, respectively (Ono *et al.*, 1983*a*). Sheep casein showed a pattern very similar to that of bovine casein. In goat casein, the first peak was strongly reduced in intensity. In the case of equine casein, only the first and the third peaks were observed; the first peak was reported to contain α_s- and β-caseins and the third peak β-casein. The composition of the different peaks was examined by gel electrophoresis. By this technique the authors were not able to detect κ-casein in equine casein. Moreover, they found a relatively high proportion (50·5%) of α_s-casein in goat casein. However, it is known that goat casein is very low in the α_s-component (Richardson & Creamer, 1975) and that equine casein does contain κ-casein (Kotts & Jenness, 1976), which throws some doubt on the identification of the different caseinate fractions.

A surface location of κ-casein was also suggested in the case of camel and buffalo casein micelles. In experiments with camel casein micelles, immobilized chymosin and neuraminidase induced a release of a macropeptide and sialic acid at rates comparable to those observed with the soluble enzymes (Mehaia, 1987*a*,*b*). In the case of buffalo casein micelles an increasing sialic acid content with decreasing micellar size was observed (Hewday *et al.*, 1989).

Sood *et al.* (1979) studied the protein, calcium and phosphate contents of milks from nine species and determined the voluminosity of the casein micelles. The protein content was lowest for human, ass, mare and camel milks and relatively high for sheep and sow milks. Appreciable variations in the voluminosity of the casein micelles were observed. Human micelles showed the highest voluminosity —nearly 8 ml/g. The lowest value (2·8 ml/g) was found for sow micelles; the value for bovine micelles was approximatley 4 ml/g. The voluminosity of the different casein micelles appears to be correlated to the casein content of the milk; the milks with the lowest casein content show the highest values for the voluminosity of the micelles.

The size distribution and the structure of the casein micelles of different species have been the subject of a number of electron microscopical studies (Richardson *et al.*, 1974; Ali & Robinson, 1985; Welsch *et al.*, 1988; Buchheim *et al.*, 1989; Farah & Rüegg, 1989). For the micelles of the 19 species studied, the same fine structure was observed as for bovine micelles. The size of the micelles is very variable amongst the different species. The smallest micelles are observed in human milk, the largest in lama milk. The width of the size distribution also shows a considerable species dependence. Very narrow size distributions are found for

TABLE 3
Interspecies Comparison of Some Micellar Properties. Protein Concentration and Voluminosity were Taken from Sood *et al.* (1979), Data on Micellar Sizes from Buchheim *et al.* (1989) and Farah & Rüegg (1989)

Species	*Protein concentration (g/100 ml)*	*Voluminosity (ml/g)*	*Number-average diameter (nm)*	*Weight-average diameter (nm)*
Cow	2·6	4·1	47	182
Buffalo	3·6	3·6	104	167
Goat	2·7	4·1	52	282
Sheep	4·3	3·8	59	122
Human	0·7	7·9	33	64
Sow	3·3	2·7	65	133
Mare	1·2	5·2	140	255
Ass	0·8	5·3	187	293
Camel	1·6	4·4	27	243

human and sheep milks whereas goat milk shows a very broad distribution. Some micellar parameters for nine species are summarized in Table 3.

Although studies on the properties of non-bovine caseins and casein micelles are scarce and disperse, it appears that the data available can be fitted within the principles and mechanisms established for the properties and behaviour of bovine caseins and casein micelles.

Acknowledgements

The author is grateful to Drs R.J. Siezen, D.G. Schmidt and H.J. Vreeman for their critical remarks and useful suggestions and to Dr M. Kalab for his generous gift of a number of electron micrographs of casein micelles.

REFERENCES

Ali, M.Z. & Robinson, R.K. (1985). Size distribution of casein micelles in camels' milk. *J. Dairy Res.*, **52**, 303–7.

Andrews, A.L., Atkinson, D., Evans, M.T.A., Finer, E.G., Green, J.P., Phillips, M.C. & Robertson, R.N. (1979). The conformation and aggregration of bovine β-casein A. I. Molecular aspects of thermal aggregation. *Biopolymers*, **18**, 1105–21.

Aoki, T. (1989). Incorporation of individual casein constituents into casein aggregates cross-linked by colloidal calcium phosphate in artificial casein micelles. *J. Dairy Res.*, **56**, 613–8.

Aoki, T., Toyooka, K. & Kako, Y. (1985). Role of phosphate groups in the calcium sensitivity of α_{s2}-casein. *J. Dairy Sci.*, **68**, 1624–9.

Aoki, T., Kako, Y. & Imamura, T. (1986). Separation of casein aggregates cross-linked by colloidal calcium phosphate from bovine casein micelles by high performance gel chromatography in the presence of urea. *J. Dairy Res.*, **53**, 53–9.

Aoki, T., Yamada, N., Tomita, I., Kako, Y. & Imamura, T. (1987). Caseins are cross-linked through their ester phosphate groups by colloidal calcium phosphate. *Biochim. Biophys. Acta*, **911**, 238–43.

Azuma, N., Kaminogawa, S. & Yamauchi, K. (1984). Molecular weight and conformation of human κ-casein and its interaction with other human milk proteins. *Agric. Biol. Chem.*, **48**, 771–6.
Azuma, N., Kaminogawa, S. & Yamauchi, K. (1985). Reconstitution of human casein micelle and its properties. *Agric. Biol. Chem.*, **49**, 2655–60.
Azuma, N., Aizawa, M. & Yamauchi, K. (1989). Effect of bound phosphate on the calcium-binding ability and calcium-dependent precipitability of human β-casein. *Agric. Biol. Chem.*, **53**, 1037–41.
Baumy, J.J. & Brule, G. (1988). Effect of pH and ionic strength on the binding of bivalent cations to β-casein. *Le Lait*, **68**, 409–18.
Berry, G.P. & Creamer, L.K. (1975). The association of bovine β-casein. The importance of the C-terminal region. *Biochemistry*, **14**, 3542–5.
Britten, M., Boulet, M. & Paquin, P. (1986). Composition and interfacial properties of casein micelle fractions. *J. Dairy Res.*, **53**, 573–84.
Buchheim, W. & Schmidt, D.G. (1979). On the size of monomers and polymers of β-casein. *J. Dairy Res.*, **46**, 277–80.
Buchheim, W., Lund, S. & Scholtissek, J. (1989). Vergleichende Untersuchungen zur Struktur and Größe von Caseinmicellen in der Milch verchiedener Species. *Kieler Milchwirtschaftliche Forschungsberichte*, **41**, 253–66.
Byler, D.M. & Farrell Jr., H.M. (1989). Infrared spectroscopic evidence for calcium ion interaction with carboxylate groups of casein. *J. Dairy Sci.*, **72**, 1719–23.
Calapaj, G.G. (1968). An electron microscope study of the ultra structure of bovine and human casein micelles in fresh and acidified milk. *J. Dairy Res.*, **35**, 1–6.
Carroll, R.J. & Farrell Jr., H.M. (1983). Immunological approach to location of κ-casein in the casein micelle by electron microscopy. *J. Dairy Sci.*, **66**, 679–86.
Chaplin, L.C. (1984). Studies on micellar calcium phosphate: composition and apparent solubility product in milk over a wide pH range. *J. Dairy Res.*, **51**, 251–7.
Choate, W.L., Heckman, F.A. & Ford, T.F. (1959). Size memory of casein colloid particles. *J. Dairy Sci.*, **42**, 761–6.
Creamer, L.K., Wheelock, J.V. & Samuel, D. (1973). The distribution of glyco-κ-casein and carbohydrate-free κ-casein between large and small bovine casein micelles, and its implication in micelle structure. *Biochim. Biophys. Acta*, **317**, 202–6.
Dalgleish, D.G. (1985). Glycosylated κ-caseins and the sizes of bovine casein micelles. Analysis of the different forms of κ-casein. *Biochim. Biophys. Acta*, **830**, 213–5.
Dalgleish, D.G. & Law, A.J.R. (1988). pH-Induced dissociation of bovine casein micelles. I. Analysis of liberated caseins. *J. Dairy Res.*, **55**, 529–38.
Dalgleish, D.G. & Law, A.J.R. (1989). pH-Induced dissociation of bovine casein micelles. II. Mineral solubilization and its relation to casein release. *J. Dairy Res.*, **56**, 727–35.
Dalgleish, D.G. & Parker, T.G. (1980). Binding of calcium ions to bovine α_{s1}-casein and precipitability of the protein-calcium ion complexes. *J. Dairy Res.*, **47**, 113–22.
Dalgleish, D.G., Horne, D.S. & Law, A.J.R. (1989). Size-related differences in bovine casein micelles. *Biochim. Biophys. Acta*, **991**, 383–7.
Davies, D.T. & Law, A.J.R. (1983). Variation in the protein composition of bovine casein micelles and serum casein in relation to micellar size and milk temperature. *J. Dairy Res.*, **50**, 67–75.
Dickson, I.R. & Perkins, D.J. (1971). Studies on the interactions between purified bovine caseins and alkaline-earth-metal ions. *Biochem. J.*, **124**, 235–40.
Donnelly, W.J., McNeill, G.P., Buchheim, W. & McGann, T.C.A. (1984). A comprehensive study of the relationship between size and protein composition in natural bovine casein micelles. *Biochim. Biophys. Acta*, **789**, 136–43.
Evans, M.T.A., Phillips, M.C. & Jones, M.N. (1979). The conformation and aggregation of bovine β-casein A. II. Thermodynamics of thermal association and the effects of changes in polar and apolar interactions on micellization. *Biopolymers*, **18**, 1123–40.

Farah, Z. & Rüegg, M.W. (1989). The size distribution of casein micelles in camel milk. *Food Microstructure*, **8**, 211–6.

Farrell Jr., H.M. (1988). Physical equilibria: proteins. In *Fundamentals of Dairy Chemistry*, ed. N.P. Wong, R. Jenness, M. Keeney & E.H. Marth. Van Nostrand Reinhold Company, New York, pp. 461–510.

Garnier, J. (1973). Models of casein micelle structure. *Neth. Milk Dairy J.*, **27**, 240–8.

Garnier, J. & Ribadeau-Dumas, B. (1970). Structure of the casein micelle. A proposed model. *J. Dairy Res.*, **37**, 493–504.

Green, M.L. (1971). The specificity for κ-casein as the stabilizer of α_s-casein and β-casein. I. Replacement of κ-casein by other proteins. *J. Dairy Res.*, **38**, 9–23.

Griffin, M.C.A. & Roberts, G.C.K. (1985). A ^{1}H-n.m.r. study of casein micelles. *Biochem. J.*, **228**, 273–6.

Griffin, M.C.A., Lyster, R.L.J. & Price, J.C. (1988). The disaggregation of calcium-depleted casein micelles. *Eur. J. Biochem.*, **174**, 339–43.

Groves, M.L. & Gordon, W.G. (1970). The major component of human casein: A protein phosphorylated at different levels. *Arch. Biochem. Biophys.*, **140**, 47–51.

Heth, A.A. & Swaisgood, H.E. (1982). Examination of casein micelle structure by a method of reversible covalent immobilization. *J. Dairy Sci.*, **65**, 2047–54.

Hewdey, M.M., Nassib, T.A., Zin El-Din, M.M. & El-Sokkary, E.S. (1989). Studies on Buffaloes casein. I. Composition of casein micelles. *Egyptian J. Dairy Sci.*, **17**, 87–92.

Hill, R.J. & Wake, R.G. (1969). Amphiphile nature of κ-casein as the basis for its micelle stabilizing property. *Nature*, **221**, 635–9.

Holt, C. (1975). The stability of bovine casein micelles. In *Proceedings International Conference on Colloid and Interface Science*, ed. E. Wolfram. Akademia Kiado, Budapest, pp. 641–4.

Holt, C. (1982). Inorganic constituents of milk. III. The colloidal calcium phosphate of cow's milk. *J. Dairy Res.*, **49**, 29–38.

Holt, C. (1985). The milk salts: Their secretion, concentrations and physical chemistry. In *Developments in Dairy Chemistry—3*, ed. P.F. Fox. Elsevier Applied Science Publishers, London, pp. 143–81.

Holt, C., Hasnain, S.S. & Hukins, D.W.L. (1982). Structure of bovine milk calcium phosphate determined by X-ray absorption spectroscopy. *Biochim. Biophys. Acta*, **719**, 299–303.

Holt, C., Davies, D.T. & Law, A.J.R. (1986). Effects of colloidal calcium phosphate content and free calcium ion concentration in the milk serum on the dissociation of bovine casein micelles. *J. Dairy Res.*, **53**, 557–72.

Holt, C., van Kemenade, M.J.J.M., Nelson Jr. L.S., Sawyer, L., Harries, J.E., Bailey, R.T. & Hukins, D.W.L. (1989). Composition and structure of micellar calcium phosphate. *J. Dairy Res.*, **56**, 411–6.

van Hooydonk, A.C.M., Hagedoorn, H.G. & Boerrigter, I.J. (1986). pH-induced physico-chemical changes of casein micelles in milk and their effect on rennetting. 1. Effect of acidification on physico-chemical properties. *Neth. Milk Dairy, J.*, **40**, 281–96.

Horisberger, M. & Rouvet-Vauthey, M. (1984). Localization of glycosylated κ-casein on thin sections of casein micelles by lectin-labelled gold markers. *Histochemistry*, **80**, 523–6.

Horisberger M. & Vauthey, M. (1984). Localization of κ-casein in thin sections of casein micelles by the gold method. *Histochemistry*, **80**, 9–12.

Horisberger, M. & Vonlanthen, M. (1980). Localization of glycosylated κ-casein in bovine casein micelles by lectin-labelled gold granules. *J. Dairy Res.*, **47**, 185–91.

Jang, H.D. & Swaisgood, H.E. (1990). Characteristics of the interaction of calcium with casein submicelles as determined by analytical affinity chromatography. *Arch. Biochem. Biophys.*, **283**, 318–28.

Jenness, R. (1982). Inter-species comparison of milk proteins. In *Developments in Dairy Chemistry—1*, ed. P.F. Fox. Applied Science Publishers, London, pp. 87–114.

Kajiwara, K., Niki, R., Urakawa, H., Hiragi, Y., Donkai, N. & Nagura, M. (1988). Micellar structure of β-casein observed by small-angle X-ray scattering. *Biochim. Biophys. Acta*, **955**, 128–34.

Kakalis, L.T., Kumosinski, T.F. & Farrell Jr., H.M. (1990). A multinuclear high-resolution NMR study of bovine casein micelles and submicelles. *Biophys. Chem.*, **38**, 87–98.

Kalab, M., Phibbs-Todd, B.E. & Allan-Wojtas, P. (1982). Milk gel structure XIII. Rotary shadowing of casein micelles for electron microscopy. *Milchwissenschaft*, **37**, 513–8.

Kaminogawa, S., Yamauchi, K. & Yoon, C.-H. (1980). Calcium insensitivity and other properties of α_{s1}-I casein. *J. Dairy Sci.*, **63**, 223–7.

van Kemenade, M.J.J.M. & de Bruyn, P.L. (1989*a*). The influence of casein on the kinetics of hydroxyapetite precipitation. *J. Colloid Interface Sci.*, **129**, 1–14.

van Kemenade, M.J.J.M. & de Bruyn, P.L. (1989*b*). The influence of casein on the precipitation of brushite and octacalcium phosphate. *Colloids and Surfaces*, **36**, 359–68.

Knoop, A.M., Knoop, E. & Wiechen, A. (1973). Electron microscopical investigations on the structure of the casein micelles. *Neth. Milk Dairy J.*, **27**, 121–7.

Kotts, C. & Jenness, R. (1976). Isolation of κ-casein-like proteins from milks of various species. *J. Dairy Sci.*, **59**, 816–22.

Kudo, S., Iwata, S. & Mada, M. (1979). An electron microscopic study of the location of κ-casein in casein micelles by periodic acid-silver methamine staining. *J. Dairy Sci.*, **62**, 916–20.

Kumosinski, T.F., Pessen, H., Farrell Jr., H.M. & Brumberger, H. (1988). Determination of the quaternary structural states of bovine casein by small-angle X-ray scattering: submicellar and micellar forms. *Arch. Biochem. Biophys.*, **266**, 548–61.

Lin, S.H.C., Leong, S.L., Dewan, R.K., Bloomfield, V.A. & Morr, C.V. (1972). Effect of calcium ion on the structure of native bovine casein micelles. *Biochemistry*, **11**, 1818–21.

Lyster, R.L.J., Mann, S., Parker, S.B. & Williams, R.J.P. (1984). Nature of micellar calcium phosphate in cows' milk as studied by high-resolution electron microscopy. *Biochim. Biophys. Acta*, **801**, 315–7.

McGann, T.C.A., Donnelly, W.J., Kearney, R.D. & Buchheim, W. (1980). Composition and size distribution of bovine casein micelles. *Biochim. Biophys. Acta*, **630**, 261–70.

McGann, T.C.A., Buchheim, W., Kearney, R.D. & Richardson, T. (1983). Composition and ultrastructure of calcium phosphate-citrate complexes in bovine milk systems. *Biochim. Biophys. Acta*, **760**, 415–20.

McMahon, D.J. & Brown, R.J. (1984). Composition, structure and integrity of casein micelles: a review. *J. Dairy Sci.*, **67**, 499–512.

Mehaia, M.A. (1987*a*). Studies on camel milk casein micelles. Treatment with soluble and immobilized chymosin. *Milchwissenschaft*, **42**, 706–8.

Mehaia, M.A. (1987*b*). Studies on camel casein micelles: Treatment with soluble and immobilized neuraminidase. *Carbohydrate Polymers*, **7**, 361–9.

Morr, C.V. (1967). Effect of oxalate and urea upon ultracentrifugation properties of raw and heated skim milk casein micelles. *J. Dairy Sci.*, **50**, 1744–51.

Ono, T. & Creamer, L.K. (1986). Structure of goat casein micelles. *N. Z. J. Dairy Sci. Technol.*, **21**, 57–64.

Ono, T. & Obata, T. (1989). A model for the assembly of bovine casein micelles from F2 and F3 subunits. *J. Dairy Res.*, **56**, 453–61.

Ono, T. & Takagi, T. (1986). Molecular weights of subunit components of bovine casein micelles. *J. Dairy Res.*, **53**, 547–55.

Ono, T., Yahagi, M. & Odagiri, S. (1980). The binding of calcium to κ-casein and para κ-casein. *Agric. Biol. Chem.*, **44**, 1499–503.

Ono, T., Odagiri, S. & Takagi, T. (1983*a*). Separation of the submicelles from micellar casein by high performance gel chromatography on a TSK-GEL G4000SW column. *J. Dairy Res.*, **50**, 37–44.

Ono, T., Furuyama, T. & Odagiri, S. (1983*b*). Formation of artificial casein micelles. *Agric. Biol. Chem.*, **47**, 221–6.

Ono, T., Kohno, H., Odagiri, S. & Takagi, T. (1989). Subunit components of casein micelles from bovine, ovine, caprine and equine milks. *J. Dairy Res.*, **56**, 61–8.

Paquin, P., Britten, M., Laliberté, M.-F. & Boulet, M. (1987). Interfacial properties of milk casein proteins. In *Proteins at Interfaces*, ed. J.L. Brash & T.A. Horbett. American Chemical Society, Washington D.C. pp. 677–86.

Parker, T.G. & Dalgleish, D.G. (1981). Binding of calcium ions to bovine β-casein. *J. Dairy Res.*, **48**, 71–6.

Parry Jr., R.M. & Carroll, R.J. (1969). Location of κ-casein in milk micelles. *Biochim. Biophys. Acta*, **194**, 138–50.

Payens, T.A.J. (1966). Association of caseins and their possible relation to structure of the casein micelle. *J. Dairy Sci.*, **49**, 1317–24.

Payens, T.A.J. (1979). Casein micelles: the colloid-chemical approach. *J. Dairy Res.*, **46**, 291–306.

Payens, T.A.J. & van Markwijk, B.W. (1963). Some features of the association of β-casein. *Biochim. Biophys. Acta*, **71**, 517–30.

Payens, T.A.J. & Schmidt, D.G. (1966). Boundary spreading of rapidly polymerizing α_{s1}-casein B and C during sedimentation. *Arch. Biochem. Biophys.*, **115**, 136–45.

Payens, T.A.J. & Vreeman, H.J. (1982). Casein micelles and micelles of κ- and β-casein. In *Solution Behavior of Surfactants*, Vol. 1, ed. K.L. Mittal & E.J. Fendler. Plenum Publishing, New York, pp. 543–71.

Payens, T.A.J., Brinkhuis, J.A. & van Markwijk, B.W. (1969). Self-association in non-ideal systems. Combined light scattering and sedimentation measurements in β-casein solutions. *Biochim. Biophys. Acta*, **175**, 434–7.

Richardson, B.C. & Creamer, L.K. (1975). Comparative micellar structure. IV. The similarity between caprine α_s-casein and bovine α_{s3}-casein. *Biochim. Biophys. Acta*, **393**, 37–47.

Richardson, B.C., Creamer, L.K., Pearce, K.N. & Munford, R.E. (1974). Comparative micelle structure. II. Structure and composition of casein micelles in ovine and caprine milk as compared with those in bovine milk. *J. Dairy Res.*, **41**, 239–47.

Roefs, S.P.F.M., Walstra, P., Dalgleish, D.G. & Horne, D.S. (1985). Preliminary note on the change in casein micelles caused by acidification. *Neth. Milk Dairy J.*, **39**, 119–22.

Rollema, H.S. & Brinkhuis, J.A. (1989). A ^{1}H-NMR study of bovine casein micelles; influence of pH, temperature and calcium ions on micellar structure. *J. Dairy Res.*, **56**, 417–25.

Rollema, H.S., Brinkhuis, J.A. & Vreeman, H.J. (1988). ^{1}H-NMR studies of bovine κ-casein and casein micelles. *Neth. Milk Dairy J.*, **42**, 233–48.

Rose, D. (1968). Relation between micellar and serum casein in bovine milk *J. Dairy Sci.*, **51**, 1897–902.

Rose, D. (1969). A proposed model of micelle structure in bovine milk. *Dairy Sci. Abstr.*, **31**, 171–5.

Ruettimann, K.W. & Ladisch, M.R. (1987). Casein micelles: structure, properties and enzymic coagulation. *Enzyme Microb. Technol.*, **9**, 578–89.

Schmidt, D.G. (1970*a*). The association of α_{s1}-casein B at pH 6·6. *Biochim. Biophys. Acta*, **207**, 130–8.

Schmidt, D.G. (1970*b*). Differences between the association of the genetic variants B, C and D of α_{s1}-casein. *Biochim. Biophys. Acta*, **221**, 140–2.

Schmidt, D.G. (1980). Colloidal aspects of casein. *Neth. Milk Dairy J.*, **34**, 42–64.

Schmidt, D.G. (1982). Association of caseins and casein micelle structure. In *Developments in Dairy Chemistry—1*, ed. P.F. Fox. Applied Science Publishers, London, pp. 61–86.

Schmidt, D.G. & Both, P. (1982). Location of α_{s1}-, β- and κ-casein in artificial casein micelles. *Milchwissenschaft*, **37**, 336–7.

Schmidt, D.G. & Buchheim, W. (1970). Elektronenmikroskopische Untersuchung der Feinstruktur von Caseinmicellen in Kuhmilch. *Milchwissenschaft*, **25**, 596–600.

Schmidt, D.G. & Koops, J. (1977). Properties of artificial casein micelles. 2. Stability

towards ethanol, dialysis, pressure and heat in relation to casein composition. *Neth. Milk Dairy J.*, **31**, 342–51.
Schmidt, D.G. & Payens, T.A.J. (1972). The evaluation of positive and negative contributions to the second virial coefficient of some milk proteins. *J. Colloid Interface Sci.*, **39**, 655–62.
Schmidt, D.G. & Payens, T.A.J. (1976). Micellar aspects of casein. In *Surface and Colloid Science*, Vol. 9, ed. E. Matijević. John Wiley & Sons, New York, pp. 165–229.
Schmidt, D.G. & Poll, J.K. (1989). Properties of artificial casein micelles 4. Influence of dephosphorylation and phosphorylation of the casein. *Neth. Milk Dairy J.*, **43**, 53–62.
Schmidt, D.G., van der Spek, C.A., Buchheim, W. & Hinz, A. (1974). On the formation of artificial casein micelles. *Milchwissenschaft*, **29**, 455–9.
Schmidt, D.G., Koops, J. & Westerbeek, D. (1977). Properties of artificial micelles. 1. Preparation, size distribution and composition. *Neth. Milk Dairy J.*, **31**, 328–41.
Schmidt, D.G., Both, P & Koops, J. (1979). Properties of artificial casein micelles. 3. Relationship between salt composition, size and stability towards ethanol, dialysis and heat. *Neth. Milk Dairy J.*, **33**, 40–8.
Shimmin, P.D. & Hill, R.D. (1964). An electron microscope study of the internal structure of casein micelles. *J. Dairy Res.*, **31**, 121–3.
Slattery, C.W. (1976). Casein micelle structure; An examination of models. *J. Dairy Sci.*, **59**, 1547–56.
Slattery, C.W. (1978). Variation in the glycosylation pattern of bovine κ-casein with micelle size and its relationship to a micelle model. *Biochemistry*, **17**, 1100–4.
Slattery, C.W. (1979). A phosphate-induced sub-micelle-micelle equilibrium in reconstituted casein micelle systems. *J. Dairy Res.*, **46**, 253–8.
Slattery, C.W. & Evard, R. (1973). A model for the formation and structure of casein micelles from subunits of variable composition. *Biochim. Biophys. Acta*, **317**, 529–38.
Slattery, C.W., Sood, S.M. & Chang, P. (1989). Hydrophobic interactions in human casein micelle formation: β-casein aggregation. *J. Dairy Res.*, **56**, 427–33.
Snoeren, T.H.M., van Markwijk, B. & van Montfoort, R. (1980). Some physicochemical properties of bovine α_{s2}-casein. *Biochim. Biophys. Acta*, **622**, 268–76.
Snoeren, T.H.M., Klok, H.J., van Hooydonk, A.C.M. & Damman, A.J. (1984). The voluminosity of casein micelles. *Milchwissenschaft*, **39**, 461–3.
Sood, S.M., Sidhu, K.S. & Dewan, R.K. (1979). Heat stability and the voluminosity and hydration of casein micelles of different species. *N. Z. J. Dairy Sci. Technol.*, **14**, 217–25.
Sood, S.M., Chang, P. & Slattery, C.W. (1985). Interactions in human casein systems: self-association of fully phosphorylated human β-casein. *Arch. Biochem. Biophys.*, **242**, 355–64.
Sood, S.M., Chang, P. & Slattery, C.W. (1988). Interactions in human casein systems: self-association of nonphosphorylated human β-casein. *Arch. Biochem. Biophys.*, **264**, 574–83.
Sood, S.M., Chang, P. & Slattery, C.W. (1990). Interactions of triply phosphorylated human β-casein: monomer characterization and hydrodynamic studies of self-association. *Arch. Biochim. Biophys.*, **277**, 415–21.
Stothart, P.H. (1989). Subunit structure of casein micelles from small-angle neutron-scattering. *J. Mol. Biol.*, **208**, 635–8.
Stothart, P.H. & Cebula, D.J. (1982). Small-angle neutron scattering study of bovine casein micelles and sub-micelles. *J. Mol. Biol.*, **160**, 391–5.
Swaisgood, H.E. & Timasheff, S.N. (1968). Association of α_s-casein C in the alkaline pH range. *Arch. Biochem. Biophys.*, **125**, 344–61.
Swaisgood, H.E., Brunner, J.R. & Lillevik, H.A. (1964). Physical parameters of κ-casein from cow's milk. *Biochemistry*, **3**, 1616–23.
Takase, K., Niki, R. & Arima, S. (1980). A sedimentation equilibrium study of the temperature-dependent association of bovine β-casein. *Biochim. Biophys. Acta*, **622**, 1–8.

Talbot, B. & Waugh, D.F. (1970). Micelle-forming characteristics of monomeric and covalent polymeric κ-caseins. *Biochemistry*, **9**, 2807–13.

Thompson, M.P., Gordon, W.G., Boswell, R.T. & Farrell Jr., H.M. (1969). Solubility solvation, and stabilization of α_{s1}- and β-caseins. *J. Dairy Sci.*, **52**, 1166–73.

Thurn, A., Burchard, W. & Niki, R. (1987*a*). Structure of casein micelles. II. α_{s1}-casein. *Colloid & Polymer Sci.*, **265**, 897–902.

Thurn, A., Burchard, W. & Niki, R. (1987*b*). Structure of casein micelles. I. Small angle neutron scattering and light scattering from β- and κ-casein. *Colloid & Polymer Sci.*, **265**, 653–66.

Toma, S.J. & Nakai, S. (1973). Calcium sensitivity and molecular weight of α_{s5}-casein. *J. Dairy Sci.*, **56**, 1559–62.

Visser, J., Minihan, A., Smits, P., Tjan, S.B. & Heertje, I. (1986). Effects of pH and temperature on the milk salt system. *Neth. Milk Dairy J.*, **40**, 351–68.

Vreeman, H.J. (1979). The association of bovine SH-κ-casein at pH 7.0. *J. Dairy Res.*, **46**, 271–6.

Vreeman, H.J., Both, P., Brinkhuis, J.A. & van der Spek, C. (1977). Purification and some physicochemical properties of bovine κ-casein. *Biochim. Biophys. Acta*, **491**, 93–103.

Vreeman, H.J., Brinkhuis, J.A. & van der Spek, C.A. (1981). Some association properties of bovine SH-κ-casein. *Biophys. Chem.*, **14**, 185–93.

Vreeman, H.J., van Markwijk, B.W. & Both, P. (1989). The structure of casein micelles between pH 5.5 and 6.7 as determined by light-scattering, electron microscopy and voluminosity experiments. *J. Dairy Res.*, **56**, 463–70.

Walstra, P. (1979). The voluminosity of bovine casein micelles and some of its implications. *J. Dairy Res.*, **46**, 317–23.

Walstra, P. (1990). On the stability of casein micelles. *J. Dairy Res.*, **73**, 1965–79.

Walstra, P. & Jenness, R. (1984). *Dairy Chemistry and Physics*, John Wiley, New York.

Waugh, D.F. & Noble Jr., R.W. (1965). Casein micelles. Formation and structure II. *J. Am. Chem. Soc.*, **87**, 2246–57.

Waugh, D.F., Slattery, C.W. & Creamer, L.K. (1971). Binding of cations to caseins. Site binding, Donnan binding, and system characteristics. *Biochemistry*, **10**, 817–23.

Welsch, U., Buchheim, W., Schumacher, U., Schinko, I. & Patton, S. (1988). Structural, histochemical and biochemical observations on horse milk-fat-globule membranes and casein micelles. *Histochemistry*, **88**, 357–65.

Whitney, R.McL. (1988). Proteins of milk. In *Fundamentals of Dairy Chemistry*, ed. N.P. Wong, R. Jenness, M. Keeney & E.H. Marth. Van Nostrand Reinhold Company, New York, pp. 81–169.

4

β-LACTOGLOBULIN

SUSAN G. HAMBLING, ALAN S. MCALPINE & LINDSAY SAWYER*

Department of Biochemistry, University of Edinburgh Medical School, Hugh Robson Building, George Square, Edinburgh EH8 9XD, UK

1. INTRODUCTION

β-Lactoglobulin (BLG) is the major whey protein secreted in the milk of ruminants like the cow or sheep, and monogastrics like the pig, horse, dog or cat. It is absent from the milk of humans and rodents. Bovine BLG can be readily isolated and, since its isolation from milk by Palmer in 1934, it has been extensively studied by essentially every technique available. Despite this, however, its biological function remains uncertain.

The name β-lactoglobulin was first proposed by Cannan *et al.* (1942) for the fraction crystallised by Palmer, which Pedersen (1936*a*) had identified with the 'β' component of whey in ultracentrifugation studies. Despite its having been called 'β-lactalbumin' by Svedberg (1938), the name β-lactoglobulin is that by which the protein is now known and it is *quite distinct* from the other whey protein α-lactalbumin, the subject of another chapter in this volume!

A number of reviews on the physico-chemical properties of BLG have been published over the years (e.g. Tilley, 1960; Townend *et al.*, 1969; McKenzie, 1971; Green *et al.*, 1979) together with several others on milk proteins in general which contain substantial sections on the properties of the protein (e.g. McKenzie, 1967; Lyster, 1972; Jenness, 1974; Thompson & Farrell, 1974; Whitney *et al.*, 1976; Jenness, 1979; Swaisgood, 1982; Jenness, 1985). It is the purpose of this present review to bring together various strands of the work on β-lactoglobulin over the last 55 years and to discuss the various clues that these give as to the function of BLG.

For example, binding studies carried out on the bovine protein *in vitro* have shown that it can bind a variety of ligands, most of which are small, hydrophobic molecules like fatty acids or vitamin A (Futterman & Heller, 1972; Puyol *et al.*, 1991). ORD studies on BLG have shown that it undergoes several conformational changes between pH 2 and pH 9, possibly the most important of which is the N $\leftrightarrow$ R transition that occurs in the physiological pH range (Tanford *et al.*, 1959).

*To whom correspondence should be addressed.

Sedimentation studies reveal the bovine protein to be a dimer under physiological conditions whereas the BLG from pigs is a monomer. Crystal structures of BLG at both pH 6·5 and pH 7·8 have recently becomê available and are being improved so that it should soon be possible to offer a detailed molecular explanation for these and other solution properties.

The primary structure of BLG shows some homology to human serum retinol binding protein (RBP) (Godovac-Zimmerman *et al.*, 1985*a*; Pervaiz & Brew, 1985). More striking, however, is the similarity in tertiary structure (Sawyer *et al.*, 1985; Papiz *et al.*, 1986) which also extends to insecticyanin (ICN) (Sawyer, 1987; Sawyer & Richardson, 1991). Both of these, RBP and ICN, carry small insoluble molecules (the former retinol, the latter biliverdin) within a hydrophobic cavity, which is characteristic of all three of these proteins. Further similarities, at least at the sequence level, have been identified in a number of other proteins giving rise to the likelihood that there is a family of proteins, the lipocalycins (cf. lipocalins; Pervaiz & Brew, 1987), which transport small, insoluble or labile molecules. That they may all have evolved from a common ancestor is supported by recent work on the gene sequences which show a similar intron/exon pattern for the several members of the family for which these data are available.

In what follows, it is not the authors' intention to include any of the extensive literature on the food science aspects of the protein since this will be covered elsewhere. Rather, the chapter concentrates upon the studies that have been carried out under what are basically physiological conditions and which provide information about BLG as an intriguing biological macromolecule.

2. Biosynthesis and Secretion

Major lactoproteins, including BLG, are biosynthesised within the secretory epithelial cells of the mammary gland under the control of prolactin (Larson, 1979). mRNA coding for BLG which is specific to the mammary tissue, is translated to yield a 180 amino acid pre-BLG. Pre-BLGs have been isolated from the mammary glands of cow (Yoshikawa *et al.*, 1978), sheep and pig, using cell-free translation systems (Mercier & Gaye, 1982). The signal peptides consist of 16 or 18 highly conserved amino acids, which are hydrophobic, and possibly α-helical, with a basic residue near the N-terminus.

Cow	Met–Lys–Cys–Leu–Leu–Leu–Ala–Leu–Ala–Leu– –Thr–Cys–Gly–Ala–Gln–Ala–
Sheep	Met–Lys–Cys–Leu–Leu–Leu–Ala–Leu–Gly–Leu–Ala–Leu–Ala–Cys–Gly–Val–Gln–Ala–
Pig	Met–Arg–Cys–Leu–Leu–Leu–Thr–Leu–Gly–Leu–Ala–Leu–Leu–Cys–Gly–Val–Gln–Ala–

A membrane-bound proteinase cleaves off the signal peptide on the C-terminal side of Ala, to generate the mature 162 amino acid protein which can then undergo post-translational modifications that begin in the cisternal space of the rough endoplasmic reticulum. Transfer of an oligosaccharide from membrane lipid to Asn in the protein sequence –Asn–X–Ser/Thr– has been reported only for BLG from the Australian Droughtmaster breed of cattle (Bell *et al.*, 1970). It is likely that a disulphide isomerase ensures correct folding (Freedman *et al.*, 1989) before transport to the Golgi apparatus and subsequent incorporation into secretory vesicles. The native protein is then secreted into the lumen, where it accumulates in the milk before removal by the suckling young.

3. Distribution

The composition of milk varies considerably from species to species and also with the time since parturition (Larson, 1979). There are also seasonal variations, which presumably reflect the dietary habits of the animal. A quantitative partition of the albumin fraction of whey proteins useful for detecting the presence of BLG, has been devised by Davies (1974).

Since the initial preparation of BLG from bovine (*Bos taurus*) milk (Palmer, 1934), dimeric BLGs have been isolated from the milks of a number of other ruminants, and monomeric BLGs have been purified from the milks of several non-ruminant livestock species (Table 1). BLG has also been detected in milks of other species, but the state of its association in these cases is uncertain. For instance, the molecular weight of kangaroo BLG has been quoted as 17 to 19 kDa (McKenzie *et al.*, 1983), suggesting a monomeric structure but a subsequent report declared that it was dimeric, despite its low (32%) sequence homology with bovine BLG (Godovac-Zimmerman & Shaw, 1987).

It seems likely that any BLGs detected by bovine BLG antiserum are dimeric, as there is known to be little cross-reactivity between this antiserum and monomeric BLGs. Hence, the lack of cross-reactivity does not rule out the presence of a monomeric BLG. Thus, in two cases, donkey (Godovac-Zimmerman *et al.*, 1988) and peccary (Lyster *et al.*, 1966; Kalan *et al.*, 1971), there was no cross-reactivity with bovine BLG antiserum whereas there was when antiserum to monomeric BLG was used. The existence of camel BLG is still in doubt; its absence was stated (quoted by Kessler & Brew, 1970), and it is known that camel milk does not cross-react with antiserum to bovine BLG (Lyster *et al.*, 1966). However, higher titres of antibodies to bovine BLG have subsequently been shown to cross-react with both camel milk and colostrum (Liberatori *et al.*, 1979*c*). This may indicate that there is a monomeric BLG present. Preliminary studies with feline milk revealed there to be no cross-reactivity between it and bovine BLG antiserum, but recently a protein thought to be BLG has been isolated and is being sequenced (Halliday *et al.*, 1991). It is probable, that this BLG will turn out to be monomeric.

It is believed that there is no BLG in human milk (Bell & McKenzie, 1964) despite the isolation of a 13·5 kDa whey protein with a pl of 5·1 (Conti *et al.*, 1980), and the characterisation of a protein from human colostrum with a pl of 6·1, a molecular weight around 18 kDa and an amino acid composition similar to that of pig BLG (Conti *et al.*, 1982). Both of these proteins were detected by immunoprcipitation with bovine BLG antiserum, and were most likely fragments of human lactoferrin which is known to cross-react with the antiserum (Brignon *et al.*, 1985). Radioimmunoassays have suggested that low concentrations of bovine BLG (5–

TABLE 1
Distribution of BLG in the Milks of Various Species

Species	*Cross-reaction with antibovine antisera*	*State of association*
Cow (*Bos taurus*)	+	Di[a]
Oxen (*Bos javanicus*)	+	Di[b]
Yak (*Bos grunniens*)	+	Di
Zebu (*Bos indicus*)	+[c]	?Di*
Waterbuffalo (*Bubalus arnee*)	+	Di
Buffalo (*Bubalis bubalis*)	+[c]	?Di*
Bison (*Bison bison*)	+[c]	?Di*
Musk ox (*Ovibos moschatus*)	+[c]	?Di*
Eland (*Taurotragus oryx*)	+[c]	?Di*
Goat (*Capra hircus*)	+	Di
Sheep (*Ovis aries*)	+	Di[d]
Mouflon (*Ovis ammon musimon*)	+	Di[e]
Red deer (*Cervus elaphus* L.)	+[c]	?Di*
European elk (*Alces alces* L.)	+[c]	?Di*
Reindeer (*Rangifer tarandus* L.)	+[c]	?Di*
White-tailed deer (*Odocoileus virginianus*)	+[c]	?Di*
Fallow deer (*Dama dama*)	+[c]	?Di*
Caribou (*Rangifer arcticus*)	+[c]	?Di*
Giraffe (*Giraffa camelopardalis*)	+[c]	?Di*
Okapi (*Okapia johnstoni*)	+[c]	?Di*
Prong-horn antelope (*Antilocapra americana*)	+[c]	?Di*
Camel (*Camelus dromedarius*)	±[f]	?Mono*
Llama (*Lama glaba* L.)	−[g]	none[g]
Peccary (*Pecari tajacu*)	−*[c]	Mono
Pig (*Sus scrofa domestica*)	±[h]	Mono
Horse (*Equus caballus*)	±[h,i]	Mono
Zebra (*Equus quagga*)	−[c]	?none/Mono*
Rhinoceros (*Diceros bicornis*)	−[c]	?none/Mono*
Donkey (*Equinus asinus*)	−*[j]	Mono[j]
Mouse (*Mus musculus*)	−	none[k]
Rat (*Ratus norvegicus*)	nm	none[l]
Guinea-pig (*Cavia porcellus*)	nm	none[m]
Dolphin (*Tursiops truncatus*)	nm	Mono[n]
Manatee (*Trichechus manatus latirostris*)	nm	Mono[n]
Beagle (*Canis familiaris*)	nm	Mono[n]
Cat (*Felis catus*)	−[o]	?Mono*
Grey kangaroo (*Macropus giganteus*)	nm	Mono*[p]
Red kangaroo (*Macropus rufus*)	nm	Mono[p]
Northern fur seal (*Callorhinus ursinus*)	nm	unknown[q]
Human (*Homo sapiens*)	±[r]	none*[s]
Rabbit (*Oryctolagus cuniculus*)	−[c]	none

+, BLG detected by antibovine BLG antisera; −, BLG not detected by antibovine BLG antisera; ±, BLG detected by antibovine BLG antisera, but only at higher titres; nm, cross-reactivity to antibovine BLG antisera not measured; Di, dimeric BLG detected; Mono, monomeric BLG detected; None, no BLG present; unknown, BLG present, but its state of association unknown; * cross reactivity to antibovine BG antisera is assumed to indicate a dimeric form of the protein.

[a]Bull *et al.* (1946a, b). [b]Bell *et al.* (1981a, b). [c]Lyster *et al.* (1966). [d]Bell & McKenzie (1967b). [e]Godovac-Zimmerman *et al.* (1987). [f]Liberatori *et al.* (1979). [g]Fernandez & Oliver (1988). [h]Liberatori *et al.* (1979a). [i]Bell *et al.* (1981). [j]Godovac-Zimmerman *et al.* (1988). [k]Clark *et al.* (1987). [l]Hennighausen & Sippel (1982). [m]Brew & Campbell (1967). [n]Pervaiz & Brew (1986). [o]Halliday *et al.*, 1991. [p]McKenzie *et al.* (1983). [q]Ashworth *et al.* (1966). [r]Conti *et al.* (1979). [s]Bell & McKenzie (1964).

800 μg/litre), or at least antigenic peptides therefrom, are present in human milk, having been absorbed from the diet (Axelsson *et al.*, 1986; Monti *et al.*, 1989).

BLG is also absent from the milk of rodents. However, both rat and mouse milks contain a major whey acidic protein (WAP) which has a similar molecular weight to, but is not, a BLG. There is no evidence of hybridisation between cDNA from a rat mammary library and cDNA for sheep BLG (Ali, 1989), and mouse WAP shows no primary sequence homology to bovine BLG (Hennighausen & Sippel, 1982). Despite a report that the major whey protein in murine milk (Lockwood *et al.*, 1966) was BLG, it failed to cross-react with bovine BLG antisera (Lyster *et al.*, 1966) and was most likely WAP. A recent publication has confirmed both the abundance of WAP (2 g/litre) and the absence of BLG (Clark *et al.*, 1987). The 'mouse BLG' that has become commercially available is probably WAP, despite the successful expression of sheep BLG in transgenic mice (Simons *et al.*, 1987).

4. Isolation

The isolation of BLG from milk is a simple procedure, involving just four stages: the removal of fats, the removal of the caseins, the fractionation of the whey proteins, and the purification of BLG. Since each stage can be carried out in a number of ways, a variety of published methods is available.

The original method of Palmer (1934) for the isolation of bovine BLG from skim milk was eventually superceded by the method of Aschaffenburg and Drewry (1957), which involved an acid precipitation step at pH 2. Mailliart and Ribadeau-Dumas (1988) have devised a similar salting out procedure suitable for large scale use. Later, it was suggested that the harshness of this step might cause irreversible changes to BLG and the method was modified by Armstrong *et al.* (1967) such that the whey proteins were fractionated by acid precipitation at pH 3·5. More recently, Monaco *et al.* (1987) separated the whey proteins by DEAE-cellulose chromatography without the pH going below 6·6. Crystals grown from this BLG diffracted to a higher resolution than those grown from BLG purified by the method of Aschaffenburg and Drewry, suggesting that acid precipitation at pH 2 does indeed damage the protein in some undefined way. Despite this, most methods for the commercial preparation of bovine BLG use the procedure of Aschaffenburg and Drewry. A summary of these methods is presented in Table 2.

Other dimeric BLGs have also been prepared by these methods, or slight variations thereof. Yak BLG (Grosclaude *et al.*, 1976) has been isolated by the procedure of Aschaffenburg and Drewry (1957), whereas red deer (McDougall & Stewart, 1976), waterbuffalo (Braunitzer *et al.*, 1979), sheep (Maubois *et al.*, 1965) and goat BLGs (Kalan & Basch, 1969) have been obtained by a modified method. More recently, both goat (Preaux *et al.*, 1979) and ovine (Godovac-Zimmerman *et al.*, 1987) BLGs have been isolated by a non-salt precipitation method in which the fat and caseins were removed by centrifugation, and then the whey protein mixture was dialysed, lyophilised, and separated by gel filtration (cf. Davies, 1974). A procedure for the isolation of BLG from both cow and goat milk based upon

TABLE 2
Some Isolation Procedures for Bovine β-Lactoglobulins

Method	*Removal of fat*	*Removal of caseins*	*Fractionation of whey proteins*	*Purification*
Palmer (1934)	Skim milk used	Acid precipitation	Na_2SO_4 precipitation at up to 30°C	Dialysis, recrystallization
Aschaffenburg & Drewry (1957)	With Na_2SO_4 precipitation at 40°C		Acid precipitation at pH 2 to remove α-La and SA, $(NH_4)_2SO_4$ precipitation of BLG at pH 6	Dialysis, recrystallization
Armstrong *et al.* (1967)	$(NH_4)_2SO_4$ precipitation at 20°C		Acid precipitation at pH 3·5 to remove α-La and SA, $(NH_4)_2SO_4$ precipitation of BLG at pH 6	Dialysis, recrystallization
Monaco *et al.* (1987)	Centrifugation	$CaCl_2$ precipitation at pH 6·6	Dialysis, DEAE-cellulose chromatography using NaCl to elute BLG	Gel filtration, crystallization

α-La = α-Lactalbumin.
SA = Serum albumin.

ethanol fractionation has been reported (Bain & Deutsch, 1948) but neither isolate crystallised, suggesting that some form of denaturation has occurred.

The isolation of monomeric BLGs from the milk of non-ruminants can be carried out with equal ease using procedures similar to those described above. Pig BLG has been isolated and purified by three different methods, which are summarised in Table 3. No comparison appears to have been published on the advantages of these various methods. The procedure of Kalan *et al.* (1971) uses many precipitation steps which probably cause the low yield (25%) obtained. The method of Kessler and Brew (1970), on the other hand, has been successfully used to isolate monomeric BLGs from the milks of dolphin, manatee, beagle (Pervaiz & Brew, 1986) and horse (Godovac-Zimmerman *et al.*, 1985*b*).

The amount of BLG obtained by the various methods depends upon both the procedures used and the quantity of BLG in the initial milk, which is known to vary with season and time since parturition. The quantities of BLG isolated from the milks of a few species are given in Table 4.

5. Genetic Variants and Primary Structure

Although Li (1946) demonstrated by electrophoresis that BLG was heterogeneous, and Polis *et al.* (1950) separated two components of BLG by fractional crystallisation, it was Aschaffenburg and Drewry (1955) who showed that there were two components in bovine milk whose origin was probably genetically determined, a fact that has been proved (Aschaffenburg & Drewry, 1957) and reviewed (Aschaffenburg, 1968; see also Chapter 10). The presence of genetic variants of BLG has also been shown in other species. The original phenotyping procedure used paper electrophoresis at pH 8·6 but a more convenient method uses gel electrophoreis on undiluted, whole milk samples (Aschaffenburg, 1964; Aschaffenburg & Thymann, 1965; Thompson, 1970; Kiddy *et al.*, 1972; see Chapter 10).

The original fractionation of the whey proteins of bovine milk produced a heterogeneous preparation of BLG. This was due to the presence of two genetic variants, BLGs A and B, which have been separated (Piez *et al.*, 1961), sequenced (Braunitzer *et al.*, 1972), and are the ones most often present in commercial preparations. Other genetic variants of bovine BLG have also been detected, isolated and sequenced. Cow (*Bos taurus*) is currently believed to have six variants: BLG-A and BLG-B as mentioned above; BLG-C which has been positively identified only in the milk of Jersey cows (Bell & McKenzie, 1967*a*), but may also be present in South African Nguni cattle (Osterhoff & Pretorius, 1966); BLG-D (Brignon *et al.*, 1969; Brignon & Ribadeau-Dumas, 1973); BLG-H (Conti *et al.*, 1988) and BLG-W which is thought to exhibit only neutral amino acid substitutions when compared to BLG-B (Godovac-Zimmerman *et al.*, 1990). A crossbreed between *Bos taurus* and *Bos indicus*, Australian Droughtmaster, was shown to produce variants BLG-A, BLG-B and a new variant BLG-Dr which is the only BLG known to be N-glycosylated (Bell *et al.*, 1970). Bali cattle, *Bos javanicus*, have three genetic variants. It has recently been suggested that BLG-E is identical to yak

TABLE 3
Isolation Procedures for Pig Porcine β-Lactoglobulin

Method	*Removal of fat*	*Removal of caseins*	*Fractionation of whey proteins*	*Purification*
Kessler & Brew (1970)		Acidification	Gel filtration	Ion-exchange chromatography, gel filtration
Kalan *et al.* (1971)	Separator at 40°C	By acid precipitation	By Na_2SO_4 precipitation to removeα-La, acid precipitation of BLG at pH 3	pH fractionation, $(NH_4)_2SO_4$ fractionation, crystallization
Bell *et al.* (1981a)	Centrifugation	By $(NH_4)_2SO_4$ precipitation	$(NH_4)_2SO_4$ precipitation of whey proteins, gel filtration	Ion-exchange chromatography

α-La = α-Lactalbumin.

TABLE 4
Quantities of β-Lactoglobulin in the Milk of Various Species

Species	*Quantity of BLG (mg/ml)*	*References*
Bovine	1·8–5·0	Palmer (1934), Butler (1974)
Goat	1·4	Kalan & Basch (1969)
Sheep	2·8	Simons *et al.* (1987)
Red deer	2·8–3·0	McDougall & Stewart (1976)
Pig	0·6[a]	Kalan *et al.* (1971)
Dolphin	16·2	Pervaiz & Brew (1986)
Manatee	14·1	Pervaiz & Brew (1986)
Beagle	10·1	Pervaiz & Brew (1986)
Human	0·000 005–0·000 8[b]	Axelsson *et al.* (1986)
Mouse	0	Simons *et al.* (1987)

[a]This value is an underestimate due to the poor fractionation of the porcine whey proteins at pH 3, and the numerous precipitation steps used subsequently.
[b]It is unlikely that this result refers to human BLG. Radioimmunoassays using antibovine BLG antisera have detected both fragments of lactoferrin and bovine BLG in human milk (Monti *et al.*, 1989).

(*Bos grunniens*) BLG, as position 11 is now thought to be Asp (Bell *et al.*, 1981*a*) and not Asn (Grosclaude *et al.*, 1976). These data imply, not unexpectedly, a close relationship between bovine subgenera.

Other ruminant BLGs also exhibit genetic polymorphism. Red deer possess two variants (McDougall & Stewart, 1976), whereas sheep have variants A, B (Bell & McKenzie, 1967*b*; Preaux *et al.*, 1980; Kolde & Braunitzer, 1983) and C (Erhardt *et al.*, 1989). Mouflon BLG is homogeneous, and identical in sequence to ovine BLG-B (Godovac-Zimmerman *et al.*, 1987). Waterbuffalo (Sen & Sinha, 1961; Braunitzer *et al.*, 1979) and goat (Sen & Chaudhuri, 1962; Preaux *et al.*, 1979) BLGs are both homogeneous by electrophoresis, but the latter (unlike any of the above proteins), contains an amino acid substitution which cannot be the result of a single point mutation, namely, Asp-130 (GAU/C) in bovine BLG-A becomes Lys-130 (AAA/G) in goat.

Monomeric BLGs from kangaroo, dolphin, horse and pig have also been purified to homogeneity and sequenced. Their sequences diverge considerably from that of bovine BLG-A and from one another. Many of the substitutions observed cannot have arisen from single point mutations. Kangaroo BLGs are homogeneous (Godovac-Zimmerman & Shaw, 1987), whereas dolphin, donkey, horse and pig show genetic polymorphism. Two dolphin BLGs have been detected and partially sequenced (Pervaiz & Brew, 1985, 1986). Of the two donkey BLGs isolated, one has been fully sequenced (Godovac-Zimmerman *et al.*, 1988). Equine BLG has been shown to be heterogeneous (Conti *et al.*, 1982*a*), despite an earlier report to the contrary (Bell *et al.*, 1981*b*). Two variants have been purified and sequenced—the complete primary structure of BLG-I (Conti *et al.*, 1984) agrees

TABLE 5
Amino Acid Compositions of Pig β-Lactoglobulin

Amino Acid	*Kalan*[a]		*Bell*[b]		*Conti*[c]	*Conti*[d]			*This work**
	A	*B*	*A*	*C*	*I*	*1*	*2*	*3*	
Asx	17	17	17	18	17	16	18	18	17
Thr	10	10	10	9	10	10	9	10	10
Ser	10	10	9	9	8	7	9	7	7
Glx	23	23	24	22	24	22	21	24	24
Pro	8	8	8	8	7	7	6	7	8
Gly	3	3	4	4	3	7	5	4	3
Ala	13	14	14	13	13	13	12	13	13
Cys	4	4	4	4	5	4	6	5	nd
Val	13	12	13	13	13	15	10	12	12
Met	4	4	4	4	4	nd	2	3	3
Ile	6	6	6	6	6	6	5	6	6
Leu	24	24	25	24	24	22	27	25	25
Tyr	2	2	2	2	2	nd	5	3	2
Phe	3	3	3	3	3	3	6	3	3
Lys	11	11	11	11	11	14	11	11	12
His	3	3	3	4	3	4	4	4	4
Arg	5	5	5	5	5	5	4	5	6
Trp	1	1	1	1	1	nd	nd	nd	nd

nd = Amino acid residue not determined; * = this amino acid composition was obtained from pig BLG kindly supplied by Dr Carl Holt.
[a]Kalan *et al.* (1971).
[b]Bell *et al.* (1981*c*).
[c]Conti *et al.* (1986*b*).
[d]Conti *et al.* (1986*a*).

with its partial sequence determined earlier (Bell *et al.*, 1981*b*), but reveals only 70% identity to BLG-II (Godovac-Zimmerman *et al.*, 1985*a*).

The heterogeneity of pig BLG was first reported in 1969 (Kraeling & Gerrits, 1969) and the variants named BLG-A and BLG-B (Kalan *et al.*, 1971). A later study also isolated two variants: one was believed to be BLG-A, and the other, distinct from BLG-B, was named BLG-C (Bell *et al.*, 1981*c*). Recently, BLG has been purified from sow's milk, and studied by two methods (Conti *et al.*, 1986*a*): chromatofocusing yielded three peaks of which only the first at pH 4·7, was homogeneous. This was sequenced as pig BLG-I (Conti *et al.*, 1986*b*). On the other hand, isoelectricfocusing with an immobilised pH gradient gave three bands, BLG-1 BLG-2 and BLG-3, with pls of 4·4, 4·65 and 4·9, respectively. The amino acid compositions of all these proteins are given in Table 5 together with a composition from this laboratory from a pig believed to be BB. It is unclear which variants are the same. Homology between the first 50 amino acids of BLG-A (Bell *et al.*, 1981*c*) and BLG-I (Conti *et al.*, 1986*b*) suggests that these two may be equivalent. However, it is apparent from the published work on the pig BLGs that inadequate genetic phenotyping has been performed before the biochemical analysis.

Now that primary sequences are routinely obtainable, reviewing the literature on amino acid composition and the various N- and C-terminal analyses is considered unnecessary. The interested reader should refer to the full discussion by McKenzie (1971). One point worth reporting, however, is the removal of the two carboxy-terminal residues by carboxypeptidase A treatment of bovine variants A, B and C (Kalan *et al.*, 1965). The rate of removal of the residues from variant C is significantly slower than that from variants A or B. Of course the disulphide bridge in the native sequence terminates any further proteolysis so that a homogeneous product can be obtained. This has been crystallised (Greenberg & Kalan, 1965) and a low resolution crystal structure has been determined (Green *et al.*, 1979).

Although an incomplete sequence was published by Frank and Braunitzer (1967), the first full sequence was reported by Braunitzer *et al.* (1972). The positions of the disulphide bridges have given rise to some controversy. Whilst the cystine 66–160 has been unambiguously identified by a variety of methods (Mainferme *et al.*, 1971; Martial *et al.*, 1971; Perez-Gomez *et al.*, 1971 with revised numbering reported by Preaux & Lontie, 1972), McKenzie *et al.* (1972) have shown that equimolar amounts of 106–119 and 106–121 were obtained when ^{14}C-iodoacetamide was covalently bound to the free thiol in the presence of 8M urea, the disulphide bridges reduced, and then unlabelled iodoacetamide reacted with the newly generated sulphydryls. Subsequent peptide digestion and analysis showed that the label was evenly distributed between Cys-119 and Cys-121. However, McKenzie and Shaw (1972) report that there is no appreciable reaction between the free thiol and iodoacetamide in the absence of 8M urea, and that 8M urea can disrupt disulphide bridges (McKenzie *et al.*, 1972). Thus the interchange observed could be explained by the effects of the urea. This result conflicts with those of Preaux and Lontie (1972), and the crystal structures at medium resolution (Papiz *et al.*, 1986; Monaco *et al.*, 1987; Yewdall, 1988) show clearly that there is but one disulphide residue 106–119. Indeed, the separation of the Cys residues 119 and 121 is some 10 Å so that a significant rearrangement is required to accomplish the interchange. What is not ruled out by the crystallography is the selection by crystallisation of the cystine 106–119 isomer, although the two forms which grow together from the same solution at pH 7·8, both have this cystine. Mutant proteins have been produced with either 119 or 121 replaced by Ser but they have not yet been fully analysed (Paterson, G., Fothergill-Gilmore, L.A. & Sawyer, L., unpublished; Batt, C.A. 1990, pers. comm.).

Not only have there been a number of protein sequence determinations on a wide variety of BLGs (Godovac-Zimmerman & Braunitzer, 1987; Godovac-Zimmerman, 1988) but there are now a number of partial and complete DNA sequences available for mRNA-derived cDNA (Willis *et al.*, 1982; Mercier *et al.*, 1985; Gaye *et al.*, 1986; Jamieson *et al.*, 1987; Ivanov *et al.*, 1988; Anderson *et al.*, 1989) and the gene (Ali & Clark, 1988) which reveals the presence of introns.

The genetic variants have usually been detected because they have a charge difference detectable on electrophoresis (but see Krause *et al.*, 1988). However, now that DNA sequencing techniques are available, mutations that are not reflected in altered mobility will perhaps be found more commonly. For example, the DNA sequence of bovine BLG determined by Anderson *et al.* (1989) implies the change Phe105Val (TTC → GTC). As no comment is made about this, however, an error of some kind cannot be ruled out.

Examination of the primary structure of BLG reveals no obviously repetitive or unusual stretches of sequence. The ruminant proteins as purified are not modified post-translationally, with the exception of the BLG-Dr whose Asp/Asn mutation has already been referred to. It is interesting to note, however, that they all possess a sequence Asn–Pro–Thr at 152 which does not appear to be glycosylated. From the crystal structure at pH 7·6 (Papiz *et al.*, 1986), this can be explained on steric grounds since the Asn residue lies in-between the two subunits. Since assembly is assumed to occur as the last stage in the biosynthesis, this observation could mean that there is a penultimate *de*glycosylation process.

The BLG variants from ruminants show greater than 90% sequence identity between one another within the same species, and other dimeric BLGs (Table 6). They are synthesised under the control of codominant alleles. The non-ruminant variants are monomeric, show much lower primary sequence homology to other monomeric and dimeric BLGs (typically 30–70% identity), and may originate from gene duplication, as suggested for the horse BLGs (Godovac-Zimmerman *et al.*, 1985*b*). Table 7 gives the sequences of the various non-ruminant species with cow BLG-A as the representative ruminant protein.

6. Solution Studies

Probably because of its abundance, ease of preparation and relative stability, numerous solution studies have been carried out on this protein using a wide variety of techniques. These experiments have elucidated the physico-chemical behaviour of the protein, provided an insight into the importance of specific amino acids and yielded information on the binding of ligands to BLG *in vitro*. The vast majority of these studies have been carried out with the bovine protein and, unless specifically mentioned, it is this protein that has been used. A number of physico-chemical properties are summarised in Table 8.

6.1. Solubility

Tilley (1960) reviewed the considerable variation in the solubility of BLG as reported in the literature up to 1960. In this laboratory, protein is taken up into solution in dilute buffer from which it can be precipitated and even crystallised by dialysis against distilled water (Senti & Warner, 1948). It is clear, however, that the addition of salt or buffer increases the solubility of BLG quite dramatically: Polis *et al.* (1950) dissolved 0·18 g/100 ml water at the pI compared to 1·65 g/100 ml at pH 5·2 in 0·2 M NaCl—a 10-fold increase. Similar increases were observed by Treece *et al.* (1964) who also showed BLG-B to be about five times more soluble than BLG-A. A solubility minimum always exists at the pI, and so different absolute values are obtained at other pH values. Arakawa and Timasheff (1987) have examined the behaviour of BLG with NaCl and conclude that because the solubil-

TABLE 6
Comparison of the Amino Acid Sequences of Ruminant β-Lactoglobulins

β-Lactoglobulin	*Variable amino acid positions*																		
	1	*11*	*20*	*28*	*45*	*50*	*53*	*59*	*64*	*78*	*84*	*87*	*118*	*129*	*130*	*148*	*150*	*158*	*162*
Bovine A	V	D	Y	D	E	P	D	Q	D	I	I	L	V	D	D	R	S	E	I
Bovine B	V	D	Y	D	E	P	D	Q	G	I	I	L	A	D	D	R	S	E	I
Bovine C	V	D	Y	D	E	P	D	H	G	I	I	L	A	D	D	R	S	E	I
Bovine D	V	D	Y	D	Q	P	D	Q	G	I	I	L	A	D	D	R	S	E	I
Bovine E	V	D	Y	D	E	P	D	Q	G	I	I	L	A	D	D	R	S	G	I
Bovine F	V	D	Y	D	E	S	D	Q	G	I	I	L	A	D/Y[a]		R	S	G	I
Bovine G	V	D	Y	D	E	P	D	Q	G	M	I	L	A	D	D	R	S	G	I
Bovine Dr	V	D	Y	N[b]	E	P	D	Q	D	I	I	L	V	D	D	R	S	E	I
Yak	V	N/D[c]	Y	D	E	P	D	Q	G	I	L	I	A	D	D	R	S	G	I
Water buffalo	I	D	Y	D	E	P	D	Q	D	I	I	L	A	D	D	R	S	E	V
Goat	I	D	Y	D	E	P	N	Q	D	I	I	L	A	D	K	R	A	G	V
Sheep A	I	D	Y	D	E	P	N	Q	D	I	I	L	A	D	N	R	A	G	V
Sheep B	I	D	H	D	E	P	N	Q	D	I	I	L	A	D	N	R	A	G	V
Sheep C	I	D	Y	D	E	P	N	Q	G	I	I	L	A	D	N	Q	A	G	V

[a]It is unclear which residue is at which position since the changes in E, F and G are inferred (Bell *et al.*, 1981b).
[b]This Asn is glycosylated in the Droughtmaster breed.
[c]This position was originally believed to be Asn (Grosclaude *et al.*, 1976), but has also been reported as Asp (Bell *et al.*, 1981).

TABLE 7
Comparison of the Primary Sequences of Non-ruminant β-Lactoglobulin and Bovine β-Lactoglobulin-A

BOVINE A	LIVTQTMKGL	DIQKVAGTWY	SLAMAASDIS	LLDAQSAPLR	-VYVEELKPT	PEGDLEI-LL
KANGAROO	VENIRSKNDL	GVEKFVGSWY	LRE-AAKTME	F----SIPLF	DMDIKEVNLT	PEGNLELVLL
DOLPHIN I	VSVIRTMEDL	DIQRVAGTWH	SVAMAASDIS	LLDTEEAPLR	-VNVEELRPT	PQGDLEI-FL
DONKEY	TNIPQTMQDL	DLQEVAGKWH	SVAMAASDIS	LLDSEEAPLR	-VYIEKERPT	PEDNLEI-IL
HORSE I	TNIPQTMQDL	DLQEVAGKWH	SVAMAASDIS	LLDSEEAPLR	-VYIEKERPT	PEDNLEI-IL
HORSE II	TNIPQTMQDL	DLQEVAGRWH	SVAMVASDIS	LLDSEEAPLR	-VYVEELRPT	PEGNLEI-IL
PIG I	VEVTPIMTEL	DTQKVAGTWH	TVAMAVSDVS	LLDAKSSPLK	-AYVEGLKPT	PEGDLEI-LL
FELINE II	ATLPPTMEDL	DIRQVAGTWH	SMAMAASDIS	LLDSETAPLR	-VYVQELRPT	PRDNLEI-IL
BOVINE A	QKWENDECAQ	KKIIAEKTKI	PAVFKIDALN	ENKVL--VLD	TDYKKYLLFC	MENSAEPEQS
KANGAROO	EKTDR--CVE	KKLLLKKTKK	PTEFEIYISS	ESSYTFCVME	TDYDSYFLFC	LYNISDREK-
DOLPHIN I	QKRDKNGCVK	EKIIAEKTEI	PAVFKINFLN	ENKIF--VLD	SDYTNYLLFC	MENTADPERS
DONKEY	REGENKGCAE	KKIFAEKTES	PAEFKINYLD	EDTVF--ALD	TDYKNYLFLC	MKNAATPGQS
HORSE I	REGENKGCAE	KKIFAEKTQS	PAQFKINALD	EDTVF--YLD	TDYKNYLFLC	MKNAATPGQS
HORSE II	REGANHACVE	RNIVAQKTED	PAVFTVNYQG	ERKIS--VLD	TDYANYMFFC	VGPPLPSAEH
PIG I	QKRENDKCAQ	EVLLAKKTDI	PAVFKINALD	ENQLF--LLD	TDYDSHLLLC	MENSAPQEHS
FELINE II	RKRENRACIE	GNIMAQRTED	PAVFMVDYQG	ERKIS--VLD	TDYTNYMFFC	MEAPAPGTEN
BOVINE A	-LVCQCL	VRTPEVDDEA	LEKFDKALK-	ALPMHIRLSF	NPTQLEEQCH	I
KANGAROO	-MACAHY	VRRIEENKGM	NE-FKKILR-	TLAMPYTVI-	-EVRTRDMCH	V
DOLPHIN I	-LTCAYL	ARTLQVDDGV	MEKFNKAIKP	ALPMHIRL-F	SPTQLEEQCH	V
DONKEY	-LVCNYL	ARTQMVDEEI	MEKFRRALQ-	PLPGRVQILP	DLTRMAERCR	I
HORSE I	-LVC-YL	ARTQMVDEEI	MEKFRRALQ-	PLPGRVQIVP	DLTRMAERCR	I
HORSE II	GMVCQYL	ARTQKVDEEV	MEKFSRALQ-	PLPGRVQIVQ	DPSGGQERCG	F
PIG I	-LVCQSL	ARTLEVDDQI	REKFEDALK-	TLSVPMRIL-	-PAQLEEQCR	V
FELINE II	GMMCQYL	ARTLKADNEV	MEKFDRALQ-	TLPVHIRIIL	DLTQGKEQCR	V

-, Gap inserted to help alignment.

TABLE 8
Selected Physicochemical Parameters for β-Lactoglobulin

Parameter	*Value*	*Reference*
Isoelectric point	5·18	Cannan *et al.* (1942)
Molecular mass (Daltons)	36 000	Townend & Timasheff (1957)
Sedimentation coefficient ($S^{o}_{20w} \times 10^{13}\ s^{-1}$)	2·83	Cecil & Ogston (1949)
Diffusion coefficient ($cm^2\ s^{-1} \times 10^{-7}$)	7·70 ± 0·09	Cecil & Ogston (1949)
Crystal density for salt-free Lattice Q ($g\,cm^{-3}$)	1·147 ± 0·003	Green *et al.* (1956)
Intrinsic viscosity [η] ($ml\,g^{-1}$)	2·9	Townend *et al.* (1960*b*)
Partial specific volume ($cm^3\ g^{-1}$)	0·751	Swaisgood (1982)
Hydration (%)	35–40	Pantaloni (1965)
Axial ratio	2:1	Green & Aschaffenburg (1959)
Stokes radius (nm)	2·68	Swaisgood (1982)
Dipole moment (Debye)	730	Ferry & Oncley (1941)
Refractive Index (at 589 nm)	1·594	Swaisgood (1982)
R_G (nm)	2·17	Witz *et al.* (1964)
Extinction coefficient (litres $g^{-1}\ cm^{-1}$)	0·96	Townend *et al.* (1060*b*)
Charge at pH 6·6	− 10·0	Basch & Timasheff (1967)

ity is anomolous: there is a considerable binding of electrolytes resulting from the unique distribution of surface charge at neutral pH, which results in the large observed dipole moment of 730 Debye (Ferry & Oncley, 1941). Salting-out from solutions containing well in excess of 50 mg/ml protein is the standard way of growing crystals for X-ray studies (Aschaffenburg *et al.*, 1965).

6.2. Molecular Size

Ultracentrifugation studies by Pederson (1936*b*) found variation of the sedimentation coefficient with pH and, in sedimentation equilibrium studies, found the M_w to increase from 37 600 at pH 6·4 to 50 200 at pH 9·8. This indicated a slow aggregation not observed during his sedimentation velocity measurements. The molecular weight determined by osmotic pressure (Gutfreund, 1945; Bull & Currie, 1946), light scattering (Heller & Klevens, 1946; Halwer *et al.*, 1951), sedimentation and diffusion (Pederson 1936*b*; Cecil & Ogston, 1949), and X-ray diffraction (McMeekin & Warner, 1942; Green *et al.*, 1956) gave values around 35 600. However, Bull (1946*a*,*b*) found a value of 17 100 using a surface film technique, which further revealed a value of 34 300 when carried out in the presence of Cu^{2+}. Subsequently, SDS-PAGE in reducing conditions, and amino acid analysis provided values of about 18 500 (Piez *et al.*, 1961). Townend *et al.* (1960*b*) found a molecular weight of 18 000 at pH values below 3·5. These results are consistent with the normal form of ruminant BLG being a dimer of subunit molecular weight around 18 200, and this is entirely in keeping with the crystal structure work (Green & Aschaffenburg, 1959; Aschaffenburg *et al.*, 1965) and the value of 18 200 calculated from the sequence. However, in the decade following 1957, a large number of studies were performed on the association of BLG as a function of pH, protein concentration and temperature. These studies are discussed fully elsewhere (McKenzie, 1967, 1971; Swaisgood, 1982). Here only the main conclusions are reported and updated as necessary.

As has been mentioned, below pH 3·5 dissociation into monomers occurs with little change to the ORD (Timasheff *et al.*, 1966*a*), CD (Townend *et al.*, 1967; Teller *et al.*, 1979) or tryptophan fluorescence (Townend *et al.*, 1967). Timasheff and Townend (1961) examined the sedimentation and light scattering properties of the A and B variants and found dissociation constants at 15°C and pH 2·7 of $7{\cdot}5 \times 10^{-5}$ and $2{\cdot}4 \times 10^{-5}$ mol/litre, respectively. These values are not greatly affected by ionic strength (Townend *et al.*, 1960*a*, *b*) or genetic variant (Townend *et al.*, 1964) and correspond to a free energy of association of about −25 kJ/mol. In keeping with this, Mills and Creamer (1975) conclude that there is a rapid dissociation to monomer followed by a slow conformational change when the pH is dropped from 4·5 to 2·0, from both intrinsic and extrinsic fluorescence measurements.

Between pH 3·7 and 6·5, bovine BLG-A, and mixtures of A and B can reversibly octamerise to produce aggregates of 144 kDa (Townend & Timasheff, 1960; Timasheff & Townend, 1961; Pessen *et al.*, 1985). Neither tetrameric nor hexameric intermediates are present in significant amounts. The interactions within the octamer of BLG-A are maximal around pH 4·5 and near 0°C. The association constant at pH 4·4, 4·5°C is $2{\cdot}8 \times 10^{11}\,dm^3 mol^{-3}$, corresponding to a free energy of $-60{\cdot}8\,kJ\,mol^{-1}$. The packing of subunits to form a compact octamer with 422 symmetry (Timasheff & Townend, 1964) is accompanied by an increase in hydration which proton NMR estimates to be 5–6 extra water molecules/dimer. Aromatic residues are believed to not be involved as their UV spectra are unaltered by association. However, carboxyl groups have been implicated in this aggregation (Armstrong & McKenzie, 1967). Up to 30% modification of the carboxyls by carbodiimide has no effect on the ORD spectrum of the protein, but does reduce the protein's ability to octamerise. Two carboxyl residues likely to be involved are Asp-64, which is present in BLG-A but not BLG-B, the latter being unable to form an octamer alone, and Asp-28, which is the only amino acid substituted by Asn in BLG-Dr. Glycosylation at Asn-28 prevents octamerisation of BLG-Dr, presumably by steric hindrance since the location of residues 28 and 64 in the structure shows them to be 24 Å apart.

Ruminant BLGs are dimeric over the pH range 5 to 8, although there is evidence that a monomer–dimer equilibrium exists in cow and red deer BLGs at pH 6·5 (McDougall & Stewart, 1977). The decrease in sedimentation coefficient observed across this pH range for bovine BLG-A and BLG-B most probably corresponds to an increase in the dissociation of the dimer (Georges *et al.*, 1962; McKenzie *et al.*, 1967). Around pH 5·3, which is near the isoelectric point for all bovine variants, the free energy of association is about −40 kJ/mol (Timasheff & Townend, 1961;

Reithel & Kelly, 1971). Zimmerman *et al.* (1970) found values for the bovine A and B variants at 20°C of −26·3 and −28·8 kJ/mol, respectively, for the free energies of association at pH 6·9, in fair agreement with the values of −23·6 and −28·6 kJ/mol found at pH 7·5 by McKenzie *et al.* (1967) in ultracentrifugation studies and −23·6 kJ/mol by light scattering (Georges *et al.*, 1962) for variant B.

Over the pH range 8 to 9·5, slow time-dependent changes occur in BLG. At pH values above 8·5 reversible dissociation occurs (Georges & Guinand, 1960; Townend *et al.*, 1960*b*), and at above pH 9 the optical rotation, circular dichroism and solubility alter with time as the protein denatures irreversibly and aggregates (Groves *et al.*, 1951; Christensen, 1952; Herskovits *et al.*, 1964). The addition of a thiol-blocking group can inhibit aggregation, implicating thiol oxidation and/or thiol/disulphide exchange in the formation of the heavier components (Roels *et al.*, 1966).

Preliminary crystallographic studies on the dimeric form of bovine BLG have suggested that the interface between subunits may involve both hydrophobic interactions between Ile-29 and Ile-147, and the stacking of the symmetry-related His-146 residues (Papiz *et al.*, 1986).

Little, if any, information about the aggregation of the monomeric BLGs has been published.

7. Denaturation

The denaturation of bovine BLG involves the dissociation of dimer to monomer, a major change in the conformation of the polypeptide chain, and aggregation. The denaturant can be alkali, heat, organic compounds or heavy metal ions, and it is not clear how each of these acts to yield insoluble aggregates. Although it is improbable that there is a common mechanism throughout, the initial stages of unfolding by heat at neutral pH and by alkali do appear to follow a similar path (Casal *et al.*, 1988).

7.1. Alkali Denaturation

Alkali denaturation of bovine BLG becomes significant above pH 8 (Macheboeuf & Robert, 1952). As the pH rises from 8 to 9, the sedimentation coefficient decreases to a value consistent with a monomeric structure (Townend *et al.*, 1960*b*), and it remains at this value up to pH 10. Above this, further structural changes occur. ORD studies have shown that around pH 11 the α-helix remains intact whilst the β-sheet is converted to a random conformation (Timasheff *et al.*, 1966*a*). In complete contrast, infrared spectroscopy shows that it is the helix which unfolds first in the initial stage of alkali denaturation (Casal *et al.*, 1988). FTIR reveals that at pH 12 the structure is completely random (Purcell & Susi, 1984). These observations were obtained with freshly-made solutions since alkaline denaturation is time-dependent. The rate of denaturation rises rapidly with pH and leads eventually to the formation of large aggregates (Groves *et al.*, 1951; Hin Bon Hoa *et al.*, 1973).

7.2. Thermal Denaturation

Although the thermal denaturation of BLG is unlikely to be connected with its function *in vivo*, studies have been carried out which provide information about the environments of specific amino acids. Information about the interactions between BLG and other milk components, which are of interest to industry will be dealt with elsewhere.

Thermal denaturation of bovine BLG *in vitro* has suggested that upon increasing the temperature from 30 to 55°C, the dimer dissociates to monomer (Dupont, 1965; Sawyer, 1969). At higher temperatures, unfolding occurs concomitant with increased activity and oxidation of the thiol group (Larson & Jenness, 1952). Differential scanning calorimetry (DSC) reveals a denaturation peak at about 80°C, pH 6·5, which is pH dependent, and moves to about 60°C at pH 8 (de Wit & Klarenbeek, 1981). Park and Lund (1984), however, observed no such decrease over the same pH range under nearly the same experimental conditions, except that 50 mM buffer was present. A shift in the denaturation peak to around 60°C is also observed when the DSC experiment is carried out in the presence of β-mercaptoethanol. The two Trp residues become exposed, one fully and the other partially (Mills, 1976). It is likely that Trp-61, which is located on a flexible external loop, is the one that becomes fully exposed, whilst Trp-19, normally fully buried at the foot of the rigid β-barrel, becomes only partially exposed. The effect of pH on the conversion of the monomers to a denatured form has indicated that the two imidazoles present per subunit are involved (Dupont, 1965). As His-146 is close to the α-helix, this implies some destruction of the helical structure which is not inconsistent with the large change in optical rotation observed, although there is still some residual secondary structure after heat treatment at 90°C (Ananthanarayanan *et al.*, 1977). This is also consistent with the infrared spectroscopic results which show the initial unfolding of helical structure (Casal *et al.*, 1988). In contrast, lowering the pH has been shown to increase the thermal stability of the protein and it is suggested that extra hydrogen bonding is responsible (Kella & Kinsella, 1988).

The effect of heating BLG in the presence of other milk components has also been investigated. Studies *in vitro* have shown that lactose stabilises BLG against thermal denaturation (Park & Lund, 1984), by forming a browning complex which is believed to be antigenic (Otani *et al.*, 1985*a*). κ-Casein, on the other hand, destabilises BLG, the enhanced rate of its unfolding being entropy-driven, and indicative of hydrophobic residues becoming exposed (Park & Lund, 1984). The interaction between κ-casein and BLG is believed to involve the free thiol, the disulphide bridges and Ca^{2+} (Sawyer, 1969).

7.3. Organic Compounds

BLG can be denatured by compounds like urea, guanidinium hydrochloride, alcohol and detergents. Urea denaturation is believed to occur via a two-step process. At pH 3·5, where BLG tends to be monomeric, the first step is the first-order, reversible unfolding of the polypeptide chain to a random-coiled conformation (Greene & Pace, 1974). However, at pH 5·2, this step follows more complex kinetics which, upon addition of NEM, which dissociates dimeric BLG, revert to first-order (McKenzie & Ralston, 1971, 1973). This indicates that the first stage involves both dissociation and unfolding. The second stage involves the unfolded protein undergoing –SH/SS exchange reactions (and possibly –SH oxidation), to yield irreversible aggregates. Creighton (1980) has used urea-gradient electrophoresis to study the kinetics of the refolding of urea-denatured BLG, and Alexander and Pace (1971) have shown that the midpoint concentrations for the urea denaturation of bovine A, B and goat BLG at 25°C and pH 3 are 5·3, 5·19 and 4·68 M. These values show that BLG-A is 1 kJ/mol more stable than BLG-B, and 4 kJ/mol more stable than goat.

The effect of ethanol on the structure of bovine BLG has also been examined. Ethanol unfolds the protein at pH 3 in a similar manner to urea (Tanford & De, 1961; Dufour & Haertlé, 1990) although unlike urea, the reaction is not entirely reversible with the unfolded protein refolding to a more α-helical conformation. Timasheff *et al.* (1966*a*) observed a similar increase in helix content in MeOH. It has also been observed that the optical rotation does not appear to vary for concentrations of EtOH less than 10%, although Hemley *et al.* (1979) advise against using more than 5% EtOH when wishing to study the native conformation of BLG in an ethanolic solution.

Sodium dodecyl sulphate (SDS) can also affect the conformation of bovine BLG, but only at high concentrations. BLG binds one mole of the dodecyl sulphate anion per subunit, without any change in the secondary or quaternary structure (Seibles, 1969). However, at high concentrations, an 'all-or-none' complex forms with 22 moles of SDS adsorbed onto the protein surface (Jones & Wilkinson, 1976; Hilquist *et al.*, 1982), but with negligible change to the secondary structure. Continued addition of SDS causes further associations and a change in secondary structure as the amount of α-helical structure increases. The CD spectrum in the presence of a 50-fold molar excess does not alter appreciably in the near UV region but the far UV spectrum is considerably altered, especially at low pH (Su & Jirgensons, 1977). Urea denaturation has been used by Cupo and Pace (1983) to monitor the effect of small changes to the environment of the free thiol at pH 2·8; adding —SCH_2CH_2X, where X = —COOH, —CH_3, —OH, —NH_2, progressively destabilises the native structure by amounts varying from 4·5 to 30 kJ mol^{-1} with respect to the native protein.

7.4. Heavy Metals

The addition of heavy metal compounds to BLG in solution can induce an irreversible conformational change, which is prevented if the protein is in crystalline form. The crystalline array clearly stabilises the native conformation (Dunnill *et al.*, 1966; Green *et al.*, 1979), enabling the isomorphous heavy atom derivatives necessary for X-ray crystallographic studies to be obtained by soaking protein crystals with heavy metal compounds. However, in some cases, even in the solid phase, conformational changes appear to occur (Sawyer & Green, 1979).

The denaturation of BLG by heavy metals above pH 6 has been studied using mercurial compounds, silver(I) and copper salts. Ag(I) and some Hg(II) salts, for example, *p*-chloromercuribenzoate (PCMB), cause an increase in laevorotation up to a ratio of 2 moles salt per dimer BLG (Pantaloni, 1965) by binding to the free thiol and enhancing dissociation. Hg(II) acetate also shows similar behaviour, although at pH 6·8 the binding of only one mole of Hg(II) per dimer of BLG also shows an increase in laevorotation consistent with a mercurial bridge between the free thiols of each subunit (Lontie & Preaux, 1966). For this to occur, some rearrangement of the protein is necessary as the thiol groups are about 25 Å apart in the crystallised native protein (Green *et al.*, 1979). The binding of one Hg(II) per dimer has been studied (Pantaloni, 1964): its reaction is first-order, specific for the free thiol, and involves a Trp, although no direct interaction between these residues has been proved. The positions of the Trp residues in the crystal structure are 9 and 31 Å from Cys-121. The denaturation of BLG by Cu(II) occurs by two consecutive, first-order reactions. The first involves the binding of one Cu(II) per subunit (Pantaloni, 1965), possibly at an amide or anomalous carboxyl site, whilst the second stage involves a conformational change of the complex (Pantaloni, 1962). The subsequent loss of copper, and the low concentrations used, suggest that it might in some way enhance the Tanford transition.

8. Secondary Structure

The secondary structure of BLG in solution has been studied by a variety of techniques. Raman spectroscopy showed a strong band at 1242 cm^{-1} corresponding to *β*-sheet and random coil conformations, and a weak band at 938 cm^{-1} from which it was deduced that the protein possessed about 10% α-helix (Frushour & Koenig, 1975). ORD studies have confirmed an α-helical content of about 10%, and suggested that the remaining 90% was divided evenly between *β*-sheet and random coil (Timasheff *et al.*, 1966*a*), whilst CD experiments gave values for each conformation of 10% α-helix, 43% anti-parallel *β*-sheet and 47% unordered (Townend *et al.*, 1967). These values agree with data obtained from IR (Timasheff & Susi, 1966; Rüegg *et al.*, 1975; Casal *et al.*, 1988), FTIR (Byler & Susi, 1986), from some secondary structure prediction algorithms (Deckmyn & Preaux, 1978; Creamer *et al.*, 1983), and from the 2·8 Å resolution crystal structure, which revealed 51% anti-parallel *β*-strands and 7% α-helix (Papiz *et al.*, 1986). Byler *et al.* (1983) have reported a laser-Raman study of the cystine groups in the 500–700 cm^{-1} spectral region that reveals that both disulphides adopt a *gauche-gauche–gauche* conformation. As had been noted above, extremes of pH and modification of the properties of the solvent alter these values markedly.

9. Conformational Changes

Circular dichroism, and the related technique of optical rotatory dispersion, provide an excellent means of monitoring conformational change. ORD, or even

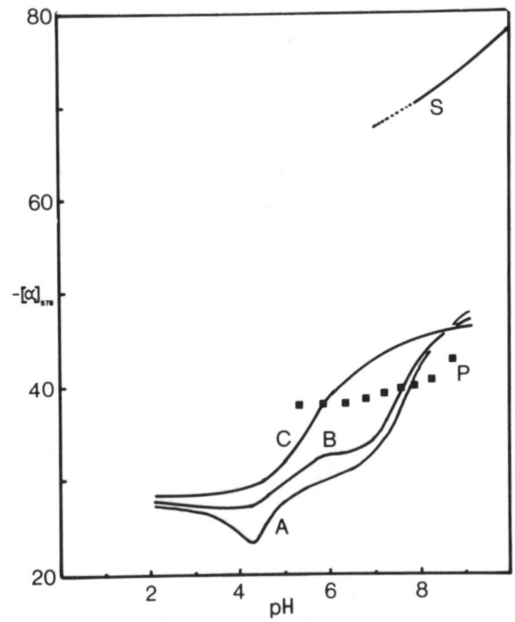

Fig. 1. Specific rotation as a function of pH showing the conformational changes for bovine β-lactoglobulin variants A, B and C (data from McKenzie & Sawyer, 1967) and for pig (■—P, S.G. Hambling, unpublished results). The curve S corresponds to the denatured state (Schellman, 1958).

optical rotation, and proton-binding experiments (titration curves) have shown that bovine BLG undergoes three pH-dependent conformational transitions which can be summarised as:

$$Q \leftrightarrow N \leftrightarrow R \rightarrow S$$

Figure 1 shows the curves obtained by McKenzie and Sawyer (1967), together with the curve from the authors' laboratory for pig BLG, and the values typical of denatured protein.

The first transition (Q ↔ N) between pH 4 and 6 is reversible, and has been described for the bovine BLG variants A, B and C (Timasheff *et al.*, 1966*b*; McKenzie & Sawyer, 1967). The increase in sedimentation coefficient observed, with increasing pH, correlates with the contraction of the protein. BLG-A undergoes a two-group ionisation whereas BLG-B follows a single proton transition, suggesting that the additional carboxyl in BLG-A, Asp-64, may be one of the groups being ionised. Also, the pH at which octamers form most readily with BLG-A corresponds to a significant dip in the curve for that variant alone. BLG-C also follows a single proton transition, but unlike BLG-B, one extra cationic residue per subunit, possibly the additional His-59, is exposed upon increasing the pH. From ORD studies, Timasheff *et al.* (1966*a*) report that no aromatic residues are involved in this transition, and Townend *et al.* (1969) also came to this conclusion using solvent perturbation difference spectroscopy. The latter is slightly surprising as His -59, Trp-61 and Asp-64 are located in the same region of the protein. No change is detected over this region in the IR spectrum (Casal *et al.*, 1988).

Between pH 6·5 and 7·8, the second reversible conformational change (N↔R), often called the Tanford transition, of observed (Tanford *et al.*, 1959). In bovine BLG, this transition can be detected by a change in optical rotation ($[\alpha]_D$ of $-25°$ at pH 6, but $-48°$ at pH 8), by a decrease in the sedimentation coefficient (3·2 to 2·6 *s*) or possibly even by a change in the position of the thermal denaturation peak observed by DSC (de Wit & Klarenbeek, 1981). The change in sedimentation coefficient may be due to an expansion in the volume of the protein, a variation in the shape of the protein (Tanford *et al.*, 1959, 1960; Timasheff *et al.*, 1966 *b*) or the increased dissociation of dimer to monomer (Georges *et al.*, 1962; Pantaloni, 1965; McKenzie & Sawyer, 1967).

Which amino acids are responsible for this conformational change is still unclear, although the effect of this transition on some residues is known. Upon increasing the pH, one buried carboxyl per subunit becomes exposed and ionised, with a positive enthalpy, which suggests that the buried carboxyl was originally hydrogen-bonded (Tanford & Taggart, 1961). The anomolous carboxyl is a consequence of the native structure since no such observation is made when titrating urea-denatured BLG. The pK_a of this anomalous carboxyl is around 7·3, similar to that of a His. Goat BLG also has an anomolous carboxyl group (Ghose *et al.*, 1968). Cu(II), which is known to alter the Tanford transition (Pantaloni, 1962, 1965), can bind to both carboxyl and imidazole groups, thus making it unclear whether His is involved. Absorption spectroscopy has shown that a Tyr is involved —it is believed that a hindered Tyr becomes partially exposed (Pantaloni, 1965). The free sulphydryl also becomes more accessible when the pH is raised from 6 to 8, as demonstrated by an increase in the second-order rate constants for the binding of pCMB (Lyster, 1964; Dunnill & Green, 1965) or tetracyanoaurate(III) (Sawyer & Green, 1979) to bovine BLG. The Tanford transition is dramatic when observed by optical rotation but much less so by other techniques. In particular, a large change in the fold in the polypeptide is not revealed, by measurement change at 222 nm (Su & Jirgensons, 1977), or by the crystal structures at pH 6·5 and 7·6. Rather, a subtle rearrangement of side chains is observed (Yewdall, 1988). Further, it is not yet clear that either change is important for the function of the protein although the pH range in which they occur coincides with that in which the protein can be found *in vivo*.

The third, irreversible, conformational change is the alkali denaturation of BLG, which has been discussed previously.

10. Amino Acid Environments

A large number of protein chemical investigations have been carried out to determine the environment of specific amino acids within the protein, many of which were summarised by Townend *et al.* (1969). These studies have mostly been performed with bovine BLG although a few have used the caprine or ovine protein. Within the ruminant proteins, however, the results appear to be broadly similar. Far less is known about the monomeric BLGs.

An obvious target for modification is the free cysteine, shown to be Cys-121 by both crystallographic studies (Papiz *et al.*, 1986; Monaco *et al.*, 1987; Yewdall, 1988) and a series of enzymatic digestion experiments (Mainferme *et al.*, 1971; Martial *et al.*, 1971; Perez-Gomez *et al.*, 1971). However, the possibility of disulphide interchange between positions Cys-119 and Cys-121 (McKenzie & Shaw, 1972; McKenzie *et al.*, 1972) has not been eliminated. The environment of the free thiol has been investigated using a variety of compounds. The rate of binding of pCMB increases above pH 6 by a factor greater than can be accounted for by the ionisation of the –SH, indicating that the thiol becomes more exposed at higher pH (Dunnill & Green, 1965). It is known that when NEM binds to the free sulphydryl, the protein is completely dissociated into subunits (Lontie & Preaux, 1966), implying a location near the site of association. However, dissocation upon sulphydryl modification is by no means invariable (Cunningham & Nuenke, 1960; Townend *et al.*, 1969; Stone & Wishnia, 1978; Cupo & Pace, 1983). Thus,whether the pH-dependent reactivity of the Cys residue results from the N $\leftrightarrow$ R conformational change or dissociation (or, indeed, both) is still unclear. From a comparison of the dimensions of the different substituents, Townend *et al.* (1969) suggested that the free thiol and dimer association site must be at least 6 Å apart, and that the contact area for octamer formation must be in the same vicinity. Examination of the relative positions of the free thiol and the region of subunit association in the crystal structure shows that this is correct: the helix at the top of the molecule runs between Cys-121 and the molecular dyad axis.

There are four tyrosine residues per monomer of bovine BLG-A and their accessibility has been investigated by reactions with N-acetylimidazole and cyanuric fluoride (Gorbunoff, 1967; Townend *et al.*, 1969): two of the tyrosines are readily available at pH 9·3 where one can be acetylated with no change in the CD spectrum; the third is partially hindered and only becomes reactive at pH 10·8 after some conformational change of the protein; the fourth is unavailable, titrating with an apparent pK_a of 12·3 (Tanford *et al.*, 1959). In the case of the goat protein, the reactivity of the third tyrosine is slightly greater. In both cases, these residues are not involved in subunit association, although one or more are affected by the Tanford transition (Pantaloni, 1965).

Bovine BLG contains two tryptophans per monomer, Trp-19 and Trp-61. Both solvent perturbation spectroscopy (Townend *et al.*, 1969) and Trp fluorescence studies (Mills, 1976) have revealed that about half of these residues are buried, but neither distinguished between one buried and one exposed residue, or both partially buried. The structure reveals that Trp-19 is buried and Trp-61 is in an undefined region which is most probably mobile. In contrast, the modification of BLG by 2-hydroxy-5-nitrobenzyl bromide (HNBB) at pH 2, where the protein is a monomer, revealed that both Trp residues were accessible. Further, the presence of retinol had no effect on the Trp modification although the converse is not true

(Fugate & Song, 1980). These workers also found that neither Trp is affected over the pH range 2–7·5. Townend *et al.* (1969) have excluded the involvement of either Trp residue in the association/dissociation of the dimer or the conformational transition between pH 4 and 6. As electron transfer between Trp and Tyr amino acids is predominantly initiated by the electron-deficient Trp-61, and its rate is independent of pH over the range 6·1–7·9, this is in keeping with neither Trp residue being involved in the Tanford transition (Prutz *et al.*, 1980).

Phosphorylation of BLG by $POCl_3$ in CCl_4 produced no phosphoserines, the major species being N-phospholysine and/or N-phosphohistidine (Woo *et al.*, 1982). The extent of phosphorylation increases with pH, but never exceeds a stoichiometry of 14 mol phosphate/mol BLG. This is less than the total number of Lys and His sites (15 + 2), indicating that some are not fully exposed. Most of the Lys residues in bovine BLG are accessible for reductive alkylation although those in the α-helix (Lys-139, -138, -141) are only partially modified, and Lys-47 remains unaltered (Brown *et al.*, 1988). Cleavage of the penultimate His and C-terminal Ile of bovine BLGs has been carried out using carboxypeptidase A (Greenberg & Kalan, 1965).

The carboxyl groups in BLG have also been studied. Treatment of bovine BLG with carbodiimide has shown that modifying 7–8 of the carboxyls prevents octamerisation of BLG-A (Armstrong & McKenzie, 1967). Esterification with MeOH, EtOH and BuOH affects 23, 19 and 12 carboxyls, respectively, out of the 25 per monomer (Halpin & Richardson, 1985). Of the two hindered carboxyls not modified by MeOH, one is likely to be the anomalous carboxyl associated with the Tanford transition. Mattarella and Richardson (1983) have shown that the carboxyl modification leads to significant changes in the character of the protein. A detailed comparison of the crystal structures at either side of this transition will enable the anomalous carboxyl to be identified unambigously.

11. Binding Studies

A wide variety of compounds is known to bind to BLG, some of which are listed together with their association constants in Table 9. However, it is still unclear which regions of the protein are involved, and whether any of the ligands which bind are important for the biological function of this protein.

11.1. Aliphatic Molecules

Probably the first binding was shown by the surface area measurements of Bull (1946*b*) who found 8 molecules of sodium lauryl sulphate (SLS) associated with the BLG dimer. Since continued addition of SLS led to a sharp change at 40 mol/dimer, Bull considered the effect to be one of titration rather than association. It is not surprising, therefore, that SDS has since been shown to bind to bovine BLG. The first stage involves the tight binding of two anions/dimer to yield a complex that has been crystallised (McMeekin *et al.*, 1949), but not yet studied by X-ray

TABLE 9
Parameters for Ligand Binding to Bovine BLG

Ligand	*Number bound per dimer*	*Association constant* $(M^{-1})^a$	*Reference*
Retinol	2	5×10^7	Fugate & Song (1980)
Stearate	2	$1{\cdot}7 \times 10^5$	Spector & Fletcher (1970)
Palmitate	2	$6{\cdot}8 \times 10^5$	Spector & Fletcher (1970)
Laurate	2	$0{\cdot}5 \times 10^5$	Spector & Fletcher (1970)
Oleate	2	$0{\cdot}4 \times 10^5$	Spector & Fletcher (1970)
Heptane	2	$0{\cdot}48 \times 10^6$	Mohammedzadah-K. *et al.* (1969)
Butane	2	$1{\cdot}7 \times 10^3$ $(5{\cdot}8 \times 10^2)$	Wishnia & Pinder (1966)
Pentane	2	$7{\cdot}1 \times 10^3$ $(6{\cdot}2 \times 10^3)$	Wishnia & Pinder (1966)
Iodobutane	2	2.8×10^3	Wishnia & Pinder (1966)
SDS	2	$3{\cdot}1 \times 10^5$	O'Neill & Kinsella (1987)
2,6-MANS[b]	2	$3{\cdot}4 \times 10^5$	Lovrien & Anderson (1969)
Methyl orange	2	$0{\cdot}2 \times 10^4$	O'Neill & Kinsella (1987)
n-Octylbenzene-*p*-sulphonate	3	$6{\cdot}3 \times 10^4$	O'Neill & Kinsella (1987)
p-Nitrophenol	2	$1{\cdot}9 \times 10^4$	Farrell *et al.* (1987)
p-Nitrophenylacetate	2	$3{\cdot}1 \times 10^4$	Farrell *et al.* (1987)
p-Nitrophenyl-β-glucuronide	2	$1{\cdot}6 \times 10^4$	Farrell *et al.* (1987)
p-Nitrophenyl sulphate	2	$2{\cdot}0 \times 10^3$	Farrell *et al.* (1987)
p-Nitrophenyl pyridoxal phosphate	2	$3{\cdot}1 \times 10^3$	Farrell *et al.* (1987)
2-Heptanone	2	$0{\cdot}15 \times 10^3$	O'Neill & Kinsella (1987)
2-Octanone	2	$0{\cdot}50 \times 10^3$	O'Neill & Kinsella (1987)
2-Nonanone	2	$2{\cdot}44 \times 10^3$	O'Neill & Kinsella (1987)
Toluene		$4{\cdot}5 \times 10^2$ $(5{\cdot}9 \times 10^1)$	Robillard & Wishnia (1972*a*)
Trifluorotoluene		$4{\cdot}2 \times 10^2$ $(3{\cdot}1 \times 10^1)$	Robillard & Wishnia (1972*a*,*b*)
Hexafluorobenzene		$1{\cdot}6 \times 10^3$	Robillard & Wishnia (1972*a*,*b*)

[a] Figures in parentheses represent the association constant for the binding of a second ligand at a second site.
[b] 2,6-MANS = *N*-methyl-2-anilino-6-naphthelene sulphonate.

crystallography. Chemical modifications of BLG before complexation revealed that neither Trp, Cys nor His-161 were essential for binding, but did show that the hindered His, His-146, was important for the formation of the BLG + SDS complex (Seibles, 1969). Higher concentrations of SDS caused a conformational change associated with the binding of 22 anions. This complex is in equilibrium with that to which two SDS anions are bound (Hillquist *et al.*, 1982; see above).

Free fatty acids (FFA) can also bind to bovine BLG in a manner similar to SDS. They reversibly bind at a single strong binding site per subunit with an association constant of about $10^5\,\text{M}^{-1}$, inducing a conformational change. A further binding at up to 24 weaker secondary sites, presumably similar to those for SDS, has been reported (Spector & Fletcher, 1970). Binding at the primary site depends upon the nature of the ligand, decreasing in the order: palmitic(C16:0) > stearic (C18:0) > oleic(C18:1) > lauric(C12:0). Further studies using palmitate and bovine BLG have indicated that the binding affinity increases over the pH range 6·5–8·5, suggesting a possible correlation with the Tanford transition. Spin-labelled stearate derivatives, e.g. 12-doxyl stearic acid, have been studied by ESR and show some weak binding to BLG but do not form 1:1 complexes (Keith, E.O., 1987, pers. comm.). This suggests that the alkyl chain binding site has little spare capacity. Interestingly, the FFAs with C > 14 which bind to BLG, are those most prevalent in milk. Whether this is coincidence, or because BLG has a role as a general FFA carrier in the lipid metabolism of the mammary gland is still unclear (Diaz de Villegas *et al.*, 1987; Puyol *et al.*, 1991).

Triglycerides bind to BLG (Smith *et al.*, 1983), but no details of the binding are given. Phosphatidylcholines appear unable to bind to native BLG (Brown *et al.*, 1983) but in the presence of a helix-inducing solvent, BLG can refold to a structure containing about 50% α-helix, which is then able to bind dipalmitoyl phosphatidylcholine. The vesicles formed on sonication of the complex revealed that the ratio of lipid : BLG was 20 : 1, in keeping with both the SDS and FFA complexes. Further, CD revealed that BLG now contained 30% α-helix. BLG in this form also exhibited a 10% increase in Trp fluorescence, compared to 8% when a FFA is bound, although the hydrophobicity of the residues remained unaltered, as indicated by a reduction in quenching (Brown *et al.*, 1983). The ability of partially unfolded BLG to bind to lipids may be of importance for its transport across a membrane. More recently, Cornell and Patterson (1989) showed that BLG can adsorb to an artificial phospholipid membrane surface at pH 4·4 though not at pH 7·0. CD measurements showed no change in conformation on adsorption so that an electrostatic mechanism has been proposed. This is consistent with the large dipole moment of the molecule and also with the adsorption to colloidal gold (Horisberger & Vauthey, 1984).

Many alkanes bind to BLG, the amount bound decreasing as the chain length increases (Mohammadzadeh-K. *et al.*, 1969). Butane, pentane and iodobutane can all bind to a single hydrophobic region per subunit of bovine BLG (Wishnia & Pinder, 1966). The site, believed to be a hydrophobic pocket going from the surface towards the interior of the protein, is unaffected by the state of association of the protein. This pocket can bind either two butanes equally, two pentanes unequally, possibly for steric reasons or because of exposure of a hydrocarbon tail, or one iodobutane. The position of the latter, soaked into BLG crystals and determined by X-ray crystallography, locates this alkane-binding site inside the hydrophobic cavity (Papiz, 1982).

Aliphatic ketones, e.g. 2-heptanone, also bind at a single hydrophobic site in each subunit of bovine BLG, with association constants of the order of 10^3M^{-1}. As these are of the same magnitude as those for alkane binding, these ketones probably occupy the alkane binding cavity (O'Neill & Kinsella, 1987).

11.2. Conjugated Compounds

Both fused and single-ring aromatic compounds are known to bind to BLG. The binding of N-methyl-2-anilino-6-naphthalenesulphonate at pH 8 has been studied by calorimetry, equilibrium dialysis and fluorescence spectrometry, and reveals that bovine BLG-B possesses two anionic binding sites, one of which binds more strongly than the other (Lovrien & Anderson, 1969). Both of these sites are hydrophobic and one is likely to correspond to that above, although as has been seen the alkyl binding site is apparently restricted in its capacity. The rather feeble binding to a single site at pH 6 found by Lovrien and Anderson (1969) is probably that for 8-anilino-naphthalene-1-sulphonate examined by Mills and Creamer (1975) between pH 2 and 6·5.

The hydrolysis of *p*-nitrophenyl phosphate (PNPP) by bovine spleen phosphoprotein phosphatase was partially inhibited by BLG, suggesting that BLG could bind PNPP (Thompson & Farrell, 1974). The binding of several *p*-nitrophenyl compounds to bovine BLG-A was confirmed by Trp fluorescence studies (Farrell *et al.*, 1987), implicating Trp in the binding. In addition, the dissociation constant for the 1 : 1 complex of bovine BLG and PNPP was independent of pH over the range 4 to 7·5, indicating little dependence on conformational state.

The binding of toluene, trifluorotoluene and hexafluorobenzene to bovine BLG has been investigated, and reveals two distinct forms of binding: strong association at a single hydrophobic site per subunit with an accessible volume of around 220 ml/mol, which is believed to be the alkane binding site, and weaker interactions at one or more other sites. Neither type of association is affected by dimerisation of the protein (Robillard & Wishnia, 1972*a,b*), although the octamerisation of BLG-A is abolished.

By what may have been a chance experiment, Futterman and Heller (1972) showed that BLG enhanced the fluorescence lifetime of retinol from 3 to 10 ns and reduced it susceptability to oxidation. UV absorption spectroscopy has shown that BLG and retinol form a complex (Hemley *et al.*, 1979), whilst CD has revealed that one retinol is bound per subunit, with a dissociation constant of 0·02μM (Fugate & Song, 1980) compared with 0·2μM for the retinol–RBP complex. The location and nature of the retinol binding site in BLG is uncertain. Modification of the Trp residues with 2-hydroxy-5-nitrobenzyl bromide (HNBB) caused a blue shift in the absorption spectrum of the BLG–retinol complex to that of free retinol (Fugate & Song, 1980), implying that bonding interactions have been destroyed, and implicating a Trp in the binding site. Surprisingly, retinol still binds to the protein in 8M urea or SDS. The binding is independent of pH in the range 2–7·5, suggesting that either a non-ionisable group is involved or the binding site is hydrophobic. Since HNBB treatment can also affect free sulphydryls, involvement of Cys-121 has not been ruled out. Crystallographic studies of the trigonal lattice Z, grown at pH 7·8 from a solution containing bovine BLG and excess retinol, have shown that retinol apparently binds in an external hydrophobic channel, adjacent to the α-helix and Cys-121, and not in the hydrophobic cavity near Trp-19 (Monaco *et al.*, 1987).

11.3. Ions

There is evidence for the binding of both monovalent and divalent cations to bovine BLG (Baumy & Brule, 1988). Na^+ is able to bind to carboxyls and imidazoles in this protein above its pI value, the number of ions bound per subunit increasing from 0 to 2 as the pH is raised from 6 to 9 (Baker & Saroff, 1965). The binding curve is of a similar shape to that observed by optical rotation over the Tanford transition, although slightly displaced to higher pH. Various heavy metal cations have also been shown to bind Ag^+ (Pantaloni, 1965), Cu^{2+} (Pantaloni, 1962) and mercurial salts (Pantaloni, 1964), and these can form complexes with one cation bound per free sulphydryl. However, above pH 6·8 Hg(II) acetate can also form a complex in which there is only one cation bound per dimer of BLG. It is believed that the Hg^{2+} forms a bridge between the free thiol in each subunit (Lonti & Preaux, 1966), although for this to be the case considerable rearrangement and/or disulphide interchange of the protein must occur since the free thiols are about 25Å apart in the native structure (Green *et al.*, 1979).

One type of anion which is known to bind to BLG is the iodide ion. I_3^- reacts specifically with the free thiols of bovine BLG, producing a sulphenyl iodide complex. This is less stable than the native protein and is able to react with different mercaptans to yield derivatives with varying abilities to octamerise (Cunningham & Nuenke, 1960). Stone and Wishnia (1978) have shown that there is a single site for binding HgI_3^- on BLG when the free thiol group is blocked with mercaptoethanol. This site most probably is that identified by Papiz *et al.* (1986) as the major site for the 'HgI_4^{2-}' derivative, close to the molecular dyad axis. This site is not that for alkane binding either, though when native BLG is used, the binding of mercuriiodide complexes in solution is less straightforward.

These binding studies indicate that BLG has the ability to bind hydrophobic or amphipathic molecules to at least two distinct sites. Neither site interferes with or is affected by the monomer or dimer state of the protein, but higher association is interrupted by one of them. Binding at one site involves changes in the sulphydryl environment which may lead to dissociation whereas binding at the other site appears in some cases to involve a tryptophan residue. Surprisingly, denaturation does not appear to abolish the binding of some ligands. Examination of the native structures reveals a deep, calyx-like cavity which by analogy with the homologous retinol-binding protein, is able to accommodate retinol but which is quite restricted in its lateral dimensions. In both structures, a Trp residue is at the foot of the calyx.

Further, the horse-shoe shaped bilin molecule binds at the entrance to this cavity in insecticyanin (Holden *et al.*, 1987) and bilin-binding protein (Huber *et al.*, 1987), both of whose structures show the same polypeptide fold as BLG. Tempting though it is to claim that this internal site is that for vitamin A, electron density not inconsistent with vitamin A has been found on the outer surface of the calyx, close to the helix, in one of the high pH crystal forms co-crystallised with the ligand (Monaco *et al.*, 1987). This site also appears to be hydrophobic in nature and is less restricted since it is a crevice on the outer surface of the molecule. Crystal soaking experiments are hampered by the instability of the crystals in solvents which can dissolve ligands like vitamin A or alkanes so that it has proved difficult to obtain convincing direct evidence on the detailed nature of the binding sites.

12. Macromolecule Binding

12.1. Non-milk Proteins

Cytochrome-c has been reported to interact with BLG. The initial binding to subunits of bovine BLG produces a 1 : 1 complex. If the pH is above 7·5, the conformation of BLG in the complex is altered, thus enabling the free sulphydryl of BLG to reduce the iron in cytochrome-c (Brown & Farrell, 1978). A non-reduced complex has also been observed between cytochrome-c and pig BLG since the latter has no free thiol. It is unclear at present if it is the N ↔ R conformational change that leads to this interaction, or which amino acids are involved. BLGs A, B, C all form a similar complex with cytochrome-c, although the rate of iron reduction depends on the variant. Other proteins have also been shown to associate with BLG, mostly through inhibition of enzyme activity. For instance, the inhibition of human renin by bovine BLG-B has been reported (Workman *et al.*, 1974). It is non-competitive, and requires a leucine residue near a hydrophobic region in BLG. Similarly, Farrell and Thompson (1971) have investigated the effects of BLG on phenylphosphate hydrolysis by bovine spleen phosphoprotein phosphatase as mentioned above.

12.2. Milk Proteins

There are numerous studies of the interactions between milk proteins, many resulting from the milk processing and food industries, which are outwith the scope of this review. Bovine BLG is known to interact with α-lactalbumin (Hunziker & Tarassuk, 1965), and several of the caseins. For example, it can react with κ-casein to form a 3 : 1 complex involving hydrophobic interactions. This complex is subsequently stabilised by covalent bonding, and a conformational change which makes the disulphide bridges less susceptible to attack (Haque *et al.*, 1987), possibly therefore as a result of disulphide interchange.

12.3. Immunoglobulins

Antibodies raised against bovine BLG-B cross-react with ruminant BLGs to different extents. This appears to be dependent upon the immunochemical method used to detect the cross-reactivity as well as the species from which the BLG originates (Jenness *et al.*, 1967). It is therefore possible that the few amino acid substitutions between ruminant BLGs occur in the antigenic regions of the protein. Cross-reaction with antisera against bovine BLG and monomeric BLG from pig, horse (Liberatori *et al.*, 1979*b*) or camel (Liberatori *et al.*, 1979*a*) is low, indicating some similarity in the two types of protein, though with the possibility that the antigenic region(s) involve both subunits.

Antibodies towards bovine BLG have been shown to cross-react with human milk, although the presence of bovine BLG in human milk is most likely to be a transfer of the protein or at least antigenic peptides, from the mother's diet (Axelsson *et al.*, 1986). Indeed, on removal of milk proteins from the mother's diet the levels fall to non-detectable amounts (Jakobsson *et al.*, 1985). Similarly, antibody studies have revealed the presence of bovine BLG in cows' urine after cessation of milking for prolonged periods (Lyster & Wheelock, 1967) and in the urine of new-born suckled calves (Pierce, 1960).

As well as using antibodies to bovine BLG to detect the presence of BLG in milks and other fluids, they are also useful in monitoring the presence of BLG in the gastro-intestinal (GI) tract, and for identifying the allergenic and antigenic sites of BLG.

Koritz *et al.* (1987) used antibodies towards bovine BLG, raised within a rabbit, to show that the concentration of BLG decreases upon its progression down the GI tract owing to its increased exposure to proteinases and intestinal absorption. Levels remain high enough to detect up to two days after removal of BLG from the diet, after which the levels drop (Suzuki *et al.*, 1987). These antibodies were tentatively identified as secretory IgA (sIgA) and have also been detected at low levels of BLG in the saliva of infants, although their origin is unclear (Frick & Rieger, 1987).

Increasing the amount of cows' milk ingested has been shown to cause an increase in the amount of IgG in serum (Fallstrom *et al.*, 1978). Challenges with isolated milk proteins confirm that BLG is often the offending agent in infants with an allergy to cows' milk (Lebenthal *et al.*, 1970). It is thought that the immature nature of the neonate intestinal surface enables BLG to be absorbed, raising IgG levels and complexing with them (Robertson *et al.*, 1982). Why only some infants are susceptible is unclear, although the nature of the ingested BLG may be important. Huang *et al.* (1985) have investigated tryptic and cyanogen bromide fragments for both allergenicity and antigenicity. A monoclonal IgE towards bovine BLG showed reactivity with both native and aggregated forms of the protein; however, only the latter induced positive cutaneous anaphylaxis (Granato & Piguet, 1986).

Ingestion by guinea-pigs of bovine BLGs A and B leads to higher levels of antibody being raised towards the A variant. However, injection into sensitised guinea-pigs induces anaphylactic shock when challenged with either variant, indicating the same allergenic site in both variants distinct from the antigenic sites which bind to IgG. It is possible that the antigenic site involves one or both of the two substituted amino acids, Asp-64 and Val-118, in the A variant (Malik *et al.*, 1988). IgE and IgG antibody binding, to reduced and carboxymethylated BLG, indicates that the disulphide bridges are unimportant in allergenicity, though important in the antigenic region (Otani & Hosono, 1987).

Various fragments and derivatives of bovine BLG have been exposed to IgE and IgG antibodies raised against native BLG, to try to locate the antigenic sites. Tryptic, chymotryptic and peptic digests of BLG show no cross-reaction with IgE raised towards native protein, suggesting that the antigenic sites are discontinuous (Kurisaki *et al.*, 1982). Kurisaki *et al.* (1982) also indicated the importance of the disulphide bridges but noted that modifications of the free thiol, either Trp, two of the three Arg residues or the amino groups did not inhibit cross-reaction on incubation with IgE. Four antigenic sites have been proposed for native BLG which are present at both pH 6 and 8·5. This indicates that neither the Tanford transition, nor the monomer–dimer state of association significantly affects the antigenic nature of the protein (Kurisaki *et al.*, 1985). Identical modifications to those performed by Kurisaki *et al.* (1982) gave similar results with IgG antibodies except that Trp and amino group modification did affect the IgG interaction. This may indicate the importance of these sites in antigenicity but not allergenicity. Additional modifications have also implicated histidine, carboxyl and possibly tyrosine sidechains in the antigenic sites. However, CD measurements on chemically modified BLGs indicate that the structures may no longer be in their native conformation (Otani *et al.*, 1985*b*).

Some recent work may give a better indication of the position of an antigenic region within bovine BLG. Monti *et al.* (1989) showed a reaction between antibodies towards bovine BLG and milk from humans fed on a milk-free diet. Recognition was attributed to a fragment at the N-terminus of human lactoferrin. Note that the cross-reactivity of lactoferrin with an anti-bovine BLG antibody is an alternative explanation for the reports of a human BLG. Further, sequence comparisons between β_2-microglobulin and BLG, initiated because of cross-reactivity with some antibodies to BLG, implicate a region between residues 121 and 140 in BLG as being responsible (Conti & Godovac-Zimmerman, 1990). This is identical to that proposed by Monti *et al.* (1989) and corresponds to the main stretch of α-helix.

13. Crystallographic Studies

Many of the methods used to isolate BLG from milk produce an impure protein, which can be purified by recrystallisation. Microcrystals of BLG have been obtained during the preparations of BLG from the milks of ruminants—cow (Palmer, 1934; Aschaffenburg & Drewry, 1957), goat (Askonas, 1954) and sheep (Maubois *et al.*, 1965)—and of non-ruminants such as pigs (Kalan *et al.*,

TABLE 10
Results of Crystallographic Studies on Three Crystal Forms of Bovine BLG-A

		$a/Å$	$b/Å$	$c/Å$	α	β	γ	
Lattice X Triclinic	P1	37·8	49·6	56·6	123·42°	97·28°	103·66°	pH 6·5
Lattice Y Orthorhombic	$B22_12$	55·7	67·2	81·7	90·0°	90·0°	90·0°	pH 7·6
Lattice Z Trigonal	$P3_221$	54·4	54·4	113·1	90·0°	90·0°	120·0°	pH 7·6

1971). These were often too small or too disordered to be suitable for X-ray crystallography but crystals suitable for this type of structural body have been grown from bovine and buffalo BLGs, and characterised (Crowfoot & Riley, 1938; Senti & Warner, 1948; Green *et al.*, 1956; Aschaffenburg *et al.*, 1965; Bolognesi *et al.*, 1979). These papers reveal that there are at least 12 crystal forms of the protein grown under various conditions of pH, salt concentration and the presence of ligands. Of these, the salted-out forms have proved to be the most satisfactory for X-ray crystallography.

BLG was one of the first proteins to be subjected to X-ray analysis. Work on both wet and air-dried crystals grown without salt, revealed the dimeric nature of the protein although this was not recognised at the time (Crowfoot & Riley, 1938; Crowfoot, 1941). The dimeric nature of the protein in these and other crystal forms was discussed by Green and Aschaffenburg (1959) and an account of the early X-ray work is given by Hodgkin and Riley (1968).

The structures of four distinct crystal forms, lattices K, X, Y and Z, have been obtained by X-ray crystallography using the method of isomorphous replacement (Green *et al.*, 1979). All four structures show features suggestive of an α-helix and some β-sheet, but no distinct differences are evident. Unlike lattices X, Y and Z, lattice K was produced from protein which had the C-terminal –His–Ile residues removed with carboxypeptidase. However, at this low resolution, no indication of the position of the C-terminus could be obtained. A comparison of the positions of the three heavy-atom sites in common amongst the lattices X, Y and Z indicated a possible movement of the free sulphydryl group in keeping with the expected conformational transition between pH 6·5 (lattice X) and 7·8 (lattices Y and Z).

Detailed X-ray crystallographic studies have been carried out on three crystal forms of bovine BLG-A, which were obtained by salting-out with ammonium sulphate (Table 10). Of the two higher pH forms, lattice Z crystals usually appear first, and then convert to lattice Y, providing that the protein concentration is not too high (Papiz, 1982). Each of these forms has so far given rise to an interpretation of the electron density map at 2·8 Å resolution or better (Papiz *et al.*, 1986; Monaco *et al.*, 1987; Yewdall, 1988). Work on lattices X and Y is being extended to 2·0 and 1·8 Å resolution, respectively. As mentioned above, the structure of the lattice Z protein has also been obtained with retinol bound (Monaco *et al.*, 1987).

The first structure of bovine BLG to be obtained was that of the orthorhombic lattice Y at 2·8 Å resolution (Papiz *et al.*, 1986). The structure, which is shown in Fig. 2, consists of nine strands of anti-parallel β-sheet, eight of which wrap round to create a flattened, conical barrel, or calyx, closed at one end. There is a three-turn α-helix on the outer surface of the calyx. The positions of two important residues, Trp-19 and Cys-121, and the interface between subunits, have been located. Trp-19 is situated at the foot of the β-barrel, on the inner surface; modelling retinol into the cavity revealed that the β-ionone ring is 10 Å away from the Trp, whilst its hydroxyl is near Lys-70. The free thiol, Cys-121, is located near the surface, in a hydrophobic channel, bordered on one side by the α-helix. The interface between subunits involves hydrophobic interactions between Ile-29, Ile-147 and the stacking of the His-146, and their two-fold related equivalents. Unfortunately, there is a considerable amount of disorder in the larger, outer loops which means that it has not been possible to locate all of the side chains with certainty. However, it is tempting to equate Tyr-102 with the buried one, Tyr-20 with that part buried and associated with the Tanford transition, and Trp-19 as that mostly buried (Townend *et al.*, 1969). The anomalous carboxyl group is harder to assign since several could be considered as being in atypical environments.

BLG and serum RBP show only low amino acid similarity (about 25% identity) (Godovac-Zimmerman *et al.*, 1985*a*; Pervais & Brew, 1985), but tertiary structures show a high degree of similarity (Sawyer *et al.*, 1985; Papiz *et al.*, 1986). The eight-stranded β-barrel, α-helix and disulphide bridges are conserved in both proteins, with the insertions in RBP relative to BLG appearing in external loops. The structure of the isolated RBP–retinol complex revealed that the chromophore was located within the hydrophobic cavity and prevented from penetration to the foot of the calyx by a ring of five Phe residues (Newcomer *et al.*, 1984). Removal of the chromophore produced apo-RBP, for which a crystal structure has not yet been obtained, although its structure has been modelled (Åqvist *et al.*, 1986). RBP *in vitro* can bind a variety of retinoids (Horwitz & Heller, 1974), suggesting that the binding site is not an exact match to retinol. An external binding site has been proposed for the binding of β-ionylidene ethylbromoacetate to RBP which is located near the carboxy-end of the α-helix (Gawinowicz & Goodman, 1985). This position appears to be similar to the hydrophobic channel to which retinol binds when co-crystallised with BLG (Monaco *et al.*, 1987).

The same β-barrel with 8-strands and an α-helix near their C-terminus is found in two insect proteins which bind bilin or biliverdin, insecticyanin (ICN) from *Manduca sexta* L., and bilin binding protein (BBP) from *Pieris brassicae*. Both the ICN structure at 2·6 Å resolution (Holden *et al.*, 1987) and the BBP structure at 2·0 Å (Huber *et al.*, 1987) are similar to BLG, RBP and to one another. They bind biliverdin IXg (in a cyclic conformation) at the same position within, but near the top of, the hydrophobic cavity where retinol is located in RBP. The orientation of the chromophore differs between these two biliproteins. Despite the structural homology between ICN, BBP and BLG, it is unlikely that BLG binds biliverdin IXg *in vivo*. Although biliverdin is found in buffaloes' milk, it is only present at low concentrations (0·6 μg/ml), is associated with the casein fraction, and exists in a different isomeric form (Kamur *et al.*, 1987).

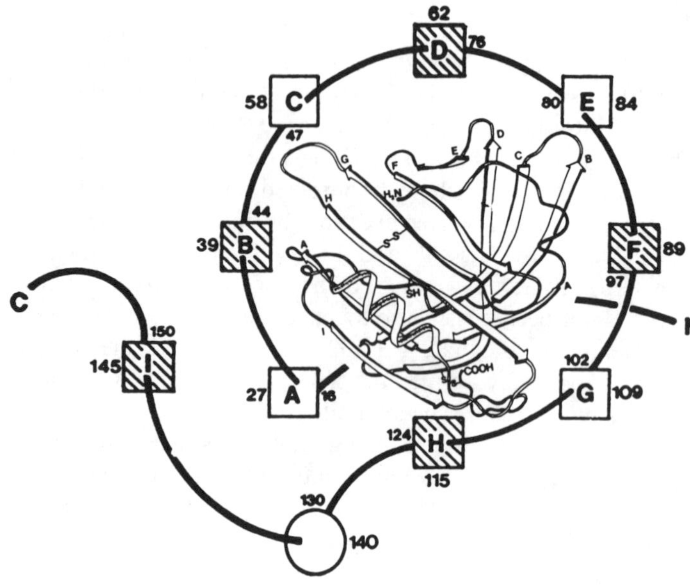

(a)

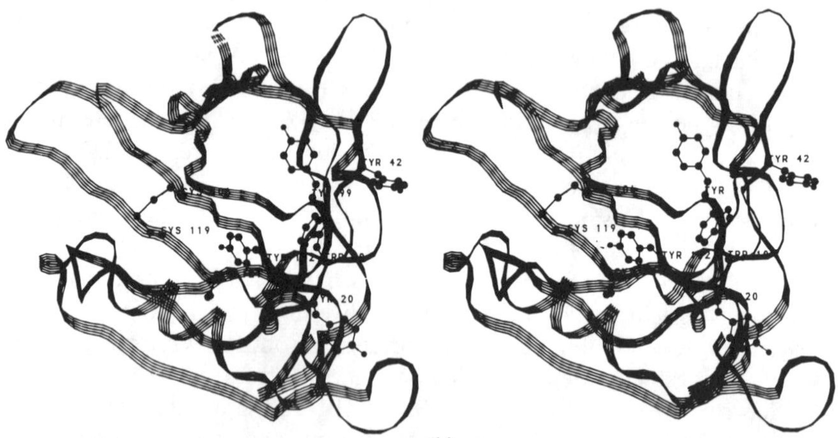

(b)

Fig. 2. (a) A cartoon representation of the backbone together with the strand identification of bovine β-lactoglobulin, based upon a figure from Papiz *et al.* (1986). (Adapted by permission from *Nature*, **324**, 383–5. Copyright (C) 1986 MacMillan Magazines Ltd.). (b) A stereodiagram of the backbone trace of the lattice Y crystal form of bovine β-lactoglobulin showing the positions of the disulphide bridge 106–119 and the free thiol at 121, the Trp-19 and the Tyr side chains. The view has the entrance to the calyx on the left and the helix-sheet putative binding site in the centre around Cys-121. (c) A stereodiagram showing the dimer looking approximately along the two-fold axis of symmetry. The subunits are arranged one on top of the other on the page.

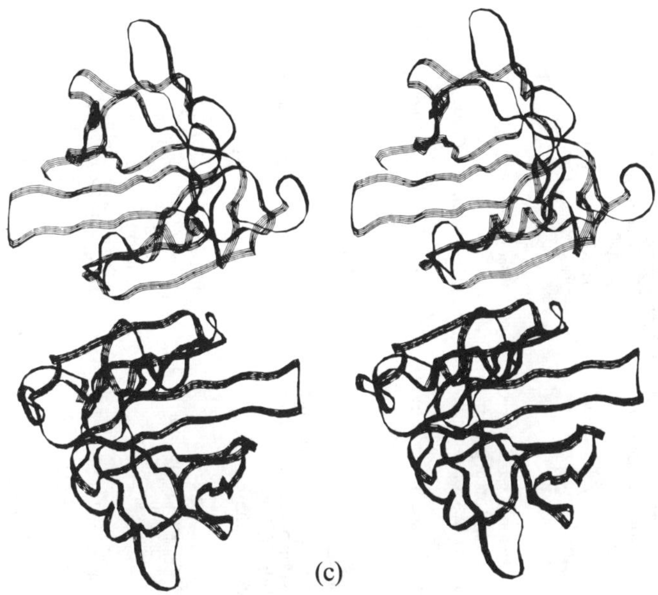

Fig. 2. (Continued).

14. Homologous Proteins

There has emerged over the past decade a series of proteins whose structures are probably closely related to that of BLG. This family has been called the lipocalycin superfamily (Sawyer & Richardson, 1991). Sawyer (1987) pointed out that the necessary requirements for this family were a short sequence near the N-terminus which contained the residues –Gly–X–Trp–Tyr–, another stretch –Thr–Asp–Tyr–, a conserved basic residue and a disulphide bridge between the C-terminus and a residue around 60. These requirements have been further refined (North, 1989*a*, *b*) to produce a consensus pattern for the two short stretches of sequence:

–*h*–(D,N,E,S,T)–X–X–*u*–(L,I,V,F,Y)–X–G–X–W–(W,Y,F,H,R,K)–X–(L,I,V,M)–*h*–

–*h*–*h*–X–T–D–Y–X–X–*y*–*h*–

where *h* is hydrophobic, *u* is often basic and *y* is aromatic.

Although these common amino acids represent only a small proportion of the sequence, North (1989*a*, *b*) has found them to be grouped spatially at the base of the calyx furthest from its entrance. He further proposes a receptor recognition function for this clustering which implies that there are subtle modifications, analogous to those found in antibodies, between the members, allowing their quite distinct interaction with their receptors.

Table 11 lists the various members of this family, together with their functions where these are known and the molecules that are known to bind to them. All of the members have molecular weights around 18 kDa, and appear to bind and

TABLE 11
Some characteristics of the β-Lactoglobulin Superfamily Proteins

Protein	*Number of residues*	*Location*	*Ligand*	*Function*	*References*
Bovine BLG	162	Milk	Retinol, FFA?	Transport/transfer in gut of young?	Godovac-Zimmermann & Braunitzer (1987)
PP14 (α-PEG)	162	Amniotic fluid	—	—	Julkunen *et al.* (1988),
Human RBP	182	Serum	Retinol	Retinol transport in complex with transthyretin	Rask *et al.* (1979)
Purpurin	175	Neural retina	Retinol	Transport retinol across interphotoreceptor matrix	Berman *et al.* (1987)
α_1-Microglobulin	167	Serum, urine, cerebrospinal fluid	Brown-yellow chromophore	Binds IgA or albumin	Åkerström & Lögdberg (1990)
Protein HC	181				
Frog BG protein	159	Nasal mucosa	Pyrazines, other odorants	Presentation of odorants to receptors on neurons	Pervaiz & Brew (1985)
Cow nasal protein	159	Nasal mucosa			Lee *et al.* (1987)
Odorant BP	172	Nasal mucosa			Tirindelli *et al.* (1987) Pevsner *et al.* (1988)
Aphrodisin	151	Vaginal discharge	—	Aphrodisiac	Henzel *et al.* (1988)
Apolipoprotein-D	169	Serum	Cholesteryl esters?	Part of LCAT complex? Transports cholesterol or bilin?	Drayna *et al.* (1986) Peitsch & Boguski (1990)
α_1-Acid glycoprotein	181	Serum, urine	Steroids, unknown lipid	Acute phase protein	Cooper *et al.* (1987)
Rat ESP	166	Epididymal luminal fluid	—	Binds to sperm membrane, and aids maturation	Brooks *et al.* (1986)
α_{2u}-Globulin	162	Urine, serum	Pheromones	—	Unterman *et al.* (1981)
MUP	162	Urine, serum	Pheromones	—	Clark *et al.* (1984)
Complement C8γ	182	Serum	—	Inessential bit of complement complex that lyses cells	Haefliger *et al.* (1987)
Crustacyanin A	168	Carapace	Astaxanthin	Colouration	Keen *et al.* (1990)
Crustacyanin C	175				
Insecticyanin	189	Hemolymph	Biliverdin IX	Camouflage	Riley *et al.* (1984)
Bilin BP	173				Sutor *et al.* (1988)

α-PEG, Pregnancy-associated endometrial α-globulin; BG, Bowman's gland protein; BP, binding protein; ESP, epididymal protein; MUP, mouse major urinary protein.

transport a hydrophobic/labile/insoluble ligand. The relatedness of these proteins is further supported by the gene homology between five members of the superfamily (Ali & Clark, 1988).

15. Function

To date, no biological function for BLG has been discovered. The amino acid composition is such that the protein is of high nutritional value but the molecular properties, particularly the acid stability, lead to the supposition that some other, more specific function exists. It might be supposed, too, that some telling correlation between the presence of the protein in the milk of a species and some specific physiological or biochemical function would help direct research towards the true function. However, as yet no such compelling correlation has been derived. Several have been proposed without much evidence (or conviction!) by Thompson and Farrell (1974), including an involvement with phosphorylation–dephosphorylation in the mammary gland or as a transferase. Interestingly, the protein in the lipocalycin family whose sequence is closest is the human placental protein 14, or pregnancy-associated endometrial glycoprotein (PEG). On the other hand, it is unlikely that the same requirement for a particular developmental function in the first trimester of human pregnancy can be met from an ingested milk protein in the neonate calf. However, the similarity of BLG to the wider family of transport proteins, together with its resistance to low pH and gastric proteolysis (Miranda & Pelissier, 1983; McAlpine & Sawyer, 1990), point towards a transport or transfer function in the intestinal tract where specific receptors have been detected in the neonate calf (Papiz *et al.*, 1986). Further, Said *et al.* (1989) have shown that BLG enhances retinol uptake in the jejunum and ileum of suckling rats but whether retinol, a fatty acid or some other ligand is the primary substrate is not yet known. Indeed, the question as to whether BLG is loaded with retinol or something else in colostrum or milk has not yet been satisfactorily answered although indications have been obtained that both retinol (Garrick, 1986) and fatty acids (Perez *et al.*, 1989; Puyol *et al.*, 1991) are bound to the protein in milk.

The disappearance of specific receptors fairly early on in the life of the neonate calf is certainly consistent with the transfer of something derived exclusively from colostrum or possibly milk. While a transport function can be proposed on the basis of circumstantial evidence and a few experiments, an involvement with the transfer of a passive immunity as has been proposed by Jenness (1985) must remain a possibility. Superficially, the protein fold is not unlike an immunoglobulin domain (both proteins are mostly β-sheet), but a detailed examination shows them to be quite distinct.

It can be pointed out, too, that the protein family which includes the cellular RBP, p2 myelin, photoreactive yellow protein and fatty acid binding protein appears to be distinct from that of BLG although it is based upon a 10-stranded β-barrel formed from a polypeptide chain which is some 30 amino acids shorter than BLG (Jones *et al.*, 1988; Sacchettini *et al.*, 1989; McRee *et al.*, 1989). The consensus pattern at the N-terminus is:

G–X–(W,Y,F)–X–(I,V,L,M)–X–X–X–N–(F,Y)–(D,E)–

the first few residues of which are reminiscent of the first pattern in the BLG family.

16. Conclusion

From the above, it can be seen that BLG is a molecule about which much is known. It is capable of binding small molecules and of interacting with large ones, including itself. Now that the protein is available in over-expressed forms in both yeast and *E. coli*, modifications which can affect these interactions should be fairly straightforward since they can be based upon the three-dimensional structure. Such changes to the protein will provide an excellent opportunity to explain fully the conformational, binding and self-association properties of the protein and may even lead to a proper description of its biological function.

References

Åkerström, B. & Lögdberg L. (1990). An intriguing member of the lipocalin family: α_1-microglobulin. *TIBS*, **15**, 240–243.

Alexander, S.S. & Pace, C.N. (1971). A comparison of the denaturation of bovine β-lactoglobulins A and B, and goat β-lactoglobulin. *Biochem.*, **10**, 2738–43.

Ali, S. (1989). Structure and expression of the gene encoding ovine β-lactoglobulin. PhD Thesis, University of Edinburgh.

Ali, S. & Clark, A.J. (1988). Characterization of the gene encoding ovine β-lactoglobulin. Similarity to the genes for retinol binding protein and other secretory proteins. *J. Mol. Biol.*, **199**, 415–26.

Ananthanarayanan, V.S., Ahmad, F. & Bigelow, C.C. (1977). The denaturation of β-lactoglobulin-A at pH 2. *Biochim. Biophys. Acta*, **492**, 194–203.

Anderson, L.J., Hayes, G., Pearse, M.J., Stewart, A.F., Willis, T.M. & Mckinlay, A.G. (1989). Complete sequence of bovine β-lactoglobulin cDNA. *Nucleic Acid Research*, **17**, 6739.

Åqvist, J., Sandblom, P., Jones, T.A., Newcomer, M.E., van Gunsteren, W.F. & Tapia, O. (1986). Molecular dynamics simulations of the holo and apo forms of retinol binding protein. Structural and dynamical changes induced by retinol removal. *J. Mol. Biol.*, **192**, 593–604.

Arakawa, T. & Timasheff, S.N. (1987). Abnormal solubility behaviour of β-lactoglobulin: Salting-in by glycine and NaCl. *Biochem.*, **26**, 5147–53.

Armstrong, J.McD. & McKenzie, H.A. (1967). A method for modification of carboxyl groups in proteins: its application to the association of bovine β-lactoglobulin A. *Biochim. Biophys. Acta*, **147**, 93–9.

Armstrong, J.McD., McKenzie, H.A. & Sawyer, W.H. (1967). On the fractionation of β-lactoglobulin and α-lactalbumin. *Biochim. Biophys. Acta*, **147**, 60–72.

Aschaffenburg, R. (1964). Protein phenotyping by direct polyacrylamide-gel electrophoresis of whole milk. *Biochim. Biophys. Acta*, **147**, 60–72.

Aschaffenburg, R. (1968). Genetic varients of milk proteins: their breed distribution. *J. Dairy Res.*, **35**, 447–60.

Aschaffenburg, R. & Drewry, J. (1955). Occurrence of different β-lactoglobulins in cow's milk. *Nature*, **176**, 218–219.

Aschaffenburg, R. & Drewry, J. (1957). Improved method for the preparation of crystalline β-lactoglobulin and α-lactalbumin from cow's milk. *Biochem. J.*, **65**, 273–7.

Aschaffenburg, R. & Thymann, M. (1965). Simultaneous phenotyping process for the principal proteins of cows' milk. *J. Dairy Sci.*, **46**, 1524–6.

Aschaffenburg, R., Green, D.W. & Simmons, R.M. (1965). Crystal forms of β-lactoglobulin. *J. Mol. Biol.*, **13**, 194–201.

Ashworth, U.S., Ramaiah, G.D. & Keyes, M.C. (1966). Species difference in the composition of milk with special reference to the Northern fur seal. *J. Dairy Sci.*, **49**, 1206–11.

Askonas, B.A. (1954). Crystallization of goat β-lactoglobulin. *Biochem. J.*, **58**, 332–6.

Axelsson, I., Jakobsson, I., Lindberg, T. & Benediktsson, B. (1986). Bovine β-lactoglobulin in the human milk. *Acta Paed. Scand.*, **75**, 702–7.

Bain, J.A. & Deutsch, H.F. (1948). Studies on lactoglobulins. *Arch. Biochem.*, **16**, 221–9.

Baker, H.P. & Saroff, H.A. (1965). Binding of sodium ions to β-lactoglobulin. *Biochemistry*, **4**, 1670–7.

Basch, J.J. & Timasheff, S.N. (1967). Hydrogen ion equilibria of the genetic variants of bovine β-lactoglobulin. *Archs. Biochem. Biophys.*, **118**, 37–47.

Baumy, J.J. & Brule, G. (1988). Binding of bivalent cations to α-lactalbumin and β-lactoglobulin: effect of pH and ionic strength. *Le Lait*, **68**, 33–48.

Bell, K. & McKenzie, H.A. (1964). β-Lactoglobulins. *Nature*, **204**, 1275–9.

Bell, K. & McKenzie, H.A. (1967*a*). The isolation and properties of bovine β-lactoglobulin C. *Biochim. Biophys. Acta*, **147**, 109–22.

Bell, K. & McKenzie, H.A. (1967*b*). The whey proteins of ovine milk: β-Lactoglobulins A and B. *Biochim. Biophys. Acta*, **147**, 123–4.

Bell, K., McKenzie, H.A., Murphy, W.H. & Shaw, D.C. (1970). β-Lactoglobulin droughtmaster: a unique protein variant. *Biochim. Biophys. Acta*, **214**, 427–36.

Bell, K., McKenzie, H.A. & Shaw, D.C. (1981*a*). Bovine β-lactoglobulin E, F and G of Bali (Banteng) Cattle, *Bos (Bibos) javanicus*. *Aust. J. Biol. Sci.*, **34**, 133–47.

Bell, K., McKenzie, H.A., Muller, V., Rogers, C. & Shaw, D.C. (1981*b*). Equine whey proteins. *Comp. Biochem. Physiol.*, **68B**, 225–36.

Bell, K., McKenzie, H.A. & Shaw, D.C. (1981*c*). Porcine β-lactoglobulin A and C. *Mol. Cell. Biochem.*, **35**, 103–11.

Berman, P., Gray, P., Chen, E., Keyser, K., Ehrlich, D., Karten, H., LaCorbiere, M., Esch, F. & Schubert, D. (1987). Sequence analysis, cellular localisation, and expression of a neuroretina adhesion and cell survival molecule, *Cell*, **51**, 135–142.

Bolognesi, M., Liberatori, J., Oberti, R. & Ungaretti, L. (1979). Preliminary crystallographic data on buffalo β-lactoglobulin. *J. Mol. Biol.*, **131**, 411–3.

Braunitzer, G., Chen. R., Schrank, B. & Stangl, A. (1972). Die sequenzanalyse des β-lactoglobulins. *Hoppe-Seyler's Z. Physiol. Chem.*, **354**, 867–78.

Braunitzer, G., Liberatori, J. & Kolde, H.-J. (1979). The primary structure of the β-lactoglobulin of the waterbuffalo (*Bubalus arnee*). *Z. Naturforsch.*, **34c**, 880–1.

Brew, K. & Campbell, P.N. (1967). The characterization of the whey proteins of guinea-pig milk. *Biochem. J.*, **102**, 258–64.

Brignon, G. & Ribadeau-Dumas, B. (1973). Localisation dans la chaine peptidique de la β-lactoglobuline bovine de la substitution Glu/Gln differenciant les varients genetiques B et D. *FEBS Lett.*, **33**, 73–6.

Brignon, G., Ribadeau-Dumas, B., Garnier, J., Pantaloni, D., Guinard, S., Basch, J.J. & Timasheff, S.N. (1969). Chemical and physico-chemical characterization of genetic variant D of bovine β-lactoglobulin D. *Arch. Biochem. Biophys.*, **129**, 720–7.

Brignon, G., Chtourou, A. & Ribadeau-Dumas, B. (1985). Does β-lactoglobulin occur in human milk? *J. Dairy Res.*, **52**, 249–54.

Brooks, D.E., Means, A.R., Wright, E.J., Singh, S.P. & Tiver, K.K. (1986). Molecular cloning of the cDNA for two major androgen-dependent secretory proteins of 18.5 kilodaltons synthesised by the rate epididymis. *J. Biol. Chem.*, **261**, 4956–61.

Brown, E.M. & Farrell, H.M. (1978). Interaction of β-lactoglobulin and cytochrome c: complex formation and iron reduction. *Arch. Biochem. Biophys.*, **185**, 156–64.

Brown, E.M., Carroll, R.J., Pfeffer, P.E. & Sampugna, J. (1983). Complex formation in sonicated mixtures of β-lactoglobulin and phosphatidylcholine. *Lipids*, **18**, 111–8.

Brown, E.M., Pfeffer, P.E., Kumosinski, T.F. & Greenberg, R. (1988). Accessibility and mobility of lysine residues in β-lactoglobulin. *Biochem.*, **27**, 5601–10.

Bull, H.B. (1946*a*). Monolayers of β-lactoglobulin. I. Film molecular weight. *J. Am. Chem. Soc.*, **68**, 745–7.

Bull, H.B. (1946*b*). Monolayers of β-lactoglobulin. II. Mixed monolayers of β-lactoglobulin and sodium lauryl sulphate. *J. Am. Chem. Soc.*, **68**, 747–8.

Bull, H.B. & Currie, B.T. (1946). Osmotic pressure of β-lactoglobulin solutions. *J. Am. Chem. Soc.*, **68**, 742–5.

Butler, J.E. (1974). Immunoglobulins of mammary secretions. *Lactation*: A comprehensive treatise, **III**, 217–256.

Byler, D.M. & Susi, H. (1986). Examination of the secondary structure of proteins by deconvolved FTIR spectra. *Biopolymers*, **25**, 469–487.

Byler, D.M., Susi, H. & Farrell (Jr), H.M. (1983). Laser-Raman spectra, sulphydryl groups, and conformation of the cystine linkages of β-lactoglobulin. *Biopolymers*, **22**, 2507–11.

Cannan, R.K., Palmer, A.H. & Kibrick, A.C. (1942). Hydrogen ion dissociation curve of β-lactoglobulin. *J. Biol. Chem.*, **142**, 803–33.

Casal, H.L., Kohler, U. & Mantsch, H.N. (1988). Structural and conformational changes of β-lactoglobulin B: an infrared spectroscopic study of the effect of pH and temperature. *Biochim. Biophys. Acta*, **957**, 11–20.

Cecil, R. & Ogston, A.G. (1949). The sedimentation constant, diffusion constant and molecular weight of lactoglobulin. *Biochem. J.*, **44**, 33–5.

Christensen, L.K. (1952). Denaturation and enzymic hydrolysis of lactoglobulin. *C. R. Lab. Carlsberg (Ser. Chim.)*, **28**, 37–174.

Clark, A.J., Clissold, P.M., Shawi, A., Beattie, P. & Bishop, J. (1984) The structure of mouse major urinary protein genes: differing splicing configurations in the 3′-non-coding region. *EMBO J.* **3** 1045–1052.

Clark, A.J., Simons, P., Wilmut, I. & Lathe, R. (1987). Pharmaceuticals from transgenic livestock. *TIBTech.*, **5**, 20–4.

Conti, A. & Godovac-Zimmermann, J. (1990). Antibodies raised against bovine β-lactoglobulin react with β_2-microglobulin. *Biol. Chem. Hoppe-Seyler*, **371**, 261–3.

Conti, A., Liberatori, J. & Napolitano, L. (1980). Isolation and preliminary physico-chemical characterization of human β-lactoglobulin. *Milchwissenschaft*, **35**, 65–8.

Conti, A., Liberatori, J. & Napolitano, L. (1982*a*). Heterogeneity of equine β-lactoglobulins. *Milchwissenschaft*, **37**, 338–40.

Conti, A., Napolitano, L. & Liberatori, J. (1982*b*). Characterisation of β-lactoglobulins from human colostrum. Some physico-chemical properties and preliminary amino-acid evaluation. *Dairy Industries International*, **47**, 163–4.

Conti, A., Godovac-Zimmermann, J., Liberatori, J. & Braunitzer, G. (1984). The primary structure of monomeric β-lactoglobulin I from horse colostrum (*Equus caballus*, Perissodactyla). *Z. Physiol. Chem.*, **366**, 1393–401.

Conti, A., Napolitano, L., Godovac-Zimmermann, J., Liberatori, J. & Braunitzer, G. (1986*a*). Preparative separation of pig polymorphic β-lactoglobulins by IPG and non-IPG. *Protides Biol. Fluids Proc. Colloq.*, **34**, 907–10.

Conti, A., Godovac-Zimmermann, J. Pirchner, F., Liberatori, J. & Braunitzer, G. (1986*b*). Pig β-lactoglobulin I (*Sus scrofa domestica*, Artiodactyla). *Biol. Chem. Hoppe-Seyler*, **367**, 871–8.

Conti, A., Napolitano, L., Cantisani, A.M., Davoli, R. & Dall'Olio, S. (1988). Bovine β-lactoglobulin H: isolation by preparative isoelectric focusing in immobilized pH gradients and preliminary characterization. *J. Biochem. Biophys. Meth.*, **16**, 205–14.

Cooper, R., Eckley, D.M. & Papaconstantinou, J. (1987). Nucleotide sequence of the mouse α_1-acid glycoprotein gene 1. *Biochemistry*, **26**, 5244–50.

Cornell, D.G. & Patterson, D.L. (1989). Interaction of phospholipids in monolayers with β-lactoglobulin adsorbed from solution. *J. Agric. Food Chem.*, **37**, 1455–9.

Creamer, L.K., Parry, D.A.D. & Malcolm, G.N. (1983). Secondary structure of bovine β-lactoglobulin B. *Arch. Biochem. Biophys.*, **227**, 98–105.
Creighton, T.E. (1980). Kinetic study of protein unfolding and refolding using urea gradient electrophoresis. *J. Mol. Biol.*, **137**, 61–80.
Crowfoot, D. (1941). A recent review of some recent X-ray work on protein crystals. *Chem. Rev.*, **28**, 215–28.
Crowfoot, D.M. & Riley, D.P. (1938). An x-ray study of Palmer's lactoglobulin. *Nature*, **141**, 521–2.
Cunningham, L.W. & Nuenke, B.J. (1960). Analysis of modified β-lactoglobulins and ovalbumins prepared from the sulfenyl iodide intermediates. *J. Biol. Chem.*, **235**, 1711–5.
Cupo, J.F. & Pace, C.N. (1983). Conformational stability of mixed disulphide derivatives of β-lactoglobulin B. *Biochemistry*, **22**, 2654–8.
Davies, D.T. (1974). The quantitative partition of the albumin fraction of milk serum proteins by gel chromotography. *J. Dairy Res.*, **41**, 217–28.
de Wit & Klarenbeek, G. (1981). A differential scanning calorimetric study of the thermal behaviour of bovine β-lactoglobulin at temperatures up to 160°C. *J. Dairy Res.*, **48**, 293–302.
Deckmyn, H. & Préaux, G. (1978). Chain folding prediction of the bovine β-lactoglobulins. *Arch. Int. Physiol. Biochim.*, **86**, 938–9.
Diaz de Villegas, M.C., Oria, R., Sala, F.J. & Calvo, M. (1987). Lipid binding by β-lactoglobulin of cow milk. *Milchwissenschaft*, **42**, 357–8.
Drayna, D., Fielding, C., McLean, J., Baer, B., Casro, G., Chen, E., Comstock, L., Henzel, W., Kohr, W., Rhee, L., Wion, K. & Lawn, R. (1986). Cloning and expression of human apolipoprotein D cDNA. *J. Biol. Chem.*, **261**, 16535–9.
Dufour, E & Haertlé (1990). Alcohol-induced changes of β-lactoglobulin-retinol-binding stoichiometry. *Protein Engineering*, **4**, 185–190.
Dunnill, P. & Green, D.W. (1965). Sulphydryl groups and the N-R conformational change in β-lactoglobulin. *J. Mol. Biol.*, **15**, 147–51.
Dunnill, P., Green, D.W. & Simmons, R.M. (1966). Heavy-atom sulphydryl derivatives of oxy haemoglobin and β-lactoglobulin; factors affecting isomorphism of native and derivative crystals. *J. Mol. Biol.*, **22**, 135–144.
Dupont, M. (1965). Etude d'une etape reversible dans la thermo-denaturation de la β-lactoglobuline bovine A. *Biochim. Biophys. Acta*, **102**, 500–13.
Erhardt, G., Godovac-Zimmerman, J. & Conti, A. (1989). Isolation and complete primary sequence of a new ovine wild-type β-lactoglobulin C. *Biol. Chem. Hoppe-Seyler*, **370**, 757–62.
Fallstrom, S.P., Ahlstedt, S. & Hanson, L.A. (1978). Specific antibodies in infants with gastro-intestinal intolerance to cow's milk protein. *Int. Arch. All. Appl. Immun.*, **56**, 97–105.
Farrell, H.M. (Jr) & Thompson, M.P. (1971). Biological significance of milk protein polymorphism. *J. Dairy Sci.*, **54**, 1219–28.
Farrell, H.M., Bede, M.J. & Enyeart, J.A. (1987). Binding of *p*-nitrophenyl phosphate and other aromatics by β-lactoglobulin. *J. Dairy Sci.*, **70**, 252–8.
Fernandez, F.M. & Oliver, G. (1988). Proteins present in llama milk. I. Quantitative aspects and general characteristics. *Milchwissenschaft*, **43**, 299–302.
Ferry, J.D. & Oncley, J.L. (1941). Studies on the dielectric properties of protein solutions. III. Lactoglobulin. *J. Amer. Chem. Soc.*, **63**, 272–8.
Frank, G. & Braunitzer, G. (1967). Der primarstruktur der β-lactoglobuline. *Hoppe Seyler's Z. Physiol. Chem.*, **348**, 1691–2.
Freedman, R.B., Bullied, N.J., Hawkins, H.C. & Paver, J.L. (1989). Role of protein disulphide-isomerase in the expression of native proteins. *Biochem. Soc. Symp.*, **55**, 167–92.
Frick, M. & Reiger, C.H.L. (1987). Local antibodies to α-casein and β-lactoglobulin in saliva of infants. *Pediat. Res.*, **22**, 399–401.
Frushour, B.G. & Koenig, L.J. (1975). Raman studies of the crystalline, solution and alkaline-denatured states of β-lactoglobulin. *Biopolymers*, **14**, 649–62.

Fugate, R.D. & Song, P.-S (1980). Spectroscopic characterization of β-lactoglobulin–retinol complex. *Biochim. Biophys. Acta*, **625**, 28–42.
Futterman, S. & Heller, J. (1972). The enhancement of fluorescence and the decreased susceptibility to enzymatic oxidation of retinol complexed with bovine serum albumin, β-lactoglobulin and the retinol-binding protein of human plasma. *J. Biol. Chem.*, **247**, 5168–72.
Garrick, J.D. (1986). Is the role of β-lactoglobulin to serve as a retinol carrier in milk? BSc Honours Project Thesis, University of Aberdeen.
Gawinowicz, M.A. & Goodman, D.S. (1985). The site of a retinoid affinity label to plasma retinol-binding protein. *J. Prot. Chem.*, **4**, 199–213.
Gaye, P., Hue-Delahaie, D., Mercier, J.-C., Soulier, S., Vilotte, J.L. & Furet, J.P. (1986). Ovine β-lactoglobulin messenger RNA: nucleotide sequence and mRNA levels during functional differentiation of the mammary gland. *Biochimie*, **68**, 1097–107.
Georges, C. & Guinand, S. (1960). Sur la dissociation reversible de la β_2-lactoglobuline, a des pH superieurs a 5·5. 1. Etude par la diffusion de la lumiere. *J. Chem. Phys.*, **57**, 606–14.
Georges, C., Guinand, S. & Tonnelat, J. (1962). Etude thermodynamique de la dissociation reversible de la β-lactoglobuline B pour des pH superieurs a 5·5. *Biochim. Biophys. Acta*, **59**, 737–9.
Ghose, A.C., Chaudhuri, S. & Sen, A. (1968). Hydrogen ion equilibria and sedimentation behaviour of goat β-lactoglobulins. *Arch. Biochem. Biophys.*, **126**, 232–43.
Godovac-Zimmermann, J. (1988). The structural motif of β-lactoglobulin and retinol binding protein: a basic framework for binding and transport of small hydrophobic molecules? *TIBS*, **13**, 64–6.
Godovac-Zimmermann, J. & Braunitzer, G. (1987). Modern aspects of the primary structure and function of β-lactoglobulins. *Milchwissenschaft*, **42**, 294–7.
Godovac-Zimmermann, J. & Shaw, D. (1987). The primary structure, binding site and possible function of β-lactoglobulin from Eastern grey kangaroo (*Macropus giganteus*). *Biol. Chem. Hoppe-Seyler*, **368**, 879–86.
Godovac-Zimmermann, J., Conti, A., Liberatori, J. & Braunitzer, G. (1985*a*). Homology between the primary structures of β-lactoglobulins and human retinol binding protein: evidence for a similar biological function? *Biol. Chem. Hoppe-Seyler*, **366**, 431–4.
Godovac-Zimmermann, J., Conti, A., Liberatori, J. & Braunitzer, G. (1985*b*). The amino-acid sequence of β-lactoglobulin II from horse colostrum (*Equus caballus*, Perissodactyla): β-lactoglobulins are retinol-binding proteins. *Biol. Chem. Hoppe-Seyler*, **366**, 601–8.
Godovac-Zimmermann, J., Conti, A. & Napolitano, L. (1987). The complete amino-acid sequence of dimeric β-lactoglobulin from mouflon (*Ovis ammon musimon*) milk. *Biol. Chem. Hoppe-Seyler*, **368**, 1313–9.
Godovac-Zimmermann, J., Conti, A., James, L. & Napolitano, L. (1988). Micro-analysis of the amino-acid sequence of monomeric β-lactoglobulin I from donkey (*Equus asinus*) milk. *Biol. Chem. Hoppe-Seyler*, **369**, 171–9.
Gorbunoff, M.J. (1967). Exposure of tyrosine residues in proteins. Reaction of cyanuric fluoride with ribonuclease, α-lactalbumin, and β-lactoglobulin. *Biochemistry*, **6**, 1606–15.
Granato, D.A. & Piguet, P.F. (1986). A mouse monoclonal IgE antibody anti-bovine milk β-lactoglobulin allows studies of allergy in gastrointestinal tract. *Clin. Exp. Immun.*, **63**, 703–10.
Green, D.W. & Aschaffenburg, R. (1959). Twofold symmetry of the β-lactoglobulin molecule in crystals. *J. Mol. Biol.*, **1**, 54–64.
Green, D.W., North, A.C.T. & Aschaffenburg, R. (1956). Crystallography of β-lactoglobulin from cow's milk. *Biochim. Biophys. Acta*, **21**, 583–5.
Green, D.W., Aschaffenburg, R., Camerman, A., Coppola, J.C., Diamand, R.D., Dunnill, P., Simmons, R.M., Komorowski, E.S., Sawyer, L., Turner, E.M.C. & Woods, K.F. (1979). Structure of bovine β-lactoglobulin at 6 Å resolution. *J. Mol. Biol.*, **131**, 375–97.
Greenberg, R. & Kalan, E.B. (1965). Studies on β-lactoglobulins A, B and C. II. Preparation of modified proteins by treatment with carboxypeptidase A. *Biochemistry*, **4**, 1660–6.

Greene, R.F. & Pace, C.N. (1974). Urea and guanidine hydrochloride denaturation of ribonuclease, lysozyme, α-chymotrypsin and β-lactoglobulin. *J. Biol. Chem.*, **249**, 5388–93.

Grosclaude, F., Mahe, M.-F., Mercier, J.-C., Bonnemarie, J. & Teissier, J.H. (1976). Polymorphisme des lactoproteines de bovines nepalais. *Ann. Genet. Sel. Anim.*, **8**, 461–79.

Groves, M.L., Hipp, N.J. & McMeekin, T.L. (1951). Effect of pH on the denaturation of β-lactoglobulin and its dodecyl sulphate derivative. *J. Am. Chem. Soc.*, **73**, 2790–3.

Gutfreund, H. (1945). The molecular weight of Palmer's β-lactoglobulin. *Nature*, **155**, 237.

Haefliger, J.-A., Jenne, D., Stanley, K.K. & Tschopp, J. (1987). Structural homology of human complement component C8γ and plasma protein HC: identity of the cysteine bond pattern. *Biochem. Biophys. Res. Comm.*, **149**, 750–4.

Halliday, J.A., Bell, K. & Shaw, D.C. (1991). The complete amino acid sequence of feline β-lactoglobulin II and a partial revision of the equine β-lactoglobuli II sequence. *Biochim. Biophys. Acta*, **1077**, 25–30.

Halpin, M.I. & Richardson, T. (1985). Selected functionality changes of β-lactoglobulin upon esterification of side chain carboxyl groups. *J. Dairy. Sci.*, **68**, 3189–98.

Halwer, M., Nutting, G.C. & Brice, B.A. (1951). Molecular weight of lactoglobulin, ovalbumin, lysozyme and serum albumin by light scattering. *J. Am. Chem. Soc.*, **73**, 2786–90.

Hambling, S.G. (1990). Crystallographic and solution studies on bovine β-lactoglobulin. PhD Thesis, University of Edinburgh.

Haque, Z., Kristjannson, M.M. & Kinsella, J.E. (1987). Interaction between κ-casein and β-lactoglobulin: possible mechanism. *J. Agri. Food Chem.*, **35**, 644–9.

Heller, W. & Klevens, H.B. (1946). Molecular weights of proteins as determined by light scattering. *Phys. Rev.*, **67**, 61.

Hemley, R., Kohler, B.E. & Siviski, P. (1979). Absorption spectra for the complexes formed from vitamin-A and β-lactoglobulin. *Biophysics J.*, **28**, 447–55.

Hennighausen, L.G. & Sippel, A.E. (1982). Mouse whey acidic protein is a novel member of the family of 'four-disulphide core' proteins. *Nucl. Acid Res.*, **10**, 2677–84.

Henzel, W.J., Rodriguez, H., Singer, A.G., Stults, J.T., Macrides, F., Agosta, W.C. & Niall, H. (1988). The primary structure of aphrodisin. *J. Biol. Chem.*, **263**, 16682–7.

Hilerskovits, T.T., Townend, R. & Timasheff, S.N. (1964). Molecular interactions in β-lactoglobulin. IX. Optical rotatory dispersion of the genetic variants in different states of association. *J. Am. Chem. Soc.*, **86**, 4445–52.

Hillquist, A.J. Dan, mo & Kresheck, G.C. (1982). Influence of surfactants on the conformation of β-lactoglobulin B using circular dichroism. *Biopolymers*, **21**, 895–908.

Hin Bon Hoa, G., Douzou, P. & Pantaloni, C. (1973). Transformations alcalines de la β-lactoglobulines en milieu hydro-alcoolique a basses temperatures. *Biochemie*, **55**, 269–76.

Hodgkin, D.C. & Riley, D.P. (1968). Some ancient history of protein X-ray analysis. In *Structural Chemistry and Molecular Biology*, ed. A. Rich & N. Davidson. Freeman, New York and London, pp. 15–28.

Holden, H.M., Rypiewski, W.R., Law, J.H. & Rayment, I. (1987). The molecular structure of insecticyanin from the tobacco hornworm (*Manduca sexta L.*) at 2·6 Å resolution. *EMBO J.*, **6**, 1565–70.

Holzwarth, G. (1972). Ultraviolet spectroscopy of biological membranes. In *Membrane Molecular Biology*, ed. C.F. Fox & A.D. Keith. Stanford Conn: Sinauer Assoc., pp. 228–86.

Horisberger, M. & Vauthey, M. (1984). Labelling of colloidal gold with protein: A quantitative study using β-lactoglobulin. *Histochemistry*, **22**123.

Horwitz, J. & Heller, J. (1974). Properties of the chromophore binding site of retinol-binding protein from human plasma. *J. Biol. Chem.*, **249**, 4712–9.

Huang, Q., Coleman, J.W. & Stanworth, D.R. (1985). Investigation of the allergenicity of β-lactoglobulin and its cleavage fragments. *Int. Arch. All. Appl. Immun.*, **78**, 337–45.

Huber, R., Schneider, M., Mayer, I., Muller, R., Deutzmann, R., Suter, F., Zuber, H., Falk, H. & Kayser, H. (1987). Molecular structure of the bilin binding protein (BBP) from *Pieris brassicae* after refinement at 2·0 Å. *J. Mol. Biol.*, **198**, 499–513.

Huhtala, M.-L., Seppala, M., Narvanen, A., Palomaki, P., Mulkunen, J. & Hohn, B. (1987). Amino acid sequence homology between human placental protein 14 and β-lactoglobulins from various species. *Endocrinol.*, **120**, 2620–2.

Hunziker, H.G. & Tarassuk, N.P. (1965). Chromatographic evidence for heat induced interaction of α-lactalbumin and β-lactoglobulin. *J. Dairy Sci.*, **48**, 733–44.

Ivanov, V.N., Judinkova, E.S. & Gorodetsky, S.I. (1988). Molecular cloning of bovine β-lactoglobulin cDNA. *Biol. Chem. Hoppe-Seyler*, **369**, 425-9.

Jakobsson, I., Lindberg, T., Benediktsson, B. & Hansson, B.-G. (1985). Dietary bovine β-lactoglobulin is transferred to human milk. *Acta Paed. Scand.*, **74**, 342–5.

Jamieson, A.C., Vandeyar, M.A., Kang, Y.C., Kinsella, J.E. & Batt, A.C. (1987). Cloning and nucleotide sequence of the bovine β-lactoglobulin gene. *Gene*, **61**, 85–90.

Jenness, R. (1974). The composition of milk. In *Lactation*, vol. III, eds. B.L. Larson & V.R. Smith. Academic Press, New York, pp. 3–107.

Jenness, R. (1979). Comparative aspects of milk proteins. *J. Dairy Res.*, **46**, 197–210.

Jenness, R. (1985). Biochemical and nutritional aspects of milk and colostrum. In *Lactation*, ed. B.L. Larson. Iowa State University Press, pp. 164–197.

Jenness, R., Phillip, N.I. & Kalan, E.B. (1967). Immunochemical comparison of β-lactoglobulins. *Fed. Proc.*, **26**, 340.

Jones, M.N. & Wilkinson, A. (1976). The interaction between β-lactoglobulin and sodium *n*-dodecyl sulphate. *Biochem. J.*, **153**, 713–8.

Jones, T.A., Bergfors, T., Sedzik, J. & Unge, T. (1988). The three-dimensional structure of P2 myelin protein. *EMBO J.*, **7**, 1597–604.

Julkunen, M., Seppala, M. & Janne, O.A. (1988). Complete amino acid sequence of human placental protein 14: A progesterone-regulated uterine protein homologous to β-lactoglobulins. *Proc. Natl. Acad. Sci.*, **85**, 8845–9.

Kalan, E.B. & Basch, J.J. (1969). Preparation of goat β-lactoglobulin. *J. Dairy Sci.*, **49**, 406–9.

Kalan, E.B., Greenberg, R. & Walter, M. (1965). β-Lactoglobulins A, B and C. I. Comparison of chemical properties. *Biochem.*, **4**, 991–7.

Kalan, E.B., Kraeling, R.R. & Gerrits, R.J. (1971). Isolation and partial characterization of a polymorphic swine whey protein. *Int. J. Biochem.*, **2**, 232–44.

Kamur, M., Dastur, N.N. & Bhatt, M.V. (1987). Studies on the pigments of buffalo milk. I. Identification of biliverdin IXa isomer in fresh buffalo milk. *Milchwissenschaft*, **39**, 29–31.

Keen, J.N., Caceres, I., Eliopolous, E.E., Zagalsky, P.F. & Findlay, J.B.C. (1990). Complete sequence and model for the A_2-subunit of the carotenoid pigment complex, crusticyanin. *Eur. J. Biochem.*, **197**, 407–17.

Kella, N.K.D. & Kinsella, J.E. (1988). Enhanced thermodynamic stability of β-lactoglobulin at low pH. *Biochem. J.*, **255**, 113–8.

Kessler, E. & Brew, K. (1970). The whey proteins of pig's milk isolation and characterization of a β-lactoglobulin. *Biochim. Biophys. Acta*, **200**, 449–58.

Kiddy, C.A., Rollins, R.E. & Zikakis, J.P. (1972). Discontinuous polyacrylamide electrophoresis for β-lactoglobulin of cow milk. *J. Dairy Sci.*, **55**, 1506–7.

Kinsella, J.E. & Whitehead, D.M. (1987). Proteins in whey: chemical, physical, and functional properties. *Advances in Food and Nutrition Research*, **33**, 343–438.

Kolde, H.-J. & Braunitzer, G. (1983). The primary structure of ovine β-lactoglobulin. 1. Isolation of the peptides and sequence. *Milchwissenschaft*, **38**, 18–20.

Koritz, T.N., Suzuki, S. & Coomb, R.R.A. (1987). Antigenic stimulation with proteins of cows milk via the oral route in guinea-pigs and rats. 1. Measurement of antigenically intact β-lactoglobulin and casein in the gastrointestinal contents of duodenum, jejunum and ileum. *Int. Arch. All. Appl. Immun.*, **82**, 72–5.

Kraeling, R.R. & Gerrits, R.J. (1969). Polymorphism of a protein of sows' whey. *J. Dairy Sci.*, **52**, 2036–8.

Krause, I., Buchberger, J., Weiss, G., Pfugler, M. & Klostermeyer, H. (1988). Isoelectric focusing in immobilised pH gradients with carrier ampholytes added for high-resolution phenotyping of bovine β-lactoglobulins: Characterisation of a new genetic varient. *Electrophoresis*, **9**, 609–13.

Kurisaki, J., Nakamura, S., Kaminogawa, S. & Yamauchi, K. (1982). The antigenic properties of β-lactoglobulin examined with mouse IgE antibody. *Agri. Biol. Chem.*, **46**, 2069–75.

Kurisaki, J., Nakamura, S., Kaminogawa, S., Yamauchi, K., Watanabe, S., Hotta, K. & Hattori, M. (1985). Antigenicity of modified β-lactoglobulin examined by three different assays. *Agri. Biol. Chem.*, **49**, 1733–7.

Larson, B.L. (1979). Biosynthesis and secretion of milk proteins: a review. *J. Dairy Sci.*, **46**, 161–74.

Larson, B.L. & Jenness, R. (1952). Characterization of the sulphydryl groups and the kinetics of the heat denaturation of crystalline β-lactoglobulin. *J. Am. Chem. Soc.*, **74**, 3090–3.

Lebenthal, E., Laor, J., Lewitus, Z., Matoth, Y. & Freier, S. (1970). Gastrointestinal protein loss in allergy to cows milk β-lactoglobulin. *Isr. J. Med. Sci.*, **6**, 506–10.

Lee, K.-H., Wells, R.G. & Reed, R.R. (1987). Isolation of an olfactory cDNA: similarity to retinol-binding protein suggests a role in olfaction. *Science*, **235**, 1053–6.

Li, C.H. (1946). Electrophoretic inhomogeneity of crystalline β-lactoglobulin. *J. Amer. Chem. Soc.*, **68**, 2746–7.

Liberatori, J., Morisio, L. Guetti, L.M. Conti, A. & Napolitano, L. (1979*a*). β-Lactoglobulins in the mammary secretions of camel (*Camelus dromedarius*) and she-ass. Immunological detection and preliminary physico-chemical characterization. *Boll. Soc. It. Biol. Sper.*, **55**, 1369–73.

Liberatori, J., Guidetti, L.M. & Conti, A. (1979*a*). Immunological evidence of β-lactoglobulins in human colostrum and milk. *Boll. Soc. It. Biol. Sper.*, **55**, 822–25.

Liberatori, J., Guidetti, L.M. & Conti, A. (1979*b*). Immunochemical studies on β-lactoglobulins. Precipitin reactions of sow's and mare's mammary secretions against anti-bovine β-lactoglobulin antiserum. *Boll. Soc. It. Biol. Sper.*, **55**, 815–21.

Lockwood, D.H., Turkington, R.W. & Topper, Y.J. (1966). Hormone-dependent development of milk protein synthesis in mammary gland in vitro. *Biochim. Biophys. Acta*, **139**, 493–501.

Lontie, R. & Preaux, G. (1966). Polarimetric investigation of β-lactoglobulin A and B and of the reactivity of their thiol groups. *Protides Biol. Fluids*, **14**, 475–82.

Lovrien, R. & Anderson, W.F. (1969). Resolution of binding sites in β-lactoglobulin. *Arch. Biochem. Biophys.*, **131**, 139–44.

Lyster, R.L.J. (1964). The reactivity of the –SH groups of β-lactoglobulin subunits. *Abst. 6th Int. Congress Biochem.*

Lyster, R.L.J. (1972). Reviews of the progress of dairy science. Section C. Chemistry of milk proteins. *J. Dairy Res.*, **39**, 279–318.

Lyster, R.L.J. & Wheelock, J.V. (1967). Occurrence of milk proteins in urine of cows during an extended milking interval. *J. Dairy Res.*, **34**, 27–30.

Lyster, R.L.J., Jenness, R., Phillips, N.I. & Sloan, R.E. (1966). Comparative biochemical studies of milks—III. Immuno-electrophoretic comparisons of milk proteins of the Artiodactyla. *Comp. Biochem. Physiol.*, **17**, 967–71.

Macheboeuf, M. & Robert, B. (1953). Etude sur la denaturation des proteines 1. Autoxydation des proteines en milieu alcalin. *Bull. Soc. Chim. Biol.*, **35**, 399–411.

Maillert, P. & Ribadeau-Dumas, B. (1988). Preparation of β-lactoglobulin and β-lactoglobulin-free proteins from whey retenate by NaCl salting out at low pH. *J. Food Sci.*, **53**, 743–5.

Mainferme, F., Préaux, G. & Lontie, R. (1971). Location of the disulphide bridges in the sequence of β-lactoglobulin A with CNBr and thermolysin. *Arch. Int. Physiol. Biochem.*, **79**, 840–1.

Malik, Z., Bottomley, R. & Austen, B. (1988). Allergenic properties of the genetic variants A and B of bovine β-lactoglobulin. *Int. Arch. All. Appl. Immun.*, **86**, 245–8.

Martial, J., Préaux, G. & Lontie, R. (1971). Location of the thiol group and of the disulphide bridges in the sequence of β-lactoglobulin B. *Arch. Int. Physiol. Biochem.*, **79**, 842–3.

Mattarella, N.L. & Richardson, T.L. (1983). Physicochemical and functional properties of

positively charged derivatives of bovine β-lactoglobulin. *J. Agric. Food Chem.*, **31**, 972–8.
Maubois, J.L., Pion, R. & Ribadeau-Dumas, B. (1965). Preparation et etude de la β-lactoglobulin de brebis crystallisee. *Biochim. Biophys. Acta*, **107**, 501–10.
McAlpine, A.S. & Sawyer, L. (1990). β-Lactoglobulin: a protein drug carrier? *Biochem. Soc. Trans.*, **197,** 407–17.
McDougall, E.I. & Stewart, J.C. (1976). The whey proteins of the milk of red deer (*Cervus elaphus* L.). *Biochem. J.*, **153**, 647–55.
McDougall, E.I. & Stewart, J.C. (1977). The state of aggregation of red deer (*Cervus elaphus* L.) β-lactoglobulin preparations near neutral pH. *Biochem. J.*, **167**, 45–51.
McKenzie, H.A. (1967). Milk Proteins. *Adv. Prot. Chem.*, **22**, 147–234.
McKenzie, H.A. (1971). β-Lactoglobulins. In *Milk Proteins—II*, ed. H.A. McKenzie. Academic Press, New York, pp. 257–330.
McKenzie, H.A. & Ralston, G.B. (1971). The denaturation of proteins: two states? reversible or irreversible? *Experientia*, **27**, 617–24.
McKenzie, H.A. & Ralston, G.B. (1973). Nature of products formed by the action of urea on bovine β-lactoglobulins. *Aust. J. Biol. Sci.*, **26**, 859–70.
McKenzie, H.A. & Sawyer, W.H. (1967). The effect of pH on β-lactoglobulins. *Nature*, **214,** 1101–1109.
McKenzie, H.A. & Shaw, D.C. (1972). Alternative positions for the sulphydryl group in β-lactoglobulin: the significance for sulphydryl location in other proteins. *Nature*, **238**, 147–9.
McKenzie, H.A., Sawyer, W.H. & Smith, M.B. (1967). Optical rotatory dispersion and sedimentation of assocation–dissociation: bovine β-lactoglobulins near pH 5. *Biochim. Biophys. Acta*, **147**, 73–92.
McKenzie, H.A., Ralston, G.B. & Shaw, D.C. (1972). Location of sulphydryl and disulphide groups in bovine β-lactoglobulins and effects of urea. *Biochemistry*, **11**, 4539–47.
McKenzie, H.A., Muller, V.J. & Treacy, G.B. (1983). 'Whey' proteins of milk of the red (*Macropus rufus*) and Eastern grey (*Macropus giganteus*) kangaroo. *Comp. Biochem. Physiol.*, **74B**, 259–71.
McMeekin, T.L. & Warner, R.C. (1942) The hydration of β-lactoglobulin crystals. *J. Am. Chem. Soc.* **64** 2393–8.
McMeekin, T.L., Polis, B.D., Della-Monica, E.S. & Custer, J.H. (1949). a crystalline compound of β-lactoglobulin with dodecyl sulphate. *J. Am. Chem. Soc.*, **71**, 3606–9.
McRee, D.E., Tainer, J.A., Meyer, T.E., van Beeuman, J., Cusanovich, M.A. & Getzoff, E.D. (1989). Crystallographic structure of a photoreceptor protein at 2.4 Å resolution. *Proc. Nat. Acad. Sci.*, **86** , 6533–7.
Mercier, J.C. & Gaye, P. (1982). Early events in secretion of main milk proteins: occurrence of precursors. *J. Dairy Sci.*, **65**, 299–316.
Mercier, J.C., Gaye, P., Soulier, S., Hue-Delahaie, D. & Vilotte, J.L. (1985). Construction and identification of recombinant plasmids carrying cDNAs coding for ovine α_{s1}-, α_{s2}-, β-, κ-casein and β-lactoglobulin. Nucleotide sequence of α_{s1}-casaein cDNA. *Biochimie*, **67**, 959–71.
Mills, O.E. (1976). Effect of temperature on tryptophan fluorescence of β-lactoglobulin B. *Biochim. Biophys. Acta*, **434**, 324–32.
Mills, O.E. & Creamer, L.K. (1975). Conformational changes of bovine β-lactoglobulin at low pH. *Biochim. Biophys. Acta*, **379**, 618–26.
Miranda, G. & Pelissier, J.-P. (1983). Kinetic studies of *in vivo* digestion of bovine unheated skim-milk proteins in rat stomach. *J. Dairy Res.*, **50**, 27–36.
Mohammadzadeh-K., A., Feeney, R.E. & Smith, L.M. (1969). Hydrophobic binding of hydrocarbons by proteins. I. Relationship of hydrocarbon structure. *Biochim. Biophys. Acta*, **194**, 246–55.
Monaco, H.L., Zanotti, G., Spadon, P., Bolognesi, M., Sawyer, L. & Eliopoulos, E. (1987). Crystal structure of the trigonal form of bovine β-lactoglobulin and its complex with retinol at 2.5 Å resolution. *J. Mol. Biol.*, **197**, 695–706.
Monti, J.C., Mermoud, A.-F. & Jolles, P. (1989). Anti-bovine β-lactoglobulin antibodies react with a human lactoferrin fragment and bovine β-lactoglobulin present in human

milk. *Experentia*, **45**, 178–80.
Newcomer, M.E., Jones, T.A., Åqvist, J., Sundelin, J., Erikkson, U., Rask, L. & P. Peterson, A. (1984). The three dimensional structure of retinol-binding protein. *EMBO J.*, **3**, 1451–4.
North, A.C.T. (1989*a*). Three-dimensional arrangement of conserved amino acid residues in a superfamily of specific ligand-binding proteins. *Int. J. Biol. Macromol.*, **11**, 56–8.
North, A.C.T. (1989*b*). Application of molecular graphics for the study of recognition. *J. Mol. Graphics*, **7**, 67–70.
Novotny, J. & Franek, F. (1975). Different degrees of interspecies homology in immunoglobulin lambda chain constant domain correlated with three-dimensional structure. *Nature*, **258**, 641–3.
O'Neill, T.E. & Kinsella, J.E. (1987). Binding of alkanone flavors to β-lactoglobulin: effects of conformational and chemical modification. *J. Agri. Food Chem.*, **35**, 770–4.
Osterhoff, D.R. & Pretorius, A.M.G. (1966). Inhereited biochemical polymorphism. *Proc. S. Afri. Soc. Anim. Prod.*, **5**, 166–73.
Otani, H. & Hosono, A. (1987). Antigenic reactive regions of S-carboxymethylated β-lactoglobulin. *Agri. Biol. Chem.*, **51**, 531–6.
Otani, H., Morita, S. & Tokita, F. (1985*a*). Studies on the antigenicity of the browning product between β-lactoglobulin and lactose. *Jap J. Zootech. Sci.*, **56**, 67–74.
Otani, H., Uchio, T. & Tokita, F. (1985*b*). Antigenic reactivities of chemically modified β-lactoglobulins with antiserum to bovine β-lactoglobulin. *Agri. Biol. Chem.*, **49**, 2531–6.
Palmer, A.H. (1934). The preparation of a crystalline globulin from the albumin fraction of cow's milk. *J. Biol. Chem.*, **104**, 359–72.
Pantaloni, D. (1962). Role des groupes SH dans la structure de la β-lactoglobuline. *Compte. Rend.*, **254**, 1884–6.
Pantaloni, D. (1964). Etude de la transition R-S de la β-lactoglobuline par spectropolarimetrie et par spectrophotometrie de differences. *Compte. Rend. Seanc. Acad. Sci. Paris*, **258**, 5753–6.
Pantaloni, D. (1965). Structure et changements de conformations de la β-lactoglobuline en solution. Doctural Thesis, University of Paris.
Papiz, M.Z. (1982). Crystallographic studies of β-lactoglobulin at 2·8 Å resolution. Doctoral Thesis, Napier College, Edinburgh.
Papiz, M.Z., Sawyer, L., Eliopoulos, E.E., North, A.C.T., Findlay, J.B.C., Sivaprasadarao, R., Jones, T.A., Newcomer, M.E. & Kraulis, P.J. (1986). The structure of β-lactoglobulin and its similarity to plasma retinol-binding protein. *Nature*, **324**, 383–5.
Park, K.H. & Lund, D.B. (1984). Calorimetric study of the thermal denaturation of β-lactoglobulin. *J. Dairy Sci.*, **67**, 1699–706.
Pederson, K.O. (1936*a*). Introduction and preliminary results with fractionation of skim milk. *Biochem. J.*, **30**, 948–60.
Pederson, K.O. (1936*b*). The lactoglobulin of Palmer: Detected changes in sedimentation coefficient at pH 5·0 and 7·5. *Biochem. J.*, **30**, 961–70.
Peitsch, M.C. & Boguski, M.S. (1990). Is apolipoprotein D a mammalian bilin-binding protein? *The New Biologist*, **2**, 197–206.
Perez, M.D., Diaz de Villegas, C., Sanchez, L., Arando, P., Ena, J.M. & Calvo, M. (1989). Interaction of fatty acids with β-lactoglobulin and albumin from ruminant milk. *J. Biochem.*, **106**, 1094–7.
Perez-Gomez, G., Préaux, G. & Lontie, R. (1971). Location of the disulphide bridges in β-lactoglobulin B with pepsin and trypsin after CNBr cleavage. *Arch. Int. Physiol. Biochem.*, **79**, 843–4.
Pervaiz, S. & Brew, K. (1985). Homology of β-lactoglobulin, serum retinol-binding protein, and protein HC. *Science*, **228**, 335–7.
Pervaiz, S. & Brew, K. (1986). Purification and characterization of the major whey proteins from the milks of the Bottlenose dolphin (*Tursiops truncatus*), the Florida manatee (*Trichechus manatus latirostris*) and the beagle (*Canis familiaris*). *Arch. Biochem. Biophys.*, **246**, 846–54.

Pervaiz, S. & Brew, K. (1987). Homology and structure function correlations between α_1-acid glycoprotein and serum retinol-binding protein and its relatives. *FASEB*, **1**, 209–14.

Pessen, H. Purcell, J.M. & Farrell, H.M. (1985). Proton relaxation rates of water in dilute solutions of β-lactoglobulin. Determination of cross relaxation and correlation with structural changes by the use of two genetic variants of a self-associating globular protein. *Biochim. Biophys. Acta*, **828**, 1–12.

Pevsner, J., Reed, R.R., Feinstein, P.G. & Snyder, S.H. (1988). Molecular cloning of odorant-binding protein: member of a ligand carrier family. *FEBS Lett.*, **212**, 225–8.

Pierce, A.E. (1960). β-Lactoglobulins in the urine of the new-born suckled calf. *Nature*, **186**, 940–1.

Piez, K.A., Davie, E.W., Folk, J.E. & Gladner, J.A. (1961). β-Lactoglobulins A and B. *J. Biol. Chem.*, **236**, 2912–6.

Polis, P.D., Schmuckler, H.W., Custer, J.H. & McMeekin, T.L. (1950). Isolation of an electrophoretically homogeneous crystalline component of β-lactoglobulin. *J. Amer. Chem. Soc.*, **72**, 4965–8.

Préaux, G. & Lontie, R. (1972). Revised number of the free cysteine residues and of the disulphide bridges in the sequence of β-lactoglobulins A and B. *Arch. Int. Physiol. Biochim.*, **80**, 980–1.

Préaux, G., Braunitzer, G., Schrank, B. & Stangl, A. (1979). The amino acid sequence of goat β-lactoglobulin. *Hoppe Seyler's Z. Physiol. Chem.*, **360**, 1595–604.

Préaux, G., Braunitzer, G. & Kolde, H.-J. (1980). Primary structure of ovine β-lactoglobulin. *Arch. Int. Physiol. Biochem.*, **88**, (B45) 35–6.

Prutz, W.A., Butler, J., Land, E.J. & Swallow, A.J. (1980). Direct demonstration of electron transfer between tryptophan and tyrosine in proteins. *Biochem. Biophys. Res. Comm.*, **96**, 408–14.

Purcell, J.M. & Susi, H. (1984). Solvent denaturation of proteins as observed by resolution-enhanced Fourier transform infrared spectroscopy. *J. Biochem. Biophys. Meth.*, **9**, 193–9.

Puyol, P., Pérez, M.D., Ena, J.M. & Calvo, M. (1991). Interaction of bovine β-lactoglobulin and other bovine and human whey proteins with retinol and fatty acids. *Agric. Biol. Chem.*, **55**, 2515–20.

Rask, L., Anundi, H. & Peterson, P.A. (1979). The primary structure of human retinol binding protein. *FEBS. Lett.*, **104**, 55–8.

Reithel, F.J. & Kelly, M.J. (1971). Thermodynamic analysis of the monomer–dimer association of β-lactoglobulin A at the isoelectric point. *Biochem.*, **10**, 2639–44.

Riley, C.T., Barbeau, B.K., Keim, P.S., Kezdy, F.J., Heinrikson, R.L. & Law, J.H. (1984). The covalent protein structure of insecticyanin, a blue biliprotein from the hemolymph of the tobacco hornworm, *Manduca sexta* L. *J. Biol. Chem.*, **259**, 13159–65.

Robertson, D.M., Paganelli, R., Dinwiddie, R. & Levinsky, R.J. (1982). Milk antigen absorption in the preterm and term neonates. *Arch. Dis. Child.*, **57**, 369–72.

Robillard, K.A. & Wishnia, A. (1972*a*). Aromatic hydrophobes and β-lactoglobulin A. Thermodynamics of binding. *Biochemistry*, **11**, 3835–40.

Robillard, K.A. & Wishnia, A. (1972*b*). Aromatic hydrophobes and β-lactoglobulin A. Kinetics of binding by nuclear magnetic resonance. *Biochemistry*, **11**, 3841–5.

Roels, H., Préaux, G. & Lontie, R. (1966). Stabilization of β-lactoglobulin A and B at pH 8.9 by blocking the thiol groups. *Arch. Int. Biochem.*, **74**, 522–3.

Ruegg, M., Moor, U. & Blanc, B. (1975). Hydration and thermal denaturation of β-lactoglobulin. A calorimetric study. *Biochim. Biophys. Acta*, **400**, 534–42.

Sacchettini, J.C., Gordon, J.I. & Banaszak, L.J. (1989). The structure of crystalline *Escherichia coli*-derived rat intestinal fatty acid-binding protein at 2.5 Å resolution. *J. Biol. Chem.*, **363**, 5815–9.

Said, H.M., Ong, D.E. & Shingleton, J.L. (1989). Intestinal uptake of retinol: enhancement by bovine milk β-lactoglobulin. *Am. J. Clin. Nutr.*, **49**, 690–4.

Sawyer, L. (1987). One fold among many. *Nature*, **2327**, 659.
Sawyer, L. & Green, D.W. (1979). The reaction of cow β-lactoglobulin with tetracyanoaurate (III). *Biochim. Biophys. Acta*, **579**, 234–9.
Sawyer, L. & Richardson, J. (1991). Letter in *TIBS*, **16**, 11 only.
Sawyer, L., Papiz, M.Z., North, A.C.T. & Eliopoulos, E.E. (1955). Structure and function of bovine β-lactoglobulin. *Biochem. Soc. Trans.*, **13**, 265–6.
Sawyer, W.H. (1969). Complex between β-lactoglobulin and κ-casein. A review. *J. Dairy Sci.*, **52**, 1347–55.
Schellman, J.A. (1958). Rotatory properties of β-lactoglobulin. *Compt. Rend. Trav. Lab. Carlsberg Ser. Chim.*, **30**, 395–414.
Seibles, T.S. (1969). Interaction of dodecyl sulphate with native and modified β-lactoglobulin. *Biochemistry*, **8**, 2949–54.
Senti, F.R. & Warner, R.C. (1948). X-ray molecular weight of β-lactoglobulin. *J. Amer. Chem. Soc.*, **70**, 3318–20.
Sen, A. & Chaudhuri, S. (1962). Non-casein proteins of goat's milk. *Nature*, **195**, 286–7.
Sen, A. & Sinha, N.K. (1961). Comparison of the β-lactoglobulins of buffalo's and cow's milk. *Nature*, **190**, 343–4.
Simons, J.P., McClenaghan, M. & Clark, A.J. (1987). Alteration of the quality of milk by expression of sheep β-lactoglobulin in transgenic mice. *Nature*, **328**, 530–2.
Smith, L.M., Fantozzi, P. & Creveling, R.K. (1983). Study of triglyceride-protein interaction using a microemulsion-filtration method. *J. Am. Oil Chem. Soc.*, **60**, 960–7.
Spector, A.A. & Fletcher, J.E. (1970). Binding of long chain fatty acids to β-lactoglobulin. *Lipids*, **5**, 403–11.
Stone, W.L. & Wishnia, A. (1978). Binding of iodomercurates to sulphydryl-blocked β-lactoglobulin-A, -B and -C. *Bio. Inor. Chem.*, **8**, 517–29.
Su, Y.T. & Jirgensons, B. (1977). Further studies on detergent-induced conformational transitions in proteins. Circular dichroism of ovalbumin, bacterial α-amylase, papain, and β-lactoglobulin at various pH values. *Arch. Biochem. Biophys.*, **181**, 137–46.
Sutor, F., Kayser, H. & Zuber, H. (1988). The complete amino-acid sequence of the bilin-binding protein from *Pieris brassicae* and its similarity to a family of serum transport proteins like retinol-binding proteins. *Biol. Chem. Hoppe-Seyler*, **369**, 497–505.
Suzuki, S., Koritz, T.N. & Coombs, R.R. (1987). Antigenic stimulation with proteins of cows milk via the oral route in guinea-pigs and rates. 2. Antibodies to β-lactoglobulin secreted into the alimentary canal and serum. *Int. Arch. All. Appl. Immun.*, **82**, 76–82.
Svedberg, T. (1938). The ultracentrifuge and its field of research. *Ind. Eng. Chem.*, **10**, 113–29.
Swaisgood, H.E. (1982). Chemistry of milk protein. *Developments in Dairy Chemistry—1*, ed. P.F. Fox. Applied Science Publishers, London, pp. 1–59.
Tanford, C. & De, P.K. (1961). The unfolding of β-lactoglobulin at pH 3·0 by urea, formamide, and other organic substances. *J. Biol. Chem.*, **236**, 1711–5.
Tanford, C. & Taggart, V.G. (1961). Ionization-linked changes in protein conformation. II. The N-R transition in β-lactoglobulin. *J. Am. Chem. Soc.*, **83**, 1634–1638.
Tanford, C., Bunville, L.G. & Nozaki, Y. (1959). The reversible trans-formation of β-lactoglobulin at pH 7·5. *J. Am. Chem. Soc.*, **81**, 4032–6.
Tanford, C., De, P.K. & Taggart, V.G. (1960). The role of the α-helix in the structure of proteins. Optical rotatory dispersion of β-lactoglobulin. *J. Am. Chem. Soc.*, **82**, 6028–34.
Teller, D.C., Swanson, E. & DeHaen, C. (1979). The translational frictional coefficient of proteins. *Meth. in Enzymol.*, **61**, 103–24.
Thompson, M.P. (1970). Phenotyping milk proteins: A review. *J. Dairy Sci.*, **53**, 1341–8.
Thompson, M.P. & Farrell, H.M. (1974). Genetic variants of milk proteins. In *Lactation: A Comprehensive Treatise*, Vol. III, eds. B.L. Larson & V.R. Smith. Academic Press, New York and London, pp. 109–34.
Tilley, J.M.A. (1960). The chemical and physical properties bovine β-lactoglobulin. *Dairy Sci. Abstr.*, **22**, 111–25.
Timasheff, S.N. & Susi, H. (1966). Infrared investigation of the secondary structure of β-lactoglobulins. *J. Biol. Chem.*, **241**, 249–51.

Timasheff, S.N. & Townend, R. (1961). Molecular interactions in β-lactoglobulin. V. The association of the genetic species of β-lactoglobulin below the isoelectric point. *J. Am. Chem. Soc.*, **83**, 464–9.

Timasheff, S.N. & Townend, R. (1964). Structure of the β-lactoglobulin tetramer. *Nature*, **203**, 517–9.

Timasheff, S.N., Townend, R. & Mescanti, L. (1966*a*). The optical rotatory dispersion of β-lactoglobulins. *J. Biol. Chem.*, **241**, 1863–70.

Timasheff, S.N., Mescanti, L., Basch, J.J. & Townend, R. (1966*b*). Conformational transitions of bovine β-lactoglobulins A, B and C. *J. Biol. Chem.*, **241**, 2496–501.

Tirindelli, R., Keen, J.N., Cavaggioni, A., Eliopolous, E.E. & Findlay J.B.C. (1989). Complete amino acid sequence of pyrazine-binding protein from cow nasal mucosa. *Eur. J. Biochem.*, **185**, 569–72.

Townsend, R. & Timasheff, S.N. (1957). The molecular weight of β-lactoglobulin. *J. Am. Chem. Soc.*, **79**, 3613–3614.

Townend, R. & Timasheff, S.N. (1960). Molecular interactions in β-lactoglobulin III. Light scattering investigation of the stoichiometry of the association between pH 3·7 and 5·2. *J. Am. Chem. Soc.*, **82**, 3168–74.

Townend, R., Winterbottom, R.J. & Timasheff, S.N. (1960*a*). Molecular interactions in β-lactoglobulin. II. Ultracentrifugal and electrophoretic studies of the association of β-lactoglobulin below its isoelectric point. *J. Am. Chem. Soc.*, **82**, 3161–8.

Townend, R., Weinberger, L. & Timasheff, S.N. (1960*b*). Molecular interactions in β-lactoglobulin. IV. The dissociation of β-lactoglobulin below pH 3·5. *J. Am. Chem. Soc.*, **82**, 3175–9.

Townend, R., Herskovits, T.T., Swaisgood, H.E. & Timasheff, S.N. (1964). Solution properties of β-lactoglobulin C. *J. Biol. Chem.*, **239**, 4196–201.

Townend, R., Kumosinski, T.F. & Timasheff, S.N. (1967). The circular dichroism of variants of β-lactoglobulin. *J. Biol. Chem.*, **242**, 4538–45.

Townend, R., Herskovits, T.T., Timasheff, S.N. & Gorbunoff, M.J. (1969). The state of amino acid residues in β-lactoglobulin. *Arch. Biochem. Biophys.*, **129**, 567–80.

Treece, J.M., Sheinson, R.S. & McMeekin, T.L. (1964). The solubilities of β-lactoglobulins A, B, and AB. *Arch. Biochem. Biophys.*, **108**, 99–108.

Unterman, R.D., Lynch, K.R., Nakhasi, H.L., Dolan, K.P., Hamilton, J.W., Cohn, D.V. & Feigelson, P. (1981). Cloning and sequence of several α_{2u}-globulin cDNAs. *Proc. Nat. Acad. Sci.*, **78**, 3478–82.

Whitney, R.McL., Brunner, J.R., Ebner, K.E., Farrell, H.M., Jr., Josephson, R.V., Morr, C.V. & Swaisgood, H.E. (1976). Nomenclature of the proteins of cows milk: fourth revision. *J. Dairy Sci.*, **59**, 798–815.

Willis, I.M., Stewart, A.F., Caputo, A., Thompson, A.R. & Mackinlay, A.G. (1982). Construction and identification by partial nucleotide sequence analysis of bovine casein and β-lactoglobulin cDNA clones. *DNA*, **1**, 375–86.

Wishnia, A. & Pinder, T.W., Jr. (1966). Hydrophobic interactions in proteins. The alkane binding site of β-lactoglobulins A and B. *Biochemistry*, **5**, 1534–42.

Witz, J., Timasheff, S.N. & Luzzati, V. (1964). Molecular interaction in β-lactoglobulin. VIII. Small-angle X-ray scattering investigation of the geometry of β-lactoglobulin. A tetramerisation. *J. Am. Chem. Soc.*, **86**, 168–73.

Woo, S.L., Creamer, L.K. & Richardson, T. (1982). Chemical phosphorylation of bovine β-lactoglobulin. *J. Agri. Food Chem.*, **30**, 65–70.

Workman, R.J., McKown, M.M. & Gregerman, R.I. (1974). Renin: inhibition by proteins and peptides. *Biochemistry*, **13**, 3029–35.

Yang, J.T., Wu, C.-S.C. & Martinez, H.M. (1986). Calculation of protein conformation from circular dichroism. *Meths. Enzymol.*, **130**, 208–69.

Yewdall, S.J. (1988). Structural studies on β-lactoglobulin. PhD Thesis, University of Leeds.

Yoshikawa, M., Mizukami, T., Sasaki, R. & Chiba, H. (1978). Pre-β-lactoglobulin synthesis by mRNA from bovine mammary gland. *Agri. Biol. Chem.*, **42**, 2185–6.

Zimmerman, J.K., Barlow, G.H. & Klotz, I.M. (1970). Dissociation of β-lactoglobulin near neutral pH. *Arch. Biochem. Biophys.*, **138**, 101–9.

5

α-LACTALBUMIN

KEITH BREW & JAY A. GROBLER
Department of Biochemistry and Molecular Biology, University of Miami School of Medicine, Miami, Florida 33101, USA

1. INTRODUCTION

Since the ability to produce milk is a characteristic that defines the mammals as a group, it is reasonable to believe that lactation arose during an early phase of mammalian evolution. Despite this relatively recent origin (in terms of biochemical evolution), milk production involves the biosynthesis of a considerable number of milk-specific proteins, as well as distinctive metabolic pathways and is associated with quite complex regulatory mechanisms. Lacto-proteins and enzyme activities that are uniquely associated with milk production are therefore of considerable interest from a biochemical and evolutionary perspective, as well as having practical and economic significance in relation to the production of milk.

The subject of this review, α-lactalbumin (LA), is a low molecular weight (14·2 kDa) globular whey protein that was first crystallized in the nineteenth century, and was the subject of many studies during the early years of protein chemistry (see McKenzie (1967) or Gordon (1971) for reviews of this earlier work). Despite this extended history, the results of investigations of LA have continued to produce unexpected findings, e.g. the discovery of its role as a specificity regulator in lactose synthase (Brodbeck & Ebner, 1966; Brodbeck *et al.*, 1967; Brew *et al.*, 1968), the demonstration of its homology with the type-c lysozymes (Brew & Campbell, 1967; Brew *et al.*, 1968) and, most recently, the realization that it is a calcium metalloprotein (Hiroaka *et al.*, 1980).

Earlier reviews have emphasized the function of LA in lactose biosynthesis and its relationship with lysozyme (Brew, 1970; Hill & Brew, 1975) and the molecular genetics and the regulation of the expression of LA (Hall & Campbell, 1986). A recent extensive review by Kronman (1989) discusses the metal-binding and physico-chemical properties of LA in considerable depth. To avoid repetition of material in these previous accounts, the present authors have provided a relatively brief overview of earlier studies with a more detailed discussion of areas that have been illuminated by recent advances. These relate to the 3D structure of LA, and its relationship to its role in lactose synthase, the complex evolutionary relationship between LA and different subgroups of the type-c lysozymes and the structure and function of the Ca^{2+}-binding site.

2. Overview of Earlier Work

2.1. Role in Lactose Synthase

Lactose synthase consists of two distinct protein components: the catalytic component, galactosyltransferase and the regulatory component, LA. Galactosyltransferase (UDP-galactose N-acetylglucosamine β-4 galactosyltransferase), is found in a wide range of secretory cells in addition to the mammary epithelial cell and in most situations acts in isolation in catalyzing the transfer of galactose from UDP-galactose into a β-linkage with the 4-hydroxy group of β-linked N-acetylglucosamine in N-linked oligosaccharides in glycoproteins:

$$\text{UDP-Gal} + \text{GlcNAc } \beta\text{1-Mann} \ldots \text{Asn} \rightarrow \text{Gal } \beta\text{1-4GlcNAc } \beta\text{1-Mann} \ldots \text{Asn} + \text{UDP} \qquad (1)$$

This represents a later processing step which occurs after the transfer of a core oligosaccharide from a dolichol-pyrophosphoryl carrier to an asparagine in the glycoprotein, followed by a trimming of sugars (glucose and mannose) and the enzymatic addition of N-acetylglucosamine. During these events, the glycoprotein is transported from the rough endoplasmic reticulum to the golgi apparatus, where galactosyltransferase is an intrinsic component of the membranes of the trans region (see Strous, 1986). Galactosyltransferase can be conveniently assayed using the free monosaccharide, N-acetylglucosamine (GlcNAc), as a model substrate. A proportion of cellular galactosyltransferase is present on the cell surface, where it has been proposed to function in certain cell–cell interactions (see Shur, 1984), and some is secreted, following partial proteolysis, into milk and serum. Most studies of galactosyltransferase and lactose synthase have utilized the soluble enzyme form isolated from milk (see Hill & Brew (1975) for a discussion of purification procedures for soluble galactosyltransferase).

Galactosyltransferase in isolation does not catalyze lactose synthesis (reaction 2) effectively because of its very poor affinity for glucose (reflected in a K_m of about 2 M).

$$\text{UDP-Gal} + \text{Glucose} \rightarrow \underset{\text{(Lactose)}}{\text{Gal } \beta\text{1-4Glucose}} + \text{UDP} \qquad (2)$$

The function of LA in lactose synthase is that of promoting the binding of glucose to galactosyltransferase, reducing the K_m by up to 1 000-fold, so that lactose is synthesized at physiological concentrations of glucose. At the same time, LA acts as a competitive inhibitor with respect to glycoprotein or oligosaccharide substrates of galactosyltransferase (Powell & Brew, 1976; Bell *et al.*, 1976). With free

N-acetylglucosamine as substrate, LA reduces the K_m and V_m so that activation is obtained at low substrate concentrations and inhibition at higher concentrations. As an inhibitor of this reaction, LA appears to bind to complexes containing N-acetylglucosamine and has a much lower K_i than that associated with its inhibition of galactose transfer to glycoproteins. The K_i for the last reaction is equivalent to the K_d associated with LA activation of glucose binding for lactose synthesis (see Powell & Brew, 1976).

2.2. LA Secretion and the Organization and Regulation of Lactose Synthase

LA is produced specifically in rough endoplasmic reticulum of the lactating mammary gland together with other milk proteins as a pre-protein with a hydrophobic leader sequence (see Hall & Campbell, 1986). After translocation across the ER membrane and proteolytic removal of the leader sequence, the LA molecule is transported to the golgi apparatus, where galactosyltransferase is located. In this subcellular compartment, it performs its role as a K_m modulator of galactosyltransferase through reversible binding with the catalytic domain, which projects into the lumen of the golgi (see Section 5.2). Lactose, being synthesized in the lumen of the golgi, is too large to traverse the golgi membrane to the cytosol and accumulates, generating a flow of cellular water into the golgi to maintain osmotic equilibrium. Since lactose is the principal osmole in most milks, this process is thought to generate a major part of the aqueous phase of milk (see Brew, 1970; Linzell & Peaker, 1971). LA interacts only temporarily with galactosyltransferase and is secreted into milk, together with lactose, by exocytosis in secretory vesicles formed from the golgi membranes. This arrangement represents a mechanism through which the hormonally-regulated synthesis of LA, and other milk proteins, is coupled with lactose biosynthesis since, if the production of active LA molecules ceases, the remaining protein in the golgi lumen will be secreted and lactose production terminated (Brew, 1970).

While the organization of lactose synthase has intrinsic regulatory implications, it is likely that other factors, such as glucose transport into the golgi (White *et al.*, 1981), will influence the rate of production of lactose *in vivo*.

2.3. Relationship with the Type-c Lysozymes

The close evolutionary relationship of LA with chicken lysozyme and its avian and mammalian homologues was first demonstrated by sequence studies of bovine LA (Brew *et al.*, 1967) and has been confirmed by the protein sequences of other LAs (see Shewale *et al.*, 1984), the cDNA sequences of LAs and lysozymes (Dandekar & Qasba, 1981; Hall *et al.*, 1982), as well as the gene structures of LAs and lysozymes (see Section 3.1).

Because the type-c lysozymes are widely distributed in the animal kingdom, being found in birds, mammals, reptiles and even insects (see Jolles & Jolles, 1984), whereas LA and lactose are uniquely mammalian products, it is reasonable to think that the ancestor of the lysozyme–LA gene superfamily was a lysozyme structural gene. The simplest scheme for explaining the origin of LA involves the duplication of a lysozyme structural gene following which one copy continues to encode a functional lysozyme, while the second gene copy diverges in structure to eventually produce LA and consequently the ability of mammals to synthesize lactose. Since galactosyltransferase is found in non-mammalian groups (Hill & Brew, 1975), the evolutionary development of lactose synthesis was dependent on the acquisition of LA activity. Additional LA and lysozyme protein and cDNA sequences are now available together with improved methods for generating phylogenetic trees from sequence differences. Recent analyses indicate that the scheme described above is probably incorrect and that the LA and lysozyme gene lines diverged considerably earlier than the divergence of birds and mammals (Prager & Wilson, 1988). Since lactose synthase activity is found only in mammals, this conclusion implies that a protein with a distinct function was formed during the development of LA from lysozyme. This putative intermediate may have been identified through the discovery of the Ca^{2+}-binding lysozymes (see Section 3.2), a group of type-c lysozymes, found in the milks of some mammals, which are clearly the products of a paralogous lysozyme gene line of considerable antiquity, and are more similar to the LAs than are the better-studied type-c lysozymes.

Despite their similarity in structure, LA and lysozyme do not show any significant crossover in biochemical activity. An exception to this is described in a report by Hopper and McKenzie (1974) regarding a protein fraction from the milk of the echidna, a monotreme, which was found to have both lysozyme activity and a low level of LA activity with bovine galactosyltransferase. However, this protein was isolated as a drop-through fraction from a DEAE-Sephadex column at pH 7·9 and does not appear to have been further purified. Similar fractions from the milks of other individual echidnas showed lysozyme but not LA activity. As far as the authors are aware, there have been no further reports regarding the structure or activity of this interesting protein. Based on present knowledge of structure–function relationships in LA and lysozyme, it is difficult to see how both activities could residue in a single protein. However, the possibility of the existence of such a protein, representing an intermediate in LA evolution, cannot be excluded.

2.4. Mechanism of Galactosyltransferase and Lactose Synthase

Current views of the mechanism of galactosyltransferase-catalyzed reactions and the mechanism of action of LA have been derived from steady-state kinetic studies of the action of galactosyltransferase in the absence and presence of LA (Morrison & Ebner, 1971*a*, *b*; Khatra *et al.*, 1974; Bell *et al.*, 1976; Powell & Brew, 1976) as well as the properties of covalently crosslinked LA-galactosyltransferase complexes (Brew *et al.*, 1975; Bell *et al.*, 1976) and studies of complexes formed between galactosyltransferase and substrates, inhibitors and LA. From this work, it can be concluded that in the absence of ligands (substrates and metal ions), LA and galactosyltransferase do not form detectable complexes, whereas the addition of Mn^{2+}, UDP-derivatives and monosaccharides (glucose or GlcNAc) progressively strengthens the formation of 1:1 molecular complexes of the two proteins. In conjunction with kinetic studies, these results lead to the conclusion that the effects of LA are related to interactions involving acceptor substrates (mono- or oligo-saccharides) and LA with galactosyltransferase. With monosaccharide substrates there is synergism between sugar and LA binding, so that the binding of LA enhances sugar binding and sugar binding enhances LA binding. With oligosaccharide (e.g. diacetylchitobiose) or glycoprotein substrates, there is mutually exclusive binding of substrate and LA, so that LA is a competitive inhibitor of the transferase reaction.

The nature of the effects of LA were used to develop a model for the arrangement of binding sites for acceptor substrates and LA on galactosyltransferase (Brew *et al.*, 1979) that does not require any allosteric regulatory effects from LA binding; such effects would be inconsistent with the kinetic effects of LA, which indicate rapid equilibrium binding of LA to enzyme complexes during the catalytic cycle (Khatra *et al.*, 1974; Bell *et al.*, 1976; Powell & Brew, 1976). This model is illustrated in Fig. 1. A closely similar view of the mechanism of LA action has since been proposed by Lambright *et al.* (1985). The acceptor binding site of galactosyltransferase is viewed as containing at least two subsites that are capable of binding monosaccharides, one of which can bind free N-acetylglucosamine or glucose (with low affinity) substrates (S_1) while the second (S_2) binds an additional sugar in extended substrates. This arrangement is consistent with the observation that oligosaccharides with non-reducing terminal GlcNAc have lower K_m values than the monosaccharide GlcNAc (Prieels *et al.*, 1976). The binding site for LA overlaps the S_2 site, so that LA and oligosaccharides bind in a mutually exclusive manner. In the model, the binding of LA at this site in galactosyltransferase, adjacent to the monosaccharide substrate binding site, brings a neighboring site in LA close to S_1. This site in LA, (S^{LA}), which is proposed to be a region corresponding to part of the lysozyme active site cleft, forms favorable, stabilizing interactions with a monosaccharide bound to S_1. This three-way interaction readily explains the synergism between monosaccharides and LA for complex formation with galactosyltransferase, and also provides a functional correlation between LA and lysozyme. The proximity of the binding site for LA to the monosaccharide binding site is further supported by the finding that extension of a monosaccharide by a few carbons at the 1- or 2-position of the pyranose ring results in competitive inhibition by LA (Bell *et al.*, 1976; Lambright *et al.*, 1985).

3. Structure

3.1. Primary Structure

Complete amino acid sequences have been determined for LAs from 10 mammalian species by direct protein sequencing or translation from cDNA sequences or a combination of the two approaches. Only one of these is from a non-eutherian mammal, the red-necked wallaby (see Shewale *et al.* (1984) and Nitta & Sugai (1989) for references).

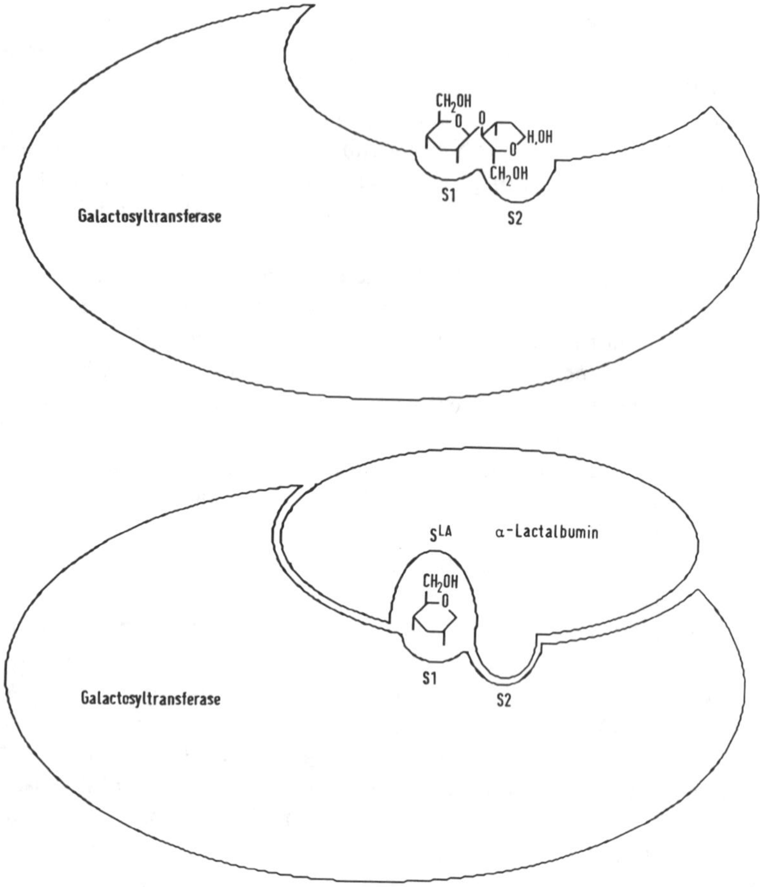

Fig. 1. A model for the molecular basis of the effects of α-lactalbumin on the acceptor substrate specificity of galactosyltransferase (see text for an explanation).

The currently known complete LA sequences are displayed in Fig. 2 in alignment with six selected lysozyme sequences, of which three bind Ca^{2+} and three are of the 'conventional' or non-Ca^{2+}-binding type (see Section 1). One additional Ca^{2+}-binding lysozyme has been sequenced, from the donkey (Godovac-Zimmerman *et al.*, 1988), but since this differs from the horse milk protein at only three positions it is not included here. Figure 2 also indicates sites that are conserved in all known LA sequences. It can be seen that these residues can be subdivided into: a group (printed in boldface in the LA and lysozyme sequences) which is also conserved in all known lysozymes, a further group which is conserved in both LAs and lysozymes but are different in the two groups (italics), residues that are conserved in all of the Ca^{2+}-binding proteins (boldfaced and underlined), and residues that are conserved in LAs but not lysozymes (printed in boldface in the LA sequences only). While it is possible that some of these amino acids may vary in other presently-unknown LA sequences, their high level of conservation suggests that

```
ALPHA-LACTALBUMINS
           1         10            20           30          40            50          60          70
           !          !             !            !           !             !           !           !
Bovine     EQLTKCEVFRELK--DLKGYGGVSLPEWVCTTFHTSGYDTQAIVQNN--DSTEYGLFQINNKIWCKDDQNPHSSN
Sheep      EQLTKCEAFQKLK--DLKDYGGVSLPEWVCTAFHTSGYDTQAIVQNN--DSTEYGLFQINNKIWCKDDQNPHSRN
Goat       EQLTKCEVFQKLK--DLKDYGGVSLPEWVCTAFHTSGYDTQAIVQNN--DSTEYGLFQINNKIWCKDDQNPHSRN
Camel      KQFTKCKLSDELK--DMNGHGGITLAEWICIIFHMSGYDTETVVSNN--GNREYGLFQINNKIWCRDNENLQSRN
Horse      KQFTKCQLSQVLK--SMDGYKGVTLPEWICTIFHNSGYDTQTIVKNN--GKTEYGLFQINNKMWCRDNQILPSRN
Human      KQFTKCELSQLLK--DIDGYGGIALPELICTMFHTSGYDTQAIVENN--ESTEYGLFQISNKLWCKSSQVPQSRN
Guinea pig KQLTKCALSHELN--DLAGYRDITLPEWLCIIFHISGYDTQAIVKNS--DHKEYGLFQINDKDFCESSTTVQSRN
Rat        TEFTKCEVSHAIE--DMDGYQGISLLEWTCVLFHTSGYDSQAIVKNN--GSTEYGLFQISNRNWCKSSEFPESEN
Rabbit     TQLTRCELTEKLK--ELDGYRDISMSEWICTLFHTSGLDTKITVNNN--GSTEYGIFQINSKLWCVSKQNPQSKN
Macropus   IDYRKCQASQILKEHGMD--KVIPLPELVCTMFHISGLSTQAEVNNH--SNKEYGIFQISNDGWCAEKQEDVANS
LYSOZYMES
Calcium-binding
Dog        KIFSKCELARKLKSMGMDGFHGYSLANWVCMAEYESNFNTQAFNGRNSNGSSDYGIFQLNSKWWCKSNSHS-SAN
Horse      KVFSKCELAHKLKAQEMDGFGGYSLANWVCMAEYESNFNTRAFNGKNANGSSDYGLFQLNNKWWCKDNKRS-SSN
Pigeon     KDIPRCELVKILRRHGFEGFVGKTVANWVCLVKHESGYRTTAFNNNG-PNSRDYGIFQINSKYWCNDGKTRGSKN
Conventional
Human      KVFERCELARTLKRLGMDGYRGISLANWMCLAKWESGYNTRATNYNAGDRSTDYGIFQINSRYWCNDGKTPGAVN
Chicken    KVFGRCELAAAMKRHGLDNYRGYSLGNWVCAAKFESNFNTQATNRNT-DGSTDYGILQINSRWWCNDGRTPGSRN
Moth       KRFTRCGLVQELRRLGFD---ETLMSNWVCLVENESGRFTDKIGKVNKNGSRDYGLFQINDKYWCSKGSTPG-KD

ALPHA-LACTALBUMINS
                    80          90         100         110           120    124
                     !           !           !           !             !      !
Bovine     ICNISCDKFLDDDLTDDIMCVKKIL-DKVGINYWLAHKALCSEK-LDQWL--CEKL
Sheep      ICNISCDKFLDDDLTDDIVCAKKIL-DKVGINYWLAHKALCSEK-LDQWL--CEKL
Goat       ICNISCDKFLDDDLTDDIVCAKKIL-DKVGINYWLAHKALCSEK-LDQWL--CEKL
Camel      ICDISCDKFLDDDLTDDKMCAKKIL-DKEGIDYWLAHKPLCSEK-LEQWQ--CEKW
Horse      ICGISCNKFLDDDLTDDVMCAKKIL-DSEGIDYWLAHKPLCSEK-LEQWL--CEEL
Human      ICDISCDKFLDDDITDDIMCAKKIL-DIKGIDYWLAHKALCTEK-LEQWL--CEKL
Guinea pig ICDISCDKLLDDDLTDDIMCVKKIL-DIKGIDYWLAHKPLCSDK-LEQWY--CEAQ
Rat        ICDISCDKFLDDELADDIVCAKKIV-AIKGIDYWKAHKPMCSEK-LEQWR--CEKPGAPALVVPALNSETPVP
Rabbit     ICDTPCENFLDDNLTDDVKCAMKIL-DKEGIDHWLAHKPLCSEN-LEQWV--CKK
Macropus   VCGILCSKFLDDDITDDIECAKKILQLPEGLGYWKAHETFCLED-LDQWR--C
LYSOZYMES
Calcium-binding
Dog        ACNIMCSKFLDDNIDDDIACSKRVVKDPNGMSAWVAWVKHCKGKDLSKYLASCNL
Horse      ACNIMCSKLLDENIDDDISCAKRVVRDPKGMSAWKAWVKHCKDKDLSEYLASCNL
Pigeon     ACNINCSKLRDDNIADDIQCAKKIAREARGLTPWVAWKKYCQGKDLSSYVRGC
Conventional
Human      ACHLSCSALLQDNIADAVACAKRVVRDPQGIRAWVAWRNRCQNRDVRQYVQGCGV
Chicken    LCNIPCSALLSSDITASVNCAKKIVSDGNGMNAWVAWRNRCKGTDVQAWIRGCRL
Moth       -CNVTCNQLLTDDISVAATCAKKIY-KRHKFDAWYGWKNHCQHG-LPDISD-C
```

Fig. 2. Sequence alignment of the α-lactalbumin (LA) and representative calcium binding and non-calcium binding lysozymes. Residues that are conserved in both LA and lysozyme sequences are printed in boldface. Residues conserved only in the LAs are also printed in boldface but in the LA sequences only. Residues that are conserved in both LAs and lysozymes but are different in the two groups are printed in italics, and residues conserved in the calcium binding proteins are boldfaced and underlined.

most of them are important for the structure and/or action of LA. Their roles in the 3D structure of LA and lysozyme and functional roles are further discussed in Sections 3.2 and 5.1.

It should be noted that the marsupial LA has a residue at position 97 that is not present in the eutherian proteins and also differs in the location of a gap near the amino terminal end of the sequence that is required to optimally align the LAs with the lysozymes. The LAs also differ in size at the C-terminus where although most have three residues following the cysteine at position 121, the rabbit protein has two and the marsupial protein, none. Rat LA has a C-terminal extension of 20 residues following the corresponding cysteine. A similar 'tail' was also present in the translated sequence from a cDNA clone for guinea-pig LA (Hall *et al.*, 1982), but in this case it appears to represent a cloning artifact since protein sequencing (Brew, 1972) clearly shows that guinea-pig milk LA terminates at glutamine 123 (position 124 in Fig. 1). Bovine, sheep, goat, rabbit, human, rat and wallaby LAs have a consensus N-glycosylation sequence (Asn–X–Ser) at residues 45–47. In bovine LA (and probably in the other ruminants), a small proportion (around 10%) of molecules are glycosylated, whereas rabbit and rat LAs appear to be uniformly glycosylated at this site (Hopp & Woods, 1979; Prasad *et al.*, 1982). This difference has been attributed to the sequence at this site in the rabbit and rat, Asn–Gly–Ser, being a more effective substrate for the N-glycosylation system than the corresponding ruminant sequence, Asn–Asp–Ser (Prasad *et al.*, 1982). However, peptides derived from this region of bovine LA have been found to be good substrates for the dolichol pyrophosphoryl oligosaccharide protein oligosaccharide transferase *in vitro* (Struck *et al.*, 1978). The other LAs do not appear to be significantly glycosylated.

3.1.1. Evolutionary Relationship with Different Lysozyme Groups

As discussed in Section 1, the identification of the Ca^{2+}-binding lysozymes has required a new evaluation of the evolutionary origins of LA. The discovery of these enzymes originated from the observation by Stuart *et al.* (1986) that the sequence of horse milk lysozyme contained the aspartyl residues identified as functioning in Ca^{2+} liganding in the Ca^{2+}-binding site of baboon LA. Horse lysozyme was earlier noted to be highly divergent in sequence from other mammalian and avian lysozymes (McKenzie & Shaw, 1985). Subsequently, horse milk and pigeon egg lysozymes (Rodruiges *et al.*, 1985) have been shown to bind Ca^{2+} with binding constants of 2×10^6 and $1{\cdot}6 \times 10^7 \text{M}^{-1}$, respectively, comparable to the binding constant of LA measured under the same conditions (Nitta *et al.*, 1988). Ca^{2+}-binding lysozymes have now been obtained from donkey milk and from the milks of the dog and hooded seal which are similar in sequence to the equine protein but very divergent from the other mammalian lysozymes (K.R. Rao, J.A. Grobler & K. Brew, unpublished). It can be seen from Fig. 2 that the mammalian Ca^{2+}-binding lysozymes have several regions of characteristic sequence that distinguish them from the 'conventional' lysozymes in addition to the presence of aspartyl residues that act in Ca^{2+}-liganding.

Present information therefore indicates that the Ca^{2+}-binding mammalian lysozymes are major components of the milk whey fraction (> 1 mg/ml) of at least two mammalian orders, the Carnivores and the Perissodactyls. The biological significance of the unusually high milk lysozyme levels in these species is unclear and it is also not known if similar proteins are expressed in other mammals or in tissues other than the lactating mammary gland. Pigeon lysozyme does not readily group with any of the other proteins in the alignment, being highly divergent in sequence from the mammalian Ca^{2+}-binding lysozymes as well as from the other known avian lysozymes.

Nitta and Sugai (1989) have re-examined the relationship between the LAs and lysozymes using all the protein sequences known at that time. Consequently, their analysis did not include the Ca^{2+}-binding canine lysozyme, but did include both the horse and pigeon enzymes. They quantified differences between all pairs of proteins using the genetic code matrix (minimal number of nucleotide base changes required to interconvert sequences), and used these scores to generate the most parsimonious phylogenetic trees (Fitch & Margoliash, 1967). The three best trees based on this criterion were, however, rejected by these authors to minimize the number of gene duplications implied by the arrangement of sequences on the tree and the number of acquisitions and losses of the Ca^{2+}-binding function. Their preferred evolutionary sequence is shown in Fig. 3(a). In this sequence, a gene for a non Ca^{2+}-binding lysozyme is duplicated to generate ancestral genes for the non Ca^{2+}-binding 'conventional' lysozymes found in birds and mammals, and the Ca^{2+}-binding lysozyme/LA group. Ca^{2+}-binding activity was acquired in the later line prior to a gene duplication event that generated the ancestral gene for LA. The authors (Grobler, J.G., Rao, K.R., Pervaiz, S. & Brew, K., unpublished results) have generated phylogenetic trees for the lysozymes and LAs using a somewhat different approach. Sequence differences were calculated using a different matrix (Mutation Data Matrix; Dayhoff *et al.*, 1983) and were corrected for the level of similarity expected for random sequences of the same composition as well as for the probability of multiple substitutions at the same site as described by Feng and Doolittle (1987). The tree was also generated using a procedure in which the tree was built starting with the most similar pair of sequences (Farris, 1972). Despite the differences in methodology, the result was identical in topology to the most parsimonious tree generated by Nitta and Sugai (1989), which had been rejected by these workers. It can be argued that the basis for rejecting this tree was not valid since multiple gene duplications are clearly common in the lysozyme lineage (e.g. Dobson *et al.*, 1984) and losses of Ca^{2+}-binding function have been documented in other protein families. The evolutionary scheme implied by this phylogenetic tree is indicated in Fig. 3(b). Assuming that the point at which the insect lysozyme joins the tree represents the 'root', it appears that pigeon lysozyme represents a gene line that diverged at a point earlier than the divergence of the LA and conventional lysozyme lines. From a comparison of the sequence of pigeon and other lysozymes, Rodruigez *et al.* (1987) also reached the conclusion that it represents a distinct gene line from the 'conventional' and equine lysozymes. However, they did not include LA sequences in their comparisons.

In the scheme shown in Fig. 3(b), a later gene duplication represents the point at which the 'conventional' lysozyme and LA lines diverged. The least certain aspect of the phylogenetic tree is the location of the branch point leading to the Ca^{2+}-binding lysozyme line. We find that when other scoring methods are used to quantify the degree of divergence of the sequences, it appears to branch from the LA line whereas when the procedure described above is used, it appears to diverge from the 'conventional' lysozyme line. In both cases, it diverges at a point close to the divergence point of the LA and 'conventional' lysozyme lines. It can be concluded that more protein sequence data and perhaps gene structures are required to resolve this question.

Despite this uncertainty, these results do lead to the conclusion that Ca^{2+}-binding activity is not a recent acquisition of the protein family since it is present in the earlier-diverging pigeon lysozyme line. The nearly-ideal bipyramidal arrangement of Ca^{2+}-liganding groups in the binding site and the high affinity for Ca^{2+} are

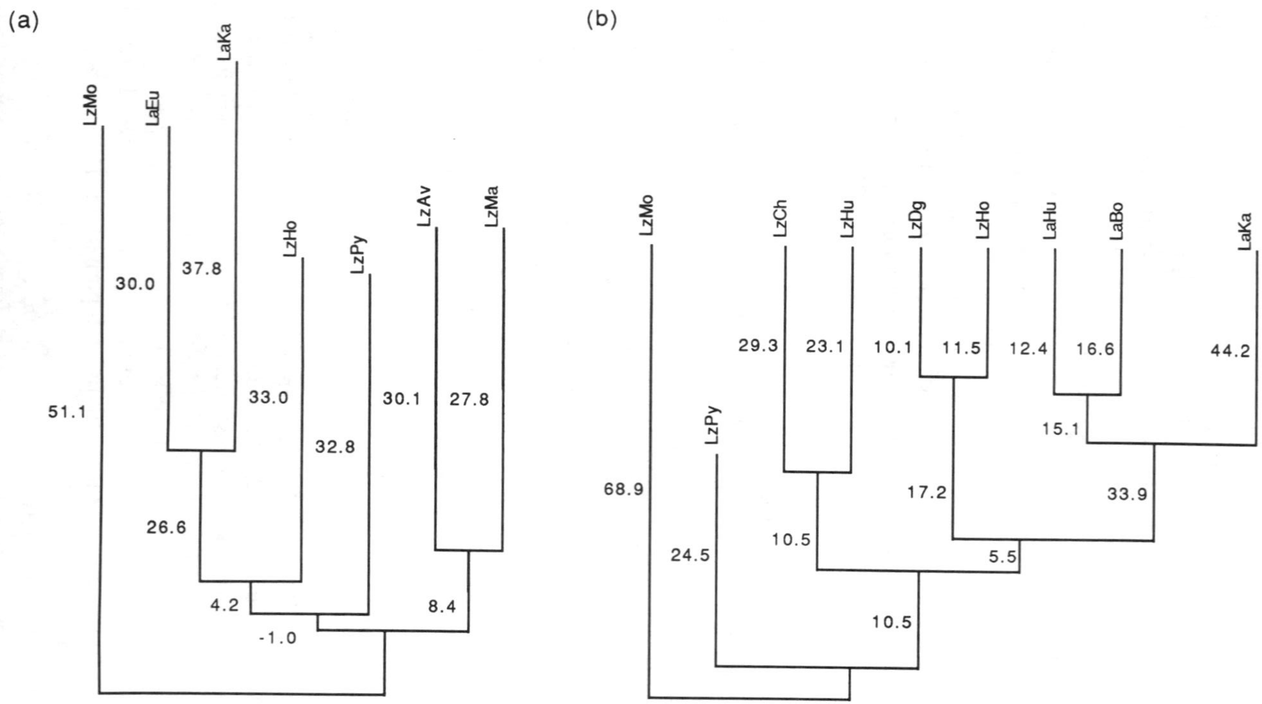

Fig. 3. Phylogenetic trees for α-lactalbumin (LA) and lysozyme based on sequence differences. The tree on the left is that selected by Nitta and Sugai (1989), while that on the right was constructed as described in the text. Abbreviations for (a) are LzMo, Hyalophora crecopia lysozyme; LaEu, eutherian mammal LA (average); LaWa, Wallaby LA; LzHo, horse lysozyme; LzPy, pigeon egg lysozyme; LzAv, avian egg lysozyme (average); LzMa, 'conventional' mammalian lysozyme (average). In (b), the additional abbreviations are: LzCh, chicken egg white lysozyme; LzHu, human lysozyme; LzDg, dog milk lysozyme; LaHu, human LA; LaBo, bovine LA.

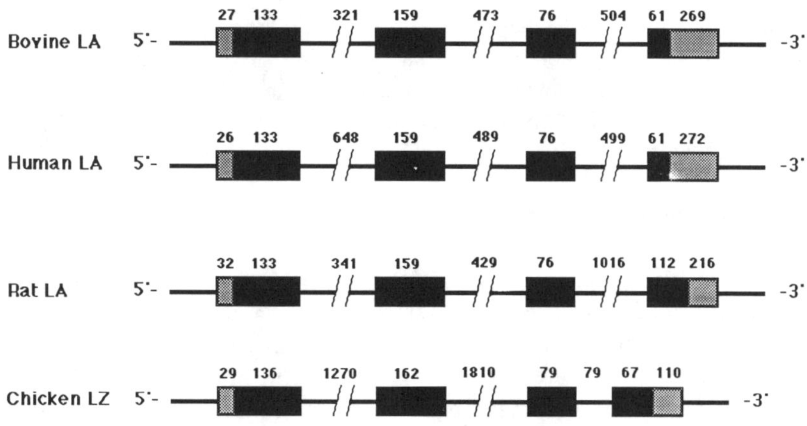

Fig. 4. Schematic structures of the genes for α-lactalbumins and lysozymes. The open blocks correspond to coding sequences (exons). The numbers above each diagram indicate the number of bases in each intron or exon.

features that also suggest a highly adapted structure of considerable ancestry. The model for the evolution of the superfamily also indicates that Ca^{2+}-binding activity has been lost from the non Ca^{2+}-binding mammalian and avian lysozymes. This is consistent with the presence of aspartates at locations that act in Ca^{2+}-binding in the Ca^{2+}-binding proteins in some species variants of this line. In a recent study, Kuroki *et al.* (1989) introduced Ca^{2+}-binding activity into a recombinant human lysozyme molecule by altering residues corresponding to LA residues 82 and 88 to aspartates. The ease with which this functional change was produced is consistent with the idea that it represented the 'restoration' of activity at a site that had performed this function in an ancestral molecule.

3.1.2. Gene Structure

The structures of the genes for rat, human and bovine LAs have been determined (Qasba & Safaya, 1984; Hall *et al.*, 1987; Vilotte *et al.*, 1987). As shown schematically in Fig. 4, the coding sequence is divided by three intervening sequences (introns) into four exons. The exon/intron boundaries are at precisely corresponding sites in the coding region in the three LA genes and two known lysozyme genes. While the exon sizes are quite similar, intron sizes vary within the LAs as well as lysozymes. In relation to the mature protein sequences shown in Fig. 2, introns are located between the first and second bases in the codon for Trp 26, between the first and second bases in the codon for Lys 79 and between the second and third bases of the codon for Trp 105 of LA. The correspondence of the intron locations in the two protein groups provides further evidence of their common ancestry which is independent of amino acid and cDNA sequence data. Acharya *et al.* (1989) have discussed the relationship between the exon/intron organization and the structure and activity of LA and lysozyme. In lysozyme, exon 2 encodes the principal residues that function in substrate binding and catalysis while exon 3 in the LA gene (and presumably the gene for the Ca^{2+}-binding lysozymes) encodes the aspartyl residues that are involved in Ca^{2+}-binding. Exon 4 represents the region where LA and lysozyme deviate most in sequence and 3D structure and contains a number of residues that appear to be important for the interaction of LA with galactosyltransferase (see Section 5).

3.2. Three-dimensional Structure

The similarity in sequence and disulfide bond arrangements between LA and lysozyme prompted Browne *et al.* (1969) to construct a model of LA, based on the main chain conformation established for lysozyme (Blake *et al.*, 1967). It was found that deletions indicated by the alignment of the protein sequences could be accommodated with little disruption of the overall conformation of the polypeptide backbone. Changes in side chains that are buried in the lysozyme structure were coordinated with respect to total side chain volume and also tended to preserve the hydrophobic nature of the interior of the folded molecule. Differences were noted in the model of LA that are dictated by sequence changes. In particular, the blocking of the upper region of the lysozyme active site cleft region by the side chain of Tyr 103 (Ala in lysozyme), which appeared to form a new hydrophobic core through interaction with Trp 104 and Trp 60. Also, deletions and substitutions near the C-terminus of LA suggested a significant 3D structural change in this region and it proved difficult to build a satisfactory structure for this region.

A model was later proposed by Warme *et al.* (1974) which was formulated by replacing the lysozyme sequence in its crystal structure by that of LA, and computing a structure with minimized free energy. This structure was generally similar to the earlier model, and again, it proved difficult to define a structure for the C-terminal region. Three alternative conformations with similar free energies were proposed for this region.

The inability to obtain either suitable crystals or their isomorphous heavy atom derivatives delayed early attempts to determine the crystallographic structure of LA. A low resolution structure (6 Å and 4·5 Å) was eventually reported by Smith *et al.* (1987) for baboon LA, which provided satisfactory crystals. However, to obtain a useful isomorphous derivative, it proved necessary to chemically modify the protein by opening the disulphide bond between cysteines 6 and 120 to provide a specific binding site for a mercury ion. Since then, a high resolution structure has been reported for the Ca^{2+}-binding site by Stuart *et al.* (1986) and, most recently, a refined structure for the whole protein at 1·7 Å (Acharya *et al.*, 1989).

As predicted in the earlier modeling studies, the polypeptide backbone of LA (dimensions 23 Å × 26 Å × 40 Å) is practically superimposible on that of lysozyme with significant deviations found only at the C-terminal ends (see Fig. 5). As in lysozyme, the compact globular structure contains several regions of regular secondary structure including four α-helices, several regions of 3_{10} helix, and an antiparallel β-pleated sheet separated by irregular β-turns (Table 1 and Fig. 5). The Ca^{2+}-binding elbow, the cleft region (corresponding to the active site of lysozyme) and a region proposed to be the site of interaction with galactosyltransferase (Sinha

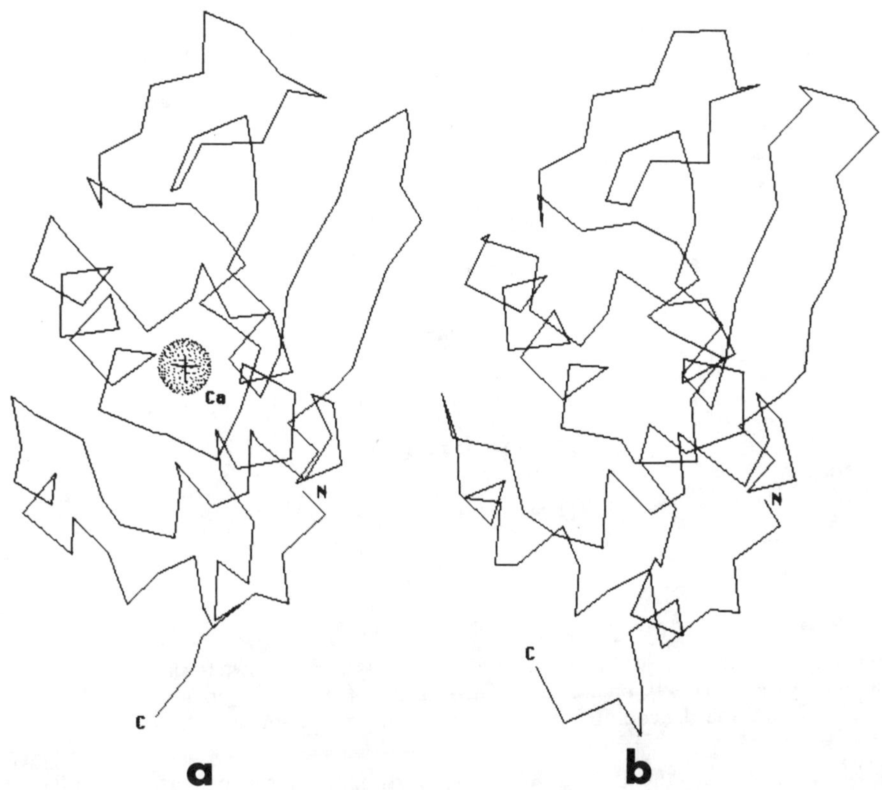

Fig. 5. Comparison of the polypeptide backbone structure of (a) baboon α-lactalbumin (LA) and (b) chicken egg white lysozyme. The location of the coordinated calcium ion in the structure of LA is indicated (Ca). The structural coordinates for lysozyme were obtained from the Brookhaven Protein Structure Bank, while those for baboon LA were kindly provided by Professor Sir David Phillips. The structures were visualized on a Silicon Graphics IRIS Molecular Graphics System and the images were printed with a CH5300 Seiko Thermal Color Printer.

TABLE 1
Locations of Secondary Structure Elements in the Sequence of Baboon α-Lactalbumin

Structure	*α-Helix*	3_{10} *Helix*	*β-Structure*
Sequence position	5–11	12–16	40–43
	23–34	17–21	47–50
	86–99	76–82	
	105–109	101–104	
		115–119	

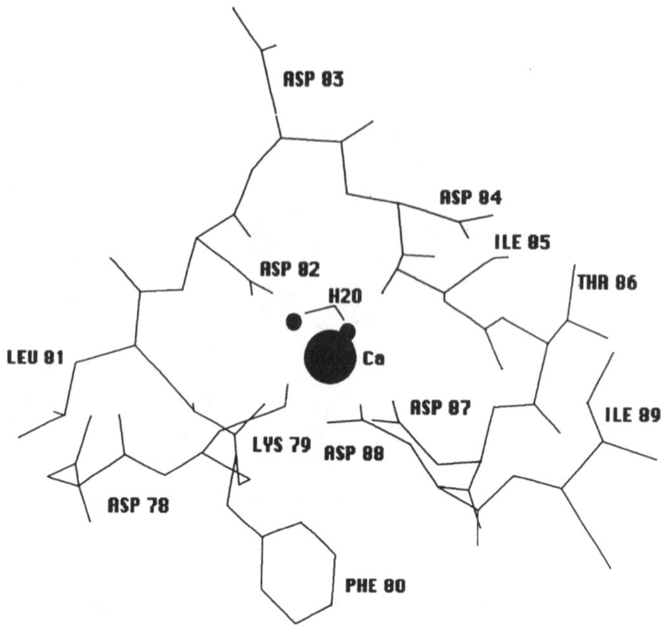

Fig. 6. Calcium binding site on α-lactalbumin. This has been redrawn from a structure given by Acharya *et al.* (1989), Figure 13(a). The location of the Ca^{2+} is indicated by the larger circle and the coordinating water molecules by the smaller filled circles.

& Brew, 1981; Shewale *et al.*, 1984) are of particular interest because they differ in properties and function from the corresponding regions of lysozyme. These differences do not result from any significant changes in secondary or tertiary structure but from specific residue side chain substitutions.

3.2.1. The Calcium Binding 'Elbow'

The Ca^{2+}-coordinating oxygens in LA are all located within a 10-residue region of sequence, which has been described as an 'elbow' structure. This forms part of a loop that is bounded by a disulfide bridge linking residues 73 and 91. The structure around the elbow is analogous to a helix–turn–helix motif found in calmodulin and related proteins with residues 76–82 forming the first helix (a 3_{10} helix) and residues 85–93, the second (an α-helix). Aspartate 82 acts as a spacer between the two helices as part of an irregular β-turn. Coordinating oxygens are provided by the side chain carboxyls of aspartates 82, 87 and 88 and by the peptide carbonyl groups of lysine 79 and aspartate 84. Two water molecules provide two additional coordinating oxygens for the calcium ion which has an apparent coordination number of 7. The coordination structure around the calcium is a slightly distorted pentagonal bipyramid (see Fig. 6) with relatively short calcium–oxygen distances (2·3–2·5 Å). The complex is therefore compact, and the Ca^{2+}-binding loop is the most ordered part of the LA molecule (as indicated by the low temperature factors). In all of the LAs and Ca^{2+}-binding lysozymes, the Ca^{2+}-coordinating aspartates are conserved.

It is interesting that glutamate, which could function similarly in Ca^{2+}-liganding as it does in the Ca^{2+}-binding sites of calmodulin and troponin C (Strynadka & James, 1989) is not found at these sites in any of these proteins. This may reflect incompatability of the longer side chain of glutamate with the constrained fold in this region of structure.

While the K_a for Ca^{2+} with apo-LA is constant between pH 6 and 8, it decreases sharply below pH 4 (see Kronman, 1989). The exposure of native LA to more acidic pH values results in a conformational transition to a form with markedly different physical properties (see below for a discussion of the A and N conformers). This conformational change is believed to be associated with the dissociation of the bound Ca^{2+}, produced by the protonation of the coordinating aspartyl β-carboxyl groups. The pH dependence of this transition is compatible with the involvement of three carboxylates with abnormally low pK values, probably resulting from their coordination with the bound Ca^{2+} (see Kronman, 1989).

The similarity between the Ca^{2+}-binding elbow of LA with the 'E–F' hand structure found in a variety of intracellular Ca^{2+}-binding proteins has been discussed by Stuart *et al.* (1986) and Acharya *et al.* (1989). In both cases, the Ca^{2+}-binding site is a continuous stretch of sequence, rather than being composed, as in a large proportion of protein calcium-binding sites, of residues from well-separated regions of sequence (see Strynadka & James, 1989). The E–F hand and elbow structures are flanked by two nearly orthogonal helixes separated by a loop, and utilize similar residues as ligands to the Ca^{2+}. The structures are also similar in utilizing Asx turns to stabilize the loop (one in the elbow and several in the E–F hand) and a hydrophobic residue (Ile 85 in LA) to stabilize the overall structure. Besides the lack of any overall sequence correlation between E–F hand proteins and LA, several other considerations indicate that the apparent similarity reflects convergence rather than divergence from a common ancestor. Thus, the E–F hand is larger than the LA elbow, consisting of 12 residues rather than 10, and the patterns of coordination within the sequences of these loops are different; the E–F hand provides 6 Ca^{2+}-coordinating groups rather than 5, as in LA, and the arrangement of coordinating groups around the Ca^{2+} in the E–F hand is a distorted octahedron, unlike the pentagonal bipyramid found in LA. The elbow structure contains only a single turn, as compared with two in the E–F hand and the helix preceding the elbow is a 3_{10} rather than an α-helix. It seems possible that a loop flanked by two regions with defined secondary structures represents a particularly satisfactory structure for translating Ca^{2+}-binding into a conformational transition.

While three aspartyl residues participate in Ca^{2+}-binding through their side chains, at least five aspartates are found in the Ca^{2+}-binding elbow of the LAs and mammalian Ca^{2+}-binding lysozymes. The roles of the additional aspartyl residues (83 and 84 in bovine LA), which are directed away from the Ca^{2+}-binding site towards the surface of the molecule, will be considered below.

3.2.2. Binding Site for Mn^{2+}

Mn^{2+} binds with a relatively high affinity to apo-LA and stabilizes a conformer that resembles the native form. However, the rate of change in conformation is slow, taking about 3 min to reach completion, in contrast with the essentially instantaneous Ca^{2+}-induced change (Bratcher & Kronman, 1984). The state stabilized by Mn^{2+} is therefore not identical with the true native state. The high-affinity Mn^{2+} site on LA is also different from the Ca^{2+} binding site since there is not a simple competition between the two ions for binding (see Kronman, 1989). NMR spectroscopy of ^{13}C-methylated bovine LA has suggested the location of the Mn^{2+} binding site (Gerken, 1984) which is primarily based on the distance of the coordinated Mn^{2+} from the N-terminal α-amino group. The site adjoins the Ca^{2+}-binding elbow and involves coordination of the Mn^{2+} with the side chain carbonyl oxygens of glutamates 1, 7 and 11, peptide carbonyls of Glu 2, Thr 4 and Asp 83 and the hydroxyl oxygen of Thr 4. Asp 83 and Leu 85, whose peptide bond and side chain lie directly above the putative binding site, are also components of the Ca^{2+}-binding elbow; stabilization of an N-like state through Mn^{2+} binding may therefore directly reflect the participation of these two residues. Glutamates 1, 7 and 11 are not conserved in all known LA sequences as would be expected if this site is present in all LAs. The nature of residues found at these positions varies widely in different LAs, raising the question of whether this is the true structure of the Mn^{2+} site. Certainly, the data indicating a site located close to the amino terminus are quite convincing but a final conclusion regarding the structure of this site requires additional crystallographic or NMR studies.

3.2.3. The Cleft Region

In LA, the cleft region of lysozyme that divides the molecule into two wings is retained, although with significant changes associated with amino acid substitutions between the two proteins. Changes in both residues involved in catalysis in lysozyme—Glu 35 to Thr (residue 33) and Asp 52 to Glu (residue 49)—in bovine LA appear to be responsible for the complete loss of hydrolytic activity in LA. The report by McKenzie and White (1987) that LA has a low level of activity in cleaving bacterial cell walls (about 10^{-6} of that of lysozyme) was based on very prolonged assays times and requires further investigation since the nature of the chemical reaction occurring was not defined (hydrolysis versus aggregation associated with LA binding to membranes). Based on the essential nature of Glu 35 in lysozyme (Malcolm *et al.*, 1989), a true hydrolytic activity in LA would seem to be extremely improbable.

While a number of the residues in lysozyme that act in substrate binding are conserved in all known LAs, one particular substitution, Tyr 103 (LA) for Ala 107 in lysozyme, has the effect of blocking the upper region of the cleft containing the A and B monosaccharide binding sites in lysozyme. The interaction of this residue with other hydrophobic residues in this area (Trp 104, Ile 95 and Trp 60) results in the formation of a new hydrophobic center, as anticipated by Browne *et al.* (1969). The existence of this hydrophobic region, termed a 'hydrophobic box', has also been indicated by NMR studies (Koga & Berliner, 1985). Although these workers suggested that this represents a hydrophobic site responsible for the interaction of LA with galactosyltransferase, other data do not support this view (see Section 5.1).

3.2.4. The N and A Conformers

Earlier studies of conformational states of LA revealed that exposure of the native protein (N conformer) to acidic conditions (pH lower than 5) results in a transition to a conformer designated as the A (or acid) state. Later work has shown that the same transition can be effected by other environmental changes including moderately elevated temperatures, alkaline pH values, low concentrations of denaturants and the removal of Ca^{2+} or addition of certain metal ions, notably Zn^{2+} (see Kronman (1989) for a discussion of this work). The classification of a conformer as N or A is based on the nature of its near-UV CD spectrum or its tryptophan fluorescence emission spectrum. Typically, an A conformer lacks the complex CD spectrum associated with aromatic residues that is found in N forms of LA, although the far UV CD spectrum characteristic of secondary structure is largely unchanged. A states also have increased tryptophan fluorescence quantum yields and fluorescence emission maxima at longer wavelengths. A conformers produced by the treatments indicated above all show these types of change but differ slightly in character. Multiple N-type conformers also exist, e.g. the true native Ca^{2+} holo-protein and the Mn^{2+}-bound form. Hence, neither the N conformer nor the A conformer is unique, and should be regarded as categories of the 'N-like' and 'A-like' forms.

While spectroscopic studies of A conformers of LA initially suggested that they were denatured forms, as did the increased propensity for aggregation, the far UV CD spectra of these forms indicate that they retain the secondary structure of the native protein. Increased exposure of tryptophan residues is also not observed during the N to A transition, although the loss of the near UV CD spectra for these groups indicates that they have increased mobility (Kronman, 1989). The overall general topology of the molecule is also preserved as indicated by minor changes in shape and the preservation of the four disulfide bonds. As discussed below, these features of A states are characteristic of the 'molten globule model', a type of structure that has been proposed to represent an intermediate state in the folding of proteins (see Baum *et al.*, 1989).

4. Functional Aspects of Metal Binding

4.1. Binding of Different Metal Ions

Since the initial report of the presence of a tightly bound Ca^{2+} in bovine LA, a large number of papers have been published reporting studies of the interaction of various metal ions, including Ca^{2+}, with LAs and the effects of such interactions on the physico-chemical and functional properties of the protein. The stoichiometries of binding of different metal cations and their effects on the conforma-

TABLE 2
Stoichiometry of Binding of Different Metal Ions to α-Lactalbumin (LA) and Their Effects on the Molecular Conformation. Binding Site 1 for Each Metal Represents the Highest Affinity Binding Site for That Metal. These Sites for the Different Metals Do Not Correspond to Equivalent Sites (See Text for Details)

Metal	*Binding site*	*Conformer stabilized*	*Comments*
Calcium	1	N	
	2	N	Binding at this site inhibited by potassium ion
Manganese	1	N	High affinity binding site distinct from calcium site. The two ions do not bind competitively
	2–3 (or 4)	N or A	N conformation promoted by binding of ion to highest affinity sites
Terbium	1	N	
	2	N	Binding at second site reduces LA affinity for Tb at first site
	3	A	Shift to A state upon binding of third Tb is time dependent
Zinc	1	A	Binds simultaneously with Ca
	2	A	Dissociation of Ca takes place. Conformer of LA upon binding second zinc is described as an 'expanded' A state
Sodium	1	N	Probably binds at high affinity Ca binding site
Potassium	1	N	Inhibits binding of second Ca

tional state of LA are summarized in Table 2. Since this work has been recently reviewed in detail by Kronman (1989), the present authors have chosen to focus on Ca^{2+} binding since this represents the ion that is bound to LA under physiological conditions.

4.2. Affinity for Ca^{2+}

The association equilibrium constant for Ca^{2+}-binding with apo-LA has been investigated by a number of different laboratories using physico-chemical parameters to monitor the conformational change in LA associated with Ca^{2+} binding (Permyakov *et al.*, 1981; Murakami *et al.*, 1982; Segawa & Sugai, 1983), ^{45}Ca in direct binding measurements (Kronman *et al.*, 1981; Bratcher & Kronman, 1984; Bryant & Andrews, 1984; Schaer *et al.*, 1985) and, most recently, Ca^{2+}-sensitive electrodes or fluorescent probes to determine free Ca^{2+} levels during Ca^{2+} titrations (Hamano *et al.*, 1986; Nitta *et al.*, 1988). These studies have given values for the association constant (K_a^{app}) of Ca^{2+} with apo-LA ranging from $2{\cdot}5 \times 10^6$ to 10^{10}. The lowest values represent those determined by direct binding measurements with $^{45}Ca^{2+}$ in the absence of chelating agents by four different laboratories; a similar binding constant has been obtained from measurements of fluorescence changes during Ca^{2+}-titration in the absence of chelators (Kronman, 1989). The higher values represent those determined by indirect measurements and using EGTA to buffer free Ca^{2+} levels.

Recent values for K_a^{app} close to 10^7 have been obtained using fluorescent Ca^{2+} indicators and by calculation from the difference in the ΔG of unfolding of LA between 0 and 1 mM Ca^{2+} (Kuwajima *et al.*, 1986; Mitani *et al.*, 1986; Harushima & Sugai, 1989). Van Cauwelaert also obtained values in the 10^6 M^{-1} range (personal communication cited by Kronman, 1989). Various authors have attempted to resolve the discrepancy between the varying values reported by different groups. A suggestion that higher values could reflect the direct binding of chelators to apo-LA (Bratcher & Kronman, 1984) has been discounted. However, the presence of counter-ions (particularly Na^+) in chelators such as EDTA and EGTA may explain some discrepancies since such ions bind to LA and stabilize 'native-like' (N) states (vide infra). Other explanations involve the leaching of ions from chelating resins (Chelex) that have been used by some workers in preparing apo-LA. A very high value reported by Murakami *et al.* (1982) (5×10^9 M^{-1}) has been attributed to an incorrect value assumed for the stability constant for the Ca^{2+}–EGTA complex (Bryant & Andrews, 1984; Mitani *et al.*, 1986). Having personally conducted direct binding measurements using $^{45}Ca^{2+}$, which consistently support the lower value obtained also by other workers using similar methods, one author of this review (KB) wishes to affirm strongly that this represents the correct value under this particular set of experimental conditions (20 mM tris buffer pH 7·4; 25°C). However, it is likely that some of the higher values for the apparent affinity constant are also correct with respect to the experimental conditions under which they were determined. Thus, as discussed above, LA molecules in aqueous solution exist in an equilibrium between two groups of conformational states designated A and N. There is good agreement that only molecules in N states can bind Ca^{2+} (Ikeguchi *et al.*, 1986b; Kronman, 1989) with consequent effects on the conformational equilibrium. The equilibrium is also affected by other molecules that preferentially interact with either N or A conformers and by external conditions (ionic strength, pH, etc.). It is clear that a variety of cationic molecules do interact with LA, including organic bases such as tris and glucosamine (Rawitch & Gleason, 1971), as well as a range of inorganic cations including Na^+ and K^+. An additional possibility discussed by Kronman (1989) is the binding of an EGTA–Ca^{2+} complex with apo-LA. As summarized in Table 2, several ions bind in greater than 1 : 1 stoichiometries (including Ca^{2+} in the absence of KCl) adding further complexities, since binding at one site may influence the affinity of a second site for the ion; also some metals bind competitively with Ca^{2+} (e.g. NaCl) while others have a complex interrelationship with Ca^{2+} (e.g. Mn^{2+}). The N and A states stabilized by different ions can also be distinguished kinetically and by subtle differences in physico-chemical properties. The propensity of LA for self-association is a further influence on discrepancies between experimental measurements of its affinity for metals. This has been used to explain the discrepancies between K_a values reported for the binding of Mn^{2+} to its high-affinity site on apo-LA, which range from $6{\cdot}3 \times 10^5$ at 4 μM LA to $3{\cdot}2 \times 10^4$ at millimolar protein concentrations (Kronman, 1989).

When correction is made for the magnitude of the $N \rightleftharpoons A$ conformational equilibrium, to give a K_a for Ca^{2+} binding to LA molecules in an N state, both Hamano *et al.* (1986) and Kronman (1989) agree that this may be in the order of 10^9 to 10^{10}, reflecting a very high affinity for a hypothetical Ca^{2+}-free N conformer.

4.3. Effects of Metals on Structure and Activity

As mentioned in Section 1, the binding of Ca^{2+} to apo-LA or decalcification of holo-LA has pronounced effects on its stability to heat and chemical denaturants as well as on a number of physico-chemical properties. The generation of the 'A' state in LA by acidification to pH values below 4 results in equivalent changes; a phenomenon which when linked with the pH-dependence of Ca^{2+}-binding, indicates that the low pH 'A' state results from the dissociation of Ca^{2+}. The strong conservation of the LA Ca^{2+}-binding site in the protein from different species suggests that it has a significant biochemical role. However, stabilization of LA to denaturation is unlikely to represent the function of the bound Ca^{2+} because, as discussed in Section 1, the action of LA occurs at an intracellular site, and is transient in nature since the protein is subsequently secreted into milk. Because the catalytic component of lactose synthase, galactosyltransferase, also binds Mn^{2+} and Ca^{2+} which appear to have roles in catalysis and UDP-galactose binding, a role for metals bound to LA, or for the LA metal-binding sites in lactose synthase is an intriguing possibility. At the same time, the metal-dependence of galactosyltransferase activity makes this difficult to examine experimentally. Two groups have investigated these using somewhat different approaches. Kronman *et al.* (1981) used Zn^{2+} and Co^{2+} to activate galactosyltransferase; the former binds to LA, stabilizing an A-like state, whereas Co^{2+} was shown to not compete with Ca^{2+} for binding with apo-LA. Increasing concentrations of holo- and apo-LA were added to galactosyltransferase and lactose synthase activity determined. It was found that Zn^{2+}-LA (A state) gave higher activity than Ca^{2+}-LA with Zn^{2+}-activated galactosyltransferase, while apo-LA and Ca^{2+}-LA gave closely similar activity profiles with Co^{2+}-activated galactosyltransferase. These results therefore indicated that binding of Ca^{2+} to LA is not essential for its action in lactose synthase and that the Zn^{2+}-bound protein (in an A-like conformation) was more active than the holo-protein. Similar conclusions were reached in later reports by Berliner and co-workers (Murakami & Berliner, 1983; Musci & Berliner, 1985) which, surprisingly, make no reference to the earlier work of Kronman *et al.* (1981). These workers used low concentrations of Mn^{2+}, which would only partially saturate apo-LA for activating galactosyltransferase. They suggest that Zn^{2+} may act in the lactating mammary epithelial cell to shift LA into an A-like state which appears to be more active for lactose synthesis. However, Kronman has pointed out that holo-LA would not be converted to an A conformer at the concentrations of Zn^{2+} used in their study. Also, LA, as isolated from milk, is in a Ca^{2+}-bound state (Kronman, 1989), and would be expected to be in this form in the golgi apparatus, where lactose is synthesized.

4.4. Binding of LA to phospholipids and membranes

In apo-LA, the side chains of groups of hydrophobic amino acids are more accessible to solvent as indicated by the efficient binding of apo-LA to phenyl-Sepharose and its elution with Ca^{2+} (Lindahl & Vogel, 1984), the interaction of LA with phospholipid vesicles under acidic conditions (Hanssens *et al.*, 1980, 1983, 1985; Herreman *et al.*, 1981), and the increasing ability of LA to induce fusion of phosphatidyl serine/phosphatidyl ethanolamine vesicles as the acidity of the medium is increased below pH 6 (Kim & Kim, 1986). Berliner and Koga (1987) subsequently showed that apo-LA binds to phospholipid vesicles at pH 7·4 and slowly dissociates on incubation with Ca^{2+}. Since proteolysis studies with LA bound to such vesicles indicate that a region (residues 80 to 108) is associated with the bilayer (Kim & Kim, 1986), the slow release by Ca^{2+} may be attributable to the necessity for Ca^{2+} to diffuse into the membrane to interact with the LA Ca^{2+}-binding site prior to release. Based on these findings, Berliner and Koga (1987) propose an alternative view of the role of Ca^{2+} in LA; namely, that Ca^{2+} could modulate the release of LA from smooth endoplasmic reticulum membranes for transport to the golgi and secretion from the mammary cell. Although these authors had a misconception regarding the role of the smooth endoplasmic reticulum since the biosynthesis and initial processing of proteins occurs in the rough endoplasmic reticulum and not the smooth endoplasmic reticulum, the idea that apo-LA is retained at its site of biosynthesis (the rough endoplasmic reticulum), represents an element of what appears to be the true function of Ca^{2+}.

4.5. Function of Ca^{2+} in the Formation of Native Structure

Although there have been a large number of analyses of the thermodynamics and physico-chemical changes associated with the denaturation and refolding of LA (see Kuwajima *et al.* (1986), Ikeguchi *et al.* (1986*b*), Baum *et al.* (1989) for recent examples, and Kronman (1989) for a detailed review), these have involved denatured states, or partially denatured states, in which the disulfide crosslinks are intact. The four disulfide bonds of LA, which link cysteins 6 and 120, 28 and 111, 61 and 77, and 73 and 91, can be expected to place considerable restraints on the polypeptide chain, even in the presence of high concentrations of denaturants. It is therefore questionable as to whether results obtained for refolding from such a state reflect the mechanism of folding of LA *in vivo*, where the initial state is the reduced form of the protein.

Rao and Brew (1989) investigated the conditions required for refolding of bovine LA from the reduced state. As previously found for refolding and disulfide-bond formation in other disulfide-containing proteins such as chicken lysozyme (Saxena & Wetlaufer, 1970) or bovine pancreatic trypsin inhibitor (Creighton, 1980), alkaline pH values and thiol or disulfide compounds or thiol/disulfide mixtures, such as reduced and oxidized glutathione, are required for the oxidation and disulfide exchange reactions that occur during generation of the native protein. However, as shown in Fig. 7, LA differs from previously studied proteins in

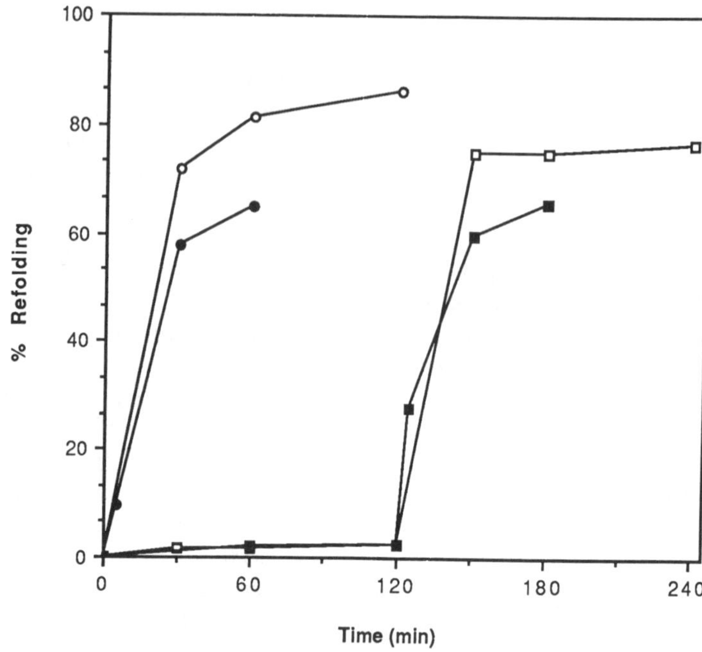

Fig. 7. Time course of refolding of reduced denatured α-lactalbumin (LA). The reaction mixtures contained reduced LA (100 μg/ml) in 0·02M tris HCl pH 9·5, 9 mM reduced glutathione, 1 mM oxidized glutathione, and either $CaCl_2$ (100 μM) or EGTA (2 mM). Samples were taken at time intervals and the quantity of refolded LA determined by assay of activity (○, $CaCl_2$; □, EGTA) or by HPLC (●, $CaCl_2$; ■, EGTA). After 120 min, excess $CaCl_2$ (2·1 mM) was added to the EGTA-containing system and the incubation continued as shown (Reproduced from Rao and Brew, 1989).

requiring Ca^{2+} for the formation of the native protein. When a chelator (EGTA) is added to bind endogenous Ca^{2+}, only 2% (or less) of active LA is regenerated, as compared with as much as 90–100% in the presence of 100 μM Ca^{2+}. When reduced protein preparation and reagents were depleted of Ca^{2+}, a low level of active LA was regenerated in the absence of chelators, which was further reduced to about 2% when 2 mM EGTA was added, and a dependence of renaturation levels on exogenous Ca^{2+} was obtained, indicating that the Ca^{2+} dependence was associated with Ca^{2+} binding to a high affinity site with a 1 : 1 stoichiometry (Fig. 8). The level of refolding obtained with reduced LA in the absence of Ca^{2+} is similar to that obtained when ribonuclease, another protein with four disulfide bonds, is reoxidized from the denatured reduced state under denaturing conditions (Hillson *et al.*, 1984). Peptide mapping studies also indicated that LA when reoxidized in the presence of EGTA contained randomly arranged disulfide bonds.

Since Ca^{2+} increases the stability of native (disulfide crosslinked) LA, it is reasonable to question whether the effect of Ca^{2+} on the folding and disulfide-bond formation in the denatured reduced protein simply reflects a difference in stability between the apo- and holo-protein forms under the conditions used for refolding.

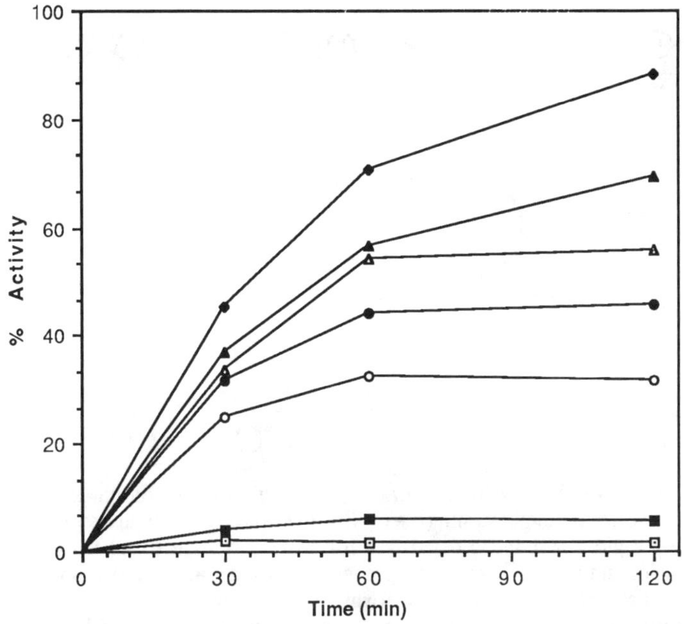

Fig. 8. Effect of different total Ca^{2+} concentrations on the renaturation of reduced denatured α-lactalbumin (LA). The experiment was conducted as in Fig. 7, except that the reduced LA concentration was 280 μg/ml and the reagents used had been treated to deplete the endogenous Ca^{2+}. The levels of refolding were determined by activity assays of 5 μl aliquots. The additions were: ⊡, 2 mM EGTA; ■, none; ○, 5 μM $CaCl^2$; ●, 10 μM $CaCl_2$; △, 15 μM $CaCl_2$; ▲, 20 μM $CaCl_2$; ◆, 50 μM $CaCl_2$. (Reproduced from Rao and Brew, 1989).

However, a similar Ca^{2+}-dependence was observed when refolding and reoxidation was carried out over a range of pH, temperatures, ionic strength and redox conditions. Thus, it appears that Ca^{2+} is an essential part of the structure of the LA molecule under all conditions so that the binding of Ca^{2+} is coupled with native folding and disulfide bond formation. The essential structural role of Ca^{2+} in LA indicated by these results is consistent with the large negative entropy change associated with Ca^{2+}-binding to the apo-protein (Schaer *et al.*, 1985; Kuwajima *et al.*, 1986).

The requirement by LA for Ca^{2+} for native folding might appear to be contrary to the paradigm that the amino acid sequence of a protein determines its 3D structure and disulfide-bond arrangements. However, this 'central dogma' regarding protein folding requires some qualification since in a number of proteins, the native structure is influenced in a major way by a ligand (e.g. a dominant prosthetic group such as heme, in the case of a globin) or an extraneous structure (e.g. a phospholipid bilayer in the case of a membrane protein). Nevertheless, the Ca^{2+} effect on LA refolding is unusual, and possibly unique when considered in relation to other comparable globular proteins. For example, in the case of trypsinogen, another protein that is stabilized by Ca^{2+} binding, the addition of very high levels

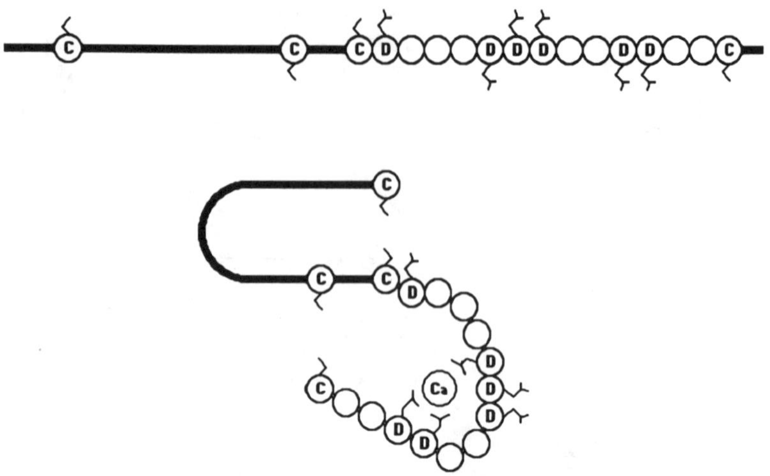

Fig. 9. A model for the role of calcium in the regulation of folding and disulfide bond formation in reduced α-lactalbumin (LA). The upper part of the figure represents the central disulfide loop region of reduced denatured (or nascent) LA in the absence of Ca^{2+}, where charge repulsion between the aspartates prevents bending and disulfide bond formation. The lower part represents an intermediate form that is stabilized by Ca^{2+}-binding, with an orientation of cysteinyl residues that favors native disulfide bond arrangement.

of Ca^{2+} (50 mM) during refolding gives only a modest increase of 5–10% in the final yield of native protein (Sinha & Light, 1975). Perhaps more significantly, chicken egg white lysozyme which, as discussed earlier, has a three-dimensional structure that is essentially superimposible on that of LA, folds efficiently in the presence of chelating agents (Saxena & Wetlaufer, 1970).

The effect of Ca^{2+} on LA refolding and the absence of this phenomenon in lysozyme, despite a close correspondence in their 3D structures, led Rao and Brew (1989) to propose that in the absence of Ca^{2+}, the region of sequence in LA that constitutes the Ca^{2+}-binding site becomes a feature that blocks the formation of the correct three-dimensional structure. As discussed previously, the Ca^{2+}-binding region contains five aspartyl residues in a stretch of seven consecutive amino acids, and has to form a tight bend or elbow so that flanking cysteines (residues 77 and 91) can join in disulfide linkages with their partners (residues 61 and 73, respectively). In the absence of Ca^{2+}, charge repulsion between the carboxylates on these residues is proposed to prevent the formation of the bend at residue 82 that is needed for native disulfide bond pairing. In contrast, the presence of Ca^{2+} will stabilize folding intermediates with bends that favor native disulfide pairing and suitable Ca^{2+}-liganding arrangements. A scheme depicting this proposal is shown in Fig. 9, where it can be seen that the disulfide bond between cysteines 73 and 91 would be expected to be highly Ca^{2+}-dependent. The observation by Segawa *et al.* (1981) that a single disulfide bond in apo-LA is rapidly reduced by dithioerythritol at pH 8 in the absence of denaturatants may reflect the Ca^{2+}-dependence of this bond. The identity of the rapidly reduced bond was not determined, although the authors assumed it to be the Cys 6—Cys 120 bond which had been previously shown by Schechter *et al.* (1974) to be susceptible to reduction in the absence of denaturants. However, Schechter *et al.* (1974) used holo-LA rather than apo-LA in their study; based on the considerations discussed above, it is quite likely that the relative stabilities of the different disulfide bonds in LA are different in the apo and holo states. Thus, the 73–91 bond would be predicted to be highly stable in the presence of Ca^{2+} and very susceptible to reduction in its absence and it may be that it is this bond that is rapidly reduced in apo-LA.

The idea that charge repulsion between the apartates in the Ca^{2+}-binding site may initiate the global change in LA that occurs on removal of Ca^{2+} was suggested by Kronman (1989) to account for the difference in properties of the N and A forms of LA, prior to the publication of the studies of LA refolding. As discussed in detail in his review, the mutual repulsion of these residues would be expected to produce an increased separation of helices 76–82 and 85–93 which would be transmitted to a large part of the LA molecule, thereby disrupting the tertiary structure over a large part of the protein.

4.5.1. Implications for the Regulation of Lactose Biosynthesis

The characteristics of the LA Ca^{2+}-binding site imply that its function *in vivo* is regulatory, and that its role is that of imposing Ca^{2+} regulation on the folding of nascent LA after its biosynthesis. Like other secreted proteins, LA is synthesized on ribosomes of the endoplasmic reticulum as a pre-protein with a hydrophobic leader sequence. After translocation to the lumen of the endoplasmic reticulum and proteolytic removal of the leader sequence, folding and disulfide bond formation occurs in this subcellular compartment. Consequently, the formation of native LA will be dependent on the availability of Ca^{2+} in the lumen of the ER. For this to be accepted as a plausible mechanism of regulation, two questions require answering. Firstly, is there evidence for the regulation of protein production *in vivo* at the level of folding? Secondly, do Ca^{2+} levels in the lumen of the rough endoplasmic reticulum have regulatory effects?

Affirmative answers can be given to both of these questions. With regard to the first, although Ca^{2+}-modulated folding of a secreted protein has not been previously reported, there is strong evidence that a mechanism is present in the rough endoplasmic reticulum for selectively degrading incorrectly folded proteins, including mutant proteins, and unassembled protein subunits. The term 'architectural editing' has been applied to this process, which is thought to play a particularly important role in regulating the production and assembly of polypeptide components of complex hetero-oligomeric proteins, such as plasma membrane receptors, that are produced in the endoplasmic reticulum (Klausner, 1989). The processes associated with the generation of native secretory (and membrane) proteins in the endoplasmic reticulum are quite complex and involve the action of residential lumen proteins including BiP—a soluble protein that binds to mutant or abnormally folded proteins, including unassembled subunits, and is thought to facilitate subunit assembly and folding. Protein disulfide isomerase is another lumenal protein that is important in catalyzing disulfide exchange reactions during the folding of nascent proteins, as well as being involved in other types of protein processing (Pelham, 1989).

Recently acquired evidence also indicates that Ca^{2+} plays an important regulatory role in the endoplasmic reticulum. The endoplasmic reticulum contains a Ca^{2+}-ATPase that acts in sequestering Ca^{2+} from the cytosol; the activity of this pump is modulated by c-AMP and calmodulin and the release of Ca^{2+} from the compartment by phosphoinositides (Carafoli, 1987; Sambrook, 1990). Several endoplasmic reticulum lumenal proteins bind Ca^{2+} (Sambrook, 1990) so that although in most cells the endoplasmic reticulum represents the main intracellular Ca^{2+}-storage compartment, the intra-lumenal concentration of free Ca^{2+} may not be high. Ca^{2+} have now been shown to regulate the transport of secretory proteins from the endoplasmic reticulum to the golgi and other cell compartments (Sambrook, 1990), a finding that adds circumstantial support for a role for Ca^{2+} in regulating the production of native LA. Since, as discussed previously, the production of lactose is dependent on native LA production, previously published data that indicate a regulatory role for Ca^{2+} in lactose synthesis (Munoz *et al.*, 1985) also indirectly support this scheme.

Ca^{2+} metabolism in the lactating mammary gland may be quite different from other cells since there is clearly an enormous flow of Ca^{2+} through lactating mammary epithelial cells. The total Ca^{2+} level (free and bound) in milk is around 30 mM; all of the Ca^{2+} secreted in milk originate in the blood stream and pass through the secreting mammary cells (Linzell & Peaker, 1971). Cytosolic levels of Ca^{2+} are generally submicromolar (Carafoli, 1987) so that Ca^{2+} entering the mammary cells must be rapidly sequestered in some compartment of the secretory pathway of the cell (i.e. rough ER, golgi or secretory vesicles) for secretion into milk (Baumrucker, 1978). Current evidence indicates that the golgi membranes are a major site of calcium sequestration in the mammary epithelial cell. Thus, Ca^{2+} uptake by the ER membranes may contribute a minor fraction of the total Ca^{2+} found in milk, and a site at which lactose synthesis is indirectly regulated *in vivo*.

4.5.2. The A Conformer as a Model for the Molten Globule State

As discussed previously, the A conformer of LA, generated by acid or other treatments, has the characteristics of a 'molten globule' conformation (Baum *et al.*, 1989), a structural state that has been proposed as a possible intermediate stage in protein folding. The physico-chemical properties of the A state, which indicate a similarity in secondary structure and shape to the native protein but a high level of amino acid side chain mobility are those expected of the molten globule (Baum *et al.*, 1989).

The availability of a stable conformer of LA with these intermediate characteristics between a native and denatured state has aroused some interest because of its possible utility as a model as an intermediate in protein folding. However, the requirement of LA for Ca^{2+} for folding from the reduced denatured state puts the A conformer in a different perspective, since it is not an intermediate in the Ca^{2+}-modulated folding of reduced LA but is better described as a denatured form that is constrained by disulfide crosslinks. Its suitability as a model for a protein folding intermediate is therefore questionable.

5. Structural Basis of Action in Lactose Synthase

5.1. Location of Functional Sites in LA

As discussed in Section 2, the action of LA in lactose synthase suggests that a site through which LA interacts with galactosyltransferase is adjacent to a region that can provide stabilizing interactions with a bound monosaccharide in the lactose synthase complex. The conservation in LA of the general structure of the cleft region and several residues which function directly in carbohydrate binding in lysozyme raises the possibility that this region of LA may also mediate carbohydrate binding in lactose synthase.

The ideal approach for determining the structural basis of LA action would be to elucidate the structure of an LA–galactosyltransferase complex by X-ray crystallography. Unfortunately, the difficulty of determining the structure of LA in isolation, the low solubility and lability of galactosyltransferase and the relatively weak stability of LA–galactosyltransferase complexes suggest that this will not be accomplished for a long time. Present information relating to the locations of functional sites on LA is consequently indirect in nature, being based on chemical modification studies and the nature of the conserved amino acids in LAs and lysozymes.

The results of chemical modification studies of LA with group-specific reagents have been reviewed by Hill and Brew (1975), Richardson and Brew (1980) and Kronman (1989). Of the four tryptophans of bovine LA (26, 60, 104 and 118), residues 60 and 118 are the most susceptible to modification and reaction at either site reduces its affinity for galactosyltransferase or causes complete inactivation, depending on the nature of the modification (Schechter *et al.*, 1974; Bell *et al.*, 1975). Tyrosines 103 and 18 of human LA are susceptible to nitration with tetranitromethane (Tyr 103 > Tyr 18) with essentially no effect on its activity. However, more extensive modification results in the additional modification of Trp 104 with extensive loss of activity and some perturbation of the conformation (Prieels *et al.*, 1975). Modification of the amino groups of LA by maleylation (Kitchen & Barman, 1973) or acetylation (Sinha & Brew, 1981) had little effect on activity. Modification of LA carboxyl groups by coupling with amines resulted in major activity losses associated with the reaction of three groups (Lin, 1970; Brew & Sinha, 1980). The most specific and potentially informative effects come from chemical modification studies of LA histidyl residues. In human LA, treatment with diethylpyrocarbonate resulted in the ethoxyformylation of His 32, but no reaction with His 108. This modification resulted in complete loss of activity (Schindler *et al.*, 1976). Modification of bovine LA, which has an additional histidine (residue 68), at moderate protein concentrations (1 mg/ml) with the same reagent resulted in loss of activity in association with the modification of two residues (32 and 68). At high protein concentrations (46 mg/ml), where LA is largely dimerized (Kronman, 1989), only His 32 is modified with a concomitant elimination of activity. In contrast, carboxymethylation of His 32 produces some, but not complete, loss of activity (Prieels *et al.*, 1979). The two conclusions to be drawn from this are that (a) the N1 of His 32 of LA (modified by ethoxyformylation) but not N3 (modified by carboxymethylation) is directly involved in its association with galactosyltransferase, and (b) His 68, but not His 32, is protected on dimer formation.

The effects of complex formation with galactosyltransferase on the reactivities of the 13 individual amino groups of LA has been examined by differential trace acetylation with [^{3}H]-acetic anhydride (Richardson & Brew, 1980). In this procedure, modification levels of less than 1 group per molecule are used, avoiding problems resulting from modification-induced conformational changes that are intrinsic in traditional chemical modification methods. It was found that the reactivity of the epsilon amino group of lysine 5 was reduced in reactivity by a factor of three and the reactivity of lysine 114 was increased two-fold on complex formation. The other 11 amino groups were insignificantly affected by the interaction. These reactivity changes could potentially result from changes in either accessibility, pK or nucleophilicity, and may reflect the direct effects of the protein–protein contact or localized conformational changes. Nevertheless, it was noted that the specific changes in these residues, which are quite close together on the surface of the LA molecule, and the lack of effects on other residues, indicates the involvement of a region of LA that includes a number of conserved residues, including His 32, in the interaction with galactosyltransferase. A label-selection procedure in which trace-acetylated LA (<1 group/mol) was crosslinked with galactosyltransferase using bisimidoesters, was later used by Brew and Sinha (1980) to determine the amino group in LA that is crosslinked with galactosyltransferase by these reagents (see Brew *et al.*, 1975). This study indicated that lysine 108 is the main crosslinking site, but also suggested specific affinity effects associated with the acetylation of lysines 5 and 114, thus providing further support for the results of Richardson and Brew (1980).

Examination of the results of chemical modification and trace-labeling studies in relation to the nature of the residues that are conserved in all known LA sequences, as well as the structures of Ca^{2+}-binding and non-Ca^{2+}-binding lysozymes shows a high degree of consistency and provides the basis for reasonable hypotheses regarding the location of functional regions in LA (see Shewale *et al.*, (1984) for a similar analysis that does not include the Ca^{2+}-binding lysozymes). As summarized in Table 3, a group of residues conserved in both LAs and lysozymes have apparently essential roles in the structures of both groups. Another group of residues, present in the cleft region, are also conserved in both LAs and lysozymes, supporting the view that this region of LA has a functional role. Some of these are the same in LAs and lysozymes, while others are different in the LAs and lysozymes, although being conserved in each group. An example of this is Glu 49 in LA, corresponding to the catalytic site residue, Asp 52, in lysozyme. Changes in such critical residues can be expected to be associated with the functional differences in

TABLE 3
Residues Conserved in α-Lactalbumins and/or Lysozymes

1. RESIDUES CONSERVED IN BOTH LA AND LYSOZYME			
Essential structural roles		*Directly or indirectly involved in substrate binding in lysozyme*	
CYS	6, 28, 73, 77, 91, 111, 120	ASN	44
SER	34	GLN	53
TYR	50	ILE	55
GLY	51, 100	TRP	104
		ALA	106
2. RESIDUES CONSERVED IN LA AND CALCIUM-BINDING LYSOZYMES			
Components of Ca^{2+}-binding site			
LEU	8		
ASP	82, 83, 87, 88		
3. RESIDUES CONSERVED ONLY IN LA			
Components of the GT binding site		*Residues with unknown function*	
HIS	32	GLU	25
GLY	35	LYS	94
LEU	115	ILE	95
GLN	117		
TRP	118		

the two proteins. Another group of residues are conserved in the LAs but not in the lysozymes; their location in a region between lysines 5 and 114 on the surface of LA, proximity to lysine 108 and inclusion of amino acids whose modification eliminates or reduces the affinity of La for galactosyltransferase (His 32, Trp 118) indicates that they are components of the region through which LA interacts with the enzyme. Unfortunately, the structure of this region of LA was not discussed in detail in the report of the high resolution structure by Acharya *et al.* (1989). In Fig. 10, the location of this region in an earlier lower resolution structure of LA is shown (coordinates kindly provided by D.C. Phillips). It should be noted that it includes both hydrophilic and hydrophobic side chains (His 32, Gln 117 cf. Phe 31, Leu 115, Trp 118) and appears to consist of two ridges of amino acid side chains separated by a groove or pocket. However, it is possible that the structure differs in detail in the later high resolution LA structure. As shown in Figs. 10 and 11, the region adjoins the cleft, a feature that is consistent with the model for the action of LA discussed in Section 2. It should be noted that this site is distinct from the 'hydrophobic box' discussed by others as a possible galactosyltransferase binding site. The fact that the specific residues in the latter structure are not conserved and that some can be modified without effect on the activity of LA (Tyr 103) indicates that it does not have a direct functional role but instead is associated with the blockage of the upper region of the cleft in LA.

The technique of site-directed mutagenesis is a powerful approach to relating structure to function in proteins. In the case of LA, the availability of a high

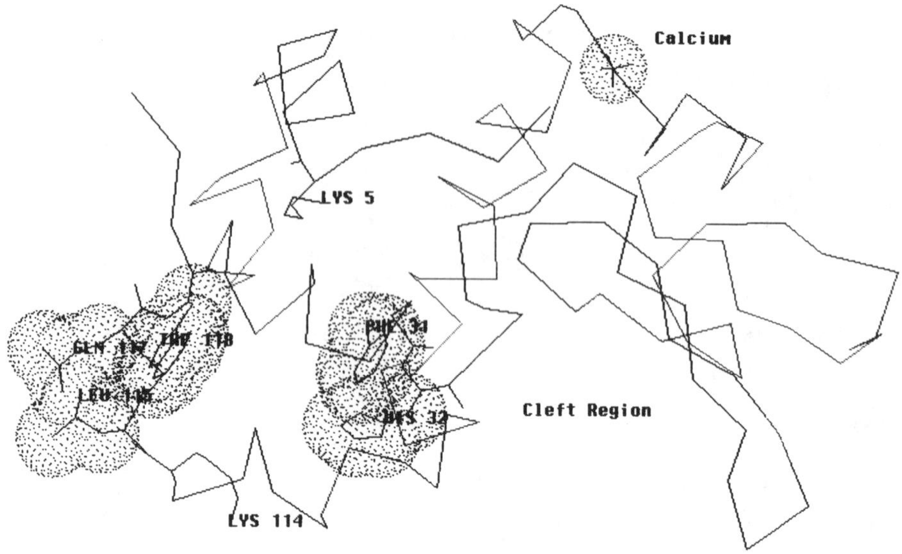

Fig. 10. Location of the proposed galactosyltransferase binding site on α-lactalbumin. Van der Waals radii of the component conserved residues are shown to indicate the surface of the site. The location of the coordinated calcium ion on the opposite face of the molecule is also shown. The image was obtained as described in the legend for Fig. 5.

resolution structure, hypotheses regarding the mechanism of action and its structural basis as well as the role of Ca^{2+} provides an ideal background for designing informative mutagenesis experiments. Recently, the cDNA for bovine LA has been cloned and expressed in *Escherichia coli* in high yield as a fusion protein with a fragment of human cathepsin D. The recombinant protein was produced as an insoluble aggregate (inclusion body) in the bacterium, but was readily solubilized with denaturants and largely purified by gel filtration under denaturing conditions. The product was shown to fold, with the generation of LA activity, in a Ca^{2+}-dependent manner as for bovine milk LA; however, the maximum activity of the fusion protein was only 25% of that of the milk protein. Proteolysis of the folded protein with trypsin resulted in the cleavage of a Lys–Glu bond at the junction of the cathepsin D-derived and LA-derived sequences. A recombinant bovine LA molecule was purified from this cleavage mixture with N-terminal sequence and activity indistinguishable from native LA (Wang *et al.*, 1989). It can be concluded that an N-terminal extension on LA reduces but does not eliminate its affinity for galactosyltransferase, and does not impinge on the binding of Ca^{2+}. Using this system, an investigation of the structural and functional effects of specific changes in a number of the amino acid residues listed in Table 3 is currently in progress (M. Wang, J.A. Grobler & K. Brew, unpublished).

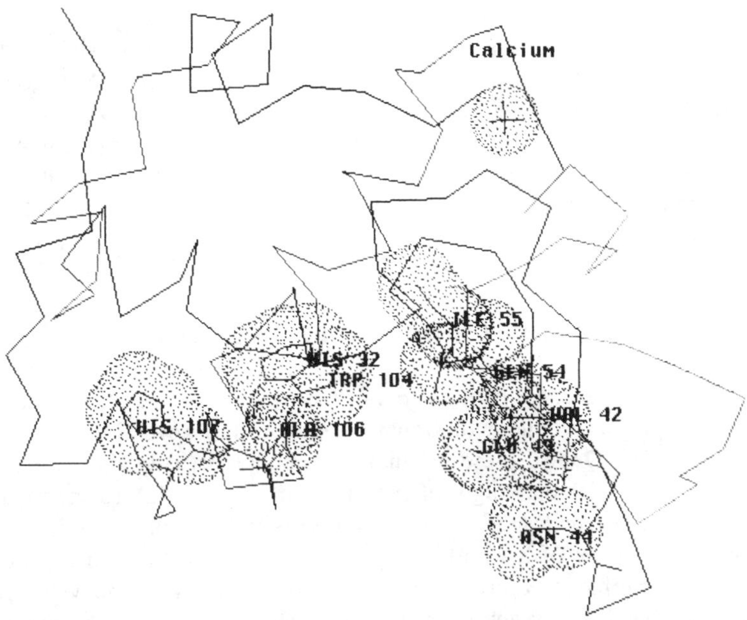

Fig. 11. The cleft region of α-lactalbumin. Van der Waals radii are shown for the conserved residues. Histidine 32 is also shown to demonstrate its proximity to the putative galactosyltransferase binding site (see text for details). The image obtained as described in the legend for Fig. 5.

5.2. Interaction Site of Galactosyltransferase for LA

A complete understanding of the mechanism of action of LA and its structural basis will require the determination of the structure of the active site and LA binding site of galactosyltransferase. Although cDNA and partial protein sequences have been determined for bovine, human and mouse galactosyltransferases (Appert *et al.* 1986; Narimatsu *et al.*, 1986; Shaper *et al.*, 1986, 1988), at present little is known about the 3D structure and functional sites. From the amino acid sequence of GT, it can be deduced that, as for two other golgi-located glycosyltransferases, CMP-sialic acid α-2–6 transferase and UDP-galactose α-3 galactosyltransferase, the enzyme contains an N-terminal short cytoplasmic domain, a hydrophobic trans-membrane domain, and a region that projects into the lumen of the golgi consisting of a stem region of about 60 residues and a catalytic domain of about 300 residues (see Paulson & Colley (1989) for a review). The presence of a small region of similar sequence in the β-4 and α-3 galactosyltransferases has been noted (residues 342 to 354 in the β-transferase) and suggested to represent a common region that is involved in the binding of UDP-galactose (Joziasse *et al.*, 1989; Larsen *et al.*, 1989). Chemical modification studies have confirmed the role of this region in the binding of UDP-galactose and suggest that the C-terminal 150 residues of galactosyltransferase may be a UDP-galactose-binding domain (Yadav

& Brew, 1990). Other chemical modification and affinity labeling studies indicate that the acceptor binding site of galactosyltransferase is localized in a distinct region from that involved in UDP-galactose binding (Lee *et al.*, 1983). Because of the synergistic or competitive interaction between the binding LA and different acceptor substrates, the LA binding region of galactosyltransferase is likely to be located in the amino-terminal region of the catalytic domain (residues 96 to 278 of the translated cDNA sequence). This, however, is an area that requires additional, more definitive studies.

6. Conclusion: Future Prospects

Despite significant recent advances in our knowledge of the structure, action, metal-binding properties and evolutionary origins of LA, several major questions remain unresolved. One of these originates from the virtually simultaneous discoveries, more than 20 years ago, of the biological role of LA as the regulatory component of lactose synthase, and its homology with lysozyme. The close similarity in the 3D structures of LA and lysozyme, now confirmed by a high resolution X-ray crystallographic structure, remains to be reconciled with the wide gulf that separates their biochemical activities as a regulator of carbohydrate binding with galactosyltransferase and an anti-microbial glycosidase. The model for LA action discussed in this review provides a slight functional correlation between the two proteins. However, a direct role for LA in glucose binding in the lactose synthase complex has not yet been demonstrated. It is to be hoped that with a bacterial expression system available that provides high yields of recombinant bovine LA, site-directed mutagenesis can be used to further illuminate the mechanism of action of LA and the relationship of its structure and function. Studies of this type can also determine the roles of the cleft region in facilitating glucose binding, the importance of individual residues in the interaction with galactosyltransferase and whether the 'hydrophobic box' plays any significant role in the activity of LA.

While the requirement of Ca^{2+} for refolding of reduced LA and the nature of the Ca^{2+}-binding site both imply a role for Ca^{2+} in regulating the production of native LA *in vivo*, this requires testing by determining if alteration of the aspartates in the binding site can eliminate Ca^{2+} binding and uncouple Ca^{2+} binding from native folding. It is also important to establish the in-vivo significance of the effect of Ca^{2+} by determining if Ca^{2+} availability can affect the production of native LA in lactating mammary cells or in a mammalian cell expression system.

The existence of the mammalian Ca^{2+}-binding lysozymes provides confirmation that the LA Ca^{2+} binding site is not directly linked with the activity of LA in lactose synthase. At present these lysozymes have only been found in mammalian milks, further linking Ca^{2+} with milk production. However, the tissue expression of these proteins and their distribution requires clarification as well as their evolutionary interrelationship with LA, the 'conventional' lysozymes and the anomalous pigeon lysozyme. Studies of the gene structures of the Ca^{2+}-binding lysozymes might be expected to provide information relevant to the origins of LA, and hence lactose biosynthesis and lactation. Clarification of the nature of proteins involved in lactose biosynthesis in the more 'primitive' monotremes and their structural analysis would be of great interest in this respect.

The recent insertion of the gene for bovine LA into transgenic mice, and the demonstration of its specific expression at high levels in some lines (Vilotte *et al.*, 1989) opens up the exciting future prospect of manipulating the composition of milk in cows using similar technology. This would provide a practical application for knowledge of LA structure–function relationships if these can be used to generate a mutant LA with altered functional properties.

A full understanding of the role of LA in lactose production requires more knowledge of the structure and action of galactosyltransferase. The cDNA for this enzyme has been expressed recently in a mammalian cell line (Masibay & Qasba, 1989), opening the way to investigations of structure–function relationships, including the nature and location of its binding site for LA.

Finally, the possibility that LA has biological functions in addition to its role in lactose biosynthesis is suggested by the presence of proteins with LA-like characteristics and activities in male reproductive systems (Hamilton, 1981; Jones & Brown, 1982; Byers *et al.*, 1984; Limpaseni & Chulavatnatol, 1986). Although these proteins have been isolated, no sequence data have been reported from which their structural relationship with milk LA can be evaluated. In view of proposed roles for galactosyltransferase in cellular interactions, the natures and properties of these proteins merit careful investigation. If any general principle is to be learned from studies of LA during the past 25 years, it is to never conclude that the final word has been uttered on any aspect of the structure or activity of this unusual protein.

Acknowledgement

Work of the authors that is described in this review has been supported by grant number GM 21363 from the National Institutes of Health.

References

Acharya, K.R., Stuart, D.I., Walker, N.P.C., Lewis, M. & Phillips, D.C. (1989). Refined structure of baboon alpha-lactalbumin at 1·7 Å resolution. Comparison with c-type lysozyme. *J. Mol. Biol.*, **208**, 99–127.

Andrews, P., Kitchen, B.J. & Winzor, F.J. (1973). Use of affinity chromatography for the quantitative study of acceptor-ligand interactions: the lactose synthase system. *Biochem. J.*, **135**, 897–900.

Appert, H.E., Rutherford, T.J., Tarr, G.E., Wiest, J.S., Thomford, N.R. & McCorquodale, D.J. (1986). Isolation of a cDNA coding for human galactosyltransferase. *Biochem. Biophys. Res. Commun.*, **139**, 163–8.

Baum, J., Dobson, C.M., Evans, P.A. & Hanley, C. (1989). Characterization of a partly folded protein by NMR methods: studies on the molten globule state of guinea-pig alpha-lactalbumin. *Biochemistry*, **28**, 7–13.

Baumrucker, C.R. (1978). Calcium transport in lactation. In *Lactation: A Comprehensive Treatise*, Vol. IV, ed. B.L. Larson. Academic Press, New York, pp. 463–74.

Bell, J.E., Castellino, F.J., Trayer, I.P. & Hill, R.L. (1975). Modification of alpha-lactalbumin with N-bromosuccinimide and 2-hydroxy-5-nitrobenzylbromide. *J. Biol. Chem.*, **250**, 7579–85.

Bell, J.E., Beyer, T.A. & Hill, R.L. (1976). The kinetic mechanism of bovine milk galactosyltransferase. *J. Biol. Chem.*, **251**, 3003–13.

Berliner, L. & Koga, K. (1987). Alpha-Lactalbumin binding to membranes: evidence for a partially buried protein. *Biochemistry*, **26**, 3006–9.

Blake, C.C.R., Mair, G.A., North, A.C.T., Phillips, D.C. & Sarma, V.R. (1967). On the conformation of the hen egg-white lysozyme molecule. *Proc. Roy. Soc. B*, **167**, 365–77.

Bratcher, S.C. & Kronman, M.J. (1984). Inter- and intramolecular interactions of alpha-lactalbumin. XIV. Metal ion binding to the N and A conformers of bovine alpha-lactalbumin. *J. Biol. Chem.*, **259**, 10875–86.

Brew, K. (1970). Lactose synthetase: Evolutionary origins, structure and control. In *Essays in Biochemistry*, ed. P.N. Campbell & F. Dickens. Academic Press, New York, pp. 93–118.

Brew, K. (1972). The complete amino acid sequence of guinea-pig alpha-lactalbumin. *Eur. J. Biochem.*, **27**, 341–53.

Brew, K. & Campbell, P.N. (1967). The characterization of the whey proteins of guinea-pig milk. *Biochem. J.*, **102**, 258–64.

Brew, K. & Sinha, S.K. (1980). α-Lactalbumin: studies of the binding site for galactosyltransferase. In *Protides of the Biological Fluids*, Vol. 28, ed. H. Peeters. Pergamon Press, Oxford, pp. 149–52.

Brew, K., Vanaman, T.C. & Hill, R.L. (1967). Comparison of the amino acid sequences of bovine alpha-lactalbumin and hen's egg white lysozyme. *J. Biol. Chem.*, **242**, 3747–9.

Brew, K., Vanaman, T.C. & Hill, R.L. (1968). The role of alpha-lactalbumin and the A protein in lactose synthetase: a unique mechanism for the control of a biological reaction. *Proc. Natl. Acad. Sci. USA*, **59**, 491–7.

Brew, K., Shaper, J.H., Olsen, K.W., Trayer, I.P. & Hill, R.L. (1975). Cross-linking of the components of lactose synthetase with dimethylpimelimidate. *J. Biol. Chem.*, **250**, 1434–44.

Brew, K., Richardson, R.H. & Sinha, S.K. (1979). Structural basis of the regulation of galactosyltransferase. In *From Gene to Protein: Information Transfer in Normal and Abnormal Cells*, ed. T.R. Russell, K. Brew, H. Faber & J. Schultz. Miami Winter Symposia, Academic Press, New York. **16**, pp. 433–47.

Brodbeck, U. & Ebner, K.E. (1966). Resolution of a soluble lactose synthetase into two protein components and solubilization of microsomal lactose synthetase. *J. Biol. Chem.*, **241**, 762–4.

Brodbeck, U., Denton, W.L., Tanahashi, N. & Ebner, K.E. (1967). The isolation and identification of the B protein of lactose synthetase as alpha-lactalbumin. *J. Biol. Chem.*, **242**, 1391–7.

Browne, W.J., North, A.C.T., Phillips, D.C., Brew, K., Vanaman, T.C. & Hill, R.L. (1969). A possible three-dimensional structure of bovine alpha-lactalbumin based on that of hen's egg-white lysozyme. *J. Mol. Biol.*, **42**, 65–86.

Bryant, D.T.W. & Andrews, P. (1984). High-affinity binding of calcium to bovine alpha-lactalbumin in the absence and presence of EGTA. *Biochem. J.*, **220**, 617–28.

Byers, S.W., Dym, M., Hewlett, I.K. & Qasba, P.K. (1984). Alpha-lactalbumin-like proteins in the male reproductive track. *Ann. N.Y. Acad. Sci.*, **438**, 8–17.

Carafoli, E. (1987). Intracellular calcium homeostasis. *Ann. Rev. Biochem.*, **56**, 395–433.

Creighton, T.E. (1980). Role of the environment in the refolding of reduced pancreatic trypsin inhibitor. *J. Mol. Biol.*, **144**, 521–50.

Dandekar, A.M. & Qasba, P.K. (1981). Rat alpha-lactalbumin has a 17-residue long COOH-terminal extension as judged by sequence analysis of the c-DNA clones. *Proc. Natl. Acad. Sci. USA*, **78**, 4853–7.

Dayhoff, M.O., Barker, W.C. & Hunt, L.T. (1983). Establishing homologies in protein sequences. *Meth. Enzymol.*, **91**, 524–45.
Dobson, D.E., Prager, E.M. & Wilson, A.C. (1984). Stomach lysozymes of ruminants. I. Distribution and catalytic properties. *J. Biol. Chem.*, **259**, 11607–16.
Farris, J.S. (1972). Estimating phylogenetic trees from distance matrices. *Am. Nat.*, **106**, 645–68.
Feng, D.F. & Doolittle, R.F. (1987). Progressive sequence alignment as a prerequisite to correct phylogenetic trees. *J. Mol. Evol.*, **25**, 351–60.
Fitch, W.M. & Margoliash, E. (1967). Construction of phylogenetic trees. *Science*, **155**, 279–84.
Gerken, T.A. (1984). Amino group environments and metal-binding properties of carbon-13 reductively methylated bovine alpha-lactalbumin. *Biochemstry*, **23**, 4688–97.
Godovac-Zimmerman, J., Conti, A. & Napolitano, L. (1988). The primary structure of donkey (*Equus asinus*) lysozyme contains the Ca(II) binding site of alpha-lactalbumin. *Biol. Chem. Hoppe-Seyler*, **269**, 1109–15.
Gordon, W.G. (1971). Alpha-lactalbumin. In *Milk Proteins*, Vol. II, ed. H.A. McKenzie. Academic Press, New York, pp. 331–65.
Hall, L. & Campbell, P.N. (1986). Alpha-lactalbumin and related proteins: a versatile gene family with an interesting parentage. In *Essays in Biochemistry*, vol. 22, ed. F.D. Marshall & K.F. Tipton. Academic Press, New York, pp. 1–26.
Hall, L., Craig, R.K., Edbrooke, M.R. & Campbell, P.N. (1982). Comparison of the nucleotide sequence of cloned human and guinea-pig pre-alpha-lactalbumin cDNA with that of chick pre-lysozyme cDNA suggests evolution from a common ancestral gene. *Nucleic Acids Res.*, **10**, 3503–15.
Hall, L, Emery, D.C., Davies, M.S., Parker, D. & Craig, R.K. (1987). Organization and sequence of the human alpha-lactalbumin gene. *Biochem. J.*, **242**, 735–42.
Hamano, M., Nitta, K., Kuwajima, K. & Sugai, S. (1986). Binding strength of calcium ion to bovine alpha-lactalbumin. *J. Biochem. (Tokyo)*, **100**, 1617–22.
Hamilton, D.W. (1981). Evidence for alpha-lactalbumin-like activity in reproductive fluid of the male rat. *Biol. Reprod.*, **25**, 385–92.
Hanssens, I., Houthuys, C., Herreman, W. & Van Cauwelaert, F.H. (1980). Interaction of alpha-lactalbumin with dimyristoyl phosphatidylcholine vesicles. I. A micro-calorimetric and fluorescence study. *Biochim. Biophys. Acta*, **602**, 539–57.
Hanssens, I., Herreman, W., Van Ceunebroeck, J-C., Dangreau, H, Gielens, C., Preaux, G. & Van Cauwelaert, F.H. (1983). Interaction of alpha-lactalbumin with dimyristoyl phosphatidylcholine vesicles. III. Influence of the temperature and of the lipid to protein molar ratio on the complex formation. *Biochim. Biophys. Acta*, **728**, 293–304.
Hanssens, I., Van Ceunebroeck, J-C., Pottel, H., Preaux, G. & Van Cauwelaert, F.H. (1985). Influence of the protein conformation on the interaction between alpha-lactalbumin and dimyristoyl phosphatidylcholine vesicles. *Biochim. Biophys. Acta*, **817**, 154–64.
Harushima, Y. & Sugai, S. (1989). Hydrogen exchange of the tryptophan residues in bovine, goat, guinea pig and human alpha-lactalbumin. *Biochemistry*, **28**, 8568–76.
Herreman, W., Van Tornout, P., Van Cauwelaert, F.H. & Hanssens, I. (1981). Interaction of alpha-lactalbumin with dimyristoyl phosphatidylcholine vesicles. II. A fluorescence polarization study. *Biochim. Biophys. Acta*, **640**, 418–29.
Hill, R.L. & Brew, K. (1975). Lactose synthetase. *Adv. Enzymol. Relat. Areas Mol. Biol.*, **43**, 411–90.
Hillson, D.A., Lambert, N. & Freedman, R.B. (1984). Formation and isomerization of disulfide bonds in proteins: protein disulfide isomerase. *Methods Enzymol.*, **107**, 281–94.
Hiroaka, Y., Segawa, T., Kuwajima, K., Sugai, S. & Murai, N. (1980). Alpha-lactalbumin: a calcium metallo-protein. *Biochem. Biophys. Res. Commun.*, **93**, 1098–104.
Hopp, T.P. & Woods, K.R. (1979). Primary structure of rabbit alpha-lactalbumin. *Biochemistry*, **18**, 5182–91.

Hopper, K.E. & McKenzie, H.A. (1974). Comparative studies of alpha-lactalbumin and lysozyme: echidna lysozyme. *Mol. Cell. Biochem.*, **3**, 93–108.

Ikeguchi, M., Kuwajima, K., Mitani, M. & Sugai, S. (1986*a*). Evidence for identity between the equilibrium unfolding intermediate and transient folding intermediate: a comparative study of the folding reactions of alpha-lactalbumin and lysozyme. *Biochemistry*, **25**, 6965–72.

Ikeguchi, M., Kuwajima, K. & Sugai, S. (1986*b*). Calcium-induced alteration in the unfolding behaviour of alpha-lactalbumin. *J. Biochem. (Tokyo)*, **99**, 1191–201.

Jolles, P. & Jolles, J. (1984). What's new in lysozyme research? *Mol. Cell. Biochem.*, **63**, 165–89.

Jones, R. & Brown, C.R. (1982). Association of epididymal secretory proteins showing alpha-lactalbumin-like activity with the plasma membrane of rat spermatazoa. *Biochem. J.*, **206**, 161–4.

Joziasse, D.H., Shaper, J.H., Van den Eijnden, D.H., Van Tunen, A.J. & Shaper, N.L. (1989). Bovine alpha 1-3 galactosyltransferase: isolation and characterization of a cDNA clone. Identification of homologous sequences in human genomic DNA. *J. Biol. Chem.*, **264**, 14290–7.

Khatra, B.S., Herries, D.G. & Brew, K. (1974). Some kinetic properties of human-milk galactosyltransferase. *Eur. J. Biochem.*, **44**, 537–60.

Kim, J. & Kim, H. (1986). Fusion of phospholipid vesicles induced by alpha-lactalbumin at acid pH. *Biochemistry*, **25**, 7867–74.

Kitchen, B. & Barman, T.E. (1973). Maleyl α-lactalbumin as lactose synthase specifier protein. *Biochim. Biophyta. Acta*, **298**, 861–7.

Klausner, R.D. (1989). Architectural editing: determining the fate of newly synthesized membrane proteins. *The New Biologist*, **1**, 3–8.

Koga, K. & Berliner, L.J. (1985). Structural elucidation of a hydrophobic box in bovine alpha-lactalbumin by NMR: nuclear Overhauser effects. *Biochemistry*, **24**, 7257–62.

Kronman, M.J. (1989). Metal-ion binding and the molecular conformational properties of alpha-lactalbumin. *C.R.C. Crit. Rev. Biochem. and Mol. Biol.*, **24**, 565–667.

Kronman, M.J., Sinha, S.K. & Brew, K. (1981). Characteristics of the binding of Ca^{2+} and other divalent metal ions to bovine alpha-lactalbumin. *J. Biol. Chem.*, **256**, 8582–7.

Kuroki, R., Taniyama, Y., Seko, C., Nakamura, H., Kikuchi, M. & Ikehara, M. (1989). Design and creation of a Ca^{2+} binding site in human lysozyme to enhance stability. *Proc. Natl. Acad. Sci. USA*, **86**, 6903–7.

Kuwajima, K., Harushima, Y. & Sugai, S. (1986). Influence of calcium-binding on the structure and stability of bovine alpha-lactalbumin studied by circular diochroism and nuclear magnetic resonance spectra. *Int. J. Pept. Protein Res.*, **27**, 18–27.

Lambright, D.G., Lee, T.K. & Wong, S.W. (1985). Association–dissociation modulation of enzyme activity: case of lactose synthase. *Biochemistry*, **24**, 910–4.

Larsen, R.D., Rajan, V.P., Ruff, M.M., Kukowska-Latallo, J., Cummings, R.D. & Lowe, J.B. (1989). Isolation of a cDNA encoding a murine UDP galactose: beta-D-galactosyl-1, 4-N-acetyl-D-glucosaminide alpha-1, 3-galactosyltransferase: expression cloning by gene transfer. *Proc. Natl. Acad. Sci. USA*, **86**, 8227–321.

Lee, T.K., Wong, L-J.C. & Wong, S.S. (1983). Photoaffinity labeling of lactose synthase with a UDP-galactose analogue. *J. Biol. Chem.*, **258**, 13166–71.

Limpaseni, T. & Chulavatnatol, M. (1986). A rat epidydimal sialoglycoprotein with alpha-lactalbumin activity. *Biochem. Int.*, **13**, 41–9.

Lin, T. (1970). Lactose synthase. Modification of carboxyl groups in α-lactalbumin. *Biochemistry*, **9**, 984–95.

Linzell, J.L. & Peaker, M. (1971). Mechanism of milk secretion. *Physiol.Rev.*, **51**, 564–97.

Lindahl, L. & Vogel, H.J. (1984). Metal-ion dependent hydrophobic-interaction chromatography of alpha-lactalbumin. *Anal. Biochem.*, **140**, 394–402.

Malcolm, B.A., Rosenberg, S., Corey, M.J., Allen, J.S., De Baestelier, A. & Kirsch, J.F.

(1989). Site-directed mutagenesis of the catlytic residues Asp-52 and Glu-35 of chicken egg white lysozyme. *Proc. Natl. Acad. Sci. USA*, **86**, 133–7.
Masibay, A.S. & Qasba, P.K. (1989). Expression of bovine beta-1, 4-galactosyltransferase cDNA in COS-7 cells. *Proc. Natl. Acad. Sci. USA*, **86**, 5733–7.
McKenzie, H.A. (1967). Milk proteins. *Adv. Prot. Chem.*, **22**, 56–234.
McKenzie, H.A. & Shaw, D.C. (1985). The amino acid sequence of equine milk lysozyme. *Biochem. Int.*, **11**, 841–3.
McKenzie, H.A. & White, F.H. (1987). Studies of a trace lytic activity associated with alpha-lactalbumin. *Biochem. Int.*, **14**, 347–56.
Mitani, M., Harushima, Y., Kuwajima, K., Ikeguchi, M. & Sugai, S. (1986). Innocuous character of [ethylene bis (oxyethylenenitrilo)] tetraacetic acid and EDTA as metal ion buffers in studying Ca^{2+} binding by alpha-lactalbumin. *J. Biol. Chem.*, **261**, 8824–9.
Morrison, J.F. & Ebner, K.E. (1971*a*). Studies on galactosyltransferase. Kinetic investigations with glucose as the galactosyl group acceptor. *J. Biol. Chem.*, **246**, 3985–91.
Morrison, J.F. & Ebner, K.E. (1971*b*). Studies on galactosyltransferase. Kinetic effects of alpha-lactalbumin with N-acetylglucosamine and glucose as galactosyl group acceptors. *J. Biol. Chem.*, **246**, 3992–8.
Munoz, A., Lavandero, S., Donoso, E., Puente, J. & Sapag-Hagar, M. (1985). A role for adrenaline and calmodulin in modulating cyclic AMP levels during the lactogenic cycle. *FEBS Lett.*, **187**, 173–6.
Murakami, K. & Berliner, L.J. (1983). A distinct zinc-binding site in the alpha-lactalbumins regulates calcium binding. Is there a physiological role for this control? *Biochemistry*, **22**, 3370–4.
Murakami, K., Andree, P.J. & Berliner, L.J. (1982). Metal ion binding to alpha-lactalbumin species. *Biochemistry*, **21**, 5488–94.
Musci, G. & Berliner, L.J. (1985). Physiological roles of zinc and calcium binding to alpha-lactalbumin in lactose biosynthesis. *Biochemistry*, **24**, 6945–8.
Narimatsu, H., Sinha, S., Brew, K., Okayama, H. & Qasba, P.K. (1986). Cloning and sequencing of cDNA of bovine N-acetylglucosamine (beta 1-4) galactosyltransferase. *Proc. Natl. Acad. Sci. USA*, **83**, 4720–4.
Nitta, K. & Sugai, S. (1989). The evolution of lysozyme and alpha-lactalbumin. *Eur. J. Biochem.*, **151**, 111–18.
Nitta, K., Tsuge, H., Shimazaki, K. & Sugai, S. (1988). Calcium-binding lysozymes. *Biol. Chem. Hoppe-Seyler*, **369**, 1109–15.
Paulson, J.C. & Colley, K.J. (1989). Glycosyltransferases. Structure, localization, and control of cell type-specific glycosylation. *J. Biol. Chem.*, **264**, 17615–8.
Pelham, H.R.B. (1989). Control of protein exit from the endoplasmic reticulum. *Ann. Rev. Cell Biol.*, **5**, 1–23.
Permyakov, E.A., Yarmolenko, V.V., Kalinichenko, L.P., Morozova, L.A. & Burshtein, E.A. (1981). Calcium binding to alpha-lactalbumin: structural rearrangement and association constant evaluation by means of intrinsic protein fluorescence changes. *Biochem. Biophys. Res. Commun.*, **100**, 191–7.
Powell, J.T. & Brew, K. (1976). A comparison of the interactions of galactosyltransferase with a glycoprotein substrate (ovalbumin) and with alpha-lactalbumin. *J. Biol. Chem.*, **251**, 3653–63.
Prager E.M. & Wilson, A.C. (1988). Ancient origin of alpha-lactalbumin from lysozyme: analysis of DNA and amino acid sequences. *J. Mol. Evol.*, **27**, 326–35.
Prasad, R.V., Butkowski, R.J., Hamilton, J.W. & Ebner, K.E. (1982). Amino acid sequence of rat alpha-lactalbumin. A unique alpha-lactalbumin. *Biochemistry*, **21**, 1479–82.
Prieels, J.P., Dolmans, N., Leonis, J. & Brew, K. (1975). Nitration of tyrosyl residues in human α-lactalbumin. Effect on lactose synthase specifier activity. *Eur. J. Biochem.*, **60**, 533–9.
Prieels, J.P., Dolmans, M., Schindler, M. & Sharon, N. (1976). The binding of glycoconjugates to human-milk D-galactosyltransferase. *Eur. J. Biochem.*, **66**, 579–82.

Prieels, J.P., Bell, J.E., Schindler, M., Castellino, F.J. & Hill, R.L. (1979). Involvement of histidine-32 in the biological activity of alpha-lactalbumin. *Biochemistry*, **18**, 1771–6.

Qasba, P.K. & Safaya, S.K. (1984). Similarity of the nucleotide sequences of rat alpha-lactalbumin and chicken lysozyme genes. *Nature*, **308**, 377–80.

Rao, K.R. & Brew, K. (1989). Calcium regulates folding and disulfide-bond formation in alpha-lactalbumin. *Biochem. Biophys. Res. Commun.*, **163**, 1390–6.

Rawitch, A.B. & Gleason, M. (1971). The interaction of alpha-lactalbumin with tris cations and other cationic hydroxamines. *Biochem. Biophys. Res. Commun.*, **45**, 590–7.

Richardson, R.H. & Brew, K. (1980). Lactose synthase. An investigation of the interaction site of alpha-lactalbumin for galactosyltransferase by differential kinetic labeling. *J. Biol. Chem.*, **255**, 3377–85.

Rodruigez, R., Menendez-Arias, L., Gonzales de Buitrago, G. & Gavilanes, J.G. (1985). Amino acid sequence of pigeon egg-white lysozyme. *Biochem. Int.*, **11**, 841–3.

Rodruigez, R., Menedez-Arias, L., Gonzalez de Buitrago, G. & Gavilanes, J.G. (1987). Structure of pigeon lysozyme and its relationship with other type c lysozymes. *Comp. Biochem. Physiol.*, **88B**, 791–6.

Sambrook, J.F. (1990). The involvement of calcium in transport of secretory proteins from the rough endoplasmic reticulum. *Cell*, **61**, 197–9.

Saxena, V.P. & Wetlaufer, D.B. (1970). Formation of three-dimensional structure in proteins. I. Rapid nonenzymatic reactivation of reduced lysozyme. *Biochemistry*, **9**, 5015–23.

Schaer, J.J., Milos, M. & Cox, J.A. (1985). Thermodynamics of the binding of calcium and strontium to bovine alpha-lactalbumin. *FEBS Lett.*, **190**, 77–80.

Schechter, Y., Patchornik, A. & Burstein, Y. (1974). Selective sulfenylation of tryptophan residues in alpha-lactalbumin of bovine milk. *J. Biol. Chem.*, **249**, 413–9.

Schindler, M., Sharon, N. & Prieels, J.P. (1976). Reversible inactivation of lactose synthase by the modification of His 32 in human alpha-lactalbumin. *Biochem. Biophys. Res. Commun.*, **69**, 167–73.

Segawa, T. & Sugai, S. (1983). Interactions of divalent metal ions with bovine, human and goat alpha-lactalbumins. *J. Biochem. (Tokyo)*, **93**, 1321–8.

Segawa, T., Kuwajima, K. & Sugai, S. (1981). Kinetic correlations between the disulfide bond reduction and the induced conformational change of proteins. *Biochim. Biophys. Acta*, **668**, 89–97.

Shaper, N.L., Shaper, J.H., Meuth, J.L., Fox, J.L., Chang, H., Kirsch, I.R. & Hollis, G.F. (1986). Bovine galactosyltransferase: identification of a clone by direct immunological screening of a cDNA expression library. *Proc. Natl. Acad. Sci. USA*, **83**, 1573–7.

Shaper, N.L., Hollis, G.F., Douglas, J.G., Kirsch, I.R. & Shaper, J.H. (1988). Characterization of the full length cDNA for murine beta 1-4 galactosyltransferase. Novel features at the 5'-end predict two translational start sites at two in-frame AUGs. *J. Biol. Chem.*, **253**, 10420–8.

Shewale, J.G., Sinha, S.K. & Brew, K. (1984). Evolution of alpha-lactalbumins. The complete amino acid sequence of the alpha-lactalbumin from a marsupial (Macropus rufogriseus) and corrections to regions of sequence in bovine and goat alpha-lactalbumins. *J. Biol. Chem.*, **259**, 4974–56.

Shur, B.D. (1984). The receptor function of galactosyltransferase during cellular interactions. *Mol. Cell. Biochem.*, **61**, 143–58.

Sinha, S.K. & Brew, K. (1981). A label selection procedure for determining the location of protein–protein interaction sites by cross-linking with bisimidoesters. Application to lactose synthase. *J. Biol. Chem.*, **256**, 4193–204.

Sinha, N.K. & Light, A. (1975). Refolding of reduced, denatured trypsinogen and trypsin immobilized on agarose beads. *J. Biol. Chem.*, **250**, 8624–9.

Smith, S.G., Lewis, M., Aschaffenburg, R., Fenna, R.E., Wilson, I.A., Sundaralingam, M., Stuart, D.I. & Phillips, D.C. (1987). Crystallographic analysis of the three-dimensional structure of baboon alpha-lactalbumin at low resolution. Homology with lysozyme. *Biochem. J.*, **242**, 353–60.

Strous, G.J. (1986). Golgi and secreted galactosyltransferase. *C.R.C. Crit. Rev. Biochem.*, **21**, 119–51.

Struck, D.K., Lennarz, W.J. & Brew, K. (1978). Primary structural requirements for the enzymatic formation of the N-glycosidic bond in glycoproteins. *J. Biol. Chem.*, **253**, 5786–94.

Strynadka, N.C.J. & James, M.N.G. (1989). Crystal structures of the helix-loop-helix calcium-binding proteins. *Annu. Rev. Biochem.*, **58**, 951–98.

Stuart, D.I., Acharya, K.R., Walker, N.P.C., Smith, S.G., Lewis, M. & Phillips, D.C. (1986). Alpha-Lactalbumin possesses a novel calcium binding loop. *Nature*, **324**, 84–7.

Vilotte, J-L., Soulier, S., Mercier, J-C., Gaye, P., Hue-Delahaie, D. & Furet, J.R. (1987). Complete nucleotide sequence of bovine alpha-lactalbumin gene: comparison with its rat counterpart. *Biochimie (Paris)*, **69**, 609–20.

Vilotte, J-L., Soulier, S., Stinnakre, M-G., Massoud, M. & Mercier, J-C. (1989). Efficient tissue-specific expression of bovine alpha-lactalbumin in transgenic mice. *Eur. J. Biochem.*, **186**, 43–8.

Wang, M., Scott, W.A., Rao, K.R., Udey, J., Conner, G.E. & Brew, K. (1989). Recombinant bovine alpha-lactalbumin obtained by limited proteolysis of a fusion protein expressed at high levels in Escherichia coli. *J. Biol. Chem.*, **264**, 21116–21.

Warme, P.K., Momany, F.A., Rumball, S.V., Tuttle, R.W. & Scheraga, H.A. (1974). Computation of structures of homologous proteins. Alpha-Lactalbumin from lysozyme. *Biochemistry*, **13**, 768–81.

White, M.D., Kuhn, N.J. & Ward, S. (1981). Mannitol and glucose movement across the Golgi membrane of lactating-rat mammary gland. *Biochem. J.*, **194**, 173–7.

Yadav, S.P. & Brew, K. (1990). Identification of a region of UDP-galactose N-acetylglucosamine beta-4 galactosyltransferase involved in UDP-galactose binding by differential labeling. *J. Biol. Chem.*, **265**, 14163–9.

6

IMMUNOGLOBULINS OF THE MAMMARY SECRETIONS

BRUCE L. LARSON

Department of Animal Sciences, University of Illinois, Urbana, Illinois 61801, USA

1. INTRODUCTION

1.1. Historical, Scope and Definitions

The apparent presence of a factor that prevented disease in the early lactation mammary secretions of certain species is one of the earliest immunological observations. Successful herders understood even in ancient times that neonate survival and healthy development were dependent upon early ingestion of sufficient amounts of the early maternal secretions (colostrum). This was particularly true in the ruminants but also to some extent in the horse and pig. This situation contrasted to the human where neonate survival was not dependent upon colostrum ingestion.

Recognition that agents with immune properties were present in animal systems developed in the 1800s (Ehrlich, 1892; Fischl & von Wunscheim, 1895; Brambell, 1970). By the early 1900s, investigators were isolating and characterizing proteins (globulins) with immuno-properties (antibodies) from blood serum and other sources, such as colostrum, rich in these proteins (Crowther & Raistrick, 1916). The similarity of proteins isolated from these sources, and the appearance of immunoglobulins (Igs) in the bloodstream of the neonate of certain species upon ingestion of the maternal colostrum (Howe, 1921, 1924; Orcutt & Howe, 1922; Smith & Little, 1922; Smith, 1946, 1948; Smith & Holm, 1948; Hansen & Phillips, 1949; Larson & Kendall, 1957; Larson, 1958; Murphy *et al.*, 1964; Porter, 1969; Witty *et al.*, 1969), made it appear likely that the Igs of the mammary secretions were strongly related to those of blood. Definitive data on their origin, as well as the origin of the other milk proteins, did not become available until the mid-1900s with the development of sophisticated protein separation and identification procedures and radioactive tracer techniques (Askonas *et al.*, 1954; Larson & Gillespie, 1957; Dixon *et al.*, 1961). Recognition of the Igs as large molecular weight glycoproteins that originate in the cells of the reticuloendothelial system, and elucidation of the molecular structures of the several classes, opened the door to research on the molecular aspects of their functional roles.

Research continues on the Igs to define in more detail their associated prosthetic structures, three-dimensional configurations, pathways and mechanisms associated with their biological functions, the complex genetic mechanisms associated with their biological functions, and the complex genetic mechanisms involved in their synthesis. Studies on growth factors and other materials that may be present in the mammary secretions in small but biologically important amounts, some of which may be immunologically related, is also an active area.

This chapter will discuss the component Igs of the mammary secretions. These include the colostrum (first few milkings after parturition), transition milk, and normal milk as it is secreted during the course of the lactation period. Species differ widely in the rate and time at which Igs, as well as the colostral secretions in total, appear after parturition. The definition of colostrum is somewhat loose and it is not always possible to determine whether a reported investigation on colostrum refers specifically to the first secretions or milk samples obtained at some nonspecific time early in the lactation period.

Some previous reviews of particular interest are those concerned with the milk protein system including Igs (Jenness, 1974, 1979, 1982; Larson, 1979; Larson & Jorgensen, 1974; Blanc, 1981; Whitney, 1988), the ruminant immune system (Butler, 1974, 1981, 1983, 1986; Lascelles & McDowell, 1974; Lascelles, 1979; Guidry, 1985), and general understanding of Igs and the immune system (Newby *et al.*, 1982; Tizard, 1982; Smith *et al.*, 1983; Darnell *et al.*, 1986). General information drawn from these sources will be considered in many aspects of this review. Even though specific reference on some matter is made to one or two of these review sources, the others also may contain significant information.

1.2. Occurrence and Significance of Immunoglobulins

Igs are present in the mammary secretions of all mammalian species as part of the immune defense systems of the animal. There are large differences between species in the composition and total amounts. The species differences are largely a reflection of the two mechanisms, *in utero* or *via colostrum*, by which passive immunity by maternal Igs is transferred to the offspring. In general, those mammalian species that transfer passive immunity to the fetus *in utero* prior to birth (human, rabbit) demonstrate lower total amounts and different ratios of Ig classes in their mammary secretions compared to those species that transfer passive immunity to the neonate only *via colostrum* after birth (cow, horse, pig) where ingestion of Igs from the colostrum is essential for healthy survival.

Differences between species and their Ig relationships will be considered but the emphasis will be placed on the species that are the major producers of milk consumed by humans. The domesticated ruminant species (family Bovidae), which includes the dairy cow, goat, sheep, buffalo and yak, comprise the major producers of commercial milk in the world, and of these, the dairy cow (*Bos taurus*) is probably the most highly developed for milk production. Significantly, by coincidence or evolutionary design, the mammary glands of these species are also the largest producer of mammary Igs. It is these ruminant and closely related species that are the most dependent upon the colostral transport of Igs to the neonate for their healthy survival.

The significance and quantitative rationale for the presence of Igs in the mammary secretions requires some background knowledge and understanding of the immune defense systems of the animal.

2. Animal Immune Systems

The immune capability of a mammalian species is composed of two major systems, cell-mediated immunity and humoral immunity (Butler, 1974, 1983; Smith *et al.*, 1983; Guidry, 1985; Darnell *et al.*, 1986). Each has the capability of responding to some degree with protective mechanisms when stimulated by foreign antigens such as bacteria, viruses, toxins, foreign tissues and tumors. Both systems are derived from lymphoblasts produced in the bone marrow, but each takes a different path in processing (differentiation) which results in two types of lymphocytes: T-lymphocytes and B-lymphocytes (Fig. 1).

2.1. Cell-Mediated Immune (CMI) System

Lymphocytes differentiated in the thymus gland to T-lymphocytes form the CMI system. T-lymphocytes respond to foreign body stimulation by producing a host of different types of activated cells which include cytotoxic killer, helper, supressor and memory cells and cells that produce specific active proteins called lymphokines. Lymphokines are involved in many effector and supressor functions such as inhibition or stimulation of macrophages, enhancement or inhibition of antibody production in lymphocytes, and growth inhibition and cytotoxic effects in various other cells. Transfer factor, involved with the transfer of hypersensitivity from one animal to another, and interferon, an inhibitor of viral and bacterial replication, also are produced by T-lymphocytes.

The response of the CMI system to foreign bodies is strong and is particularly involved with mycotic agents, cancer cells and foreign tissues such as grafts and transplants.

2.2. Humoral Immune System

Humoral immunity involves antibodies and the complement system. Antibodies are produced by plasma cells derived (evolved) from B-lymphocytes after contact and stimulation with a foreign antigen. The differentiation or maturation of avian B-lymphocytes takes place in an organ called the bursa but, in mammalian species, a specific organ as such has not been found. Nevertheless, this processing does occur in the bone marrow. The multiple production of daughter plasma cells derived from stimulated B-lymphocytes is a function of the strength of the antigen-

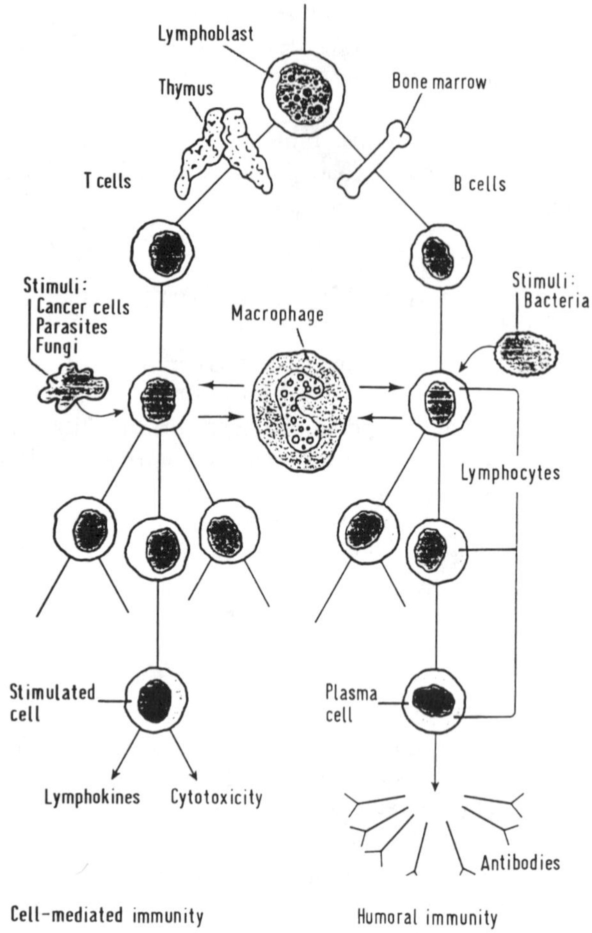

Fig. 1. The two-component immune system. Cell-mediated immunity: lymphoblasts from the bone marrow differentiate in the thymus to form T-lymphocytes. Upon stimulation by a foreign antigen, the T-cells divide to form clones that recognize the antigen. Further stimulation results in more functional cells and products to combat the foreign antigen. Humoral immunity: lymphoblasts from bone marrow differentiate into B-lymphocytes which, upon stimulation by a foreign antigen, multiply to produce clones of plasma cells, each producing specific antibodies to the antigen. Macrophages are surveillance cells that recognize antigens and present them to T- and B-cells plus other functions. (From Sherman & Johnston (1982), reproduced by permission of *Illinois Research*, the University of Illinois).

ic response, and even years after a specific antigen (disease) has been overcome, the body still retains a few B-cells which can reproduce rapidly if the foreign antigen should again be encountered.

The humoral immune system is especially effective against bacteria and their toxic products. The antibodies (immunoglobulins) are synthesized in plasma cells located throughout the body, including some in the mammary gland. Large amounts of Igs are present in the blood and in significant amounts in various body tissues, including the submucosa of ducts and tracts, and secretory and mucosal tissues.

3. Structure and Properties of Immunoglobulins

3.1. Immunoglobulin Classes

There are five classes (or isotypes) of Igs (Butler, 1974, 1981, 1983; Smith *et al.*, 1983; Guidry, 1985; Darnell *et al.*, 1986). All have a similar basic structure, but there are differences in the organizational structure as well as the amino acid sequences and carbohydrate groups present. Certain portions of the molecules are identical within a class or subclass. Differences relate both to the structures that permit specific cellular-related functions of the class and subclass and to the antigen-binding portions of the functional molecules, which differ widely in amino acid sequence depending upon the immunological specificity to a particular antigen. Research on the structure and function of Igs has progressed furthest with the human species from which the nomenclature has been derived (Cepellini, 1964). Studies with other species show gross class similarities, but differences occur in amino acid chain composition, the possible absence of one class (IgD) and in the occurrence of subclasses.

3.1.1. Basic Structure

All Igs have a similar basic structure composed of four subunits with a molecular weight that varies around 160 000 and a sedimentation constant around 7*S* (Fig. 2). IgG is the predominant Ig class in serum and always exists in monomeric form. Using IgG as an example, the intact molecule contains two identical light chains (molecular weight each about 25 000) and two identical heavy chains (molecular weight each about 55 000). There are two types of light chains (κ and λ) which differ in amino acid chain structure but have homologous sequences. The light chains contain a constant region (C_L) and a variable region (V_L) associated with immunological specificity. The heavy chains differ for each class (γ-chain for IgG). They contain constant (C_H) regions as well as a variable (V_H) region associated with immunological specificity. Each light chain is attached to the heavy chain by a disulfide bond and the two heavy chains are held together by disulfide bonds near the hinge region which imparts conformational flexibility to the complex molecule in antibody–antigen interactions. The upper (diagramatic) variable region domains of the molecule (Fig. 2) provide two sites where antigen-specific binding can occur. Models of antibody (Ig) molecules derived from X-ray crystallographic data clearly show the Y-shape with two antigen binding sites (Silverton *et al.*, 1977; Darnell *et al.*, 1986).

Treatment of the IgG molecule with a reducing agent (β-mercaptoethanol) breaks the disulfide bonds (Fig. 3) resulting in separation into the individual light and heavy γ-chains. Digestion with papain splits the Ig molecule near the hinge

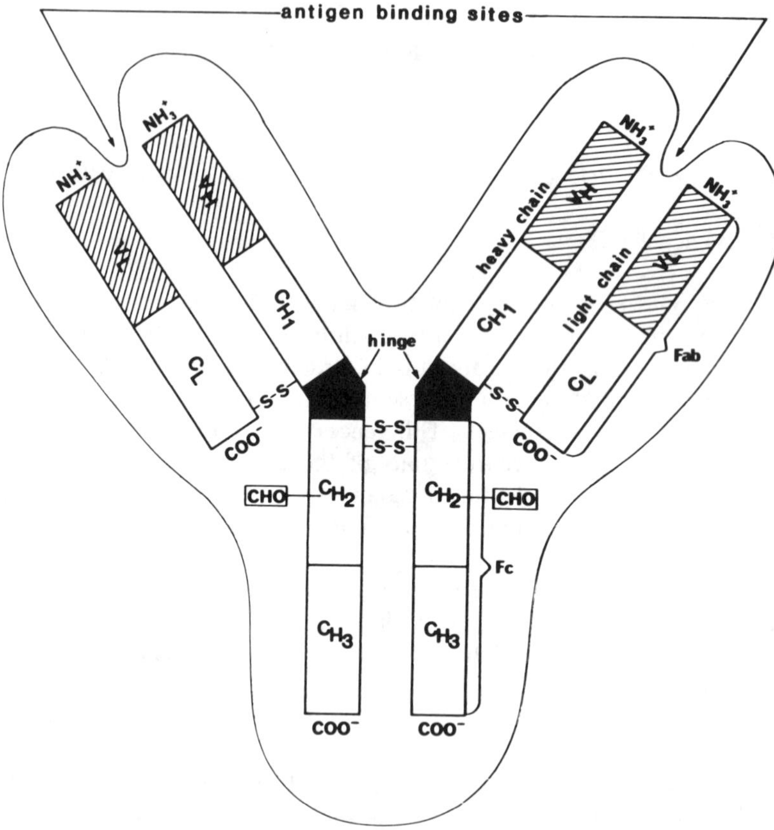

Fig. 2. Schematic diagram of a basic 7*S* immunoglobulin (Ig) molecule showing two heavy and two light chains joined by disulfide bonds: V = variable region; C = constant region; L = light chain; H = heavy chain; 1, 2, and 3 subscripts refer to the three constant regions of the heavy chains; CHO = carbohydrate groups; Fab refers to the (top) antigen-specific portion of the Ig molecule; Fc refers to the cell-binding effector portion of the Ig molecule. (From Guidry (1985), reprinted by permission from *Lactation*, edited by B.L. Larson, 1985 by Iowa State University Press, Ames.)

region into a lower portion (Fc) and two upper segments (Fab), each of which is composed of a light chain and heavy chain constant and a variable region. Pepsin digestion of the Ig molecule breaks up the Fc portion but leaves the two Fab segments attached as $F(ab')_2$, held together by disulfide bonds.

The Fc portion of the Ig molecule contains two constant regions of the heavy chains with most of the segment(s) that impart class and subclass specificity (Figs. 2 and 3). Carbohydrate groups (IgG contains about 3% carbohydrate) are located here as well as the cell-binding effector chain segments recognized by receptors on cells involved in transport and other specific effector functions of the Ig molecule that can include complement activation, opsonization, and antibody-dependent toxicity that is cell-associated.

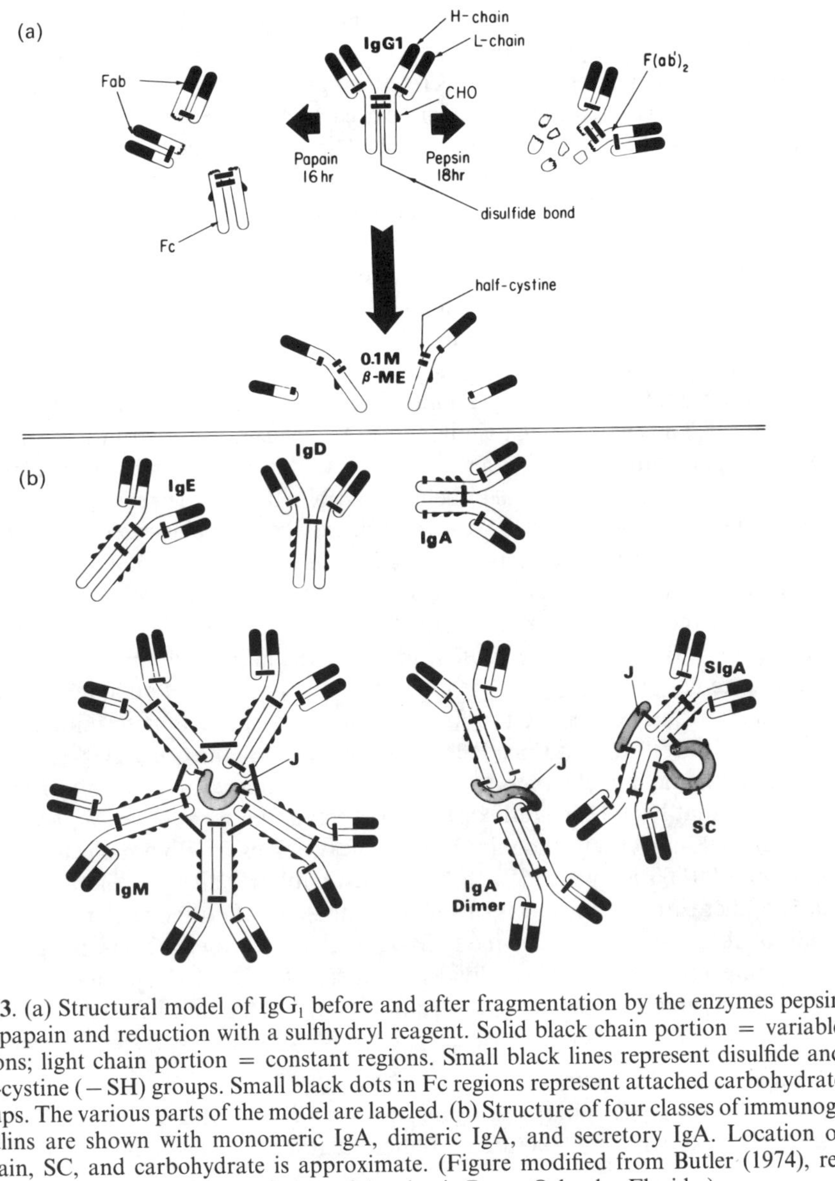

Fig. 3. (a) Structural model of IgG_1 before and after fragmentation by the enzymes pepsin and papain and reduction with a sulfhydryl reagent. Solid black chain portion = variable regions; light chain portion = constant regions. Small black lines represent disulfide and half-cystine (– SH) groups. Small black dots in Fc regions represent attached carbohydrate groups. The various parts of the model are labeled. (b) Structure of four classes of immunoglobulins are shown with monomeric IgA, dimeric IgA, and secretory IgA. Location of J-chain, SC, and carbohydrate is approximate. (Figure modified from Butler (1974), reproduced by permission of Academic Press, Orlando, Florida.)

3.1.2. Class Differences

IgA and IgM are similar to IgG in their basic monomeric structure with the exception that there is an additional 18-amino acid extension at the C-terminal end of the heavy chains; α for IgA and μ for IgM (Butler, 1974, 1981; Guidry, 1985; Darnell *et al.*, 1986). Both contain approximately 10% carbohydrate. IgA may occur free or in complexed forms. Two monomeric 7*S* IgA molecules bind to

J-chain, a protein of about 16 000 molecular weight, to form a dimer of about 370 000 molecular weight with a sedimentation constant of about 9*S* (Fig. 3). The dimeric form is synthesized and secreted by plasma cells into the submucosa of various tractal tissues and in the stromal area adjacent to the secretory epithelium of the mammary and other exocrine glands. Dimeric IgA, facilitated by the J-chain, binds to the secretory component (SC), a protein of about 75 000 molecular weight synthesized in the epithelial cells to form secretory IgA (SIgA) with a sedimentation constant of about 11*S* (Bradtzaeg, 1981: Solari & Kraehenbuhl, 1984; Darnell *et al.*, 1986; Kraehenbuhl *et al.*, 1988). Free secretory component (FSC) also is present in colostrum and milk (Butler, 1974, 1983; Pringnitz *et al.*, 1985*a*).

IgM is a more complex Ig molecule with a molecular weight of about one million and a sedimentation constant of about 19*S*. It consists of five basic 7*S* molecules bound in a circular manner (Fig. 3) to J-chain. IgM is synthesized in plasma cells and binds to SC, similar to IgA, when transported intact across the secretory epithelium. IgM in the mammary secretions comes both from plasma cells adjacent to the secretory epithelium and blood serum (Frenyo *et al.*, 1987*a*). This and other transfer system pathways for the Igs will be discussed in Section 4.

IgG is the dominant Ig class comprising most of the Igs of blood serum in probably all mammalian species. IgG is the carrier of passive immunity in all species, whether transferred *in utero* before or *via colostrum* to the neonate after birth. Both IgA and IgM are present in blood serum in small amounts as well as in the mammary secretions. In the initial immune response of an animal to a new antigen, IgM antibodies are the first produced and predominate until sufficient plasma cells are formed to produce larger amounts of the other classes.

Diagrammatic model structures for all of the Ig classes are shown in Fig. 3. In all species that have been studied, IgG, IgA (monomeric, dimeric and secretory), IgM and IgE have been recognized. IgE (ε heavy chains) with a molecular weight of about 190 000 is involved with hypersensitivity and allergic reactions. In the human, there are variant forms of IgA (A_1 and A_2), IgD (function unknown) and additional IgG subclasses. Bovine IgG_2 exists in two further subclasses, IgG_{2a} and IgG_{2b}, and IgG_{2a} is further subdivided (Butler, 1983, 1986; Heyerman & Butler, 1987). Undoubtedly considerable variation exists within and between species.

3.1.3. Antigenic Determinants and Genetic Markers

There are three types of major antigenic determinants located on Ig molecules that serve as markers in the structure and immune responses of the Igs (Butler, 1981, 1986).

—*Isotype* denotes differences in determinants that distinguish the class and subclasses of the Igs. Isotypes reflect differences in amino acids or carbohydrates located in the constant regions of the heavy and light chains.

—*Allotype* denotes genetic variants among Igs of similar isotype in an animal population. These reflect amino acid differences from mutations in the DNA coding the constant regions (usually) of either the heavy or light chains.

—*Idiotype* denotes the antigenic determinants that distinguish different antigen combining sites. These reflect amino acid differences in the variable regions of the heavy and light chains.

3.2. Composition, Structure and Properties

3.2.1. Clonal Selection and Diversity of Antibodies

Idiotypes with different conformational specificities reflect amino acid differences in the variable regions. How the body could recognize and, after a short lag period, begin to produce specific antibodies in large numbers to a new antigen was a mystery for many years. It is now understood that the body is constantly elaborating many new B-lymphocytes, each of which has surface Ig molecules that may differ widely in specificity between cells, but are the same for each individual cell. When a new antigen is encountered that combines with an appropriate Ig configuration on a particular B-cell, that cell is triggered into differentiating into many (new) plasma cells which now produce large numbers of Ig molecules of the same specificity.

In the first encounter with a new antigen, IgM is produced predominantly. A second encounter produces IgA and then IgG. After a new antigen is overcome and no longer present, a few memory cells persist which quickly multiply and produce Igs if the antigen is ever encountered again.

The diversity of antibodies reflects rearrangements of DNA in the genes that generate the variable regions of the Ig molecule. In the mouse, the coding regions for the heavy and light chains are encoded in three genetic loci on three different chromosomes, with each locus having multiple variable region genes and a few constant region genes. It has been estimated that 10^{11} different antibody molecules could be produced by the B-cells in the mouse; a number probably larger than the animal could produce in a lifetime (Darnell *et al.*, 1986). The Igs are synthesized in the rough endoplasmic reticulum and either are secreted or reside in the cell surface membrane.

3.2.2. Amino Acid Content (Nutritional Value)

The amino acid composition of the Igs can be of nutritional importance when they make up a significant portion of the diet, such as with bovine colostrum. The amino acid sequence and composition can vary with the isotype, idiotype and allotype, but these differences are not normally large.

Typical data for the amino acid content of bovine serum and colostral IgG_1 and IgG_2 are summarized from two sources in Table 1. Data on IgG from some other species (human, rabbit and horse) show close similarities to those in Table 1 (Kickhofen *et al.*, 1968).

Other sources of amino acid data of value on mammary Igs include those on the 'euglobulin' and 'pseudoglobulin' fractions isolated from bovine colostrum and milk in the era before the current class recognition and nomenclature were developed (Hansen *et al.*, 1947; Smith, 1948). These fractions represent a composite of the several Ig classes present, but mainly IgG_1, since it usually comprises 75%

TABLE 1
Amino Acid Composition of Bovine Serum and Colostral IgG_1 and IgG_2[a]

Amino acid	*Colostral IgG_1*		*Serum IgG_1*		*Colostral IgG_2*	*Serum IgG_2*	
Lys	5·75	(5·16)	6·17	(5·58)	8·75	8·08	(5·61)
His	1·14	(1·48)	1·74	(1·70)	1·99	2·08	(1·61)
Arg	3·47	(3·16)	4·68	(3·93)	4·20	4·40	(3·68)
Asp	7·13	(7·55)	7·59	(7·94)	7·55	7·51	(7·70)
Thr	9·42	(9·90)	9·63	(9·40)	7·40	8·38	(9·73)
Ser	13·57	(13·76)	13·89	(14·17)	10·41	11·04	(13·98)
Glu	8·20	(8·48)	8·03	(8·69)	8·65	7·44	(8·38)
Pro	8·36	(8·16)	7·07	(7·13)	10·95	7·83	(7·30)
Gly	6·78	(7·47)	6·32	(6·94)	6·83	7·09	(7·18)
Ala	5·76	(5·18)	5·55	(4·97)	5·95	5·98	(4·98)
Cys	2·28	(3·25)	2·94	(2·46)	2·27	3·04	(2·65)
Val	8·59	(8·95)	8·96	(9·42)	5·74	9·35	(9·64)
Met	0·71	(0·65)	1·21	(0·91)	1·22	0·61	(0·81)
Ile	4·94	(2·34)	2·53	(2·51)	3·04	2·36	(2·55)
Leu	6·87	(6·50)	6·90	(6·28)	7·55	7·30	(6·31)
Tyr	4·10	(3·68)	3·98	(3·60)	4·04	4·47	(3·53)
Phe	2·93	(2·50)	2·91	(2·59)	3·46	3·04	(2·58)
Trp	—	(1·83)	—	(1·79)	—	—	(1·78)

[a]Data from Lisowski *et al.* (1975) and, in parentheses, Kickhofen *et al.* (1968). The results are expressed as the percentage of amino acids recovered (residues of amino acid per 100 residues).

or more of the total Igs of bovine colostrum. The classical 'euglobulin' fraction probably contains most or all of the SIgA and IgM.

3.2.3. Physical Properties

A summary is shown in Table 2 of data on the molecular components and physical characteristics of bovine mammary secretion Igs. Physical data on Igs for different species are not plentiful and most of those available come from studies with the human, mouse and bovine. The footnote relating to the designation of these Igs by their electrophoretic mobility is important for relating the older literature to the current nomenclature.

The concentrated Igs in bovine colostrum store well at refrigerator temperatures for several weeks and are stable on freezing or lyophilization for long periods. Storage is common since some dairy cows produce large amounts of excess colostrum containing high levels of Igs. Other cows, often first-calf heifers, produce insufficient amounts for the effective passive immunization of their calf (see Sections 4.1 and 5.2). If excess colostrum turns acid from fermentation, practical dairymen will continue to utilize the still active Igs present by making it palatable with added sodium bicarbonate.

The Igs are the most thermolabile of the major milk proteins (Larson & Rolleri, 1955; Lyster, 1970; Parry, 1974; de Wit & Klarenbeek, 1984). For example, 30 min at 70°C will denature 89% of the Igs in bovine milk but only about 52% or less of the other major milk proteins (Larson & Rolleri, 1955).

TABLE 2
Properties of Bovine Mammary Secretion Immunoglobulins[a]

Property	*Immunoglobulin or component*						
	IgG$_1$	*IgG*$_2$	*SIgA*	*IgM*	*L-chain*[b]	*J-chain*	*Secretory component*
Electrophoretic fraction[c]	γ_1 and β_2	γ or γ_2	β_2	β_2	n/a	n/a	β_2
Molecular weight × 10^3:							
Intact	146–163	146–154	385–430	1 030	22–27	16·5	70–96
Heavy chain	56–59	54–59	61–63	62–76	n/a	n/a	n/a
Heavy chain type	γ_1	γ_2	α	μ	n/a	n/a	n/a
Carbohydrate:							
Total	2·8–3·1	2·6–3·0	6–10	10–12	0·5	—	3·1–5·9
Hexose[d]	1–1·5	1–1·5	5·6	5–5·5	—	—	2·6–2·9
Hexosamine	0·9–1·7	0·8–1·6	—	2–3	—	—	2·5–3
Fucose	0·1–0·3	0·2–0·3	—	1·25	—	—	0·2–0·8
Sialic acid	0·3	0·2	—	1·4	—	—	0·4
$S_{w,20}$[e]	6·3–7	6·5–7·1	10·8–11	18–20	—	—	4–4·9
Isoelectric point	5·5–6·8	7·5–8·3	—	—	—	—	—
$A_{1\%,1\,cm}$[f]	12·1, 13·5	12·0, 12·1	—	11·8	—	—	—

[a] Adapted from data compiled by Butler (1983, 1986) and Eigel *et al.* (1984). n/a = Not applicable. See text on IgE.
[b] L-chains (κ and λ) predominantly of the λ type.
[c] Relative location in classical electrophoretic pattern of blood serum proteins (serum albumin; α-, $\beta(\beta_1 + \beta_2)$- and $\gamma(\gamma_1 + \gamma_2)$-globulins) in order of decreasing mobility; γ_1 was also called 'Fast γ'.
[d] Hexoses of IgG_1 and IgG_2 are mannose, galactose and glucose, present in approximately similar amounts; hexoses of SC are predominantly mannose.
[e] Sedimentation constant in Svedberg units ($\times 10^{-13}$ s).
[f] Absorptivity of a 1% solution in a 1 cm light path at 280 nm.

The Igs, and particularly the IgM in the classical 'euglobulin' fraction, are related to the creaming phenomenon or fat globule clustering in milk with cold agglutination (Farrell & Thompson, 1974; Brown, 1988). Other clustering phenomena involving bacteria also have been ascribed. IgM and SIgA are more prevalent in the fat layer than in whey and significant amounts of IgM and IgG_2 are associated with the casein pellet on centrifugation (Frenyo *et al.*, 1987*b*). Colostrum and milk contain many intact cells and cellular debris as well as bacteria, and some of the Igs present are associated with these materials (Larson, 1979; Paape *et al.*, 1982; Weaver *et al.*, 1982).

3.2.4. Related Immune System Proteins

Complement is a complex group of proteins present in blood which assists the function of antibodies. Among the functions of complement are the enhancement of vascular permeability and attraction of phagocytic leukocytes, facilitation of the ingestion and destruction of antigen by the phagocytes, and the lysis of certain types of organisms. There are at least 11 proteins in the complement group, of which some are undoubtedly present in milk but in extremely low amounts (mg/dl and less range) related to the number and type of leukocytes present, the degree of any inflammation, and other factors (Guidry, 1985).

Closely related to the functions of the immune systems are a group of membrane-bound glycoproteins on the surface of most cells that are products of the HLA multigene system (Smith *et al.*, 1983; Darnell *et al.*, 1986). The antigens in this system are called the major histocompatibility complex (MHC) antigens. The HLA system, delineated first in humans as the 'human leukocyte system A', encodes for histocompatability antigens involved in the rejection of transplanted tissues; it is the BoLA system in the bovine. Functions of the HLA or BoLA system include the presentation of antigen to T-lymphocytes, response by helper T-lymphocytes to help induce B-lymphocytes to proliferate and become plasma cells, and other T- and B-cell functions.

β-Microglobulin is a protein that associates with MHC gene products (Simister & Mostov, 1989). This protein of unknown function, when isolated many years ago from bovine milk (Groves *et al.*, 1963, 1985), was named 'lactollin' and then subsequently recognized to be the same as β_2-microglobulin of bovine blood serum (Jenness, 1982; Whitney, 1988). It has a molecular weight of about 12 000, a structure homologous to the constant regions of the light and heavy chains of IgG, and it exists in milk as a tetramer. Human colostrum contains about 0·1 mg/ml and normal milk about 0·01 mg/ml. Bovine milk and colostrum contain about 0·006 mg/ml. The β_2-microglobulin in milk appears to arise from proteolytic-dependent degradation of the cellular fraction, probably monocytes (Pringnitz *et al.*, 1985*a*,*b*). It recently has been shown (Simister & Mostov, 1989) that β_2-microglobulin is a component of the Fc receptor involved in IgG transport in the rat intestine (see section 4.2).

4. Transfer of Passive Immunity

4.1. Pathways and Species Differences

Igs are transported from the bloodstream of the mammalian mother to the neonate by one or both of two major pathways (Brambell, 1970; Butler, 1974). In the *in utero* pathway, IgG passes from the maternal blood across the yolk sac or placental membrane into the fetus (Dancis, 1960; Butler, 1974, 1981; Lin, 1980). In the *via colostrum* pathway through the mammary gland and intestinal route, IgG is first transferred from the bloodstream across the mammary secretory cells into the colostrum (Smith, 1946, 1948; Larson, 1958; Dixon *et al.*, 1961; Murphy *et al.*, 1964; Butler, 1983). Upon its ingestion by the neonate, the IgG is transferred from the gut across the intestinal cells into the blood (Hansen & Phillips, 1949; Deutsch & Smith, 1957; Lecce & Morgan, 1962; Witty *et al.*, 1969; Brandon & Lascelles, 1971; Rodewald, 1973; Campbell *et al.*, 1977). The passage of maternal IgG across the yolk sac, placental membrane, or the mammary gland secretory cell layer is a highly specific transport pathway. The passage across the neonate's intestinal cells

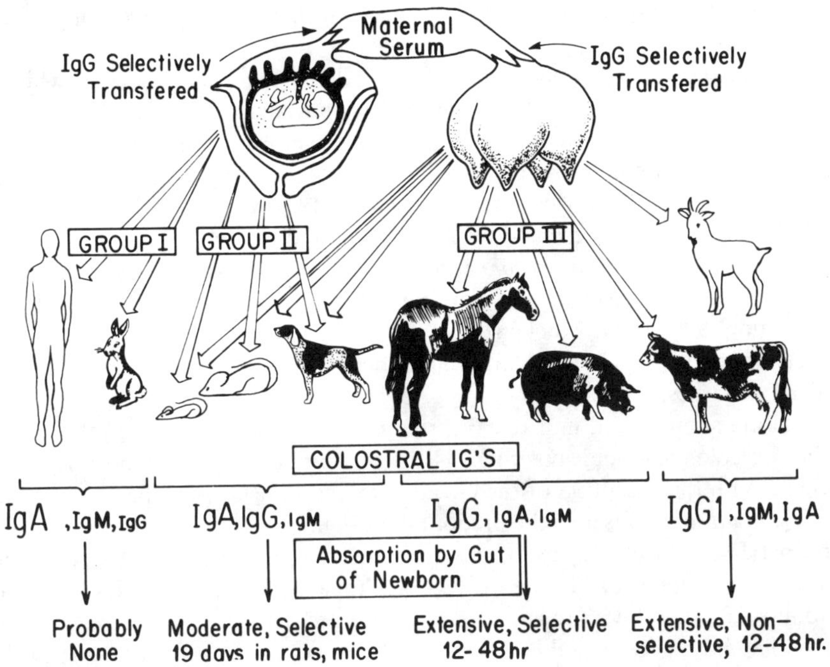

Fig. 4. The transfer of maternal immunoglobulins to the newborn of representative mammalian species. Group I species transfer *in utero* before birth. Group II species transfer both *in utero* before birth and *via colostrum* after birth. Group III species transfer only *via colostrum* after birth. The size of the immunoglobulin notation (IgA, IgM, IgG, IgG_1) denotes the relative percentage composition of the immunoglobulins in the colostrum. Species in group II may have IgG as the predominant Ig in colostrum. Significant IgG_2 also may be present in the colostrum of some Group III species. See text for comments. The relative absorption of immunoglobulins in the gut of the neonate is also shown. (From Butler (1974), reproduced by permission of Academic Press, Orlando, Florida.)

can be specific or non-specific depending on the species, and it occurs only for a specified period after birth until the intestinal cells undergo closure.

These pathways are summarized in Figure 4 with the species divided into three major groups. At the left side of the figure in Group I are the species (human, rabbit, guinea pig) that only transport IgG prior to birth *in utero* by a selective transport pathway. The colostrum of these species does contain Igs which have a beneficial therapeutic effect to counteract infective intestinal organisms, but the Igs do not pass across the intestinal barrier into the neonate's circulation. IgA (SIgA) is the predominant Ig in the milk and colostrum of these species.

At the far right of Fig. 4 in the Group III species are the ruminants (cow, goat, sheep) which only transport IgG *via colostrum* after birth. The mammary transport process is highly selective for the IgG_1 subclass which predominates in the colostrum with smaller amounts of IgG_2, IgM and IgA. The absorption process in the intestine of the neonate is non-selective and occurs only for about a day (Deutsch & Smith, 1957; Lecce & Morgan, 1962; Brambell, 1970; Brandon & Lascelles, 1971;

Butler, 1974; Campbell *et al.*, 1977). IgG continues as the predominant Ig in normal milk.

Close to the ruminant species in Group III are the horse and pig, which are similar in that a subspecies of IgG is the chief Ig transported, but different in that the absorption process in the neonate may or may not be selective and SIgA becomes the predominant Ig in normal milk (Lecce & Morgan, 1962; Porter, 1969; Rouse & Ingram, 1970; McGuire & Crawford, 1972; Butler, 1981).

Group II species, represented by the dog, cat and rodents, transport IgG to the neonate both *in utero* and *via colostrum* and milk after birth. Intestinal absorption in the neonate is selective and continues for up to 3 weeks (Brambell, 1970; Rodewald, 1973, 1980; Mackenzie, 1984). IgG is usually the predominant Ig in colostrum (carnivores in particular) but either IgG or IgA may predominate in the milk of a species.

Data are summarized in Table 3 for some species on the concentrations of Igs in blood, colostrum and milk. IgG is the predominant Ig of blood serum in all species. With the exception of the Group I species that transmit passive immunity entirely *in utero*, IgG is also the predominant Ig in colostrum. In the Group II and Group III species that transmit passive immunity, an IgG subclass is probably always involved, such as in the rat (IgG_{2a}), horse (IgT), and cow (IgG_1). Smaller amounts of other IgG subclass(es) probably are also present in the mammary secretions; however, available data are limited to only a few species. It must be recognized that large variations probably occur in all species, and in particular those that transport large amounts of Igs through the colostrum. Some of the values in Table 3 represent limited sample numbers and should be taken as representative approximate amounts.

Comment is appropriate for some of the species data in Table 3. With regard to the values for IgG and IgA in dog colostrum, it should be noted that the research and summarized data (Heddle & Rowley, 1975) showing that the Igs of dog colostrum are about 80% IgG and only 20% IgA have been misinterpreted in at least one text source (Tizard, 1982) to indicate the reverse.

The data on the rat are inconsistent and may represent differences due to several factors, including sampling time and specificity of antisera for the various forms of IgA (monomeric, dimeric, secretory) used in the analyses. Ten-fold differences (0·13–1·42 mg/ml) in the colostral level of IgA have been reported, with the maximum about 4 days after parturition. Analyses on colostrum, and particularly milk, for IgG have focused on the subclass IgG_{2a}, largely ignoring other IgG subclasses that may be present. The rat data in Table 3 represent average values from several studies. Even though a value is lacking for the total IgG content of rat milk, the amount of IgG_{2a} subclass alone is greater than that of IgA. In colostrum, the total IgG is larger even when compared to the highest reported value for IgA. It must be concluded that IgG is the predominant Ig in rat serum, colostrum and milk.

Comment concerning the amount of IgG_1 that can be present in bovine colostrum and the ratio of IgG_1 to IgG_2 in the mammary secretions is pertinent. The amount of IgG_1 leaving the blood and accumulating in the mammary gland of a

TABLE 3
Concentration of Immunoglobulins and Percentage of Major Component in Serum and Mammary Secretions of Several Species[a]

Species	*Immunoglobulin*	*Concentration* (mg/ml)			*% Major component immunoglobulin*		
		Blood serum	*Colostrum*	*Milk*	*Serum*	*Colostrum*	*Milk*
Human	IgG	12·1	0·43	0·04	78		
	IgA	2·5	17·35	1·00		90	87
	IgM	0·93	1·59	0·10			
	FSC		2·09	0·02			
Rat[b]	IgG-total	24·6	2·6	—[c]	96	76	—[c]
	IgG_{2a}	8·0	0·9	1·53			
	IgA	0·15	0·8	0·59			
	IgM	0·77	ND[d]	ND[d]			
Dog[b]	IgG	11·1	23·4	0·24	81	68	
	IgA	0·7	9·8	2·63			85
	IgM	1·7	0·8	0·22			
Pig	IgG	21·5	58·7	3·0	89	80	
	IgA	1·8	10·7	7·7			70
	IgM	1·1	3·2	0·3			
Horse	IgG-total	21·9	113·4	0·39	89	88	
	IgG(T)	8·2	15·2	0·09			
	IgA	1·5	10·7	0·48			43
	IgM	1·2	5·4	0·03			
Cow (*Bos taurus*)[b]	IgG-total	25·0	32–212	0·72	88	85	66
	IgG_1	14·0	20–200	0·6			
	IgG_2	11·0	12·0	0·12			
	IgA	0·4	3·5	0·13			
	IgM	3·1	8·7	0·04			
	FSC		0·5	0·2			

[a] Approximate values, some from limited observations. Data compiled and calculated from: human and pig (Butler, 1974); rat (Stechschulte & Austen, 1970; Bazin *et al.*, 1974; McGhee *et al.*, 1975; Michalek *et al.*, 1975; Rousseaux & Bazin, 1979); dog (Vaerman & Heremans, 1969; Heddle & Rowley, 1975); horse (Rouse & Ingram, 1970; Vaerman *et al.*, 1971; McGuire & Crawford, 1972); cow (Butler, 1981, 1983; Devery-Pocius & Larson, 1983). Certain IgG subclasses for rat, horse and cow are shown and included in the total IgG. Where subclasses in other species were reported, they are grouped in the total for the class.
[b] See text for comments on rat and dog secretions, amounts of bovine IgG_1, and ratio of bovine IgG_1 to IgG_2.
[c] Data not available but total IgG must be > 1·53 mg/ml and 72%.
[d] ND = Not consistently detected.

dairy cow around parturition can be less than 100 g to more than 2 kg; the latter being an amount that can significantly reduce the blood serum level (Larson & Kendall, 1957; Larson, 1958). An increase in the blood serum level of IgG with age (Larson & Touchberry, 1959; Williams & Millar, 1978) makes this more apparent in older animals beyond the first or second lactation. In a study involving 87 dairy cows (Devery-Pocius & Larson, 1983), the median amount of total Igs recovered in the colostrum (first four milkings) was over 800 g with IgG_1 accounting for about 78% of the total. Individual samples of first colostrum have been observed with over 200 mg/ml IgG_1. The largest amounts transported were found with older cows in their second and subsequent lactations. An increase of colostral IgG with age beyond the first lactation has also been noted in ewes (Gilbert *et al.*, 1988) as have Ig concentrations exceeding 150 mg/ml in ewe colostrum (Campbell *et al.*, 1977).

Review data (Butler, 1974, 1983) generally indicate that the IgG_1/IgG_2 ratio in colostrum and milk normally exceeds 10 and may be as high as 30. However, in the study noted above (Devery-Pocius & Larson, 1983) with cows representing the major dairy breeds but predominantly (75%) Hostein–Friesian, the IgG_1/IgG_2 ratios for individual animals did range as high as 16 for the composited first four milkings and were much higher in some individual first colostrum samples, but the overall average ratio for all cows in the composited first four milkings was only 5. In fact, the amount of IgG_1 was so small in two cows that it did not exceed that of IgG_2. Data are presented in another source (Butler, 1974) with six Holstein–Friesian cows showing an IgG_1/IgG_2 ratio in colostrum of about 10 at parturition, but only averaging about 2 in milk during the next month. These observations emphasise the caution necessary with limited data, where large variation occurs. For these reasons, the bovine data in Table 3 are composites from several sources which this writer believes represent a more general average estimate for high producing dairy cattle in the United States in their second and subsequent lactations.

4.2. Transport Mechanisms

The transport of IgA through epithelial cells has been extensively studied and the mechanism in the mammary secretory cell appears to be similar to other secretory tissue sources (Brandtzaeg, 1981; Mostov *et al.*, 1984; Solari & Kraehenbuhl, 1984; Darnell *et al.*, 1986; Kraehenbuhl *et al.*, 1988). Secretory component (SC) synthesized in the secretory cell, moves into position on the basolateral membrane as a surface receptor where it picks up dimeric IgA with attached J-chain synthesized in an adjacent plasma cell. The complex moves into (endocytosis) and through the mammary secretory cell to the apical surface where it passes (exocytosis) into the lumen as SIgA, with a segment of the SC clipped off in the release process. It is likely that IgM is transported by a similar process, but pieces of blood origin IgM also appear in the mammary secretions (Frenyo *et al.*, 1987*a*). The poly-Ig receptor for polymeric IgA and IgM is structurally related to the Igs (Mostov *et al.*, 1984).

The mechanism for the transport of IgG_1 and IgG_2 through the mammary secretory cell has not been well studied. However, the presence of specific receptor groups with a high affinity for IgG_1 at the time of maximum Ig transport, as well as additional lower affinity receptors for both $IgG_1$and IgG_2 during lactation, corresponds in general to the observed amounts of these subclasses in colostrum and milk (Hammer *et al.*, 1969; Sasaki *et al.*, 1977; Guidry, 1985). Bovine mammary cells at the time of maximum transport rapidly take up IgG_1 at their basolateral membrane surfaces and large amounts of IgG_1 can be observed both in the cells and accumulated in the lumen at the time of maximum transport just preceding parturition (Leary *et al.*, 1982; Larson, 1985). This transport appears to take place through a Fc receptor-mediated endocytosis (Lascelles, 1979; Leary & Larson, 1982). However, in contrast to IgA transport, the receptor does not remain with the IgG molecule in the secretions. Whether the receptor is then digested or recycled back to the surface of the cell has not been defined, with several mechanisms being possible (Pearse & Bretscher, 1981; Farquhar, 1983; Hopkins, 1983; Pastan & Willingham, 1983). The IgG_1 appears to be carried through the cell in vesicles that are not associated with the pathway of any other major milk constituent. The IgG_1 is secreted (exocytosis) into the lumen as part of the colostrum or milk as an intact molecule carrying no apparent attachments from the transport process.

Fc receptors for IgG subclasses are involved also in the *in utero* transport of IgG in Group I species and the intestinal IgG transport in Group II species (Dancis, 1960; Lin, 1980; Rodewald, 1980; Mackenzie, 1984). They also are found on many types of blood cells involved with the immune system (Darnell *et al.*, 1986; Simmons & Seed, 1988). Fc receptors in complex tissues have been difficult to isolate and characterize because of their diverse complexity, lability and multi-component nature. The Fc receptor for IgG transport in the intestine of the neonatal rat contains two components with molecular weights of 41 000–50 000 and 15 000 with the functional unit about 110 000 (Simister & Rees, 1985). The smaller component has been identified as β_2-microglobulin (Simister & Mostov, 1989) (see last paragraph in Section 3.2.3). It has not yet been determined if β_2-microglobulin serves a similar role in other IgG receptors.

5. Mammary Secretions and Some Immunological Considerations

5.1. Origin of Bovine Mammary Immunoglobulins

The vast majority of total Igs present in the mammary secretions of ruminant species (bovine) are transported there from the blood. These blood origin Igs are overwhelmingly IgG, specifically the IgG_1 subclass, and result from a highly specific transport system. The large amount of IgG in the mammary secretions of ruminants is different from the situation in many other species, and even from other secretory tissues in the ruminants themselves. The predominant Ig of the other external secretions is IgA which is derived from plasma cells lying adjacent to the cells of the secretory epithelium. IgG is predominant in the colostral secretions of some other species also (Table 3) but IgA predominates in milk. The question arises as to what explanation can be given for the low concentration of IgA and IgM derived from local synthesis in plasma cells of the ruminant mammary secretions.

The ruminant (bovine) mammary gland does contain plasma cells adjacent to the secretory epithelium where synthesis and direct transfer of Igs occur (Lee & Lascelles, 1970; Butler, 1981; Frenyo *et al.*, 1987*a*). This is mainly SIgA, but plasma cells producing IgM and IgG are also present. Compared to most other species, however, the ruminant mammary gland contains few plasma cells and most of those present are producers of IgA and IgM. In contrast to the IgG_1 derived almost exclusively from the blood, only part of the IgA and IgM present in ruminant milk and colostrum comes from the blood, with the remainder being derived from local synthesis by plasma cells and transferred directly to the mammary secretory cells. This is probably also true for part of the IgG_2 which is supported by the amino acid data (Table 1) which show significant differences between IgG_2 isolated from blood and colostrum, suggesting that the majority of subgroup IgG_2 found in colostrum and milk differs from that present in the blood serum (Lisowski *et al.*, 1975). Also, it should not be overlooked that although the concentration of IgA in bovine milk is low, the large size of the mammary gland and the huge volume of milk produced, results in a significant total production of IgA; probably the largest of any species (Butler, 1981; Mach & Pahud, 1971).

5.2. Altering the Mammary Immune System

Since the mammary transfer of passive immunity is of great importance in the domestic ruminant species, there have been many attempts to take advantage of it by enhancing the production of specific antibodies. For example, any increase in the antibody level to specific bacteria that cause mastitis or enteric infections should be of benefit. Enhancement of the CMI system also might be of benefit to the neonate by increasing the numbers of stimulated lymphocytes and other cells in the colostrum. The availability of large amounts of specific antibodies also might have further applications within and beyond the ruminant species such as in the human for use in the prevention and treatment of enteric pathogens.

An area of interest for years in all species (especially human) following the recognition of the CMI system has been to assess the significance of and potential to increase the transfer of activated T-lymphocytes and other cells to the neonate through the mammary secretions. For example, macrophages carrying IgA are a significant source of Igs in human colostrum. Many studies in this area have been conducted and, although the transfer of some CMI has been shown (Lascelles & McDowell, 1974; Head & Beer, 1978; Schultz, 1982; Weaver *et al.*, 1982), the significance of this transfer route to the newborn is elusive and its importance not well established.

There have been many attempts to stimulate and increase in the ruminant mammary secretions antibodies to specific organisms, especially those involved with mammary and intestinal infections. Mammary infusions of foreign antigens have given variable results. Use of a coliform antigen infusate into the mammary gland early in the dry period prior to parturition is not uncommon and may in some cases provide a significant benefit for the calf. Some promise has been shown in several species by oral administration of antigen, which stimulates sensitive areas in the small intestine called Peyers patches (Lee & Lascelles, 1970; Butler, 1974,

1981; Lascelles & McDowell, 1974; Newby *et al.*, 1982; Brandtzaeg, 1983). These patches contain M cells that transfer the antigen across the mucosal epithelia to reach the lymphoid tissue (Kraehenbuhl *et al.*, 1988). The oral route may result in more specific antigen-stimulated B-lymphocytes, which migrate to other parts of the body, including the mammary gland, and there evolve into plasma cells producing antibodies to the specific antigen. Obstacles to this approach in the bovine have been low antigenic response, difficulties in enhancing IgG_1 antibody production, and the small numbers of plasma cells in the mammary gland. Thus, the results of studies to enhance mammary secretion Igs have been variable with claims and the development of commercial products that often have been of questionable value.

Several commercial products that purport to be of benefit to the calf as colostrum supplements contain preparations of Igs isolated from bovine colostrum or milk or possibly from some other source. Analyses of these products (Hunt *et al.*, 1988; Haines *et al.*, 1990) have shown that their Ig contents are either negligible, or if present, considerably lower than the amount required by the newborn calf for the effective passive transfer of immunity. The products are not colostrum replacers although some benefit as a colostrum supplement may be provided. There have been test marketings of some products for humans that purport to contain bovine antibodies against enteric pathogens. Time will be the judge to determine if products are developed that contain effective amounts of specific antibodies or if the advertising claims will outstrip the rate of product development as has happened in the bovine field.

5.3. Conclusion

The immunoglobulins are a complex group of proteins that make a significant contribution to the protein content as well as the physical properties of the mammary secretions. They are present because milk is a secretory product and in many species they play an important biological role in the transfer of passive immunity to the newborn. This is especially true in the ruminant species that are the primary sources of milk for human consumption.

Even though some of the earliest observations were made with other species (see Section 1), the development of understanding of the molecular structures and biological functions has oriented around the human species. Since the human transmits passive immunity only *in utero*, understanding of the role of specific immunoglobulins in the transmission of passive immunity *via colostrum* has proceeded at a slower pace and has been influenced by knowledge from the human system.

The effects of this have been shown in several ways. For example, one can still find general statements in texts that IgA is the primary immunoglobulin of colostrum and milk. While this is true for a few species, including the human, it is certainly not true for many other species. Another result is that resources for research have supported and allowed studies on the molecular mechanisms of IgA transport to proceed at a much faster rate than that for IgG transport. As the human system has been studied, there has been a growing appreciation of the suitability of colostrum and milk for the human neonate. Perhaps this may be viewed as a positive result.

References

Askonas, B.A., Campbell, P.N., Humphrey, J.H. & Work, T.S. (1954). The source of antibody globulin of rabbit milk and goat colostrum. *Biochem. J.*, **56**, 597–601.

Bazin, H., Beckers, A. & Querinjean, P. (1974). Three classes and four (sub)classes of rat immunoglobulins: IgM, IgA and IgG_1, IgG_{2a}, IgG_{2b}, IgG_{2c}. *Eur. J. Immunol.*, **4**, 44–8.

Blanc, B. (1981). Biochemical aspects of human milk—comparison with bovine milk. *Wld Rev. Nutri. Diet*, **36**, 1–89.

Brambell, F.W.R. (1970). *The Transmission of Passive Immunity from Mother to Young*, North-Holland Publishing, Amsterdam.

Brandon, M.R. & Lascelles, A.K. (1971). Relative efficiency of absorption of IgG_1, IgG_2, IgA and IgM in the newborn calf. *Aust. J. Exp. Biol. Med. Sci.*, **49**, 629–33.

Brandtzaeg, P. (1981). Transport models for secretory IgA and secretory IgM. *Clin. Exp. Immunol.*, **44**, 221–32.

Brandtzaeg, P. (1983). The secretory immune system of lactating human mammary glands compared with other exocrine organs. *Ann. New York Acad. Sci.*, **409**, 353–81.

Brown, R.J. (1988). Milk coagulation and protein denaturation. In *Fundamentals of Dairy Science*, 3rd edn, N.P. Wong, R. Jenness, M. Keeney & E.H. Marth. AVI Publishing, Westport, Connecticut, pp. 583–607.

Butler, J.E. (1974). Immunoglobulins of the mammary secretions. In *Lactation: A Comprehensive Treatise*, vol. 3, ed. B.L. Larson & V.R. Smith. Academic Press, New York, pp. 217–55.

Butler, J.E. (1981). Concept of humoral immunity among ruminants. In *The Ruminant Immune System*, Volume 137 in Series Advances in Experimental Biology and Medicine, ed. J.E. Butler. Plenum Press, New York, pp. 3–55.

Butler, J.E. (1983). Bovine immunoglobulins: An augmented review. *Vet. Immunol. Immunopath.*, **4**, 43–152.

Butler, J.E. (1986). Biochemistry and biology of ruminant immunoglobulins. In *Progress in Veterinary Microbiology and Immunology*, ed. R. Pandey. Karger AG, Basel, Switzerland, pp. 1–53.

Campbell, S.D., Siegel, M.J. & Knowlton, E.J. (1977). Sheep immunoglobulins and their transmission to the neonatal lamb. *New Zealand Vet. J.*, **25**, 361–5.

Cepellini, R. (1964). Nomenclature for human immunoglobulins. *Bull. World Hlth. Org.*, **30**, 447.

Crowther, C. & Raistrick, H. (1916). A comparative study of the proteins of the colostrum and milk of the cows and their relations to serum proteins. *Biochem. J.*, **10**, 434–52.

Dancis, J. (1960). Transfer of proteins across the human placenta. In *The Placenta and Fetal Membranes*, ed. C.A. Vilee. Williams and Wilkens, Baltimore, pp. 185–7.

Darnell, J., Lodish, H. & Baltimore, D. (1986). *Molecular Cell Biology*, W.H.Freeman, New York.

Deutsch, H.F. & Smith, V.R. (1957). Intestinal permeability to proteins in the newborn herbivore. *Am. J. Physiol.*, **191**, 271–6.

Devery-Pocius, J.E. & Larson, B.L. (1983). Age and previous lactations as factors in the amount of colostral immunoglobulins. *J. Dairy Sci.*, **66**, 221–6.

de Wit, J.N. & Klarenbeek, J.N. (1984). Effects of heat treatments on structure and solubility of whey proteins. *J. Dairy Sci.*, **67**, 2701–10.

Dixon, F.J., Weigle, W.O. & Vazquez, J.J. (1961). Metabolism and mammary secretion of serum proteins in the cow. *Lab. Invest.*, **10**, 216–37.

Ehrlich, P. (1892). Ueber immunitat durch verbung und saugung. *Hyg. Infectionkrankh*, **12**, 183–203.

Eigel, W.N., Butler, J.E., Ernstrom, C.A., Farrell, H.M., Jr., Harwalkar, V.R., Jenness, R. & Whitney, R.McL. (1984). Amer. Dairy Sci. Assoc. Comm.—Nomenclature of Proteins of Cow's Milk: Fifth Revision. *J. Dairy Sci.*, **67**, 1599–631.

Farquhar, M.G. (1983). Multiple pathways of exocytosis, endocytosis and membrane recycling: Validation of a Golgi route. *Fed. Proc.*, **42**, 2407–13.

Farrell, H.M., Jr. & Thompson, M.P. (1974). Physical equilibria: Proteins. In *Fundamentals of Dairy Science*, 2nd edn, ed. B.H. Webb, A.H. Johnson & J.A. Alford. AVI Publishing, Westport, Connecticut, pp. 442–73.

Fischl, R. & von Wunscheim, O.Z. (1895). Ueber Schutzkorper im Blute des Neugeborenen: das Verhalten des Blutserums des Neugeborenen gegen Diptheriebacillen und Diptheriegift nebst kritischen Bermerkungen zur humoralen Immunitatstheorie. *Heilk*, **16**, 429–82.

Frenyo, V.L., Butler, J.E., Wilson, R.A. & Kavanagh, L. (1987*a*). The transport and metabolism of bovine IgM. *Immunol.*, **24**, 207–19.

Frenyo, V.L., Butler, J.E. & Guidry, A.J. (1987*b*). The association of extrinsic bovine IgG_1, IgG_2, SIgA and IgM with the major fractions of milk. *Vet. Immunol. Immunopath.*, **13**, 239–54.

Gilbert, R.P., Gaskins, C.T., Hillers, J.K., Parker, C.F. & McGuire, T.C. (1988). Genetic and environmental factors affecting immunoglobulin G1 concentrations in ewe colostrum and lamb serum. *J. Animal Sci.*, **66**, 855–63.

Groves, M.L., Basch, J.J. & Gordon, W.G. (1963). Isolation, characterization and amino acid composition of a new crystalline protein, lactollin, from milk. *Biochem.*, **2**, 814–17.

Groves, M.L., Greenberg, R. & Farrell, H.M. Jr. (1985). Isolation and properties of goat β_2-microglobulin. *Comp. Biochem. Physiol.*, **81(B)**, 621–7.

Guidry, A.J. (1985). Mastitis and the immune system of the mammary gland. In *Lactation*, ed. B.L. Larson. Iowa State University Press, Ames, pp. 229–62.

Guidry, A.J., Butler, J.E., Pearson, R.E. & Weinland, B.T. (1980). IgA, IgG_1, IgG_2, IgM and BSA in serum and mammary secretion throughout lactation. *Vet. Immunol. Immunopath.*, **1**, 329–41.

Haines, D.M., Chelak, B.J. & Naylor, J.M. (1990). Immunoglobulin concentrations in commercially available colostrum supplements for calves. *Can. Vet. J.*, **31**, 36–7.

Hammer, D.K., Kickhofen, B. & Malchow, H. (1969). Preferential adsorption of a single IgG type by isolated epithelial cells of the mammary gland. In *Protides of the Biological Fluids*, ed. H. Peters. Pergamon Press, New York, p. 663.

Hansen, R.G. & Phillips, P.H. (1949). Studies on proteins from bovine colostrum: III. The homologous and heterologous transfer of ingested protein to the blood stream of the young animal. *J. Biol. Chem.*, **179**, 523–7.

Hansen, R.G., Potter, R.L. & Phillips, P.H. (1947). Studies on proteins from bovine colostrum. II. Some amino acid analyses of a purified colostrum pseudoglobulin. *J. Biol. Chem.*, **171**, 229–32.

Head, J. & Beer, A.E. (1978). The immunologic role of viable leukocytic cells in mammary exosecretions. In *Lactation: A Comprehensive Treatise*, vol. 4, ed. B.L. Larson. Academic Press, New York, pp. 337–64.

Heddle, R.J. & Rowley, D. (1975). Dog immunoglobulins. I. Immunochemical characterization of dog serum, parotid saliva, colostrum, milk and small bowel fluid. *Immunol.*, **29**, 185–95.

Heyerman, H. & Butler, J.E. (1987). The heterogeneity of bovine IgG2—IV. Structural differences between IgG2a molecules of the A1 and A2 allotype. *Mol. Immunol.*, **24**, 1327–34.

Hopkins, C.R. (1983). The importance of the endosome in intracellular traffic. *Nature*, **304**, 684–5.

Howe, P.E. (1921). An effect of the ingestion of colostrum upon the composition of blood of new-born calves. *J. Biol. Chem.*, **49**, 115–8.

Howe, P.E. (1924). The relation between the ingestion of colostrum or blood serum and the appearance of globulin and albumin in the blood and urine of the new-born calf. *J. Exp. Med.*, **39**, 313–20.

Hunt, E.H., Van Camp, S.D. & Flemming, S. (1988). Developing technologies for prevention of bovine failure of passive transfer of antibody (FPTA). *Bovine Practitioner*, **23**, 131–4.

Jenness, R. (1974). Composition of milk. In *Lactation: A Comprehensive Treatise*, vol. 3, ed. B.L. Larson & V.R. Smith. Academic Press, New York, pp. 3–107.

Jenness, R. (1979). The composition of human milk. *Seminars in Perinatology*, **3**, 225–39.

Jenness, R. (1982). Interspecies comparison of milk proteins. In *Developments in Dairy Chemistry*, vol. 1, ed. P.F. Fox. Applied Science Publishers, London, pp. 87–114.

Kickhofen, B., Hammer, D.K. & Scheel, D. (1968). Isolation and characterisation of γG type immunoglobulins from bovine serum and colostrum. *Hoppe-Seyler's Z. Physiol. Chem.*, **349**, 1755–73.

Kraehenbuhl, J.-P., Weltzin, R., Schaerer, E. & Solari, R. (1988). Mucosal immunity and the mammary gland. In *Biology of Human Milk*, vol. 15, ed. L.A. Hanson. Nestle Nutrition Workshop Series, Vevey/Raven Press, New York, pp. 185–94.

Larson, B.L. (1958). Transfer of specific blood serum proteins to the lacteal secretions near parturition. *J. Dairy Sci.*, **41**, 1033–44.

Larson, B.L. (1979). Biosynthesis and secretion of milk proteins: A review. *J. Dairy Res.*, **46**, 161–74.

Larson, B.L. (1985). Biosynthesis and cellular secretion of milk. In *Lactation*, ed. B.L. Larson. Iowa State University Press, Ames, pp. 155–63.

Larson, B.L. & Gillespie, D.C. (1957). Origin of the major specific proteins in milk. *J. Biol. Chem.*, **227**, 565–73.

Larson, B.L. & Jorgensen, G.N. (1974). Biosynthesis of the milk proteins. In *Lactation: A Comprehensive Treatise*, vol. 2, ed. B.L. Larson & V.R. Smith. Academic Press, New York, pp. 115–46.

Larson, B.L. & Kendall, K.A. (1957). Changes in specific blood serum protein levels associated with parturition in the bovine. *J. Dairy Sci.*, **40**, 659–66.

Larson, B.L. & Rolleri, G.D. (1955). Heat denaturation of the specific serum proteins in milk. *J. Dairy Sci.*, **38**, 351–60.

Larson, B.L. & Touchberry, R.W. (1959). Blood serum protein level as a function of age. *J. Animal Sci.*, **18**, 983–90.

Lascelles, A.K. (1979). The immune system of the ruminant mammary gland and its role in the control of mastitis. *J. Dairy Sci.*, **62**, 154–60.

Lascelles, A.K. & McDowell, G.H. (1974). Localized humoral immunity with particular reference to ruminants. *Transplant Rev.*, **19**, 170–208.

Leary, H.L., Jr. & Larson, B.L. (1982). Intracellular transport of IgG_1 and IgG_2 through the prepartum mammary alveolar cell. *J. Dairy Sci.*, **65**, (Suppl. 1), 82–3.

Leary, H.L., Jr., Larson, B.L. & Nelson, D.R. (1982). Immunochemical localization of IgG1 and IgG2 in prepartum and lactating bovine mammary tissue. *Vet. Immunol. Immunopath.*, **3**, 509–14.

Lecce, J. & Morgan, D.O. (1962). Effect of dietary regimen on cessation of intestinal absorption of large molecules (closure) in neonatal pigs. *J. Nutr.*, **78**, 263–8.

Lee, C.S. & Lascelles, A.K. (1970). Antibody producing cells in antigenetically stimulated mammary glands and in the gastrointestinal tract of sheep. *Austral. J. Exp. Biol. Med. Sci.*, **48**, 525–35.

Lin, C-T. (1980). Immunoelectron microscopic location of immunoglobulin G in human placenta. *J. Histochem. Cytochem.*, **28**, 339–46.

Lisowski, J., Janusz, M., Tyran, B., Morawiecki, A., Golab, S. & Bialkowska, H. (1975). Immunoglobulins of colostrum—IV: Comparative studies of bovine serum and colostral IgG1 and IgG2. *Immunochem.*, **12**, 167–72.

Lyster, R.L. (1970). The denaturation of α-lactalbumin and β-lactoglobulin in heated milk. *J. Dairy Res.*, **37**, 233–43.

Mach, J.P. & Pahud, J.J. (1971). Secretory IgA, a major immunoglobulin in most bovine external secretions. *J. Immunol.*, **106**, 552–63.

Mackenzie, N. (1984). Fc receptor-mediated transport of immunoglobulin across the intestinal epithelium of the neonatal rodent. *Immunol. Today*, **5**, 364–6.

McGhee, J.R., Michalek, S.M. & Ghanta, V.K. (1975). Rat immunoglobulins in serum and secretions: Purification of rat IgM, IgA and IgG and their quantitation in serum, colostrum, milk and saliva. *Immunochem.*, **12**, 817–23.

McGuire, T.C. & Crawford, T.B. (1972). Identification and quantitation of equine serum and secretory immunoglobulin A. *Infect. and Immunol.*, **6**, 610–15.

Michalek, S.M., Rahman, A.F.R. & McGhee, J.R. (1975). Rat immunoglobulins in serum and colostrum: Comparison of IgM, IgA and IgG in serum, colostrum, milk and saliva of protein malnourished and normal rats. *Proc. Soc. Exp. Biol. Med.*, **148**, 1114–8.

Mostov, K.E., Friedlander, M. & Blobel, G. (1984). The receptor for transepithelial transport of IgA and IgM contains multiple immunoglobulin-like domains. *Nature (London)*, **308**, 37–43.

Murphy, F.A., Aalund, O., Osebold, J.W. & Carrow, E.J. (1964). Gamma globulins of bovine lacteal secretions. *Arch. Biochem. Biophys.*, **108**, 230–9.

Newby, T.J., Stokes, C.R. & Bourne, F. (1982). Immunological activities of milk. *J. Vet. Immunol. Immunopath.*, **3**, 67–94.

Orcutt, M.L. & Howe, P.E. (1922). The relation between the accumulation of globulins and the appearance of agglutinins in the blood of new-born calves. *J. Exp. Med.*, **36**, 291–308.

Paape, M.J., Wergin, W.D., Guidry, A.J. & Schultz, W.D. (1982). Phagocytic defense of the ruminant mammary gland. In *The Ruminant Immune System* (*Adv. Exptl. Med. Biol.*, **137**), ed. J.E. Butler. Plenum Press, New York, pp. 555–78.

Parry, R.M., Jr. (1974). Milk coagulation and protein denaturation. In *Fundamentals of Dairy Science*, ed. B.H. Webb, A.H. Johnson & J.A. Alford, 2nd edn. AVI Publishing, Westport, Connecticut, pp. 603–61.

Pastan, I. & Willingham, M.C. (1983). Receptor-mediated endocytosis: Coated pits, receptosomes and the Golgi. *Trends Biochem. Sci.*, **8**, 50–4.

Pearse, B.M.F. & Bretscher, M.S. (1981). Membrane recycling by coated vesicles. *Ann. Rev. Biochem.*, **50**, 85–101.

Porter, P.H. (1969). Transfer of immunoglobulins IgG, IgA and IgM to lacteal secretions in the parturient sow and their absorption by the neonatal piglet. *Biochim. Biophys. Acta*, **181**, 381–92.

Pringnitz, D.J., Butler, J.E. & Guidry, A.J. (1985*a*). *In vivo* proteolytic activity of the mammary gland. Contribution to the origin of secretory component, β_2-microglobulin, and bovine-associated mucoprotein (BAMP) in cows milk. *Vet. Immunol. Immunopath.*, **9**, 143–60.

Pringnitz, D.J., Butler, J.E. & Guidry, A.J. (1985*b*). Quantitation of bovine β_2-microglobulin: Occurrence in body fluids, on milk fat globules and origin in milk. *Molecular Immunol.*, **22**, 779–86.

Rodewald, R. (1973). Intestinal transport of antibodies in the newborn rat. *J. Cell Biol.*, **58**, 189–211.

Rodewald, R. (1980). Distribution of immunoglobulin G receptors in the small intestine of the rat. *J. Cell Biol.*, **85**, 18–32.

Rouse, E.T. & Ingram, D.G. (1970). The total protein and immunoglobulin profile of equine colostrum and milk. *Immunol.*, **19**, 901–7.

Rousseaux, J. & Bazin, H. (1979). Rat immunoglobulins. *Vet. Immunol. Immunopath.*, **1**, 61–78.

Sasaki, M., Larson, B.L. & Nelson, D.R. (1977). Kinetic analysis of the binding of immunoglobulins IgG1 and IgG2 to bovine mammary cells. *Biochim. Biophys. Acta*, **497**, 160–70.

Schultz, R.D. (1982). The role of cell-mediated immunity in infectious diseases of cattle. In *The Ruminant Immune System* (*Adv. Exptl. Med. Biol.*, **137**), ed. J.E. Butler. Plenum Press, New York, pp. 57–90.

Sherman, A. & Johnston, P. (1982). Nutrition and immunity. *Illinois Research*, **24**, 4–7 (University of Illinois Agricultural Experiment Station, Urbana).

Silverton, E.W., Navia, M.A. & Davies, D.R. (1977). Three-dimensional structure of an intact human immunoglobulin. *Proc. Nat. Acad. Sci. (USA)*, **74**, 5140–4.

Simister, N.E. & Mostov, K.E. (1989). An Fc receptor structurally related to MHC class I antigens. *Nature (London)*, **337**, 184–7.

Simister, N.E. & Rees, A.R. (1985). Isolation and characterization of an Fc receptor from neonatal small rat intestine. *Eur. J. Immunol.* , **15**, 733–8.

Simmons, D. & Seed, B. (1988). The Fcγ receptor of natural killer cells is a phosphate-linked membrane protein. *Nature*, **333**, 568–70.

Smith, E.L. (1946). The immune proteins of bovine colostrum and plasma. *J. Biol. Chem.*, **164**, 345–58.

Smith, E.L. (1948). The isolation and properties of the immune proteins of bovine milk and colostrum and their role in immunity: A review. *J. Dairy Sci.*, **31**, 127–38.

Smith, E.L. & Holm, A. (1948). The transfer of immunity to the new-born calf from colostrum. *J. Biol. Chem.*, **175**, 349–57.

Smith, E.L., Hill, R.L., Lehman, I.R., Lefkowitz, R.J., Handler, P. & White, A. (1983). *Principles of Biochemistry: Mammalian Biochemistry*, 7th edn. McGraw-Hill, New York.

Smith, T. & Little, R.B. (1922). The significance of colostrum to the new-born calf. *J. Exp. Med.*, **36**, 181–98.

Solari, R. & Kraehenbuhl, J.-P. (1984). Biosynthesis of the IgA antibody receptor: A model for the transepithelial sorting of membrane glycoprotein. *Cell*, **36**, 61–71.

Stechschulte, D.J. & Austen, K.F. (1970). Immunoglobulins of rat colostrum. *J. Immunol.*, **104**, 1052–62.

Tizard, I.R. (1982). *An Introduction to Veterinary Immunology*, 2nd edn. Saunders, Philadelphia.

Vaerman, J.-P. & Heremans, J.F. (1969). The immunoglobulins of the dog—II. The immunoglobulins of canine secretions. *Immunochem.*, **6**, 779–86.

Vaerman, J.-P., Querinjean, P. & Heremans, J.F. (1971). Studies on the IgA system of the horse. *J. Immunol.*, **21**, 443–54.

Weaver, E.A., Tsuda, H., Goldblum, R.M., Goldman, A.S. & Davis, C. P. (1982). Relationship between phagocytosis and immunoglobulin release from human colostral macrophages. *Infect. Immun.*, **38**, 1073–7.

Whitney, R.McL. (1988). Proteins of milk. In *Fundamentals of Dairy Science*, 3rd edn, ed. N.P. Wong, R. Jenness, M. Keeney & E.M. Marth. AVI Publishing, Westport, Connecticut, pp. 81–169.

Williams, M.R. & Millar, R (1978). Changes in immunoglobulin G-2 levels with age in British cattle. *Res. Vet. Sci.*, **25**, 82–5.

Witty, R., Brown, P. & Smith, M.W. (1969). The transport of various immunoglobulins by the new-born pig intestine. *Experientia*, **25**, 310–12.

7

BIOLOGICAL PROPERTIES OF MILK PROTEINS

P.F. FOX[a] & A. FLYNN[b]

[a]*Department of Food Chemistry,* [b]*Department of Nutrition, University College, Cork, Republic of Ireland*

1. INTRODUCTION

Milks are secreted by mammals (> 4000 species) for the primary function of providing the complete nutritional requirements of the neonate of the species. However, by accident (e.g. mechanism of excretion), but more probably by design, milk serves other functions in addition to being strictly nutritional. Most of the non-nutritional functions of milk are served by proteins and peptides which include immunoglobulins, enzymes and enzyme inhibitors, binding or carrier proteins, growth factors and antibacterial agents. Because the nutritional and physiological requirements of the neonate of each species are at least to some extent unique, the milk of each species is more or less unique (Widdowson, 1984). Of the > 4000 species of mammal, the milks of only 176 have been analysed for fat, protein and carbohydrate and of these, the data for only ~ 50 species are considered to be reliable (Oftedal, 1980).

In this review, we will consider primarily bovine and human milks, which are rather well characterized. The proteins of bovine milk are probably the best characterized food protein system. Although the principal proteins of human milk are also well characterized, there are several minor proteins in human milk that have not yet been studied in detail.

When the protein chemist or food scientist thinks of milk proteins, she/he normally considers their chemical, physical or functional/technological properties. Although the food scientist/technologist has learned to exploit many of the unique functional properties of milk proteins, these are not the natural functions of these proteins. Even the nutritionist normally considers milk proteins in terms of supplying an adequate level of essential amino acids, but many proteins are capable of doing this. Many of the unique nutritional, physiological, biochemical, immunological and other biological properties of these proteins are ignored.

Human milk has been studied much more thoroughly in this regard than bovine or other milks. For obvious reasons, this work has been done mainly by paediatricians whose work is not frequently referred to by dairy or food scientists. The presence of these biologically-active peptides in bovine milk has generally been investigated also by paediatricians with a view to infant nutrition. Generally, when they have been sought, similar biologically active proteins/peptides have been found in the milks of both species, although usually at different, sometimes very marked, concentrations. Presumably, they are common to the milks of most species. Books and reviews on human milk and lactation include: Jensen and Neville (1985), Atkinson and Lonnerdal (1989), Goldman *et al.* (1987), Blanc (1981), Koldovsky (1989), Schaub (1985), Goldman *et al.* (1985), Goldman (1988) and Gaull *et al.* (1982).

The principal objective of this chapter is to review the less familiar unique biological properties of milk proteins. Some of the proteins and peptides reviewed herein are also discussed briefly in Chapter 11 from the nutritional viewpoint.

2. Enzymes and Enzyme Inhibitors

Milk contains up to 60 indigenous enzymes, many of which, because of their technological significance, have been studied extensively. The literature is reviewed in Chapter 8. The indigenous enzymes in human milk are described by Hamos *et al.* (1985) and Hernell and Lonnerdal (1989).

α-Lactalbumin, one of the principal whey proteins in all milks, is an enzyme modifier in lactose synthetase. The extensive literature on α-lactalbumin is reviewed in Chapter 5.

3. Binding Proteins

Most, perhaps all, milk proteins exhibit some form of binding property. These range from essentially non-specific metal binding by the caseins to rather highly specific binding.

3.1. Metal Binding

3.1.1. Casein

All the caseins are phosphoproteins: bovine α_{s1}-, α_{s2}-, β- and κ-caseins contain 8–9, 10–13, 4–5 and 1–2 mole P/mole protein, respectively. The phosphate residues, which exist as monoesters of serine, occur mainly in clusters. They are capable of binding metal ions strongly which results in charge neutralization, and in the case of α_{s1}-, α_{s2}- and β-caseins, in precipitation or, as in milk, in micelle formation. Calcium is the principal metal involved but a substantial portion of the zinc in milk is also chelated by the phosphate residues of casein. Additional calcium and zinc are associated with the colloidal calcium phosphate (CCP) in the casein micelles.

Both the casein-bound calcium and the colloidal calcium phosphate are of major significance in many technologically important properties of milk, e.g. rennet coagulation, heat stability, alcohol stability, and are probably nutritionally significant also. It is generally believed that casein phosphopeptides facilitate the absorption of metal ions and are being promoted as nutritionally-desirable supplements. However, although bovine milk contains more Zn than human milk, it is commonly reported that Zn is less readily available from bovine than from human milk, a feature commonly attributed to the higher degree of phosphorylation of the former. The distribution of Zn differs markedly between bovine and human milks. In the former, ~95% of total Zn is associated with the casein micelles, ~40% of which is bound by the casein, presumably via phosphoserine residues, with 60% being a constituent of CCP (Singh *et al.*, 1989). In human milk, casein binds ~14% of Zn, and whey proteins and citrate each bind $\sim\frac{1}{3}$ of total Zn (Lonnerdal *et al.*, 1982).

The reduced bioavailability of Zn from bovine milk may be due to its high content of CCP which is low or zero in human milk. Reduction of the CCP content of bovine milk by acidification/dialysis increases the availability of Zn from this milk (Kiely *et al.*, 1988). Recent studies on suckling rats (Brennan, 1990) have shown that Zn is almost fully absorbed from both bovine and human milks although absorption is somewhat faster from the latter.

Phosphopeptides derived from casein stimulate the absorption of Ca (Sato *et al.*, 1986). Such peptides are resistant to proteolysis due to their high density of negative charges; they have been detected in the small intestine of the rat and may pass intact through the intestinal wall. Stimulation of Ca absorption by phosphopeptides is proportional to the degree of phosphorylation.

Iron is also bound by phosphoserine residues although the influence of this binding on the bioavailability of Fe is ambiguous. Casein phosphopeptide–Fe complexes have been proposed as useful products for supplementation of diets with Fe. Methods for the production of caseinophosphopeptides for nutritional and/or medical applications have been developed (Maubois & Léonil, 1989).

The serine phosphate groups of casein may have a specific metabolic function (Williams & Sanger, 1959). One of the phosphate groups of β-casein is released faster than the others (West & Dalgleish, 1976) and one may be transferred to ADP by a specific AMP-independent casein kinase to form ATP, suggesting that it may be a type of high-energy phosphate (Cochet *et al.*, 1983). The structure and function of the phosphorylated residues of casein were reviewed by West (1986).

3.1.2. α-Lactalbumin and Serum Albumin

α-Lactalbumin binds calcium strongly (see Chapter 5) but this is probably not of nutritional significance. Blood serum albumin (BSA) also has strong metal-binding sites (see Peters, 1985) but this is probably not of specific significance in milk, which contains little BSA.

3.1.3. Lactotransferrin

Probably, the most significant of the specific metal-binding proteins are the transferrins, a group of evolutionarily related iron-binding proteins, the best characterized members of which are:

—serotransferrin (present in blood plasma and other extracellular fluids, e.g. milk, spinal fluid, semen).
—ovotransferrin (conalbumin; present in avian and reptile egg whites).
—lactotransferrin (milk, pancreatic juice, tears, leucocytes).

Related but less well-characterized proteins occur in melanoma cells, some lymphomas and the blood of crustaceans and prochordates. Various aspects of iron-binding proteins are reviewed in the text edited by Spik *et al.* (1985).

Human colostrum and milk contain 6–8 mg/ml and 2–4 mg/ml lactotransferrin, respectively; concentrations in bovine colostrum and milk are ~1 and 0·02–0·35 mg/ml (Reiter, 1985*a*,*b*). The concentration of lactotransferrin in human milk decreases slightly during lactation (Harzer & Bindels, 1985). Lactotransferrin was first isolated from human milk by Montreuil *et al.* (1960) and Johansson (1960), while Gordon *et al.* (1962) and Groves (1965) isolated lactotransferrin from bovine milk. Lactotransferrins were quantified in the milks of guinea pig, goat, cow, pig, mare and mouse but milks of the rat, rabbit and dog are apparently devoid of lactotransferrin (Masson & Heremans, 1971). Because of the apparent physiological and nutritional significance of lactotransferrins, several improved methods for their isolation, many using various forms of affinity chromatography, have been published; some are designed for commercial-scale operation (Law & Reiter, 1977; Wang *et al.*, 1984; Davidson & Lonnerdal, 1985; Bezwoda & Mansoor, 1986; Kawakami *et al.*, 1987; Al-Mashikhi *et al.*, 1988; Shimazaki & Nishio, 1991).

Human lactotransferrin is the best characterized of the lactotransferrins (Montreuil *et al.*, 1985). Its complete amino acid sequence has been published (Metz-Boutigue *et al.*, 1984). Based on the amino acid sequence and the attached carbohydrate, the molecular weight is calculated to be 82 000 ± 40. It contains 16 disulphides, 15 of which also occur in serotransferrin. The molecule exists as two domains which contain 125 residues at corresponding positions (37% homology), suggesting gene duplication. There are 59% and 49% homologies between human lactotransferrin and human serotransferrin and chicken ovotransferrin, respectively. Each domain binds 1 mole of iron. The molecule has a pI of 5·6–6·1 (Bezwoda & Mansoor, 1986).

The amino acid composition of bovine lactotransferrin was reported by Castellino *et al.* (1970) and its N-terminal sequence by Wang *et al.* (1984). The isolation amino acid composition and N-terminal sequence of equine lactotransferrin were reported by Jolles *et al.* (1984). Two molecular forms of bovine lactotransferrin have been reported (Tsuji *et al.*, 1989); differences in the degree of glycosylation are partially responsible for the microheterogeneity.

Lactotransferrin binds iron very strongly, which indicates two roles for this protein: iron absorption and protection against enteric infection in the noeonate; the literature on this area has been reviewed extensively (Arnold *et al.*, 1977; Spik *et al.*, 1985; Reiter, 1985*a*,*b*; Lonnerdal, 1985). Because the concentration of lactotransferrin in human milk is considerably higher than that in bovine milk, there is considerable interest in supplementing bovine milk-based infant formulae with lactotransferrin. The concentration of lactotransferrin in milk increases during mastitic infections, suggesting that it may have a protective role in the mammary gland (Harmon *et al.*, 1975, 1976).

Milks of the various species studied also contain serotransferrin. These have been reviewed by Williams (1985). The heterogeneity of bovine serotransferrin was studied by Maeda *et al.* (1980) and the properties of lactotransferrin and serotransferrin compared by Aisen and Leibman (1972).

3.2. Vitamin-binding Proteins

Milks of several species contain proteins capable of binding retinol (β-lactoglobulin), folic acid, riboflavin or Vitamin B_{12}. There is evidence that these proteins facilitate the absorption of the appropriate vitamin and presumably have a specific function.

3.2.1. β-Lactoglobulin

β-Lactoglobulin is the principal whey protein in the milks of domesticated dairy animals and a similar protein occurs in the milks of several species but is absent from human and rodent milks. The extensive literature on this very well characterized protein is reviewed in Chapter 4.

A biological function for β-lg has been sought for some time. The protein is capable of binding several hydrophobic molecules, including retinol, and it has been proposed that *in vivo* β-lg functions to protect retinol against oxidation and to transport it through the stomach to the small intestine. The resistance of β-lg to peptic hydrolysis is probably significant in this regard (see Chapter 4).

3.2.2. Vitamin B_{12}-binding Protein

Gregory (1954) showed that Vitamin B_{12} in the milks of cow, goat, sheep, sow, woman and rat was almost entirely bound to a whey protein which was partially purified from porcine milk and characterized by Gregory and Holdsworth (1955*a*,*b*). The B_{12}-binding protein was isolated from human milk by Burger and Allen (1974) using affinity chromatography on B_{12}-Sepharose. The protein, which was an R-type B_{12}-binding protein, had a molecular weight of 61 000–63 000 and contained ~35% carbohydrate. More recent work by Sandberg *et al.* (1981) indicates that there are two Vitamin B_{12}-binding proteins in human milk, the principal one being the R-type to which most or all of the vitamin B_{12} is bound, and the other being transcobalamin II, which is present at a low concentration and is devoid of cobalamin.

Ford *et al.* (1977) reported that porcine milk is rich in Vitamin B_{12}-binding protein which Trugo (1988) isolated by affinity chromatography on B_{12}-Sepharose and characterized it with respect to Mr (~62 000), carbohydrate content (24%), amino acid profile and binding affinity for B_{12}. The Vitamin B_{12}-binding protein of rat milk was partly purified by Raaberg *et al.* (1989) and some of its properties determined.

The possible nutritional significance of this protein was investigated by Ford (1974). The protein probably plays two roles: aids absorption of B_{12} and acts as a bacteriocidal agent. Animal experiments indicate that protein-bound Vitamin B_{12} is efficiently absorbed from the intestine, both before and for some time after cessation of transport of intact protein across the neonatal gut epithelium (Ford *et al.*, 1975; Trugo *et al.*, 1985). However, Sandberg *et al.* (1981) suggest that cobalamin must be released from its binding protein by proteolysis and then combined with an intrinsic factor for absorption. Trugo and Turvey (1987) showed that Vitamin B_{12}-binding protein attaches to the brush border membrane where it stimulates absorption of vitamin B_{12}. Human milk has the capacity to bind ~43 ng cyanocobalamin/ml but in milk most of the protein is unsaturated (Ford *et al.*, 1977). The binding capacity is reduced by heat treatments to a minimum at 75°C × 15 min but binding capacity increases again after more severe heat treatments (Ford *et al.*, 1977). Human milk inhibits the uptake of cyanocobalamin by intestinal bacteria (Ford *et al.*, 1977). This inhibitory effect is progressively lost on heating up to 75°C × 15 min but less so at higher temperatures (Ford *et al.*, 1977). It therefore appears likely that this protein protects Vitamin B_{12} against bacterial uptake in the gut and may strongly influence the ecology of the intestinal microflora and the vitamin nutrition of the suckling infant (Ford, 1974). However, Cole *et al.* (1983) could not confirm this hypothesis for vitamin B_{12}-binding protein, folate-binding protein and lactotransferrin in the case of rabbits.

Ford *et al.* (1977) suggest the minimum stability of Vitamin B_{12}-binding protein at ~75°C may be due to the action of a milk enzyme (presumably proteinase) which is itself denatured at higher temperatures. (Plasmin is in fact activated by mild heat treatments, presumably due to inactivation of inhibitors of plasminogen activators (see Chapter 8)). Ford *et al.* (1977) argue that if this is so, then refrigerated storage should also cause considerable loss; they estimate that storage at 10°C for 16 h would be as damaging as heating at 70°C for 15 min. As far as the authors are aware, this possibility has not been investigated.

3.2.3. Folate-binding Proteins (FBPs)

A number of proteins from a variety of tissues have been characterized as folate-binding proteins. Such proteins fall into two molecular weight ranges: 100 000–200 000 and 25 000–40 000. FBPs may be classified as extracellular, i.e. in fluids such as milk, blood serum, cerebrospinal fluid, or intracellular, which may be sub-divided into those that are membrane-bound and those that are intracellular but soluble. The membrane-associated folate-binding proteins can be solubilized by detergents and are believed to participate in the transport of folate derivatives. Reviews on folate-binding proteins include Waxman (1975), Colman and Herbert (1980) and Wagner (1982).

FBP was detected in milk by Ghitis (1966, 1967), Ghitis *et al.* (1969) and Metz *et al.* (1968). Low MW FBPs have been isolated from cow's milk (Ford *et al.*, 1969; Salter *et al.*, 1972, 1981; Svendsen *et al.*, 1979), goat's milk (Selhub & Grosowicz, 1973; Rubinoff *et al.*, 1977) and human milk, which also contains a high MW FBP (Antony *et al.*, 1982; Hansen *et al.*, 1977*a*; Waxman & Schreiber, 1975*a*). FBP appears to accompany bovine β-lactoglobulin in commercial preparations (Waxman & Schreiber, 1975*b*). ELISA methods have been developed for the FBPs of bovine (Høier-Madsen *et al.*, 1986) and human (Høier-Madsen *et al.*, 1987) milks; the FBPs from these milks show no cross-reactivity. The proteins are glycoproteins with pIs in the neutral range. The protein–folate complex dissociates at acidic pH values and reforms on neutralization.

The complete amino acid sequence of FBP from bovine milk has been determined by Svendsen *et al.* (1979, 1982, 1984). The amino acid composition of human milk FBP was reported by Antony *et al.* (1982). A generally similar composition was reported by Svendsen *et al.* (1982) who also determined the sequence of 60 residues from the N-terminal. The five N-terminal residues of the bovine protein were missing from the human protein but otherwise the two proteins showed a high degree of homology. The complete amino acid sequence (222 residues) of FBP from bovine milk was reportd by Svendsen *et al.* (1984). The protein contains ~10% carbohydrate, attached to Arg residues, probably 49 and 141, and had a molecular weight of ~30 000, including carbohydrate (Svendsen *et al.*, 1984). Antony *et al.* (1982) proposed that the high and low molecular FBP proteins were the products of gene duplication but Svendsen *et al.* (1982) propose that the low MW form is the result of proteolytic cleavage of the larger protein. FBP undergoes extensive aggregation, up to 30 monomers, depending on pH and protein concentration (Pedersen *et al.*, 1980). The mechanism of folate binding by FBP from milk has been studied by Hansen *et al.* (1977*b*, 1978) and Holm *et al.* (1978).

The physiological significance of folate-binding proteins in milk has been reviewed by Salter and Mowhem (1983) and the bioavailability of folate in general by Bailey (1988). Protein-bound folate is efficiently absorbed from the intestine (Ford *et al.*, 1972; Izak *et al.*, 1972; Ford, 1974; Waxman & Schreiber, 1975*a*; Salter & Blakeborough, 1988). Breast-fed babies have higher plasma and red blood cell folate concentrations than their mothers or adult references, suggesting that milk contains a factor that facilitates folate uptake by intestinal cells (Ek, 1987). In vitro studies on intestinal cells showed (Colman *et al.*, 1981) that FBP-bound folate was absorbed more effectively than free folate and the beneficial effect of FBP on folate absorption increased throughout the length of the small intestine while absorption of free folate was independent of intestinal segment. Absorption of folate from FBP was not influenced by glucose or dilantin in contrast to free folate; Ca or EDTA had little effect on free folate uptake whereas 1 mM $CaCl_2$ approximately doubled uptake of FBP-bound folate. Uptake of FBP-bound folate was not affected by temperature in the range 4–37°C or oxygen concentration between 20 and 95%. The demonstration (Colman & Herbert, 1980) that the intestine contains FBP receptor sites supports this function of FBP in milk. However, Said *et al.* (1986) found that free folate was absorbed faster than FBP-bound folate in the jejunum and at about the same rate in the ileum; they suggested that FBP regulates the nutritional bioavailability of folate. Tani *et al.* (1983) and Swiatlo and Picciano (1988) also reported that FBP reduces the rate of absorption of folate in perfused rat intestine or the bioavailability of folate to the rat, respectively.

FBP reduces the availability of folate to bacteria in the intestine and thus may have an antibacterial effect (Ford, 1974; Ford *et al.*, 1977) although Cole *et al.* (1983) could not confirm this using rabbits. The folate binding properties of FBP, and hence its antibacterial efficacy, are reduced by heat tretment although the extent of change is slight for commercial pasteurization, e.g. ~10% for low-temperature long-time pasteurization (Ford *et al.*, 1977). A detailed study by Gregory (1982) showed that the folate binding properties of FBP in heated milks were reduced in comparison with raw milk.

3.2.4. Riboflavin Binding

The white and yolk of egg contain almost identical riboflavin-binding proteins which are required for the transport of riboflavin across the vitellin membrane from the blood serum, which contains a similar protein, into the white and yolk. These proteins have eight phosphoseryl residues, somewhat analogous to casein; these are essential for the transport function of the protein, apparently not at the vitamin-binding site but presumably as a recognition site for binding of the protein at the membrane (Miller *et al.*, 1982, 1984).

Riboflavin-binding proteins have been reviewed by White and Merrill (1988).

Kanno and Kanehara (1985) reported that raw bovine milk contains a riboflavin-binding protein (RfBP). This was partially purified by Toyosaki *et al.* (1987) using DEAE cellulose, gel filtration and gel electrophoresis; the molecular weight was estimated to be 38 000. Based on amino acid, carbohydrate and phosphate composition, the M_r of RfBP of egg white is 29 200, although higher values (30 000–36 000) were found using sedimentation equilibrium and SDS-gel electrophoresis (see White & Merrill, 1988). Since the RfBP of egg white and yolk are derived from blood plasma, it is likely that RfBP in milk also originates in the blood. Riboflavin bound to RfBP of milk or egg white has good antioxidant properties (Toyosaki *et al.*, 1987; Toyosaki & Mineshita, 1988). RfBP plays a major nutritional role in the egg; a similar role for RfBP in milk has not been demonstrated.

West (1986) argues that although the caseins do not function as riboflavin-binding proteins, the similarity of the clustered serine phosphates in these two unrelated proteins suggests the possibility that the casein serine phosphates may act as a membrane recognition signal.

4. Bifidus Factors

It has been recognized for many years that breast-fed babies are more resistant to gastroenteritis than bottle-fed babies (see Bullen & Willis (1971) and Rasic & Kurmann (1983) for references). This is undoubtedly a multi-factorial phenomenon including better hygiene, more appropriate milk composition, several anti-bacterial systems, especially immunoglobulins, lysozyme, lactotransferrin, vitamin-binding proteins, lactoperoxidase, which are discussed in this and other chapters, and a lower intestinal pH. Bullen and Willis (1971) reported that the mean pH of the faeces of breast-fed babies was 5·1 while that of bottle-fed babies was 6·4. These authors attributed the low pH of the faeces of breast-fed infants to the lower buffering capacity of human milk compared to bovine milk due to its lower content of protein and phosphate. While buffering capacity is undoubtedly important, differences in the intestinal microflora of breast-fed and bottle-fed infants are also of major significance.

Bifidobactria represent about 99% of the faecal microflora of breast-fed infants (see Rasic & Kurmann, 1983). These bacteria also represent a high proportion of the microflora of bottle-fed infants but several other genera, e.g. *Bacteroides*, *Clostridium*, *Coliforms*, also occur in high numbers. Furthermore, the predominant species of *Bifidobacterium* in breast-fed infants is *B. bifidum* with lesser numbers of *B. longum*; the faecal microflora of bottle-fed infants is dominated by *B. longum* with lower numbers of *B. bifidum*, *B. infantis*, *B. adolescentis* and *B. breve* (Beerens *et al.*, 1980).

The preponderance of *B. bifidum* in the faeces of breast-fed infants is due to the presence of stimulatory factors in human milk. Several factors stimulate the growth of *Bifidobacteria* (see Rasic & Kurmann, 1983):

(1) Lactulose and some of its reduction products stimulate the growth of *Bifidobacteria*, *Lactobacillus* and several other genera. It is an unnatural sugar but is produced from lactose by relatively mild heat treatments, such as used in the manufacture of UHT milk products and sterilized infant formulae (Andrews, 1989). Inclusion of 1–1·5 g lactulose per kg body-weight in infant formulae stimulates the growth of *Bifidobacteria*; the inclusion of lactulose in the diets of calves and piglets is recommended (see Modler *et al.*, 1990).
(2) Pantethine: some strains of *Bifidobacteria*, including *B. bifidum*, require pantethine since they cannot utilize pathotenic acid.
(3) N-Acetylglucosamine-containing saccharides (bifidus factor 1).

György (1953) and György *et al.* (1954*a*,*b*,*c*) reported that N-acetylglucosamine-containing saccharides are essential for the growth (cell wall synthesis) of *B. bifidum*. This factor, referred to as the Bifidus factor, is present at high concentrations in human milk and colostrum and in bovine colostrum but at very low levels in the milks of cows, goats and sheep. Several such saccharides are present in human milk, especially colostrum, but not in bovine milk (see Barker & Stacey (1963); Smith *et al.* (1978) and Blanc (1981) for reviews on milk oligosaccharides).

Most (40–75%) of the bifidus-promoting factors in human milk are non-dialysable. These appear to be a mixture of glycoproteins and higher oligosaccharides. The dialysable (or ethanol soluble) bifidus factors in human milk have been resolved into two components by HPLC (Ashoor & Monte, 1983); the fractions were not characterized.

György *et al.* (1974) fractionated the non-dialysable growth promoters into seven fractions, one of which (I) contained glycoproteins, the others being mixtures of oligosaccharides. Four of the fractions (IV–VII) contained N-acetylneuraminic acid (NANA) and had no bifidus activity unless the NANA residue(s) was removed by treatment with neuraminidase.

—*Glycopeptides* (Bifidus factor 2): the non-dialysable 'supplementary' bifidus factor was studied by György and Rose (1955), Raynaud *et al.* (1959) and

Raynaud and Bizzini (1971) and shown to be a protein; it is referred to as Bifidus factor 2.

—*Glycoproteins*: bovine and human colostra and milks contain several glycoproteins which include N-acetylglucosamine in the carbohydrate moiety. Glycoproteins are found in both the casein and whey fractions.

4.1. Whey Glycoproteins

Bezkorovainy (1965) isolated several acidic glycoproteins from bovine colostrum and whey. The most acidic of these, M-1, was further characterized by Bezkorovainy (1967). Although homogeneous by several criteria, the protein showed several N-terminal residues. It had a molecular weight of ~10 000 and contained NANA, galactose and hexosamine in the ratio of 2:3:3. Five glycopeptides with bifidus activity were isolated from human colostrum by Hirano *et al.* (1968).

Nichols and Bezkorovainy (1973) showed that the M-1 acid glycoprotein fraction of human colostrum contained at least six glycoproteins, one of which was isolated to apparent homogeneity. The protein had a molecular weight of ~31 000 and contained 27% galactose, 22% hexosamine, 8% fucose and 11% NANA. Obviously, this protein differed markedly from the M-1 glycoprotein of bovine colostrum in regard to MW, a higher carbohydrate content and the presence of fucose.

A total of seven M-1 glycoprotein fractions with bifidus activity were isolated from human colostral whey by Nichols *et al.* (1974), five of which were purified by Nichols *et al.* (1975). They appeared to represent a closely-related family of glycoproteins with molecular weights of 26 000–35 000 and carbohydrate contents of 50–80%. Treatment with neuramindase was not necesssary to activate the bifidus promoting activity of these proteins.

Mature human milk (whey) also contains glycoproteins with bifidus activity (Bezkorovainy & Nichols, 1976). The proteose peptone fraction of human milk also has bifidus activity (Bezkorovainy *et al.*, 1976) as do tryptic/chymotryptic digests of human casein (Bezkorovainy *et al.*, 1979).

A high molecular weight, mucin-like glycoprotein from the fat globule membrane of human milk was isolated by Shimizu & Yamauchi (1982) and shown to have bifidus-promoting properties.

Beerens *et al.* (1980) showed that bovine, ovine and porcine milks did not stimulate the growth of *B. bifidum* but did contain factors that stimulated *B. infantis* and *B. longum*; the nature of the growth factors was not studied.

4.2. Caseins

κ-Casein is the only glycosylated member of the casein group of proteins. Although other sugars have been reported in bovine κ-casein, the general view is that it contains galactose, N-acetylgalactosamine and NANA, which occur as trisaccharides or tetrasaccharides containing one extra NANA residue. κ-Casein contains 0–5 oligosaccharide moieties (Vreeman *et al.*, 1986). Apparently, κ-casein does not stimulate *B. bifidum* but mild acid treatment releases bifidus growth factors (Kehagias & Hansen, 1976).

Human κ-casein is also a glycoprotein (40% carbohydrate); it has a molecular weight of ~33 000 (~20 000 carbohydrate-free) and contains galactose (15%), fucose (3%), N-acetylglucosamine (10%), N-acetylgalactosamine (5%) and NANA (5%) (Yamauchi *et al.*, 1981). Brignon *et al.* (1985), who determined the sequence of human κ-casein, estimated that it may contain ~52% carbohydrate; they reported a molecular weight of ~37 000.

A second highly glycosylated phosphoprotein (HGPP) was isolated from human milk by Azuma and Yamauchi (1987). HGPP, which had a molecular weight of ~41 000, contained 1·6% P and ~38% carbohydrate (7·5% fucose, 15% galactose, ~7% N-acetylglucosamine, ~2% N-acetylgalactosamine and ~7·5% NANA). On SDS-PAGE, HGPP has an electrophoretic mobility between κ-casein and lactotransferrin. It was present in an approximate ratio of 1 : 4 with κ-casein with which it associated strongly.

Bezkorovainy *et al.* (1979) isolated a glycopeptide with bifidus-promoting activity from tryptic and chymotryptic digests of human casein (two further minor peptides were demonstrated and a minor peptide was found in digests of bovine casein; intact human casein had little bifidus activity). The peptide contained galactose, N-acetylgalactosamine, N-acetylglucosamine, fucose and NANA, and had a molecular weight of ~30 000. The above authors suggest that the peptide is derived from κ-casein (presumably, the value for the MW is incorrect since the MW of κ-casein is 33 000–37 000) and that it corresponds to the glycoprotein isolated from human whey (see above); they speculate that the glycoprotein in whey may be produced from casein by the action of indigenous proteinase (which is now known to have little activity on κ-casein). Enzymatic digests of bovine casein were also shown by Poch and Bezkorovainy (1988) to have bifidus activity.

5. Growth Factors

'Growth factor' is a term applied to a group of highly potent hormone-like polypeptides which play a critical role in regulation and differentiation of a variety of cells, acting through cell-membrane receptors. Milks of several species have been found to stimulate in-vivo growth of gastrointestinal tissues, to promote in-vitro replication of cultured cells and to contain assayable growth factors. In general, such growth-promoting characteristics are more pronounced in colostrum than in mature milk.

At least 16 distinct growth factors have been detected in the milk of human and other species. These include epidermal growth factor, insulin, insulin-like growth factors 1 and 2 (IGFI, 2), three human milk growth factors ($TGF_{\alpha 1}$, $TGF_{\alpha 2}$ and TGF_{β}), two mammary-derived growth factors (MDGF I, II), colony stimulating factor, nerve growth factor, platelet-derived growth factor (PDGF) and bombesin. Some of these can be grouped together on the basis of similarity in biological activity (Kidwell & Solomon, 1989).

The source of these polypeptides may be blood plasma, mammary gland or both. The biological significance of these growth-promoting activities in colostrum and mature milks is not yet clear. In terms of possible physiological significance, two potential targets may be considered, i.e. the mammary gland or the neonate. In general, most attention has focussed on the latter. It is not known whether the factors in milk that possess the capacity to promote cell proferation (a) influence growth of mammary tissue, (b) promote growth of cells within the intestines of the recipient neonate, or (c) are absorbed in a biologically-active form and exert an effect on enteric or other target organs.

The original work in this area was carried out by Widdowson and her colleagues who observed striking increases in the mass of intestinal mucosa of suckled newborn piglets (Widdowson *et al.*, 1976) and rabbits (Hall & Widdowson, 1979) compared to litter-mates that were fasted for the first 24 h after birth. Subsequently, Heird *et al.* (1984) showed that small intestinal mucosal mass was increased by 75%, and mucosal DNA and protein contents by 65% and 93%, respectively, in beagle puppies suckled for the first 24 h after birth compared with litter-mates fed artificial formula. Berseth *et al.* (1983) reported similar findings in rats and Mirand *et al.* (1990) showed that feeding newborn lambs with cow or ewe colostrum stimulated intestinal protein synthesis more than cows' milk over the first 18 h of life.

Klagsbrun (1978) showed that human milk contains growth factor activity, and is capable of stimulating the growth of fibroblasts in culture. Bovine milk contains growth-promoting factors that stimulate DNA synthesis and cell division in cultured Balb/c 3T3 cells (Klagsbrun & Neumann, 1979). The mitogenic activity of colostrum is higher than that of mature milk from several species, including human, cow, sheep and goat (Klagsbrun, 1978; Klagsbrun & Neumann, 1979; Brown & Blakely, 1983).

Epidermal growth factor (EGF) belongs to a family of growth factors that act through the EGF receptors on the cell membrane. Five such factors have been detected in human milk: EGF, MDGFII, $TGF_{\alpha 1}$, $TGF_{\alpha 2}$ and HMGIII.

EGF contains 53 amino acids (M_r: 6045; pI: 4·4) and is derived from a precursor of M_r 130 000 (Scott *et al.*, 1983) that is synthesized in the salivary glands and other tissues (Cohen, 1962). Human milk and colostrum contain considerable amounts of EGF, as shown by radioimmunoassays with specific antibodies (Carpenter, 1980). The concentrations are highest in colostrum (23 to 125 nmol/1) (Read *et al.*, 1984) and fall shortly after delivery to about 5 nmol/1. About 70% of the growth factor activity of human milk can be neutralized by antibodies against human EGF (Carpenter, 1980), suggesting that EGF is the major growth factor in human milk. EGF has been purified to homogeneity from human milk and sequenced (Petrides *et al.*, 1985).

EGF has been shown to be present also in the milk of the mouse (Beardmore & Richards, 1983) and rat (Thornburg *et al.*, 1984) and EGF-like activity has been detected in bovine colostrum (Klagsbrun & Neumann, 1979) and milk (Yagi *et al.*, 1986), but not in cows' milk formulae (Carpenter, 1980; Yagi *et al.*, 1986). Klagsbrun and Neumann (1979) showed that the EGF-like activity of bovine milk is markedly dependent on the stage of lactation. Activity is highest in colostrum and samples of milk obtained 32 and 50 h after birth are 20% and 1%, respectively, as active as a sample from the same cow 8 h after parturation. Fractionation of colostrum by molecular sieve chromatography gives one major peak of activity with an M_r of 30 000–45 000. Upon isoelectric focusing, this peak is resolved into two peaks with PI values of 6·5 and 9·0. These results suggest that these EGF-like factors are molecularly distinct from EGF.

Klagsbrun and co-workers isolated and characterized three polypeptides from human colostrum and milk which stimulate DNA synthesis and proliferation of cultured fibroblasts in vitro and which they called human milk growth factors I, II and III (Klagsbrun, 1978; Shing & Klagsbrun, 1984). HMGF III, which constituted 75–95% of human milk growth factor activity, has a M_r of 5 000–6 000 and a pI of 4·4–4·7, all of which suggest that it may be identical to EGF. HMGF I (M_r 34 000–38 000; pI 4·2–4·5) and HMGF II (M_r 10 000–20 000; pI 3·2–4·8) represent about 5% and 20%, respectively, of the fibroblast growth factor activity in human milk.

Zwiebel *et al.* (1986) isolated a factor from human milk with EGF-like activity which they called mammary-derived growth factor II (MDGF II). It has a M_r of 17 000 and a pI of 4·0, which suggests that it is molecularly distinct from EGF.

Human milk contains growth factor activity that resembles transforming growth factor α (TGF_α), a factor which has the capacity to transform anchorage-dependent cells into cells that grow in an anchorage-independent manner (Zwiebel *et al.*, 1986). The TGF_α activity of human milk can be resolved into two components by isoelectric focusing with pI values of 6·5 ($TGF_{\alpha 1}$) and 7·0 ($TGF_{\alpha 2}$).

Members of the insulin family include insulin itself and two insulin-like growth factors, IGF1 and IGF2. Insulin and IGF1 have been detected in human milk (Baxter *et al.*, 1984; Read *et al.*, 1984). While IGF2 has not been measured in human milk, IGF2 activity has been detected (Read *et al.*, 1983).

The concentration (3·75 nmol/1) of insulin (M_r 5 796; pI 5·7) in human colostrum is 30 times that in blood serum but falls to 15–20% of the original level within a few days (Read *et al.*, 1984). While Read *et al.* reported that the concentration of insulin in human colostrum and milk was less than that required to stimulate significant increases in protein and DNA synthesis in several cell lines in vitro, Ballard *et al.* (1982) have claimed that insulin accounts for a substantial amount of the mitogenic activity of human colostrum, as shown by the selective removal of insulin by immunoaffinity chromatography. This apparent discrepancy may be due to the different cell lines used by the two groups. Ballard *et al.* (1982) reported that the concentration of insulin in bovine colostrum (2–3 nmol/1) exceeded that in blood serum by about 100-fold.

IGF1 (M_r 7 649; pI 8·4; Rindernecht & Humbel, 1976) is largely bound to a carrier protein in human milk (Baxter *et al.*, 1984). Suikkari (1989) reported that the concentration of IGF1 in human milk decreased from 49 μg/litre at 4 h post-partum to 29 μg/litre at 92 h post-partum. Frankenne *et al.* (1989) purified IGF1 from bovine colostrum and showed it to have the same molecular weight as IGF1 from human milk. Campbell and Baumrucker (1989) showed that IGF1 is mainly associated with a 45 kDa binding protein in bovine milk, with a small amount unbound. The binding protein is unsaturated and the complex is readily dissociated by acid treatment.

Prosser *et al.* (1989) reported that treatment of cows at 35–47 weeks of lactation with recombinantly-derived bovine growth hormone (rbGH) for 7 days increased the concentration of IGF1 in milk from 0·44 to 1·6 nmol/litre. However, the peak IGF1 concentration during rbGH treatment was lower than that in milk collected during early lactation and equivalent to that in mature human milk. They showed that about 80% of IGF1 in bovine milk is associated with proteins of molecular weight 40 000–150 000.

Mammary-derived growth factor I (MDGF I) is a polypeptide which stimulates mouse mammary cells and rat kidney cells in vitro to synthesize extracellular proteins, including collagen. It has been purified from human milk by Bano *et al.* (1985) and shown to be a polypeptide with a M_r of 62 000 and a pI of 4·8.

Platelet-derived growth factor (PDGF), a polypeptide with a M_r of 28 000–35 000 and a pI of 9·9, has been shown to be present in colostrum and mature milk from cows, sheep and goats (Brown & Blakely, 1984). Transforming growth factor β (TGF_β), which acts synergistically with TGF_α and IGF1 in promoting cell growth in culture, has been detected in colostrum and milk from human (Noda *et al.*, 1984) and goat (Brown & Blakeley, 1984). Sinha and Yunis (1983) partially purified a protein from human milk with colony stimulating factor (CSF) activity; it had a M_r of 240 000–250 000 and a pI of 4·4–4·9. Jahnke and Lazarus (1984) detected a bombesin immunoactive peptide with a M_r of 3200 in bovine milk while Ekman *et al.* (1985) showed that human milk contains 60–430 μg/ml of bombesin-immunoactive material.

Nerve growth factor (NGF), which is known to play a role in the survival and differentiation of selected neurons of the nervous system, has been shown to be present in the milk of mouse (Grueters *et al.*, 1985) and human (Wright & Gaull, 1983).

It is not known if the gastrointestinal mucosal proliferation observed on feeding colostrum to newborn animals is attributable to any of the growth factors described above. To date, most attention has focussed on EGF which appears to be the major growth factor in human milk. Feeding EGF in pharmacological doses to neonatal rats (Berseth, 1985) or 3-day-old rabbits (O'Loughlin *et al.*, 1985) resulted in increased DNA in intestinal tissues. Falconer (1987) showed that stomach growth in neonatal rats during the first 24 h after delivery was faster in neonates fed formula containing EGF than in suckled neonates. Feeding EGF at physiological levels to neonatal rats resulted in increased intestinal length (Moore *et al.*, 1986). Neonatal rats given artificial formula containing EGF for 39 h had greater DNA synthesis and content in the intestine than formula-fed rats without EGF (Berseth, 1987). Rats given pooled milk with antibodies to EGF for 39 h had lower intestinal weight and less DNA in the intestine than animals fed rat milk without antibodies.

There is also evidence that orally-administered EGF can affect tissues other than the intestine. Berseth and Go (1988) showed that ingestion of a formula containing added EGF by newborn rats for 5 days increased liver DNA and RNA and heart and kidney weight compared to formula-fed controls. These results suggest that oral ingestion of EGF is associated with growth of non-intestinal organs. For EGF consumed orally to exert such effects requires absorption of intact EGF or one of its metabolites and distribution to target tissues. Britton *et al.* (1988) showed that little degradation of EGF occurred in the rat stomach during the suckling and weanling periods, and degradation by luminal fluid from the small intestine of 12-day-old rats was lower than in weanling (31-day-old) rats. Absorption of intact EGF has been demonstrated in suckling, weanling and adult rats (Thornburg *et al.*, 1984, 1987). Gale *et al.* (1989) reported that urinary excretion of EGF was higher in infants who were breast fed than in infants receiving cows' milk-based formula or total parenteral nutrition. They suggested that these results were consistent with the hypothesis that EGF ingested in human milk crosses the gastrointestinal wall to enter the general circulation in the suckling infant.

6. Cholesterol Effect of Milk Proteins

Since the first decade of this century it has been known that dietary proteins may increase serum cholesterol in several animal species (see Terpstra *et al.* (1983) for a review). Particularly in rabbits, feeding casein, skim milk protein or lactalbumun, in cholesterol-free or cholesterol-containing semi-purified diets, results in much higher levels of plasma cholesterol and more marked artherosclerosis than feeding similar diets containing soy protein. Guinea pigs, rats, hamsters, pigs and monkeys show a similar but less marked response to the type of dietary protein, and only if the diet contains cholesterol, while chickens, calves and mice appear to be rather insensitive to this dietary manipulation (van der Meer & Beynen, 1987). In strictly controlled human studies, casein did not increase serum cholesterol relative to soy protein (Sacks *et al.*, 1983).

In most studies on dietary protein and serum cholesterol, the effects of casein and soy protein have been compared, and three hypotheses have been advanced to explain the species dependence of the cholesterolaemic effects of casein and soy protein:

(1) It is related to the high digestibility of casein relative to soy protein (Carroll, 1981; West *et al.*, 1984). It is proposed that incompletely hydrolysed soy proteins would bind bile acids and thus interrupt the enterohepatic cycle of steroids, thereby lowering serum cholesterol levels.
(2) It is due to the high lysine to arginine ratio in casein (Kritchevsky, 1979; Park & Liepa, 1982; Kritchevsky *et al.*, 1987). This ratio is more than twice as high in casein as in soy protein. It has been suggested that the lysine : arginine ratio influences cholesterol metabolism by affecting the synthesis of apoprotein E (Kritchevsky *et al.*, 1982) or by affecting the ratio of insulin to glucagon in plasma (Sugano *et al.*, 1982; Nagata *et al.*, 1981).
(3) It is due to the high degree of phosphorylation of casein (van der Meer, 1983, 1988; van der Meer & Beynen, 1987). In this hypothesis it is assumed that

in the lumen of the small intestine, bile acids and bilary micelles may bind to insoluble calcium phosphate and that this binding is inhibited by dietary casein and phosphopeptides derived from it by digestion. Consequently, casein would increase the availability of bile acids for reabsorption which would result in enhanced lipid, cholesterol and bile acid absorption and infux to the liver, leading eventually to an increased concentration of plasma cholesterol. Thus, it is postulated that casein interferes with normal excretion of bile acids and cholesterol into faeces, which is the main route for removal of cholesterol from the body. This is the only hypothesis that addresses the species dependence of the cholesterolaemic effects of casein and soy protein. Species dependence is postulated to be related to the nature of the bile acid conjugates and the activity of intestinal alkaline phosphatase. Glycine-conjugated (but not taurine-conjugated) bile acids, which are predominant in the rabbit, bind readily to insoluble calcium phosphte at pH > 5·5 and their availability for reabsorption is thereby reduced. However, phosphopeptides derived from casein, which can persist in the small intestine of the rabbit due to its low intestinal alkaline phosphatase activity, can prevent binding of bile acids to insoluble calcium phosphate and increase their availability for absorption. In rats, bile acids are predominantly conjugated to glycine and intestinal alkaline phosphatase activity is high. Thus, phosphopeptides may not persist in the small intestine and there would be no interference with binding of bile acids to insoluble calcium phosphate, and dietary casein would, therefore, not increase bile acids and cholesterol absorption.

It is clear from the many studies on the cholesterolaemic effects of proteins that none of these hypotheses is consistent with all of the experimental findings. For example, the similarity in cholesterolaemic response in animals to casein and other animal proteins (e.g. lactalbumin, whole egg protein, whole milk protein) (Krichevsky *et al.*, 1987) which have much different amino acid compositions and much lower degrees of phosphorylation than casein is not explained. Up to now most attention has focussed on identifying a hypercholesterolaemic factor in casein. An equally plausible interpretation of the data would be the presence of a hypocholesterolaemic factor in soy protein.

This interpretation is supported by the results of recent research by Sugano *et al.* (1988, 1990) who found that the undigested insoluble peptide fraction obtained after microbial protease treatment of soybean protein significantly lowered serum cholesterol in rats fed diets enriched with cholesterol and cholate when compared to the parent soybean protein. This fraction markedly stimulated faecal excretion of both neutral and acidic steroids and also bound conjugated bile acids in vitro, suggesting that the hypocholesteralaemic effect was due to interference with cholesterol and bile acid reabsorption from the intestine. The undigested fraction obtained after pepsin digestin exerted a similar hypocholesterolaemic effect as that obtained after digestion with microbial proteases.

While these findings may help to explain the hypocholesterolaemic effect of soybean protein relative to animal proteins, such as casein, in animals, the absence of a significant hypocholesterolaemic effect of soybean protein in humans (Sacks *et al.*, 1983) suggests that this mechanism may not be operative in all species.

7. Biologically Active Peptides from Casein Hydrolysates

Peptides with various types of biological activity have been isolated from the hydrolysates of casein and whey proteins (Meisel *et al.*, 1989; Maubois & Léonil, 1989). It has been demonstrated that at least some of these peptides are produced in vivo and therefore may play a role in vivo. The presence of these latent peptides led Migliore-Samour and Jolles (1988) to refer to casein as a 'prohormone'.

7.1. Caseinomorphins

The existence of several endogenous opioid peptides (endorphins) has been demonstrated in various mammalian tissues (see Teschemacher, 1978); their functions have been comprehensively reviewed by Kromer (1988). Although these peptides range in size from 5 to 31 residues, their N-terminal sequences are, invariably, Try–Gly–Gly–Phe (see Chiba & Yoshikawa, 1986). The peptides exert their activity by binding to specific receptors of target cells which are classified according to the specific physiological effects.

Peptides with opioid activity have been identified in partial enzymatic digests of proteins derived from foods and have been called exorphins. Several peptides with opioid activity have been isolated from digests of milk proteins; the literature has been reviewed by Chiba and Yoshikawa (1986) and Meisel and Schlimme (1990).

Peptides with opioid activity were detected at considerable concentrations in certain batches of baby foods, casein digests and cows' milk by Brantl and Teschemacher (1979). These peptides were isolated from enzymatic digests of casein and characterized as a family of peptides containing 4–7 amino acids with a common N—terminal sequence, H–Tyr–Pro–Phe–Pro–, and 0–3 additional residues (Gly, Pro, Ile) (Brantl *et al.*, 1979; Henschen *et al.*, 1979).

The peptides represent residues 60–63/6 of β-casein and were called β-caseinomorphins 4 to 7, respectively; they possess μ-type opiate receptor activity. Obviously, since their amino acid sequence differs from that of the endorphins, the caseinomorphins (CM) represent a new class of opioid pepides. β-CM-5 was the most effective of the β-CMs which are 300–4 000 times less effective than morphine (Brantl *et al.*, 1981). The amide of β-CM-4 is more active than β-CM-5 (Chang *et al.*, 1981, 1985). Varius physiological effects of β-CMs were investigated by Brantl's group and others (see Hautefeuille *et al.* (1986) and Brantl & Neubert (1986) for references). β-CMs are very resistant to enzymes of the GIT and appear in the contents of the small intestine following ingestion of milk (see Brantl & Neubert, 1986). Meisel (1986) isolated a new β-caseinomorphin, β-CM-11 (β-CN f60–70), from the duodenal chyme of minipigs. β-CM-11 is less effective than β-CM-5 but Meisel (1986) suggests that β-CM-11 may be hydrolysed to smaller, more active β-CMs by the action of peptidases in the brush border. Further studies on β-CM-11 and its comparison with β-CM-5 were reported by Meisel and Frister (1989).

Teschemacher *et al.* (1986) found no evidence for β-CM in human plasma after ingestion of cows' milk and milk products but Teschemacher (1987) reported that β-CM precursor occurs in the plasma of new-born calves and infants and of pregnant, lactating and non-lactating females. Certain proteolytic bacteria growing in milk produce β-casomorphins (Hamel *et al.*, 1985).

Thus, while β-casomorphins appear to have physiological effects in the intestine (Tome *et al.*, 1987), rather high concentrations are required to affect sleep patterns and concentrations reaching the plasma or brain are probably not sufficiently high (Taira *et al.*, 1990).

The sequence 51–57 of human β-casein, Tyr–Pro–Phe–Val–Glu–Pro–Ile, corresponds to bovine β-casein 60–66 (i.e. Tyr–Pro–Phe–Pro–Gly–Pro–Ile). The opioid activity of synthetic hβ-CM-4 and -5 was considerably less than that of bovine β-CM-4 and β-CM-5 (Brantl, 1984). Several peptides with sequences corresponding to segments of human β-casein 51–57 were synthesized by Yoshikawa *et al.* (1986*a*) and found to have opioid activity. Peptides corresponding to human β-casein 41–44 and 59–63 were also synthesized and found to have weak opioid activity, as did the peptides Tyr–Gly–Leu–Phe, which occurs in α-lactalbumins from several species, and Tyr–Leu–Leu–Phe, which is found in bovine β-lactoglobulin (see also Chiba & Yoshikawa (1986) and Yoshikawa *et al.* (1986*a*) for other references).

A delta-sleep-inducing peptide (DSIT) has been demonstrated in human milk and colostrum (Graf *et al.*, 1984). This was confirmed by Ernst *et al.* (1985) who showed that bovine milk contains similar peptides. The identity and characteristics of DSIP, which appears not to be a β-casomorphin, have not been established.

Exorphins have also been isolated from peptic hydrolysates of α-casein (Zioudrou *et al.*, 1979) and shown to be a mixture of two peptides, α-casein f90–95 and f90–96 (Arg–Tyr–Leu–Gly–Tyr–Leu (Glu)); the latter was the more effective and the N-terminal Arg appeared to be important for activity (Loukas *et al.*, 1983). Therefore, the α-casein-derived exorphins have a considerably different structure from the β-caseinomorphins.

Opiate peptides are also released from α-lactalbumin (α-lactorphin, residues 50–53, Tyr–Gly–Leu–Phe) and β-lactoglobulin (β-lactorphin, residues 102–105, Tyr–Leu–Leu–Phe) on hydrolysis by pepsin and trypsin *in vitro* (Antila *et al.*, 1991).

Several sequences, especially 35–41, 57–60 and 25–34, of bovine and human κ-caseins were found to have opoid activity; these peptides were called casoxins A, B and C, respectively (Yoshikawa *et al.*, 1986*a*,*b*; Chiba *et al.*, 1989).

Three peptides with opioid properties were also isolated from peptic digests of lactotransferrin: Tyr–Leu–Gly–Ser–Gly–Tyr (lactoferroxin A), Arg–Tyr–Tyr–Gly–Tyr (lactoferroxin B), Lys–Tyr–Leu–Gly–Pro–Gln–Tyr (lactoferroxin C) (Chiba *et al.*, 1989). The activity of the three peptides was markedly increased by methylation of the α-carboxyl group.

Thus, some milk products contain preformed pepides with opioid properties while all the major milk proteins contain sequences which, when liberated by gastrointestinal proteinases, possess opioid activity. These peptides are very resistant to protolysis by gastrointestinal proteinases and because of their high hydrohobicity can be absorbed intact from the intestine. They have been shown to have physiological activity *in vitro* but their activity *in vivo* is as yet uncertain.

7.2. Immunomodulating Peptides

Migliore-Samour *et al.* (1989) and Gattegno *et al.* (1988) reported that enzymatic digest of human caseins contained immunomodulating peptides which stimulated the phagocytic activity of human macrophages *in vitro* and exerted a protective effect in vivo in mice against *Klebsiella preumoniae* infection. Two of the peptides were characterized as Val–Glu–Pro–Ile–Pro–Tyr (residues 54–59 of β-casein) and Gly–Leu–Phe (origin not identified).

7.3. Platelet-modifying Peptide

The undecapepide, Met–Ala–Ile–Pro–Pro–Lys–Lys–Asn–Gln–Asp–Lys (residues 106–116 of bovine κ-casein) inhibited the aggregation of, and the binding of fibrinogen to, ADP-treated platelets; its behaviour was similar to that of the structurally related C-terminal dodecapeptide (residues 400–411) of human fibrinogen γ-chain (Jolles *et al.*, 1986). This peptide was produced from the caseino(glyco) macropeptide, residues 106–169 of κ-casein produced by the action of chymosin on κ-casein or whole casein. Shorter peptides, 106–112 and 113–116, were also produced; these had similar effects on platelet aggregation but were less effective than the undecapeptide.

A peptide with similar properties was isolated from a hydrolysate of lactotransferrin (Mazoyer *et al.*, 1990).

7.4. Angiotensin Converting Enzyme (ACE) Inhibitor

ACE is a peptidyl-dipeptidase (EC 3.4.15.1) which cleaves dipeptides from the carboxyl terminal of peptides. It converts angiotensin I to the potent vasoconstrictor, angiotensin II, and inactivates the vasodilator bradykinin. Maruyama and Suzuki (1982) isolated the dodecapeptide, Phe–Phe–Val–Ala–Pro–Phe–Pro–Glu–Val–Phe–Gly–Lys, from tryptic hydrolysates of casein; this corresponded to the sequence 23–34 of α_{s1}-casein B. Other inhibitory peptides were later isolated, e.g. Thr–Thr–Met–Pro–Leu–Trp, the C-terminal sequence of α_{s1}-casein (Maruyama *et al.*, 1987). Kohmura *et al.* (1989) synthesized 69 peptides corresponding to sequences of human β-casein; several peptides had inhibitory activity, especially those with sequences within the region β-CN 39–52, the most active being Ser–Phe–Gln–Pro–Gln–Pro–Leu–Ile–Tyr–Pro (β-CN 43–52).

References

Aisen, P. & Leibman, A. (1972). Lactoferrin and transferrin: A comparative study. *Biochim. Biophys. Acta*, **257**, 314–23.

Al-Mashikhi, S.A., Li-Chan, E. & Nakai, S. (1988). Separation of immunoglobulins and lactoferrin from cheese whey by chelating chormatography. *J. Dairy Sci.*, **71**, 1747–55.

Andrews, G. (1989). Lactulose in heated milk. In *Heat-Induced Changes in Milk*, ed. P.F. Fox. International Dairy Federation Bulletin 238, Brussels, pp. 45–52.

Antila, P., Paakkari, I., Järvinen, A., Mattila, M.J., Laukkanen, M., Pihlanto-Leppälä, A., Mäntsälä, P. & Hellman, J. (1991). Opioid peptides derived from *in vitro* proteolysis of whey proteins. *Intern. Dairy J.*, **1**, 215–29.

Antony, A.C., Utley, C.S., Marcell, P.D. & Kolhouse, J.F. (1982). Isolation, characterization and comparison of the solubilized particulate and soluble folate binding proteins from human milk. *J. Biol. Chem.*, **257**, 10081–9.

Arnold, R.R., Cole, M.F. & McGhee, J.R. (1977). A bacteriocidal effect for human lactoferrin. *Science*, **197**, 263–5.

Ashoor, S.H. & Monte, W.C. (1983). High performance liquid chromatographic determination of *Bifidobacterium bifidum* growth factors in human milk. *J. Assoc. Off. Anal. Chem.*, **66**, 135–9.

Atkinson, S.A. & Lonnerdal, B. (ed.) (1989). *Protein and Non-protein Nitrogen in Human Milk*. CRC Press, Boca Raton, FL.

Azuma, N. & Yamauchi, K. (1987). A glycoprotein from human milk. *J. Dairy Res.*, **54**, 199–205.

Bailey, L.B. (1988). Factors affecting folate bioavailability. *Food Technol.*, **24** (10), 206–210, 212, 238.

Ballard, F.J., Nield, M.K., Francis, G.L., Dahlenberg, G.W. & Wallace, J.C. (1982). The relationship between the insulin content and inhibitory effects of bovine colostrum on protein breakdown in cultured cells. *J. Cell Physiol.*, **110**, 249–54.

Bano, M., Salomon, D.S. & Kidwell, W.R. (1985). Purification of a mammary derived growth factor from human milk and human mammary tumors. *J. Biol. Chem.*, **260**, 5745–52.

Barker, S.A. & Stacey, M. (1963). The carbohydrates of milk and colostrum. *Dairy Sci. Abstr.*, **25**, 445–50.

Baxter, R.C., Zaltsman, Z. & Turtle, J. (1984). Immunoreactive somatomedin C/IGF1 and its binding protein in human milk. *J. Clin. Endocrinol. Metab.*, **58**, 955–9.

Beardmore, J.M. & Richards, R.C. (1983). Concentrations of epidermal growth factor in mouse milk throughout lactation. *J. Endocrinol.*, **96**, 287–92.

Beerens, H., Romond, C. & Neut, C. (1980). Influence of breast-feeding on the bifid flora of the newborn intestine. *Am. J. Clin. Nutr.*, **33**, 2434–9.

Berseth, C.L. (1985). EGF-mediated breast milk-enhanced intestinal growth in neonatal rats. *Pediatr. Res.*, **19**, 213A.

Berseth, C.L. (1987). Enhancement of intestinal growth in neonatal rats by epidermal growth factor in milk. *Am. J. Physiol.*, **253**, G662–5.

Berseth, C.L. & Go, V.L.W. (1988). Enhancement of neonatal somatic and hepatic growth by orally administered epidermal growth factor. *J. Pediatr. Gastroenterol. Nutr.*, **7**, 889–93.

Berseth, C.L., Lichtenberger, L.M. & Morriss, F.H. (1983). Comparison of the intestinal growth promoting effects of rat colostrum and mature rat milk in newborn rats in vivo. *Am. J. Clin. Nutr.*, **37**, 52–60.

Bezkorovainy, A. (1965). Comparative study of the acid glycoproteins isolated from bovine serum, colostrum, and milk whey. *Arch. Biochem. Biophys.*, **110**, 558–67.

Bezkorovainy, A. (1967). Physical and chemical properties of bovine milk and colostrum whey M-1 glycoprotein. *J. Dairy Sci.*, **50**, 1368–75.

Bezkorovainy, A. & Nichols, J.H. (1976). Glycoproteins from mature human milk whey. *Pediat. Res.*, **10**, 1–5.

Bezkorovainy, A., Nichols, J.H. & Sly, D.A. (1976). Proteose-peptone fractions of human and bovine milk. *Int. J. Biochem.*, **7**, 639–45.

Bezkorovainy, A., Grohlich, D. & Nichols, J.H. (1979). Isolation of a glycopeptide fraction with *Lactobacillus bifidus* subspecies *pennsylvanius* growth-promoting activity from whole human milk casein. *Am. J. Clin. Nutr.*, **32**, 1428–32.

Bezwoda, W.R. & Mansoor, N. (1986). Isolation and characterization of lactoferrin separated from human whey by adsorption chromatography using Cibacron Blue F3 G-A linked affinity adsorbent. *Clin. Chim. Acta*, **157**, 89–94.

Blanc, B. (1981). Biochemical aspects of human milk—comparison with bovine milk. *Wld Rev. Nutr. Diet.*, **36**, 1–89.

Brantl, V. (1984). Novel opioid peptides derived from human β-casein: human β-casomorphins. *Eur. J. Pharmacol.*, **106**, 213–14.

Brantl, V. & Neubert, K. (1986). Opioid peptides derived from food proteins. *Trends Pharmacol. Sci.*, **7**, 6–7.

Brantl, V. & Teschemacher, H. (1979). A material with opioid activity in bovine milk and milk products. *Naunyn-Schmildeberg's Arch. Pharmacol.*, **306**, 301–4.

Brantl, V., Teschemacher, H., Henschen, A. & Lottspeich, F. (1979). Novel opioid peptides derived from casein (β-casomorphins). I. Isolation from bovine casein peptone. *Hoppe-Seyler's, Z. Physiol. Chem.*, **360**, 1211–16.

Brantl, V., Teschemacher, H., Blasig, J., Henschen, A. & Lottspeich, F. (1981). Opioid activities of β-casomorphins. *Life Sci.*, **28**, 1903–9.

Brennan, M. (1990). Absorption of iron and zinc from milks and infant formulae in suckling rats. M.Sc. Thesis, National University of Ireland, Dublin.

Brignon, G., Chtourou, A. & Ribadeau Dumas, B. (1985). Preparation and amino acid sequence of human κ-casein. *FEBS Lett.*, **188**, 48–54.

Britton, J.R., George-Nasciento, C. & Koldovsky, O. (1988). Luminal hydrolysis of recombinant human epidermal growth factor in the rat gastrointestinal tract: segmental and developmental differences. *Life Sci.*, **43**, 1339–47.

Brown, K.D. & Blakely, D.M. (1983). Cell growth—promoting activity in mammary secretions of the goat, cow and sheep. *Br. Vet. J.*, **139**, 68–78.

Brown, K.D. & Blakely, D.M. (1984). Partial purification and characterization of a growth factor from goat colostrum. *Biochem. J.*, **219**, 609–17.

Bullen, C.L. & Willis, A.T. (1971). Resistance of the breast-fed infant to gastroenteritis. *Br. Med. J.*, **3**, 338–43.

Burger, R.L. & Allen, R.H. (1974). Characterization of Vitamin B_{12}-binding proteins isolated from human milk and saliva by affinity chromatography. *J. Biol. Chem.*, **249**, 7220–7.

Campbell, P.G. & Baumrucker, C.R. (1989). Insulin-like growth factor-1 and its association with binding proteins in bovine milk. *J. Endocrinol.*, **120**, 21–9.

Carpenter, G. (1980). Epidermal growth factor is the major growth promoting agent in human milk. *Science*, **210**, 198–9.

Carroll, K.K. (1981). Soya protein and atherosclerosis. *J. Am. Oil. Chem. Soc.*, **58**, 416–19.

Castellino, F.J., Fish, W.W. & Mann, K.G. (1970). Structural studies on bovine lactoferrin. *J. Biol. Chem.*, **245**, 4269–75.

Chang, K.-J., Killian, A., Hazum, E. & Cautrecasas, P. (1981). Morphiceptin (NH_4 Tyr.-Pro.Phe.Pro-$CONH_2$): A potent and specific antagonist for morphine (μ) receptors. *Science*, **212**, 75–7.

Chang, K.-J., Su, Y.F., Brent, D.A. & Chang, J.F. (1985). Isolation of a specific μ-opiate receptor peptide, morphiceptin, from an enzymatic digest of milk proteins. *J. Biol. Chem.*, **260**, 9706–12.

Chiba, H. & Yoshikawa, M. (1986). Biologically functional peptides from food proteins:

new opioid peptides from milk proteins. In *Protein Tailoring for Food and Medical Uses*, ed. R.E. Feeney & J.R. Whitaker. Marcel Dekker, New York, pp. 123–53.

Chiba, H., Tani, F. & Yoshikawa, M. (1989). Opioid antagonist peptides derived from κ-casein. *J. Dairy Res.*, **56**, 363–6.

Cochet, C., Feige, J.-J. & Chambaz, E.M. (1983). Reversibility of the phosphate transfer between ATP and phosphoproteins catalysed by a cyclic nucleotide independent (G type) casein kinase. *Biochim. Biophys. Acta*, **744**, 147–54.

Cohen, S. (1962). Isolation of a mouse submaxillary gland protein accelerating incisor eruption and eyelid opening in the newborn animal. *J. Biol. Chem.*, **237**, 1555–62.

Cole, C.B., Scott, K.J., Henschel, M.J., Coates, M.E., Ford, J.E. & Fuller, R. (1983). Trace-nutrient-binding proteins in milk and the growth of bacteria in the gut of infant rabbits. *Br. J. Nutr.*, **49**, 231–40.

Colman, N. & Herbert, V. (1980). Folate binding proteins. *Ann. Rev. Med.*, **31**, 433–9.

Colman, N., Hettiarachchy, N. & Herbert, V. (1981). Detection of a milk factor that facilitates folate uptake by intestinal cells. *Science*, **211**, 1427–9.

Davidson, L.A., Lonnerdal, B. (1985). Isolation and characterization of monkey milk lactoferrin and identification of a specific brush border receptor. In *Proteins of Iron Storage and Transport*, ed. G. Spik, J. Montreuil, R.R. Crichton & J. Mazurier. Elsevier Science Publishers, Amsterdam, pp. 275–8.

Ek, J. (1987). Folate and human milk. In *Human Lactation—3. The Effects of Human Milk on the Recipient Infant*, ed. A.S. Goldman, A. Atkinson & L.A. Hanson. Plenum Press, New York, pp. 99–104.

Ekman, R., Ivarsson, S. & Jansson, L. (1985). Bombesin, neurotensin and pro-gamma-metatotropin immunoreactants in human milk. *Regul. Peptides*, **10**, 99–105.

Ernst, A., Monti, J.C. & Schoenenberger, G.A. (1985). Delta-sleep-inducing peptide (DSIP) in human and cow milk. In *Sleep '84, Proc. 7th Eur. Congr. on Sleep Res.*, ed. W.P. Koella, E. Ruther & H. Schulz. Gustav Fischer Verlag, Stuttgart, pp. 215–17.

Falconer, J. (1987). Oral epidermal growth factor in trophic for the stomach in the neonatal rat. *Biol. Neonate*, **52**, 347–50.

Ford, J.E. (1974). Some observations on the possible nutritional significance of vitamin B_{12} and folate-binding proteins in milk. *Br. J. Nutr.*, **31**, 243–57.

Ford, J.E., Salter, D.N. & Scott, K.J. (1969). The folate-binding protein in milk. *J. Dairy Res.*, **36**, 435–48.

Ford, J.E., Knaggs, G.S., Salter, D.N. & Scott, K.J. (1972). Folate nutrition in the kid. *Br. J. Nutr.*, **27**, 571–83.

Ford, J.E., Scott, K.J., Sansom, B.F. & Taylor, P.J. (1975). Some observations on the possible nutritional significance of Vitamin B_{12}- and folate-binding proteins in milk. Absorption of [^{58}Co] cyanocobalamin by suckling piglets. *Br. J. Nutr.*, **34**, 469–92.

Ford, J.E., Law, B.A., Marshall, M.E. & Reiter, B. (1977). Influence of the heat treatment of human milk on some of its protective constituents. *J. Pediatrics*, **90**, 29–35.

Frankenne, F., Marcotty, C., Duyckaerts, A., Beeumen, J. & Maghuin-Rogister, G. (1989). Purification and characterization of insulin-like growth factor-1 from cow colostrum. *Biochem. Soc. Trans.*, **17**, 602.

Gale, S.M., Read, L.C., George-Nasciento, C., Wallace, J.C. & Ballard, F.J. (1989). Is dietary epidermal growth factor absorbed in premature human infants? *Biol. Neonate*, **55**, 104–10.

Gattegno, L., Migliore-Samour, D., Saffar, L. & Jolles, P. (1988). Enhancement of phagocytic activity of human monocytic-macrophagic cells by immunostimulating peptides from human casein. *Immunol. Lett.*, **18**, 27–32.

Gaull, G.E., Jensen, R.G., Rassin, D.K. & Malloy, M.H. (1982). Human milk as food. In *Advances in Perinatal Medicine*, Vol. 2, ed. A. Milunsky, E.A. Friedman & L. Gluck. Plenum Medical Book Co., New York, pp. 47–120.

Ghitis, J. (1966). The labile folate of milk. *Am. J. Clin. Nutr.*, **18**, 452–7.

Ghitis, J. (1967). The folate binding in milk. *Am. J. Clin. Nutr.*, **20**, 1–4.

Ghitis, J., Mandelbaum-Shavit, F. & Grossowicz, N. (1969). Binding of folic acid and derivatives by milk. *Am. J. Clin. Nutr.*, **22**, 156–62.

Goldman, A.S. (1988). Immunologic supplementation of cows' milk formulations. Special Address, International Dairy Federation, Annual Sessions, Budapest.

Goldman, A.S., Ham Pong, A.J. & Goldblum, R.M. (1985). Host defences: development and maternal contribution. *Adv. Pediatrics*, **32**, 71–100.

Goldman, A.S., Atkinson, S.A. & Hanson, L.A. (ed.) (1987). *Human Lactation—3. The Effects of Human Milk on the Recipient Infant*. Plenum Press, New York.

Gordon, W.G., Ziegler, J. & Basch, J.J. (1962). Isolation of an iron-binding protein from cows' milk. *Biochim. Biophys. Acta*, **60**, 410–11.

Graf, M.V., Hunter, C.A. & Kastin, A.J. (1984). Presence of delta-sleep-inducing peptide-like material in human milk. *J. Clin. Endocrinol. Metabolism*, **59**, 127–32.

Gregory, M.E. (1954). The microbiological assay of Vitamin B_{12} in the milk of different animal species. *Br. J. Nutr.*, **8**, 340–7.

Gregory, J.F. (1982). Denaturation of the folacin-binding protein in pasteurized milk products. *J. Nutr.*, **112**, 1329–38.

Gregory, M.E. & Holdsworth, E.S. (1955*a*). The occurrence of a cyanobalamin-binding protein in milk and the isolation of a cyanobalamin-protein complex from sow's milk. *Biochem. J.*, **59**, 329–34.

Gregory, M.E. & Holdsworth, E.S. (1955*b*). Some properties of the cyanobalamin-protein complex from sow's milk, and the mode of linkage of cyanobalamin with protein. *Biochem. J.*, **59**, 335–40.

Groves, M.L. (1965). Preparation of some iron-binding proteins and α-lactalbumin from bovine milk. *Biochim. Biophys. Acta*, **100**, 154–62.

Grueters, A., Lakshmanan, J., Tarris, R., Alm, J. & Fisher, D.A. (1985). Nerve growth factor in mouse milk during early lactation: lack of dependency on submandibular glands. *Pediatr. Res.*, **19**, 934–7.

György, P. (1953). A hitherto unrecognized biochemical difference between human milk and cow's milk. *Pediatrics*, **11**, 98–108.

György, P. & Rose, C.S. (1955). Further observations on the metabolic requirements of *Lactobacillus bifidus* var *pennsylvanicus*. *J. Bact.*, **69**, 483–90.

György, P., Norris, R.F. & Rose, C.S. (1954*a*). Bifidus factor. I. A variant of *Lactobacillus bifidus* requiring a special growth factor. *Arch. Biochem. Biophys.*, **48**, 193–201.

György, P., Kuhn, R., Rose, C.S. & Zilliken, F. (1954*b*). Bifidus factor, II. Its occurrence in milk from different species and in other natural products. *Arch. Biochem. Biophys.*, **48**, 202–8.

György, P., Hoover, J.R.E., Kuhn, R. & Rose, C.S. (1954*c*). Bifidus factor. III. The rate of dialysis. *Arch. Biochem. Biophys.*, **48**, 209–13.

György, P., Jeanloz, R.W., von Nicolai, H. & Zilliken, F. (1974). Undialyzable growth factor for *Lactobacillus bifidus* var. *pennsylvanicus*. *Eur. J. Biochem.*, **43**, 29–33.

Hall, R.A. & Widdowson, E.M. (1979). Response of the organs of rabbits to feeding during their first days after birth. *Biol. Neonate*, **35**, 131–9.

Hamel, U., Killwein, G. & Teschemacher, H. (1985). *β*-Casomorphin immunoreactive materials in cows' milk incubated with various bacterial species. *J. Dairy Res.*, **52**, 139–48.

Hamos, M., Freed, L.M., Jones, J.B., Berkow, S.E., Bitman, J., Mehta, N.R., Happ, B. & Hamos, P. (1985). Enzymes in human milk. In *Human Lactation: Milk Components and Methodologies*, ed. R.G. Jensen & M.C. Neville. Plenum Press, New York, pp. 251–66.

Hansen, S.I., Holm, J. & Lyngbye, J. (1977*a*). Folate binding by human milk protein. *Scand. J. Clin. Lab. Invest.*, **37**, 363–7.

Hansen, S.I., Holm, J. & Lyngbye, J. (1977*b*). Mechanism of folate binding in milk. *Clin. Chim. Acta*, **75**, 321–4.

Hansen, S.I., Holm, J. & Lyngbye, J. (1978). Cooperative binding of folate to a protein isolated from cow's whey. *Biochim. Biophys. Acta*, **535**, 309–18.

Harmon, R.J., Schanbacher, F.L., Ferguson, L.C. & Smith, K.L. (1975). Concentration of lactoferrin in milk of normal lactating cows and changes occurring during mastitis. *Am. J. Vet. Res.*, **36**, 1001–7.

Harmon, R.J., Schanbacher, F.L., Ferguson, L.C. & Smith, K.L. (1976). Changes in lactoferrin, immunoglobulin G, bovine serum albumin, and α-lactalbumin during acute experimental and natural coliform mastitis in cows. *Infect. Immun.*, **13**, 533–42.

Harzer, G. & Bindels, J.G. (1985). Changes in human milk innunoglobulin A and lactoferrin during early lactation. In *Composition and Physiological Properties of Human Milk*, ed. J. Schaub. Elsevier Science Publishers, Amsterdam, pp. 285–95.

Hautefeuille, M., Brantl, V., Dumontier, A.-M. & Desjeux, J.-F. (1986). In vitro effects of β-casomorphins on ion transport in rabbit ileum. *Am. J. Physiol.*, **250**, G92–7.

Heird, W.C., Schwarz, S.M. & Hansen, I.H. (1984). Colostrum induced enteric mucosal growth in beagle puppies. *Pediatr. Res.*, **18**, 512–15.

Henschen, A., Lottspeich, F., Brantl, V. & Teschemacher, H. (1979). Novel opioid peptides derived from casein (β-casomorphins). II. Structure of active components from bovine casein peptone. *Hoppe Seyler's, Z. Physiol. Chem.*, **360**, 1217–24.

Hernell, O. & Lonnerdal, B. (1989). Enzymes in human milk. In *Protein and Non-protein Nitrogen in Human Milk*, ed. S.A. Atkinson & B. Lonnerdal. CRC Press, Boca Raton, FL, pp. 67–75.

Hirano, S., Hayashi, H., Terabayashi, T., Onodera, K., Iseki, S., Kochibe, N., Nagai, Y., Yagi, N., Nakagaki, T. & Imagawa, T. (1968). Biologically active glycopeptides in human colostrum. *J. Biochem.*, **64**, 563–5.

Høier-Madsen, M., Hansen, S.I. & Holm, J. (1986). Rabbit antibodies against the folate binding protein from cow's milk. Production, characterization and use for development of an enzyme-linked immunosorbent assay (ELISA). *Biosciences Report*, **6**, 895–900.

Høier-Madsen, M., Hansen, S.I. & Holm, J. (1987). Rabbit antibodies against the low molecular weight folate binding protein from human milk. Use for immunological characterization of human folate binding proteins in an enzyme-linked immunosorbent assay (ELISA). *Bioscience Report*, **7**, 553–7.

Holm, J., Hansen, S.I. & Lyngbye, J. (1978). Effect of methotrexate on folate binding to a folate binding protein in cow's milk. *Acta Pharmacol. Toxicol.*, **42**, 77–80.

Izak, G., Galewski, K., Rachmilewitz, M. & Grossowicz, N. (1972). The absorption of milk-bound pteroglutamic acid from small intestinal segments. *Proc. Soc. Expt. Biol. Med.*, **140**, 248–50.

Jahnke, G.D. & Lazarus, L.H. (1984). A bombesin immunoreactive peptide in milk. *Proc. Natl. Acad. Sci., USA*, **81**, 578–82.

Jensen, R.G. & Neville, M.C. (ed.) (1985). *Human Lactation: Milk Components and Methologies*. Plenum Press, New York.

Johansson, B.G. (1960). Isolation of an iron-containing red protein from human milk. *Acta Chem. Scand.*, **14**, 510–12.

Jolles, J., Donda, A., Amiguet, P. & Jolles, P. (1984). Mare lactotransferrin: purification, analysis and N-terminal sequence determination. *FEBS Lett.*, **176**, 185–8.

Jolles, P., Levy-Toledano, S., Fiat, A.-M., Soria, C., Gillessen, D., Thomaidis, A., Dunn, F.W. & Caen, J.P. (1986). Analogy between fibrinogen and casein. Effect of an undecapeptide isolated from κ-casein on platelet function. *Eur. J. Biochem.*, **158**, 379–82.

Kanno, C. & Kanehara, N. (1985). Influence of riboflavin-binding protein in raw cow milk. *Proc. Ann. Meeting Soc. Heric. Chem. Jpn*, p. 718 (Abstr.), cited from Toyosaki *et al.* (1987).

Kawakami, H., Shinmoto, H., Dosako, S.I. & Sogo, Y. (1987). One-step isolation of lactoferrin using immobilized monoclonal antibodies. *J. Dairy Sci.*, **70**, 752–9.

Kehagias, C. & Hansen, P.M.T. (1976). Isolation of sialic acid from casein and reclamation of residual casein. *J. Dairy Sci.*, **59**, 1718–26.

Kidwell, W.R. & Solomon, D.S. (1989). Growth factors in human milk: sources and potential physiological roles. In *Protein and Non-protein Nitrogen in Human Milk*, ed. S.A. Atkinson & B. Lonnerdal. CRC Press, Boca Raton, FL, pp. 77–91.

Kiely, J., Flynn, A., Singh, H. & Fox, P.F. (1988). Improved zinc bioavailability from colloidal calcium phosphate-free cow's milk. In *Trace Elements in Man and Animals—6*, ed. L.S. Hurley, C.L. Keen, B. Lonnerdal & R.B. Rucker. Plenum Press, New York, pp. 499–500.

Klagsbrun, M. (1978). Human milk stimulates DNA synthesis in cultural fibroblasts. *Proc. Natl. Acad. Sci.*, **75**, 5057–61.

Klagsbrun, M. & Neuman, J. (1979). The serum-free growth of Balb/C 3T3 cells in medium supplemented with bovine colostrum. *J. Supramol. Struct.*, **11**, 349–59.

Kohmura, M., Nio, N., Kubo, K., Minoshima, Y., Munekata, E. & Ariyoshi, Y. (1989). Inhibition of angiotensin-converting enzyme by synthetic peptides of human β-casein. *Agric. Biol. Chem.*, **53**, 2107–14.

Koldovsky, O. (1989). Search for role of milk-borne biologically active peptides for the suckling. *J. Nutr.*, **119**, 1543–51.

Kritchevsky, D. (1979). Vegetable protein and atherosclerosis. *J. Am. Oil Chem. Soc.*, **56**, 135–40.

Kritchevsky, D., Tepper, S., Czarnecki, S.K. & Klurfeld, D.M. (1982). Atherogenicity of animal and vegetable protein. *Atherosclerosis*, **41**, 429–31.

Kritchevsky, D., Tepper, S. & Klurfeld, D.M. (1987). Dietary protein and atherosclerosis. *J. Am. Oil Chem. Soc.*, **64**, 1167–71.

Kromer, W. (1988). Endogenous and exogenous opioids in the control of gastrointestinal mobility and secretion. *Pharmacological Revs.*, **40**, 121–62.

Law, B.A. & Reiter, B. (1977). The isolation and bacteriostatic properties of lactoferrin from bovine milk whey. *J. Dairy Res.*, **44**, 595–9.

Lonnerdal, B. (1985). Biochemistry and physiological function of human milk proteins. *Am. J. Clin. Nutr.*, **42**, 1299–317.

Lonnerdal, B., Hoffman, B. & Hurley, L.S. (1982). Zinc and copper binding proteins in human milk. *Am. J. Clin. Nutr.*, **36**, 1170–6.

Loukas, S., Varoucha, D., Zioudrou, C., Streaty, R.A. & Klee, W.A. (1983). Opioid activities and structures of α-casein derived exorphins. *Biochemistry*, **22**, 4567–73.

Maeda, K., McKenzie, H.A. & Shaw, D.C. (1980). Nature of the heterogeneity within genetic variants of bovine serum transferrin. *Anim. Blood Grps. Biochim. Genet.*, **11**, 63–75.

Maruyama, S. & Suzuki, H. (1982). A peptide inhibitor of angiotensin I-converting enzyme in the tryptic hydrolyzate of casein. *Agric. Biol. Chem.*, **46**, 1393–4.

Maruyama, S., Mitachi, H., Awaya, J., Kurono, M., Tonizuka, N. & Suzuki, H. (1987). Angiotensin I-converting enzyme inhibitory activity of the C-terminal hexapeptide of α_{s1}-casein. *Agric. Biol. Chem.*, **51**, 2557–61.

Masson, P.L. & Heremans, J.F. (1971). Lactoferrin in milk from different species. *Comp. Biochem. Physiol.*, **39B**, 119–29.

Maubois, J.L. & Léonil, J. (1989). Peptides du lait à activité biologique. *Lait*, **69**, 245–69.

Mazoyer, E., Levy-Toledano, S., Rendu, F., Hermant, L., Lu, H., Fiat, A.-M., Jolles, P. & Caen, J. (1990). KRDS, a new peptide derived from lactotransferrin, inhibits platelet aggregation and release reaction. *Eur. J. Biochem.*, **194**, 43–9.

Meisel, H. (1986). Chemical characterization and opioid activity of an exorphin isolated from *in vivo* digests. *FEBS Lett.*, **196**, 223–7.

Meisel, H. & Frister, H. (1989). Chemical characterization of bioactive peptides from *in vivo* digests of casein. *J. Dairy Res.*, **56**, 343–9.

Meisel, H. & Schlimme, E. (1990). Milk proteins: precursors of bioactive peptides. *Trends Food Sci. Technol.*, **1**, 41–3.

Meisel, H., Frister, H. & Schlimme, E. (1989). Biologically active peptides in milk proteins. *Z. Ernährungswiss.*, **28**, 267–78.

Metz, J., Zalusky, R. & Herbert, V. (1968). Folic acid binding by serum and milk. *Am. J. Clin. Nutr.*, **21**, 289–97.

Metz-Boutigue, M.-H., Jollcs, J., Mazurier, J., Schoentgen, F., Legrand, D., Spik, G.,

Montreuil, J. & Jolles, P. (1984). Human lactoferrin: amino acid sequence and structural comparisons with other transferrins. *Eur. J. Biochem.*, **145**, 659–76.
Migliore-Samour, D. & Jolles, P. (1988). Casein, a prohormone with an immunomodulating role for the newborn? *Experientia*, **44**, 188–93.
Migliore-Samour, D., Floc'h, F. & Jolles, P. (1989). Biologically active casein peptides implicated in immunomodulation. *J. Dairy Res.*, **56**, 357–62.
Miller, M.S., Benore-Parsons, M. & White, H.B. (1982). Dephosphorylation of chicken riboflavin-binding protein and phosvitin decreases their uptake by oocytes. *J. Biol. Chem.*, **257**, 6818–24.
Miller, M.S., Mas, M.T. & White, M.S. (1984). Highly phosphorylated region of chicken riboflabvin-binding protein: chemical characterization and ^{31}P NMR-studies. *Biochem.*, **23**, 569–76.
Mirand, P.P., Mosini, L., Levieux, D., Attaix, D., Bayle, G. & Bonnet, Y. (1990). Effect of colostrum on protein metabolism in small intestine of newborn lambs. *Biol. Neonate*, **57**, 30–6.
Modler, H.W., McKellar, R.C. & Yaguchi, M. (1990). Bifidobacteria and bifidogenic factors. *Can. Inst. Food Sci. Technol. J.*, **23**, 29–41.
Montreuil, J., Tonnelat, J. & Mullet, S. (1960). Preparation et propietes de la lactosiderophiline (lactotransferrine) due lait de femme. *Biochim. Biophys. Acta*, **45**, 413–21.
Montreuil, J., Mazurier, J., Legrand, D. & Spik, G. (1985). Human lactotransferrin; structure and function. In *Proteins of Iron Storage and Transport*, ed. G. Spik, J. Montreuil, R.R. Crichton & J. Mazurier. Elsevier Science Publishers, Amsterdam, pp. 25–38.
Moore, M.C., Greene, H.L., Said, H.M., Ghishan, F.K. & Orth, D.N. (1986). Effect of epidermal growth factor and artificial feeding in suckling rats. *Pediatr. Res.*, **20**, 1248–51.
Nagata, Y., Tanaka, K. & Sugano, M. (1981). Further studies on the hypocholesterolemic effect of soy-bean protein in rats. *Br. J. Nutr.*, **45**, 233–41.
Nichols, J.H. & Bezkorovainy, A. (1973). Isolation and characterization of a glycoprotein from human colostrum. *Biochem. J.*, **135**, 875–80.
Nichols, J.H., Bezkorovainy, A. & Landau, W. (1974). Human colostral whey M-1 glycoprotein and their *L. bifidus* var *penn.* growth promoting activity. *Life Sci.*, **14**, 967–76.
Nichols, J.H., Bezkorovainy, A. & Paque, R. (1975). Isolation and characterization of several glycoproteins from human colostrum whey. *Biochim. Biophys. Acta*, **412**, 99–108.
Noda, K., Umeda, M. & Ono, T. (1984). Transforming growth factor activity in human colostrum. *Gann*, **75**, 109–12.
Oftedal, O.T. (1980). Milk composition and formula selection for hand-rearing young mammals. In *Nutrition of Captive Wild Animals*, ed. E.R. Maschgan, M.E. Allen & L.E. Fisher. Proc. First Annual Dr. Scholl Nutrition Conference, pp. 67–83.
O'Loughlin, E.V., Chung, M., Hollenberg, M., Hayden, J., Zahavi, I. & Gall, D.G. (1985). Effect of epidermal growth factor on ontogeny of the gastrointestinal tract. *Am. J. Physiol.*, **249**, G674–8.
Park, M.-S. & Liepa, G.U. (1982). Effects of dietary protein and amino acids on the metabolism of cholesterol-carrying lipoproteins in rats. *J. Nutr.*, **112**, 1892–8.
Pedersen, T.G., Svendsen, I., Hansen, S.I., Holm, J. & Lyngbye, J. (1980). Aggregation of a folate-binding protein from cow's milk. *Carlsberg Res. Commun.*, **45**, 161–6.
Peters, T., Jr. (1985). Serum albumin. *Adv. Protein Chem.*, **37**, 161–245.
Petrides, P.E., Hosang, M., Shooter, E., Esch, F. & Bohlen, P. (1985). Isolation and characterization of epidermal growth factor from human milk. *FEBS Lett.*, **187**, 89–95.
Poch, M. & Bezkorovainy, A. (1988). Growth-enhancing supplements for various species of the genus *Bifidobacterium*. *J. Dairy Sci.*, **71**, 3214–21.
Prosser, C.G., Fleet, I.R. & Corps, A. (1989). Increased secretion of insulin-like growth factor I into milk of cows treated with recombinantly derived bovine growth hormone. *J. Dairy Res.*, **56**, 17–26.
Raaberg, L., Nexø, E., Poulsen, S.S. & Tollund, L. (1989). Cobalamin and its binding protein in rat milk. *Scand. J. Lab. Invest.*, **49**, 529–35.

Rasic, J.L. & Kurmann, J.A. (1983). *Bifidobacteria and their Role*. Birkhäuser Verlag, Basel.

Raynaud, M. & Bizzini, B. (1971). Purification et proprietés du facteur bifidus 2. *Ann. Nutr. Alim.*, **25**, A209–23.

Raynaud, M., Maté-Levy, J. & Rouillon, N. (1959). Le facteur bifidus 2. *Sem. Hosp. Paris*, **35**, 240–55.

Read, L.C., Francis, G.L., Wallace, J.C. & Ballard, F.J. (1983). Characterization of growth factors in milk. *Proc. Aust. Biochem. Soc.*, **15**, 38.

Read, L.C., Upton, F.J., Francis, G.L., Wallace, J.C., Dahlenberg, G.W. & Ballard, F.J. (1984). Growth promoting activity of human milk during lactation. *Pediatr. Res.*, **18**, 133–9.

Reiter, B. (1985*a*). Protective proteins in milk—Biological significance and exploitation. *Intern. Dairy Fed. Bull.*, **191**, 1–35.

Reiter, B. (1985*b*). The biological significance of the non-immunoglobulin protective proteins in milk: lysozyme, lactoferrin, lactoperoxidase. In *Developments in Dairy Chemistry—3*, ed. P.F. Fox. Elsevier Applied Science Publishers, London, pp. 281–336.

Rindernecht, E. & Humbel, R.E. (1976). The amino acid sequence of human insulin-like growth factor I and its structural homology with proinsulin. *J. Biol. Chem.*, **253**, 2769–76.

Rubinoff, M., Schreiber, C. & Waxman, S. (1977). The isolation and characterization of the folate binding protein from goat milk. *FEBS Lett.*, **75**, 244–8.

Sacks, F.M., Breslow, J.L., Wood, P.G. & Kass, E.H. (1983). Lack of an effect of dairy protein (casein) and soy protein on plasma cholesterol in strict vegetarians. An experiment and a critical review. *J. Lipid Res.*, **24**, 1012–20.

Said, H.M., Horne, D.W. & Wagner, C. (1986). Effect of human milk folate binding protein on folate intestinal transport. *Arch. Biochem. Biophys.*, **251**, 114–20.

Salter, D.N. & Blakeborough, P. (1988). Influence of goat's-milk folate-binding protein on transport of 5-methyltetrahydrofolate in neonatal-goat small intestinal brush-border membrane visicles. *Br. J. Nutr.*, **59**, 497–507.

Salter, D.N. & Mowhem, A. (1983). Neonatal role of milk folate-binding protein: studies on the course of digestion of goat's milk folate binder in the 6-d-old kid. *Br. J. Nutr.*, **50**, 589–96.

Salter, D.N., Ford, J.E., Scott, K.J. & Andrews, P. (1972). Isolation of the folate-binding protein from cows' milk by the use of affinity chromatography. *FEBS Lett.*, **20**, 302–6.

Salter, D.N., Scott, K.J., Slade, H. & Andrews, P. (1981). The preparation and properties of folate-binding protein from cow's milk. *Biochem. J.*, **193**, 469–76.

Sandberg, D.P., Begley, J.A. & Hall, C.A. (1981). The content, binding, and forms of Vitamin B_{12} in milk. *Am. J. Clin. Nutr.*, **34**, 1717–24.

Sato, R., Noguchi, T. & Naito, H. (1986). Casein phosphopeptide (CPP) enhances calcium absorption from the ligated segment of rat small intestine. *J. Nutr. Sci. Vitaminol.*, **32**, 67–76.

Schaub, J. (ed.) (1985). *Composition and Physiological Properties of Human Milk*. Elsevier Science Publishers (Biomedical Division), Amsterdam.

Scott, J., Urdea, M., Quiroga, M., Sanchez-Pescador, R., Fong, N., Selby, M., Rutter, W. & Bell, G.I. (1983). Structure of a mouse submaxillary gland mRNA coding for EGF and seven other related peptides. *Science*, **221**, 236–40.

Selhub, J. & Grossowicz, N. (1973). Chemical fixation of folate binding protein to activated Sepharose. *FEBS Lett.*, **35**, 76–8.

Shimazaki, K.-I. & Nishio, N. (1991). Interacting properties of bovine lactoferrin with immobilized Cibacron Blue F3GA in colum chromatography. *J. Dairy Sci.*, **74**, 404–8.

Shimizu, M. & Yamauchi, K. (1982). Isolation and characterization of mucin-like glycoprotein in human milk fat globule membrane. *J. Biochem.*, **91**, 515–24.

Shing, V.W. & Klagsbrun, M. (1984). Human and bovine milk contain different sets of growth factors. *Endocrinol.*, **115**, 273–82.

Singh, H., Flynn, A. & Fox, P.F. (1989). Zinc binding in bovine milk. *J. Dairy Res.*, **56**, 249–63.

Sinha, S.K. & Yunis, A.A. (1983). Isolation of a colony stimulating factor from human milk. *Biochem. Biophys. Res. Commun.*, **114**, 797–803.

Smith, D.F., Zopf, D.A. & Ginsburg, V. (1978). Fractionation of sialyl oligosaccharides of human milk by ion exchange chromatography. *Anal. Biochem.*, **85**, 602–8.

Spik, G., Montreuil, J., Crichton, R.R. & Mazurier, J. (ed.) (1985). *Proteins of Iron Storage and Transport*. Elsevier Science Publishers, Amsterdam.

Sugano, M., Ishiwaki, N., Nagata, Y. & Imaizumi, K. (1982). Effects of arginine and lysine addition to casein and soya-bean protein on serum lipids, apolipoproteins and glucagon in rats. *Br. J. Nutr.*, **48**, 211–21.

Sugano, M., Yamada, Y., Hashimoto, Y., Matsuo, T. & Kimoto, M. (1988). The hypocholesterolemic action of the undigested fraction of soybean protein in rats. *Atherosclerosis*, **72**, 115–22.

Sugano, M., Goto, S., Yamada, Y., Yoshida, K., Hashimoto, Y., Matsuo, T. & Kimoto, M. (1990). Cholesterol-lowering activity of various undigested fractions of soybean protein in rats. *J. Nutr.*, **120**, 977–85.

Suikkari, A.M. (1989). Insulin-like growth factor (IGF-1) and its low molecular weight binding protein in human milk. *Eur. J. Obstet. Gynecol. Reprod. Biol.*, **30**, 19–25.

Svendsen, I., Martin, B., Pedersen, T.G., Hansen, S.I., Holm, H.J. & Lyngbye, J. (1979). Isolation and characterization of the folate-binding protein from cows' milk. *Carlsberg Res. Commun.*, **44**, 89–99.

Svendsen, I., Hansen, S.I., Holm, J.. & Lyngbye, J. (1982). Amino acid sequence homology between human and bovine low molecular weight folate binding protein isolated from milk. *Carlsberg Res. Commun.*, **47**, 371–6.

Svendsen, I., Hansen, S.I., Holm, J. & Lyngbye, J. (1984). The complete amino acid sequence of the folate-binding protein from cows' milk. *Carlsberg Res. Commun.*, **49**, 123–31.

Swiatlo, N.L. & Picciano, M.F. (1988). Relative folate bioavailability from human, bovine and goat milk containing diets. *FASEB J.*, **2**, 4600 (Abstr.).

Taira, T., Hilakivi, L.A., Aalto, J. & Hilakivi, I. (1990). Effect of Beta-casomorphin on neonatal sleep in rats. *Peptides*, **11**, 1–4.

Tani, M., Fushiki, T. & Iwai, K. (1983). Influence of folate-binding protein from bovine milk on the absorption of folate in gastrointestinal tract of rat. *Biochim. Biophys. Acta*, **757**, 274–81.

Terpstra, A.H.M., Hermus, R.J.J. & West, C.E. (1983). The role of dietary protein in cholesterol metabolism. *Wld Rev. Nutr. Diet.*, **42**, 1–55.

Teschemacher, H. (1978). Endogenous ligands of opiate receptors (endorphins). In *Developments in Opiate Research*, ed. A. Herz. Marcel Dekker, New York, pp. 67–151.

Teschemacher, H. (1987). β-Casomorphins: do they have physiological significance? In *Human Lactation. 3. The Effects of Human Milk on the Recipient Infant*, ed. A.S. Goldman, S.A. Atkinson & L.A. Hanson. Plenum Press, New York, pp. 213–25.

Teschemacher, H., Umbach, M., Hamel, U., Prectorius, K., Ahnert-Hilger, G., Brantl, V., Lottspeich, F. & Henschen, A. (1986). No evidence for the presence of β-casomorphins in human plasma after ingestion of cows' milk or milk products. *J. Dairy Res.*, **53**, 135–8.

Thornburg, W., Matrisian, L., Magun, B. & Koldosisky, O. (1984). Gastrointestinal absorption of epidermal growth factor in suckling rats. *Am. J. Physiol.*, **246**, G80–5.

Thornburg, W., Rao, R.K., Matrisian, L.M., Magun, B.E. & Koldovsky, O. (1987). Effects of maturation on gastrointestinal absorption of epidermal growth factor in rats. *Am. J. Physiol.*, **253**, G68–71.

Tome, D., Dumontier, A.-M., Hautefeuille, M. & Desjeux, J.-F. (1987). Opiate activity and transepithelial passage of intact β-casomorphins in rabbit ileum. *Am. J. Physiol.*, **253**, G737–44.

Toyosaki, T. & Mineshita, T. (1988). Antioxidant effects of protein-bound riboflavin and free riboflavin. *J. Food Sci.*, **53**, 1851–3.

Toyosaki, T., Yamamoto, A. & Mineshita, T. (1987). Partial purification of an antioxidizing component in raw cow milk. *J. Food Sci.*, **52**, 88–90.

Trugo, N.M.F. (1988). Characterization of the Vitamin B_{12}-binding protein isolated from sow's milk and its affinity for cyanocobalamin and other corrinoids. *Brazilian J. Med. Biol. Res.*, **21**, 883–94.

Trugo, N.M.F. & Turvey, A. (1987). Immunofluorescent localization of Vitamin B^{12}-binding protein from sow's milk in the ileal mucosa of piglets. *Brazilian J. Med. Biol. Res.*, **20**, 285–8.

Trugo, N.M.F., Ford, J.E. & Salter, D.N. (1985). Vitamin B_{12} absorption in the neonatal piglet. 3. Influence of Vitamin B_{12}-binding protein from sows' milk on uptake of Vitamin B_{12} by microvillus membrane visicles prepared from small intestine of the piglet. *Br. J. Nutr.*, **54**, 269–83.

Tsuji, S., Hirata, Y. & Matsuko, K. (1989). Two molecular forms of bovine lactoferrin. *J. Dairy Sci.*, **72**, 1130–6.

van der Meer, R. (1983). Is the hypercholesterolaemic effect of dietary casein related to its phosphorylation state? *Atherosclerosis*, **49**, 339–41.

van der Meer, R. (1988). Species-dependent effects of dietary casein and calcium on cholesterol metabolism. *International Dairy Federation Bulletin*, **222**, 24–7.

van der Meer, R. & Beynen, A.C. (1987). Species-dependent responsiveness of serum cholestrol to dietary prteins. *J. Am. Oil Chem. Soc.*, **64**, 1172–7.

Vreeman, H.J., Visser, S., Slangen, C.J. and van Riel, J.A.M. (1986). Characterization of bovine κ-casein fractions and the kinetics of chymosin induced macropeptide release from carbohydrate-free and carbohydrate-containing fractions determined by high performance gel-permeation chromatography. *Biochem. J.*, **240**, 87–97.

Wagner, C. (1982). Cellular folate binding proteins; function and significance. *Ann. Rev. Nutr.*, **2**, 229–48.

Wang, C.S., Chan, W.Y. & Kloer, H.U. (1984). Comparative studies on the chemical and immulological properties of human milk, human pancreatic juice and bovine milk lactotransferrin. *Comp. Biochem. Physiol. B.*, **78**, 575–80.

Waxman, S. (1975). Folate binding proteins. *Br. J. Haematol.*, **29**, 23–9.

Waxman, S. & Schreiber, C. (1975*a*). The purification and characterization of the low molecular weight human folate binding protein using affinity chromatography. *Biochemistry*, **14**, 5422–8.

Waxman, S. & Schreiber, C. (1975*b*). The isolation of the folate binding protein from commercially purified bovine beta-lactoglobulin. *FEBS Lett.*, **55**, 128–30.

West, C.F., Beynen, A.C., Scholtz, K.E., Terpstra, A.H.M., Schutte, J.B., Deuring, K. & van Gils, L.G.M. (1984). Treatment of dietary casein with formaldehyde reduces its hypercholesterolaemic effect in rabbits. *J. Nutr.*, **14**, 17–25.

West, D.W. (1986). Structure and function of the phosphorylated residues of casein. *J. Dairy Res.*, **53**, 333–52.

West, D.W. & Dalgleish, D.G. (1976). A kinetic analysis of the dephosphorylation, by bovine spleen phosphoprotein phosphatase (EC 3.1.3.16) of a phosphopeptide derived from β-casein. *Biochim. Biophys. Acta*, **438**, 169–75.

White, H.B. & Merrill, A.H. (1988). Riboflavin-binding proteins. *Ann. Rev. Nutr.*, **8**, 279–99.

Widdowson, E.M. (1984). Milk and the newborn animal. *Proc. Nutr. Soc.*, **43**, 87–100.

Widdowson, E.M., Colombo, V.E. & Artavanis, C.A. (1976). Changes in the organs of pigs in response to feeding for the first 24 h after birth. II. The digestive tract. *Biol. Neonate*, **28**, 272–81.

Williams, J. (1985). The structure of transferrins. In *Proteins of Iron Storage and Transport*, ed. G. Spik, J. Montreuil, R.R. Crichton & J. Mazurier. Elsevier Science Publishers, Amsterdam, pp. 13–23.

Williams, J. & Sanger, F. (1959). The grouping of serine phosphate residues in phosvitin and casein. *Biochim. Biophys. Acta*, **33**, 294–6.

Wright, C.E. & Gaull, G.E. (1983). Nerve growth factor is present in human milk. *Pediatr. Res.*, **17**, 144A.

Yagi, H., Suzuki, S., Noji, T., Nagashima, K. & Kurouma, T. (1986). Epidermal growth factor in cows' milk and milk formulae. *Acta Paediatr. Scand.*, **75**, 233–5.

Yamauchi, K., Azuma, N., Kobayashi, H. & Kaminogawa, S. (1981). Isolation and properties of human κ-casein. *J. Biochem.*, **90**, 1005–12.

Yoshikawa, M., Tani, F., Yoshimura, T. & Chiba, H. (1986*a*). Opioid peptides from milk proteins. *Agric. Biol. Chem.*, **50**, 2419–21.

Yoshikawa, M., Tani, F., Ashikaga, T., Yoshimura, T. & Chiba, H. (1986*b*). Purification and chracterization of an opioid antagonist from a peptic digest of bovine κ-casein. *Agric. Biol. Chem.*, **50**, 2951.

Zioudrou, C., Streaty, R.A. & Klee, W.A. (1979). *J. Biol. Chem.*, **254**, 2446–9.

Zwiebel, J.A., Bano, M., Nexo, E., Salomon, D.S. & Kidwell, W.R. (1986). Partial purification of transforming growth factors from human milk. *Cancer Res.*, **46**, 933–9.

8

INDIGENOUS ENZYMES IN MILK

Editor's Comment

This chapter is divided into six more-or-less independent self-contained sections, covering the principal indigenous enzymes in milk.

I. General Introduction

A.T. ANDREWS

Biotechnology Group, Department of Food Science and Technology, University of Reading, Whiteknights, PO Box 226, Reading RG6 2AP, UK

II. Lipases

T. OLIVECRONA, S. VILARÓ & G. BENGTSSON-OLIVECRONA

Department of Medical Biochemistry and Biophysics, University of Umea, S-90187 Umea, Sweden

III. Proteinases

P.F. FOX

Department of Food Chemistry, University College, Cork, Republic of Ireland

IV. Phosphatases

A.T. ANDREWS

Biotechnology Group, Department of Food Science and Technology, University of Reading, Whiteknights, PO Box 226, Reading RG6 2AP, UK

V. Lactoperoxidase

L. BJORCK

Department of Food Science, Swedish University of Agricultural Sciences, Box 7051, S-750 07, Uppsala, Sweden

VI. Other Enzymes

N.Y. FARKYE

Products Technology Center, California Polytechnic State University, San Luis Obispo, California 93407, USA

I. General Introduction

Bovine milk contains approximately 32 g/litre of protein of which the four caseins (α_{s1}-, α_{s2}-, β- and κ-) represent 24–25g/litre and the major whey proteins, β-lactoglobulin and α-lactalbumin, are present at about 4 and 2 g/litre, respectively. By far the most important of the remaining 1–2 g of whey proteins per litre are serum albumin and the immunoglobulins which unlike the above are not unique to milk but are passively or actively transferred into it from blood. The proteose-peptone fraction, which is somewhat variable in amount but may be as much as 1–3 g/litre, is often regarded as part of the whey protein (non-casein) fraction, but in fact is composed mainly of peptides produced from caseins (Andrews & Alichanidis, 1983) through the action of proteinases, particularly plasmin, the principal indigenous proteinase in good quality milk (Kaminogawa *et al.*, 1972; de Rham & Andrews, 1982*a,b*; Andrews, 1983*a*). If we exclude all these proteins present in relatively large amounts, all the other proteins in milk represent much less than 1% of the total protein, and most of them are present at less than 1 mg/litre.

Over 60 different enzymes have been reported to be indigenous to milk (Got, 1971; Blanc, 1982). Only a relatively small number have been isolated and characterised in detail and doubts must remain about the truly indigenous nature of many of those that have been described only in terms of identifiable activities or where only limited studies of their properties have been reported. Many so-called milk enzymes may actually originate from contaminating microorganisms.

The first enzyme reported in bovine milk was lactoperoxidase (Arnold, 1881) and although milk is not a particularly rich source of enzymes, the number reported to be indigenous to milk slowly increased to 19 by 1966 (Shahani, 1966). The advent of readily-available low molecular weight radioactive and chromogenic substrates greatly improved the sensitivity of detection and facilitated kinetic studies so that by 1971 this number had increased to 55–60 (Got, 1971). Since that time there has been little further increase, probably because all the enzymes present in milk at reasonably high levels have already been detected but also because merely detecting the presence of trace amounts of hitherto undetected enzymes is no longer considered to be an adequate research objective per se.

Thus, rather than hunting for new enzymes, more recent studies on milk enzymes are concerned with their origin, their biological or physiological significance as indicators of performance or state of health, or with their relevance in food processing, nutrition, and product quality or stability. Such considerations preclude most of the enzymes reported in milk from being of any real practical importance, often because enzyme levels are too low or because suitable substrates are either not present or are unimportant to milk quality.

1.1. Occurrence and Distribution of Indigenous Milk Enzymes

Bovine milk has been far more intensively studied than that of other species but all milks contain enzymes and it is now known that although there may be marked quantitative differences, the complement of enzymes, and their properties, in the milks of all mammals are very similar. This is not surprising when it is remembered that the enzymes in all milks originate via almost identical pathways, i.e. either from mammary tissue cells, blood plasma or leucocytes. Interspecies differentiation between mammals is a recent evolutionary event so genetic differences in their complements of proteins and enzymes are largely confined to relatively conservative amino acid substitutions, deletions or insertions.

The distribution of enzymes in milk is often not homogeneous so that while catalase, lactoperoxidase, ribonuclease, γ-glutamyltransferase, *N*-acetyl-β-D-glucosaminidase, α-mannosidase, etc., are found principally in the whey (serum) phase, others are associated with the casein micelles (e.g. proteinases, lipases) or with the cream phase (xanthine oxidase). Also, the distribution of enzymes is often variable and difficult to reconcile with their origin. For example, plasmin is of blood serum origin but rather than being found in the soluble (whey) fraction it is associated principally with the casein micelles. Lysosomal enzymes, such as acid phosphatase and catalase, are found largely in the milk serum rather than in any particulate fraction, although up to 25% of the catalase activity has been reported to be in the cream phase (Kitchen *et al.*, 1970). Many enzymes (e.g. catalase) are found predominantly in one of the phases of milk, but are not confined exclusively to it. This may result from a post-secretory redistribution of the enzyme, but it often reflects differences in the behaviour of isoenzymes or even of totally distinct enzymes having similar activity but from different sources. For example, proteinases can originate from both blood serum and leucocytes, although the properties of the two groups of enzymes are very different. Catalases are present in both cytoplasm and lysosomal particles and also in bacteria. Various milk processing operations, such as homogenisation, heating, churning, etc., can also bring about some redistribution of enzymes between the different phases in milk.

1.2. Factors that Influence Enzyme Activities in Milk

Neither the enzyme complement in milk nor the enzyme distribution are constant but are influenced by several factors which adds further complications. These factors include not only the type of animal from which the milk originates but also its breed, age, lactation number, diet and nutritional status, stage of lactation (especially very early or very late) and health, particularly with regard to mammary gland infections (mastitis). Qualitatively, most milks have a rather similar complement of enzymes, but there can be considerable quantitative differences. For example, the level of lysozyme in human milk (400 μg/ml) is over 1000-fold higher than in bovine, ovine or caprine milks (0·1–0·25 μg/ml), whereas bovine milk contains more ribonuclease and lipase than the others (Chandan *et al.*, 1968). Caution must be exercised when considering this sort of comparative data, however, because in some reports only a small number of animals were considered and it is important that all other factors that may influence enzyme levels are comparable.

Non-enzyme proteins may show considerable interspecies variation (e.g. human milk contains no β-lactoglobulin or α_{s1}-casein which are major proteins in bovine, buffalo, caprine or ovine milks) so it is reasonable that even when all influencing factors are taken into account there will still be considerable interspecies differences in enzyme levels. Interbreed differences within a single species are very much smaller, but may still be significant. For example, the milk of Jersey cows contains higher levels of superoxide dismutase (1·27 U/ml) than that of Holstein milk (0·92 U/ml) and xanthine oxidase levels are also slightly higher (0·078 and 0·065 U/ml, respectively) (Holbrook & Hicks, 1978).

Dietary factors, generally, have only minor effects on milk enzyme profiles and levels, although there are reports that in undernourished women, supplementing the diet with protein or fat increases levels of lipase, esterase, xanthine oxidase and acid and alkaline phosphatases in human milk (Karmarkar & Ramakrishnan, 1959; Belavady, 1960, 1962).

Lactational effects on milk enzyme levels are far greater than breed or nutritional effects. Colostrum is very different in composition from mature mid-term milk in almost all its major constituents, and consequently also differs markedly with respect to enzymes, both in total amounts and in relative proportions. The activities of many enzymes, including α-amylase, catalase, lactoperoxidase and acid phosphatase, are higher in colostrum than in mature milk; only lipase levels are lower (Wuthrich *et al.*, 1963). The extent and time span of these differences are very variable, however. For example, β-glucuronidase levels are more than 10-fold higher than normal at parturition but decrease to normal values after 3 days (Kiermeier & Gull, 1968) while lactoperoxidase increases to a maximum after 3–5 days before gradually decreasing again (Kiermeier & Kuhlmann, 1972). Acid and alkaline phosphatases are both elevated in colostrum, but decline to normal values at very different rates; 10 days for alkaline phosphatase but 40 days for acid phosphatase (Kiermeier & Meinl, 1968), which may reflect the different origins of the enzymes. Alkaline phosphatase probably originates in the plasmalemma of mammary gland cells whereas acid phosphatase is a lysosomal enzyme. Alternatively, it may reflect variations in isoenzyme behaviour, with some isoenzymes returning to normal faster than others.

Once the enzyme composition has reached that typical of mature milk there is comparatively little further change until very late in lactation when the somatic cell content of milk gradually increases, so that those enzymes associated with somatic cells also increase slightly (e.g. acid phosphatase, acid proteinase, *N*-acetyl-β-D-glucosaminidase, catalase, etc). The permeability of mammary gland cell walls also increases slightly in late lactation and consequently the content of soluble components of blood serum origin increases, including that of enzymes such as plasmin. In some respects, the enzyme profile of very late lactation milk approaches that of colostrum. Generally, cell wall permeability and the concentrations in milk of both soluble and cellular constituents of blood origin, including the levels of enzymes from those sources, increase with lactation number (and hence usually age). This is shown by the fact that the somatic cell count of mid-lactation milk from a healthy cow in its first lactation is approximately 2–4 $\times$ 10^4 cells/ml but by the seventh lactation, 2–4 $\times$ 10^5 cells/ml is more typical.

Apart from the marked lactational influences discussed above, the most striking changes in milk composition within a single species or animal are caused by disease. The occurrence of mastitis is accompanied by very pronounced changes in the concentrations of nearly all milk components (for reviews see Haenlein *et al.*, 1973; Newstead, 1973; Newbold, 1974; Randolph *et al.*, 1974; Kitchen, 1981). Mastitis is accompanied by a marked increase in the permeability of mammary gland cell walls so that the concentration of virtually all blood constituents increases in the milk. The whey protein fraction of milk increases sharply due to the influx of blood proteins, especially serum albumin and the immunoglobulins, while there is little change in the concentrations of α-lactalbumin and β-lactoglobulin (Anderson & Andrews, 1977). High levels of plasminogen, the inactive precursor of plasmin, also enter the milk and this is rapidly converted to the active enzyme (de Rham & Andrews, 1982; Schaar & Funke, 1986). Other proteinases, including thrombin and cellular proteinases, also increase greatly so proteolysis can be very rapid and extensive in severely mastitic milk, which obviously leads to further changes in milk protein composition (Barry & Donnelly, 1981; de Rham & Andrews, 1982*a*, *b*; Andrews, 1983*b*). The influence of mastitis on milk composition of course varies with the severity and stage of the disease, being much greater during severe and clinical infections than in relatively mild or subclinical cases.

1.3. Benefits of Indigenous Milk Enzyme Activities

Changes in the levels, or the appearance in milk, of enzymes of blood origin can be used in the diagnosis of mastitis. Catalase, arylesterase and various glycosidase activities have all been suggested for this purpose but the most successful so far is *N*-acetyl-β-D-glucosaminidase activity (Kitchen *et al.*, 1978, 1980). There have been reports of the possible beneficial roles of milk enzymes in the development of desirable gut flora and in enhancing immune defence mechanisms in the newborn (both human and other animals). Interest in these areas appears to have waned recently, perhaps because they are now considered to be of minor importance in this context. Meanwhile, studies in cell biology and in biosynthetic pathways within the mammary gland, including the synthesis, transport and secretion of enzymes, have gained in popularity. The most frequent rationale for investigating milk enzymes, however, still includes an appraisal of their technological significance. Flavour, texture or stability in dairy products are generally the prime considerations but the use of enzymes as bacteriocidal agents or as indices of thermal treatments (Andrews *et al.*, 1987) are also of technological interest. Table 1 summarises the most important enzymes from these viewpoints.

The enzymes in bovine milk have been studied intensively and the literature reviewed at regular intervals (Whitney, 1958; Shahani, 1966; Kitchen *et al.*, 1970; Got, 1971; Groves, 1971; Jodar, 1972; Dwivedi, 1973; Shahani *et al.*, 1973; Teadlecova & Hanus, 1973; Johnson, 1974; Fox & Morrissey, 1981; Kitchen, 1985). Some of the enzymes in human milk have also been thoroughly investigated (see Shahani *et al.* (1980), Hamosh *et al.* (1985) and Hernell & Lonnerdal (1989) for reviews).

TABLE 1
The Most Important Enzyme Activities in Bovine Milk

Enzyme	*EC Number*	*Action*
Proteinase (plasmin, thrombin, cathespin D)	3.4.21.7	Stability and textural defects in milk and dairy products Bitter peptide formation
Lipase	3.1.1.3	Off-flavour production and rancidity
Alkaline phosphatase	3.1.3.1	Index of pasteurisation
Acid phosphatase	3.1.3.2	
Lactose synthetase	2.4.1.22	Lactose biosynthesis
γ-Glutamyltranspeptidase	2.3.1.1	Amino-acid uptake in protein synthesis
Superoxide dimutase	1.15.1.1	Antioxidant Free radical inhibitor
Xanthine oxidase	1.2.3.2	Pro-oxidation
Catalase	1.11.1.6	Pro-oxidation Mastitis diagnosis
Lactoperoxidase	1.11.1.7	Antibacterial agent Pro-oxidation
N-Acetylglucosaminidase	3.2.1.30	Mastitis diagnosis
Lysozyme	3.2.1.17	Antibacterial agent
Ribonuclease	3.1.27.5	
Leucine aminopeptidase	3.4.11.1	
Rhodanase	2.8.1.1	
Sulphydryl oxidase	1.8.3	Reduction of cooked flavours

REFERENCES

Anderson, M. & Andrews, A.T. (1977). Progressive changes in individual milk protein concentrations associated with high somatic cell counts. *J. Dairy Res.*, **44**, 223–35.

Andrews, A.T. (1983*a*). Proteinases in normal bovine milk and their action on caseins. *J. Dairy Res.*, **50**, 45–55.

Andrews, A.T. (1983*b*). Breakdown of caseins by proteinases in bovine milks with high somatic cell counts arising from mastitis or infusion with bacterial endotoxin. *J. Dairy Res.*, **50**, 57–66.

Andrews, A.T. & Alichanidis, E. (1983). Proteolysis of caseins and the proteosexpeptone fraction of bovine milk. *J. Dairy Res.*, **50**, 275–90.

Andrews, A.T., Anderson, M. & Goodenough, P.W. (1987). A study of the heat stabilities of a number of indigenous milk enzymes. *J. Dairy Res.*, **54**, 237–46.

Arnold, C. (1881). Einige neue Reactionen der Milch. *Arch. Pharm.*, **219**, 41–62.

Barry, J.G. & Donnelly, W.J. (1981). Casein compositional studies. II. The effect of secretory disturbance on casein composition in freshly drawn and aged bovine milks. *J. Dairy Res.*, **48**, 437–46.

Belavady, B. (1960). Xanthine oxidase and alkaline phosphatase activity in the breast milk of poor Indian women. *Indian J. Med. Res.*, **48**, 654–60.

Belavady, B. (1962). Effect of protein supplementation to nursing mothers on some milk constituents: casein and xanthine oxidase. *Indian J. Med. Res.*, **50**, 100–3.

Blanc, D. (1982). Les protèines du lait à activitè enzymatique et hormonale. *Le Lait*, **62**, 350–95.

Chandan, R.C., Parry, R.M. Jr. & Shahani, K.M. (1968). Lysozyme, lipase and ribonuclease in milk of various species. *J. Dairy Sci.*, **51**, 606–7.

de Rham, O. & Andrews, A.T. (1982*a*). The roles of native milk proteinase and its zymogen during proteolysis in normal bovine milk. *J. Dairy Res.*, **49**, 577–85.

de Rham, O. & Andrews, A.T. (1982*b*). Qualitative and quantitative determination of proteolysis in mastitic milk. *J. Dairy Res.*, **49**, 587–96.

Dwivedi, B.K. (1973). The role of enzymes in food flavours: Part 1, dairy products. *CRC Crit. Rev. Food Technol.*, **3**, 457–78.

Fox, P.F. & Morrissey, P.A. (1981). Indigenous enzymes of bovine milk. In *Enzymes and Food Processing*, ed. G.C. Birch, N. Blakeborough & K.J. Parker. Applied Science Publishers, London, pp. 213–38.

Got, R. (1971). Les enzymes de laits. *Ann. Nutr. Alim.*, **25**, A291–A311.

Groves, M.L. (1971). Minor milk proteins and enzymes. In *Milk Proteins: Chemistry and Molecular Biology*, Vol. II, ed. H.A. McKenzie. Academic Press, New York, pp. 367–418.

Haenlein, G.F.W., Schultz, L.H. & Zikakis, J.P. (1973). Composition of proteins in milk with varying leucocyte contents. *J. Dairy Sci.*, **56**, 1017–24.

Hamosh, M., Freed, L.M., Jones, J.B., Berkow, S.E., Bitman, J., Mehta, N.R., Happ, B. & Hamosh, P. (1985). In *Human Lactation: Milk Components and Methodologies*, ed. R.G. Jensen & M.C. Neville. Plenum Press, New York, pp. 251–66.

Hernell, O. & Lonnerdal, B. (1989). Enzymes in human milk. *Protein and Non-Protein Nitrogen in Human Milk*, ed. S.A. Atkinson & B. Lonnerdal. CRC Press, Boca Raton, FL, pp. 67–75.

Holbrook, J. & Hicks, C.L. (1978). Variation of superoxide dismutase in bovine milk. *J. Dairy Sci.*, **61**, 1072–77.

Jodar, L.G. (1972). Principales enzimes de la leche: Importancia y detección. *Rev. Esp. Lech.*, **83**, 3–18.

Johnson, H.A. (1974). The composition of milk. In *Fundamentals of Dairy Chemistry*, 2nd edn, ed. B.H.Webb, A.H. Johnson & J.A. Alford. AVI Publishing Co, Wesport, CT, pp. 1–57.

Kaminogawa, S., Mizobuchi, H. & Yamauchi, K. (1972). Comparison of bovine milk protease with plasmin. *Agr. Biol. Chem.*, **36**, 2163–7.

Karmarker, M.G. & Ramakrishnan, C.V. (1959). Relation between dietary fat, fat content of milk and concentration of certain enzymes in human milk. *J. Nutr.*, **69**, 274–6.

Kiermeier, F. & Gull, K. (1968). Biologische Einflüsse auf die Pb-glucuronidase-akticität de milch. *Z. Lebensm. Unters. Forsch.*, 205–8.

Kiermeier, F. & Meinl, E. (1968). Zur kenntnis de milchphosphatasen. II. Verhalten der sauren phosphatase neber der alkalishen in kuhmilch. *Z. Lebensm. Unters. Forsch.*, **114**, 189–202.

Kiermeier, F. & Kuhlmann, H. (1972). Laktoperoxidase-aktivität in human- and kuhmilch. *Munchner Medizinische Woehenschrift*, **114**(49), 2144–6 (cited in *Dairy Science Abst.*, **35** (1973), 5338.

Kitchen, B.J. (1981). Bovine mastitis: milk compositional changes and related diagnostic tests. *J. Dairy Res.*, **48**, 167–88.

Kitchen, B.J. (1985). Indigenous milk enzymes. In *Developments in Dairy Chemistry—3*, ed. P.F. Fox. Elsevier Applied Science Publishers, London, pp. 239–79.

Kitchen, B.J., Taylor, G.C. & White, I.C. (1970). Milk enzymes—their distribution and activity. *J. Dairy Res.*, **37**, 279–88.

Kitchen, B.J., Middleton, G. & Salmon, M.C. (1978). Bovine milk *N*-acetyl-β-D-glucosaminidase and its significance in the detection of abnormal udder secretions. *J. Dairy Res.*, **45**, 15–20.

Kitchen, B.J., Middleton, G., Durward, I.G., Andrews, R.J. & Salmon, M.C. (1980). Mastitis diagnostic tests to estimate mammary gland epithelial cell damage. *J. Dairy Res.*, **63**, 978–83.

Newbold, F.S.H. (1974). Microbial diseases of the mammary gland. In *Lactation*, ed. B.L. Larson & V.R. Smith. vol. 2, Academic Press, New York, pp. 269–88.

Newstead, D.F. (1973). Effects of mastitis on milk composition and properties. *N.Z.J. Dairy Sci. Technol.*, **8**, 52–5.
Randolph, H.E., Erwin, R.E. & Richter, R.L. (1974). Influence of mastitis on properties of milk. VII. Distribution of milk proteins. *J. Dairy Sci.*, **57**, 15–18.
Schaar, J. & Funke, H. (1986). Effect of subclinical mastitis on milk plasminogen and plasmin compared with that on sodium, antitrypsin and *N*-acetyl-*β*-D-glucosaminidase. *J. Dairy Res.*, **53**, 515–28.
Shahani, K.M. (1966). Milk enzymes: their role and significance. *J. Dairy Sci.*, **49**, 907–20.
Shahani, K.M., Harper, W.J., Jensen, R.G., Parry, R.M., Jr & Zittle, C.A. (1973). Enzymes in bovine milk: a review. *J. Dairy Sci.*, **56**, 531–43.
Shahani, K.M., Kwan, A.J. & Friend, B.A. (1980). Role and significance of enzymes in human milk. *Am. J. Clin. Nutr.*, **33**, 1861–8.
Teadlecova, M. & Hanus, J. (1973). Das corkommen von enzymen in kuhmilch. *Die Nahrung*, **17**, 565–77.
Whitney, R.McL. (1958). The minor proteins of bovine milk. *J. Dairy Sci.*, **41**, 1303–23.
Wuthrich, S., Richterich, R. & Hostettler, H. (1963). Untersuchungen über milchenzyme: I. Enzyme in kuhmilch und frauenmilch. *Z. Lebensm. Unters. Forsch.*, **124**, 336–48.

II. Lipases in Milk

Two endogenous lipases have been found in milk (Olivecrona & Bengtsson, 1983). They are present in quite different amounts in different species. One of the enzymes, the bile salt-stimulated lipase, accounts for about 1% of the protein in human milk and has a role in digestion of milk lipids in the offspring. This enzyme is, however, not present in milk from the domestic milk-producing animals. Since milk from these species is the focus for the present volume, we will not review the interesting biochemical and physiological properties of the bile-salt stimulated lipase here. The reader is referred to recent papers on its structure (Nilsson *et al.*, 1990) and reviews on its importance for growth and development of human infants (Hernell & Bläckberg, 1991).

The other lipase (LPL), lipoprotein lipase, is present in milk from all mammals, albeit in different amounts (Hamosh & Scow 1970; Wallinder, *et al.*, 1982; Olivecrona & Bengtsson, 1983; Chilliard *et al.*, 1984; Chilliard & Doreau, 1985; Olivecrona & Bengtsson-Olivecrona, 1991). It has an important role for milk production in the mammary gland (Scow *et al.*, 1987), but thus far, no demonstrated role for milk utilisation in the offspring. It is responsible for development of lipolytic rancidity in bovine milk, and thus has great importance in dairy science. It is this enzyme that we will discuss here.

Evolutionary Aspects

LPL is a member of a family of lipases which catalyze steps in the digestion and transport of triglycerides (Datta *et al.*, 1988; Kirchgessner *et al.*, 1989; Persson *et al.*, 1989). Kirchgessner *et al.* have deduced the evolution outlined in Fig. 1 from an analysis of the intron/exon organisation of the genes. The human LPL gene has 10 exons and spans about 30 kb. The gene for hepatic lipase is similar in structure (Ameis *et al.*, 1990), wheareas the gene for pancreatic lipase has 13 exons and 12

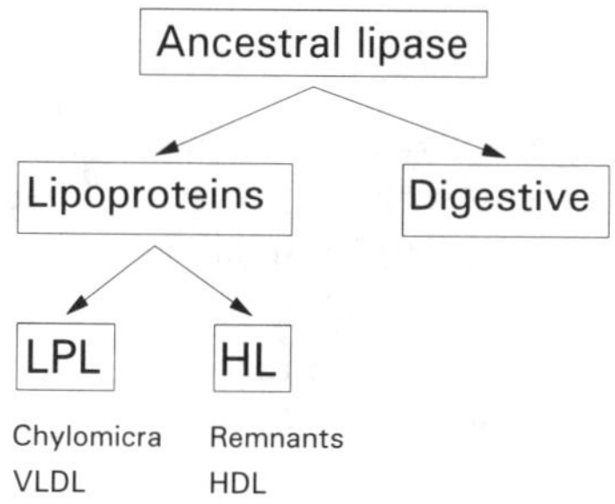

Fig. 1. Evolutionary relationships within the lipase gene family.

introns (Mickel *et al.*, 1989), of which four introns have no counterparts in the genes for lipoprotein and hepatic lipases. The hypothesis states that the gene for ancestral lipase was duplicated. This allowed specialisation into a digestive lipase, now recognised as pancreatic lipase, and a lipoprotein-metabolising lipase. A second gene duplication allowed this latter enzyme to specialise further into LPL and hepatic lipase. These two enzymes catalyze overlapping, but different steps in metabolism of triglyceride-rich lipoproteins (Olivecrona & Bengtsson-Olivecrona, 1990). LPL is specialised for rapid hydrolysis of lipoprotein triglycerides in peripheral tissues. This accomplishes two things. It enables the tissues to take up and utilise fatty acids from the lipoproteins for energy production, synthesis of tissue components or storage (Eckel, 1989) and it transforms the large triglyceride-rich lipoproteins (chylomicrons and very low density lipoproteins, VLDL) into cholesterol-rich remnant lipoproteins (Eisenberg, 1990). This process is relatively rapid and is completed within minutes to a few hours after the lipoprotein has entered circulation. Some of the remnants are rapidly removed from the circulation, but some are transformed into low density lipoproteins (LDL) and high density lipoproteins (HDL). These remnant-derived particles are the main lipoproteins in fasting plasma. Hepatic lipase is involved in remodelling of the remnant lipoproteins (Olivecrona & Bengtsson-Olivecrona, 1990). It is not present in milk.

The lipase gene family has a fourth member, the vitellogenins (Baker, 1988; Kirchgessner *et al.*, 1989). These yolk proteins have a central region of about 100 amino acid residues which show extensive sequence similarities to a region around the active site in the lipases. The vitellogenins lack the active site serine, and have no catalytic activity. They function as lipid transport molecules and are recognised by surface receptors on target cells. Interestingly, recent data indicate that LPL may also function as a receptor ligand, independent of its catalytic activity (Beisiegel *et al.*, 1991). Hence, the evolutionary history may have to be reconsidered. Was the original molecule a receptor ligand that transported lipids, which later extended its repertoire with a hydrolytic function, or was it a lipase which acquired the ability to bind to cell surfaces?

Function in Lipoprotein Metabolism

LPL has been found in high activity in all mammals, birds and fishes studied (Cryer, 1987). This is in keeping with its role in delivery of triglycerides to peripheral tissues, which is important for animals whose diet occasionally contains lipids. Hepatic lipase, on the other hand, is present in widely varying amounts in different animals. Humans and rats have high activity (Olivecrona & Bengtsson-Olivecrona, 1990), while dogs (Muller *et al.*, 1985) and rabbits (Clay *et al.*, 1989) have relatively low activity. Calves have very little or perhaps no hepatic lipase (Etienne *et al.*, 1981). These differences probably relate to the large variation in how different species handle remnant lipoproteins, LDL and HDL.

LPL has a directive effect on the metabolism of plasma triglycerides (Scow *et al.*, 1977). The activity is acutely and profoundly regulated by changes in energy metabolism (Eckel, 1989; Bensadoun, 1991). For instance, the activity is high in adipose tissue in well-fed animals, but drops to low levels during fasting (Eckel, 1987). In the mammary gland, LPL activity is low before and during pregnancy, increases shortly before parturition and remains high throughout lactation (Scow *et al.*, 1987). This is the mechanism by which blood triglycerides are diverted to the mammary gland for synthesis of milk lipids.

Structure

The catalytically active form of (bovine) LPL is a noncovalent homodimer (Iverius & Östlund-Lindqvuist, 1976; Olivecrona *et al.*, 1985; Osborne *et al.*, 1985) of a subunit with 450 amino acid residues, five disulphide bridges and two oligosaccharide chains (Yang *et al.*, 1989). The primary structure of the monomer is shown in Fig. 2. Physical studies indicate that the dimer is in rapid reversible equilibrium with monomers, but these are prone to undergo an essentially irreversable change of conformation with loss of catalytic activity. This has been alluded to as a 'built in mechanism to self-destruct'. Other related lipases, e.g. pancreatic lipase, are apparently stable also in a monomeric state (Chapus *et al.*, 1988). This suggests that the dimeric state of LPL may not be necessary for catalytic activity as such, but maintains the subunits in their active conformation. On the other hand, there are well-documented examples of lipases which are monomeric in solution, but attain a dimeric state at lipid-water interfaces (Hazlett *et al.*, 1985).

Recently, the 3-dimensional structure for human pancreatic lipase was resolved by X-ray crystallography (Winkler *et al.*, 1990). This molecule has two folding units, a major N-terminal domain, and a smaller C-terminal domain (Fig. 3). Inspection of sequences for the other members of the family reveals a region with extensive homology, corresponding to residues 125–230 in pancreatic lipase (Komaromy & Schotz, 1987; Datta *et al.*, 1988; Persson *et al.*, 1991). This segment, the central homology region, is located in the center of the major folding unit in pancreatic lipase and encompasses the active site, with its triad of residues, Ser-152, Asp-176 and His-263. These residues are conserved in LPL and hepatic lipase.

```
bLPL                        DRITGGKDFRDIESKFAIRTPEDTAEDTCHLIPGVTESVA   40
hHL               GQSLKPEPFGRRAQAVETNKTLHEMKTRFLL-FGE-TNQ-GCQIRINHPDTLQ   50
pPL   SEVCFPRLGCFSDDAPWAGIVQRPLKILPPDKDVDTRFLLYT--NQNQNNYQELVADPSTIT   60

bLPL  NCHFNHSSKTFVVIHGWTVTGMYESWVPKLVAALYKREPDS-NVIVVDWLSRAQQHYPVSAG  101
hHL   ECGFNSSLPLVMIIHGWSVDGVLENWIWQMVAALKSQPAQPVNVGLVDWITLAHDHYTIAVR  112
pPL   NSNFRMDRKTRFIIHGFIDKG-EEDWLSNICKNLFKVE--SVNCICVDWKGGSRTGYTQASQ  119

bLPL  YTKLVGQDVAKFMNWMADEFNYPLGNVHLLGYSLGAHAAGIAGSLTNK--KVNRITGLDPAG  161
hHL   NTRLVGKEVAALLRWLEESVQLSRSHVHLIGYSLGAHVSGFAGSSIGGTHKIGRITGLDAAG  174
pPL   NIRIVGAEVAYFVEVLKSSLGYSPSNVHVIGHSLGSHAAGEAGRRTNG--TIERITGLDPAE  179

bLPL  PNFEYAEAPSRLSPDDADFVDVLHTFTRGS-PGRSIGIQKPVGHVDIYPNGGTFQPGCNIGE  222
hHL   PLFEGSAPSNRLSPDDASFVDAIHTFTREH-MGLSVGIKQPIGHYDFYPNGGSFQPGCHFLE  235
pPL   PCFQGTPELVRLDPSDAKFVDVIHTDAAPIIPNLGFGMSQTVGHLDFFPNGGKQMPGCQKNI  241

bLPL  ALR-VIAERGLGDVDQLVKCSHERSVHLFIDSLLNEENPSKAYRCNSKEAFEKGLCLSCRKN  283
hHL   LYR-HIAQHGFNAITQTIKCSHERSVHLFIDSLLHAGTQSMAYPCGDMNSFSQGLCLSCKKG  296
pPL   LSQIVDIDGIWEGTRDFVACNHLRSYKYYADSILNPDGFA-GFPCDSYNVFTANKCFPCPSE  302

bLPL  RCNNMGYEINKVRAKR--SSKMYLKTRSQMPYKVFHYQVKIHFSGTESNTYTNQAFEISLYG  343
hHL   RCNTLGYHVRQEPGSK--SKRLFLVTRAQSPFKVYHYQLKIQFI-NQTETPIQTTFTMSLLG  355
pPL   GCPQMGHYADRFPGKTNGVSQVFYLNTGDASNFARWRYKVSVTLSGKKVTGHILVSL---FG  361

bLPL  TVAESENIPFTLPE-VS-TNKTYSFLLYTEVDIGELLMLKLKW-IS----------DSYFSW  392
hHL   TKEKMQKIPITLGKGIA-SNKTYSFLITLDVDIGELIMIKFKWENSAVWANVWDTVQTIIPW  416
pPL   NEGNSR--QYEIYKGTLQPDNTHSDEFDSDVEVGDLQKVKFIWYNNNV--------INPTL--  412

bLPL  SNWWSSPGFDIGKIRVKAGETQKKVIFCSREK--MSYLQKGKSPVIFVKCHDKSLNRKSG    450
hHL   STGPRHSGLVLRTIRVKAGETQQRMTFCSENT--DDLLLRPTQEKIFVKCEIKSKTSKRKIR  476
pPL   ------PRVGASKITVERNDGK-VYDFCSQETVREEVLLTLNP------C              449
```

Fig. 2. Sequence homologies in the lipase family. This figure shows an alignment of the sequences of bovine lipoprotein lipase (bLPL, Yang, *et al*., 1989), human hepatic lipase (hHL, Datta *et al*., 1988) and porcine pancreatic lipase, (deCaro *et al*., 1981). There are disulfide bridges in bLPL which corespond to disulfide bridges between residues 237–261 and 433–449 in pPL. The cysteins are in disulfide bridges 285–296, 299–304 in pPL, but these appear to have changed partners in bLPL to give bridges corresponding to 285–304 and 296–299 (Yang *et al*., 1989). Cysteins are present in these positions also in hHL but the arrangement has not been determined. Cysteins forming a disulfide bridge 29–42 in bLPL are also present in hHL, but not in pPL, which, on the other hand, has two unique disulfides.

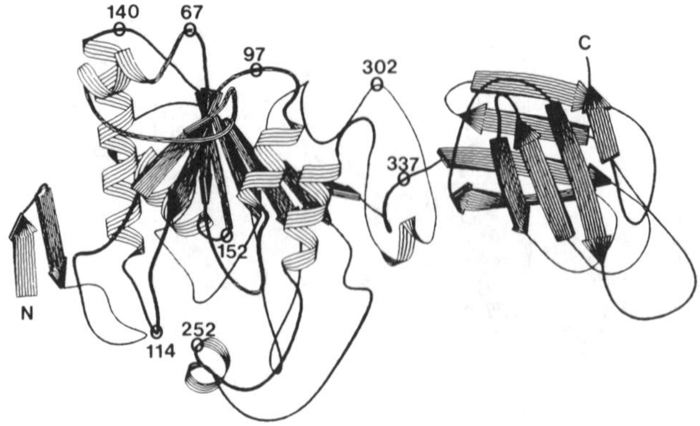

Fig. 3. Three-dimensional structure of pancreatic lipase. Courtesy of F. Winkler.

Site-directed mutagenesis of the serine residue (Ser-132 in human LPL) has confirmed that it is required for catalytic activity (Faustinella *et al.*, 1991).

It seems highly likely that lipoprotein and hepatic lipases are folded similarly to pancreatic lipase and have similar active sites (Derewenda & Derewenda, 1991; Persson *et al.*, 1991). Shortly after the central homology region in pancreatic lipase there is a surface loop with a short one-turn alpha helix (between disulphide-bridged Cys-237 and Cys-261). This loop, the so-called flap, covers the entrance to the active site. It is assumed that it moves away when the lipase adsorbs to a lipid-water interface. The presence of such a surface loop in LPL also is supported by the demonstration of a protease-sensitive site, corresponding to the tip of the putative flap (Persson *et al.*, 1991). This cleavage has profound effects on the kinetic properties of the lipase (Bengtsson & Olivecrona, 1981). There is not a single residue with identity across all lipases in the flap segment. In contrast, eight of the 14 residues immediately before the flap are identical in all three lipases, and 10 of the 14 residues after the flap are identical in hepatic lipase and LPL. Some of the functional differences between the three members of the family may relate to differences in the flap.

In the segment corresponding to the C-terminal folding unit of pancreatic lipase, residues 336–449, there are only 10 residue identities between pancreatic lipase and hepatic lipase or LPL. A further indication of divergence is that several gaps must be introduced to align the sequences, some of them fairly long (9 and 11 residues). There are more identities between hepatic lipase and LPL in this segment. One function of the C-terminal domain in pancreatic lipase is to interact with colipase (Abousalham *et al.*, 1991). It was recently reported that a chimeric lipase constructed from the N-terminal domain of LPL and the C-terminal domain of hepatic lipase was stimulated by apo CII in a similar way as LPL (Dichek *et al.*, 1992). This indicates that apo CII binds to the N-terminal domain in LPL, in this respect it is of interest that colipase and apo CII have evolved separately and from different ancestors (Persson *et al.*, 1989). Apo CII is clearly a member of the apolipoprotein gene family, whereas colipase bears no resemblance to this family. The segment between the two folding units in pancreatic lipase is cleaved by chymotrypsin (Abousalham *et al.*, 1991). In contrast, chymotrypsin cleaves LPL in the middle of the putative C-terminal domain (Bengtsson-Olivecrona *et al.*, 1986). Thus, several bits of evidence indicate that the C-terminal parts of the enzymes differ in both structure and function.

Eleven point mutations have been reported in the human LPL gene, excluding those caused by premature stop codons (Derewenda & Cambillau, 1991; Ma *et al.*, 1992). Two of the mutations have no effect whereas nine result in inactive LPL. Only two directly affect residues in the catalytic triad (Ma *et al.*, 1992). All but one of the mutants that cause loss of activity are located in the putative N-terminal domain, and most of them locate to presumably internal regions. From a theoretical analysis Derewenda and Cambillou conclude that most of the mutations probably cause loss of activity through unfolding rather than through direct effects on critical residues. This agrees with the physical evidence that the LPL molecule is labile and prone to undergo conformational changes with loss of catalytic activity (Olivecrona & Bengtsson-Olivecrona, 1987).

LPL is a glycoprotein. The bovine (Yang *et al.*, 1989) and the mouse (Masuno *et al.*, 1991) enzymes have two N-linked oligosaccharides, the guinea pig enzyme may have three oligosaccharides (Semb & Olivecrona, 1989). It has been found in several laboratories that LPL synthesised in tunicamycin-treated cells has no catalytic activity (Ailhaud, 1990; Masuno *et al.*, 1991). Tunicamycin inhibits the addition of oligosaccharide chains, and the results have therefore led to the view that the oligosaccharides are required for the formation of active LPL. The glycosylation sites in human LPL are Asn-43 and Asn-359 (Yang *et al.*, 1989). Substitution by mutagenesis of Asn-43 (but not of other asparagines) by Ala resulted in a catalytically inactive protein (Semenkovich *et al.*, 1990). Mice with the cld/cld mutation synthesise a form of LPL which has the correct amino acid sequence (Oka *et al.*, 1991), but incorrect oligosaccharides, and is catalytically inactive (Masuno *et al.*, 1990). The need for glycosylation is likely a conformational effect, without the correct oligosaccharides the subunits may not assemble into correctly folded dimers.

Catalytic Properties

As discussed above, the structures in and around the active site are probably similar in pancreatic lipase, LPL and hepatic lipase. All three enzymes hydrolyse the 1,3-positions in tri-, di- and monoglycerides and the 1-position in glycerophospholipids (Olivecrona & Bengtsson-Olivecrona, 1987). The enzymes do not hydrolyse cholesterol esters or sphingolipids at significant rates.

The action of lipases involves two separate steps (Verger & De Haas, 1976). First, the lipase adsorbs to the lipid/water interface. Then, the enzyme seeks out a single substrate molecule at the interface, aligns this molecule into the active site and hydrolyses it. Thermodynamic considerations and experimental evidence (Komaromy & Reed, 1991) suggests that the 'interface recognition site' is separate from the active site. The energy available from substrate-lipase interactions within the active site is used to split the substrate, not to bind the lipase to the lipid droplet. A recent study suggests that for LPL the preferred route for substrate entry into the active site is from below (i.e. from the core of the lipid droplet) rather than from the side (i.e. along the interface) (Rojas *et al.*, 1991). This is in accord with the picture of catalysis that emerges from the three dimensional structure of pancreatic lipase.

During lipolysis of milk lipids, short and medium chain fatty acids are preferentially released. In this respect, a recent study is of interest (Deckelbaum, 1990). They compared hydrolysis of phospholipid-stabilised emulsions of medium-chain versus long-chain triglycerides. Hydrolysis was at least 2-fold more rapid with the medium-chain triglycerides. NMR studies using ^{13}C-enriched triglycerides showed that the maximum solubility of trioctanoin in phospholipid bilayers is much higher than that for triolein, 11 mol% vs 3 mol%. The authors conclude that shorter chain triglycerides are more readily hydrolysed by LPL than long-chain triglycerides because of greater solubility and mobility at the emulsion-water interface. Hence, the apparent preference for shorter chain fatty acids is based on the physical properties of the substrate molecules rather than on specific properties of the active site. Nilsson and his colleages have noted that LPL has relatively low activity against ester bonds involving fatty acids with double bonds close to the carbonyl (Ekstrom *et al.*, 1989). This is probably due to steric hindrance in the active site.

Earlier reports stated that LPL shows a bell-shaped profile for activity versus pH, with optimum between 8 and 9. More recent studies indicate, however, that the activity continues to rise with pH to at least 10 (Rapp & Olivecrona, 1978; Bengtsson & Olivecrona, 1982; Bengtsson & Olivecrona, 1983). The implication is that the decrease in activity earlier observed above pH 9 was due to some factor other than the active site reaction. This was explored using milk fat globules as substrate (Bengtsson & Olivecrona, 1983). Binding of LPL to the globules was found to be a pH-dependent process with an optimum around pH 8, and the activity curve was a composite of the binding step and the pH effect on the active site reaction. This is one explanation for bell-shaped pH curves. Another is that LPL is often not stable under conditions used for assay, and the rate of inactivation increases with pH. If this is not kept under control one will underestimate the activity at high pH.

Activator Proteins

For optimal activity, LPL requires apo CII (Smith & Pownall, 1984). *In vivo*, apo CII is supplied as an integral part of the substrate lipoproteins, chylomicrons and VLDL, and its interaction with LPL presumably occurs with or after binding of the lipoprotein particle to the enzyme. There is no evidence that activation of the lipase by soluble apo CII would precede binding of a lipoprotein particle or that apo CII would remain with the enzyme when the lipoprotein detaches after a round of lipolysis. As the core of the lipoprotein is depleted of triglyceride, excess surface material is shed from the particle and taken up into the HDL fraction. These lipoproteins serve as a reservoir of apo CII from which the activator can move to

Species	Residue number (−20 … −10 … 1 … 10 … 20)	Residue number (21 … 30 … 40 … 50 … 60 … 70 … 79)
Rat	MGSRFFLALFLALLVLGNEVQGTEE---DDPGSSALLDTVQE	HLFSYWNSAKAAAGELYQKTYLTSVDEKLRDMYSKSSAAMTTYAGIFTDQLLTLLKGE
Human	MGTRLLPALFLVLLVLGFEVQGTQQPQQDEMPSPTFLTQVKE	SLSSYWESAKTAAQNLYEKTYLPAVDEKLRDLYSKSTAAMSTYTGIFTDQVLSVLKGEE
Monkey	MGTRFLLALCLVLLVLGFEVQGAQLPQQDEPPSPALLSRVQE	SLSSYWESAKAAAQKLYEKTYLPAVDEKLRDLYSKSTAAMSTYTGIFTDQVLSVLKGEE
Canine	MGTRYLLVLLLVLLVLGFEVQGAHESQQDETTSSALLTQMQE	SLYSYWGTARSAAEDLYKKAYPTTMDEKIRDIYSKSTAAVSTYAGIFTDQLLSMLKGDS
Bovine	AHVPQQDEASSPALLTQVQE	SLLGYWDTAKAAAQKLYKKTYLPAVDEKIRDIYSKSTAAVTTYAGIITDQVFSVLSGKD
Guinea Pig	MDARSLLLLWLLLPLLLLLGCEVQGAHLTQQDEPTSPDLL----E	TLSTYWDSAKAAAQGLYNNTYLPAVDETIRDIYSKGSAAISTYTGILTDQILTMLQGKQ

Fig. 4. Sequence homologies between apo CII from different mammals. The sequences are deduced from cloned cDNA for rat (Andersson *et al.*, 1991*b*), human (Wion *et al.*, 1987), monkey (Whitted *et al.*, 1989), canine (Datta *et al.*, 1987) and guinea pig (Andersson *et al.*, 1991*a*, *b*), and from protein sequencing for bovine (Bengtsson-Olivecrona & Sletten, 1990) apo CII.

newly secreted chylomicrons/VLDL (Eisenberg, 1990). This system may not operate in guinea pigs (Andersson *et al.*, 1991*a*) that have very low levels of HDL.

Human apo CII contains 79 amino acid residues (Myklebost *et al.*, 1984). Studies with peptide fragments have demonstrated that the carboxyterminal third of the molecule contains the structures needed for interaction with LPL and for activation, while the amino-terminal part probably forms amphipathic helices that anchor the protein to the lipid particle (Smith & Pownall, 1984). Sequences are known for apo CII from five other mammals, including the cow (Fig. 4). Sequence conservation is high in the C-terminal part in accord with the view that the interactions between the lipase and the activator occur here. The solution structure of fragment 50–79 was recently studied by 2D proton NMR (Lycksell *et al.*, 1992). The peptide showed a strong tendency for helix formation spanning residues 55–77 but interrupted by a segment with less well defined conformation around residues 60–65. This segment could be flexible and have an important role for interaction with the lipase. This is in concert with studies with fragments of apo CII which indicate that the region around Tyr-63 is crucial for activation of the lipase (Smith & Pownall, 1984).

Interaction with Polyanions

LPL binds strongly to heparin and related polysaccharides (Olivecrona & Bengtsson-Olivecrona, 1989). *In vivo*, this serves to anchor the enzyme to heparan sulphate proteoglycans at the vascular endothelium. The heparin binding region of bovine LPL has been shown by limited proteolysis to be located in the segment 230–390 (Persson *et al.*, 1989). From Arg-264 to Lys-321 there are 15 basic residues (Arg or Lys) which are conserved in all LPLs, and there are only three acidic residues of which one is not conserved in avian LPL. The corresponding segment in pancreatic lipase, residues 283 to 342, contains 6 to 7 basic and 3 to 5 acidic residues. This region is located between the two major folding units and contains two disulphides which have changed partners in LPL compared to pancreatic lipase (Yang *et al.*, 1989). Charged residues within this region should be available for binding to a heparin chain. If one considers the opening into the active site pocket as the face side of LPL, the groove that could accommodate a heparin molecule is located on the back (Persson *et al.*, 1991). This is in concert with observations that binding of heparin is fully compatible with enzymatic activity, as implied by the proposed mode of binding to endothelium *in vivo* (Olivecrona & Bengtsson-Olivecrona, 1989). There is evidence that the heparin chain can bind simultaneously to binding sites on both subunits in the active dimeric molecule (Clarke *et al.*, 1983).

Product Inhibition

LPL is subject to strong product inhibition (Olivecrona & Bengtsson-Olivecrona, 1987). A mechanism shared with many other lipases is that fatty acids can act as substrates, and di- and mono-glycerides can act as acyl acceptors. If the fatty acids are not removed or complexed, the lipase will spend much of its time in the non-productive pursuit of hydrolysing and then re-forming glycerides. A mechanism more specific for LPL is that the enzyme binds fatty acids. This has at least two effects relevant to product inhibition. 1) Fatty acids interfere with binding of the enzyme to the lipid droplets, and the enzyme which remains with the droplets can be sequestered in complexes with fatty acids in the surface film on the particle or in bilayer extensions of the surface film. 2) Fatty acids abolish the effects of the activator protein (Bengtsson & Olivecrona, 1980; Saxena & Goldberg, 1990).

Product inhibition is probably an important mechanism for control of LPL action *in vivo*, so that products are not formed more rapidly than they can be taken up by the tissue (Olivecrona *et al.*, 1989). In contrast, the availability of fatty acid acceptors does not seem to be an important parameter for regulation of lipolytic rancidity in bovine milk. The fatty acids most relevant to rancidity are short-chain and relatively water-soluble. They probably do not accumulate around sites of lipolysis on the fat droplets. Furthermore, there are ample Ca^{2+} to complex the fatty acids and prevent re-esterification.

Synthesis and Secretion into Milk

LPL activity in mammary gland is greatly increased during lactation in guinea pigs (McBride & Korn, 1963; Robinson, 1963; Wallinder *et al.*, 1982), rats (Hamosh & Scow, 1970), mice (Jensen *et al.*, 1991), goats (Chilliard *et al.*, 1985), and cows (Jensen & Pitas, 1976). The high level of LPL activity in breast milk suggests that human mammary gland also has high LPL activity during lactation (Hernell & Bläckberg, 1985; Freed *et al.*, 1989). Studies on rats and guinea pigs have shown that the mammary gland has low LPL activity during pregnancy, but activity

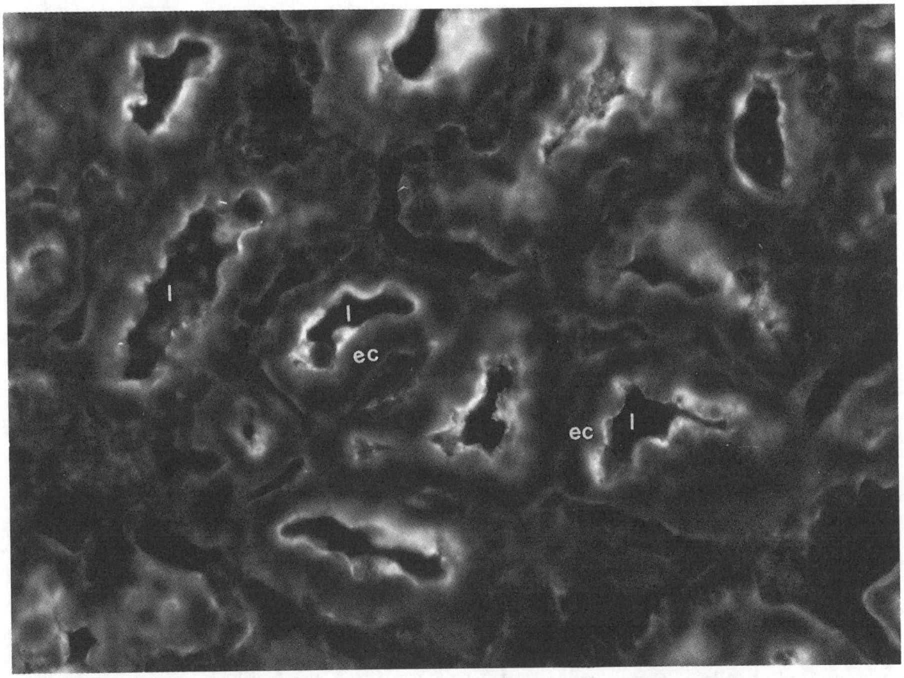

Fig. 5. Localisation by immunofluorescence of LPL in cryosections of mammary gland from a lactating guinea pig. The tissue was fixed by 4% paraformaldehyde, cryoprotected and 6 μm cryosections prepared. The primary antibodies were from a rabbit immunised with LPL purified from guinea pig milk. The second antibody was goat anti-rabbit IgG conjugated with fluorescein. Note that most of the label is present in the apical cytoplasm of epithelial cells (ec) that line the alveoli of the mammary gland. By electron microscopy (not shown) LPL could also be visualised at the basolateral membrane of the epithelial cells, and associated with endothelial cells.

increases sharply about 2 days before parturition and remains high throughout lactation. At the same time, LPL activity in adipose tissue decreases and remains low during lactation. Hormonal regulation of the synthesis of LPL in mammary gland during lactation has been reviewed in depth (Scow & Chernick, 1987). It has generally been assumed that the main source of LPL in the mammary gland are the milk-producing alveolar epithelial cells. This view was supported by a study in guinea pigs, where lipase mRNA and protein was demonstrated in this cell type by *in situ* hybridisation and immunofluorescence (Camps *et al.*, 1990). Further studies in guinea pigs have confirmed that the main localisation of LPL is the apical cytoplasm of epithelial cells (Fig. 5). Electron microscopy showed LPL within vesicles along the secretory path, i.e. Golgi, condensing vesicles, mature intracellular vesicles, and secretory vesicles. In the earlier part of this passage, LPL was associated with the membranes. When the casein micelles are assembled, LPL appears to transfer to them, and was found decorating the periphery of the micelles (Fig. 6). No LPL was seen associated with the milk fat globules within the cells.

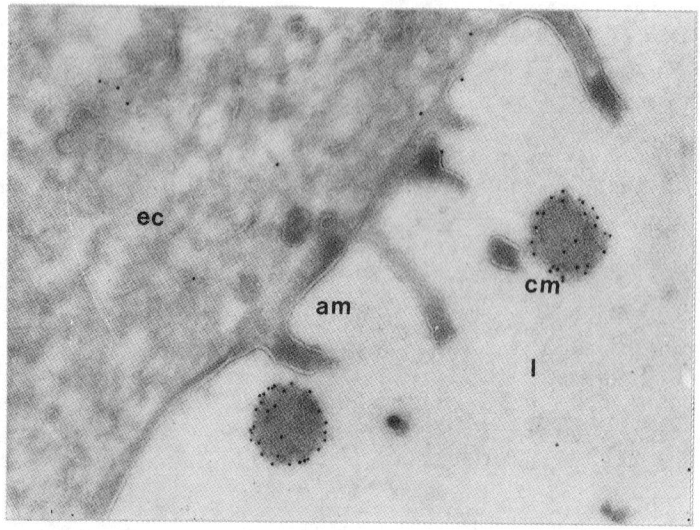

Fig. 6. Localisation by immuno-electron microscopy of LPL associated to casein micelles in the lumen of mammary alveoli of guinea pigs. The tissue was fixed by 8% paraformaldehyde, cryoprotected and sectioned. The first antibody was the same as in Fig. 5. The antibodies were visualised with a protein A-gold (9 nm) conjugate. LPL was detected inside the epithelial cells in Golgi stacks and in condensing vesicles (not shown in this picture). Casein micelles (cm) secreted into the lumen of the alveoli (l) are decorated by LPL at their periphery (arrows). am: apical membrane.

Apparently, the enzyme is secreted with the casein micelles. Some transfer to the milk fat globules may occur in the mammary gland, but this association may occur largely during collection and storage of the milk (see below).

Jensen *et al.* (1991) have studied the localisation of LPL in mouse mammary gland using *in situ* hybridization and immunofluorescence. They concluded that the main sources of LPL were interstitial cells located between epithelial structures. These cells are probably regressed, lipid-depleted mammary adipocytes. Their results indicate a quite different cellular origin of LPL in mouse compared to guinea pig mammary gland, suggesting a species difference. LPL activity in milk differs widely between species. It is extremely high in guinea pigs, but has been reported to be low in rat milk (Hamosh & Scow, 1970). An interesting possibility is that species which produce the lipase in the mammary epithelial cells also release it into milk, whereas other species produce the lipase in a different cell type, and release much less of the enzyme into milk.

It has been suggested that LPL leaks out from the mammary gland, because of tissue damage, or because the mammary cells are unable to sort adequately between proteins destined for secretion into milk and those which are destined for secretion through the basal aspect of the cell towards the blood vessels (Hamosh & Scow, 1970). In guinea-pigs, LPL accounts for about 0·1% of the protein in milk (Wallinder *et al.*, 1982), and for about 0·1% of the proteins synthesised in the mammary gland (Semb & Olivecrona, 1986). This suggests that LPL is transferred into milk as efficiently as other milk proteins are. The immunohistochemical studies in guinea pigs showed more LPL immunoreaction over milk within secretory vesicles, alveoli and ducts, than at the vascular endothelium (Camps *et al.*, 1990). This is in accord with the early studies of McBride and Korn (1963) and Robinson (1963) who found that most of the LPL activity in mammary glands is associated with the milk.

State of LPL in Milk

In bovine and in guinea pig milk more than 80% of the LPL is associated with casein micelles (Fox *et al.*, 1967; Downey & Murphy, 1975; Hohe *et al.*, 1985). The binding is primarily electrostatic, between negatively charged phosphates on the caseins and positively charged amino acid residues in the 'heparin binding site' on the enzyme. In accord with this, the enzyme can be released from the micelles by addition of heparin (Sundheim & Bengtsson-Olivecrona, 1985). In human milk a larger fraction of the LPL is associated with the fat globules (Hernell & Olivecrona, 1974; Neville *et al.*, 1991). This may relate to the much lower amount of casein in human than in bovine milk. Cooling or freezing of milk results in transfer of LPL from the skim milk fraction to the milk fat globules (Sundheim & Bengtsson-Olivecrona, 1985, 1987*a*; Neville *et al.*, 1991). This is probably one of the reasons for so-called cold-induced lipolysis in bovine milk.

It is hard to reconcile the data on synthesis and secretion of LPL with the view that its appearance in milk is a mistake, and that the enzyme has no useful purpose in milk. There are in fact several possible functional roles for the enzyme:

—It was recently shown that heparin-like molecules in the intestinal mucosa bind pancreatic lipase and cholesterol esterase/non-specific lipase (Bosner *et al.*, 1989). The implication is that intestinal lipolysis occurs mainly at the brush border. This could facilitate absorption of the lipolysis products. If so, the process would be similar to lipolysis of lipoprotein triglycerides at the surface of blood capillaries. Since LPL binds strongly to heparin-like molecules it is tempting to suggest that the enzyme in milk might attach to the intestinal wall and help bind the milk fat globules to the mucosa.

—It has been demonstrated that LPL can catalyse transfer of lipids (e.g. cholesteryl esters, tocopherol) into cells (Stein *et al.*, 1984; Traber *et al.*, 1985). This transfer activity requires that the lipase binds to the cell surface, but is not dependent on the hydrolytic activity of the lipase. Perhaps milk LPL has such a role in the intestine.

—Milk fat globules are not efficiently attacked by pancreatic lipase (Hernell & Bläckberg, 1991). To initiate lipolysis it is necessary that some other lipase hydrolyses a small fraction of the triglycerides. This function is usually attributed to the gastric lipase. The possibility that LPL can also initiate intestinal lipolysis has not been explored.

—Hernell and Bläckberg (1985); Hernell & Bläckberg *et al.* (1986) have shown that fatty acids generated in human milk by LPL have a powerful antiparasitic function.

Lipolysis in Milk

Freshly drawn bovine milk contains only small amounts of free fatty acids. If the milking and storage procedures are appropriate, the milk can be stored for several days with little further development of free fatty acids. If this were not so, the modern dairy industry would not exist (Downey, 1980). Hydrolysis of as little as 1–2% of the milk triglycerides to fatty acids gives a rancid or 'lipolysed' flavor to the milk (Pillay *et al.*, 1980). This makes the milk unpalatable for consumption. Concentration of the reaction products in the cream may in turn impair the flavor quality of products manufactured from milk.

It is now generally agreed that lipolysis is largely due to the LPL indigenous to the milk (Castberg *et al.*, 1975; Deeth & Fitz-Gerald, 1976; Jensen & Pitas, 1976; Chilliard, 1980; Andersson, 1982). It is only after prolonged storage or after poor hygienic handling that bacterial lipases become important. Virtually all bovine milks, with the possible exception of colostrum (Driessen, 1976), contain substantial amounts of lipolytic activity, but there seems to be only a weak correlation between the (potential) lipase activity of the milk and its susceptibility to lipolysis (Deeth & Fitz-Gerald, 1976; Jensen & Pitas, 1976; Jellema, 1980). If the conditions in milk were optimal for lipase activity, hydrolysis would proceed rapidly. A typical lipase activity for bovine milk at 37°C, pH 7, with activator added is 100 nmol fatty acid released per min per ml milk. At this rate, the milk would become unpalatable in a few minutes. Thus, the question is not why some samples of milk become lipolysed, but rather why most milk samples do not.

Lypolysis as it occurs in dairy practice is often differentiated into spontaneous and induced (Deeth & Fitz-Gerald, 1976). Spontaneous lipolysis is initiated in some milk samples simply by prompt cooling of the fresh milk (Jellema, 1980). In contrast, lipolysis can be induced in all milk samples by rough mechanical treatment (Fleming, 1980). If the lipase was not destroyed by prior pasteurisation, the conventional homogenisation of the milk would result in rapid lipolysis and render the milk unusable. Induced lipolysis was not a big problem when milk was collected by hand milking and transported to the dairy the same day. It has become more of a problem with the modern pipeline milking machines and holding of the milk for several days at the farms before transportation to a central dairy.

At the biochemical level, induced lipolysis is presumably equivalent to disruption of the organised structure of the fat globule membrane; this allows lipase to bind to the fat globules and hydrolise their lipids. Spontaneous lipolysis is less well understood, but again lipolysis correlates to binding of lipase to the fat globules (Fleming, 1980; Sundheim & Bengtsson-Olivecrona, 1985; Sundheim, 1988). It should be emphasized again that there is an abundance of lipase in milk; the question is why most milks are so resistant to lipolysis. It has been demonstrated that this resistance of the milk fat is partly due to factors related to the structure of the milk fat globule itself (Sundheim & Bengtsson-Olivecrona, 1986, 1987*a*) partly due to factors in the skim milk fraction (Sundheim & Bengtsson-Olivecrona, 1987 *b,c*). Methods to assess these factors have been worked out in two laboratories (Cartier & Chilliard, 1986; Sundheim, 1988).

References

Ailhaud, G. (1990). Cellular and secreted lipoprotein lipase revisited. *Clin. Biochem.*, **23**, 343–7.

Ameis, D., Stahnke, G., Kobayashi, J., McLean, J., Lee, G., Buscher, M., Schotz, M.C. & Will, H. (1990). Isolation and characterisation of the human hepatic lipase gene. *J. Biol. Chem.*, **265**, 6552–5.

Andersson, M. (1982). Factors affecting the distribution of lipoprotein lipase activity between serum and casein micelles in bovine milk. *J. Dairy Res.*, **49**, 51–9.

Andersson, Y., Thelander, L. & Bengtsson-Olivecrona, G. (1991*a*). Demonstration of apolipoprotein CII in guinea pigs. Functional characteristics, cDNA sequence and tissue expression. *J. Biol. Chem.*, **266**, 4074–80.

Andersson, Y., Thelander, L. & Bengtsson-Olivecrona, G. (1991*b*). Rat apolipoprotein C-II lacks the conserved site for proteolytic cleavage of the pro-form. *J. Lipid Res.*, **32**, 1805–9.

Baker, M.E. (1988). Is vitellogenin an ancestor of apolipoprotein B-100 of human low-density lipoprotein and human lipoprotein lipase. *Biochem. J.*, **255**, 1057–60.

Beisiegel, U., Weber, W. & Bengtsson-Olivecrona, G. (1991). Lipoprotein lipase enhances the binding of chylomicrons to low density lipoprotein receptor-related protein. *Proc. Natl. Acad. Sci. USA.*, **88**, 8342–6.

Bengtsson, G. & Olivecrona, T. (1980). Lipoprotein lipase: mechanism of product inhibition. *Eur. J. Biochem.*, **106**, 557–62.

Bengtsson, G. & Olivecrona, T. (1981). Lipoprotein lipase: modification of its kinetic properties by mild tryptic digestion. *Eur. J. Biochem.*, **113**, 547–54.

Bengtsson, G. & Olivecrona, T. (1982). On the pH dependency of lipoprotein lipase activity. *Biochem. Biophys. Acta.*, **712**, 196–199.

Bengtsson, G. & Olivecrona, T. (1983). The effects of pH and salt on the lipid binding and enzyme activity of lipoprotein lipase. *Biochim. Biophys. Acta.*, **751**, 254–259.

Bengtsson-Olivecrona, G. & Sletten, K. (1990). Primary structure of the bovine analogues to human apolipoproteins CII and CIII. Studies on isoforms and evidence for proteolytic processing. *Eur. J. Biochem.*, **192**, 51–21.

Bengtsson-Olivecrona, G., Olivecrona, T. & Jörnvall, H. (1986). Lipoprotein lipase from cow, guinea-pig and man. Structural characterisation and identification of protease-sensitive internal regions. *Eur. J. Biochem.*, **161**, 281–8.

Bensadoun, A. (1991). Lipoprotein lipase. *Annu. Rev. Nutr.*, **11**, 297–306.

Bosner, M.S.T., Gulick, D.J.S., Riley, D.J.S., Spilburg, C.A. & Lange, L.G. (1989). Heparin-modulated binding of pancreatic lipase and uptake of hydrolized triglycerides in the intestine. *J. Biol. Chem.*, **264**, 20261–4.

Camps, L., Reina, M., Llobera, M., Vilaró. S. & Olivecrona, T. (1990). Lipoprotein lipase: cellular origin and functional distribution. *Am. J. Physiol.*, **258**, C673–81.

Cartier, O. & Chilliard, Y. (1986). Effects of different skim milk fractions on activity of cow milk purified lipoprotein lipase. *J. Dairy Sci.*, **69**, 951–5.

Castberg, H., Egelrud, T. & Olivecrona, T. (1975). Lipase in bovine milk and the relationship between the lipoprotein lipase and the tributyrate hydrolyzing activities in cream and skim milk. *J. Dairy Res.*, **42**, 255–66.

Abousalham, A., Chaillain, C., Kerfelec, B., Foglizzo, E. & Chapus, C. (1991). Uncoupling of catalysis and colipase binding in pancreatic lipase by limited proteolysis. *Prot. Engin.*, **4**.

Chapus, C., Rovery, M., Sarda, L. & Verger, R. (1988). Minireview on pancreatic lipase and colipase. *Biochimie* **70**, 1223–34.

Chilliard, Y. (1980). Variations physiologiques des activités lipasiques et de la lipolyse spontanée dans les laits de vache, de chèvre et de femme: revue bibliographique. *Lait* **60**, 1–60.

Chilliard, Y. & Doreau, M. (1985). Characterisation of lipase in mare milk. *J. Dairy Sci.*, **68**, 37–9.

Chilliard, Y., Selselet-Attou, G., Bas, P. & Morand-Fehr, P. (1984). Characteristics of lipopolytic system in goat milk. *J. Dairy Sci.*, **67**, 2216–23.

Clarke, A.R., Luscombe, M. & Holbrook, J.J. (1983). The effect of the chain length of heparin on its interaction with lipoprotein lipase. *Biochim. Biophys. Acta.*, **747** 130–7.

Clay, M.A., Hopkins, G.J., Ehnholm, C. & Barter, P.J. (1989). The rabbit as an animal model of hepatic lipase deficiency. *Biochim. Biophys. Acta.*, **1002**, 173–81.

Cryer, A. (1987). Comparative biochemistry and physiology of lipoprotein lipase. In *Lipoprotein lipase*. ed. J. Borensztajn. Evener Press, Chicago. 277–327.

Datta, S.W., Li, -H., Ghosh, I., Luo, C.-C. & Chan, L. (1987). Structure and expression of dog apolipoprotein C-II and C-III mRNAs. *J. Biol. Chem.*, **262**, 10588–93.

Datta, S., Luo, C.-C., Li, W.-H., VanTuinen, P., Ledbetter, D.H., Brown, M.A., Chen, S.-H., Lui, S.-W. & Chan, L. (1988). Human hepatic lipase. Cloned cDNA sequence, restriction fragment length polymorphisms, chromosomal localisation, and evolutionary relationships with lipoprotein lipase and pancreatic lipase. *J. Biol. Chem.*, **263**, 1107–10.

De Caro, J., Boudouard, M., Bonicel, J., Guidoni, A.A., Desnuelle, P. & Rovery, M. (1981). Porcine pancreatic lipase. Completion of the primary structure. *Biochim. Biophys. Acta.*, **671**, 129–38.

Deckelbaum, R.J., Hamilton, J.A., Moser, A., Bengtsson-Olivecrona, G., Butbul, E., Carpentier, Y.A., Gutman, A. & Olivecrona, T. (1990). Medium-chain vs long-chain triacylglycerol emulsion hydrolysis by lipoprotein lipase and hepatic lipase: Implications for the mechanism of lipase action. *Biochemistry*, **29**, 1136–42.

Deeth, H.C. & Fitz-Gerald, C.H. (1976). Lipolysis in dairy products: A review. *Aust. J. Dairy. Tech.*, **31**, 53–64.

Derewenda, Z.S. & Cambillau, C. (1991). Effects of gene mutations in lipoprotein and hepatic lipases as interpreted by a molecular model of the pancreatic triglyceride lipase. *J. Biol. Chem.*, **266**, 23112–19.

Derewenda, Z.S. & Derewenda, U. (1991). Relationships among serine hydrolases: Evidence for a common structural motif in triacylglyceride lipases and esterases. *Biochem. Cell Biol.*, **69**, 842–51.

Dichek, H.L., Parrot, C., Brewer, H.B., Jr. & Santamarina-Fojo, S. (1992). Functional characterization of a chimeric protein genetically engineered from human lipoprotein lipase and human hepatic lipase. [Abstract]. *Clin. Res. J.*, **40,** 321.

Downey, W.K. (1980). Risks from pre- and post-manufacture lipolysis. *Bull. Int. Dairy Fed. Doc.*, **118**, 4–18.

Downey, W.K. & Murphy, R.F. (1975). Partitioning of the lipolytic enzymes in bovine milk. *Bull. Int. Dairy Fed. Doc.*, **86**, 19–23.

Driessen, F.M. (1976). A comparative study of the lipase in bovine colostrum and in bovine milk. *Neth. Milk Dairy J.*, **30**, 186–96.

Eckel, R.H. (1987). Adipose tissue lipoprotein lipase. In *Lipoprotein Lipase*, ed. J. Borensztajn. Evener Publishers, Chicago, pp. 79–132.

Eckel, R.H. (1989). Lipoprotein lipase: a multifunctional enzyme relevant to common metabolic diseases. *N. Engl. J. Med.*, **320**, 1060–8.

Eisenberg, S. (1990). Metabolism of apolipoproteins and lipoproteins. *Curr. Opin. Lipidol.*, **1**, 205–21.

Ekstrom, B., Nilson, A. & Akesson, B. (1989). Lipolysis of polyenoic fatty acid esters of human chylomicrons by lipoprotein lipase. *Eur. J. Clin. Invest.*, **19**, 259–64.

Etienne, J., Noé, L., Rossignol, M., Dosne, A.-M. & Debray, J. (1981). Post-heparin lipolytic activity with no hepatic triacylglycerol lipase involved in a mammalian species. *Biochim. Biophys. Acta.*, **663**, 516–23.

Faustinella, F., Smith, L.C., Semenkovich, C.F. & Chan, L. (1991). Structural and functional roles of highly conserved serines in human lipoprotein lipase. Evidence that serine 132 is essential for enzyme catalysis. *J. Biol. Chem.*, **266**, 9481–5.

Fleming, M.G. (1980). Mechanical factors associated with milk lipolysis in cows's milk. *Bull. Int. Dairy Fed. Doc.*, **118**, 41–52.

Fox, P.F., Yaguchi, M. & Tarassuk, N.P. (1967). Distribution of lipase milk proteins. II. Dissociation from k-casein with dimethylformamide. *J. Dairy Sci.*, **50**, 307–12.

Freed, L.M., Berkow, S.E., Hamosh, P., York, C.M., Mehta, N.R. & Hamosh, M. (1989). Lipases in human milk: Effect of gestational age and length of lactation on enzyme activity. *J. Am. Coll. Nutrit.*, **8**, 143–50.

Hamosh, M. & Scow, R.O. (1970). Lipoprotein lipase activity in guinea pig and rat milk. *Biochim. Biophys. Acta.*, **231**, 283–9.

Hazlett, T.L. & Dennis, E.A. (1988). Lipid-induced aggregation of phospholipase A2: Sucrose density gradient ultracentrifugation and cross linking studies. *Biochim. Biophys. Acta.*, **961**, 22–9.

Hernell, O. & Bläckberg, L. (1985). Lipolysis in human milk: causes and consequences. In *Composition and Physical Properties of Human Milk*, ed. J. Schaub. Elsevier Science Publishers, Amsterdam. pp. 165–75.

Hernell, O. & Bläckberg (1991). Digestion and absorption of human milk lipids. In *Encyclopaedia of Human Biology*, Vol. 3, ed. R. Dulbecco. Academic Press, New York, pp. 47–56.

Hernell, O. & Olivecrona, T. (1974). Human milk lipases. I. Serum-stimulated lipase. *J. Lipid Res.*, **15**, 367–74.

Hernell, O., Ward, H., Bläckberg, L. & Periera, M.E.A. (1986). Killing of *Giardia Lamblia* by human milk lipases: an effect mediated by lipolysis of milk lipids. *J. Inf. Dis.*, **153**, 715–20.

Hohe, K.A., Dimick, P.S. & Kilara, A. (1985). Milk lipoprotein distribution in the major fractions of bovine milk. *J. Dairy Sci.*, **68**, 1067–73.

Iverius, P.-H. & Östlund-Lindqvuist, A.M. (1976). Lipoprotein lipase from bovine milk. Isolation procedure, chemical charaterisation and molecular weight analysis. *J. Biol. Chem.*, **251**, 7791–5.

Jellema, A. (1980). Physiological factors associated with lipolytic activity in cow's milk. *Bull. Int. Dairy Fed. Doc.*, **118**, 33–40.

Jensen, R.G. & Pitas, R.E. (1976). Milk lipoprotein lipase: a review. *J. Dairy Sci.*, **59**, 1203–14.

Jensen, D.R., Bessesen, D.H., Etienne, J., Eckel, R.H. & Neville, M.C. (1991). Distribution and source of lipoprotein lipase in mouse mammary gland. *J. Lipid Res.*, **32**, 733–42.

Kirchgessner, T.G., Chaut, J.C., Heinzmann, C., Etienne, J., Guilhot, K., Svenson, K., Ameis, D., Pilon, C., D'Auriol, L., Andalibi, A., Schotz, M.C., Galibert, F. & Lusis, A.J. (1989). Organisation of the human lipoprotein lipase gene and evolution of the lipase gene family. *Proc. Natl. Acad. Sci. USA.*, **86**, 9647–51.

Komaromy, M.C. & Schotz, M.C. (1987). Cloning of rat hepatic lipase cDNA: Evidence for a lipase gene family. *Proc. Natl. Acad. Sci. USA.*, **84**, 1526–30.

Komaromy, M.C. & Reed, M. (1991). Expression of rat hepatic lipase in heterologous systems: Evidence for different sites for interface binding and catalysis. *J. Lipid Res.*, **32**, 963–75.

Lycksell, P.O., Öhman, A., Bengtsson-Olivecrona, G., Johansson, L.B.-Å., Wijmenga, S.S., Wernic, D. & Gräslund, A. (1992). Sequence specific 1H NMR assignments and secondary structure of a carboxy terminal functional fragment of apolipoprotein CII. *Eur. J. Biochem.* **205**, 223–31.

Ma, Y., Bruin, T., Tuzgol, S., Wilson, B.I., Roederer, G., Liu, M.-S., Davignon, J., Kastelein, J.J.P., Brunzell, J.D. & Hayden, M.R. (1992). Two naturally occurring mutations at the first and second bases of codon asparic acid 156 in the proposed catalytic triad of human lipoprotein lipase. *In vivo* evidence that aspartic acid 156 is essential for catalysis. *J. Biol. Chem.*, **267**, 1918–23.

Masuno, H., Blanchette-Mackie, E.J., Chernick, S.S. & Scow, R.O. (1990). Synthesis of inactive nonsecretable high mannose-type lipoprotein lipase by cultured brown adipocytes of combined lipase-deficient *cld/cld* mice. *J. Biol. Chem.*, **265**, 1628–38.

Masuno, H., Schultz, C.J., Park, J.-W., Blanchette-Mackie, E.J., Mateo, C. & Scow, R.O. (1991). Glycosylation, activity and secretion of lipoprotein lipase in cultured brown adipocytes of newborn mice. Effect of tunicamycin, monensin, 1-deoxymannojirimycin and swainsonine. *Biochem. J.*, **277**, 801–9.

McBride, O.W. & Korn, E.D. (1963). The lipoprotein lipase of mammary gland and the correlation of its activity to lactation. *J. Lipid Res.*, **4**, 17–23.

Mickel, F.S., Weidenbach, F., Swarovsky, B., LaForge, K.S. & Scheele, G.A. (1989). Structure of the canine pancreatic lipase gene. *J. Biol. Chem.*, **264**, 12895–901.

Muller, D.L., Saudek, C.D. & Applebaum-Bowden, D. (1985). Hepatic triglyceride lipase in diabetic dogs. *Metabolism*, **34**, 251–4.

Myklebost, O., Williamson, B., Markham, A.F., Myklebost, S.R., Rogers, J., Woods, D.E. & Humphries, S.E. (1984). The isolation and characterisation of cDNA clones for human apolipoprotein CII. *J. Biol. Chem.*, **259**, 4401–4.

Neville, M.C., Waxman, L.C., Jensen, D.R. & Eckel, R.H. (1991). Lipoprotein lipase in human milk: compartmentalisation and effect of fasting, insulin, and glucose. *J. Lipid Res.*, **32**, 251–7.

Nilsson, J., Bläckberg, L., Carlsson, P., Enerbäck, S., Hernell, O. & Bjursell, G. (1990). cDNA cloning of human-milk bile-salt-stimulated lipase and evidence for its identity to pancreatic carboxylic ester hydrolase. *Eur. J. Biochem.*, **192**, 543–50.

Oka, K., Nakano, T., Tkalcevic, G.T., Scow, R.O. & Brown, W.V. (1991). Molecular cloning of mouse hepatic triacylglycerol lipase: Gene expression in combined lipase-deficient (*cld/cld*) mice. *Biochim. Biophys. Acta Gene Struct. Expression*, **1089**, 13–20.

Olivecrona, T. & Bengtsson, G. (1983). Lipases in milk. In *Lipases*, ed. B. Borgström & H. Brockman. Elsevier, Amsterdam, pp. 206–61.

Olivecrona, T. & Bengtsson-Olivecrona, G. (1987). Lipoprotein lipase from milk—the model enzyme in lipoprotein lipase research. In *Lipoprotein Lipase*, ed. J. Borensztajn. Elsevier, Chicago, pp. 15–58.

Olivecrona, T. & Bengtsson-Olivecrona, G. (1989). Heparin and lipases. In *Heparin*, ed. D. Lane and U. Lindahl. Edward Arnold, London, pp. 335–361.

Olivecrona, T. & Bengtsson-Olivecrona, G. (1990). Lipoprotein lipase and hepatic lipase. *Curr. Opin. Lipidol.*, **1**, 222–30.

Olivecrona, T. & Bengtsson-Olivecrona, G. (1991). Indigenous enzymes in milk II, Lipase. In *Food enzymology*. ed. P.F. Fox. Elsevier Applied Science, London. 62–78.

Olivecrona, T., Bengtsson-Olivecrona, G., Osborne, J.C., Jr. & Kempner, E.S. (1985). Molecular size of bovine lipoprotein lipase as determined by radiation inactivation. *J. Biol. Chem.*, **260**, 6888–91.

Olivecrona, T., Bengtsson-Olivecrona, G., Deckelbaum, R.J. & Carpentier, Y. (1989). Do fatty acids control lipoprotein lipase action? In *Intestinal Lipid Lipoprotein Metabolism*, ed. E. Windler and H. Greten. Zuckschwert Verlag, München, Germany, pp. 174–9.

Osborne, J.C., Jr., Bengtsson-Olivecrona, G., Lee, N. & Olivecrona, T. (1985). Studies on inactivation of lipoprotein lipase. Role of dimer to monomer dissociation. *Biochemistry*, **24**, 5606–11.

Persson, B., Bengtsson-Olivecrona, G., Enerbäck, S., Olivecrona, T. & Jörnvall, H. (1989). Structural features of lipoprotein lipase. Lipase family relationships, binding interactions, non-equivalence of lipase cofactors, vitellogenin similarities, and functional subdivisions of lipoprotein lipase. *Eur. J. Biochem.*, **179**, 39–45.

Persson, B., Jörnvall, H., Olivecrona, T. & Bengtsson-Olivecrona, G. (1991). Lipoprotein lipases and vitellogenins in relation to the known three-dimensional structure of pancreatic lipase. *FEBS Lett.*, 33–6.

Pillay, V.T., Myhr, A.N. & Gray, J.I. (1980). Lipolysis in milk. Determination of free fatty acid and threshold for lipolysed flavor detection. *J. Dairy Sci.*, **63**, 1213–18.

Rapp, D. & Olivecrona, T. (1978). Kinetics of milk lipoprotein lipase. Studies with tributyrin. *Eur. J. Biochem.*, **91**, 379–85.

Robinson, D.S. (1963). Changes in the lipolytic activity of the guinea pig mammary gland at parturition. *J. Lipid Res.*, **4**, 21–3.

Rojas, C., Olivecrona, T. & Bengtsson-Olivecrona, G. (1991). Comparison of the action of lipoprotein lipase on triacylglycerols and phospholipids when presented in mixed liposomes or in emulsion particles. *Eur. J. Biochem.*, **197**, 315–21.

Saxena, U. & Goldberg, I.J. (1990). Interaction of lipoprotein lipase with glycosaminoglycans and apolipoprotein C-II: Effects of free-fatty-acids. *Biochim. Biophys. Acta.*, **1043**, 161–8.

Scow, R.O. & Chernick, S.S. (1987). Role of lipoprotein lipase during lactation. In *Lipoprotein Lipase*, ed. J. Borensztajn. Evener Publishers, Chicago. 149–85.

Scow, R.O., Chernick, S.S. & Fleck, T.R. (1977). Lipoprotein lipase and uptake of triacylglycerol, cholesterol and phophatidylcholine from chylomicrons by mammary and adipose tissue of lactating rats *in vivo*. *Biochim. Biophys. Acta.*, **487**, 297–306.

Semb, H. & Olivecrona, T. (1986). Lipoprotein lipase in guinea pig tissues. Molecular size and rates of synthesis. *Biochim. Biophys. Acta.*, **878**, 330–7.

Semb, H. & Olivecrona, T. (1989). The relation between glycosylation and activity of guinea pig lipoprotein lipase. *J. Biol. Chem.*, **264**, 4195–4200.

Semenkovich, C.F., Luo, C.-C., Nakanishi, M.K., Chen, S.-H., Smith, L.C. & Chan, L. (1990). *In vitro* expression and site-specific mutagenesis of the cloned human lipoprotein lipase gene. Potential N-linked glycosylation site asparagine 43 is important for both enzyme activity and secretion. *J. Biol. Chem.*, **265**, 5429–33.

Smith, L.C. & Pownall, H.J. (1984). Lipoprotein lipase. In *Lipases*, ed. B. Borgström and H. Brockman. Elsevier, Amsterdam, pp. 263–305.

Stein, O., Halperin, G., Leitersdorf, E., Olivecrona, T. & Stein, Y. (1984). Lipoprotein lipase mediated uptake of non-degradable ether analogues of phosphatidylcholine and cholesteryl ester by cultured cells. *Biochim. Biophys. Acta.*, **795**, 47–59.

Sundheim, G. (1988). Spontaneous lipolysis in bovine milk: Combined effects of cream, skim milk and lipoprotein lipase activity. *J. Dairy Sci.*, **71**, 620–6.

Sundheim, G. & Bengtsson-Olivecrona, G. (1985). Lipolysis in milk introduced by cooling or by heparin: comparisons of amount of lipoprotein lipase in the cream fraction and degree of lipolysis. *J. Dairy Sci.*, **68**, 589–93.

Sundheim, G. & Bengtsson-Olivecrona, G. (1986). Hydrolysis of bovine milk fat globules by lipoprotein lipase: Inhibition by proteins extracted from milk fat globule membrane. *J. Dairy Sci.*, **70**, 1815–21.

Sundheim, G. & Bengtsson-Olivecrona, G. (1987*a*). Isolated milk fat globules as substrate for lipoprotein lipase: Study of factors relevant to spontaneous lipolysis in milk. *J. Dairy Sci.*, **70**, 499–505.

Sundheim, G. & Bengtsson-Olivecrona, G. (1987*b*). Hydrolysis of bovine and caprine milk fat globules by lipoprotein lipase. Effects of heparin and of skim milk on lipase distribution and on lipolysis. *J. Dairy Sci.*, **70**, 2467–75.

Sundheim, G. & Bengtsson-Olivecrona, G. (1987*c*). Methods to access the propensity of milk fat globules toward lipolysis and the ability of skim milk to inhibit lipolysis. *J. Dairy Sci.*, **70**, 2040–5.

Traber, M.G., Olivecrona, T. & Kayden, H.J. (1985). Bovine milk lipoprotein lipase transfers tocopherol to human fibroblasts during triglyceride hydrolysis *in vitro*. *J. Clin. Invest.*, **75**, 1729–34.

Verger, R. & deHaas, G.H. (1976). Interfacial enzyme kinetics of lipolysis. *Annu. Rev. Biophys. Bioeng.*, **5**, 77–117.

Wallinder, L., Bengtsson, G. & Olivecrona, T. (1982). Purification and properties of lipoprotein lipase in guinea pig milk. *Biochim. Biophys. Acta.*, **711**, 107–13.

Whitted, B.E., Castle, C.K., Polites, H.G., Melchior, G.W. & Marotti, K.R. (1989). Purification, cloning and nucleotide sequence determination of cynomolgus monkey apolipoprotein C-II: comparison to the human sequence. *Mol. Cell. Biochem.*, **90**, 69–79.
Winkler, F.K., D'Arcy, A. & Hunziker, W. (1990). Structure of human pancreatic lipase. *Nature*, **343**, 771–4.
Wion, K.L., Kirchgessner, T.G., Lusis, A.J., Schotz, M.C. & Lawn, R.M. (1987). Human lipoprotein lipase complementary DNA sequence. *Science*, **235**, 1638–41.
Wong, H., Davies, R.C., Nikazy, J., Seebart, K.E. & Schotz, M.C. (1991). Domain exchange: Characterization of a chimeric lipase of hepatic lipase and lipoprotein lipase. *Proc. Natl. Acad. Sci. USA* **88**, 11290–4.
Yang, C.-Y., Gu. Z.-W., Yang, H.-X, Rohde M.F., Gotto, A.M. Jr. & Pownall, H.J. (1989). Structure of bovine milk lipoprotein lipase. *J. Biol. Chem.*, **264**, 16822–7.

III. Proteinases

3.1. Introduction

Bovine milk contains a number of indigenous proteinases (Reimerdes, 1983). The most thoroughly studied is an alkaline serine proteinase, which has been shown to be plasmin (EC 3.4.21.7). The isolation, properties and significance of plasmin in milk have been reviewed (Humbert & Alais, 1979; Fox, 1981; Visser, 1981; Grufferty & Fox, 1988a). Proteolytic activity has also been reported in human (see Hernell & Lonnerdal, 1989), buffalo (Mathur & Sharma, 1979; Madkor & Fox, 1991), goat (Varshney & Mathur, 1979) and equine (Visser, 1981) milks. Most ($\sim 90\%$) of the plasmin in bovine milk exists as its precursor, plasminogen (Kaminogawa *et al.*, 1972; Eigel *et al.*, 1979; de Rham & Andrews, 1982*a*; Korycka-Dahl *et al.*, 1983; Richardson, 1983*a*).

Plasmin and plasminogen in milk are associated with the casein micelles (de Rham & Andrews, 1982*a*; Korycka-Dahl *et al.*, 1983; Richardson, 1983*a*) and are concentrated in rennet casein (Richardson & Elston, 1984), cheese (Farkye & Fox, 1990) and casein micelles prepared by centrifugation (Reimerdes & Klostermeyer, 1974). Both are dissociated from the micelles by 0·1 M with lysine or 6-aminohexanoic acid, suggesting that they associate with casein via lysine residues, ethanol (Warner & Polis, 1945), 1 M NaCl (Grufferty & Fox, 1988*b*) or by acidification to pH 4·0 (McMeekin *et al.*, 1959). According to Richardson & Elston (1984), plasmin dissociates from the micelles at pH values $< 5{\cdot}7$ but Grufferty and Fox (1988*b*) and Madkor and Fox (1991) found no dissociation at pH values $\geqslant 4{\cdot}8$ on holding for up to 4 h; dissociation was complete at pH values $< 4{\cdot}6$.

The milk fat globule membrane also prossesses proteolytic activity which appears to be mainly plasmin (Hofmann *et al.*, 1979).

3.2. Isolation of Plasmin from Milk

Plasmin has been purified from milk to varying degrees (up to 17 000-fold) by acid extraction and $(NH_4)_2SO_4$ precipitation (Warner & Polis, 1945; Zittle, 1965; Yamauchi *et al.*, 1969) and chromatography on one or more of the following:

DEAE cellulose, CM cellulose, DEAE Sepharose 6B, CM Sephadex, lysine–Sepharose 4B, Sephadex G-100 or G-200 (Kaminogawa *et al.*, 1971; Dulley, 1972; Reimerdes & Klostermeyer, 1974; Halpaap *et al.*, 1977; Snoeren & van Riel, 1979).

Plasmin is a serine proteinase with trypsin-like activity; it is inhibited by DFP, Cu, Zn, Hg, soyabean trypsin inhibitor and tosyllysine chloromethylketone, partly by Mg and Ca but not by EDTA, α,α'-dipyridyl, ρ-CMB or iodoacetate (Kaminogawa *et al.*, 1969, 1971, 1972; Chen & Ledford, 1971; Andrews, 1983*a*).

Kaminogawa *et al.* (1971, 1972) reported that the molecular weight of the alkaline proteinase from milk, determined by gel filtration, is 100 000 but based on the amino acid sequence of plasmin, its molecular weight is 83 200 or 81 000 (Castellino & Powell, 1981). Plasmin is maximally active at ~pH 7·5 and 37°C (Warner & Polis, 1945; Chen & Ledford, 1971; Dulley, 1972; Kaminogawa *et al.*, 1972) and is stable over the pH range 1·5–9·0 (Chen & Ledford, 1971) or 4–9 (Kaminogawa *et al.*, 1972). As will be discussed later, it is quite heat stable.

3.3. The Plasmin System in Milk

The fibrinolytic system in blood consists of five main components: plasminogen, plasmin, plasminogen activators, plasmin inhibitors and inhibitors of plasminogen activators. All components appear to be present in milk (see Grufferty & Fox, 1988*a*).

The level of plasmin in freshly drawn milk is reported (de Rham & Andrews, 1982*a*) to be negligible but increases during storage owing to activation at plasminogen (Richardson, 1983*a*). The plasminogen activators in bovine milk are associated with the casein micelles, while plasmin inhibitors occur in the serum (Reimerdes *et al.*, 1976; Korycka-Dahl *et al.*, 1983). Richardson (1983*a*) estimated that the plasminogen activator activity in milk is equivalent to 0·5 Ploug units of urokinase/ml. The activator is quite heat stable: it retained full activity after heating for 10 min at 70°C and ~50% after 10 min at 80°C (Richardson, 1983*a*). There may be three plasminogen activators in milk and colostrum with M_r = 13 000, ~65 000 and ~700 000; the last two probably correspond to α_2-antiplasmin (the principal plasmin inhibitor in blood) and α_2-macroglobulin, respectively (Honkanen-Buzalski & Sandholm, 1981).

The plasmin activity in milk is increased by 30–40% by pasterurisation (Noomen, 1975) or cooking rennet casein or cheese curd at temperatures > 50°C (Farkye & Fox, 1990). Noomen (1975) suggested that increased activity is due to inactivation of the plasmin inhibitor. However, the heat stabilities of plasmin, plasminogen and trypsin inhibitor in milk are similar (Richardson, 1983*a*). Since the activation of plasminogen in milk is increased by pasteurisation (de Rham & Andrews, 1982*a*; Richardson, 1983*a*), it is likely that the increased plasmin activity in heated milk is due to inactivation of an inhibitor of plasminogen activator, rather than of a plasmin inhibitor (Richardson, 1983*a*).

The plasmin inhibitor in milk strongly reduces the activity of the enzyme in milk, as is indicated by the markedly greater proteolysis observed in redispersed ultracentrifugally-sedimented micelles compared with milk (Grufferty & Fox, 1988*a,c,d*; Madkor & Fox, 1991).

3.4. Proteolysis of Milk Proteins by Plasmin

β-Casein is hydrolysed more quickly than α_{s1}-casein in milk, Na-caseinate or isolated form by indigenous milk proteinase or added plasmin (Chen & Ledford, 1971; Kaminogawa *et al.*, 1972; Noomen, 1975; Eigel, 1977*a,b*; de Rham & Andrews, 1982*a*). It is hydrolised to γ-caseins (Kaminogawa & Yamauchi, 1972*a*, 1974; Eigel, 1977*a*; Snoeren & van Riel, 1979) and three members of the proteose peptone (PP) fraction (Andrews, 1978*a,b*; Eigel & Keenan, 1979; Andrews & Alichanidis, 1983). The γ_1-, γ_2- and γ_3-caseins consist of residues 29–209, 106–209 and 108–209 of β-casein, respectively (Groves *et al.*, 1972, 1973). PP5 consists of two peptides, β-casein f1–105 and 1–107, while PP8-fast represents β-casein f1–28 (Andrews, 1978*a,b*). Eigel and Keenan (1979) suggested that PP8-slow is β-casein f29–105/107 but band 12 of the PP fraction corresponds to these sequences (Andrews & Alichanidis, 1983).

The PP fraction of milk contains at least 38 peptides, 25 of which appear to be produced from the caseins by plasmin (Andrews & Alichanidis, 1983); 52% of the total PP originates from β-casein, 29% from α_{s1}-casein, 9% from α_{s2}-casein and 4% from κ-casein; the remaining 6% are attributed to the action of other indigenous proteinases. However, Le Bars and Gripon (1989) reported that most of the pH 4·6-soluble N produced by plasmin from whole casein is derived from α_{s2}-casein. There is a statistically significant relationship between plasmin activity and the PP content (Schaar, 1985) and the γ-casein content (Donnelly & Barry, 1983) of milk.

Plasmin produces several peptides from α_{s1}-casein (Eigel, 1977*b*; Andrews & Alichanidis, 1983), some of which appear to be λ-casein (Aimutis & Eigel, 1982). α_{s2}-Casein is rapidly hydrolysed by plasmin to several peptides (Snoeren & van Riel, 1979; Andrews & Alichanidis, 1983); eight of the bonds hydrolysed have been identified (Le Bars & Gripon, 1989). Chen and Ledford (1971 reported that κ-casein is resistant to proteolysis by indigenous milk proteinase but Andrews and Alichanidis (1983) found it to be hydrolysed quite rapidly by plasmin.

β-Lactoglobulin (β-lg) and α-lactalbumin (α-la) are not hydrolysed by plasmin (Chen & Ledford, 1971; Yamauchi & Kaminogawa, 1972); in fact, denatured β-lg inhibits plasmin (Snoeren *et al.*, 1980) probably via thiol-disulphide interchange between the activated sulphydryl group of denatured β-lg and the disulphide bonds of plasmin (Grufferty & Fox, 1986).

3.5. Factors Affecting Plasmin Activity in Milk

3.5.1. Stage of Lactation and Age of Cow

The concentrations of plasmin and especially plasminogen in milk increase in late lactation (Korycka-Dahl *et al.*, 1983; Richardson, 1983*b*; Schaar, 1985). The elevated plasmin activity in late lactation milks is reflected in reduced levels of β-casein with concomitant increases in γ-casein, which may represent 30% of β-casein (Davies & Law, 1977; Barry & Donnelly, 1980; Donnelly & Barry, 1983).

3.5.2. Breed

The increase in γ-casein, with the concomitant decrease in β-casein, in late lactation is especially marked in milks from older cows (Davies & Law, 1977) and plasmin activity in milk increases with lactation number (Schaar, 1985).

Plasmin activity appears to be higher in some breeds, e.g. Friesians and Swedish Red and White, than others, e.g. Jersey and Ayrshire (Davies & Law, 1977; Barry & Donnelly, 1980; Richardson, 1983*b*; Schaar, 1985).

3.5.3. Mastitis

The levels of plasmin and plasminogen in milk increase during mastitis (Barry & Donnelly, 1981; de Rham & Andrews, 1982*b*; Andrews, 1983:b; Schaar & Funke, 1986) and there is a partial transfer of both from the casein micelles to whey (Schaar & Funke, 1986). Barry and Donnelly (1981) suggest that most of the proteolytic activity in mastitic milk is due to plasmin, with proteinases from leucocytes representing only a small proportion but de Rham and Andrews (1982*b*) and Andrews (1983*b*) presented evidence showing that somatic cells are major contributors to the proteolytic activity of mastitic milk. Leucocytes contain plasminogen activators (Christman *et al.*, 1977) which may be partly responsible for the increased plasmin activity in mastitic and late lactation milks.

3.5.4. Low Temperature Storage

Reimerdes and Herlitz (1979) reported that more γ-casein is formed in milk stored at 4°C than at 10 or 26°C and attributed this to the dissociation of both β-casein and plasmin from the micelles. However, Donnelly and Barry (1983) showed that the conversion of β-casein to γ-casein in milk was 20 times faster at 37°C than at 4°C and the plasmin activity in casein micelles prepared from milks stored at 4 or 20°C for 16 h was the same (Grufferty & Fox, 1988*b*).

3.6. Heat Stability of Plasmin in Milk and Buffer Systems

Plasmin isolated from milk and dissolved in buffer, pH 7·0, was completely inactivated in 10 min at 80°C (Kaminogawa *et al.*, 1972). The heat stability of plasmin in buffers (Dulley, 1972) and in casein systems (Grufferty & Fox, 1988*d*) decreases with increasing pH in the range 3·5–9·2.

Pasteurisation of milk at 72°C for 15 s reduces plasmin activated by 10% (Korycka-Dahl *et al.*, 1983) or 17% (Richardson, 1983*a*) but leads to increased activity during subsequent storage (Noomen, 1975; de Rham & Andrews, 1982*a*; Andrews, 1983*a*; Andrews & Alichanidis, 1983; Korycka-Dahl *et al.*, 1983; Richardson, 1983*a*), probably owing to the inactivation of the inhibitors of plasminogen activator (Richardson, 1983*a*). The heat treatments necessary to prevent proteolysis in UHT and in bottle-sterilised milks during storage were 142°C for 18 s or 120°C for 15 min, respectively (Driessen & van der Waals, 1978).

The heat stability of porcine plasmin in phosphate buffer, pH 7·0, was increased when casein was added at 25 mg/ml but addition of skim milk, to a similar casein concentration, had no effect (Alichanidis *et al.*, 1986). The heat stability of indigenous plasmin was lower in milk than in a micellar casein dispersion (Grufferty & Fox, 1988*d*; Rollema & Poll, 1986), probably due to sulphydryl-disulphide interaction between denatured β-lg and plasmin.

The heat stability of indigenous plasmin was higher in a micellar than in a non-micellar casein system and the heat stability of plasmin in buffer was increased by adding lysine, suggesting that the binding of plasmin to casein via lysine residues increases its heat stability (Grufferty & Fox, 1988*d*).

3.7. Significance of Plasmin Activity in Dairy Products

Until recently, indigenous plasmin activity in milk was considered not to be very significant from the technological viewpoint. However, owing to relatively new developments in dairy practices and products (e.g. improved microbiological quality, long-term cold storage, UHT-sterilised products), it has assumed considerable importance.

3.7.1. Viscosity

Proteolysis by plasmin has been reported to reduce the viscosity of some dairy products, e.g. UHT-sterilised custard (Driessen *et al.*, 1981) and caseinates (Warner & Polis, 1945; Richardson, 1982; Grufferty & Fox, 1988*c*) especially from late lactation milk (Southward & Dolby, 1968), presumably due to elevated plasmin activity (although most of the plasmin in milk should be removed in the whey at pH 4·6), commercial acid casein contains considerable plasmin activity (Richardson & Elston, 1984). Most of the plasmin in milk is present in rennet casein (Richardson & Elston, 1984) and may cause proteolysis and textural changes in cheese analogues during storage.

3.7.2. Non-Casein Nitrogen

Indigenous plasmin forms pH 4·6-soluble N in Na caseinate (Warner & Polis, 1945) or micellar casein dispersions (Grufferty & Fox, 1988*c*) during storage. The N content of cheese whey increased during storage of milk, presumably due to plasmin activity; pasteurisation of the milk before storage increased the loss (Noomen, 1975). The proteose fraction, which is soluble at pH 4·6, is probably mainly responsible for these increases.

3.7.3. Ethanol Stability

Even extensive proteolysis of a micellar casein dispersion by indigenous plasmin did not affect its ethanol stability (Grufferty & Fox, 1988*c*). The low ethanol stability of late lactation milk is probably due mainly to a high concentration of calcium (Donnelly & Horne, 1986) rather than high plasmin activity.

3.7.4. Heat Stability

The considerable seasonal variation in the heat stability of milk (see Fox, 1982; van Boekel *et al.*, 1989; Chapter 15), appears to be due mainly to changes in the concentrations of urea and soluble calcium; plasmin activity is probably of little consequence, although very limited data are available.

3.7.5. Cheesemaking Characteristics

The cheesemaking properties of milk (rennet coagulation time (RCT), curd strength and curd syneresis) deteriorate with advancing lactation and the resulting cheese has a high moisture content (White & Davies, 1958; O'Keeffe *et al.*, 1982; O'Keefe, 1984). The stretchability and meltability of Mozzarella cheese made from late lactation milk were inferior to those of mid-lactation controls and cheese could not be made at all from very late lactation milk (Kindstedt *et al.*, 1991). In many respects, the cheesemaking properties of mastitic milk are similar to those of late lactation milk.

Many changes occur with advancing lactation which could contribute to these changes (e.g. increased pH, altered Ca concentration, modified micelle structure) but increased plasmin activity in both late lactation and mastitic milks is considered by many to be a major contributor. However, the available data are inconclusive. Pearse *et al.* (1986) and Grufferty and Fox (1988*c*) showed that the RCT of micellar casein systems increased only after extensive proteolysis by added or indigenous plasmin. Late lactation milks with poor curd-forming properties had higher contents of γ-casein and other casein fragments and lower levels of α_{s1}-, β- and κ-caseins than milks with good curd-forming properties (Okigbo *et al.*, 1985). However, the rennet coagulability of the former could be improved by acidification or addition of $CaCl_2$, suggesting that high plasmin activity is not the only factor responsible for the poor curd-forming properties of late lactation milk. Kindstedt *et al.* (1991) found that although the RCT of late lactation milk could be restored by acidification to the normal value of milk, other properties of the cheese were not restored; the significance of plasmin action in these changes has not yet been studied directly.

Donnelly *et al.* (1984) reported that the moisture content of Cheddar cheeses was directly related to the extent of pre-hydrolysis of cheesemilk by trypsin (which has similar specificity to plasmin) and concluded that proteolysis of caseins by indigenous plasmin contributes significantly to the high moisture levels in cheese made from late lactation milk. However, according to Pearse *et al.* (1986), normal syneresis occurred in a model milk in which $\sim 50\%$ of the β-casein had been hydrolysed by plasmin.

3.7.6. Cheese Ripening

Swiss and Cheddar cheeses contain 6–13 and 3–4·6 μg plasmin/g, respectively, and there are higher levels of γ-casein in 1-month-old Swiss cheese than in 3-month-old Cheddar, indicating that plasmin is very active in Swiss cheese during ripening (Richardson & Pearce, 1981). The higher cooking temperature used in Swiss cheese appears to increase plasmin activity (Farkyc & Fox, 1990). However, a study on cheese containing the plasmin inhibitor, 6-amino hexanoic acid, showed that plasmin makes a significant contribution to the formation of water soluble N in Cheddar (Farkye & Fox, 1991). β-Casein is hydrolysed to γ-caseins more quickly in Gouda than in Cheddar, possibly because its higher pH and moisture content favour plasmin activity (Creamer, 1974). An extensive study by Visser (1977) indicated that plasmin contributes less to the formation of soluble N in Gouda than rennet or starter proteinases and it did not contribute to bitter flavour development. Stadhouders (1960) considered that plasmin plays no role in the ripening of Edam but Reiter *et al.* (1969) hold the opposite view. Proteolysis by plasmin in Romano-type cheese was indicated by gel electrophoresis (Guinee & Fox, 1984); this type of cheese contains considerable plasmin activity, probably due to the high temperature used (Farkye & Fox, 1990).

In Meshanger-type cheese, plasmin activity was found to be dependent on pH, NaCl concentration and ripening temperature and was normally slight (Noomen, 1978).

Over 90% of exogenous plasmin added to cheese milk was incorporated into the cheese and accelerated ripening (Farkye & Fox, 1992).

3.7.7. Age Gelation of UHT Milks

This subject is reviewed in Chapter 17. While Nakai *et al.* (1964) and Cheng and Gelda (1974) failed to demonstrate a proteinase in UHT milks or creams, Snoeren *et al.* (1980) isolated plasmin from UHT milk by affinity chromatography. Several studies (e.g. Bjorck, 1973; Corradini, 1975; Snoeren *et al.*, 1979) have shown that proteolysis does occur in UHT milk during storage and that β-casein is hydrolysed more quickly than α_{s1}-casein with the formation of γ-caseins, suggesting plasmin activity.

When inhibitors of plasmin were added to UHT milk, no proteolysis or gelation occurred after 9 months at 20°C although extensive proteolysis occurred in the control which gelled after ~60 days (de Koning *et al.*, 1985). However, in concentrated UHT milk in which no plasmin or plasminogen was detected, gelation occurred after 41 days at 30°C. It was concluded that gelation of unconcentrated UHT-sterilised milks is due to proteolysis, but that concentrated UHT milks gelled via a non-enzymic, physicochemical mechanism. Milk sterlised by a direct UHT process gelled during storage whereas indirectly heated UHT milk did not (Manji *et al.*, 1986); although no correlation was observed between the shelf-life of the directly sterilised milk and the extent of proteolysis, this milk contained higher levels of plasmin and plasminogen than indirectly sterilised milk. Plasminogen was activated during storage and was probably mainly responsible for proteolysis. In a later study, Manji and Kakuda (1988) showed that UHT-treated concentrated milk gelled via a physicochemical mechanism, supporting the hypothesis of de Koning *et al.* (1985).

3.8. Other Indigenous Proteinases in Milk

Slight differences in the breakdown pattern of caseins in milk (with and without added plasmin) during storage at 37°C, the failure of soybean trypsin inhibitor to totally inhibit proteolysis in milk (Andrews, 1983*a*) and the fact that only 25 of the 38 peptides in the PP fraction could be attributed to the action of plasmin on individual caseins (Andrews & Alichanidis, 1983), all suggest that other proteinases besides plasmin occur in milk.

3.8.1. Acid Milk Proteinase

Kaminogawa and Yamauchi (1972*b*) isolated an acid proteinase from milk with a yield of 10% and a 2100-fold increase in specific activity. The enzyme had a pH optimum of 4·0 and a mol. wt. ~36 000; it was not inhibited by EDTA, α,α'-dipyridyl, iodoacetic acid or DFP and only ~10% by p-CMB. Its properties were similar to those of cathepsin D. α_{s1}-Casein was more susceptible than β-casein and its specificity on individual caseins was generally similar to that of chymosin (Kaminogawa *et al.*, 1980). The acid proteinase is associated with the casein in milk and may be active during cheese ripening.

3.8.2. Thrombin

Reimerdes (1983) described the isolation of two serine proteinases from milk by affinity chromatography on lysine-Sepharose; one enzyme was similar to plasmin while the properties of the other suggested that it might be thrombin.

3.8.3. Aminopeptidases

Using 4-nitroanilides of free and substituted amino acids, Reimerdes (1983) reported that there are aminopeptidases in milk serum specific for the 4-nitroanilides of alanine, leucine and lysine. Milk and blood sera showed similar peptidase profiles, suggesting that the aminopeptidases in milk originate from blood.

3.8.4. Proteinases from Leucocytes in Milk

At the levels found in milk, plasmin was 2–8 times more proteolytic than leucocytes isolated from blood and added to caseinate at a level of 10^6/ml (Grieve & Kitchen, 1985). Pasteurisation (65°C for 30 min) of milk completely inactivated the leucocyte enzyme(s). Gel electrophoresis showed that the leucocytes from blood hydrolysed the individual caseins in the order α_{s1}- > β- > κ-, in contrast to plasmin which hydrolised β- > α_{s1}- > κ-. Verdi *et al.* (1987) reported that the proteinase activity associated with elevated somatic cell counts increased the breakdown of both α_{s1}- and β-caseins during storage.

REFERENCES

Aimutis, W.R. & Eigel, W.N. (1982). Identification of λ-casein as plasmin-derived fragments of bovine α_{s1}-casein. *J. Dairy Sci.*, **65**, 175–81.

Alichanidis, E., Wrathal, J.H.M. & Andrews, A.T. (1986). Heat stability of plasmin (milk proteinase) and plasminogen. *J. Dairy Res.*, **53**, 259–69.

Andrews, A.T. (1978*a*). The composition, structure and origin of proteose–peptone component 5 of bovine milk. *Eur. J. Biochem.*, **90**, 59–65.
Andrews, A.T. (1978*b*). The composition, structure and origin of proteose–peptone component 8F of bovine milk. *Eur. J. Biochem.*, **90**, 67–71.
Andrews, A.T. (1983*a*). Proteinases in normal bovine milk and their action on caseins. *J. Dairy Res.*, **50**, 45–55.
Andrews, A.T. (1983*b*). Breakdown of caseins by proteinases in bovine milks with high somatic cell counts arising from mastitis or infusion with bacterial endotoxin. *J. Dairy Res.*, **50**, 57–66.
Andrews, A.T. & Alichanidis, E. (1983). Proteolysis of caseins and the proteose–peptone fraction of bovine milk. *J. Dairy Res.*, **50**, 275–90.
Barry, J.G. & Donnelly, W.J. (1980). Casein compositional studies. I. The composition of casein from Friesian herd milks. *J. Dairy Res.*, **47**, 71–82.
Barry, J.G. & Donnelly, W.J. (1981). Casein compositional studies. II. The effect of secretory disturbance on casein composition in freshly drawn and aged bovine milks. *J. Dairy Res.*, **48**, 437–46.
Bjorck, L. (1973). Modification of casein in ultra-high-temperature treated milk during storage. *Milchwissenshaft*, **28**, 291–3.
Castellino, F.J. & Powell, J.R. (1981). Human plasminogen. *Meth. Enzymol.*, **80**, 365–78.
Chen, J.H. & Ledford, R.A. (1971). Purification and characterisation of milk protease. *J. Dairy Sci.*, **54**, 763 (Abstr.).
Cheng, W.S. & Gelda, C.S. (1974). Proteolysis in ultra-high-temperature treated and canned ten percent cream. *J. Dairy Sci.*, **57**, 1502–4.
Christman, J.K., Silverstein, S.C. & Acs, G. (1977). Plasminogen activators. In *Proteinases in Mammalian Cells and Tissues*, ed. A.J. Barrett. North Holland Publishing Company, New York, pp. 91–149.
Corradini, C. (1975). Gel formation behaviour in UHT sterilised milk. *Milchwissenschaft*, **30**, 413–16.
Creamer, L.K. (1974). β-Casein degradation in Gouda and Cheddar cheese. *J. Dairy Sci.*, **58**, 287–92.
Davies, D.T. & Law, A.J.R. (1977). The composition of whole casein from the milk of Ayrshire cows. *J. Dairy Res.*, **44**, 447–54.
De Koning, P.J., Kaper, J., Rollema, H.S. & Driessen, F.M. (1985). Age-thinning and gelation in unconcentrated and concentrated UHT-sterilised skim milk. Effect of native milk proteinase. *Neth. Milk Dairy J.*, **39**, 71–87.
De Rham, O. & Andrews, A.T. (1982*a*). The roles of native milk proteinase and its zymogen during proteolysis in normal bovine milk. *J. Dairy Res.*, **49**, 577–85.
De Rham, O. & Andrews, A.T. (1982*b*). Qualitative and quantitative determination of proteolysis in mastitic milks. *J. Dairy Res.*, **49**, 587–96.
Donnelly, W.J. & Barry, J.G. (1983). Casein compositional studies. III. Changes in Irish milk for manufacturing and role of milk proteinase. *J. Dairy Res.*, **50**, 433–41.
Donnelly, W.J. & Horne, D.S. (1986). Relationship between ethanol stability of bovine milk and natural variation in milk composition. *J. Dairy Res.*, **53**, 23–33.
Donnelly, W.J., Barry, J.G. & Buchheim, W. (1984). Casein micelle composition and syneretic properties of late lactation milk. *Ir. J. Food Sci. Technol.*, **8**, 121–30.
Driessen, F.M. & van der Waals, C.B. (1978). Inactivation of native milk proteinase by heat treatment. *Neth. Milk Dairy J.*, **32**, 245–54.
Driessen, F.M., van Hooydonk, A., Streuper, A. & Kingma, F. (1981). Influence of native milk proteinase on the viscosity of UHT-sterilized custard during storage. *Neth. Milk Dairy J.*, **35**, 121–31.
Dulley, J.R. (1972). Bovine milk protease. *J. Dairy Res.*, **39**, 1–9.
Eigel, W.N. (1977*a*). Formation of γ_1-A^2, γ_2-A^2 and γ_3-A^2 caseins by *in vitro* proteolysis of β-casein A^2 with bovine plasmin. *Intern. J. Biochem.*, **8**, 187–92.

Eigel, W.N. (1977*b*). Effect of bovine plasma on α_{s1}-β, and κ-A caseins. *J. Dairy Sci.*, **60**, 1399–403.

Eigel, W.N. & Keenan, T.W. (1979). Identification of proteose peptone component 8 slow as plasmin derived fragment of bovine β-casein. *Intern. J. Biochem.*, **10**, 529–35.

Eigel, W.N., Hofman, C.J., Chibber, B.A.K., Tomich, J.M., Keenan, T.W. & Mertz, E.T. (1979). Plasmin-mediated proteolysis of casein in bovine milk. *Proc. Natl Acad. Sci. (USA)*, **76**, 2244–8.

Farkye, N.Y. & Fox, P.F. (1990). Observations on plasmin activity in cheese. *J. Dairy Res.*, **57**, 413–18.

Farkye, N.Y. & Fox, P.F. (1991). Preliminary study on the contribution of plasmin to proteolysis in Cheddar cheese: cheese containing plasmin inhibitor, 6-aminohexanoic acid. *J. Agric. Chem.*, **39**, 786–8.

Farkye, N.Y. & Fox, P.F. (1992). On the contribution of plasmin to Cheddar cheese ripening. Effect of added plasmin. *J. Dairy Res.*, **59**, 209–16.

Fox, P.F. (1981). Proteinases in dairy technology. *Neth. Milk Dairy J.*, **35**, 233–53.

Fox, P.F. (1982). Heat induced coagulation of milk. In *Developments in Dairy Chemistry —1: Proteins*, ed. P.F. Fox. Applied Science Publishers, London, pp. 189–228.

Grieve, P.A. & Kitchen, B.J. (1985). Proteolysis in milk: the significance of proteinases originating from milk leucocytes and a comparison of the action of leucocyte, bacterial and natural milk proteinases on casein. *J. Dairy Res.*, **52**, 101–12.

Groves, M.L., Gordon, W.G., Kalan, E.B. & Jones, S.B. (1972). Composition of bovine γ-caseins A^1 and A^3, and further evidence for a relationship in biosynthesis of γ- and β-caseins. *J. Dairy Sci.*, **55**, 1041–9.

Groves, M.L., Gordon, W.G., Kalan, E.B. & Jones, S.B. (1973). TS-A^2, TS-B, R-, and S-caseins: Their isolation, composition, and relationship to the β- and γ-casein polymorphs A^2 and B. *J. Dairy Sci.*, **56**, 558–68.

Grufferty, M.B. & Fox, P.F. (1986). Potassium iodate-induced proteolysis in ultra heat treated milk during storage: the role of β-lactoglobulin and plasmin. *J. Dairy Res.*, **53**, 601–13.

Grufferty, M.B. & Fox, P.F. (1988*a*). Milk alkaline proteinase. *J. Dairy Res.*, **55**, 609–30.

Grufferty, M.B. & Fox, P.F. (1988*b*). Factors affecting the release of plasmin activity from casein micelles. *N.Z. J. Dairy Sci. Technol.*, **23**, 153–63.

Grufferty, M.B. & Fox, P.F. (1988*c*). Functional properties of casein hydrolysed by alkaline milk proteinase. *N.Z. J. Dairy Sci. Technol.*, **23**, 95–108.

Grufferty, M.B. & Fox, P.F. (1988*d*). Heat stability of the plasmin system in milk and casein systems. *N.Z. J. Dairy Sci. Technol.*, **23**, 143–52.

Guinee, T.P. & Fox, P.F. (1984). Studies on Romano-type cheese: General proteolysis. *Ir. J. Food Sci. Technol.*, **8**, 105–14.

Halpaap, I., Reimerdes, E.H. & Klostermeyer, H. (1977). Milk proteinases. VI. Comparative isolation of plasminogen from bovine blood and proteinase from cows milk. *Milchwissenschaft*, **32**, 341–6.

Hernell, O. & Lonnerdal, B. (1989). Enzymes in human milk. In *Protein and Non-protein Nitrogen in Human Milk*, ed. S.A. Atkinson & B. Lonnerdal. CRC Press, Boca Raton, FL, pp. 67–75.

Hofmann, C.J., Keenan, T.W. & Eigel, W.N. (1979). Association of plasminogen with bovine milk fat globule membrane. *Intern. J. Biochem.*, **10**, 909–17.

Honkanen-Buzalski, T. & Sandholm, M. (1981). Trypsin-inhibitors in mastitic milk and colostrum: correlation between trypsin-inhibitor capacity, bovine serum albumin and somatic cell contents. *J. Dairy Res.*, **48**, 213–23.

Humbert, G. & Alais, C. (1979). Reviews of the progress of dairy science: The milk proteinase system. *J. Dairy Res.*, **46**, 559–71.

Kaminogawa, S. & Yamauchi, K. (1972*a*). Decomposition of β-casein by milk protease. *Agric. Biol. Chem.*, **36**, 255–60.

Kaminogawa, S. & Yamauchi, K. (1972*b*). Acid protease of bovine milk. *Agric. Biol. Chem.*, **36**, 2351–6.

Kaminogawa, S. & Yamauchi, K. (1974). Similarity between γ-casein and a product of β-casein hydrolysis by milk protease. *Agric. Biol. Chem.*, **38**, 2343–8.

Kaminogawa, S., Yamauchi, K. & Tsugo, T. (1969). Properties of milk protease concentrated from acid precipitated casein. *Jap. J. Zootech. Sci.*, **40**, 559–65.

Kaminogawa, S., Mizobuchi, H. & Yamauchi, K. (1972). Comparison of bovine milk protease with plasmin. *Agric. Biol. Chem.*, **36**, 2163–7.

Kaminogawa, S., Sato, F. & Yamauchi, K. (1971). Purification and some properties of milk protease. *Agric. Biol. Chem.*, **35**, 1465–7.

Kaminogawa, S., Yamauchi, K., Miyazawa, S. & Koga, Y. (1980). Degradation of casein components by acid protease of bovine milk. *J. Dairy Sci.*, **63**, 701–4.

Korycka-Dahl, M., Ribadeau-Dumas, B., Chene, N. & Martal, J. (1983). Plasmin activity in milk. *J. Dairy Sci.*, **66**, 704–11.

Le Bars, D. & Gripon, J.C. (1989). Specificity of plasmin towards α_{s2}-casein. *J. Dairy Res.*, **56**, 817–21.

Manji, B. & Kakuda, Y. (1988). The role of protein denaturation, extent of proteolysis, and storage temperature on the mechanism of age gelation in a model system. *J. Dairy Sci.*, **71**, 1455–63.

Manji, B., Kakuda, Y. & Arnott, D.R. (1986). Effect of storage temperature on age gelation of ultra-high temperature milk processed by direct and indirect heating systems. *J. Dairy Sci.*, **69**, 2994–3001.

Madkor, S.A. & Fox, P.F. (1991). Plasmin in buffalo milk. *Food Chemistry*, **39**, 139–56.

Mathur, M.P. & Sharma, R.K. (1979). Studies on buffalo milk protease. *Indian J. Dairy Sci.*, **32**, 177–81.

McMeekin, T.L., Hipp, N.J. & Groves, M.L. (1959). The separation of components of α-casein. I. The preparation of α_{s1}-casein. *Arch. Biochem. Biophys.*, **83**, 35–43.

Nakai, S., Wilson, H.K. & Herreid, E.O. (1964). Assaying sterile concentrated milk for native proteolytic enzymes. *J. Dairy Sci.*, **47**, 754–7.

Noomen, A. (1975). Proteolytic activity of milk protease in raw and pasteurised cow's milk. *Neth. Milk Dairy J.*, **29**, 153–61.

Noomen, A. (1978). Activity of proteolytic enzymes in simulated soft cheeses (Meshanger type). 1. Activity of milk protease. *Neth. Milk Dairy J.*, **32**, 26–48.

O'Keeffe, A.M. (1984). Seasonal and lactational influences on moisture content of Cheddar cheese. *Ir. J. Food Sci. Technol.*, **8**, 27–37.

O'Keeffe, A.M., Phelan, J.A., Keogh, K. & Kelly, P. (1982). Studies on milk composition and its relationship to some processing criteria. IV. Factors influencing the renneting properties of a seasonal milk supply. *Ir. J. Food Sci. Technol.*, **6**, 39–48.

Okigbo, L.M., Richardson, G.H., Brown, R.J. & Ernstrom, C.A. (1985). Casein composition of cow's milk of different chymosin coagulation properties. *J. Dairy Sci.*, **68**, 1887–92.

Pearse, M.J., Linklater, P.M., Hall, R.J. & MacKinlay, A.G. (1986). Extensive degradation of casein by plasmin does not impede subsequent curd formation and syneresis. *J. Dairy Res.*, **53**, 477–80.

Reimerdes, E.H. (1983). New aspects of naturally occurring proteases in bovine milk. *J. Dairy Sci.*, **66**, 1591–600.

Reimerdes, E.H. & Herlitz, E. (1979). The formation of γ-casein during cooling of raw milk. *J. Dairy Res.*, **46**, 219–21.

Reimerdes, E.H. & Klostermeyer, H. (1974). Milk Proteases. 1. Micelle associated protease-purification and hydrolysis of caseins. *Milchwissenschaft*, **29**, 517–23.

Reimerdes, E.H., Klostermeyer, H. & Sayk, E. (1976). Milk proteinases. 7. Fractionation of components of the proteinase inhibitor system in milk. *Milchwissenschaft*, **31**, 329–34.

Reiter, B., Sorokin, Y., Pickering, A. & Hall, A.J. (1969). Hydrolysis of fat and protein in small cheeses made under aseptic conditions. *J. Dairy Res.*, **36**, 65–76.

Richardson, B.C. (1982). The effect of storage on the viscosity of some casein solutions. *N.Z. J. Dairy Sci. Technol.*, **17**, 277–82.

Richardson, B.C. (1983*a*). The proteinases of bovine milk and the effect of pasteurisation on their activity. *N.Z. J. Dairy Sci. Technol.*, **18**, 233–45.

Richardson, B.C. (1983*b*). Variation of the concentration of plasmin and plasminogen in bovine milk with lactation. *N.Z. J. Dairy Sci. Technol.*, **18**, 247–52.

Richardson, B.C. & Elston, P.D. (1984). Plasmin activity in commercial caseins and caseinates. *N.Z. J. Dairy Sci. Technol.*, **19**, 63–7.

Richardson, B.C. & Pearce, K.N. (1981). The determination of plasmin in dairy products. *N.Z. J. Dairy Sci. Technol.*, **16**, 209–20.

Rollema, H.S. & Poll, J.K. (1986). The alkaline milk proteinase system: kinetics and mechanism of heat-inactivation. *Milchwissenschaft*, **41**, 536–40.

Schaar, J. (1985). Plasmin activity and proteose peptone content of individual milks. *J. Dairy Res.*, **52**, 369–78.

Schaar, J. & Funke, H. (1986). Effect of subclinical mastitis on milk plasminogen and plasmin compared with that on sodium, antitrypsin and *N*-acetyl-β-D-glucosaminidase. *J. Dairy Res.*, **53**, 515–28.

Snoeren, T.H.M. & van Riel, J.A.M. (1979). Milk proteinase, its isolation and action on α_{s2}- and β-casein. *Milchwissenschaft*, **34**, 528–31.

Snoeren, T.H.M., van der Spek, C.A., Dekker, R. & Both, P. (1979). Proteolysis during the storage of UHT sterilised wholemilk. 1. Experiments with milk heated by the direct system for 4 seconds at 142°C. *Neth. Milk Dairy J.*, **33**, 31–9.

Snoeren, T.H.M., van Riel, J.A.M. & Both, P. (1980). Enkele eigensahappen van het uit UHT-gesteriliseerde melk geisoleerde melkproteinase. *Zuivelzicht*, **72**, 42–3.

Southward, C.R. & Dolby, R.M. (1968). Seasonal variation in the viscosity index and adhesive strength of casein from the milk of individual cows. *J. Dairy Res.*, **35**, 25–30.

Stadhouders, J. (1960). The hydrolysis of protein during the ripening of Dutch cheese. The enzyme and bacteria involved. *Neth. Milk Dairy J.*, **14**, 83–110.

Van Boekel, M.A.J.S., Nieuwenhuijse, J.A. & Walstra, P. (1989). The heat coagulation of milk. 1. Mechanisms. *Neth. Milk Dairy J.*, **43**, 97–127.

Varshney, G.C. & Mathur, M.P. (1979). Purification and properties of goat milk protease. *Indian J. Biochem. Biophys.*, **16**, 375–8.

Verdi, R.J., Barbano, D.M., Dellavalle, M.E. & Senyk, G.F. (1987). Variability in true protein, casein, nonprotein nitrogen and proteolysis in high and low somatic cell milks. *J. Dairy Sci.*, **70**, 230–42.

Visser, F.M.W. (1977). Contribution of enzymes from rennet, starter bacteria and milk to proteolysis and flavour development in Gouda cheese. 5. Some observations on bitter extracts from aseptically made cheese. *Neth. Milk Dairy J.*, **31**, 265–76.

Visser, S. (1981). Proteolytic enzymes and their action on milk proteins—A review. *Neth. Milk Dairy J.*, **35**, 65–88.

Warner, R.C. & Polis, E. (1945). On the presence of a proteolytic enzyme in casein. *J. Am. Chem. Soc.*, **67**, 529–32.

White, J.C.D. & Davies, D.T. (1958). The relation between the chemical composition of milk and the stability of the caseinate complex. III. Coagulation of rennet. *J. Dairy Res.*, **25**, 267–80.

Yamauchi, K. & Kaminogawa, S. (1972). Decomposition of milk proteins by milk protease. *Agric. Biol. Chem.*, **36**, 249–54.

Yamauchi, K., Kaminogawa, S. & Tsugo, T. (1969). Separation of milk protease fraction from casein. *Jap. J. Zootech. Sci.*, **40**, 551–8.

Zittle, C.A. (1965). Purification of protease in cows milk. *J. Dairy Sci.*, **48**, 771 (Abstr.).

IV. Phosphatases in Milk

4.1. Alkaline Phosphatase (E.C.3.1.3.1)

4.1.1. Technological Significance

Indigenous milk alkaline phosphatase is important from the dairy technology viewpoint and for the safety of dairy foods because of the widespread use of the phosphatase test as an index of HTST pasteurisation efficiency (Kay & Graham, 1935; Aschaffenburg & Mullen, 1949). Alkaline phosphatase is always present in raw milk at easily measured levels and its heat stability profile closely follows that required for adequate pasteurisation. A wide range of pathogenic bacteria, including *Salmonella* (Anon., 1984), *Campilobacter* (Anon., 1985) and *Listeria* (Hayes *et al.*, 1986) can proliferate in milk so the test is crucial to public health in terms both of assuring adequate heat treatments and for establishing whether raw milk has been added to heated or unheated products. False positive readings, which can be attributed to a partial reactivation of the enzyme, have been reported (Lyster & Aschaffenburg, 1962; Peereboom, 1969), but in general the test has proved to be both useful and reliable. Reactivation tends to occur when milk or cream is exposed briefly to high temperatures (80–180°C for a few seconds) followed by storage at 5–40°C (optimum about 30°C). Lower temperature treatments such as those typical of pasterisation are seldom followed by reactivation. It is more common in cream than in milk but it never exceeds 1% of the initial activity. If the milk is preheated and Mg^{2+}, glycerophosphate or low levels of β-lactoglobulin added to promote reactivation, as much as 30% of the activity has been restored (Lyster & Aschaffenburg, 1962). When reactivation occurs, enzyme activity first becomes detectable after about 6 h and then progressively increases. To avoid concluding unjustifiably that the heat treatment had not been adequate for pasteurisation, enzyme assays should be performed as soon as possible after the heating (Linden, 1979).

Alkaline phosphatase occurs in all milks, although the levels present may vary considerably between species. A modified phosphatasc test for pasteurisation has been applied successfully to caprine milk (Williams, 1986). Many variations of the test have been applied to bovine milk, most of the alterations being to improve speed or to facilitate its application to particular products. The Scharer Rapid Phosphatase Test is used widely in the USA and when combined with pasteurisation can be used to distinguish mammalian alkaline phosphatases from the more thermostable bacterial isoenzymes (Pratt-Lowe *et al.*, 1987). There are some indications that alkaline phosphatase activity is influenced by lactose and protein content, being lower at high protein and low lactose levels, at least in ultrafiltered milks (Mistry, 1989), so the phosphatase test may require modification under such conditions. There are also reports (Lorient & Linden, 1976; Jasinska *et al.*, 1985) that β-lactoglobulin has an inhibitory effect on alkaline phosphatase activity but it is small and probably of no technological significance. The different genetic variants of β-lactoglobulin appeared to vary in inhibitory power (Jasinska *et al.*, 1985).

4.1.2. Occurrence and Distribution

The milks of individual animals (cows) can differ by as much as 40-fold in alkaline phophatase content and even bulk supplies can show a seasonal 2-fold variation in levels (Haab & Smith, 1956). This variation is most probably due to lactational effects rather than to breed or diet, since there are very high levels in colostrum which fall to a mininum 1–2 weeks after parturition and then they rise again, reaching an approximately constant level about 25 weeks post-partum (Jenness & Patton, 1959; Kiermeier & Meinl, 1961). There is an inverse corelation between alkaline phosphatase and milk yield but there does not seem to be any consistent relationship to fat content. Mastitis causes considerable elevations in milk alkaline phosphatase activity and this probably also contributes to observed fluctuations in enzyme levels (Bogin & Ziv, 1973; Singh & Ganguli, 1975). Alkaline phosphatase is associated with microsomal particles in milk and these are adsorbed onto fat globules so about 30–50% of the enzyme is also found in the cream phase.

4.1.3. Purification

Most extractions begin by churning the milk or cream and separating the buttermilk from which the enzyme is recovered by a butanol extraction step (Morton, 1953; Linden *et al.*, 1974) (the enzyme is stable to the organic solvent which dissociates the protein–fat complexes). Gel filtration (Sephadex G-200) and ion-exchange chromatography (DEAE-cellulose) are most frequently used for further purification (Linden *et al.*, 1974; Linden & Alais, 1976; Kuzuya *et al.*, 1980) but affinity chromatography on concanavalin A-Sepharose 4B has also been used in a study giving apparently homogeneous enzyme (Kuzuya *et al.*, 1982). The overall purification factor was almost 7200 and the final product has an activity of 43·1 units/mg of protein. Other reports (Linden *et al.*, 1974) achieved a similar purification factor (7440) and claimed a homogeneous enzyme preparation with a specific activity of 96 units/mg, even without an affinity chromatography step.

There is a report (Peereboom, 1968) that there are a number of alkaline phosphatase isoenzymes, but most workers have found little evidence for this. They may represent the same enzyme in different states of aggregation or enzyme bound to membrane fragments, other proteins or lipids. The possible glycoprotein nature of the enzyme is still not entirely clear, however. Much of the more thorough analytical work on the enzyme (Morton, 1953; Linden *et al.*, 1974; Linden & Alais, 1976) did not report the presence of sialic acid or any other carbohydrate, while others claim that it is a sialoglycoprotein (Sharma & Ganguli, 1969) and that the electrophoretic mobility of the larger 'isoenzyme' was altered by neuraminidase treatment (Peereboom, 1968). The latter could also be due to the liberation of sialic acid from any accompanying bound material rather than from the enzyme itself. However, the usefulness of concanavalin-A-Sepharose as a purification step (Kuzuya *et al.*, 1982) strongly supports a glycoprotein structure for the enzyme.

TABLE 1
Molecular Parameters of Bovine Milk Alkaline Phosphate

Parameters	*Value*
Native M_R	160 000–190 000
Subunit M_R	85 000
Sedimentation constant ($S_{20,w}$)	6·0
pH optimum	9·8–9·9
Zn^{2+} content	4·9 ± 0·6 g atom/mol
Specific activity	43–96 IU/mg
pH stability range	> 7·0–10·0
Activators	Cd^{2+}, Co^{2+}, Mg^{2+}, Mn^{2+}, Ni^{2+}
Inhibitors	Be^{2+}, Ca^{2+}, Cu^{2+}
	Chelating and oxidizing agents, orthophosphate (K_i = 3·0 mM); L-phenylalanine (K_i = 3·7 mM); L-leucine (K_i = 0·2 mM)
K_m(*p*-nitrophenylphosphate, pH 9, 37°C)	0·5–0·7 mM
V_{max}(*p*-nitrophenylphosphate, pH 9, 37°C)	210 μmol min^{-1} mg^{-1}
Thermal stability:	
D-value at 60°C, pH 9	27·2 min
D-value at 63°C, pH 9	8·3 min

4.1.4. Characterisation

Alkaline phosphatase is a phosphomonoesterase with optimum activity in the alkaline pH range (9–10·5). Like many other phosphomonoesterases, it is active against a wide range of substrates and hydrolyses most phosphate ester bonds, including AMP, glycerophosphate, phosphates of glucose, serine and threonine and also phosphoproteins, but it will not cleave ADP, ATP, pyrophosphates or polyphosphates. The optimum pH for hydrolysis is influenced by the nature of the substrate. For example, phosphoserine is hydrolysed most rapidly at pH 8·5–9·0, but pH 6·5–7·0 is best for cleaving the phosphate groups off phosphoproteins such as casein (Zittle & Bingham, 1959; Linden & Alais, 1976; Lorient & Linden, 1976). Catalysis proceeds via a transient phosphoryl-enzyme intermediate involving a reactive serine hydroxyl group. The κ_m value when *p*-nitrophenylphosphate is used as substrate (Barman & Gutfreund, 1966; Linden & Alais, 1976; Kuzuya *et al.*, 1980) at pH 10·0 and 25°C, is about 5×10^{-4} M. Inorganic orthophosphate is a strong competitive inhibitor of *p*-nitrophenylphosphate hydrolysis but κ_i values reported (Barman & Gutfreund, 1966; Kuzuya *et al.*, 1980) differ by a factor of 10 from 0·3 to 3·0 mM.

Molecular properties of bovine milk alkaline phosphatase are summarised in Table 1 from which it is apparent that the enzyme is a dimer, containing 4 or 5 Zn atoms per molecule. It is stable and active at alkaline pH values but various specific activity values from 43 to 96 U/mg have been reported for supposedly homogeneous pure enzyme samples. Zn^{2+} are required for activity but Mg^{2+} is also a strong activator so 1 mM Mg^{2+} is usually added to all buffers before assay. Other metal ions shown as activators in Table 1 are much less effective. Chelating agents

TABLE 2
Molecular Parameters of Bovine Milk Acid Phosphatase

Parameters	*Value*
Native M_R	42 500
Subunit M_R	41 500
Sedimentation constant ($S_{20,w}$)	2·8
Partial specific volume ($\bar{v}$) (calculated)	0·728
Extinction coefficient ($E^{1\%}_{1\,cm}$) at 280 nm	9·4
Isoelectric point (pI)	7·9
Specific activity	30·2 IU/mg
pH optimum	4·9
pH stability range	3·0–9·0
Activators	Ascorbic acid, β-mercaptoethanol, dithiothreitol, L-cysteine thioglycollic acid
Inhibitors	Ag^+, Al^{3+}, Cu^{2+}, F^-, Fe^{3+}, Hg^{2+}, Sn^{2+}, Zn^{2+}, oxidizing agents
K_m (*p*-nitrophenylphosphate, pH 4·8, 37°C)	0·81 mM
K_m (*p*-nitrophenylphosphate, pH 5·2, 37°C)	1·41 mM
K_m (total casein, pH 5·2, 37°C)	2·84 mM
Thermal stability:	
D-value at 85°C, pH 4·9	25 min
D-value at 89°C, pH 4·9	6·5 min
D-value at 75°C, pH 6·7	10 min

such as EDTA, EGTA, etc., are powerful inhibitors, as expected for a metallo-enzyme. The purified enzyme has high contents of serine, glutamic acid, alanine and lysine and low amounts of aromatic and hydrophobic amino acids (Linden *et al.*, 1974).

4.1.5. Similarity to Mammary Gland Enzyme

Kitchen (1985) pointed out that bovine milk alkaline phosphatase has many similarities to the mammary gland enzyme (O'Keeffe & Kinsella, 1979). Both enzymes are dimers, consisting of two large polypeptide chains, each having molecular weights of about 80 000, and K_m values (0·5–0·7) for *p*-nitrophenyl-phosphate hydrolysis are very similar. The pH optimum of the mammary enzyme varies with substrate concentration, being close to 10·0 at 2 mM *p*-nitrophenyl-phosphate and 9·0 at 0·1 mM substrate. This variation has not been reported for the milk enzyme, but the pH optima of the two enzymes are in the same range. The mammary enzyme is a glycoprotein with a carbohydrate content of almost 30% and it binds strongly to concanavalin A-bound to agarose beads. Although the milk enzyme has not been proven to be a glycoprotein, affinity chromatography on the same immobilised ligand has been used successfully for its purification (Kuzuya *et al.*, 1982). Therefore, it seems that the mammary gland itself is probably the source of the milk enzyme and that these two enzymes are the same.

4.2. Acid Phosphatase (E.C.3.1.3.2)

4.2.1. Technological Significance

Milk acid phosphatase is also a phosphomonoesterase, and, like the alkaline phosphatase, it is distinguished in operational terms by the pH range over which it is active. The levels of acid phosphatase in milk are lower than those of the alkaline enzyme but nevertheless it may be of some technological importance in dairy products for three reasons. First, it is relatively stable to heat so that in pasteurized products it is the only active indigenous phosphatase. Second, many food products have a pH (about 5) close to its activity maximum and third, phosphoproteins are excellent substrates. In milk the caseins, being phosphoproteins, are good subtrates for acid phosphatase and as such should be competitive inhibitors of *p*-nitrophenylphosphate hydrolysis. This was indeed found to be the case (Andrews, 1974), their inhibitory activity being directly proportional to their molecular content of phosphate. α_{s1}-Casein ($8\,PO_4$ groups mol^{-1}) is a more powerful competitive inhibitor than β-casein ($5\,PO_4\,mol^{-1}$) and κ-casein ($1\,PO_4\,mol^{-1}$) is least effective. In milk the caseins occur as multimolecular aggregates, containing all four caseins (α_{s1}-, α_{s2}-, β- and κ) which associate further into micelles containing 5×10^4 or more casein molecules. Both hydrophobic and electrostatic bonding contribute to micelle formation and stability but interchain bridging by Ca^{2+}, linking phosphate groups on different casein polypeptide chains is also important. The dramatic change in milk transparency on addition of a chelating agent such as EDTA which binds the calcium ions and causes the micelles to disintegrate with an accompanying drop in light scattering power is good evidence for the importance of calcium and phosphate in maintaining micelle integrity. The resulting calcium-free caseins still remain in an aggregated form (hydrophobic and electrostatic bonds remain) but as much smaller particles with reduced light scattering ability. The action of acid phosphatase in cleaving phosphate groups off the serine residues of the caseins has similar consequences and also disrupts calcium phosphate bridging. Thus, phosphatase activity reduces the ability of caseins to form stable complexes (Pepper & Thompson, 1963; Bingham *et al.*, 1972, 1976; Knoop & Peters, 1975). Micellar-type aggregates may be formed but compared to unmodified casein these are both fewer in number and larger with dephosphorylated casein, suggesting weaker intermolecular bonding and a 'looser' structure.

The presence of casein micelles and the maintenance of their integrity are vital to the technological properties of milk so it is not surprising that acid phosphatase has been investigated in relation to the gelling of UHT milk (Andrews & Pallavicini, 1973), curd structure and strength (Ohmiya *et al.*, 1983; Pearce *et al.*, 1986) and in cheese maturation and flavour development (Schormuller *et al.*, 1960; Andrews & Alichanidis, 1975*a*). Extensive proteolysis of the caseins occurs during cheese maturation, producing phosphopeptides among the products formed and it has been suggested (Schormuller *et al.*, 1960) that the synergistic action of proteinases and phosphatases is required to hydrolyse these further, to avoid bitter peptides and ensure correct flavour development. Starter organisms also contribute

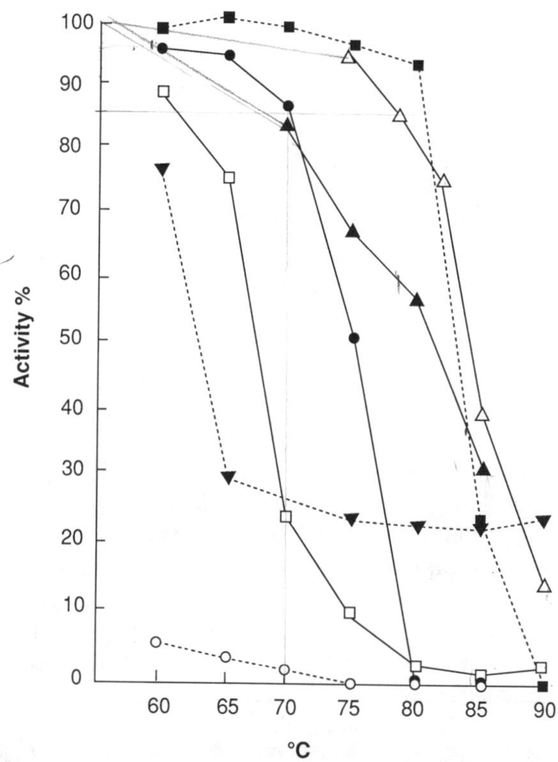

Fig. 1. Heat stability of bovine milk enzymes. Whole milk samples (50–500 μl) sealed in thin-walled glass capillary tubes were heated for 15 s at the indicated temperatures before cooling and assay of residual enzyme activity. Lipoprotein lipase, ○; α-L-fucosidase, ▼; N-acetyl-β-glucosaminidase, □; γ-glutamyl transpeptidase, ●; xanthine oxidase, ▲; α-D-mannosidase, ■; acid phosphatase. △ (reproduced with permission from Journal of Dairy Research).

to phosphatase levels in cheese but the amounts depend on starter strain and may be quite small, so the predominant phosphatase activity in a typical hard cheese can be attributed to milk acid phosphatase. The pH of most cheeses is close to the optimum for the milk acid phosphatase so that dephosphorylation of caseins and phosphopeptides certainly occurs, although possibly only slowly (Dulley & Kitchen, 1973) and the importance, if any, of the enzyme to cheese flavour profiles has not yet been established.

When first reported (Mullen, 1950), acid phosphatase was identified as the most heat-stable indigenous milk enzyme although it is now known that plasmin is probably even more heat stable (Alichanidis *et al.*, 1986). Figure 1 shows the heat stability of bovine milk acid phosphatase together with that of a number of other indigenous milk enzymes which may have value as indicators of various heat treatments given to milk and dairy products (Andrews *et al.*, 1987). Acid phosphatase survives normal pasteurisation conditions and heating at pH 6·7 for nearly

5 s at 100°C or 20 s at 90°C is required for 90% inactivation (Andrews, 1974), while 15–20% of the initial activity may survive even UHT processing (Andrews & Pallavicini, 1973). pH greatly affects heat stability, however, and the acid phosphatase is even more stable at acid pH values (e.g. 4·9–5·2) which are typical of many dairy products. Thermal denaturation follows first-order kinetics and is unaffected by the addition of casein substrate (Andrews, 1974). The acid phosphatase may be quite stable to heat but it has been reported (Mullen 1950) that visible light (but not UV) causes quite extensive inactivation.

4.2.2. Occurrence and Distribution

Acid phosphatase levels in normal bovine milk range from 0·0026 to 0·0037 IU/ml when *p*-nitrophenylphosphate is used as substrate at pH 4·0 (Kitchen *et al.*, 1970). At the pH optimum (pH 4·9), activity levels are about double this. While other workers have reported higher (Bingham & Zittle, 1963) or lower (Andrews & Palllavicini, 1973) values, it is apparent that acid phosphatase levels are about two orders of magnitude lower than those of alkaline phosphatase (Kitchen *et al.*, 1970). The acid phosphatase is much more active against phosphoprotein substrates than the alkaline phosphatase, so in this case the difference is rather less.

In milk, the distribution of acid phosphatase is similar to that of alkaline phosphatase and likewise, although most of the enzyme is in the skim phase, its specific activity is greater in cream (Bingham *et al.*, 1961; Kitchen *et al.*, 1970). All the acid phosphatase is rapidly bound to an ion-exchange resin when it is stirred into whole milk, so the distribution equilibrium is obviously quite readily shifted towards the aqueous phase from which the enzyme is extracted (Bingham *et al.*, 1972; Andrews & Pallavicini, 1973).

4.2.3. Purification

Ion-exchange extraction provides an extemely efficient first step in most purification protocols and often gives a 300-fold increase in specific activity (IU/mg protein) in a single step. Gel filtration then gives a further 30-fold increase in specific activity. If these are followed by other steps, such as ion-exchange, cellulose phosphate and affinity chromatography, a 10^{-6}-fold overall purification can be achieved, giving apparently homogeneous pure enzyme with a specific activity of over 30 IU/mg (Andrews, 1976*a*).

4.2.4. Characterisation

Bovine milk acid phosphatase is a phosphomonoesterase with broad specificity. It is capable of hydrolysing aromatic phosphomonoesters (e.g. *p*-nitrophenylphosphate, phenolphthalein phosphate, etc.), pyrophosphates (e.g. ADP and ATP), polyphosphates and the phosphoserine residues of phosphoproteins. It does not hydrolyse aliphatic phosphomonoesters such as AMP, *β*-glycerophosphate or free serine phosphate (Bingham & Zittle, 1963). Its substrate specificity is, therefore, different from that of milk alkaline phosphatase. During the hydrolysis of *p*-nitrophenylphosphate, an enzyme–substrate complex is formed rapidly and then hydrolysed, first releasing *p*-nitrophenol and then the phosphate moiety (Andrews

& Pallavicini, 1973). From the influence of pH on K_m, and the fact that the enzyme is inhibited by iodoacetate but not by other more specific thiol modifying reagents, it was concluded that the active site contains a histidine residue.

The molecular properties of the acid phosphatase are shown in Table 2. The enzyme is a single polypeptide-chain molecule with a molecular weight of about 42 000, determined under both denaturing and non-denaturing conditions. The enzyme is activated slightly by reducing agents, unaffected by metal chelating agents, tartrate (which inhibits some phosphatases) or thiol blocking reagents, and it is inhibited by oxidising agents and heavy metal ions. F^- (K_i = 0·3 mM) is a powerful non-competitive inhibitor of *p*-nitrophenylphosphate hydrolysis. Al^{3+} (K_i = 0·12 mM), orthophosphate (K_i = 2·2 nM), sodium pyrophosphate (K_i = 1·1 mM) and sodium hexametaphosphate (K_i = 1·9 mM) are all competitive inhibitors.

The purified enzyme is a glycoprotein containing 2 mol/mol of galactose, 2 mol/mol of mannose and 4 mol/mol of *N*-acetylglucosamine. The enzyme contains low levels of proline and histidine and no methionine, but, as might be expected for an enzyme with a basic pI, there are relatively large amounts of lysine and arginine.

4.2.5. Relationship to Other Acid Phosphatases

Almost one-third of the acid phosphatase in normal milk is associated with the cream phase (Bingham & Zittle, 1963; Kitchen *et al.*, 1970), where it is probably bound to the milk fat globule membrane. This enzyme is identical to that in the skim phase. Ion-exchange resin treatment extracts all the acid phosphatase from good quality bovine milk and there is no evidence in subsequent purification steps for the presence of more than one isoenzyme. This shows that there is an interaction between part of the enzyme in the skim phase and the milk fat globule membrane and that this interaction is readily reversible. Polyacrylamide gel electrophoresis of extracts of enzyme from skim and from the cream phase, followed by staining for activity, show only one band and hence there appears to be only one isoenzyme present (Andrews & Pallavicini, 1973).

In the milk of animals with mastitis, total levels of acid phosphatase activity are elevated to an extent depending upon the stage and severity of the infection. This increase, which may be as much as 5 to 10-fold, is due to the appearance in the milk of isoenzymes of leucocyte origin (Anderson *et al.*, 1975; Andrews & Alichanidis, 1975*b*). A large proportion of this isoenzyme activity is bound to particles and cellular fragments so that only a small proportion is extractable with ion-exchange resin. The extractable enzyme appears to be identical to a blood serum acid phosphatase. All these isoenzymes have properties different from the normal milk enzyme and are largely inactivated by pasteurisation so they are unlikely to be of much technological significance (Andrews, 1976*b*).

References

Alichanidis, E., Wrathall, J.H.M. & Andrews, A.T. (1986). Heat stability of plasmin (milk proteinase) and plasminogen. *J. Dairy Res.*, **53**, 257–69.

Anderson, M., Brooker, B.E., Andrews, A.T. & Alichanidis, E. (1975). Membrane material in bovine skim-milk from udder quarters infused with endotoxin and pathogenic organisms. *J. Dairy Res.*, **42**, 401–17.

Andrews, A.T. (1974). Bovine milk acid phosphatase. II. Binding to casein substrates and heat-inactivation studies. *J. Dairy Res.*, **41**, 229–37.

Andrews, A.T. (1976*a*). Bovine milk acid phosphatase. III. Purification and characterisation of the enzyme. *Biochim. Biophys. Acta*, **434**, 345–53.

Andrews, A.T. (1976*b*). Further studies on acid phosphatase of leucocyte origin in normal and mastitic bovine milks. *J. Dairy Res.*, **43**, 127–31.

Andrews, A.T. & Alichanidis, E. (1975*a*). Acid phosphatase activity in cheese and starters. *J. Dairy Res.*, **42**, 327–39.

Andrews, A.T. & Alichanidis (1975*b*). The acid phosphatases of bovine leucocytes, plasma and the milk of healthy and mastitic cows. *J. Dairy Res.*, **42**, 391–400.

Andrews, A.T. & Pallavicini, C. (1973). Bovine milk acid phosphatase. I. Some kinetic studies and other properties using a partially purified preparation. *Biochim. Biophys. Acta*, **321**, 197–209.

Andrews, A.T., Anderson, M. & Goodenough, P.W. (1987). A study of the heat stabilities of a number of indigenous milk enzymes. *J. Dairy Res.*, **54**, 237–46.

Anon. (1984). Salmonellosis from inadequately pasteurised milk. *Center for Disease Control, Morbid Mortal. Weekly Rep.*, **33**, 505–7.

Anon. (1985). *Campylobacter* outbreak associated with raw milk provided on dairy tour. California. *Center for Disease Control, Morbid Mortal. Weekly Rep.*, **35**, 311–14.

Aschaffenburg, R. & Mullen, J.E.C. (1949). A rapid and simple phosphatase test for milk. *J. Dairy Res.*, **16**, 58–67.

Barman, T.E. & Gutfreund, H. (1966). The catalytic-centre activity and kinetic properties of bovine milk alkaline phosphatase. *Biochem. J.*, **101**, 460–6.

Bingham, E.A. & Zittle, C.A. (1963). Purification and properties of acid phosphatase in bovine milk. *Arch. Biochem. Biophys.*, **101**, 471–7.

Bingham, E.W., Jasewicz, L. & Zittle, C.A. (1961). Acid phosphatase in cream and in skimmilk. *J. Dairy Sci.*, **44**, 1247–56.

Bingham, E.A., Farrell, H.M. & Carroll, R.J. (1972). Properties of dephosphorylated α_{s1}-casein system. *J. Dairy Sci.*, **46**, 764–7.

Bingham, E.A., Farrell, H.M. & Dahl, K.J. (1976). Removal of phosphate groups from casein with potato acid phosphatase. *Biochim. Biophys. Acta*, **429**, 448–60.

Bogin, E. & Ziv, G. (1973). Enzymes and minerals in normal and mastitic milk. *Cornell Vet.*, **63**, 666–8.

Dulley, J.R. & Kitchen, B.J. (1973). The acid phospatase of Cheddar cheese. *Aust. J. Dairy Technol.*, **28**, 114–6.

Haab, W. & Smith, L.M. (1956). Variations in alkaline phosphatase activity of milk. *J. Dairy Sci.*, **39**, 1644–50.

Hayes, P.S., Feeley, J.C., Graves, L.M., Ajello, G.W. & Fleming, D.W. (1986). Isolation of *Listeria monocytogenes* from raw milk. *Appl. Environ. Microbiol.*, **51**, 438–44.

Jasinska, B., Kleczkowski, K. & Michalak, W. (1985). Influence of β-lactoglobulin on milk alkaline phosphatase activity toward the main milk caseins. *J. Dairy Sci.*, **68**, 2172–5.

Jenness, R. & Patton, S. (1959). Milk enzymes, In *Principles of Dairy Chemistry*, Wiley & Sons, New York, pp. 189–203.

Kay, H.D. & Graham, W.R. (1935). The phosphatase test for pasteurised milk. *J. Dairy Res.*, **6**, 191–203.

Kiermeier, F. & Meinl, E. (1961). Zur kenntnis de milchphophatasen. II. Verhalten der sauren phosphatase neber der alkalischen in kuhmilch. *Z. Lebensm. Unters. Forsch.*, **114**, 189–202.

Kitchen, B.J. (1985). Indigenous milk enzymes. In *Developments in Dairy Chemistry*, vol. 3, ed. P.F. Fox. Elsevier Applied Science, London, pp. 239–79.

Kitchen, B.J., Taylor, G.C. & White, I.C. (1970). Milk enzymes—their distribution and activity. *J. Dairy Res.*, **37**, 279–88.

Knoop, A.M. & Peters, R.H. (1975). Phosphataseaktivität in Sauermilch. *Milchwissenschaft*, **30**, 680–5.

Kuzuya, Y., Kanamaru, Y. & Tanahashi, T. (1980). Partial purification and some properties of bovine milk alkaline phophatase. *Jap. J. Zootech. Sci.*, **51**, 501–10.

Kuzuya, Y., Kanamaru, Y. & Tanahashi, T. (1982). Purification of bovine milk alkaline phosphatase with affinity chromatography and effect of flavanoids and saccharides on the enzyme activities. *Jap. J. Zootech. Sci.*, **53**, 45–9.

Linden, G. (1979). Biochemical study of some aspects of milk alkaline phosphatase reactivation. *Milchwissenschaft*, **34**329–32.

Linden, G. & Alais, C. (1976). Phosphatase alcaline du lait de vache. II. Structure sous-unitaire, nature métalloprotéique et paramètres cinétiques. *Biochim. Biophys. Acta*, **429**, 205–13.

Linden, G., Mazeron, P., Michalowski, J.B. & Alais, C. (1974). Phosphatase alcaline du lait de vache. I. Purification et propriétés moléculaires. *Biochim. Biophys. Acta*, **358**, 82–90.

Lorient, D. & Linden, G. (1976). Dephosphorylation of bovine casein by milk alkaline phosphatase, *J. Dairy Res.*, **43**, 19–26.

Lyster, R.L.J. & Aschaffenburg, R. (1962). The reactivation of milk alkaline phosphatase after heat treatment. *J. Dairy Res.*, **29**, 21–35.

Mistry, V.V. (1989). Thermal inactivation characteristics of alkaline phosphatase in unfiltered milk. *J. Dairy Sci.*, **43**, 19–26.

Morton, R.K. (1953). Alkaline phosphatase of milk. *Biochem. J.*, **55**, 785–800.

Mullen, J.E.C. (1950). The acid phosphatase of cow's milk. *J. Dairy Res.*, **17**, 288–305.

Ohmiya, K., Sugano, S., Yun, S-E. & Shimizu, S. (1983). Immobilisation of acid phosphatase and its use for dephosphorylation of casein. *Agric. Biol. Chem.*, **47**, 535–42.

O'Keeffe, R.B. & Kinsella, J.E. (1979). Alkaline phosphatase from bovine mammary tissue: Purification and some molecular and catalytic properties. *Int. J. Biochem.*, **10**, 125–31.

Pearce, M.J., Linklater, P.M., Hall, R.J. & Mackinley, A.G. (1986). Extensive degradation of casein by plasmin does not impede subsequent curd formation and syneresis. *J. Dairy Res.*, **53**, 477–80.

Peereboom, J.W.C. (1968). Studies on alkaline milk phosphatase. II. Occurrence of various phosphatase isozymes in dairy products. *Neth. Milk & Dairy J.*, **22**, 137–52.

Peereboom, J.W.C. (1969). Theory on the renaturation of alkaline milk phosphatase reactivation. *Milchwissenschaft*, **24**, 266–9.

Pepper, L. & Thompson, M.P. (1963). Dephosphorylation of α_s- and κ-caseins and its effect on micelle stability in the κ- α_{s1}-casein system. *J. Dairy Sci.*, **46**, 764–7.

Pratt-Lowe, E.L., Richardson, T. & Geiger, R.M. (1987). Inactivation of microbial alkaline phosphatase in Hispanic-style cheeses with the Scharer Rapid Phosphatase Test. *J. Dairy Sci.*, **70**, 1804–6.

Schormuller, J., Lehmann, K. & Belitz, H.D. (1960). Phosphate and morpavic phosphatase compounds in foods. VI separation and characterisation of phosphopeptides from enzymic casein hydrolysates. *Z. Lebensm.-Forsch.*, **111**, 180–90.

Sharma, R.S. & Ganguli, N.C. (1969). Alkaline phosphatase of milk: a sialoprotein. *Indian J. Dairy Sci.*, **22**, 272–4.

Singh, L.N. & Ganguli, N.C. (1975). A comparative appraisal of certain enzymes in abnormal bovine milk. *Indian J. Dairy Sci.*, **28**, 67–8.

Williams, D.J. (1986). A modification to the Aschaffenburg and Mullen alkaline phosphatase test suitable for goat's milk. *Aust. J. Dairy Technol.*, **41**, 28–30.

Zittle, C.A. (1964), Dynamic state of milk: xanthine oxidase, acid phosphatase, ribonuclease and protease. *J. Dairy Sci.*, **47**, 202–5.

Zittle, C.A. & Bingham, E.A. (1959). Action of purified milk phosphatase on phosphoserine and on casein. *J. Dairy Sci.*, **42**, 1772–80.

V. Lactoperoxidase

5.1. Introduction

Peroxidases are widely distributed in nature, in plants and animals. Horseradish peroxidase is the best characterized member of the group. Peroxidases catalyse the following general reaction:

$$AH_2 + H_2O_2 \rightarrow A + H_2O$$

where AH_2 and A are reduced and oxidised forms of a suitable electron donor, respectively. Several organic and inorganic compounds can serve as electron donors but substrate specifically varies among the different types of peroxidase.

Various mammalian tissues contain different peroxidases, i.e. gluthathione peroxidase, thyroid peroxidase, eosinophil peroxidase, myeloperoxidase and lactoperoxidase. Peroxidase activity in bovine milk was observed by Hansen (1924), who also demonstrated a correlation between the antibacterial activity of bovine milk and peroxidase activity. Although Thurlow (1925) separated milk peroxidase from oxidase activity by fractionation with $(NH_4)_2SO_4$, lactoperoxidase was first isolated, crystallised and characterised by Theorell and Åkesson (1943). Agner (1941) isolated a peroxidase from *empyema* from patients suffering from pulminary tuberculosis which was initially called verdoperoxidase but was later renamed myeloperoxidase when it was discovered that lactoperoxidase was also greenish.

5.2. Distibution of Lactoperoxidase

Lactoperoxidase has been demonstrated in milks from many species, e.g. bovine, porcine, caprine, guinea pig, murine and human. Bovine milk is rich in lactoperoxidase, i.e. ~30 mg/litre (Polis & Shmukler, 1953) but guinea pig milk, the richest source of lactoperoxidase, contains 10–15 times as much (Stephens *et al.*, 1979). However, inter-study comparison of activity is difficult because different assay methods have been used.

The type of peroxidase in human milk has been disputed. Peroxidase activity in human milk was first studied by Gothefors and Marklund (1975) who reported that the lactoperoxidase activity was only about 5% of that normally present in bovine milk. Moldoveanu *et al.* (1982) concluded that human milk peroxidase is derived from leucocytes but this has been refuted by Langbakk and Flatmark (1984), who isolated a peroxidase from human colostrum with chromatographic and immunoreactive properties similar to those of bovine lactoperoxidase. These results were confirmed by Langbakk and Flatmark (1989) who reported that lactoperoxidase represents 0·004% of the total protein in human colostrum.

Lactoperoxidase activity has also been demonstrated in other human glands/secretions: salivary glands (Morrison & Allen, 1963), cervical mucus (Shindler *et al.*, 1976), nasal glands (Wanatabe, 1978) and tears (van Haeringen *et al.*, 1979). Lactoperoxidase is also present in the salivary glands of pigs (Morrison & Steele,

TABLE 1
Amino Acid and Carbohydrate Composition of Lactoperoxidase B1, Expressed to the Nearest Integral per 78 000 g of Protein (Carlström, 1969*a*).

Aspartic acid	71
Glutamic acid	61
Amide nitrogen	62
Theronine	32
Serine	33
Proline	42
Glycine	40
Alanine	37
Valine	28
Isoleucine	27
Leucine	68
Methionine	11
Cysteine	0
Cystine	9
Tyrosine	15
Phenylalanine	30
Tryptophan	15
Lysine	34
Histidine	14
Arginine	37
Mannose	26
N-Acetylglucosamine	14
N-Ac-galactosamine	4
Sialic acid	0

1968), monkeys (Mäkinen *et al.*, 1978), rats (Mäkinen *et al.*, 1980), guinea pigs and hamsters (Kumlien, 1972).

The nomenclature of these peroxidases is somewhat confused. Although the peroxidases present in saliva and uterine mucus are very similar to bovine lactoperoxidase it has been suggested that, until further chemical characterisations have been made, they should be referred to as salivary and uterine peroxidase, respectively (Pruitt & Tenovuo, 1985).

5.3. Chemical Stucture of Lactoperoxidase

Lactoperoxidase is a glycoprotein with a molecular weight of ~78 kDa and contains one heme group (Theorell & Pedersen, 1944). The iron content is 0·0680–0·0709% and the carbohydrate content 9·9–10·2% (Carlström, 1969*a*). The amino acid composition is given in Table 1.

The inability of acidic acetone to extract the heme was taken as evidence for a covalent heme–protein bond, which was tentatively identified as an ester or amide bond (Hultqvist & Morrison, 1963; Morell & Clezy, 1963). However, Sievers (1979) demonstrated that no covalent heme–protein bonds exist and that the prosthetic group is protoheme IX.

Dumonet and Rousset (1983) reported the isolation of a non-heme lactoperoxidase from bovine milk by ion exchange chromatography. This molecule was reported to have the same apparent molecular weight, but it did not absorb at 412 nm and had no enzymic activity; it showed partial imunochemical identity with lactoperoxidase. However, Ekstrand (unpublished) was unable to confirm these results, but found similarities between this compound and lactoferrin. Thus, the existence of an apo-form of lactoperoxidase in milk remains to be proven.

Several different forms of lactoperoxidase have been demonstrated in bovine milk. Polis & Schmukler (1953) separated milk lactoperoxidase into two fractions, A and B, the former being the more acidic. A total of 10 different forms of lactoperoxidase, designated A1, A2/I, A2/II, A3, B1, B2/I, B2/II, B3, B4 and B5, were subsequently identified (Carlström, 1969*b*). Based on chemical analysis of the different fractions. Paul and Ohlsson (1985) proposed the following relationships:

$$\begin{array}{ccccccccccc} \mathrm{B\ 1} & \rightarrow & \mathrm{B\ 2/I} & \rightarrow & \mathrm{B\ 2/II} & \rightarrow & \mathrm{B\ 3} & \rightarrow & \mathrm{B\ 4} & \rightarrow & \mathrm{B\ 5} \\ \downarrow & & \downarrow & & \downarrow & & \downarrow & & & & \\ \mathrm{A\ 1} & & \mathrm{A\ 2/I} & & \mathrm{A\ 2/II} & & \mathrm{A\ 3} & & & & \end{array}$$

B1 represents the intact, or least degraded, form of lactoperoxidase. Its amino acid composition gives a molecular weight of 78 431 Da and an iron content of 0·071%. The horizontal changes within both groups occur under conditions known to favour deamination of aspargine or glutamine residues while each of the vertical shifts (B → A) is due to the loss of 4 glucosamine, 6 mannose and 1 galactosamine residues.

The primary structure of lactoperoxidase is unknown but information on the secondary structure indicates that 65% of the molecule exists as β-structures, 23% as α-structures and 12% as uncoordinated structures (Sievers, 1979). Some information is also available regarding the tertiary structure: electrophoretic studies on lactoperoxidase after pre-treatment with 1% SDS and 1% mercaptoethanol showed a single component, indicating that the molecule consists of a single peptide chain; 8 disulphide bonds contribute to molecular rigidity (Sievers, 1980).

5.4. Isolation and Analysis of Lactoperoxidase

Lactoperoxidase can be easily isolated from skim milk or whey by cation exchangers such as DEAE–Sephadex CG-50-NH_4. Paul *et al.* (1980) described a method by which the enzyme was adsorbed first on an ion exchanger, type CG-50-NH_4 (20 g/litre), and then further purified on phenyl–Sepharose. A large-scale process using ion exchange chromatography has been developed for the isolation of lactoperoxidase on a commercial scale (Prieels & Peiffer, 1986).

Lactoperoxidase activity is traditionally measured by a spectrophotometric technique using a chromogen as electron donor. Originally, various types of benizidine derivatives were used, e.g. dianisidine or *o*-toluidine, but owing to the toxicity of these substances, they are seldom used today. 2,2^{1}-Azino-di-(3-ethylbenzthiazoline-6-sufonic acid) (ABTS, Boeringer Mannheim, Mannheim, Germany) has been introduced as an ideal reagent for peroxidase assays; it is relatively non-toxic, is available in purified form and has a high absorbance in the oxidized form. A method for assaying lactoperoxidase activity using ABTS was published by Shindler *et al.* (1976). Lactoperoxidase can also be determined by an FPLC-technique (Ekstrand & Björck, 1986).

5.5. Biological Significance of Lactoperoxidase

Lactoperoxidase has attracted much interest owing to its antibacterial activity in various secretions which has been reviewed in detail by Pruitt and Tenovuo (1985). This phenomenon has been studied especially in human saliva and bovine milk. In the following, only a brief summary of the antibacterial mechanism and its biological significance will be given.

Although it was observed at the beginning of this century that lactoperoxidase activity could be correlated with the antibacterial properties of raw milk (Hansen, 1924), this phenomenon was not elucidated in detail until the mid 1960s. Portman and Auclair (1959) demonstrated that lactoperoxidase was involved in the inhibition of certain types of starter cultures in milk. Jago and Morrison (1962) demonstrated that H_2O_2 was necessary for the antibacterial effect and Reiter *et al.* (1964) showed that the hydrogen donor involved was thiocyanate ions (SCN^-). The exact nature of the antibacterial agent was obscure until Hoogendoorn *et al.* (1977) and Aune and Thomas (1977) independently showed that hypothiocyanite ($OSCN^-$) was the antibacterial agent formed by the lactoperoxidase-catalysed oxidation of SCN^-. More recent investigations indicate that hypothiocyanite may not be the only antibacterial agent formed. Björck and Claesson (1980) demonstrated that non-enzymatically formed hypothiocyanite did not have the same antibacterial effect as an optimal mixture of lactoperoxidase, SCN^- and H_2O_2. They concluded that in addition to hypothiocyanite, other short-lived oxidation products are formed, as originally proposed by Hogg and Jago (1970). Further support for this hypothesis has been presented by Pruitt *et al.* (1982) who, based on polographic studies of the oxidation products, concluded that the higher oxyacids of thiocyanate are also involved in the antibacterial effect. Thus, although hypothiocyanite is, unquestionably, formed in the lactoperoxidase-catalysed oxidation of SCN^- and has antibacterial activity, there is now much indirect evidence that some higher oxyacids, i.e. HO_2SCN and/or HO_3SCN, are involved in the antibacterial mechanism.

The biological significance of the lactoperoxidase/SCN^-/HO_2O_2 system is today only partially known. Bovine milk contains an array of antibacterial agents, which provide the neonate with vital protection against various infectious agents before its own defence systems are developed (Reiter, 1981, 1985). The lactoperoxidase system is one of these agents. This system may be involved in the protection of the mammary gland (Reiter, 1980).

It is now well established that the lactoperoxidase system can inhibit certain starter cultures, i.e. those strains that accumulate H_2O_2 through their oxygen metabolism. Most of these strains have now been identified and are no longer in use. It is less well known that some strains of *Lactococcus* are, in fact, stimulated by the presence of lactoperoxidase and SCN^-. It has been suggested that this effect is due to the removal of H_2O_2 and alleviation of the direct inhibitory effect of H_2O_2 (Stadhouders, J., 1986, pers. comm.).

Many bacteria show a longer lag time and multiply more slowly in raw milk than in pasteurised milk (Langeveld *et al.*, 1973; Stadhouders, 1982). Whether this effect can be attributed to the antibacterial systems in milk is still uncertain. However, many milk-spoiling Gram-negative organisms are sensitive to the lactoperoxidase system (Björck *et al.*, 1975; Reiter & Härnulv, 1984). Beumer *et al.* (1985) demonstrated that *Campylobacter jenuni* is killed in raw milk and that this effect can be attributed to the lactoperoxidase system.

The more detailed knowledge of the antibacterial mechanism of the lactoperoxidase system has led to different practical applications. One such is the activation of the system in raw milk, by supplementation with thiocyanate and H_2O_2, to prevent bacterial deterioration of non-refrigerated milk during collection and storage in developing countries (Björck *et al.*, 1979; International Dairy Federation, 1988).

Another application is in calf milk replacers as a substitute for low-dose antibiotics. Improved performance of calves by the activated lactoperoxidase system has been demonstrated (Waterhouse & Mullan, 1980*a*,*b*; Reiter *et al.*, 1981; Still *et al.*, 1990). Consequently, calf milk replacers containing an activated lactoperoxidase system are marketed in several countries in which the use of feed antibiotics is restricted, e.g. Denmark, Sweden, Finland and Norway.

The availability of lactoperoxidase, through large-scale isolation from skim milk or whey, has opened up several applications for the antibacterial effects of the lactoperoxidase system. There is great interest in several dental products presently being developed.

REFERENCES

Agner, K. (1941). Verdoperoxidase—a ferment isolated from leucocytes. *Acta. Physiol. Scant.*, **2**, (Suppl. 8), 8–62.

Aune, T & Thomas, E. (1977). Accumulation of hypothiocyanate ion during peroxidase catalysed oxidation of thiocyanate ion. *Eur. J. Biochem.*, **80**, 209–14.

Beumer, R., Noomen, A., Marijs, J. & Kampelmacher, E. (1985). Antibacterial action of the lactoperoxidase on *Campylobacter jejuni* in cow's milk. *Neth. Milk Dairy J.*, **39**, 107–14.

Björck, L. & Claesson, O. (1980). Correlation between concentration of hypothiocyanate and antibacterial effect of the lactoperoxidase system against *Escherichia coli*. *J. Dairy Sci.*, **63**, 919–22.

Björck, L., Rosén, C.G., Marshall, V. & Reiter, B. (1975). Antibacterial activity of the lactoperoxidase system in milk against pseudomonads and other Gram-negative bacteria. *Appl. Microbiol.*, **30**, 199–204.

Björck, L., Claesson, O. & Shulthess, W. (1979). The lactoperoxidase/thiocyanate/hydrogen peroxide system as a temporary preservative for raw milk in developing countries. *Milchwissenschaft*, **34**, 726–9.

Carlström, A. (1969*a*). Lactoperoxidase. Identification of multiple molecular forms and their interrelationships. *Acta. Chem. Scand.*, **23**, 171–84.

Carlström, A. (1969*b*). Physical and compositional investigations of the subfractions of lactoperoxidase. *Acta. Chem. Scand.*, **23**, 185–202.

Dumonet, C. & Rousset, B. (1983). Identification, purification and characterisation of a non-heme lactoperoxidase in bovine milk. *J. Biol. Chem.*, **258**, 14166–72.

Ekstrand, B. & Björck, L. (1986). Fast protein liquid chromatography of antibacterial components in milk; lactoperoxidase, lactoferrin and lysozyme. *J. Chromatography*, **382**, 429–33.

Gothefors, L. & Marklund, S. (1975). Lactoperoxidase activity in human milk and in saliva of newborn infants. *Infect. Immun.*, **11**, 1210–15.

Hansen, F.S. (1924). Bactericidal properties of milk. *Br. J. Exp. Pathol.*, **5**, 271–80.

Hogg, D.M.C. & Jago, G. (1970). The antibacterial action of lactoperoxidase. The nature of the bactericidal inhibitor. *Biochem. J.*, **117**, 779–90.

Hoogendoorn, H., Piessens, J., Scholtes, W. & Stoddard, L. (1977). Hypothiocyanite ion; the inhibitor formed by the system lactoperoxidase–thiocyanate–hydrogen peroxide. *Caries Res.*, **11**, 77–84.

Hultqvist, D.E. & Morrison, M. (1963). Lactoperoxidase I. The prosthetic group of lactoperoxidase. *J. Biol. Chem.*, **238**, 2843–6.

International Dairy Federation (1988). Code of practice for preservation of raw milk by lactoperoxidase system. *International Dairy Federation*, Brussels, Bulletin No. 234.

Jago, G. & Morrison, M. (1962). Anti-streptococcal activity of lactoperoxidase III. *Proc. Soc. Exp. Biol. Med.*, **111**, 585–8.

Kumlien, A. (1972). Iodine metabolism and peroxide activity in salivary glands. *Acta. Endocrinol.*, **70**, 239–46.

Langbakk, B, & Flatmark, T. (1984). Demonstration and partial purification of lactoperoxidase from human colostrum. *FEBS Letters*, **174**, 300–3.

Langbakk, B. & Flatmark, T. (1989). Lactoperoxidase from human colostrum. *Biochem. J.*, **259**, 627–31.

Langeveld, L., Cuperus, F. & Stadhouders, J. (1973). Bacteriological aspects of the keeping quality at 5°C of reinfected and non-reinfected pasteurised milk. *Neth. Milk Dairy J.*, **27**, 54–65.

Mäkinen, K., Bowden, W., Dalgard, D. & Fitzgerald, G. (1978). Effect or peroral administration of xylitol on the exocrine secretions of monkeys. *J. Nutr.*, **108**, 779–89.

Mäkinen, K., Kölling, D. & Söderberg, E. (1980). Exocrine gland enzyme and sorbitol on the rat. *Intern. J. Vit. Nutr. Res.*, **50**, 79–83.

Moldoveanu, Z., Tenovuo, J., Mestecky, J. & Pruitt, K. (1982). Human milk peroxidase is derived from milk leucocytes. *Biochim. Biophys. Acta*, **718**, 103–8.

Morrell, D. & Clezy, P. (1963). The haematin prosthetic groups of some animal peroxidases. I. The preparation and some properties of an ether soluble haematin from milk peroxidase. *Biochim. Biophys. Acta*, **71**, 157–64.

Morrison, M. & Allen, P.Z. (1963). The identification and isolation of lactoperoxidase from salivary gland. *Biochem. Biophys. Res. Comm.*, **13**, 490–4.

Morrison, M. & Steele, W. (1968). Lactoperoxidase, the peroxidase of the salivary gland. In *Biology of the Mouth*, ed. P. Person. *American Association for the Advancement of Science*, Washington, DC, pp. 89–110.

Paul, K. & Ohlsson, P. (1985). Chemical structure of lactoperoxidase. In: *The Lactoperoxidase System: Chemistry and Biological Significance*, ed. K. M. Pruitt & J. Tenovuo. Marcel Dekker Inc., New York, pp. 15–29.

Paul, K., Ohlsson, P. & Henriksson, A. (1980). The isolation of some liganding properties of lactoperoxidase. *FEBS Lett.*, **110**, 200–4.

Polis, B. & Shmukler, H. (1953). Crystalline lactoperoxidase. I. Isolation by displacement chromatography. *J. Biol. Chem.* **201**, 475–500.

Portman, A. & Auclair, J. (1959). Relation entre la lactenine L2 et la lactoperoxidase. *Le Lait*, **39**, 147–58.

Prieels, J.P. & Peiffer, R. (1986). Process for the purification of proteins from a liquid such as milk. UK Patent Application, BD 2 171 102 A1.

Pruitt, K.M. & Tenovuo, J. (ed.) (1985). *The Lactoperoxidase System: Chemistry and Biological Significance*. Marcel Dekker Inc., New York.

Pruitt, K., Tenovuo, J., Andrews, R. & McKane, T. (1982). Lactoperoxidase catalysed oxidation of thiocyanate—polargraphic study of the oxidation product. *Biochemistry*, **21**, 52–7.

Reiter, B. (1980). Do the antibacterial factors in the milk protect the mammary gland? In *Resistance Factors and Genetic Aspects of Mastitis Control*, ed. L. Basselik-Chabielska. Ossolinium, Jablonna, Poland, pp. 351–92.

Reiter, B. (1981). The contribution of milk to resistance to intestinal infection in the newborn. In *Immunological Aspects in the Foetus and Newborn*, ed. H.P. Lamberet and C.B.S. Woods. Academic Press, London, pp. 155–95.

Reiter, B. (1985). Protective proteins in milk—biological significance and exploitation. *International Dairy Federation, Brussels, Bulletin*, No. 191.

Reiter, B. & Härnulv, G. (1984). Lactoperoxidase antibacterial system: Natural occurrence, biological functions and practical applications. *J. Food Protection*, **47**, 724–32.

Reiter, B., Pickering, A. & Oram, J. (1964). An inhibitory system-lactoperoxidase/thiocyanate/hydrogen peroxide—in raw milk. In *Microbial Inhibitors in Food*, ed. N. Molin. Alqvist & Wiksell, Stockholm, pp. 297–305.

Reiter, B., Fulford, R. & Marshall, V. (1981). An evaluation of the growth promoting effect of the lactoperoxidase system in newborn calves. *Animal Prod.*, **32**, 297–306.

Shindler, J., Childs, R. & Bardsley, W. (1976). Peroxidase from human cervical mucus. *Eur. J. Biochem.*, **65**, 325–31.

Sievers, G. (1979). The prosthetic group of milk lactoperoxidase is proteheme IX. *Biochim. Biophys. Acta*, **579**, 181–90.

Sievers, G. (1980). Structure of milk lactoperoxidase. A study using circular dichroism and difference absorption spectroscopy. *Biochim. Biophys. Acta*, **624**, 249–59.

Stadhouders, J. (1982). Cooling and therminisation as a means to extend the keeping quality of raw milk. *Kieler. Milchw. Forshungsber*, **34**, 19–28.

Stephens, S., Harkness, R.A. & Cockle, S.M. (1979). Lactoperoxidase activity in guinea-pig milk and saliva: correlation in milk of lactoperoxidase with bactericidal activity against *Escherichia coli*. *Br. J. Exp. Pathol.*, **60**, 252–8.

Still, J., Delabut, P., Coppe, P., Kaeckenbeeck, A. & Perraudin, J. (1990). Treatment of induced enterotoxigenic colibacellocis (scours) by the lactoperoxidase system and lactoreffin. *Ann. Rech. Vet.*, **21**, 143–52.

Theorell, H. & Åkesson, Å. (1943). Highly purified milk peroxidase. *Arkiv för Kemi. Mineralogi ock Geologi.*, **17B**, 1–6.

Theorell, H. & Pedersen, K. (1944). The molecular weight and light absorption of crystallised lactoperoxidase. In *The Svedberg*, Almqvist & Wiksell, Stockholm, pp. 523–9.

Thurlow, S. (1925). XXV. Studies on xanhine oxidase. IV. Relation of xanthine oxidase and similar oxidising systems to Bach's oxygenase. *Biochem. J.*, **XIX**, 175–87.

van Haeringen, N., Ensink, F. & Glasius, E. (1979). The peroxidase–thiocyanate–hydrogen peroxide system in tear fluid and saliva of different species. *Exp. Eye Res.*, **28**, 343–7.

Wanatabe, K. (1978). Localisation of peroxidase activity in the human nasal gland. Comparison with patients with a house dust allergy to persons with normal glands. *Acta Histochem. Cytochem.*, **11**, 151–9.

Waterhouse, A. & Mullen, W. (1980*a*). Addition of an active lactoperoxidase system to a milk substitute diet for calves. *Irish J. Food Sci. Technol.*, **4**, 69–70.

Waterhouse, A. & Mullan, W. (1980*b*). Re-inclusion of an active lactoperoxidase system in a milk substitute diet for calves. *Animal Prod.*, **30**, 458 (Abstr.).

TABLE 1
Activity and Distribution of LDH in the Milk of Some Species

Species	*Activity (U/ml)*	*Distribution*				
		LDH1	*LDH2*	*LDH3*	*LDH4*	*LDH5*
Bovine[a]						
Normal	0·28 ± 0·15	93·5	6·5	—	—	—
Mastitic	4·54 ± 1·21	34·8	22·3	16·7	15·4	7·3
Human[b]						
Colostrum	0·90 ± 0·41	low				high
Normal	0·39 ± 0·17	high				low
Porcine[c]	1·31 ± 0·28					

[a]Bogin & Ziv (1973); Bogin *et al.* (1977).
[b]Patil & Rangnekar (1983).
[c]Grigor & Hartman (1985).

VI. OTHER ENZYMES

6.1. Introduction

Trace amounts of several enzymes have been identified or purified from milks of various species. Some of these 'trace' enzymes have an established function in milk, essential for the development of the newborn whose primary food is milk. The activities of others may reflect the physiological changes and pathological state of the mammary gland during lactation, while some have been recognised for their importance in the processing of dairy products. The following is a review of some of the milk enzymes not considered in the previous sections of this chapter.

6.2. L–Lactate Dehydrogenase (EC 1.1.1.27)

Lactate dehydrogenase (LHD) plays an important role in carbohydrate metabolism by catalysing the reversible oxidation of lactate to pyruvate:

$$\text{L-lactate} + NAD^+ \rightleftharpoons \text{pyruvate} + NADH + H^+$$

The conversion of lactate to pyruvate occurs under aerobic conditions while the reverse reaction is anaerobic.

LDH is a tetrameric molecule with a molecular weight of 140 kDa. Each molecule consists of two different polypeptide chains (H and M), which are so named according to their origins (Everse & Kaplan, 1973; Holbrook *et al.*, 1975). Type H originates from aerobic tissues such as heart muscle (H) while type M is from anaerobic tissues like skeletal muscle (M). Five LDH isoenzymes (H_4, H_3M, H_2M_2, HM_3 and M_4), which result from permutations of the two polypeptide chains, occur in milk. The isoenzymes are called LDH 1, 2, 3, 4 and 5, respectively.

The enzyme has not yet been purified from milk. Normal bovine and human milks contain similar levels of LDH (Table 1).

LDH activities are higher in mastitic milk and colostrum than in corresponding normal milks. LDH enters milk from blood granulocytes and lymphocytes (Kato *et al.*, 1989). In colostrum, the concentration of LDH isoenzymes decreases from LDH5 to LDH1 (Kjellberg & Karlsson, 1967; Patil & Rangnekar, 1983) but the order reverses as lactation progresses, such that about 95% of the LDH activity in bovine milk is due to LDH1 (Bogin *et al.*, 1977).

According to Sommer *et al.* (1986), elevated LDH activity in milk is a useful indicator of mastitis. The significance of LDH in dairy processing has not been demonstrated although it is conceivable that it may play a role in dairy fermentations.

6.3. Catalase (EC 1.11.1.6)

Catalase ($H_2O_2 : H_2O_2$ oxidoreductase) is a hemo-protein which catalyses the decomposition of hydrogen peroxide:

$$2H_2O_2 \rightleftharpoons 2H_2O + O_2$$

It also catalyses peroxide-dependent oxidation of acids (hydrazoic, formic, nitrous), lower aliphatic alcohols and hydroxylaminine (Schonbaum & Chance, 1976).

The enzyme consists of oligomeric subunits associated by hydrophobic interactions (Schonbaum & Chance, 1976). It is present in various living tissues and cells. It enters milk in association with somatic cells. Thus, catalase activity in raw milk is directly proportional ($r = 0{\cdot}7$) to the somatic cell count (Read *et al.*, 1969; Kitchen, 1976) and may be an indicator of mastitis since the mammary gland cell wall becomes increasingly permeable to blood constituents (somatic cells) during mastitis.

Catalase activity in milk varies with the stage of lactation; its activity is high in colostrum, decreases during lactation and increases in late lactation milk (Wuthrich *et al.*, 1964). About 22 and 73% of the catalase activity in bovine (Kitchen *et al.*, 1980) and caprine (Ito & Akuzawa, 1982) milks are present in the cream and skim milk fractions, respectively; the remainder (5%) is probably associated with the separator sludge. About 56% of the activity in cream is firmly bound to insoluble lipoprotein of the milk fat globule membrane (Ito, 1971).

Catalase has been purified (23,000-fold) and crystallised (0·1 mg/kg) from bovine milk (Ito & Akuzawa, 1983*a*). The purification of catalase from buttermilk involves *n*-butanol extraction, $(NH_4)_2SO_4$ (35–55% saturation) fractionation, chloroform : methanol extraction and chromatography of the extract on DEAE-Sephacel and Sephacryl S-300 (Ito & Akuzawa, 1983*a*).

Milk contains at least two catalase isoenzymes (Ito & Akuzawa, 1983*b*) which are immunochemically similar but have different pH optima, i.e. 7 and 8 (Ito *et al.*, 1984). The former is more active in the pH range 5–8. Inhibitors of catalase include Cu^{2+}, Fe^{2+}, Hg^{2+}, Sn^{2+}, CN^- and N_3^- (Ito, 1974).

Milk catalase is a pentameric molecule with an estimated molecular weight of 225 kDa (Ito & Akuzawa, 1983*a*). When treated with SDS (1% × 48 h), the native enzyme dissociates into five subunits of mol. wt. 55, 40, 34, 24 and 11 kDa. All but the 55 kDa subunit are inactive (Ito & Akuzawa, 1983*c*).

Catalase crystallised from milk is stable in NaCl (> 90% activity survives in 10% NaCl), but is inactivated at temperatures below its optimum of 20°C and when heated (70°C × 30 min) in buffer at pH 8·0 (Ito & Akuzawa, 1983*a*). When milk is heated at 77–83°C for 20 s, 88–90°C for 3 min or 91–95°C without holding, its catalase activity is reduced by 63, 74 and 84%, respectively (Shidlovskaya, 1982). However, some reactivation of catalase occurs during refrigerated storage of heated milk (Shidlovskaya, 1982; Griffiths, 1986), presumably due to the protection of the enzyme present in microorganisms in milk that survive heating of milk and the subsequent release of the enzyme into milk. Therefore, Phillips and Griffiths (1987) suggested that catalase activity in pasteurised milk may be a good indicator of bacterial growth within 25·5 h of processing.

Catalase activity in milk is a good index of mastitis (Kitchen *et al.*, 1970). A possible role of catalase in lipid oxidation was suggested by Fox and Morrissey (1981) because the enzyme contains a prosthetic ferric group. However, this role has not been demonstrated experimentally.

6.4. Glutathione Peroxidase (EC 1.11.1.9)

Glutathione peroxidase (GPx) is a selenium (Se)-containing protein which catalyses the oxidation of glutathione:

$$2\ GSH + H_2O_2 \rightleftharpoons GSSH + 2\ H_2O$$

Human and bovine milks contain about 1·35 and 0·71 μg GPx per ml: respectively (Hojo, 1982). In defatted human milk, the average GPx activity is 31·3 ± 1·8 mU/ml: where 1 mU GPx oxidizes 1 nmol DPH per min (Bhattacharya *et al.*, 1988). GPx activity decreases during human lactation (Bhattacharya *et al.*, 1988). In Japan, Hojo (1982) found > 1·5 times more GPx activity in bovine milk samples collected in June than in those collected in November, suggesting seasonal variation in GPx activity of milk. The variation in GPx activity of bovine milk is attributable to the Se content of the feed which may be a result of geographic differences.

The enzyme is a source of biologically active Se, providing about 12% (Hojo, 1982) and 22–27% (Hojo, 1982; Bhattacharya *et al.*, 1988) of the total Se in bovine and human milks, respectively. GPx also contributes to xenobiotic metabolism through the decomposition of hydroperoxides.

GPx has been purified (4,500-fold; Bhattacharya *et al.*, 1988) from human milk by acetone precipitation and repeated ion-exchange (DEAE Sephadex A-50) and gel filtration (Sephadex G-150) chromatography. GPx is a tetrameric molecule with a molecular weight of 92 kDa. The active enzyme consists of four 23 kDa subunits and contain 0·31% Se (Bhattacharya *et al.*, 1988)

Laboratory heating (80°C × 10 min) and commercial processing of milk destroys its GPx activity (Hojo, 1982), raising questions on the bioavailability of Se from pasteurised and sterilised milk and dairy products.

6.5. Superoxide Dismutase (EC 1.15.1.1)

Superoxide dismutase (SOD), properly known as superoxide: superoxide oxido-reductase, catalyses the dismutation of superoxide anion, $O_2^{\bar{\cdot}}$:

$$2O_2^{\bar{\cdot}} + 2H^+ \rightleftharpoons H_2O_2 + O_2$$

The $O_2^{\bar{\cdot}}$ radical, which is potentially dangerous to living matter, is a product of the oxidation of xanthine by xanthine oxidase at high pH, high O_2 and low xanthine concentrations (Fridovich, 1970). The superoxide ion is also formed during autoxidation of hemoglobin (Misra & Fridovich, 1972).

There are 3 distinct types of SOD: (1) Mn-containing SOD in prokaryotes; (2) Cu/Zn-containing SOD in eukaryotes, and (3) Fe-containing SOD found in some microorganisms and plants (Fridovich, 1986). This review is limited to the Cu/Zn-SOD—the type found in milk.

Cu/Zn-SOD is a dimer (mol. wt. 31–33 kDa) with chemically identical subunits (Evans *et al.*, 1974) which are non-covalently associated, but stabilised by intra-chain disulphide bonds (Abernethy *et al.*, 1974). Each subunit contains one Cu^{2+} and one Zn^{2+}. The native enzyme is resistant to thiol-modifying reagents but complete removal of the Zn exposes the SH groups (Rotilio *et al.*, 1972), suggesting that a Zn–S complex is essential for stability while Cu^{2+} is responsible for electron transfer (Fridovich, 1974) during action of the enzyme.

SOD is characterised by a broad, visible absorption band with a maximum wavelength at about 680 nm. Thus, concentrated solutions of the enzyme are bluish-green in colour (Malmstrom *et al.*, 1975).

SOD is present in milk serum. Its concentration in bovine milk ranges from 0·15 to 2·5 mg/litre (Hill, 1975; Asada, 1976; Korycka-Dahl *et al.*, 1979). Milk from Jersey cows contains about 38% more SOD activity than Holstein milk (Holbrook & Hicks, 1978) while SOD activity in human milk is lower (Keen *et al.*, 1980) than that in bovine milk.

Bovine (Hill, 1975) and human (Keen *et al.*, 1980) milk SOD have eletrophoretic and chromatographic properties identical to SOD purified from their corresponding erythrocytes, suggesting that the enzyme originates from blood plasma. However, according to Holbrook and Hicks (1978), the activity of SOD in milk is not dependent on the stage of lactation, age of cow or mastitis—all of which affect the permeability of the mammary gland cell membrane.

SOD was purified from acid whey (Hill, 1975) by a sequence of chloroform-ethanol fractionation, gel filtration (Sephadex G-25) of the ethanolic phase (which contains SOD activity), followed by ion exchange chromatography on DE32. Alternatively, SOD may be isolated from skim milk by ultrafiltration (using 100 and 10 kDa MW cut-off membranes), followed by gel filtration on Sephadex G-100 (Keen *et al.*, 1980).

Although the Cu/Zn-SOD predominates in milk, it has been suggested (Hicks, 1980) that some Mn-SOD may be present. This is questionable because bovine and human milks lose their SOD activities in the presence of CN^- (Keen *et al.*, 1980), which strongly binds to copper, thus inhibiting Cu/Zn-SOD. Therefore, the occurrence of Mn-SOD in milk may be due to contaminating bacteria.

Normal pasteurisation (72°C × 15 s) has little effect on the SOD activity of milk, but only 25% of its activity remains in milk heated at 75°C for 20 min (Hicks *et al.*, 1979).

Since high SOD activity results in low levels of oxidative rancidity in milk (Holbrooks & Hicks, 1978), SOD may be important in maintaining the oxidative stability of milk by scavenging the superoxide ion produced by xanthine oxidase during oxidation of its substrate.

6.6. Xanthine Oxidase (EC 1.1.3.22)

Xanthine oxidase (XO) is one of the most extensively studied indigenous enzymes in milk. It is a non-specific oxido-reductase which catalyses the oxidation of aldehydes, purines and pterins. It also catalyses the oxidation of hypoxanthine to xanthine, and xanthine to uric acid, with the concomitant reduction of O_2 to H_2O_2. At alkaline pH and high O_2 concentration, limited oxidation of xanthine results in the production a very reactive superoxide anion, O_2^{-} (Fridovich, 1970), which is a substrate for SOD activity (see above).

XO is a dimeric metallo-flavoprotein with a molecular weight of about 300 kDa (Bray, 1975). The enzyme contains flavin adenine dinucleotide (FAD), molybdenum, iron and acid-labile sulphur in the ratio 1 : 1 : 4 : 4 (Massey *et al.*, 1969) plus one mole of phosphorus per active centre molybdenum (Davis *et al.*, 1980). XO is present in microorganisms, various mammalian organs, tissues and milk. Compared to caprine (Modi *et al.*, 1959) and human (Zikakis *et al.*, 1976) milks, bovine milk is a rich source of XO—containing about 35 mg/litre (Walstra & Jenness, 1980) or even more. In some organs (e.g. rat liver) the enzyme is present as xanthine dehydrogenase EC 1.1.1.204), XD (Stirpe & Della Corte, 1969), which can be converted to XO by treatment with sulphydryl oxidase (Clare *et al.*, 1981). XD can be converted to XO by partial proteolysis or treatment with SH-modifying reagents (Waud & Rajagopalan, 1976).

There are at least five genetic variants of XO with different isoelectric pH values ranging from 6·9 to 7·6 (Sullivan *et al.*, 1982). Therefore, XO activity in bovine milk was thought to be determined genetically because three different activities of XO were found in Guernsey and Holstein milk (Zikakis & Treece, 1971). This observation has not been confirmed. However, Bray (1975) found that milk contains both enzymatically active XO and inactive (desulpho-XO and demolybdo-XO) forms of the enzyme. The relative amounts of the active and inactive enzyme in milk of individual cows are influenced by nutrition (Bray, 1975).

Active XO differs from the inactive desulpho-XO and demolybdo-XO by the substitution of terminal sulphur for oxygen in desulpho-XO (Gutteridge *et al.*, 1978; Bordas *et al.*, 1980) and the substitution of a pterin-like moiety for molybdenum in demolybdo-XO (Ventom *et al.*, 1988). Treatment of the active enzyme with CN^- (Massey & Edmondson, 1970) or H_2O_2 (Ventom *et al.*, 1988) converts it to desulpho-XO.

XO is a component of the milk fat globule membrane (MFGM), where it is attached to the inner (cytoplasmic) surface of the membrane by butyrophilin (Valivullah & Keenan, 1989), a phospholipid-binding protein (Freundenstein *et al.*, 1979). Therefore, the enzyme must be first released from the MFGM by proteolysis before purification. There are several published methods for purifying XO from human (Graham *et al.*, 1989) and bovine (Gilbert & Bergel, 1964; Edmondson *et al.*, 1972; Mangino & Brunner, 1977; Nathans & Hade, 1978; Nishino *et al.*, 1981; Sullivan *et al.*, 1982; Cheng *et al.*, 1988) milks. The purified enzyme can be immobilized on *n*-octyl-sepharose 4B (Tramper, 1987).

Protease treatment of the MFGM prior to purification of XO causes partial proteolysis of the enzyme (Waud *et al.*, 1975; Nathans & Hade, 1978; Cheng *et al.*, 1988) since purified milk XO is hydrolysed by chymotrypsin, papain or subtilisin (Battelli *et al.*, 1973). However, XO is resistant to plasmin (indigenous milk protease) activity (Cheng *et al.*, 1988), suggesting that milk XO is not hydrolysed *in situ*. Addition of dithioerythritol to milk prior to purification results in an enzyme with xanthine dehydrogenase activity (Nakamura & Yamazaki, 1982).

The optimum pH for activity of milk XO is about 8·3 and is dependent on the substrate and experimental conditions (Bray, 1975). Its K_m for oxygen at pH 8·5–10·0 is in the range 5–8 $\times$ $10^{-5}\,M$ (Bray, 1975). Inhibitors of XO include allopurinol, ammeline, purine-6-aldehyde (Bray, 1975), *o*-iodosobenzoate and iodoacetamide (Green & O'Brien, 1967).

The role of XO in the development of spontaneous oxidised flavour in milk is not yet clear. While Aurand *et al.*, (1967, 1977) attribute oxidative rancidity in milk, in part, to XO activity, others (Smith & Dunkley, 1960; Bruder *et al.*, 1982) disagree.

Commercial processing (pasteurisation and homogenisation) of milk reduces its XO activity to < 30% (Cerbulis & Farrell, 1977; Zikakis & Wooters, 1980). On the contrary, laboratory heating of milk for 15 s at 72°C (Greenbank & Pallansch, 1962), 80°C (Andrews *et al.*, 1987) or 90°C (Greenbank & Pallansch, 1962) destroys < 20, 40 or 100%, respectively, of milk XO activity. These differences in the heat stability of milk XO are attributable to renaturation since XO undergoes conformational changes at 61·4 and 77·6°C (Ruegg *et al.*, 1977).

Questions on the function of milk XO in health and disease aroused the attention of the dairy industry in the early 1970s when Oster (1971) postulated that individuals who drink homogenised milk are prone to atherosclerosis because XO causes depletion of plasmalogen in cell membranes. Following extensive investigation of XO activity in dairy products (Cerbulis & Farrell, 1977; Zikakis & Wooters, 1980), in addition to nutritional and epidemioligical studies (American Heart Association, 1981; Clifford *et al.*, 1983; Deeth, 1983), the association of XO activity with atherosclerosis was disclaimed.

6.7. Sulphydryl Oxidase

Sulphydryl oxidase (SOX) functions in the formation of the three-dimensional structure of proteins (Swaisgood & Horton, 1980) by catalysing the oxidation of all groups of cysteine and SH-containing peptides and proteins to their corresponding disulphides using molecular oxygen as electron acceptor:

$$2RSH + O_2 \rightleftharpoons RSSR + H_2O_2$$

At present, no EC number has been assigned to SOX, and therefore it is often confused with thiol oxidase (EC 1.8.3.2). Nevertheless, mammalian milk SOX is distinctly different from thiol oxidase, both in molecular and biochemical properties (Janalino & Swaisgood, 1975). The mammalian enzyme also differs from SOX from *Aspergillus niger* (Swaisgood, H.E., 1991, pers. comm). Unlike bovine milk SOX, the microbial enzyme is a soluble flavoprotein with a subunit molecular weight of 47 kDa; it does not bind to cysteinylsuccinamidopropyl-glass nor catalyse the formation of disulphide bonds in proteins (Swaisgood, H.E., 1991, pers. comm).

SOX is present in colostrum (Isaacs, 1985) and in milk from various species (Kiermeier & Petz, 1967*a*,*b*,*c*; Clare *et al.*, 1984; Isaacs *et al.*, 1984). Bovine milk contains about 10 mg SOX per litre (Swaisgood & Abraham, 1980), most of which is bound to the skim milk membrane (Sliwkowski *et al.*, 1982). Similarly, as much as 95% of human milk SOX is associated with skim milk membrane material (Isaacs, 1985).

SOX has been purified (to different degrees) from skim milk or sweet whey ($pH > 6$) by a variety of methods (Swaisgood, 1980; Swaisgood *et al.*, 1982). However, purification by transient covalent affinity chromatography on cysteinylsuccinamidopropyl-glass (Sliwkowski *et al.*, 1982) yields a relatively pure enzyme for sequencing (Janalino *et al.*, 1990).

SOX is a highly aggregated Fe/Cu-containing glycoprotein (Janalino & Swaisgood, 1975) with a blocked N-terminus (Janalino *et al.*, 1990). Its high degree of aggregation makes it difficult to estimate accurately the molecular weight of the native enzyme. However, it has a subunit molecular weight of about 89 kDa (Janalino & Swaisgood, 1975; Janalino *et al.*, 1980). SOX contains two chemically reactive SH groups, of which one (Song *et al.*, 1986), in addition to the Fe (Janalino & Swaisgood, 1975), is essential for activity. The carbohydrate moiety increases the solubility of the enzyme (Swaisgood & Horton, 1980).

Bovine milk SOX is acid-labile (Janalino & Swaisgood, 1975). However, Isaacs *et al.* (1984) reported that human milk SOX partially survives passage through the gastrointestinal (GI) tract, losing about 50% of its activity after 1 h at pH 2·5. SOX activity is optimal at pH 7·0 and 37°C (Janalino & Swaisgood, 1975).

Commercially processed milk contains about 40% of the SOX activity of raw milk (Swaisgood, 1980), suggesting that the enzyme partially survives pasteurisation. Low temperature (4 or −20°C) storage of human milk has no effect on its SOX activity (Isaacs, 1985). EDTA (1 nM) inhibits SOX activity (Janalino & Swaisgood, 1975).

The H_2O_2 generated from SOX activity may be involved in the lactoperoxidase system of milk (Swaisgood & Abraham, 1980). The purified enzyme may be added to UHT milk to reduce the cooked flavour defect (Swaisgood *et al.*, 1987).

6.8. γ-Glutamyl Transferase (EC 2.3.2.2)

γ-Glutamyl transferase or γ-glutamyl transpeptidase (GGT) catalyses the transfer of γ-glutamyl residues from γ-glutamyl-containing peptides to an acceptor L-amino acid (except proline) or its peptide (Tate & Meister, 1974):

$$\text{(5-L-Glutamyl)-peptide} + \text{amino acid} \rightleftharpoons \text{peptide} + \text{5-L-glutamyl-amino acid}$$

GGT is a secretory enzyme present in mammalian tissues (Meister & Tate, 1976) and fluids (Majumder & Ganguli, 1972; Binkley *et al.*, 1975), as well as in plants and microorganisms (Yasumoto *et al.*, 1978).

In milk, GGT is bound to the membranes of skim milk and fat globules (Kitchen, 1974; Baumrucker, 1979). Its activity in both human (Binkley *et al.*, 1975; Patil & Rangnekar, 1982) and bovine (Barcelos & Weight, 1986) milks is influenced by the stage of lactation, with activity highest in colostrum.

Active GGT is a glycoprotein (mol. wt. 80 kDa), consisting of two subunits with molecular weights of 57 and 25·5 kDa (Yasumoto *et al.*, 1978). It has a pI of 3·85 (Yasumoto *et al.*, 1978). Depending on the acceptor amino acid or peptide (Yasumoto *et al.*, 1978), its optimum pH for activity is in the range 8·0–9·0 (Sobiech *et al.*, 1974; Baumrucker, 1980). Inhibitors of GGT activity include iodoacetamide, diisopropyl fluorophosphate and metal ions like Cu^{2+} and Fe^{3+} (Sobiech *et al.*, 1974).

Bovine milk GGT is stable at 20°C (Sobiech *et al.*, 1974) but its activity is reduced by > 80% in pasteurised (72–75°C × 15 s) milk (Majumder & Ganguli, 1972; Patel & Wilbey, 1989). Therefore, its activity may be useful in detecting milk heated to 70–80°C (Andrews *et al.*, 1987).

GGT plays a role in the transport of amino acid from blood into the mammary gland for use in milk protein synthesis (Baumrucker, 1980; Pocuis *et al.*, 1987; Vina *et al.*, 1985). GGT is readily absorbed in the GI tract (Braun *et al.*, 1984), resulting in high serum levels of GGT in newborns fed colostrum or early breast milk (Patil & Rangnekar, 1982). Since heat treatment of infant formulae destroys its GGT activity, the activity of this enzyme in the sera of newborns can be used to distinguish between infants fed breastmilk or a milk-based formula.

6.9. Ribonuclease (EC 3.1.27.5)

Milk ribonuclease (RNase) is identical to pancreatic RNase which catalyses the endonucleolytic cleavage of 3′-phosphomononucleotides and 3′-phosphooligonucleotides ending in polycytidylic (Cp) or polyuridylic (Up) acids with 2′, 3′-cyclic phosphate intermediates.

RNase is present in bovine, caprine, human, porcine (Chandan *et al.*, 1968) and buffalo (Gupta & Mathur, 1989) milks. Bovine milk contains about 25 mg RNase per litre (Shahini *et al.*, 1973), most of which is concentrated in the serum phase of milk (Meyer *et al.*, 1987*a*). RNase activity in bovine milk is > 3-fold higher than in human milk (Chandan *et al.*, 1968) and in one instance (Meyer *et al.*, 1987*b*), as much as 100 times more RNase was found in the former. Generally, RNase activity is higher in colostrum (3–5 ×) and mastitic milk (2 ×) than in normal milk (Meyer *et al.*, 1987*b*).

The purification of RNase from milk (acid whey) involves chromatography on the carboxylic acid resin, IRC-50 (Bingham & Zittle, 1964; Gupta & Mathur, 1989), Sephadex G-50 (Dalaly *et al.*, 1970) or phosphocellulose (Meyer *et al.*, 1987*a*). Earlier research (Bingham & Kalan, 1967) showed two RNase isoenzymes (A & B) in bovine milk. Recently, three isoenzymes (A, B & II-1) in the ratio 72 : 27 : 1 were purified from bovine milk (Meyer *et al.*, 1987*a*). The major differences between the three isoenzymes are that RNase B is a glycoprotein containing 3·0% glucosamine, 1·2% galactosamine and 5·7% mannose. RNase A contains no carbohydrate (Bingham & Kalan, 1967), while RNase II-1 is unable to hydrolyse phosphonucleotides ending in Cp (Meyer *et al.*, 1987*a*).

The molecular weight of pancreatic RNase is in the range 12·7–15·5 kDa (see Scheraga & Rupley, 1962) and that of bovine milk RNase is 13·6 kDa (Bingham & Zittle, 1964) but recently, Gupta and Mathur (1989) reported a molecular weight of 29 kDa for caprine milk RNase. Bovine milk RNase exhibits optimal activity at 37°C (Shahani *et al.*, 1973) and pH 7·5 (Bingham & Zittle, 1962) compared to 50°C and pH 9·0, respectively, for caprine milk RNase (Gupta & Mathur, 1989). Inhibitors of RNase activity include Cu^{2+}, Ag^{+}, Fe^{2+}, Ca^{2+}, Hg^{2+}, Zn^{2+} and Mn^{2+} (Gupta & Mathur, 1989).

Less than 0·3% of bovine milk RNase survives sterilisation (121°C × 10 s) of milk (Meyer *et al.*, 1987*a*) while about 60% survives in milk heated to 80°C × 15 s (Griffiths, 1986) or 72°C × 2 min (Meyer *et al.*, 1987*a*). However, the enzyme retains its activity when heated (90°C × 20 min) at pH 3·5 (Bingham & Zittle, 1962), suggesting that RNase is stable at low pH.

RNase may protect milk from viruses since it is reported (McCormick *et al.*, 1974) to inhibit the activity of RNA-dependent DNA-polymerase, an enzyme involved in the replication of viruses. A bacteriocidal role of RNase has also been suggested (Meyer *et al.*, 1987*b*).

6.10. Amylases

Three types of amylases (α-, EC 3.2.1.1; β-, EC 3.2.1.2; and γ-, EC 3.2.1.3) occur in nature. In bovine milk, α- and β-amalyses are present (Guy & Jenness, 1958; Griffiths, 1986) but the former predominates and is reviewed herein.

α-Amylase hydrolyses 1,4-α-D-glucosidic linkages in polysaccharides containing at least three 1,4-α-linked D-glucose units. Bovine and human milk α-amylases are salivary-type amylases (Stejskal *et al.*, 1981). α-Amylase activity varies (1000–8000 U/litre) in human milk, but is generally about 25-fold higher than that in bovine milk (Shahani *et al.*, 1980). Human milk contains 10–25 times more α-amylase than blood (Stejskal *et al.*, 1981), suggesting that the enzyme does not enter milk from blood, but is ectopically produced in the mammary gland, in response to a hormonal stimulus (Stejskal *et al.*, 1981).

α-Amylase has been purified from milk by chromatography on Sephadex G-100 (Goussault *et al.*, 1966) or G-150 (Stejskal *et al.*, 1981). It shows optimal activity in the pH range 6·5–7·5 (Goussault *et al.*, 1966; Guy & Jenness, 1958) and at 44°C (Guy & Jenness, 1958). The enzyme is stable in human milk frozen at −20 or −70°C (Hamosh *et al.*, 1985) but bovine milk loses <45% of its activity on pasteurisation (75°C × 15 s) (Griffiths, 1986).

The purified enzyme is stable at pH > 3 (Lindberg & Skude, 1982; Jones *et al.*, 1982) and is resistant to proteolysis by pepsin (Heitlinger *et al.*, 1983), suggesting that it can survive passage through the GI tract after ingestion of milk. α-Amylase activity increases in the presence of Ca^{2+} (Stein & Fischer, 1958), Cl^- (Guy & Jenness, 1958) and bovine serum albumin (Heitlinger *et al.*, 1983), but decreases in the presence of I^- (Jenness & Patton, 1959).

The presence of α-amylase in milk is perhaps surprising because milk is thought to lack a suitable substrate for the enzyme. However, recent HPLC analysis (Newburg *et al.*, 1986) shows that human milk and colostrum contain about 120–150 mg oligosaccharides (mostly penta-to tetradecasaccharides) per litre, suggesting that milk may contain substrates for the enzyme. Since pancreatic α-amylase is not developed at birth (Hamosh, 1988), breastmilk appears to be the immediate source of the enzyme for premature infants who may require high-calorie starch-containing supplements in their diets. In light of this, research on amylase activity in bovine milk and infant formulae is needed.

6.11. Lysozyme (EC 3.2.1.17)

Lysozyme cleaves β-(1 → 4) glycosidic bonds between *N*-acetylmuramic acid and *N*-acetyl-D-glucosamine residues in peptidoglycan, a constituent of bacterial cell walls.

There are two types of lysozymes: those found in hen egg-whites (HEWL), also known as chick-type (c) lysozyme and those found in Embden goose egg-whites or goose-type (g) lysozyme (Arnheim *et al.*, 1973). Lysozymes c and g differ in their amino acid sequence, molecular weight and extinction coefficients (E_{280nm}) (Table 2).

Lysozyme g is heat labile, and contains half as much cystine (3) and tryptophan (4) residues as lysozyme c (Dianoux & Jolles, 1969). The concentration of lysozyme in mammalian milks varies from <0·3 mg/100 ml in bovine milk (BML) to about 79 mg/100 ml in equine milk (EML). Generally, lysozyme activity in precolostrum and colostrum is higher than in corresponding human (Hyslop *et al.*, 1974; Lewis-Jones & Reynolds, 1983) and bovine (Panfil-Kuncewicz & Kisza, 1976) milks. In bovine milk, lysozyme activity increases with somatic cell count and mastitis (Sokolova *et al.*, 1983; Carlsson *et al.*, 1989).

HML, CML and EML are considered type c lysozymes because their molecular weights are identical to HEWL. However, it is unclear whether BML is a type c or g lysozyme. Both types c and g lysozymes are present in bovine stomach tissues and fluids (Dobson *et al.*, 1984), raising the possibility that both types of lysozyme are present in bovine milk. Recent studies (White *et al.*, 1988) show that BML consists of two components which differ by seven amino acids in the first 39 residues.

Although HML and hen egg-white lysozyme (HEWL) are structurally homolo-

TABLE 2
Comparison of Lysozyme from Milk of Different Species

Source	*Concentration (mg/100 ml)*	*Molecular weight (daltons)*	$E_{280\,nm}$
Bovine (BML)	< 0·3[a]	18 000[b]	7·1[c]
Human (HML)	10–12[d]	15 000[d,e]	
Equine (EML)	79[f]	14 400[h]	
Caprine (CPL)	0·5[g]	15 000[g]	
Lysozyme c		15 000	27
Lysozyme g		20 500	15

[a]Chandan *et al.* (1968).
[b]Eitenmiller *et al.* (1971).
[c]Chandan *et al.* (1965).
[d]Parry *et al.* (1969).
[e]Jolles & Jolles (1967).
[f]Jauregui-Adell (1975).
[g]Duhaiman (1988).
[h]Jauregui-Adell *et al.* (1972).

gous, they differ in their amino acid composition: HEWL has 129 amino acid residues compared to 130 for HML, which has an extra glycine residue (Canfield *et al.*, 1974). HML is antigenically and serologically different from BML (Eitenmiller *et al.*, 1974).

There are similarities between milk lysozyme and α-lactalbumin in 3-dimensional structure (Brew & Campbell, 1967; Browne *et al.*, 1969) and intron–exon organisation of their genes (Phillips *et al.*, 1987), suggesting similar genetic lineage.

Lysozyme is purified from milk or acid whey by affinity chromatography on heparin-Sepharose followed by gel filtration on Sepharose 4B (Wang & Kloer, 1984), Sephadex-G50 (Boesman-Finkelstein & Finkelstein, 1982) and/or hydroxyapatite chromatography (Duhaiman, 1988).

BML and HML exhibit optimal activities at pH 7·5 (McKenzie & White, 1986) and 6·35 (Parry *et al.*, 1969), respectively. Both enzymes are most stable in the pH range 3–4 (Jauregui-Adell, 1975; Eitenmiller *et al.*, 1976) and retain > 75% of their activity in milk heated to 75°C × 15 min (Friend *et al.*, 1983) or 80°C × 15 s (Griffiths, 1986).

When the intrachain disulphide bonds in lysozyme are reduced, enzymatic activity is lost (Friend *et al.*, 1972) but reoxidation of the reduced enzyme restores its native conformation. However, renaturation rates differ among lysozymes; HEWL renatures more rapidly than HML as a result of the greater hydrophobicity of the latter (Dubois *et al.*, 1982). Also reoxidation of reduced HML and BML regenerates 84 and 328% of their original activities, respectively (Friend *et al.*, 1972).

In association with lactoferrin, lysozyme functions as a bacteriocidal agent in milk (Carlsson *et al.*, 1989). HML, BML (Repaske, 1956) and CML (Duhaiman, 1988) lyse *Escherichia coli*. Therefore, feeding human precolostrum, which contains high levels of lysozyme, to premature infants can help prevent gastrointestinal invasion by pathogenic bacteria (Lewis-Jones & Reynolds, 1983).

6.12. *N*-acetyl-*β*-D-glucosaminidase (EC 3.2.1.30)

N-Acetyl-*β*-D-glucosaminidase (NAGase) hydrolyses terminal, non-reducing *N*-acetyl-*β*-D-glucosamine residues in glycoproteins. NAGase is a lysosomal enzyme present in various mammalian tissues and secretions. Its activity in bovine milk was first detected by Mellors (1968). NAGase is also active in human (Oberkotter *et al.*, 1982), caprine (Timms & Schultz, 1985*a*; Maisi & Riipinen, 1988) and porcine (Hurley & Grieve, 1988) milks.

Most of the NAGase in milk originates from somatic cells (macrophages and neutrophile granulocytes) and mammary gland epithelial cells (Kitchen & Middleton, 1976; Kitchen *et al.*, 1980; Dulin *et al.*, 1985; Fox *et al.*, 1988), resulting in a correlation of 0·73 and 0·54 between NAGase activity and somatic cells in bovine and caprine milk, respectively (Timms & Schultz, 1985*a*). Thus, NAGase activity increases ($r > 0{\cdot}62$) with mastitis (Kitchen *et al.*, 1978; Obara & Komatsu, 1984; Miller & Paape, 1988). There are variations in NAGase activity between milkings (Marschke *et al.*, 1987) and during lactation (Miller & Paape, 1988). Dry cow secretions contain about 10-fold more NAGase activity than normal milk (Timms & Schultz, 1985*b*; Miller & Paape, 1988).

In fresh milk, NAGase is associated with the cells from which it originates (Kaartinen *et al.*, 1988) but is released as 'free' enzyme during storage at temperatures >4°C, freeze–thawing or detergent treatment (Kaartinen *et al.*, 1988) as a result of damage to these cells. Consequently, there are variations in the literature on the percentage (94%, Mellors (1968); 91%. Kitchen *et al.* (1978) or 63%, Fox *et al.* 1988)) of whole milk NAGase found in skim milk.

Bovine milk and mammary gland NAGase have identical properties (Kitchen *et al.*, 1979). Both consist of 2 dimeric isoenzymes (A and B) which differ in their molecular weights and heat stability. Isoenzyme A has a molecular weight of 120 kDa and is more heat stable than B, which has a molecular weight of 230 kDa (Kitchen *et al.*, 1979). Similarly, human milk contains two NAGase isoenzymes (Oberkotter *et al.*, 1982).

NAGase exhibits optimal activity at 50°C (Kitchen *et al.*, 1978) and at pH 4·2 (Mellors, 1968). Milk loses its NAGase activity when heated to 70–71°C for 15–18 s (Griffiths, 1986; Andrews *et al.*, 1987).

NAGase activity in unheated quarter foremilks, composite (Kitchen *et al.*, 1984*a*; Mattilla *et al.*, 1986) or bulk (Kitchen *et al.*, 1984*b*) samples is useful in the accurate detection of mastitis in quarters, cows or herds, respectively, provided the time of milking (Marschke *et al.*, 1987) and milk handling prior to testing (Kaartinen *et al.*, 1988) are standardised. NAGase activity may also be used to detect teat canal inflammations (Kaartinen & Jensen, 1988) and to monitor damage to NAGase-containing cells in milk (Kaartinen *et al.*, 1988).

6.13. α-Mannosidase (EC 3.2.1.24)

α-Mannosidase is a Zn-containing enzyme which catalyses the hydrolysis of terminal, non-reducing residues in α-D-mannosides. Its activity has been demonstrated only in bovine milk, where it is associated with β-casein (Mellors & Harwalkar, 1968).

The enzyme has been partially purified (12-fold) from bovine skim milk by ethanol precipitation and $(NH_4)_2SO_4$ fractionation. Attempts to purify it further by chromatography destroyed its activity (Mellors & Harwalkar, 1968).

α-Mannosidase is stable over a broad pH range (3·4–10). The enzyme is inhibited by Cu^{2+}, EDTA and dibasic amino acids (Harwalkar & Mellors, 1968). Freezing (−10°C for 5 days) of milk reduces its α-mannosidase activity by about 50% (Mellors & Harwalker, 1968). Pasteurisation (72°C × 15 s) of milk has no effect on its α-mannosidase activity but heating milk at 90°C for 15 s completely inactivates the enzyme (Andrews *et al.*, 1987).

The role of α-mannosidase in milk is unknown. However, its activity may be used to monitor the heat treatment of milk between 80 and 90°C (Andrews *et al.*, 1987).

6.14. β-Glucuronidase (EC 3.2.1.31)

β-Glucuronidase catalyses the hydrolysis of β-D-glucuronoside:

$$\beta\text{-D-glucuronoside} + H_2O \rightleftharpoons \text{alcohol} + \text{D-glucuronic acid}$$

β-Glucuronidase activity has been detected (Kiermeier & Gull, 1968) in human and bovine milks, but the enzyme has not yet been purified from milk. β-Glucuronidase activity is high (10 × normal) in colostrum and decreases to normal values between the 2nd and 10th days of lactation (Kiermeier & Gull, 1968). Since β-glucuronidase is a lysosomal enzyme, its activity in milk increases proportionally with somatic cell count up to 2×10^6 cells/ml (Kiermeier & Gull, 1968; Perdigon *et al.*, 1986).

Pasteurisation (72°C × 15 s) of milk destroys its β-glucuronidase activity although some activity survives UHT treatment (Kiermeier & Doruk, 1972*a*). Cu^{2+} is an uncompetitive inhibitor of the enzyme. Ascorbic acid increases the inhibitory effect of Cu (Kiermeier & Doruk, 1972*b*).

β-Glucuronidase activity in milk may be used as a complementary diagnostic test for mastitis (Perdigon *et al.*, 1986).

6.15. 5′-Nucleotidase (EC 3.1.3.5)

5′-Nucleotidase catalyses the hydrolysis of 5′-nucleotides:

$$5'\text{-ribonucleotide} + H_2O \rightleftharpoons \text{ribonucleoside} + P_i$$

The enzyme, which is a component of the MFGM (Patton & Trams, 1971; Snow *et al.*, 1980), has been purified from acid whey (Caulini *et al.*, 1972) and from MFGM (Huang & Keenan, 1972*a*).

TABLE 3
Some Properties[a] of 5′-Nucleotidases Isolated from MFGM

Property	*Fraction V*	*Fraction VI*
Activation energy (cal/mol)	10 400	9 600
Phospholipid (μg/mg)	8·2	30·5
Hydrolysis at 5 mM	CMP > IMP > GMP > AMP	GMP > CMP > IMP > AMP
Inhibition by ATP	competitive ($K_i = 5 \times 10^{-6}$ M)	non-competitive ($K_i = 4{\cdot}76 \times 10^{-5}$ M)
K_m for AMP	0·94 mM	5·0 mM
K_m for CMP	2·5 mM	3·95 mM
Activity loss due to 5 mM $MnCl_2$	15	60

[a]Huang & Keenan (1972*a*).

Purification of 5′-nucleotidase from MFGM involves detergent treatment, $(NH_4)_2SO_4$ fractionation, heat treatment, sonication and chromatography on Sepharose-4B. This yields two fractions (designated V and VI) with 5′-nucleotidase activity (Huang & Keenan, 1972*a*). The two isoenzymes differ in phospholipid content, substrate specificity and kinetic properties (Table 3). Although Fraction V has a lower phospholipid content than Fraction VI, it is rich in sphingomylin, which forms at least 78% of its phospholipid.

5′-Nucleotidase does not require a metal cofactor (Caulini *et al.*, 1972; Huang & Keenan, 1972*a*); its activity is optimal at pH 7·0–7·5 (Caulini *et al.*, 1972; Huang & Keenan, 1972*a*) and 69°C (Huang & Keenan, 1972*a*). However, in the presence of 4·5 mM Mg^{2+} with AMP as substrate, there is a second optimum at pH 10·0 (Huang & Keenan, 1972*a*). Heating at 60°C × 30 min (Huang & Keenan, 1972*a*), and 72 or 80°C for 15 s (Andrews *et al.*, 1987) reduces its activity by 20, 40 or 97%, respectively.

Although milk contains substrates (nucleotides) for 5′-nucleotidase, it is unknown whether the nucleotides are hydrolysed in milk. 5′-Nucleotidase activity is a useful 'marker' for studying secretory mechanisms of the milk fat globules and biogenesis of MFGM material (Kitchen, 1983).

6.16. Adenosine Triphosphatase (EC 3.6.1.3)

Adenosine triphosphatase (ATPase) catalyses the dephosphorylation of adenosine triphosphate, ATP:

$$ATP + H_2O \rightleftharpoons ADP + P_i$$

Unlike many ATPases involved in the active transport of Na^+ across membranes, bovine milk ATPases are not Na^+-dependent (Patton & Trams, 1971; Huang & Keenan, 1972*b*).

Bovine milk ATPases are associated with the MFGM. They are stimulated by Mg^{2+} and K^+ (Patton & Trams, 1971; Huang & Keenan, 1972*b*) and show

TABLE 4
Partial List of Minor Enzymes in Milk

EC No.	*Enzyme*	*Reaction catalysed*	*Source*	*Distribution in milk*	*Reference*
1.1.1.14	L-Iditol dehydrogenase	L-Iditol + NAD^+ ⇌ L-sorbose + NADH	—	SM	Neuman *et al.* (1965)
1.1.1.37	Malate dehydrogenase	Malate + NAD^+ ⇌ oxaloacetate + NADH	Mammary gland	SM	Grigor & Hartmann (1985); Andrews *et al.* (1987)
1.1.1.40	Malic enzyme	Malate + $NADP^+$ ⇌ pyruvate + CO_2 + NADH	Mammary gland	SM	Deodhar *et al.* (1964); Karlsson & Kjellberg (1966); Grigor & Hartmann (1985)
1.1.1.42	Isocitrate dehydrogenase	Isocitrate + $NADP^+$ ⇌ 2-oxogluterate + CO_2 + NADH	Mammary gland	SM	Grigor & Hartmann (1985)
1.1.1.44	Phosphoglucuronate dehydrogenase (decarboxylating)	6-Phospho-D-gluconate + $NADP^+$ ⇌ D-ribose-5 phosphate + CO_2 + NADPH	Mammary gland	SM	Grigor & Hartmann (1985)
1.1.1.49	Glucose-6-phosphate dehydrogenase	D-Glucose-D-gluconate + $NADP^+$ ⇌ D-glucono-1,5-lactone-6-phosphate + NADPH	Mammary gland	SM	Sklavunu-Zurukzoglu *et al.* (1965); Grigor & Hartmann (1985);
1.4.3.6	Amine oxidase (Cu-containing)	$RCH_2NH_2 + H_2O + O_2 \rightleftharpoons RCHO + NH_3 + H_2O_2$	—	SM	Patterson *et al.* (1969)
—	Polyamine oxidase	Spermine → spermidine → putrescine	—	SM	Morgan & Toothill (1985)
—	Fucosyltransferase	Catalyses the transfer of fucose from GDP L-fucose to specific oligosaccharides and glycoproteins	—	SM	Jarkovsky *et al.* (1970); Prieels *et al.* (1977)

TABLE 4 Continued

EC No.	*Enzyme*	*Reaction catalysed*	*Source*	*Distribution in milk*	*Reference*
1.6.99.3	NADH dehydrogenase	NADH + acceptor ⇌ NAD^+ + reduced acceptor	—	FGM	Martel-Pradol & Got (1972); Jarasch *et al.* (1977)
1.8.1.4	Dihydrolipoamide dehydrogenase (Diaphorase)	Dihydrolipoamide + NAD^+ ⇌ lipoamide + NADH	—	SM/FGM	Kitchen (1974)
2.4.1.22	Lactose synthetase A protein UDP-galactose: D-glucose, 1-galactosyltransferase B protein: α-lactalbumin	UDP galactose + D-glucose ⇌ UDP + lactose	Golgi apparatus	SM	Babad & Hassid (1964); Brodbeck & Ebner (1966); Bushway & Keenan (1979)
2.4.1.38	Glycoprotein 4-β-galactosyltransferase	UDP galactose + *N*-acetyl-D-glucosaminyl-glycopeptide ⇌ UDP + 4,β-D-galactosyl-*N*-acetyl-D-glucosaminyl glycopeptide	—	FGM	Blaken *et al.* (1982)
2.4.1.90	*N*-Acetyllactosamine synthase	UDP galactose + *N*-acetyl-D-glucosamine ⇌ UDP *N*-acetyllactosamine	Golgi apparatus		Powell *et al.* (1977)
2.4.99.6	CMP-*N*-acetyl-*N*-acetyl-lactosaminide α-2, 3-sialytransferase	CMP-*N*-acetylneuraminate + β-D-galactosyl 1,4-*N*-acetyl-D-glucosaminyl glycoprotein ⇌ CMP + α-*N*-acetylneuraminyl 1-2, 3-β-D-galactasyl-1,4-*N*-acetyl-D-glucosaminyl-glycoprotein	—	SM	Hesford & Berger (1983)

2.5.1.3	Thiamin-phosphate pyrophosphorylase	2-Methyl-4-amino-5-hydroxymethyl/pyrimidine diphosphate + 4-methyl-5-(2-phosphonooxyethyl)-thiazole $\rightleftharpoons$ pyrophosphate + thiamine monophosphate	—	FGM	Martel-Pradal & Got (1972)
2.6.1.1	Aspartate aminotransferase	L-aspartate + 2-oxogluterate $\rightleftharpoons$ oxaloacetate + L-glutamate	Blood	SM	Heyndrickx (1962); Neuman *et al.* (1965)
2.6.1.2	Alanine aminotransferase	L-alanine + 2-oxogluterate $\rightleftharpoons$ pyruvate + L-glutamate	Blood	SM	Heyndrickx (1962); Neuman *et al.* (1965)
2.7.7.49	RNA-directed DNA polymerase	*n* Deoxynucleoside triphosphate $\rightleftharpoons$ *n* pyrophosphate + DNA_n	—	SM	Schlom *et al.* (1972); Gerwin *et al.* (1973); Roy-Burman *et al.* (1973)
2.8.1.1	Thiosulphate sulphurtransferase	Thiosulphate + cyanide $\rightleftharpoons$ sulphite + thiocyanate	—	SM	Garcia Alfonso & Costella Bertran (1953)
3.1.1.8	Cholinesterase	An acylcholine + H_2O $\rightleftharpoons$ choline + a carboxylic acid anion	Blood	FGM	Augustinsson, K.-B. (1960); Montoure & Forster (1960)
3.1.3.9	Glucose-6-phosphatase	D-Glucose 6-phosphate + H_2O $\rightleftharpoons$ D-glucose + Pi		FGM	Dowben *et al.* (1967); Martel-Pradal & Got (1972)
3.1.6.1.	Arylsulphatase	A phenol sulphate + H_2O $\rightleftharpoons$ a phenol + sulphate	—	—	Poczekaj & Wenclewski (1963); Kitchen (1976)
3.2.1.21	β-Glucosidase	Hyrolysis of terminal non-reducing β-D-glucose residues	Lysosomes	FGM	Anderson (1977)
3.2.1.23	β-Galactosidase	Hydrolysis of terminal non-reducing β-D-galactose residues in β-D-galactosidase	Lysosomes	FGM	Anderson & Cheeseman (1975); Griffiths (1986)
3.2.1.51	α-Fucosidase	An α-L-fucoside + H_2O $\rightleftharpoons$ an alchohol + L-fucose	Lysosomes	—	Kitchen (1976); Andrews *et al.* (1987)

TABLE 4 Continued

EC No.	Enzyme	Reaction catalysed	Source	Distribution in milk	Reference
3.4.11.1	Cytosol aminopeptidase (Leucine aminopeptidase)	Aminoacyl-peptide + $H_2O \rightleftharpoons$ amino acid + peptide	—	SM	Serban *et al.* (1969); Eichel (1981)
3.4.11.3	Cystyl-aminopeptide (Oxytocinase)	Cystyl-peptidase + $H_2O \rightleftharpoons$ amino acid + peptide	—	SM	Brouet-Yager *et al.* (1974)
3.4.21.4	Trypsin	Hydrolyses peptide bond, preferentially Arg-, Lys-	—	SM	Mizakorimov & Tashmatova (1971); Borulf *et al.* (1987)
3.6.1.1	Inorganic pyrophosphatase	Pyrophosphate + $H_2O \rightleftharpoons$ 2 orthophosphate	—	SM/FGM	Heyndrickx (1963); Kitchen (1974)
3.6.1.9	Nucleotide pyrophosphate	A dinucleotide + $H_2O \rightleftharpoons$ 2 mononucleotides	—	SM/FGM	Patton & Trams (1971); Kitchen (1974)
4.2.1.1	Carbonate dehydratase	$H_2CO_3 \rightleftharpoons CO_2 + H_2O$	—	SM	Garcia Alfonson & Costella Vertran (1953)
5.3.1.9	Glucose-6-phosphate isomerase	D-Glucose-6-phosphate $\rightleftharpoons$ fructose 6-phosphate	—	SM	Heyndrickx (1962)
6.4.1.2	Acetyl-CoA carboxylase	ATP + acetyl-CoA + $HCO_3 \rightleftharpoons$ ADP orthophosphate + malonyl-CoA	—	FGM	Shriver *et al.* (1989)

SM = Skim milk; FGM = fat globule membrane.

maximum activity at 55°C and pH 8·5 (Huang & Keenan, 1972*b*). The enzyme has an activation energy of 10 200 cal/mole and its K_m for ATP is $5{\cdot}68 \times 10^{-5}$ M (Huang & Keenan, 1972*b*).

The observation that milk lacks Na^+-ATPase supports the view that the apical plasma membrane of the mammary gland secretory cell lacks the sodium pump and that Na^+ ions are passively distributed across the apical plasma membrane (Linzell & Peaker, 1971).

6.17. Fructose-Bisphosphate Aldolase (EC 4.1.2.13)

Fructose-bisphosphate aldolase (aldolase) catalyses the reaction described by:

$$\text{D-Fructose 1,6-bisphosphate} \rightleftharpoons \text{glycerol phosphate} + \text{D-glyceraldehyde-3-phosphate}$$

Milk aldolase originates from blood (leucocytes), resulting in the elevation of its activity during mastitis (Kitchen *et al.*, 1970). Most (66%) of the milk aldolase activity is in the skim milk (Kitchen *et al.*, 1970). In cream, the enzyme is associated with the MFGM (Erwin & Randolph, 1975).

Acidification (to pH 4·6) or warming (37°C) of milk destroys its aldolase activity (Polis & Shmukler, 1950) but the enzyme remains active after $(NH_4)_2SO_4$ precipitation of caseins and whey proteins (Polis & Shmukler, 1950; Kitchen *et al.*, 1970).

According to Dwivedi (1973), aldolase may play a role in the flavour development of dairy products.

6.18. Other enzymes

Table 4 is a partial list of other enzymes that have been detected in milk but not described above because of the limited information on their biochemical and molecular properties in milk.

6.19. Conclusions

Although enzymes represent < 1% of the total protein in milk, they are present in catalytic amounts, enough to cause changes in the quality of milk and milk products. Research on milk enzymes has focused on their use in the determination of milk quality (mastitis) prior to processing and to monitor heat treatment of milk and milk products. Very little attention has been paid to the value of milk enzymes, particularly those in bovine milk, in infant nutrition. The use of somatotropin to increase milk yield, if it becomes standard and acceptable practice, may affect the activities of some enzymes. As shown by Politis *et al.* (1990), milk from cows injected with somatotropin has lower than normal plasmin activity but activity increases rapidly to normal levels on cessation of somatotropin injection.

Acknowledgement

The preparation of this review was assisted by funds from the California Dairy Research Foundation.

References

Abernethy, J.L., Steinman, H.M. & Hill, R.L. (1974). Bovine erythrocyte superoxide dismutase. Subunit structure and sequence location of the intrasubunit disulphide bond. *J. Biol. Chem.*, **249**, 7339–47.

American Heart Association (AHA). (1981). Nutrition Program Committee. Advisory statement concerning the claims that consumption of homogenised milk increases the risk of heart disease. AHA, Dallas, TX.

Anderson, M. (1977). Source and significance of lysosomal enzymes in bovine milk fat globule membrane. *J. Dairy Sci.*, **60**, 1217–22.

Anderson, M. & Cheeseman, G.C. (1975). Stability of the fat globule membrane. Proceedings of the Lipolysis Symposium, Cork, Ireland. *I.D.F. Bulletin, Document*, **86**, 11–8.

Andrews, A.T., Anderson, M. & Goodenough, P.W. (1987). A study of the heat stabilities of a number of indigenous milk enzymes. *J. Dairy Res.*, **54**, 237–46.

Arnheim, N., Hindenburg, A., Begg, G.S. & Morgan, F.J. (1973). Multiple genes of lysozyme in birds. Studies on black swan egg white lysozyme. *J. Biol. Chem.*, **248**, 8036–42.

Asada, K. (1976). Occurrence of superoxide dismutase in bovine milk. *Agric. Biol. Chem.*, **40**, 1659–60.

Augustinsson, K.-B. (1960). The esterase activity of dog's colostrum. *Experientia*, **16**, 411–22.

Aurand, L.W., Chu, T.M., Singleton, J.A. & Shen, R. (1967). Xanthine oxidase activity and development of spontaneously oxidized flavor in milk. *J. Dairy Soc.*, **50**, 465–71.

Aurand, L.W., Bone, N.H. & Gidding, G.G. (1977). Superoxide and singlet oxygen in milk lipid peroxidation. *J. Dairy Sci.*, **60**, 363–9.

Babad, H. & Hassid, W.Z. (1964). A soluble lactose synthesising enzyme from bovine milk. *J. Biol Chem.*, **239**, PC946–7.

Barcelos, J. de A. & Weight, U. (1986). Gamma-glutamyl transferase activity in cow's milk during the cycle of udder function, in relation to changes in mammary gland tissue. *J. Vet. Med.*, **33**, 308–19.

Battelli, M.G., Lorenzoni, E. & Stirpe, F. (1973). Milk xanthine oxidase type D (Dehydrogenase) and type O (Oxidase). *Biochem. J.*, **131**, 191–8.

Baumrucker, C.R. (1979). γ-Glutamyl transpeptidase of bovine milk membranes: distribution and characterisation. *J. Dairy Sci.*, **62**, 253–8.

Baumrucker, C.R. (1980). Purification and identification of γ-glutamyl transpeptidase of milk membranes. *J. Dairy Sci.*, **63**, 49–54.

Bhattacharya, I.D., Picciano, M.F. & Milner, J.A. (1988). Characteristics of human milk glutathione peroxidase. *Biol. Trace Element Res.*, **18**, 59–70.

Bingham, E.W. & Kalan, E.B. (1967). Ribonuclease B of bovine milk. *Arch. Biochem. Biophys.*, **121**, 317–24.

Bingham, E.W. & Zittle, C.A. (1962). Ribonuclease in bovine milk. *Biochem. Biophys. Res. Commun.*, **7**, 408–13.

Bingham, E.W. & Zittle, C.A. (1964). Ribonuclease of bovine milk: purification and properties. *Arch. Biochem. Biophys.*, **106**, 235–9.

Binkley, F., Weisemann, M.L., Groth, D.P. & Powell, R.W. (1975). γ-Glutamyl transferase: a secretory enzyme. *FEBS Lett.*, **51**, 168–70.

Blaken, W.M., Hooghwinkel, G.J.M. & van den Eijnden, D.H. (1982). Biosynthesis of blood-group I and i substances. Specificity of bovine colostrum β-N-acetyl-D-glucosaminide β1-4 galactosyltransferase. *Eur. J. Biochem.*, **127**, 547–52.

Boesman-Finkelstein, M. & Finkelstein, R.A. (1982). Sequential purification of lactoferrin, lysozyme and secretory immunoglobulin A from human milk. *FEBS Lett.*, **144**, 1–5.

Bogin, E. & Ziv, G. (1973). Enzymes and minerals in normal and mastitic milk. *Cornell Vet.*, **63**, 666–76.

Bogin, E., Ziv, G., Avidar, J., Rivetz, B., Gordin, S. & Saran, A. (1977). Distribution of lactate dehydrogenase isoenzymes in normal and inflamed bovine udders and milk. *Res. Vet. Sci.*, **22**, 198–200.

Bordas, J., Bray, R.C., Garner, C.D. Gutteridge, S. & Hasnain, S.S. (1980). X-ray absorption spectroscopy of xanthine oxidase. The molybdenum centres of the functional and desulpho forms. *Biochem. J.*, **191**, 499–508.

Borulf, S., Lindberg, T. & Mansson, M. (1987). Immunoreactive anionic trypsin and anionic elastase in human milk. *Acta Pediatric Scand.*, **76**, 11–15.

Braun, J.P., Tainturier, D., Bezille, P., Raviart, I. & Rico, A.G. (1984). Transfer of gamma-glutamyltransferase from mother colostrum to newborn goat and foal. *Enzyme*, **31**, 193–6.

Bray, R.C. (1975). Molybdenum iron-sulphur flavin hydroxylases and related enzymes. In *The Enzymes*, vol. XIII, 3rd edn, ed. P.D. Boyer. Academic Press, New York, pp. 303–88.

Brew, K. & Campbell, P.M. (1967). The characterisation of whey proteins of guinea-pig milk. The isolation and properties of α-lactalbumin. *Biochem. J.*, **102**, 258–64.

Browne, W.J., North, A.C.T., Phillips, D.C., Brew, K., Vanaman, T.C. & Hill, R.L. (1969). A possible three-dimensional structure of bovine α-lactalbumin based on that of hen's egg-white lysozyme. *J. Mol. Biol.*, **42**, 65–86.

Brodbeck, U. & Ebner, K.E. (1966). Resolution of a soluble lactose synthetase into two protein components and solubilisation of microsomal lactose synthetase. *J. Biol. Chem.*, **241**, 762–4.

Brouet-Yager, M., Kleiner, H., Bogaert, E. & van Guff, G.L.A. (1974). L-cystinyl-di-α-naphthylamide hydrolase ('oxytocinase') in human colostrum and early milk plasma. *Clin. Chim. Acta.*, **54**, 387–9.

Bruder, G., Heid, H., Jorasch, E.-D., Keenan, T.W. & Mather, I.H. (1982). Characteristics of membrane-bound and soluble forms of xanthine oxidase from milk and endothelial cells of capillaries. *Biochim. Biophys. Acta*, **701**, 357–69.

Bushway, A.A. & Keenan, T.W. (1979). Characterisation of a soluble glycolipid galactosyltransferase which occurs in bovine milk. *Biochim. Biophys. Acta*, **572**, 146–52.

Canfield, R.E., Collins, J.C. & Sobel, J.H. (1974). Human leukemia lysozyme. In *Lysozymes*, ed. E.F. Osserman, R.E. Canfield & S. Beychock. Academic Press, New York, pp. 63–79.

Carlsson, A., Bjorck, L. & Persson, K. (1989). Lactoferrin and lysozyme in milk during acute mastitis and their inhibitory effect on delvotest P. *J. Dairy Sci.*, **72**, 3166–75.

Caulini, G., Binotti, I., Allessandrini, M., Ipata, P.L. & Magni, G. (1972). Preliminary results on 5′-nucleotidase in milk. *Bull. Soc. Italiana Biol. Sperimentale*, **48**, 890–3.

Cerbulis, J. & Farrell, H.M. (1977). Xanthine oxidase activity in dairy products. *J. Dairy Sci.*, **60**, 170–6.

Chandan, R.C., Parry, R.M. Jr. & Shahani, K.M. (1965). Purification and some properties of bovine milk lysozyme. *Biochim. Biophys. Acta*, **110**, 389–98.

Chandan, R.C., Parry, R.M. Jr. & Shahani, K.M. (1968). Lysozyme lipase and ribonuclease in milk of various species. *J. Dairy Sci.*, **51**, 606–7.

Cheng, S.G., Koch, U. & Brunner, J.R. (1988). Characteristics of purified cows' milk xanthine oxidase and its submolecular characteristics. *J. Dairy Sci.*, **71**, 901–16.

Clare, D.A., Blakistone, B.A., Swaisgood, H.E. & Horton, H.E. (1981). Sulphydryl oxidase-catalyzed conversion of xanthine dehydrogenase to xanthine oxidase. *Arch. Biochem. Biophys.*, **211**, 44–7.

Clare, D.A., Horton, H.R., Stabel, T.J., Swaisgood, H.E. & Lecce, J.G. (1984). Tissue distribution of mammalian sulphydryl oxidase. *Arch. Biochem. Biophys.*, **230**, 138–45.

Clifford, A.J., Ho, C.Y. & Swenerton, H. (1983). Homogenised bovine milk xanthine oxidase: a critique of the hypothesis relating to plasmalogen depletion and cardiovascular disease. *Am. J. Clin. Nutr.*, **38**, 327–32.

Dalaly, B.K., Eitenmiller, R.R., Vakil, J.R. & Shahani, K.M. (1970). Simultaneous isolation of human milk ribonuclease and lysozyme. *Anal. Biochem.*, **37**, 208–11.

Davis, M.D., Edmondson, D.E. & Muller, F. (1984). ^{31}P nuclear magnetic resonance and chemical studies of the phosphorus residues in bovine milk xanthine oxidase. *Eur. J. Biochem.*, **145**, 237–43.

Deeth, H.C. (1983). Homogenised milk and atherosclerotic disease: a review. *J. Dairy Sci.*, **66**, 1419–35.

Deodhar, A.D., Rajalakshimi, R. & Ramakrishnan, C.V. (1964). Studies on human lactation. Part IV. Lactic and malic dehydrogenases and xanthine oxidase in breast milk and their variations with progess of lactation and dietary vitamin supplementation. *Acta Pediatr. Stockh.*, **53**, 101–4. (Cited from *Dairy Sci. Abst.*, **26**, (1968) 602.).

Dianoux, A.-C. & Jolles, P. (1969). Differences in the behaviour of egg-white lysozymes of hens and geese with regards to *Micrococcus lysodeikticus. Bull. Soc. Chim. Biol.*, **51**, 1559–64.

Dobson, D.E., Prager, E.M. & Wilson, A.C. (1984). Stomach lysozyme of rumminants. I. Distribution and catalytic properties. *J. Biol. Chem.*, **259**, 11607–16.

Dowben, R.M., Brunner, J.R. & Philpott, D.E. (1967). Studies on milk fat globule membranes. *Biochim. Biophys. Acta*, **135**, 1–10.

Dubois, T., Guillard, R., Prieels, J.-P. & Perraudin J.-P. (1982). Comparison between the folding of reduced hen egg white lysozyme and that of reduced human milk lysozyme. *Biochemistry*, **21**, 6516–23.

Duhaiman, A.S. (1988). Purification of camel milk lysozyme and its lytic effect on *Escherichia coli* and *Micrococcus lysodeikticus. Comp. Biochem. Physiol.*, **91B**, 793–6.

Dulin, A.M., Paape, M.J. & Miller, R.H. (1985). NAGase activity in neutrophils, macrophages and lymphocytes. *J. Dairy Sci.*, **68** (Suppl. 1), 207.

Dwivedi, B.K. (1973). The role of enzymes in flavour development. Part I. Dairy products. *CRC Crit. Rev. Food Technol.*, **3**, 457–78.

Edmondson, D., Massey, V., Palmer, G., Beacham, L.M. & Elion, G.B. (1972). The resolution of active and inactive xanthine oxidase by affinity chromatography. *J. Biol. Chem.*, **247**, 1597–604.

Eichel, V. (1981). Leucine aminopeptidase activity in cow's milk. *Arch. Expt. Vet.*, **35**, 235–44. (Cited from *Dairy Sci. Abstr.*, **44**, (1982) 478.)

Eitenmiller, R.R., Friend, B.A. & Shahani, K.M. (1971). Comparison of bovine and human milk lysozymes. *J. Dairy Sci.*, **54**, 762 (abst).

Eitenmiller, R.R., Friend, B.A. & Shahani, K.M. (1974). Immunochemical studies on human and bovine milk lysozymes. *J. Food Sci.*, **39**, 930–3.

Eitenmiller, R.R., Friend, B.A. & Shahani, K.M. (1976). Relationship between composition and stability of bovine milk lysozyme. *J. Dairy Sci.*, **59**, 834–9.

Erwin, R.W. & Randolph, H.E. (1975). Influence of mastitis on the properties of milk. XI. Fat globule membrane. *J. Dairy Sci.*, **58**, 9–12.

Evans, H.J., Steinman, H.M. & Hill, R.L. (1974). Bovine erythrocyte superoxide dismutase. Isolation and characterization of tryptic, cyanogen bromide and maleylated tryptic peptide. *J. Biol. Chem.*, **249**, 7315–25.

Everse, J. & Kaplan, N.O. (1973). Lactate dehydrogenases: structure and function. *Adv. Enzymol.*, **37**, 61–133.

Fox, L.K., Hancock, D.D., McDonald, J.S. & Gaskins, C.T. (1988). *N*-Acetyl-*β*-D-glucosaminidase activity in whole milk and milk fractions. *J. Dairy Sci.*, **71**, 2915–22.

Fox, P.F. & Morrissey, P.M. (1981). Indigenous enzymes of bovine milk. In *Enzymes and*

Food Processing, ed. G.C. Birch, N. Blakenbrough & K.J. Parker. Applied Science Publishers, London, pp. 213–38.

Freundenstein, C., Keenan, T.W., Eigel, W.N., Sasaki, M., Sadler, J. & Franke, W. (1979). Preparation and characterisation of the inner coat material associated with milk fat globule membrane from bovine and human milk. *Expt. Cell Res.*, **118**, 277–94.

Fridovich, I. (1970). Quantitative aspects of the production of superoxide anion radical by milk xanthine oxidase. *J. Biol. Chem.*, **245**, 4053–7.

Fridovich, I. (1974). Superoxide dismutases. *Adv. Enzymol.*, **41**, 35–97.

Fridovich, I. (1986). Superoxide dismutases. *Adv. Enzymol,*, **58**, 61–97.

Friend, B.A., Eitenmiller, R.R. & Shahani, K.M. (1972). Reduction and reactivation of human and bovine milk lysozyme. *Arch. Biochem. Biophys.*, **149**, 435–40.

Friend, B.A., Shahani, K.M., Long, C.A. & Agel, E.N. (1983). Evaluation of freeze-drying, pasteurisation, high-temperature heating and storage in selected enzymes, B-vitamins and lipids in mature human milk. *J. Food Prot.*, **46**, 330–4.

Garcia Alfonso, C. & Costella Bertran, E. (1953). Composition of milk. Carbonic anhydrase, rhodonase and alkaline phosphomonoesterase in milk of cow, goat and ewe. *Le Lait*, **33**, 386–93.

Gerwin, B.I., Ebert, P.S., Chopra, H.C., Smith, S.G., Kverdar, J.P., Albert, S. & Brennan, M.J. (1973). DNA polymerase activities in human milk. *Science*, **180**, 198–201.

Gilbert, D.A. & Bergel, F. (1964). The chemistry of xanthine oxidase. *Biochem. J.*, **90**, 350–3.

Graham, K., Fleming, J.E., Young, R. & Bensch, K. (1989). Preparation of antibodies against xanthine oxidase from human milk. *Int. J. Biochem.*, **21**, 715–22.

Green, R.C. & O'Brien, P.J. (1967). Xanthine oxidase inactivation by reagents that modify thiol groups. *Biochem. J.*, **105**, 585–9.

Greenbank, G.R. & Pallansch, M.J. (1962). Inactivation and reactivation of xanthine oxidase in dairy products. *J. Dairy Sci.*, **45**, 958–61.

Griffiths, M.W. (1986). Use of milk enzymes as indices of heat treatment. *J. Food Prot.*, **49**, 696–705.

Grigor, M.R. & Hartman, P.E. (1985). NADP-linked dehydrogenases in secreted milk. *J. Dairy Sci.*, **52**, 501–6.

Gupta, N. & Mathur, M.P. (1989). Ribonuclease in goat milk. *Neth. Milk Dairy J.*, **43**, 495.

Gutteridge, S., Tanner, S.J. & Bray, R.C. (1978). Comparison of the molybdenum centres of native and desulpho xanthine oxidase. *Biochem. J.*, **175**, 887–97.

Guy, E.J. & Jenness, R. (1958). Separation, concentration and properties of alpha-amylase from cow's milk. *J. Dairy Sci.*, **41**, 13–27.

Hamosh, M. (1988). Enzymes in milk: Their function in the mammary gland, in milk, and in infant. In *Biology of Human Milk*, vol. 15, ed. L.A. Hanson. Nestle Nutrition Workshop Series, Raven Press, New York, pp. 45–61.

Hamosh, M., Freed, L.M., Jones, J.B., Berkow, S.E., Bitman, J., Mehta, N.R., Happ, B. & Hamosh, P. (1985). Enzymes in human milk. In *Human Lactation. Milk Components and Methodoligies*, ed. R.G. Jensen & M.C. Neville. Plenum Press, New York, pp. 251–66.

Heitlinger, L.A., Lee, P.C., Dillion, W.P. & Lebenthal, E. (1983). Mammary amylase: a possible alternate pathway of carbohydrate digestion in infancy. *Pediatr. Res.*, **17**, 15–18.

Hesford, F.J. & Berger, E.G. (1983). Purification of bovine colostrum sialyltransferase to homogeneity by affinity chromatography *Experientia*, **39**, 648 (abstr).

Heyndrickx, G.V. (1962). Investigation on enzymes in human milk. *Ann. Pediatr.*, **198**, 365–2.

Heyndrickx, G.V. (1963). Further investigations on the enzymes in human milk. *Pediatrics*, **31**, 1019–23.

Hicks, C.L. (1980). Occurrence and consequence of superoxide in milk products. *J. Dairy Sci.*, **63**, 1199–204.

Hicks, C.L., Bucy, J. & Stofer, W. (1979). Heat inactivation of superoxide dismutase in bovine milk. *J. Dairy Sci.*, **62**, 529–32.

Hill, R.D. (1975). Superoxide dismutase activity in bovine milk. *Aust. J. Dairy Technol.*, **30**, 26–8.

Hojo, Y. (1982). Selenium concentration and glutathione peroxidase activity in cow's milk. *Biol. Trace Element Res.*, **4**, 233–9.

Holbrook, J. & Hicks, C.L. (1978). Variation of superoxide dismutase in bovine milk. *J. Dairy Sci.*, **61**, 1072–7.

Holbrook, J.J., Liljas, A., Steindel, S.J. & Rossmann, M.C. (1975). Lactate dehydrogenase. In *The Enzymes*, vol. XI, 3rd edn. ed. P.D. Boyer. Academic Press, New York, pp. 191–292.

Huang, C.M. & Keenan, T.W. (1972*a*). Preparation and properties of 5′-nucleotidases from bovine milk fat globule membranes. *Biochim. Biophys. Acta*, **274**, 246–57.

Huang, C.M. & Keenan, T.W. (1972*b*). Adenosine triphosphatase activity of bovine milk fat globule membranes. *Comp. Biochem. Physiol.*, **43B**, 277–82.

Hurley, W.L. & Grieve, R.J.C. (1988). Total and differential cell counts and *N*-acetyl-β-D-glucosaminidase activity in sow milk during lactation. *Vet. Res. Comm.*, **12**, 149–53.

Hyslop, N.E. Jr., Kern, K.C. & Walker, W.A. (1974). Lysozyme in human colostrum and breast milk. In *Lysozyme*, ed. E.F. Osserman, R.E. Canfield & S. Beychock. Academic Press, New York, pp. 449–62.

Isaacs, C.E. (1985). Milk enzyme function: effects of compartmentation and storage conditions on sulphydryl oxidase and glutamyl transpeptidase. In *Human Lactation: Milk Components and Methodologies*, ed. R.G. Jensen and M.C. Neville. Plenum Press, New York, pp. 277–82.

Isaacs, C.E., Pascal, T., Wright, C.E. & Gaull, G.E. (1984). Sulphydryl oxidase in human milk: stability of milk enzymes in the gastrointestinal tract. *Pediatr. Res.*, **18**, 532–5.

Ito, O. (1971). Studies on catalase in milk. 4. Relationship between catalase activity and lipoprotein in buttermilk. *Bull. Nippon Vet. Zootech. Coll.*, **20**, 56–61.

Ito, O. (1974). Studies on catalase in milk. XII. Physicochemical properties of catalase (3). *Bull. Nippon Vet. Zootech. Coll.*, **23**, 45–8.

Ito, O. & Akuzawa, R. (1982). Distribution of catalase in goat milk. *Bull. Nippon Vet. Zootech. Coll.*, **31**, 198–202.

Ito, O. & Akuzawa, R. (1983*a*). Purification, crystallisation, and properties of bovine milk catalase. *J. Dairy Sci.*, **66**, 967–73.

Ito, O. & Akuzawa, R. (1983*b*). Isozymes of bovine milk catalase. *J. Dairy Sci.*, **66**, 2468–73.

Ito, O. & Akuzawa, R. (1983*c*). Dissociation of cow's milk catalase by sodium dodecyl sulphate. *Jpn. J. Zootech. Sci.*, **54**, 39–45.

Ito, O., Akuzawa, R. & Kamata, S.I. (1984). Immunochemical differences between two types of catalase from cream. *Jpn. J. Zootech. Sci.*, **55**, 722–7.

Janalino, V.G. & Swaisgood, H.E. (1985). Isolation and characterisation of sulphydryl oxidase from bovine milk. *J. Biol. Chem.*, **250**, 2532–8.

Janalino, V.G., Barnes, C.S. & Swaisgood, H.E. (1980). Dissociation and unfolding of sulphydryl oxidase in solutions of guanidinium chloride. *J. Dairy Sci.*, **63**, 1969–74.

Janalino, V.G., Morrison-Rowe, S.J. & Swaisgood, H.E. (1990). Confirmation of blocked amino terminus of sulphydryl oxidase. *J. Dairy Sci.*, **73**, 2287–91.

Jarasch, E.-D., Brudder, G., Keenan, T.W. & Franke, W.W. (1977). Redox constituents in milk fat globule membranes and rough endoplasmic reticulum from lactating mammary gland. *J. Cell Biol.*, **73**, 223–41.

Jarkovsky, Z., Marcus, D.M. & Gollman, A.P. (1970). Fucosyltransferase found in human milk. Product of the Lewis blood group gene. *Biochemistry*, **9**, 1123–8.

Jauregui-Adell, J. (1975). Heat stability and reactivation of mare milk lysozyme. *J. Dairy Sci.*, **58**, 835–8.

Jauregui-Adell, J., Cladell, G., Ferraz-Pina, C. & Rech, J. (1972). Isolation and partial characterisation of mare milk lysozyme. *Arch. Biochem. Biophys.*, **151**, 353–5.

Jenness, R. & Patton, S. (1959). *Principles of Dairy Chemistry*. John Wiley and Sons, New York, pp. 192–3.

Jolles, J. & Jolles, P. (1967). Human tear and human milk lysozyme. *Biochemistry*, **6**, 411–17.
Jones, J.B., Metha, N.R. & Hamosh, M. (1982). α-Amylase in preterm milk. *J. Pediatr. Gastroenterol. Nutr.*, **1**, 43–8.
Kaartinen, L. & Jensen, N.E. (1988). Use of *N*-acetyl-β-glucosaminidase to detect teat canal inflammations. *J. Dairy Res.*, **55**, 603–7.
Kaartinen, L., Kuosa, P.L., Veijalainen, K. & Sandholm, M. (1988). Compartmentalisation of milk *N*-acetyl-β-D-glucosaminidase (NAGase). *J. Vet. Med.*, **35**, 408–14.
Karlsson, B. & Kjellberg, B. (1966). Changes in the lactic dehydrogenase isoenzyme pattern of porcine milk during the lactation period. *Naturwissenschaften*, **53**, 90.
Kato, K., Mori, K. & Katoh, N. (1989). Contribution of leucocytes of the origin of lactate dehydrogenase isozymes in milk of bovine mastitis. *Jpn. J. Vet. Sci.*, **51**, 530–9.
Keen, C.L., Lonnerdal, B., Stein, T.N. & Hurley, L.S. (1980). Superoxide dismutase isoenzymes in bovine and human milk. *Biol. Trace Element Res.*, **2**, 221–7.
Kiermeier, F. & Doruk, M. (1972*a*). Inactivation of β-glucuronidase in milk by heat. *Z. Lebensm. Unters-Forsch.*, **150**, 220–4.
Kiermeier, F. & Doruk, M. (1972*b*). On the influence of copper on the β-glucuronidase in milk. *Z. Lebensm. Unters-Forsch.*, **150**, 274–6.
Kiermeier, F. & Gull, J. (1968). Biologische einflusse auf die β-glucuronidase-aktivitat der milch. *Z. Lebensm. Unters-Forsch.*, **138**, 205–8.
Kiermeier, F. & Petz, E. (1967*a*). Uber ein sulfhydrylgruppen oxidierrendes enzym in der milch. I. Isolierung und cherakterisierung des enzyms. *Z. Lebensm. Unters-Forsch.*, **132**, 342–52.
Kiermeier, F. & Petz, E. (1967*b*). Uber ein sulfhydrylgruppen oxydierendes enzym in der milch. II. Einflub auf die sulfhydrylgruppen erhitzter milch und molke. *Z. Lebensm. Unters-Forsch.*, **134**, 97–102.
Kiermeier, F. & Patz, E. (1967*c*). Uber ein sulfhydrylgruppenoxydierendes enzym in der milch. III. Einflub von erhitzungstemperatur und lagerzeit. *Z. Lebensm. Unters-Forsch.*, **134**, 149–56.
Kitchen, B.J., Middleton, G., Durward, I., Andrews, R.J. & Salmon, M.C. (1980). Mastitis diagnostic tests to estimate mammary gland epithelial cell damage. *J. Dairy Sci.*, **63**, 978–83.
Kitchen, B.J., Kwee, W.S., Middleton, G. & Andrews, R.J. (1984*a*). Relationship between the levels of *N*-acetyl-β-D-glucosaminidase (NAGase) in bovine milk and the presence of mastitis pathogens. *J. Dairy Res.*, **51**, 11–16.
Kitchen, B.J., Middleton, G., Kwee, W.S. & Andrews, R.J. (1984*b*). *N*-Acetyl-β-D-glucosaminidase (NAGase) in bulk herd milk. *J. Dairy Res.*, **51**, 227–32.
Kjellberg, B. & Karlsson, B.W. (1967). Comparative analyses of lactic and malic dehydrogenases and their multiple molecular forms in milk from animal species and man. *Comp. Biochem. Physiol.*, **22**, 397–413.
Korycka-Dahl, M., Richardson, T. & Hicks, C.L. (1979). Superoxide dismutase activity in bovine milk serum. *J. Food Prot.*, **42**, 867–71.
Lewis-Jones, D.I. & Reynolds, G.J. (1983). A suggested role for precolostrum in preterm and sick newborn infants. *Acta Pediatr. Scand.*, **72**, 13–17.
Lindberg, T. & Skude, G. (1982). Amylase in human milk. *Pediatrics*, **70**, 235–8.
Linzell, J.L. & Peaker, M. (1971). Intracellular concentrations of sodium, potassium and chloride in the lactating mammary gland and their relation to the secretory mechanism. *J. Physiol.*, **216**, 683–700.
Maisi, P. & Riipinen, I. (1988). Use of California mastitis test, *N*-acetyl-β-D-glucosaminidase, and antitrypsin to diagnose caprine subclinical mastitis. *J. Dairy Res.*, **55**, 309–14.
Majumder, G.C. & Ganguli, N.C. (1972). γ-Glutamyl transpeptidase in milk. I. Certain parameters influencing its activity. *Milchwissenschaft*, **27**, 296–9.
Malmstrom, B,G, Andreasson, L.-E. & Heinhammar, B. (1975). Copper-containing oxidases and superoxide dismutase. In *The Enzymes*, vol. XII, 3rd edn, ed. P.D. Boyer. Academic Press, New York, pp. 507–79.

Mangino, M.E. & Brunner, J.R. (1977). Isolation and partial characterisation of xanthine oxidase associated with the milk fat globule membrane of cows' milk. *J. Dairy Sci.*, **60**, 841–50.

Marschke, R.J., Roberts, R. & Kitchen, B.J. (1987). The effect of sampling time on *N*-acetyl-β-D-glucosaminidase (NAGase) levels in bovine milk and its relevance to mastitis diagnosis. *Aust. J. Dairy Technol.*, **42**, 3–6, 10.

Martel-Pradel, M.B. & Got, R. (1972). Presence d'enzymes marqueurs des membranes plasmiques, de l'appareil de golgi et du recticulum endoplasmique dans les membranes de globules lipidique de lait maternel. *FEBS Lett.*, **21**, 220–2.

Massey, V. & Edmondson, D. (1970). On the mechanism of inactivation of xanthine oxidase by cyanide. *J. Biol. Chem.*, **245**, 6595–8.

Massey, V., Brumby, P.E., Komai, H. & Palmer, G. (1969). Studies on milk xanthine oxidase. Some spectral and kinetic properties. *J. Biol. Chem.*, **244**, 1682–91.

Mattila, T., Pyorala, S. & Sandholk, M. (1986). Comparison of milk antitrypsin albumin, *N*-acetyl-β-D-glucosaminidase, somatic cells and biological analysis as indicators of bovine subclinical mastitis. *Vet. Res. Commun.*, **10**, 113–24.

McCormick, J.J., Larson, L.J. & Rich, M.A. (1974). RNAse inhibition of reverse transcriptase activity in human milk. *Nature*, **251**, 737–40.

McKenzie, H.A. & White, F.H. Jr. (1986). Determination of lysozyme activity at low levels with emphasis on the milk enzyme. *Anal. Biochem.*, **157**, 367–74.

Meister, A. & Tate, S.S. (1976). Glutathione and related γ-glutamyl compounds: biosynthesis and utilisation. *Ann. Rev. Biochem.*, **45**, 559–604.

Mellors, A. (1968). *β-N*-acetylglucosaminase in bovine milk. *Can. J. Biochem.*, **46**, 451–5.

Mellors, A. & Harwalkar, V.R. (1968). Glycosidases in bovine milk: α-mannosidase and its inhibition by zwitterions. *Can. J. Biochem.*, **46**, 1351–6.

Meyer, D.H., Kunin, A.S., Maddalena, J. & Meyer, .L. (1987*a*). Ribonuclease activity and isozymes in raw and processed cows' milk and infant formulas. *J. Dairy Sci.*, **70**, 1797–803.

Meyer, W.L., Meyer, D.H., Kunin, A.S., Capeless, E.L., Simmons, K.R. & Pankey, J.W. (1987*b*). Ribonuclease II isoenzymes in milk. Their possible role and use as diagnostic tools. *J. Dairy Sci.*, **70** (Suppl. 1), 266 (abstr).

Miller, R.H. & Paape, M.J. (1988). Effects of parity, bacteriological status, stage of lactation, and dry period on *N*-acetyl-β-D-glucosaminidase activity of milk and dry secretions. *J. Dairy Sci.*, **71**, 2508–12.

Misra, H.P. & Fridovich, I. (1972). The role of superoxide anion in the autoxidation of epinephrine and a simple assay for superoxide dismutase. *J. Biol. Chem.*, **247**, 3170–5.

Mizakorimov, U.M. & Tashmatova, K.Y. (1971). Hydrolytic enzymes of human colostrum. *Med. Zh. Uzbeck.*, **5**, 45–6. (Cited from *Dairy Sci. Abstr.*, 34 (1972) 571.).

Modi, U.V., Owen. E.C. & Proudfoot, R. (1959). Species differences in the occurrence of xanthine oxidase in milk. *Proc. Nutr. Soc.*, **18**, i(abstr).

Montoure, J.E. & Forster, T.L. (1960). Observations on the cholinesterase activity of bovine plasma and colostrum. *J. Dairy Sci.*, **43**, 878–9.

Morgan, D.M.L. & Toothill, V.J. (1985). Polyamine oxidase activity in human and bovine milk. *Biochem. Soc. Trans.*, **13**, 352–3.

Nakamura, M. & Yamazaki, I. (1982). Preparation of bovine milk xanthine oxidase as a dehydrogenase form. *J. Biochem.*, **92**, 1979–86.

Nathans, G.R. & Hade, E.P.K. (1978). Bovine milk xanthine oxidase. Purification by ultrafiltration and conventional methods which omit addition of proteases. Some criteria for homogeneity of native xanthine oxidase. *Biochim. Biophys. Acta*, **526**, 328–34.

Nishino, T., Nishino, T. & Tsushima, K. (1981). Purification of highly active milk xanthine oxidase by affinity chromatography on Sepharose 4B/folate gels. *FEBS Lett.*, **131**, 369–72.

Neuman, V., Kudelka, E. & Sidelarova, K. (1965). Transaminases and sorbitol dehydrogenase in cows' milk. *Vet. Med. Praha*, **10**, 349–54. (Cited from *Dairy Sci. Abstr.*, **27**, (1965) 536.)

Newburg, D.S., Daniel, P.F., O'Neill, N.E. & McCluer, R.H. (1986). High performance liquid chromatography of neutral and acidic oligosaccharides from human milk and colostrum. In *Human Lactation: Maternal-Environmental Factors*, ed. M. Hamosh & A.S. Goldman. Plenum Press, New York, pp. 581–8.

Obara, Y. & Komatsu, M. (1984). Relationship between *N*-acetyl-β-*D*-glucosaminidase activity and cell count, lactose, or lactoferrin in cow milk. *J. Dairy Sci.*, **67**, 1043–6.

Oberkotter, L.V., Koldovsky, O. & Tenore, A. (1982). *N*-acetyl-β-hexosaminidase activity in human breast milk. *Int. J. Biochem.*, **14**, 151–4.

Oster, K.A. (1971). Plasmalogen diseases: A new concept of the etiology of the atherosclerotic process. *Am. J. Clin. Res.*, **2**, 30–5.

Panfil-Kuncewicz, H. & Kisza, J. (1976). Study of lysozyme content in human and cows' milk. *Zeszyty Naukowe Acad. Rolniczo-Technioznej w Olsztynie Technol. Zwynosci.*, **151**, 105–12. (Cited from *Dairy Sci. Abstr.*, **39**, (1977) 119.)

Parry, R.M. Jr., Chandan, R.C. & Shahani, K.M. (1969). Isolation and characterisation of human milk lysozymes. *Arch. Biochem. Biophys.*, **130**, 59–65.

Patel, S.S. & Wilbey, R.A. (1989). Heat exchanger performance: γ-glutamyl transpeptidase assay as a heat treatment indicator for dairy products. *J. Soc. Dairy Technol.*, **42**, 79–80.

Patil, K.P. & Rangnekar, N.R. (1982). γ-Glutamyltransferase activity in human milk. *Clin. Chem.*, **28**, 1724–5.

Patil, K.P. & Rangnekar, N.R. (1983). Lactate dehydrogenase and its isoenzymes in human milk. *Clin. Chem.*, **29**, 1568–70.

Patterson, D.S.P., Berrett, S. & Cullen, G.A. (1969). Oxidative enzymes in milk and their value as indicators of subclinical mastitis in the cow. *Vet. Record*, **85**, 708–13.

Patton, S. & Trams, E.G. (1971). The presence of plasma membrane enzymes on the surface of bovine milk fat globules. *FEBS Lett.*, **14**, 230–2.

Perdigon, G. & Pesce de Ruiz Hogado, A.A. (1986). Significance of β-glucuronidase in mastitis detection. *J. Dairy Sci.*, **69**, 27–31.

Phillips, D.C., Acharya, K.R., Handoll, H.H.G. & Stuart, O.I. (1987). From lysozyme to α-lactalbumin: protein engineering and evolution. *Biochem. Soc. Trans.*, **15**, 737–44.

Phillips, J.D. & Griffiths, M.W. (1987). The use of the catalase metre in assessing the quality of milk. *J. Appl. Bact.*, **62**, 223–6.

Pocius, P.A., Baumrucker, C.R., McNamara, J.P. & Bauman, D.E. (1987). γ-Glutamyl transpeptidase in rat mammary tissue. *Biochem. J.*, **188**, 565–8.

Poczekaj, J. & Wenclewski, A. (1963). The activity of arylsulphatase of human milk. *Ginek. Polska.*, **34**, 577–82.

Polis, B.D. & Smukler, H.W. (1950). Aldolase in bovine milk. *J. Dairy Sci.*, **33**, 619–22.

Politis, I., Block, E. & Turner, J.D. (1990). Effect of somatotropin on the plasminogen and plasmin system in the mammary gland: Proposed mechanism of action for somatotropin on the mamary gland. *J. Dairy Sci.*, **73**, 1494–9.

Powell, J.T., Jarlfors, U. & Brew, K. (1977). Enzymic characteristics of fat globule membranes from bovine colostrum and bovine milk. *J. Cell. Biol.*, **72**, 617–27.

Prieels, J.-P., Bayer, T. & Hill, R.L. (1977). Human milk fucosyltransferases. *Biochem. Soc. Trans.*, **5**, 838–9.

Read, R.B. Jr., Reyes, A.L., Bradshaw, J.G. & Peelre, J.T. (1969). Evaluation of seven procedures for detecting abnormal milk due to mastitis. *J. Dairy Sci.*, **52**, 1359–67.

Repaske, R. (1956). Lysis of Gram-negative bacteria by lysozyme. *Biochim. Biophys. Acta*, **22**, 189–91.

Rotilio, G., Calabrese, L., Bossa, F., Barra, D., Finazzi Agro, H. & Mondovi, B. (1972). Properties of the apoprotein and role of copper and zinc in protein conformation and enzyme activity of bovine superoxide dismutase. *Biochemistry*, **11**, 2182–92.

Roy-Burman, P., Rongey, R.W., Henderson, B.E. & Gardner, M.B. (1973). Attempts to detect RNA tumor virus in human milk. *Nature New Biol.*, **244**, 146.

Ruegg, M., Moor, U. & Blanc, B. (1977). A calorimetric study of the thermal denaturation of whey proteins in simulated milk infiltrate. *J. Dairy Res.*, **44**, 509–20.

Scheraga, H.A. & Rupley, J.A. (1962). Structure and function of ribonuclease. *Adv. Enzymol.*, **24**, 161–261.

Schlom, J., Spiegelman, S. & Moore, D.H. (1972). Detection of high molecular-weight RNA in particles from human milk. *Science*, **172**, 542–4.

Schonbaum, G.R. & Chance, B. (1976). Catalase. In *The Enzymes*, vol. XIII, 3rd edn. ed. P.D. Boyer. Academic Press, New York, pp. 363–408.

Serban, A.M.D., Biener, J. & Diaconescu, C. (1969). Enzyme contents of human colostrum, milk and abnormal mammary secretions. *Studii Cerc. Endocr.*, **20**, 147–52. (Cited from *Dairy Sci. Abstr.*, **33**, (1971) 563.)

Shahani, K.M., Harper, W.J., Jensen, R.G., Parry, R.M. Jr. & Zittle, C.A. (1973). Enzymes in bovine milk: a review. *J. Dairy Sci.*, **56**, 53--43.

Shahani, K.M., Kwam, A.J. & Friend, B.A. (1980). Role and significance of enzymes in human milk. *Am. J. Clin. Nutr.*, **33**, 1861–8.

Shidlovskaya, V.P. (1982). Activity of enzymes and prediction of milk stability. *Proc. 21st Int. Dairy Congr. (Moscow)*, **1**(2), 520–1.

Shriver, B.J., Allred, J.B. & Roman-Lopez, C.R. (1989). Bovine milk-fat-globule-membrane contains an enzymically inactive form of acetyl-CoA carboxylase. *Biochem. J.*, **257**, 925–7.

Sklavunu-Zurukzoglu, S., Mameletziz, C. & Katriu, D. (1965). Observations on the glucose-6-phosphate dehydrogenase of the breast milk. *Helv. Paediat. Acta*, **20**, 193–6. (Cited from *Dairy Sci. Abstr.*, **51**, (1965) 40.)

Sliwkowski, M.B., Swaisgood, H.E. & Horton, H.R. (1982). Solubilisation of sulphydryl oxidase, a bovine skim milk membrane enzyme. *J. Dairy Sci.*, **65**, 1681–7.

Smith, G.J. & Dunkley, W.L. (1960). Xanthine oxidase and incidence of spontaneous oxidised flavor in milk. *J. Dairy Sci.*, **43**, 278–80.

Snow, L.D., Doss, R.C. & Carraway, K.L. (1980). Cooperativity of the concanavalin A inhibition of bovine milk fat globule membrane 5′-nucleotidase. Response to the extraction of nucleotides and of the putative cytoplasmic surface coat components. *Biochim. Biophys. Acta*, **611**, 333–41.

Sobiech, K.A., Ziomek, E. & Szewczuk, A. (1974). Purification and some properties of γ-glutamyl transpeptidase from cow's milk. *Arch. Immun. Ther. Exp.*, **22**, 645–56. (Cited from *Dairy Sci. Abstr.*, **38**, (1976) 410.)

Sokolova, A.P., Sokolova, I.V. & Lyubimor, I.A. (1983). Lysozyme activity of milk and the functional state of the mammary gland. *Soviet Agric. Sci.*, **1**, 34–5. (Cited from *Dairy Sci. Abstr.*, **46**, (1984) 272.)

Sommer, H., Hambitzer, R. & Aydin, I. (1986). LDH-activity and correlation to pathogenic microorganisms in dairy-milk. *Israel J. Vet. Med.*, **42**, 372–5.

Song, K.B., Swaisgood, H.E. & Horton, H.R. (1986). Requirement for a sulphyfryl group for sulphydryl oxidase activity. *J. Dairy Sci.*, **69**, 2589–92.

Stein, E.A. (1987). Extraction and purification of molybdenum cofactor from milk xanthine oxidase. *Eur. J. Biochem.*, **169**, 349–52.

Stejskal, K., Kamaryt, J. & Muchova, M. (1981). The alpha-amylase activity in breast milk. *Refrigeration Sci. Technol.*, **2**, 85–9.

Stirpe, F. & Della Corte, E. (1969). The regulation of rat liver xanthine oxidase. Conversion *in vitro* of the enzyme activity from dehydrogenase (Type D) to oxidase (Type O). *J. Biol. Chem.*, **244**, 3855–63.

Sullivan, C.H., Mather, I.H., Greenwalt, D.E. & Madara, P.J. (1982). Purification of xanthine oxidase from the fat globule membrane of bovine milk by electrofocusing. Determination of isoelectric points and preparation of specific antibodies to the enzyme. *Mol. Cell. Biochem.*, **44**, 13–22.

Swaisgood, H.E. (1980). Sulphyldryl oxidase: properties and applications. *Enzyme Microb. Technol.*, **2**, 265–72.

Swaisgood, H.E. & Abraham, P. (1980). Oxygen activation by sulphydryl oxidase and the enzyme's interaction with peroxidase. *J. Dairy Sci.*, **63**, 1205–10.

Swaisgood, H.E. & Horton, H.R. (1980). Sulphur in biology: oxidation of sulphydryl groups and the formation of three-dimensional structure in proteins. In *The Biology of Sulphur*, CIBA Foundation Symposium No. 72, pp. 205–22.

Swaisgood, H.E., Sliwkowski, M.X., Skudder, P.J. & Janalino, V.G. (1982). Sulphyldryl oxidase: characterisation and application for flavour modification of UHT milk. In *Use of Enzymes in Food Technology*, ed. P. Dupuy. Technique et Documentation Lavoisier, Paris, pp. 229–35.

Swaisgood, H.E., Janalino, V.G. & Skudder, P.J. (1987). Continuous treatment of ultra-high-temperature sterilised milk using immobilised sulphyldryl oxidase. In *Methods in Enzymology* vol. 136, ed. K. Mosbach, pp. 423–31. Academic Press, London, pp. 423–31.

Tate, S.S. & Meister, A. (1974). Interaction of γ-glutamyl transpeptidase with amino acids, dipeptides, and derivatives and anologs of glutathione. *J. Biol. Chem.*, **249**, 7593–602.

Timms, L.L. & Schultz, L.H. (1985*a*). *N*-Acetyl-β-D-glucosaminidase activity and somatic cells in goat milk. *J. Dairy Sci.*, **68**, 3363–6.

Timms, L.L. & Schultz, L.H. (1985*b*). *N*-Acetyl-β-D-glucosaminidase in milk and blood plasma during dry and early post-partum periods. *J. Dairy Sci.*, **68**, 3368–70.

Tramper, J. (1987). Preparation of immobilised milk xanthine oxidase and application in organic syntheses. In *Methods in Enzymology*, vol. 136, ed. K. Mosbach. Academic Press, London, pp. 254–62.

Valivullah, M.H. & Keenan, T.W. (1989). Butyrophilin of lipid globule membrane contains N-linked carbohydrates and cross-links with xanthine oxidase. *Int. J. Biochem*,, **21**, 103–7.

Ventom, A.M., Diestung, J. & Bray, R.C. (1988). The isolation of demolybdo xanthine oxidase from bovine milk. *Biochem. J.*, **255**, 949–56.

Vina, J.R., Puertes, I.R., Montoro, J.B., Saez, G.T. & Vina, J. (1985). Gamma-glutamyl-amino acids as signals for the hormonal regulation of amino acid uptake by the mammary gland of the lactating rat. *Biology of the Neonate*, **48**, 250–6.

Walstra, P. & Jenness, R. (1980). *Dairy Chemistry and Physics*. John Wiley and Sons, New York, pp. 128.

Wang, C.S. & Kloer, H.U. (1984). Purification of human lysozyme from milk pancreatic juice. *Anal. Biochem.*, **139**, 224–7.

Waud, W.R. & Rajagopalan, K.V. (1976). A mechanism of the conversion of liver xanthine dehydrogenase from an NAD^+-dependent form (Type D) to an O_2-dependent form (Type O). *Arch. Biochem. Biophys.*, **172**, 365–79.

Waud, W.R., Bray, F.D., Wiley, R.D. & Rajagopalan, K.V. (1975). A new purification procedure for bovine milk xanthine oxidase: effect of proteolysis on the subunit structure. *Arch. Biochem. Biophys.*, **169**, 695–701.

White, F.H. Jr., McKenzie, H.A. & Shaw, D.C. (1988). Studies on a partially purified milk lysozyme. *Biochem. Int.*, **16**, 521–8.

Wuthrich, S., Richterrich, R. & Hosteller, H. Studies of milk enzymes. II. Enzymes in centrifuged and heat treated cows' milk. *Z. Lebensm. Unters-Forsch.*, **124**, 345–8.

Yasumoto, K., Iwami, K., Fushiki, T. & Mitsuda, H. (1978). Purification and enzymatic properties of γ-glutamyltransferase from bovine colostrum. *J. Biochem. (Tokyo)*, **84**, 1227–36.

Zikakis, J.P. & Treece, J.M. (1971). Xanthine oxidase polymorphism in bovine milk. *J. Dairy Sci.*, **54**, 648–54.

Zikakis, J.P. & Wooters, S.C. (1980). Activity of xanthine oxidase in dairy products. *J. Dairy Sci.*, **63**, 893–904.

Zikakis, J.P., Dougherty, T.M. & Biasoto, N.O. (1976). The presence and some properties of xanthine oxidase in human milk and colostrum. *J. Food Sci.*, **41**, 1408–12.

9

PRODUCTION, FUNCTIONAL PROPERTIES AND UTILIZATION OF MILK PROTEIN PRODUCTS

D.M. MULVIHILL

Department of Food Chemistry, University College, Cork, Republic of Ireland

1. INTRODUCTION

Bovine milk contains ~ 13% solids which include fat, lactose, protein, and organic and inorganic salts. Normal milk contains ~ 3·5 g total protein per 100 ml which falls into two main categories based on solubility at pH 4·6 at > ~ 8°C. Under these conditions ~ 80% of the total nitrogen precipitates (this fraction is referred to as casein), while 20% remains soluble in the serum or whey; ~ 15% being whey proteins with the remainder being non-protein nitrogeneous components. As discussed in Chapters 2–6, both the casein and non-casein fractions are heterogeneous; casein includes four principal primary proteins (gene products), α_{s1}-, α_{s2}-, β-, κ- and several minor proteins, while the non-casein fraction includes β-lactoglobulin, α-lactalbumin, blood serum albumin, immunoglobulins, casein-derived proteose peptones, several minor proteins, including lactotransferrin, and several enzymes. The characteristics of the caseins and whey proteins differ very significantly (Table 1):

(1) The caseins are insoluble at their isoelectric points (~ pH 4·6) at temperatures > ~ 8°C. At very low ionic strength, most of the whey proteins are also insoluble at their isoelectric points (~ pH 5) but they are soluble in this pH range in the ionic environment of milk.
(2) Coagulation of the caseins can be induced by limited proteolysis using crude proteinase preparations, known as rennets, but the whey proteins remain soluble.
(3) Caseins are extremely heat-stable proteins while the whey proteins are heat labile: sodium caseinate dissolved in water does not coagulate in 60 min at 140°C, while the whey proteins are dentured on heating at temperatures > 70°C. At its normal pH (~ 6·7), milk may be heated at 140°C for 20 min before coagulation occurs; however, on heating milk at > ~ 72°C, whey proteins become denatured and interact with casein to form a complex.
(4) In milk, the caseins occur as large aggregates, micelles, which can be separ-

TABLE 1
Principal Differences between Casein and Whey Proteins

Characteristic	*Casein*	*Whey*
Solubility at pH 4·6	No	Yes
Rennet coagulation	Yes	No
Heat stability	High	Low
Particle size	Large (micelles; mol.wt $\sim 10^8$)	Small (molecules; mol.wt $\sim 1{\cdot}5$–$7{\cdot}0 \times 10^4$)

ated from the molecularly dispersed whey proteins by ultracentrifugation (e.g. 100 000 g for 1 h).

Some of these differences between caseins and whey proteins are exploited in industrial methods for casein and whey protein isolation; however, caseins and whey proteins can also be isolated together in various high-protein products, referred to as co-precipitates. Because of the heterogeneity of both protein systems, methods with industrial scale-up posssibilities are now being developed to effect fractionation to individual proteins.

Because of their source and their nutritional value, dehydrated milk protein-enriched products are 'high esteem' food ingredients. As well as contributing to the nutritional status of foods, the physico-chemical and functional properties of these protein-enriched products are exploited to modify or enhance the textural and rheological characteristics of foods. These proteins bind and emulsify fat, bind and entrap water, and entrap and stabilize air in food products, thus contributing to product stability and sensory appeal.

This chapter will provide an overview of methods for the production of dehydrated milk protein-enriched products, the functional properties of these products and their applications in foods.

2. PRODUCTION OF MILK PROTEIN PRODUCTS

2.1. Production of Caseins

Casein has been produced commercially for at least 70 years. Initially, casein was used for industrial purposes, e.g. glues, paper glazing and synthetic fibres. It was not until the 1960s that isolated casein became an important food protein, due mainly to pioneering work in Australia and New Zealand. Today, casein, produced by acid or rennet coagulation, is one of the principal functional food proteins with an annual world production of ~250 000 tonnes. It has some rather unique properties and cannot be replaced by other proteins in certain food applications. Methods for the manufacture of casein and caseinates have been reviewed by Muller (1971, 1982); Mulvihill (1989) and Fox and Mulvihill (1990).

The first step in the isolation of the casein fraction from milk is removal of fat by centrifugation to yield a skim milk from which the casein is isolated after destabilizing it and rendering it insoluble. The use of skim milk ensures that the fat

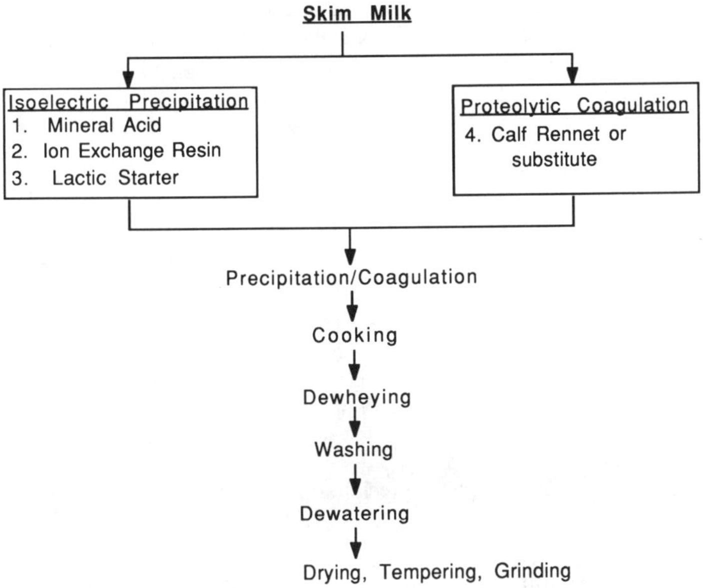

Fig. 1. Casein manufacture.

content of the casein is low enough to minimize flavour defects arising from deterioration of lipids in the dried casein products. Following destabilization, the insoluble casein is separated from the soluble whey proteins, lactose and salts, washed to remove residual soluble solids and then dried (Fig. 1).

2.1.1. Destabilization/Precipitation

(a) In the manufacture of mineral acid casein, precipitation is accomplished by spraying dilute (1–2 M) mineral acid, usually HCl, under pressure into milk (preheated to 25–30°C) flowing in the opposite direction to give a precipitation pH of ~4·6. Steam is then injected to heat the acidified milk to the required precipitation temperature (~50°C) and a holding or acidulation tube is used to ensure complete coagulation and agglomeration of the curd prior to separation of the curd and whey (Fig. 2).

(b) The pH of skim milk can also be reduced to the isoelectric point of casein by mixing skim milk at < 10°C with a cation exchange resin in the hydrogen form in a reaction column; this replaces cations in the milk by H^+ to give a pH of ~2·2. The deionized, acidified milk is then mixed with untreated milk to give the final desired precipitation pH of ~4·6. The mixture is then heated to the coagulation temperature by direct steam injection (Fig. 2).

(c) Precipitation in the manufacture of lactic casein is accomplished by inoculating pasteurized skim milk with a mixed or multiple defined-strain starter and incubating at 22–26°C. During an incubation period of 14–16 h, the starter slowly ferments some of the lactose to lactic acid and a casein gel network or coagulum, with good water holding capacity, is formed as the pH of the milk

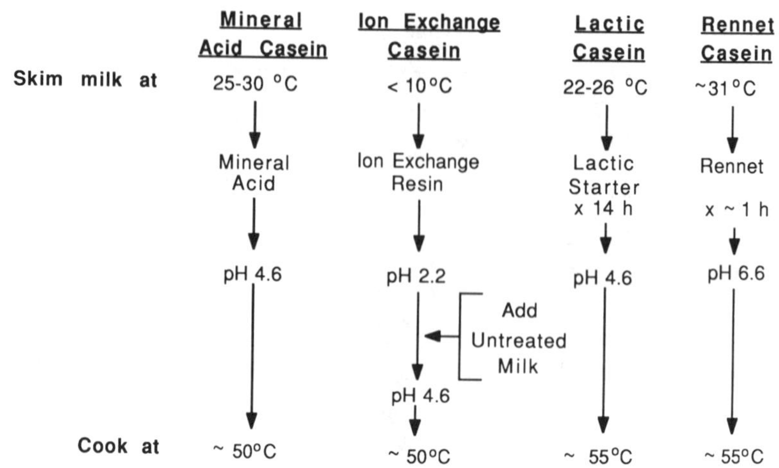

Fig. 2. Casein precipitation conditions.

falls slowly under quiescent conditions to the isoelectric pH of the casein. Following coagulation, the coagulum is pumped from the coagulation vats and cooked by direct steam injection. To permit the curd particles to agglomerate and to initiate syneresis, a period of contact with whey, termed acidulation, either in a holding pipe or vat, is allowed prior to curd and whey separation (Fig. 2).

(d) Any of a number of proteinases can coagulate milk at its natural pH ($\sim$6·7) in a two-stage process: the first stage involves the specific hydrolysis of κ-casein to yield para-κ-casein and (glyco)macropeptides, while the second stage involves coagulation of the rennet-altered casein micelles by Ca^{2+} at temperatures above 20°C. When a coagulum of this nature is formed from skim milk, it can be further processed to yield rennet casein in a manner similar to that used for the manufacture of lactic casein following quiescent acid coagulation. Rennet casein has a high ash content, especially colloidal calcium phosphate (CCP), calcium and phosphate. The proteinases traditionally used, referred to as rennets, are crude preparations of gastric proteinases prepared from calf vells. However, the supply of calf rennet has been inadequate for many years and rennet substitutes are now widely used. The primary and secondary phases of rennet coagulation, including the mechanism of gel assembly and the factors influencing gel strength, are now well understood (see Chapter 14). In the traditional method for rennet casein manufacture, skim milk at pH $\sim$ 6·7, is 'set' with rennet (1 : 4500) in large jacketed vats at $\sim$31°C, in a manner similar to that practised in cheesemaking. When coagulation has progressed to the desired stage, the gel is pumped from the coagulation vat to a cooking pipe where steam is injected to raise the temperature to 50–55°C for $\sim$45 s before separation of curds and whey (Fig. 2).

2.1.2. Dewheying

Following destabilization of the casein, the curd is separated from the whey prior to washing. The efficiency of the 'dewheying' step, which is of the utmost importance in determining the volume of whey recovered for further processing, the efficiency of the washing operation and the quality of the final casein produced, depends on the pH and temperature of precipitation and on the equipment used to achieve separation.

2.1.3. Washing

During washing, residual whey constituents (lactose, whey proteins, salts) and free acid are removed from the dewheyed curd to a limited extent by washing of the surface of the curd particles and to a much greater extent by diffusion from within the curd particles. The rate of diffusion depends on the size and permeability of the curd particles, the difference in the concentration of the constituents between the interior of the particles and the surrounding wash water and on the amount, temperature and movement of the wash water.

Washing systems used include multi-stage counter-current systems and counter-current tower washing systems in which the curd falls through an ascending column of water.

It is normal to use a gradient of wash water temperatures during the washing operation. A typical temperature profile for washing acid curd in a four-stage washing system is 55, 65, 75 and 35°C for the first to the fourth stage, respectively.

2.1.4. Dewatering

When washing is complete, casein curd is mechanically dewatered to produce a curd of minimum moisture content to minimize the quantity of water to be evaporated and thus minimize the energy required during the subsequent thermal drying operation. The properties of the casein curd following washing should be such as to allow for maximum dewatering under the conditions of operation of the dewatering machine while at the same time maintaining the curd in a suitable condition for subsequent drying. Mechanical dewatering devices include roller and screw presses and decanting centrifuges.

2.1.5. Drying, Tempering and Grinding

To produce a stable, long-life casein that meets the internationally recognized compositional standards for edible-grade product, the casein curd is dried to $< 12\%$ moisture in any one of a variety of drier types. Traditional driers used are of a semi-fluidized, vibrating type in which casein curd passes along vibrating perforated stainless steel conveyors while warm air is forced up through the perforations, partially fluidizing the curd as it is dried.

Pneumatic ring driers, which consist essentially of a large, stainless steel, ring-shaped duct through which high-velocity, heated air and moist, disintegrated casein curd are circulated continuously, are now widely used to dry casein.

A drying technique, referred to as 'attrition drying', based on the principle of grinding and drying in a single operation, is also widely used in casein plants since it allows the production of a casein product closely resembling spray dried casein. The drier consists of a fast-revolving, multi-chambered rotor and a stator with a serrated surface. Turbulence, vortices and cavitation effects in the drier result in highly efficient grinding, which pulverizes the curd into very small particles with a large total surface area. These particles are simultaneously dried in a hot air stream that passes through the drier concurrently with the curd. The dried casein is very fine with an overall average particle size of $\sim 100\,\mu m$. The particles have good wettability and dispersability in water because they are irregularly shaped and many contain cavities due to the rapid evaporative process.

Dried casein is relatively hot as it emerges from the drier and the moisture content of individual particles varies. Therefore, it is necessary to temper and blend the dried curd to achieve a cooled final product of uniform moisture content. This is usually achieved by pneumatic circulation of the curd between a number of holding bins.

Following drying, tempering and blending, the casein is ground in roller or pin-disc mills to produce the small-sized particles required by users of casein. Milled material is separated on screens to products of desired particle-size range and oversized material is re-cycled for further milling.

2.2. Production of Caseinates

Acid caseins are insoluble in water but will dissolve in alkali under suitable conditions to yield water soluble caseinates that may be spray or roller dried.

2.2.1. Sodium Caseinate

Sodium caseinate, usually prepared by solubilizing acid casein with NaOH, is the water-soluble casein most commonly used in foods.

The steps involved in its manufacture (outlined in Fig. 3) are as follows:

(1) Casein curd from a dewatering device ($\sim 45\%$ solids) is minced to disintegrate the curd which is then mixed with water at 40°C to give a solids content of $\sim 25\%$, before passing it through a colloid mill.

(2) NaOH (2·5 M) is pumped into the casein slurry, emerging from the mill at < 45°C and with the consistency of 'toothpaste', to give a final caseinate pH of 6·6 to 6·8. The NaOH and slurry must be efficiently mixed with a mixer capable of coping with the high viscosity.

(3) The mixture is transferred to a vat where solubilization occurs as the mixture is vigorously agitated and heated. The slurry is recirculated and/or transferred by pump to a second vat where solubilization is completed as the solution temperature is raised to ~ 75°C. An in-line pH meter is used to indicate whether the correct volume of NaOH solution has been added to the curd and to regulate the addition.

(4) The caseinate solution is pumped to a balance tank through a heat exchanger in which the temperature is increased to ~ 95°C. A second in-line pH meter is used to control further addition of NaOH, if necessary, to give a caseinate of the desired pH.

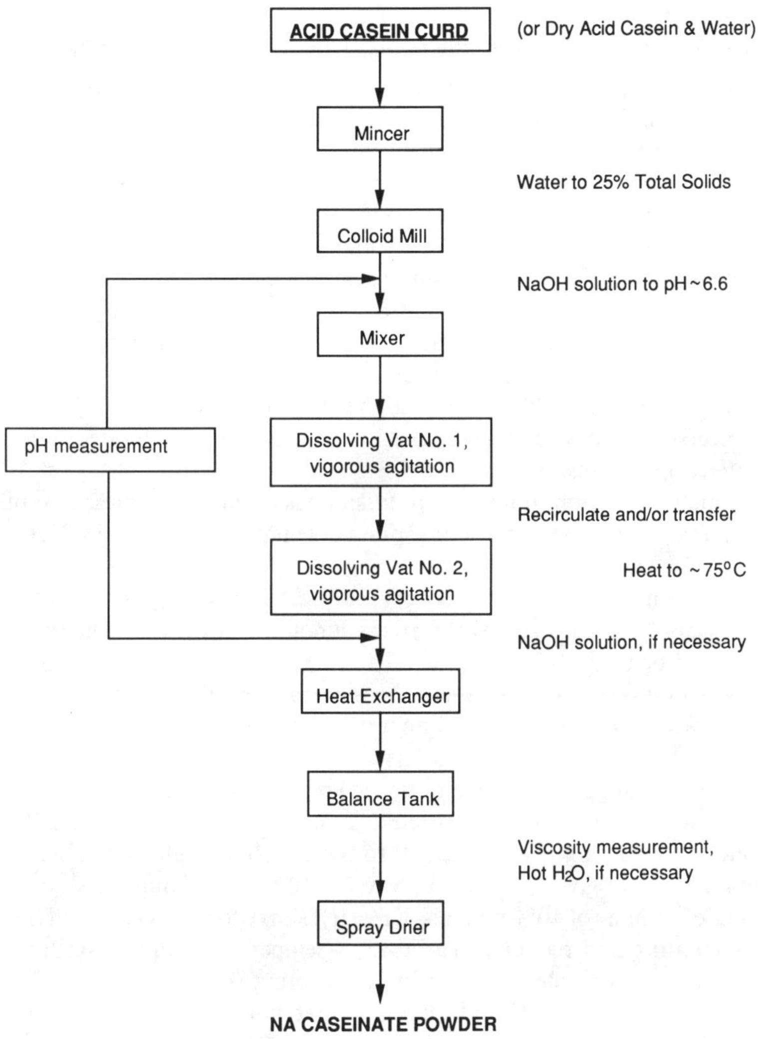

Fig. 3. Sodium caseinate manufacture.

(5) The solution is pumped from the balance tank to the spray drier via an in-line viscometer which regulates hot water addition to control viscosity and ensure efficient atomization of the solution in the drier.

During caseinate production, care must be taken to minimize: (a) the time for which the caseinate solution is held at high temperatures, since brown colouration may occur due to reactions between the protein and residual lactose; (b) the time for which the casein is exposed to high pH during dissolving, as this may lead to the production of lysinoalanine and the development of off-flavours.

2.2.2. Other Caseinates

Other methods used to produce different caseinate types include (Fig. 4):

(1) Production of roller-dried sodium caseinate by feeding a mixture of curd (50–65% moisture) and an alkaline sodium salt (Na_2CO_3 or $NaHCO_3$) onto the drying drum of a roller-drier.

(2) Production of granular sodium caseinate by lowering the moisture content of acid casein curd to < 40%, reacting the curd with Na_2CO_3, with agitation, for up to 60 min and drying the resultant caseinate in a pneumtic ring drier or a fluidized bed drier. The resulting caseinate has a higher bulk density and improved dispersibility compared to spray- and roller-dried products.

(3) Drying a mixture of acid casein curd (45% dry matter) and Na_2CO_3 in an attrition drier to produce a product that looks like spray-dried sodium caseinate but which has a much higher bulk density.

(4) Conversion of casein to caseinate in the presence of a limited amount of water using extrusion techniques.

(5) Production of ammonium and potassium caseinates in a manner similar to that used for the production of sodium caseinate by substituting NH_4OH or KOH for NaOH.

(6) Production of granular ammonium caseinate by exposing dry acid casein to ammonia (gas) and removing excess ammonia with a stream of air in a fluidized bed degassing system.

(7) Production of citrated caseinate by a method similar to that used for the preparation of spray-dried sodium caseinate by using a mixture of trisodium citrate and tripotassium citrate instead of NaOH.

(8) Production of calcium caseinate by: (1) passing 'soft' casein curd through a mixer to give evenly-sized particles; (2) mixing with water to ~25% total solids; (3) passing the mixture through a colloid mill and adjusting the temperature to give a milled slurry at 35–40°C; (4) mixing the slurry with a metered volume of 10% aqueous $Ca(OH)_2$ slurry to give the desired final pH; (5) agitating and recirculating in a low-temperature conversion tank until conversion is complete (> 10 min); (6) heating the dispersion in a tubular heat exchanger to 70°C and pumping directly to a spray drier.

2.3. Miscellaneous Methods of Casein and Co-precipitate Isolation

In addition to the 'traditional' methods described above, several new alternative methods for the preparation of casein or co-precipitates have been reported, some of which may have commercial potential.

One such method involves precipitation of milk proteins by addition of ethanol to reduce the dielectric constant of the mixture (Hewedi *et al.*, 1985). Addition of an equal volume of 60% (v/v) ethanol to pasteurized skim milk or to skim milk that had been heated at 90°C × 10 min, both adjusted to pH 6·3, caused precipitation of ~82% and ~90% of total nitrogen, respectively.

Another method uses selective solubilization of lactose from nonfat-dry-milk using ethanol; the insoluble residue, which may be regarded a a milk protein

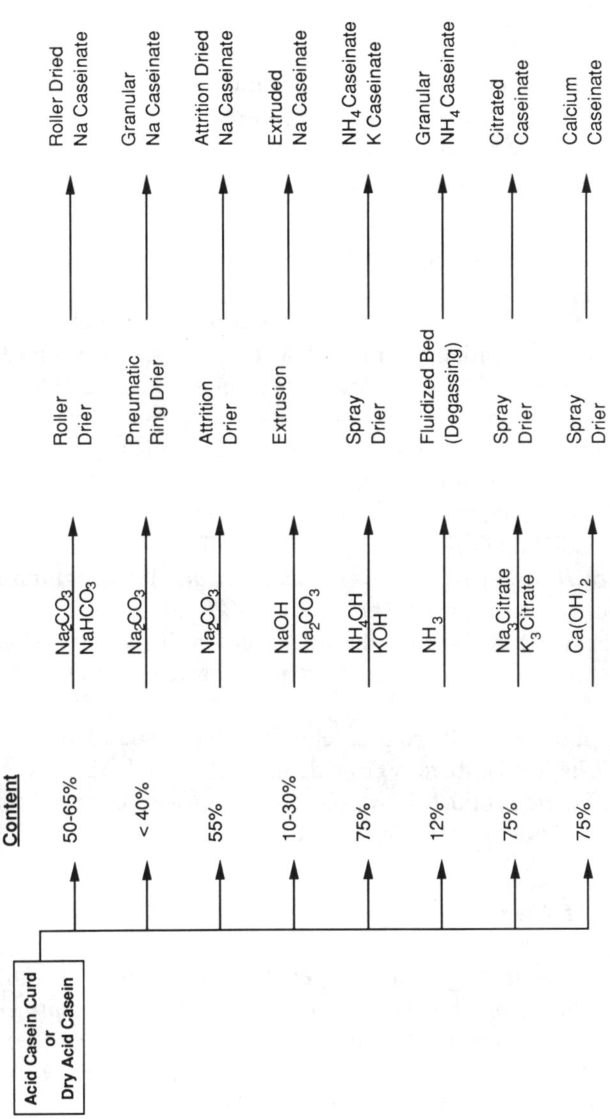

Fig. 4. Methods for the manufacture of different caseinate types.

co-precipitate, had a lactose content of 2–4% using a one-step extraction process under optimum conditions (Hoff *et al.*, 1987).

Ultrafiltration of skim milk to a 4 or 6 volume concentration ratio (VCR = initial volume of milk/volume of retentate) to remove lactose, followed by storage of the retentate at −8°C for 1–4 weeks leads to cryo-destabilization of casein and some whey proteins which are sedimentable by centrifugation at 5000 g for 10 min at 0–5°C (Lonergan, 1983).

Membrane filtration methods for the separation of casein micelles (phosphocaseinate) from all the other constituents of skim milk, including the whey proteins, are reported to have been developed at laboratory level but have not as yet been scaled up to industrial level (Maubois, 1990).

2.4. Industrial Scale Fractionation of Caseins

A number of methods for fractionating casein into β-casein-rich and α_s-/κ-casein-rich fractions on a potentially industrial scale have been developed. These are based on the association characteristics of the caseins which are dependent on ionic strength and/or temperature. At low temperatures, β-casein exists in solution as monomers (Payens & van Markwijk, 1963), a characteristic exploited by Allen *et al.* (1985) to prepare β-casein by renneting calcium caseinate at 4°C; under these conditions, β-casein remains soluble while α_s- and para-κ-caseins coagulate. A method for the isolation of β-casein by microfiltration of calcium caseinate at 5°C was reported by Terre *et al.* (1986). Famelart *et al.* (1989) optimized the same technique to purify β-casein from whole casein at 4°C and pH 4·2–4·6. Murphy and Fox (1990) reported a method for the fractionation of a dilute sodium caseinate solution by ultrafiltration into a β-casein-rich permeate and an α_s-/κ-casein-rich retentate (Fig. 5).

As many peptides derived from caseins have been shown to have biological functions (see Chapter 7), there is growing interest in methods for the production and isolation of these peptides. However, commercial-scale production and isolation methods have not as yet been reported.

2.5. Production of Whey Protein-enriched Products

Whey is the liquid remaining after removal of fat and casein from milk during the manufacture of cheese or acid and rennet casein. There are two principal types of whey; sweet whey (minimum pH 5·6) from the manufacture of cheese or rennet casein, and acid whey (maximum pH 5·1) from the manufacture of acid casein. Average compositons of some whey types are shown in Table 2. Acid whey has a higher mineral/ash content and if the acid has been produced by the action of starter bacteria, the lactose concentration is reduced. Whey proteins represent only 10% of the total solids of whey and on drying wheys the resulting powders have low protein contents. However, a number of processes (Fig. 6) have been developed, and are now being exploited commercially, to recover the whey proteins

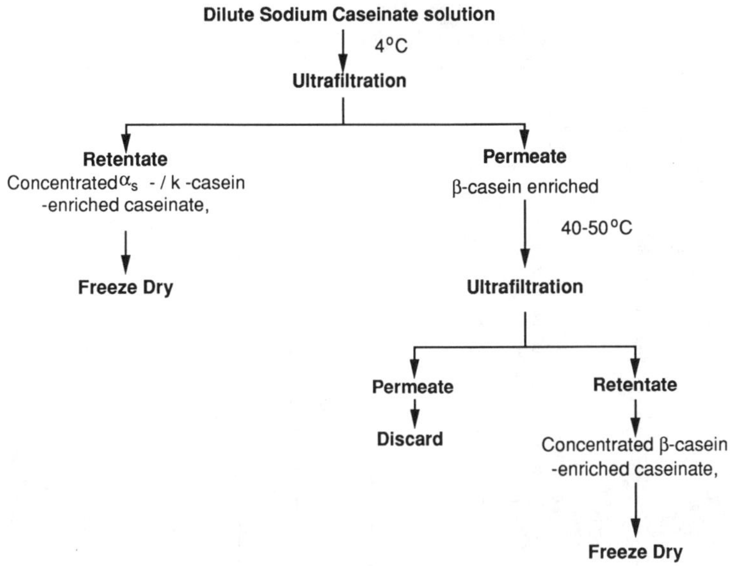

Fig. 5. Method for preparing α_s-/κ- and β-casein enriched caseins (Murphy & Fox, 1990).

in more concentrated forms. These processes have been reviewed by Marshall (1982), Matthews (1984), IDF (1987) and Morr (1989).

Whey and whey protein-enriched solutions are usually pasteurized using minimum temperature and holding times and maintained at low temperature to minimize microbial and physico-chemical deterioration of the proteins and other

TABLE 2
Average Composition and pH of Sweet (Rennet Casein and Cheddar Cheese) and Acid (Lactic and Mineral Acid) Wheys

Component	*Composition (g/litre)*			
	Sweet wheys		*Acid wheys*	
	Rennet casein	*Cheddar cheese*	*Lactic acid casein*	*Mineral acid casein*
Total solids	66·0	67·0	64·0	63·0
Total protein (N × 6·38)	6·6	6·5	6·2	6·1
Non-protein nitrogen (NPN)	0·37	0·27	0·40	0·30
Lactose	52·0	52·0	44·0	47·0
Milk fat	0·20	0·20	0·30	0·30
Minerals (ash)	5·0	5·2	7·5	7·9
Calcium	0·50	0·40	1·6	1·4
Phosphate	1·0	0·50	2·0	2·0
Sodium	0·53	0·50	0·51	0·50
Lactate	—	2·0	6·4	—
pH	6·4	5·9	4·6	4·7

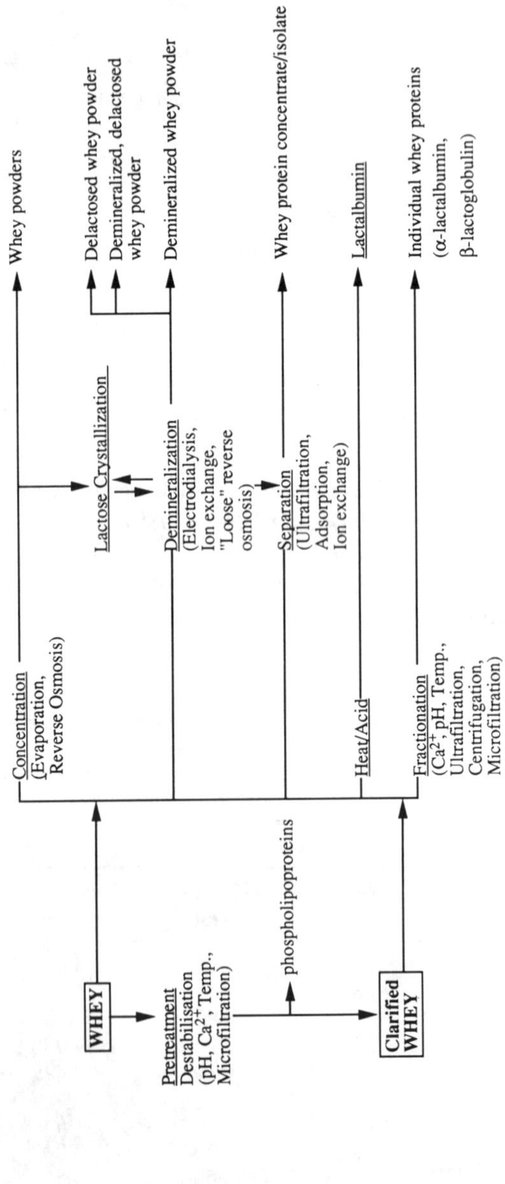

Fig. 6. Industrial isolation of protein products from whey.

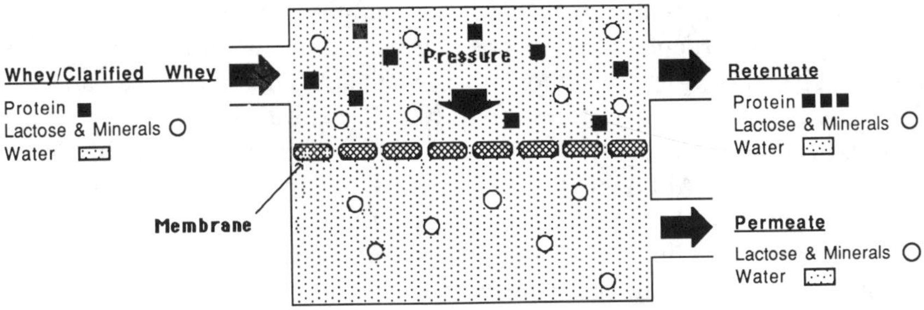

Fig. 7. Production of WPC by ultrafiltration.

whey constituents that would adversely alter functional and organoleptic properties of the resulting protein-enriched products.

2.5.1. Whey Powders/Modified Whey Powders

Whole whey powders containing less than 15% protein are produced by concentrating whey by evaporation or a combination of reverse osmosis and evaporation, followed by spray drying. Demineralization by 'loose' reverse osmosis, electrodialysis or ion-exchange and/or lactose crystallization are used commercially to reduce the lactose and/or mineral concentration of whey and produce modified whey powders such as demineralized and demineralized-delactosed whey powders which contain ~15–35% protein.

Lactose is traditionally crystallized from whey by: concentration of the whey to 58–62% total solids using a multiple-effect falling-film evaporator at a maximum product temperature of ~70°C; controlled cooling of the concentrate and seeding to induce nucleation and crystal growth; separation of the crystals from the mother liquor by decanter centrifugation, followed by washing (if desired) and crystal drying. The mother liquor may be further concentrated and spray dried as a whey protein concentrate powder containing ~30% protein.

2.5.2. Whey Protein Concentrate Production by Ultrafiltration–Diafiltration

Ultrafiltration (UF) is a pressure membrane filtration process that facilitates the selective separation of whey proteins from lactose, salts and water under mild conditions of temperature and pH. It is a physico-chemical separation technique in which a pressurized solution flows over a porous membrane that allows the passage of only relatively small molecules. The retained solution (retentate) flows over the membrane, while, under the influence of pressure, water flows through the membrane, together with the low molecular weight solutes. The protein is retained by the membrane and is concentrated relative to the other solutes in the retentate (Fig. 7). Fat globules and suspended solids are also retained.

The membranes used in UF are asymmetric microporous structures, the effective layers of which appear to contain pores with diameters ranging from 1 to 20 nm. Commonly used membrane configurations include tubular, spiral-wound, plate and frame and hollow-fibre, with each configuration offering advantages and disadvantages for particular applications. The membranes are manufactured from synthetic polymers (e.g. polysulphone or polyamide). They are characterized by high resistance to high temperatures (up to 100°C), can withstand a wide pH range (1–13) and can be cleaned with agents normally used in the dairy industry (e.g. HNO_3 and NaOH). Although UF is currently the method of choice for the commercial manufacture of WPC of varying protein concentrations, it has several major problems which limit its operational performance. These problems include: high capital and operating costs; membrane fouling, with concomitant loss of permeate flux rate; incomplete removal of low molecular weight solutes unless diafiltration (dilution of retentate with water and repeated UF) is used; cleaning, sanitation and related microbial problems; disposal of large volumes of permeate.

Prior to processing, whey is commonly pre-treated by methods involving pH and/or temperature adjustments, addition of calcium or calcium complexing agents and either quiescent standing, centrifugation or microfiltration to dissolve colloidal calcium phosphate and/or to remove insoluble cheese curd or casein fines, milkfat and calcium lipophosphoprotein complexes (Hayes *et al.*, 1974; Breslau *et al.*, 1975; De Wit & De Boer, 1975; Lee & Merson, 1976; De Wit *et al.*, 1978; Matthews *et al.*, 1978; Muller & Harper, 1979; Maubois *et al.*, 1987).

These pre-treatments increase flux during ultrafiltration, prevent fouling of the membranes and modify the properties of the whey protein concentrates.

The limit for whey concentration by UF on modern plants is ~24% total solids, with a protein/total solids ratio limit of ~0·72 : 1. Difiltration is employed to achieve a higher ratio of protein/total solids (~0·80 : 1) and a total solid content of ~28%.

2.5.3. *Production of Whey Protein Isolate (WPI) by Ion Exchange Adsorption*

Whey proteins are amphoteric molecules and may be considered as ions. At pH values lower than their isoelectric point (~pH 4·6), whey proteins have a net positive charge and behave as cations which can be adsorbed on cation exchangers. At pH values above their isoelectric point, the proteins have a net negative charge and behave as anions which can be adsorbed on anion exchangers. Media with suitable pore sizes and surface characteristics have been developed specifically for the recovery of proteins from dilute solutions, depending upon the pH of the medium. Two major ion exchange fractionation processes have been commercialized for the manufacture of WPI (Fig. 8).

The 'Vistec' process uses a cellulose-based exchanger in a stirred tank reactor (Burgess & Kelly, 1979; Palmer, 1982). The process involves a series of steps that are performed as a fractionation cycle: (1) whey is adjusted to pH $< 4·6$ with acid, pumped into a tank reactor and stirred to allow protein adsorption onto the ion exchanger; (2) lactose and other unadsorbed material are filtered off with the water; (3) the resin is resuspended in water and the pH adjusted to $> 5·5$ with alkali to release the proteins from the ion exchanger; (4) the aqueous solution of proteins is separated from the resin by filtration in the tank reactor, concentrated by ultrafiltration and evaporation, and spray dried as WPI containing ~95% protein.

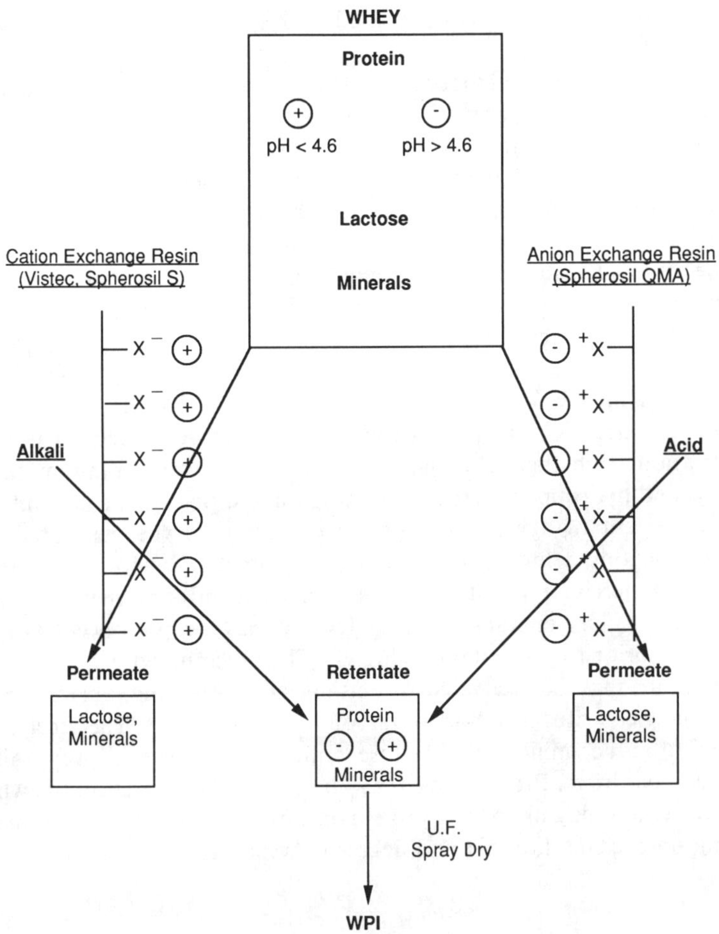

Fig. 8. Production of whey protein isolate (WPI) by ion exchange adsorption.

UF treatment of the protein-rich eluate fraction is essential for purification and concentration of the protein.

The 'Spherosil' processes (Mirabel, 1978; Kaczmarek, 1980) use either cationic Spherosil S or anionic Spherosil QMA ion exchangers, and fractionation is accomplished in fixed-bed column reactors. Acidified whey at pH $<$ 4·6 is applied to the Spherosil S column reactor to allow protein adsorption. After lactose and other unadsorbed solutes have been eluted with water, the pH is raised by addition of alkali to elute adsorbed proteins from the reactor. The protein-rich eluate fraction is concentrated by UF and evaporation and spray dried to produce WPI. Sweet whey at pH $>$ 5·5 is applied to the Spherosil QMA column reactor to permit negatively charged protein molecules to adsorb onto the ion exchanger. After elution of non-protein materials, the proteins are released by lowering the pH with acid. Released proteins are concentrated and spray dried as WPI, as for the Spherosil S process.

These adsorption processes recover ~85% of the protein under ideal operating conditions. The recovered concentrates are characterized by high protein and low lactose and lipid concentrations and have good functionality. However, several major problems are associated with these ion exchange processes, including: (1) production of large volumes of rinse water, chemical solutions and deproteinized whey which must be processed or disposed of; (2) the need to concentrate and purify the dilute protein-containing eluate by UF, evaporation and drying; (3) the excessive time requirement for conducting each fractionation cycle; (4) microbial contamination of the reactor.

2.5.4. Lactalbumin Production

Whey proteins are globular proteins and are readily denatured on heating. On transformation from their globular conformations to more random structures, sulphydryl and hydrophobic groups are exposed and protein–protein interactions occur. The extent of aggregation and precipitation of the denatured proteins depend on heating temperature and holding time, pH and concentration of calcium. Commercial precipitation conditions depend on whey type and the desired final product characteristics and whey may be preconcentrated and/or dimineralized prior to precipitation (Fig. 9). The precipitated protein, referred to as lactalbumin, may be recovered by settling and decanting, vacuum filtration, self-desludging centrifuges or horizontal solid-bowl decanters. The precipitate may be washed to reduce mineral and lactose contents and dried in spray, roller, ring or fluidized bed driers. Protein yields may be up to 80% of that in the whey, and lactalbumin containing up to 90% protein on a dry weight basis may be recovered, depending on precipitation pH and degree of washing.

2.5.5. Fractionation of Whey Proteins

A number of methods have been developed (Amundson *et al.*, 1982; Slack *et al.*, 1986; Maubois *et al.*, 1987; Pearse, 1987), that may have commercial scale potential, to fractionate the major whey protein components, β-lactoglobulin and α-lactalbumin, and produce whey protein concentrates enriched in these fractions. These methods (Fig. 10) depend on either mild heat treatments of a whey concentrate or a clarified whey under controlled pH and ionic conditions or demineralization of whey concentrate under controlled pH conditions, to effect selective reversible precipitation of α-lactalbumin or β-lactoglobulin enriched fractions and the separation of the precipitate from β-lactoglobulin or α-lactalbumin enriched solutions. The precipitate is resolubilized by water addition and pH adjustment and then dried while the soluble protein is further concentrated by ultrafiltration/diafiltraton prior to drying.

There is also considerable interest in the isolation of biologically-active proteins, such as lactoferrin and lactoperoxidase, and biologically- or functionally-active peptides from whey. Since biologically-active proteins and peptides are reviewed in Chapter 7, they will not be considered further here.

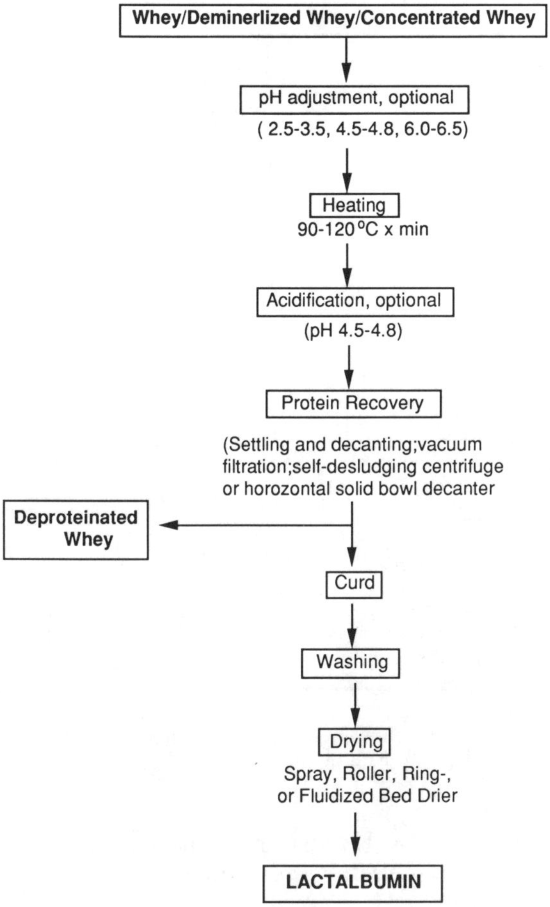

Fig. 9. Production of lactalbumin.

2.6. Co-precipitate Production

Following precipitation of caseins from milk by acidification or renneting, the whey proteins remain soluble in the whey. However, these can be co-precipitated with the casein by first heating milk, at its natural pH, to temperatures that denature the whey proteins and induce their complexation with casein, followed by precipitation of the milk protein complex by acidification to pH 4·6 or by a combination of added $CaCl_2$ and acidification (Fig. 11). Products produced in this manner are referred to as casein–whey protein co-precipitates. Yields of 92–95% of total milk protein are obtained, compared to < 80% for acid or rennet caseins. Co-precipitates produced by these methods have poor solubility properties; however, processes for the manufacture of similar products with good solubility have been described (Fig. 12). These processes involve pre-adjusting milk to

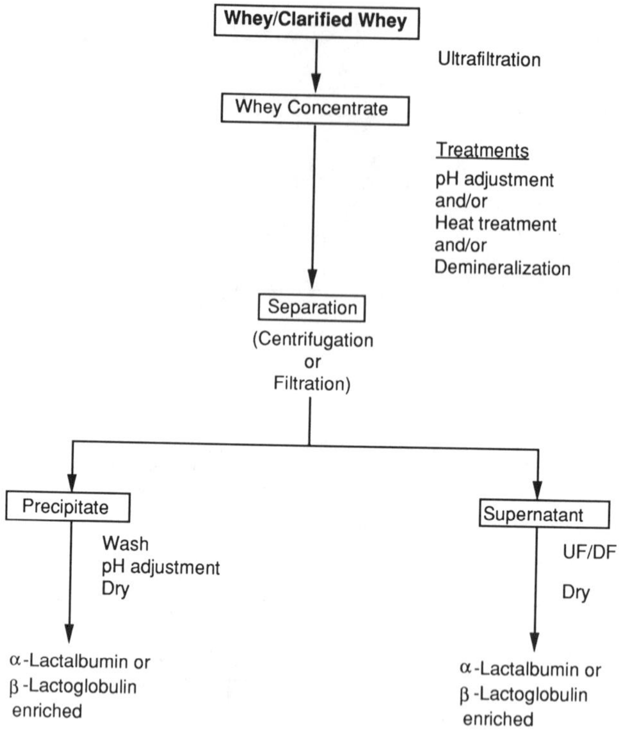

Fig. 10. Fractionation of whey proteins.

pH 7·0–7·5, heating at 90°C × 15 min or pre-adjusting milk to pH 10 and heating at 60°C × 3 min prior to isoelectric precipitation.

2.7. Production of Milk Protein Concentrate

Skim milk may also be processed directly by ultrafiltration/difiltration to yield milk protein concentrates that contain a range of protein contents up to ~80% and in which the casein is in a similar, micellar, form to that found in milk while the whey proteins are also reported to be in their native form. The products have a relatively high ash content, since protein-bound minerals are retained.

3. Functional Properties of Milk Protein Products

The utilization of milk protein products as food ingredients is dependent on their physico-chemical and functional properties. Therefore, a brief overview of some of the important functional characteristics is included here; more extensive reviews on these topics include Fox & Mulvihill (1983), Kinsella (1984), De Wit (1989*a*) and Mulvihill & Fox (1989).

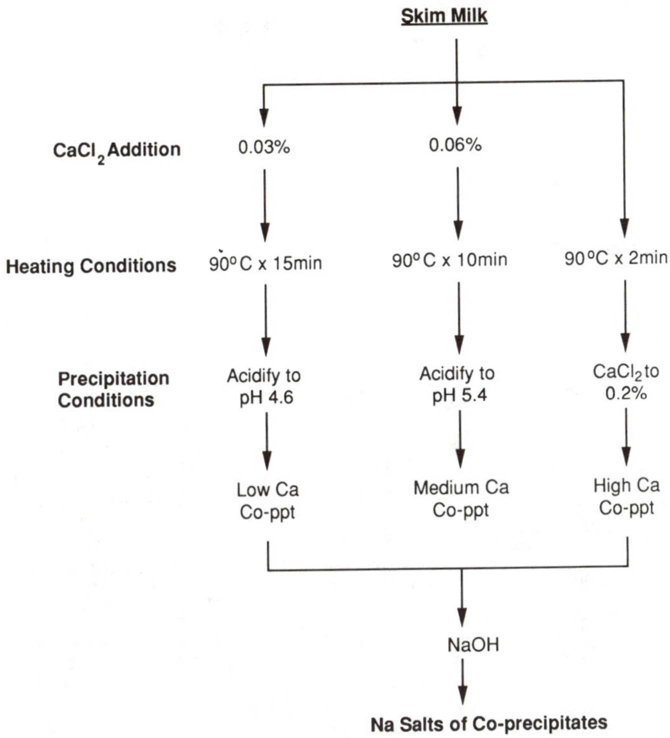

Fig. 11. Manufacture of conventional co-precipitates.

3.1. Solubility

A typical solubility–pH profile of casein shows that close to its isoelectric pH, i.e. pH 4·0–5·0, the acid form of casein is completely insoluble, while at pH values > 5·5, it is converted to cationic salt (Na, K, NH_3) and is completely soluble. Solutions containing 10–15% of these latter caseinates can be readily prepared at pH 6·0–7·0. At pH values < 3·5, casein is also soluble but at this pH it is more viscous than at neutral pH values and gel-like systems are formed.

When rennet casein is dispersed in water, the natural pH of the dispersion is ~7·0; however, rennet casein is insoluble because of its high calcium content, but can be solubilized by either raising the pH above ~9·0 or by adding calcium chelators, normally food-grade polyphosphates and/or citrates. Sodium forms of conventional 'co-precipitates' are also somewhat insoluble at pH 6·0–7·0, but the sodium forms of casein–whey protein co-precipitates prepared from milk heated at alkaline pH values (i.e. SLP and TMP) have solubility characteristics similar to caseinates.

Calcium caseinates and medium and high calcium co-precipitates form coarse colloidal dispersions rather than solutions.

Na-, NH_3- and K-caseinates are remarkably heat stable, e.g. a 3% (w/v) solution

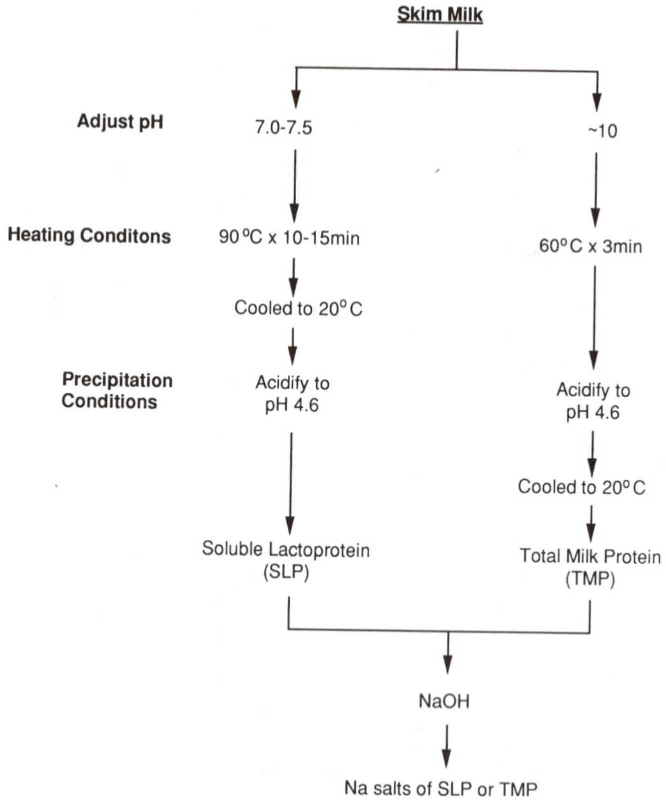

Fig. 12. Methods for the manufacture of soluble lactoprotein and total milk protein.

of sodium caseinate, pH 7·0, may be heated at 140°C × 60 min without coagulating. Calcium caseinate is less stable and a 1% (w/v) solution gels at 50–60°C. Whey proteins are unique among the milk proteins used in food applications since in their native conformation they are soluble at low ionic strength over the entire pH range required for food applications. However, being globular proteins, their solubility decreases at high salt concentrations due to salting out and they are susceptible to thermal denaturation at temperatures > ~70°C. Solubility at pH 4·6 is widely used as an index of the extent of denaturation caused by processing and storage of protein-rich whey products. The level of denaturation and subsequent insolubility at pH 4·6 depends on heating temperature and time, whey pH and ionic calcium concentration (Donovan & Mulvihill, 1987; Mulvihill & Donovan, 1987).

3.2. Gelation and Coagulation

Gels are systems in which a small proportion of solid is dispersed in a relatively large proportion of liquid but which have the property of mechanical rigidity or the ability to support shearing stress at rest (properties of solids). Milk undergoes gelation when subjected to one of several treatments and usually casein is the gelling component involved (Fox & Mulvihill, 1990).

Gelation or coagulation occurs when milk is subjected to limited proteolysis by acid proteinases, e.g. rennets, which hydrolyse the micelle-stabilizing κ-casein, producing para-κ-casein-containing micelles which coagulate at the concentration of Ca^{2+} in the milk serum. This phenomenon forms the basis for the manufacture of rennet casein and most cheese varieties. On mixing equal volumes of milk and 80% (v/v) ethanol, the casein micelles are destabilized and gels or precipitates are formed.

Acid (isoelectric) gelation/coagulation of milk is exploited in the manufacture of fermented milks, acid cheeses (Cottage and Quark) and acid caseins. Depending on the pre-heat treatment of the milk, acid gels may or may not expel whey (synerese): pre-heating to temperatures greater than the denaturation temperature of the whey proteins (usually > 85°C) is exploited to reduce syneresis in fermented milks while milk used for the production of acid cheeses and caseins is heated as little as possible to promote syneresis or whey expulsion.

The viscosity of casein is much higher at low pH (~2·5–3·5) than at neutral pH and gel-like structures are formed at > 5% protein at temperatures < 40°C, which may be exploited in the preparation of milk protein-containing fruit gels.

Concentrated Ca-caseinate dispersions (> 15% protein) gel on heating to 50–60°C. Gelation temperature increases with protein concentration from 15 to 20% and with pH in the range 5·2–6·0. The gel liquefies slowly on cooling but reforms on heating; calcium caseinate is the only milk protein system reported to exhibit reversible thermal gelation properties.

Although thermal sensitivity is undesirable when one is seeking to prepare a soluble whey protein-enriched product, this property can be exploited in the production of thermal gels from whey proteins, which have excellent thermal gelling properties. The minimum protein concentration and heating regime required for gelation and gel characteristics, such as opacity, strength and elasticity or brittleness, depend on characteristics of the whey protein product, such as method of production, contents of protein, total ash, selected minerals and other non-protein components, the extent of whey protein denaturation during manufacture of the whey protein product, solution conditions such as pH, ionic species present, other non-protein components added and the presence of reducing agents (Mulvihill & Kinsella, 1987, 1988: Mulvihill *et al.*, 1990, 1991). Whey protein concentrates and isolates with a range of gelling properties can be produced by selection of whey type and variations in processing conditions during manufacture.

3.3. Hydration Properties

Many of the functional food applications of dairy proteins depend on their ability to hydrate and thus bind or entrap water. In this context, water binding or hydration is defined as the grams of water associated with or occluded by 1 g dry protein. Hydration values for casein micelles, calculated from voluminosity data, range from 1·4 to 6·4 g H_2O/g. Hydration values for 68 casein and caseinate samples calculated from viscosity data ranged from 0·7 to 3·8 g H_2O/g. Hydration of acid casein was reported to be relatively independent of temperature, while hydration of Na caseinate decreased from 3 g/g at 25°C to ~0 at 80°C.

The level of hydration of proteins is strongly influenced by the level of available water and it is common to relate the degree of hydration to the relative humidity of the environment to which the protein is exposed. A plot of bound (sorbed) water as a function of relative humidity, P/P_o or a_w, yields a water sorption isotherm which gives useful information on the water binding or hydration characteristics of proteins. Isotherms for sodium caseinate, acid casein and micellar casein show that the hydration of acid casein is higher than that of ultracentrifugal (micellar) casein; the differences are small when $a_w < 0{\cdot}6$ but at $a_w > 0{\cdot}6$, acid casein sorbs much more water than micellar casein. High hydration values for Na caseinate at high a_w values reflect swelling and solubilization.

Hydration values determined for individual native whey proteins range from 0·32 to 0·60 g H_2O/g, depending on the assay method used. However, when whey protein solutions of sufficient protein content and suitable solution conditions (pH, ions, etc.) are heated, gels are formed and the water holding capacity of such gels make significant contributions to the texture and rheology of a number of procesed foods.

Using a method based on water uptake by flour doughs to which various milk protein products were added, the water sorption capacity of several milk protein products was determined by Knightsbridge and Goldman (1975); the value reported ranged from 0·96 to 3·45 g H_2O per g product.

3.4. Viscosity

Owing to hydration, swelling and polymer–polymer interactions, caseinates form highly viscous solutions at concentrations > ~15% and even at high temperatures, the viscosity of solutions containing > ~20% protein is so high as to make them difficult to process, e.g. spray-dried sodium caseinate has a low bulk density.

The effects of solution conditions on the viscosity of caseins/caseinates have been extensively investigated. The viscosity of sodium caseinate is strongly dependent on pH, with a minimum at ~pH 7·0. The viscosity of casein is much higher at low pH (~2·5–3·5) than at neutral pH, and, as already noted, gel-like structures are formed at > 5% protein at temperatures < 40°C. The viscosity of sodium caseinate is logarithmically related to concentration, while there is a linear relationship between log viscosity and the reciprocal of absolute temperature. Caseinates exhibit pseudoplastic rheological behaviour and are thixotropic at high shear rates. The cation present has a significant effect on the viscosity of caseinates, but this in turn depends on pH, temperature and protein concentration.

Limited proteolysis by indigenous milk proteinase reduces the viscosity of caseinate solutions and may explain the low viscosity of caseinates produced from late lactation milk, which has a high level of indigenous proteinase. The viscosity of caseinates can also be reduced by treatment with disulphide-reducing and/or sulphydryl blocking agents.

Calcium concentration influences the viscosity of caseinate; the viscosity of caseinate containing 1% Ca was reported to decrease sharply in a curvilinear fashion from 30 to 38°C, then remain constant up to ~57°C, above which the solution gelled at pH 5·4 but not at higher pH values. The shape of the viscosity–temperature curve was strongly dependent on protein concentration, pH and $[Ca^{2+}]$. Low levels of Ca increase the viscosity of Na caseinate above pH 7·0 but below pH 7·0 viscosity decreases due to micelle formation.

The effects of various manufacturing conditions on the viscosity of casein/caseinates have also been studied. Excessive heating of milk prior to casein manufacture or of casein curd during drying, leads to increased viscosity of the caseinates prepared from such caseins. Precipitation at lower than normal pH values (e.g. ~3·8) and especially at higher pH values (e.g. ~5·05) also increases the viscosity of caseinates. The viscosity of roller-dried caseinate is higher than that of spray-dried caseinate. Solubilized conventional co-precipitates are more viscous than sodium caseinate and their viscosity increases with increasing calcium concentration. Solutions of total milk protein have viscosities intermediate between those of sodium caseinate and conventional co-precipitates.

Due to their compact globular shapes, solutions of undenatured whey proteins are much less viscous than caseinate solutions. They exhibit minimum viscosity around the isoelectric point (pH 4·5) and relative to water, their viscosity decreases between 30 and 65°C, but increases thereafter owing to protein denaturation. Solutions of WPC containing 4–12% (w/v) protein were reported to exhibit Newtonian flow while at higher concentrations flow became more pseudoplastic and at 18–20% yield values were observed (Pradipasena & Rha, 1977; Rha & Pradipasena, 1986).

3.5. Surface Active Properties

The strongly amphipathic nature of proteins, arising from the mixture of polar and non-polar amino acid residues, causes them to concentrate at interfaces. Because milk proteins have been available in pure form and have good surface activity, the surfactant properties have been extensively studied (for a review, see Mulvihill & Fox, 1989).

Sodium caseinate is a more effective interfacial tension depressor than whey protein, blood plasma, gelatin or soy protein; it diffuses more quickly to an interface and on reaching the interface adsorbs more quickly than the other proteins, probably because of direct and rapid anchoring of freely-available hydrophobic segments. The order of surface activity reported for the individual milk proteins is: β-casein > monodispersed casein micelles > serum albumin > α-lactalbumin > α_s-casein = κ-casein > β-lactoglobulin euglobulins (see Mulvihill & Fox, 1989).

The effectiveness of whey proteins as surface active agents is enhanced by partial heat denaturation. The surface activity of whole and individual caseins may be modified enzymatically. Dephosphorylation or treatment of sodium caseinate with plasmin (to produce γ-caseins and proteose peptones) greatly increases its surface activity. γ_2/γ_3-Caseins are small, very hydrophobic peptides and thus are very surface active.

Surface films of Na caseinate or β-casein are much more flexible and less viscoelastic at both oil/water and air/water interfaces than films of β-lactoglobulin, α-lactalbumin or bovine serum albumin.

3.6. Emulsifying and Foaming Properties

In studies on the emulsifying and foaming properties of milk proteins, as reported in the literature, a wide range of apparatus type and environmental conditions (pH, ionic strength, temperature, protein concentration) have been used to prepare emulsions and foams in model and pilot scale studies. Also, different terms were used to express the results of these studies. Terms commonly used in relation to the emulsifying properties of food proteins are:

—emulsifying capacity (g oil emulsified/g protein),
—emulsion stability (rate of creaming, globule coalescence, flocculation),
—emulsifying activity index (area of interface stabilized per unit weight of protein, m^2/g),
—interfacial area (m^2/ml emulsion),
—protein load (mg/m^2).

Soyabean oil emulsions prepared in a valve homogenizer and stabilized by sodium caseinate were found to have lower creaming stabilities than similar emulsions stabilized by either whey protein concentrate or soy isolate. The emulsifying properties of highly dispersed sodium, ammonium and low-calcium caseinates and more aggregated high-calcium caseinate and ethanol-precipitated and ultracentrifugal (micellar) caseins have been studied (Mulvihill & Murphy, 1991). Although the highly dispersed caseinates had higher emulsifying capacities than the more aggregated caseins/caseinates, emulsions formed using the latter were more stable than those stabilized by the highly dispersed caseinates. Fat surface area formed on emulsification increased (i.e. globule size decreased) as the power input during emulsification was increased for all the proteins and the extent of the increase was inversely related to the degree of aggregation of the emulsifying caseins/caseinates. The protein loads (mg/m^2) of the emulsions formed using aggregated caseins/caseinates were greater than for the dispersed caseinates and protein load was directly related to emulsion stability. In general, milk protein products, especially caseinates, are very good fat emulsifiers and are widely used in emulsifying applications in foods.

In relation to foaming properties, important characteristics are foam volume (overrun) and foam stability. Caseinates generally give higher foam overruns but produce less stable foams than egg white or whey protein concentrates. Whey protein-enriched products are widely used in foaming appications in foods and

TABLE 3
Applications of Milk Protein Products in Bakery Products

Caseins/caseinates/co-precipitates	
Used in:	Bread, biscuits/cookies, breakfast cereals, cake mixes, pastries, frozen cakes and pastries, pastry glaze
Effect:	Nutritional, sensory, emulsifier, dough consistency, texture, volume/yield
Whey proteins	
Used in:	Bread, cakes, muffins, croissants
Effect:	Nutritional, emulsifier, egg replacer

factors such as protein concentration, level of denaturation, ionic environment, pre-heat treatment and the presence of lipids all influence whipping properties.

4. Food Uses of Milk Protein Products

Details of many of the food uses of milk protein products are proprietary to food processors and are not reported in the literature. However, reviews on the food uses of milk proteins include: Southward & Goldman (1978), International Dairy Federation (1982), Southward & Walker (1982), Zadow (1986), Hugunin (1987), De Wit (1989*b*) and Southward (1986, 1989). The following are brief outlines of some reported food applications of dairy protein products.

4.1. Bakery Products

Milk proteins do not have properties close enough to those of wheat gluten to enable them to replace the latter protein to any great extent in bakery products; however, their use as a nutritional supplement and for functional effects in cereal-based products has considerable potential (Table 3). The limiting amino acid in most cereal proteins is lysine and since caseins are particularly rich in lysine, they make excellent supplements for cereals. Only about 4% casein in a casein–wheat flour mixture is required to increase the lysine content by ~60%. The protein efficiency ration (PER) of white wheat flour is only 1·1 compared with 2·5 for casein and on blending casein and wheat flour to give a mixture containing 75% wheat protein and 25% casein, the PER is increased to ~1·8. Another important functional characteristic of milk protein products in bakery applications is water binding which affects dough consistency.

Casein/caseinates are added to breakfast cereals, milk biscuits, protein-enriched bread and biscuits, high-protein bread and cookies as a nutritional supplement and to frozen baked cakes and cookies as an emulsifier and to improve texture. The type of casein/caseinate must be carefully chosen to be compatible with the particular bakery application. Co-precipitates are used in pastry glaze to improve colour; in milk biscuits and cookies as a nutritional supplement and in fortified bread to improve dough consistency, sensoric properties and to increase volume and yield.

TABLE 4
Applications of Milk Protein Products in Dairy-Type Products

Caseins/caseinates/co-precipitates	
Used in:	Imitation cheeses (vegetable oil, caseins/caseinates, salts and water)
Effect:	Fat and water binding, texture enhancing, melting properties, stringiness and shredding properties
Used in:	Coffee creamers (vegetable fat, carbohydrate, sodium caseinate, stabilizers and emulsifiers)
Effect:	Emulsifier, whitener, gives body and texture, promotes resistance to feathering, sensory properties
Used in:	Cultured milk products, e.g. yoghurt
Effect:	Increase gel firmness, reduces syneresis
Used in:	Milk beverages, imitation milk, liquid milk fortification, milk shakes
Effect:	Nutritional, emulsifier, foaming properties
Used in:	High fat powders, shortening, whipped toppings and butter-like spread
Effect:	Emulsifier, texture enhancing, sensory properties
Whey proteins	
Used in:	Yoghurt, Quarg cheese, Ricotta cheese
Effect:	Yield, nutritional, consistency, curd cohesiveness
Used in:	Cream cheeses, cream cheese spreads, sliceable/squeezable cheeses, cheese fillings and dips
Effect:	Emulsifier, gelling, sensory properties

Whole whey protein products generally have a loaf volume depressing effect which has been associated with proteose peptone components. When whey is concentrated by ultrafiltration, the depressant appears to be removed since dough fortified with 1% UF-WPC caused only little loaf volume depression. The concentration of whey lipids during UF also contributes to good baking characteristics.

Replacement of eggs by whey protein in cake manufacture would have economic and nutritional advantages. However, simply replacing whole eggs by WPC in madeira-type cakes results in poor quality cakes but much better results are obtained when the fat and WPC are pre-emulsified.

Various types of WPC have been used in convenience breakfast bakery products, like muffins and croissants, to increase their nutritional value.

4.2. Dairy Products

Milk protein products are widely used to supplement the protein content and, therefore, enhance sensory characteristics of conventionally processed dairy products and are also used in the production of a range of imitation dairy products (Table 4). Imitation cheeses (cheese analogues) are made from vegetable fat, caseins, salts and water and are used in pizza, lasagne and sauces and on burgers, grilled sandwiches, macaroni, etc., at a significant cost-saving compared to the use of natural cheese. The functional properties of casein which favour their use in imitation cheese include fat and water binding, texture enhancing, melting properties, stringiness and shredding ability. While caseins (both acid and rennet) and caseinates have been used most commonly for cheese analogues, co-precipitates also have potential in this area.

Sodium caseinate is used in powdered coffee creamers, which also contain vegetable fat, a carbohydrate source and added emulsifier and stabilizers. These creamers are cheaper, have a longer shelf-life and are more convenient to use (e.g. they require no refrigeration) than fresh coffee creams. In these products, sodium caseinate acts as an emulsifier/fat encapsulator and whitener, imparts body and flavour and promotes resistance to feathering (i.e. coagulation of cream in hot coffee solutions).

Sodium caseinate is used to increase gel firmness and reduce syneresis in yoghurts, and is added to milk shakes for its emulsifying and foaming properties. In the manufacture of imitation milks, the principal ingredients are caseins/caseinates, vegetable fat and carbohydrate, e.g. corn syrup. The main advantages associated with imitation milk products are low cost and the absence of lactose to which some people are intolerant. There is also interest in the fortification of liquid milk with casein products such as sodium, potassium and calcium caseinates and co-precipitates.

Sodium caseinate is used as an emulsifying and fat encapsulating agent in the manufacture of high-fat powders for use as shortenings in baking or cooking. Dry whipping fats or whipping creams contain casein products. A number of butter-like dairy spreads are manufactured using milk and/or vegetable fat and various casein products. In these applications, casein acts mainly as an emulsifier and in the case of dairy spreads, it also enhances texture and flavour.

Whey protein products are widely used in yoghurts and various cheeses to improve the yield, nutritional value and consistency. Up to 20% of the casein in Quark can be replaced by thermally-modified WPC, resulting in an increase in yield and nutritional value. The use of sweet UF-WPC in Ricotta cheese manufacture increases the cohesiveness of the curd. Emulsions prepared using heat-denatured whey proteins and fat are used as a protein base for formulated cream cheeses and cream cheese spreads.

The viscosity and stability of yoghurts are improved by fortification with WPC to replace skim milk solids. Sliceable and squeezeable cheese-type products, based on the emulsifying and gelling properties of whey proteins, have been produced by heat treatment of skim milk and WPC solids dispersed in an emulsion of milk fat in WPC. Whey protein concentrates are also used in cheese filling and dips as they tend to complement cheese flavour and produce a soft end product.

4.3. Beverages

Casein products are used as stabilizers or for their whipping and foaming properties in drinking chocolate, fizzy drinks and fruit beverages (Table 5). There is also a large market for sodium caseinate as an emulsifier in cream liqueurs and to a

TABLE 5
Applications of Milk Protein Products in Beverages

Caseins/caseinates/co-precipitates	
Used in:	Drinking chocolate, fizzy drinks and fruit beverages
Effect:	Stabilizer, whipping and foaming properties
Used in:	Cream liqueurs, wine apertifs
Effect:	Emulsifier
Used in:	Wine and beer industry
Effect:	Fines removal, clarification, reduce colour and astringency
Whey proteins	
Used in:	Soft drinks, fruit juices, powdered or frozen orange beverages
Effect:	Nutritional
Used in:	Milk-based flavoured beverages
Effect:	Viscosity, colloidal stability

lesser extent in wine aperitifs. Cream liqueurs typically contain 16% (w/w) milk fat, 3·3% sodium caseinate, 19% added sugar and 14% ethanol. Trisodium citrate is also added to prevent calcium-induced age gelation. Casein products have also been used in the wine and beer industries as fining agents, to decrease colour and astrigency and to aid in clarification.

Fruit juices, soft drinks or milk-based beverages supplemented with whey protein concentrates are highly nutritious products. For use in soft drinks, defatted WPC with a low ash content, good solubility at pH 3·0 and a bland flavour are required. The WPC must also be resistant to physical deterioration or flavour changes on storage of the product and it must not mask the typical flavour of the soft drink via protein–flavour component interactions. WPCs are added to milk-like flavoured drinks to impart viscosity, body and colloidal stability and they have been included as protein supplements in powdered orange beverages and in frozen orange juice concentrates.

4.4. Dessert-Type Products

Sodium caseinate is used in ice cream substitutes and frozen desserts to improve whipping properties, body and texture and to act as a stabilizer (Table 6). It is also used extensively in mousses, instant puddings and whipped toppings for similar reasons and also because it acts as an emulsifier and film-former. The basic ingredients of whipped toppings are vegetable fat, sugar, protein (sodium caseinate), emulsifier, stabilizers and water. After blending the ingredients at 38–46°C, the mixture is pasteurized and homogenized and then either cooled rapidly to below freezing point or spray dried.

In the manufacture of ice cream, up to 10% of the skim milk solids can be replaced by whey powder. A higher level of replacement (up to 25%) may be possible by using delactosed, demineralized whey powder or UF-WPC with no adverse effect on flavour, texture or appearance. WPC has also been used in frozen

TABLE 6
Applications of Milk Protein Products in Dessert Type Products

Caseins/caseinates/co-precipitates	
Used in:	Ice cream, frozen desserts
Effect:	Whipping properties, body and texture
Used in:	Mousses, instant pudding, whipped topping
Effect:	Whipping properties, film former, emulsifier, imparts body and flavour
Whey proteins	
Used in:	Ice cream, frozen juice bars, frozen dessert coatings
Effect:	Skim milk solid replacement, whipping properties, emulsifying, body/texture

juice bars and in compound coatings, especially chocolate coatings, for frozen desserts.

4.5. Pasta Products

Milk protein products are often incorporated into the flour base for pasta manufacture to enhance nutritional quality and also to improve texture (Table 7). Products fortified by the addition of sodium or calcium caseinate, low calcium co-precipitate or WPC prior to extrusion include maraconi and pasta.

Enrichment of pasta flours with undenatured whey protein products produces firmer cooked noodles which are also more freeze–thaw stable and suitable for microwave cooking. 'Imitation' or 'synthetic' pasta-type products containing a substantial proportion of milk protein have also been manufactured.

4.6. Confectionery

Caseins are used in toffee, caramel, fudge and other confections as they form a firm, resilient, chewy matrix on heating and they contribute water binding and aid emulsification (Table 8). WPCs have limited use in these products as they result in a softer coagulum and the high lactose content tends to cause crystallization during storage. However, whey proteins are very suitable for use in aerated candy mixtures and are incorporated as a frappé, which is a highly aerated sugar syrup containing the whipping protein. Casein hydrolysates are used as foaming agents in place of egg albumen in marshmallow and nougat as they confer stability to high cooking temperatures and good flavour and browning properties. Replacement of egg white by WPC in the manufacture of meringues results in acceptable products only when defatted WPCs are used, while the manufacture of acceptable sponge cakes requires fat-containing rather than defatted WPCs.

TABLE 7
Applications of Milk Protein Products in Pasta Products

Used in:	Macaroni, pasta, imitation pasta
Effect:	Nutritional, texture, freeze–thaw stability, microwaveable

TABLE 8
Applications of Milk Protein Products in Confectionary

Caseins/caseinates/co-precipitates	
Used in:	Toffee, caramel, fudges
Effect:	Confers firm, resilient, chewy texture; water binding, emulsifier
Used in:	Marshmalow and nougat
Effect.	Foaming, high temperature stability, improve flavour and brown colour
Whey proteins	
Used in:	Aerated candy mixes, meringues, sponge cakes
Effect:	Whipping properties, emulsifier

4.7. Meat Products

In the meat industry, milk proteins are used mainly in comminuted meat products rather than prime cuts; however, they are also used in injection brines for uncomminuted products like cooked hams (Table 9). Caseins in comminuted meat products contribute to fat emulsification, water binding and improved consistency as they release meat proteins for gel formation and water binding. While sodium caseinate is the preferred additive in meat applications, various types of co-precipitates have also been used.

In frankfurters and luncheon rolls, up to 20% of the meat protein may be replaced by whey proteins. In these systems, whey proteins are used to prepare pre-emulsions of part of the fat and to support network formation, via gelation, during subsequent cooking.

Soluble, low viscosity whey protein concentrates are suitable for use in injection brines for fortification of whole meat products. Injection of fresh and cured meats with whey protein solution ($\geqslant$ 10%, w/v, protein) may increase the yield by as much as 30%.

TABLE 9
Applications of Milk Protein Products in Meat Products

Caseins/caseinates/co-precipitates	
Used in:	Comminuted meat products
Effect:	Emulsifier, water binding, improves consistency, releases meat proteins for gel formation and water binding
Whey proteins	
Used in:	Frankfurters, luncheon rolls
Effect:	Pre-emulsion, gelation
Used in:	Injection brine for fortification of whole meat products
Effect:	Gelation, yield

TABLE 10
Applications of Milk Protein Products in Dietary Pharmaceutical and Medical Products

Used in:	*Special dietary preparations*
	— Ill or convalenscent patients
	— Dieting patients/people
	— Athletes
	— Astronauts
Used in:	*Infant foods*
	— Nutritional fortification
	— 'Humanized' infant formulae
	— Low-lactose infant formulae
	— Specific mineral balance infant foods
	— Casein hydrolysates: used for infants suffering from diarrhoea, gastroenteritis, galactosemia, malabsorption, phenylketonuria
	— Whey protein hydrolysates used in hypoallergic formulae preparation
	— Nutritional fortification
Used in:	*Intraveneous feeds*
	— Patients suffering from metabolic disorders, intenstinal disorders or for post-operative patients
Used in:	*Special food preparations*
	— Patients suffering from cancer, pancreatic disorders or anaemia
Used in:	*Specific drug preparations*
	— β-casomorphins used in sleep or hunger regulation or insulin secretion
	— Sulphonated glycopeptides used in treatment of gastric ulcers
Used in:	*Miscellaneous products*
	— Toothpastes
	— Cosmetics
	— Wound treating preparations

4.8. Dietary, Pharmaceutical and Medical Applications

Since milk protein products are of high nutritional quality, they are used extensively in dietary preparations (Table 10) for people who are ill or convalescing, for malnourished children in developing countries on a therapeutic diet and for people on weight-reducing diets. Caseins are used in special preparations to enhance athletic performance and have been incorporated into formula diets for space feeding.

While casein products are not generally used in infant formulae, they are used extensively in specialized preparations for infants with specific nutritional problems. Caseinates and co-precipitates are used in low-lactose formulae for lactose-intolerant infants while various types of caseinates have been used in infant foods with a specific mineral balance, e.g. low-sodium infant formulae for children with specific renal problems. Casein hydrolysates are used in specialized foods for premature infants, in formulae for infants suffering from diarrhoea, gastroenteritis,

TABLE 11
Applications of Milk Protein Products in Convenience Foods

Used in:	Gravy mixes, soup mixes, sauces, canned cream soups and sauces, dehydrated cream soups and sauces, salad dressings, microwavable foods, low lipid convenience foods
Effect:	Whitening agents, dairy flavour, flavour enhancer, emulsifier, stabilizer, viscosity controller, freeze–thaw stability, egg yolk replacement, lipid replacement

galactosaemia and malabsorption. A special casein hydrolysate, low in phenylalanine, has been prepared for use in formulae for feeding infants with phenylketonuria. Casein products are also added to various foods for children and infants and to drinks as a nutritional supplement.

Modified low-mineral whey powders are used to produce 'humanized' infant formulae which have a whey protein-to-casein ratio close to that of human milk. Hypoallergic, peptide-based formulae have been developed based on whey protein hydrolysates. Selected individual caseins and whey proteins have been proposed as possible ingredients for the 'next generation' of 'more humanized' infant formulae.

Milk protein hydrolysates are used for intravenous nutrition for patients suffering from intestinal disorders, protein metabolism disorders and for post-operative patients. Special casein preparations have been used as food for patients suffering from cancer, pancreatic disorders or anaemia.

Specific drugs have been produced from casein; β-casein is used as raw material for production of β-casomorphins, tetra- to heptapeptides which can regulate sleep, hunger or insulin secretion. Sulphonated glycopeptides prepared from casein have been used for the treatment of gastric ulcers. It is claimed that the use of casein in toothpaste prevents dental caries, in cosmetics it conceals facial wrinkles and in special therapeutic creams it heals wounds.

4.9. Convenience Foods

Milk protein products are widely used in convenience foods, i.e. foods which require a minimum of preparation by the consumer (Table 11). Either skim milk powder or whey/caseinate blends are used as whitening agents in gravy mixes. Whey solids are used in dehydrated soup mixes and sauces to impart a milky or dairy flavour, as flavour enhancers and to provide emulsifying and stabilizing effects. Caseinates are used as emulsifying agents and viscosity controllers in canned cream soups and sauces and for the preparation of dry emulsions for use in dehydrated cream soups and sauces. Sauces and gravies containing whey proteins are reported to be less prone to cook-on to utensil walls, require minimum agitation and have stability to freeze–thaw cycling. In some convenience foods, caseinate–whey protein blends are used as cheap replacements for skim milk powders. Whey protein products have potential as replacements for egg yolk in salad dressing and modified whey protein-based products, with potential to replace lipids in a variety of convenience foods, have been developed. Milk protein

TABLE 12
Applications of Milk Protein Products in Textured Products

Used in:	Puffed snack foods, protein-enriched snack-type products, meat extenders
Effect:	Structuring, texturing, nutritional

products have been proposed as texture, stability and flavour enhancers in microwaveable foods.

4.10. Textured Products

Textured milk protein-based foods, in the form of cheeses, have been manufactured from milk for thousands of years. However, milk protein-enriched products have been used in the production of textured foods only recently (Table 12). Rewetted acid caseins or acidified rennet casein or co-precipitate have been mixed with carbonates or bicarbonates of alkali metals or alkali earth metals and extruded to produce puffed snack foods while caseinates have been co-extruded with wheat flour to produce protein-enriched snack-type food products.

Fibrous meat-like structures, formed from caseins by fibre spinning techniques have been produced for use as extenders in comminuted meats. Whey proteins may be co-spun with the casein to produce stronger fibres than those containing casein alone. Meat-like structures can also be formed from casein or co-precipitates by renneting followed by thermoplastic extension which involves a combination of heat treatment and extrusion or working. Microwave heating of whey protein solutions results in simultaneous expansion and gelation to give textured products with possible applications in comminuted meats.

5. FUTURE DEVELOPMENTS

As outlined above, a large range of milk protein products is presently recovered from milk. It is likely that this range will be extended in the future as methods developed to effect separation of individual caseins and whey proteins become commercial. It is likely that enzymatic and physical modification methods will be applied to confer new physico-chemical and functional properties to these milk protein products.

While the main area of application of milk proteins will be as functional ingredients in formulated foods, it is likely that there will be increased use of milk protein-based products in dietary, pharmaceutical and medical products.

REFERENCES

Allen, I.M., McAuliffe, A.G. & Donnelly, W.T. (1985). Simplified approaches to casein fractionation. *Irish Journal of Food Science and Technology*, **9**, 85 (Abstract).

Amundson, C.H., Watanawanichakorn, S. & Hill, C.G. (1982). Production of enriched protein fractions of β-lactoglobulin and α-lactalbumin from cheese whey. *Journal of Food Processing and Preservation*, **6**, 55–71.

Breslau, B.R., Cross, R.A. & Goulet, J. (1975). Production of a crystal clear, bland tasting, protein solution from cheese whey. *Journal of Dairy Science*, **58**, 782 (Abstract).
Burgess, K.J. & Kelly, J. (1979). Technical note: selected functional properties of a whey protein isolate. *Journal of Food Technology*, **14**, 325–9.
De Wit, J.N. (1989*a*). Functional properties of whey proteins. In *Developments in Dairy Chemistry—4*, ed. P.F. Fox. Elsevier Applied Science Publishers, London, pp. 285–322.
De Wit, J.N. (1989*b*). The use of whey protein products. In *Developments in Dairy Chemistry—4*, ed. P.F. Fox. Elsevier Applied Science Publishers, London, pp. 323–46.
De Wit, J.N. & De Boer, R. (1975). Ultrafiltration of cheese whey and some functional properties of the resulting whey protein concentrate. *Netherlands Milk and Dairy Journal*, **29**, 198–211.
De Wit, J.N., Klarenbeek, G. & De Boer, R. (1978). A simple method for the clarification of whey. In *Brief Communications, Proceedings of the International Dairy Congress*, Paris, Congrilait Publishers, Paris, France, pp. 919–20.
Donovan, M. & Mulvihill, D.M. (1987). Thermal denaturation and aggregation of whey proteins. *Irish Journal of Food Science and Technology*, **11**, 87–100.
Famelart, M.H., Hardy, C. & Brule, G. (1989). Optimisation of the preparation of β-casein-enriched solution. *Le Lait*, **69**, 47–57.
Fox, P.F. & Mulvihill, D.M. (1983). Functional properties of caseins, caseinates and co-precipitates. In *Physico-chemical Aspects of Dehydrated Protein-Rich Milk Products.* Proceedings of IDF Symposium, Helsingor, Denmark, International Dairy Federation, Brussels, pp. 188–259.
Fox, P.F. & Mulvihill, D.M. (1990). Casein. In *Food Gels*, ed. P. Harris. Elsevier Applied Science Publishers, London, pp. 121–74.
Hayes, J.F., Dunkerley, J.A., Muller, L.L. & Griffin, A.T. (1974). Studies on whey processing by ultrafiltration. II. Improving permeation rates by preventing fouling. *Australian Journal of Dairy Technology*, **29**, 132–40.
Hewedi, M.M., Mulvihill, D.M. & Fox, P.P. (1985). Recovery of milk protein by ethanol precipitation. *Irish Journal of Food Science and Technology*, **9**, 11–23.
Hoff, J.E., Nielsen, S.S., Peng, I.C. & Chambers, J.V. (1987). Ethanol extraction of lactose from non-fat dry milk: production of protein raffinate. *Journal of Dairy Science*, **70**, 1785–96.
Hugunin, A.G. (1987). Applications of UF whey protein: developing new markets. International Dairy Federation Bulletin 212, Brussels, Belgium, pp. 135–44.
International Dairy Federation (1982). Dairy Ingredients in Food Products. IDF Bulletin 147, Brussels, Belgium.
International Dairy Federation (1987). Trends in Whey Utilization. IDF Bulletin 212, Brussels, Belgium.
Kaczmarke, J. (1980). Whey protein separation and processing. In *Proceedings 1980 Whey Production Conference*, Chicago, IL. Whey Products Institute, USDA Philadelphia, PA, pp. 68–80.
Kinsella, J.E. (1984). Milk proteins: physicochemical and functional properties. *CRC Critical Reviews in Food Science and Nutrition*, **21**, 197–262.
Knightsbridge, J.P. & Goldman, P.A. (1975). Water absorptive capacity of dried milk products. *New Zealand Journal of Dairy Science and Technology*, **10**, 152–7.
Lee, D.N. & Merson, R.L. (1976). Chemical treatments of cottage cheese whey to reduce fouling of ultrafiltration membranes. *Journal of Food Science*, **41**, 778–86.
Lonergan, D.A. (1983). Isolation of casein by ultrafiltration and cryodestabilization. *Journal of Food Science*, **48**, 1817–21.
Marshall, K.R. (1982). Industrial isolation of milk proteins: whey protein. In *Developments in Dairy Chemistry—1*, ed. P.F. Fox. Applied Science Publishers, London, pp. 339–73.
Matthews, M.E. (1984). Whey protein recovery processes and products. *Journal of Dairy Science*, **67**, 2680–92.
Matthews, M.E., Doughty, R.K. & Short, J.L. (1978). Pretreatment of acid casein whey to

improve processing rates in ultrafiltration. *New Zealand Journal of Dairy Science and Technology*, **13**, 216–20.
Maubois, J.L. (1990). Applications of membrane techniques in the dairy industry—proposal for a new IDF group of experts. International Dairy Federation, Bulletin No. 244, Brussels, Belgium, pp. 20–9.
Maubois, J.L., Pierre, A., Fauquant, J. & Piat, M. (1987). Industrial fractionation of main whey proteins. International Dairy Federation Bulletin 212, Brussels, Belgium, pp. 154–9.
Mirabel, B. (1978). Nouveau procédé d'extraction des protéins du lactosérum. *Annales de la Nutrition et de l'Alimentation*, **23**, 243–53.
Morr, C.V. (1989). Whey proteins: manufacture. In *Developments in Dairy Chemistry—4*, ed. P.F. Fox. Elsevier Applied Science Publishers, London, pp. 245–84.
Muller, L.L. (1971). Manufacture and uses of casein and co-precipitate. *Dairy Science Abstracts*, **33**, 659–74.
Muller, L.L. (1982). Manufacture of caseins, caseinates and co-precipitates. In *Developments in Dairy Chemistry—1*, ed. P.F. Fox. Applied Science Publishers, London, pp. 315–37.
Muller, L.L. & Harper, W.J. (1979). Effects on membrane processing of pretreatments of whey. *Journal of Agricultural and Food Chemistry*, **27**, 662–4.
Mulvihill, D.M. (1989). Caseins and caseinates: manufacture. In *Developments in Dairy Chemistry—4*, ed. P.F. Fox. Elsevier Applied Science Publishers, London, pp. 97–130.
Mulvihill, D.M. & Donovan, M. (1987). Whey proteins and their thermal denaturation. A review. *Irish Journal of Food Science and Technology*, **11**, 43–75.
Mulvihill, D.M. & Fox, P.F. (1989). Physicochemical and functional properties of milk proteins. In *Developments in Dairy Chemistry—4*, ed. P.F. Fox. Elsevier Applied Science Publishers, London, pp. 131–72.
Mulvihill, D.M. & Kinsella, J.E. (1987). Gelation characteristics of whey proteins and β-lactoglobulin. *Food Technology*, **41**, 102–11.
Mulvihill, D.M. & Kinsella, J.E. (1988). Gelation of β-lactoglobulin: effects of sodium chloride and calcium chloride on the rheological and structural properties of gels. *Journal of Food Science*, **53**, 231–6.
Mulvihill, D.M. & Murphy, P.C. (1991). Surface active and emulsifying properties of caseins/caseinates as influenced by state of aggregation. *International Dairy Journal*, **1**, 13–37.
Mulvihill, D.M., Rector, D. & Kinsella, J.E. (1990). Effects of structuring and destructuring anionic ions on the rheological properties of thermally induced β-lactoglobulin gels. *Food Hydrocolloids*, **4**, 267–76.
Mulvihill, D.M., Rector, D. & Kinsella, J.E. (1991). Mercaptoethanol, N-ethylmaleimide, propylene glycol and urea effects on the rheological properties of thermally induced β-lactoglobulin gels at alkaline pH. *Journal of Food Science*, **56**, 1338–41.
Murphy, J.M. & Fox, P.F. (1990). Fractionation of sodium caseinate by ultrafiltration. *Food Chemistry*, **39**, 27–38.
Palmer, D.E. (1982). Recovery of proteins from food factory waste by ion exchange. In *Food Proteins*, eds. P.F. Fox & J.J. Condon. Applied Science Publishers, London, pp. 341–52.
Payens, T.A.J. & van Markwijk, B.W. (1963). Some features of the association of β-casein. *Biochimica et Biophysica Acta*, **71**, 517–30.
Pearse, R.J. (1987). Fractionation of whey protein. In *Trends in Whey Utilization*, International Dairy Federation Bulletin 212, Brussels, Belgium, pp. 150–3.
Pradipasena, P. & Rha, C.K. (1977). Effect of concentration on apparent viscosity of a globular protein. *Polymer, Engineering and Science*, **17**, 861–4.
Rha, C.K. & Pradipasena, P. (1986). Viscosity of proteins. In *Functional Properties of Food Macromolecules*, eds. J.R. Mitchell & D.A. Ledward. Elsevier Applied Science Publishers, London, pp. 79–120.
Slack, A.W., Amundson, C.H. & Hill, C.G. (1986). Nitrogen solubilities of β-lactoglobulin and α-lactalbumin enriched fractions derived from ultrafiltered cheese whey retentate.

Journal of Food Processing and Preservation, **10**, 19–29.

Southward, C.R. (1986). Utilization of milk components: casein. In *Modern Dairy Technology, Volume 1. Advances in Milk Processing*, ed. R.K. Robinson. Elsevier Applied Science Publishers, London, pp. 317–68.

Southward, C.R. (1989). Use of casein and caseinates. In *Developments in Dairy Chemistry—4*, ed. P.F. Fox. Elsevier Applied Science Publishers, London, pp. 173–244.

Southward, C.R. & Goldman, A. (1978). Co-precipitates and their application in food products. II. Some properties and applications. *New Zealand Journal of Dairy Science and Technology*, **13**, 97–105.

Southward, C.R. & Walker, N.J. (1982). Casein, caseinates and milk protein co-precipitates. In *CRC Handbook of Processing and Utilization in Agriculture, Volume 1. Animal Products*, ed. I.A. Wolf. CRC Press Inc., Boca Raton, FL, pp. 445–552.

Terre, E., Maubois, J.L., Brule, G. & Pierre, A. (1986). Procede d'obtention d'une matiere enrichie en caseine beta, appareillage pour la mise en oeuvre de ce procede, et application des produits obtenus par ce procede comme aliments, complements alimentaires ou additifs en industrie alimentaire et pharmaceutique ou dans la preparation de peptides a activite physiologique. French Patent No. 2 592 769.

Zadow, J.G. (1986). Utilization of milk components: whey. In *Modern Dairy Technology, Volume 1. Advances in Milk Processing*, ed. R.K. Robinson. Elsevier Applied Science Publishers, London, pp. 273–316.

10

Genetic Polymorphism of Milk Proteins

K.F. Ng-Kwai-Hang

Department of Animal Science, Macdonald College of McGill University, 21,111 Lakeshore Road, Ste Anne de Bellevue, P.Q., Canada H9X 1CO

&

F. Grosclaude

Laboratoire de Génétique Biochimique, Institut National de la Recherche Agronomique, 78350 Jouy-en-Josas, France

1. Introduction

Extensive studies on the qualitative and quantitative aspects of milk proteins in more than 100 mammalian species have demonstrated that the protein contents vary from 1 to 20% between different species and within the same species of different genetic backgrounds under different environmental conditions. All the milks so far analysed contain an acid precipitable fraction, commonly known as casein, and an acid soluble fraction known as the whey protein or milk serum protein. Gel electrophoretic techniques have been used to reveal the identity of several types of caseins and whey proteins and to establish the presence of homologous proteins across several species. The discovery of two electrophoretically distinct forms of β-lactoglobulin by Aschaffenburg and Drewry in 1955 resulted in the initiation of very active research in the field of genetic polymorphism of milk proteins in several countries of the world. Although various investigators have reported on the occurrence of polymorphisms of the major milk proteins, distinction must be made between genetic polymorphism, which is due to a mutation resulting in a change in the amino acid sequence of the protein, and polymorphism due to post-transcriptional modification such as different degrees of phosphorylation and glycosylation of the protein.

The purpose of this chapter is to present an overall picture of our present knowledge on the genetic polymorphisms of milk proteins in the following areas: methods of identification, number of variants already identified, differences in amino acid sequence, genetic basis, geographical and breed distribution, relationships with production traits, milk composition and milk quality. For obvious economic reasons, most of the research has been directed toward the bovine and to a lesser extent to the caprine species. Hence, most of the information given in this chapter pertains to milk of those two species.

2. METHODS OF DETECTING GENETIC POLYMORPHISM

2.1. Electophoretic Techniques

It has been 35 years since the occurrence of genetic polymorphisms in milk protein was first reported by Aschaffenburg and Drewry (1955). This event was marked by the observation that when milk samples from individual cows were subjected to electrophoresis on filter paper in barbitone buffer at pH 8·6 and under 230 V for 16 h, these samples produced either one of two or a mixture of both electrophoretically distinct β-lactoglobulin bands which were denoted as β_1- and β_2-lactoglobulin in order of decreasing mobility. This nomenclature was later replaced by A and B, respectively, when it was discovered (Aschaffenburg & Drewry, 1957) that the synthesis of these two types of β-lactoglobulin was under genetic control, determined by two allelic autosomal genes. The discovery of Aschaffenburg and Drewry was followed by a very active search for the existence of genetic polymorphisms in the different caseins and whey proteins. Electrophoretic methods for the detection of polymorphs by far exceed any other methods used because of their simplicity and ease of application for large numbers of samples. The original paper electrophoretic procedure has been replaced by several other electrophoretic methods which have been developed, modified or adapted in order to improve the resolving power for the purpose of phenotyping of different milk proteins from various species of animal.

The paper electrophoretic technique of Aschaffenburg and Drewry (1955) was applied by Blumberg and Tombs (1958) for the detection of genetic variants of α-lactalbumin in Nigerian White Fulani cattle. Paper electrophoresis of milk proteins was limited to the serum proteins because of the difficulties encountered in the separation of the native caseins which tend to associate and form micelles. However, by the inclusion of 6·0 M urea as dissociating agent in a pH 7·5 citrate–phosphate buffer, Aschaffenburg (1961) was able to overcome this limitation by demonstrating the presence of three polymorphs labelled A, B and C in order of decreasing mobility for β-casein in the milk of Jersey and Guernsey cows.

Thompson *et al.* (1962) introduced the use of starch gel urea electrophoresis at pH 8·6 for the resolution of three α_{s1}-casein variants known as A, B and C, in Holstein, Brown Swiss and Ayrshire breeds. The resolution of β-lactoglobulin A, B and C was achieved by Bell (1962) through the use of a pH 8·5 borate buffer without urea in a starch gel electrophoresis system. The method of Wake and Baldwin (1961) was modified (Grosclaude *et al.*, 1966*a*) by the inclusion of 2-mercaptoethanol in the gel for the separation of A, B and D variants of β-lactoglobulin. The inclusion of small quantities of 2-mercaptoethanol in the sample buffer or electrophoretic gel contributed to the characterisation of two bands corresponding to κ-casein by three independent research groups (Neelin, 1964; Schmidt, 1964; Woychik, 1964). Genetic studies conducted by Grosclaude *et al.* (1965) demonstrated that these electrophoretic bands represented two genetic variants of κ-casein. Starch gel electrophorphesis was used by Thompson *et al.*

(1964) and Michalak (1967) for the detection of variants A, B and C in β-casein. Evidence of genetic polymorphism in α_{s2}-casein was first demonstrated by Grosclaude *et al.* (1976*a*) in a study involving milk from Nepalese yak, yak × zebu crosses, and bovine × yak crosses. The separation of three α_{s2}-casein variants was achieved by electrophoresis in starch gel slabs containing 2-mercaptoethanol and 7 M urea under alkaline conditions.

Several methods have been applied for the phenotyping of genetic polymorphism of β-lactoglobulin, α_{s1}-casein, and β-casein under alkaline conditions in polyacrylamide gels (Aschaffenburg, 1964; Thompson *et al.* 1964; Aschaffenburg & Thymann, 1965). The fastest single moving band of variant A for β-casein, which was not resolved under several alkaline conditions, was later found by Peterson and Kopfler (1966) to consist of three variants now known as A^1, A^2, and A^3 which could be separated by electrophoresis at pH 2·8 in polyacrylamide gel containing 4·5 M urea. Resolution of these three distinct types of β-casein A was later confirmed by Grosclaude *et al.* (1966*b*) and Arave (1967) with starch gel electrophoresis at low pH. The use of polyacrylamide gels as supporting electrophoretic medium instead of starch gels has some advantages in that the former are easier to prepare and manipulate because they can be formed without heat treatment and can be stained without slicing.

Other supporting media for electrophoresis include agar gels which have been used by Aschaffenburg (1965) for the phenotyping of four polymorphs of β-lactoglobulin at pH 8·6 in veronal buffer. Bell and Stone (1978) obtained good resolution of β-lactoglobulin A and B by electrophoresis in cellulose acetate strips under alkaline conditions. The same support, in combination with tris-glycine buffer at pH 9·0 and containing 6·0 M urea in the presence of 2-mercaptoethanol, was employed by Davoli (1981) for the phenotyping of variants of α_{s1}-casein, κ-casein, and α-lactalbumin. Because β-lactoglobulin B had the same mobility as β-casein A, the author recommended separate runs for caseins and whey proteins. Cellulose acetate electrophoresis requires 4 h for separation and hence is much faster than the traditional starch gel electrophoresis which needs 20–22 h.

The application of different zone electrophoretic techniques for the separation of major milk proteins has contributed enormously to the study of genetic polymorphism in this particular group of proteins. Thompson (1970) presented a comprehensive review of different electrophoretic techniques that could be used for the phenotyping of milk proteins. It has to be recognised that the above listing of different methods utilised for identification of genetic variants of α_{s1}-casein, α_{s2}-casein, β-casein, κ-casein, β-lactoglobulin, and α-lactalbumin is far from being complete. A search of the literature will reveal a bewildering array of electrophoretic procedures that have been modified for the phenotyping of major milk proteins in the different breeds of cattle and in different species of mammals. A more complete and thorough survey of the different milk protein polymorphs that have been identified will be addressed in a later section of this chapter. For a new researcher in the field of milk protein polymorphism, the problem is not so much finding a suitable method for phenotyping, but rather deciding which method to choose from the hundreds that have been published. In spite of this difficulty, for practical purposes, the choice can be narrowed down to either the use of starch gel electrophoresis or polyacrylamide gel electrophoresis. The selection between these two gel supporting media depends on availability of facilities because both types of electrophoresis have been developed along parallel lines, and provide good resolution of the caseins when used in gel buffers containing urea and 2-mercaptoethanol at alkaline pH. Better separation of β-lactoglobulin and α-lactalbumin polymorphs is achieved in the absence of urea and 2-mercaptoethanol under alkaline conditions. Because alkaline gels cannot be used to resolve β-casein A into its components A^1, A^2 and A^3, another electrophoretic run is required for their resolution. Three separate electrophoretic runs under different sets of conditions would be needed in order to establish the whole spectrum of genetic polymorphism for α_{s1}-casein, α_{s2}-casein, β-casein, κ-casein, β-lactoglobulin and α-lactalbumin.

2.2. Isoelectric Focusing Techniques

Isoelectric focusing can be considered as a special type of gel electrophoresis whereby each protein migrates to a specific point across a pH gradient, that is its isoelectric point. Longer running times are required for isoelectric focusing than for polyacrylamide gel electrophoresis because the heating effect of current interferes with the formation of a pH gradient on the gel, and the mobility of the protein decreases as it approaches its isoelectric point. Whole casein, when subjected to electrophoresis across a pH gradient of 3·0 to 10·0 in 5% polyacrylamide containing urea, has been shown to produce as many as 20 bands. This separation method has the capability of resolving proteins differing by only 0·02 pH unit in their isoelectric points (Peterson, 1969). This technique could separate variants A^1, A^2, A^3, B and C of β-casein and Peterson claimed its usefulness for the separation of β-lactoglobulin and α-lactalbumin variants. Josephson (1972) reported successful identification of κ-casein A and B through the use of pH gradient electrophoresis of bovine milk caseins. Isoelectric focusing at a pH range of 3·5 to 5·0 in 5% polyacrylamide containing 6 M urea was used by Pearce and Zadow (1978) for separation of the A and B variants of β-lactoglobulin. A procedure for simultaneously phenotyping the four major caseins by isoelectric focusing was described by Addeo *et al.* (1983). Later, the same group (Di Luccia *et al.*, 1988) reported that κ-casein C had an isoelectric pH between the A and B variants. The use of agarose gel containing 7 M urea as support medium in an isoelectric system with a pH gradient of 2·5 to 7·0 could also produce good results for the phenotyping of α_{s1}-casein, β-casein, κ-casein and β-lactoglobulin (Bech & Munk, 1988). Seibert *et al.* (1985) described a rapid method for the simultaneous phenotyping of these four proteins by isoelectric focusing in ultra-thin (100μm thick) 4·5% polyacrylamide gels containing 8 M urea and 2-mercaptoethanol. With this method they (Seibert *et al.*, 1987) were able to detect a new variant of κ-casein in the milk of German Simmental cows, known as variant D, which had the same mobility as the A variant in polyacrylamide or starch gel electrophoresis. The isoelectric point of variant D is 5·76 compared to 5·62 for A and 5·83 for B. More recently, Erhardt (1989*a*) reported on another newly discovered type of κ-casein denoted as the E variant. Application of isoelectric focusing has also revealed the existence of three β-lactoglobulin variants in sheep (Erhardt *et al.*, 1989).

The use of isoelectric focusing for the phenotyping of milk proteins can be considered as an alternative to starch gel or polyacrylamide gel electrophoresis. This method seems attractive due to the fact that a single run is adequate for phenotyping of the seven known variants of β-casein. Close attention should be paid to isoelectric procedures for the purpose of phenotyping of milk proteins in view of the recently discovered variants of κ-casein.

2.3. Column Chromatographic Techniques

During the past three decades, many chromatographic techniques have been developed and adapted for the study of milk proteins. Yaguchi and Rose (1971) prepared an extensive review covering the application of column chromatographic techniques for the separation and purification of milk proteins. The most commonly used stationary phase in those methods is based on the principle of either ion exchange or size exclusion chromatography. For the purpose of phenotyping milk proteins, size exclusion chromatography is of limited value because the differences in molecular weights between the several variants of a specific protein are not sufficient to result in differences in the elution profile. Although ion exchange chromatography may have some potential application as a method of detecting genetic polymorphism of milk proteins, its use would be impractical due to the fact that the process of equilibrating the column, doing the actual chromatographic separation, and regenerating the column for the next sample may take several days. Another limitation pertains to the lack of discrimination between genetic variants. Nevertheless, it is interesting to note the possibility of separating the β-casein variant A from C in a DEAE–cellulose system containing urea (Thompson & Pepper, 1964), and by QAE–cellulose ion exchange chromatography (Ng-Kwai-Hang & Pélissier, 1989). Chromatography on a DEAE–cellulose column using imidazole–HCl–urea at pH 7·0 (Thompson, 1966) could resolve variants A and B of κ-casein.

With the development of high performance liquid chromatography it became possible to use chromatographic columns of considerably smaller dimensions, for example, 50 mm length and 5 mm diameter, and hence significantly reduce the analysis time for milk proteins. The HPLC system described by Pearce (1983) was able to resolve α-lactalbumin and the variants A and B of β-lactoglobulin within 30 min. Similarly, Humphrey and Newsome (1984) and Andrews *et al.* (1985) reported on the resolution of β-lactoglobulin A and B by HPLC on an anion exchange column. The only other known genetic variants that have satisfactorily been separated by HPLC are κ-caseins A and B (Dalgleish, 1986; Visser *et al.*, 1986; Guillou *et al.*, 1987). The technique used by Guillou *et al.* (1987) could resolve neither the variants B and C of α_{s1}-casein, nor the variants A^1, A^2 and B of β-casein. In the process of fractionating bovine caseins by HPLC on a reversed phase column, Carles (1986) discovered a new variant of β-casein which is more hydrophobic than β-casein A^1 by having a leucine instead of a proline in the 114-169 region of β-casein. This newly found variant is an example of the so called 'silent variants' because the substitution of an amino acid in this case would not result in a change in the net electrical charge of the protein and hence such variants would not be resolved by electrophoresis or ion exchange chromatography. With this as background information, the search for other polymorphs of milk proteins by HPLC techniques opens up a new area of research as discussed in the next section.

2.4. HPLC techniques for silent variants

Until now, the detection of genetic polymorphism of milk proteins has been achieved through various electrophoretic techniques and some of the identified variants have been ascertained by chromatographic methods. These methods are restricted to the resolution of proteins with differing net charges caused by the substitution of one amino acid for another. In theory, it has been estimated that mutations in proteins due to amino acid substitutions that do not lead to a change in the net charge on proteins should occur three times more frequently than those resulting in an alteration in net charge. The search for methods other than electrophoresis for the identification of further genetic variants of the major milk proteins should open up a new area in the biochemistry of milk proteins.

Considering that the total number of amino acid residues varies from 123 in α-lactalbumin to 209 in β-casein, changes in the net hydrophobic nature of the whole protein due to mutation caused by the substitution of one amino acid for another may be too small for detection by the reversed phase HPLC technique described by Carles (1986). Complete hydrolysis of the protein followed by peptide mapping, which has been successfully used for determination of mutants in hemoglobin (Schroeder *et al.*, 1982), could be adapted for identification of the silent variants of milk proteins. Basically, this method involves digesting the purified protein of interest with trypsin. The resulting mixture of peptides is then subjected to HPLC in a reversed phase column and its elution profile is compared with that of a standard protein of known primary structure. The eluent fraction corresponding to a peak with different retention time from the standard elution profile is collected for further analysis. Often, amino acid analysis of the aberrant peptide is sufficient to determine the type and position of the substitution in the variant. Should several genetic variants be discovered by application of this technique, then one will be faced with the problem of classification and nomenclature of the different milk proteins (Eigel *et al.*, 1984) which so far has been based on electrophoretic mobility. The feasibility of phenotyping milk proteins by peptide mapping will depend on availability of equipment and facilities. This method is more time consuming and considerably more involved than electrophoresis and hence would not be as practical as the latter method for large scale routine screening of genetic polymorphism of milk proteins in animal populations.

2.5. Genetic Variants at the DNA Level

Because milk production is a sex-limited trait, the determination of genetic polymorphism in milk proteins has to await the initiation of lactation in the female of the species. The possibility of phenotyping female animals prior to lactation, right at birth for example, and of phenotyping male animals for specific milk protein genes would have obvious profound advantages. With the development in molecular biology, techniques applicable for the identification of variants of some known milk proteins at the DNA level have been tested and perfected.

Genetic polymorphism observed in the milk proteins is a consequence of mutations which result in changes in the nucleotide sequence of the particular gene involved and hence different amino acid sequence will result. Prior to the use of several techniques available for the phenotyping of milk protein in genes, information is required regarding the primary structure and the position along the peptide chain where amino acid substitutions occur for the coded protein. Based on these observations, nucleotide sequence consisting of a relatively small number of bases, 15 for example, and corresponding to the sequence in the region where the mutation occurs could be constructed. These synthetic primers are then used to amplify the region of interests in an in-vitro system containing DNA extract isolated from various cells (blood or semen samples) through the polymerase chain reaction. Amplification is followed by digestion with restriction enzymes and analysis of the resulting fragments by electrophoresis. The number of electrophoresis bands, obtained from specific restriction endonucleases, identify the variant in question. Recently, Zadworny *et al.* (1990) isolated DNA from semen samples from bulls for the determination of κ-casein genotypes by PCR.

A second method exploits the principle of restriction fragment length polymorphism. Here, the DNA sequence of interest is cut by restriction endonucleases and the resulting fragments are separated by electrophoresis on an agarose gel slab, followed by transfer to nitrocellulose. The bands obtained are identified through visualisation after hybridisation with specific cDNA probes which have been labelled with P^{32}.

3. Occurrence of Genetic Polymorphism in Bovine Species

Most of the work on the genetic polymorphism of milk proteins has been performed on cattle. The different genetic variants of major milk proteins so far identified are represented schematically in Fig. 1. Presently, five genetic variants of α_{s1}-casein, designated as A, D, B, C and E in order of decreasing electrophoretic mobility in alkaline gels containing urea and 2-mercaptoethanol, have been reported. The variants A, B and C were discovered by Thompson *et al.* (1962) while variant D was first reported in French Flamande cattle (Grosclaude *et al.*, 1966*a*). Ten years elapsed before the discovery of the fifth variant, E, in Nepalese Yak's milk (Grosclaude *et al.*, 1976*b*). Evidence of a new variant of α_{s1}-casein with slower mobility than the C variant was reported in milk from Bali cattle, *Bos javanicus*

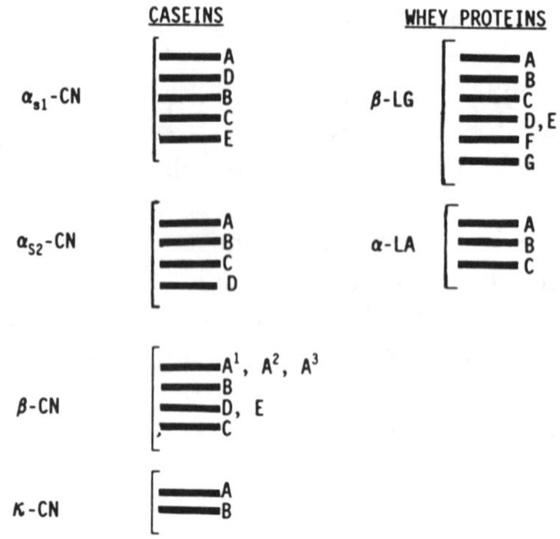

Fig. 1. Schematic representation of zone electrophoresis at alkaline pH of major milk proteins to show relative migrations of genetic variants. The positions shown for the different proteins are approximate and are only for illustration purposes. Migrations of other variants described in the text (β-casein A′, B^2, A^4, $A^3_{Mongolie}$; κ-casein C, D, E; β-lactoglobulin Dr) are not shown.

(Bell *et al.*, 1981*c*). Before assigning another letter to this particular variant, more information is required regarding its primary structure; this unique variant might be identical to either the D or E variant which have previously been reported.

The first report on genetic polymorphism of α_{s2}-casein was by Grosclaude *et al.* (1976*a*) who used the designation A for the already known variant common in European breeds and assigned the letter B to a newly discovered variant in cattle (*Bos taurus*) and Zebus (*Bos indicus*), and the letter C to another variant observed in yaks (*Bos grunniens*) from the Nepalese valley. Among the European breeds of *Bos taurus* in France, a fourth variant, D, was later identified in the Vosgienne and Montbéliarde breeds (Grosclaude *et al.*, 1979).

In search of genetic polymorphism of milk proteins in several dairy cattle breeds, Aschaffenburg (1961) discovered three variants, denoted as A, B and C, of β-casein. The A variant could be resolved into A′, A^2 and A^3 by gel electrophoresis under acidic conditions (Peterson & Kopfler, 1966). Evidence for a rare variant, designated D, in the Deshis breed of India, and the Borans of Kenya was reported by Aschaffenburg *et al.* (1968). A seventh variant, E, of β-casein was discovered by Voglino (1972) in the Piedmont breed of Italy. In alkaline gels containing urea, the decreasing order of electrophoretic mobility for the seven genetic variants of β-casein is: $A^1 = A^2 = A^3 > B > D, E > C$ (Kiddy, 1975). At least four additional variants of rare occurrence have been described. Abe *et al.* (1975) detected β-casein A′, which had a significantly lower mobility than A^1, A^2 or A^3 at pH 1·7, among Japanese Brown cows. In the same year, the presence of variant B^2 in two cows of unidentified breed was reported in New Zealand (Creamer & Richardson, 1975). A new variant of β-casein, provisionally denoted as A^4, was observed in the Bali (Bell *et al.*, 1981*a*) and Korean (Han *et al.*, 1983, 1984) cattle. The identification of a new variant among the Mongolian cattle led to the designation of β-casein $A^3{}_{Mongolie}$ (Grosclaude *et al.*, 1982).

The inclusion of 2-mercaptoethanol in urea-containing gels disclosed the presence of genetic polymorphism in κ-casein (Grosclaude *et al.*, 1965). Each of the two variants, A and B, of κ-casein revealed multiple bands when subjected to alkaline electrophoresis due to different degrees of glycosylation (Pujolle *et al.*, 1966; Doi *et al.*, 1979). κ-Casein has long been considered as diallelic, consisting of variants A and B. A third variant C has been described by Di Stasio and Merlin (1979) and two additional variants, D and E, were detected by Seibert *et al.* (1987) and Erhardt (1989*a*) through the techniques of isoelectric focusing. More recently, results of amino acid sequencing of variants C, D and E revealed that variants C and D are one and the same (Erhardt, G., pers. comm. 1989).

Genetic polymorphism of milk proteins was first recognised after the discovery of β-lactoglobulin of A and B by Aschaffenburg and Drewry (1955). The presence of variant C (Bell, 1962) was later detected in the milk of Jersey cows. A fourth variant, D, of β-lactoglobulin was identified in the French Montbéliarde breed (Grosclaude *et al.*, 1966a). Another variant of β-lactoglobulin, designated Dr, unique to the Droughtmaster cattle was described by Bell *et al.* (1970*a*). The Dr variant is identical to the A variant except for the presence of a covalently-bound carbohydrate moiety. This glycosylated form of β-lactoglobulin is not a true genetic variant but rather the product of post-translational modification. However, Bell *et al.* (1981*a*) indicated that β-lactoglobulin in Dr differs from the A variant by the substitution of asparagine for aspartic acid at position 28, the site of attachment for the carbohydrate moiety. If this point mutation could be confirmed, then β-lactoglobulin Dr could be considered as a true genetic variant. Yak milk has been shown to contain only one type of β-lactoglobulin, D_{yak} (Grosclaude *et al.*, 1976*a*, 1982), which has the same electrophoretic mobility as bovine β-lactoglobulin D. To avoid the use of subscript and superscript in milk protein nomenclature, Eigel *et al.* (1984) suggested that the variant in yak be labelled as E due to its order of discovery. There is strong evidence for the presence of two additional variants, F and G, among Banteng cattle (Bell *et al.*, 1981*b*).

Although Aschaffenburg and Drewry (1957) suspected the possible occurrence of genetic polymorphism in α-lactalbumin, they were unable to confirm it because all the individual milk samples analysed at that time contained only variant B. It was Blumberg and Tombs (1958) who discovered the presence of α-lactalbumin A among the Zebu cattle in Nigeria. A third variant of α-lactalbumin with slower mobility than α-lactalbumin B has been provisionally designated as C, pending amino acid sequencing (Bell *et al.*, 1981*a*). This latest variant of α-lactalbumin was identified in Bali cattle.

4. Molecular Basis for Genetic Polymorphism in Bovine Species

Genetic polymorphism in the milk proteins is due to either substitutions of amino acids, or deletion of a certain amino acid sequence along the peptide chain as a consequence of mutations causing changes in the sequence of base pairs of the DNA molecule which constitute the protein gene. For the 30 genetic variants identified in six types of bovine milk protein, mutations leading to deletion of amino acids occur in only two cases (α_{s1}-casein A, α_{s2}-casein D) whereas the more frequent mutations leading to amino acid substitutions occur for the other genetic variants. Determination of the primary structure of a protein is a prerequisite for pinpointing the exact location where mutation has occurred and thus resulting in genetic polymorphism. The first milk protein to be totally sequenced was the relatively small α-lactalbumin which contains 123 amino acids (Brew *et al.*, 1970). The complete amino acid sequence of the second whey protein, β-lactoglobulin, with 162 amino acids, is now known (Braunitzer & Chen, 1972; Braunitzer *et al.*, 1972) after several attempts at partial sequencing (Frank & Braunitzer, 1967; McKenzie, 1971). Determination of the amino acid sequences of different variants in the family of caseins constitutes one of the major contributions to milk protein biochemistry by the research group at Jouy-en-Josas, France. A summary of the differences in amino acid sequences giving rise to genetic variants for the milk protein is presented in Table 1. Only the positions of the altered amino acids for a particular genetic variant are shown. For a more complete sequence of each of the milk proteins, the reader is referred to either the original publications, the last review on milk proteins nomenclature (Eigel *et al.*, 1984) or appropriate chapters in this book.

4.1. Bovine α_{s1}-casein

α_{s1}-Casein was resolved into two distinct electrophoretic bands which were characterised as α_{s1}-casein for the major component and α_{s0}-casein for the faster moving minor component. Determination of the primary sequence of the two components of α_{s1}-casein (Mercier *et al.* 1971; Grosclaude *et al.*, 1973*a*) revealed identical amino acid sequences. α_{s0}-Casein has a faster electrophoretic mobility than α_{s1}-casein due to the presence of an additional phosphate group attached at the serine residue in position 41 (Manson *et al.*, 1976, 1977). Post-translational modification giving rise to a component with a different electrophoretic mobility could not be considered as genetic variation because the amino acid sequence specified by the gene is identical for both components. From the primary structure of α_{s1}-casein B, containing 199 amino acid residues, which was established by Mercier *et al.* (1971) and Grosclaude *et al.* (1973*a*), the same group of researchers were able to deduce the sites of mutation giving rise to the genetic variants A, C, D and E of α_{s1}-casein (Grosclaude *et al.*, 1972, 1976*b*). When compared to variant B, there is a deletion of a sequence of 13 amino acids, residues 14–26, for variant A. Variants C and D differ from B by having a glycine instead of a glutamic acid at position 192, and

TABLE 1
Positions and Amino Acid Differences in Genetic Variants of Milk Proteins

Protein[a]	*Variant*	*Position and amino acid in the protein*							
		14–26		53		59		192	
α_{s1}-CN (199)	A	Deleted							
	B			Ala		Gln		Glu	
	C							Gly	
	D			Thr					
	E					Lys		Gly	
		33	47		50–58		130		
α_{s2}-CN (207)	A	Glu	Ala				Thr		
	B								
	C	Gly	Thr				Ile		
	D				Deleted				
		18	35		36	37	67	106	122
β-CN (209)	A[1]						His		
	A[2]	SerP	SerP		Glu	Glu	Pro	His	Ser
	A[3]							Gln	
	B						His		Arg
	C		Ser			Lys	His		
	D	Lys							
	E				Lys				
				97		136		148	155
κ-CN (169)	A					Thr		Asp	Ser
	B			Arg		Ile		Ala	
	C			His					
	E								Gly
		45	50	59	64	78	118	129 or 130	158
β-LG (162)	A				Asp		Val		
	B	Glu	Pro	Gln	Gly	Ile	Ala	Asp	Glu
	C			His					
	D	Gln							
	E								Gly
	F		Ser					Tyr	Gly
	G					Met			Gly
			10			?			
α-LA (123)	A		Gln						
	B		Arg			Asp			
	C					Asn			

[a]Numbers in parentheses indicate the total number of amino acid residues in the protein.

a threonine instead of an alanine at position 53, respectively. The substitution of glutamine for lysine at position 59 and of glutamic acid for glycine at position 192 of α_{s1}-casein B gives α_{s1}-casein E.

4.2. Bovine α_{s2}-casein

Gel electrophoretic separation of casein revealed a group of several fast-moving protein bands ahead of the β-casein zones which were labelled as: α_{s0}-, α_{s1}-, α_{s2}-, α_{s3}-, α_{s4}-, α_{s5}- and α_{s6}-casein in order of decreasing mobility (Annan & Manson, 1969; Hoagland *et al.*, 1971). The structures of the first two, α_{s0}-casein and α_{s1}-casein were described in the previous section. When Brignon *et al.* (1976, 1977) conclusively showed that the remaining α_s-caseins had the same amino acid sequence, but different degrees of phosphorylation, it was logically proposed to collectively denote this group as α_{s2}-casein. α_{s5}-Casein is a dimer consisting of α_{s2}-casein and α_{s4}-casein, linked by a disulphide bond. The complete sequence of the 207 amino acids in α_{s2}-casein variant A was elucidated by Brignon *et al.* (1977). Variant D of α_{s2}-casein is another example of a mutation where there is a deletion of several amino acids in sequence. Here, nine amino acids are absent in one of the three following sequences: 50–58, 51–59 or 52–60 (Grosclaude *et al.*, 1978, 1979). The difference between the primary structures of variants A and C of α_{s2}-casein is due to the substitution of glutamic acid, alanine and threonine at positions 33, 47 and 130, for glycine, threonine and isoleucine, respectively (Mahé & Grosclaude, 1982). The substitution at position 130 affects the phosphorylation of serine residues at positions 129 and 131: in variant A, only serine 129 can be partially phosphorylated, whereas in variant C both serine 129 and serine 131 can be phosphorylated.

4.3. Bovine β-casein

Within β-casein, the A^2 variant was the first to be completely sequenced (Ribadeau-Dumas *et al.*, 1972). The locations along this polypeptide chain of 209 amino acids where mutations have occurred to produce several genetic variants were later elucidated by the same group of researchers (Grosclaude *et al.*, 1972, 1974*a*, *b*). Using the sequence of β-casein A^2 as a reference, variants A′, B and C differ by having a histidine instead of a proline at position 67. Variant B also differs from A^2 by having an arginine instead of a serine at position 122. β-Casein C also differs from β-casein A^2 by a glutamic acid to lysine substitution at position 37 and the absence of glutamic acid at position 37 leads to a non-phosphorylated serine at position 35 in variant C. Substitution of histidine in variant A^2 at position 106 for a glutamine leads to the rare variant A^3. Replacement of serine at position 18 of A^2 by a lysine gives variant D of β-casein. Substitution of lysine for glutamic acid in variant A^2 at position 36 leads to variant E. The primary structures of variants A′, B^2, A^4 and $A^3{}_{\text{Mongolie}}$ are not yet known.

4.4. Bovine κ-casein

The primary structures of two genetic variants, A and B, of κ-casein were established by Grosclaude *et al.* (1972*b*) and Mercier *et al.* (1973). κ-Casein A differs from the B variant by substitution of a threonoine for isoleucine at position 136 and of an aspartic acid for alanine at position 148. Both amino acid substitutions occur in the macropeptide region, residues 106 to 169, resulting from the hydrolysis by chymosin. The difference between κ-casein A and B in Zebu involves similar amino acid substitutions at positions 136 and 148, and a third substitution at position 135, isoleucine (κ-A) → threonine (κ-B) (Grosclaude *et al.*, 1974*a*). Substitution of arginine in variants B for histidine at position 97 gives variant C. At position 155, variant A has serine while variant E has glycine.

4.5. Bovine β-lactoglobulin

To date, seven genetic variants of β-lactoglobulin, which contains 162 amino acids, are known. The primary structure of β-lactoglobulin A and B was determined by Braunitzer and Chen (1972), Braunitzer *et al.* (1972) and Kolde and Braunitzer (1983). The substitution of aspartic acid at position 64 and valine at position 118 by glycine and alanine, respectively, constitute the difference between variants A and B. Substitution of glutamine at position 59 of variant B by histidine gives variant C, while replacement of glutamic acid at position 45 of variant B by a glutamine results in variant D (Brignon & Ribadeau-Dumas, 1973). Variant E, which was discovered in yak's milk (Grosclaude *et al.*, 1976*b*), differs from variant B by having glycine instead of glutamic acid at position 158. The complete amino acid sequence of β-lactoglobulins F and G, which are restricted to Bali cows (Bell *et al.*, 1981*b*), have yet to be fully established. However, it has been suggested that variants F and G, like variant E, have glycine at position 158 replacing glutamic acid in variant B. In addition, β-lactoglobulin G has methionine at position 78 instead of isoleucine as in variant B. Variant F also differs from B by substitution of serine for proline at position 50 and tyrosine for aspartic acid either at position 129 or 130. The exact locations of amino acid substitutions in β-lactoglobulin F and G have to be ascertained after the determination of the primary structures of these two variants.

4.6. Bovine α-lactalbumin

The complete amino acid sequences of α-lactalbumins A and B were determined by Brew *et al.* (1970). Variant A differs from B by a single substitution of glutamine for arginine at position 10. The exact amino acid substitutions involved that differentiate variant C, observed in *Bos javanicus* (Bell *et al.*, 1981*a*), from A or B have not yet been established. It was tentatively proposed that variant C differs from B by an amide group, most probably by a substitution of asparagine for aspartic acid.

5. Occurrence and Molecular Basis of Genetic Polymorphism in Species other than Bovine

5.1. Caprine Species

Investigations on the polymorphism of milk proteins were less extensive in the goat than in the cow. Although a limited number of studies were carried out on a small proportion of the existing breeds, a remarkable diversity of polymorphism of α_{s1}-casein in goat's milk has been discovered and analysed. Richardson and Creamer (1975) concluded from an analytical study of a group of caprine caseins that there was no protein corresponding to α_{s1}-casein. In contrast, Boulanger *et al.* (1984) not only identified goat α_{s1}-casein, but also showed that, at least in the Saanen and Alpine dairy breeds, the electrophoretic polymorphism of α_{s1}-casein was significantly associated with its quantitative variations. Grosclaude *et al.* (1987) and Mahé and Grosclaude (1989) established the existence, in those breeds, of at least seven alleles, associated with various amounts of α_{s1}-casein in the milk. While alleles α_{s1}-Cn^A, α_{s1}-Cn^B and α_{s1}-Cn^C each contributed approximately 3·6 g casein per litre of milk, α_{s1}-Cn^E contributed only 1·6 g, α_{s1}-Cn^F and α_{s1}-Cn^D 0·6 g, and α_{s1}-Cn^0 appeared to be a null allele. Quantitative variations of α_{s1}-caseins were also observed in Italian breeds (Russo *et al.*, 1977; Ciafarone & Addeo, 1984; Russo *et al.*, 1986; Addeo *et al.*, 1988; Ambrosoli *et al.*, 1988).

It is not possible to study the product of allele α_{s1}-Cn^0 because it is associated with the absence of α_{s1}-casein. In contrast, the variants corresponding to six other alleles of α_{s1}-Cn were characterized by Brignon *et al.* (1989, 1990). Because variants A and B show the closest homology with the ovine and bovine sequences, it may be assumed that the higher amount of α_{s1}-casein associated with alleles α_{s1}-Cn^A, α_{s1}-Cn^B and α_{s1}-Cn^C is the normal amount for the species, the other alleles thus being defective mutants. Brignon *et al.* (in press) suggested that the deletions observed in variants D and F were due to mutations inducing abnormal splicing of the mRNA rather than to deletions within the gene. This suggestion is strongly supported by results on nucleotide sequences obtained by Leroux and Martin (pers. comm., 1989).

As shown by Grosclaude *et al.* (1987), the amount of total casein in caprine milk is positively correlated with the amount of α_{s1}-casein. Milks possessing alleles α_{s1}-Cn^A, α_{s1}-Cn^B or α_{s1}-Cn^C contain significantly more total casein than milk without those alleles. Interestingly, the total frequency of the three normal alleles in the two investigated dairy breeds does not exceed 0·2, the defective alleles, α_{s1}-Cn^E and α_{s1}-Cn^F being far more frequent. Thus, there is a great potential for improving the casein content of milk by selection in favour of alleles α_{s1}-Cn^A, $\alpha_{s1}Cn^B$ and $\alpha_{s1}Cn^C$.

In addition to the polymorphism of α_{s1}-casein, Boulanger *et al.* (1984) described a polymorphism of α_{s2}-casein, with two alleles, $\alpha_{s2}Cn^A$ and α_{s2}-Cn^B, the former being predominant in the Alpine (0·85) and Saanen (0·87) breeds. This polymorphism is probably widely distributed (Russo *et al.*, 1986; Grosclaude F., & Mahé,

M.-F., unpublished). In general, β-casein, κ-casein, β-lactoglobulin and α-lactalbumin have been considered to be monomorphic in the caprine species. However, Boulanger (1976) observed an additional, slower β-lactoglobulin band in 5 Alpine or Saanen samples. Russo *et al.* (1986) found two bands of κ-casein in three Alpine samples. Regarding β-casein, Mácha (1981) makes reference to four alleles in the Czech white short-haired breed, but no description was given of this polymorphism. In the Italian Garganica breed, Dall'olio *et al.* (1989) found one sample with no visible β-casein band. In fact, the existence of a null β-casein allele had already been established in 1986 by Grosclaude and Mahé (unpublished results) in the Criollo breed of Guadeloupe. Two additional variants of β-casein were also observable in that population—one slower and the other faster than the normal type. Future work on different breeds may thus be expected to confirm the existence of polymorphism of β-casein.

5.2. Ovine Species

The best documented polymorphism in the sheep is that of β-lactoglobulin, discovered by Bell and McKenzie (1964) and King (1969). Two variants, β-Lg A and β-Lg B, were found in all investigated breeds, β-Lg A being normally predominant (Chiofalo & Micari, 1987). They differ by a substitution of Tyr in variant A for His in variant B at position 20 (Gaye *et al.*, 1986). Such a difference involving histidine versus a neutral amino acid explains why the polymorphism may be seen in SGE at pH 7·2 (King, 1969) but not at pH 8·6 (Chiofalo *et al.*, 1986). Owing to lack of description, a third variant, reported in the Tajik breed of USSR and denoted as β-Lg C (Aliev & Koloteva, 1975), is only putative. On the other hand, the same designation, β-Lg C, was used by Erhardt *et al.* (1989) for an additional variant observed only, but at a high frequency (0·175), in the German Merinoland breed. Because variant β-Lg C differs from β-LgA by one single substitution (Gln for Arg at position 148) allele $\beta\text{-Lg}^{C}$ may be considered as a mutant of $\beta\text{-Lg}^{A}$. Attempts to establish whether $\beta\text{-Lg}^{A}$ or $\beta\text{-Lg}^{B}$ is the original type in sheep by comparing the amino acid at position 20 in related species resulted in controversial conclusions. For example, with His at position 20, ovine β-Lg B compares to β-lactoglobulin of the wild Sardinian mouflon, while, with Tyr at the same position, ovine β-LgA compares to caprine and bovine β-lactoglobulins (Godovac-Zimmermann *et al.*, 1987). In fact, Sardinian mouflon, which is now considered as a feral population (Poplin, 1979) does not represent the original wild type of sheep. In addition, the absence of $\beta\text{-Lg}^{A}$ in this population is by no means certain. Because of its similarity with the goat and bovine proteins, and its higher frequency, $\beta\text{-Lg}^{A}$ is probably the original form of ovine β-lactoglobulin.

Except for κ-casein, the other main proteins were also found, or suspected, to be polymorphic. In addition to the common α-lactalbumin type, α-La A, the rare α-La B variant was observed in three Italian breeds (f = 0·004–0·012) by Chiofalo and Micari (1987), and also reported in USSR (Stambekov *et al.*, 1977*a*, *b*) and Western Germany (Erhardt, 1989*a*). As early as 1966, King (1966) called attention to variations in the electrophoretic patterns of 'α-casein', which is now known to constitute α_{s1}- and α_{s2}-casein fractions, and of β-casein. To account for the observation in the Welsh Mountain, Clun Forest and Suffolk breeds of a pattern with two additional bands of 'α-casein', King (1966) assumed the existence of a variant called 'Welsh'. The occurrence of a minimum of six different electrophoretic patterns in the 'α-casein' group is confirmed by data from Arave *et al.* (1973), Di Stasio (1983), Micari *et al.* (1986), Chiofalo and Micari (1987) and Dall'olio *et al.* (1989), all of whom identified the 'Welsh' variant in various breeds, without establishing whether it was α_{s1} or α_{s2}-casein. This variant is in fact α_{s1}-casein (Mahé M.-F. & Grosclaude, F., unpublished; Mauriello *et al.*, 1990). King (1966) also showed the existence of a β-casein variant, accounting for a pattern with an additional faster-migrating band. Micari *et al.* (1986) and Chiofalo and Micari (1987) observed a β-casein pattern with an additional slower band. A genetic analysis of the polymorphism of ovine caseins remains to be undertaken.

5.3. Porcine Species

The first descriptions of polymorphism of porcine milk proteins were made even before those proteins had been identified and as a consequence, the designations used turned out to be inappropriate. Glasnak (1966, 1968*a*, *b*) reported on a polymorphism of the so-called 'β_1- and β_2-caseins', as well as of the 'X- and Z-proteins'. Likewise, Kraeling and Gerrits (1969) and Althen and Gerrits (1972) described the polymorphism of 'Whey$_1$' and 'Whey$_2$' proteins, and Gerrits *et al.* (1969) studied a casein fraction called 'Cn$_3$'. In fact, 'β_1-casein' and 'Cn$_3$' were later identified as β-casein while 'Whey$_1$' was shown to be β-lactoglobulin (Kalan *et al.*, 1971). 'Whey$_2$' and 'X' were apparently the same protein which according to Althen and Gerrits (1972) behaved electrophoretically like serum transferrin, although it was not this protein.

At present, the best characterised polymorphisms of porcine milk proteins are those of β-lactoglobulin, the A and B variants were identified at alkaline pH by Kraeling and Gerrits (1969) working at that time on 'Whey$_1$'. Variant A was predominant in Duroc pigs (0·95) and variant B in Yorkshire pigs (0·75). Because the A and B variants were shown to differ by an Ala (B)/Val (A) substitution, differences in conformation and/or ion binding resulting from compositional differences were considered to be responsible for the difference in electrophoretic mobility of the polymorphs (Kalan *et al.*, 1971). A third variant, C, was found by Bell *et al.* (1981*d*) who used SGE at pH 7·8. The overall frequency of allele β-Lg^{C} in the six breeds or crosses investigated was about 0·06. Interestingly, a quantitative variation was superimposed on this polymorphism with about two-thirds of the heterozygous animal having a much reduced amount of variant C. Variant C differs from variant A by two amino acid substitutions (Glu for Asp at position 9 and His for Gln at position 69). Recently, Erhardt and Senft (1987) and Erhardt (1989*b*) using acid PAGE confirmed the existence of three β-lactoglobulin variants —A, B and C. Their identity with the previously described A, B and C, although not rigorously established, is probable. In particular, the C variant described by Erhardt (1989*b*) cannot be distinguished from A under alkaline conditions, a behaviour expected from the amino acid substitution (His/Gln) found by Bell *et al.* (1981*d*) for their C variant.

In the case of β-casein, the A and B variants recently considered by Erhardt and Senft (1987) and Erhardt (1989*c*) in three German breeds are probably the same as the A and B variants described by Gerrits *et al.* (1969), and the B and C variants discovered earlier by Glasnak (1966, 1968*b*). Variant A (= B for Glasnak) appears to be predominant in all breeds investigated.

In addition, particular situations were observed for porcine α-lactalbumin (Schmidt and Ebner, 1972) and α_{s2}-casein. Using semi-discontinuous buffers, Bell *et al.* (1981*e*) discovered two variants of α-lactalbumin, called A and B. Variant A was found to differ from variant B by a substitution of Arg for Lys. This particular difference explains why this polymorphism cannot be detected under the usual conditions. The polymorphism of α_{s2}-casein discovered by Erhardt (1989*c*) consists of a fast migrating band, in the presence or absence of a minor satellite. The present authors suggest that the latter band is the consequence of the incomplete saturation of a phosphorylation site of the protein, as in the case of bovine α_{s0}- and α_{s1}-caseins. The absence of this minor band would thus be due to a straightforward non-phosphorylation of the site, induced by an amino acid substitution affecting the coding structure proposed by Mercier (1981). The designation α_{s2}-Cn^0, which suggests the existence of a null allele would, under this hypothesis, be inappropriate.

5.4. Equine Species

The most clearly established polymorphism of equine milk protein is that of a minor whey protein (Bell *et al.*, 1970*b*), called protein of zone 1 by Bell *et al.* (1981*a*) and Whey$_1$ protein by Kingsbury and Gaunt (1977). The first group distinguished five variants (A, B, C, D and E), while the second group showed four variants (A, B, C and D). Bell *et al.* (1981*a*) were unable to identify this protein as a known whey protein from its molecular weight and amino acid composition. Using alkaline starch gel electrophoresis, Mácha and Novackova (1975) observed, in Czechoslovakian breeds, polymorphisms of 'α-casein', β-casein and unidentified 'X3' protein, with, in each case, three alleles called A, B and C. They also found, at acid pH, two variants of β-lactoglobulin, called A and B. In contrast, Chiofalo *et al.* (1983) observed variations only in the patterns of one of the casein zones among Italian breeds.

From the colostrum of a single mare, Godovac-Zimmermann *et al.* (1985) isolated two forms of β-lactoglobulin, called I and II. Likewise, they also discovered (Godovac-Zimmermann *et al.*, 1987) two forms of α-lactalbumin, called B and C, which differed in their amino acid sequence from the A type studied by Kaminogawa *et al.* (1984). It is not clear whether those forms are true genetic variants. According to the amino acid sequence data published by Godovac-Zimmermann *et al.* (1985, 1987), the forms of both proteins differ in their net charge at alkaline pH and should therefore be distinguishable by electrophoresis. However, the indications for heterogeneity of α-lactalbumin or β-lactoglobulin after electrophoresis at alkaline pH were apparently limited to a single mare having two α-lactalbumin bands (Kingsbury & Gaunt, 1977).

5.5. Other Species

Only fragmentary data are available for genetic polymorphism of milk proteins in other species. In water buffalo (*Bubalus bubalis*), α_{s2}-casein (Addeo *et al.*, 1977) and κ-casein (Chiofalo *et al.*, 1982*a*) were suggested to be polymorphic. In a more recent publication, Chiofalo *et al.* (1984) reported that α-lactalbumin, β-lactoglobulin and serum albumin were monomorphic; α_{s1}-casein and β-casein were dimorphic. From a population of 30 buffaloes, 24 had three electrophoretic bands (B, C and D), while the remainder had four bands (A, B, C and D).

Analysis of casein and whey protein fractions from Somalian camels' milk by Di Stasio *et al.* (1983) revealed the occurrence of polymorphism in the β-lactoglobulin, but not in the other proteins. Two variants, A and B, of α-lactalbumin were detected by Conti *et al.* (1985) in Somalian camel (*Camelus dromedarius*) milk. These two forms of α-lactalbumin had similar mobility in PAGE at pH 8·3, but differed in pI due to a difference in amino acid composition at the N terminal.

Voglino *et al.* (1975) reported two types of β-casein in human milk. They suggested that substitution of a charged amino acid in the protein resulted in the difference in electrophoretic mobility observed in acidic buffer. Casein samples isolated from milk collected from different women populations (Turin, Sardinian, Kikuyu) showed polymorphism in the α- and β-casein systems (Voglino & Ponzone, 1972; Ponzone *et al.*, 1975; Ponzone & Voglino, 1976). Amino acid sequence analysis (Greenberg *et al.*, 1984) indicated the existence of genetic polymorphism involving uncharged residues in human β-casein.

Polymorphisms of milk proteins have also been identified in the milks of laboratory animals. For example, rat milk contains two variants of α-lactalbumin (McKenzie & Larson, 1978). Variants have been reported in the caseins and whey proteins of mouse milk (Piletz & Ganschow, 1981). Analysis of feline milk protein (Halliday *et al.*, 1990) revealed the presence of β-lactoglobulin-like proteins with complex electrophoretic patterns resulting from three distinct loci. A range of two to five variants are expressed by each locus.

6. Linkage Between Milk Protein Genes

One of the important features in milk protein genetics is the close linkage of the genes encoding the four caseins, a situation that was discovered and analysed in cattle. Grosclaude *et al.* (1964) presented the first evidence of a close linkage between loci α_{s1}-Cn and β-Cn. Soon afterwards, having established the genetic determinism of the κ-caseins, they also concluded the linkage of locus κ-Cn with α_{s1}-Cn and β-Cn (Grosclaude *et al.*, 1965). Those results were confirmed by data of Larsen and Thymann (1966*a*), who then introduced the concept 'pleiotropy' instead of linkage, an interpretation that raised controversy (see the discussion of

TABLE 2
Number of Informative Dam–Daughter Pairs Collected for Analysing Linkage between Loci α_{s1}-Casein, β-Casein, α_{s2}-Casein. No Recombinant was Found—a Result Indicating Very Close linkage

Linkage investigated	*No. of informative dam–daughter pairs*	*References*
α_{s1}-Casein and β-casein	78	Grosclaude *et al.* (1964)
	45	Larsen & Thymann (1966)
	25	Grosclaude (unpublished)
α_{s1}- or β-Casein and κ-Casein	108	Grosclaude *et al.* (1965)
	143	Larsen & Thymann (1966)
α_{21}-, β-, κ-Casein and α_{s2}-casein	49	Grosclaude *et al.* (1978)

the communication by Larsen and Thymann (1966*b*)) and was indeed wrong. With regard to α_{s2}-Cn, no genetic study was possible until polymorphism was found. This was achieved by Grosclaude *et al.* (1978) who showed that locus α_{s2}-Cn was also linked to the three other loci.

Table 2 recapitulates the data gathered by Grosclaude and coworkers and by Larsen and Thymann. As can be seen, no recombination was observed over 448 possibilities within the casein gene cluster a result which gives evidence of the close linkage between the members of this cluster. Hines *et al.* (1969), as well as Macha and Mullerova (1968, 1969), produced data that would indicate, on the contrary, a rather frequent occurrence of recombination. Grosclaude *et al.* (1973) suggested that these authors probably used bovine families with the wrong progenies. In other species, linkage of casein genes was found whenever it was investigated; loci α_{s1}-Cn and α_{s2}-Cn are linked in the goat (Grosclaude *et al.*, 1987) and loci α_{s1}-Cn, α_{s2}-Cn, β-Cn and κ-Cn in the sheep (Lévéziel *et al.*, 1991). In contrast, the locus of β-lactoglobulin is genetically independent of the casein cluster in the bovine species (Larsen, 1971). In addition, population data suggest that loci of β-lactoglobulin and α-lactalbumin are not closely linked (Grosclaude *et al.*, 1974*a*; Mariani & Leoni, 1982).

The non-independent inheritance of casein alleles has observable consequences in the genotypic structure of populations. King *et al.* (1965) were the first to report the apparent absence of particular combinations of α_{s1}- and β-casein genotypes. In the Jersey breed, for example, the expected genotypes comprising α_{s1}-casein BC and β-casein B, as well as α_{s1}-casein C and β-casein AB or B, were missing. The authors postulated the following 'chromosomes', or 'combinations': α_{s1}-Cn^{B}, β-Cn^{A}; α_{s1}Cn^{B}, β-Cn^{B} and α_{s1}-Cn^{C}, β-Cn^{A}. Since the α_{s1}-Cn^{B}, β-Cn^{A} combination prevailed in all examined breeds, the other combinations were assumed to be the result of mutations in either α_{s1}-Cn or β-Cn. The combination α_{s1}-Cn^{C}, β-Cn^{B} was considered as a recombinant one, and its apparent absence was interpreted as a consequence of a very close linkage between loci α_{s1}-Cn and β-Cn. Such an interpretation, dividing the combinations into an original type, mutant derivatives

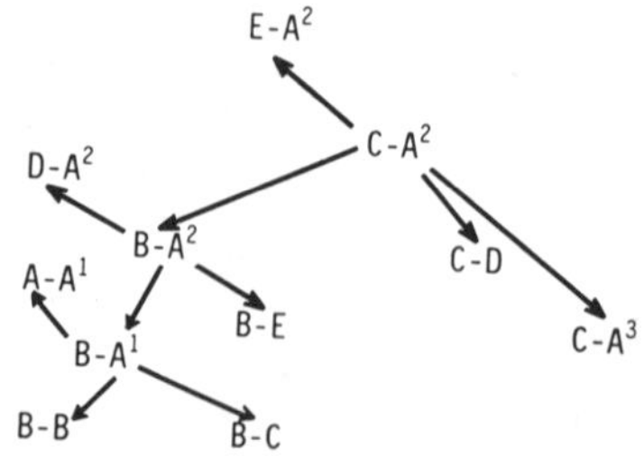

Fig. 2. Phylogeny of the alleles of the α_{s1}-casein, β-casein cluster. The α_{s1}-casein C–β-casein A^2 allele (C–A^2) in the figure may be considered as the original one from which the others have derived by successive mutations either at the α_{s1}-casein or at the β-casein locus.

and recombinant types, was developed independently by Grosclaude *et al.* (1966*b*, 1973*b*, 1974*a*) and Grosclaude (1974) who positioned the additional variants as they were discovered.

In the early interpretations of King *et al.* (1965) and Grosclaude *et al.* (1966*b*), the combination α_{s1}-Cn^B, β-Cn^A—later α_{s1}-Cn^B, β-Cn^{A2}—was considered as the original type because it prevailed in all western European breeds (*Bos taurus*). However, because the combination α_{s1}-Cn^C, β-Cn^{A2} strongly predominates in zebu (*Bos indicus*) and yak (*Bos grunniens*) populations (α_{s1}-Cn^B, β-Cn^{A2} being absent even in the yak), Grosclaude (1979) and Grosclaude *et al.* (1982) suggested this combination to be actually the original type. Their conclusion is substantiated by the fact that the amino acid residue found in the caprine as well as in the ovine α_{s1}-casein sequence at the position homologous to bovine 192 is Gly as in α_{s2}-CnC, and not Glu, as in α_{s1}-CnB (Brignon *et al.*, 1989). Figure 2 shows the postulated relationships between the original α_{s1}-Cn^C, β-Cn^{A2} combination and its mutant derivatives. The term 'haplotype', coined in human histocompatibility studies, will be used in place of 'combination' to designate a combination of alleles of closely linked genes.

The non-independent occurrence of casein genotypes in populations is an example of the phenomenon termed 'linkage disequilibrium'. Statistical tests may be applied to estimate the strength of linkage disequilibrium between alleles of two loci. In the case of bovine caseins, such tests, carried out in 32 breeds, were summarized by Graml *et al.* (1986*b*). On the whole, linkage disequilibrium between alleles of the α_{s1}-Cn and β-Cn loci is statistically significant in 18 of 32 breeds, and, except in one case, always of the same sign. This means that the same haplotypes are either present or absent in nearly all breeds. This is not the case when alleles of either α_{s1}-Cn and κ-Cn, or β-Cn and κ-Cn, are being considered. However, linkage disequilibrium between alleles of β-Cn and κ-Cn is significant in 16 of 32 breeds; between alleles of α_{s1}-Cn and κ-Cn, it is only significant in 11 of 32 breeds. These results confirm the early suggestion by Grosclaude *et al.* (1965) that the strongest linkage disequilibrium occurred between alleles of the α_{s1}-Cn and β-Cn loci. Based on these observations, Grosclaude (1979) proposed a model in which the β-Cn and α-$_{s1}$-Cn loci were very close to each other, in contrast to the κ-Cn locus which was further apart. Such a model is now confirmed (Ferretti *et al.*, 1990) and moreover, it is known that the α_{s2}-locus is situated between the β-Cn and κ-Cn loci. Because of such a close linkage between casein loci it is logical to consider, as units of inheritance, not only the alleles of the individual loci, but also the haplotypes. This is particularly true for the haplotypes of the α_{s1}-Cn, β-Cn pair of loci, which are so closely linked. Their occurrence and frequencies in the breeds will be discussed in the next section.

7. Frequency Distribution of Genetic Variants in Cattle Species

Data available on the occurrence and frequencies of milk protein variants in the three main species of the genus *Bos*—*Bos taurus* (taurines), *Bos indicus* (zebus) and *Bos grunniens* (yaks)—, although numerous, are far from being complete. First, the proportion of investigated populations is higher for taurines than for zebus or yaks, but, even for taurines, important gaps exist. In Europe, for example, while Italian and French breeds have been extensively investigated, nothing is known on native Spanish and Portuguese cattle. Moreover, breeds in Switzerland and Great Britain, two countries that have played an important role in modern cattle history, were only poorly investigated in their native areas. Fortunately, due to their international distribution, they were studied outside their settlements in Europe and elsewhere.

Other reasons contributing to data heterogeneity include differences in the electrophoretic techniques used in the course of time or even contemporarily, and also in the degree of attention given to particular polymorphisms. As discussed earlier, the last point is well exemplified by the polymorphism of α_{s2}-casein, which is certainly less restricted than the literature would indicate.

When considering data about breed distribution of milk protein variants, it is of interest to know, for each locus, the relative appearance in time of different alleles, a process which can be represented by a phylogenetic tree. This implies the identification of the original variant, which can be based on two different criteria. First, the original variant should normally give the closest sequence alignment with the same protein or related species. Second, the corresponding allele may be expected to have, usually, the highest frequency over a large number of populations. Although this rule cannot be considered as absolute (Watterson & Guess, 1977) the availability of population data in three different species of the genus *Bos* makes this criterion fairly convincing in the particular case of milk protein polymorphism. Finally, when trying to infer conclusions from population data on protein polymorphism, attention should be paid to the risk of cryptic heterogeneity. This does not refer to the well-known existence of 'silent variants', which cannot be distinguished by electrophoresis because their differences in structure have no effect on the net charge of the protein. It concerns the possibility that different amino acid substitutions, resulting from distinct mutations, have the same effect on the net charge of the protein, and thus give rise to variants indistinguishable by electrophoresis. Without a biochemical analysis, those variants would thus be considered as one and the same. As discussed earlier, variant β-CnA[3] provides an example of such heterogeneity. In the same way, the common type of β-lactoglobulin of the Yak and β-Lg D of European breeds are identical on electrophoresis although, compared to β-Lg B, they have different amino acid substitutions. Other similar situations may not have been yet discovered.

7.1. α-Lactalbumin

α-Lactalbumin was found to be polymorphic in all Zebu populations investigated so far, allele α-La^{B} being always predominant over α-La^{A}. As already pointed out by Aschaffenburg (1968), the frequency of α-La^{A} appears to be higher in Indian (0·22 to 0·40) than in African (0·03 to 0·15) breeds. In Zebu, the most extensive survey was carried out in Madagascar where, in a sample of 778 cows, the frequency of α-La^{A} was 0·166, with no regional heterogeneity (Grosclaude *et al.*, 1974*a*). As far as *Bos taurus* is concerned, α-lactalbumin appears to be monomorphic in all northern European breeds, while allele α-La^{A} is clearly present in southern Europe (Fig. 3). In this area, systematic investigations were carried out only in Italy where as many as 12 breeds were found to be polymorphic. Those breeds are essentially Podolic, or breeds known to have been crossed with Podolic cattle (Bettini & Masina, 1972; Mariani & Russo, 1977; Chianese *et al.*, 1988). This is also the case for the few polymorphic Romanian and Russian breeds. Complementary investigations are desirable to establish a more precise localisation of the region for this allele. A verification of the biochemical uniqueness of variant α-La A found in different locations would also be important. The variant of the north-eastern France Vosgienne breed, for example, may possibly be different from the Zebu variant (Grosclaude F., & Mahé, M.-F., unpublished). The frequency criterion unequivocally designates α-La B as the original type of the genus *Bos*, but curiously, when amino acid residues at position 10 of α-lactalbumins are compared, variant B, with Arg, is identical with α-lactalbumins of the Italian water buffalo, *Bubalus arnee*, (Addeo *et al.*, 1976) and of the Banteng, *Bos (Bibos) javanicus* (Bell *et al.*, 1981*a*) while variant A, with Gln, is identical with goat and sheep proteins (McGillivray *et al.*, 1979; Gaye *et al.*, 1987). Whether this puzzling situation is only a consequence of chance remains to be established.

7.2. β-Lactoglobulin

The two common variants of β-lactoglobulin, β-Lg A and B, are probably universally distributed in bovine and zebu populations; β-Lg B is predominant in all Zebu, as well as in about 75% of all taurine breeds so far investigated. Because this variant also gives the best alignment with β-lactoglobulins of other species (Godovac-Zimmermann & Braunitzer, 1987), it is clearly the original type for *Bos taurus* and *Bos indicus*. The Yak has its own β-lactoglobulin type, β-Lg D_{yak} (β-Lg E according to Bell *et al.*, 1981*b*) which, interestingly, appears to be derived from β-Lg B (Grosclaude *et al.*, 1976*b*). Discovered by Bell (1962) in Australian Jersey,

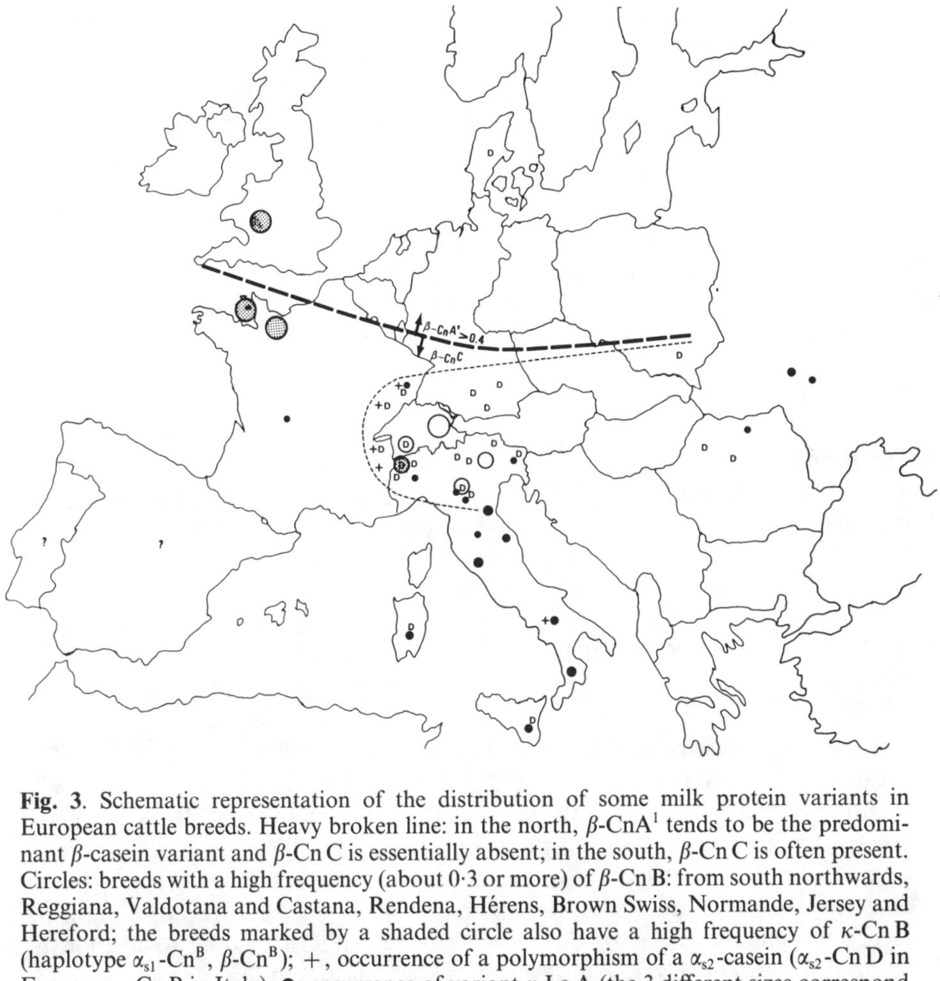

Fig. 3. Schematic representation of the distribution of some milk protein variants in European cattle breeds. Heavy broken line: in the north, β-CnA[1] tends to be the predominant β-casein variant and β-Cn C is essentially absent; in the south, β-Cn C is often present. Circles: breeds with a high frequency (about 0·3 or more) of β-Cn B: from south northwards, Reggiana, Valdotana and Castana, Rendena, Hérens, Brown Swiss, Normande, Jersey and Hereford; the breeds marked by a shaded circle also have a high frequency of κ-Cn B (haplotype α_{s1}-Cn^{B}, β-Cn^{B}); +, occurrence of a polymorphism of a α_{s2}-casein (α_{s2}-Cn D in France, α_{s2}-Cn B in Italy); ●: occurrence of variant α-La A (the 3 different sizes correspond to the following frequencies: less than 0·05, between 0·05 and 0·10, more than 0·10 and up to 0·23); D: occurrence of β-Lg D; the main extension area is surrounded by a light dotted line.

variant C is present in various settlements of that breed (Aschaffenburg, 1968; Bech & Kristiansen, 1990). Its occurrence in other populations remains uncertain because, as already pointed out by Aschaffenburg (1968), a number of techniques used for β-lactoglobulin phenotyping fail to detect this variant. This is well exemplified by the case of the Normande, a breed having probable connections with the Jersey, where β-Lg C was only recently identified (Grosclaude, F. & Mahé, M.-F., unpublished) using isoelectric focusing instead of conventional techniques. On the other hand, the β-lactoglobulin C reported in a few central European breeds is probably β-Lg D. This is at least the case for the data of Seibert *et al.* (1985), as confirmed by Seibert (pers. comm. 1989). First observed in the Montbéliarde, a breed of French Jura (Grosclaude *et al.*, 1966*a*), β-Lg D is usually considered as a rare variant (Baker & Manwell, 1980). However, although its frequency never exceeds 0·04, this variant occurs (Fig. 3) in most breeds of the Alpes, Vosges and Jura mountains, like Simmental, Pinzgau, Brown Swiss, Tarine, Vosgienne and others (Russo & Mariani, 1978; Merlin & Di Stasio, 1982; Raimondi, 1982; Felenczak, 1983; Graml *et al.*, 1984*a*, *b*; Buchberger *et al.*, 1986; Grosclaude, 1988). Its occurrence was also mentioned outside of this area, e.g. in Danish Jersey (Larsen & Thymann, 1966*a*, *b*) and in the Sicilian Modicana and in Sardinian cattle (Raimondi, 1982). This may be due to migrations and/or to cryptic heterogeneity. As a matter of fact, any new mutation of the common β-Lg^{B} allele that induces the loss of one charge in the protein would produce a D type variant.

7.3. α_{s1}-Casein

The B and C variants are present in all *Bos taurus* and *Bos indicus* populations investigated so far, but, while the variant C is most frequent in all Zebu breeds, variant B was always found to be predominant in taurine breeds, except in two cases, the Jersey on the Isle of Jersey (Larsen *et al.*, 1974) and the Italian Calabrian (Bettini & Masina, 1972), a Podolic breed included by Baker and Manwell (1980) in the *Primigenius* group. As a matter of fact, frequencies of allele α_{s1}-Cn^{C} higher than 0·25 are in the Channel Island (Jersey, Guernsey, Normande) and in the *Primigenius* group. Because of the reversed frequencies of alleles α_{s1}-Cn^{B} and α_{s1}-Cn^{C} in *Bos taurus* as compared to *Bos indicus*, the identification of the original allele at locus α_{s1}-Cn was not unmistakable (Grosclaude *et al.*, 1966*b*, 1973). Because α_{s1}-Cn^{C} is predominant in the Yak, which apparently lacks α_{s1}-Cn^{B}, α_{s1}-Cn^{C} was finally retained as the original allele (Grosclaude, 1979). This conclusion is supported by the fact that variant α_{s1}-Cn C gives the closest alignment with α_{s1}-caseins of other, related species, such as sheep and goat (Brignon *et al.*, 1989). Thus, while in yaks and zebus the original allele α_{s1}-Cn^{C} still prevails, in taurine breeds the mutant allele, α_{s1}-Cn^{B}, predominates, almost tending to fixation in North European breeds. In contrast to B and C, the A and D variants, which are derived from B and are thus more recent, are far less common. Variant A is very rare, both in terms of breed distribution and of frequencies within breeds. Originally identified in the American Holstein (Thompson *et al.*, 1962), it appeared to be confined to a single blood line originating in the state of Michigan (Aschaffenburg, 1968). However, later on, it was found in three Red Danish cows (Larsen & Thymann, 1966*a*,*b*). Aschaffenburg (1968) suggested that α_{s1}-Cn A represents a comparatively recent mutation. Because of the occurrence of this variant in both Holstein and Danish Red, Farrell and Thompson (1971) suggested a more ancient ancestry, while Baker and Manwell (1980) suggested early crosses between those two breeds in the USA. Variant A differs from variant B by the deletion of residues 14 to 26 (Grosclaude *et al.*, 1970). This deletion could be the result of abnormal splicing of the mRNA sequence corresponding to one exon of the gene, a mechanism already suggested by Brignon *et al.* (in press) for the three other deletions occurring in caseins (caprine α_{s1}-Cn D and F, bovine α_{s2}-Cn D). The unexpected isolation of a cDNA clone for variant α_{s1}-Cn A from mammary tissue of a cow homozygous for allele α_{s1}-Cn^B (McKnight *et al.*, 1989) suggests this sequence as being prone to instability. In this situation the possibility of independent, recurrent-like mutations, giving rise to the same α_{s1}-Cn A variant, should not be categorically rejected. Variant D, first observed in the Flamande, a Red breed of Northern France (Grosclaude *et al.*, 1966*a*), is usually considered as a rare variant. However, although it occurs at low frequencies (0·002 to 0·08) it has now been identified in a total of 21 French, Dutch, Italian, Polish and Canadian breeds belonging to different phylogenetic groups (Corradini, 1969; Michalak, 1969; Russo & Mariani, 1978; Merlin & Di Stasio, 1982; Raimondi, 1982; Di Stasio & Dupont, 1983; Chianese *et al.*, 1988; Grosclaude, 1988).

7.4. α_{s2}-Casein

While the existence of genetic polymorphism of the three other caseins was established within a short period (1961–1965), the first example of polymorphism of the somewhat neglected α_{s2}-casein was discovered much later (Grosclaude *et al.*, 1976*a*) in Napalese yaks and crosses between yaks and local taurines or zebus. Besides the common type, α_{s2}-Cn A, two variants were suggested to occur, α_{s2}-Cn B and α_{s2}-Cn C, the latter apparently specific to the Yak. Later, the same variants, α_{s2}-Cn B and C, were observed in Mongolian yaks with frequencies of 0·09 and 0·11, respectively (Grosclaude *et al.*, 1982). This confirmed that α_{s2}-casein was normally polymorphic in the Yak. In the meantime, the same group had discovered polymorphism of α_{s2}-casein in two breeds of eastern France, the Vosgienne and the Montbéliarde. Both shared an additional variant, α_{s2}-Cn D, with a frequency of only 0·01 in the Montbéliarde, but 0·09 in the Vosgienne (Grosclaude *et al.*, 1978). Later, this variant was also found in two French alpine breeds, the Abondance and the Tarine (Nuyts & Mahé, unpublished). It thus emerges that, in France, variant α_{s2}-Cn D occurs in the same breed groups as β-Lg ;D (Fig. 3). It would be surprising not to find this variant in related neighbour breeds, perhaps in the whole β-Lg D area; a search with appropriate techniques would be desirable. Recently, Chianese *et al.* (1988) reported on the occurrence of α_{s2}-Cn B at low frequency (0·004) in the Italian Podolic breed. Again, a search for this allele in other breeds, especially those of the podolic group, but also in Zebu, would certainly be worthwhile. One should also mention here the unexplained absence of α_{s2}-caseins, apparently balanced by more κ-casein, in the milks of 4 Red Danish cows in Poland (Michalak, 1969), which could possibly be due to the existence of a null α_{s2}casein allele. On the whole, it can be assumed that much of the polymorphism of α_{s2}-casein in cattle still remains to be discovered.

7.5. β-Casein

Subtyping of 'β-casin A' into A^1, A^2 and A^3 by acid gel electrophoresis (Peterson & Kopfler, 1966) was performed in less than half of the published breed investigations. Despite this, there is enough evidence to conclude that β-CnA^1 and

β-CnA^2 are present in all *Bos taurus* and *Bos indicus* populations, while *Bos grunniens* may only have β-CnA^2 (Grosclaude *et al.*, 1982). Variant A^2 is predominant in all zebu populations investigated so far and in the majority of taurine breeds, and may thus be considered as the original type of the genus *Bos*. Among the international breeds, the Guernsey has, by far, the highest frequency of allele β-Cn^{A2} (0·8 to 0·95 according to Rausch *et al.*, 1967; Kiddy *et al.*, 1968; Zikakis *et al.*, 1974). In the present state of knowledge on European cattle, the highest concentration of breeds with a high frequency of β-CnA^2 (more than 0·7) is found in the group of essentially beef breeds of central and southern France, including the Charolais, Limousin, Blonde d'Aquitaine, Salers, Aubrac, Ferrandaise (Grosclaude, 1988). On the other hand, variant β-CnA^1 occurs at high frequencies (0·4 to 0·7) in breeds of North European location or origin (Fig. 3) like the Ayrshire (Kiddy *et al.*, 1968; Li & Gaunt, 1972; Lin *et al.*, 1986), the British Shorthorn (Grosclaude, F. & Mahé, M.-F., unpublished), Black and White Lowland breeds including the North-American Holstein (Kiddy *et al.*, 1968; Szmelik *et al.*, 1971; Li & Gaunt, 1972; McLean *et al.*, 1984; Ng-Kwai-Hang *et al.*, 1984; Bech & Kristiansen, 1990), European Red breeds such as Danish Red, Polish Red and the French Flamande (Szmelik *et al.*, 1971; Grosclaude, 1988; Bech & Kristiansen, 1990), but also in a few Italian breeds such as the Chianina and the Burlina (Merlin & Di Stasio, 1982; Raimondi, 1982) and in the Japanese Brown (Abe *et al.*, 1975; Komatsu & Abe, 1980). The notation B_Z was introduced for the B variant of Zebu (Aschaffenburg *et al.*, 1968) following the finding of differences in composition and chymotryptic peptide maps between β-caseins B prepared from the milk of only one Indian Zebu and one Jersey cow (see also Thompson *et al.* (1969*b*) and Thompson (1971)). Although the occurrence of silent differences within the B variant is of interest, the introduction of B_Z was an error, and will not be retained here. The reasons are, first, that extrapolation from one single cow to the whole of *Bos indicus* is not justified and, more fundamentally, that silent polymorphisms should not be taken into consideration because, precisely, their occurrence and frequencies cannot be evaluated at the level of populations. Variant B, derived from A^1 and thus less ancient than A^2 and A^1, was found in nearly all *Bos taurus* and *Bos indicus* populations in the world. Considering only breed samples of more than 50 animals, the few exceptions were the Italian Calabrian (71 cows), the Japanese Brown (231) and the Madagascar Zebu (586) (Bettini & Masina, 1972; Grosclaude *et al.*, 1974*a*; Komatsu & Abe, 1980). The observation on Madagascar Zebu may be significant because of the fairly large sample size. In a majority of breeds, the frequency of allele β-Cn^B was less than 0·1. On the other hand, the highest frequencies were observed (Fig. 3) in two distinct breed groups in Europe. The first group includes Brown Swiss and Brown Swiss-related European Brown breeds (0·2 to 0·4 according to: Corradini, 1969; Horvath & Mészaros, 1972; Russo & Mariani, 1978; Raimondi, 1982; Samarieanu *et al.*, 1986; Graml *et al.*, 1984*b*; Grosclaude, 1988) as well as several rare Swiss and Italian breeds located close to the South of the Brown Swiss area; more particularly, Valdotana Black-Pied (0·4), Hérens and Castana (0·3), Reggiano and Rendena (0·25) (Russo & Mariani, 1978; Merlin & Di Stasio, 1982; Raimondi, 1982; Bolla *et al.*, 1983; Di

Stasio & Dupont, 1983; Pagnacco & Caroli, 1987). The second group located in north-western Europe, includes the various Jersey populations (approximately 0·3 according to: Michalak, 1969; Szmelik *et al.*, 1971; Li & Gaunt, 1972; Larsen *et al.*, 1974; McLean *et al.*, 1984; Bech & Kristiansen, 1990), the French Normande (0·47 according to Grosclaude, 1974) which is the continental breed closest to the Jersey, and the Hereford (0·25 to 0·45 in rather small samples studied by: Aschaffenburg, 1968; Caldwell *et al.*, 1971; Komatsu & Abe, 1980). Variant C, also derived from A^1, has not been so far observed in Zebu, and thus seems to be restricted to *Bos taurus*. Until the yet unexplored breeds are investigated, it may be considered as essentially absent from North European breeds. On the other hand, it is normally present in Mid- and Southern European populations (Fig. 3). It was also found in Mongolian cattle (Grosclaude *et al.*, 1965), in Korean cattle (Han *et al.*, 1983) and in the Egyptian Baladi cattle (Graml *et al.*, 1986*c*). While its breed frequency is often less than 0·05, the highest values (0·10–0·23) were found, from east–westwards, in Polish Simmental (Michalak, 1969), in Hungarian Simmental and Spotted breeds (Horvath & Mészaros, 1972), in the Italian Chianina (Bettini & Masina, 1972; Russo & Mariani, 1974), in the French Tarine or Italian Tarina (Raimondi, 1982; Grosclaude, 1988) and in the Bretonne Pie-Noire, in extreme west France (Grosclaude, 1988). Whenever present, variant A^3, derived from A^2, has a frequency less than 0·04. According to Baker and Manwell (1980), this variant is possibly indicative of some gene flow from Pied Lowlands into the Channel Island breeds. This variant indeed occurs in North European cattle (Ayrshire, Shorthorn, Holstein–Friesian) as well as in the Guernsey and its continental neighbour, the Normande (Grosclaude *et al.*, 1966*b*; Rausch *et al.*, 1967; Kiddy *et al.*, 1968; Li & Gaunt, 1972; Zikakis *et al.*, 1974; Hines *et al.*, 1977; Grosclaude, F. and Mahé, M.-F., unpublished). However, its occurrence was also reported in Polish Jersey (Szmelik *et al.*, 1971), German Simmental (Seibert *et al.*, 1987), Italian Burlina, Grigio Alpina, Piemontese, and Rendena (Merlin & Di Stasio, 1982; Raimondi, 1982), as well as in Mongolian (Grosclaude *et al.*, 1982) and Korean native cattle (Han *et al.*, 1984). In fact, variant A^3 provides an example of cryptic heterogeneity because the A^3 variant of Mongolian cattle was shown to be different from the A^3 variant characterised from a Normande milk. Interestingly, the difference was suspected because A^3 was not in the same haplotype in the two populations (Grosclaude *et al.*, 1982). Further investigations are needed to appreciate the extent of this phenomenon.

7.6. κ-Casein

The two common variants of κ-casein, A and B, were found in all investigated populations, including those of yaks (Grosclaude, *et al.*, 1976*b*, 1982). Variant A, which is predominant in a large majority of them, may thus be considered as the original type on the basis of the frequency criterion. On the other hand, sequence data (Mercier *et al.*, 1976) revealed a puzzling situation; while at position 136, variant A is identical with κ-caseins of other species (water buffalo, goat, sheep, pig, man), all having Thr, on the contrary, at position 148, it is variant B that is identical with the same proteins. Mercier *et al.* (1976) suggested that the original κ-casein type of the genus *Bos* was in fact an intermediate between A and B, having Thr at position 136 like A, and Ala at position 148 like B. On electrophoresis, this hypothetical ancestral variant would migrate like B. The reasons for the divergence in the conclusions of the frequency and of the sequence criteria are unclear. A possible conciliation would be to suggest that an ancestral form of type B, as proposed by Mercier *et al.* (1976), was very early supplanted by a mutant of type A. Until further developments, the present authors are inclined to consider variant A as the original type, thus favouring the frequency criterion. Curiously, as already noted by Baker and Manwell (1980), the highest frequencies of κ-Cn^{A} may be observed in two quite distant groups, e.g. *Bos Indicus*, on one side, and on the other side, breeds of Northern Europe, the same also having high frequencies of α_{s1}-Cn^{B} and β-Cn$^{A^1}$ (Ayrshire, Shorthorn, Holstein-Friesian, Flamande, Danish Red, etc.). On the other hand, the highest values for κ-Cn^{B} are found in the various Jersey extensions (usually 0·7 or more, and up to 0·95 on the Isle of Jersey according to Larsen *et al.* (1974)) and in its continental neighbour, the Normande (0·66 in 1972, according to Grosclaude (1974); increased to 0·74, according to Balland (1987), but also in several breeds of the Italian podolic group (Calabrese, Maremmana, Marchigiana, Romagnola, Chianina and Modicana) which have frequencies close to 0·7 or higher (Russo & Mariani, 1978). The distribution of the more recently discovered variants remains to be established.

7.7. Overall considerations for frequency distribution

According to Epstein (1971), all present humpless and humped cattle have a common origin, in south-west Asia, where their early ancestors were selected from domestic cattle derived from the wild auroch, *Bos primigenius*. The existence of sets of variants shared by *Bos taurus* and *Bos indicus* (α-La A and B, β-Lg A and B, α_{s1}-Cn B and C, β-Cn A^1, A^2 and possibly B, κ-Cn A and B) is consistent with the hypothesis of a common origin, and suggests that the corresponding mutations occurred before the separation of these two branches of the genus *Bos*. Note that Grosclaude *et al.* (1974*a*) ascertained that the amino acid substitutions between α_{s1}-C and B, β-Cn A^1, A^2 and B, as well as κ-Cn A and B, were the same in Zebu as in western humpless cattle. As discussed before, the original variants were α-La B, β-Lg B, α_{s1}-Cn C, α_{s2}-Cn A, β-Cn A^2 and possibly κ-Cn A. Interestingly, all those original variants are still predominant in *Bos indicus* while important shifts have occurred in *Bos taurus*, especially at the α_{s1}-Cn locus, but also at the β-Cn locus. Those shifts are best illustrated by taking into account the frequencies of haplotypes formed by the alleles of this pair of very closely linked loci (Grosclaude *et al.*, 1966*b*, 1973). Table 3 gives the frequencies of such haplotypes in breeds representative of different phylogenetic groups. Clearly, Zebu has retained a high frequency of the original haplotype CA^2 (for α_{s1}-Cn^{C}, β-Cn$^{A^2}$) while North European breeds have nearly lost this haplotype, which is replaced, by BA^2, and above all, BA^1; the latter reaches very high frequencies. Note that in this group, the rare α_{s1}-Cn^{C} allele tends to be associated with the rare β-Cn$^{A^3}$.

TABLE 3
Frequency Distribution of α_{s1}-Cn, β-Cn Haplotypes for Different Breeds in Various Countries

Breeds and Country	*α_{s1}-Cn, β-Cn Haplotypes*								*References*
	AA^1	*BA^1*	*BA^2*	*BB*	*BC*	*CA^1*	*CA^2*	*CA^3*	
Zebu Madasgascar	—	0·027	0·139	—	—	0·072	0·762	—	Grosclaude *et al.* (1974*a*)
Ayshire Canada	—	0·554	0·440	0·003	—	—	—	0·003	Lin *et al.* (1986)
Shorthorn Great Britain	—	0·671	0·321	0·004	—	—	—	0·004	Grosclaude & Mahé (1984, unpublished)
Holstein Canada	0·003	0·561	0·405	0·007	—	—	0·016	0·011	Ng-Kwai-Hang *et al.* (1984)
Black & White Denmark	—	0·55	0·39	0·03	—	—	< 0·01	0·03	Bech & Kristiansen (1990)
Red Danish Denmark	< 0·01	0·71	0·23	0·06	—	—	< 0·01	—	Bech & Kristiansen (1990)
Normande France	—	0·19	0·11	0·47	0·01	—	0·18	0·04	Grosclaude (1988)
Jersey USA	—	0·09	0·28	0·37	0·03	—	0·24	—	Li & Gaunt (1972)
Brown Swiss Italy	—	0·28	0·42	0·30	0·02	< 0·01	0·07	—	Raimondi (1982)
Simmental France	—	0·26	0·55	0·03	0·06	0·02	0·08	—	Grosclaude (1988)
Charolaise France	—	0·10	0·68	0·13	0·01	—	0·08	—	Grosclaude (1988)
Podolica Italy	—	0·28	0·32	0·04	—	0·03	0·33	—	Ferrara *et al.* (1986)

The taurine breed groups having kept the highest frequencies of the original haplotype, CA^2, are the Channel Island and Podolic groups. Remarkably, as discussed before, Podolic breeds are, like *Bos indicus*, also polymorphic at the α-lactalbumin locus. This may be the result of introgression from *Bos indicus*, as suggested by Bettini and Masina (1972) for the Modicana breed in which about one-third of the animals have bifid processes to the last thoracic vertebrae, as usually observed in the Zebu. However, those observations are also consistent with the concept that both Zebu and Podolic cattle have retained, more than the other groups, genes of the common ancestor of all cattle populations. Important additional work is needed to obtain a complete picture of the distribution of milk protein variants in cattle populations all over the world. Research on the polymorphism at the DNA level is also expected to provide data of considerable interest.

8. Milk Protein Polymorphism as Marker Genes

Several aspects emerging from the study of genetic polymorphism of milk proteins have been proposed to have practical applications for the purpose of improving the overall efficiency in different sectors of the dairy industry. Because the different polymorphic forms of the major milk proteins are controlled by autosomal genes which are inherited in accordance with Mendelian inheritance, selection of cows for a specific milk protein type is feasible, and it is possible to breed for desirable specific gentic variants. Consequently, researchers around the world have shown considerable interest in using milk protein genes as genetic markers for increasing milk production, improving production-related traits, and altering milk composition. It has to be emphasised that any effects a certain genetic variant of a specific milk protein would have on production traits are in the strictest sense statistical effects only, i.e. any observed differences are mere associations and in most cases little is known about the cause-and-effect relationship. In contrast, the effect of a genetic variant on protein composition of milk may be considered as a more direct effect because mutation of the protein in question is located in the coding section of its gene and hence could alter its rate of biosynthesis.

8.1. Milk Production

To give a summary of the extensive literature that deals with possible relationships between genetic polymorphism of milk proteins and milk production would be a rather difficult task because of differing trends and at times, conflicting results that have been reported. In many cases, results from different studies are not comparable for various reasons, including population size, breeds of animal, frequency of genetic variants under consideration, methods of measuring production traits (test-day or entire lactation yield), and most importantly, the rigor of statistical analysis to adjust for other important factors contributing to milk production such as: age of cow, season, stage of lactation, health status, and effects of other genetic variants. Before any conclusions could be drawn concerning the association between milk protein polymorphism and milk yield, all the above mentioned factors have to be considered. The relationships between genetic variants of β-lactoglobulin and milk production serve as an example of apparent confusion existing in the literature. Several studies (Brum *et al.*, 1968; Cerbulis & Farrell, 1975; McLean *et al.*, 1984; Ng-Kwai-Hang *et al.*, 1984, 1990*a*; Lin *et al.*, 1986; Gonyon *et al.*, 1987; Haenlein *et al.*, 1987) have reported no relationships between β-lactoglobulin variants and milk production, whereas others (Janicki, 1978; Yebroski & Komissarenko, 1982; Jairam & Nair, 1983) showed superior production for cows with phenotype BB. There are also reports of cows bearing the homozygous A for β-lactoglobulin being higher milk producers (Comberg *et al.*, 1964; Konevtzova, 1978; Atroshi *et al.*, 1982; Babukov *et al.*, 1982; Ng-Kwai-Hang *et al.*, 1986). The association of β-lactoglobulin AB with higher milk yield has also been reported (Pupkova, 1980).

The effects of casein variants on milk production are also not well established and the information obtained so far is inconclusive and depends on the nature of the study. No significant associations were observed between milk yield and genetic variants of α_{s1}-casein, β-casein and κ-casein (McLean *et al.*, 1984; Gonyon *et al.*, 1987; Haenlein *et al.*, 1987). At the α_{s1}-casein locus, the superiority of phenotype BB over AB or BC for milk yield has been reported (Pupkova, 1980; Yebrovski & Komissarenko, 1982; Ng-Kwai-Hang *et al.*, 1984, 1986; Lin *et al.*, 1986). High milk yield is usually associated with the A variant of β-casein (Babukov *et al.*, 1982; Yebrovski & Komissarenko, 1982; Matyukov, 1983; Ng-Kwai-Hang *et al.*, 1984, 1986; Lin *et al.*, 1986), more specifically the A^2 and A^3 (Matyukov, 1983; Ng-Kwai-Hang *et al.*, 1986, 1990*a*). No definite conclusion could be drawn concerning the association between milk production and κ-casein variants. Some studies indicate no relationship (McLean *et al.*, 1984; Ng-Kwai-Hang *et al.*, 1984, 1990*a*), while others suggest that κ-casein AA (Gonyon *et al.*, 1987; Babukov *et al.*, 1982) or κ-casein AB (Yebrovski & Komissarenko, 1982; Ng-Kwai-Hang *et al.*, 1986), or κ-casein BB (Pupkova, 1980) is positively associated with milk production.

8.2. Milk composition

The absolute contents and relative proportions of various constituents present in the milk will dictate the overall value of this farm product in terms of economics, nutrition and technological properties. There are many reports that relate genetic polymorphism of milk proteins to milk content of fat and protein, and to a lesser extent of minerals and other organic ions that have a profound effect on the manufacturing properties of the milk. It appears that reports on the associations of genetic variants with different milk components is not as controversial as those with milk production.

8.2.1. Milk fat

Although some studies (Cerbulis & Farrell, 1975; Graml *et al.*, 1985; Lin *et al.*, 1986) indicate no significant correlation between genetic variants of β-lactoglobulin and the fat content of milk, others tend to show the contrary (Comberg *et al.*, 1964; Hoogendoorn *et al.*, 1969; Janicki, 1978; McLean *et al.*, 1984; Ng-Kwai-Hang *et al.*, 1984, 1986; Graml *et al.*, 1986*a*). In most cases, β-lactoglobulin B, either in the homozygote or heterozygote form, AB, favours a higher fat percentage in the milk. For example, results of 25 000 test-day milk samples from Holstein cows averaged 3·72 and 3·67% fat for β-lactoglobulin BB and AA, respectively. This difference of 0·05% in fat content may seem small but it is quite significant at the production level when one considers the volume of milk produced.

At the α_{s1}-casein loci, some reports (Zebrovskii *et al.*, 1977; Munro, 1978; Ng-Kwai-Hang *et al.*, 1986) show a higher fat content to be associated with the BC variant. Unfortunately, in all those studies, there was an absence of homozygous CC for α_{s1}-casein so that it was not possible to disclose whether variant C would be even more advantageous for fat content when it is present in the homozygous form. Highest milk fat is associated with β-casein B (Babukov, 1978; McLean *et al.*, 1984; Ng-Kwai-Hang *et al.*, 1986). For the different possible phenotypes of β-casein that were present, the fat content of the milk was in the following descending order according to genetic variants: A^1B, A^1A^1, A^1A^2, A^2A^2, A^1A^3, A^2A^3, A^2B (Ng-Kwai-Hang *et al.*, 1986). Higher fat content has also been correlated to κ-casein BB with 3·76% compared to 3·67% for κ-casein AA (Ng-Kwai-Hang *et al.*, 1986).

8.2.2. Milk Protein

The original discovery by Aschaffenburg and Drewry (1957) that β-lactoglobulin AA milk contains more protein than the AB or BB milks has been confirmed by many other researchers (Moustgaard *et al.*, 1960; Feagan *et al.*, 1972; Cerbulis & Farrell, 1975; Janicki, 1978; Mariani *et al.*, 1979*b*; McLean *et al.*, 1984; Ng-Kwai-Hang *et al.*, 1984, 1986, 1990*a*). The difference in protein content is largely due to altered rate of synthesis of β-lactoglobulin (Aschaffenburg & Drewry, 1957; Cerbulis & Farrell, 1975; McLean *et al.*, 1984; Ng-Kwai-Hang *et al.*, 1987). The higher β-lactoglobulin content associated with the AA phenotype will result in a higher whey protein fraction in milk, and if the rate of synthesis of casein is not altered by the same extent, there will be an alteration in the ratio of casein to total protein. This ratio, which is often known as 'casein number', is an important measure of casein or cheese yielding capacity of milk. Although a few reports indicate no influence of β-lactoglobulin polymorphism on total casein content (Cerbulis & Farrell, 1975; Mityukov, 1982), most work confirms that β-lactoglobulin AA is associated with less casein than the BB genotype (Mariani *et al.*, 1979*a*; Buchberger *et al.*, 1982; McLean *et al.*, 1984; Mariani, 1985; Ng-Kwai-Hang *et al.*, 1986, 1987; Rampilli *et al.*, 1988). The higher content of whey protein mainly due to β-lactoglobulin and the lower content of casein associated with the AA genotype contributes to the lower casein number observed with this phenotype (Schaar *et al.*, 1985; Ng-Kwai-Hang *et al.*, 1986). Studies on the effects of β-lactoglobulin polymorphism on the contents of individual caseins and whey proteins (Aaltonen & Antila, 1987; Ng-Kwai-Hang *et al.*, 1987) indicated that β-lactoglobulin AA is associated with lower contents of α_s-casein, α-lactalbumin, serum albumin and immunoglobulin, but a higher content of β-lactoglobulin. From a study involving eight breeds of cattle, Komatsu *et al.* (1977) estimated that the mean content of β-lactoglobulin in milk is 2·58 for allele A compared to 1·70 mg/ml for allele B. Such a considerable difference in β-lactoglobulin content associated with the two types of this protein suggests that the structural gene for β-lactoglobulin is a major gene which determines the amount of this particular protein being synthesised. Genetic polymorphisms of α_{s1}-casein, β-casein and κ-casein are associated with either the amount of the corresponding casein, or with other casein or whey protein fractions so that there is an alteration in total casein or protein content of milk. Phenotype BC of α_{s1}-casein is associated with higher contents of α_{s1}-casein, total casein and total protein, but lower β-lactoglobulin and whey protein than the BB or AB types (McLean *et al.*, 1984; Kroeker *et al.*, 1985; Ng-Kwai-Hang *et al.*, 1986, 1987). For the polymorphic forms of β-casein, the A^1 and A^2 types are associated with a higher content of α_S-casein, but lower β-casein and κ-casein than the B type. The phenotype A^1B is associated with the highest protein (3·48%) and casein (2·78%) contents (Ng-Kwai-Hang *et al.*, 1986) whereas A^2A^2 is lowest in both components, 3·36 and 2·66%, respectively. Several reports have consistently shown that the B variant of κ-casein (Mariani *et al.*, 1976; McLean *et al.*, 1984; Kroeker *et al.*, 1985; Ng-Kwai-Hang *et al.*, 1986, 1987; Rampilli *et al.*, 1988) is associated with a high amount of κ-casein in milk. Variant B is also associated with high total casein and protein, a relationship which has generated great interest within the cheese industry for the genetic polymorphism of κ-casein.

8.2.3. Other Milk Components

Most of the research has concentrated on finding associations between protein polymorphisms and the two major constituents of milk, fat and protein. However, there also exist fewer reports relating polymorphisms with minor constituents. Mariani *et al.* (1979*b*) showed that the citric acid content of milk averaged 165 mg/100 ml for κ-casein A as compared to 135 mg/100 ml for the B variant. The same group of workers (Mariani *et al.*, 1984) reported a lower sialic acid content of total casein containing κ-casein A compared to κ-casein B. Comparison of milk composition for different types of α-lactalbumin is rare because only one variant, B, is known to exist in the common dairy breeds. Research on Zebu cows showed that α-lactalbumin AB was associated with higher potassium concentration in milk than the BB genotype (Perez-Beato & Ponce, 1985), but there were no significant differences in lactose and sodium for the two types of α-lactalbumin.

8.3. Dairy production-related traits

In addition to the search for associations between genetic polymorphs of milk proteins and milk production and milk composition, there are also some studies that are aimed at finding relationships with production-related traits. The rationale is that milk production is dependent on such factors as growth rate of the heifer, reproductive performance and health status of the cow.

8.3.1. Growth rate and body weight

Initial growth rates of calves from birth to production (Kriventsov, 1978; Singh *et al.*, 1981) were faster for those fed on β-lactoglobulin AA milk than those fed on β-lactoglobulin BB milk. One possible explanation for this observation is the fact that β-lactoglobulin AA milk has more protein, as discussed in the previous section, and hence will be more nutritious. Other studies (Pokalov, 1975; Lin *et al.*, 1987) have shown that heifers with the gene for β-lactoglobulin A were heavier than those with the gene for β-lactoglobulin B. In this case, the difference observed is due to genetics and not nutrition because both groups of heifers were raised under the same conditions, including nutritional regime.

8.3.2. Reproductive Performance

The age of heifers at first calving is related to the variant of α-lactalbumin, β-lactoglobulin and κ-casein (Jairam & Nair, 1983; Ronda *et al.*, 1984). Heifers with BB α-lactalbumin and heterozygote AB for β-lactoglobulin or κ-casein are usually younger than heifers of other phenotypes at the start of the first lactation, and β-lactoglobulin AB heifers have a longer gestation period than either the AA or BB heifers (Lin *et al.*, 1987). β-Lactoglobulin AB heifers are younger at first conception and require fewer days from first service to conception. In a study involving more than 6000 cows and 700 sires, Hargrove *et al.* (1980) found no significant differences between polymorphs of β-lactoglobulin, α_{s1}-casein, β-casein and κ-casein, and conception rate, average days open and the number of cows that conceived by the third service. From another study based on 4819 Holstein cows, Ng-Kwai-Hang *et al.* (1990*b*) concluded that there were no associations between genetic variants of the above four proteins and number of: days to first breeding, days open and services per conception.

8.3.3. Mastitis

Mastitis is a disease caused by inflammation of the mammary gland and is of great economic importance to the dairy industry. Besides affecting the health of the cow, this condition drastically affects milk production and milk quality. From the literature, no definite conclusion can be drawn regarding the association of genetic polymorphism with the incidence of mastitis. In a study involving more than 2000 cows over two lactations (Jansen *et al.*, 1981), no difference was observed between the incidence of mastitis and polymorphs of β-lactoglobulin, β-casein and κ-casein. Similar conclusions were reached by Ng-Kwai-Hang, Monardes and Hayes (unpublished) who observed no association between lactational somatic cell count and genetic variants of α_{s1}-casein, β-casein, κ-casein and β-lactoglobulin over three lactations for 2900 cows. A lower incidence of mastitis was associated with β-lactoglobulin AB (Giesecke & Osterhoff, 1975; Stur *et al.*, 1976) but there was no relationship between the polymorphs of α_{s1}-casein, β-casein and κ-casein and the incidence of mastitis. While Leonhard-Kluz and Kaminska (1981) and Atroshi *et al.* (1982) reported that a higher resistance to mastitis was associated with β-lactoglobulin BB, Han *et al.* (1986) reported the opposite. With an increased infection of the mammary gland, it is known that the concentration of immunoglobulins in milk increases at the expense of β-lactoglobulin. Therefore, one would expect mastitis to be associated with protein variants which alter either the β-lactoglobulin or immunoglobulin content in milk. From other studies (for example, Ng-Kwai-Hang *et al.*(1987)), α_{s1}-casein BC, β-casein A^1B, κ-casein BB and β-lactoglobulin BB, which are associated with a high immunoglobulin concentration, could also be associated with a high incidence of mastitis.

9. Genetic Polymorphism and Properties of Milk

The properties of milk and dairy products are influenced to a large extent by the amounts and relative proportions of each of the milk constituents. The protein fraction, because of its diversity in concentration and type, contributes significantly in this regard. Polymorphism of a milk protein caused by deletion or substitution of amino acids is bound to alter the properties of this particular protein which depend on the linear array of amino acids endowing it with a unique primary structure. Differences in amino acid composition and sequence for genetic variants could partially explain changes in the properties of the molecules through a combination of a series of modification including net charge, hydrophobicity, degrees of phosphorylation and glycosylation, all of which contribute to the behaviour of milk proteins and hence the overall manufacturing properties of the milk. Here, we encounter the direct effect of genetic polymorphism since there are modifications in protein structure due to mutations which lead to alterations in the overall properties of the protein.

9.1. Physico-Chemical Properties of Genetic Variants

Although there is a paucity of research that compares the physico-chemical properties of α_{s1}-casein polymorphs due to the very high frequency of the B variant and the lack of A or C variants in most of the dairy cattle population, the few observations that have been made are worth noting. The quality of cheese made from milk containing α_{s1}-casein A is inferior to that from milk with the variant B (Sadler *et al.*, 1968) and the former milk is more difficult to process (Thompson *et al.*, 1969*a*). The deletion of 13 amino acid residues in α_{s1}-casein A reduces the hydrophobic nature of the protein compared to α_{s1}-casein B (Creamer *et al.*, 1982), leading to a change in the rate of proteolysis during cheese ripening. Casein adhesives manufactured from casein containing α_{s1}-casein A are less viscous and have better adhesive strength than those containing either the B or C variants (Mountfort, 1984). This desirable attribute of α_{s1}-casein A has prompted New Zealand researchers to establish a dairy herd consisting of cows homozygote for the very rare A for α_{s1}-casein. The stability of casein micelles in the presence of calcium, lactate, citrate and phosphate is dependent on the genetic variants of α_{s1}-casein, β-casein and κ-casein present (El-Negoumy, 1974; Imafidon, 1990; Ng-Kwai-Hang & Imafidon, 1990). Micelles containing the combination BB–AB–

AB of the three respective caseins were found to be more stable than micelles containing the combination BB–AA–BB. This difference in casein micelle stability may be due to differences in hydrophobicity between the polymorphs of β-casein (Eigel & Randolph, 1976). The B variant of β-casein is also different from the A variant by having a lower water-binding capacity (Kirchmeir *et al.*, 1983). Differences in physico-chemical properties between variants A and B of κ-casein include higher heat stability (Kirchmeier *et al.*, 1983; McLean *et al.*, 1987) and smaller casein micelles (Morini *et al.*, 1975) in milk containing the B variant. The heat stability of milk is affected by β-lactoglobulin variant (Feagan, 1979; Hillier *et al.*, 1979; McLean *et al.*, 1987; Imafidon, 1990) as well as temperature and pH. Hillier *et al.* (1979) reported that at temperatures below 90°C, β-lactoglobulin A was more heat stable than β-lactoglobulin B, but at temperatures above 90°C, the situation was reversed. At pH 4 and 9, β-lactoglobulin A was less stable than β-lactoglobulin B over the entire range of temperature. Differential scanning calorimetric measurements in phosphate buffer at pH 6·8 (Imafidon, 1990; Ma *et al.*, 1990) indicated that β-lactoglobulin BB had a higher denaturation temperature than either the AB or AA phenotype. Three genetic variants of β-lactoglobulin were tested for their ability to inhibit the hydrolysis of p-nitrophenyl phosphate by phosphoprotein phosphatase (Farrell & Thompson, 1971); the results indicated the inhibitory effect of β-lactoglobulin was in the order A $>$ B $>$ C. These differences are related to structural changes in the β-lactoglobulin variants.

9.2. Coagulating Properties of Milk

The rate and extent of coagulation of milk by rennet and the characteristics of the resulting curd are very important factors which have to be considered in the process of cheesemaking. The first indication of effects of genetic polymorphism on coagulating properties (Sherbon *et al.*, 1967) was through the following observations: β-lactoglobulin B milk had higher curd tension than milk containing the A variant; milk containing α_{s1}-casein BC was different from BB milk in terms of clotting time and consistency of curd, and κ-casein AA milk took longer to coagulate and produced a softer curd than either AB or BB milk. To date, several laboratories have confirmed these original findings and most of the reports have concentrated on the effects of κ-casein polymorphism on: rennet clotting time, rate of curd formation, and curd firmness as determined by a formagraph. In general, most of the reports (Losi *et al.*, 1973, 1975; Mariani *et al.*, 1976; Tervala *et al.*, 1983, 1985; Schaar, 1984; Mariani & Leoni, 1985; Jacob & Puhan, 1986; Marziali & Ng-Kwai-Hang, 1986*a*; Aaltonen & Antila, 1987; Rampilli *et al.*, 1988) are consistent in that milk containing κ-casein B is better than κ-casein A milk by having shorter coagulation time, faster rate of firming, and producing a firmer curd. This effect of κ-casein B may be due partially to higher fat and casein contents in milk containing this variant. However, Marziali and Ng-Kwai-Hang (1986*a*) demonstrated the direct effect of κ-casein variant on coagulation properties after adjustments were made for the effects of milk composition. In terms of genetic polymorphism in other caseins, it appears that α_{s1}-casein BC is better than BB (Mariani *et *al.*, 1988) and β-casein BB is better than the AA type (Mariani *et al.*, 1986). Differences also exist in the coagulation behaviour between milks containing β-lactoglobulin A or β-lactoglobulin B (Marziali & Ng-Kwai-Hang, 1986*a*).

9.3. Cheesemaking

Several studies have demonstrated that the effects of genetic polymorphism of milk proteins on coagulating properties are also reflected in cheese yield and cheese composition which is a determinant of cheese quality. In the manufacture of Parmesan cheese, Morini *et al.* (1979) obtained approximately 8% higher yield with κ-casein B than with κ-casein A milk. This yield differential in favour of the B variant was about half in the case of Cheddar-type cheese (Marziali & Ng-Kwai-Hang, 1986*b*). However, no differences were observed in the yield of Svecia cheese (Schaar, 1986) between the two variants of κ-casein, possibly due to differences in cheesemaking processes. Higher yield for κ-casein B milk is also associated with higher fat and protein contents in the cheese (Marziali & Ng-Kwai-Hang, 1986*c*). According to Morini *et al.* (1979), Parmesan cheese made from κ-casein BB milk is more desirable than that from κ-casein AA milk because of better organoleptic and physicochemical properties. Cheese from κ-casein BB milk underwent greater syneresis during the first 24h, lost less water during ripening and contained more fat with a higher fat to protein ratio. Comparison of cheese yield from milks containing β-casein A^1A^1 or β-casein A^1A^2 revealed the superiority of the former (Marziali & Ng-Kwai-Hang, 1986*b*) due to greater retention of fat and protein in the cheese curd and lower losses of milk components in the whey (Marziali & Ng-Kwai-Hang, 1986*c*). Although no significant differences were observed in cheese yields for phenotypes AA, AB and BB of β-lactoglobulin, recovery of fat and protein in cheese was higher for the BB type due to lower losses in the whey (Marziali & Ng-Kwai-Hang, 1986*b*, *c*). Based on the results of a preliminary study, Graham *et al.* (1984) suggested that the combination of AB, BB, BB for κ-casein, β-casein and β-lactoglobulin, respectively, would be more favourable than AA, AA, AA in terms of casein content, coagulating properties and cheese yield.

10. Concluding Remarks

The first report of genetic polymorphism in β-lactoglobulin approximately 35 years ago stimulated active research efforts in the areas of biochemistry and genetics of milk proteins. Initially, most research was focussed on developing methods for identification and on studying the mode of inheritance and frequency distribution of genetic variants of different milk proteins in various bovine breeds. Detection of genetic variants by electrophoresis has been possible because the mutants so far identified involve a change in net charge of the protein. In future, it will be possible to adapt HPLC techniques for the purpose of peptide mapping of milk proteins and thus characterise the silent variants. With the developments in molecular biology, methods are becoming available for typing milk protein genes at the DNA level and it will be practical to phenotype cows before initiation of lactation, as well as sires located in the artificial insemination centres. Because milk protein genes are simply inherited in the Mendelian fashion, there is considerable interest in using them as markers for production and milk composition traits. From the literature, reports relating bovine milk protein variants with various aspects of dairy production are inconclusive and at times contradictory, mainly due to differences in the rigor of statistical analysis among the various studies. However, it has been consistently shown that the B variant of κ-casein and the A variant of β-lactoglobulin are associated with a higher protein content in the milk. The ratio of casein to whey protein is also altered by the variants of β-lactoglobulin and κ-casein present and this probably has a profound effect on the manufacturing properties of the milk. Studies on genetic polymorphism of caprine α_{s1}-casein have shown convincing evidence that certain alleles of this protein are associated with higher concentrations of casein in milk. Before recommendations can be made regarding the selection for specific variants of milk proteins, it would be wise to ascertain that these variants are not associated with factors which are detrimental to the overall dairy industry. Reports showing differences in the physicochemical and coagulating properties of milk due to the presence of certain protein variants have to be confirmed by studies comparing isolated and pure forms of individual proteins without the interference of other milk components. Hence, this opens up a new area of research in the biochemistry of milk proteins.

References

Aaltonen, M.L. & Antila, V. (1987). Milk renneting properties and the genetic variants of proteins. *Milchwissenschaft*, **42**, 490–2.

Abe, T., Komatsu, M., Oishi, T. & Kageyama, A. (1975). Genetic polymorphism of milk proteins in Japanese and European cattle breeds in Japan. *Jap. J. Zootech. Sci*, **46**, 591–9.

Addeo, F., Mercier, J.-C. & Ribadeau-Dumas, B. (1976). Eléments de structure primaire de l'α-lactalbumine et de la β-lactoglobuline de buffle. *FEBS Lett.*, **63**, 255–9.

Addeo, F., Mercier, J.-C. & Ribadeau-Dumas, B. (1977). The caseins of buffalo milk. *J. Dairy Res.*, **44**, 455–68.

Addeo, F., Chianese, L., Di Luccia, A., Petrilli, P., Mauriello, R. & Anelli, G. (1983). Identification of bovine casein variants by gel isoelectric focusing. *Milchwissenschaft*, **38**, 586–8.

Addeo, F., Mauriello, R. & Di Luccia, A. (1988). A gel electrophoretic study of caprine casein. *J. Dairy Res.*, **55**, 413–21.

Aliev, G.A. & Koloteva, R.S. (1975). Genetic variations in the β-lactoglobulin in the milk of sheep. *Anim. Breed. Abstr.*, **43**, 3440.

Althen, T.G. & Gerrits, R.J. (1972). Polymorphism of $whey_2$ protein of sow's milk. *J. Dairy Sci.*, **55**, 331–3.

Ambrosoli, R., Di Stasio, L. & Mazzocco, P. (1988). Content of α_{s1}-casein and coagulation properties in goat's milk. *J. Dairy Sci.*, **71**, 24–8.

Annan, W.D. & Manson, W. (1969). A fractionation of the α_s-casein complex of bovine milk. *J. Dairy Res.*, **36**, 259–68.

Andrews, A.T., Taylor, M.D. & Owen, A.J. (1985). Rapid analysis of bovine milk proteins by fast protein liquid chromatography. *J. Chrom.*, **348**, 177–85.

Arave, C.W. (1967). Procedure for simultaneous phenotyping of β-casein and β-lactoglobulin variants in cow's milk. *J. Dairy Sci.*, **50**, 1320–2.
Arave, C.W., Gillet, T.A., Price, D.A. & Matthews, D.H. (1973). Polymorphisms in casein of sheep milk. *J. Anim. Sci.*, **36**, 241–4.
Aschaffenburg, R. (1961). Inherited casein variants in cow's milk. *Nature*, **192**, 431–2.
Aschaffenburg, R. (1964). Protein phenotyping by direct polyacrylamide gel electrophoresis of whole milk. *Biochim. Biophys. Acta*, **82**, 188–91.
Aschaffenburg, R. (1965). Variants of milk proteins and their pattern of inheritance. *J. Dairy Sci.*, **48**, 128–32.
Aschaffenburg, R. (1968). Genetic variants of milk proteins: their breed distribution. *J. Dairy Res.*, **35**, 447–60.
Aschaffenburg, R. & Drewry, J. (1955). Occurrence of different beta-lactoglobulins in cow's milk. *Nature*, **176**, 218–19.
Aschaffenburg, R. & Drewry, J. (1957). Genetics of the β-lactoglobulins of cow's milk. *Nature*, **180**, 376–8.
Aschaffenburg, R. & Thymann, M. (1965). Simultaneous phenotyping procedure for the principal proteins of cow's milk. *J. Dairy Sci.*, **48**, 1524–6.
Aschaffenburg, R., Sen A. & Thompson, M.P. (1968). Genetic variants of casein in Indian and African Zebu cattle. *Comp. Biochem. Physiol.*, **25**, 177–84.
Atroshi, F., Kangasniemi, R., Honkanon-Buzalski, T. & Sandholm, M. (1982). Beta-lactoglobulin phenotypes in Finnish Ayrshire and Friesian cattle with special reference to mastitis indicators. *Acta Vet. Scand.*, **23**, 135–43.
Babukov, A.V. (1978). Intrapopulation variability of the beta-casein locus in Black Pied cattle of the USSR and its effect on realisation of productive qualities. *Nau. Tru. Lenin. Sel'sko. Inst.*, **342**, 43–4.
Babukov, A.V., Mityutko, V.E. & Komissarenko, A.D. (1982). Use of genofond of polymorphic milk proteins for improvement of Black-and White cattle. *Proc. XXI Intern. Dairy Congr. Moscow*, Vol. 1, Book 1, p.39.
Baker, C.M.A. & Manwell, C. (1980). Chemical classification of cattle. 1. Breed groups. *Anim. Blood Grps. Biochem. Genet.*, **11**, 127–50.
Balland, O. (1987). Analyse et évolution du polymorphisme électrophorétique des lactoprotéines des races bovines Montbéliarde et Normande. (Mimeographed report).
Bech, A.M. & Kristiansen, K.R. (1990). Milk protein polymorphism in Danish dairy cattle and the influence of genetic variants on milk yield. *J. Dairy Res.*, **57**, 53–62.
Bech, A.M. & Munk, K.S. (1988). Studies on bovine milk protein polymorphism by electrofocusing in agarose gels containing 7M urea. *Milchwissenschaft*, **43**, 230–2.
Bell, J.W. & Stone W.K. (1978). Rapid separation of whey proteins by cellulose acetate electrophoresis. *J. Dairy Sci*, **62,** 502–4.
Bell, K. (1962). One-dimensional starch gel electrophoresis of bovine skim milk. *Nature*, **195**, 705–6.
Bell, K. & McKenzie, H.A. (1964). β-Lactoglobulins. *Nature*, **204,** 1275–9.
Bell, K., McKenzie, H.A., Murphy, W.H. & Shaw, D.C. (1970*a*). β-lactoglobulin Droughtmaster: a unique protein variant. *Biochim. Biophys. Acta*, **214**, 427–36.
Bell, K., McKenzie, H.A. & Ralston, G.B. (1970*b*). Do β-lactoglobulins occur in equine and porcine secretions? *Proc. Aust. Biochem. Soc.*, **3**, 82.
Bell, K., McKenzie, H.A., Muller, U., Rogers, C. & Shaw, D.C. (1981*a*). Equine whey proteins. *Comp. Biochem. Physiol.*, **68B**, 225–36.
Bell, K., McKenzie, H.A. & Shaw, D.C. (1981*b*). Bovine β-lactoglobulin E, F, and G of Bali (Banteng) cattle, *Bos* (Bibos) *Javanicus. Aust. J. Biol. Sci.*, **34**, 133–47.
Bell, K., Hopper, K.E. & McKenzie, H.A. (1981*c*). Bovine α-lactalbumin C and α_{s1}-, β- and κ-caseins of Bali (Banteng) cattle, *Bos* (Bibos) *javanicus. Aust. J. Biol. Sci.*, **34**, 149–59.
Bell, K., McKenzie, H.A. & Shaw, D.C. (1981*d*). Porcine β-lactoglobulin A and C. Occurrence, isolation and chemical properties. *Mol. Cell. Biochem.*, **35**, 103–11.

Bell, K., McKenzie, H.A. & Shaw, D.C. (1981*e*). Porcine α-lactalbumin A and B. *Mol. Cell. Biochem.*, **35**, 113–19.

Bettini, T.M. & Masina, P. (1972). Proteine e polimorfismo proteico del latte vaccino. *Produz. Anim.*, **11**, 107–26.

Blumberg, B.S. & Tombs, M.P. (1958). Possible polymorphism of bovine α-lactalbumin. *Nature*, **181**, 683–4.

Bolla, P., Caroli, A. & Ceriotti, G. (1983). Analisi genetica delle proteine del latte nella razza bovina Rendera. *Atti Soc. Ital. Sci. Vet.*, **37**, 432–6.

Boulanger, A. (1976). Etude biochimique et génétique des protéines du lait de chèvre (*Capra hircus*). Thesis, University of Paris VII.

Boulanger, A., Grosclaude, F. & Mahé, M.F. (1984). Polymorphisme des caséines α_{s1} et α_{s2} de la chèvre (*Capra hircus*). *Génét. Sél. Evol.*, **16**, 157–76.

Braunitzer, G. & Chen, R. (1972). Die spaltung des β-lactoglobulins AB mit bromcyan. *Hoppe-Seyler's Z. Physiol. Chem.*, **353**, 674–6.

Braunitzer, G., Chen, R., Schrank, B. & Strangl, A. (1972). Automatische sequenzanalyse eines proteins (β-lactoglobulin AB). *Hoppe-Seyler's S. Physiol. Chem.*, **353**, 832–4.

Brew, K., Castellino, F.J., Vanaman, T.C. & Hill, R.L. (1970). The complete amino acid sequence of bovine α-lactalbumin. *J. Biol. Chem.*, **245**, 4570–82.

Brignon, G. & Ribadeau-Dumas, B. (1973). Localisation dans la chaine peptidique de la β-lactoglobuline bovine de la substitution Glu/Gln différenciant les variants génétique B et D. *FEBS Lett.*, **33**, 73–6.

Brignon, G., Ribadeau-Dumas, B. & Mercier, J.C. (1976). Premiers elements de structure primaire des caséines α_{s2} bovines. *FEBS Lett.*, **71**, 111–16.

Brignon, G., Ribadeau-Dumas, B., Mercier, J.C. & Pélissier, J.P. (1977). Complete amino acid sequence of bovine α_{s2}-casein. *FEBS Lett.*, **76**, 274–9.

Brignon, G., Mahé, M.F., Grosclaude, F. & Ribadeau-Dumas, B. (1989). Sequence of caprine α_{s1}-casein and characterization of those of its genetic variants which are synthesized at a high level, α_{s1}-Cn A, B and C. *Protein Seq. Data Anal.*, **2**, 181–8.

Brignon, G., Mahé, M.F., Ribadeau-Dumas, B., Mercier, J.C. & Grosclaude, F. (1990). Two of the three genetic variants of goat α_{s1}-casein which are synthesized at a reduced level have an internal deletion possibly due to altered RNA splicing. *Eur. J. Biochem.*, **193**, 237–41.

Brum, E.W., Rausch, W.H., Hines, H & Ludwick, T.M. (1968). Association between milk and blood polymorphism types and lactation traits of Holstein cattle. *J. Dairy Sci.*, **51**, 1031–8.

Buchberger, J., Kiermeier, F., Kirchmeier, O., Graml, R. & Pirchner, F. (1982). Effect of genetic variants of milk proteins on milk composition. *Proc. XXI Intern. Dairy Congr. Moscow*, Vol. 1., Book 1, P40–1.

Buchberger, J., Graml, R. & Klostermeyer, H. (1986). Genfrequenzen der Milchproteine beim bayerischen gelbvieh, beim pinzgauer und beim murnau-wordonfelser rind. *Bayer. Landwirtsch. Jahrb.*, **63**, 267–72.

Caldwell, J., Wesseli, D.F. & Cartwright, T.C. (1971). Occurrence of α_{s1}- and β-casein types in five breeds of beef cattle. *J. Anim. Sci.*, **32**, 601–4.

Carles, C. (1986). Fractionation of bovine caseins by reverse phase high performance liquid chromatography: identification of a genetic variant. *J. Dairy Res.*, **53**, 35–41.

Cerbulis, J. & Farrell, H.M. (1975). Composition of milks of dairy cattle. I. Protein, lactose, and fat contents and distribution of protein fraction. *J. Dairy Sci.*, **58**, 817–27.

Chianese, L., Di Luccia, A., Maurielle, R., Ferrara, L., Zehender, G. & Addeo, F. (1988). Polimorfismo biochimico delle proteine del latte in bovine di razza Podolica. *Zootec. Nutr. Anim.*, **14**, 189–97.

Chiofalo, L. & Micari, P. (1987). Attuali conoscenze sulle varianti delle proteine del latte nelle popolazioni ovine allevate in Sicilla. Osservazioni sperimentali. *Sci. Tec. Latt.-Casearia.*, **38**, 104–14.

Chiofalo, L., Micari, R. & Sturniolo, G. (1982*a*). Indagini sulle varianti genetiche delle proteine nel latte de bufale allevate in Sicilla (Italia). *Ann. Fac. Med. Vet. Messina.*, **19**, 1-10.

Chiofalo, L., Micari, P. & Sturniolo, G. (1982*b*). Polimorfismo delle caseine a pH acido nel cavallo Sanfratellano e in altre populazini cavaline allevate in Sicilia. *Ann. Fac. Med. Vet. Messina.*, **19**, 109–19.

Chiofalo, L., Micari, P. & Sturniolo, G. (1983). Polimorfismo delle proteine del latte in popolazioni caralline allevate in Sicilia. *Zoot. Nutr. Anim.*, **9**, 311–8.

Chiofalo, L., Micari, P. & Sturniolo, G. (1984). Genetic variants of milk proteins of buffalos reared in Sicily. *Ann. della. Fac. Med. Vet.*, **19**, 49–58.

Chiofalo, L., Micari, P. & Girmenia, A.M. (1986). Polimorfismo genetico del locus β-lattoglobulina nella razza ovina Comisana allevata in Sicilla. Zootec. *Nutr. Anim.*, **12**, 73–80.

Ciafarone, N. & Addeo, F. (1984). Compositione della caseina et propieta del latte di capra. *Il Vergaro*, **11**, 17–24.

Comberg, G., Meyer, H. & Growing, M. (1964). Correlations between beta-lactoglobulin types in cattle and age at first calving, milk yield and fat contents. *Zuchtungskunde*, **36**, 248–55.

Conti, A., Godovac-Zimmermann, J., Napolitano, L. & Liberatori, J. (1985). Identification and characterization of two alpha-lactalbumin from Somali camel milk (*Camelus dromedarius*). *Milchwissenschaft*, **40**, 673–5.

Corradini, C. (1969). Distribution of the genetic variants of α_{s1}-, β-, and κ-caseins in milk from Jersey cows in the Netherlands. *Neth. Milk Dairy J.*, **23**, 79-83.

Creamer, L.K. & Richardson, B.C. (1975). A new genetic variant of beta-casein. *N.Z. J. Dairy Sci. Technol.*, **10**, 170–1.

Creamer, L.K., Zoerb, H.F., Olson, N.F. & Richardson, T. (1982). Surface hydrophobicity of alpha s_1-I, alpha s_1-casein A and B and its implication in cheese structure. *J. Dairy Sci.*, **65**, 902–6.

Dalgleish, D. (1986). Analysis by fast protein liquid chromatography of variants of κ-casein and their relevance to micellar structure and renneting. *J. Dairy Res.*, **53**, 43–51.

Dall'olio, S., Davoli, R. & Russo, V. (1989). Una nuova variante di β-casearia caprina. *Sci. Tec. Latt.-Casearia.*, **40**, 24–8.

Davoli, R. (1981). Detection of genetic variants of milk proteins by electrophoresis on cellulose acetate. *Riv. Zoo. Vet.*, **9**, 96–100.

Di Luccia, A., Addeo, F., Corradini, C. & Mariani, P. (1988). Bovine kappa-casein C. an electrophoretic study. *Sci. Tecn. Latt-Cas.*, **39**, 413–22.

Di Stasio, L. (1983). New phenotypes of α_s-casein in sheep. *Anim. Blood Grps. Biochem. Genet.*, **14**, 229–32.

Di Stasio, L. & Merlin, P. (1979). Polimorfisimo biochimici del latte nella razza bovina Grigio Alpina. *Riv. Zootec. Vet.*, **2**, 64–7.

Di Stasio, L. & Dupont, E. (1983). La popolazione bovina 'castana': affinita genetiche con le razze Valdostana P.N. e Herens. *Atti Soc. Ital. Sci. Vet.*, **37**, 505–7.

Di Stasio, L., Cristofori, F. & Sartore, G. (1983). Phenotypic variations in blood and milk of the Somali camel. *Anim. blood. Grps. Biochem. Genet.*, **14**, 225–8.

Doi, H., Ibuki, F. & Kanamori, M. (1979). Heterogeneity of reduced bovine κ-casein. *J. Dairy Sci.*, **62**, 195–203.

Eigel, W.N. & Randolph, H.E. (1976). Comparison of calcium sensitivities of alpha s_1-B, beta-A2, and gamma-A2 caseins and their stabilization by kappa-casein A. *J. Dairy Sci.*, **59**, 203–6.

Eigel, W.N., Butler, J.E., Ernstrom, C.A., Farrell, H.M. Jr., Harwalkar, V.R., Jenness, R. & Whitney, R. McL (1984). Nomenclature of proteins of cow's milk. Fifth revision. *J. Dairy Sci.*, **67**, 1599–631.

El-Negoumy, A.M. (1974). Effect of polymorphism on casein stability in salt solutions of varying complexity before and after mixing. *J. Dairy Sci.*, **57**, 1170–6.

Epstein, H. (1971). *The Origin of the Domestic Animals of Africa*, Vol. 1. Africana Publishing Corporation, New York.

Erhardt, G. (1989*a*). κ-Casein in bovine milk—evidence of a further allele (κ-Cn^E) in different breeds. *J. Anim. Breed. Genet.*, **106**, 225–31.
Erhardt, G. (1989*b*). Isolierung und charakterisierung von caseinfraktionen sowie deren genetische varianten in schweinemilch. *Milchwissenschaft*, **44**, 17–20.
Erhardt, G. (1989*c*). Isolierung und charakterisierung der molkenproteine sowie deren genetische varianten in schweinemilch. *Milchwissenschaft*, **44**, 145–9.
Erhardt, G. & Senft, B. (1987). Protein polymorphisms in porcine milk. Abstract, Proc. 20th Int. Conf. Anim. Blood Grps. Biochem. Polym. (Helsinki, 1986). Anim. Genet. **18** (supp. **1**), 56–7.
Erhardt, G., Godovac-Zimmermann, J. & Conti, A. (1989). Isolation and complete primary sequence of a new ovine wild-type β-lactoglobulin C. *Biol. Chem. Hoppe-Seyler.*, **370**, 757–62.
Farrell, H.M. & Thompson, M.P. (1971). Biological significance of milk protein polymorphism. *J. Dairy Sci.*, **54**, 1219–28.
Feagan, J.T. (1979). Factors affecting protein composition of milk and their significance to dairy processing. *Aust. J. Dairy Technol.*, **34**, 77–81.
Feagan, J.T., Bailey, L.A., Hehir, A.F., McLean, D.M. & Ellis, N.J.S. (1972). Coagulation of milk proteins. I. Effect of genetic variants of milk proteins on rennet coagulation and heat stability of normal milk. *Aust. J. Dairy Technol.*, **27**, 129–34.
Felenczak, A. (1983). I. Genetic polymorphism of milk proteins. *Zesz. Nauk. Akad. Rol. Krak., Zootech.*, **22**, 175–91.
Ferrara, L., DiLuccia, A., Astolfi, P., Mauriello, R. & Addeo, F. (1986). Polimorfismo delle proteine del latte in bovini di razza podolica. II. Aspetti genetici. In *Convegno su l'allavamento del boviho Podolico nel Mezzogiorno d'Italia*, pp.299–308.
Ferretti, L., Leone, P., Rognoni, G. & Sgaramella, V. (1990). Linkage of the four bovine casein genes as demonstrated by pulsed gene electrophoresis. *Proc. 4th World Congr. Genet. Appl. Livestock Prod.*, **XIII**, 75–8.
Frank, G. & Braunitzer, G. (1967). Zur primarstruktar der β-lactoglobuline. *Hoppe-Seyler's Z. Physiol. Chem.*, **348**, 1691–2.
Gaye, P., Hue-Delahaie, D., Mercier, J.C., Soulier, S., Vilotte, J.L. & Furet, J.P. (1986). Ovine β-lactoglobulin messenger RNA: nucleotide sequence and mRNA levels during fractional differentiation of the mammary gland. *Biochimie*, **68**, 1097–107.
Gaye, P., Hue-Delahaie, D., Mercier, J.C., Soulier, S., Vilotte, J.L. & Furet, J.P. (1987). Complete sequence of ovine α-lactalbumin mRNA. *Biochimie*, **69**, 601–8.
Gerrits, R.J., Kraeling, R.R. & Kincaid, C.M. (1969). Polymorphism in a casein fraction of cow's milk. *Biochem. Genet.*, **3**, 355–8.
Giesecke, W.H. & Osterhoff, D.R. (1975). Milk protein phenotyping as a practicable genetic means of augmenting conventional methods of mastitis control and prevention. *Ann. Bull. Int. Dairy Fed.*, **85**, 223–33.
Glasnak, V. (1966). Protein polymorphism in cow's milk. Research note. In *Polymorphismes biochimiques des animaux*. Proc. Xth Eur. Conf. Anim. Blood Grps Biochem. Polym. (Paris 1966), pp.433–5.
Glasnak, V. (1968*a*). Polymorphism of β_1-caseins in sow's milk. *Folia Biol.*, **14**, 70–3.
Glasnak, V. (1968*b*). Inter- and intraspecific differences in milk proteins of cattle and swine. *Comp. Biochem. Physiol.*, **25**, 355–7.
Gonyon, D.S., Mather, R.E., Hines, H.C., Haenlein, G.F.W., Arave, C.W. & Gaunt, S.N. (1987). Association of bovine blood and milk polymorphisms with lactation traits: Holsteins. *J. Dairy Sci.*, **70**, 2585–98.
Graham, E.R.B., McLean, D.M. & Zviedrans, P. (1984). The effect of milk protein genotypes on the cheesemaking properties of milk and on the yield of cheese. *4th Conf. Aust. Ass. Anim. Breed. Genet.*, Adelaide, 4–6 June, pp. 136–7.
Graml, R., Buchberger, J., Kirchmeier, O., Kiermeier, F. & Pirchner, F. (1984*a*). Genfrequenschatzung bei Milchproteinen des bayerischen Fleckviehs. *Züchtungskunde*, **56**, 73–87.

Graml, R., Buchberger, J., Klostermeyer, H. & Pirchner, F. (1984*b*). Untersuchungen über die genfrequenzen der caseine and β-lactoglobuline bei der bayerischen braunviehpopulation. *Züchtungskunde*, **56**, 221–30.

Graml, R., Buchberger, J., Klostermeyer, H. & Pirchner, F. (1985). Pleiotrope wirkungen von β-lactoglobulin-und-casein genotypen auf milchinhaltsstoffe der bayerischen Fleckviehs und Braunviehs. *Z. Tierzucht. Zuch Biol.*, **102**, 353–70.

Graml, R., Buchberger, J., Klostermeyer, H. & Pirchner, F. (1986*a*). Pleiotrope wirkungen von β-lactoglobulin-und caseine genotypen auf milchfett-und milchproteinmengen der bayerischen Fleckviehs und Braunviels. *Z. Tierzucht. Zuch Biol.*, **103**, 33–45.

Graml, R., Buchberger, J. & Pirchner, F. (1986*b*). Genetic disequilibria between the α_{s1}-, β-, κ-casein and the β-lactoglobulin loci of the Bavarian Brown and Bavarian Simmental cattle. *Génét. Sél. Evol.*, **18**, 1-10.

Graml, R., Ohmayer, G., Pirchner, F., Erhardt, L., Buchberger, J. & Mostageer, A. (1986*c*). Biochemical polymorphism in Egyptian Baladi cattle and their relationship with other breeds. *Anim. Genet.*, **17**, 61–76.

Godovac-Zimmermann, J. & Braunitzer, G. (1987). Modern aspects of the primary structure and function of β-lactoglobulins. *Milchwissenschaft*, **42**, 294–7.

Godovac-Zimmermann, J., Conti, A., Liberatori, J. & Braunitzer, G. (1985). The amino-acid sequence of β-lactoglobulin II from horse colostrum (*Equus caballus*, Perissodactyla): β-lactoglobulins are retinol-binding proteins. *Biol. Chem. Hoppe-Seyler*, **366**, 601–8.

Godovac-Zimmermann, J., Shaw, D., Conti, A. & McKenzie, H. (1987). Identification and the primary structure of equine α-lactalbumin B and C (*Equus caballus*, Perissodactyla). *Biol. Chem. Hoppe-Seyler*, **368**, 427–33.

Greenberg, R., Groves, M.L. & Dower, H.J. (1984). Human beta-casein. Amino acid sequence and identification of phosphorylation sites. *J. Biol. Chem.*, **259**, 5132–8.

Grosclaude, F. (1974). Analyse génétique et biochimique du polymorphisme électrophorétique des caséines α_{s1}, β et κ chez les bovines (*Bos taurus*) et les zébus (*Bos indicus*). Thesis, University of Paris VII.

Grosclaude, F. (1979). Polymorphism of milk proteins: some biochemical and genetical aspects. Proc. XVIth Int. Conf. Anim. Blood Grps Biochem. Polym., Leningrad, 1978, vol. 1, pp. 54–92.

Grosclaude, F. (1988). Le polymorphisme génétique des principales lactoprotéines bovines. Relations avec la quantité, la composition et les aptitudes fromagères du lait. *INRA Prod. Anim.*, **1**, 5–17.

Grosclaude, F., Garnier, J., Ribadeau-Dumas, B. & Jeunet, R. (1964). Etroite dépendance des loci contrôlant le polymorphisme des caséines α_s et β. *C.R. Hebd. Séanc. Acad. Sci. Paris*, **259**, 1569–71.

Grosclaude, F., Pujolle, J., Garnier, J. & Ribadeau-Dumas, B. (1965). Déterminisme génétique des caséines κ du lait de vache; étroite liaison du locus κ-Cn avec les loci α_{s1}-Cn et β-Cn. *C.R. Hebd. Séances Acad. Sci., Paris*, **261**, 5229–32.

Grosclaude, F., Pujolle, J. Garnier, J. & Ribadeau-Dumas, B. (1966*a*). Mise en évidence de deux variants supplémentaires des protéines du lait de vache: α_{s1}-Cn^D et Lg^D. *Ann. Biol. Anim. Biochim. Biophys.*, **6**, 215–22.

Grosclaude, F., Pujolle, J., Ribadeau-Dumas, B. & Garnier. J. (1966*b*). Analyse génétique du groupe de loci de structure synthétisant les casèines bovines. In *Polymorphismes Biochimiques des Animaux*. INRA, Paris, pp. 415–20.

Grosclaude, F., Mahé, M.F., Mercier, J.C. & Ribadeau-Dumas, B. (1970). Localisation, dans la partie NH_2-terminale de la caséine α_{s1} bovine, d'une délétion de 13 acides aminés différenciant le variant A des variants B et C. *FEBS Lett.*, **11**, 109–12.

Grosclaude, F., Mahé, M.F., Mercier, J.C. & Ribadeau-Dumas, B. (1972*a*). Charactérisation des variants génétiques des caséines α_{s1} et β bovines. *Eur. J. Biochem.*, **26**, 328–37.

Grosclaude, F., Mahé, M.F., Mercier, J.C. & Ribadeau-Dumas, B. (1972*b*). Localisation des substitutions d'acides aminés différenciant les variants A et B de la caséine κ bovine. *Ann. Génét. Sél. Anim.*, **4**, 515–21.

Grosclaude, F., Mahé, M.F. & Ribadeau-Dumas, B. (1973*a*). Structure primaire de la caséine α_{s1}, et de la caséine β-bovine. *Eur. J. Biochem.*, **40**, 323–4.

Grosclaude, F., Mercier, J.C. & Ribadeau-Dumas, B. (1973*b*). Genetic aspects of cattle casein research. *Neth. Milk Dairy J.*, **27**, 328–40.

Grosclaude, F., Mahé, M.F. & Mercier, J.C. (1974*a*). A comparaison du polymorphisme génétique des lactoprotéines du Zébu et des bovins. *Ann. Génét. Sél. Anim.*, **6**, 305–29.

Grosclaude, F., Mahé, M.F. & Voglino, G.F. (1974*b*). Le variant βE et le code de phosphorylation des caséines bovines. *FEBS Lett.*, **45**, 3–5.

Grosclaude, F., Mahé, M.F., Mercier, J.C., Bonnemaire, J. & Teissier, J.H. (1976*a*). Polymorphisme des lactoprotéines de bovines Néplais. II. Polymorphisme des caséine 'α_s-mineurs'; le locus α_{s2}-Cn est il lié aux loci α_{s1}-Cn, β-Cn et κ-Cn? *Ann. Génét. Sél. Anim.*, **8**, 481–91.

Grosclaude, F., Mahé, M.F., Mercier, J.C., Bonnemaire, J. & Teissier, J.H. (1976*b*). Polymorphisme des lactoprotéines de bovines Népalais. I. Mise en évidence, chez le yak, et caractérisation biochimique de deux nouveaux variants: β-lactoglobuline D (yak) et caséine α_{s1}-E. *Ann. Génét. Sél. Anim.*, **8**, 461–79.

Grosclaude, F., Joudrier, P. & Mahé, M.F. (1978). Polymorphisme de la caséine α_{s2} bovine: étroite liaison du locus α_{s2}-Cn avec les loci α-Cn, β-Cn et κ-Cn; mise en évidence d'une délétion dans le variant α_{s2}-Cn D. *Ann. Génét. Sél. Anim.*, **10**, 313–27.

Grosclaude, F., Joudrier, P. & Mahé, M.F. (1979). A genetic and biochemical analysis of a polymorphism of bovine α_{s2}-casein. *J. Dairy Res.*, **46**, 211–13.

Grosclaude, F., Mahé, M.F. & Accolas, J.P. (1982). Note sur le polymorphisme génétique des lactoprotéines de bovins et de yaks Mongols. *Ann. Génét. Sél. Anim.*, **14**, 545–50.

Grosclaude, F., Mahé, M.F., Brignon, G., Di Stasio, L. & Jeunet, R. (1987). A Mendelian polymorphism underlying quantitative variations of goat α_{s1}-casein. *Génét. Sél. Evol.*, **19**, 399–412.

Guillou, H., Miranda, G. & Pelissier, J.P. (1987). Analyse quantitative des caséines dans le lait de vache par chromatographie liquide rapide d'échange d'ions (FPLC). *Le Lait*, **67**, 135–48.

Haenlein, G.F.W., Gonyon, D.S., Mather, R.E. & Hines, H.C. (1987). Associations of bovine blood and milk polymorphisms with lactation traits. Guernseys. *J. Dairy Sci.*, **70**, 2599–609.

Halliday, J.M., Bell, K., McKenzie, H.A. & Shaw, D.C. (1990). Feline whey proteins: identification, isolation and initial characterization of α-lactalbumin, β-lactoglobulin and lysozyme. *Comp. Biochem. Physiol.*, **95B**, 773–9.

Han, S.K., Chung, E.Y. & Lee, K.M. (1983). Studies on the genetic polymorphism of milk proteins in Korean cattle. *Proc. 5th World Conf. Anim. Prod.*, **2**, 51–2.

Han, S.K., Lee, K.M., Chung, E.Y. & Jang, K.J. (1984). Studies on the genetic polymorphism of milk proteins. II. Genetic variants of β-casein and β-lactoglobulin. *Korean J. Anim. Sci.*, **26**, 217–24.

Han, S.K., Chang, K.J. & Chung, E.Y. (1986). Relationship between milk protein polymorphisms and bacterial mastitis in Holstein-Friesian cattle. *Kor. J. Anim. Sci.*, **28**, 37–42.

Hargrove, G.L., Kiddy, C.A., Young, C.W., Hunter, A.G., Trimberger, G.W. & Mather, R.E. (1980). Genetic polymorphisms of blood and milk and reproduction in Holstein cattle. *J. Dairy Sci.*, **63**, 1154–66.

Hillier, R.M., Lyster, R.L.J. & Cheeseman, G.C. (1979). Thermal denaturation of alpha-lactalbumin and beta-lactoglobulin in cheese whey: effect of total solids concentration and pH. *J. Dairy Res.*, **46**, 103–11.

Hines, H.C., Kiddy, C.A., Brum, E.W. & Arave, C.W. (1969). Linkage among cattle blood and milk polymorphisms. *Genetics*, **62**, 401–12.

Hines, H.C., Haenlein, G.F.W., Zikakis, J.P. & Dickey, H.C. (1977). Blood antigen, serum protein and milk protein gene frequencies and genetic interrelationships in Holstein cattle. *J. Dairy Sci.*, **60**, 1143–51.

Hoagland, P.D., Thompson, M.P. & Kalan, E.B. (1971). Amino acid composition of α_{s3}-, α_{s4}, and α_{s5}-caseins. *J. Dairy Sci.*, **54**, 1103–10.

Hoogendorn, M.P., Moxley, J.E. Hawes, R.O. & MacRae, H.F. (1969). Separation and gene frequencies of blood serum transferrin, casein and beta-lactoglobulin loci of dairy cattle and their effects on certain production traits. *Can. J. Anim. Sci.*, **49**, 331–41.

Horvath, I. & Mészaros, I. (1972). Casein types in Hungarian spotted cattle and other breeds playing a role in its development. In *Proc. XIIIth Europ. Conf. Anim. Blood Grps Biochem. Polym.*, Budapest, 1970. Dr. W. Junk N.V., The Hague (ed.) pp. 217–23.

Humphrey, R.S. & Newsome, L.J. (1984). High performance ion-exchange chromatography of the major bovine milk proteins. *N.Z. J. Dairy Sci. Technol.*, **19**, 197–204.

Imafidon, G.I. (1990). Genetic polymorphism and physico-chemical properties of milk proteins. PhD Thesis, McGill University.

Jacob, E. & Puhan, Z. (1986). The significance of genetic variants of kappa-casein for rennetability of milk. *Deut. Molk. Zeit.*, **107**, 833–4.

Jairam, B.T. & Nair, P.G. (1983). Genetic polymorphisms of milk proteins and economic characters in dairy animals. *Ind. J. Anim. Sci.*, **53**, 1–8.

Janicki, C. (1978). Beta-lactoglobulin and some milk production characteristics of Holstein–Friesian cows in Poland imported from USA and Canada. *Rocz. Akad. Roln. Poz.*, **25**, 169–75.

Jansen, N.E., Klastrup, O., Madsen, P.S., Madsen, P. & Nielsen, S.M. (1981). The influence of milk protein types on mastitis. 32nd Ann. Meeting. EAAP III, P6.

Josephson, R.V. (1972). Isoelectric focusing of bovine milk caseins. *J. Dairy Sci.*, **55**, 1535–43.

Kalan, E.B., Draeling, R.R. & Gerrits, R.J. (1971). Isolation and partial characterization of a polymorphic swine whey protein. *Int. J. Biochem.*, **2**, 232–44.

Kaminogawa, S., McKenzie, H.A. & Shaw, D.C. (1984). The amino-acid sequence of equine α-lactalbumin. *Biochem. Int.*, **9**, 539–46.

Kemmer, B. (1969). β-Lactoglobulin typen in der sauenmilch. *Züchtungskunde*, **41**, 331–4.

Kiddy, C.A. (1975). Gel electrophoresis in vertical polyacrylamide beds. In *Methods of Gel Electrophoresis of Milk Proteins*, ed. H.E. Swaisgood, B.L. Larson, E.B. Kalan, J.R. Brunner, C.V. Morr & P.M.T. Hansen. American Dairy Science Association, Champaign, Illinois, p.14.

Kiddy, C.A., McCann, R.E. & Thatcher, W.W. (1968). Gene frequencies in milk protein polymorphisms in dairy cattle. *Immunogenetics Letter.*, **5**, 150–2.

King, J.W.B. (1966). The caseins of sheep's milk. In *Polymorphismes Biochimiques des Animaux*. Proc. Xth Eur. Conf. Anim. Blood Grps Biochem. Polym., Paris, 427–31.

King, J.W.B. (1969). The distribution of sheep β-lactoglobulins. *Anim. Prod.*, **11**, 53–7.

King, J.W.B., Aschaffenburg, R., Kiddy, C.A. & Thompson, M.P. (1965). Non-independent occurrence of α_{s1}- and β-casein variants of cow's milk. *Nature*, **206**, 424.

Kingsbury, E.T. & Gaunt, S.N. (1977). Heterogeneity in whey proteins of mare's milk. *J. Dairy Sci.*, **60**, 274–7.

Kirchmeier, O., Mehana, A., Graml, R. Buchberger, J. & Pirchner, J. (1983). Studies on physico-chemical properties of casein: genetic and seasonal effects. *Milchwissenschaft*, **38**, 589–91.

Kolde, H.J. & Braunitzer, G. (1983). The primary structure of ovine β-lactoglobulin 1. Isolation of the peptides and sequence *Milchwissenchaft*, **38**, 18–20.

Komatsu, M. & Abe, T. (1980). Milk casein haplotypes in nine Japanese and European cattle breeds. *Jap. J. Zootech. Sci.*, **51**, 799–802.

Komatsu, M., Abe, T. & Oishi, T. (1977). Relationships between beta-lactoglobulin types and the concentrations of beta-lactoglobulin and alpha-lactalbumin in milk. *Jap. J. Zootech. Sci.*, **48**, 237–42.

Konevtsova, M.I. (1978). Milk protein polymorphism in Red steppe cows and its connection with milk production indices. *Nau. Tru. Lenin. Selsko. Inst.*, **342**, 46–8.

Kraeling, R.R. & Gerrits, R.J. (1969). Polymorphism of a protein of cow's whey. *J. Dairy Sci.*, **52**, 2036–8.

Kriventsov, Y.M. (1978). Relationship between the preweaning growth of crossbred heifers and the beta-lactoglobulin type of the milk of nurse cows. *Sel'skok. Nauk. Lenina.*, **7**, 29–30.

Kroeker, E.M., Ng-Kwai-Hang, K.F., Hayes, J.F. & Moxley, J.E. (1985). Effects of environmental factors and milk protein polymorphism on composition of casein fraction in bovine milk. *J. Dairy Sci.*, **68**, 1752–7.

Larsen, B. (1971). Blood groups and polymorphic protein in cattle and swine. *Ann. Génét. Sél. Anim.*, **3**, 54–70.

Larsen, B. & Thymann, W. (1966*a*). Studies on milk protein polymorphism in Danish cattle and the interaction of the controlling genes. *Acta Vet. Scand.*, **7**, 189–205.

Larsen, B. & Thymann, M. (1966*b*). Studies on milk protein polymorphism and the interaction of the controlling genes. In *Polymorphismes Biochimiques des Animaux*. Proc. Xth Eur. Conf. Anim. Blood Grps Biochem. Polym., Paris, pp.421–5.

Larsen, B. Gruchy, C.L. & Moustgaad, J. (1974). Studies on blood groups and polymorphic protein systems in Jersey cattle on the Isle of Jersey. *Acta Agric. Scand.*, **24**, 99–110.

Leonhard-Kluz, I. & Kaminska, B. (1981). Beta-lactoglobulin polymorphism and udder health in cows. *Dairy Sci. Abstr.*, **44**, 2356.

Levéziel, H., Méténier, L., Guérin, G., Cullen, P., Provot, C., Bertaud, M. & Mercier, J.-C. (1991). Restriction fragment length polymorphism of ovine casein genes. Close linkage between the α_{s1}-, α_{s2}-, β- and κ-Cn loci. *Anim. Genet.*, **22**, 1–10.

Li, F.H.F. & Gaunt, S.N. (1972). Study of genetic polymorphisms of milk β-lactoglobulin, α_{s1}-casein, β-casein and κ-casein in five dairy breeds. *Bioch. Genet.*, **6**, 9-20.

Lin, C.Y., McAllister, A.J., Ng-Kwai-Hang, K.F. & Hayes. J.F. (1986). Effects of milk protein loci on first lactation production in dairy cattle. *J. Dairy Sci.*, **69**, 704–12.

Lin, C.Y., McAllister, A.J., Ng-Kwai-Hang, K.F., Hayes, J.F., Batra, T.R., Lee, A.J., Roy, G.L., Vesely, J.A., Wauthy, J.M. & Winter, K.A. (1987). Association of milk protein types with growth and reproductive performance of dairy heifers. *J. Dairy Sci.*, **70**, 29–39.

Losi, G., Capella, P., Castagnetti, G.B., Grazia, L., Zambonelli, C., Mariani, P. & Russo, V. (1973). Influenza delle variants genetiche della caseine κ sulla formazione e sulle caracteristiche della cagliata. *Scien. Tecn. Alimenti.*, **3**, 373–6.

Losi, G., Castagnetti, G.B. & Morini, D. (1975). Importanza delle varianti genetiche delle proteine del latte per l'industria lattiero-casearia. *Mondo del Latte.*, **29**, 727–39.

Ma, C.Y., Ng-Kwai-Hang, K.F., Harwalkar, V.R. & Imafidon, G.I. (1990). Effect of genetic variants of β-lactoglobulin and κ-casein on thermostability of β-lactoglobulin in model systems. *23rd Int. Dairy Congr.* 2: No. 679.

Mácha, J. (1981). Genetical polymorphism of protein in the milk of goats and performance of goats. *Acta Univ. Agric. Fac. Agron.*, **29**, 233–7.

Mácha, J. & Müllerova, Z. (1968). Milk protein polymorphism of cattle breeds in Czechoslovakia. *Proc. 2nd Eur. Conf. Anim. Blood Grps Biochem. Polym.* DR. W. Junk, M.V. (ed) The Hague, pp.247–52.

Mácha, J. & Müllerova, Z. (1969). A contribution to the study of associations between loci controlling cow's milk protein synthesis. *Acta Univ. Agric. Fac. Agron.*, **17**, 139–47.

Mácha, J. & Novackova (1975). Polymorphisms bilkovin Mléce klisen. *Zivoc. Vyroba.*, **20**, 73–8.

Mahé, M.F. & Grosclaude, F. (1982). Polymorphisme de la caséine α_{s2} des bovines: caractérisation de variant C du yak (Bos grunniens). *Ann. Génét. Sél. Anim.*, **14**, 401–16.

Mahé, M.-F. & Grosclaude, F. (1989). α_{s1}-Cn^{D}, another allele associated with a decreased synthesis rate at the caprine α_{s1}-casein locus. *Génét. Sél. Evol.*, **21**, 127–9.

Manson, W., Annan, W.D. & Barnes, G.K. (1976). α_{s0}-Casein: Its preparation and characterization. *J. Dairy Res.*, **43**, 133–6.

Manson, W., Carolan, T. & Annan, W.D. (1977). Bovine α_{s0}-casein: a phosphorylated homologue of α_{s1}-casein. *Eur. J. Biochem.*, **78**, 411–17.

Mariani, P. (1985). Casein content in Friesian cow's milk. *Scien. Tech. Latt.-Casear.*, **36**, 191–209.

Mariani, P. & Leoni, M. (1982). Relazione genetica tra i loci α-lattalbumina e β-lattoglobulina nei bovine. *Ann. Fac. Med. Vet. Parma.*, **2**, 209–16.

Mariani, P. & Leoni, M. (1985). Il tempo di coagulazione del latte in rapporto alla varianti genetiche delle casein β et κ. *Ann. Fac. med. Vet. Univ. Parma*, **5**, 185–95.

Mariani, P. & Russo, V. (1977). Polimorfismo genetico della α-lattalbumina nelle razze bovine. *Riv. Zootec. Vet.*, **5**, 603–13.

Mariani, P., Losi, G., Russo, V., Castagnetti, G.B., Grazia, L., Morini, D. & Fossa, E. (1976). Caseification tests made with milk characterised by Variants A and B of κ-casein in the production of Parmigiano-Reggiano cheese. *Scien. Tecn. Latt.-Casear.*, **27**, 208–27.

Mariani, P., Morini, D., Losi, G., Castagnetti, G.B., Fossa, E. & Russo, V. (1979*a*). Ripartizione delle frazioni azotate del latte in vacche caratterizatte da genetipo diverso nel locus β-lactoglobulina. *Scien. Tecn. Latt.-Casear.*, **30**, 153–76.

Mariani, P., Losi, G., Morini, D. & Castagnetti, G.B. (1979*b*). Citric acid content in milk of cows with different kappa-casein genotypes. *Scien. Tecn. Latt.-Casear.*, **30**, 375–84.

Mariani, P., Resmi, P. & Losi, G. (1984). Sialic acid content of total casein in relation to cow genotype at the kappa-Cn locus. *Att. Soc. Ital. Sci. Vet.*, **38**, 410–12.

Mariani, P., Meazza, M., Resmini, P., Pagani, M.A., Pecorari, M. & Fossa, E. (1986). Observations on beta-casein types and renneting properties of milk. *Ind. del Latte.* , **22**, 35–58.

Mariani, P., Bonatti, P. & Pecorari, M. (1988). Rennet coagulation properties of cow milk in relation to alpha s_1-casein genotype. *Scien. Tecn. Latt.-Casear.*, **39**, 431–8.

Marziali, A.S. & Ng-Kwai-Hang, K.F. (1986*a*). Effects of milk composition and genetic polymorphism on coagulation properties of milk. *J. Dairy Sci.*, **69**, 1793–8.

Marziali, A.S. & Ng-Kwai-Hang, K.F. (1986*b*). Relationships between milk protein polymorphisms and cheese yielding capacity. *J. Dairy Sci.* , **69**, 1193–201.

Marziali, A.S. & Ng-Kwai-Hang, K.F. (1986*c*). Effects of milk composition and genetic polymorphism on cheese composition. *J. Dairy Sci.*, **69**, 2533–42.

Matyukov, V.S. (1983). The value of β-casein polymorphism for selection in cattle. *Selsk. Khoz. Biol.*, **18**, 73–8.

Mauriello, R., Pieragostini, E., Bufano, G. & Addeo, F. in press. Polimorfismo delle caseine in pecore di razza Altamurana. *Sci. Tec. Latt.—Casear.*

McGillivray, R.T.A., Brew, K. & Barnes, K. (1979). The amino-acid sequence of goat α-lactalbumin. *Arch. Biochem. Biophys*, **197**, 404–14.

McKenzie, H.A., (1971). In *Milk Proteins: Chemistry and Molecular Biology*, ed. H.A. McKenzie Academic Press, New York, p.257.

McKenzie, H.A. & Larson, B.L. 1978. Purification and characterization of rat alpha-lactalbumin: apparent genetic variants. *J. Dairy Sci.*, **61**, 714–22.

McKnight, R.A., Jimenes-Flores, R., Kang, Y. Creamer, L.K. & Richardson, T. (1989). Cloning and sequencing of a complementary deoxyribonucleic acid coding for a bovine α_{s1}-casein A from mammary tissue of a homozygous B variant cow. *J. Dairy Sci.*, **72**, 2464–73.

McLean, D.M., Graham, E.R.B., Ponzoni, R.W. & McKenzie, H.A. (1984). Effects of milk protein genetic variants on milk yield and composition. *J. Dairy Res.*, **51**, 531–46.

McLean, D.M., Graham, E.R.B., Ponzoni, R.W. & McKenzie, H.A. (1987). Effects of milk protein genetic variants and composition on heat stability of milk *J. Dairy Res.*, **54**, 219–35.

Mercier, J.-C. (1981). Phosphorylation of caseins, present evidence for an amino acid triplet code posttranslationally recognized by specific kinases. *Biochimie*, **63**, 1–17.

Mercier, J.C., Grosclaude, F. & Ribadeau-Dumas, B. (1971). Structure primaire de la caséine α_{s1}-bovine. *Eur. J. Biochem.*, **23**, 41–51.

Mercier, J.C., Brignon, G. & Ribadeau-Dumas, B. (1973). Structure primaire de la caséine k B bovine. Séquence complète. *Eur. J. Biochem.*, **35** 222–35.

Mercier, J.C., Chobert, J.M. & Addeo, F. (1976). Comparative study of the amino acid sequences of the caseino macropeptides from seven species. *FEBS Lett.*, **72**, 208–14.

Merlin, P. & Di Stasio, L. (1982). Study on milk proteins loci in some decreasing Italian cattle breeds. *Ann. Génét. Sél. Anim.*, **14**, 17–28.

Micari, P., Chiofalo, L. & Michailidis, J. (1986). Studio elettroforetico dei loci caseinici e delle siero-proteine nel latte della pecora di Chios (Grecia): raffronti con la pecora Barbaresca-Siciliana. *Atti. Soc. Ital. Vet.*, **40**, 578–81.

Michalak, W. (1967). Anomalous electrophoretic pattern of milk proteins. *J. Dairy Sci.*, **50**, 1319–20.

Michalak, W. (1969). Hereditary polymorphism of milk proteins in some breeds of cattle raised in Poland. Part II: Biul. *Zaki. Hodowli Dosw. Zwierzat PAN*, **15**, 89–111.

Mityukov, A.S. (1982). The content of genetically determined polymorphous proteins in cow's milk. *Proc. XXI Intern. Dairy Congr (Moscow)*., vol. 1, book 1, p.204.

Morini, D., Losi, G., Castagnetti, G.B., Beneveilli, M., Resmini, P. & Volonterio, G. (1975). The influence of genetic variants of kappa-casein on the size of casein micelles. *Scien. Tecn. Latt.-Casear.*, **26**, 437–44.

Morini, D., Losi, G., Castagnetti, G.B. & Mariani, P. (1979). Cheesemaking experiments with milk characterized by kappa-casein variants A and B: characteristics of the ripened cheese. *Scien. Tecn. Latt-Casear.*, **30**, 243–62.

Mountfort, M. (1984). Patient project building herd of rare casein type-A cows. *Dairy Exporter*, December 1984, 29–31.

Moustgaard, J., Moller, I. & Sorensen, P.H. (1960). Polymorphism of bovine β-lactoglobulin. *Aarsberetn, Inst. Sterilitetsforskn. K. Vet. Landbohojsk. Kbh.*, 111–23.

Munro, G.L. (1978). Effect of genetic variants of milk proteins on yield and composition of milk. *Proc. XX Intern. Dairy Congr. Paris*, p10.

Neelin, J.M. (1964). Variants of κ-casein revealed by improved starch gel electrophoresis. *J. Dairy Sci.*, **47**, 506–9.

Ng-Kwai-Hang, K.F. & Imafidon, G.I. (1990). Effect of genetic variants of κ-casein and β-casein on stability of calcium-caseinate model systems. *23rd Int. Dairy Congr.* 2: No. 688.

Ng-Kwai-Hang, K.F. & Pélissier, J.P. (1989). Rapid separation of bovine caseins by mass ion exchange chromatography. *J. Dairy Res.*, **56**, 391–7.

Ng-Kwai-Hang, K.F., Hayes, J.F., Moxley, J.E. & Monardes, H.G. (1984). Association of genetic variants of casein and milk serum proteins with milk, fat, and protein production by dairy cattle. *J. Dairy Sci.*, **67**, 835–40.

Ng-Kwai-Hang, K.F., Hayes, J.F., Moxley, J.E. & Monardes, H.G. (1986). Relationships between milk protein polymorphisms and major milk constituents in Holstein–Friesian cows. *J. Dairy Sci.*, **69**, 22–6.

Ng-Kwai-Hang, K.F., Hayes, J.F., Moxley, J.E. & Monardes, H.G. (1987). Variation in milk protein concentration associated with genetic polymorphism and environmental factors. *J. Dairy Sci.*, **70**, 563–70.

Ng-Kwai-Hang, K.F., Monardes, H.G. & Hayes, J.F. (1990*a*). Association between genetic polymorphism of milk proteins and production traits during three lactations. *J. Dairy Sci.*, **73**, 3414–20.

Ng-Kwai-Hang, K.F., Monardes, H.G. & Hayes, J.F. (1990*b*). Association between genetic polymorphism of milk proteins and parameters of reproduction in Holstein cows. Research Report, Department of Animal Science, Macdonald College, McGill University, pp.6–7.

Pagnacco, G. & Caroli, A. (1987). Effect of casein and β-lactoglobulin genotypes on renneting properties of milks. *J. Dairy Res.*, **54**, 479–85.

Pearce, R.J. (1983). Analysis of whey proteins by high performance liquid chromatography. *Aust. J. Dairy Technol.*, **38**, 114–17.

Pearce, R.J. & J.G. Zadow. (1978). Isoelectric focusing of milk proteins. *Proc. XX Intern. Dairy Congr. (Paris)*, pp. 217–18.

Perez-Beato, O. & Ponce, P. (1985). Concentrations of lactose, sodium and potassium in the milk of Zebu cows in Cuba and their relationship with genotype at the alpha-lactalbumin locus. *Rev. Sal. Anim.*, **7**, 211–15.

Peterson, R.F. (1969). Electrofocusing of milk proteins. *J. Dairy Sci.*, **52**, 903 (abstr.).

Peterson, R.F. & Kopfler, F.C. (1966). Detection of new types of β-casein by polyacrylamide gel electrophoresis at acid pH: a proposed nomenclature. *Biochem. Biophys. Res. Commun.*, **22**, 388–92.

Piletz, J.E. & Ganschow, R.E. (1981). Genetic variation of milk proteins in mice. *Biochem. Genet.*, **19**, 1023–30.

Pokalov, V.P. (1975). Relationship of blood serum and milk proteins with body weight of cows. *Anim. Breeding Abstr.*, **44**, 4147.

Ponzone, A. & Voglino, G.F. (1976). Milk casein polymorphism in man. *Helv. Paeddiat. Acta*, **31**, 47–52.

Ponzone, A., Voglino, G.F. & Tognolo, A. (1975). Milk casein polymorphism in the Kikuyu population. *Ann. Genet.*, **18**, 203–5.

Poplin, F. (1979). Origine du mouflon de Corse dans une nouvelle perspective paléontologique: par marronage. *Ann. Génét. Sél. Anim.*, **11**, 133–43.

Pujolle, J., Ribadeau-Dumas, B., Garnier, J. & Pion, R. (1966). A study of κ-casein components. *Biochem. Biophys. Res. Commun.*, **25**, 285–90.

Pupkova, G.V. (1980). Milk protein polymorphism and milk production of Estonian Black Pied cows. *Dairy Sci. Abstr.*, **45**, 6620.

Raimondi, R. (1982). Marcatori genetici in popolazioni animali scarsamente valorizzate o in via di estinzione. Progetto finalizzato Difese delle risorse genetiche delle popolazioni animali; obiettivi e risultati. Milano, pp. 120–38.

Rampilli, M., Caroli, A., Bolla, P. & Pirlo, G. (1988). Relationships of milk protein genotypes with casein composition and rennet ability of milk during lactation. *Scien. Tecn. Latt-Casear.*, **39**, 262–79.

Rausch, W.H., Hines, H.C., Treece, J.M. & Tyrrell, H.F. (1967). A preliminary report on polymorphism frequencies of 15 genetic systems in Guernseys. *Immunogenetics Letter.*, **5**, 87–92.

Ribadeau-Dumas, B., Brignon, G., Grosclaude, F. & Mercier, J.C. (1972). Structure primaire de la caséine β bovine. Séquence complète. *Eur. J. Biochem.*, **25**, 505–14.

Richardson, B.C. & Creamer, L.K. (1975). Comparative micelle structure. IV. The similarity between caprine α_s-casein and bovine α_{s3}-casein. *Biochim. Biophys. Acta*, **393**, 37–47.

Ronda, R., Perez-Beato, O. & Granado, A. (1984). Relationship between milk proteins and quantitative parameters in $\frac{3}{4}$ Holstein-Friesian × $\frac{1}{4}$ Zebu cows. *Rev. Sol. Anim.*, **6**, 593–603.

Russo, V. & Mariani, P. (1974). Polymorfismo delle proteine del latte nelle razze bovine da carne. III. Chianina. *Riv. Zootec. Vet.*, **2**, 139–43.

Russo, V. & Mariani, P. (1978). Polimorfismo delle proteine de latte e relazioni tra varianti genetiche e caratteristiche di interesse zootecnico, technologico e caseario. *Riv. Zootec. Vet.*, **6**, 365–79.

Russo, V., Chiofalo, L. & Micari, R. (1977). Esame elettroforetico delle proteine del latte in popolazioni caprine allevate in Sicilia. *Zoot, Nutr. Anim.*, **3**, 247–53.

Russo, V., Davoli, R., Dall'olio, S. & Tedeschi, M. (1986). Ricerche sul polimorfismo del latte caprino. *Zoot. Nutr. Anim.*, **12**, 55–62.

Sadler, A.M., Kiddy, C.A., McCann, R.E. & Mattingly, W.A. (1968). Acid production and curd toughness in milks of different α_{s1}-casein types. *J. Dairy Sci.*, **51**, 28–30.

Samarieanu, M., Stamazescu, E., Granciu, I. & Sotu, E. (1963). Genetic variations of β-caseins in cattle. *Anim. Breed. Abstr.* (1986) Abst. No. 2767, **54**, 362.

Schaar, J. (1984). Effects of κ-casein genetic variants and lactation number on the renneting properties of individual milks. *J. Dairy Res.*, **51**, 397–406.

Schaar, J. (1986). Variation in milk protein composition. Studies on κ-casein and β-lactoglobulin genetic polymorphism and on milk plasmin. Thesis, Department of Animal Breeding & Genetics, Swedish University of Agricultural Sciences, Report No. 71.

Schaar, J., Hansson, B. & Petterson, H.E. (1985). Effects of genetic variants of κ-casein and β-lactoglobulin on cheesemaking. *J. Dairy Res.*, 429–37.

Schmidt, D.G. (1964). Starch gel electrophoresis of κ-casein. *Biochim. Biophys. Acta*, **90**, 411–14.

Schmidt, D.V. & Ebner, K.E. (1972). Multiple forms of pig, sheep and goat α-lactalbumin. *Biochim. Biophys. Acta*, **263**, 714–20.

Schroeder, W.A., Shelton, J.B., Shelton, J.R., Powars, D., Friedman, S., Baker, J., Finklestein, J.Z., Miller, B., Johnson, C.S., Sharpsteen, J.R., Sieger, L. & Kawaoka, E. (1982). Identification of eleven human haemoglobin variants by high-performance liquid chromatography: additional data on functional properties and clinical expression. *Biochem. Genet.*, **20**, 133–52.

Seibert, B., Erhardt, G. & Senft, B. (1985). Procedure for simultaneous phenotyping of genetic variants in cow's milk by isolectric focusing. *Anim. Blood Grps and Biochem. Genet.*, **16**, 183–91.

Seibert, B., Erhardt, G. & Senft, B. (1987). Detection of a new κ-casein variant in cow's milk. *Anim. Genet.*, **18**, 269–72.

Sen, A. & Chaudhuri, S. (1962). Non-casein proteins of goat's milk. *Nature*, **195**, 286–7.

Sherbon, J.W., Ledford, R.A. & Regenstein, J. (1967). Variants of milk proteins and their possible relation to milk properties. *J. Dairy Sci.*, **50**, 951.

Singh, H., Bhat, P.N. & Singh, R. (1981). Association of protein polymorphic genotypes with certain performance traits in crossbred cattle. *Ind. J. Anim. Sci.*, **51**, 5–10.

Stambekov, S. Zh., Shapiro, Y.U.O., Mandrusova; E.E. & Romanyuk, N.A. (1977*a*) Polymorphism of milk proteins in Latvian Darkheaded ewes. *Dairy Sci. Abstr.*, **39** no. 1000.

Stambekov, S. Zh., Shapiro, Y.U.O. & Mandrusova, E.E. (1977*b*). Milk protein polymorphism in Soviet Merino and Latvian Darkheaded ewes and its connection with economically valuable characteristics. *Dairy Sci. Abstr.*, **39** no. 6873.

Stur, I., Neumeister, E., Mayr, B., Glawischnig, E. & Schleger, W. (1976). Relationships between mastitis resistance, degree of heterozygosis and milk proteins in Austrian Brown cattle. *Wien. Tier: Monat.*, **63**, 352–6.

Szmelik, L., Zagulski, T. & Michalak, W. (1971). Investigations on genetic polymorphism of β-casein in cattle bred in Poland. *Biul. Inst. Genet. Hodowli Zwierzat Pol. Akad. Nauk.*, **23**, 51–65.

Tanev, G. (1967). Thin layer starch gel electrophoretic separation of the principal proteins of ewe's milk. *Zhivotnovdi Nauki.*, **41**, 23–9.

Tervala, H.L., Antila, V., Syvaejaervi, J. & Lindstroem, U.B. (1983). Variations in the renneting properties of milk. *Meij. Aika.*, **41**, 24–33.

Tervala, H.L., Antila, V. & Syvaejaevi, J. (1985). Factors affecting the renneting properties of milk. *Meij. Aika.*, **43**, 16–25.

Thompson, M.P. (1966). DEAE-cellulose-urea chromatography of casein in the presence of 2-mercaptoethanol. *J. Dairy Sci.*, **49**, 792–5.

Thompson, M.P. (1970). Phenotyping milk proteins: A review. *J. Dairy Sci.*, **53**, 1341–8.

Thompson, M.P. (1971). α_s and β-caseins. In *Milk Proteins*, vol. II, ed. H.A. McKenzie. Academic Press, New York and London, pp.117–73.

Thompson, M.P. & Pepper, L. (1964). Genetic polymorphism in caseins of cow's milk. IV. Isolation and properties of β-casein A, B, and C. *J. Dairy Sci.*, **47**, 633–6.

Thompson, M.P., Kiddy, C.A., Pepper, L. & Zittle, C.A. (1962). Variations in the α_s-casein fraction of individual cow's milk. *Nature*, **195**, 1001–2.

Thompson, M.P., Kiddy, C.A., Johnston, J.O. & Weinberg, R.M. (1964). Genetic polymorphism in caseins of cow's milk. II. Confirmation of the genetic control of β-casein variation. *J. Dairy Sci.*, **47**, 378–81.

Thompson, M.P., Gordon, W.G., Boswell, R.T. & Farrell, H.M. (1969*a*). Solubility, solvation, and stabilization of α_{s1}- and β-caseins. *J. Dairy Sci.*, **52**, 1166–73.

Thompson, M.P., Gordon, W.G., Pepper, L. & Greenberg, R. (1969*b*). Amino acid composition of β-caseins from the milks of *Bos indicus* and *Bos taurus* cows: a comparative study. *Comp. Biochem. Physiol.*, **30**, 91–8.

Visser, S., Jenness, R. & Mulin, R.J. (1982). Isolation and characterization of β- and γ-caseins from horse milk. *Biochem. J. Mol. Asp.*, **203** 131–9.

Visser, S., Slangen, K.J. & Rollema, H.S. (1986). High performance liquid chromatography of bovine caseins with the application of various stationary phases. *Milchwissenschaft*, **41**, 559–62.

Voglino, G.F. (1972). A new β-casein variant in Piedmont cattle. *Anim. Blood Grps Biochem. Genet.*, **3**, 61–2.

Voglino, G.F. & Ponzone, A. (1972). Polymorphism in human casein. *Nature New Biol.*, **238**, 149–50.

Voglino, G.F., Caone, M. & Ponzone, A. (1975). A new beta-casein variant in human milk. *Biomedicine*, **32**, 342–5.

Wake, R.G. & Baldwin, R.L. (1961). Analysis of casein fractions by zone electrophoresis in concentrated urea. *Biochim. Biophys. Acta*, **47**, 225–39.

Watterson, G.A. & Guess, H.A. (1977). Is the most frequent allele the oldest? *Theor. Popul. Biol.*, **11**, 141–60.

Woychik, J.H. (1964). Polymorphism in κ-casein of cow's milk. *Biochem. Biophys. Res. Commun.*, **16**, 267–71.

Yaguchi, M. & Rose, D. (1971). Chromatographic separation of milk proteins: A review. *J. Dairy Sci.*, **54**, 1725–43.

Yebrovski, L.S. & Komissarenko, A. (1982). Effect of genotype on intra-breed variation in the amino acid content of milk proteins. *Proc. XXI Intern. Dairy Congr. (Moscow)*. vol. 1, book 1, p.46.

Zadworny, D., Kuhnlein, U. & Ng-Kwai-Hang, K.F. (1990). Determination of kappa-casein alleles in Holstein dairy cows and bulls using the polymerase chain reaction. *Proc. 4th World Congr. Genet. Appl. Livestock Prod. XIV:251–4.*

Zebrovskii, L.S., Babukov, A.V. & Mityutko, V.E. (1977). The relationship of polymorphic proteins with productivity in Black Pied cattle. *Zhivotnovodstvo*, **7**, 25–7.

Zikakis, J.P., Haenlein, G.F.W., Hines, H.C., Mather, R.E. & Tung. S. (1974). Gene frequencies of electrophoretically determined polymorphisms in Guernsey blood and milk. *J. Dairy Sci.*, **57**, 405–10.

11

NUTRITIONAL ASPECTS OF MILK PROTEINS

LEIF HAMBRÆUS

Department of Nutrition, University of Uppsala, Dag Hammarskjölds väg 21, S-752 37 Uppsala, Sweden

1. INTRODUCTION

The nutritional significance of milk proteins in the human diet can essentially be divided into two sections: a *macronutrient aspect* which essentially comprises the role of milk proteins as a source of dietary protein, and a *micronutrient or specific physiological aspect* in which the specific physiological functions of certain protein fractions are taken into consideration.

The *macronutrient aspect* comprises, essentially, the content of amino acids, especially the essential amino acids, and nitrogen in milk proteins in order to meet the requirements of protein synthesis.

The *specific physiological aspect* of certain milk proteins comprises their potential role in the defence against infections as well as for the transport of certain nutrients (e.g. trace elements) or for certain metabolic implications, e.g. casein and hypercholesterolaemia or endocrine functions, e.g. epithelial growth factor and caseinomorphines.

From a strictly nutritional point of view, only their roles for the transport of certain nutrients as well as for the metabolism of other metabolites of nutritional interest should be considered in addition to the macronutrient aspects. Their role in the defence mechanisms against infections is, however, not essentially a nutritional problem although there is a close relationship between malnutrition and infection. A strict nutritional approach to the role of milk proteins versus the specific physiological role of certain milk protein fractions will, however, lead to slight confusion in our evaluation of the nutritional role of milk proteins in the diet. As will be commented upon later in this review, this is especially obvious when the true protein content of human milk is discussed in relation to that which is nutritionally available as a source of amino acids (Woelderen, 1987).

1.2. Milk—An Exclusive Source of Nutrients

One of the major characteristics of the nutritional situation during the neonatal period in mammals is the fact that the offspring has to rely upon one single source of nutrients—milk. Consequently, never in life is there less possibility to compensate for any lack of nutrients in the diet. Furthermore, there is a very pronounced growth and maturation of tissues during the neonatal period which creates an increased and specific demand for essential nutrients. At the same time there is a reduced tolerance for deviations in food intake due to the immaturity of the liver and kidneys, both of which play essential roles in the metabolism and excretion of metabolites. Consequently, it has rightly or wrongly been assumed that the composition of the milk of any mammalian species is the definitive indicator of the nutrient requirements of the offspring during the neonatal period.

The composition of milk obtained from various species shows great variations (Jenness, 1974) (Table 1). It is therefore logical to assume on a teleological basis that the composition of milks of different species represents the optimum composition of nutrients required during the neonatal period of that species. Several authors have anticipated that there would be a relationship between the rate of growth of the offspring and nutrient density, i.e. the amount of essential nutrients in relation to its energy content, e.g. g protein per 10 MJ or 1000 kcal.

2. The Role of Dairy Products in the Human Diet

In the industrialized world, dairy products constitute an essential part in the staple diet, especially in northern Europe and North America. In these areas milk products constitute 20–30% of the total dietary protein intake and about 65% of the animal protein intake. The type of dairy products differ; in some countries liquid milk products, in the form of more or less standardized consumer's milk with various fat contents, dominate while in other countries dairy products are essentially consumed in the form of cheese or fermented products, e.g. yoghurt.

In the developing world, however, milk and dairy products constitute only a minor component, 1–2% of the protein intake. They are usually only available for privileged people. Furthermore, it is almost impossible to transport fresh milk, and consequently fermented milk products and cheese dominate. In India and Pakistan, however, it is estimated that 8–10% of the protein intake in the diet comes from dairy products. As a matter of fact, India ranks third in the world as milk producer after the Soviet Union and the United States, and the availability of milk per caput increased from 39 kg per person in 1966 to 57 kg per person in 1987.

Another important factor is that milk products are usually consumed by those who are in special need of a nutritionally well-balanced diet, i.e. the vulnerable groups comprising infants and pre-school children. The various milk products, i.e. fresh milk, fermented products, as well as cheese, all represent very valuable supplements in the human diet. Cheese manufacture also represents a method of

TABLE 1
Composition of Milks Obtained from Different Mammals and Growth Rate of Their Offspring

Species	*Total protein*[a]	*Casein (% total protein)*	*Fat*[a]	*Lactose*[a]	*Ash*[a]	*Days required to double birth weight*
Man	0·9	30	3·8	7·0	0·2	180
Goat	2·9	86	4·5	4·1	0·8	12
Cow	3·4	80	3·7	4·8	0·7	47
Water buffalo	3·8	84	7·4	4·8	0·8	
Indian elephant	4·9	39	11·6	4·7	0·7	
Sheep	5·5	84	7·4	4·8	1·0	10
Rat	8·4	76	10·3	2·6	1·3	6
Northern fur seal	8·9	52	53·3	0·1	0·5	
Blue whale	10·9	66	42·3	1·3	1·4	10

Source: Jenness (1974) with some slight revision.
[a]Expressed as g per 100 g whole milk.

preserving the nutritionally high-value proteins in milk, which itself cannot be stored readily.

Milk represents one of the most complete single food items within the human diet. However, milk and dairy products as an ingredient in the diet are rarely used as a single food item but in combinations with other food components. This means that their nutritional role in the diet is essentially as a supplement of nutrients in the diet for man at all ages from the weaning period up to old age. Thus, most interest should be devoted to the components in which dairy products are especially rich or have a specific nutritional value as supplements to a conventional diet. Dairy products can also be used as a component in the diet of small children, e.g. in formulas and gruels. Although it is obvious that it is possible to formulate high quality breast-milk substitutes to be used when the mother for one or other reason is unable to breast-feed her child, this might lead to a conflict between formula-feeding and breast-feeding, which led to the WHO international code for marketing breast-milk substitutes (WHO, 1981).

The use of milk proteins in non-dairy products is of increasing interest as this is often based on the exploitation of the functional characteristics of certain milk proteins but could be further supported as an alternative to other animal or vegetable proteins due to their high nutritional value.

Milk may also contain certain components which play a specific physiological role in the body. Whether or not this should only be of significance during infancy and childhood is not known. There is reason to believe that the specific physiologic or metabolic roles of certain milk protein fractions, e.g. lactotransferrin, or minor proteins or biologically active peptides, e.g. epidermal growth factor, are valid also in the diets of adults.

When discussing the nutritional significance of milk, a number of questions arise:

(1) Is there a difference with respect to the various components of the milk, i.e. are some constituents more specific for the species than others? If so, does this mean that they are also more essential from the nutritional point of view not only for the development of the offspring?
(2) What evidence do we have that milk is the most relevant source of nutrients? During what age period does it seem to be the optimum food? Are milk products always valuable constituents of the human diet?
(3) What do we know regarding the stability of the composition of milk? What do the physiological variations in milk composition mean nutritionally?
(4) What do we know about the significance of various environmental factors on milk composition, i.e. maternal diet, drugs, contaminants, environmental pollution?

Based on observed differences in the composition of milk from various species, Jenness (1974) divided milk components into the following groups based on whether they were specifically secreted in the mammary gland or not and whether they occurred only in certain species:

(1) organ and species specific;

(2) organ specific but not species specific;
(3) not organ specific but species specific; and
(4) neither organ nor species specific.

It can be assumed that the milk components that are both organ and species specific or organ specific but not species specific are of greatest interest when nutritional and metabolic aspects of dairy products are discussed. This hypothesis seems to be valid also in the discussion of the nutritional aspects of milk proteins.

3. Composition of Milk Proteins

All milks analysed to date contain considerable amounts of protein although the concentration varies between less than 1% up to 10%, the protein content of human milk being the lowest at 0·8 g per 100 ml, and those of the blue whale and reindeer at about 11 g per 100 ml, being the highest. There is also a very pronounced variation in the number and type of proteins present in milks from different species.

Milk proteins have been intensively studied throughout the years by protein biochemists (McKenzie, 1971). Thus, the origin of the word protein dates to 1838 when the Dutch Chemist Mulder, who devised one of the first acid precipitation methods for the separation of bovine casein, fibrin and egg white, accepted to call them proteins as was suggested by the Swedish chemist Berzelius in a letter to him dated July 10 1838 (Vickery, 1950). Originally, casein was assumed to be the only or major specific milk protein while most of the less abundant milk proteins found were thought to be derived from blood. Later, extensive studies have shown that many of the non-casein proteins in milk are also synthesized in the mammary gland and must be considered as milk specific (Kon & Cowie, 1961; McKenzie, 1971). The proteins of bovine milk have, for obvious reasons, been more widely studied than those of any other species, man included.

Studies on composition and the specific nutritional and physiological function of bovine milk proteins have been intensive during the last two decades (see Galesloot & Tinbergen, 1985; Barth & Schlimme, 1989). Increased interest has also been devoted to the composition of human milk and its specific nutritional and metabolic implications in relation to cows' milk (Blanc, 1981; Hambraeus, 1984; Lönnerdal, 1985). The nomenclature of milk proteins is, however, still essentially based on the characterization of bovine milk proteins (Whitney *et al.*, 1976).

Milk contains a heterogeneous mixture of proteins and consequently a variety of methods has been used to separate and characterize the various milk protein fractions since acid precipitation of casein was introduced at the beginning of the 19th century. Isolation and characterization of milk proteins is complicated by their considerable tendency to associate and form complexes which makes the definition of their specific physiological, nutritional and immunological characteristics difficult.

As dairy products are essentially based on bovine milk, most nutritional studies on milk proteins deal with bovine milk proteins. In countries where nutritional

TABLE 2
Protein Composition in Mature Human Milk and Cows' Milk

Component	*Human milk (mg/ml)*		*Cows' milk (mg/ml)*[a]
Total nitrogen	1·54	0·16[b]	5·2
Whey nitrogen	1·20	0·27[b]	0·7
Non-protein nitrogen	0·31	0·04[b]	0·3
Casein	2·33	1·69[c]	27
Whey proteins			
α-Lactalbumin	1·73	0·25[b]	0·9
β-Lactoglobulin	—		3·0
Lactoferrin	1·48	0·19[b]	0·001 2
Secretory IgA	1·15	0·62[b]	
Lysozyme	0·07	0·05[b]	0·000 1
Serum albumin	0·38	0·15[b]	0·3

[a]Compiled data from literature and author's data.
[b]Woodhouse & Lönnerdal (1988).
[c]Lönnerdal & Forsum (1985).

problems may be more serious than in the industrialized world, the nutritional implications of differences in milk protein composition may be potentiated. It should, in this respect, be remembered that in less developed countries milks from other species are consumed, e.g. buffalo, camel and goat, making the specific characteristics of their milk protein composition worth consideration. The principal nutritional problems are, however, the same, e.g. the potential hazards and benefits of consuming non-species-specific milk proteins. As most data are available regarding the composition of bovine milk, this review will essentially deal with the nutritional aspects of the differences between bovine and human milk proteins.

When analysing the specific nutritional value of milk proteins in the diet it is of special interest first to compare the protein composition of bovine milk with that of human milk and analyse the points of resemblance and differences and what nutritional implications these might have.

As indicated in Table 2, there are two major differences:

(1) there is an almost inverse relationship between the contents of caseins and whey proteins in the two milks;
(2) there are obvious differences in the composition of the whey proteins.

Unfortunately we still do not fully understand the physiological implications, neither nutritionally nor metabolically, of these differences. This of course hampers our complete understanding of the specific nutritional role of milk proteins.

4. PROTEIN REQUIREMENT IN MAN

The macronutrient aspect, or nutritive value, of milk protein is related to its amino acid content as well as composition and to the availability of these amino acids.

The nutritive value thus has two dimensions: quantitative and qualitative. The content of essential amino acids, e.g. those that cannot be synthesized within the body and consequently must be supplied through the diet, is of special concern. This should then be related to the protein requirement of man. When discussing the nutritive value of a protein we must furthermore consider that the protein requirement can be divided into two categories: (1) growth requirement and (2) maintenance requirement. The first refers particularly to the situation in infants and children but also to pregnant and lactating women as well as to adults during convalescence, physical training or body-building, when there is increased protein synthesis. The maintenance requirement, on the other hand, refers to the need to cover the daily turnover of about 400 g of body protein of which about 25% comes from the diet.

It is still an open question whether there is a qualitative difference with regard to the need for essential amino acids to meet these two requirements. In formulating the 1985 edition of protein requirements, the FAO/WHO/UNU expert consultation followed the same principle as the 1973 FAO/WHO committee where they establish two principle reference points: the maintenance protein requirement of the young child and the requirement of the young male adult. However, the 1985 FAO/WHO/UNU consultation concluded that few normal diets provide insufficient amounts of essential amino acids, except for infants and pre-school children.

5. What is meant by the Nutritive Value of a Protein?

Milk protein, as well as egg protein, have often been used as references for evaluation of the nutritive value of food proteins (FAO, 1970; FAO/WHO, 1973). This is essentially due to the fact that they seem to be the only ones originally intended to function as a single source of nutrients for the offspring. However, it is obvious that there is quite a difference in the protein composition, and thereby also the amino acid composition, not only between milk and egg but also between milks obtained from various species (FAO, 1970). Consequently, it seems more logical to use human milk composition as reference for the optimal protein and amino acid composition, at least during the neonatal period. This is also accepted for infants in the latest edition of the FAO/WHO/UNU protein requirement (1985). Above the age of 12, however, the FAO/WHO/UNU do not find it necessary to correct the amino acid score when a mixed diet is consumed, as most natural diets provide sufficient amounts of essential amino acids. However, in the 1985 edition of the FAO/WHO/UNU protein recommendations, the expert consultation stresses that in calculations of the nutritive value of dietary proteins more attention should be given to the *digestibility* of the proteins in a mixed diet as the availability of dietary protein for all age groups can be significantly affected by digestibility. True digestibility is defined as the per cent of protein or nitrogen absorbed from a given protein (nitrogen) intake

$$100 \times (\text{nitrogen intake} - \text{faecal nitrogen}) \div \text{nitrogen intake}.$$

This is considered to be the most important factor for the nutritional value of dietary proteins for adults (FAO/WHO/UNU, 1985). Interestingly this was almost neglected in earlier reports from expert committees dealing with the protein requirement of man. The relatively low relevance of the amino acid composition for the nutritive value of dietary proteins for adults means that the superiority of milk proteins, owing to their relatively high content of essential amino acids is less valid for adults than for children. However, in comparison with vegetable proteins, the high digestibility of milk proteins is still a nutritional advantage.

6. How do we Estimate the Nutritional Value of Food Proteins?

From a biochemical point of view, protein is defined essentially as polypeptides containing more than about 50 amino acids and/or a molecular weight above 6000. However, from a nutritional viewpoint, proteins are of interest as a source of amino acids. Consequently, the size of the protein molecule is of less importance and small peptides are also relevant. This obviously has an impact when comparing results obtained by various methods used to estimate the nutritional value of milk proteins.

Assessment of the nutritive value of food proteins involves several types of analyses (Table 3). Nutritionally, the protein content and composition of milk must be analysed from at least two aspects: first, as a source of amino acids for protein synthesis and maintenance and secondly, the various protein components *per se* must be analysed from their potential specific physiological roles with impact on the nutritional status.

6.1. Biochemical Evaluations

Biochemical analysis involves determination of the *total protein content*, usually based on assays of total nitrogen by the Kjeldahl method, as well as determination of *specific protein components*, usually by immunochemical methods or gel filtration. *Amino acid analysis* after acid hydrolysis or enzymatic digestion *in vitro* of the proteins will also give information on the nutritive value of the various protein fractions.

6.2. Total Protein Analysis

There is no absolute method for the determination of total protein content in a complex mixture of proteins in solutions containing peptides, amino acids and other nitrogen-containing components, e.g. milk. In food tables, the protein content of various food items is almost always based on indirect calculations derived from nitrogen analyses by the Kjeldahl method. The Kjeldahl method is an adequate reference method when the correct digestion temperature and catalyst are chosen. The nitrogen value derived from the analysis is then multiplied by a

TABLE 3
Methods for Evaluation of Nutritive Value of Protein

In-vitro methods
Biochemical methods
Nitrogen analysis
Amino acid analysis
Microbiological methods
Amino acid analysis
Digestion by proteolytic microorganisms
Streptococcus zymogenes
Tetrahymena pyriformis
Enzymatic methods
Enzyme preparations
Proteolytic microorganisms
In-vivo methods
Biological tests with experimental animals
Screening methods (eg. PER)
Nitrogen balance studies (NPU, BV NDpCal%)
Indirect nitrogen analysis
Toxicological studies
Human tests (in adults and infants)
Acceptability tests
Nitrogen balance studies

conversion factor, usually 6·25 since most proteins have a nitrogen content of 16%. However, it is quite obvious that the nitrogen content of various amino acids differs (Table 4). In addition, since total nitrogen comprises protein nitrogen as well as non-protein nitrogen, the conversion factor is substantially dependent not only on the amino acid composition of the protein but also on the ratio between non-protein nitrogen and protein nitrogen. Consequently, various conversion factors must be used for different proteins, as illustrated in Table 5. However, in the interest of standardization, the protein content of food items calculated using a conversion factor of 6·25 should be given as 'crude protein'.

Since the Kjeldahl method is not as simple and rapid as would be needed when large numbers of samples are to be analysed for protein content, other analytical approaches have been tested, e.g. dye-binding methods, spectrophotometric methods and infra-red spectroscopy. Results obtained with different methods are, however, not directly comparable and may lead to different nutritional conclusions.

Interestingly, human and bovine milks contain almost the same amount of non-protein nitrogen (40–50 mg per 100 ml) (Table 2). However, the total amount of nitrogen is much lower in the former, being only about 200 mg per 100 ml versus 540 mg per 100 ml in bovine milk. The protein content of human milk, calculated on the basis of total nitrogen determinations using 6·38 as conversion factor, has

TABLE 4
Nitrogen Content of Amino Acids (%)

Amino acid	*Nitrogen content* (%)
Less than 16·0%	
Tyrosine	7·7
Phenylalanine	8·5
Methionine	9·4
Glutamic acid	9·5
Aspartic acid	10·5
Hydroxyproline	10·7
Isoleucine	10·7
Leucine	10·7
Cystine	11·7
Threonine	11·8
Valine	12·0
Proline	12·2
Serine	13·3
Tryptophan	13·7
Alanine	15·7
More than 16·0%	
Asparagine	18·7
Glycine	18·7
Glutamine	19·2
Lysine	19·2
Histidine	27·1
Arginine	32·2

TABLE 5
Nitrogen-Protein Conversion Factors

Wheat (whole grain)	5·83
Maize	6·25
Rice	5·95
Potato	6·25
Bean (*Phaseolus vulgaris*)	6·25
Soybean	5·71
Sunflower	5·30
Beef	6·25
Egg	6·25
Cow's milk	6·38
Human milk	6·38[a]

[a] If corrected for non-protein nitrogen content based on data from Lönnerdal *et al.* (1976*a*), the conversion factor would be 5·18.

been reported to be 1·1–1·2 g per 100 ml (FAO, 1970; Fomon 1974). However, in bovine milk only 5% of the nitrogen occurs as non-protein nitrogen while as much as 25% of the nitrogen in human milk is in a non-protein form. The value of crude protein consequently represents an overestimation by about 20% in human milk. Analysis of total nitrogen minus non-protein nitrogen, as well as calculations based on amino acid analysis of true proteins, have shown that the protein content is only 0·8–0·9 per 100 ml in milks obtained from well-nourished as well as from malnourished, lactating women (Lindblad and Rahimtoola, 1974; Lönnerdal *et al.*, 1976*a*,*b*; Hambraeus *et al.*, 1978) (Table 6). Such a low protein content in human milk was reported by Macy *et al.* (1931), but their data seem to have been neglected for many years.

6.3. Amino Acid Analysis

Amino acid analysis of milk samples are usually performed by ion exchange chromatography using amino acid analysers. Although this analysis does not differentiate between free and protein-bound amino acids, this might be of little practical nutritional significance. However, some amino acids are destroyed during analysis, e.g. tryptophan, or may react to form unavailable components, e.g. lysine, due to Maillard reactions during heat treatment in the dairy industry, i.e. roller drying, or during food preparation (Abrahamsson *et al.*, 1977).

The amino acid content of a food protein is often expressed in relation to that of a reference protein, using the term 'chemical score', first introduced by Mitchell and Block (1946). The amino acid that occurs in the lowest concentration in relation to the amount in the reference protein is called the 'first limiting amino acid' and the percentage represents the 'chemical score' or 'amino acid score'. However, one serious limitation of all amino acid reference patterns used to date is the fact that sulphur amino acids, methionine and cysteine/cystine, are added together, as are the aromatic amino acids, tyrosine and phenylalanine. This has some nutritional implications, especially when dietary proteins for use during early infancy are evaluated as nutritive sources. Another disadvantage is the fact that the score value is influenced only by the content of the limiting amino acid but it gives no information whatsoever regarding any changes in the content of other essential amino acids as long as there is no change in the first limiting amino acid. Thus, any destruction of lysine in milk products as a result of the Maillard reaction will not affect the chemical score value until lysine has become the first limiting amino acid. Finally, the usefulness of amino acid score as an index of the nutritive value of a protein is of course completely dependent on the selection of a relevant reference protein.

Various amino acid scoring systems have been used throughout the years by international groups (FAO, 1957; FAO/WHO, 1965,1973; FAO/WHO/UNU, 1985). This is illustrated in Table 7 where the egg protein reference pattern recommended by FAO in 1965, and the provisional amino acid score patterns recommended by FAO in 1957, 1973 and 1985 are shown in relation to the essential amino acid content of casein and skim milk protein. It shows that with the egg

TABLE 6

Essential Amino Acid Content in Skim Milk Powder, Whey Proteins and Casein, and Amino Acid Pattern in FAO Egg Reference (1965) and FAO Provisional Amino Acid Scoring Pattern (1973)[a]

Amino acid	*Skim milk powder (FAO, 1970)*	*Casein (FAO, 1970)*	*Whey protein (Forsum* et al., *1973)*	*Human milk*		*FAO/WHO (1965)*	*FAO/WHO (1973)*	*FAO/WHO/UNU (1985)*
				FAO (1970)	*Lönnerdal* et al. *(1976a)*			
Isoleucine	52	54	76	40	49	66	40	46
Leucine	97	95	118	86	91	88	70	93
Lysine	71	81	113	67	65	64	55	66
Methionine + cysteine	34	32	52	29	37	55	35	42
Phenylalanine + tyrosine	96	111	70	66	76	100	60	72
Threonine	41	47	84	44	44	51	40	43
Tryptophan	14	16	24	NA[b]	NA[b]	16	10	17
Valine	63	75	72	45	52	73	50	55
Total essential amino acids	468	511	609	377	414[c]	513	360	424

[a]The values refer to mg amino acid per g protein.
[b]NA, not analysed.
[c]Tryptophan not included.

TABLE 7
Nutritionally Available Proteins in Human Milk (in g per 100 ml)

Based on Kjeldahl analysis of total N	
(a) using conversion factor 6·38	1·1–1·2
(b) using conversion factor 5·18[a]	0·8–0·9
Based on true protein content	
(a) total protein[b]	0·8–0·9
(b) with correction for nutritionally not available proteins	0·5–0·7

[a]Assessing 25% of total N to be derived from non-protein nitrogen (Hambræus *et al.*, 1984).
[b]Based on the sum of different major protein fractions analysed by specific methods.

protein pattern as a reference, the sulphur-containing amino acids are the limiting ones, giving a chemical score of 62 and 58, respectively, while the scores are 97 and 91 with the FAO reference pattern. The amino acid scoring systems have developed expectations that they should not only predict the potential nutritional value of a food protein for humans but also correlate directly to the results obtained by means of biological assays. As the amino acid content is considered to be of little importance in relation to the digestibility of dietary protein, the concept of chemical score and amino acid score no longer has the same nutritional priority in the evaluation of dietary proteins.

6.4. Biological Evaluations

The biochemical analyses are often supplemented by the use of some biological methods in order to establish whether the amino acids analysed are available for digestion and absorption in the gastrointestinal tract. These studies are obviously more relevant since the 1985 edition of the international protein requirements (FAO/WHO/UNU, 1985) places special emphasis on the digestibility of proteins. Various biological methods can be used (Table 3) including in-vitro methods using enzymatic digestion or microbiological assays using *Streptococcus zymogenes* or *Tetrahymena pyriformis*. However, most biological methods for assessing protein quality are based on some type of feeding study with experimental animals in metabolic cages, usually rats. Since we are usually interested in the nutritive value of proteins for *human* consumption it should always be remembered that each species has its own requirements for protein and amino acids. The final proof of the nutritional value of a protein in the human diet can, consequently, only be based on metabolic studies on humans. It is, for obvious reasons, not possible to perform these studies to any large extent. The need of a perfect method for quality evaluation of dietary proteins is obvious, but has somewhat decreased since most natural diets seem to provide the needed amounts for protein synthesis and turn-over.

7. Milk Protein Composition

Milk proteins are often subdivided into two major groups: the *caseins* and the *whey proteins*. This separation is based on the fact that casein is precipitated in the dairy industry during cheese manufacture, which leaves whey as a by-product. Maximum precipitation of bovine casein is obtained at pH 4·6 and this is also the most commonly used definition of casein. However, the conditions for maximum precipitation of casein vary for milks from different species, e.g. pH 4·0 is reported to be optimal for casein precipitation in rat milk (Glass, 1956) while human casein is very difficult to precipitate (Mellander & Fölsch, 1972; Lönnerdal & Forsum, 1985). Furthermore, there is often a co-precipitation of whey proteins or non-casein proteins with the caseins. In the manufacture of some cheeses this is considered very desirable in order to get optimal recovery from milk. From a nutritional point of view, this is also an advantage since the whey proteins have a higher nutritional quality than the caseins (Forsum, 1973).

Since dairy products (e.g. fresh or fermented milks, hard and soft cheeses) contain the different milk proteins in varying amounts, which will influence the protein nutritive value of the product, the specific nutritional aspects on the milk protein components will be commented on.

7.1. The Caseins

The caseins comprise a group of proteins with a unique molecular structure (Swaisgood, 1973). The caseins occur as micelles containing calcium and inorganic phosphate and represent some of the few naturally occurring phosphoproteins. The caseins are milk-specific proteins but the types of casein show inter-species variation. On the other hand, some of the inter-species similarities may have nutritional implications, e.g. a high content of ester-bound phosphate, a high proline content, a low content of sulphur-containing amino acids, especially of cystine, and a low solubility at pH 4–5. The caseins constitute the dominant protein fraction in bovine milk, where it represents about 80% of the total protein. In human milk, however, casein constitutes only 20–30% of total protein (Lönnerdal & Forsum, 1985) (Table 2). Caseins are not easily quantified due to their heterogeneity and micellar composition. They are, however, classically determined by acid precipitation at pH 4·6, which is the isoelectric point of casein. However, as the casein profile differs in milks from various species, this also influences their isoelectric points. Human casein does not have the same distinct isoelectric point as bovine casein and other major protein components precipitate between pH 4 and 5 (Lönnerdal & Forsum, 1980). Furthermore, certain whey proteins form complexes, some of them with casein at this pH. Consequently, acid precipitation overestimates the casein content of human milk (Lönnerdal & Forsum, 1985). By means of isoelectric precipitation and ultracentrifugation methods, as well as indirect quantitation of casein by analysis of total proteins and subtracting the values for the major whey proteins determined by immunochemical methods, less than 30% of the protein in human milk is casein.

α-Casein is the dominant casein in ruminant milks, but does not occur in human milk, where β-casein is the dominant form (Jenness & Sloan, 1970). The γ-caseins represent the C-terminal segments of β-casein from which the acidic phosphopeptide is missing; γ-caseins are, consequently, no longer referred to as a separate casein group. κ-Casein is unique since it remains soluble in calcium solution under conditions that precipitate all other casein components and it can consequently also occur in the whey fraction. It stabilizes the other caseins in the presence of calcium and plays a key role in the formation of calcium-containing casein micelles. κ-Casein also has carbohydrate side chains, the carbohydrate content of human κ-casein being higher than that of bovine κ-casein (Yamauchi *et al.*, 1984). When chymosin or pepsin act on κ-casein, a glycomacropeptide is formed which is suggested to be part of the so-called bifidus factor (György *et al.*, 1953) which plays a significant role during infancy as it inhibits gut colonization by *Escherichia coli* and promotes growth and colonization of *Bifidobacteria* in the gut (Bezkorovainy *et al.*, 1979). This initiated the production of special bifidus-factor-containing infant formulae.

It has been suggested that the primary function of the caseins is the nutrition of the newborn. This is due not only to their role as a source of amino acids but also to their role as sources of calcium and inorganic phosphate. Calcium-containing casein micelles make the content of calcium and inorganic phosphate in milk far higher than would be expected from their physicochemical solubility in milk.

Curd formation may also be of physiological interest for the function of the digestive tract and may enable the gastric enzymes to start protein digestion in the newborn but such functions are difficult to evaluate. There are, however, physicochemical differences between human and bovine caseins resulting in different curd formation characteristics in the stomach. The casein micelles of human milk are smaller than those of bovine milk. Casein is very heat-resistant but is precipitated at low pH values, resulting in curd formation; it is also precipitated by a special enzyme, chymosin, which occurs in the stomach. The coagulum from human milk is soft and flocculant whilst bovine milk coagulum is resistant and firmer, the physiological significance of which is still not fully understood.

Being a very specific phosphoprotein, casein probably plays a role in the phosphorus balance in adults also. From the nutritional point of view, this means that the caseins not only function as a protein source but also as a source of calcium and phosphorus. A milk-free diet will consequently also have an impact on the calcium intake of a population at all ages. Thus, in Sweden 75% of the dietary calcium intake is derived from dairy products.

From the nutritional point of view it is also of interest to comment on the *uniqueness of the amino acid composition* of the caseins. The physiological role of casein in the human diet is still not very well known. It is consequently somewhat difficult to establish whether it can be substituted for, nutritionally, by other dietary proteins. This is illustrated in Table 8. It is seen that the very low cystine content of the caseins results in a methionine/cystine ratio that is 2–3 times higher than that in other animal proteins and more than 7 times higher than that in human milk. Human milk, as a matter of fact, seems to be the only animal protein source

TABLE 8
Amino Acid Content in Casein and Whey Protein (mg Amino Acid per g Total Nitrogen)

	Human breast-milk	*Casein*[a]	*Whey protein*[b]	*Egg*[a]	*Beef*[a]	*Wheat (whole grain)*[a]	*Soybean*[a]
Isoleucine	254	345	476	393	301	204	284
Leucine	471	607	736	551	507	417	486
Lysine	337	518	704	436	556	179	399
Methionine	78	178	151	210	169	94	79
Cystine	114	23	174	152	80	159	83
Phenylalanine	171	334	224	358	275	282	309
Tyrosine	223	371	214	260	225	187	196
Threonine	228	297	527	320	287	183	241
Tryptophan		103	147	93	70	68	80
Valine	296	430	449	428	313	276	300
Arginine	171	239	175	381	395	288	452
Histidine	114	186	144	151	213	163	158
Alanine	166	196	341	370	365	226	266
Aspartic acid	451	455	766	601	562	308	731
Glutamic acid	1 000	1 406	1 231	796	955	1 866	1 169
Glycine	98	126	126	207	304	245	261
Proline	513	738	450	260	236	621	343
Serine	228	385	374	478	252	287	320

[a]FAO (1970).
[b]Forsum *et al.* (1973).

that has a methionine/cystine ratio below or close to 1·0. This is probably due to its high content of whey proteins and concomitant low casein content. This is of considerable metabolic and nutritional interest since it is known that the premature infant lacks the enzymatic capacity to convert methionine to cysteine which makes cysteine or cystine an essential amino acid (Sturman *et al.*, 1970). The lack of cystine in casein is consequently a nutritional disadvantage. Interestingly, the cystine content of milk is almost exclusively derived from β-lactoglobulin and α-lactalbumin. The use of casein-rich infant formulae therefore warrants further consideration from the nutritional and metabolic point of view.

Casein has been used as a reference protein for many years in the evaluation of the nutritive value of proteins, especially in biological assays in rats. It is, however, well known that casein is by no means the optimal protein from a nutritional point of view since several animal proteins (e.g. egg protein, whey proteins) as well as a few vegetable proteins (e.g. soy protein, rapeseed proteins), show a higher nutritive value (FAO, 1970). As the sulphur-containing amino acids are the limiting ones, casein is sometimes supplemented with methionine when used as a reference protein.

7.1.1. β-Caseinomorphines

β-Caseinomorphines are sequences in the caseins which have been shown to exert an opoid, morphine-like activity (Brantl & Teschemacher, 1982). They have further-

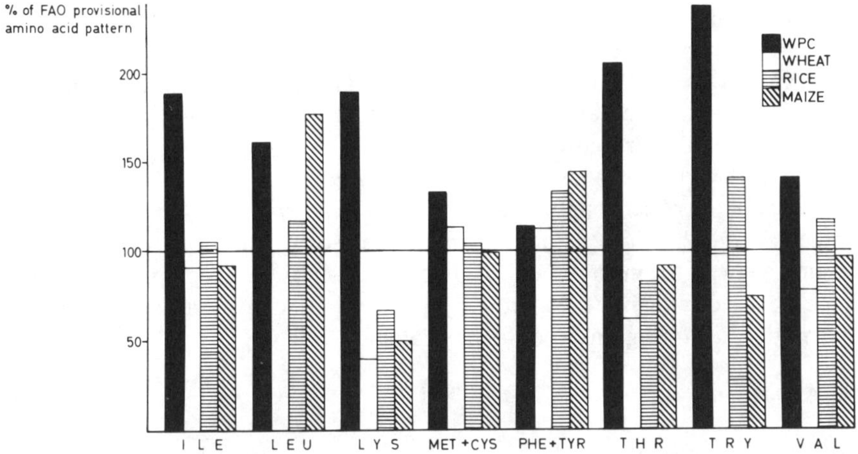

Fig. 1. Essential amino acid contents of a whey protein concentrate (WPC), wheat, corn and rice illustrated as a percentage of the FAO provisional amino acid reference pattern of 1973.

more been shown to stimulate insulin release postprandially but they do not seem to have a hypercholesterolemic effect (Pfeuffer & Barth, 1988).

7.2. The Whey Proteins

Whey is a highly nutritious by-product of cheese and casein manufacture and contains about 0·6% protein. The whey proteins represent the non-casein proteins as well as fragments of the caseins which remain soluble when the caseins have been precipitated enzymatically by chymosin or isoelectrically by acid precipitation at their isoelectric point, pH 4·6. The whey proteins represent a very heterogeneous group of proteins (Woodhouse & Lönnerdal, 1988) which share very few common characteristics except that of being soluble under conditions that precipitate the caseins. Milks from carnivorous mammals contain a considerable amount of milk-specific non-casein proteins while milks from other species, especially rodents, contain very little. One of the characteristics of the amino acid pattern of the whey protein fraction is the relatively low content of aromatic amino acids, phenylalanine and tyrosine.

One important function of the whey proteins is their contribution to the nutritive value of milk protein. Thus, the nutritive value of casein, which does not seem to be optimal from the human point of view, is inferior to that of whey proteins. It has also been postulated that casein is counterbalanced by whey proteins in human milk in order to balance the supply of essential amino acids to the newborn. The high nutritive value of the whey proteins is mainly due to their high content of essential amino acids (Forsum & Hambraeus, 1974). This is illustrated in Fig. 1 which shows the essential amino acid content of whey protein concentrate, WPC, as well as of some cereal proteins (wheat, maize and rice) expressed as a percentage of the 1973 FAO reference pattern. It is seen that the whey protein concentrate has a surplus of all essential amino acids. Of special interest are the high contents of isoleucine, leucine, threonine and tryptophan. Wheat is low in lysine and threonine, while maize is low in lysine and tryptophan. Rice, on the other hand, has a relatively well-balanced amino acid composition although the contents of lysine and threonine are fairly low.

Some of the whey proteins seem to have distinct physiological and biochemical roles, e.g. *lactoferrin* strongly binds iron, *α-lactalbumin* is a constituent of lactose synthetase and *lysozyme* is an enzyme that destroys the bacterial cell wall, to mention a few of the most discussed whey proteins. With respect to *β-lactoglobulin*, which is the dominant whey protein in bovine milk, we still lack unequivocal information regarding any specific physiological role.

7.3. β-Lactoglobulin

β-Lactoglobulin is the dominant whey protein in bovine milk. It is relatively heat-labile and occurs in the milks of all ruminants studied but has been reported to be absent from the milks of the camel and guinea pig (Bell & McKenzie, 1964) as well as human milk—although minor amounts have been reported to occur in human milk (Conti *et al.*, 1980). Although the physiological role of β-lactoglobulin is still not fully understood, it has been speculated that it plays a regulatory role in phosphorus metabolism in the mammary gland (Farrell & Thompson, 1971) and retinol binding (see Chapter 4). Some of the changes in the properties of milk that occur on heating are due to denaturation and aggregation of denatured β-lactoglobulin with other proteins such as the formation of a disulphide bond between β-lactoglobulin and κ-casein. It has been postulated that β-lactoglobulin may be partly responsible for cows' milk protein intolerance observed during infancy in those fed formula during the neonatal period.

7.4. α-Lactalbumin

α-Lactalbumin is a well-defined whey protein and one of the three dominant whey protein components in human milk, second only to β-lactoglobulin in bovine whey (Blanc, 1981). The reader should be aware of the fact that there is still some confusion as to nomenclature. The term 'α-lactalbumin' is not equivalent to 'lactalbumin' which was earlier used as a common name of the total whey protein fraction and still can be used in the nutritional or pediatric literature. α-Lactalbumin occurs in the milks of all mammals although it is present at very low levels in milks containing little or no lactose, i.e. milk from the northern fur seal and sea-lion. α-Lactalbumin has a stable configuration between pH 5·4 and 9 and it is also a relatively most heat-stable whey protein. It is part of the enzyme *lactose synthetase* (EC 2.4.1.22) and the biosynthesis of lactose seems to be regulated by the production of α-lactalbumin. Consequently there seems to be a relationship between the content of lactose and that of α-lactalbumin when milks obtained from various mammals are compared. There is however no direct correlation between the lactose and α-lactalbumin contents of human milk according to transversal and

TABLE 9
Iron-Binding Proteins in Milk from Various Species

μg/ml	*Iron*	*Lactoferrin*	*Transferrin*
0·3–0·6	Human, cow, goat		
1·4–2·4	Sow		
2–4	Rabbit		
3–14	Rat, dog		
< 50		Rat, rabbit, dog	Dog, human
20–200		Cow, goat, sow	Sow, cow, goat, mare
200–2000		Guinea-pig, mouse, mare	Guinea-pig, mouse
> 2000		Human	Rabbit, rat

longitudinal studies, which showed that the lactose concentration increases whilst α-lactalbumin content decreases throughout lactation (Lönnerdal *et al.*, 1976*a*,*b*). Although the primary role of α-lactalbumin seems to be its enzymatic function, it is obvious that it also plays a significant nutritional role as it has a very high nutritive value (Forsum, 1973). Its amino acid composition seems to be optimal for the requirements of the human infant, and it has a high digestibility.

α-Lactalbumin has a metal-ion binding site which normally binds calcium. It has therefore been suggested to have a key role in the biosynthesis of casein in the mammary gland (Hiraoka *et al.*, 1980; Lönnerdal & Glazier, 1985).

7.5. Iron-binding proteins

Milk contains two kinds of iron-binding proteins: transferrin, which seems to be almost identical to the transferrin in blood, and lactoferrin, which is closely related to transferrin but characteristic of milk and exocrine glands. All mammals seem to be able to produce these two iron-binding proteins although they do not occur to the same extent in all milks (Table 9). Lactoferrin has thus been identified, but at a much lower concentration than in human milk, in milks from cow, goat, horse, pig, mouse and guinea-pig. It seems to be absent in milk from rabbit, rat and dog where transferrin is the dominant iron-binding protein.

Lactoferrin was first isolated by Johansson (1960) as a red-coloured iron-binding protein in milk. It is synthesized in the mammary gland but also in other exocrine glands and occurs in saliva, tears and semen (Masson, 1970); the concentration in human milk is extremely high. Each molecule of lactoferrin can bind two ferric ions with the incorporation of bicarbonate. The two iron-binding sites are not equivalent, one being more acid-labile than the other.

The potential physiological role of lactoferrin was first described as a bacteriostatic one since lactoferrin is present in the human milk in a highly unsaturated form. Lactoferrin could then effectively compete with the siderophores of iron-demanding bacteria. Lactoferrin was consequently considered to be at least partly responsible for the defence against gastrointestinal infections during the neonatal period (Bullen *et al.*, 1972).

However, of more nutritional interest is a second possible function of lactoferrin:

to facilitate iron absorption. It is a well-known fact that although the iron content in human milk is very low, the bioavailability of the iron is high (Saarinen *et al.*, 1977). This has been suggested to be due to lactoferrin which should function as an iron transporter. Further evidence for such a role has been gained in recent years with the demonstration that a lactoferrin receptor exists in the brush-border membrane of the small intestine in several species (Cox *et al.*, 1979; Mazurier *et al.*, 1985; Davidson & Lönnerdal, 1988). Thus, it has been suggested that the receptor-mediated uptake of lactoferrin iron is responsible for the reported high bioavailability of breast milk iron. Lactoferrin in human milk is very unsaturated with respect to iron (3–5% of its total iron-binding capacity (Fransson & Lönnerdal, 1980)); therefore it can compete effectively with bacterial siderophores, making iron unavailable to bacteria. Thus, only a minor fraction of the iron in human milk is actually bound to lactoferrin whilst about one-third is bound to the fat fraction (fat globule membrane) and another one-third to low-molecular components (Fransson & Lönnerdal, 1980). Saturation of lactoferrin has been found to reverse its bacteriostatic effect in nature (Reiter, 1978). However it is possible that iron present in other compartments may be released and transferred to lactoferrin during digestion. Interestingly, iron-saturated lactoferrin is more resistant to proteolysis than the apo-protein (Brock *et al.*, 1976).

The specific role of lactoferrin both with respect to the defence against gastrointestinal infections and as promotor of iron transport in the intestinal tract depends on the fact that it is not digested in the gastrointestinal tract which also means that it is not nutritionally available as a source of amino acids.

7.6. Immunoglobulins

The milk protein fraction also contains a number of immunoglobulins which represent a very heterogeneous group of proteins. Immunoglobulin A, IgA, seems to be most specific for milk and occurs primarily in certain species, i.e. human milk and milk from primates, while it is a minor component in milk from ruminants where the less milk-specific immunoglobulins, IgG and IgM, dominate. Immunoglobulin A occurs as secretory IgA which is synthesized in the mammary gland by binding two molecules of IgA with two other components, a joint chain (J-chain) and a secretory component. The secretory IgA is resistant to proteolysis in the gastrointestinal tract, which is probably necessary for it to exert its physiological action as part of the defence mechanism against infections during the neonatal period.

From a strictly nutritional point of view, the immunoglobulins are of less interest since their role is not primarily as nutrients. It is however of interest that if they have an essential role in the defence mechanism against gastrointestinal infections they are not nutritionally available as sources of amino acids. This has a practical implication in evaluation of the nutritional value of those milks that contain immunoglobulins.

7.7. Enzymes

Milk contains a number of enzymes; not less than 44 different enzymes have been detected in bovine, human and some other milks and have been listed by Jenness (1974). Some of them seem essentially to be involved in the defence mechanism against bacteria in the gastrointestinal tract (i.e. lysozyme, lactoperoxidase) and are therefore of limited interest from strictly nutritional view point. Others (e.g. lipase, α-amylase, protease and protease inbibitors) may be of nutritional significance as they will affect the availability of other nutrients. Lysozyme (EC 3.2.1.17) appears in especially high amounts in human milk and is suggested to have a bacteriostatic function in the gastrointestinal tract since it destroys the cell wall structure of Gram-positive bacteria. The digestion of lipids is enhanced in breast-fed infants by a bile salt-stimulated lipase. It has been postulated that α-amylase which occurs in human and bovine milks, survives passage through the stomach and consequently helps in the digestion of starch and other oligosaccharides. The role of proteases and protease inhibitors in milk remains unknown although the nutritional significance seems to be limited.

7.8. Vitamin-binding Proteins

Two vitamin-binding proteins that occur in milks are of special nutritional interest: folate-binding proteins (Ford *et al.*, 1969) and cobalamin-binding proteins (Sandberg *et al.*, 1981). Both occur in minor amounts but there is evidence that they facilitate the uptake of the corresponding vitamin from milk in the intestinal tract as they survive low pH and some of the proteolytic enzymes (Salter & Mowlem, 1983; Andersen & von der Lippe, 1979) and also lead to the accumulation of these vitamins in the milk during secretion in the mammary gland.

7.9. Hormone-binding Proteins

7.9.1. Growth Factors

Several growth-promoting factors have been reported to occur in milk, e.g. epidermal growth factor, insulin, somatomedin-C and transforming growth factor (Koldovsky, 1989). They occur at very low concentrations and since they essentially act as hormones, with very little nutritional value, they are beyond the scope of this review.

7.9.2. Milk Fat Globule Membrane Proteins

The fat globules in the milk are surrounded by a membrane formed from the epithelial cells in the mammary gland (Freudenstein *et al.*, 1979). The membranes also seem to contain a number of enzymes, among them *xanthine oxidase* and *alkaline phosphatase*. We still know very little about the specific physiological and nutritional roles of these proteins. However, they may be of nutritional significance since they may influence the bioavailability of minerals such as iron, zinc and manganese which occur in significant amounts in the milk fat globule membrane protein fraction. This might explain the findings of trace elements, i.e. iron, zinc and manganese, in human milk fat. It has already been mentioned that a substantial amount of the iron in milk is found in the lipid fractions where it occurs in the fat globule membrane proteins (Fransson & Lönnerdal, 1980).

8. Nutritional versus Physiological Role of Milk Proteins

Some of the proteins in human milk are considered to have a specific physiological role and will withstand hydrolysis in the gastrointestinal tract, e.g. secretory IgA, lactoferrin and lysozyme. Since secretory IgA and lactoferrin constitute a considerable amount of the whey protein fraction in human milk, this illustrates that the macronutrient aspect of milk proteins must take any specific physiological role of certain protein fractions into consideration.

There is obviously a conflict between a nutritional approach to the role of milk proteins and that of a holistic or complete physiological approach to their role. If specific proteins exert their physiological role only if they are not digested and absorbed in the gastrointestinal tract, they should not be included in calculations of nutritionally available proteins in the milk. The practical implication of this is illustrated in the case of human milk which may have a content of nutritionally available protein as low as 0·5–0·6 g per 100 ml (Hambraeus *et al.*, 1984) (Table 7). Human milk thus seems to have the lowest protein content of all milks and the protein fraction constitutes only 4% or even less of the total energy content. A protein determination based only on nitrogen analysis might consequently be greatly misleading when the nutritive value of various milk proteins is analysed without compensation for proteins of specific physiological importance which are not digested in the gastrointestinal tract (Hambraeus *et al.*, 1984; Woelderen, 1987).

9. Can the Nutritional Value of Milk Proteins be Damaged?

Most milk products are consumed after being subjected to some form of heat treatment. This might lead to a number of chemical reactions which may interfere with their nutritional quality. In the case of milk proteins, any heat treatment will lead to some denaturation and the protein fractions can react with each other as well as with other milk constituents, even if the heat treatment is mild. One of the essential amino acids which, furthermore, is often the limiting one in most diets of vegetable origin, lysine, can react in essentially two ways which reduce its bioavailability.

First, the ε-amino group of lysyl residues can react with aldehydes from carbohydrates according to the so-called *Maillard reaction* (Eriksson *et al.*, 1981). This reaction, which is complicated and not fully understood in its details, leads to an essential nutritional problem as lysine will no longer be available for absorption and utilization (Abrahamsson *et al.*, 1977). Trypsin can no longer cleave the peptide bond between the modified lysyl residue and the next amino acid and hence protein digestibility is impaired and abnormal peptides are formed which, it has been suggested, are toxic or harmful.

Lysine can also form a peptide, *lysinoalanine*, which occurs when casein is heated in dilute alkali leading to the elimination of phosphate from phosphoseryl with the formation of dehydroalanine which reacts with the ε-group of lysine (Bohak, 1964). Processes that include heating in dilute alkali, e.g. the manufacture of sodium caseinate and the preparation of textured proteins, may lead to the formation of lysinoalanine. Since this peptide improves the physical properties of textured casein products, it may be desirable, but the nutritional value is concomitantly reduced since lysine will be less available. It has also been suggested that lysinoalanine may be potentially toxic, but this is still not completely verified.

Treatment of milk proteins, especially casein, with proteolytic enzymes in order to produce soluble peptides which may be more readily absorbed or may have reduced antigenic characteristics, may give rise to a bitter taste, probably caused by peptides from β-casein (Clegg *et al.*, 1974).

10. The Role of Milk Proteins in the Human Diet

10.1. How much Milk is Needed?

The role of milk products for the nutritional status of the population varies. In some industrialized countries, i.e. in northern Europe and the USA, milk products constitute about 30% of the protein intake and 75% of the calcium intake. Since there is no protein deficiency in these countries, neither qualitatively nor quantitatively, there is no minimal quantity of milk needed to cover the protein needs. The necessary amount of dairy products in the diet is therefore essentially determined by their role as a source of calcium since there are few other calcium sources in the Western diet. This leads to a recommended intake for adults of slightly less than 500 ml milk per person per day.

10.2. Supplementary Effect

As illustrated in Table 6, the amount of essential amino acids in milk protein is close to what is expected to be the optimal amino acid requirement and composition for the human being. Consequently, the best application of milk proteins seems to be as a supplement for cheaper vegetable proteins, thus taking advantage of its abundance of essential amino acids which enhance the nutritive value of a mixed diet as it nutritionally supplements inferior proteins (Table 8). This enhancing effect is most obvious when small amounts of milk are used to supplement cereals. This is also of significant international importance with respect to global food scarcity since by far the larger part of the world's population relies on cereals as their staple food. The new approach to amino acid scoring by the 1985 FAO/

WHO/UNU expert commission, however, reduces the nutritional impact of milk protein supplementation of diets.

To date, the whey proteins have not been used to any great extent in the human diet in ways that recognize their high nutritive value; when they *are* used it is mainly for their functional properties in beverages, creams, etc. Most of the whey proteins are still used in animal feed since the production of whey still represents a serious pollution problem for most dairy industries. So far, the US seems to be the country where the whey proteins are used to the greatest extent in the human diet.

10.3. Formulae for Breast-milk Substitutes and Weaning

The adequacy of milk proteins as a source of dietary proteins for the growing infant was shown by Mellander *et al.* (1959), when they compared babies fed on artificial cows' milk preparations as an alternative to breast-feeding. If anything, those children grew somewhat faster. Although modern adapted infant formula, or so-called humanized infant formula, might show the same ratio between whey proteins and casein as in human milk, there are still significant differences since the whey protein composition differs markedly in bovine and human milks. This is illustrated in Fig. 2, which shows the protein composition of bovine milk, human milk and an adapted formula, expressed as grammes of nitrogen per litre. The nutritional significance of these differences is still not known.

During the last few years growing interest has been devoted to any differences in the metabolic effect and nutritive value of various proteins in infant feeding. It has then been shown that the casein-dominated formulae lead to certain metabolic effects and changes in the plasma amino acid pattern in pre-term as well as term infants. Whether or not these observations are also relevant for the use of milk protein-based products in clinical dietetics of adults has to be further elucidated. No doubt a general recommendation should be to reduce the casein content and increase the use of high nutritionally valued whey proteins in such dietetic products.

Räihä and his collaborators have been able to show the effect of protein quality and quantity in formulae based on whey-predominant or casein-predominant milk proteins in relation to isocaloric amounts of human breast-milk on several biochemical parameters, e.g. blood urea nitrogen, urine osmolality and plasma amino acid pattern (Lönnerdal & Chen, 1987; Räihä *et al.*, 1976, 1986*a,b*).

11. Milk Proteins in Clinical Dietetics

The benefits of dairy products are due to their high content of nutritionally valuable proteins as well as of calcium and a number of vitamins. The negative factors in a milk-dominated diet are the low fibre and iron contents which often lead to constipation and iron deficiency anemia. Another disadvantage is the milk fat, which constitutes a much discussed and dominating source of saturated fatty acids in many industrialized countries.

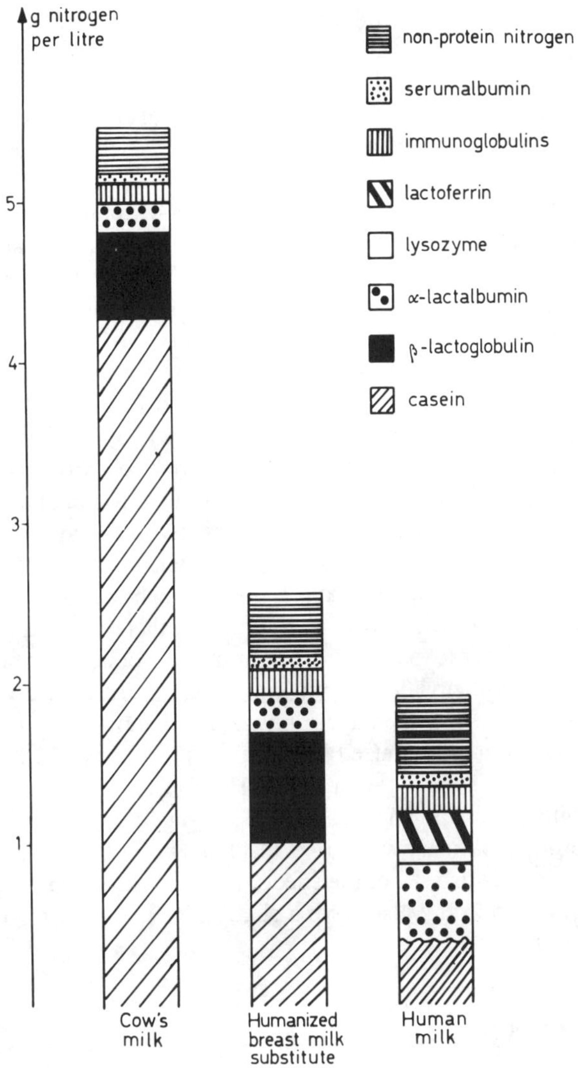

Fig. 2. Protein composition of cows' milk, humanized infant formula (adapted formula) and human milk. Nitrogen derived from the various proteins and non-protein nitrogen are given as grammes of nitrogen per litre (from Hambraeus *et al.*, 1977).

Milk proteins are still considered by some to be the nutritionally best valued food proteins due to their content of essential amino acids and a high digestibility. Dairy products are, therefore, often used in the treatment of malnutrition in children in various health programmes in developing as well as developed countries; the use of milk and dairy products in school feeding programmes in many countries being one example.

Dairy products are used in specific preparations for clinical dietetics as a major ingredient in formulae due to their high nutritive value and high acceptability. Milk and milk products are often metabolically well accepted and easy to take in small quantities. They are therefore suitable for use in the formulation of well-balanced small meals. In the case of advanced gastrointestinal dysfunction, however, the lactose content must be reduced as such diseases may include a secondary lactase deficiency.

The increased interest in enteral feeding of patients with nutritional problems pre- and post-operatively as an alternative or supplement to parenteral nutrition has also increased the interest for high-quality, easily digested proteins.

Dairy products have also for a long time been used in the diets of patients with peptic ulcers. This was motivated by their high protein and phosphate content which gave them a good buffering capacity. However, modern treatment with antacid preparations has reduced the need for diets with a high buffer capacity. The ingestion of large quantities of milk and dairy products by these patients often led to a high intake of saturated fat and an increased risk for coronary heart disease.

The use of partly hydrolysed milk protein products for patients with reduced digestive capacity, e.g. patients with cystic fibrosis, has also been tested but is hampered by the formation of bitter-tasting products.

The low content of phenylalanine and tyrosine in whey protein has made them suitable as a protein source in formulae used in the dietetic treatment of patients with disturbed metabolic capacity for these amino acids, i.e. phenylketonuria, tyrosinemia and hyperphenylalaninemia. Using such a formula it was possible to obtain a more positive nitrogen balance and a more pronounced nitrogen retention than with a conventional humanized breast-milk substitute, which illustrated the high nutritional value of the whey proteins (Hambraeus *et al.*, 1974).

Fermented milk products, e.g. products with *Lactobacilllus acidophilus*, have been used in geriatric patients with good results. The nutritive value of the milk protein has not been disturbed but the effect on the intestinal flora has reduced the risk of constipation and may also have a slight antibiotic effect on the intestinal microflora (Nahaisi, 1986).

12. Milk for High-energy Consumers

High energy consumers such as athletes, especially those involved in weight lifting and body-building, but also those engaged in endurance sports, have been using milk products as a source of high-valued dietary protein supplements. Some reports are presented where they try to verify the positive effect of high protein intake on their performance. However, there is still a lot to be learnt about any potential benefit of a high protein intake as well as whether or not there is an increased demand for protein in athletes.

Protein does not contribute to muscular energy to any great extent according to studies on urea excretion in the urine. However, this does not take into account the nitrogen that is lost through sweat and increased amino acid oxidation. During endurance exercise, decreased protein synthesis (Millward *et al.*, 1982) and increased protein degradation (Lemon & Nagle, 1981; Young & Torun, 1981) have been observed. The contribution of protein to energy expenditure may also be related to the intensity and length of exercise (Lemon, 1985).

Exercise-induced hypertrophy of the muscles depends on increased protein synthesis which may result in an increased protein requirement. Protein intake in excess may, however, affects performance by decreasing carbohydrate intake and stressing the liver and kidneys. Thus, a carbohydrate-rich diet, constituting 60–70% of the energy intake, and a protein intake corresponding to 2 g per kg per day has been recommended for athletes. In order to reduce any waste of protein due to a low quality, animal protein sources are usually recommended due to their high digestibility and complete amino acid composition. On these premises, milk protein should be a favourable source of dietary protein.

The high energy consumption by athletes, however, indirectly leads to a concomitant high protein intake when they consume a conventional mixed diet. Thus, an athlete with a daily energy turnover of 21 000 kJ (or 5000 kcal) on a conventional diet containing protein at 12% of energy will consume 150 g of protein. This is more than twice their protein requirement (0·8 g per kg body weight for a 75-kg man). This will consequently meet any increased demand of extra protein requirement due to muscle hypertrophy or compensate for an increased turnover of muscle protein. In addition, most athletes consume a diet which has more than 12% of energy from protein. Furthermore, the higher daily intake of protein reduces the need for high-valued protein, e.g. content of essential amino acids, since there is an efficient re-utilization of amino acids, especially of essential amino acids, in the body. Thus, there seems to be very limited scientific evidence so far that milk protein is especially needed by athletes for nutritional reasons. However, most dairy products are well tolerated and could be consumed as convenient foods. Especially, high-energy consumers and athletes in intensive training have problems to meet their energy needs during training periods and need easily-digested highly-nutritive meals. This makes dairy products especially well suited for athletes.

13. Can Milk Proteins be Hazardous

13.1. Milk intolerance

Intolerance to milk, essentially cows' milk as this is most commonly used in the diet, has been reported and will of course interfere with the nutritional value of dairy products. The intolerance could be due to many various mechanisms and thereby be related to different constituents in the milk. Perhaps the most common syndrome is due to lactose intolerance but quite a number of syndromes are also due to milk protein components and should be commented on here as they interfere with the nutritional value of milk protein. Almost all known mechanisms of food intolerance have been reported to be caused by milk or milk products. One of the practical problems is the fact that milk or milk components are often used in combinations with other nutrients in processed foods, which will have implications in clinical dietetics for patients with milk protein intolerance.

Food intolerance can include various forms of adverse reactions from psychological reasons, which often are referred to as food aversion, to allergic reactions due to an abnormal immunological reaction. In addition, adverse reactions could be due to a lack of certain enzymes, e.g. in lactose intolerance, or could be due to a pharmacological effect due to certain compounds, irritation of mucous membranes, leading to gastritis or diarrhea, or increased gas production and colic due to fermentation of unabsorbed food residues. It can often be difficult to differentiate between adverse effects to milk proteins due to food intolerance, food allergy or food aversion. In most cases the clinical diagnosis of food intolerance can be established only if the symptoms and signs disappear when an elimination diet is administered.

Small intestinal mucosa damage has been reported in children leading to malnutrition as a result of cows' milk protein intolerance and irritable bowel syndrome has been experienced in some adults and considered to be related to milk protein intolerance. However, the incidence and prevalence of milk protein intolerance is still not known.

Hypersensitivity reactions to cows' milk are confined to milk protein components as antigenic reactions do not occur to fat or carbohydrate components. Consequently, all different milk protein fractions, i.e. any of the four caseins as well as various whey proteins, may give rise to an allergenic reaction. Interestingly, it has been reported that the capacity to induce an immune response can be markedly increased by the Maillard reaction. Digestion of milk proteins may also give rise to new and quite distinct patterns of immune response. Thus, partial digestion of milk proteins may not necessarily result in a hypoallergic milk product. Small amounts of cows' milk protein may lead to an increased risk of anaphylactic sensitization of infants when compared to full cows' milk protein diet. Since there are reasons to believe that the problem in milk allergy is more likely to be due to the response in the infant than by the food itself, the concept of hypoallergenic milks as a result of partly digested or reduced cows' milk protein content is not considered scientifically sound (Ferguson & Watret, 1988). Pasteurization does not seem to reduce the potential allergenic effect of milk proteins in sensitized individuals but intensive heat treatment and extensively hydrolysed casein or whey products do. Such treatment may however destroy the nutritional value of other essential components in the milk, e.g. vitamins. In the evaluation of the nutritional significance of the use of hypoallergenic milk formulae the immunogenicity of the products must be related to the nutritional quality.

Atopic eczema has been reported to occur as result of cows' milk protein intolerance in children and has been considered to occur also in exclusively breastfed infants due to carry-over of cows' milk protein antigen in the mothers' milk. However, cows' milk protein intolerance does not seem to be an important component in adult atopic eczema (Barnetson *et al.*, 1981).

Intolerance to cows' milk protein leading to *diarrhoea* and *failure to thrive* has been reported in a number of studies since the 1960s but these cases must be differentiated diagnostically from coeliac disease as well as lactose intolerance (Kuitunen *et al.*, 1975). *Rectal bleeding* resulting in loose mucousy and bloody stools has also been described in infants under the age of one year as a result of cows' milk protein intolerance. During the last decade, increased interest has been devoted to the appearance of infantile colic in breast-fed infants as a result of exposure to antigens of cows' milk protein which had to be eliminated from the mothers' diet. In their studies Lothe and his collaborators found that cows' milk whey proteins can elicit symptoms of infantile colic in colicky infants and that infants with infantile colic had significantly higher macromolecular absorption. Their studies indicated that the gut mucosa is affected in infantile colic which could be due to damage mucosa or immature intestinal function (Jakobsson & Lindberg, 1978; Lothe *et al.*, 1982; Lothe & Lindberg, 1989).

13.2. Are Milk Proteins Atherogenic?

In the nutritional debate on the relationship of diet and health there has been an ongoing discussion on the potential risks of homogenized milk (Oster, 1971). This has been related to the role of xanthine oxidase in the milk fat globule membrane as homogenization might increase the effect of xanthine oxidase which would then oxidize plasmalogenes and increase the risk of heart infarcts. This is still being discussed, although it has been much criticized and is not generally accepted (Oster & Ross, 1982; Clifford *et al.*, 1983). Of special nutritional concern is the discussion of the role of casein as a promotor of atherosclerosis (Terpstra *et al.*, 1983). As early as the 1920s, rabbits fed casein were shown to develop arterial lesions. In the 1950s increased levels of serum cholesterol were reported in rats fed casein-based diets. However, in strictly controlled human studies, casein was not found to increase serum cholesterol (Sacks *et al.*, 1983). Although such effects have been shown to occur in experimental animals, usually rabbits and rats, no clear differential effect of casein and soy protein has been demonstrated on plasma cholesterol in man (West & Beynen, 1985). Thus, there seems to exist a species-dependent effect of dietary casein on cholesterol metabolism.

The hypothesis regarding the hypercholesterolemic effect of casein has been divided in three groups: (1) an effect of the amino acid composition of the dietary protein on the endocrine response; (2) an increased intestinal absorption of cholesterol; (3) reduced bile acid synthesis. It has been suggested that the casein-induced hypercholesterolemia is primarily caused by the effect of casein on the enterohepatic cycle of steroids. The hypercholesterolemic effect of dietary casein depends on a low intestinal phosphatase activity in combination with a high glycine-conjugated/taurine-conjugated bile acid ratio. The intestinal phosphatase activity in man, however, seems to counteract any casein-induced hypercholesterolemia. Of nutritional interest is the fact that dietary calcium phosphate may have a favourable effect on cholesterol metabolism. Thus, as dairy products are the main sources of calcium phosphate as well as of casein, more research is needed before the potential hazard effect of casein on atherosclerosis is understood.

14. Conclusion

Milk certainly represents one of the most complete single foods and the different composition of milks from various species might be an indicator by Nature of the specific nutrient requirements of that species. The use of various milks in the human diet without doubt takes advantage of their content of essential nutrients and milk products function as high-valued supplements to many diets in different regions of the world. To date, most interest has been devoted to the macronutrient aspects of the role of milk proteins in the human diet. There is, however, a conflict between the strict nutritional aspects which essentially take into account the milk proteins as a source of amino acids for protein synthesis, and the more specific physiological role of certain protein components in the milk which exert well-defined metabolic effects. There are also a number of very potentially active proteins and peptides which may have a significant physiological role during the neonatal period and which may also have an effect in the adult. There are, consequently, still many problems to be solved regarding the physiological implications of the various characteristics of the protein constituents of milks and their relation to the nutritional and metabolic effects in man. More studies are required and it is possible that various specific milk protein components could be used in the future for specific nutritional and pharmacological purposes. This might be one of the most economic ways for the utilization of whey proteins, which today still represent a problem for the dairy industry and at the same time take advantage of advanced dairy technology.

References

Abrahamsson, L., Bengtsson, O., Hambraeus, L. & Holm, H. (1977). Protein quality of milk-cereal based foods for infants and children in relation to processing methods and composition of the products. *J. Food Technol.*, **14**, 429–40.

Andersen, K.J. & von der Lippe, G. (1979). The effect of proteolytic enzymes on the vitamin B_{12}-binding proteins of human gastric juice and saliva. *Scand. J. Gastroent.*, **14**, 833–8.

Barnetson, R. St.C., Merrett, T.G. & Ferguson, A. (1981). Studies on hyperimmunoglobulinaemia E in atopic diseases with particular reference to food allergens. *Clin. Exper. Immunol.*, **46**, 54–60.

Barth, C.A. & Schlimme, E. (1989). *Milk Proteins—Nutritional, Clinical, Functional and Technological Aspects*. Springer Verlag, New York.

Bell, K. & McKenzie, H.A. (1964). β-Lactoglobulins. *Nature*, **204**, 1275–79.

Bezkorovainy, A., Grohlich, D. & Nichols, J.H. (1979). Isolation of a glycopolypeptide fraction with *Lactobacillus bifidus* subspecies *pennsylvanicus* growth-promoting activity from whole human milk casein. *Am. J. Clin. Nutr.*, **32**, 1428–32.

Blanc, B. (1981). Biochemical aspects of human milk—comparison with bovine milk. *Wld Rev. Nutr. Diet.*, **36**, 1–89.

Bohak, Z. (1964). N-epsilon-(DL-2-amino-2-carboxyethyl)-L-lysine, a new amino acid formed on alkaline treatment of proteins. *J. Biol. Chem.*, **239**, 2878.

Brantl, V. & Teschemacher, H. (1982). Opiartig wirkende Stoffe in Milch und Milchprodukten. 2. Isolierung, Strukturaufklärung und biologische Wirkung der β-casomorphine. *Milchwissenschaft*, **37**, 641–4.

Brock, J.H., Arzabe, F., Lampreave, F. & Pineiro, A. (1976). The effect of trypsin on bovine transferrin and lactoferrin. *Biochim. Biophys. Acta*, **446**, 214–25.

Bullen, J.J., Rogers, J.J. & Leigh, L. (1972). Iron-binding proteins in milk and resistance to *Escherichia coli* infection in infants. *Brit. Med. J.*, **1**, 69.

Clegg, K.M., Lim, C.L. & Manson, W. (1974). The structure of a bitter peptide derived from casein by digestion with papain. *J. Dairy Res.*, **41**, 283–7.

Clifford, A.J., Ho, C.Y. & Swenerton, H. (1983). Homogenized bovine milk xanthine oxidase: a critique of the hypothesis relating to plasmalogen depletion and cardiovascular disease. *Am. J. Clin. Nutr.*, **38**, 327–32.

Conti, A., Liberatori, J. & Napolitano, L. (1980). Isolation and preliminary physico-chemical characterization of human β-lactoglobulin. *Milchwissenschaft*, **35**, 65–8.

Cox, T.M., Mazurier, J., Spik, G., Montreuil, J. & Peters, T.J. (1979). Iron binding proteins and influx of iron across the duodenal brush border. Evidence for specific lactoferrin receptors in the human intestine. *Biochim. Biophys. Acta*, **588**, 120–8.

Davidson, L. & Lönnerdal, B. (1988). Specific binding of lactoferrin to brush border membrane: ontogeny and effect of glycan chain. *Am. J. Physiol.*, **254**, G580–5.

Eriksson, C., Asp. N.G. & Theander, O. (ed.) (1981). Maillard reactions in food: chemical, physiological and technological aspects. *Proc. Food Nutr. Sci.*, **5**, 1–501.

FAO (1957). Protein requirements. *FAO Nutritional Studies*, No. 16. FAO, Rome.

FAO (1970). Amino acid content of foods. *FAO Nutritional Studies*, No. 24. FAO, Rome.

FAO/WHO (1965). Protein requirements. *WHO Technical Report Series*, No. 301. WHO, Geneva.

FAO/WHO (1973). Protein and energy requirements. *WHO Technical Report Series*, No. 522. WHO, Geneva.

FAO/WHO/UNU Expert Consultation (1985). Energy and protein requirements. *WHO Technical Report Series*, No. 724. WHO, Geneva.

Farrell, H.M.,Jr & Thompson, M.P. (1971). Biological significance of milk protein polymorphism. *J. Dairy Sci.*, **54**, 1219–28.

Ferguson, A. & Watret, K.C. (1988). Cows' milk intolerance. *Nutr. Rep. Rev.*, **1**, 1–22.

Fomon, S. (1974). *Infant Nutrition*, 2nd edn. WB Saunders, Philadelphia.

Ford, J.E., Salter D.N. & Scott, K.J. (1969). The folate-binding protein in milk. *J. Dairy Res.*, **36**, 435–46.

Forsum, E. (1973). Nutritional evaluation of whey protein concentrates and their fractions. *J. Dairy Sci.*, **57**, 665–70.

Forsum, E. & Hambraeus, L. (1974). Utilization of whey proteins in human nutrition. In *Joint IUPAC/IUFoST Symposium—The Contribution of Chemistry to Food Supplies*, ed. I. Morton & D.N. Rhodes. Butterworths, London, pp., 375–83.

Forsum, E., Hambraeus, L. & Siddiqi, I.H. (1973). Fortification of wheat by whey protein concentrate, dried skim milk, fish protein concentrate and lysine monohydrochloride. *Nutr. Rep. Int.*, **8**, 39–48.

Fransson, G.B. & Lönnerdal, B. (1980). Iron in human milk. *J. Pediat.*, **96**, 380–4.

Freudenstein, C., Keenan, T.W., Eigel, W.N., Sasaki, M., Stadler, J. & Franke, W.W. (1979). Preparation and characterization of the inner coat material associated with fat globule membranes from bovine and human milk. *Exp. Cell. Res.*, **118**, 227–94.

Galesloot, T.E. & Tinbergen, B.J. (ed.) (1985). *Milk Proteins '84*. Proc. Intern. Cong. Milk Proteins. Agricultural Publishing and Documentation, Wageningen.

György, P., Norris, R.F. & Rose, C.S. (1953). Bifidus factor. I. A variant of *Lactobacillus bifidus* requiring a special growth factor. *Arch Biochem. Biophys.*, **48**, 193–201.

Hambraeus, L. (1984). Human milk composition. *Nutr. Abstr. Rev.*, **54**, 219–36.

Hambraeus, L., Hardell, L.I., Forsum, E. & Lorentsson, R. (1974). Use of a formula based on a whey protein concentrate in the feeding of an infant with hyperphenylalaninemia. *Nutr. Metabol.*, **17**, 84–90.

Hambraeus, L., Forsum, E., & Lönnerdal, B. (1977). Nutritional aspects of breast milk versus cow's milk formula. In *Food and Immunology, Symposia of the Swedish Nutrition*

Foundation XIII, ed. L. Hambraeus, L.-Å. Hanson, & H. McFarlane. Almqvist & Wiksell, Uppsala, pp. 116–24.

Hambraeus, L., Lönnerdal, B., Forsum, E. & Gebre-Medhin, M. (1978). Nitrogen and protein components of human milk. *Acta Paed. Scand.*, **67**, 561–5.

Hambraeus, L., Fransson, G.-B. & Lönnerdal, B. (1984). Nutritional availability of breast milk protein. *Lancet*, **ii** 167–8.

Hiraoka, Y., Segawa, T., Kuwajima, K., Sugai, S. & Murai, N. (1980). α-Lactalbumin: a calcium metalloprotein. *Biochem. Biophys. Res. Comm.*, **95**, 1098–104.

Jakobsson, I. & Lindberg, T. (1978). Cow's milk as a cause of infantile colic in breast-fed infants. *Lancet*, **ii**, 437–40.

Jenness, R. (1974). The composition of milk. In *Lactation—A Comprehensive Treatise*, Vol. 3, ed. B. Larson & V.R. Smith. Academic Press, New York. pp. 3–105.

Jenness, R. & Sloan, R.E. (1970). The composition of milks of various species: a review. *Dairy Sci. Abstr.*, **32**, 599–612.

Johansson, B.G. (1960). Isolation of an iron-containing red-protein from human milk. *Acta Chem. Scand.*, **14**, 510–12.

Koldovsky, O. (1989). Search for role of milk-borne biologically active peptides for the suckling. *J. Nutr.*, **119**, 1543–51.

Kon, S.K. & Cowie, A.T. (1961). *Milk—The Mammary Gland and Its Secretion*, Vol. 1 & 2. Academic Press, New York.

Kuitunen, P., Visakorpi, J.K., Savilahti, E. & Pelkonen, P. (1975). Malabsorption syndrome with cow's milk intolerance. Clinical findings and course in 54 cases. *Arch. Dis. Childh.*, **50**, 351–6.

Lemon, P.W.R. (1985). Metabolic aspects of proteins for top sportsmen. In *Milk Proteins '84*, Proc. Int. Congr. Milk Proteins, ed. T.E. Galesloot & B.J. Tinbergen. Agricultural Publishing and Documentation, Wageningen, pp. 88–98.

Lemon, P.W.R. & Nagle, F.J. (1981). Effects of exercise on protein and amino acid metabolism. *Med. & Sci. in Sports & Exercise*, **13**, 141–3.

Lindblad, B.S. & Rahimtoola, R.J. (1974). A pilot study of the quality of human milk in a lower socio-economic group in Karachi, Pakistan. *Acta Paed. Scand.*, **63**, 125.

Lönnerdal, B. (1985). Biochemistry and physiological function of human milk proteins. *Amer. J. Clin. Nutr.*, **42**, 1299–317.

Lönnerdal, B. & Chen, C.-L. (1987). Effect of protein level and casein: whey ratio of infant formula on growth, plasma amino acids and trace elements. *Amer. J. Clin. Nutr.*, **45**, 824 (abstr.).

Lönnerdal, B. & Forsum. E. (1980). Casein content of human milk. In *Nutrition in Europe, Proc. of the Third European Nutrition Conf.*, ed. L. Hambraeus. Almqvist & Wiksell International, Stockholm, pp. 139

Lönnerdal, B. & Forsum, E. (1985). Casein content of human milk. *Amer. J. Clin. Nutr.*, **41**, 113–20.

Lönnerdal, B. & Glazier, C. (1985). Calcium binding by α-lactalbumin in human milk. *J. Nutr.*, **115**, 1209–16.

Lönnerdal, B., Forsum, E. & Hambraeus, L. (1976*a*). The protein content of human milk. I. A transversal study of Swedish normal material. *Nutr. Rep. Int.*, **13** 125–34.

Lönnerdal, B., Forsum, E. & Hambraeus, L. (1976*b*). A longitudinal study of the protein, nitrogen and lactose contents of human milk from Swedish well-nourished mothers. *Amer. J. Clin. Nutr.*, **29**, 1127–33.

Lothe, L. & Lindberg, T. (1989). Cow's milk whey protein elicits symptoms of infantile colic in colicky formula-fed infants: A double-blind cross-over study. *Pediatrics*, **83**, 262–6.

Lothe, L., Lindberg, T. & Jakobsson, I. (1982). Cow's milk formula as a cause of infantile colic: a double blind study. *Pediatrics*, **70**, 7–10.

Macy, J.G., Nims, B., Brown, M. & Hunscher, H.A. (1931). Human milk studies VII. Chemical analysis of milk representative of the entire first and last halves of the nursing period. *Am. J. Dis. Child.*, **42**, 569–89.

Masson, P. (1970). *La Lactoferrine*. Editions Arsica, Brussels, 232.
Mazurier, J., Montreuil, J. & Spik, G. (1985). Visualization of lactoferrin brush-border receptors by ligand-blotting. *Biochim. Biophys. Acta*, **821**, 453–60.
McKenzie, H. A. (ed.) (1971). *Milk Proteins, Chemistry and Molecular Biology*, Vol. 1 & 2. Academic Press, New York.
Mellander, O. & Fölsch, G. (1972). Enzyme resistance and metal binding of phosphorylated peptides. In *Protein and Amino Acid Functions*, ed. E.J. Bigwood. Pergamon Press, Oxford, pp. 569–79.
Mellander, O., Vahlquist, B. & Mellbin, T. (1959). Breast feeding and artificial feeding. A clinical serological, and biochemical study of 402 infants with a survey of the literature. The Norrbotten Study. *Acta Paediatrica*, **48**(Suppl. 116), 1–99.
Millward, D.J., Davies, C.T.M., Halliday, D., Wolman, S.L., Matthews, D. & Rennie, M. (1982). Effect of exercise on protein metabolism in humans as explored with stable isotopes. *Fed. Proc.*, **41**, 2686–91.
Mitchell, H.H. & Block, R.J. (1946). Some relationships between the amino acid contents of proteins and their nutritive values for the rat. *J. Biol. Chem.*, **163**, 599–630.
Nahaisi, M.H. (1986). *Lactobacillus acidophilus*: therapeutic properties, products and enumeration. *Develop. Food Microbiol.*, **2**, 153–78.
Oster, K.A. (1971). Plasmalogen diseases: a new concept of the etiology of the atherosclerotic process. *Am. J. Clin. Res.*, **2**, 30–5.
Oster, K.A. & Ross, D.J. (1982). Flawed objectives to the bovine milk xanthine oxidase theory. *J. Am. Coll. Nutr.*, **1**, 420.
Pfeuffer, M. & Barth, C.A. (1988). Influences of casomorphines on plasma lipid levels and lipid secretion rates. In *Milk Proteins, Nutritional, Clinical, Functional and Technological Aspects*, ed. C.A. Barth & E. Schlimme. Springer Verlag, New York, pp. 291–2.
Räihä, N.C.R., Heinonen, K., Rassin, D.K. & Gaull, G.E. (1976). Milk protein quantity and quality in low birth weight infants. I. Metabolic responses and effect on growth. *Pediatrics*, **57**, 659–74.
Räihä, N.C.R., Minoli, I., Moro, G. (1986*a*). Milk protein intake in the term infant: I. Metabolic responses and effects on growth. *Acta Paediatr. Scand.*, **75**, 881–6.
Räihä, N.C.R., Minoli, I., Moro, G. & Bremer, H. (1986*b*). *Milk protein intake in the term infant: II. Effects on plasma amino acid concentrations. Acta Pediatr. Scand.*, **75**, 887–92.
Reiter, B. (1978). Review of the progress of dairy science: antimicrobial systems in milk. *J. Dairy Res.*, **45**, 131.
Saarinen, U.M., Siimes, M. A. & Dalmann, P.R. (1977). Iron absorption in infants. High bioavailability of breast milk iron as indicated by extrinsic tag method of iron absorption and by the concentration of serum ferritin. *J. Pediat.*, **91**, 36–9.
Sacks, F.M., Breslow, J.L., Wood, P. G. & Kass, E. H. (1983). Lack of an effect of dairy protein (casein) and soy protein on plasma cholesterol of strict vegetarians. An experiment and a critical review. *J. Lipid Res.*, **24**, 1012–20.
Salter, D.N. & Mowlem, A. (1983). Neonatal role of milk folate-binding proteins. Studies on the course of digestions of goat's milk folate binder in the 6-d child. *Brit. J. Nutr.*, **50**, 589–96.
Sandberg, D.P., Begley, J.A. & Hall, C.A. (1981). The content, binding and forms of vitamin B_{12} in milk. *Am. J. Clin. Nutr.*, **34**, 1717–24.
Sturman, J.A., Gaull, G.E. & Räihä, N.C.R. (1970). Absence of cystathionase in human fetal liver: Is cystine essential? *Science*, **169** 74–5.
Swaisgood, H.E. (1973). The caseins. *CRC Critical Reviews in Food Technology*, **3**, 375–414.
Terpstra, A.H.H., Hermus, R.J.J. & West, C.E. (1983). The role of dietary protein in cholesterol metabolism. *World Rev. Nutr. Diet.*, **42**, 1–55.
West, C.E. & Beynen, A.C. (1985). Milk proteins in contrast to plant proteins: effects on plasma cholesterol. In *Milk Proteins '84*. Proc. Int. Congr. Milk Protein, ed. T.E. Galeshot & B.J. Tinbergen. Agricultural Publishing and Documentation, Wageningen, pp. 80–7.

Whitney, R. McL., Brunner, J.R., Ebner, K.E., Farrell, H.M.,Jr, Josephson, R.V., Morr, C.V. & Swaisgood, H.E. (1976). Nomenclature of the proteins of cow's milk: Fourth revision. *J. Dairy Sci.*, **59**, 795–815.

Woelderen, B.F., van (1987). Changing insight into milk proteins: some implications. *Nutr. Abstr. Rev. (ser. A)*, **57**, 129–34.

Woodhouse, L.R. & Lönnerdal, B. (1988). Quantitation of the major whey proteins in human milk, and the development of a technique to isolate minor whey proteins. *Nutr. Res.*, **8**, 853–64.

World Health Organization (1981). *International Code of Marketing Breast-milk Substitute.* WHO, Geneva.

Yamauchi, K., Azuma, N., Kobayashi, H. & Kaminogawa, S. (1984). Isolation and properties of human K-casein. *Agric. Biol. Chem.*, **48**, 771–6.

Young, V.R. & Torun, B. (1981). Physical activity: impact on protein and amino acid metabolism and implications for nutritional requirements. In *Nutrition in Health and Disease and International Development*, ed. A.E. Harper & G.K. Davis. A.R. Liss Inc., New York, pp. 57–85.

12

BIOSYNTHESIS OF MILK PROTEIN

T.B. MEPHAM

Department of Physiology and Environmental Science, University of Nottingham, Sutton Bonington, Loughborough, Leicestershire LE12 5RD, UK

&

P. GAYE,[a] P. MARTIN[b] & J.C. MERCIER[b]

[a]Laboratoire de Biologie Cellulaire et Moléculaire, [b]Laboratoire de Génétique Biochimique, Institut National de la Recherche Agronomique, INRA-CRJ, 78350 Jouy-en-Josas, France

1. INTRODUCTION

Synthesis and secretion of large quantities of specific milk proteins are major functions of mammary epithelial cells during lactation. The process, which is under control of peptide and steroid hormones, requires the concomitant activation of both milk protein and 'house keeping' genes. The full expression of the former genes is obviously largely dependent on efficient transcriptional, translational and secretory machineries, and on an optimal supply of amino acids to mammary cells.

In the previous edition (Fox, 1982), this chapter was devoted to six main topics: (1) uptake, transport and metabolism of amino acids in mammary epithelial cells, (2) hormonal induction of RNA messengers encoding milk proteins, (3) assembly of polypeptide chains on ribosomes, (4) intracellular transport and exocytosis of milk proteins, (5) co- and post-translational covalent modifications of milk proteins, and (6) secretion of non-specific milk proteins. There has since been a rapid growth of information in the field of milk protein synthesis, regarding the structure and function of mammalian mRNAs and genes, and the intracellular routing and sorting of proteins.

The purpose of this revised version is to summarize the previous knowledge of the biosynthesis of milk proteins and to present the most pertinent information provided by current research in the field. Attention will be focused on milk protein mRNAs and genes and on the intracellular transport of proteins, with special emphasis on early stages studied in various types of eukaryotic cells which are assumed to be similar in mammary epithelial cells. The practical applications of these studies will not be developed here and only brief comment will be made about their obvious interest for the dairy field.

2. Structure and Expression of Milk Protein mRNAs and Genes

Before presenting an account of the current knowledge of milk protein mRNAs and genes, a brief survey of what is currently known in higher eukaryotes is given.

2.1. From Gene to mRNAs in Eukaryotes

2.1.1. Structure and Expression of Genes

All the genetic information relevant to species is stored in DNA, of which only a tenth or so is transcribed into proteins. Genes or sets of genes are scattered through large masses of apparently non-coding DNA, termed 'selfish'. Inactive genes are tightly packaged with histones into chromatin and their activation requires decondensation of this structure so that transcriptional control sequences become available to regulatory proteins.

All but histone and interferon eukaryotic nuclear genes (Breathnach & Chambon, 1981) contain intervening sequences or introns (Gilbert, 1978) which are transcribed but not translated. Introns, which generally account for the most part of pre-mRNA, are excised through formation of a loop or lariat (reviewed in Sharp, 1987). Although sequences of introns are frequently conserved, their biological and evolutionary significance is unclear (Crick, 1979; Sharp, 1985). It has been proposed that they provide a mechanism for generating diversity during evolution (Gilbert, 1985); in agreement with such a view, many introns separate exons encoding functional and/or structural protein domains or modules (Gilbert, 1985; Gilbert *et al.*, 1986). Another selective pressure for the retention of introns is the flexibility that they give to biological systems by regulating the expression of relevant proteins (Sharp, 1985): the same gene can generate different products from the same primary transcript by alternative splicing. It is one amongst several strategies that eukaryotic cells use to diversify their gene products, as discussed below.

Transcription of eukaryotic genes is a highly regulated process, mediated by RNA polymerase II (Pol II) and miscellaneous transcription factors. The enzyme, made up of 8–10 subunits, is able to polymerize ribonucleotides into mRNA when moving along a $3' \rightarrow 5'$ DNA template. Unfortunately, neither the complete credits of the 'transcription' play nor the detailed scenario are well-known for any mammalian gene, including milk protein genes. Transcription of a mammalian protein-encoding gene depends on the recognition of several constitutive DNA sequences and/or local DNA conformations (DNA *cis*-elements identified by gel retardation and footprinting assays) by miscellaneous effectors (*trans*-acting elements) (Fig. 1).

Cis-acting elements comprise:

(i) promoter elements which function in a 5′ position-dependent fashion, either proximal to the transcription unit (the TATA box of non-house-keeping genes which determines the site of initiation) or more distal (e.g. CCAAT

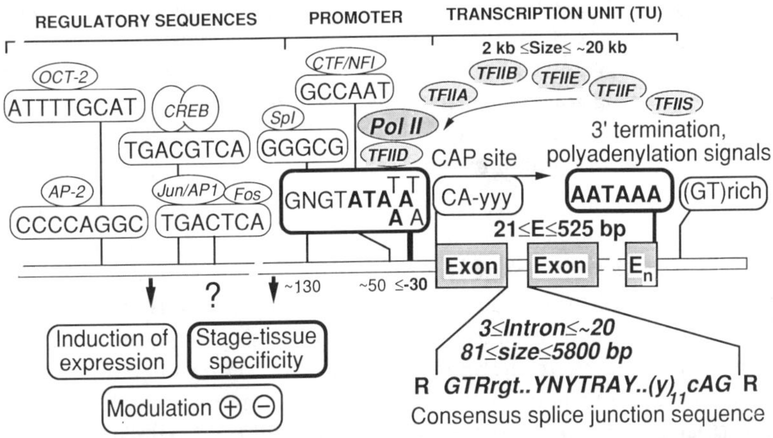

Fig. 1. Schematic representation of a hypothetical milk protein gene. The figure is not meant to imply that all *cis*-acting motifs described here exist in any gene or occur necessarily at that position (see text): bold-faced boxes refer to strong *cis*-acting signals (Bucher & Trifonov, 1986); a lowercase letter refers to the nucleotide most frequently found at the position; r and y refer to purine and pyrimidine, respectively; N refers to any nucleotide; Pol.II = RNA polymerase II; TFII = factors of the general transcription machinery; other circled proteins are various nuclear DNA-binding proteins.

box and GGGCG motif in some promoters, functional in both orientations and required for efficient initiation). Mutations in the promoter region can generate a non-functional gene, termed pseudogene.

(ii) enhancers, whose activity is independent of their position and orientation as they can occur 5′, 3′ or within the transcription unit; however, promoter responsiveness to enhancer can depend on the sequence of the TATA box (Wefald *et al.*, 1990). DCR elements (Dominant Control Region, also named Nuclear DNA attachment element and Matrix Attachment Region, MAR) identified in immunoglobulin, insulin, elastase (quoted in Jones *et al.*, 1988) and lysozyme genes (Stief *et al.*, 1989; Phi-Van *et al.*, 1990) are particular enhancers which confer tissue-specificity to spatially close promoters. Silencer elements (Renkawitz, 1990), the negative enhancer counterparts, have been identified in a few genes, including that encoding lysozyme.

Trans-acting protein factors fall into two classes:

(i) factors of the general transcriptional machinery which are required for accurate basal transcription by Pol II. Initiation requires the stepwise involvement of at least 5 factors: TFIIA and TATA-binding TFIID, then TFIIB, TFIIE, and TFIIF, which is also involved in elongation together with TFIIS. The four later factors bind to Pol II.

(ii) factors, such as hormone–receptor complexes, nuclear proteins CREB, Jun, CTF/NFI, Spl, OCT-1 and 2, AP-2, etc., able to induce or modulate transcription through binding to specific DNA sites, and responsible for the topological, temporal and optimal transcription of a given gene. Some factors, such as Sp1, appear to be ubiquitous, whereas others might be specific to a given type of cell. Therefore, selective stage and tissue-specific optimal expression of a gene is likely to depend on both the nature and organization of its *cis*-acting elements and the occurrence of the proper combination of recognition factors in the host cell at a given time.

Structural motifs that specify termination of transcription are still ill-defined. In addition to the well-known AATAAA sequence, at least one downstream signal, either a GT- or T-rich nucleotide stretch, may be involved (Proudfoot & Whitelaw, 1988; Wickens, 1990). A 3′ end formation of mRNA results from the combination of both transcriptional termination and RNA processing by an endonuclease, since polyadenylation occurs on the shortened 3′ end of the primary transcript, about 12–20 nucleotides downstream from the AAUAAA motif.

For further details on the regulation of transcription, the reader is referred to recent reviews (Jones *et al.*, 1988; La Thangue & Rigby, 1988; Hatzopoulos *et al.*, 1988; Levine & Manley, 1989; Mitchell & Tjian, 1989; Saltzman & Weinmann, 1989).

2.1.2. Structure and Processing of RNA Molecules

In prokaryotes, transcription and translation are coupled, as the mRNA being synthesized can bind to ribosomes thus allowing concomitant protein synthesis.

In eukaryotes, synthesis of translatable mRNAs (mature mRNAs) involves several steps (Fig. 2) occurring in the nucleus (synthesis of pre-mRNA, capping, 3′ end cleavage and polyadenylation, protein binding, methylation of selected nucleotides and splicing), termed 'RNA processing' (reviewed in Nevins, 1983). It is conceivable, therefore, that alterations in any of these processes could lead to an altered expression of the relevant gene. Recently, it has been demonstrated that mature mRNAs can be, in some cases, submitted to modifications of the primary sequences of coding regions ('RNA editing'), which provides an additional means of modulating genetic expression (Simpson & Shaw, 1989).

Shortly after the beginning of transcription, the 5′ end of pre-mRNA is 'capped' by addition of 7-methylguanosine, which is linked by an unusual 5′–5′ triphosphate bond to the first transcribed base. The cap structure occurring in most, if not all, eukaryotic mRNAs confers a higher translational activity and stability on these molecules (reviewed by Banerjee (1980) and Rhoads (1985)). Poly(A) sequence is added to the 3′ end of RNA primary transcript by a multiprotein complex (reviewed by Humphrey and Proudfoot (1988)) responsible for both cleavage of transcript 3′ end, 12–20 nucleotides downstream from the AAUAAA polyadenylation signal, and subsequent stepwise addition of up to 200 adenylate residues (Moore & Sharp, 1985). In some immunoglobulin genes, alternative poly(A) site usage generates mRNAs encoding membrane-bound or secreted forms of antibodies (Early *et al.*, 1980). Both the integrity and position of the AAUAAA sequence are essential, as mutations altering this sequence prevent 3′ end formation and accumulation of mature mRNAs (Wickens & Stephenson, 1984). Poly(A) sequence appears to play an important role in both transport of mRNAs from the nucleus

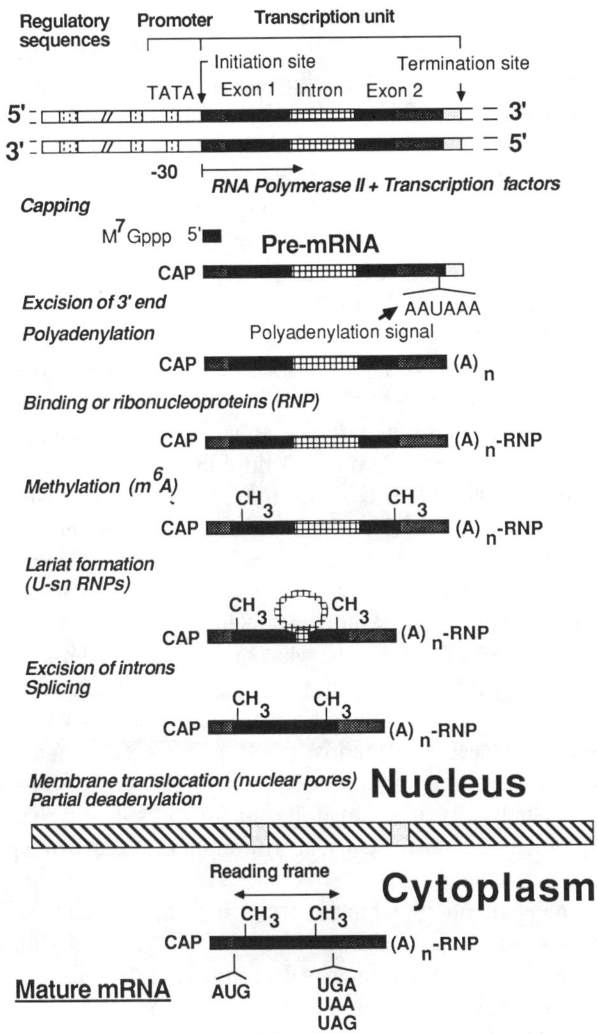

Fig. 2. Stepwise maturation of pre-mRNA within the nucleus. The primary RNA transcript undergoes an intricate processing which generates the mature mRNA, as outlined in this scheme. A 7-methylguanosine residue is first linked to the 5′ end of the growing pre-mRNA through a 5′–5′ triphosphate bridge (capping). Stepwise addition of 150 to 200 adenosine residues to the 3′ end is carried out by a poly-A polymerase. Other modifications include binding of proteins, methylation and RNA splicing, as described in Fig. 3.

to the cytoplasm and mRNA stability, in conjunction with its associated binding protein (PABP) (Bernstein *et al.*, 1989). Accordingly, accumulation of vasopressin mRNA correlates with increased poly(A) tract size (Robinson *et al.*, 1988). The 3′ non-coding regions of a variety of unstable mRNAs, e.g. those encoding c-*myc*, c-*fos* and lymphokines, contain AU-rich stretches including one or several AUUUA motifs that may be responsible for their instability (Shaw & Kamen,

1986). According to Bernstein and Ross (1989), the destabilizing effect of AU-rich sequences could be due to their interaction with the poly(A) protein complex.

In addition to methylation and binding to specific ribonucleic proteins, pre-mRNA undergoes removal of introns and subsequent splicing, which involves primarily the recognition by several effectors of specific intronic sequences close to the actual splice site (Fig. 3). Removal of introns proceeds by a two-step mechanism: (i) pre-mRNA is cleaved at the 5′ splice site and the 5′-terminal residue (guanosine) of the intron is linked by a 2′, 5′-phosphodiester bond (lariat formation) to a specific adenosine residue (the branch point) usually located 20 to 50 nucleotides upstream from the 3′ splice site; (ii) cleavage at the 3′ splice site is followed by ligation of the exons. Splicing events take place in large multicomponent complexes called spliceosomes, which are composed of the small nuclear ribonucleoprotein particles (snRNPs U1, U2, U4/U6, U5) and other splicing factors. The U snRNPs subunits of the spliceosome are themselves complex structures composed of a set of up to 10 different proteins and of one or two snRNAs (Steitz *et al.*, 1988). This complex nuclear machinery ensures an accurate selection of splice sites in pre-mRNA. The only conserved sequences known to play a role in splicing are those found at both 5′ and 3′ intronic ends and at the branch site. Consensus motifs for those 3 sites have been derived from compilations of sequences (Padgett *et al.*, 1986; Shapiro & Senapathy, 1987). U1 snRNP may recognize and bind to the 5′ splice site via complementary base pairing and U2 snRNPs may interact with the branch point sequence (BPS), within which even single-base changes can severely decrease splicing efficiency without preventing spliceosome assembly (Reed & Maniatis, 1988). U5 snRNP is an absolute requirement for the second step of splicing (cleavage at the 3′ splice-site and ligation of the exons) and is needed for efficient initiation of the splicing reaction (Winkelmann *et al.*, 1989). The role of U4/U6, although essential for splicing, is still ill-defined.

Although the mechanism by which specific interactions between snRNPs and pre-mRNA lead to site-specific cleavage and ligation is not fully understood, it is now clear that snRNPs are crucial factors for sequence recognition during splicing. Changes in such factors could regulate splicing patterns observed in different cell types or physiological states (Leff *et al.*, 1986). Alternative splicing of pre-mRNA offers a rapid and reversible response in terminally differentiated cells that have lost the power to synthesize and rearrange DNA. This phenomenon is quite widespread since more than 50 genes, in species from Drosophila to Man, are known to generate protein diversity through alternative splicing (Breitbart *et al.*, 1987). One of the best examples is provided by the calcitonin gene. Calcitonin and calcitonin gene related peptide (CGRP) are both produced by using different exons in different tissues (Rosenfeld *et al.*, 1983). Thus, in the brain very little calcitonin is produced whereas in the thyroid the calcitonin/CGRP ratio is about 95 : 1. Principles governing splice site selection remain obscure: a single 5′ splice site can be joined to one of several 3′ splice sites or vice versa. A combination of these events can occur, giving rise to exon skipping. Whereas alternative splicing can produce two or more functional forms of mRNA, recent studies indicate that splicing can

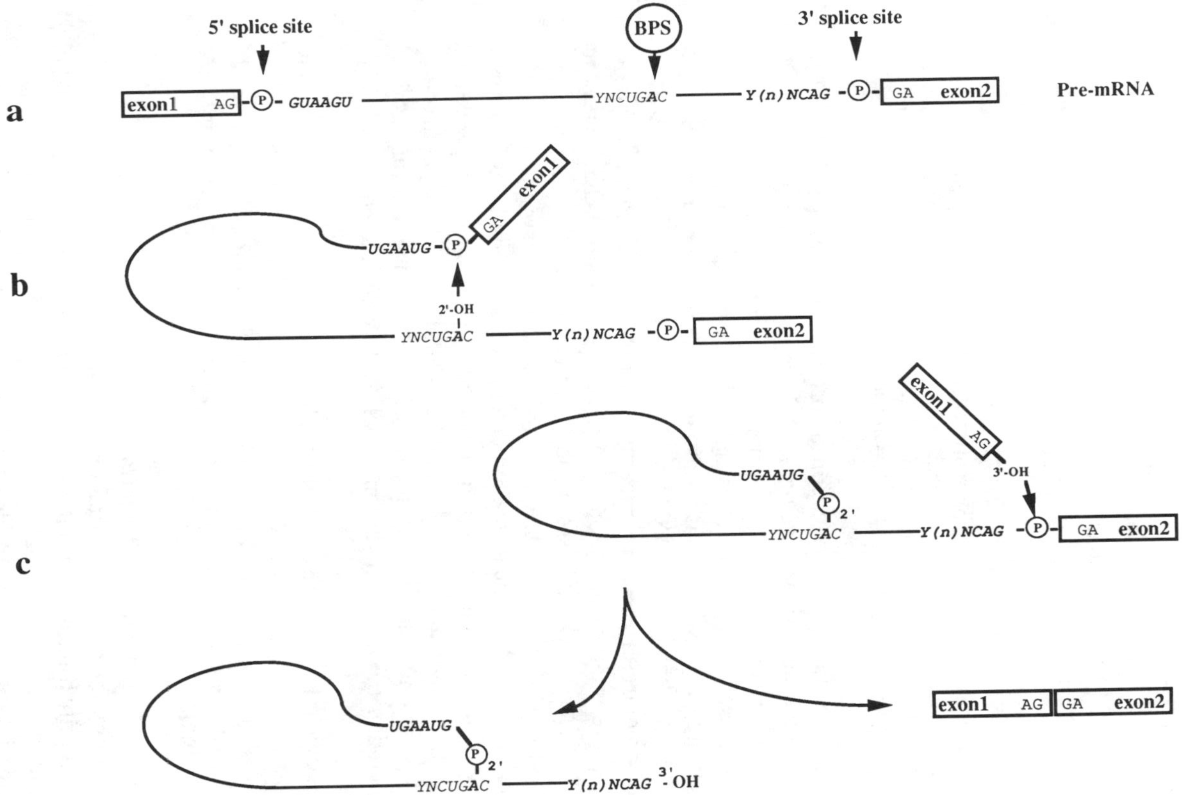

Fig. 3. The pre-mRNA splicing mechanism. A schematic representation of a precursor mRNA splicing pathway is shown. (a) The intervening sequence or intron, spanning from the 5′ to the 3′ splice sites, is represented as a line, flanked by two boxes representing exons 1 and 2. Consensus sequences (Y = pyrimidine, N = any base) at both splice sites and branch point (BPS) are indicated. The excised intron is released as a lariat which is generated through: (b) a cleavage at the 5′ splice site, leading to the formation of a 2′–5′ phosphodiester bond between the 5′-P terminal gunanosine residue and the 2′-OH adenosine residue within the branch site; (c) a second cleavage at the 3′ splice site, followed by ligation of the two exons.

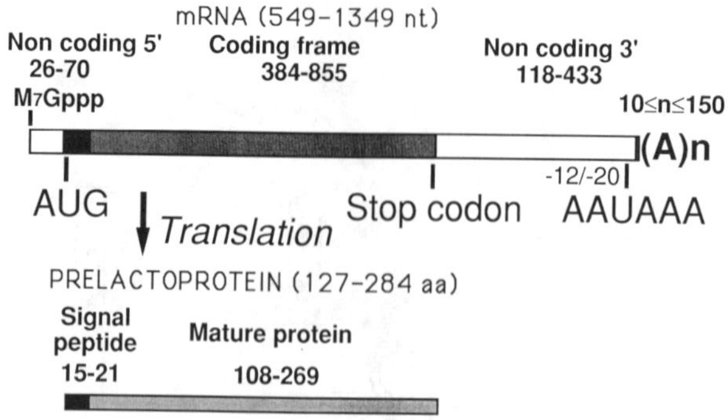

Fig. 4. Schematic representation of mammary mRNAs encoding the major milk proteins: 4 caseins, β-lactoglobulin, α-lactalbumin and whey acidic protein (WAP). Numbers indicate the extreme sizes of both mRNAs and the relevant proteins, and of constitutive regions, in the eight species investigated so far (nt = nucleotides; aa = amino acids).

also be used to turn on/off the expression of protein encoding genes. The best characterized examples are some developmentally regulated genes in Drosophila (Bingham *et al.*, 1988). Another alternative mechanism for the production of two proteins from a single DNA fragment has been recently recognized in eukaryotes: the opposite strand of gonadotrophin releasing hormone gene is transcribed in a functional RNA (Adelman *et al.*, 1987).

The typical structure of eukaryotic mature mRNA is schematized in Fig. 4. Between the cap site and the translation initiation codon, usually AUG, there is an untranslated sequence of variable size which precedes the coding sequence terminated by a stop codon (UAG, UAA or UGA). Loss of termination codon (frameshift mutation) leads to the formation of a longer polypeptide, as illustrated by the carboxy terminal extension of human chorionic gonadotrophin β subunit, which arises from a readthrough event (Talmadge *et al.*, 1984). The terminal codon is followed by a second untranslated region of variable length (3′ non-coding sequence) containing the polyadenylation signal AAUAAA and ending with the poly(A) tail.

Although specific mRNAs have characteristic half-lives, a variety of endogenous and exogenous stimuli can trigger changes in rates of mRNA turnover. For example, oestrogens increase 30-fold the half-life of vitellogenin mRNA (Brock & Shapiro, 1983) whereas dexamethasone destabilizes type I collagen mRNA (Raghow *et al.*, 1986).

2.1.3. Translation of RNA Molecules into Protein

The process whereby the genetic information encoded in the nucleotide sequence of mRNA is faithfully translated into the amino acid sequence of a protein can be divided into four major stages: formation of aminoacyl-tRNAs, initiation, elongation and termination of the polypeptide chain: these have been well described in current books of biochemistry. Therefore, translation will not be discussed here and the reader is referred to those books or reviews such as that of Rhoads (1988).

Only a brief comment will be made on regulation of translation. There is much evidence that it occurs early, at the initiation step, in response to various stimuli, such as hormones and nutritional conditions. Regulatory factors fall into two classes: (i) extrinsic factors, e.g. those controlling the translation process of all mRNAs, for example, the availability of rate-limiting initiation factors and their transient activation through covalent modifications such as phosphorylation; (ii) intrinsic factors relevant to the structure of each mRNA (the speed of translation depends in particular on the accessibility of the cap, the number of secondary structures occurring in the 5′ untranslated region and the flanking sequences of the initiation codon).

2.2. Structure of Milk Protein mRNAs and Genes

Previous structural and genetic studies of milk proteins provided an indirect knowledge of the relevant mRNAs and genes. Intra- and inter-species comparisons of amino acid sequences showed: (i) internal homologies within αs1- and αs2-casein, suggesting that the cognate genes evolved through intragenic duplications; (ii) the ubiquity of the major phosphorylation site and the striking homology of the signal peptides of calcium-sensitive αs1-, αs2- and β-casein, indicating a possible common origin; (iii) the structural relationship between α-lactalbumin and lysozyme, and to a lesser extent between β-lactoglobulin, retinol binding proteins (RBP), α-microglobulins and human placental protein 14, and between κ-casein and fibrinogens. Whey-acidic protein (WAP), a major protein found in rodents' and camels' milks, was identified as a member of the 'four disulphide core' family which includes neurophysins, wheat germ agglutinin, protease inhibitors, venom toxins, etc. (Hennighausen & Sippel, 1982); (iv) the rapid evolutionary rate of milk proteins. Furthermore, genetic variants were used as markers to demonstrate the close linkage of the four casein genes in the bovine species.

Structures and expression of milk protein mRNAs and genes have now been investigated in at least eight species, including ruminants, and nucleotide sequences of many mRNAs and genes have been reported (Fig. 5).

2.2.1. Milk Protein mRNAs

The rate of synthesis of milk proteins primarily depends upon the concentration of the relevant mRNAs, whose accumulation results from both an increase of the transcription rate (up to 4-fold) and a better stability (half-life × 25). Their level increases steadily from late pregnancy, and at mid-lactation they amount to up to 100 000 copies per cell, accounting for 60% of total mRNA in rat (Guyette *et al.*, 1979; Rosen *et al.*, 1981).

The general organization of milk protein mRNAs is depicted in Fig. 4. Sizes of the 30 or so mRNAs investigated range from 549 nucleotides for rabbit WAP to 1349 nucleotides for rat α-casein (αs1 type), excluding the poly(A) tail. The 5′ and 3′ untranslated regions amount to 28–44% of the mRNA and the polyadenylation recognition signal AAUAAA occurs 12–19 nucleotides upstream from the poly(A) tail. Both bovine αs1- and αs2-mRNA contain intra-duplicated regions (data not shown) as predicted from previous studies of the relevant proteins: a 24-nucleotide

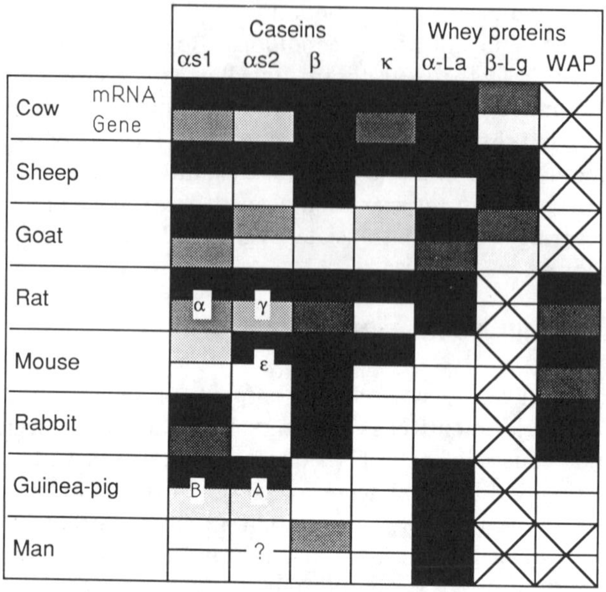

Fig. 5. Milk protein mRNAs and genes sequenced either completely (dark) or partially (grey) in various species. Dark-grey means that most of the sequence is known and ? that the protein may not exist. Sources are referenced in Mercier *et al.* (1990) or in the text for the most recent data. A few sequences have not been published yet.

repeat (codons 70–77/110–117) in the former and two long regions with 59% homology (codons 63–123/145–207) in the latter.

Inter-species comparison of homologous mRNAs (Fig. 6) confirmed the high rate of evolution of the open reading frames whose conserved domains are those encoding the major phosphorylation sites of the calcium-sensitive caseins and the signal peptide (see Figs 9 and 14 in Sections 2.2.2.1 and 3.3.3, respectively). The faster evolutionary rate of αs1 and αs2 casein coding frames, as compared to the 5′ and 3′ flanking untranslated regions, suggests that the latter have undergone an evolutionary constraint for maintaining local conformations involved in the high rate of translation and/or the stability of the mRNA.

Sequence similarities observed between milk proteins and the aforementioned proteins have been confirmed at the cDNA level, for example, coding frames of ovine β-Lg cDNA and human placental protein 14 share 67% homology (Julkunen *et al.*, 1988).

2.2.2. Milk Protein Genes

Since the first report on the partial structure of a casein gene by Yu-Lee & Rosen (1983), 15 or so nucleotide sequences of milk protein genes have been established (Fig. 5). These mosaic genes have the canonical organization described for tissue-specific eukaryotic genes (see Section 2.1.1 and Fig. 1).

The transcription unit scattered with several introns contains from 3′ to 5′, the polyadenylation signal AATAAA, several consensus splice junction sequences,

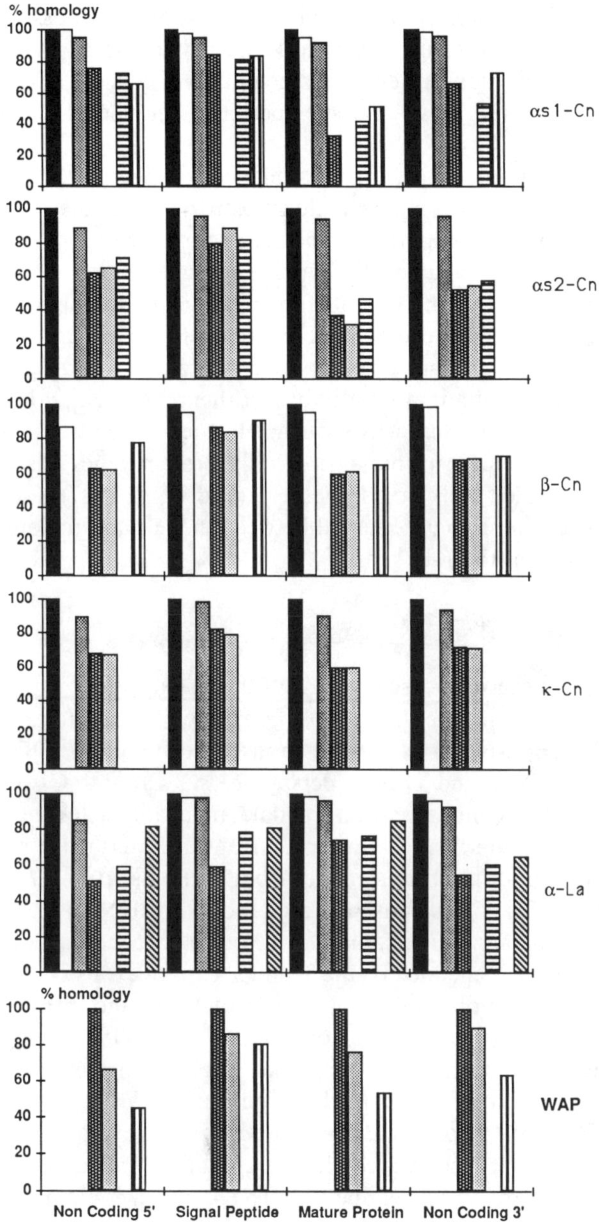

Fig. 6. Interspecies comparisons of homologous milk protein cDNAs. The sequence of reference is that of sheep for all but one type of cDNAs ■ Sheep □ Goat ▩ Cow ▩ Rat ▩ Mouse ▤ Guinea pig ▥ Rabbit ▧ Man. Except for the signal peptide encoding region which is well-conserved, the reading frame appears to have evolved at least as rapidly as the 5′ and 3′ untranslated regions.

and the cap site. Upstream from it, occur the TATA box and possible CAAT boxes which signal the start of transcription to RNA polymerase II. Activation of these genes requires, in addition, the effector recognition of consensus sequences or local conformations involved in stage-tissue specificity, induction and modulation of expression.

Potential recognition sites for miscellaneous effectors were identified by computer search primarily in the region upstream from the transcription unit. Intra-and inter-species comparisons of milk protein genes sequenced so far have corroborated previous phylogenetic hypotheses and provided additional interesting information: thus (i) the overall organization of investigated genes is conserved during evolution, as exemplified by those of β-casein and α-la (see Sections 2.2.2.1 and 2.2.2.2); the main changes concern sizes of introns, where repetitive sequences have been found, at least in one of them, in casein, α-la and β-lg genes; (ii) the occurrence of a consensus sequence located a hundred nucleotides or so upstream from the transcription start of milk protein genes has been suggested. However, the so-called 'milk box' hypothesis (Hall *et al.*, 1987) is still controversial owing to the weak homology observed between the relevant regions of some genes and its absence from the β-lg gene.

2.2.2.1. Casein genes (Fig. 7).

2.2.2.1.1. Calcium-sensitive casein genes. With the exception of the β-casein gene, which has been sequenced either almost completely in rat (Jones *et al.*, 1985) or entirely in cow (Bonsing *et al.*, 1988), mouse (Yoshimura & Oka, 1989), rabbit (Thépot *et al.*, 1990) and sheep (Persuy, M.A., Provot, C. & Mercier, J.C. unpublished), only incomplete structural data are available for the genes encoding the two other calcium-sensitive caseins, αs1 and αs2, termed α and γ in the rat.

The β-casein transcription unit ranges from 7·2 kb in rat to 8·7 kb in sheep, and contains 9 exons. The 5′ untranslated region of the mRNA is divided into exons I and II, and all splicing in the coding frame occurs between codons. The multiple phosphorylation site is encoded by the 3′ and 5′ ends of exons IV and V, respectively. On the basis of the distribution and sizes of exons and introns, and sequence similarity existing between some exons, Jones *et al.* (1985) suggested that the β-casein gene evolved through intragenic duplications: exon III might have generated exon IV and the set exon III–intron 3–exon IV further duplicated to give the set exon V–intron 5–exon VI, or reciprocally. The overall organization of the β-casein gene has been conserved during evolution and differences in sizes of transcription units are primarily due to intron 4, which contains repeated sequences in cows and sheep.

Estimated sizes of αs1 casein transcription units are 15 kb in rat and 20 kb in ruminant and rabbit species. Partial organization of the first five exons, comprising only 16% of rat mRNA (Yu-Lee *et al.*, 1986; Rosen, 1987), as well as sequence data relevant to the region upstream from exon II in the bovine and rabbit species, were reported. In goat, the αs1 casein gene is divided into 19 small exons (Leroux *et al.*, 1992).

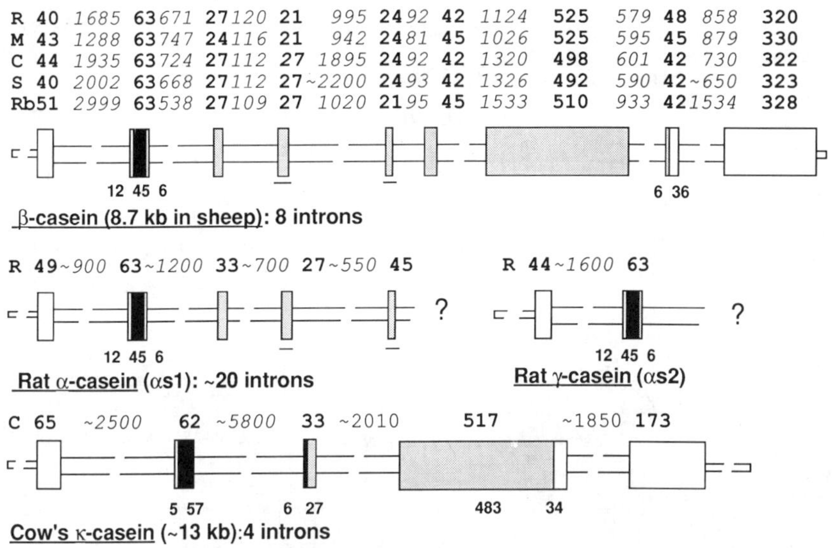

Fig. 7. Organization of the casein encoding genes: R, rat; M, mouse; C, cow; S, sheep; Rb, rabbit. Exons are represented by large boxes: black and grey regions encode the signal peptide and the mature protein, respectively. Broken lines represent regions encoding a multiple phosphorylation site. Sizes of exons and introns (italicized) are indicated. The drawing is based on data reported by Jones *et al.* (1985), Yoshimura & Oka (1989), Bonsing *et al.* (1988), Persuy, M.A., Provot, C. & Mercier, J.C. (unpublished) and Thépot *et al.* (1991) for the β-casein gene and Alexander *et al.* (1988) for the bovine κ-casein gene. Only fragmentary data for the rat species (Rosen, 1987) are available on the αs1- and αs2-casein genes.

The overall organization of the αs2-casein gene is still unknown. In rats, the 15-kb αs2-like casein gene may contain at least 9 exons, and many repeated sequences occur within the transcription unit and the 5′ flanking region. Sequence data are available only for the region upstream from exon II.

As first observed by Rosen (1987), the genes encoding the 'calcium-sensitive' caseins have a similar organization at the 5′ end of the transcription unit (Fig. 7) and share common structural patterns upstream from it (Fig. 8(a)), which strongly supports the hypothesis of a common ancestor. Exon II always comprises the end of the 5′ untranslated region and 17 codons, including those encoding the 15 amino acid residue long signal peptide; the ubiquitous major phosphorylation site is encoded by a sequence arising from the splicing of two exons in the calcium-sensitive casein genes, which is advantageous as there is a double selection pressure for maintaining both the splicing consensus sequence and the encoded major phosphorylation recognition site (Fig. 9). The 200 bp 5′ flanking region of the transcription unit contains several well conserved structural motifs (Fig. 8(a,b)) which are likely to play a role in the regulation of transcription. For example, the 3′ end of the invariant 18 nucleotide stretch close to the TATA box, is very similar to the octamer ATTTGCAT occurring in many genes and recognized by transcriptional

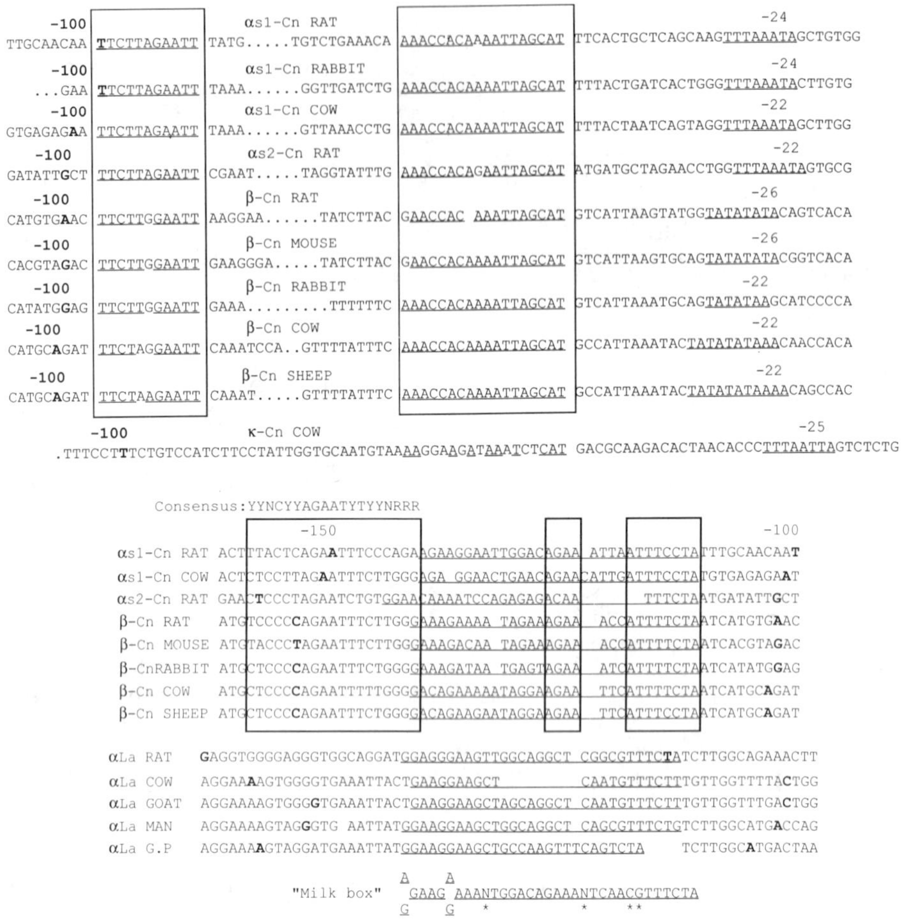

Fig. 8. (a) Specific structural motifs (boxed) occurring upstream from the TATA box (underlined) of the genes encoding the calcium-sensitive caseins αs1, αs2 and *β*. *κ*-casein and whey protein genes (data not shown) lack these sequences. Sequences were aligned to get the best inter-species homology for a given gene. Bold-faced nucleotides occur at position -100 with respect to the site of transcription initiation. (b) Structural motifs (bold-faced boxes) common to the genes encoding the calcium-sensitive caseins αs1, αs2 and *β*, and consensus sequences (underlined 'Milk box') shared with the α-La (Hall *et al.*, 1987; Laird *et al.*, 1988) gene. Sequences were aligned to get the best inter-species homology for a given gene. Bold-faced nucleotides occur at positions -150 and -100 with respect to the site of transcription initiation; * indicates a possible gap in the consensus sequence.

factors OCT-1 and/or OCT-II. However, the precise functional role of these motifs is still unknown.

2.2.2.1.2. *κ*-Casein gene. The *κ*-casein gene, partially sequenced in the bovine species, is quite different structurally from the three other casein genes, with a 13-kb long transcription unit split into only 5 exons (Fig. 7), amounting to 850 nucleotides. All introns have a large size and both the second, which divides the

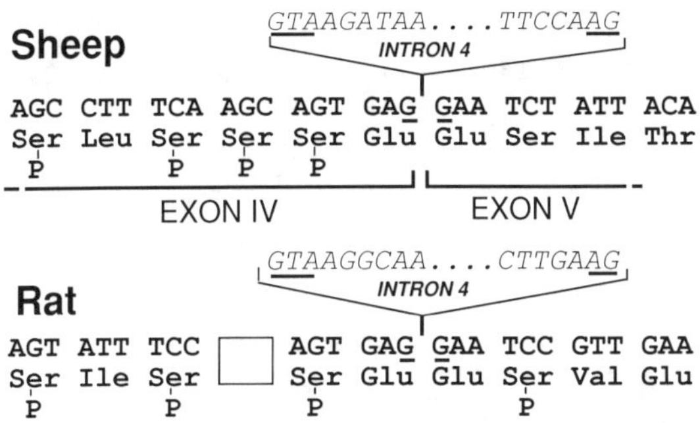

Fig. 9. The RNA stretch encoding the major phosphorylation site of ovine and rat β-caseins results from a splicing event (Jones *et al.*, 1985; Persuy, M.A., Provot, C. & Mercier, J.C. unpublished results). Phosphorylation requires the enzymatic recognition of the tripeptide sequence Ser–X–A, where A is an acidic residue, and X any amino acid residue (Mercier, 1981).

signal peptide encoding region, and the third contain many Alu-type repetitive sequences. Furthermore, the 5′ flanking region of the transcription unit lacks the structural motifs shared by the other casein genes.

κ-Casein and fibrinogens are functionally similar, as their limited proteolytic cleavage triggers milk and blood clotting, respectively. The most striking sequence similarity, observed between the proteins and the relevant cDNAs, respectively, involves a nucleotide stretch corresponding to the 5′ end of κ-casein exon IV and the 3′ end of γ-fibrinogen exon II (Alexander *et al.*, 1988), which suggests an evolutionary relationship.

2.2.2.1.3. Chromosomal localization. Electrophoretic analyses of dam–daughter's bovine caseins (reviewed by Grosclaude, (1979)) and of dam–offspring's enzymatic digests of ovine genomic DNA (RFLP study with casein cDNA probes (Levéziel *et al.*, 1991)) showed the unique transmission of parental haplotypes by heterozygous males, thus demonstrating the close linkage of casein genes in both bovine and ovine species. Clustering of casein genes has also been evidenced by somatic cell and/or in-situ chromosomal hybridization studies using casein cDNA probes, in the mouse (Gupta *et al.*, 1982; Geissler *et al.*, 1988), rabbit (Gellin *et al.*, 1985; Dalens & Gellin, 1986) and bovine (Womack *et al.*, 1989) species. Casein genes were assigned to chromosomes 5 in the W region, 12 and 6, respectively. Threadgill and Womack (1990) as well as Italian workers (Ferretti *et al.*, 1990) have recently localized the cluster of bovine casein genes in a rather short genomic DNA fragment with sizes of 185 and 250 kb, respectively. Casein genes may occur as a single-copy per haploid chromosome set, as no related sequences have so far been reported.

2.2.2.2. Whey protein genes (Fig. 10). The gene coding for α-lactalbumin, the calcium metalloprotein involved in lactose synthesis, has been completely sequenced in rat (Qasba & Safaya, 1984), bovine (Vilotte *et al.*, 1987), human (Hall *et al.*, 1987), guinea-pig (Laird *et al.*, 1988) and goat (Vilotte *et al.*, 1991) species.

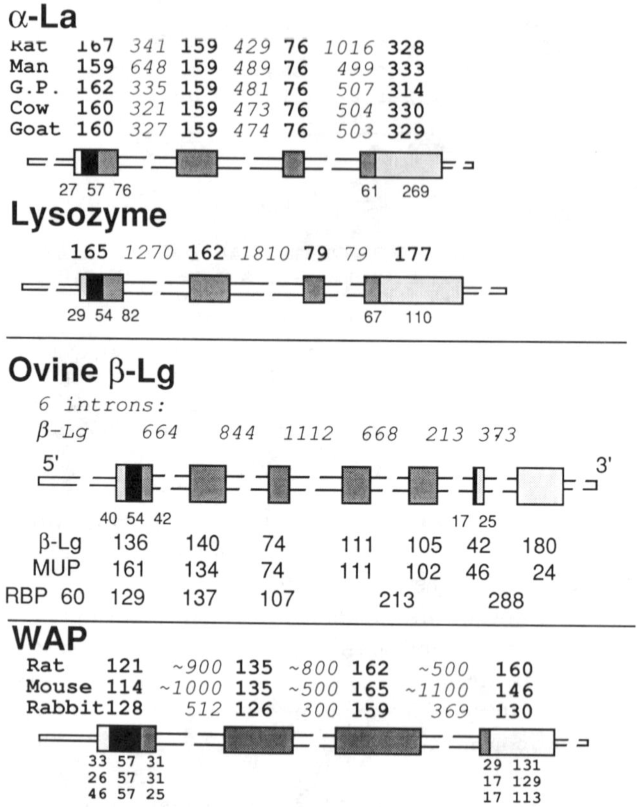

Fig. 10. Organization of the main whey protein genes and evolutionary related genes. (Same legend as in Fig. 7). α-Lactalbumin (α-La) genes: the length of the transcription unit is 2 kb for the rat (Qasba & Safaya, 1984), cow (Vilotte *et al.*, 1987), goat (Vilotte *et al.*, 1991) and guinea-pig (G.P.) (Laird *et al.*, 1988) α-La genes, 2·3 kb for the human gene (Hall *et al.*, 1987) and 3·7 kb for the chicken egg white lysozyme gene (Jung *et al.*, 1980). Sheep *β*-lactoglobulin (*β*-Lg) gene and two related genes encoding rat major urinary protein (MUP) and retinol binding protein (RBP), respectively (adapted from Ali & Clark (1988) and Harris *et al.* (1988)): the 5′ untranslated region of the RBP gene is divided into two exons, and the last two exons are equivalent to the pairs IV–V and VI–VII of the *β*-Lg gene. Whey acidic protein (WAP) genes: drawing based on the data reported in Campbell *et al.* (1984) and Thépot *et al.* (1991). Only the 5′ and 3′ ends of introns were sequenced in the rat and mouse species.

The 2 kb transcription unit with 3 exons is similar to that of lysozyme, corroborating the previous hypothesis of a common ancestor. The signal peptide and the calcium binding loop are encoded by exons I and III, respectively, whereas codons for amino acid residues essential for interaction with UDP-galactosyltransferase are located in both exons II and IV.

The *β*-lactoglobulin gene was completely sequenced in the ovine species (Ali & Clark, 1988; Harris *et al.*, 1988). The 4662 nucleotide-long transcription unit contains 6 introns and its organization is similar to that of genes encoding mouse major urinary protein (MUP) and retinol binding protein (RBP). WAP genes were sequenced in rat, mouse (Campbell *et al.*, 1984) and rabbit species (Thépot *et al.*, 1990). They contain 3 introns and the two cysteine domains are encoded by exons II and III, which may result from an intragenic duplication.

2.2.2.2.1. Chromosomal localization. Human α-La and mouse WAP loci were assigned to chromosomes 12 (Davies *et al.*, 1987) and 11 (Gupta *et al.*, 1982), respectively,by chromosomal in-situ hybridization. Analysis of panels of hamster–sheep somatic cell hybrids with an ovine α-La cDNA probe showed the synteny of the α-La locus (Imam-Ghali *et al.*, 1991) with a set of six markers already assigned to ovine chromosome 3, equivalent to bovine and caprine chromosomes 5. Several α-lactalbumin-related sequences occur in the genome of domestic ruminants (Soulier *et al.*, 1989; Vilotte *et al.*, 1991). One of them studied in both bovine and ovine species, contains a 1·5 kb region sharing 81 and 73 homology, respectively, with the bovine α-lactalbumin gene downstream from the end of exon II. One putative pseudogene has been detected for ovine β-Lg (Clark, J., 1990, pers. comm.).

2.3. Expression of Milk Protein Genes

*2.3.1. Regulatory Sequences (*cis-*and* trans-*Regulatory Elements)*

Functional studies of milk protein genes have been carried out in different ways.

(1) Computer searches have been performed to find consensus sequences functionally important according to the literature, such as those recognized by the above-mentioned effectors. Accordingly, various putative *cis*-acting DNA motifs, such as those binding hormone–receptor complexes (glucocorticoids, progesterone) or nuclear regulatory proteins (CTF/NF-1, TGGCA-protein, Spl, etc.), were localized in several milk protein genes.
(2) Computer comparisons have been performed for identifying structural motifs either common to different milk protein genes or conserved during the evolution of a given gene, such as those previously described in the 0·2-kb 5′ flanking region. A priori, those conserved regions should have a functional role not yet elucidated.
(3) Structural modifications of a gene and study of the relevant functional alterations in cell lines and/or transgenic animals.
(4) Identification of the constitutive elements of the gene which may interact with specific mammary nuclear ligands, using DNAase I protection (footprinting) and/or DNA-protein mobility shift (gel retardation) assays.
(5) Study of alleles with altered expression.

At present, comparative studies of expression of native and modified genes in transfected mammary cells or transgenic animals are rather difficult: available cell lines have partly lost their potentiality and the number of gene copies incorporated is quite difficult to assess; similarily, each transgenic animal differs by the number of copies integrated and the site of chromosomal insertion. Therefore, knowledge of the functioning of milk protein genes at thc molecular level is poor as experimental data are still fragmentary. Two recent developments in the field have provided some information:

(1) Study of expression of milk protein genes, more or less deleted upstream from the transcription start, in cell lines and transgenic animals, showed that some constitutive elements involved in expression are quite close to the beginning of the transcription unit. In particular, those responsible for temporal and stage-tissue specificity: β-Lg and α-La genes with a 5′ flanking region shortened to 0·8 kb (Harris *et al.*, 1990) and to 0·25 kb (Vilotte, J.L., unpublished), respectively, retained their temporal and specific expression in mammary epithelial cells of transgenic mice. However, a WAP-tPA (tissue Plasminogen Activator) hybrid gene with 2·6 kb of the 5′ flanking region of the WAP transcription unit was not specifically expressed in mammary epithelial cells (Pittius *et al.*, 1988). Except for ovine β-Lg gene (Simons *et al.*, 1987), expression of genes is generally lower in a transgenic animal, which suggests that the micro-injected DNA fragments lack some regulatory elements. The best results were obtained with bovine α-La gene (0·45 mg α-La /ml of mouse milk (Vilotte *et al.*, 1990), hybrid bovine αs1-casein–urokinase gene (up to 2 mg urokinase /ml mouse milk (Meade *et al.*, 1990)) and rat WAP gene (expressed as its endogenous counterpart (Bayna & Rosen, 1990).
(2) Study of constitutive elements interacting with specific mammary nuclear effectors, using DNAase I protection (footprinting) or/and DNA-protein mobility shift assays, showed the preferential binding of nuclear proteins to sites located in the region within 0·2 kb from the transcription start in both WAP (Lubon & Hennighausen, 1987) and α-La genes (Lubon & Hennighausen, 1988; Lubon *et al.* 1989).

A search for polymorphic animals differing in the rate of synthesis of a given milk protein and analysis of the structure–function relationship is a promising alternative. Such work is in progress in the caprine species. Nine alleles at the αs1—Cn locus have already been detected (Grosclaude *et al.*, 1987; Mahé & Grosclaude, 1989; Leroux *et al.*, 1990): αs1-Cn^{A},^{B} and ^{C} are associated with a high concentration of αs1-casein (3·6 g/litre per allele) while αs1-Cn^{E} and αs1-Cn^{D},^{F} and $^{F'}$ are associated with intermediate (1·6 g/litre) and low (0·6 g/litre) contents, respectively; αs1-Cn^{0} and $^{0'}$ are probably true null alleles. According to available data, a 0·5-kb insertion may be responsible for the lesser expression of allele E, and mutations altering, presumably, either the branchpoint or a 3′ splice site consensus sequence of alleles D and F might be responsible for an abnormal RNA splicing and the low αs1-casein yield (Brignon *et al.*, 1990; Leroux *et al.*, 1992).

2.4.2. Transcription and Maturation of mRNA: Differential Splicing (Deleted Caseins)

Little is known about milk protein pre-mRNA splicing pathways. Rat α-lactalbumin pre-mRNA processed *in vitro* with HeLa cell nuclear extracts showed the preferential sequential deletion of introns 1, 2 and 3 (Lang & Spritz, 1987). However, this does not imply that processing of other milk protein pre-mRNAs would also proceed linearly from one end of the transcript, either *in vitro* or *in vivo*. Other data available to date refer primarily to the intron/exon junction sequences at the two splice sites, provided by comparing milk protein cDNA and genomic sequences (Fig. 11). The search for a potential branchpoint sequence (BPS) within the known intervening sequences of milk protein encoding genes has allowed frequent identification, at a relatively conserved distance (ranging between 15 and 150 nt) from the 3′ splice site, of a putative BPS: YNYURAY, remarkably matching with the mammalian consensus sequence: YNCUGAC (Reed & Maniatis, 1988).

Aberrant splicing of primary transcripts, leading to exon-skipping, is suspected to be responsible for polymorphism of αs1- and αs2-caseins in ruminant species (Jones *et al.*, 1985; Stewart *et al.*, 1987; Brignon *et al.*, 1990). Peptide as well as DNA structural data strongly support such a hypothesis. In fact, in goats, in which the polymorphism of αs1-casein was shown to be under the control of at least 7 autosomal alleles (Grosclaude *et al.*, 1987), 2 variants (D and F) have an internal deletion (11 and 37 amino acid residues, respectively) starting at the same position in the peptide chain and both including the major phosphorylation site of the protein (Brignon *et al.*, 1990). The deletion occurring in αs1-casein D ends between two glutamate residues located immediately after the hydrophilic cluster of five phosphoseryl residues. This position is typical of an exon boundary according to Jones *et al.* (1985), who first noticed, studying the rat β-casein gene, that two glutamate codons belonging to two contiguous exons contribute to the conserved major phosphorylation site. The nucleotide sequence of the region encoding the major phosphorylation site in the ovine β-casein gene, although different from its rat counterpart, also arises from a RNA-splicing event (Persuy, M.A., Provot, C. & Mercier, J.C., unpublished data). These observations suggested that the deletion of αs1-caseins D and F could be due to a genetic event inducing an abnormal splicing of the primary transcript rather than to a deletion within the gene.

Comparison of the structural organization of a reference allele (αs1-CnA), apparently correctly spliced, to that of the αs1-CnF allele, in the region containing exons putatively outspliced, has provided evidence that substantiates this hypothesis. Leroux *et al.* (1992) have shown that most transcripts from the αs1-CnF allele are aberrantly spliced, most of them lacking 3 exons. However, correctly spliced messengers are also produced as well as transcripts in which up to 5 exons are lacking. Moreover, these data clearly demonstrate that the DNA region encoding the peptide sequence deleted in variant D does indeed correspond to an exon. Mutational events that lead to exon-skipping have been characterized but the first clues to the mechanisms involved in the phenomemon remain to be found.

Ovine αs1-casein, which is shorter (eight amino acid residues) than its caprine and bovine counterparts, might be a further example, since, although the putative 'systematic deletion' does not affect a multiphosphorylation site, the relevant bovine and caprine cDNA sequences display a typical exon structure. Another possible example of abnormal RNA-splicing is provided by a rare αs1-casein variant (A), found in the bovine species, which differs from common variants by

		EXON	INTRON	EXON
	CONSENSUS	(.r)	GTRrgt.//.yyyyyyyyyyyNcAG	(r.)
βLg	Sheep	(AG)	GTTCGA.//.AAGGTCCCTCTCCAG	(GT)
		(TG)	GTGGGC.//.GCAGCTGTCTTTCAG	(GG)
		(TG)	GTGAGT.//.CTTGACCGCGTCCAG	(CC)
		(GG)	GTGGGT.//.GTCCCTGCCCCATAG	(TC)
		(GG)	GTGAGC.//.CTCCCTCCCCCACAG	(GG)
		(GG)	GTAAGC.//.CCATGTCCATTTCAG	(GG)
WAP	Rat	(AG)	GTAGGC.//......TTTTGTACAG	(TT)
	Mouse	--	---A-- .....------G---	--
	Rabbit	--	------ CTGCCCCGT--G---	-G
	Rat	(CA)	GTAAGT.//.CCATCTCTTCCCCAG	(TT)
	Mouse	--	------ -------- ------	--
	Rabbit	-G	----CG --TGTCTC-------	-C
	Rat	(AA)	GTATGC.//.CTGTATCTCCTACAG	(AA)
	Mouse	-G	------ -CA-C--C-------	TG
	Rabbit	-G	-----T ----G------G---	-G
αLa	Cow	(AT)	GTGAGT.//.TCGTCTTTCTTTCAG	(GG)
	Goat	--	------ ---------------	--
	Rat	--	--A--- -TC-G-C--CC----	--
	G.P.	--	------ --A-A--T--C----	--
	Man	--	------ --A------CC----	TG
	Cow	(CA)	GTGAGT.//.AGAGGCTCCTGCCAG	(AG)
	Goat	--	------ T-TAA----------	--
	Rat	--	------ CTACA---- --T--	--
	G.P.	--	--A--- TACAA---T--T---	--
	Man	--	------ TGTAA----------	--
	Cow	(TG)	GTGAGT.//.CTTCTTCATGATCAG	(GT)
	Goat	--	------ -------G-------	--
	Rat	--	------ T----C-G-------	-A
	G.P.	--	------ T----C---T-C---	--
	Man	--	----A- -----C---T-----	--
κCn	Cow	(AG)	GTAATC.//.TTTAATTTAATTTAG	(CT)
		(TG)	GTGAGT.//.TCTGAATCTCAACAG	(GG)
		(TA)	GTAAGT.//.TTTATTTTTTACAG	(AG)
		(AG)	GTAAAC.//.CTGTCATTTTTATAG	(GT)
αCn	Rat	(AG)	GTATTG.//.TCCTTTTCAACACAG	(AT)
	Cow	--	-----A ?	
	Rabbit	--	------ ?	
	Rat	(CT)	GTGAGT. .TCTCTTACTTTTCAG	(AG)
	Rat	(AA)	GTAAGT. .ATTTATTCCTTGCAG	(AC)
	Rat	(AG)	GTGAGC. .TTTTAATTCCCATAG	(GA)
	Rat	(AG)	GTAAGT. . ?	
	Goat	(AG)	? TTCTCTTTCTTTTAG	(CA)
	Goat	(AG)	GTATAC TTTTTTTTATTTAAG	(GA)
	Goat	(AG)	GTAAGA TTTTGAATTTTTTAG	(CA)
	Goat	(TG)	GTAAAA ?	
	CONSENSUS	(.r)	GTRrgt.//.yyyyyyyyyyyNcAG	(r.)
βCn	Sheep	(AG)	GTAAGA.//.TCTTCCCATTCACAG	(GA)
	Cow	--	------ ------------C---	--
	Rat	--	------ ---CTT----------	--
	Mouse	--	------ --ACTT----------	--
	Rabbit	--	------ ---CT-----------	--
	Sheep	(GA)	GTAAAT.//.TTTTTCTCTCTTTAG	(CA)
	Cow	AG	------ -----T---------	-T
	Rat	AG	--GTG- -G---GCTTT-AC--	A-
	Mouse	AG	--GTG- --CC-T-TA-AAA--	AC
	Rabbit	AG	---TG- -------CT-T--C--	A-
	Sheep	(AG)	GTAAGA.//.ATTTTTCCTTTATAG	(AC)
	Cow	--	------ ---------------	-T
	Rat	--	------ -------TG----C--	--
	Mouse	--	------ TA----TG----C--	--
	Rabbit	--	------ -------TT----C--	G-
	Sheep	(AG)	GTAAGA.//.TTTATTTTTCCAAAG	(GA)
	Cow	--	------ ---------------	--
	Rat	--	-----G C--TC---CTTG---	--
	Mouse	--	-----G ---TC-----TGC--	--
	Rabbit	--	------ -CAT------TG---	--
	Sheep	(AG)	GTAAAA.//.TTTTTTTTTAACCAG	(AA)
	Cow	--	------ ---------------	--
	Rat	--	---TTT G---A---CTGG---	--
	Mouse	--	-----C A--G---A-TT-T--	C-
	Rabbit	--	----G- A-A----G-TTTT--	C-
	Sheep	(AG)	GTAATT.//.TTTCCTTCTTTCCAG	(GA)
	Cow	--	------ ---------------	--
	Rat	--	----AC --CTT--G----T--	--
	Mouse	--	----AC --CTT-----C-T--	--
	Rabbit	--	---T-- ---------------	--
	Sheep	(TT)	GTAAGT.//.TTTTTAATTTTTAAG	(GT)
	Cow	--	------ -----T---------	--
	Rat	C-	------ A-A--TGCA------	--
	Mouse	C-	------ A-G--TGCA------	--
	Rabbit	A-	------ --A--T-----A---	--
	Sheep	(TG)	? .//. ?	(AA)
	Cow	--	GTAAGC TTCTCTATTCCACAG	--
	Rat	--	-----T.//.A---T-C--T-----	--
	Mouse	--	------ ---------------	--
	Rabbit	--	------ GCTCTGTA-T-CACA	G-
γCn	Rat	(AG)	GTAATG.//. ?	

Fig. 11. Known splice junction sequences of milk protein genes. Dashes represent nucleotides identical to those of the sequence taken as reference. In the consensus sequence, a lowercase letter refers to the nucleotide most frequently found at that position. R/r and y refer to purine and pyrimidine, respectively.

a 13 amino acids internal deletion. Recently, a cDNA clone for bovine αs1-casein variant A was isolated from a mammary gland cDNA library constructed with mammary tissue of a homozygous B variant cow as starting material (McKnight *et al.*, 1989). This result suggests that such an event may occasionally occur in the course of the maturation of primary transcripts from an allele (αs1-CnB) usually correctly spliced.

In the ovine and caprine species, αs2-casein occurs as two non-allelic forms transcribed from four types of mRNAs resulting from the combination of insertion/deletion events, possibly due to aberrant splicing, affecting the 5′ untranslated and the coding regions (Boisnard & Petrissant, 1985; Boisnard *et al.*, 1991). In the bovine species, a deletion of nine residues found in αs2-casein variant D (Grosclaude *et al.*, 1978) affecting once again a major phosphorylation site, as in the goat αs1-casein, could also correspond to the loss of an exon.

Recently, it has been demonstrated that exon-skipping, which seemed up to now, restricted only to ruminants and rodents αs1- and αs2-casein mRNAs, also affects the β-casein transcripts in humans (Menon *et al.*, 1992; Martin & Leroux, 1992). Therefore, such a phenomenon can be enlarged to all three 'calcium-sensitive' casein genes.

2.4.3. Hormonal Regulation of Milk Protein Gene Expression

During the last two decades numerous studies have been devoted in several laboratories to understanding of the hormonal regulation of milk protein synthesis in mammary cells. Studies with cDNA probes and cloned genes have provided valuable new insights into the molecular mechanisms involved in the hormonal control of milk protein gene expression. This field has been extensively reviewed (Rosen & Shields, 1980; Rosen *et al.*, 1986, 1988; Houdebine *et al.*, 1985), and only a brief survey is given here.

The onset of milk protein gene expression and milk secretion at parturition and during lactation, is induced and modulated by the interplay of a complex of hormones, including steroid and polypeptide hormones, which act sequentially. While ovarian steroids and prolactin are both required for mammary gland morphogenesis, prolactin and glucocorticoids are essential for the expression of milk proteins (for more detailed information the reader is referred to the excellent reviews of Forsyth (1983, 1986)). The regulation of milk protein gene expression by lactogenic hormones has been extensively analysed at the molecular level with cDNA to casein mRNAs as probes.

Initial studies carried out *in vivo* and *in vitro* cultures of mammary glands from rabbit (Houdebine & Gaye, 1975; Devinoy *et al.*, 1978), rat (Matusik & Rosen, 1978) and mouse (Terry *et al.*, 1977; Nagaiah *et al.*, 1981) have clearly demonstrated that prolactin is responsible for the primary accumulation of casein mRNAs in the mammary gland of these species. Although the signal transduction pathway by which prolactin activates milk protein genes specifically is not known, it seems to proceed with little delay, since induction of casein mRNA occurs in rat mammary organ culture within 1 h of the addition of prolactin to the culture medium (Matusik & Rosen, 1978).

In order to determine the precise level at which prolactin exerts its effect, a comparison of mRNA levels and rate of transcription of casein mRNAs has been carried out using different experimental approaches. From pulse labelling experiments, in rat mammary gland in tissue culture, it has been shown that prolactin exerts a pleiotropic effect by increasing, on the one hand, casein gene transcription and, on the other hand, the half-life of newly synthesized mRNA molecules (Guyette *et al.*, 1979). Similar results were obtained by measuring the rate of β-casein mRNA transcription and the β-casein mRNA levels in pseudopregnant rabbits treated with prolactin (Teyssot & Houdebine, 1980). More recently, Rosen *et al.* (1986) have shown that the major effect of prolactin on the regulation of casein mRNAs appears to be exerted at the level of nuclear RNA processing.

Although the expression of milk protein genes appears to be mediated primarily by prolactin, the roles of glucocorticoids and progesterone are also clearly of major importance in the modulation of the complex events involved in this process. While glucocorticoids alone are unable to induce casein mRNA, the presence of such steroids is absolutely necessary for the full induction of casein mRNAs by prolactin in mammary tissue cultures of rabbit (Devinoy & Houdebine, 1977), rat (Matusik & Rosen, 1978) and mouse (Nagaiah *et al.*, 1981). These results, as well as those of Chomczynski *et al.* (1986), suggest that the regulation of β-casein mRNA stability requires the simultaneous presence of both prolactin and glucocorticoid, in these species. Inhibition of glucocorticoid receptor complex binding to nuclear acceptor sites has also been shown to cause a near complete suppression of casein mRNA in murine mammary gland cultured *in vitro* (Majumder *et al.*, 1983), thus confirming the essential role of this steroid. Glucocorticoids in the absence of prolactin stimulate considerably (68-fold) the level of whey acidic protein (WAP) mRNA in tissue culture preparations of mammary gland, suggesting that the expression of WAP gene is regulated independently of casein genes (Hobbs *et al.*, 1982). Progesterone always acts as a potent inhibitor of prolactin stimulation of mammary cells (reviewed by Kuhn (1983)). This steroid prevents the accumulation of casein mRNAs inducible by prolactin in the pseudopregnant rabbit mammary gland (Houdebine & Gaye, 1975; Houdebine, 1976; Teyssot & Houdebine, 1981*a*,*b*) and in ovariectomized pregnant rats (Rosen *et al.*, 1978). Similar results were also obtained in organ cultures of mammary gland from rats (Matusik & Rosen, 1978) and mice (Ganguly *et al.*, 1982). It has been proposed that progesterone may exert its inhibitory action on casein gene expression by reducing glucocorticoid binding to mammary cytosol receptors (Ganguly *et al.*, 1982). However, the mechanism by which progesterone blocks the expression of milk protein genes is still unclear; different mechanisms might be involved, some local and some systemic (Kuhn, 1983).

Although studies in organ culture have been particularly useful in defining the respective roles of prolactin and glucocorticoids, these systems suffer from limitations (for example, as a result of problems of residual hormone levels) and do not permit direct manipulation or transfer experiments of milk protein genes to identify sequence(s) involved in hormonal regulation. The development of a mammary cell culture line, designated Comma-ID, derived from pregnant Balb/c mice, cultured in floating collagen gels, has been reported to synthesize caseins in the presence of insulin, hydrocortisone and prolactin (Danielson *et al.*, 1984). Using this cell line, Rosen's group has characterized the hormonal and cell substratum requirements for the regulation of milk protein gene expression (Rosen *et al.*, 1986; Bisbee & Rosen, 1987). Analysis of β-casein mRNA levels has revealed that cell substratum interactions, as well lactogenic hormones, play a major role in the regulation of casein gene expression. This effect was primarily exerted at the post-transcriptional level, as determined by analysis of relative rates of transcription using a nuclear 'run on' transcription assay (Eisenstein & Rosen, 1988). Recent investigations, using protein synthesis inhibitors, have shown that post-transcriptional effects of lactogenic hormones on casein gene expression in Comma-ID cells are mediated by labile protein(s) exerting a selective effect on casein mRNA stability (Poyet *et al.*, 1989).

Recently, Ball *et al.* (1988) have isolated a clonal mammary epithelial cell line, HC11, derived from Comma-ID cell line, which was able to produce β-casein under the control of lactogenic hormones, but did not require cultivation in an extracellular matrix for efficient expression of the β-casein gene. This cell line has been used recently to analyse *cis*-acting sequences which are required for mediating the effect of prolactin and glucocorticoids. A strong induction of CAT activity by these two hormones was observed when HC11 cells were transfected with chimeric gene constructs containing 2300 bp of the 5′ flanking DNA sequence of rat β-casein gene fused to a chloramphenicol acetyl transferase (CAT) reporter gene. Analysis of a 5′ deletion series of the β-casein CAT gene constructs revealed a stepwise loss of hormone inducibility: 285 bp of 5′ flanking sequence was sufficient to mediate the synergistic action of lactogenic hormones on expression, but the response was only half of that with the construct containing 2300 bp (Doppler *et al.*, 1989).

The study of gene regulation in transgenic mice under conditions of appropriate germline transmission also provides a powerful approach in elucidation of the role of hormones. To investigate the relative contribution of the promoter and intragenic sequences in the hormonal regulation of β-casein gene, Lee *et al.* (1989) have used explant cultures derived from transgenic mice bearing the entire 7·2-kb rat β-casein gene and a β-casein promoter/chloramphenicol acetyl transferase (CAT) fused construct, containing either 2300 or 524 bp of 5′ flanking DNA, exon I and 490 bp of intron 1. In explant cultures derived from transgenic mice, a maximal increase of 4·5-fold in CAT activity was observed in the presence of lactogenic hormones, in contrast to a 25-fold induction of the entire rat β-casein transgene. These results are consistent with the role of intragenic sequences in the hormonal regulation of casein gene expression and suggest that, while promoter sequences contain elements important for tissue-specific expression and prolactin induced transcription, they clearly are not sufficient to elicit the full expression of casein genes.

Similar conclusions have been obtained in studies of the whey acidic protein (WAP)-tissue plasminogen activator (TPA) fusion gene and endogenous WAP gene in transgenic mice (Pittius *et al.*, 1988; Lubon *et al.*, 1989).

Taken together these data show that posttranscriptional mechanisms play a major role in the regulation of milk protein gene expression.

3. Assembly and Secretion of Milk Protein Polypeptide Chains

3.1. Blood Precursors of Milk Protein

Identification of the blood precursors of milk protein proved to be a contentious issue until the early 1960s, when a number of methodological innovations facilitated resolution of the debate (see Mepham, 1986). Thus, the incorporation of ^{14}C labelled amino acids into secreted milk protein, following their intravenous injection into goats and rabbits, provided strong circumstantial evidence for their direct uptake by mammary cells from blood plasma (see Mepham, 1971). However, such experiments did not provide conclusive proof of the immediate blood precursors of milk protein because the accumulation of isotope in the milk protein might have involved prior incorporation of label into a small fraction of the plasma proteins, subsequently selectively absorbed by the mammary glands.

3.1.1. Mammary Uptake of Amino Acids

It was the combination of improved techniques for mammary arteriovenous (AV) difference and blood flow (MBF) determination in ruminants, together with the introduction of automated chromatographic procedures for amino acid determination, that first permitted reasonably quantitative estimation of the extent to which the mammary uptake of individual free amino acids of blood plasma accounted for the output of the same amino acid residues in the milk protein. In principle, the AV difference technique, involving the determination of substrate concentrations in simultaneous samples of arterial and mammary venous blood, dates from 1906, but until the refinements introduced by Linzell (1960) it had proved capable of giving very variable results. Linzell (1966) was also responsible for greatly improving estimation of MBF, using the technique of continuous thermodilution. According to the Fick Principle, when steady state conditions apply, mammary uptake is given by the product of AV difference and MBF. Since the output of amino acids in protein can be deduced from the milk yield, milk protein concentration and its amino acid composition, it is possible to construct a 'balance sheet' of substrate uptake and product output.

Although simple in principle, the AV difference method is beset with problems for the unwary, and it ideally requires use of 'trained' animals which have been surgically prepared with exteriorized blood vessels (see Fleet & Mepham, 1983). A simplified account of the method is given by Mepham (1987).

In the first study combining AV difference determinations, MBF measurement and estimation of amino acids by ion exchange chromatography in goats, mean AV differences ranged from 0·03 mg/100 ml blood plasma for serine to 1·03 mg/100 ml for leucine (Mepham & Linzell, 1966). The variation in AV difference between animals and, as subsequently shown (Mepham & Linzell, 1974*a*), with time in the same animal, was much greater for non-essential amino acids (NEAAs) than for essential amino acids (EAAs). Frequently, and most markedly for serine, AV differences for NEAAs were negative, i.e. the mammary venous concentration was greater than the arterial. An indication of the efficiency of absorption is given by the 'percentage extraction', i.e. the AV difference expressed as a percentage of the arterial concentration. In these investigations (Mepham & Linzell, 1966), the average EAA extraction was 36% (with methionine the highest at 77%) and the average NEAA extraction was 24% (with serine the lowest at 0·1%).

The significance of these results for milk protein synthesis was estimated by comparing the output of individual amino acid residues in the protein contained in 100 ml of milk with the calculated amino acid uptakes over the period required to secrete that quantity of milk. (This comparison assumes that the contribution to total milk protein output of blood serum proteins, such as immunoglobulins, is negligible.) On this basis, it was clear that EAAs were absorbed by the mammary glands in amounts adequate to satisfy the requirements for the corresponding residues in milk protein: in some cases the amount absorbed was in excess of this requirement. By contrast, NEAAs generally showed highly variable uptakes, which were on average inadequate to supply the corresponding residues of milk protein. Despite the lack of correspondence between the uptake and output of certain amino acids, calculation of total amino acid nitrogen and carbon uptake balanced well with protein nitrogen and carbon output (Mepham & Linzell, 1966).

Subsequently, the general applicability of this distinction between the mammary utilization of EAAs and NEAAs was established in studies on other species, e.g. cows (Peeters *et al.*, 1979), sheep (Fleet & Mepham, 1985) and the perfused guinea pig mammary gland *in vitro* (Davis & Mepham, 1974). Numerous other reports have tended to confirm the generalization, but their value is often suspect because of the failure either to measure MBF or to do so reliably, e.g. by ensuring that mammary venous blood flow is not augmented by blood of non-mammary origin (see Fleet & Mepham, 1983). However, it must be admitted that, even at its best, the AV difference technique is subject to inherent disadvantages as a quantitative method. Principally, this results from the fact that while milk outputs are measured over periods of hours, uptake from blood is extrapolated from determinations on samples collected over periods of seconds. Nevertheless, the close correspondence that is evident between the uptake and output of certain EAAs (e.g. threonine in the studies of Mepham and Linzell (1966)) confirms the validity and value of the technique.

3.1.1.1. Uptake/output balances for individual amino acids. Consideration of the balance between uptake and output in milk protein of individual amino acids permits certain generalizations and raises a number of important questions. Firstly, the inadequate uptake of several NEAAs indicates that they must be provided either from a source other than plasma free amino acids or by their intramammary synthesis. Secondly, the excessive uptakes of certain EAAs suggested that they might undergo metabolism in the mammary cells and, since there was an overall balance between amino acid nitrogen uptake and milk protein nitrogen output, thereby compensate for the deficit in NEAA uptake by participation in transamination reactions (Mepham, 1971). For certain EAAs (e.g. methionine, phenylalanine and threonine in the goat; Mepham & Linzell, 1966), there was essentially stoichiometric transfer from blood plasma to milk protein; but for others, including the branched chain amino acids and, most notably, arginine, uptake was generally excessive. The former have been categorized as group I and the latter as group II (Mepham, 1982), the basis for the distinction being confirmed by studies of their intracellular metabolism (see below). Typically, in ruminant species group I EAAs show greater percentage extractions than group II EAAs or NEAAs (cows, Clark *et al.* (1978); goats, Mepham (1982); sheep, Fleet & Mepham (1985)).

However, the AV difference method is, *per se*, incapable of providing information about intracellular changes: it merely treats the mammary gland as a proverbial 'black box'. It follows that the excess uptakes of group II EAAs might not be used, or used exclusively, for NEAA synthesis. Moreover, the lability of NEAA AV differences and consequent imprecision of estimates of required rates of NEAAs synthesis, together with evidence that blood plasma AV differences for certain organs may underestimate amino acid uptake (Heitmann & Bergman, 1980), suggested that free amino acids of plasma might not be the only precursors of mammary synthesized milk proteins. Certain observations have seemed to give support to this view. Thus, Pocius *et al.* (1981) showed positive AV differences for glutathione (the tripeptide, γ-glutamyl–cysteinyl–glycine) with analysis of whole blood from lactating cows. This indicated an alternative source of those NEAAs, which are normally absorbed from plasma in inadequate amounts for milk protein synthesis. However, the suggestion that red blood cells might also provide significant quantities of other amino acids for this purpose (Baumrucker, 1985) is more questionable. The studies of Rulquin (1981) revealed lower mammary AV differences when whole blood was analysed by comparison with plasma, while Overthrow (1986) reported significant differences between AV differences for whole blood and plasma for only two amino acids in the goat.

The possibility has also been raised that plasma oligopeptides might act as precursors of milk protein (Jois, 1986). In so far as this proposal is based on determinations of amino acids in plasma samples before and after deproteinization, it depends on an assumption of complete deproteinization, which needs to be fully substantiated.

A further dimension to the problem of identifying milk protein precursors is added by the claim that there is substantial lysosomal degradation of protein prior to secretion (Mayer *et al.*, 1980). In addition, cellular proteins are subject to cytoplasmic degradation (Zak *et al.*, 1979). Recently, Oddy *et al.* (1988) estimated that in the lactating goat degradation of milk proteins accounts for about one-third of milk protein synthesis; there was also evidence of appreciable degradation of tissue proteins. Since amino acids released in these proteolytic processes contribute to the intracellular amino acid pool, they may represent a form of longer-term storage capable of augmenting the more labile source of supply from the blood.

3.1.2. Intracellular Metabolism of Amino Acids
It was noted above that neither AV difference determinations nor experiments involving intravenous injection of tracer compounds can, of themselves, provide information on the manner in which mammary cells utilize absorbed substrates. This objective requires use of isopic tracers under conditions in which the activity of the mammary gland can be isolated from that of the whole animal. There are various ways of achieving such conditions, e.g. incubation of mammary slices, perfusion of the isolated mammary gland and mammary transplantation (see Fleet & Mepham, 1983). In practice, because of its versatility and maintenance of relatively physiological conditions over periods of several hours, it is the perfused mammary gland preparation that has provided the most extensive data on the intracellular metabolism of amino acids.

Confirmation of the view derived from AV difference determinations that EAA uptake is sufficient to satisfy requirements for milk protein synthesis is provided by experiments with perfused goat and sheep glands employing [^{14}C] and [^{35}S] amino acids (Verbeke *et al.*, 1967, 1968, 1972; Roets *et al.*, 1979*a*). These experiments showed that the 'product quotient', i.e. the percentage of a milk protein amino acid residue derived from a plasma precursor, was 86% for methionine, 85% for threonine, 96% for phenylalanine, 96% for valine and 84% for arginine. Allowing for the presence of some preformed milk in the glands contributing to the output, these results are sufficiently close to 100% to strongly suggest that plasma EAAs almost entirely account for their corresponding residues in protein. Similar conclusions were drawn for valine, leucine, histidine and threonine (Davis & Mepham, 1976), tryptophan (Mepham *et al.*, 1976) and methionine, isoleucine, lysine and tyrosine (Peters, 1977) from experiments employing the isolated guinea pig mammary gland preparation.

Perfusion experiments also proved valuable in describing the fate of the 'excess' uptake of group II EAAs. Following the discovery that in goats the mammary uptake of arginine was up to four times the output of arginine residues in milk protein (Mepham & Linzell, 1966), hydrolysis of [^{14}C] arginine to ornithine and urea, and the use of arginine carbon for synthesis of milk protein proline were demonstrated in the perfused goat gland (Mepham & Linzell, 1967). Subsequent studies employing [^{14}C] and [^{15}N] isotopes in perfusion experiments (Roets *et al.*, 1974, 1979*b*) and enzyme assays on rat mammary tissue (Yip & Knox, 1972; Mezl & Knox, 1977) indicated that, with the exception of the step by which ornithine is converted to citrulline, mammary tissue possesses all the enzymes of the 'urea cycle'. The major significance of the excess arginine uptake would seem to relate to its role as a precursor of ornithine, since [^{14}C] urea production by the goat gland was consistent with all the excess uptake being hydrolysed (Mepham & Linzell, 1967). In addition to its intramammary synthesis, ornithine is appreciably absorbed from blood plasma by both ruminant (e.g. Peeters *et al.*, 1979) and non-ruminant glands (Linzell *et al.*, 1969). The uptake of this non-protein amino acid is important in at least three ways: (i) after conversion to glutamic γ-semialdehyde, it acts as a carbon precursor of proline; (ii) by irreversible transamination reactions involving the δ-amino group, it acts as a N precursor of several NEAAs, including glutamate, alanine and serine (Roets *et al.*, 1979b); (iii) via ornithine decarboxylase activity, it acts as a precursor of the polyamines, spermine and spermidine (Oka, 1983). The last compound has been shown to act as an intracellular mediator of cortisol action (Oka, 1983).

Intracellular metabolism of the branched-chain EAAs, valine, leucine and isoleucine, has been extensively investigated with isotopic tracers, e.g. in bovine mammary slices (Wohlt *et al.*, 1977), in perfused goat glands (Roets *et al.*, 1979*a*; 1983) and perfused guinea pig glands (Davis & Mepham, 1976). In all experiments, evidence was obtained for the oxidation of these amino acids. In goat glands (Roets *et al.*, 1979*a*), approximately 30% of the absorbed [U-^{14}C] valine activity was transferred to CO_2 and significant amounts were recovered in milk citrate. In bovine slice experiments with [^{14}C] valine, leucine and isoleucine, tracer was substantially transferred to β-hydroxybutyrate, isovalerate and methylmalonate, respectively (Wohlt *et al.*, 1977). Collectively, such results indicate that branched chain EAAs undergo an initial transamination reaction, with the carbon skeletons being catabolized by pathways yielding tricarboxlyic acid cycle intermediates via acetyl CoA and succinyl CoA. The role of valine in transamination reactions is demonstrated by results of goat gland perfusion experiments involving use of [U^{14}C,2,3-^{3}H] valine, which showed that about 10% of valine was reversibly transaminated during a single passage through the gland (Roets *et al.*, 1979*a*). The reversibility of the reactions was confirmed by the incorporation of ^{15}N into branched chain acids in an experiment in which DL [1-^{14}C, δ-^{15}N] ornithine was added to the perfusate (Roets *et al.*, 1979*b*). However, the importance of the excess uptake of branched chain EAAs is not limited to provision of nitrogen for NEAA synthesis and acting as an energy source. The results of experiments by Vina and Williamson (1981) on tissue from rats and by Roets *et al.* (1983) on perfused goat glands indicate significant involvement of leucine in mammary lipid synthesis.

It is apparent that group II EAAs contribute significantly to the total mammary CO_2 output and therefore to the gland's energy requirement. Since, however, there is a good balance between amino acid carbon uptake from blood plasma and protein carbon output (Mepham & Linzell, 1966), there would not seem to be a *net* transfer of carbon to CO_2. The incorporation of ^{14}C from glucose and acetate into NEAA residues of milk protein of goats (Linzell & Mepham, 1968) confirms that these compounds and those derived from amino acid catabolism enter common metabolic pools. Nevertheless, catabolism of group II EAAs seems to play an important role in mammary metabolism. In perfused guinea pig mammary gland experiments, depletion of the perfusate of a single EAA for up to 3 h significantly depressed uptake of group I EAAs, but had no significant effect on uptake of group II EAAs, for which oxidation rates were markedly increased (e.g. by a factor of over six for arginine and leucine) (Mepham *et al.*, 1979). The persistence of group II EAA uptake in conditions of suppressed protein synthesis suggests that the factors governing the partitioning of these amino acids between pathways of anabolism and catabolism may have an important bearing on the amino acid requirements of the gland.

3.1.3. Amino Acid Transport Systems

The high milk yields of dairy animals predicate an intense flux of nutrients from the blood into the mammary secretory cells, e.g. cows at peak lacatation may secrete 2 kg of milk protein daily. The point of entry is, essentially, the basolateral membrane of the secretory epithelium. Being predominantly lipid in composition, this effectively prevents entry of the hydrophilic amino acids by simple diffusion, so that cellular uptake depends on the activity of specific carrier systems located in the membrane.

Following the proposal of Meister (1973), attention was focused on the possibility that glutathione might be instrumental in amino acid transport in mammary tissue, by means of membrane–bound γ-glutamyl transpeptidase (GTP) and a series of intracellular enzymes which collectively operate the 'γ-glutamyl cycle'. This model involves the GTP using amino acids in the plasma and intracellular glutathione in an enzymatic sequence termed 'group translocation'. The result is to transfer one amino acid from the plasma to the cell interior linked to the γ-glutamyl moiety of glutathione. Four other intracellular enzymes then effect cleavage and resynthesis of glutathione, resulting in the *net* transfer of one amino acid.

Evidence that appeared to support a role for GTP in mammary amino acid transport was presented by Baumrucker and Pocius (1978). They tested the ability of 22 amino acids to serve as acceptors for GTP extracted from bovine and rat mammary tissue and claimed that the relative reactivities of the amino acids to the bovine enzyme compared favourably with the relative percentage extractions of EAAs by bovine mammary glands *in vivo*. However, the significance of GTP in this context became highly questionable with the report that the enzymatic site for glutathione of membrane-bound GTP was oriented towards the extracellular, rather than the intracellular, surface of the cell. It now seems that the significance of mammary GTP might lie chiefly in its role in the provision to the mammary cells of NEAAs produced on the hydrolysis of red cell glutathione (discussed above). The released amino acids are thought to be transported into cells by the systems discussed below. Whatever its precise function, GTP seems to be intimately connected with the lactational process, as indicated by the observation that the activity of the enzyme increases markedly at lactogenesis in rat mammary gland (Puente *et al.*, 1979) in response to increasing blood prolactin concentrations (Pocius *et al.*, 1980).

The most productive approach to seeking mechanisms of amino acid transport in mammary tissue has built on those characterized for other tissues by Christensen and colleagues (see Christensen, 1984). Several systems have been described which effect amino acid transport into cells, e.g. for neutral amino acids there are A, ASC and L systems. (The letters were originally used to denote particular affinities, but since the affinities are now known to be broader, the nomenclature retains only historical significance.) Other systems described for neutral amino acids in certain tissues include the N, T and Gly systems, while cationic amino acids are transported by the Y^+ system.

The systems are characterized by certain properties which facilitate their identification (Barker & Ellory, 1990). Thus, the A and ASC, but not the L, system are

Na^+-dependent, whereas the A system alone transports N-methylated amino acids. The L system is specifically inhibited by 2-amino-bicyclo [2.2.1] heptane 2-carboxlic acid (BCH).

Several points concerning transport systems in general merit attention, because the properties of those present in mammary tissue have often been inferred from theories derived from studies on other tissues. These may be listed as follows.

(i) Transport is mediated by processes involving membrane proteins ('carriers' or 'translocases') which exhibit properties akin to enzymes, i.e. transport shows saturation kinetics and is characterized by K_m and V_{max} data.

(ii) Generally speaking, carriers show affinity for several amino acids, resulting in various forms of interaction. Thus, uptake of one amino acid might stimulate (*cis*-stimulation) or inhibit (*cis*-inhibition) the uptake of another. Intracellular concentrations may also be important in controlling uptake, e.g. by *trans*-inhibition.

(iii) Carrier proteins experience turnover and activation or inhibition by other factors, such as hormones and ions. Consequently, transport activity might be modulated by changes in both the numbers of carrier molecules and their substrate affinity.

3.1.3.1. Studies on transport systems in mammary tissue. Neville *et al.* (1980) described three modes for the cellular uptake of aminiosobutyric acid (AIB, a non-metabolized amino acid) by diced lactating mouse mammary tissue. These were; (i) a Na^+-dependent, methyl AIB-inhibitable mode (identified as the A system); (ii) a sodium-independent, BCH-inhibitable mode (identified as the L system); and (iii) a residual mode (approximately 10% of total flux), not inhibited by methyl AIB or BCH, and sodium-independent. Baumrucker (1984), using bovine mammary slices, provided evidence for the Y^+ system, transporting arginine and lysine, but not histidine. Later work with the same slice preparation indicated the presence of A, ASC and L systems (Baumrucker, 1985).

A major disadvantage of using diced and sliced preparations of mammary cells in transport studies is that the normal spatial relations of the vascular, interstitial, intracellular and luminal compartments are disrupted. Not only are large numbers of cells damaged, but their baso-apical orientation is lost. These considerations prompted investigations employing the isolated perfused guinea pig mammary gland preparation, previously used in metabolic studies (Davis & Mepham, 1974). Use of this preparation, in which the anatomical integrity of the gland is maintained, avoids the above problems and also permits characterization of amino acid entry and exit from the cells under conditions in which interactions between different amino acids in the blood plasma can be studied (Mepham *et al.*, 1984, 1985). In these studies, the 'paired tracer' method of Yudilevich *et al.* (1979) was used. The technique involves the injection, as a bolus, of [^{3}H] sucrose (an extracellular fluid marker) and a ^{14}C-labelled amino acid into the arterial supply of the perfused gland. Since sucrose does not enter cells, the change in ^{3}H:^{14}C ratio between the injectate and the venous effluent provides a measure of cellular uptake of the amino acid. When an unlabelled amino acid is injected simultaneously, any change in uptake of the [^{14}C] amino acid indicates the extent of the shared affinity of the two amino acids for the same carrier system. For a description of the techniques employed, see Mepham (1988).

The results of these studies (Mepham *et al.*, 1984, 1985; Overthrow, 1986) point to the existence of at least four carrier systems in the lactating guinea pig mammary gland, viz. L, ASC, Y^+ and A systems. The L system was shown to transport methionine, leucine, phenylalanine and threonine, the ASC system serine and threonine, and the Y^+ system arginine, lysine and ornithine. Surprisingly, amino acids shown to be transported by the A system in other tissues (including mammary tissue of mice and cows) did not use A sites. Only for serine was a weak transport activity of the A system demonstrated. Similar disparities in relation to the A system have been reported for other tissues (see Overthrow, 1986), but the possibility remains that disaggregated tissue (such as in slice preparations) may permit transport across the apical membrane of the cell, as well as significant cytoplasmic leakage, and thus distort measurements of transport flux.

Two important points emerging from these studies on the perfused guinea pig gland, which are likely to have significance for other species, are (i) the marked but variable degree of inhibition of uptake mutually exerted by amino acids sharing the same transport sites; and (ii) the high rates of amino acid efflux (back into the blood plasma), which are also subject to *cis* and *trans* effects. The uptake of an amino acid from plasma is thus highly dependent, not only on its arterial concentration, but also on the concentration of other amino acids, intracellular and extracellular, which affect its binding to carrier sites.

Hormonal modulation of amino acid transport systems has usually been associated with the A system (see Kilberg *et al.*, 1986) and there is some evidence that this applies to mammary tissue. For example, Anderson and Rillema (1976) showed that insulin stimulated uptake by mouse mammary tissue explants of [^{14}C] AIB, but not of [^{14}C] leucine. However, hormonal effects on the L system of certain tissues have also been claimed. While some such hormonal effects are rapid, others are dependent on active protein synthesis, presumably of carrier molecules (e.g. see Kletzien *et al.*, 1976). Vina *et al.* (1985) demonstrated that γ-glutamyl amino acids (GAAs) increased mammary uptake of amino acids in rats treated with the prolactin release-inhibitor, bromocriptine. This suggested that GAAs may act as signals for the hormonal regulation of amino acid uptake, and that GTP is important in generating such signals. Apart from endocrine modulation, certain changes in activities of transport systems occur in response to substrate supply. For example, the A system of rat hepatocytes in culture is stimulated by amino acid starvation and depressed by increased supply, by processes involving gene transcription (Kilberg *et al.*, 1986). Milk accumulation in the mammary gland may also affect amino acid uptake, since Vina *et al.* (1981) reported that amino acid AV differences were significantly depressed in 5-hour unsuckled glands of rats by processes which were independent of hormonal regulation and MBF.

3.1.4. Limitation of Milk Protein Synthesis by Amino Acid Supply

At the level of the mammary secretory cell, the extent to which the genetic potential of an animal for milk protein synthesis is realized depends on either or both of two processes, viz. transcription and translation. Clearly, an important determinant of the latter process is the supply of amino acids in the appropriate intracellular pools. Identification of the source of individual intracellular amino acids and their order of rate-limitation is a matter of obvious potential application, but it must be admitted that to date no consensus on these issues has emerged.

A number of studies on cows, goats and sheep have revealed significant positive correlations between arterial plasma concentrations and AV differences of several EAAs and NEAAs in samples from different animals and from the same animal on different days (e.g. Peeters *et al.*, 1979; Mepham, 1982; Fleet & Mepham, 1985). When lactating rats are fasted for 6 or 24 h, the resulting reduction in mammary amino acid uptake is due to both reduced AV differences (and arterial concentrations) and reduced MBF (Vina *et al.*, 1987). On the basis of the observed correlations, increasing plasma amino acid concentrations by dietary supplementation or by intravascular infusion has appeared to be a plausible means of increasing milk protein yield. Ideally, this approach first requires definition of the order in which supply of the individual amino acids is rate-limiting, but with over 20 amino acids the number of possible permutations of order and degree of limitation is, effectively, infinite. One strategy for addressing the problem is to consider the amino acids in the three broad groups defined above, i.e. groups I and II EAAs, and NEAAs. In theory, amino acids in group I might be limiting because of constraints in their transport into cells, possibly implied by their high percentage extractions. Group II EAAs might be limiting because, although absorbed in amounts in excess of requirements for milk protein, they might be involved in obligatory catabolic and transamination reactions. NEAAs might be limiting because of restrictions on their rates of intramammary synthesis. Of course, in practice, rate-limitation might be exerted by the supply of any one amino acid, or several might be co-limiting.

Clark (1975) summarized studies of attempts to increase protein yield by post-ruminal infusion of proteins or by intravenous infusion of EAAs in cows. Despite important procedural and breed differences between these studies, there appeared to be strong evidence for categorizing methionine and phenylalanine (both group I) as first-limiting, while threonine and lysine might be co-limiting in some conditions. Experiments on lactating goats involving intra-arterial infusion of EAAs (Linzell & Mepham, 1974) and NEAAs (Mepham & Linzell, 1974*b*) suggested that methionine might be rate-limiting in this species, but provided no evidence for the importance of NEAAs.

Attempts to increase milk protein yield by increasing blood plasma amino acid concentrations might not, however, be soundly based. In experiments on goats (Mepham & Linzell, 1974*a*) and sheep (Davis *et al.*, 1987) no correlations were evident between arterial concentrations and AV differences of amino acids in samples taken over 8–12 h. The experiments of Henderson and Peaker (1983) point to a similar conclusion. They induced unilateral inhibition of milk secretion in goats by intraductal infusion of colchicine. This resulted in increased milk secretion in the non-infused gland, a response that was not accompanied by markedly increased arterial concentrations of EAAs.

The discovery that glutathione of red blood cells is absorbed by lactating mammary glands, together with evidence of red blood cell involvement in supplying amino acids to sheep tissues, suggested the need to re-evaluate concepts of rate-limitation (Baumrucker, 1985). The observations of Atroshi and Sandholm (1982) lent support to this view, since they discovered that, in a population of Finn sheep containing ewes with two distinct inherited red cell glutathione concentrations, higher concentrations were associated with higher milk yields. More recently, Rizzi *et al.* (1988) have demonstrated a significant correlation between glutathione levels and stage of lactation in Massese ewes.

The notion that milk protein yield might be readily controlled by manipulation of blood levels of amino acids begins to appear somewhat simplistic in the light of results of investigations of amino acid transport systems, even though such studies are not extensive. Even if it were possible to discover the ideal combination of amino acids to be supplied to the gland in the arterial blood (itself a formidable enough accomplishment), cognizance would need to be taken of the effects of their utilization by non-mammary organs and of neural and endocrine responses that might be elicited. At present, however, milk protein concentration is most reliably increased by increasing feed energy intake. Supplementation of the diet with extra protein, whatever its degradability, has not been shown to have a consistent effect on milk protein concentration (Sutton, 1989).

3.2. Translation

Assembly of polypeptide chains on ribosomes has not been investigated in detail in mammary cells, and the authors' knowledge of the field relies on experimental data obtained for other eukaryotic cells. For an overview on the topic, the reader is referred to any recent book on biochemistry and reviews such as that of Rhoads (1988).

3.3. Intracellular Transport and Co- and Post-Translational Processes of Milk Proteins

Intracellular transport and co- and post-translational processes undergone by milk proteins were described in the first edition of this book (Fox, 1982). The section covering covalent modifications is still valid as a whole whereas that dealing with the intracellular transit of proteins is partially obsolete. Consequently, in this section, attention is focused on recent developments in the latter field, with special emphasis on early stages of intracellular transport studied in various types of eukaryotic cells and which are assumed to be similar in mammary epithelial cells.

3.3.1. General Considerations

Many proteins synthesized in the cytoplasm are subsequently directed to various intracellular compartments, viz. (i) organelles surrounded with a double

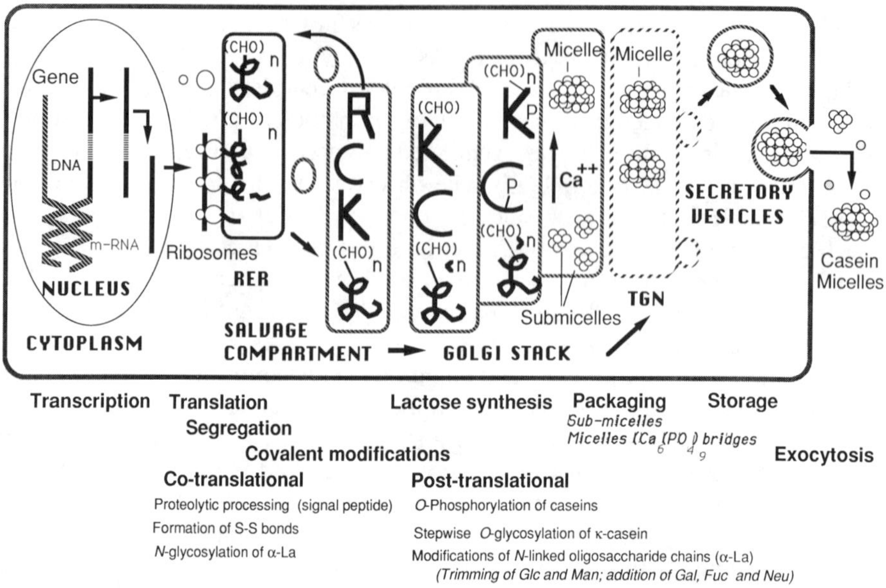

Fig. 12. Synthesis and intracellular transport of milk proteins across the vacuolar pathway of mammary epithelial cells. The polypeptide chains undergo various co- and post-translational covalent modifications along the different compartments of this pathway, as outlined in this schematic representation of the mammary epithelial cell. R, C, K and L symbolize resident proteins of the ER, calcium-sensitive caseins, κ-casein and α-lactalbumin, respectively; CHO: sugar chain; P: phosphate group; RER: rough endoplasmic reticulum; TGN: *trans*-Golgi network; Glc: glucose; Man: mannose; Gal: galactose; Fuc: fucose; Neu: neuraminic acid.

membrane: nucleus, mitochondria and (in plants) chloroplasts; (ii) organelles bounded with a single membrane of the central vacuolar system: the rough and smooth endoplasmic reticulum (ER), the *cis*-medial and *trans*-Golgi, the *trans*-Golgi network (TGN), secretory vesicles and secretory granules; (iii) Golgi-derived primary lysosomes; (iv) the plasma membrane and derived endosomes, smooth ER-derived peroxisomes (Fig. 12).

The occurrence of specific transient or permanent routing signals, directing proteins from the cytosol to the nucleus, mitochondria (Hartl & Neupert, 1990), chloroplasts (Smeekens *et al*., 1990), the ER or peroxisomes and recognized by cytosolic or/and membrane receptors, is now well documented. All classes of signals are relatively short stretches of amino acid residues, generally located at the amino terminus of the polypeptide chain, those responsible for nucleus and peroxisome targeting can occur elsewhere. There is no well-defined consensus amino acid sequence within each type of targeting signal and our ignorance about the molecular basis of their recognition by receptors reflects our present inability to predict accurately the conformation of a polypeptide chain according to its primary structure and the environment. It should be noted that in a few cases, proteins apparently lacking a conventional signal, were found to be translocated.

Translocation of proteins within or across membranes of organelles is an energy-demanding process that is still not well understood. Elucidation of the manner in which the cell solves the problem of transferring hydrophilic stretches of amino acid residues through the hydrophobic lipid phase of membranes is a major challenge in biology, and both alternatives, direct transit through the lipid bilayer or translocation through proteinaceous pores, remain open. However, possible components of the putative translocator systems have been isolated recently. It is now well recognized that translocation is not necessarily coupled to translation, as first demonstrated for transit into the ER of higher eukaryotes. Translocation into the nucleus, mitochondria, chloroplasts, lysosomes and peroxisomes is a post-translational process and in yeast, and even in mammalian cells, transport across the ER could also occur after completion of the polypeptide chain. It is generally assumed that a loosely folded, if not unfolded conformation, is a prerequisite for translocation of proteins, except perhaps for nuclear ones which could be imported in a fully-folded conformation through nuclear pores. In any case, the accessibility of the signal as well as translocation competence are vital for correct routing and transit of proteins, which suggests the possible involvement of protein helpers preventing premature folding. Accordingly, there is now some genetic and biochemical evidence that molecular chaperones such as heat-shock protein Hsp-70 and related binding protein BiP (Ellis & Hemmingsen, 1989), which prevent, through transient non-covalent association, improper interactions between potentially complementary regions, might play a key role in protein transport into the ER and mitochondria, protein assembly within the ER, and diassembly of clathrin cages.

3.3.2. Protein Transit in the Central Vacuolar System

The central vacuolar system of eukaryotic cells, through which a variety of proteins including secretory ones travel, consists of the aforementioned ordered array of discrete single membrane-bounded organelles (Fig. 12), dynamically coupled to a microtubule-based skeleton, and connected by various distinct carrier vesicles that permanently bud at the organelle surface and fuse to the next closed one, thus providing a vectorial flow of lipid and proteins. Each organelle contains specific resident membrane proteins, responsible for its identity, stability and function. This intricate system was initially termed the secretory pathway, which seems now to be inappropriate as it is not an unbroken track only followed by the secretory proteins; many of the passenger proteins will take up their residence at different stages of the course before the terminus—the apical cellular pole where secretory proteins are discharged into the alveoli lumen—is reached. The ER is the port of entry for proteins destined for export or residentship in either the lumen or the membrane of one of the organelles of the vacuolar system. Once in the ER, non-resident proteins travel vectorially to the *cis*-, medial and *trans*-Golgi cisternae and the *trans*-Golgi network (TGN), from where they are dispatched to appropriate end-stage organelles such as primary lysosomes, secretory vesicles, or the plasma membrane, via carrier vesicles that ultimately fuse with these organelles.

Increasing secretion of large amounts of proteins obviously requires the concomitant and coordinated development of the secretory apparatus. According to a recent study carried out with B lymphocytes (Wiest *et al.*, 1990), a 12-fold increase of lg secretion was correlated with a 3– to 4–fold increase in the area and volume of both the ER and Golgi stacks, and of the density of ER-bound ribosomes. Concentration of most proteins resident in either the lumen or membrane of the ER, such as BiP and ribophorins, respectively, remained unchanged but a minority (10%) was relatively more abundant in the induced cell.

3.3.2.1. Entry into the ER requires a visa, the signal peptide. Following the pioneering works of Palade (1975) and co-workers, who elucidated the intracellular pathway of secretory proteins (Fig. 12) and of Milstein *et al.* (1972), who discovered an immunoglobulin precursor with a transient amino terminal extension, Blobel and Dobberstein (1975) provided the conceptual basis known as the 'signal hypothesis' for explaining the first step in protein secretion, namely the recognition of a signal sequence by specific machinery able to transfer selectively the nascent polypeptide chain through the endoplasmic reticulum (ER) membrane via putative transient tunnels. In addition, both authors developed in-vitro systems able to carry out translation of a given mRNA and translocation of the relevant nascent protein.

According to the updated version of the 'signal concept' schematized in Fig. 13, the signal sequence emerging from the ribosomal groove interacts with the 54-kDa polypeptide of the cytosolic ribonucleoprotein complex (reviewed by Bernstein *et al.* (1989)), called the signal recognition particle (SRP), which greatly enhances (at least 10^3 times) SRP affinity for the ribosome and induces a transient elongation arrest. Then, the 68 and 72-kDa proteins and 7S RNA of the SRP may interact with a highly basic domain of the α subunit of an ER heterodimeric receptor, termed the docking protein, thus mediating the binding of the ribosome to ribophorins. Subsequently, SRP may dissociate from its GTP binding receptor and release the signal sequence which then interacts with an integral ER membrane *N*-glycosylated protein. This signal-sequence receptor (SSR) is a possible component of the putative translocation complex responsible for the transfer of the growing polypeptide chain across the ER membrane with concomitant excision of the signal sequence by a specific signal peptidase located on the luminal side. Eukaryotic signal peptidases have been purified from dog pancreas and hen oviduct (quoted in Caulfield *et al.* (1989)), as apparent hetero-pentamer and dimer, respectively. One of the components is a 23-kDA glycoprotein not functional as such. Although the activity and the exact substrate specificity of the enzyme, presumably a subtle local conformation, are not fully understood, the occurrence of amino acid residues with small chain at position -3 and -1 in the signal peptide (Von Heijne, 1983) seems to be essential for cleavage. As already mentioned, transport across membranes, including that of the ER, might depend on the maintenance of an unfolded conformation of the protein. Thus, in addition to its recognition and targeting functions, SRP may play the role of a molecular 'chaperone' (Ellis & Hemmingsen, 1989) in preserving a loose, open conformation of the nascent polypeptide chain by arresting elongation and in ensuring the

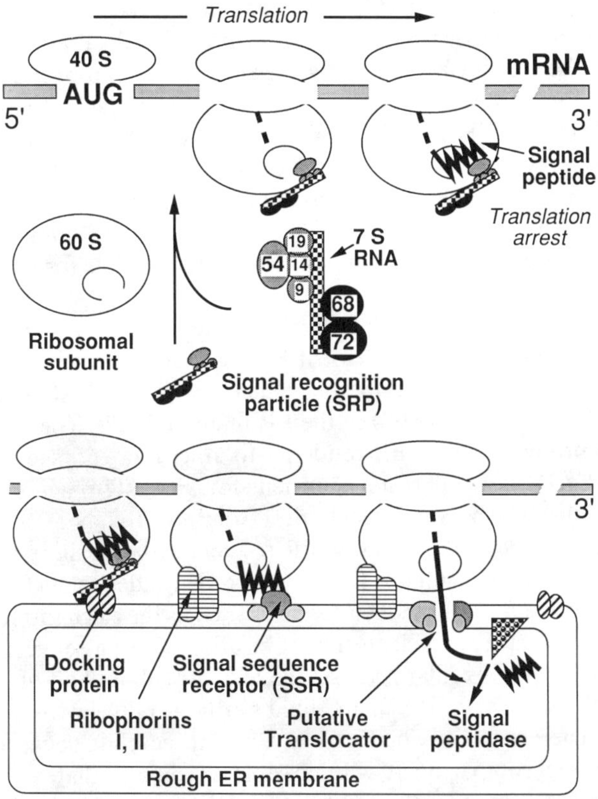

Fig. 13. Protein translocation across the endoplasmic reticulum (ER) membrane. The signal sequence emerging from the ribosomal groove interacts with the SRP, allowing the binding of the complexed SRP to an ER receptor, termed a 'docking protein'. Then, other ER proteins, such as ribophorins, bind to the ribosome, the SRP is released and the signal peptide interacts with an ER receptor, allowing the translocation of the polypeptide chain through the ER membrane, with concomitant release of the signal peptide by a signal peptidase.

accessibility of the signal sequence to the SSR (Meyer, 1988). Photocrosslinking experiments showed that the growing polypeptide chain is close to the SSR during translocation. Sequencing of both the purified SSR and the relevant cDNA in the canine species (Prehn *et al.*, 1990) showed that the preprotein contains 287 amino acid residues. The mature protein is characterized by a very acidic N-terminal end (23 negative charges in a segment of 35 residues), an internal hydrophobic stretch of about 20 residues, and a basic C-terminal region exposed to the cytosol. SSR function might be regulated by phosphorylation of the cytoplasmic tail.

3.3.2.2. ER to Golgi and intra-Golgi protein transit. Conceptually, the cell might determine the fate of any polypeptide along this transit circuit in several ways, as illustrated by Klausner's metaphor (Klausner, 1989): *You should stay* (transport incompetency); *you cannot go* (retention signal recognized by a receptor); *you can go* (transport competency); *you must go* (transport signal). Presumably, all tricks are utilized by the cell for facing the diverse situations of intracellular traffic. For example, transport by default (transport competence and lack of retention signal) might work from the ER to the Golgi until proteins reach a fork such as the *trans*-Golgi network where a sorting mechanism is needed for dispatching them to the proper endpoints, such as the lysozomes, secretory granules and plasma membrane-directed secretory vesicles. Accordingly, proteins bound to lysozomes share a common routing motif, the mannose 6-P, responsible for lysozome targeting.

Short C-terminal motifs are involved in the retention of proteins in the lumen (such as Ser–Glu–Lys–Asp–Glu–Leu (SEKDEL) found in two luminal proteins (Pelham, 1989)) and the membrane (such as Asp–Glu–Lys–Lys–Met–Pro (DEKKMP) (Nilsson *et al.*, 1989)) of the ER in animal cells. The cytoplasmic tail of integral ER proteins confers ER residency to other types of membrane proteins whereas the SEKDEL motif is not sufficient for absolute retention of chimeric secretory proteins which could reach the Golgi, although at a reduced rate (Zagouras & Rose, 1989). Presumably, ER resident proteins share other features, possibly a conformation unsuitable for transport. Only adequately folded proteins, presumably stablized by molecular chaperones such as the calcium-binding protein BiP, one of the luminal ER residents, might be allowed as passengers of ER-Golgi shuttle vesicles. In this vesicular process, the ER loses about half of its membrane every 10 min (Wieland *et al.*, 1987), and lipid synthesis in the cytosol is insufficient to replenish it. There is some evidence for a vesicular recycling between the ER and the *cis*/medial Golgi compartment, with possible receptor-mediated retrieval of ER luminal resident proteins tagged with the SEKDEL motif from a putative intermediate salvage compartment (Lippincott-Schwartz *et al.*, 1989; Pelham, 1989; Vaux *et al.*, 1990).

The vectorial transport of proteins between the ER and the Golgi and between the Golgi stacks requires many factors including ATP and GTP which provide energy, Ca^{2+} and NSF (NEM-sensitive fusion protein, whose function is inhibited by the sulphydryl-alkylating reagent *N*-ethylmaleimide). A large number of GTP-binding proteins may be involved in the sequential regulation of vesicular transport along the entire vacuolar system, each one controlling specific targetting and fusion to an acceptor compartment through GTP hydrolysis. Transiently-coated transport vesicles, mediating ER to Golgi and intra-Golgi transport, lack clathrin (Fine, 1989) and may behave as bulk protein carriers, in contrast to clathrin-coated vesicles first occuring in the TGN compartment and specialized in the concentration of specific proteins such as receptors. NSF, which is not involved in the budding process, requires both a Golgian receptor and a cytosolic component to bind to Golgi membranes. It is probably a component of a transient protein complex, ensuring efficient fusion between the uncoated transport vesicle and its target Golgian membrane. Calcium may play a major regulatory role in both early and late stages of intracellular transport. According to the oscillator model, described by Sambrook (1990), tidal calcium flux might be involved in the formation of vesicles carrying properly folded secretory proteins from the ER to the Golgi. The calcium concentration in the ER, which can be 10^3 times higher than in the cytosol, might depend on the cyclic and antagonistic activity of calcium pumps and calcium release valves: an ion-motive ATPase might transfer Ca^{2+} ions from the cytoplasm to negatively charged phospholipid head groups of the membrane and to luminal resident calcium-binding proteins able to bind up to 30 Ca^{2+}, thus forming a dense calcium–protein matrix stabilizing the underlying membrane. On the contrary, partial release of the sequestered calcium through isositol triphosphate-gated calcium channels might trigger vesiculation of the destabilized ER membrane, facilitated by loosely associated hetero-oligomeric complexes recruited from the cytosol. [Ins(1,4,5P_3)] receptors and an inositol-dependent phospholipase C-like protein, able to generate [Ins(1,4,5P_3)] from [Ins(4,5)P_2], have been identified in the ER.

3.3.3. Salient Features of Signal Peptides of Milk Proteins

As mentioned above, there is no clear consensus sequence shared by signals routing proteins to a definite location, although each class has its own distinct characteristics. Thus, signal peptides, the amino terminal extensions routing proteins to the ER, have some common features: the presence of positively charged residues close to the amino terminus, upstream from a hydrophobic stretch of amino acids with apolar side chains; the lack of negatively charged residues; the very frequent occurrence of residues with short side chains (Gly, Ala) at positions -1 and -3, and of a helix-breaking residue at position -6.

Since the first reports on both the occurrence of precursors of milk proteins and the amino acid sequences of some relevant signal peptides (Gaye *et al.*, 1977) and their striking conservation (Mercier & Gaye, 1982), 30 or so amino acid sequences of milk preproteins are now known, most of them deduced from the nucleotide sequences of cDNAs (quoted in Mercier *et al.*, 1990). These new data have confirmed the striking similarity (60–80% and 73–82% at the amino acid and nucleotide levels, respectively) and the evolutionary conservation of the primary structures of the signal peptides (interspecies homology $\geqslant$ 87%) of the calcium-sensitive caseins, αs1, αs2 and β. As illustrated in Fig. 14, most mutations are neutral, as they affect the third nucleotide of codons without altering the encoded message, whereas the others lead to rather conservative amino acid substitutions. This reflects an evolutionary constraint probably related to the necessary conservation of an optimal and definite conformation of the signal peptide which is required for efficient translocation of large amounts of casein polypeptide chains across the rough endoplasmic reticulum membrane. Invariant cysteinyl residue at position -8 in αs1-, αs2- and β-precasein might play an important role in stabilizing, through covalent bonds, the interaction of the signal peptide with either the SRP or the SSR.

```
            NON CODING 5' SIGNAL PEPTIDE
                                                                                    MATURE PROTEIN

αs1-Cn                  -15                 -10                 -5              -1  +1              +5
Cow         CTT GACAACC ATG AAA CTT CTT ATC CTC ACC TGT CTT GTG GCT GTT GCT CTT GCC AGG CCT AAA CAT CCT ATC AAG CAC
Sheep       --- ------- --- --- --- --C --- --T --- --- --- --- --- --- --- --- --- --- --- --- --- --- --- --- ---
Rat         --- AG----- --- --- --- --- --- --- --- --C --C --- --- -C- --- --- --T CT- --- -G- GC- -A- CGT -GA A-T
G.P         --C AG----- --- --G --- --- --- --- --- --C --G --- --- TC- --- G-G --- -T- --G --- TT- --C T-- -G- ---
Rabbit      ---G---C--- --- --G --- --C --- --- --T --C --- --- --- AC- --- --- --- --- -A- --- TT- -A- T-A GGA ---

Amino acid               M   K   L   L   I   L   T   C   L   V   A  V/A  A  L/V  A  R/L P/H K/R H/A P/H IRF KRG H/N
                                                                    S/T              M           F       L
αs2-Cn                  -15                 -10                 -5              -1  +1              +5
Cow         ACAAAGCAAAC ATG AAG TTC TTC ATC TTT ACC TGC CTT TTG GCT GTT GCC CTT GCA AAG AAT ACG ATG GAA CAT GTC TCC
Sheep       -TC---T---- --- --- --- --- --T --- --- --- --- --- --C --- --- --- --- --- C-- -A- --- --- --- --- ---
Rat         -TC---T--C- --- --- --- --- --- --C --- --- --G G-- --- -C- --T --G --T --- C-- G-- G-A A-G G-- AAA CCC
G.P         --TC--T--C- --- --- C-- --- --- --C --- --- --C --- --- --G --T --C --- --C C-C -A- TCA --G --A CAG ---
Mouse       --TC-TT--T- --- --- --- A-- --T C-G --T --- --- --- --C --- --T --- --- --- C-G -G- --- --G --A TA- AT-

Amino acid               M   K  F/L F/I  I  F/L  T   C   L  L/V  A  V/A  A   L   A  K/N N/H T/K M/V E/K HDQ VKQ SPI
                                                                                         Q  A/R  S           Y
β-Cn                    -15                 -10                 -5              -1  +1              +5
Cow         AATTGAGAGCC ATG AAG GTC CTC ATC CTT GCC TGC CTG GTG GCT CTG GCC CTT GCA AGA GAG CTG GAA GAA CTC AAT GTA
Sheep       ---C------- --- --- --- --- --- --- --- --T --- --- --- --- --- --- --- --- --- -A- --- --- --- --- ---
Rat         -C----C---- --- --- --- T-- --- --- --- --- --T --- --A --T --T --- --- --G --- AA- --T -C- T-- -C- --G
Rabbit      -C---GC--T- --- --- --- --- --T --- --- --- --- --- --- --C --T --- --- --G --- AA- --- C-- --- -G- --T
Mouse       -C----C---- --- --- --- T-- --- --C --- --- --T --- --C --T --T --- --- --- --- ACT AC- TTT ACT GTA TCC

Amino acid               M   K   V  L/F  I   L   A   C   L   V   A   L   A   L   A   R   E  L/Q E/D E/A LFT NTS V/S
                                                                                            K/T  T  Q/F      V
Consensus          ARYC ATG AAR

                        -15                 -10                 -5              -1
Consensus                M   K  LFV LFI  I  L/F T/A  C   L  L/V  A  VAS  A  L/V  A
                                                                    T/L
```

```
κ-Cn                    -21                     -15                 -10                 -5              -1  +1
Cow         GGAAAGGTGCA ATG ATG AAG AGT TTT TTC CTA GTT GTG ACT ATC CTG GCA TTA ACC CTG CCA TTT TTG GGT GCC CAG GAG
Sheep       ----------- --- --- --- --- --- --- --- --- --- --- --- --A --- --- --- --- --- --- --- --- --- --- ---
Rat         -TG-------- --- --- -G- -A- --- A-- G-- --- A-- -A- --- --A --- C-- --T --- --C --- --- -C- --A G-- -T-
Mouse       AT--------- --- --- -G- -A- --- A-- G-- --G A-- -A- --T --- --- --- --T --- --C --- --- -C- --A G-- ATA

Amino acid               M   M  K/R S/N  F  F/I L/V  V  V/M T/N  I   L   A   L   T   L   P   F   L  G/A  A  Q/E EVI

β-Lg                    -18         -15                 -10                 -5                  -1  +1              +5
Sheep       CAGCTGCAGCC ATG AAG TGC CTC CTG CTT GCC CTG GGC CTG GCC CTC GCC TGT GGC GTC CAG GCC ATC ATC GTC ACC CAG

Amino acid               M   K   C   L   L   L   A   L   G   L   A   L   A   C   G   V   Q   A   I   I   V   T   Q
(pig)                    -   R   -   -   -   -   T   -   -   -   -   -   T   -   -   -   -   -   V   E   -   -   P

α-La                    -19             -15                 -10                 -5              -1  +1
Cow         GGGTCACCAAA ATG ATG TCC TTT GTC TCT CTG CTC CTA GTA GGC ATC CTA TTC CAT GCC ACC CAG GCT GAA CAG TTA ACA
Sheep       ----A------ --- --- --- --- --- --- --- --- --G --- --- --- --- --- --- --- --- --- --- --- --A --- ---
Goat        ----A------ --- --- --- --- --- --- --- --- --G --- --- --- --- --- --C --- --- --- --- --- --A --- ---
Rat         -A-C-GG---- --- --- CGT --- --T C-- --- T-- --- -CG T-T --T TCG C-G -C- --- TTT --A --C AC- G-- --T ---
G.P         CA-CAG----- --- --- --- --- T-- C-- --- T-G --G --G --- --- --G --T -C- --- GTG --- --C A-G --A C-T --C
Man         ----AG----- --- -G- -T- --- --- C-- --- T-- --G --G --- --- --G --- -C- --- -T- -T- --C A-G --A --C ---

Amino acid               M  M/R SRF  F  V/F S/P  L  L/F  L  V/A G/C  I  L/S F/L H/P  A  TFV Q/L  A  ETK Q/E L/F  T

WAP                     -19             -15                 -10                 -5              -1  +1
Rat         CCGCCGACACC ATG CGC TGT TCG ATC AGC CTC GTT CTT GGC CTG CTG GCC CTG GAG GTA GCC CTT GCT CGG AAC CTA CAG
Mouse       A-A---GT--- --- --T --C CTC --- --- --T --- --- --- --- --- --- --- --- --G --- --C --- -A- --- --- G--
Rabbit      -ACCT-CC--- --- --- --- CTC --- --- --G -CC --C --- --- --C --- --- --- -CG --- --C --- -T- GC- -CC A--

Amino acid               M   R   C  S/L  I   S   L  V/A  L   G   L   L   A   L   E  V/A  A   L   A  RQL N/A L/P QEK
```

Fig. 14. Nucleotides sequences of milk protein cDNAs in the region spanning the signal peptide codons (numbered negatively). Dashes represent nucleotides identical to those of the sequence taken as reference. Amino acids are represented according to the one-letter code.

4. Conclusions

Elucidation of the nucleotide sequences of several milk protein genes from different species was a prerequisite to a better understanding of their evolution and of the mechanisms involved in their expression and regulation at the molecular level. As predicted from the amino acid sequence data, the most conserved regions of the rapidly evolving coding frames were those encoding the signal peptide of all milk proteins and the major phosphorylation site(s) of calcium-sensitive caseins, whose common origin is substantiated by the similar organization of the relevant genes.

At present, our structural knowledge is still insufficient since only a few genes have been wholly or largely sequenced in more than one species: 5 for α-lactalbumin, 3 for the WAP and 5 for β-casein. Fortunately, more sequence data will be available in the near future, allowing a more detailed and accurate analysis of conserved sequences which might play an important functional role. Nevertheless, analysis of sequence data already obtained showed the occurrence of structural motifs specific to the genes encoding calcium-sensitive caseins, as well as the identification of a consensus sequence ('milk box') shared by most, but not all, milk protein genes. However, their biological significance remains unknown.

Rapid progress in the field of milk protein gene expression and regulation will depend in part on the development of both permanent mammary cell lines and efficient cell-free transcription systems able to mimic what occurs *in vivo*. Although the application of transgenic mice for analysing gene expression has some value, it is difficult to compare different transgenic animals, as the copy number of integrated genes and their chromosomal location differ. Recent studies of *trans*-acting nuclear factors, regulating the expression of milk protein genes, are promising and these experiments should provide useful information about the fine regulation of gene expression at the molecular level. Another interesting approach for a better understanding of the structure–function relationship in milk protein genes is to search for and analyse alleles associated with specific properties as previously discussed.

In practice, in the field of RFLP studies, the very short sizes of most exons make the detection of the relevant genomic fragments with a cDNA probe difficult in the standard conditions. Knowledge of the gene sequence will allow: (i) choice and use as efficient probes, for RFLP studies, of DNA genomic fragments devoid of repetitive sequences; (ii) development of allelic specific oligonucleotide probes recognizing in-vitro amplified mutated DNA stretches, thus permitting the rapid identification of alleles of economic interest. Similarly, knowledge of the sequences of ruminant α-lactalbumin pseudogenes will allow the accurate chromosomal assignment of the authentic cognate gene by using a probe specific for the region deleted in the pseudogenes.

Quite apart from transcriptional and translational processes, milk protein synthesis is, clearly, limited by the availability of amino acids in the appropriate intracellular pools. Most amino acids are derived from the blood (plasma or cells), possibly in some cases after intracellular proteolysis of peptides, while some are synthesized intracellularly. Moreover, a substantial turnover of intracellular proteins furnishes amino acids for milk protein synthesis. A research field of recent interest centres on the mechanisms by which amino acids are translocated into cells, particular significance being attached to the shared affinities of different amino acids for common membrane carrier sites, which lead to competitive interactions.

References

Adelman, J.P., Bond, C.T., Douglas, S.J. & Herbert, E. (1987). Two mammalian genes transcribed from opposite strands of the same DNA locus. *Science*,**235**, 1514–17.

Alexander, L.J., Stewart, A.F., MacKinlay, A.G., Kapelinskaya, T.V., Tkach, T.M. & Gorodetsky, S.I. (1988). Isolation and characterization of the bovine kappa-casein gene. *Eur. J. Biochem.*, **178**, 395–401.

Ali, S. & Clark, J. (1988). Characterization of the gene encoding ovine beta-lactoglobulin. Similarity to the genes for retinol binding protein and other secretory proteins. *J. Mol. Biol.*, **199**, 415–26.

Anderson, C.D. & Rillema, J.A. (1976). Effects of hormones on protein and amino acid metabolism in mammary gland explants of mice. *Biochem. J.*, **158**, 355–9.

Atroshi, F. & Sandholm, M. (1982). Red blood cell glutathione as a marker of milk production in Finn sheep. *Research in Vet. Sci.*, **33**, 256–9.

Ball, R.K., Friis, R.R., Schoenenberger, C.A., Doppler, W. & Groner, B. (1988). Prolactin regulation of beta-casein gene expression and of a cytosolic 120-kD-d protein in a cloned mouse mammary epithelial cell line. *EMBO J.*, **7**, 2089–96.

Banerjee, A.K. (1980). 5′ Terminal cap structure in eukaryotic messenger ribonucleic acids. *Microbiol. Rev.*, **44**, 175–205.

Barker, G.A. & Ellory, J.C. (1990). The identification of neutral amino acid transport systems. *Exper. Physiol.*, **75**, 3–26.

Baumrucker, C.R. (1984). Cationic amino acid transport by the bovine mammary gland. *J. Dairy Sci.*, **67**, 2500–6.

Baumrucker, C.R. (1985). Amino acid transport systems in bovine mammary tissue. *J. Dairy Sci.*, **68**, 2436–51.

Baumrucker, C.R. & Pocius, P.A. (1978). Gamma-glutamyl transpeptidase in lactating mammary secretory tissue of cow and rat. *J. Dairy Sci.*, **61**, 309–14.

Bayna, E.M. & Rosen, J.M. (1990). Tissue-specific, high level expression of the rat whey acidic protein gene in transgenic mice. *Nucl. Acids Res.*, **18**, 2977–85.

Bernstein, P. & Ross, J. (1989). Poly(A)-poly(A) binding protein and the regulation of mRNA stability. *Trends Biochem. Sci.*, **14**, 373–7.

Bernstein, H.D., Rapoport, T.A. & Walter, P. (1989). Cytosolic protein translocation factors. Is SRP still unique?. *Cell*, **58**, 1017–19.

Bernstein, P., Peltz, S.T. & Ross, J. (1989). The poly(A)-poly(A)-binding protein complex is a major determinant of the RNA stability 'in vitro'. *Mol. Cell. Biol.*, **9**, 659–70.

Bingham, P.M., Chou, T.B., Mims, I. & Zachar, Z. (1988). On/off regulation of gene expression at the level of splicing. *Trends Genet.*, **4**, 134–8.

Bisbee, C.A. & Rosen, J.M. (1987). DNA sequence elements regulating casein gene expression. In *Transcriptional control mechanisms*, UCLA Symposium on Molecular and Cellular Biology, New Series, Vol. 52, eds D.K. Granner, M.G. Rosenfeld & S. Chang. Alan R.Liss, New York, pp. 313–23.

Blobel, G. & Dobberstein, B. (1975). Transfer of proteins across membranes 1. Presence of proteolytically processed and unprocessed nascent immunoglobulin light chains on membrane-bound ribosomes of murine myeloma. 2. Reconstitution of functional rough microsomes from heterologous components *J. Cell Biol.*, **67**, 835–851, 852–862.

Boisnard, M. & Pétrissant, G. (1985). Complete sequence of ovine alpha-s2-casein messenger RNA. *Biochimie*, **67**, 1043–51.

Boisnard, M., Hue, D., Bouniol, C., Mercier, J.C. & Gaye, P. (1991). Multiple mRNA species code for two non-allelic forms of ovine alpha-s2-casein *Eur. J. Biochem.*, **201**, 633–641.

Bonsing, J., Ring, J.M., Stewart, A.F. & MacKinlay, A.G. (1988). Complete nucleotide sequence of the bovine beta-casein gene. *Austr. J. Biol. Sci.*, **41**, 527–37.

Breathnach, R. & Chambon, P. (1981). Organisation and expression of eukaryotic split genes coding for proteins. *Ann. Rev. Biochem.*, **50**, 349–83.

Breitbart, R.E., Andreadis, A. & Nadal-Girard, B. (1987). Alternative splicing: an ubiquitous mechanism for the generation of multiple protein isoforms from single genes. *Ann. Rev. Biochem.*, **56**, 467–95.

Brignon, G., Mahé, M.F., Ribadeau-Dumas, B., Mercier, J.C. & Grosclaude, F. (1990). Two of the three genetic variants of caprine alpha s1-casein which are synthesized at a reduced level have an internal deletion possibly due to an altered RNA splicing. *Eur. J. Biochem.*, **193**, 237–41.

Brock, M.L. & Shapiro, D.J. (1983). Estrogen stabilizes vitellogenin mRNA against cytoplasmic degradation. *Cell*, **34**, 207–14.

Bucher, P. & Trifonov, E.N. (1986). Compilation and analysis of eukaryotic POL II promoter sequences. *Nucleic Acids Res.*, **14**, 10 009–26.

Campbell, S.M., Rosen, J.M., Hennighausen, L.G., Strech-Jurk, U. & Sippel, A.E. (1984). Comparison of the whey acidic protein genes of the rat and mouse. *Nucleic Acids Res.*, **12**, 8685–97.

Caulfield, M.P., Duong, Le T., Baker, R.K., Rosenblatt, M. & Lively, M.O. (1989). Synthetic substrate for Eukaryotic signal peptidase. Cleavage of a synthetic peptide analog of the precursor region of preproparathyroid hormone. *J. Biol. Chem.*, **264**, 15813–17.

Chomczynski, P., Qasba, P. & Topper, Y.J. (1986). Transcriptional and posttranscriptional roles of glucocorticoid in the expression of the rat 25,000 molecular weight casein gene. *Biochem. Biophys. Res. Commun.*, **134**, 812–18.

Christensen, H.N. (1984). Organic ion transport during seven decades: the amino acids. *Biochim. Biophys. Acta*, **779**, 255–69.

Clark, J.H. (1975). Lactational responses to postruminal administration of proteins and amino acids. *J. Dairy Sci.*, **58**, 1178–97.

Clark, J.H., Spires, H.R. & Davis, C.L. (1978). Uptake and metabolism of nitrogenous components by the lactating mammary gland. *Federation Proc.*, **37**, 1233–8.

Crick, F. (1979). Split genes and RNA splicing. *Science*, **204**, 264–71.

Dalens, M. & Gellin, J. (1986). The gene map of the rabbit. III. alpha and beta casein gene synteny. *Genet. Sel. Evol.*, **18**, 99–104.

Danielson, K.G., Oborn, C.J., Durban, E.M., Butel, J.S. & Medina, D. (1984). Epithelial mouse mammary cell line exhibiting normal morphogenesis 'in vivo' and functional differentiation 'in vitro'. *Proc. Natl. Acad. Sci. USA*, **81**, 3756–60.

Davis, S.R. & Mepham, T.B. (1974). Amino acid and glucose uptake by the isolated perfused guinea-pig mammary gland. *Quart. J. Exp. Physiol.*, **59**, 113–30.

Davis, S.R. & Mepham, T.B. (1976). Metabolism of L-[U^{14}C] phenylalanine by the isolated perfused lactating guinea-pig mammary gland. *Biochem. J.*, **156**, 553–60.

Davis, S.R., Bickerstaffe, R. & Hart, D.S. (1978). Amino acid uptake by the mammary gland of the lactating ewe. *Austr. J. Biol. Sci.*, **31**, 123–32.

Davies, M.S., West, L.F., Davis, M.B., Povey, S. & Craig, R.K. (1987). The gene for human alpha-lactalbumin is assigned to chromosome 12q13. *Ann. Hum. Genet.*, **51**, 183–8.

Devinoy, E. & Houdebine, L.M. (1977). Effects of glucocorticoids on casein gene expression in the rabbit. *Eur. J. Biochem.*, **75**, 411–16.

Devinoy, E., Houdebine, L.M. & Delouis, C. (1978). Role of prolactin and glucocorticoids in the expression of casein genes in rabbit mammary gland organ culture. Quantification of casein mRNA. *Biochim. Biophys. Acta*, **517**, 360–6.

Doppler, W., Groner, B. & Ball, R.K. (1989). Prolactin and glucocorticoid hormones synergistically induce expression of transfected rat beta-casein gene promoter constructs in a mammary epithelial cell line. *Proc. Natl. Acad. Sci. USA*, **86**, 104–8.

Early, P., Rogers, J., Davis, M., Calame, K., Bond, M., Wall, R. & Hood, L. (1980). Two mRNAs can be produced from a single immunoglobulin gene by alternative RNA processing pathways. *Cell*, **20**, 313–20.

Eisenstein, R.S. & Rosen, J.M. (1988). Both cell-substratum and hormonal regulation of milk protein gene expression are exerted primarily at the post-transcriptional level. *Mol. Cell. Biol.*, **8**, 3183–90.

Ellis, R.J. & Hemmingsen, S.M. (1989). Molecular chaperones: proteins essential for the biogenesis of some macromolecular structures. *TIBS*, **14**, 339–42.

Ferretti, L., Leone, P., Rognoni, G. & Sgaramella, V. (1990). Long range restriction analysis of the bovine casein genes. *Nucleic Acids Res.*, **18**, 6829–33.

Fine, R.E. (1989). Vesicle without clathrin: intermediates in bulk flow exocytosis. *Cell*, **58**, 609–10.

Fleet, I.R. & Mepham, T.B. (1983). Physiological methods used in the study of mammary substrate uptake in ruminants. In *Biochemistry of Lactation*, ed. T.B. Mepham. Elsevier, Amsterdam, pp. 461–91.

Fleet, I.R. & Mepham, T.B. (1985). Mammary uptake of amino acids and glucose throughout lactation in Friesland sheep. *J. Dairy Sci.*, **52**, 229–37.

Forsyth, I.A. (1983). The endocrinology of lactation. In *Biochemistry of Lactation*, ed. T.B. Mepham. Elsevier, Amsterdam, pp. 309–49.

Forsyth, I.A. (1986). Variation among species in the endocrine control of mammary growth and function: the roles of prolactin, growth hormone, and the placental lactogen. *J. Dairy Sci.*, **69**, 886–903.

Fox, P.F. (ed.) (1982). *Developments in Dairy Chemistry—1—Proteins* Applied Science Publishers, London.

Ganguly, R., Majumder, P.K., Ganguly, N. & Banerjee, M.R. (1982). The mechanism of progesterone-glucorticoid interaction in regulation of casein gene expression. *J. Biol. Chem.*, **257**, 2182–7.

Gaye, P., Gautron, J.P., Mercier, J.C. & Hazé, G. (1977). Amino terminal sequences of the precursors of ovine caseins. *Biochem. Biophys. Res. Comm.*, **79**, 903–11.

Geissler, E.N., Cheng, S.V., Gusella, J.F. & Housman, D.E. (1988). Genetic analysis of the dominant white-spotting (W) region on mouse chromosome 5: identification of clones DNA markers near W. *Proc. Natl. Acad. Sci. USA*, **85**, 9635–9.

Gellin, J., Echard, G., Yerle, M., Dalens, M., Chevalet, C. & Gillois, M. (1985). Localization of the alpha and beta casein genes to the q24 region of chromosome 12 in the rabbit (*Oryctolagus cumiculus L.*) by in situ hybridization. *Cytogenet. Cell Genet.*, **39**, 220–3.

Gilbert, W. (1978). Why genes in pieces? *Nature*, **271**, 501.

Gilbert, W. (1985). Genes-in-pieces revisited. *Science*, **228**, 823–4.

Gilbert, W., Marchionni, M. & McKnight, G. (1986). On the antiquity of introns. *Cell*, **46**, 151–4.

Grosclaude, F. (1979). Polymorphism of milk proteins: some biochemical and genetical aspects. In *Proc. 16th Int. Conf. Anim. Blood Grps Biochem. Polymorphisms*, International Society for Animal Blood Group Research, Leningrad, pp. 54–92.

Grosclaude, F., Joudrier, P. & Mahé, M.F. (1978). Polymorphisme de la caséine alpha-s2 bovine: étroite liaison du locus alpha-s2-Cn avec les loci alpha-s1-Cn, beta-Cn et kappa-Cn; mise en évidence d'une délétion dans le variant alpha-s2-Cn D. *Ann. Genet. Sel. Anim.*, **10**, 313–27.

Grosclaude, F., Mahé, M.F., Brignon, G., Di Stasio, L. & Jeunet, R. (1987). A Mendelian

polymorphism underlying quantitative variations of goat alpha-s1-casein. *Genet. Sel. Evol.*, **19**, 399–412.
Gupta, P., Rosen, J.M., D'Eustachio, P. & Ruddle, F.H. (1982). Localization of the casein gene family to a single mouse chromosome. *J. Cell Biol.*, **93**, 199–204.
Guyette, W.A., Matusik, R.J. & Rosen. J.M. (1979). Prolactin-mediated transcriptional and post-transcriptional control of casein gene expression. *Cell*, **17**, 1013–23.
Hall, L., Emery, D.C., Davies, M.S., Parker D. & Craig, R.K. (1987). Organization and sequence of the human alpha-lactalbumin gene. *Biochem. J.*, **242**, 735–42.
Harris, S., Ali, S., Anderson, S., Archibald, A.L. & Clark, A.J. (1988). Complete nucleotide sequence of the genomic ovine beta-lactoglobulin gene. *Nucleic Acids Res.*, **16**, 10 379–80.
Harris, S., McClenaghan, M., Simons, J.P., Ali, S. & Clark, A.J. (1990). Gene expression in the mammary gland. *J. Reprod. Fert.*, **88**, 707–15.
Hartl, F.U. & Neupert, W. (1990). Protein sorting to mitochondria: evolutionary conservations of folding and assembly. *Science*, **247**, 930–8.
Hatzopoulos, A.K., Schlokat, U. & Gruss, P. (1988). Enhancers and other cis-acting regulatory sequences. In *Transcription and Splicing*, ed. B.D. Hames & D.M. Glover. IRL Press, Washington, DC, pp. 43–96.
Heitmann, R.N. & Bergman, E.N. (1980). Transport of amino acids in whole blood and plasma of sheep. *Amer. J. Physiol.*, **239**, E242–7.
Henderson, A.J. & Peaker, M. (1983). Arterial plasma composition during compensatory increase in milk secretion in the goat: relation to rate limitation. *Quart. J. Exp. Physiol.*, **68**, 203–8.
Hennighausen, L.G. & Sippel, A.E. (1982). Mouse whey acidic protein is a novel member of the family of 'four-disulfide core' proteins. *Nucleic Acids Res.*, **10**, 2677–84.
Hobbs, A.A., Richards, D.A., Kessler, D.J. & Rosen, J.M. (1982). Complex hormonal regulation of rat casein gene expression. *J. Biol. Chem.*, **257**, 3598–605.
Houdebine, L.M. (1976). Effect of prolactin and progesterone on expression of casein genes. *Eur. J. Biochem.*, **68**, 219–25.
Houdebine, L.M. & Gaye, P. (1975). Regulation of casein synthesis in the rabbit mammary gland. Titration of mRNA activity for casein under prolactin and progesterone treatments. *Mol. Cell. Endocrin.*, **3**, 37–55.
Houdebine, L.M., Djiane, J., Dusauter-Fourt, I., Martel, P., Kelly, P.A., Devinoy, E. & Servely, J.L. (1985). Hormonal action controlling mammary activity. *J. Dairy Sci.*, **68**, 489–500.
Humphrey, T. & Proudfoot, N.J. (1988). A beginning to the biochemistry of polyadenylation. *Trends Genet.*, **4**, 243–5.
Imam-Ghali, M., Saidi-Mehtar, N. & Guérin, G. (1991). Sheep gene mapping. Additional DNA markers included (CASB, CASK, LALBA, IGF-1 and AMH). *Animal Genetics*, **22**, 165–72.
Jois, M. (1986). Amino acid metabolism in ruminants. PhD thesis, University of Sydney, Australia.
Jones, W.K., Yu-Lee, L.Y., Clift, S.M., Brown, T.L. & Rosen, J.M. (1985). The rat casein multigene family. Fine structure and evolution of the beta-casein gene. *J. Biol. Chem.*, **260**, 7042–50.
Jones, N.C., Rigby, P.W.J. & Ziff, E.B. (1988). Trans-acting protein factors and the regulation of eukaryotic transcription: lessons from studies on DNa tumor viruses. *Genes & Develop.*, **2**, 267–81.
Julkunen, M., Seppälä, M. & Jänne, O.A. (1988). Complete amino acid sequence of human placental protein 14: a progesterone-regulated uterine protein homologous to beta-lactoglobulins. *Proc. Natl. Acad. Sci. USA.*, **85**,8845–9.
Jung, A., Sippel, A.E., Grez, M. & Shutz, G. (1980). Exons encode functional and structural units of chicken lysozyme. *Proc. Natl Acad. Sci. USA*, **77**, 5759–63.
Kilberg, M.S., Bracy, D.S. & Handlogten, M.E. (1986). Substrate regulation of hepatic

system A transport activity after induction by substrate starvation or gluclose. *Federation Proc.*, **45**, 24328–41.

Klausner, R.D. (1989). Sorting and traffic in the central vacuolar system. *Cell*, **57**, 703–6.

Kletzien, R.F., Pariza, M.W., Becker, J.E., Potter, V.R. & Butcher, F.R. (1976). Induction of amino acid transport in primary cultures of adult rat liver parenchymal cells by insulin. *J. Biol. Chem.*, **251**, 3014–20.

Kuhn, N.J. (1983). The biochemistry of lactogenesis. In *Biochemistry of Lactation*, ed. T.B. Mepham. Elsevier, Amsterdam, pp. 351–79.

La Thangue, N.B. & Rigby, P.W.J. (1988). Trans-acting protein factors and the regulation of eukaryotic transcription. In *Transcription and Splicing*, eds. B.D. Hames & D.M. Glover. IRL Press, Washington DC, pp. 1–42.

Laird, J.E., Jack, L., Hall, L., Boulton, A., Parker, D. & Craig, R.K. (1988). Structure and expression of the guinea-pig alpha-lactalbumin gene. *Biochem. J.*, **254**, 85–94.

Lang, K.M. & Spritz, R.A. (1987). In vitro splicing pathways of pre-mRNAs containing multiple intervening sequences. *Mol. Cell. Biol.*, **7**, 3428–37.

Lee, K.F., Atiee, S.H., Henning, S.J. & Rosen, J.M. (1989). Relative contribution of promoter and intragenic sequences in the hormonal regulation of rat beta-casein transgenes. *Endocrinology*, **3**, 447–53.

Leff, S.E., Rosenfeld, M.G. & Evans, R.M. (1986). Complex transcriptional units: diversity in gene expression by alternative RNA processing. *Ann. Rev. Biochem.*, **55**, 1091–17.

Leroux, C., Martin, P., Mahé, M.F., Leveziel, H. & Mercier, J.C. (1990). Restriction fragment length polymorphisms identification of goat alpha-s1-casein alleles. A potential tool in selection of individuals carrying alleles associated with a high level protein synthesis. *Animal Genetics*. **21**, 341–51.

Leroux, C., Mazure, N. & Martin, P. (1992). Mutations away from splice site recognition sequences might cis-modulate alternative splicing of goat αs1-casein transcripts. Structural organization of the relevant gene. *J. Biol. Chem.*, **267**, 6147–57.

Levine, M. & Manley, J.L. (1989). Transcriptional repression of eukaryotic promoters. *Cell*, **59**, 405–8.

Levéziel, H., Méténier, L., Guérin, G., Cullen, P., Provot, C., Bertraud, M. & Mercier, J.C. (1991). Restriction fragment length polymorphism of ovine casein genes. Close linkage between the alpha-s1-, alpha-s2-, beta- and kappa-Cn loci. *Animal Genetics*, **22**, 1–10.

Linzell, J.L. (1960). Mammary gland blood flow and oxygen, glucose and volatile fatty acid uptake in the conscious goat. *J. Physiol., London*, **153**, 492–509.

Linzell, J.L. (1966). Measurement of venous flow by continuous thermodilution and its application to measurement of mammary blood flow in the goat. *Circul. Res.*, **18**, 745–54.

Linzell, J.L. & Mepham, T.B. (1968). Mammary synthesis of amino acids in the lactating goat.*Biochem. J.*, **107**, 18P.

Linzell, J.L. & Mepham, T.B. (1974). Effects of intramammary arterial infusion of essential amino acids in the lactating goat. *J. Dairy Res.*, **41**, 101–9.

Linzell, J.L., Mepham, T. B., Annison, E.F. & West, C.E. (1969). Mammary metabolism in lactating sows: arteriovenous differences of milk precursors and mammary metabolism of [^{14}C] glucose and [^{14}C] acetate. *Brit. J. Nutrition*, **23**, 319–32.

Lippincott-Schwartz, J., Yuan, L.C., Bonifacio, J.S. & Klausner, R.D. (1989). Rapid redistribution of Golgi proteins into the ER in cells treated with brefelding A: evidence for membrane cycling from Golgi to ER. *Cell*, **56**, 801–13.

Lubon, H. & Henninghausen, L. (1987). Nuclear proteins from lactating mammary glands bind to the promoter of a milk protein gene. *Nucleic Acids Res.*, **15**, 2103–21.

Lubon, H. & Hennighausen, L. (1988). Conserved region of the rat alpha-lactalbumin promoter is a target site for protein binding in vitro. *Biochem. J.*, **256**, 391–6.

Lubon, H., Pittius, C.W. & Hennighausen, L. (1989). In vitro transcription of the mouse whey acidic protein promoter is affected by upstream sequences. *FEBS Letters*, **251**, 173–6.

Mahé, M.F. & Grosclaude, F. (1989). Alpha-s1-CnD, another allele associated with a decreased synthesis rate at the caprine alpha-s1-casein locus. *Genet. Sel. Evol.*, **21**, 127–9.
Majumder, P.B., Joshi, J.B. & Banerjee, M.R. (1983). Correlation between nuclear glucocorticoid receptor levels and casein gene expression in murine mammary gland 'in vitro'. *J. Biol. Chem.*, **258**, 6793–7.
Martin, P. & Leroux, C. (1992). Exon-skipping is responsible for the 9 amino acid residues deletion occurring near the N-terminal of human β-casein. *Biochem. Biophys. Res. Commun.*, **183**, 750–7.
Matusik, R.J. & Rosen, J.M. (1978). Prolactin induction of casein mRNA in organ culture. A model system for studying peptide hormone regulation of gene expression. *J. Biol. Chem.*, **253**, 2343–7.
Mayer, R.J., Burgess, R.J. & Russell, S.M. (1980). Factors controlling intracellular breakdown of protein. In *Protein Deposition in Animals*, ed. P.J. Buttery & D.B. Lindsay. Butterworths, London, pp. 21–49.
McKnight, R.A., Jimenez-Flores, R., Kang, Y., Creamer, L.K. & Richardson, T. (1989). Cloning and sequencing of a complementary deoxyribonucleic acid coding for a bovine alpha-s1-casein A from mammary tissue of a homozygous B variant cow. *J. Dairy Sci.*, **72**, 2464–73.
Meade, H., Gates, L., Lacy, E. & Lonberg, N. (1990). Bovine alpha s1-casein gene sequences direct high level expression of active human urokinase in mouse milk. *Biotechnology*, **8**, 443–6.
Meister, A. (1973). On the enzymology of amino acid transport. *Science*, **180**, 33–9.
Menon, R.S., Chang, Y.F., Jeffers, K.F. & Ham, R.G. (1992). Exon-skipping in human β-casein. *Genomics*, **12**, 13–17.
Mepham, T.B. (1971). Amino acid utilisation by the lactating mammary gland. In *Lactation*, ed. I.R. Falconer. Butterworths, London, pp. 297–315.
Mepham, T.B. (1982). Amino acid utilization by the lactating mammary gland. *J. Dairy Sci.*, **65**, 287–98.
Mepham, T.B. (1986). The development of ideas on milk protein secretion. *Vlaams Diergeneeskundig Tijdschrift*, **55**, 228–38.
Mepham, T.B. (1987). In *Physiology of Lactation.* Open University Press, Milton Keynes, pp. 51–70.
Mepham, T.B. (1988). Nutrient uptake by the lactating mammary gland. In *Nutrition and Lactation in the Dairy Cow*, ed. P.C. Garnsworthy. Butterworths, London, pp. 15–31.
Mepham, T.B. & Linzell, J.L. (1966). A quantitative assessment of the contribution of individual amino acids to the synthesis of milk proteins by the goat mammary gland. *Biochem. J.*, **101**, 76–83.
Mepham. T.B. & Linzell, J.L. (1967). Urea formation by the lactating goat mammary gland. *Nature*, **214**, 507–8.
Mepham, T.B. & Linzell, J.L. (1974*a*). Hour-to-hour variation in amino acid arteriovenous concentration differences across the lactating goat mammary gland. *J. Dairy Res.*, **41**, 95–111.
Mepham, T.B. & Linzell, J.L. (1974*b*). Effects of intramammary arterial infusion of nonessential amino acids and glucose in the lactating goat. *J. Dairy Res.*, **41**, 111–21.
Mepham, T.B., Peters, A.R. & Davis, S.R. (1976). Uptake and metabolism of tryptophan by the isolated perfused guinea-pig mammary gland. *Biochem. J.*, **158**, 659–62.
Mepham, T.B., Peters, A.R. & Alexandrov, S. (1979). Effects of restriction of amino acid supply to the isolated perfused guinea-pig mammary gland. *J. Dairy Research*, **46**, 69–73.
Mepham, T.B., Overthrow, J.I., Short, A.H. & Warne, D.E. (1984). Amino acid transport interactions in lactating mammary gland of guinea-pig. *J. Physiol., London*, **257**, 73P.
Mepham, T.B., Overthrow, J.I. & Short, A.H. (1985). Epithelial cell entry and exit competition amongst amino acids in the isolated perfused mammary gland of guinea-pig. In *Carrier-Mediated Transport from Blood to Tissue*, ed. D.L. Yudilevich & G.E. Mann. Longman, London, pp. 369–72.

Mercier, J.C. & Gaye, P. (1982). Early events in secretion of main milk proteins: occurrence of precursors. *J. Dairy Sci.*, **65**, 299–316.

Mercier, J.C., Vilotte, J.L. & Provot, C. (1990). Structure and function of milk protein genes. In *Genome Analysis in Domestic Animals*, ed. H. Geldermann & F. Ellendorff. Verlag VCH, Weinheim, pp. 233–58.

Meyer, D.I. (1988). Preprotein conformation: the year's major theme in translocation studies. *TIBS*, **13**, 471–4.

Mezl, V.A. & Knox, W.E. (1977). Metabolism of arginine in lactating rat mammary gland. *Biochem.*, **164**, 105–113.

Milstein, C., Brownlee, G.G., Harrison, T.M. & Mathews, M.B. (1972). A possible precursor of immunoglobulin light chains. *Nature*, **239**, 117–20.

Mitchell, P.J. & Tjian, R. (1989). Transcriptional regulation in mammalian cells by sequence-specific DNA binding proteins. *Science*, **245**, 371–8.

Moore, C.L. & Sharp, P.A. (1985). Accurate cleavage and polyadenylation of exogenous RNA substrate. *Cell*, **41**, 845–55.

Nagaiah, K., Bolander, F.F.Jr., Nicholas, K.R., Takemoto, T. & Topper, Y.J. (1981). Prolactin-induced accumulation of casein mRNA in mouse mammary explants: a selective role of glucocorticoid.*Biochem. Biophys. Res. Comm.*, **98**, 380–7.

Neville, M.C., Lobitz, C.J., Ripoll, E.A. & Tinney, C. (1980). The sites for α-amino isobutyric acid uptake in normal mammary gland and ascites tumor cells. *J. Biol. Chem.*, **255**, 7311–16.

Nevins, J.R. (1983). The pathway of eukaryotic mRNA formation. *Ann. Rev. Biochem.*, **52**, 411–66.

Nilsson, T., Jackson, M. & Peterson, Per.A. (1989). Short cytoplasmic sequences serve as retention signals for transmembrane proteins in the endoplasmic reticulum. *Cell*, **58**, 707–18.

Oddy, V.H., Lindsay, D.B. & Fleet, I.R. (1988). Protein synthesis and degradation in the mammary gland of lactating goats.*J. Dairy Res.*, **55**, 143–54.

Oka, T. (1983). Intracellular mediators of lactogenic hormones. In *Biochemistry of Lactation*, ed. T.B. Mepham. Elsevier, Amsterdam, pp. 381–96.

Overthrow, J.I. (1986). Transmembrane transport of amino acids in the mammary gland. PhD Thesis, University of Nottingham.

Padgett, R.A., Grabowski, P.J., Konarska, M.M., Seiler, S. & Sharp, P.A. (1986). Splicing of mesenger RNA precursors. *Ann. Rev. Biochem.*, **55**, 1119–50.

Palade, G. (1975). Intracellular aspects of the process of protein synthesis. *Science*, **189**, 347–58.

Peeters, G., Houvenaghel, A., Roets, E., Massart-Leen, A.M., Verbeke, R., Dhont, G. & Verschooten, F. (1979). Electromagnetic blood flow recording and balance of nutrients in the udder of lactating cows. *J. Animal Sci.*, **48**, 1143–53.

Pelham, H.R.B. (1989). Control of protein exit from the endoplasmic reticulum. *Annu. Rev. Cell Biol.*, **5**, 1–23.

Persuy, M.A., Stinnakre, M.G., Printz, C., Mahé, M.F. & Mercier, J.C. (1992). High expression of the caprine β-casein gene in transgenic mice *Eur. J. Biochem.*, **205**, 887–93.

Peters, A.R. (1977). Control of amino acid utilization by the isolated perfused mammary gland. PhD Thesis, University of Nottingham.

Phi-Van, L., Kries, J.P.von, Ostertag, W. & Stratling, W.H. (1990). The chicken lysozyme 5′ matrix attachment region increases transcription from a heterologous promoter in heterologous cells and dampens position effects on the expression of transfected genes. *Mol. Cell. Biol.*, **10**, 2302–7.

Pittius, C.W., Henninghausen, L., Lee, E., Westphal, H., Nicols, E., Vitale, J. & Gordon, J.K. (1988). A milk protein gene promoter directs the expression of human tissue plasminogen activator cDNA to the mammary gland in transgenic mice. *Proc. Natl. Acad. Sci. USA*, **85**, 5874–8.

Pocius, P.A., Baumrucker, C.R., McNamara, J.P. & Bauman, D.E. (1980). γ-Glutamyl transpeptidase in rat mammary tissue (activity during lactogenesis and regulation by prolactin). *Biochem. J.*, **188**, 565–8.

Pocius, P.A., Clark, J.H. & Baumrucker, C.R. (1981). Glutathione in bovine blood: possible source of amino acids for milk protein synthesis. *J. Dairy Sci.*,**64**, 1551–4.

Poyet, P., Henning, S.J. & Rosen, J.M. (1989). Hormone-dependent beta-casein mRNA stabilization requires ongoing protein synthesis. *Mol. Endo.*, **3**, 1961–8.

Prehn, S., Herz, J., Hartmann, E., Kurzchalia, T.V., Frank, R., Roemisch, K., Dobberstein, B. & Rapoport, T.A. (1990). Structure and biosynthesis of the signal-sequence receptor. *Eur. J. Biochem.*, **188**, 439–45.

Proudfoot, N.J. & Whitelaw, E. (1988). Termination and 3′ end processing of eukaryotic RNA. *In Transcription and Splicing* (eds B.D. Hames, & D.M. Glover), pp 97–129. IRL Press, Oxford and Washington D.C.

Puente, J., Varas, M.A., Bechaus, F. & Sapag-Hagar, M. (1979). γ-Glutamyl transpeptidase activity and cyclic AMP levels in rat liver and mammary gland during the lactogenic cycle and in oestradiol-progesterone pseudo-induced pregnancy. *FEBS Letters*, **99**, 215–18.

Qasba, P.K. & Safaya, S.K. (1984). Similarity of the nucleotide sequences of rat alpha-lactalbumin and chicken lysozyme genes. *Nature*, **608**, 377–80.

Raghow, R., Gossage, D. & Kang, H.A. (1986). Pretranslational regulation of type I collagen, fibronectin and a 50 kDa noncollagenous extracellular protein by dexamethasone in rat fibroblasts. *J. Biol. Chem.*, **261**, 4677–84.

Reed, R. & Maniatis, T. (1988). The role of the mammalian branch point sequence in pre-mRNA splicing. *Genes & Develop.*, **2**, 1268–76.

Renkawitz, R. (1990). Transcriptional repression in eukaryotes. *TIG*, **6**, 192–7.

Rhoads, R.E. (1985). The cap structure of eukaryotic mesenger RNA and its interaction with cap-binding protein. *Mol. Subcell. Biol.*, **9**, 104–51.

Rhoads, R.E. (1988). Cap recognition and the entry of mRNA into the protein synthesis initiation cycle. *TIBS*, **13**, 52–6.

Rizzi, R., Caroli, A., Bolla, P., Acciaioli, A. & Pagnacco, G. (1988). Variability of reduced glutathione levels in Massese ewes and its effect on daily milk production. *J. Dairy Res.*, **55**, 345–53,

Robinson, B.G., Frim, D.M., Schwartz, W.J. & Majzoub, J.A. (1988). Vasopressin mRNA in the suprachiasmatic nuclei: daily regulation of polyadenylate tail length.*Science*, **241**, 342–4.

Roets, E., Verbeke, R., Massart-Leen, A.M. & Peeters, G. (1974). Metabolism of [^{14}C] citrulline in the perfused sheep and goat udder. *Biochem. J.*, **144**, 435–46.

Roets, E., Massart-Leen, A.M., Verbeke, R. & Peeters, G. (1979*a*). Metabolism of [U^{14}C;2,3^{3}H]-L-valine by the isolated perfused goat udder. *J. Dairy Res.*, **46**, 47–57.

Roets, E., Verbeke, R., Peeters, G., Axmann, H. & Proksch, G. (1979*b*). Metabolism of ornithine in perfused goat udder. *J. Dairy Sci.*, **62**, 259–69.

Roets, E., Massart-Leen, A.M., Peeters, G. & Verbeke, R. (1983). Metabolism of leucine by the isolated perfused goat udder. *J. Dairy Res.*, **50**, 413–24.

Rosen, J.M. (1987). Milk protein gene structure and expression. In *The Mammary Gland*, ed. M.C. Neville & C.W. Daniel. Plenum Publishing Corporation, London, pp. 301–22.

Rosen, J.M., O'Neal, D.L., McHugh, J.E. & Comstock, J.P. (1978). Progesterone-mediated inhibition of casein mRNA and polysomal casein synthesis in the rat mammary gland during pregnancy. *Biochemistry*, **17**, 290–7.

Rosen, J.M. & Shields, D. (1980). Posttranslational modifications of the rat mammary gland caseins: in vitro synthesis, processing, and segregation. In *Testicular Developments, Structure and Function*, ed. A. Steinberger & E. Steinberger. Raven Press, New York.

Rosen, J.M., Supowit, S.C., Gupta, P., Yu-Lee, L.Y. & Hobbs, A.A. (ed.). (1981). Regulation of casein gene expression in hormone-dependent mammary cancer. In *Hormones and Breast Cancer* Cold Spring Harbor Laboratory, pp. 397–424.

Rosen, J.M., Rodgers, J.R., Couch, C.H., Bisbee, C.A., David-Inouye, Y., Campbell, S.M.

& Yu-Lee, L.Y. (1986). Multihormonal regulation of milk protein gene expression. *Annals N.Y. Acad. Sci.*, **478**, 63–76.

Rosen, J.M., Mediona, D., Schlein, A.R., Eisenstein, R.S. & Yu-Lee, L.Y. (1988). Hormonal and cell-substratum regulation of casein gene expression at the posttranscriptional level. In *Steroid Hormone Action*, pp. 269–278. Alan R. Liss. Inc. New York.

Rosenfeld, M.G., Mermod, J.J., Attara, S.G., Swanson, L.W., Sawchenko, P.E., Rivier, J., Vaie, W.W. & Evans, R.M. (1983). Production of a novel neuropeptide encoded by the calcitonin gene via tissue-specific RNA processing. *Nature*, **304**, 129–35.

Rulquin, H. (1981). Etude méthodologique sur la mesure des differences arterio-vineuses mammaires chez vache laitiere. *Reprod. Nutr. Devel.*, **21**, 31–46.

Sambrook, J.F. (1990). The involvement of calcium in transport of secretory proteins from the endoplasmic reticulum. *Cell*, **61**, 197–9.

Shapiro, M.B. & Senepathy, P. (1987). RNA splice junctions of different classes of eukaryotes: Sequence statistics and functional implications in gene expression. *Nucleic Acids Res.*, **15**, 7155–74.

Sharp, P.A. (1985). On the origin of RNA splicing and introns. *Cell*, **42**, 397–400.

Sharp, P.A. (1987). Splicing of messenger RNA precursors. *Science*, **235**, 766–71.

Shaw, G. & Kamen, R. (1986). A conserved AU sequence from the 3′ unstransled region of GM-CSF mRNA mediates selective mRNA degradation. *Cell*, **46**, 659–67.

Simons, J.P., McClenaghabn, M. & Clark. J. (1987). Alteration of the quality of milk by expression of sheep beta-lactoglobulin in transgenic mice. *Nature*, **328**, 530–2.

Simpson, L. & Shaw, J. (1989). RNA editing and the mitochoindrial cryptogeness of kinetoplastid protozoa. *Cell*, **57**, 355–66.

Smeekens, S., Weisbeek, P. & Robinson, C. (1990). Protein transport into and within chloroplasts. *TIBS*, **15**, 73–6.

Soulier, S., Mercier, J.C., Vilotte, J.L., Anderson, J., Clark, J. & Provot, C. (1989). Characterization of genomic clones homologous to the alpha-lactalbumin gene in the bovine and ovine species. *Gene*, **83**, 331–8.

Steitz, J.A., Blade, D.L., Gerke, V., Parker, K.A., Krämer, A., Frendewey, D. & Keller, W. (1988). Functions of the abundant U-snRNPs. In *The Structure and Function of Small Nuclear Ribonucleoprotein Particules*, ed. M.L. Birnsteil, Springer-Verlag, Berlin, pp. 115–54.

Stewart, A.F., Bonsing, J., Beattie, C.W., Shah, F., Willis, I.M. & MacKinlay, A.G. (1987). Complete nucleotide sequences of bovine alpha-s2- and beta-casein cDNAs: comparison with related sequences in other species. *Mol. Biol. Evol.*, **4**, 231–41.

Stief, A., Winter, D.M., Strätling, W.H. & Sippel, A.E. (1989). A nuclear DNA attachment element mediates elevated and position-independent gene activity. *Nature*, **341**, 343–5.

Sutton, J.D. (1989). Altering milk composition by feeding. *J. Dairy Sci.*, **72**, 2801–14.

Talmadge, K., Vamvakopoulous, N.C. & Fiddes, J.C. (1984). Evolution of the genes for the beta subunits of human chorionic gonadotropin and luteinizing hormone. *Nature*, **307**, 37–40.

Terry, P.M., Banerjee, M.R. & Lui, R.M. (1977). Hormone—inducible casein messenger RNA in a serum-free organ culture of whole mammary gland. *Proc. Natl. Acad. Sci. USA*, **74**, 2441–5.

Teyssot, B. & Houdebine, L.M. (1980). Role of prolactin in the transcription of beta-casein and 28-S ribosomal genes in the rabbit mammary gland. *Eur. J. Biochem.*, **110**, 263–72.

Teyssot, B. & Houdebine, L.M. (1981*a*). Role of progesterone and glucocorticoids in the transcription of the beta-casein and 28—S ribosomal genes in the rabbit mammary gland. *Eur. J. Biochem.*, **114**, 597–608.

Teyssot, B. & Houdebine, L.M. (1981*b*). Induction of casein gene synthesis by prolactin and inhibition by progesterone in the pseudopregnant rabbit treated by colchicine without any simultaneous variations of casein mRNA concentrations. *Eur. J. Biochem.*, **117**, 563–8.

Thépot, D., Fontaine, M.L., Houdebine, L.M. & Devinoy, E. (1990). Complete sequence of the rabbit WAP gene. *Nucleic Acids Res.*, **18**, 3641.

Thépot, D., Fontaine, M.L., Houdebine, L.M. & Devinoy, E. (1991). Structure of the rabbit beta-casein gene. *Gene*, **97**, 301–6.

Threadgill, D.W. & Womack, J.E. (1990). Genomic analysis of the major bovine casein genes. *Nucleic Acids Res.*, **18**, 6935–42.

Vaux, D., Tooze, J. & Fuller, S. (1990). Identification by anti-idiotype antibodies of an intracellular membrane protein that recognizes a mammalian endoplasmic reticulum retention signal. *Nature*, **345**, 495–502.

Verbeke, R., Simeonov, S. & Peeters, G. (1967). Incorporation de la L-methionine-^{35}S dans la caséine par la glande mammaire. *Archives Intern. Physiol. et Biochim.*, **75**, 658–66.

Verbeke, R., Peeters, G., Massart-Leen, A.M. & Cocquyt, G. (1968). Incorporation of DL-[2-^{14}C] ornithine and DL-[5-^{14}C] arginine in milk constituents by the isolated lactating sheep udder. *Biochem. J.*, **106**, 719–24.

Verbeke, R., Roets, E., Massart-Leen, A.M. & Peeters, G. (1972). Metabolism of [U-^{14}C]-L-threonine and [U-^{14}C]-L-phenylalanine by the isolated perfused udder. *J. Dairy Res.*, **39**, 239–50.

Vilotte, J.L., Soulier, S., Mercier, J.C., Gaye, P., Hue-Delahaie, D. & Furet, J.P. (1987). Complete nucleotide sequence of bovine alpha-lactalbumin gene. Comparison with its rat counterpart. *Biochimie*, **69**, 609–20.

Vilotte, J.L., Soulier, S., Stinnakre, M.G., Massoud, M. & Mercier, J.C. (1990). Efficient tissue-specific expression of bovine alpha-lactalbumin in transgenic mice. *Eur. J. Biochem.*, **186**, 43–8.

Vilotte, J.L. Soulier, S., Printz, C. & Mercier, J.C. (1991). Sequence of the goat alpha-lactalbumin-encoding gene: comparison with its bovine counterpart and evidence of alpha-lactalbumin related sequences in the goat genome. *Gene*, **98**, 271–6.

Vina, J.R. & Williamson, D.H. (1981). Utilization of L-alanine and L-glutamine by lactating mammary gland of the rat. *Biochem. J.*, **196**, 757–62.

Vina, J.R., Puertes, I.R. & Vina, J. (1981). Effect of premature weaning on amino acid uptake by the lactating mammary gland of the rat. *Biochem. J.*, **200**, 705–8.

Vina, J.R., Puertes, I.R., Montoro, J.B., Saez, G.T. & Vina, J. (1985). Gamma-glutamyl amino acids as signals for the hormonal regulation of amino acid uptake by the mammary gland of the lactating rat. *Biol. of the Neonate*, **48**, 250–6.

Vina, J.R., Puertes, I.R., Rodriguez, A., Saez, G.T., & Vina, J. (1987). Effect of fasting on amino acid metabolism by lactating mammary gland: studies in women and rats. *J. Nutrition*, **117**, 533–8.

Von Heijne, G. (1983). Patterns of amino acids near signal-sequence cleavage site. *Eur. J. Biochem.*, **133**, 17–21.

Wefald, F.C., Devlin, B.H. & Williams, R.S. (1990). Functional heterogeneity of mammalian TATA-box sequences revealed by interaction with a cell-specific enhancer. *Nature*, **344**, 260–2.

Wickens, M. (1990). How the messenger got its tail: addition of poly(A) in the nucleus. *TIBS*, **15**, 277–81.

Wickens, M.P. & Stephenson, P. (1984). Role of the conserved AAUAAA sequence: four AAUAAA point mutants prevent mRNA 3′ end formation. *Science*, **226**, 1045–51.

Wieland, F.T., Gleason, M.L., Serafini, T.A. & Rothman, J.E. (1987). The rate of bulk flow from the endoplasmic reticulum to the cell surface. *Cell*, **50**, 289–300.

Wiest, D.L., Burkhardt, J.K., Hester, S., Hortsch, M., Meyer, D.I. & Argon, Y. (1990). Membrane biogenesis during B cell differentiation: most endoplasmic reticulum proteins are expressed coordinately. *J. Cell Biol.*, **110**, 1501–11.

Winkelmann, G., Bach, M. & Lührmann, R. (1989). Evidence from complementation essays in vitro that US snRNP is required for both steps of nRNA splicing. *EMBO J.*, **8**, 3105–12.

Wohlt, J.E., Clark, J.H., Derrig, R.G. & Davis, C.L. (1977). Valine, leucine and isoleucine metabolism by lactating bovine mammary tissues. *J. Dairy Sci.*, **60**, 1875–82.

Womack, J.E., Threadgill, D.W., Moll, Y.D., Faber, L.K., Foreman, M.L., Dietz, A.B.,

Tobin, T.C., Skow, L.C., Zneimer, S.M., Gallagher, D.S. & Rogers, D.S. (1989). Syntenic mapping of 37 loci in cattle. Chromosomal conservation with mouse and man. *Cytogenet. Cell Genet.*, **51**, 1109. (Abstract).

Yip, M.C.M. & Knox, W.E. (1972). Function of arginase in lactating mammary gland. *Biochem. J.*, **127**, 893–99.

Yoshimura, M. & Oka, T. (1989). Isolation and structural analysis of the mouse beta-casein gene. *Gene*, **78**, 267–75.

Yudilevich, D.L., Eaton, B.M., Short, A.H. & Leichtweiss, H.P. (1979). Glucose carriers at the maternal and fetal sides of the trophoblast in guinea pig placenta. *Amer. J. Physiol.*, **237**, C205–12.

Yu-Lee, L.Y. & Rosen, J.M. (1983). The rat casein multigene family.I.Fine structure of the gamma-casein gene. *J. Biol. Chem.*, **258**, 10794–804.

Yu-Lee, L.Y., Richter-Mann, L., Couch, C.H., Stewart, A.F., MacKinlay, R.G. & Rosen, J.M. (1986). Evolution of the casein multigene family conserved sequences in the 5′ flanking and exon regions. *Nucl. Acids Res.*, **14**, 1883–902.

Zagouras, P. & Rose, J.K. (1989). Carboxy-terminal SEKDEL sequences retard but do not retain two secretory proteins in the endoplasmic reticulum. *J. Cell. Biol.*, **109**, 2633–40.

Zak, R., Martin, A.F. & Blaugh, R. (1979). Assessment of protein turnover by use of radioisotope tracers. *Physiol. Reviews*, **59**, 407–47.

13

MOLECULAR MODELING AND GENETIC ENGINEERING OF MILK PROTEINS

TOM RICHARDSON, SANGSUK OH, RAFAEL JIMÉNEZ-FLORES

Department of Food Science and Technology, University of California, Davis, California 95616, USA

&

THOMAS F. KUMOSINSKI, ELEANOR M. BROWN & HAROLD M. FARRELL, JR

Eastern Regional Research Center, United States Department of Agriculture, Philadelphia, Pennsylvania 19118, USA

1. INTRODUCTION

The nutritional qualities of a food protein are primary considerations to dietitians, food scientists and consumers. However, in the development of novel food products, the physico-chemical or functional properties of proteins assume fundamental importance. Characteristic properties of a food are functions of its physico-chemical properties, and interactions of other food components with proteins play a crucial role in defining parameters of food character and quality. The variety, subtelty and versatility in protein properties are basic to the final structural features of a food product.

Food scientists have been constrained and limited in the development of new products by the inherent limitations of existing commodity proteins. However, research on physical, enzymatic, chemical and, more recently, genetic modifications of food proteins offers increased potential for improving the physical and functional properties of proteins in order to develop better quality foods or novel foods products.

The physical, chemical or functional characteristics of food proteins play a major role in defining the behavior of proteins during the processing, storage and consumption of a food. For example, different functional characteristics of the caseins to a large extent govern the coagulation, texture and organoleptic development in the preparation and aging processes that occur in cheeses. The stable foam formed as a result of whipping egg-white proteins is another example of how specialized protein functions can be used to prepare a variety of foods such as meringues and other whipped products. Some typical functional properties of proteins in food systems are listed in Table 1.

The inherent superiority of some proteins over others for a particular appli-

TABLE 1
Functional Properties Performed by Functional Proteins in Food Systems (Kinsella, 1982)

Functional property	*Mode of action*	*Food system*
Solubility	Protein solvation	Beverage
Water absorption and binding	Hydrogen bonding of water; entrapment of water (no drip)	Meat, sausages, breads, cakes
Viscosity	Thickening; water binding	Soups, gravies
Gelation	Protein matrix formation and setting	Meats, curds, cheese
Cohesion–adhesion	Protein acts as adhesive material	Meat, sausages, baked goods, pasta products
Elasticity	Hydrophobic bonding in gluten; disulphide links in gel	Meats, bakery
Emulsification	Formation and stabilization of fat emulsions	Sausages, bologna, soup, cakes
Fat absorption	Binding of free fat	Meats, sausages, doughnuts
Flavor-binding	Adsorption, entrapment, release	Simulated meats, bakery
Foaming	Form stable films to entrap gas	Whipped topping, chiffon desserts, angel food cakes

cation lies at the molecular level. Much remains to be learned about the mechanisms whereby the molecular structure of proteins govern their functionalities. However, since the primary amino acid sequences of proteins largely define their three-dimensional structures, it follows that alterations in the primary sequences of proteins may alter their three-dimensional characteristics while changing the functional characteristics. Well-defined, systematic structure–function studies should supply the food industry with the knowledge and tools to eventually design proteins of superior functionalities. Ultimately, of course, it is the nucleotide sequences of the DNA which define and restrict the primary sequences and thus the functionalities of proteins. Although much research has been expended on the effects of physical, chemical and enzymatic modifications of proteins to alter their functional properties, this review will focus on the potential for genetic engineering of protein functionality via manipulation of the DNA coding for the primary sequences of proteins.

In the genetic engineering of food proteins, an integrated and holistic approach is required. First, the complementary DNAs (cDNAs) coding for the appropriate protein must be isolated and cloned into appropriate microorganisms. Development of a reliable microbial expression system is a prerequisite for the production of cloned proteins allowing in-vitro studies on protein structure and function. Several well-developed systems of mutations are available for altering specific amino acids or whole domains in a protein. Mutant proteins produced in sufficient quantities in microbial systems offer a vehicle for in-vitro structure–function studies.

In conjunction with the microbial studies, the cDNAs coding for the proteins in question can be used as probes to isolate the entire structural or genomic gene containing promoters, regulatory sequences, exons and introns. Coupled to the appropriate promoter and regulatory sequences, cDNAs or genomic genes can then be introduced into immortalized bovine epithelial secretory cells which produce milk proteins and which are now becoming available (Hung *et al.*, 1991). This represents an intermediate stage between the microbial structure–function studies and the development of transgenic animals to produce genetically engineered milk.

The development of transgenic animals, the final stage in our progression, remains problematic. Much needs to be learned about the successful integration of selected genes into the genomes of receptor animals. Nevertheless, there have been a number of successes in the development of transgenic animals in recent years. The general strategy would be to initiate transgenic work in simpler, less expensive mouse systems before progressing to more cumbersome, difficult and expensive larger animal systems such as dairy cattle. It is the intent of this review to discuss the molecular aspects in the genetic engineering of the milk proteins before progressing to the more elaborate in-vivo systems.

1.1. Genetic Variants and Functional Characteristics of Proteins

From studies on the functionalities of natural genetic variants of proteins, it has become evident that a single amino acid substitution can have a marked effect on the physico-chemical properties of proteins. Perhaps the system that has been described in most detail has been the effects of genetic mutations on the functional behavior of mutant hemoglobins (Yutani *et al.*, 1985; Horbett & Brash, 1987), where effects of single amino acid substitutions on the functionality of proteins is illustrated by the extreme surface activities of hemoglobin mutants. In related studies where recombinant DNA technlogy was applied, a series of variant α-subunits of tryptophan synthase from *Escherichia coli* substituted by each of 20 amino acids at position 49 have been constructed (Kato & Yutani, 1988). The Gibbs free energy of unfolding in water of various mutants, which is a quantitative measure of the stability of protein conformation, varied between 0·72 and 1·92 times that of the wild-type protein. Kato and Yutani (1988) described the surface properties (surface tension, foaming and emulsifying properties) of tryptophan synthase α-subunits substituted by each of seven amino acids at interior position 49 and discussed the relationship between stability and functional propertes of the proteins.

In Table 2 differences in surface properties are shown between the wild-type tryptophan synthase α-subunits with Glu at interior position 49 and a mutant protein where Ile was substituted at position 49. The free energy for unfolding of the Ile mutant was approximately 8–12 kcal/mole higher than for the wild-type, indicating that the mutant does not unfold at the interface to the same extent as the wild-type under similar circumstances. This is reflected in a higher surface tension, in reduced foaming power and foam stability and decreased emulsifying activity. Thus, the hydrophobic Ile substituted in the interior of the protein results in a more stable protein when compared to the wild-type tryptophan synthase α-subunit. Here again is an example of how a single amino acid substitution can have a marked effect on the functional characteristics of a protein.

Consideration of the thermodynamics of unfolding of lysozyme (Creighton, 1983) leads to the observation that enthalpic (H) and entropic (TS) contributions to the free energies of unfolding compensate each other, so that the free energy difference between the two is relatively small. The free energies of the folded and unfolded states also differ slightly so that the observed net stability is no greater than 16 kcal/mole. This stability of protein unfolding is in the range found by Kato and Yutani (1988) for selected substitutions of single amino acids in the protein mutants that were studied. Thus, it can be expected that one or two amino acids judiciously substituted in a protein could have a marked effect on the functional characteristics of a protein.

Indeed, genetic variants of the milk proteins which differ in primary amino acid sequences by 1 to 13 amino acids behave differently in milk systems. In Table 3 are summarized some of the functional differences found in cheese and concentrated milk resulting from variations in the amino acids of the variants. It is evident that variations in charge, hydrophobicity and other physical properties as a function of amino acid substitutions can have a significant effect on the behavior of the milk system ranging from greater heat stability in concentrated milk to variations in yield and texture of cheese.

This paper is an attempt to rationalize a basis for genetic engineering of the milk system to alter its functional characteristics. Appropriate changes in ratios of milk proteins or the introduction of novel proteins into bovine milk can lead to the improvement of existing products and the development of new products. A survey of the literature indicates that, in general, the addition of 20 to 30% of a normal or modified casein based on the normal complement of a particular casein is sufficient to alter the physical behavior of milk (Jimenez-Flores & Richardson, 1988).

For example, the heat stability of milk may be enhanced by addition of κ-casein to the normal complement of that protein in milk, β-Casein added to milk can increase the firmness of curd from enzymically coagulated milk. Dephosphorylated β-casein added to artificial micelles can increase clotting time and decrease syneresis of the curd when compared with those of controls with no additions. By

TABLE 2

Surface Properties of Wild-Type Tryptophan Synthase α Subunit Compared to Mutant Protein Where Ile_{49} is substituted for Glu_{49}.

Residue at position 49	*Surface tension (dynes/cm)*		*Foaming power (ms/cm)*		*Foam stability (min)*		*Emulsifying activity (OD 500)*
	pH 7	*pH 9*	*pH 7*	*pH 9*	*pH 7*	*pH 9*	*pH 9*
Glu (wild type)	49·2 ± 0·3	45·6 ± 0·3	2·55 ± 0·07	2·94 ± 0·6	2·4 ± 0·1	10·9 ± 0·2	0·900 ± 0·04
Ile (mutant)	51·7 ± 0·6	48·1 ± 0·1	2·22 ± 0·05	2·20 ± 0·4	1·5 ± 0·1	4·2 ± 0·3	0·755 ± 0·11

TABLE 3
Consequences of Genetic Variants in Cows' Milk

Protein	*Mutant A.A.*	*Concentration in milk (g/liter)*[a]	*Effect of the genetic variant on the composition of milk*	*Functional differences in cheese*	*Functional differences in concentrated milk*	*Possible reasons*
α_{S1}-Casein-A	Deletion of 13 a.a. (14 to 26)	—		—Variant A yields a softer curd and cheese body (Creamer *et al.*, 1988)		Loss of a hydrophobic domain and of a phosphorylation site in variant A
α_{S1}-Casein-B	Glu (192)	8·9	—Correlates with higher milk yield (Ng-Kwai-Hang *et al.*, 1984)		—No significant differences among variants (McLean *et al.*, 1987)	
α_{S1}-Casein-C	Gly (192)	9·2	—Lower concentration of κ-casein (McLean *et al.*, 1984)			
β-Casein A1	His(67) Ser (192)	9·6			—No significant differences among variants (McLean *et al.*, 1987)	
β-Casein A2	Pro(67) Ser(192)	9·7				
β-Casein B	His(67) Arg(192)	10·9	—Significant increase in fat (McLean *et al.*, 1984)	—Variant B gives faster and firmer coagulum (Schaar *et al.*, 1985)		An increased quantity of the B variant in milk and/or increase in positive charge of the B variant may enhance casein interactions
κ-Casein A	Thr(136) Asp (148)	3		—Variant A gives a longer clotting time		
κ-Casein B	Ile(136) Ala(148)	3·7	—Significant increase in concentration of κ-casein and decrease in citrate (McLean *et al.*, 1984)	—Variant B gives faster and firmer coagulation, and better yield (Schaar *et al.*, 1985)	—Variant B confers greater heat stability (McLean *et al.*, 1987; Feagan *et al.*, 1972)	Variant B is present in larger amounts and this may result in smaller micelles, along with differences in composition
β-Lactoglobulin A	Asp(64) Val(118)	4·1				
β-Lactoglobulin B (for the var AB)	(Gly(64) Ala(118)	2·9 2·6	—Significant increase in fat, total solids, casein and a decrease in whey protein (McLean *et al.*, 1984)	—Significant increase in yield when κ-casein B is present	—Variant B confers greater heat stability (McLean *et al.*, 1987)	A marked decrease in the quantity of var. B may result in less interaction with other proteins upon heating

[a] McLean *et al.* (1984) and Ng-Kwai-Hang *et al.* (1987).

controlling gene dosage and the regulatory regions of the casein genes in the mammary gland, it should be possible to obtain the 20% to 30% of additional or novel caseins to modify the functionality of the milk protein system. Depending upon the rate limiting step in milk protein biosynthesis, it may be possible to affect milk protein ratios without altering total milk proteins by inserting appropriate genes into the cow's genome. This paper contains a discussion of strategies possible for the genetic engineering of milk to control its functional properties.

The genetic engineering of food proteins, in general, has been hampered by the lack of information pertaining to their three-dimensional structures. Knowledge of the three-dimensional structures of food proteins would facilitate genetic engineering by allowing precise structure–function studies in food systems. For example, X-ray diffraction analyses of crystalline enzymes have made possible alterations in the stabilities, kinetic constants and specificities of various enzymes (Jimenez-Flores *et al.*, 1986).

Although the primary sequences of all the major bovine milk proteins are known (Eigel *et al.*, 1984), there is limited information on their secondary and tertiary structures. Holt and Sawyer (1988) recently published secondary structural predictions of bovine caseins based on the best fits between combinations of predictive models. The three-dimensional structure for bovine β-lactoglobulin has been determined using X-ray diffraction techniques (Pervaiz & Brew, 1985; Sawyer *et al.*, 1985; Papiz *et al.*, 1986; Monaco *et al.*, 1987; Godovac-Zimmerman, 1988). Less is known about the three-dimensional structure for bovine α-lactalbumin, although crystallographic analysis of baboon α-lactalbumin has been performed at low resolution (Smith *et al.*, 1987).

Since the bovine caseins have not been crystallized, it has not been possible to study their three-dimensional structures using high resolution X-ray diffraction methods. Consequently, less defined techniques such as circular dichroism (Creamer *et al.*, 1981) have been employed to gain somc knowledge of the secondary structures of the caseins. However, to obtain a more refined picture of the three-dimensional structures of the bovine caseins, a multiplicity of techniques must be ultilized. Researchers at the Eastern Regional Research Center of the United States Department of Agriculture have used a variety of techniques including energy minimization methods to provide us with a more detailed description of the three-dimensional structures of bovine α_{s1}- and κ-caseins. These more refined structures will allow us to define in more detailed and rational ways the effects of genetic engineering modifications on the structure and function of caseins. The following section focuses on the details of the three-dimensional structures of α_{s1}-and κ-caseins and of α-lactalbumin and β-lactoglobulin.

2. Molecular Modeling of Bovine Milk Proteins

At the heart of the skim milk system is a unique biocolloid, the casein micelle. This colloidal complex is in dynamic equilibrium with its environment. Changes in the state of the casein micelle system occur during milk secretion and processing

(Schmidt, 1982; Farrell, 1988). Because of its innate importance to the milk system, the casein micelle has been studied extensively. From this large basic set of physical data, various investigators have attempted with varying degrees of success to formulate models which will adequately account for the properties of the casein micelle and allow for predictions of its behavior in real systems (Schmidt, 1982; Farrell, 1988). These models often fall short because of a serious gap in the data basis set.

The combination of X-ray crystallography and molecular biology has contributed greatly to our understanding of the mechanisms of action of globular proteins. While the techniques of molecular biology are now being applied to casein, crystallographic structures will probably not be realized. It is this gap in our knowledge, the reconciliation of the physical chemical data for the caseins with molecular structure, that the authors have attempted to breech with the tools of three-dimensional molecular modeling. These models are not an end in themselves; it is hoped that they will represent a new starting point in the examination and exploration of casein structure and function.

2.1. Methodology in Molecular Modeling

2.1.1. Predictions of Secondary Structures

Selection of appropriate conformational states for the individual amino acid residues was accomplished by comparing the results of sequence-based predictive techniques, primarily those of Chou and Fasman (1978), Garnier *et al.* (1978) and Cohen *et al.* (1983, 1986). Assignments of secondary structure (α-helix, β-sheet or β-turn) for the amino acid sequences were made when either predicted by more than one method, or strongly predicted by one and not predicted against by the others. In addition, because of the large number of proline residues in the caseins, proline-based turn predictions were made using the data of Benedetti *et al.* (1983) and Ananthanarayanan *et al.* (1984).

2.1.2. Molecular Modeling

Three-dimensional structures of the individual caseins were approximated using molecular modeling methods with an Evans and Sutherland PS390* interactive computer graphics display driven by Sybyl-Mendel (Tripos) software. Segments of the individual proteins ranging in length from 80 to 100 residues were constructed, and assigned ϕ and ψ angles characteristic of the respective predicted structures. All ψ angles were assigned the conventional trans configuration. In addition, aperiodic structures are in the extended rather than totally random configuration. The Sybyl-Mendel subroutine 'SCAN' was used, on the side chains only, to adjust torsion angles and relieve bad van der Waals contacts. The individual pieces were then joined together to produce the total polypeptide model.

*Mention of a brand name does not imply endorsement by the US Department of Agriculture over others of a similar type.

2.2. Results of Molecular Modeling

2.1.1. Rationale for Generation of Secondary Structure Models

Various methodologies for sequence-based secondary structural predictions are currently available (Chou & Fasman, 1978; Garnier *et al.*, 1978; Cohen *et al.*, 1983, 1986). These methods have been applied to the caseins (Loucheux-Lefebvre *et al.*, 1978; Creamer *et al.*, 1981; Raap *et al.*, 1983). In our initial approaches, a sequence-based prediction was generated from each of three basis sets for α_{s1}- and κ-caseins. The results were ambiguous; for some portions of the sequences, structures (α-helix, etc.) were consistently predicted, while in others they were not. The relatively high proline contents of the caseins pose a problem since this residue, while occasionally found in both β-sheet and α-helical structures, is generally not favorable to either. In attempting to generate a consensus sequence-based prediction, the authors first focussed upon solving the proline problem in a way commensurate with known behavior of this residue in proteins and model peptides (Benedetti *et al.*, 1983; Ananthanarayanan *et al.*, 1984; Rose *et al.*, 1985).

The proline residues of the caseins are somewhat evenly dispersed throughout the sequences, ruling out both Type I and Type II polyproline structures (Schmidt, 1982; Farrell, 1988). It has been documented, however, in model peptides that proline frequently occupies the second position of either a four residue β-turn or a three residue γ-turn (Benedetti *et al.*, 1983; Ananthanarayanan *et al.*, 1984; Rose *et al.*, 1985). Moreover, recent Raman and FTIR spectroscopic analyses of the caseins have shown that up to 35% of the residues in caseins appear to be in β- or γ-turns (Byler *et al.*, 1988; Byler & Farrell, 1989). To assess the probability that proline residues in the caseins might be located in reverse turns, the authors examined the tetrapeptides containing Pro in the second position for similarity with peptides known to form β- and γ-turns (Benedetti *et al.*, 1983; Ananthanarayanan *et al.*, 1984). Most of the proline residues were predicted to be in a turn of some type. When the models were built, normal β-turns containing proline residues generally resulted in unfavorable van der Waal's contacts with surrounding residues, leading the authors to assign most proline residues initially to the γ-turn conformation. β-Turns, other than those based on proline, were predicted by the Chou–Fasman (1978) and Cohen *et al.* (1986) methods. The total number of amino acid residues predicted to be in turns was only slightly in excess of the total predicted by spectroscopic methods (Table 4).

In a similar fashion, 'consensus' scores for α-helix and β-sheet were arrived at by choosing from those regions having the highest predicted probability of a given structure to yield values in accord with the FTIR and Raman data (Byler *et al.*, 1988; Byler & Farrell, 1989). In this case, all residues previously assigned to proline-based turns were eliminated first from consideration in α-helical or extended β-structures. The net results of these calculations for κ- and α_{s1}-caseins are compared with spectroscopic data in Table 4. Finally, all residues not included in these periodic structures were then considered to be in an extended aperiodic conformation. The initial conformational assignments for κ-casein are shown with its sequence in Fig. 1, and for α_{s1} -casein in Fig. 2. The κ-casein model does not

TABLE 4
Comparison of Adjusted Sequence Based Predictions with Spectroscopic Data

Sample		*% Helix*	*% β-Structure*	*% Turns*	*% Unspecified*
α_{s1}-Casein	Raman[a]	8–13	18–20	29–35	33–40
	Initial	15	22	45	18
κ-Casein	FTIR[b]	10	43	15	32
	Initial	16	27	37	20

[a] Byler *et al.* (1988).
[b] Preliminary data with 10% error (Byler & Farrell, 1989).

differ widely from that of Loucheux-Lefebvre *et al.* (1978), and with the exception of the β- and γ-turns, the α_{s1}-model is comparable to that of Creamer *et al.* (1981).

2.2.2. Rationale for Generation of Three-Dimensional Models
The secondary structural assignments which had been reconciled with FTIR and Raman spectroscopic data were used as a point of departure for the generation of three-dimensional models. Idealized ϕ and ψ angles assigned initially for each structural element are given in Table 5. Protein segments consisting of 80 to 100 residues were constructed, examined visually for 'bad contacts', and adjusted for unfavorable van der Waals' interactions of the side chains. The individual pieces were then joined to produce the total polypeptide, and torsional angles again adjusted to remove unfavorable van der Waals' contacts among the side chains. The initial backbone conformation was maintained throughout this procedure.

2.2.3. Three-Dimensional Molecular Model of κ-Casein
The computer model generated as described above for κ-casein is shown in Plate I. In descriptive terms, the protein can be thought of as being represented by a 'horse and rider'. The amino-terminal 110 to 120 residues represent the 'horse' while the carboxy-terminal portion represents the 'rider'. Two distinct legs are seen in the 'horse' portion of the model. These legs are generated by β-sheet regions comprising residues 10 to 25; 29 to 34; 39 to 45 and 49 to 55, which are connected by β- or γ-turns. The overall dimensions of the κ-casein monomer predicted by this model are 8·0 × 6·2 × 5·98 nm. The following section is an attempt to reconcile some known features of the chemistry of κ-casein with this molecular model. To aid in this discussion, the backbone-only model is given in Fig. 3.

2.2.4. Chemistry of κ-Casein and the Three-Dimensional Model

2.2.4.1. Chymosin (rennet). κ-Casein differs from α_s-, and β-caseins in that it is soluble over a broad range of calcium ion concentrations (Waugh & Von Hipple, 1956). It was this calcium solubility that led Waugh and Von Hipple, upon discovering the κ-fraction, to asign to it the role of casein micelle stabilization. It is also the κ-casein fraction that is most readily cleaved by chymosin (Kalan & Woychik, 1965); the resulting products are termed para-κ-casein and the macro-

```
1                                                      20
Q E Q N Q E Q P I R C E K D E R F F S D K I A K Y I P I Q Y

                  40                                        60
V L S R Y P S Y G L N Y Y Q Q K P V A L I N N Q F L P Y P Y

                                       80
Y A K P A A V R S P A Q I L Q W Q V L S D T V P A K S C Q A

                  100                                       120
Q P T T M A R H P H P H L S F M A I P P K K N Q D K T E I P

                                       140
T I N T I A S G E P T S T P T I E A V E S T V A T L E A S P
                                                        P
                  160               169
E V I E S P P E I N T V Q V T S T A V
```

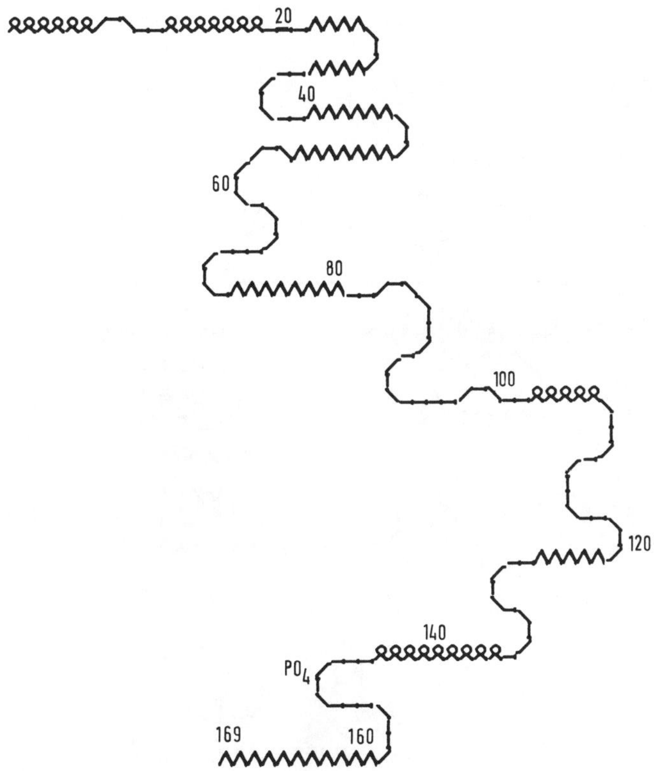

Fig. 1. Sequence of κ-casein B (top). Summary of initial secondary structural assignments made for κ-casein (bottom).

peptide. It would appear that κ-casein is the key to micelle structure in that it stabilizes the calcium insoluble α_{s1}- and β-caseins.

The action of chymosin on the micelle is primarily the hydrolysis of the highly sensitive Phe–Met peptide bond (residues 105–106) of κ-casein. Sequence data

```
1                                                        20
R  P  K  H  P  I  K  H  Q  G  L  P  Q  E  V  L  N  E  N  L  L  R  F  F  V

                                          40
A  P  F  P  Q  V  F  G  K  E  K  V  N  E  L  S  K  D  I  G  S  E  S  T  E
                                                            P     P
                           60
G  D  A  M  E  D  I  K  Q  M  E  A  E  S  I  S  S  S  E  E  I  V  P  N  S
                                       P     P  P  P                    P
               80                                                       100
V  E  Q  K  H  I  Q  K  E  D  V  P  S  E  R  Y  L  G  Y  L  E  Q  L  L  R

                                                         120
L  K  K  Y  K  V  P  Q  L  E  I  V  P  N  S  A  E  E  R  L  H  S  M  K  E
                                          P
                                             140
G  I  H  A  Q  Q  K  E  P  M  I  G  V  N  Q  E  L  A  Y  F  Y  P  E  L  F

                              160
R  Q  F  Y  Q  L  D  A  Y  P  S  G  A  W  Y  Y  V  P  L  G  T  Q  Y  T  D

               180                                                  199
A  P  S  F  S  D  I  P  N  P  I  G  S  E  N  S  E  K  T  T  M  P  L  W
```

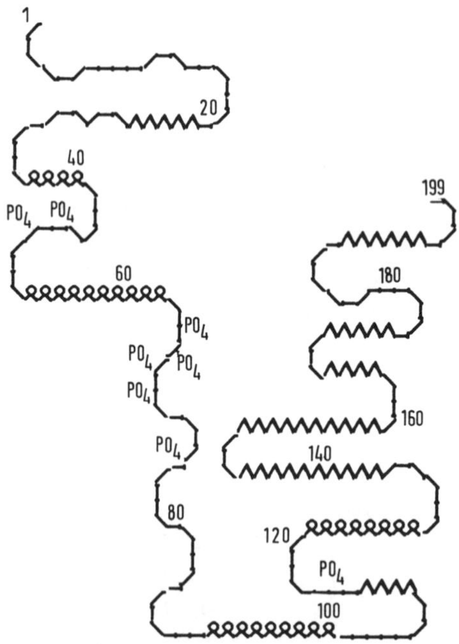

Fig. 2. Sequence of α_{s1}-casein B (top). Summary of initial secondary structural assignments made for α_{s1}-casein (bottom).

(Mercier *et al.*, 1973) show this bond to be in a proline-rich region of the molecule between Pro–His–Pro (residues 99–101) and Pro–Pro (residues 109–110), which perhaps accounts for the high susceptibility of this specific bond to chymosin. From studies of model peptides, it has been suggested that the residues lying between Pro-101 and Pro-109 occur in a β-sheet structure (Raap *et al.*, 1983). Using

TABLE 5
Dihedral (ϕ, ψ) Angles Assigned to Specific Conformational States in the Initial Structures for α_{s1}- and κ-Casein

Structures		*Angle Assignments*	
		ϕ	ψ
α-Helix		−58	−47
β-Sheet		−139	135
β-Turn	N	−139	135
	N + 1	−60	−30
	N + 2	−90	0
	N + 3	−139	135
γ-Turn	N	135	−69
	N + 1	−75	59
	N + 2	81	−126

other predictive methods, there is an equal chance that these residues could be in an α-helical conformation. In either case the Pro–X–Pro and Pro–Pro residues cause the formation of a kink which neatly presents the otherwise hydrophobic Phe–Met on the surface of the molecule. In the authors' model, the α-helix is shown, which represents the 'horse's bit' in the descriptive 'horse and rider' model. From model-building considerations, this sequence represents the minimum number of residues for a stable α-helix. Cleavage of the Phe–Met (105–106) bond would render the helical conformation untenable, the helix would unwind and a considerable amount of configurational entropy would be added to the hydrolysis reaction. The change from β-sheet to extended conformation, though significant, would be lesser in its energy contribution. In either case, helix or sheet, the Pro–Pro and Pro–X–Pro turns present this segment on the surface and make it readily accessible to chymosin. All models for casein micelle structure must in some way account for this feature of κ-casein.

2.2.4.2. Hydrophobic interactions. Earlier speculation (Hill & Wake, 1969) that κ-casein might be a linear amphiphile seems to be only partially true. The amino-terminal fifth of the molecule has a relatively high charge frequency (0·34); however, the net charge is zero and this part of the protein is also relatively hydrophobic. Residues 35 to 68 represent an exceptionally hydrophobic area with almost no charge. It is precisely within this region that the majority of the residues found in the 'legged' structures of the κ-casein molecule occur. These non-stranded, highly hydrophobic β-sheets make ideal sites for sheet–sheet interactions with other κ-casein molecules or with hydrophobic domains of α_s- and β-caseins. Indeed, the concentration-dependent reaction profile of the reduced form of purified κ-casein can be fitted with a model for polymerization with a critical micelle concentration (Vreeman *et al.*, 1981). Several investigators (Rose, 1968; Downey & Murphy, 1970; Ali *et al.*, 1980; Davies & Law, 1983) have noted that β- and κ-caseins diffuse out of the casein micelle at low temperatures. As one decreases the temperature, hydrophobic stabilization energy decreases, and κ-casein is able to

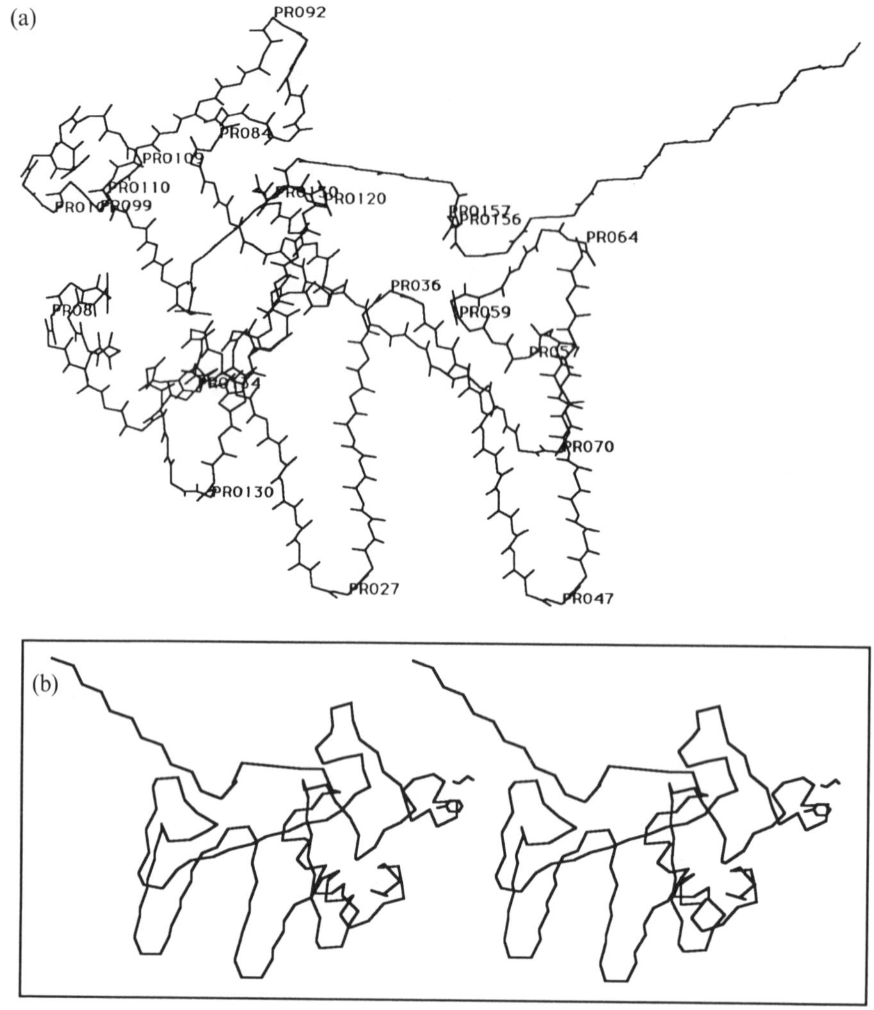

Fig. 3. (a) Chain trace of κ-casein, prolines indicated. (b) Stereo view of the three-dimensional molecular model of κ-casein, showing the backbone without side chains.

dissociate from the micelle. Finally, the importance of the hydrophobic region in casein–casein interactions can be supported by the research of Woychik and Wondolowski (1969). Of nine tyrosine residues in κ-casein, seven are located between residues 35 and 68; nitration of seven tyrosines in κ-casein severely inhibited its ability to stabilize α_{s1}-casein (Woychik & Wondolowski, 1969). As can be seen in Plate I, the 'legged' structures are constituted from this region and contain seven tyrosine residues.

2.2.4.3. Sulfhydryl-disulfide interactions. κ-Casein contains two Cys residues. Whether these can form intra- or intermolecular disulfide bonds and the effects of such bonding on micelle stabilization have not been clearly established. The occurrence of free sulfhydryl groups in the milk–protein complex has been reported by Beeby (1964), but not by others (Jolles *et al.*, 1962). Swaisgood *et al.* (1964) reported significant S–S cross-linking in purified κ-casein. However, Woychik *et al.* (1966) demonstrated that reduced and alkylated κ-casein stabilized α_{s1}-casein against calcium precipitation as well as native κ-casein. Pepper and Farrell (1982) found that in soluble whole casein, in the absence of Ca^{2+}, κ-casein occurred as a high molecular weight polydisperse complex. At low protein concentrations, this complex could be separated from the other caseins by size exclusion gel chromatography in the absence of urea. The addition of reducing agents converted the κ-casein to the sulphydryl-form which exhibited concentration-dependent associations, both with itself and with other caseins. Thus, although κ-casein represents only 13% of the casein, many κ-casein molecules must be somewhat contiguous in the micelle in order to form these disulfide-linked aggregates. It is interesting to note that Cys-11 is located between two segments predicted to be in α-helical conformations and is found on the 'rider's' left, while Cys-88 is located in a predicted β-turn on the opposite side. In the authors' model both of these residues are located near the surface of the molecule and are directed away from each other. This could account for the ability of the κ-casein molecule to form the inter-chain disulfide-bonded polymers as discussed above.

2.2.4.4. Sites for glycosylation and phosphorylation of κ-casein. Of the major components of the casein complex, only κ-casein can be glycosylated. Nearly all of the carbohydrate, as well as the phosphate, associated with κ-casein is bound to the macropeptide (Eigel *et al.*, 1984), which is the highly soluble portion formed by chymosin hydrolysis. The major site for glycosylation, Thr-133, is found in the authors' model on the back of the 'horse' and is on a β-turn. The sites of phosphorylation, Ser-149 and Thr-145, are on the back portion and are also found in β-turns.

2.2.5. Three-Dimensional Molecular Model of α_{s1}-Casein

The computer model generated for α_{s1}-casein as described above is shown in Plate II where it is displayed from carboxy- to amino-terminal (left to right). The best representation shows the molecule to be composed (right to left) of a hydrophilic amino-terminal portion, a segment of rather hydrophobic β-sheet, the phosphopeptide segment and then a very hydrophobic carboxy-terminal domain. For clarity, the backbone without side chains is shown in Fig. 4.

2.2.6. Chemistry of α_{s1}-Casein and the Three-Dimensional Model

2.2.6.1. Sites of phosphorylation in α_{s1}-casein. α_{s1}-Casein B is a single polypeptide chain with 199 amino acid residues and a molecular weight of 23 619 (Mercier *et al.*, 1971). The α_{s1}-B molecule contains eight phosphate residues, all in the form of serine monophosphates. Seven of these phosphoserine residues are clustered in an acidic portion of the molecule bounded by residues 43 and 80 (the second fifth

(a)

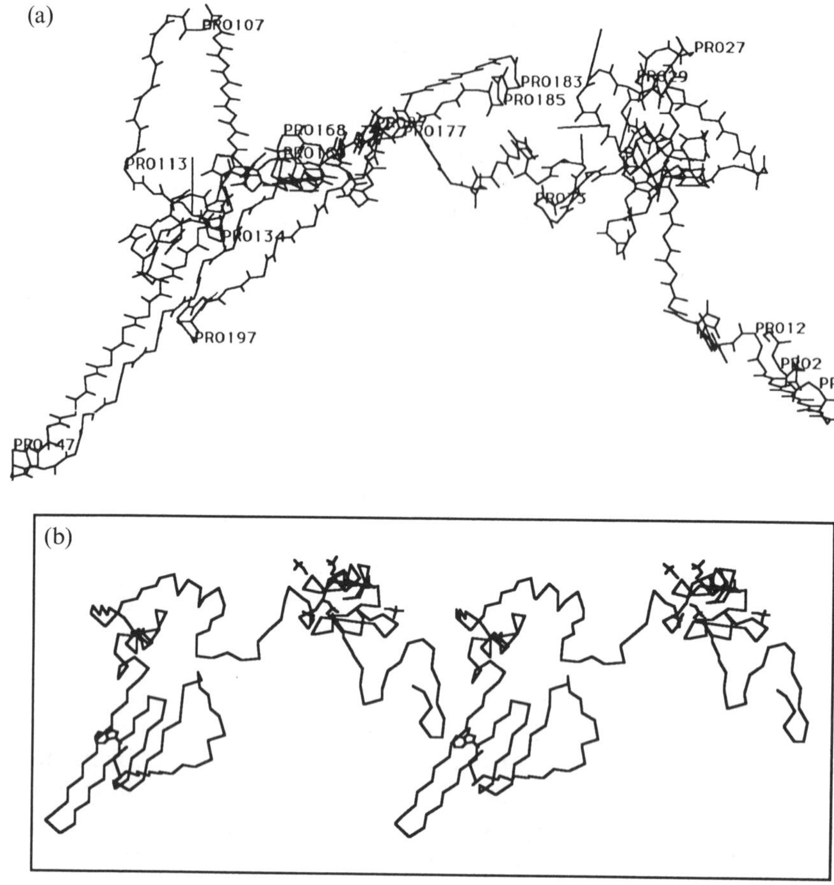

(b)

Fig. 4. (a) Chain trace of α_{s1}-casein, prolines indicated. (b) Stereo view of the three-dimensional molecular model of α_{s1}-casein, showing the backbone without side chains.

of the molecule from the amino-terminal end). This highly acidic segment contains 12 carboxylic acid groups as well as the seven phosphoserines. The model shows these residues to be located on β-turns which is compatible with other known phosphorylated residues in proteins. This cluster forms a highly hydrophilic domain on the right shoulder of the molecule.

2.2.6.2. The α_{s1}-casein A deletion and chymosin cleavage sites. The rare α_{s1}-casein A genetic variant exhibits interactions that are highly temperature dependent. The A variant is the result of the sequential deletion of 13 amino acid residues bounded by residues 13 and 27; the majority of these deleted amino acids are apolar (Farrell *et al.*, 1988) but Glu-14 and -18 and Arg-22 are also deleted. The deleted segment encompasses a region predicted to be in a β-sheet. This sheet provides a spacer-arm between the hydrophilic amino-terminal region (located at a five o'clock position in Fig. 4) and the phosphopeptide region. Deletion of this spacer-arm brings the hydrophilic section closer to the phosphate-rich shoulder. In addition, the Phe–Phe bond (residues 23 and 24) is removed; this represents a major chymosin cleavage site and its absence may account for the poor quality products prepared from α_{s1}-casein A milks (Thompson *et al.*, 1969). Additional chymosin cleavage sites are located between residues 31 and 32, 149 and 150, and 169 and 170. The site between residues 31 and 32 is relatively exposed, thus facilitating enzymatic attack. The latter two sites, 149–150 and 169–70 may be less exposed and could be involved in sheet–sheet hydrophobic interactions.

2.2.6.3. Hydrophobic interactions. For α_{s1}-casein, the high degree of hydrophobicity exhibited by the segment containing residues 100–199 is probably responsible, in part, for the pronounced self-association of the α_{s1}-casein monomer in aqueous solution (Schmidt, 1982). This self-association approaches a limiting size under conditions of lowered ionic strength; the highly charged phosphopeptide region can readily account for this phenomenon through charge repulsions. It is noteworthy that at ionic strengths $>0{\cdot}5$, α_{s1}-casein is salted out of solution at 37°C. This hydrophobic segment (residues 100–199) of α_{s1}-casein represents a hydrophobic domain and contains a segment of non-stranded β-sheet reminiscent of those found in κ-casein. This region is rich in tyrosine and as noted above for κ-casein, nitrosation of α_{s1}-casein leads to decreased stability in reconstituted micellar structures (Woychik & Wondolowski, 1969).

2.2.7. Refinement of Casein Structures

It must be stressed that the structures presented in this work represent preliminary models. They have been partially refined and poor van der Waals' side chain contacts removed, but they are not energy-minimized structures. Thus, electrostatic interactions, hydrogen bond formation, and backbone interactions are not taken into account. The refinement of these structures through the use of tools such as the Kollman Force Field will be the thrust of future work. However, it is imperative to note that even this nonrefined molecular modeling technique (when combined with predicted secondary structures and spectroscopic results) can reveal important structure–function relationships. This information can in turn be used for designing new functional proteins using site-directed mutagenesis.

2.2.8. Molecular Modeling and the Whey Proteins

We have been concerned primarily with the application of molecular modeling to the caseins. As noted in the introduction, these proteins are not amenable to crystallization and so to advance our knowledge we must rely upon predictive techniques and molecular biology. Whey proteins, on the other hand, have been crystallized and X-ray studies are in progress for both α-lactalbumin and β-lactoglobulin.

The X-ray structure for α-lactalbumin (baboon) has been refined and dimensions of the Ca^{2+} binding loop published (Stuart *et al.*, 1986). Molecular modeling holds promise of using this structure as the basis for constructing a bovine model. Force field calculations could then be used to produce an energy-minimized structure.

This structure could in turn be used to predict areas where point mutations might alter the conformation of the molecule. In addition, the wealth of chemical information developed on the interaction of this protein with galactosyl transferase (Berliner & Johnson, 1988) and its binding of metal ions could be correlated with such a molecular model.

Low resolution studies of β-lactoglobulin have been published (Papiz *et al.*, 1986). Although not complete at this time, further refinements of side chain–side chain interactions could be done by force field calculations to yield an improved model. In addition, these modeling predictions could help lead the way for refinement of the crystal structures. Such a model for the major whey protein β-lactoglobulin would open up new areas of research. The three-dimensional model could be used to predict, in advance, sites where replacement of one residue by another might produce a protein with altered physical properties. Site mutations and genetic engineering could then achieve production of the protein in, e.g. *E. coli* or yeast and the physical chemical properties of the altered proteins could be tested. Although long-term in scope, such methodologies offer the possibility of successfully producing altered proteins with tailor-made functional properties.

3. Genetic Engineering and Milk Proteins

3.1. Recombinant DNA Technology

Since the introduction of recombinant DNA technology, it is possible to alter, in a controlled manner, efficiently and systematically, the sequence of nucleotides in a segment of DNA. Genetic engineering of proteins involves the use of genetic manipulations to alter the coding sequences of a cloned gene. The modified gene product may then be expressed in a suitable system to produce essentially unlimited quantities of mutant protein.

Site-directed mutagenesis, a powerful tool for the preparation of mutant proteins, depends upon the use of synthetic nucleotides as highly specific mutagens. Synthesized oligonucleotides may be purified by gel electrophoresis or high performance liquid chromatography and used as hybridization probes (Wallace *et al.*, 1981; Simons *et al.*, 1982). Multiple amino acid substitutions can be produced by incorporating further mismatches into the oligonucleotide (Carter *et al.*, 1985; Kunkel *et al.*, 1987) or by performing sequential cycles of mutagenesis (Leatherbarrow & Fersht, 1986). The ideal site-directed mutagenesis produces a high frequency of mutants in a short time with minimum effort.

3.2. Antisense RNA Technology

Antisense RNAs are small, diffusible transcripts that pair with specific complementary target RNAs, resulting in hybridization arrest of translation, thus controlling their function or expression. Naturally-occurring antisense RNAs have been observed which control plasmid replication and conjugation (Light & Molin, 1983;

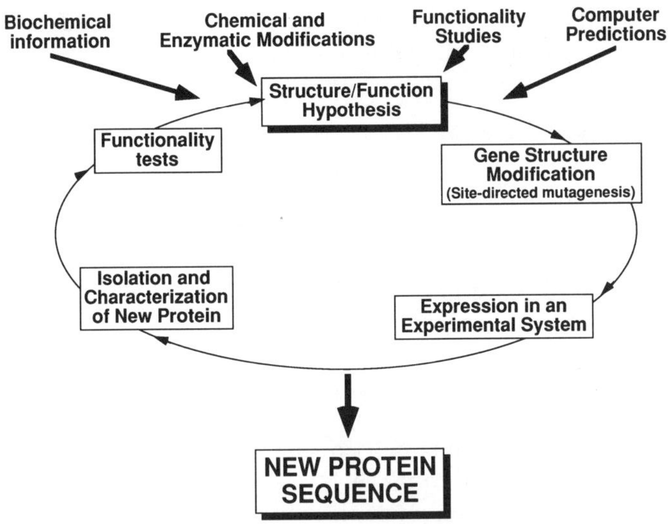

Fig. 5. Schematic diagram for protein engineering via directed mutagenesis, approach based on protein structure function (Creamer *et al.*, 1988).

Dempsey, 1987; Willetts & Skurray, 1987), transposon regulation (Simons & Kleckner, 1983), temporal control of bacteriophage development (Susskind & Youderian, 1983), and control of bacterial genes (Mizuno *et al.*, 1984). Because antisense RNA technology is based on blocking the informational flow from DNA via RNA to protein, it is possible to manipulate eukaryotic gene expression using antisense RNA technology (Nishikura & Murray, 1987). This technology may be applicable to regulating the expression of whey proteins in the mammary gland (Mercier, 1986). Absence of β-lactoglobulin or α-lactalbumin in milk could have biomedical, nutritional and technological ramifications (see Sections 3.5.4 and 3.5.5).

3.3. Strategies for Genetically Modifying Milk Proteins

The technology of genetic modification of food proteins relies on the knowledge of a relationship between the structure of the protein and its function in a food system. Figure 5 illustrates schematically a strategy for modifying food proteins which has been adapted from enzyme engineering examples. At a gross level, some relationships are known, for example, attendant calcium binding by phosphoseryl groups of β-casein affects cheese curd firmness; increased hydrophobicity of proteins decreases interfacial tension; increased pI will change protein solubility or amphiphilicity. More detailed information may allow more refined adjustments in protein structure as evidenced from the previous molecular modeling studies for α_{s1}- and κ-caseins.

Predictions of secondary structure can be used to design desired changes into the primary sequence of the proteins. Several forms of directed mutagenesis of DNA can be used to alter proteins, from changing a single base to modifying a codon, to the deletion or insertion of entire domains.

3.4. Genetic Modification of Milk Proteins

Bovine caseins are among the most widely studied food proteins. In the United States, close to 65 million kg of casein are used annually. These two factors make them good candidates for use as models for genetic modification of their functionalities (Kang & Richardson, 1985). The cDNAs coding for these caseins have been identified and sequenced.

The caseins have an open structure and are hydrated despite the high content of hydrophobic amino acids present. Some of the hydrophobic amino acids are in the outer molecular surface, as shown by the use of hydrophobic probes. These hydrophobic surfaces on the molecule are important in casein–casein interactions that are partially responsbile for the high viscosity of the casein solutions, or for the foaming and emulsification properties of caseins.

Another peculiarity of the caseins is the phosphoseryl residues. The function of milk is to provide a nutritious diet for the young; phosphorylation of the caseins can be seen as an essential biological process, being involved in the assembly of casein micelles, binding of calcium, and generally with the subsequent behavior of the caseins as functional and nutritional proteins. Particularly, changes by genetic manipulation of the phosphoseryl residues may provide a specific means to increase the mineral content of milk while retaining the colloidal stability and biological functions of the caseins.

3.5. Discussion of the Proposed Modifications on Milk Proteins

Some genetic modifications proposed for specific milk proteins are summarized in Table 6. The proposed modifications have been designed to improve the functional properties of milk during processing; each of the proposed changes are discussed in this section.

One example of a modification generalized to all caseins would be the modification or deletion of plasmin sites for hydrolysis to alter digestibility of caseins in milk (Swaisgood, 1989). The site(s) in milk proteins susceptible to attack by endogenous plasmin require two or more lysine or histidine residues in close proximity. Susceptible sites may be engineered into other parts of the protein or conversely, sites may be removed by substituting a different amino acid for a particular basic residue. Another example is the addition of genes coding for lactotransferrin, lysozyme or peroxidase to the mammary gland for their expression in milk. The particular characteristics of each of these proteins would have a desirable impact on milk from the manufacturing standpoint (Jimenez-Flores & Richardson, 1988). Of particular interest is the binding of metal ions by proteins such as lactotransferrin. This protein offers many opportunities for manipulation of metal-binding sites which may lead to in-situ antioxidants and bacteriostatic

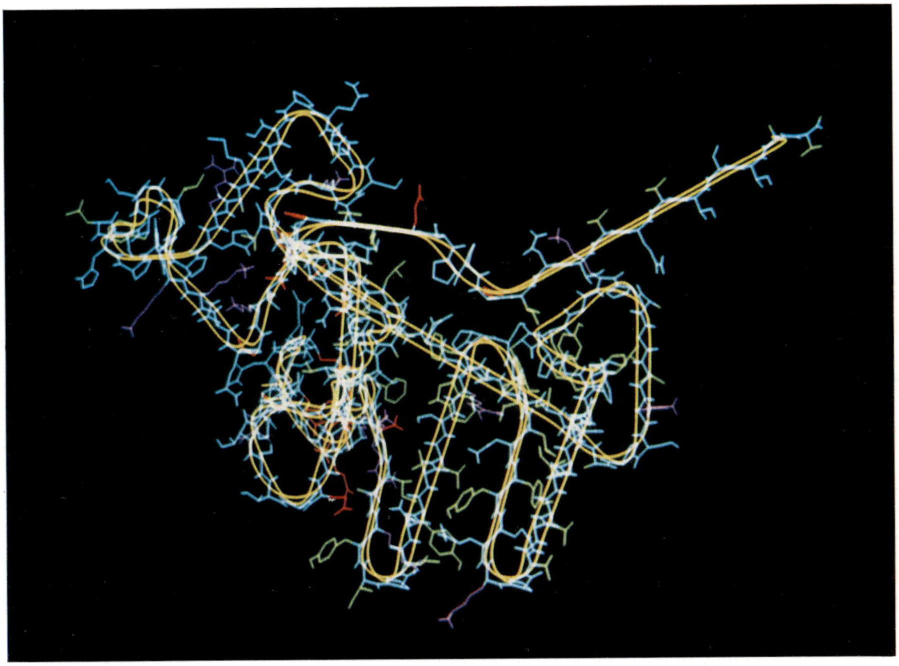

Plate I. Three-dimensional molecular model of κ-casein. The peptide backbone is coloured cyan, hydrophobic side chains green, acidic side chains red and basic side chains purple.

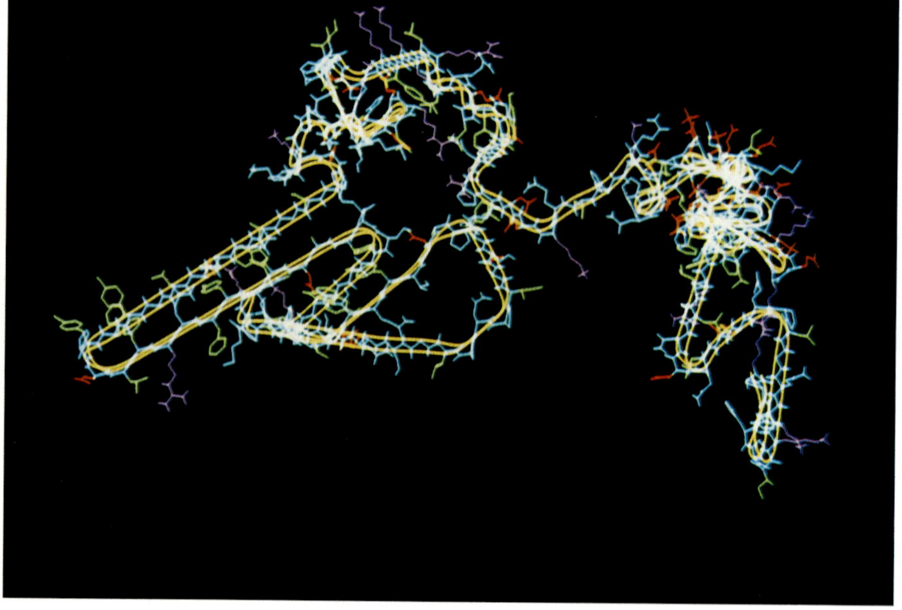

Plate II. Three-dimensional molecular model of α_{s1}-casein. The peptide backbone is coloured cyan, hydrophobic side chains green, acidic side chains red and basic side chains purple.

TABLE 6
Proposed Mutations on Milk Proteins to Alter Their Functionality

Protein	*Potential change*	*Potential Effect*
α_{s1}-Casein	—Ser-P_{115} to Ile_{115} (AGT to ATT)	—Reduce charge and increase the hydrophobicity
	—Addition of a glycosylation site in either $Asn_{17,19,74}$ or $_{114}$(all possible sites need a change of one or two bases)	—Enhance amphiphilicity and solubility —Modify calcium stability
β-Casein	—Addition of a glycosylation site in Asn_7 or Asn_{27}	—Enhance solubility, increase amphiphilicity
	—Addition of the phosphorylation domain (aa 10 to 25) at the C terminus of the protein	—Enhance binding for calcium and other divalent cations
	—Change Glu_{36} and $_{37}$ to Val (GAG to GTG and GAA to GTA)	—Enhance the hydrophobic patch on the amphipathic helix (predicted by Holt & Sawyer (1988))
	—Change $Leu_{191,192}$ to $Gln_{191,192}$ (CTG to CAG)	—Make the C terminus of β-casein into a better substrate for transglutaminase
κ-Casein	—Change Phe_{105}–Met_{106} to Phe_{105}–Phe_{106}	—Increase the hydrolysis rate by chymosin
	—Addition of a poly-Met residue at the N-terminus	—Improve the nutritional characteristics and possible enhancement of cheese flavour development
	—Change Ser-P_{149} to Glu_{149} (TCT to GAA)	—Increase calcium stability of the protein and micelle
α-Lactalbumin	—Modification of the binding site to galactosyl transferase	—Alter the efficiency of lactose synthesis
	—Anti-sense mRNA	—Reduction of α-lactalbumin concentration in milk
	—Modification of calcium binding site	
β-Lactoglobulin	—Anti-sense mRNA	—Reduction of β-lactoglobulin concentration in milk

agents. The following discussion analyses some genetic modifications in the structure of particular milk proteins.

3.5.1. α_{s1}-Casein

The phosphoseryl residue at position 115 is surrounded by hydrophobic amino acids, and according to the proposed model (Fig. 2(bottom)), it lies between well-defined regions of β-sheet and α-helical conformations. Changing this phosphoserine to an isoleucine would enhance the hydrophobic characteristic of that portion of the protein, affecting its association properties and presumably increasing its amphiphilicity.

N-linked glycosylation sites that can be added to the amino terminus of the protein (particularly residues 1 to 20 which appear to be external and on turns; Fig. 4) may enhance its hydrophilicity. The addition of glycosidic residues to the caseins increases their solubility dramatically. Further, enhancing the hydrophilicity of the amino terminus of the protein will also increase its amphipathic characteristics. Glycosylated caseins are more soluble, which should appreciably modify other functional properties such as viscosity, water holding capacity, foaming and emulsification (Courthaudon *et al.*, 1989).

3.5.2. β-Casein

The amino terminus of β-casein is hydrophilic and contains five phosphoseryl residues in close proximity; the rest of the protein is very hydrophobic. This amphipathic nature of β-casein has been correlated with excellent emulsifying, foaming and gelling characteristics. Addition of a glycosylation signal at position 7 or 27 should increase the solubility and amphipathic properties of this protein. Some proof for an increase in the solubility of β-casein after glycosylation has been reported (Courthaudon *et al.*, 1989; Jimenez-Flores *et al.*, 1990).

The role of the phosphoseryl residues in β-casein has been extensively studied (West, 1986). The phosphoseryl groups in β-casein contribute to an increase in the curd tension in cheesemaking, presumably by their interaction with calcium (Yun *et al.*, 1982*a*, *b*). The hydrophobic domain is responsible for protein associations of β-casein. The close proximity of the phosphoseryl residues in β-casein makes it simple to postulate a domain amenable to the phosphorylation of the serine residues by casein kinase (Bingham *et al.*, 1988). It is possible to add the DNA sequence corresponding to this domain, amino acids 10 to 25 in β-casein, at the amino terminus of the protein. The flexibility of the protein together with the localization of the phosphoseryl residues may lead to a protein that associates and binds calcium and other cations.

Holt and Sawyer (1988) proposed an amphipathic helix in bovine β-casein between amino acids 21 to 50. Homologous helices are predicted in the β-caseins of humans and rats. Residues 28 to 34 form the amphipathic portion of this helix in bovine β-casein which could be lengthened by the substitution of valine for glutamic acid at position 37. This change would disrupt the phosphorylation signal for serine 35, thus decreasing the probability of a β-turn in this portion of the sequence and increasing the hydrophobicity of the helix.

Transglutaminase (TGase) catalyzes a calcium-dependent acyl transfer reaction, in which one substrate is the γ-carboxamide group(s) of glutaminyl residues of proteins (acyl donors). The second substrate (acyl acceptors) is a nucleophile which could be free amino groups, such as the ε-amino group in lysine (Whitaker & Puigserver, 1982). Transglutaminase-catalyzed crosslinking of proteins has been used to form homologous and heterologous polymers of casein and other food proteins (Ikura *et al.*, 1984). Transglutaminase also catalyzes the formation of gels at sufficiently high protein concentration and at pH 7·5 (Motoki *et al.*, 1984). Change of the amino teminus of β-casein to increase its content of glutaminyl residues would provide other sites for crosslinking to form gel-like structures that may prove useful in product development.

3.5.3. κ-Casein

κ-Casein differs greatly from other caseins in many respects. It is not precipitated in the presence of Ca^{2+}, it stabilizes the casein micelle and is the main substrate of chymosin during cheesemaking. In its mature form, a portion of the κ-casein contains carbohydrate residues, and as seen from molecular modeling it has more detailed structural features than the other caseins. It is believed to have evolved independently of the other caseins, and its complete gene sequence is known (Alexander *et al.*, 1988; Kang & Richardson, 1988). This sequence shows that almost all the coding region (amino acids 10 to 169) is present in a single exon which is remarkably different from the other caseins, and simplifies site-directed mutagenesis directly on the structural gene.

In bovine κ-casein, the chymosin-sensitive bond is Phe–Met; however, the corresponding positions are occupied by Phe–Leu in rat and mouse or Phe–Ile in man and pig κ-caseins. This reflects inter-species differences in the substrate involved in the enzymatic coagulation of milk. The rennet-sensitive bond may be more susceptible if it was clamped to Phe–Phe, since many acid proteases seem to have greater affinity for this sequence (Oh & Richardson, in press).

3.5.4. α-Lactalbumin

The interaction of α-lactalbumin with UDP-galactosyltransferase (EC 2.4.1.22) is essential in the biosynthesis of lactose in the mammary gland. The structural gene for bovine α-lactalbumin has been sequenced and used to produce the protein in transgenic mice (Vilotte *et al.*, 1989). It has been suggested (Mercier, 1986; Jimenez-Flores & Richardson, 1988; Bremel *et al.*, 1989) that repression of the expression of this protein, by antisense mRNA technology (see later section), may lead to the production of reduced-lactose milk. An interesting approach would be the modification of the binding sites of α-lactalbumin and galactosyl transferase to form an irreversible complex, thus inhibiting the production of lactose but maintaining the a-lactalbumin content of milk which is desirable owing to its high levels of tryptophan.

Reduced lactose levels in milk would be desirable to minimize effects of lactose intolerance in consumers. However, since α-lactalbumin is involved in the coupling of glucose and galactose in the biosynthesis of lactose, its absence could cause problems because lactose plays an important role in maintaining the osmotic pressure of milk. The monosaccharides may not have the same osmotic effect as the disaccharide.

3.5.5. β-Lactoglubulin

Although β-lactoglobulin is the major protein component of bovine whey, its function is not clear. It is responsible for most of the allergic reactions to bovine milk in humans, and presents technological problems in the dairy industry. It would be desirable to omit the biosynthesis of this protein in the mammary gland by antisense mRNA technology.

The foregoing structural modifications of milk proteins to alter their functional characteristics are just a few of the myriad possibilities. Evaluation of the functionalities of modified proteins must await the development of suitable hosts for the production of sufficient protein.

3.6. Analysis of the Structure–Function of Modified Proteins

Microbial expression of mutant proteins offers the possibility of assessing their functionalities. Expression in eukaryotic systems, such as yeasts, are of special interest (Jimenez-Flores *et al.*, 1990), since post-translational modifications can be observed. Such changes include phosphorylation, signal processing, glycosylation and proper folding. Once the protein is purified, it can be incorporated into the system under study for analysis of its function.

Cell culture, specifically mammary cell culture, represents the ideal system for ultimately assessing any deleterious effects of the mutation on normal function. Such a system is already under investigation and development (Hung *et al.*, in press). Because of the limited quantities produced, the functionalities of cloned elaborated proteins produced by such unicellular systems must be determined using micro analytical methods.

3.7. Evidence that Novel Proteins Added to Baseline Complement Proteins will be Effective

Heretofore, we have discussed the engineering of single proteins to affect the behavior of the natural complement of milk proteins, but what about the situation where cloned proteins are added to a normal protein system, which is likely to be the case in genetic engineering? How much added protein will be necessary to alter effectively the functionality of the normal food protein system? These may become important questions in the very practical aspects of modifying the net functionality of protein mixtures that food scientists will work with in the future.

An early study in which added milk proteins altered the functionality of the milk system related to the influence of κ-casein and β-lactoglobulin on the heat stability of skim milk (Tessier & Rose, 1964). An observable enhancement in the heat stability of a type A milk between pH 6·6 and 6·8 was obtained when κ-casein was added at 13% of the normal protein level of that in milk. The heat stability minimum was totaly eliminated when the quantity of added κ-casein reached 39% of the normal complement of κ-casein in milk. Type A behaviour of milk was restored upon addition of β-lactoglobulin to the κ-casein-enriched milk. The ratio of κ-casein to β-lactoglobulin is evidently important in defining the heat stability characteristics of type A milk. Thus, the addition of only a fraction of the usual concentration of selected milk proteins can markedly influence the functional properties of the total milk system.

A second, striking example of the effect of an individual casein on milk protein functionality is the enhancement of curd firmness of enzymatically coagulated milk containing added β-casein (Yun *et al.*, 1982*a*, *b*). Fortification of skim milk with β-casein at a concentration 30% of that found in milk caused a 50% increase in curd firmness, indicating that the functional interactions of the caseins were increased, possibly by additional electrostatic and hydrophobic interactions resulting from interactions of calcium and β-casein. This is supported by the observation that the addition of 30% dephosphorylated β-casein to milk increased curd firmness by only 5%.

Added β-casein has also been shown to influence the rennet clotting time and rate of syneresis (whey expulsion) of artificial casein micelles (Pearse *et al.*, 1986; West, 1986). These are both measures of protein interactions in the cheesemaking process. Again, the phosphoseryl groups of β-casein seem to play a crucial role in these effects. The rennet coagulation and syneresis properties of artificial micelles in which completely dephosphorylated β-casein represented 20 and 30% of the total β-casein were compared with artificial micelles to which equivalent amounts of native β-casein had been added and with control micelles (no additions). The rennet coagulation time of artificial micelles with 20% of its β-casein dephosphorylated was increased by 57% and the final extent of syneresis was reduced by 18% when compared to control micelles. Adding the same amount of native β-casein, however, caused a reduction (6%) in rennet coagulation time and an increase (9%) in syneresis.

The sensitivity of rennet coagulation and syneresis to the dephosphorylation of β-casein was further investigated by incorporating partially dephosphorylated β-casein into artificial micelles. Incorporation of sufficient 25%-dephosphorylated β-casein so that 6·5 and 12·0% of the total β-casein was modified increased the rennet coagulation time of artificial micelles by 73 and 15%, respectively, and incorporating similar proportions of 50%-dephosphorylated β-casein increased the rennet coagulation time by 135 and 248%, respectively. The micelles failed to coagulate when the amount of partially dephosphorylated β-casein (25 and 50%) was increased to the point where 21% of the total β-casein was modified.

The final extent of syneresis of milks containing 25%-dephosphorylated β-casein (6·5 and 12·0%) was 12% lower than that of the control. Incorporation of 50%-dephosphorylated β-casein at 6·5 and 12·0% inhibited syneresis by 22 and 31%, respectively. It might be added that the effects of β-casein on curd firmness, rennet clotting time and syneresis are much more pronounced than those observed with added α_{s1}- and κ-caseins.

The foregoing observations indicate that a 20 to 30% change in the level of a given casein in the milk system can profoundly affect protein interactions, physical properties and behavior of milk during processing. Also, alterations in the chemical properties of caseins (dephosphorylation) to produce novel proteins can impact on the functionality of the milk system. By controlling gene structure, gene dosage and regulatory regions of the casein genes, it should be possible to obtain the 20 to 30% levels of additional or novel caseins to modify the functionality of the milk protein system.

In a recent experiment, Simons *et al* (1987) generated transgenic mice carrying

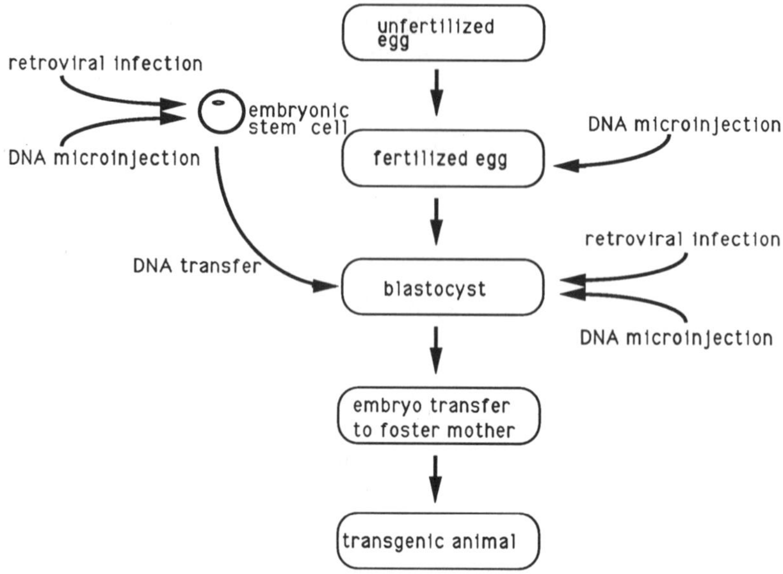

Fig. 6. Methods of gene transfer to produce transgenic animals.

the sheep β-lactoglobulin gene and showed that in these mice β-lactoglobulin is expressed at levels as high as 23 mg ml^{-1}. Because β-lactoglobulin is not normally present in mouse milk, this represents a major modification of its composition. Moreover, these levels represent five times the estimated concentration of β-lactoglobulin in sheep's milk (Simons *et al.*, 1987). All of this indicates that the protein composition in milk can be manipulated by gene transfer without any apparent adverse consequences.

3.8. Transgenic Animals

A transgenic animal is one whose genetic composition has been altered to include selected genes from other animals or species by methods other than those used in traditional breeding (Church, 1987). There have been three important advances in transgenic animal technology. One is the ability to introduce new genetic information into all somatic tissues, as well as the germ line, either by microinjection into the pronucleus of the fertilized egg (Gordon *et al.*, 1980) or by infecting embryos with retroviral vectors (Jaehner *et al.*, 1985; van der Putten *et al.*, 1985). The second is the technique of transplanting nuclei of fertilized egg into enucleated embryos. The third is the ability to establish from normal mouse embryos pluripotential stem cell lines, in culture, that can be re-integrated back into the embryo by injection into host blastocysts (Bradley *et al.*, 1984) (Fig. 6). Because transgenic animals have a potentially large impact on future agricultural products, many efforts have been made to develop them, as shown in Table 7.

Hammer *et al.* (1985) reported the insertion of the human growth hormone gene

TABLE 7
Examples of Proteins Produced in the Milk of Transgenic Animals (Bremel *et al.*, 1989)

Protein	*Gene source*	*Host*	*Promoter*
α-lactalbumin	Bovine	Mouse	α-lactalbumin
β-Lactoglobulin[a]	Sheep	Mouse	Sheep β-Lactoglobulin
Thymidine kinase	Herpes simplex virus	Mouse	Mouse metallothionein-I
Clotting factor IX[a]	Human	Sheep	Sheep β-Lactoglobulin
α-1 antitripsin[a]	Human	Sheep	Sheep β-Lactoglobulin
T-PA[a]	Human	Mouse	Rat WAP
CAT	Bacteria	Mouse	Rat β-casein
Ha-Ras	Human	Mouse	Rat WAP
Casein[a]	Rat	Mouse	Rat β-casein

[a]Secreted into the milk.

by microinjection of linearized DNA into sheep and pig ova using an MT-hGH (metallothionein-human growth hormone) fusion gene. The fused MT-hGH gene was microinjected into cow ova (Church, 1987), but the integration of MT-hGH fusion gene was low. The reason for the low integration and members of transgenic progeny is probably due to lack of knowledge of the molecular biology, physiology and reproduction of domestic animals, which makes the selection of promoters and structural sequences for microinjection constructs very difficult and empirical. Also, additional knowledge on direct integration of inserted DNA into the genome for proper expression is needed.

Most of the work on transgenic animals has been with mice because the murine system is convenient to work with in terms of economics, ease and possibility of collecting large numbers of embryos; moreover, the pronucleus is transparent which facilitates microinjection (Palmiter & Brinster, 1985). A variety of genes have been expressed in mice. The first successful transgenic animal was the classic giant mouse which was injected with metallothionein-growth hormone fusion genes (Palmiter *et al.*, 1982). Human globin genes were expressed in mice using a murine regulatory region (Kollias *et al.*, 1986). Rat pancreatic elastase 1 (Swift *et al.*, 1984) and Protamine 1 (Peschon *et al.*, 1987) have also been expressed in mice. Human tissue plasminogen activator (Pittius *et al.*, 1988) was expressed using a milk protein gene promoter (mouse whey acid protein). Rat β-casein (Lee *et al.*, 1988) and sheep β-lactoglobulin (Simons *et al.*, 1987) were expressed in mice using the respective regulatory regions. Bovine α-lactalbumin was expressed in mice using its own regulatory region (Vilotte *et al.*, 1989). In addition to the foregoing research, transgenic animals have been used to study the immune system, genetic diseases, and oncogenes (Camper, 1987).

Microinjected bovine α-lactalbumin gene with 5′ and 3′ flanking control region DNA directed tissue-specific expression to the mammary gland in transgenic mice. The expression level was close to the endogenous α-lactalbumin (Vilotte *et al.*, 1989). This suggests that the transgenic animal with an additional gene may overproduce a transgene product if the DNA construct contains all the important DNA sequences for efficient expression.

From the standpoint of dairy processing technology, transgenic animals may offer an excellent means to engineer dairy products at the raw material level. Overproduction of given milk proteins or production of novel protein mutants by transgenic animals animals may lead to improved and novel dairy products.

Because α_{s1}-casein, β-casein, and κ-casein may not be segregating independently, it may be necessary to develop homologous recombination methods of targeting milk protein genes to achieve desirable gene expression. Manipulated genes could be targeted into a desirable position by homologous recombination of DNA sequences residing in the chromosome with newly introduced DNA sequences (Mansour *et al.*, 1988). In these studies using homologous recombination it was possible to select mutants, *hprt*- (hypoxanthine-guanine phosphoribosyl transferase) and *int*-2- (proto oncogene). Many researchers are trying to develop homologous recombination methods for targeting gene insertion into the genome. With homologous gene recombination involving milk protein genes, it may be possible to develop super gene families coding for normal or mutant milk proteins with desired characteristics for dairy processing or novel diary products.

References

Alexander, L.J., Stewart, A.F., Mackinlay, A.G., Kapelinskaya, T.V., Tkach, T.M. & Gorodetsky, S.I. (1988). Isolation and characterization of the bovine κ-casein gene. *Eur J. Biochem.*, **178**, 395–401.

Ali, A.E., Andrews, A.T. & Cheeseman, G.C. (1980). Influence of storage of milk on casein distribution between micellar and soluble phases. *J. Dairy Res.*, **47**, 371–82.

Ananthanarayanan, V.S., Brahmachari, S.K. & Pattabiraman, N. (1984). Proline-containing β-turns in peptides and proteins: analysis of structual data on globular proteins. *Arch. Biochem. Biophys.*, **232**, 482–95.

Beeby, R. (1964). The presence of sulphydryl groups in κ-casein. *Biochim. Biophys. Acta*, **82**, 418–19.

Benedetti, E., Bavoso, A., Blasio, B.D., Pavone, V., Pedone, C., Toniolo, C. & Bonora, G.M. (1983). Solid-state geometry and conformation of linear, diastereosiosometric oligoprolines. *Biopolymers*,**22**, 305–17.

Berliner, L.J. & Johnson, J.D. (1988). α-Lactalbumin and calmodulin. In *Calcium Binding Proteins*, vol. VII, ed. M.P. Thompson, CRC Press, Boca Raton, FL, pp. 79–116.

Bingham, E., Parris, N. & Farrell, H.M. Jr. (1988). Phosphorylation of β-casein and α-lactalbumin by casein kinase from bovine mammary gland. *J. Dairy Sci.*, **71**, 324–36.

Bradley, A., Evans, M., Kaufman, M.H. & Robertson, E. (1984). Formation of germline chimera from embryo-derived tetatocarcinoma cell lines. *Nature*, **309**, 255–301.

Bremel, R.D., Yom, H.-C. & Bleck G.T. (1989). Alteration of milk composition using molecular genetics. *J. Dairy Sci.*, **72**, 2826–33.

Brinster, R.L., Chen, H.Y., Trumbauer, M.E., Yagle, M.K., & Palmiter, R.D. (1985). Factors affecting the efficiency of introducing foreign DNA into mice by microinjecting eggs. *Proc. Nat. Acad. Sci.*, **82**, 4438P.

Byler, D.M. & Farrell, H.M., Jr. (1989). Infrared spectroscopic evidence for calcium ion interaction with carboxylate groups of casein. *J. Dairy Sci.*, **72**, 1719–23.

Byler, D.M., Farrell, H.M., Jr. & Susi, H. (1988). Raman spectroscopic study of casein structure. *J. Dairy Sci.*, **71**, 2622–9.

Camper, S.A. (1987). Research applications of transgenic mice. *Biotechniques*, **5**, 638–50.

Carter, P., Bedouelle, H. & Winter, G. (1985). Improved oligonucleotide site-directed mutagenesis using M13 vectors. *Nucleic Acids Res.*, **13**, 4431–43.
Chou, P.Y. & Fasman, G.D. (1978). Prediction of the secondary structure of proteins from their amino acid sequence. *Adv. Enzymology*, **47**, 45–148.
Church, R.B. (1987). Embryo manipulation and gene transfer in domestic animals. *Trends Biotechnol.*, **5**, 13–19.
Cohen, F.E., Abarbanel, R.M., Kuntz, I.D. & Fletterick, R.J. (1983). Secondary structure assignment for a/b proteins by a combinatorial approach. *Biochemistry* **22**, 4894–904.
Cohen, F.E., Abarbanel, R.M., Kuntz, I.D. & Fletterick, R.J. (1986). Turn prediction in proteins using a pattern-matching approach. *Biochemistry*, **25**, 266–75.
Courthaudon, J-L., Colas, B. & Lorient, D. (1989). Covalent binding of glycosyl residues to bovine casein: effects on solubility and viscosity. *J. Agr. Food Chem.*, **37**, 32–6.
Creamer, L.K., Richardson, T. & Parry, D.A.D. (1981). Secondary structure of bovine α_{s1}-and β-casein in solution. *Arch. Biochem. Biophys.*, **211**, 689–96.
Creamer, L.K., Jimenez-Flores, R. & Richardson, T. (1988). Genetic modification of food proteins. *Trends Biotech.*, **6**, 163–9.
Creighton, T.E. (1983). *Proteins Structures and Molecular Properties*. W.H. Friedman and Co., New York, pp. 298–9.
Davies, D.T. & Law, A.J.R. (1983). Variation in the protein composition of bovine casein micelles and serum casein in relation to micellar size and milk temperature. *J. Dairy Res.*, **50**, 67–75.
Dempsey, W.B. (1987). Transcript analysis of the plasmid R100 traJ and finP genes. *Mol. Gen. Genet.*, **209**, 533–44.
Downey, W.K. & Murphy, R.F. (1970). The temperature dependent dissociation of β-casein from bovine casein micelles and complexes. *J. Dairy Res.*, **37**, 361–72.
Eigel, W.N., Butler, J.E., Ernstrom, C.A., Farrell, H.M. Jr., Harwalkar, V.R., Jenness, R. & Whitney, R. McL. (1984). Nomenclature of proteins in cow's milk: Fifth Revision. *J. Dairy Sci.*, **67**, 1599–631.
Farrell, H.M., Jr. (1988). Physical equilibria: proteins. In *Fundamentals of Dairy Chemistry*, 2nd edn, ed. N. Wong, Van Nostrand, Reinhold, New York, pp. 461–510.
Farrell, H.M., Jr., Kumosinski, T.F., Pulaski, P. & Thompson, M.P. (1988). Calcium-induced associations of the caseins: A thermodynamic linkage approach to precipitation and resolubilization. *Arch. Biochem. Biophys.*, **265**, 145–58.
Feagan, J.T., Bailey, L.F., Hehir, A.F., McLean, D.M. & Ellis, N.J.S. (1972). Coagulation of milk proteins. 1—Effect of genetic variants of milk proteins on rennet coagulation and heat stability of normal milk. *Australian J. Dairy Tech.*, **27**, 129–34.
Fersht, A.R., Ghi, J.R., Knill-Jones, J. Lowe, D.M., Wilkinson, A.J., Blow, D.M., Brick, P., Carter, P., Way, M.M.Y. & Winter, G. (1985). Hydrogen bonding and biological specificity analyzed by protein engineering. *Nature*, **314**, 235–8.
Garnier, J. Osguthorpe, D.J. & Robson, B. (1978). Analysis of the accuracy and implications of simple methods for predicting the secondary structure of globular proteins. *J. Mol. Biol.*, **120**, 97–120.
Godovac-Zimmerman, J. (1988). The structural motif of β-lactoglobulin and retinol-binding protein. *Trends Biochem. Sci.*, **13**, 64–6.
Gordon, J.W., Scangos, G.A., Plotkin, D.J., Barbosa, J.A. & Ruddle, F.H. (1980). Genetic transformation of mouse embryos by microinjection of purified DNA. *In Proc. Nat. Acad. Sci.*, **77**, 7380–9.
Hammer, R., Pursel, V.G., Rexroad, C.E., Wall, R.J., Bolt, D.J., Ebert, K.M. & Palmkiter, R.L. (1985). Production of transgenic rabbits, sheep and pigs by microinjection. *Nature*, **315**, 680–3.
Hill, R.J. & Wake, R.G. (1969). Amphiphile nature of κ-casein as the basis for its micelle stabilizing property. *Nature*, **221**, 635–9.
Holt, C. & Sawyer, L. (1988). Primary and predicted secondary structures of the caseins in relation to their biological functions. *Protein Eng.*, **2**, 251–9.

Horbett, T.A. & Brash, J.L. (1987). Proteins at interfaces: Current issues and future prospects. In *Proteins at Interfaces*, ed. J.L. Brash & T.A. Horbett. Am. Chem. Soc., Washington, DC., pp. 133.

Hung, H.T., Robitaille, G. & Turner, J.D. (1991). Establishment of bovine mammary epithelial cells (MAC-T): an in vitro model for bovine lactation. *Exp. Cell. Res.* (in press).

Ikura, J., Goto, M., Yoshikawa, M., Sasaki, R. & Chiba, H. (1984). Use of transglutaminase. reversible blocking of amino groups in substrate proteins for a high yield of specific products. *Agr. Biol. Chem.*, **48**, 2347–54.

Jaehner, D., Hasse, K., Mulligen, R. & Jaenisch R. (1985). Insertion of the bacterial gpi gene into the germline of mice by retroviral infection. *Proc. Natl Acad. Sci.*, **82**, 6927–31.

Jimenez-Flores, R. & Richardson, T. (1988). Genetic engineering of the caseins to modify the behavior of milk during processing: A review. *J. Dairy Sci.*, **71**, 2640–54.

Jimenez-Flores, R., Kang, Y.C. & Richardson, T. (1986). Genetic engineering of enzymes and food proteins. In *Protein Tailoring for Food and Medical Uses*, ed. R.E. Feeney & J.R. Whitaker, Marcel Dekker, New York, pp. 155–79.

Jimenez-Flores, R., Kang, Y.C. & Richardson, T. (1987). Cloning and sequence analysis of bovine β-casein cDNA. *Biochem. Biophys. Res. Commun.*, **142**, 617–21.

Jimenez-Flores, R., Richardson, T. & Bisson, L. (1990). Expression of bovine β-casein in *Saccaromyces cerevisiae* and characterization of the protein *in vivo*. *J. Agr. Food Chem.*, **38**, 1134–41.

Jolles, P., Alais, C. & Jolles, J. (1962). Amino acid composition of κ-casein and terminal amino acids of κ- and para κ-casein. *Arch. Biochem. Biophys.*, **98**, 56–7.

Kalan, E.B. & Woychik, J.H. (1965). Action of rennin on κ-casein, the amino acid compositions of the para-κ-casein, and glycomacropeptide fractions. *J. Dairy Sci.*, **48**, 1423–8.

Kang, Y.C. & Richardson, T. (1985). Genetic engineering of caseins. *Food Tech.*, **39**, 89–92.

Kang, Y.C. & Richardson, T. (1988). Molecular cloning and expression of bovine κ-casein in *E. coli*. *J. Dairy Sci.*, **71**, 29–40.

Kato, A. & Yutani, K. (1988). Correlation of surface properties with conformational stabilities of wild type and six mutant tryptophan synthase alpha subunits substituted at the same position. *Protein Eng.*, **2**, 153–6.

Kinsella, J.E. (1982). Relationships between structure and functional properties of food proteins. In *Food Proteins*, ed. P.F. Fox & J.J. Condon. Applied Science Publishers, London, pp. 51–103.

Kollias, G., Wrighton, N., Hurst, J. & Grosveld, F. (1986). Regulated expression of human gamma-, beta- and hybrid gamma-, beta-globin genes in transgenic mice. *Cell*, **46**, 89–94.

Kunkel, T.A., Rogerts, J.D. & Zakour, R.A. (1987). Rapid and efficient site-specific mutagenesis without phenotypic selection. *Methods Enzymol.*, **154**, 367–82.

Leatherbarrow, R.J. & Ferscht, A.R. (1986). Protein engineering. *Protein Eng.*, **1**, 7–16.

Lee, K-F., DeMayo, F.J., Atiee, S.H. & Rosen, H. (1988). Tisue-specific expression of the rat-β-casein gene in transgenic mice. *Nucleic Acid Res.*, **16**, 1027–41.

Light, J. & Molin, S. (1983). Post-transcriptional control of expression of the repA gene of plasmid R1 mediated by small RNA molecule. *EMBO J.*, **2**, 93–8.

Loucheux-Lefebvre, M.-H., Aubert, J.-P. & Jolles, P. (1978). Prediction of the conformation of cow and sheep κ-caseins. *Biophys. J.*, **23**, 323–36.

Mansour, S.L., Thomas, K.R. & Capecchi, M.R. (1988). Disruption of the proto-oncogene int-2 in mouse embryo-directed stem cells: general strategy for targeting mutations in non-selectable genes. *Nature*, **336**, 348–52.

McLean, D.M., Graham, E.R.B., Ponzoni, R.W. & McKenzie, H.A. (1984). Effects of milk protein genetic variants on milk yield and composition. *J. Dairy Res.*, **51**, 531–43.

McLean, D.M., Graham, E.R.B., Ponzoni, R.W. & McKenzie, H.A. (1987). Effects of milk protein genetic variants and composition on heat stability of milk. *J. Dairy Res.*, **54**, 219–35.

Mercier, J.C., Grosclaude, F. & Ribadeau-Dumas, B. (1971). Structure primarie de la casein α_{s1}-bovine. Sequence complete. *Eur. J. Biochem.*, **23**, 41–51.

Mercier, J.C., Brignon, G. & Ribadeau-Dumas, B. (1973). Primary structure of κ-casein B bovine. Complete sequence. *Eur. J. Biochem.*, **35**, 222–35.

Mercier, J.C. (1986). Genetic engineering applied to milk producing animals: some expectations. In *Exploiting New Technologies in Animal Breeding, Genetic Developments*, ed. C. Smith, J.W.B. King & J.C. McKay. Oxford Press, New York, pp. 122–31.

Mizuno, T., Chou, M.Y. & Inouye, M. (1984). In *Proc. Nat. Acad. Sci.*, **81**, 1966–70.

Monaco, H.L., Zonotti, G., Spadon, P., Bolognesi, M., Sawyer, L. & Eliopoulos, E.E. (1987). Crystal structure of the trigonal form of bovine β-lactoglobulin and of its complex with retinol at 2·5 A resolution. *J. Mol. Biol.*, **197**, 695–706.

Motoki, M., Nio, N. & Takanami, K. (1984). Functional properties of food proteins polymerized by transglutaminase. *Agr. Biol. Chem.*, **48**, 1257–66.

Nagao, M., Maki, M., Sasaki, R. & Chiba, H. (1984). Isolation and sequence analysis of bovine α_{s1}-casein cDNA clone. *Agr. Biol. Chem.*, **48**, 1662–7.

Nagao, M., Nakagawa, Y., Ishii, A., Sasaki, R., Tanaka, H. & Chiba, H. (1988). Expression of bovine α_{s1}-cDNA clone. *Agr. Biol. Chem.*, **48**, 1662–7.

Ng-Kwai-Hang, K.F., Hayes, J.F., Moxeley, J.E. & Monardes, H.G. (1984). Association of genetic variants of caseins and milk serum proteins with milk, fat, and protein production of dairy cattle. *J. Dairy Sci.*, **67**, 835–40.

Ng-Kwai-Hang, K.F., Hayes, J.F., Moxeley, J.E. & Monardes, H.G. (1987). Variation in milk protein concentrations associated with genetic polymorphysms and environmental factors. *J. Dairy Sci.*, **70**, 563–70.

Nishikura, K. & Murray, J.M. (1987). Antisense RNA of proto-oncogene c-fos blocks renewed growth of quiescent 3T3 cells. *Mol. Cell. Biol.*, **7**, 639–49.

Oh, S.S. & Richardson, T. (in press). Genetic engineering of bovine κ-casein to enhance its proteolysis by chymosin. In *Macromolecular Interactions and Food Colloid Stability*, ed. N. Parris & R.A. Barford. American Chemical Society, Washington, DC.

Palmiter R.D. & Brinster, R.C. (1985). Transgenic mice. *Cell*, **41**, 343–5.

Palmiter, R.D., Brinster, R.C., Hammer, R.E., Trumbauer, M.E., Rosenfeld, M.G., Birmberg, N.C. & Evans, R.M. (1982). Dramatic growth of mice that develop from eggs microinjected with metallothionein-growth hormone fusion genes. *Nature*, **300**, 611–15.

Papiz, M.Z., Sawyer, L., Eliopoulos, E.E., North, A.C.T., Findlay, J.B.C., Sivaprasadarao, R., Jones, T.A., Newcomer, M.E. & Kraulis, P.J. (1986). The structure of β-lactoglobulin and its similarity to plasma retinol-binding protein. *Nature*, **324**, 383–5.

Pearse, M.J., Linklater, P.M., Hall, R.J. & MacKinlay, A.G. (1986). Effect of casein micelle composition and casein dephosphorylation on coagulation and syneresis. *J. Dairy Res.*, **53**, 381–90.

Pepper, L. & Farrell, H.M., Jr. (1982). Interactions leading to formation of casein submicelles. *J. Dairy Sci.*, **65**, 2259–66.

Pervaiz, S. & Brew, K. (1985). Homology of β-lactoglobulin, serum retinol-binding protein, and protein HC. *Science*, **228**, 335–7.

Peschon, J.J., Behringer, R.R., Brinster, R.C. & Palmiter, R.D. (1987). Spermatid-specific expression of protamine I in transgenic mice. *Proc. Natl. Acad. Sci. USA*, **84**, 5316–19.

Pittius, C.W., Hennighausen, L., Lee, E., Westphal, H., Nicols, E., Vitale, J. & Gordon, K. (1988). A milk protein gene promoter directs the expression of human tissue plasminogen activator cDNA to the mammary gland in transgenic mice. *Proc. Natl. Acad. Sci. USA*, **85**, 5874–8.

Raap, J., Kerling, K.E.T., Vreeman, H.J. & Visser, S. (1983). Peptide substrates for chymosin: conformational studies of κ-casein and some κ-casein related peptides by circular dichroism and secondary structure prediction. *Arch. Biochem. Biophys.*, **221**, 117–24.

Rose, D. (1968). Relation between micellar and serum casein in bovine milk. *J. Dairy Sci.*, **51**, 1897–902.

Rose, G.D., Gierasch, L.M. & Smith, J.A. (1985). Turns in peptides and proteins. *Adv. Protein Chem.*, **37**, 1–109.

Sawyer, L., Papiz, M.Z., North, A.C.T. & Eliopoulos, E.E. (1985). Structure and function of bovine β-lactoglobulin. *Biochem. Soc. Trans.*, **13**, 265–6.

Schaar, J., Hansson, B. & Petterson, H.E. (1985). Effects of genetic variants of κ-casein and β-lactoglobulin on cheese making. *J. Dairy Res.*, **52**, 429–37.

Schmidt, D.G. (1982). Association of caseins and casein micelle structure. In *Developments in Dairy Chemistry—1*, ed. P.F. Fox. Applied Science Publishers, London, pp. 61–86.

Simons, G.F.M., Veeneman, G.H., Konings, R.N.H.L., van Boom, J.H. & Schoemaker, J.G.G. (1982). Oligonucleotide directed mutagenesis of gene IX of bacteriophase M13. *Nucleic Acids Res.*, **10**, 821–32.

Simons, J.P., McClennagan, M. & Clark, J. (1987). Alteration of the quality of milk by expression of sheep β-lactoglobulin in transgenic mice. *Nature*, **328**, 530–2.

Simons, R.W. & Kleckner, N. (1983). Translational control of IS10 transposition. *Cell*, **34**, 683–91.

Smith, S.G., Lewis, M., Aschaffenberg, R., Fenus, R.E., Wilson, I.A., Sundralingam, M., Stuart, D.E. & Phillips, D.C. (1987). Crystallographic analysis of the three dimensional structure of baboon α-lactalbumin at low resolution. *Biochim. J.*, **242**, 353–60.

Stuart, D.I., Acharya, K.R., Walker, N.P.C., Smith, S.G., Lewis, M. & Phillips, D.C. (1986). α-Lactalbumin possesses a novel calcium binding loop. *Nature*, **324**, 84–7.

Susskind, M.M. & Youderian, P. (1983). Bacteriophage P22 antirepressor and its control. In *Lamda II*, ed. R.W. Hendrix, J.W. Roberts, R.W. Stahl & R.A. Weisberg, Cold Spring Harbor Laboratory, Cold Spring Harbor, New York, pp. 347–63.

Swaisgood, H.E. (1982). Chemistry of milk proteins. In *Developments in Dairy Chemistry*, vol. 1, ed. P.F. Fox. Applied Science Publisher, London, pp. 1–60.

Swaisgood, H.E. (1989). Structural changes in milk proteins. In *Milk Proteins*, ed. C.A. Barth & E. Shlimme. Steinkopf verlag, Darnstadt, New York, pp. 192–210.

Swaisgood, H.E., Brunner, J.R. & Lillevik, H.A. (1964). Physical parameters of κ-casein from cow's milk. *Biochemistry*, **3**, 1616–23.

Swift, G.H., Hammer, R.E., McDonald, R.J. & Brinster, R.L. (1984). Tissue-specific expression of the rat pancreatic elastase I gene in transgenic mice. *Cell*, **38**, 639–46.

Tessier, H. & Rose, D. (1964). Influence of κ-casein and β-lactoglobulin on the heat stability of skim milk. *J. Dairy Sci.*, **47**, 1047–53.

Thompson, M.P., Gordon, W.G., Boswell, R.T. & Farrell, H.M., Jr. (1969). Solubility, solvation, and stabilization of α_{s1}- and β-caseins. *J. Dairy Sci.*, **52**, 1166–73.

Van der Putten, H., Botteri, F.M., Miller, A.D., Rosenfeld, M.G., Fan, H., Evans, R.M. & Verma, I.M. (1985). Efficient insertion of genes into the mouse germ line via retroviral vectors. *Proc. Natl. Acad. Sci.*, **82**, 6148–52.

Vilotte, J.L., Soulier, S., Stinnakre, M.G., Massoud, M. & Mercier, J.C. (1989). Efficient tissue-specific expression of bovine alpha-lactalbumin in transgenic mice. *Eur. J. Biochem*, **186**, 43–8.

Vreeman, H.J., Brinkhuis, J.A. & Van Der Spek, C.A. (1981). Some association properties of bovine SH-κ-casein. *Biophys. Chem.*, **14**, 184–93.

Wallace, R.B., Johnson, M.I., Hurose, T., Miyaka, T., Kawashima, E.H. & Itakura, K. (1981). The use of synthetic oligonucleotides as hybridization probe. *Nucleic Acids Res.*, **9**, 879–94.

Waugh, D.F. & Von Hippel, P.H. (1956). κ-Casein and the stabilization of casein micelles. *J. Am. Chem. Soc.*, **78**, 4576–82.

West, D.W. (1986). Structure and function of the phosphorylated residues of casein. *J. Dairy Res.*, **53**, 333–52.

Whitaker, J.R. & Puigserver, A.J. (1982). Fundamentals and applications of enzymatic modifications of proteins: an overview. In *Modification of Proteins: Food Nutritional and Pharmacological Aspects*, ed. R.E. Feeney & J.R. Whitaker. American Chemical Society, Washington, DC, pp. 57–94.

Willets, N. & Skurray, R. (1987). Structure and function of the F factor and mechanism of conjugation. In *Escherichia coli and Salmonella typhimurium Cellular and Molecular*

Biology, ed. F.C. Neidhardt. Am. Soc. Microbiol., Washington, DC, pp. 1110–13.
Woychik, J.H. & Wondolowski, M.V. (1969). Nitration of tyrosyl residues in κ- and α_{s1}-caseins. *J. Dairy Sci.*, **52**, 1669–72.
Woychik, J.H., Kalan, E.B. & Noelken, M.E. (1966). Chromatographic isolation and partial characterization of reduced κ-casein components. *Biochemistry*, **5**, 2276–88.
Yun, S.-E., Ohmiya, K. & Shimizu, S. (1982*a*). Role of β-casein in milk curdling. *Agr. Biol. Chem.*, **46**, 443–9.
Yun, S.-E., Ohmiya, K. & Shimizu, S. (1982*b*). Role of the phosphoryl group of β-casein in milk curdling. *Agr. Biol. Chem.*, **46**, 1505–11.
Yutani, K., Ogasahara, K. & Sugino, Y. (1985). Effect of amino acid substitutions on conformational stability of a protein. *Adv. Biophys*,**20**, 13–29.

14

THE ENZYMATIC COAGULATION OF MILK

D.G. Dalgleish
Department of Food Science, University of Guelph, Guelph, Ontario, Canada N1G 2W1

1. Introduction

After milk has been treated with rennet, there follows a stage of the reaction during which apparently little happens, followed after some time by a rapid coagulation of the milk. This phenomenon, which is the first step of cheesemaking, results from two processes. The first is the attack on κ-casein, which stabilizes the casein micelles, by the proteolytic enzymes (chymosin, pepsin or microbial proteinases) contained in the rennet; and the second is the subsequent clotting of the micelles which have been destabilized by this enzymatic attack. These processes have been described as the primary and secondary stages of the renneting reaction. This sequential description of the reaction is, however, somewhat over-simplified, since it is not necessary for the enzymatic reaction to be complete before aggregation of partly renneted micelles can occur. Indeed, the milk may begin to clot well before the enzymatic cleavage of κ-casein is complete (Green *et al.*, 1978*b*; Dalgleish, 1979; Chaplin & Green, 1980).

Since the two processes, i.e. the enzymatic hydrolysis of κ-casein and the subsequent aggregation, occur simultaneously during the later stages of the enzymic reaction, this chapter will describe both. Without the primary enzymic phase, no clotting can occur, so it is necessary to define the behaviour of the enzymes responsible for the hydrolysis of κ-casein. However, the enzymatic reaction cannot be considered in isolation from the effects that it has on the casein micelles, both in chemical and in physical and mechanistic terms, and these also will be described. There is in fact a third stage of the reaction which involves the changes in the properties and structure of the rennet curd once it has been formed. However, this chapter will describe only the mechanisms leading to coagulation (i.e. the primary and secondary stages of clotting).

To understand the coagulation reaction, it is first of all necessary to define a suitable model for the particles that are being clotted, namely the casein micelles. From the point of view of the renneting reaction, these particles, which contain

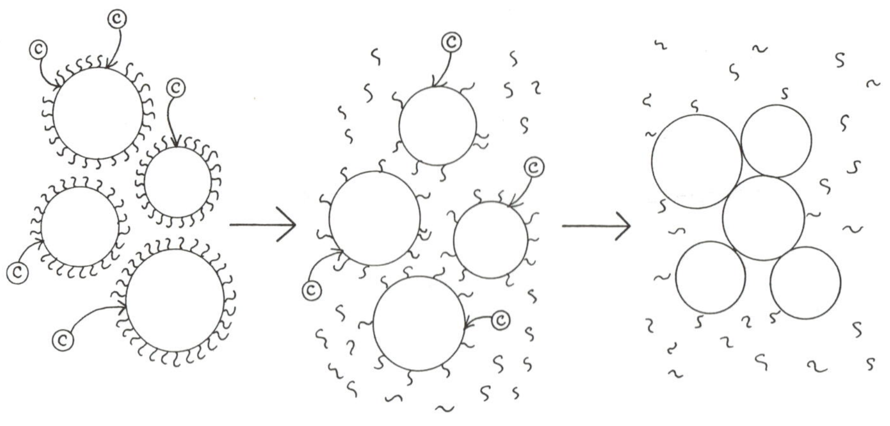

Fig. 1. Schematic diagram of the attack by chymosin (shown as C) on casein micelles. Three different points in the reaction are illustrated: (a) the κ-casein coat of the micelles is intact, and chymosin has just been added; (b) some time later, much of the κ-casein has been hydrolysed but sufficient remains to prevent aggregation; (c) at a later time still, nearly all of the κ-casein has been hydrolysed and the micelles have started to aggregate.

most of the protein in cows' milk, may be considered as being approximately spherical, composed of several thousand individual molecules of all four types of casein (α_{s1}, α_{s2}, β and κ) and containing calcium phosphate in one of its insoluble states (see Chapter 3 in this volume). The different caseins are not evenly distributed throughout the particle; in particular, κ-casein appears to be located mainly at the surface of the micelle (McGann *et al.*, 1980; Donnelly *et al.*, 1984; Rollema *et al.*, 1988), probably along with α_s-caseins (Dalgleish *et al.*, 1989), so that it can exercise a stabilizing effect on the native micelles and prevent them from coagulating. The stabilizing effect arises because κ-casein, as shown by its peptide sequence (Mercier *et al.*, 1973), is divided into two distinct regions, namely the hydrophobic para-κ-casein (residues 1–105) and the hydrophilic macropeptide or glycomacropeptide (CMP or GMP, residues 106–169). In its natural position on the surface of the micelles, κ-casein is linked to the remainder of the micelle via the para-κ-casein part of the molecule, so that the macropeptide protrudes into the surrounding solution. This hydrophilic moiety interacts with the solvent to stabilize the micelles (Holt, 1975; Walstra, 1979; Walstra *et al.*, 1981; Horne, 1984, 1986; Holt & Dalgleish, 1986; Dalgleish & Holt, 1988).

Successful milk clotting enzymes split κ-casein at the junction between the para-κ-casein and macropeptide moieties (i.e. in bovine κ-casein) at the bond between the phenylalanine residue 105 and the methionine residue 106 (Jollès *et al.*, 1968). When this occurs, the macropeptide diffuses into the serum, its stabilizing influence is lost, and the micelles can begin to coagulate once sufficient of their κ-casein has been hydrolysed (Dalgleish & Holt, 1988). These reactions are shown schematically in Fig. 1.

The enzymes that specifically cleave the Phe–Met bond of bovine κ-casein all belong to the group of acid proteinases (see Foltmann (1987) for a review). The enzymes of this class traditionally used in the manufacture of cheeses are chymosins and pepsin, the former being extracted from the stomachs of calves, kids or lambs, and the latter from adult cattle. However, in recent years, the use of enzymes of the same class from microorganisms or from other higher animals has become fairly common, for economic, religious or cultural reasons. More recently still, chymosin has been produced by microorganisms using recombinant DNA techniques, and this form of pure material is likely to become important in the near future. In the main, this chapter will deal with bovine chymosin and pepsin, since they are the best studied and the most specific for bovine milk, but some consideration will be given to the non-bovine proteinases.

Much of the literature on rennet clotting naturally relates to the activity of the proteinases. It should be pointed out that it is deceptively simple to relate the activity of the proteinases to the rate at which they clot milk. Since the clotting is a two-stage process, any variation in conditions can affect the two stages of the reaction differently. For example, milk does not clot at temperatures below 15°C (Berridge, 1942): this is caused, not by a loss of enzymatic activity, but by the low coagulation rate of renneted micelles at or below this temperature. Also, milks are found to clot differently at different pH values and concentrations of Ca^{2+} (Jakob & Puhan, 1986). In the discussion that follows, measurements of clotting time have not been used on their own as indicators of enzymatic activity, but details of the separate enzymatic and coagulation stages are considered. Similarly, the effects of additives or processing on the overall clotting time cannot be used as definite evidence of changes in the enzymatic reaction, and where possible, attempts have been made to describe their effects upon the individual steps in the clotting process.

2. Measurement of the Reaction of Acid Proteinases with κ-Casein

Since the primary stage of the reaction involves the hydrolysis of κ-casein, it is important to be able to estimate quantitatively the extent and rate of attack of the acid proteinases on κ-casein, either in isolated form in solution or in its natural state as a component of casein micelles. Unfortunately, there is no method that is simple and rapid enough to be used as a 'real-time' analysis which can be performed during the course of a renneting process.

The most obvious method for following the reaction is of course to measure the disappearance of κ-casein. This, however, can be complicated since κ-casein contains a number of components. The protein (see Chapter 10) exists as two genetic variants (a problem that can be avoided in the laboratory by using milk from individual animals), and in addition each of these genetic variants contains fractions that are glycosylated to different extents (Pujolle *et al.*, 1966; Schmidt *et al.*, 1966; Armstrong *et al.*, 1967). Thus, on anion exchange column chromatography (Vreeman *et al.*, 1977, 1986), or on electrophoresis in starch or polyacrylamide gels, κ-casein appears as a number of fractions, which complicates its estimation. Until modern methods of chromatography became available, it was generally found to be more convenient to study the kinetics by analysing the products of the reaction, namely the macropeptide or the para-κ-casein.

Insoluble para-κ-casein may be solubilized using urea, and its amount can be estimated by electrophoresis on polyacrylamide gels (Chaplin & Green, 1980) or cellulose acetate strips (Dalgleish, 1979), staining the bands and photometrically scanning the para-κ-casein band. Since all of the glycosylated residues of κ-casein are found in the macropeptide moiety (see Chapter 2), para-κ-casein appears as a single band on electrophoresis. A further advantage is that para-κ-casein is positively charged at pH values around neutrality, where the other caseins are negatively charged, so that para-κ-casein migrates in the opposite direction to the other caseins, allowing simple identification and quantification. An alternative method for quantifying para-κ-casein depends on the insolubility of the peptide in the absence of Ca^{2+}; proteolysis of milk followed by the chelation of the Ca^{2+} present has been used to assess the formation of para-κ-casein (Lawrence & Creamer, 1969). It is also possible to specifically precipitate para-κ-casein because of its higher isoelectric point (pH $>$ 5·0) compared with that of native κ-casein (pH $>$4·5; Bingham, 1975).

Quantification of the macropeptide liberated during renneting is somewhat more complex. All of the proteins in milk, with the exception of the macropeptide and lactoglobulin are precipitated by 2% trichloroacetic acid (TCA) (Wake, 1957) so that the proteinaceous material in the supernatant of renneting milk after addition of TCA to 2% is largely macropeptide. The solubility of the macropeptide in solutions of TCA depends on the degree of glycosylation; as the concentration of TCA is increased, the less glycosylated peptides are precipitated while the more highly glycosylated forms remain soluble (Armstrong *et al.*, 1967). The amount of macropeptide measured therefore depends on the concentration of TCA used to precipitate the caseins, and on the relative amounts of the differently glycosylated fractions present in the macropeptide mixture. This variation has in fact been used to differentiate between non-glycosylated and highly-glycosylated fractions, by comparing the different amounts of peptides soluble in 2% and 12% TCA (Hindle & Wheelock, 1970). Even when it has been separated, estimation of macropeptide is hampered, since it lacks aromatic amino acids, and does not absorb in the near UV. Measurement of the absorbance of the peptide at about 215 nm or determination of the nitrogen content of the TCA filtrate requires dialysis to remove traces of interfering materials.

A much more rapid estimation of the macropeptide can be achieved using high performance liquid chromatography (HPLC) (van Hooydonk & Olieman, 1982). An alternative rapid method is to use fluorescence techniques. Fluorescamine (4-phenylspiro[furan-2(3H),1-phthalan]-3,3′-dione) binds to amino groups of proteins (Udenfriend *et al.*, 1972), especially to the N-termini, and in the process becomes fluorescent. During renneting, new N-termini (of GMP) are produced, and this causes the fluorescence of samples taken during the course of the reaction and treated with fluorescamine to increase (Beeby, 1980). Treatment of the milk with TCA is still required to separate the macropeptides, since the background fluorescence arising from the other proteins in milk is otherwise too high.

HPLC can be used to estimate amounts of κ-casein and possibly the formation of para-κ-casein (Humphrey & Newsome, 1984). Another rapid method that requires no extensive prior treatment of the samples is fast protein liquid chromatography (FPLC). Using this method, it is possible to estimate, relatively rapidly, the amounts of the different forms of κ-casein in renneting samples (Dalgleish, 1986). This last method and the HPLC method of Humphrey and Newsome (1984) have the advantage that they measure the loss of the protein itself. However, there is still no method for quantifying the formation of products in real time, since all estimates depend on the analysis of samples taken during the reaction and treated with TCA or dissociating solutions.

It is possible to make approximate estimates of enzyme activity by observing the clotting time of a standardized sample of reconstituted skim milk powder under defined conditions (Berridge, 1952). An approximately linear relationship exists between the enzyme activity and the time required for the sample to clot (Foltmann, 1959; McMahon & Brown, 1983; Brown & Collinge, 1986). Such methods do not allow any detailed analysis of the kinetics of the enzymatic reaction, since the measured clotting time is compounded of the time required for a considerable proportion of κ-casein hydrolysis and the time taken by the renneted micelles to form detectable aggregates. However, under strictly defined conditions, the activities of different enzyme preparations can be compared using such measurements of clotting time.

The potential activity of the acid proteinases, rather than their specific action on κ-casein, can be studied using synthetic peptides, especially those which possess a chromophoric group whose absorbance changes when the peptide is cleaved enzymatically. The peptide, Leu–Ser–Phe(NO_2)–Nle–Ala–Leu–OMe, is a suitable peptide, the hydrolysis of which can be followed spectrophotometrically (Raymond *et al.*, 1973*b*). Other similar peptides have been proposed (Raymond & Bricas, 1979). Alternatively, peptides containing phenylalanine rather than nitrophenylalanine may be used, with the reaction being followed by estimation of the products using ninhydrin (de Koning *et al.*, 1978). These peptide substrates can also be used to measure general proteolytic activity: this can then be compared with milk-clotting activity to define the usefulness of the enzyme as a milk-clotting agent, where high clotting activity must be linked to generally low proteolytic activity.

Since commercial rennets contain varying proportions of chymosin and pepsin, it is often desirable to determine the separate amounts of the two enzymes. This has been attempted in several ways, and can be achieved by a combination of chromatographic separation of the enzymes and the subsequent analysis of the enzyme activity using the synthetic peptide described above (Collin *et al.*, 1981; Martin *et al.*, 1981). Such measurements can be related, if care is taken, to a standard milk-clotting assay, so that the assay using the reference peptide need not always be used (Collin *et al.*, 1981). It is also possible in principle to estimate the amounts of active enzymes in rennets using immuno-electrophoresis or immunodiffusion techniques (Andren & de Koning, 1982; Collin *et al.*, 1982; Garnot & Molle, 1982). Although qualitative identification of the enzymes is possible using the appropriate antisera (Collin *et al.*, 1982), the occurrence of an immunological reaction does not guarantee that the enzyme is active (Andren & de Koning, 1982; Garnot & Molle, 1982). The techniques must therefore be used with care.

3. Overall Kinetics of the Protease Reaction

Since the breakdown of the κ-casein substrate is essentially a single-step enzymic reaction, it seems reasonable to suppose that the kinetics of the proteolysis should obey the standard Michaelis-Menten formulation for the kinetics of such reactions. In this, the instantaneous rate of the reaction (i.e. the rate at which substrate $[S]$ is converted into product), v, is given by the relation:

$$v = -\mathrm{d}[S]/\mathrm{d}t = V_{\max} \cdot [S]/[K_{\mathrm{m}} + [S]] \qquad (1)$$

In this equation, $V_{\max}$ is the maximum rate at infinite concentration of substrate (i.e. it depends on the concentration of the enzyme) and K_{m} is the dissociation constant for the enzyme–substrate complex. The renneting reaction has been analysed in this way in a number of studies (Garnier, 1963; Castle & Wheelock, 1972; Dalgleish, 1979; Chaplin & Green, 1980; van Hooydonk *et al.*, 1984; Azuma *et al.*, 1984; Carles & Martin, 1985). The action of chymosin on synthetic peptides has also been shown to obey the Michaelis-Menten behaviour (Visser *et al.*, 1980). However, it should be noted that overt behaviour of this type is not always observed, because of the relation between the concentration of substrate and the constant K_{m}. It can be seen from eqn (1) that if $[S]$ is much larger than K_{m}, effectively a reaction that is zero-order with respect to the concentration of substrate will be observed, at least in the early part of the reaction. Conversely, if $K_{\mathrm{m}} \gg [S]$, the reaction becomes apparently first order, i.e.

$$-\mathrm{d}[S]/\mathrm{d}t = V_{\max} \cdot [S]/K_{\mathrm{m}} \qquad (2)$$

Various analyses of the reaction have been made using both isolated and micellar κ-casein as substrates, yielding values of K_{m} in the range 3×10^{-6} to about 5×10^{-4} mole litre^{-1}. In milk, the reaction appears to be approximately first order (van Hooydonk *et al.*, 1984), and fitting the time-course of κ-casein proteolysis gave values of K_{m} that were greater than the substrate (i.e. κ-casein) concentration, which is of the order of $1\text{–}1{\cdot}5 \times 10^{-4}$ M in unconcentrated milk. Lower values for K_{m} would not fulfil this condition, and would tend to predict a reaction with kinetics closer to zero-order. Thus, to describe the reaction that occurs in milk, it seems to be necessary to use either a first-order formulation or a Michaelis mechanism with a relatively high K_{m}. The decision as to whether or not the reaction is to be considered as a Michaelis reaction must depend on the method by which the results are analysed. Properly, the reaction should be studied during its early stages, using a variety of enzyme and substrate concentrations. However, a second method of analysing the Michaelis kinetics is to use the integrated form of eqn (1) over the time course of the reaction:

$$K_m \ln([S]_0/[S]) + ([S]_0 - [S]) = V_{max} \cdot t \quad (3)$$

where $[S]_0$ is the initial concentration of κ-casein. This has been used in some studies (Dalgleish, 1979; Chaplin & Green, 1980) and it is these that give the highest values of K_m (in milk).

There is a question as to whether the Michaelis-Menten mechanism is the correct formulation to use to describe the reaction. By using (1) and (3), an implicit assumption is made that enzyme and substrate are able to equilibrate at all times. This, in turn, implies that both enzyme and substrate are mobile throughout the solution: this assumption may also have implications for the mechanism of clotting after proteolysis (see below, Section 9). The proteinases, being relatively small in size compared with the casein micelles, are free to move through the solution, but this is hardly true of the κ-casein which forms part of the casein micelles, and which must be constrained to move through the solution as the micelles move, i.e. very slowly in comparison to the enzyme (van Hooydonk & Walstra, 1987*a*). It is possible in principle for the enzyme molecules to bind to the micelles and to create patches of para-κ-casein by attacking adjacent κ-casein molecules one after the other by a 'catch-and-razor' mechanism (Brinkhuis & Payens, 1985), rather than producing randomly distributed individual molecules of para-κ-casein, although studies of the aggregation behaviour appear to preclude this (Dalgleish, 1988*a*).

The fact that κ-casein is naturally incorporated into casein micelles suggests that the kinetics of proteolysis of micellar and soluble κ-casein might be different. It may be significant in this context that the lower values of K_m that have been measured have mainly derived from studies on isolated κ-casein or peptides (Garnier, 1963; Visser *et al.*, 1980; Azuma *et al.*, 1984; Vreeman *et al.*, 1986) while the higher values have been mainly measured using casein micelles in dispersions or in milk (Dalgleish, 1979; Chaplin & Green, 1980; van Hooydonk *et al.*, 1984). It has indeed been shown that aggregation of κ-casein increases the K_m (Vreeman *et al.*, 1986). The possibility of non-random proteolysis should, however, be borne in mind when the clotting reaction is to be studied.

In eqn (1), the value of V_{max} can be replaced by the expression $k_{cat} \cdot [E]$, where $[E]$ is the concentration of the enzyme and k_{cat} is the catalytic constant. Measured values for k_{cat} vary, as do the values for K_m. The highest value, $216\,s^{-1}$, was found for the action of chymosin on reconstituted skim milk powder (van Hooydonk *et al.*, 1984); lower values of about $60\,s^{-1}$ have been found for peptides (Visser *et al.*, 1980), and $30\,s^{-1}$ (Garnier, 1963) and 25–$40\,s^{-1}$ (Vreeman *et al.*, 1986) for isolated κ-casein. The highest value was obtained from the same experiments as those which gave high values of K_m. Since this result was obtained using milk and the others used κ-casein or peptides, once again it appears that there may be significant differences between the kinetics of chymosin action in milk and in model systems.

The inverse dependence of the clotting time of milk on the concentration of rennet is well known, and this variation is mainly attributable to the effect of enzyme concentration on the rate of proteolysis. The rate of the enzymatic reaction will increase linearly with the concentration of enzyme, which accords with either a Michaelis-Menten or a first-order mechanism (van Hooydonk *et al.*, 1984). Temperature also affects the clotting time, and although much of this variation can be attributed to the change in the rate of aggregation of renneted micelles, at least some can be attributed to the enzymatic reaction. For isolated κ-casein, there appears to be little change in K_m with temperature, and k_{cat} approximately doubles between 25 and 40°C (Garnier, 1963; Carlson *et al.*, 1987*a*). In milk, the same appears to be true for the value of k_{cat}, at least (van Hooydonk *et al.*, 1984). Certainly, the κ-casein of milk can be hydrolysed by rennet even at 4°C (Dalgleish, 1979), at which temperature the clotting reaction does not occur (Berridge, 1942).

4. Sensitivity of the Phe–Met Bond in Peptides and in κ-Casein and Factors that Affect the Rate of Hydrolysis

To clot casein micelles effectively, bovine κ-casein must be hydrolysed in the region of the Phe_{105}–Met_{106} bond (Jollès *et al.*, 1968). Cleavage at other points may, and in fact does, occur subsequently, but any general proteolysis of the caseins during the clotting reaction is a disadvantage, leading to increased losses of soluble peptides. The acid proteinases used in cheesemaking are all highly specific for this bond. This specificity may arise from factors such as the sequence of amino acids about the sensitive bond (i.e. the primary structure of the substrate), the conformation of κ-casein (i.e. its secondary and tertiary structures), and even its state of aggregation (quaternary structure). The first of these defines specific interactions between individual amino acids in the substrate and in the enzyme, and certain residues close to the sensitive bond appear to be important in this respect. However, the conformation of κ-casein must be such that the sensitive bond is located in a region of the protein readily accessible to the enzyme.

The important part of the sequence of bovine κ-casein (Chapter 2) appears to be between residues 97 and 129, and changes in this region can cause changes in the reactivity of the sensitive bond (Beeby, 1976; Hill & Hocking, 1978). The Phe–Met dipeptide is not split by chymosin (Voynick & Fruton, 1971), nor is the bond hydrolysed when it is in tri- or tetra-peptides (Hill, 1968; Schattenkerk & Kerling, 1973). However, incorporation of the bond into the pentapeptide, Ser–Leu–Phe–Met–Ala–OMethyl, allows it to be hydrolysed, and when the serine and leucine are exchanged to give the correct sequence appearing in κ-casein, the rate is enhanced (Hill, 1968, 1969). Thus, some residues in the immediate vicinity of the sensitive bond function to hold the substrate in its correct orientation in the active site of the enzyme. Further lengthening of this pentapeptide shows that amino acids added at either end, towards the N- and C-terminals of the protein, also influence the reactivity of the sensitive bond (Raymond *et al.*, 1973*b*; Schattenkerk & Kerling, 1973; Visser *et al.*, 1977). The effect of elongating the chain towards the C-terminal by three amino acids to give the peptide Ser_{104}–Lys_{111} increases the catalytic ratio, k_{cat}/K_m, and incorporation of the N-terminal Leu_{103} gives a further increase of about 600-fold in the catalytic ratio. Incorporation of His_{102} and Pro_{101}

leads to further increases in the activity of chymosin towards the peptide substrates (Visser *et al.*, 1976).

The model peptides just described have sequences identical to that of κ-casein. Other substrates have been used to determine the importance of either the phenylalanine or the methionine on the efficiency of the peptides as substrates. Norleucine is isosteric with methionine and can be used as a replacement, its incorporation increases the ratio k_{cat}/K_m by a factor of about 3 (Visser *et al.*, 1977). Conversely, the replacement of phenylalanine by *p*-nitrophenylalanine reduces the catalytic efficiency of the enzyme on the peptide substrate by about 3 (Raymond *et al.*, 1973*b*). Experiments with synthetic peptides suggest that a good substrate should possess hydrophobic residues at positions 103 and 108 (Leu and Ile in the natural protein). The hydroxyl group of Ser_{104} is strongly involved in both binding and catalysis, and the proline residues at positions 109 and 110 are also important (Visser, 1981). The shorter synthetic substrates do not show the same k_{cat} and K_m as are found for κ-casein, but a peptide derived by tryptic hydrolysis and containing residues 98–112 of κ-casein undergoes hydrolysis at a similar rate to intact κ-casein (Visser *et al.*, 1980) so that the residues at positions 98–102 (His–Pro–His–Pro–His) are also important in enzyme/substrate binding: K_m for the longer peptide is considerably smaller than that of the peptide 103–112.

The sequence 98–102 contains the only histidine residues in κ-casein, and their importance has been demonstrated by studies on the photo-oxidation of histidyl and other residues of κ-casein, which results in a much reduced sensitivity of the protein to renneting (Hill & Laing, 1965; Kaye & Jollès, 1978). It may be that only one of the histidyl residues performs an essential function and others may play a role in the aggregation of para-κ-casein. Surprisingly, iodomethylation of the methionine in intact κ-casein did not apparently reduce the sensitivity of the Phe–Met bond (Hill & Laing, 1965), so that, as found for the synthetic substrates, it appears that in κ-casein the residue in position 106 need not be methionine, so long as the sequence around it is correct. This is also confirmed by the rennetability of both human and porcine κ-caseins, which do not possess a Phe–Met bond (Chobert *et al.*, 1976). Human κ-casein has a Phe–Ile bond in the sensitive position (Fiat & Jollès, 1977; Brignon *et al.*, 1985) and is attacked by calf chymosin, although somewhat slower than bovine κ-casein. On the other hand, κ-casein from rat milk, which has a Phe–Leu bond in the sensitive region (Brignon *et al.*, 1985; Thompson *et al.*, 1985) is very resistant to hydrolysis by calf rennet (Kotts & Jenness, 1976), although the sequences 98–112 of cow and rat κ-caseins have 53% homology. The insensitivity may arise because the sequence of rat κ-casein shows a large deletion in the position occupied by residues 80–90 in the bovine protein, or because there may be a phosphorylated threonine residue close to the sensitive bond, or because lysines at positions 111 and 112 in the bovine casein are replaced by asparagine and glutamic acid in rat κ-casein (Thompson *et al.*, 1985).

The binding of the various substrates to the enzymes is significantly governed by the interaction of specific residues with particular sites on the enzyme. However, increases in the efficiency of binding will be possible if the three-dimensional structure of κ-casein presents the relevant peptides to the enzyme in an 'optimum' conformation. Many of the small peptide model substrates are structureless because they are too short to fold significantly; this is not the case in the intact protein. No detailed definite structural information is available for κ-casein, but calculations that aim to predict the conformation of proteins, have suggested that the Phe–Met bond is situated in a region of the protein that can form a projecting β-structure (Raap *et al.*, 1983), which may also interact with the enzyme by participating in β-sheet formation (Jenkins *et al.*, 1977). An alternative prediction of the secondary structure suggests that the Phe–Met bond is situated on a β-turn, which would also render it accessible to enzymatic attack (Loucheux-Lefebvre *et al.*, 1978).

There are also naturally-occurring variations in the structure of κ-casein, which may be used to assess the effect of some structural alterations remote from the Phe–Met bond (at least in terms of primary structure) on its susceptibility to hydrolysis. A number of studies has been made on the effect of glycosylation and it has been suggested that glycosylated forms of κ-casein are hydrolysed somewhat more slowly than the non-glycosylated forms (Sinkinson & Wheelock, 1970; Addeo *et al.*, 1984; van Hooydonk *et al.*, 1984; Vreeman *et al.*, 1986). The effect is not necessarily large, and other studies failed to find evidence that glycosylation had a significant effect (Chaplin & Green, 1980; Dalgleish, 1986). A second natural modification is the difference between the A and B genetic variants of κ-casein, where again the modifications consist of replacement of Ile for Thr at position 136 and Ala for Asp at position 148, both being some distance from the site of enzymatic attack. This modification seems to have a small effect on the rate of hydrolysis (Dalgleish, 1986). It is agreed that milk containing κ-casein B clots more rapidly than milk containing κ-casein A (Schaar, 1984; Aaltonen & Antila, 1987; Pagnacco & Caroli, 1987). However, there are differences in protein composition, micelle size distribution and in gel structure and syneresis (Pagnacco & Caroli, 1987; McLean & Schaar, 1989) associated with the two variants, so the specific influence of the genetic variant upon only the hydrolysis is not yet clear.

A further possible source of variation in the rate of hydrolysis of the sensitive bond may be related to the size of the micelles in which the κ-casein is located. From the measurements of clotting times on suspensions of micelles of different sizes, the clotting times of the largest and smallest micelles appeared longer than those of medium size (Ekstrand *et al.*, 1980). Since it is known that the rate of aggregation of fully-renneted micelles does not depend on micellar size (Dalgleish *et al.*, 1981), it is possible that the κ-casein in micelles of different size is hydrolysed at different rates. However, evidence for this is based on measurements of the clotting time only and may reflect factors such as the concentration of micelles in different preparations (Dalgleish, 1980*a*; Payens, 1984*a*) and the extent of aggregation required to give visible clotting, as well as possibly different rates of attack of the enzyme on κ-casein.

The proteolytic activity of the acid proteinases depends on both pH and ionic strength, which makes comparisons of rate constants difficult since in many cases measurements have been made under different conditions. Measurement of the maximal activity of the proteinases is hampered by the precipitation of the substrate caseins at pH values below 5. It has been suggested that the pH optimum for the attack of chymosin on κ-casein is in the range 5·0–5·5 (Humme, 1972), but in milk it appears that the optimum is in the region of pH 6·0 (van Hooydonk *et al.*, 1986*a*). Sufficient activity of chymosin remains at the natural pH of milk (about 6·7) to allow clotting of the milk, but some other acid proteinases lose their activity at greater or lesser rates around this pH value. The ionic strength of the renneting medium is also important in defining the activity of the proteinase. This may be because the enzyme and substrate are both negatively charged and tend to repel each other: this can be overcome by increasing the ionic strength. If the ionic strength is increased too much, however, it will interfere with specific charge interactions which are essential for enzyme activity, and consequently the activity will decrease. Activity therefore goes through a maximum as the ionic strength is increased (Payens & Both, 1980; Payens & Visser, 1981).

The concentration of specific ions, notably Ca^{2+}, has an effect on the rate of the enzymatic reaction. It is an established fact that the addition of Ca^{2+} to milk accelerates the overall clotting process, principally because of the effect on the aggregation stage of the reaction (Dalgleish, 1983). It has been reported that these ions also affect the enzymatic reaction, increasing its rate (Green & Marshall, 1977), although on the other hand, it has been claimed that concentrations of Ca^{2+} above 8 mM decrease the enzymatic activity (Bringe & Kinsella, 1986*a*). A third study has reported (van Hooydonk *et al.*, 1986*b*) that addition of $CaCl_2$ (up to 1·8 mM) to milk causes no change in the enzymatic rate. It is possible that these different observations relate to the types of experiments that were attempted, since in some the milk was diluted into a buffer solution containing Ca^{2+} (Bringe & Kinsella, 1986*a*) while in others the solution of Ca^{2+} was added to milk (van Hooydonk *et al.*, 1986*b*). It has also been noted (Bringe & Kinsella, 1986*b*) that anions affect the action of chymosin, apparently principally because of their size. Thus, inhibition increases in the order $Cl^- < Br^- < NO_3^- < SCN^-$; these ions also affect the rate of the coagulation reaction.

As described in an earlier section, the values for K_m and k_{cat} for the action of chymosin on the rennet-sensitive bond are subject to some uncertainty, although this may be partly because of different experimental conditions. Garnier (1963) quotes 33 μM for K_m, which is in good agreement with the value of 27 μM suggested by Visser *et al.* (1980), but is higher than some of the values given by Castle and Wheelock (1972) which were in some cases as low as 6 μM, and Vreeman *et al.* (1986) who found values in the range 3–9 μM, depending on the level of glycosylation. All of these values are considerably lower than the values of 500 μM reported by Dalgleish (1979) and 283 μM by Chaplin and Green (1980), and the high value defined by van Hooydonk *et al.* (1984). Values of 66 and 48 μM have been reported by Azuma *et al.*, (1984) and Carles and Martin (1985), respectively. According to Garnier (1963), the K_m was not dependent on temperature between 25 and 40°C, otherwise, some of the discrepancies might have been reconciled by postulating that K_m increased with decreasing temperature. All of the values, except those of Visser *et al.* (1980), were obtained at pH values between 6·6 and 7·0, so that the differences cannot all be explained by variation in pH. The values reported by

Garnier (1963) and Visser *et al.* (1980) for k_{cat} are in surprisingly good agreement, being $36\,s^{-1}$ at 35°C and pH 6·9 and approximately $65\,s^{-1}$ at 30°C and pH 4·7, respectively. This is reinforced by the value of $68{\cdot}5\,s^{-1}$ obtained at 30°C and pH 6·2 (Carles & Martin, 1985). These results are of interest in so far as they do not show the expected variations with temperature and pH which are shown by the overall enzyme activity (Humme, 1972; van Hooydonk *et al.*, 1986*a*).

5. Action of Acid Proteinases on Caseins other than κ-Casein

Acid proteinases that are successfully used for milk clotting are generally highly specific in their action on the bond 105–106 of κ-casein, although an enzyme from *Endothia parasitica* cleaves the bond 104–105 (Drøhse & Foltmann, 1989). Any other proteolytic reaction that occurs during clotting will render cheesemaking less efficient because soluble peptides may be released into the whey, leading to a reduced yield of cheese curd. Thus, it is important that rennets are specific, i.e. that they possess the high clotting activity and a low level of general proteolytic activity. Although the acid proteinases in general have a high specificity, they nonetheless hydrolyse caseins other than κ-casein, but considerably more slowly. This general proteolysis is too slow to be of importance during clotting, but during the ripening of cheese the residual rennet which is retained in the curd (Holmes *et al.*, 1977) plays an essential role in degrading the caseins (Grappin *et al.*, 1985; Fox, 1989). They may also, if the proteolysis is too extensive or of the wrong type, produce undesirable bitter peptides in cheese (Visser *et al.*, 1975).

Chymosin hydrolyses both α_s- and β-caseins in solution. The attack occurs at a number of specific sites in both proteins, although neither of the proteins possesses a Phe–Met bond in their structures. The primary cleavage site in β-casein is in the region of residues 189–193, being variously identified as the bond 189–190 (Pélissier *et al.*, 1974; Creamer, 1976) or 192–193 (Visser & Slangen, 1977). In fact, both bonds are susceptible to hydrolysis (Carles & Ribadeau-Dumas, 1984). The larger resultant peptide, named β-I-casein, can be further degraded to β-II casein by the hydrolysis of the bond 166–167 or 164–165, and β-II can then be cleaved at bond 139–140 to yield β-III casein (Creamer, 1976; Visser & Slangen, 1977). Splitting of the bond 127–128 is also possible on prolonged hydrolysis at low pH. It is probable that these reactions are sequential, so that it may be impossible to form β-III casein directly from β-casein without the intermediate formation of β-I and β-II peptides (Creamer, 1976): this may be a consequence of the conformation of β-casein causing the relevant bonds to be shielded from proteolysis. Alternatively, the apparently sequential formation of the peptides may arise simply because of different rates of attack on the different bonds (Visser & Slangen, 1977). β-I-Casein does not undergo temperature-dependent aggregation (Berry & Creamer, 1975) so that the differences in the rates of attack on the three susceptible bonds may arise from the dissociation of polymeric β-casein as β-I is formed by the primary enzymic attack. This dissociation may render the other bonds more accessible. In contrast to the behaviour in solution, chymosin and pepsin do not appear to hydrolyse β-casein in ripening cheese (Foster & Green, 1974), perhaps owing to low water activity (Phelan *et al.*, 1973), or perhaps as a result of the immobilized state of the β-casein.

Proteolysis of β-casein by chymosin depends on the pH and ionic strength of the solution, and upon the temperature. Decreasing the pH increases the rate of proteolysis, in accordance with the pH-dependence of the primary reaction on κ-casein (Creamer, 1976). The rates of production of β-II and β-III peptides are slowed even more than the rate of formation of β-I by increasing pH, but at low temperatures the dependence on pH is less than at 37°C. Increasing the ionic strength reduces the rate of hydrolysis of β-casein to β-I, although the rates of formation of β-II and β-III are less affected. As with the dependence on pH, the effects of ionic strength are greater at higher temperatures. These results are again consistent with the suggestion that polymeric β-casein is relatively resistant to proteolysis by chymosin, as do results showing that addition of Ca^{2+} (which increases the extent of aggregation of the protein) slows down the rate of cleavage of the first bond (Creamer *et al.*, 1971; de Jong & de Groot-Mostert, 1977).

It has been pointed out by Visser (1981) that the sites of attack of chymosin on α_s- and β-caseins appear to occur at bonds involving either phenylalanine or leucine. The primary site of attack by chymosin on α_{s1}-casein is Phe_{23}–Phe_{24} or Phe_{24}–Val_{25}, to form the peptide α_{s1}-I casein (Creamer & Richardson, 1974; Pélissier *et al.*, 1974; Carles & Ribadeau-Dumas, 1985). Further proteolysis is possible, and occurs in cheese (Creamer & Richardson, 1974) as well as in solution. The products have been identified as α_{s1}-II (residues 24/25–169 of α_{s1}-casein), α_{s1}-III and α_{s1}-IV (residues 24/25–149/150), α_{s1}V (residues 29/33–199) and α_{s1}-VII (residues 56–179) (Mulvihill & Fox, 1979*a*), but alternative positions for hydrolysis (as many as 20) have been suggested (Pélissier *et al.*, 1974). The formation of each of these peptides depends upon the particular conditions of pH used: for example at pH values above 5·8 hydrolysis follows the sequence α_{s1}I $\rightarrow$ α_{s1}-II $\rightarrow$ α_{s1}-III/IV, while at lower pH values α_{s1}-I is hydrolysed directly to α_{s1}-V (Mulvihill & Fox, 1977). The hydrolysis of the first peptide bond is, in contrast to the hydrolysis of β-casein, only slightly dependent on the ionic strength of the medium (Mulvihill & Fox, 1980), whereas the subsequent degradation of the α_{s1}-I casein is dependent on the salt concentration. As with β-casein, the state of aggregation of the casein appears to affect proteolysis, and in particular, micellar α_{s1} casein is degraded much more slowly than when it is in solution (Fox & Guiney, 1973). During cheese maturation, α_{s1}-casein is completely degraded, possibly because the proteolytic reactions are less dependent upon the concentration of salt present than is the hydrolysis of β-casein.

In all cases, these secondary reactions of chymosin on α_{s1}- and β-caseins are very much slower than the hydrolysis of κ-casein. Data available for the attack of chymosin on α_{s1}-casein in solutions give values for K_m of $4{\cdot}5 \times 10^{-4}$ and $3{\cdot}7 \times 10^{-4}$ M and k_{cat} of 3·8 and 0·66 s^{-1} (Hill *et al.*, 1974; Carles & Ribadeau-Dumas, 1985). Thus, the binding appears to be weaker than the binding of κ-casein and the catalytic constant is lower, when these values are compared with those for κ-casein on pages 589–90). Similarly, for β-casein, K_m is in the region of 7–75 μM and k_{cat} between 0·56 and 1·54 s^{-1}, depending on the aggregation state (Carles & Ribadeau-Dumas, 1984). These factors combine to make the proteolysis of α_{s1}- and β-caseins unimportant during the formation of the curd in cheesemaking, but significant as far as α_{s1}-casein is concerned during cheese ripening.

It can be argued that the chymosins and pepsins produced by a species have evolved to be specific for the clotting of the milk of the species, and that acid proteinases from other species may be less specific in their action. For example, procine pepsin is more proteolytic than calf chymosin towards bovine milk, whereas bovine pepsin is more similar to chymosin in its attack on κ-casein (Mulvihill & Fox, 1979*b*). Microbial enzymes cause degradation patterns in casein different from those produced by chymosin and pepsin (El-Shibiny & Abd El-Salam, 1976, 1977; Vamos-Vigyazo *et al.*, 1981). Comparisons between different enzymes are often difficult to make, since the reactivities of the individual acid proteinases depend differently on pH and ionic strength, as well as being affected by the aggregation state of the caseins. In respect of non-specific proteolysis, it appears that the non-bovine proteinases are generally more proteolytic than calf chymosin but that for most of the acid proteinases used in cheesemaking the primary site of attack is the Phe–Met bond in κ-casein.

6. Cloned Chymosins and Acid Proteinases other than Chymosin

Since the world supply of calf rennet is inadequate to meet the needs of cheese-makers, it has been necessary to develop the use of other acid proteinases as rennet substitutes. Only acid proteinases possess suitable activities, since, although proteinases such as trypsin cause milk to clot, they also degrade the milk clot almost as soon as it has formed. Until recently, two types of chymosin substitutes were available: the pepsins (obtained from the stomachs of ruminant and other species), and the acid proteinases of bacterial or fungal origin. However, recently the gene for chymosin has been cloned and expressed in microorganisms (e.g. yeasts or *E. coli*), to allow the production of rennets containing chymosin only (Foltmann, 1987). At the time of writing (1990), the products of cloning are becoming available and being used commercially. A number of tests have shown that, as expected, they optimize cheese yield and behave otherwise as chymosin extracted from the stomachs of calves (Hicks *et al.*, 1988; Meisel, 1988; Bines *et al.*, 1989; Pszczola, 1989; Banks, J.M., 1990, pers. comm.). It is too early to determine whether the availability of cloned chymosins will make any appreciable difference to the pattern of coagulant use, as described below, or whether the cloned chymosin will simply be used to replace veal rennet.

Among the more traditional chymosin substitutes that have been used or studied are pepsins from cow, sheep, goat, pig and chicken, and the acid proteinases of *Mucor miehei*, *M. pusillus* and *Endothia parasitica*. The most important criterion to be applied to these enzymes in relation to cheesemaking is that their general proteolytic activity should be low in comparison with their activity towards the rennet-sensitive bond of κ-casein. In addition to individual proteinases, mixtures of different proteinases can be used as rennet substitutes. The distribution of use of rennet substitutes appears to be variable worldwide, e.g. it was estimated that in 1974 about 60% of the cheese manufactured in the United States was produced using microbial proteinases (Sternberg, 1976), whereas in Europe the microbial rennets are less widely used. A number of reviews are available on the different rennets and rennet substitutes (Sternberg, 1976; Green, 1977; de Koning, 1978; Scott, 1979; Kay & Valler, 1981), and of their structures (Foltmann, 1987).

All the acid proteinases used have the ability to hydrolyse κ-casein at approximately the required position, but their properties are not identical. For example, they have different pH optima, and indeed different stabilities and pH-dependence of activity (Hofmann, 1974). Pig pepsin is unstable above pH 6·0 (Fox, 1969; Fruton, 1971), compared to chymosin (pH 6·7: Foltmann, 1966), *Mucor miehei* proteinase (pH 6·7: Alais & Lagrange, 1972) and chicken pepsin (pH 8·0: Bohak, 1969). Since pH affects the enzymes differently, it may in principle be necessary to modify the cheesemaking process to accommodate this if mixtures of proteinases are used (Jespersen & Dinesen, 1979), but differences may be sufficiently small as to render the adjustment unnecessary. A further difference between the coagulants is their heat stability, which is important if the cheese whey is to be processed. Chymosin shows maximum activity at about 40°C, and at higher temperatures shows evidence of heat-denaturation, as does pepsin, whereas for *M. miehei* proteinase inactivation occurs at temperatures about 65°C (Scott, 1979). The order of inactivation of various acid proteinases is pig pepsin, bovine pepsin, chymosin, *M. Pusillus* proteinase and *M. miehei* proteinase, and all are rendered more heat stable by decreasing the pH (Duersch & Ernstrom, 1974; Hyslop *et al.*, 1979*a*; Thunell *et al.*, 1979). All of the enzymes can be inactivated by prolonged heating, but care must be taken that the whey proteins are not also denatured by too high a temperature during this process. There are different proteinases from *M. miehei* (Lagrange *et al.*, 1980), some of which are heat-labile (Wallace, 1981), and these tend to be more widely used.

Comparisons of the proteolytic and clotting activities of the enzymes show that almost all the rennet substitutes compare unfavourably with chymosin. The clotting to proteolysis ratios of chymosin (40·9) and bovine pepsin (4·3) demonstrate the greatly enhanced general proteolysis caused by pepsin (de Koning, 1978). In a study using synthetic substrates, bovine pepsin was found to be considerably more proteolytic than chymosin; *Mucor* proteinases were somewhat less proteolytic than pepsin but considerably more so than chymosin (Martin *et al.*, 1980). However, the heat-labile *Mucor* proteinase has been reported to compare very favourably with chymosin in respect of its proteolytic activity (Ramet & Weber, 1981). Reconstituted *M. pusillus* proteinase powder, rather than enzyme extract, has been reported to be less proteolytic than chymosin (Philippos & Christ, 1976) while all other preparations tested were more proteolytic, both on κ-casein and on α_5- and β-caseins.

The products of the action of non-chymosin proteinases on the caseins have not been analysed in great detail. The enzymes have been shown to hydrolyse the B-chain of insulin with different specificities (Sternberg, 1976) and similar differences would be expected for the hydrolysis of caseins. Electrophoretic patterns of the breakdown products of β-casein are different when calf rennet and *Mucor* enzymes are used (Vanderpoorten & Weckx, 1972; El-Shibiny & Abd El-Salam, 1976) although one product from α_{S1}-casein was common to all enzymes. The peptides produced by chymosin and by *Mucor* proteinase are relatively similar, whereas proteolysis by the proteinase of *E. parasitica* gives a very different pattern (Kovacs-Proszt & Sanner, 1973; Drøhse & Foltmann, 1989), and is more proteolytic than chymosin, pepsin or *M. pusillus* protease (Tam & Whitaker, 1972). The different patterns of proteolysis are most evident during cheese ripening when bitter peptides can be formed. A good example of this is given by a number of studies on the suitability of chicken pepsin for the manufacture of Cheddar cheese: off flavours and bitterness developed much faster than when calf rennet was used (Stanley *et al.*, 1980; Green *et al.*, 1984).

It is evident that, since the proteolytic activities of non-chymosin acid proteinases are almost invariably greater than that of chymosin, there may be problems associated with long-maturing cheeses. The ideal solution is to use as coagulants mixtures of proteinases which contain enzymes less proteolytic and more proteolytic than chymosin. Of the available enzymes, pig pepsin is the only one that is potentially less proteolytic than chymosin, since it is readily denatured in the cheesemaking process (Green & Foster, 1974). Mixed proteinases have been studied (Phelan, 1973; Green & Stackpoole, 1975) and were used as the so-called Fifty-Fifty coagulants which found commercial use (Jesperson & Dinesen, 1979). In general, however, the enzymes used as coagulants are chosen so as to minimize their proteolytic activities. Most commercial calf rennets do in fact contain mixtures of chymosin and bovine pepsin in varying amounts.

7. Can Immobilized Proteinases Be Used to Clot Milk?

In economic terms, there would be clear benefits in the possession of a reusable clotting agent. Therefore, extensive research on the possibility of immobilizing chymosin or pepsin on solid supports has been performed, with the aim of using the immobilized enzyme in column reactors to perform the proteolysis of κ-casein. An ideal process would consist of cooling milk to prevent aggregation (Berridge, 1942) and passing it through a column of the immobilized proteinase, to convert its κ-casein to para-κ-casein. Warming the milk would then allow a curd to form.

There have been a number of claims that such reactors are practicable. However, there are good reasons why immobilized proteinases will be very inefficient for milk clotting. It has been demonstrated experimentally that the efficiency with which κ-casein is hydrolysed by preparations of pepsin, coupled to dextrans of different sizes, decreases with the size of the dextran–pepsin conjugate (Chaplin & Green, 1982). Extrapolation to the limit where the pepsin is completely immobilized indicates that the activity of the proteinase will be very low. Immobilized enzymes are successfully used with small, rapidly diffusing substrates, which move rapidly around the immobilized enzyme; in contrast, the casein micelles are large particles with slow rotational and translational diffusion. Normally, the clotting enzyme is small in relation to the casein micelles, and it is this which diffuses rapidly and causes rapid reaction. So, when the enzyme is immobilized, and the substrate has only slow translational and rotational diffusion, it is to be expected that proteolysis will be slow and inefficient, as has been found experimentally. This has been confirmed by an experiment which showed that only limited hydrolysis of κ-casein occurred, even when the proteinase was linked to the immobile support via a flexible link (Beeby, 1979), and it is possible that only the soluble fraction (i.e. that fraction not forming part of the casein micelles) of κ-casein was hydrolysed by this form of immobilized chymosin. Similarly, it has been shown that very prolonged treatment of milk with immobilized chymosin was required to destroy sufficient κ-casein to allow aggregation of the casein micelles (Dalgleish, 1979). Since the reactors must be run at a low temperature to prevent coagulation of the milk, and it is known that some β- and κ-caseins dissociate from the micelles into the soluble phase, this mobilization will render these caseins more susceptibile to hydrolysis by the immobilized proteinase.

Nevertheless, a number of reports has been published of apparently successful coagulation of milk by immobilized proteinases, and so, in view of the arguments given above, it must be explained how sufficient hydrolysis of κ-casein to cause coagulation of milk can occur in reactors of immobilized chymosin. The simplest explanation is that, in addition to the proteolysis of solubilized κ-casein, chymosin is released from the column and remains active in the milk (Voutsinas & Nakai, 1983). Only small amounts of proteinase dissociated in this way will eventually cause clotting when the milk is warmed. Such desorption of the proteinases from hydrophobic carriers has been demonstrated (Voutsinas & Nakai, 1983) and indeed desorption of only partly immobilized proteinases has been estimated to underlie many of the apparently successful results (Green & Crutchfield, 1969; Carlson, 1983). It is common for enzyme activity to be irreversibly lost during extensive use of the reactors (Taylor *et al.*, 1979) and it is probable that some or all of this loss of activity is attributable to desorption of the originally immobilized proteinases. The loss of enzyme activity in this way will depend partly on the method by which the enzyme is coupled to its support, but also on the materials present in the solutions being used, which may enhance the tendency to desorption.

This dissociation of the proteinases from their supports is probably the main reason why clotting has been observed in many studies. At normal temperatures and pH values, extensive proteolysis of the κ-casein is required before the micelles will aggregate (Dalgleish, 1979; Chaplin & Green, 1980), and so any experiments where coagulation seems to occur at low extents of proteolysis at normal temperature and pH should be doubted (Hicks *et al.*, 1975; Lee *et al.*, 1977). This is confirmed by the observation that apparently fully immobilized chymosin hydrolysed only limited amounts of κ-casein and no clotting was observed (Beeby, 1979). It has been shown that milk can clot when it has been mixed with ultracentrifugate that has been passed through a column of apparently immobilized pepsin

(Hicks *et al.*, 1975). This observation can most readily be explained by assuming that a quantity of the enzyme had been released into the solution, since no other mechanism for such clotting can be suggested.

Therefore, although it is in principle possible to achieve clotting via very careful use of reactors involving immobilized enzymes, the general opinion must be that this cannot be practicable in anything larger than a laboratory experiment. There are no commercial reactors available, and this must reflect the uncertainties in the mechanism, in the efficiency, and the overall practicability of the methods.

8. Effect of Pretreatment of Milk on the Enzymatic Reaction

Heating milk prior to renneting renders it difficult to clot by normal renneting procedures (Morrissey, 1969). The effects are essentially reversible until the milk has been heated to temperatures that cause the denaturation of β-lactoglobulin (Wheelock & Kirk, 1974). Several possible reasons for this loss of clotting may be advanced: the denaturation of β-lactoglobulin and its interaction with κ-casein may cause the κ-casein to be less accessible and/or susceptible to rennet action; conversely, the enzyme may hydrolyse the κ-casein but the renneted micelles may be unable to clot because of the denatured β-lactoglobulin bound to their surfaces. Alternatively, heating may cause more far-reaching changes in the micellar strucure itself such that neither of these need be considered as primary causes of the loss of rennetability of the milk. It is, for example, possible that heating causes changes in the distribution of calcium phosphate in the micelles and the serum (Fox, 1981) which render even renneted micelles less capable of aggregating (van Hooydonk *et al.*, 1987). Such a structural factor may be involved because the effect of heating (providing that it has not been too severe) can be almost totally reversed by decreasing the pH of the milk to below 6 and then readjusting it to pH 6·3 or even higher before adding rennet (Banks *et al.*, 1987; Singh *et al.*, 1988). The effect of changing the pH in this way is to partly dissociate the calcium phosphate of the micelles as the pH is lowered (Davies & White, 1960; Dalgleish & Law, 1989), followed by its reformation (and possible partial reconstitution of the micelles) as the pH is raised again, although not all of the dissociated Ca is restored to the micelles as the pH is increased (Singh *et al.*, 1988). Caseins may be lost from the micelles as the result of decreasing pH (Snoeren *et al.*, 1984), depending on the temperature (Dalgleish & Law, 1988). The major casein dissociated is β-casein, but if the milk is maintained at 30°C, very little pH-induced dissociation of the caseins occurs. The effect of heating can also be reversed, at least partially, by adding of $CaCl_2$, and this suggests that heating interferes with the coagulation mechanism of the renneted micelles, which is highly sensitive to Ca^{2+}, as well as with the enzymatic reaction.

The enzymatic reaction appears to be inhibited in milk that has been heated (Damicz & Dziuba, 1975) but inhibition does not appear to be complete, as gauged by the release of peptides by the enzyme (Wilson & Wheelock, 1972; Wheelock &

Kirk, 1974; van Hooydonk *et al.*, 1987; Reddy & Kinsella, 1990). The inhibition appears to prevent a portion (between one-third and one-half) of the κ-casein from being hydrolysed after skim milk has been heated for 1 h at 90°C or by heating casein/β-lactoglobulin mixtures for short times at 85°C (Damicz & Dziuba, 1975). The effect has been studied over a range of temperatures (van Hooydonk *et al.*, 1987; Reddy & Kinsella, 1990), and it has been demonstrated that both the rate and the extent of the hydrolysis of κ-casein are affected by heating, although different authors provide different estimates of the magnitude of the effects. It may be concluded, therefore, that the enzyme is prevented from hydrolysing sufficient κ-casein on the surfaces of the micelles to make them aggregable, since it is known (Dalgleish, 1979; Chaplin & Green, 1980) that rennet must hydrolyse a considerable proportion of the κ-casein before aggregation at a measurable rate can occur. The adjustment of pH described above may, by altering the structure of the micelles, render them more susceptible to aggregation at limited extents of hydrolysis of the κ-casein, or may, by reforming the micelles, re-establish the susceptibility of the κ-casein to rennet. Although the primary cause of inhibition is expected to be the reaction of β-lactoglobulin with κ-casein, the effect of adding reducing agents is small, and the enzyme still cannot hydrolyse all of the κ-casein (Shalabi & Wheelock, 1977). α-Lactalbumin also may play a part in the effect of heat, since it also inhibits the release of peptides by rennet (Shalabi & Wheelock, 1976; van Hooydonk *et al.*, 1987). There remains, however, much to be explained on the mechanism of the renneting reaction in heated milk.

Cooling milk before renneting tends to increase its clotting time (Qvist, 1979) and this has been shown to arise at least in part from the decrease in the rate of the renneting reaction (van Hooydonk *et al.*, 1984). Changes in the casein micelles occur during cooling, e.g. dissociation of some β and κ-caseins, and solution of some micellar calcium phosphate; which of these may be responsible for the change in clotting timc has not been established. This decrease can be partly reversed by pasteurization or by holding at high temperatures but recovery is not complete.

A further factor that may influence the renneting of milk is homogenization. After the homogenization process, casein micelles are attached to the surface of the fat globules and some of the micelles may be partly dissociated before adsorption (Oortwijn *et al.*, 1977). Change in the surroundings of κ-casein and the structures of the adsorbed casein micelles may, therefore, induce changes in behaviour during renneting. Qualitatively, this does not seem to be the case: a study where the kinetics of rennet-induced aggregation of skim and homogenized milks were compared (Robson & Dalgleish, 1984) showed that maximal rates of aggregation were attained at the same time, suggesting that rennet acted similarly on both systems, although the particles aggregated rather differently (Robson & Dalgleish, 1987). This presumably reflects the fact that the micelles are relatively intact when bound to the surfaces of the fat globules in the homogenized milk (Green *et al.*, 1983). A different situation has been obtained if κ-casein, rather than casein micelles, is bound directly to the fat globule surfaces: in this case, the action of the rennet can be considerably altered, presumably because κ-casein binds to the surface in such a conformation that the susceptible bond is not available to the enzyme (Dickinson *et al.*, 1987).

Concentration by ultrafiltration is an important factor which affects the renneting of milk. Its effect is to decrease the rate of the renneting reaction (van Hooydonk *et al.*, 1984). It is claimed that this results from the retardation of the diffusion of the enzyme because of the increased volume fraction of the casein micelles. However, the clotting of the concentrated milks is enhanced so that the overall clotting process appears faster (Garnot & Corre, 1980; Dalgleish, 1980*a*; Mehaia & Cheryan, 1983; Payens, 1984*a*).

9. The Coagulation of Renneted Casein Micelles

Before they are renneted, the casein micelles in milk show no tendency to aggregate. This stability arises from two possible mechanisms. First, the micelles carry negative charges on their surfaces, partly, but not completely, as a result of the micellar surface being coated with macropeptide regions of casein (Donnelly *et al.*, 1984; Dalgleish *et al.*, 1989). Because of the repulsion of these like charges, DLVO theory will potentially explain the micellar stability (Green & Crutchfield, 1971; Green, 1973; Payens, 1978). This principle cannot, however, completely account for all aspects of micellar stability (Payens, 1979), since the densities of surface charge appear to be rather too low to prevent aggregation. An alternative mechanism for stabilizing the micelles is that of steric stabilization, in which the macropeptide moieties of κ-casein project into the solution from the surfaces of the micelles (Fig. 1). Because the macropeptides are flexible and hydrophilic, the aggregation of the casein micelles is prevented because the 'hairy' outer layers of the micelles cannot interpenetrate. This last cause of stabilization has been favoured by a number of authors (Holt, 1975; Walstra, 1979; Walstra *et al.*, 1981; Darling & van Hooydonk, 1981; Holt & Dalgleish, 1986; Horne, 1986; Dalgleish & Holt, 1988).

During renneting, some of the charge on the surface of the casein micelles is lost. Measurements of the electrophoretic mobility of the particles (Green & Crutchfield, 1971; Pearce, 1976; Darling & Dickson, 1979; Dalgleish, 1984) show that approximately half of the charge is lost, and this in itself will make the particles coagulate more readily. Moreover, the charge decreases approximately proportionately with the extent of proteolysis of κ-casein (Holt & Dalgleish, 1986); the relationship is not linear, since the rate of change of the surface charge with the renneting is faster towards the end of the reaction. The hydrodynamic radii of the micelles decrease by approximately 5 nm during renneting (Walstra *et al.*, 1981) and this decrease also appears to be linear with the change in surface charge (Holt & Dalgleish, 1986). Because of the extensive hydration of the surface layer (Walstra, 1979), the true decrease in the radii may be as much as 12 nm (Holt & Dalgleish, 1986). It has also been shown by electron microscopy that the number average diameter of the particles decreases, although by a smaller amount than found by the other investigations which used photon correlation spectroscopy, during the early stages of renneting (Guthy *et al.*, 1989). The viscosity of the milk also decreases during this period before aggregation starts (Scott-Blair & Oosthuizen, 1961; Guthy & Novak, 1977; Guthy *et al.*, 1989), as does its turbidity (Payens, 1977). The decrease in micellar radius, which seems to have the same value irrespective of the initial radii of the micelles, is consistent with the loss of the 'hairy' macropeptide surface of the particles. However, κ-casein stabilizes even when the surface layer is not extended: the surface layer may be collapsed by treatment with ethanol, but the micelles are still stable until renneted (Horne, 1984). Therefore, the changes that are brought about in micelles by renneting involve possible losses of both charge stabilization and steric stabilization. The experimental measurements, especially the change in radius caused by renneting, strongly suggest that the κ-casein must be largely on or near to the surface of the micelle, a theory supported by other evidence (McGann *et al.*, 1980; Donnelly *et al.*, 1984; Dalgleish *et al.*, 1989).

Para-κ-casein is insoluble, and so renneting of isolated κ-casein causes precipitation almost immediately. The kinetics of the aggregation have been described (Payens, 1977; Payens *et al.*, 1977) and have been shown to describe the reaction in detail (Hyslop *et al.*, 1979*b*). However, the aggregation behaviour of isolated κ-casein during renneting does not serve as an adequate model for the clotting of casein micelles. As soon as an individual molecule of κ-casein has been hydrolysed by chymosin, it is free to aggregate; the effect of chymosin is therefore to provide a steady stream of monomeric material for a subsequent aggregation reaction (Payens *et al.*, 1977). This can indeed by seen when κ-casein is hydrolysed by pepsin (Hyslop *et al.*, 1979*b*).

In casein micelles, proteolysis of a small number of κ-casein molecules will have much less effect on the aggregation properties since the micelles contain many hundreds or even thousands of such molecules. The para-κ-casein produced in the micelles by renneting can only aggregate when the whole micelle is capable of aggregating, and it is this which causes the lag stage before aggregation is observed (Dalgleish, 1980*b*; Darling & van Hooydonk, 1981). A short lag phase is seen when isolated κ-casein is renneted, but the origin of this is different: the concentration of particles capable of aggregation is low at the start of the reaction and then increases rather rapidly (Payens *et al.*, 1977). Renneted micelles appear to be incapable of aggregating until about 60–80% of their κ-casein has been hydrolysed, after which the concentration of micelles capable of aggregating increases rapidly (Dalgleish, 1979). This behaviour can be explained either by the loss of surface charge during renneting or by loss of steric stabilization; both of the predicted curves have much the same overall shape and it is not easy to definitively conclude that only one mechanism is operative, although the model based on steric stabilization appears to provide a reasonable fit to observations (Darling & van Hooydonk, 1981; Horne, 1984; Dalgleish & Holt, 1988). The action of rennet can be seen as providing 'hot spots' via which the micelles can aggregate, these reactive areas being produced by removal of κ-casein from a sufficiently large area (Payens, 1979, 1982; Green & Morant, 1981). This area has been suggested to be the space occupied by about 20 molecules of κ-casein (Dalgleish & Holt, 1988).

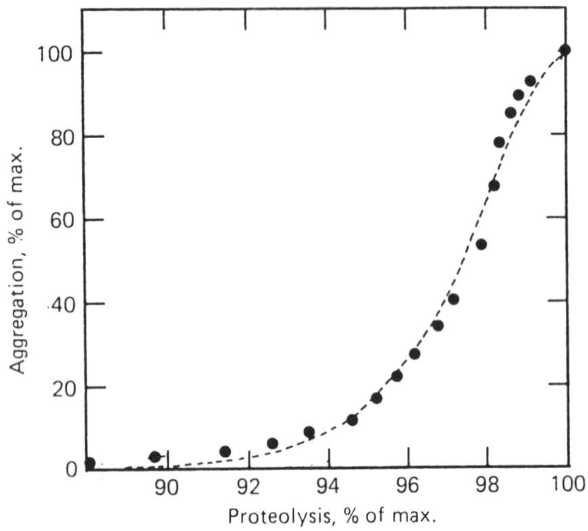

Fig. 2. The increase in aggregation of renneted casein micelles as a function of the extent of breakdown of κ-casein (Adapted from Dalgleish (1979) with permission).

During the last 20% of the proteolytic reaction both the concentration of micelles that are capable of aggregation and the rate at which they can aggregate increase rapidly, as the last of the stabilizing surface is removed (Fig. 2). Finally, when the micelles have been completely denuded of their κ-casein macropeptide a limiting rate for the aggregation is reached, and the micelles aggregate by a Smoluchowski (1917) mechanism, i.e. the growth of particle weight with time is linear (Payens *et al.*, 1977). This certainly applies in dilute solution, where most of the measurements have been made (Dalgleish *et al.*, 1981; Dalgleish, 1983); it is possible that in the more concentrated dispersion represented by milk, the aggregation mechanism may be rather different because the micelles are fairly close to each other (within one or two diameters). In such cases, Smoluchowski theory is not strictly applicable, although it may be so in practice. Most theoretical treatments of the renneting reaction (Payens *et al* ., 1977; Dalgleish, 1980*b*; Darling & van Hooydonk, 1981) have assumed that Smoluchowski kinetics apply. In milk during normal cheesemaking, aggregation is very rapid and it is demonstrable that not all of the κ-casein has been hydrolysed before clotting takes place at 30°C (Green & Marshall, 1977), so that the micelles are not aggregating at their maximum rate during clotting at the temperatures normally used for cheesemaking, nor is it necessarily true that all of the casein micelles in the milk actually participate in the formation of the initial curd.

The aggregation rate of renneted micelles is unaffected by the concentration of rennet (Brinkhuis & Payens, 1984), or by the size of the micelles (Dalgleish *et al.*, 1981). It is, however, very sensitive both to the concentration of Ca^{2+} present in the solution and to temperature (Dalgleish, 1983; Brinkhuis & Payens, 1984). Milk itself will not clot at less than about 15°C (Berridge, 1942) and this is a direct

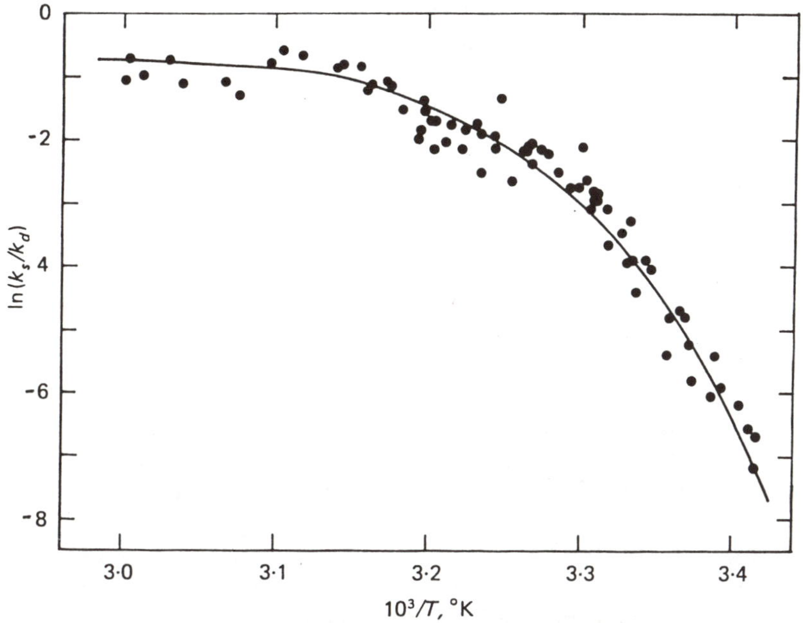

Fig. 3. The increase of the rate constant for the aggregation of completely renneted micelles, as a function of temperature, shown as an Arrhenius plot. Casein micelles were pre-renneted at 0°C before the aggregation rates were measured. k_s is the observed rate constant and k_d is the maximum possible rate constant calculated from the number of collisions. (Reproduced from Dalgleish (1983) with permission).

consequence of the slowness of the aggregation reaction at that temperature. The Arrhenius plot for the coagulation reaction is non-linear, being curved very sharply at the low-temperature end (Fig. 3). However, at temperatures of above 45°C the aggregation is very efficient indeed, and approaches the theoretical maximum rate at which particles can collide (Fig. 3). To some extent, the effect of temperature below 45°C can be overcome by increasing the concentration of Ca^{2+} present in the solution; above 45°C, the concentration of Ca^{2+} has little effect on the reaction because it is already near its maximum rate (Dalgleish, 1983). These results on temperature apply equally to measurements made when micelles are suspended in synthetic buffers (Dalgleish, 1983) or in ultrafiltrate from the same milk (Brinkhuis & Payens, 1984).

It is probable that more than one mechanism is involved in the coagulation, because a number of factors affect the rate of aggregation. In the absence of repulsive forces, van der Waals' attraction may be sufficient to hold the micelles together, but a number of other forces may act as well. From the temperature-dependence of the rate of aggregation or coagulation (Kowalchyk & Olson, 1977), it is plausible to assume that hydrophobic interactions (which become stronger with increasing temperature) may be important. However, because the rate of aggregation changes with the concentration of Ca^{2+}, these ions may play a part in the aggregation (Green & Marshall, 1977) which is incompatible with the aggregates being held together only by hydrophobic interactions (Dalgleish, 1983). Furthermore, the rate of aggregation is decreased when the ionic strength is increased, suggesting that specific ion-pair formation may be important (Knoop & Peters, 1976; Slattery, 1976; Green & Marshall, 1977; Payens, 1977).

An additional factor in the aggregation appears to be the state of the micellar calcium phosphate. It is well established that a decrease in pH leads to a decrease in the rennet coagulation time (van Hooydonk *et al.*, 1986*a*) but most of this is probably caused by the increase in enzyme activity as the pH is lowered (Humme, 1972; van Hooydonk *et al.*, 1986*a*). However, it has also been shown that pH does exert some effect on the rate of coagulation of the renneted micelles (Kowalchyk & Olson, 1977). The rate of aggregation increases as the pH decreases (Kim & Kinsella, 1989), and the extent of proteolysis required for aggregation decreases markedly (Pierre, 1983; van Hooydonk *et al.*, 1986*a*; Carlson *et al.*, 1987*b*). These results, however, may depend on the method of acidification, how long the acidified milk is stored and the effect of the buffer into which milk is diluted, since the composition of the micelles changes markedly with pH (Snoeren *et al.*, 1984; Dalgleish & Law, 1989). The increase in rate has been suggested (Pearce, 1976) to arise from the increase in the activity of Ca^{2+} as the pH is lowered, but an alternative explanation may be found by considering the effect of pH on the micellar (or colloidal) calcium phosphate (Shalabi & Fox, 1982). It has been suggested (Roefs *et al.*, 1985) that dissolution of micellar calcium phosphate may in fact lead to a decrease in the efficiency with which micelles coagulate. As the pH is lowered, more calcium phosphate dissolves, but in milk this only serves to increase the concentration of Ca^{2+} The effects may therefore tend to cancel out, to give only a small pH-dependence of the aggregation.

The fact that micellar charge is important in the clotting of the renneted micelles is confirmed by the effect of cations such as Ca^{2+} on the clotting time (Green & Marshall, 1977) and the effect of polyions. Lysozyme appears to bind to casein micelles and to catalyse clotting (Bakri & Wolfe, 1971), probably by acting as a polycation (Green, 1973). Other polycations that have similar effects are cetyltrimethylammonium bromide (Pearce, 1976) and salmine (Green, 1973). Polycations can cause milk to clot in the absence of rennet (Di Gregorio & Sisto, 1981), and so it is not surprising that subcritical amounts can enhance the rate of coagulation (Green & Marshall, 1979). The effect is produced by materials that bind to the casein micelles (Marshall & Green, 1980) and the important factor appears to be the relative charge concentration of polycations (Green, 1982*a*). Since the mean sizes and the surface potentials of the casein micelles appear to be largely unaltered by the binding of the ionic materials, it is possible that they bind in the interiors of the micelles. Binding may also occur via hydrophobic interactions and it appears that the casein, rather than the calcium phosphate, is the primary binding site (Green, 1982*b*). Additives such as proline show the importance of hydrophobic interactions (Marshall & Green, 1980), although they may also disrupt the micelles, and additives that decrease the clotting time tend to be positively charged and strongly bound to the micelles (Marshall & Green, 1980).

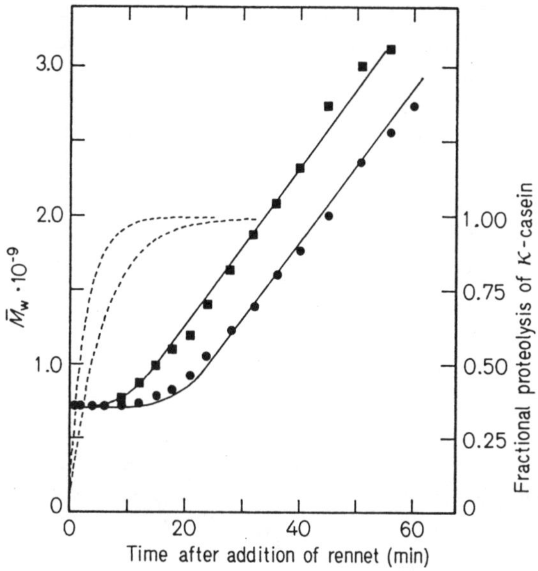

Fig. 4. Lag stage and growth of molecular weight during renneting of casein micelles. Two sets of results are shown, using different concentrations of rennet, the higher concentration giving the shorter lag time. The two broken lines show the extent of breakdown of κ-casein in the two reactions. (Reproduced from Dalgleish *et al.* (1981) with permission.)

Additives that decrease the clotting time tend to remain bound to the micelles after coagulation and most (with the notable exception of sodium dodecyl sulphate) do not affect the rate of proteolysis by rennet.

All of these results tend towards a view that the enhancement of the rate of coagulation arises from the neutralization of negative charge within the micelles, diminishing the charge repulsions and allowing hydrophobic interactions to occur once some of the stabilizing κ-casein has been hydrolysed by rennet.

10. Mechanistic Descriptions of the Overall Clotting Process

When rennet is added to milk there is no apparent action for some time, and then the milk is seen to clot rapidly. No matter what means are used to monitor the reaction, be it measurement of the visual clotting time or turbidity (Payens, 1977) or viscosity (Scott Blair & Oosthuizen, 1961) or electron microscopy (Green *et al.*, 1978*a*; Guthy *et al.*, 1989), coagulation of the milk only occurs after a period during which no aggregation occurs (Fig. 4). This is simply the result of the coagulation reaction being partly sequential on the enzymatic reaction.

In the interests of defining the rate at which milk is converted from a suspension to a gel, and thus allowing the rennet clotting of milk to be carefully controlled, several attempts to construct a usable mathematical model for the overall process, involving both enzymatic and coagulation stages, have been made. Most of the models require a certain simplification of the known mechanisms in order to provide analytical solutions, but otherwise it is possible to use computer-generated algorithms to attempt to describe the reactions in numerical terms.

The object of these models is to allow values for such factors as rennet concentration, temperature and the concentration of milk to be entered into an equation that will predict the rate of curd formation. The earliest attempt to describe the kinetics of the clotting process in this way was made in the 1870s by Storch and Segelcke (Foltmann, 1959). This simply stated that the clotting time was inversely related to the concentration of rennet used to clot the milk, i.e.

$$\mathrm{CT} = k/[E] \tag{4}$$

where k is a constant.

To a first approximation, this equation is valid (Brown & Collinge, 1986), but a further refinement was postulated by Holter (1932) and rearranged by Foltmann (1959) to give the familiar equation:

$$\mathrm{CT} = (k/[E]) + A \tag{5}$$

where k and A are constants. This preserves the observed linear dependence of the clotting time on the inverse of the enzyme concentration. In such a form, the equation is acceptable for use with a wide range of enzyme concentrations (McMahon & Brown, 1983). It may be taken as a test of any more advanced model of the reaction that it reduces to the Holter formulation, which is amply supported by empirical experimental evidence. However, the relationship given in eqn (5) is not descriptive, in the sense that the particular mechanistic significance of the two constants, A and k, is not established, and indeed they depend on the composition of the milk. More recent models have sought to describe the overall reaction by considering the details of the individual reactions involved. Although several approaches to the problem have been made, which come to somewhat different conclusions, they all possess some aspects in common.

The first of these model calculations was established by Payens and co-workers in a series of papers (Payens, 1977, 1978, 1984*a*; Payens *et al.*, 1977; Payens & Wiersma, 1980; Brinkhuis & Payens, 1984, Payens & Brinkhuis, 1986). In these calculations, it was assumed that the casein micelles became capable of aggregating in exact proportion to the amount of κ-casein hydrolysed. Thus, in the early stages of the reaction, proteolysis can only create small concentrations of micelles that can aggregate. Since the growth of the particles is a bimolecular process, aggregation can occur only slowly at this stage of the reaction. As more proteolysis occurs, the number of coagulating micelles and hence the rate of aggregate formation will increase. Thus, there will be a lag period during which no great effect of the enzyme on the average particle size will be seen. On more sensitive criteria of particle size, the size (or, more precisely, the particle weight) will decrease by a small amount because of the loss of the macropeptides from the micelles. The increase in aggregation during renneting simply arises from the increase in the number of aggregating units, as κ-casein is hydrolysed by rennet. This model described well the kinetics of coagulation of isolated κ-casein treated with pepsin (Hyslop *et al.*, 1979*b*), but also predicted that in milk a characteristic time (which may be identified as the clotting time) could be defined by the simple relation:

$$t_c = (2/k_s V)^{0 \cdot 5} \tag{6}$$

where k_s is the rate constant for flocculation and V is the enzymatic velocity (i.e. a constant rate at which the κ-casein was assumed to be destroyed and the casein micelles became capable of aggregation). This equation states that the clotting time will be inversely proportional to the square root of the enzyme concentration, rather than to the enzyme concentration itself (and is therefore not in agreement with experimental observation). The major problem with the model was that, to describe the coagulation of milk, the rate constant for the flocculation of renneted micelles had to be assumed to be very low compared with the known rates which have been measured directly (Dalgleish *et al.*, 1981; Dalgleish, 1983; Brinkhuis & Payens, 1984). This is essentially because the value of k_s which must be used in the calculations represents an average over all values and concentrations of aggregating micelles, from the low value at the start of the reaction to the high value at the end.

Considerations such as this led to a re-estimation of the model, where renneted casein micelles were considered as multi-functional particles, whose rate constants for interaction increased according to the number of functional 'hot spots' via which aggregation could occur, and in which it was possible to use a 'time-averaged' value for the rate at which proteolysis of κ-casein occurred, i.e. the rate at which functionalities were produced on micelles (Payens & Brinkhuis, 1986). Once again, the model predicted a square root dependence of the clotting time on the enzyme concentration, which arose as a consequence of the estimates of the rates at which flocculating micelles were produced.

This model of the reaction highlighted a number of important factors, among which are: (1) how much κ-casein must be hydrolysed before a micelle can aggregate; (2) whether the aggregation rate constant of micelles (i.e. k_s) is constant or variable and, if the latter, how the variation arises; (3) how the reactions of κ-casein hydrolysis and aggregation can be linked in time in such a way as to allow calculation of the overall reaction. While the first two of these are matters of the chemistry of the reactions, the third may simply be a matter of defining mathematical functions which have meaningful analytical solutions.

Having observed that micelles did not aggregate until most of their κ-casein had been hydrolyzed (Fig. 2), Dalgleish (1980*b*) developed a model that described the dependence of the aggregation tendency of micelles upon the extent of the proteolytic reaction by giving a value of zero until a defined critical extent of proteolysis had occurred on any particular micelle (i.e. micelles became capable of aggregating in a catastrophic way). In micelles where more than the critical degree of proteolysis had been achieved, k_s was given a maximal value. The proportions of micelles that had achieved the critical state at any defined extent of κ-casein breakdown could be simply computed using probability theory. The complete mathematical description of the system was relatively complex because of the necessity of incorporating the mathematics of the Michaelis mechanism into the calculation of the extent of κ-casein hydrolysis, but it was possible to employ computer simulation to show that the Holter relationship could be reproduced by the calculation.

Moreover, the calculation could be simplified by making the assumptions that (1) the translation between all micelles being stable and all being capable of aggregation could be approximated by a step function, at a defined degree of proteolysis of κ-casein, and (2) the observed coagulation time occurred at a defined degree of polymerization of the casein micelles. This gave an equation for clotting time (Dalgleish, 1980*a*):

$$t_c \sim \frac{K_m}{V_{max}} \ln \frac{1}{1 - \alpha_c} + \frac{\alpha_c \cdot s_0}{V_{max}} + \frac{1}{2k_s c_0} \frac{M_{crit}}{M_0} - 1 \tag{7}$$

where α_c is the critical extent of conversion of κ-casein, s_0 is the original concentration of κ-casein, c_0 is the number concentration of casein micelles, and M_0 and M_{crit} are average micellar particle weights of casein micelles initially and at the point where clotting is observed. A similar formulation has been used by Carlson *et al.* (1986*b*). Since for any set of conditions, most of the terms in eqn (7) are constant, the equation can be seen to reduce to the Holter eqn (5) (since of course V_{max} is proportional to the concentration of rennet).

In this mechanism, no reason was given for the sudden change in aggregation of the micelles; the model was simply based on experimental observation of the rate at which the casein micelles aggregated when renneted to different extents. However, it is possible to explain the change in the value of k_s by postulating an energy barrier that prevents aggregation of the original micelles, but that is lowered by the action of rennet (Darling & van Hooydonk, 1981). In this model, the native casein micelles were considered to be sterically stabilized, this stabilization being lost as the renneting proceeded. In this view, it was possible to define an energy function (stability ratio) for interactions between average micelles at any stage in the reaction, this stability ratio (W) being related to the extent of proteolysis at time, t, by:

$$W_t = W_0 \exp(-C_m \cdot V \cdot t) \tag{8}$$

where C_m is a constant and V is the rate of the enzymatic reaction. Like Payens's formulation, the kinetics of the enzymatic reaction were simplified to give linear proteolysis of κ-casein. At the start of the reaction, rennet has a negligible effect upon the aggregation since W_t is large at that time, and calculation shows that effective aggregation cannot take place until about 70% of the κ-casein has been hydrolysed. At this point, the ability of the micelles to aggregate increases rapidly with the extent of proteolysis, and Smoluchowski aggregation can take place at an increasing rate because the effective rate constant for the aggregation increases. This is in accordance with measurements of this rate (Green *et al.*, 1978*b*; Dalgleish, 1979; Chaplin & Green, 1980; Carlson *et al.*, 1987*b*). The description of the reaction in this way provided the first attempt to link the interaction energies of renneted micelles (Payens, 1978; Green & Crutchfield, 1971) with their changing

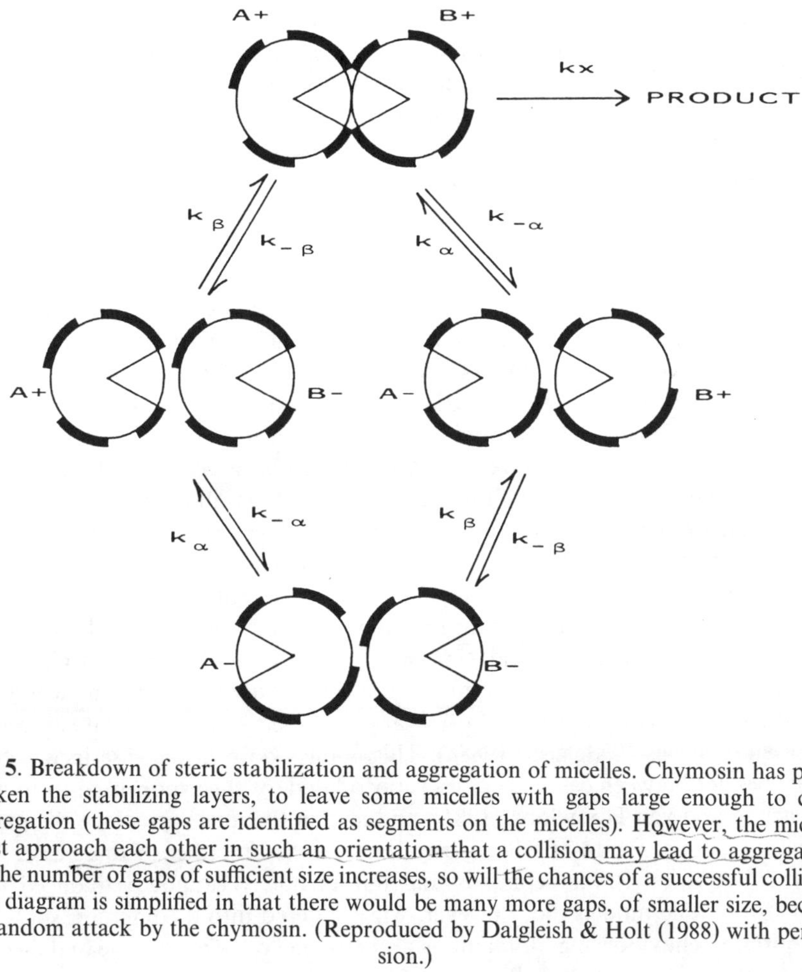

Fig. 5. Breakdown of steric stabilization and aggregation of micelles. Chymosin has partly broken the stabilizing layers, to leave some micelles with gaps large enough to cause aggregation (these gaps are identified as segments on the micelles). However, the micelles must approach each other in such an orientation that a collision may lead to aggregation. As the number of gaps of sufficient size increases, so will the chances of a successful collision. The diagram is simplified in that there would be many more gaps, of smaller size, because of random attack by the chymosin. (Reproduced by Dalgleish & Holt (1988) with permission.)

rates of aggregation during renneting. However, the difficulty with the model lay in its simplification of Michaelis kinetics to allow calculation to be made.

This model was based on the destruction of steric stabilization by rennet action, although the equation used for the stability factor (8) in fact is valid for any form of stabilizing interaction energy, whether by charge or steric means. It is, however, possible to envisage the onset of aggregation in a purely geometrical way (Dalgleish & Holt, 1988). If the micelles are stabilized by a thick layer of κ-casein and can only react when large enough gaps in this layer have been made, then from the known dimensions of the casein micelles and of their surface layers (Holt & Dalgleish, 1986), it is possible to calculate from geometric principles how large the gaps must be. A scheme of reaction showing the encounters of partially renneted micelles is shown in Fig. 5, from which it can be seen that the micelles must not only possess gaps in their surface of the appropriate size but must also collide in the proper orientation. On these purely geometrical grounds it is possible to describe the increase in aggregation rate with the extent of κ-casein hydrolysis, since as the extent of proteolysis increases, so do both the sizes of the gaps in the micellar surface and the chances of collision in a reactive orientation. This model of the reaction explains the change in aggregation of micelles with the extent of proteolysis but does not explain the changes in the critical values of α with changing temperature or pH.

In the reaction mechanisms described above, explicit analytical solutions for the coagulation time could not be calculated because of the complexity of the reaction; only by simplifying the Michaelis kinetics could analytical solutions be found. However, a calculation involving a less drastic modification of the enzyme kinetics can be made. According to van Hooydonk *et al.* (1984), the hydrolysis of κ-casein can be regarded as a first-order reaction in milk, and the high values of K_m found by Dalgleish (1979) and Chaplin and Green (1980) suggest also a pseudo-first-order reaction occurs. Van Hooydonk and Walstra (1987*a*) accepted and used this in calculating the overall reaction; however, they found that in their formulation, the exponential form of the stability factor, W, as in eqn (8), did not permit a full analytical solution. They therefore used an approximation where W was linearly related to the amount of κ-casein remaining in the micelles at any time, which is only true over a limited period of the reaction, and so the calculation did not cover the whole of the experimental time-scale from rennet addition to clotting. However, it has been found possible to calculate analytical solutions where the first-order kinetics of κ-casein hydrolysis are combined with the stability criteria of the casein micelles (Dalgleish, 1988*a*). This applies both in the case in which an energy barrier between casein micelles is postulated, of the exponential form used by Darling and van Hooydonk (1981), and also in the model in which the purely geometrical approach to micelle aggregation is used. In either of these cases, the analytical solution is very complex and unfortunately cannot be reduced to a simple form. The models did, however, when entered into a computer, show the appropriate behaviour, e.g. in relationship between coagulation time and enzyme concentration. It may be possible to use these calculations as the basis for future models.

All of the models describe the overall reaction as consisting of a lag stage followed by rapid aggregation. In two of the models, the lag time is explained as arising from the time required for aggregable micelles to be formed, giving the impression of a primary (enzymatic) stage followed by a secondary (coagulation or flocculation) stage. However, all of the models show that the onset of coagulation will occur before the exhaustion of the κ-casein substrate by the enzyme. None of the models is complete inasmuch as they cannot explain all of the changes that can be caused in the coagulation process by such factors as temperature, pH, ionic strength and concentrations of different ions, such as Ca^{2+}. What they have achieved is a rationale for the reaction in fairly general terms, within which framework more detailed investigations of the behaviour of the casein micelles and the enzyme–substrate reactions can be made.

11. Conclusion

This chapter has considered the factors that influence rennet clotting and the reactions that are believed to occur during the process. Clotting times can be measured in a variety of ways, ranging from simple observation of the formation of visible particles to more sophisticated methods (van Hooydonk, 1987). The model calculations that have been used to understand the coagulation have in fact given some insights and produced some discussion (Payens, 1987; van Hooydonk & Walstra, 1987*b*; Dalgleish, 1988*b*). However, the models have their limitations: once the aggregation of the renneted micelles is well advanced, it is almost certain that the description of the system in terms of Smoluchowski kinetics cannot be accurate and more complex theories of gel formation and properites must be used. Some attempts have been made to consider the reaction in this way (Johnston, 1984; Merin *et al.*, 1989) to describe such factors as gel strength and syneresis. These have led to partial success, and although the theories can be criticised in terms of whether they are applicable to the milk system (Payens, 1984*b*), they have made some attempt to explain gel formation and properties.

To change from considering the properties of renneting milk as those of an aqueous suspension to considering renneting milk as a gel requires changes in the type of experiments that should be done, and also the development of different theoretical backgrounds within which to discuss the results. The point of division between the two regimes is one that requires further study, since there are as yet no certain theoretical developments and very few experimental techniques (Horne & Davidson, 1990) that will allow the details of the suspension–gel transition to be investigated successfully.

References

Aaltonen, M.-L. & Antila, V. (1987). Milk renneting properties and the genetic variants of proteins. *Milchwissenschaft*, **42**, 490–2.

Addeo, F., Martin, P. & Ribadeau Dumas, B. (1984). Susceptibility of buffalo and cow κ-caseins to chymosin action. *Milchwissenschaft*, **39**, 202–5.

Alais, C. & Lagrange, A. (1972). Etude biochimique d'une protéase coagulante produite par *Mucor miehei*. 1. Activité coagulante et activité protéolytique. *Le Lait*, **52**, 407–27.

Andren, A. & de Koning, P.J. (1982). Changes in immunologically and enzymatically active centres of some milk clotting enzymes. In *Use of Enzymes in Food Technology*, ed. P. Dupuy. Technique et Documentation Lavoisier, Paris, pp. 275–80.

Armstrong, C.E., Mackinlay, A.G., Hill, R.J. & Wake, R.G. (1967). The action of rennin on κ-casein: the heterogeneity and origin of the soluble product. *Biochim. Biophys. Acta*, **140**, 123–31.

Azuma, N., Kaminogawa, S. & Yamauchi, K. (1984). Properties of glycomacropeptide and para-κ-casein derived from human κ-casein and comparison of human and bovine κ-caseins as to susceptibility to chymosin and pepsin. *Agric. Biol. Chem.*, **48**, 2025–31.

Bakri, M. & Wolfe, F.H. (1971). Destabilization of casein micelles by lysozyme. *Canad. J. Biochem.*, **49**, 882–4.

Banks, J.M., Stewart, G., Muir, D.D. & West, I.G. (1987). Increasing the yield of Cheddar cheese by the acidification of milk containing heat-denatured whey protein. *Milchwissenschaft*, **42**, 212–5.

Beeby, R. (1976). A comparison between the amino acid compositions of 2 chymosin-sensitive polypeptides and C-terminal segments of κ-casein. *J. Dairy Res.*, **43**, 37–43.

Beeby, R. (1979). The proteolysis of casein by immobilized preparations of α-chymotrypsin, chymosin and a fungal protease. *N.Z. J. Dairy Sci. Technol.*, **14**, 1–11.

Beeby, R. (1980). The use of fluorescamine at pH 6·0 to follow the action of chymosin on κ-casein and to estimate this protein in milk. *N.Z. J. Dairy Sci. Technol.*, **15**, 99–108.

Berridge, N.J. (1942) The second phase of rennet coagulation. *Nature*, **149**, 194–5.

Berridge, N.J. (1952). Some observations on the determination of the activity of rennet. *Analyst*, **77**, 57–62.

Berry, G.P. & Creamer, L.K. (1975). The association of β-casein. The importance of the C-terminal region. *Biochemistry*, **14**, 3542–5.

Bines, V.E., Young, P. & Law, B.A. (1989). Comparison of Cheddar cheese made with a recombinant calf chymosin and with standard calf rennet. *J. Dairy Res.*, **56**, 657–64.

Bingham, E.W. (1975). Action of rennin on κ-casein. *J. Dairy Sci.*, **58**, 13–17.

Bohak, Z. (1969). Purification and characterization of chicken pepsinogen and chicken pepsin. *J. Biol. Chem.*, **244**, 4638–48.

Brignon, G., Chtourou, A. & Ribadeau-Dumas, B. (1985). Preparation and amino acid sequence of human κ-casein. *FEBS Lett.*, **188**, 48–54.

Bringe, N.A. & Kinsella, J.E. (1986*a*). Influence of calcium chloride on the chymosin-initiated coagulation of casein micelles. *J. Dairy Res.*, **53**, 371–9.

Bringe, N.A. & Kinsella, J.E. (1986*b*). Inhibition of chymosin and the coagulation of para-casein micelles by anions. *J. Dairy Sci.*, **69**, 965–70.

Brinkhuis, J. & Payens, T.A. (1984). The influence of temperature on the flocculation rate of renneted casein micelles. *Biophys. Chem.* **19**, 75–81.

Brinkhuis, J. & Payens, T.A. (1985). The rennet-induced clotting of para-κ-casein revisited: inhibition experiments with pepstatin A. *Biochim. Biophys. Acta*, **832**, 331–6.

Brown, R.J. & Collinge, S.K. (1986). Actual milk coagulation time and inverse of chymosin activity. *J. Dairy Sci.*, **69**, 956–8.

Carles, C. & Martin, P. (1985). Kinetic study of the action of bovine chymosin and pepsin A on bovine κ-casein. *Arch. Biochem. Biophys.*, **242**, 411–16.

Carles, C. & Ribadeau-Dumas, B. (1984). Kinetics of action of chymosin (rennin) on some peptide bonds of bovine β-casein. *Biochemistry*, **23**, 6839–43.

Carles,C. & Ribadeau-Dumas, B. (1985). Kinetics of the action of chymosin (rennin) on a peptide bond of bovine α_{s1}-casein. Comparison of this substrate with that of β- and κ_0-caseins. *FEBS Lett.*, **185**, 282–6.

Carlson, A. (1983). The enzymatic coagulation of milk. *Diss. Abs. Int. B*, **43**, 3671.

Carlson, A., Hill, C.G. & Olson, N.F. (1987*a*). Kinetics of milk coagulation: I. The kinetics of kappa casein hydrolysis in the presence of enzyme deactivation. *Biotech. Bioeng.*, **29**, 582–9.

Carlson, A., Hill, C.G. & Olson, N.F. (1987*b*). Kinetics of milk coagulation. III. Mathematical modeling of the kinetics of curd formation following enzymatic hydrolysis of κ-casein —parameter estimation. *Biotech. Bioeng.*, **29**, 601–11.

Castle, A.V. & Wheelock, J.V. (1972). Effect of varying enzyme concentration on the action of rennin on whole milk. *J. Dairy Res.*, **39**, 15–22.

Chaplin, B. & Green, M.L. (1980). Determination of the proportion of κ-casein hydrolysed by rennet on coagulation of skim milk. *J. Dairy Res.*, **47**, 351–8.

Chaplin, B. & Green, M.L. (1982). Probing the location of casein fractions in the casein micelle using enzymes and enzyme-dextran conjugates. *J. Dairy Res.*, **49**, 631–43.

Chobert, J.-M., Mercier, J.-C., Bahy, C. & Hazé, G. (1976). Structure primaire du caséinomacropeptide des caséines κ porcine et humaine. *FEBS Lett.*, **72**, 173–8.

Collin, J.C., Martin, P., Garnot, P., Ribadeau-Dumas, B. & Mocquot, G. (1981). Deter-

mination of chymosin and bovine pepsin A in commercial bovine rennets and pepsins. *Milchwissenschaft*, **36**, 32–5.
Collin, J.-C., Muset de Retta, G. & Martin, P. (1982). Immunological identification of milk clotting enzymes. *J. Dairy Res.*, **49**, 221–30.
Creamer, L.K. (1976). A further study of the action of rennin on β-casein. *N.Z. J. Dairy Sci. Technol.*, **11**, 30–9.
Creamer, L.K. & Richardson, B.C. (1974). Identification of the primary degradation product of α_{s1}-casein in Cheddar cheese. *N.Z. J. Dairy Sci. Technol.*, **9**, 9–13.
Creamer, L.K., Mills, O.E. & Richards, E.L. (1971). The action of rennets on the caseins. I. Rennin action on β-casein B in solution. *J. Dairy Res.*, **38**, 269–80.
Dalgleish, D.G. (1979). Proteolysis and aggregation of casein micelles treated with immobilized or soluble chymosin. *J. Dairy Res.*, **46**, 643–61.
Dalgleish, D.G. (1980*a*). Effect of milk concentration on the rennet clotting time. *J. Dairy Res.*, **47**, 231–5.
Dalgleish, D.G. (1980*b*). A mechanism for the chymosin-induced flocculation of casein micelles. *Biophys. Chem.*, **11**, 147–55.
Dalgleish, D.G. (1983). Coagulation of renneted casein micelles: dependence on temperature, calcium ion concentration and ionic strength. *J. Dairy Res.*, **50**, 331–40.
Dalgleish, D.G. (1984). Measurement of electrophoretic mobilities and zeta-potentials of particles from milk using laser Doppler electrophoresis. *J. Dairy Res.*, **51**, 425–38.
Dalgleish, D.G. (1986). Analysis by Fast Protein Liquid Chromatography of different κ-caseins and their relevance to micellar structure and renneting. *J. Dairy Res.*, **53**, 43–51.
Dalgleish, D.G. (1988*a*). A new calculation of the kinetics of the renneting reaction. *J. Dairy Res.*, **55**, 521–8.
Dalgleish, D.G. (1988*b*). The coagulation of milk during renneting—further comments. *Neth. Milk Dairy J.*, **42**, 341–3.
Dalgleish, D.G. & Holt, C. (1988) A geometrical model to describe the initial aggregation of partly renneted casein micelles. *J. Colloid Interf. Sci.*, **123**, 80–4.
Dalgleish, D.G. & Law, A.J.R. (1988). pH-induced changes of bovine casein micelles. I. Analysis of liberated caseins. *J. Dairy Res.*, **55**, 529–38.
Dalgleish, D.G. & Law, A.J.R. (1989). pH-induced changes of bovine casein micelles. II. Dissociation of calcium and phosphate. *J. Dairy Res.*, **56**, 727–35.
Dalgleish, D.G., Brinkhuis, J. & Payens, T.A.J. (1981). The coagulation of differently sized casein micelles by rennet. *Eur. J. Biochem.*, **119**, 257–61.
Dalgleish, D.G., Horne, D.S. & Law, A.J.R. (1989). Size-related differences in bovine casein micelles. *Biochim. Biophys. Acta.* , **991**, 383–7.
Damicz, W. & Dziuba, J. (1975). Studies on casein proteolysis. I. Enzyatic phase of casein coagulation as influenced by heat treatment of milk proteins. *Milchwissenschaft*, **30**, 399–405.
Darling, D.F. & Dickson, J. (1979). Electrophoretic mobility of casein micelles. *J. Dairy Res.*, **46**, 441–51.
Darling, D.F. & van Hooydonk, A.C.M. (1981). Derivation of a mathematical model for the mechanism of casein micelle coagulation by rennet. *J. Dairy Res.*, **48**, 189–200.
Davies, D.T. & White, J.C.D. (1960). The use of ultrafiltration and dialysis in isolating the aqueous phase of milk and in determining the partition of milk constituents between the aqueous and disperse phases. *J. Dairy Res.*, **27**, 171–90.
de Jong, L. & de Groot-Mostert, A.E.A. (1977). The proteolytic action of rennet on different casein substrates under different conditions. *Neth. Milk Dairy J.*, **31**, 296–313.
de Koning, P.J. (1978). Coagulating enzymes in cheesemaking. *Dairy Industries International*, **43**, 7–12, 46.
de Koning, P.J., van Rooijen, P.J. & Visser, S. (1978). Application of a synthetic hexapeptide as a standard substrate for a determination of the activity of chymosin. *Neth. Milk Dairy J.*, **32**, 232–44.
Dickinson, E., Whyman, R. & Dalgleish, D.G. (1987). Colloidal properties of model

oil-in-water emulsions stabilized separately by α_{s1}-casein, β-casein and κ-casein. In *Food Emulsions and Foams*, ed. E. Dickinson. Royal Society of Chemistry Special Publication No. 58, pp. 40–51.

Di Gregorio, F. & Sisto, R. (1981). Milk coagulation by cationic polyelectrolytes. *J. Dairy Res.*, **48**, 267–71.

Donnelly, W.J., McNeill, G.P., Buchheim, W. & McGann, T.C.A. (1984). A comprehensive study of the relationship between size and protein composition in natural bovine casein micelles. *Biochim. Biophys. Acta*, **789**, 136–43.

Drøhse, H.B. & Foltmann, B. (1989). Specificity of milk-clotting enzymes towards bovine κ-casein. *Biochim. Biophys. Acta*, **995**, 221–4.

Duersch, J.W. & Ernstrom, C.A. (1974). Heat stability of residual milk clotting enzymes in whey. *J. Dairy Sci.*, **57**, 590–1 (abstract).

Ekstrand, B., Larsson-Raznikiewicz, M. & Perlmann, C. (1980). Casein micelle size and composition related to the enzymatic clotting process. *Biochim. Biophys. Acta*, **630**, 361–6.

El-Shibiny, S. & Abd El-Salam, M.H. (1976). Action of milk clotting enzymes on β-caseins from buffalo's and cow's milk. *J. Dairy Res.*, **43**, 443–8.

El-Shibiny, S. & Abd El-Salam, M.H. (1977). Action of milk clotting enzymes on α_s-caseins from buffalo's and cow's milk. *J. Dairy Sci.*, **60**, 1519–21.

Fiat, A.-M. & Jollès, P. (1977). Données structurales concernant le caséinoglycopeptide humain et comparisons avec les caséinoglycopeptides des ruminants. *C. R. Acad Sci. Paris*, **D284**, 393–5.

Foltmann, B. (1959). On the enzymatic and the coagulation stages of the renneting process. In *Proc. XV Int. Dairy Congr. (London)*, vol. 2, pp. 655–61.

Foltmann, B. (1966). A review on prorennin and rennin. *C. R. Trav. Lab. Carlsberg*, **35**, 143–231.

Foltmann, B. (1987). General and molecular aspects of rennets. In *Cheese: Chemistry, Physics and Microbiology*, vol. 1, ed. P.F. Fox. Elsevier Applied Science, London, pp. 33–61.

Foster, P.M.D. & Green, M.L. (1974). A quantitative gel-filtration method for analysis of the proteinaceous fraction of Cheddar cheese. *J. Dairy Res.*, **41**, 259–68.

Fox, K.K., Holsinger, V.H., Posati, L.P. & Pallansch, M.J. (1967). Separation of β-lactoglobulin from other milk serum proteins by trichloroacetic acid. *J. Dairy Sci.*, **50**, 1363–7.

Fox, P.F. (1969). Milk-clotting and proteolytic activities of rennet, and of bovine pepsin and porcine pepsin. *J. Dairy Res.*, **36**, 427–33.

Fox, P.F. (1981). Heat-induced changes in milk preceding coagulation. *J. Dairy Sci.*, **64**, 2127–37.

Fox, P.F. (1989). Proteolysis during cheese manufacture and ripening. *J. Dairy Sci.*, **72**, 1379–400.

Fox, P.F. & Guiney, J. (1973). Casein micelle structure: susceptibility of various casein systems to proteolysis. *J. Dairy Res.*, **40**, 229–34.

Fruton, J.S. (1971). Pepsin. In *The Enzymes*, vol 3., ed. P. Boyer. Academic Press, New York, pp. 119–64.

Garnier, J. (1963). Etude cinetique d'une protéolyse limitée: action de la présure sur la caséine κ. *Biochim. Biophys. Acta*, **66**, 366–77.

Garnot, P. & Corre, C. (1980). Influence of milk protein concentration on the gelling activity of chymosin and bovin pepsin. *J. Dairy Res.*, **47**, 103–11.

Garnot, P. & Molle, D. (1982). Influence de la nature et du taux d'inactivation sur le dosage de la chymosine et de la pepsine bovine par electro immuno-diffusion. *Le Lait*, **62**, 671–80.

Grappin, R., Rank, T.C. & Olson, N.F. (1985). Primary proteolysis of cheese proteins during ripening. A review. *J. Dairy Sci.*, **68**, 531–40.

Green, M.L. (1973). Studies on the mechanism of clotting of milk. *Neth. Milk Dairy J.*, **27**, 278–85.

Green, M.L. (1977). Review of the progress of dairy science: milk coagulants. *J. Dairy Res.*, **44**, 159–88.

Green, M.L. (1982*a*). Effect on the composition and properties of casein micelles of interaction with ionic materials. *J. Dairy Res.*, **49**, 87–98.
Green, M.L. (1982*b*). Mode of binding of ionic materials to casein micelles. *J. Dairy Res.*, **49**, 99–105.
Green, M.L. & Crutchfield, G. (1969). Studies on the preparation of water-insoluble derivatives of rennin and chymotrypsin and their use in the hydrolysis of casein and the clotting of milk. *Biochem. J.*, **115**, 183–90.
Green, M.L. & Crutchfield, G. (1971). Density-gradient electrophoresis of native and of rennet-treated casein micelles. *J. Dairy Res.*, **38**, 151–64.
Green, M.L. & Foster, P.M.D. (1974). Comparison of the rates of proteolysis during ripening of Cheddar cheeses made with calf rennet and swine pepsin as coagulants. *J. Dairy Res.*, **41**, 269–82.
Green, M.L. & Marshall, R.J. (1977). The acceleration by cationic materials of the coagulation of casein micelles by rennet. *J. Dairy Res.*, **44**, 521–31.
Green, M.L. & Marshall, R.J. (1979). Interaction of ionic materials with casein micelles. *J. Dairy Res.*, **46**, 365–7.
Green, M.L. & Morant, S.V. (1981). Mechanism of aggregation of casein micelles in rennet-treated milk. *J. Dairy Res.*, **48**, 57–63.
Green, M.L. & Stackpoole, A. (1975). The preparation and assessment of a suitable *Mucor pucillus* Lindt proteinase-swine pepsin mixture for Cheddar cheese-making. *J. Dairy Res.*, **42**, 297–312.
Green, M.L., Marshall, R.J. & Glover, F.A. (1983). Influence of homogenization of concentrated milks on the structure and properties of rennet curds. *J. Dairy Res.*, **50**, 341–8.
Green, M.L., Valler, M.J. & Kay, J. (1984). Assessement of the suitability for Cheddar cheesemaking of purified and commercial chicken pepsin preparations. *J. Dairy Res.*, **51**, 334–40.
Green, M.L., Hobbs, D.G. & Morant, S.V. (1978*a*). Intermicellar relationships in rennet-treated separated milk. I. Preparation of representative electron micrographs. *J. Dairy Res.*, **45**, 405–11.
Green, M.L., Hobbs, D.G., Morant, S.V. & Hill, V.A. (1978*b*). Intermicellar relationships in rennet-treated separated milk. II. Process of gel assembly. *J. Dairy Res.*, **45**, 413–22.
Guthy, K. & Novak, G. (1977). Observations of the primary phase of milk coagulation by rennet under standardized conditions. *J. Dairy Res.*, **44**, 363–6.
Guthy, K., Auerswald, D. & Buchheim, W. (1989). Electron microscopic and viscosimetric studies of the primary phase of rennet coagulation of milk. *Milchwissenschaft*, **44**, 560–3.
Hicks, C.L., Ferrier, L.K., Olson, N.F. & Richardson, T. (1975). Immobilized pepsin treatment of skim milk and skim milk fractions. *J. Dairy Sci.*, **58**, 19–24.
Hicks, C.L., O'Leary, J. & Bucy, J. (1988). Use of recombinant chymosin in the manufacture of Cheddar and Colby cheese. *J. Dairy Sci.*, **71**, 1127–31.
Hill, R.D. (1968). The nature of the rennin-sensitive bond in casein and its probable relation to sensitive bonds in other proteins. *Biochem. Biophys. Res. Commun.*, **33**, 659–63.
Hill, R.D. (1969). Synthetic peptide and ester substrates for rennin. *J. Dairy Res.*, **36**, 409–15.
Hill, R.D. & Hocking, V.M. (1978). A study of structural factors that affect the reactivity of the chymosin-sensitive bond in κ-casein. *N. Z. J. Dairy Sci. Technol.*, **13**, 195–201.
Hill, R.D., Lahav, E. & Givol, D. (1974). A rennin-sensitive bond in α_{s1} B-casein. *J. Dairy Res.*, **41**, 147–53.
Hill, R.D. & Laing, R.R. (1965). The action of rennin on casein: the effect of modifying functional groups on the casein. *J. Dairy Res.*, **32**, 193–201.
Hindle, E.J. & Wheelock, J.V. (1970). The primary phase of rennin action in heat-sterilized milk. *J. Dairy Res.*, **37**, 389–96.
Hofmann, T. (1974). Structure, function and evolution of acid proteases. In *Food-Related Enzymes*, ed. J.R. Whitaker, *ACS Adv. Chem. Ser.*, **136**, 146–85.
Holmes, D.G., Duersch, J.W. & Ernstrom, C.A. (1977). Distribution of milk clotting

enzymes between curd and whey and their survival during cheesemaking. *J. Dairy Sci.*, **60**, 862–9.

Holt, C. (1975). The stability of bovine casein micelles. In *Proceedings of the Conference on Colloid and Surface Science, Budapest*, ed. E. Wolfram. Akademiai Kiado, Budapest, pp. 641–4.

Holt, C. & Dalgleish, D.G. (1986). Electrophoretic and hydrodynamic properties of bovine casein micelles interpreted in terms of particles with an outer hairy layer. *J. Colloid Interf. Sci.*, **114**, 513–24.

Holter, H. (1932). Rennet activity. *Biochem. Z.*, **255**, 160–8.

Horne, D.S. (1984). Steric effects in the coagulation of casein micelles by ethanol. *Biopolymers*, **23**, 989–93.

Horne, D.S. (1986). Steric stabilization and casein micelle stability. *J. Colloid Interf. Sci.*, **111**, 250–60.

Horne, D.S. & Davidson, C. (1990). The use of dynamic light scattering in monitoring rennet curd formation. *Milchwissenschaft*, **45**, 712–15.

Humme, H.E. (1972). The optimum pH for the limited specific proteolysis of κ-casein by rennin (primary phase of milk clotting). *Neth. Milk Dairy J.*, **26**, 180–5.

Humphrey, R.S. & Newsome, L.J. (1984). High-performance ion-exchange chromatography of the major bovine milk proteins. *N.Z. J. Dairy Sci. Technol.*, **19**, 197–204.

Hyslop, D.B., Swanson, A.M. & Lund, D.B. (1979*a*). Heat inactivation of milk-clotting enzymes at different pH. *J. Dairy Sci.*, **62**, 1227–32.

Hyslop, D.B., Richardson, T. & Ryan, D.S. (1979*b*). Kinetics of pepsin-initiated coagulation of κ-casein. *Biochim. Biophys. Acta*, **566**, 390–6.

Jakob, E. & Puhan, Z. (1986). Unterschiede zwischen Labträger und normal-gerinnender Milch unter besonderer Berücksichtigung des Caseinfraktion. *Schwiez. Milchw. Forschung*, **15**, 27–35.

Jenkins, J.A., Tickle, I., Sewell, T., Ungaretti, L., Wollmer, A. & Blundell, T.L. (1977). X-ray analysis and circular dichroism of the acid protease from *Endothia parasitica* and chymosin. *Adv. Exp. Med. Biol.*, **95**, 43–60.

Jespersen, N.J.T. & Dinesen, V. (1979). Milk coagulants. *J. Soc. Dairy Technol.*, **32**, 194–7.

Johnston, D.E. (1984). Application of polymer cross-linking theory to rennet induced milk gels. *J. Dairy Res.*, **51**, 91–101.

Jollès, J., Alais, C. & Jollès, P. (1968). The tryptic peptide with the rennin-sensitive linkage of cow's κ-casein. *Biochim. Biophys. Acta*, **168**, 591–3.

Kay, J. & Valler, M.J. (1981). The aspartate proteinases used as rennet substitutes. *Neth. Milk Dairy J.*, **35**, 281–6.

Kaye, N.M.C. & Jollès, P. (1978). The involvement of one of the three histidine residues of cow κ-casein in the chymosin-initiated milk clotting process. *Biochim. Biophys. Acta*, **536**, 329–40.

Kim, B.Y. & Kinsella, J.E. (1989). Effect of temperature and pH on the coagulation of casein. *Milchwissenschaft*, **44**, 622–5.

Knoop, A.-M. & Peters, K.-H. (1976). Die Natur der bei Lab- und Säuregallertenbildung wirksamen Kräfte und die Rolle des Calciums, Phosphats und Citrats bei der Gallertenbildung. *Milchwissenschaft*, **31**, 338–345.

Kotts, C. & Jenness, R. (1976). Isolation of κ-casein-like proteins from milks of various species. *J. Dairy Sci.*, **59**, 816–22.

Kovacs-Proszt, G. & Sanner, T. (1973). Comparison of the specificity and kinetic properties of 3 milk-clotting enzymes. *J. Dairy Res.*, **40**, 263–72.

Kowalchyk, A.W. & Olson, N.F. (1977). Effects of pH and temperature on the secondary phase of milk clotting by rennet. *J. Dairy Sci.*, **60**, 1256–9.

Lagrange, A., Paquet, D. & Alais, C. (1980). Comparative study of two *Mucor meihei* acid proteinases. Purification and some molecular properties. *Int. J. Biochem.*, **11**, 347–52.

Lawrence, R.C. & Creamer, L.K. (1969). The action of calf rennet and other proteolytic enzymes on κ-casein. *J. Dairy Res.*, **36**, 11–20.

Lee, H.J., Olson, N.F. & Richardson, T. (1977). Peptide release from milk during treatment with immobilized pepsin and totality of clotting of treated milk. *J. Dairy Sci.*, **60**, 1683–8.
Loucheux-Lefebvre, M.-H., Aubert, J.-P. & Jollès, P. (1978). Prediction of the conformation of the cow and sheep κ-caseins. *Biophys. J.*, **23**, 323–36.
Marshall, R.J. & Green, M.L. (1980). The effect of the chemical structure of additives on the coagulation of casein micelle suspensions by rennet. *J. Dairy Res.*, **47**, 359–69.
Martin, P., Raymond, M.-N., Bricas, E. & Ribaudeau-Dumas, B. (1980). Kinetic studies on the action of *Mucor pusillus, Mucor miehei* acid proteases and chymosins A and B on a synthetic chromophoric hexapeptide. *Biochim. Biopys. Acta*, **612**, 410–20.
Martin, P., Collin, J.-C., Garnot, P., Ribadeau-Dumas, B. & Mocquot, G. (1981). Evaluation of bovine rennets in terms of absolute concentrations of chymosin and pepsin A. *J. Dairy Res.*, **48**, 447–56.
McGann, T.C.A., Donnelly, W.J., Kearney, R.D. & Buchheim, W. (1980). Composition and size distribution of bovine casein micelles. *Biochim. Biophys. Acta*, **630**, 261–70.
McLean, D.M. & Schaar, J. (1989). Effects of β-lactoglobulin and κ-casein variants and concentrations on syneresis of gels from renneted heated milk. *J. Dairy Res.*, **56**, 297–301.
McMahon, D.J. & Brown, R.J. (1983). Milk coagulation time: linear relationship with inverse of rennet activity. *J. Dairy Sci.*, **66**, 341–4.
Mehaia, M.A. & Cheryan, M. (1983). Coagulation studies on ultrafiltration-concentrated skimmilk. *Milchwissenschaft*, **38**, 708–10.
Meisel, H. (1988). Charakterisierung von gentechnologisch gewonnen Labpräparaten im vergleich zu Kalberlab. 2. Ergebnisse. *Milchwissenschaft*, **43**, 71–5.
Mercier, J.-C., Brignon, G. & Ribadeau-Dumas, B. (1973). Structure primaire de la caséiene κB bovine: séquence complete. *Eur. J. Biochem.*, **35**, 222–35.
Merin, U., Talpaz, H. & Fishman, S. (1989). A mathematical model for the description of chymosin action on casein micelles. *J. Dairy Res.*, **56**, 79–86.
Morrisey, P.A. (1969). The rennet hysteresis of heated milk. *J. Dairy Res.*, **36**, 333–41.
Mulvihill, D.M. & Fox, P.F. (1977). Proteolysis of α_{s1}-casein by chymosin: influence of pH and urea. *J. Dairy Res.*, **44**, 533–40.
Mulvihill, D.M. & Fox, P.F. (1979*a*). Proteolytic specificity of chymosin on bovine α_{s1}casein. *J. Dairy Res.*, **46**, 641–51.
Mulvihill, D.M. & Fox, P.F. (1979*b*). Proteolytic specificity of chymosins and pepsins on β-caseins. *Milchwissenschaft*, **34**, 680–3.
Mulvihill, D.M. & Fox, P.F. (1980). Proteolysis of α_{s1}-casein by chymosin in dilute NaCl solutions and in cheddar cheese. *Ir. J. Fd. Sci. Technol.*, **4**, 13–23.
Oortwijn, H., Walstra, P. & Mulder, H. (1977). The membranes of recombined fat globules. 1. Electron microscopy. *Neth. Milk Dairy J.*, **31**, 134–47.
Pagnacco, G. & Caroli, A. (1987). Effect of casein and β-lactoglobulin genotypes on renneting properties of milks. *J. Dairy Res.*, **54**, 479–85.
Payens, T.A.J. (1977). On enzymatic clotting processes. II. The colloidal instability of chymosin-treated casein micelles. *Biophys. Chem.*, **6**, 263–70.
Payens, T.A.J. (1978). On enzymatic clotting processes. Part 3—Flocculation rate constants for paracasein and fibrin. *Farad. Disc. Chem. Soc.*, **65**, 164–74.
Payens, T.A.J. (1979). Casein micelles: the colloid-chemical approach. *J. Dairy Res.*, **46**, 291–306.
Payens, T.A.J. (1982). Stable and unstable casein micelles. *J. Dairy Sci.*, **65**, 1863–73.
Payens, T.A. (1984*a*). The relationship between milk concentration and rennet clotting time. *J. Appl. Biochem.*, **6**, 232–9.
Payens, T.A.J. (1984*b*). On the gelpoint in the enzyme—triggered clotting of casein micelles: a comment. *Neth. Milk Dairy J.*, **38**, 195–7.
Payens, T.A. (1987). Concerning consistent and inconsistent theories of milk clotting: a comment. *Neth. Milk Dairy J.*, **41**, 289–91.
Payens, T.A.J. & Both, P. (1980). Enzymic clotting process. IV. The chymosin-triggered

clotting of para-κ-casein. In *Biochemistry: Ions, Surfaces, Membranes*, ed. M. Blank, *ACS Adv. Chem. Ser.*, **188**, 129–41.
Payens, T.A. & Brinkuis, J. (1986). Mean field kinetics of the enzyme-triggered gelation of casein micelles. *Colloids Surf.*, **20**, 37–50.
Payens, T.A.J. & Visser, S. (1981). What determines the specificity of chymosin towards κ-casein? *Neth. Milk Dairy J.*, **35**, 387–9.
Payens, T.A.J. & Wiersma, A.K. (1980). On enzymatic clotting processes—V. Rate equations for the case of arbitrary rate of production of the clotting species. *Biophys. Chem.*, **11**, 137–46.
Payens, T.A.J., Wiersma, A.K. & Brinkhuis, J. (1977). On enzymatic clotting processes. I. Kinetics of enzyme-triggered coagulation reactions. *Biophys. Chem.*, **6**, 253–61.
Pearce, K.N. (1976). Moving boundary electrophoresis of native and rennet-treated casein micelles. *J. Dairy Res.*, **43**, 27–36.
Pélissier, J.-P., Mercier, J.-C. & Ribadeau Dumas, B. (1974). Proteolysis of α_{s1} and β-caseins by rennin. Specificity of action. Bitter peptides released. *Ann. Biol. Animale, Biochim., Biophys.*, **14**, 343–62.
Phelan, J.A. (1973). Laboratory and field tests on new milk coagulants. *Dairy Industries*, **38**, 418–23.
Phelan, J.A., Guiney, J. & Fox, P.F. (1973). Proteolysis of β-casein in Cheddar cheese. *J. Dairy Res.*, **40**, 105–12.
Philippos, Sh.G. & Christ, W. (1976). Uber einige, fur die Käseherstellung vorgeschlagene mikrobielle, milchgerinnende Enzyme und deren Einwirkungen auf das Casein der Kuhmilch. 1. Wirkung milchgerinnender Enzyme auf Kuhmilchasein. *Milchwissenschaft*, **31**, 349–54.
Pierre, A. (1983). Influence de la modification de la charge des micelles de caséine sur le taux de caséinomacropeptide libéré par la présure au moment de la coagulation du lait. *Le Lait*, **63**, 217–29.
Pszczola, D.E. (1989). Rennet containing 100% chymosin increases cheese quality and yield. *Food Technol.*, **43**, 84, 88–9.
Pujolle, J., Ribadeau Dumas, B., Garnier, J. & Pion, B. (1966). A study of κ-casein components. 1. Preparation. Evidence for a common C-terminal sequence. *Biochem. Biophys. Res. Commun.*, **25**, 285–90.
Qvist, K.B. (1979). Reestablishment of the original rennetability of milk after cooling. 1. The effect of cooling and LTST pasteurization of milk and renneting. *Milchwissenschaft*, **34**, 467–70.
Raap, J., Kerling, K.E.T., Vreeman, H.J. & Visser, S. (1983). Peptide substrates for chymosin (rennin): conformational studies of κ-casein and some κ-casein-related oligopeptides by circular dichroism and secondary structure prediction. *Arch. Biochem. Biophys.*, **221**, 117–24.
Ramet, J.P. & Weber, F. (1981). Etude de l'aptitude fromagère d'une préparation enzymatique coagulante thermolabile extraite de *Mucor miehei. Le Lait*, **61**, 458–64.
Raymond, M.N. & Bricas, E. (1979). New chromophoric substrates for chymosin (rennin). *J. Dairy Sci.*, **62**, 1719–25.
Raymond, M.N., Bricas, E., Salesse, R., Garnier, J., Garnot, P. & Ribadeau-Dumas, B. (1973*a*). A proteolytic unit for chymosin (rennin) activity based on a synthetic reference peptide. *J. Dairy Sci.*, **56**, 419–22.
Raymond, M.N., Bricas, E. & Mercier, J.-C. (1973*b*). Synthesis of new oligopeptides substrates of chymosin (rennin) and kinetic parameters of their hydrolyses. *Neth. Milk Dairy J.*, **27**, 298–303.
Reddy, I.M. & Kinsella, J.E. (1990). Interaction of β-lactoglobulin with κ-casein in micelles as assessed by chymosin hydrolysis: effect of temperature, heating time, β-lactoglobulin concentration and pH. *J. Agric. Food Chem.*, **38**, 50–8.
Robson, E.W. & Dalgleish, D.G. (1984). Coagulation of homogenized milk particles by rennet. *J. Dairy Res.*, **51**, 417–24.
Robson, E.W. & Dalgleish, D.G. (1987). Aggregation rates and electrophoretic mobilities

of homogenized milk fractions treated with rennet. In *Food Emulsion and Foams*, ed. E. Dickinson. Royal Society of Chemistry Special Publication No. 58, pp. 64–71.

Roefs, S.P.F.M., Walstra, P., Dalgleish, D.G. & Horne, D.S. (1985). Preliminary note on the change in casein micelles caused by acidification. *Neth. Milk Dairy J.*, **39**, 119–22.

Rollema, H.S., Brinkhuis, J.A. & Vreeman, H.J. (1988). ^{1}H-NMR studies of bovine κ-casein and casein micelles. *Neth. Milk Dairy J.*, **42**, 233–48.

Schaar, J. (1984). Effects of κ-casein variants and lactation number on the renneting properties of individual milks. *J. Dairy Res.*, **51**, 397–406.

Schattenkerk, C. & Keerling, K.E.T. (1973). Relation between structure and capacity to function as a rennin substrate. *Neth. Milk Dairy J.*, **27**, 286–7.

Schmidt, D.G., Both, P. & de Koning, P.J. (1966). Fractionation and some properties of κ-casein variants. *J. Dairy Sci.*, **49**, 776–82.

Scott, R. (1979). 'Rennets' and cheese. In *Topics in Enzyme and Fermentation Biotechnology*, vol. 3, Applied Sciences Publishers, London, pp. 103–69.

Scott Blair, G.W. & Oosthuizen, J.C. (1961). A viscometric study of the breakdown of casein in milk by rennin and rennet. *J. Dairy Res.*, **28**, 165–73.

Shalabi, S.I. & Fox, P.F. (1982). Influence of pH on the rennet coagulation of milk. *J. Dairy Res.*, **49**, 153–7.

Shalabi, S.I. & Wheelock, J.V. (1976). The role of α-lactalbumin in the primary phase of chymosin action on heated micelles. *J. Dairy Res.*, **43**, 331–5.

Shalabi, S.I. & Wheelock, J.V. (1977). Effect of sulphydryl-blocking agents on the primary phase of chymosin action on heated casein micelles and heated milk. *J. Dairy Res.*, **44**, 351–5.

Singh, H., Shalabi, S.I., Fox, P.F., Flynn, A. & Barry, A. (1988). Rennet coagulation of heated milk: influence of pH adjustment before or after heating. *J. Dairy Res.*, **55**, 205–15.

Sinkinson, G. & Wheelock, J.V. (1970). Carbohydrates of the glycopeptides released by the action of rennin on whole milk. *Biochim. Biophys. Acta*, **215**, 517–21.

Slattery, C.W. (1976). Casein micelles structure: an examination of models. *J. Dairy Sci.*, **59**, 1547–56.

Smoluchowski, M. von (1917). Versuch einer mathematischen Theorie der Koagulationskinetik kolloider Losungen. *Z. Physik. Chem.*, **92**, 129–68.

Snoeren, T.H.M., Klok, H.J., van Hooydonk, A.C.M. & Damman, A.J. (1984). The voluminosity of casein micelles. *Milchwissenschaft*, **39**, 461–3.

Stanley, D.W., Emmons, D.B., Modler, H.W. & Irvine, D.M. (1980). Cheddar cheese made with chicken pepsin. *Can. Inst. Food Sci. Technol. J.*, **13**, 97–102.

Sternberg, M. (1976). Microbial rennets. *Adv. Appl. Microbiol.*, **20**, 135–57.

Tam, J.J. & Whitaker, J.R. (1972). Rates and extents of hydrolysis of several caseins by pepsin, rennin, *Endothia parasitica* protease and *Mucor pusillus* protease. *J. Dairy Sci.*, **55**, 1523–31.

Taylor, M.J., Olson, N.F. & Richardson, T. (1979). Coagulation of skimmilk by immobilised proteases. *Process Biochem.*, **14**(2), 10–16.

Thompson, M.D., Dave, J.R. & Nakhasi, H.L. (1985). Molecular cloning of mouse mammary gland κ-casein: comparison with rat κ-casein and rat and human γ-fibrinogen. *DNA*, **4**, 263–71.

Thunell, R.K., Duersch, J.W. & Ernstrom, C.A. (1979). Thermal inactivation of residual milk clotting enzymes in whey. *J. Dairy Sci.*, **62**, 373–7.

Udenfriend, S., Stein, S., Bohlen, P., Dairman, W., Leimgruber, W. & Weigle, M. (1972). Fluorescamine: a reagent for assay of amino acids, peptides, proteins and primary amines in the picomole range. *Science*, **178**, 871–2.

Vamos-Vigyazo, L., El-Hawary, M. & Kiss, E. (1981). Action of chymosin and of a microbial milk clotting enzyme preparation on β-casein. *Acta Alimentaria*, **10**, 278–80.

Vanderpoorten, R. & Weckx, M. (1972). Breakdown of casein by rennet and microbial milk-clotting enzymes. *Neth. Milk Dairy J.*, **26**, 47–59.

van Hooydonk, A.C.M. (1987). The renneting of milk. Thesis, Agricultural University of Wageningen, pp. 109–26.
van Hooydonk, A.C.M. & Olieman, C. (1982). A rapid and sensitive high-performance liquid chromatography method of following the action of chymosin in milk. *Neth. Milk Dairy J.*, **36**, 153–8.
van Hooydonk, A.C.M. & Walstra, P. (1987*a*). Interpretations of the kinetics of the renneting reaction in milk. *Neth. Milk Dairy J.*, **41**, 19–47.
van Hooydonk, A.C.M. & Walstra, P. (1987*b*). A consistent model for the clotting of milk by rennet: a reply to a comment. *Neth. Milk Dairy J.*, **41**, 293–5.
van Hooydonk, A.C.M., Olieman, C. & Hagedoorn, H.G. (1984). Kinetics of the chymosin-catalysed proteolysis of κ-casein in milk. *Neth. Milk Dairy J.*, **37**, 207–22.
van Hooydonk, A.C.M., Boerrigter, I.J. & Hagedoorn, H.G. (1986*a*). pH-induced physico-chemical changes of casein micelles in milk and their effect on renneting. 2. Effect of pH on renneting of milk. *Neth. Milk Dairy J.*, **40**, 297–313.
van Hooydonk, A.C.M., Hagedoorn, H.G. & Boerrigter, I.J. (1986*b*). The effect of various cations on the rendering of milk. *Neth. Milk Dairy J.*, **40**, 369–90.
van Hooydonk, A.C.M., de Koster, P.G. & Boerrigter, I.J. (1987). The renneting properties of heated milk. *Neth. Milk Dairy J.*, **41**, 3–18.
Visser, S. (1981). Proteolytic enzymes and their action on milk proteins. A review. *Neth. Milk Dairy J.*, **35**, 65–88.
Visser, S. & Slangen, K.J. (1977). On the specificity of chymosin (rennin) in its action on bovine β-casein. *Neth. Milk Dairy J.*, **31**, 16–30.
Visser, S., Slangen, K.J. & Hup, G. (1975). Some bitter peptides from rennet-treated casein. A method for their purification, utilizing chromatographic separation on silica gel. *Neth. Milk Dairy J.*, **29**, 319–34.
Visser, S., van Rooijen, P.J., Schattenkerk, C. & Kerling, K.E.T. (1976). Peptide substrates for chymosin (rennin). Kinetic studies with peptides of different chain length including parts of the sequence 1011–112 of bovine κ-casein. *Biochim. Biophys. Acta*, **438**, 265–72.
Visser, S., van Rooijen, P.J., Schattenkerk, C. & Kerling, K.E.T. (1977). Peptide substrates for chymosin (rennin). Kinetic studies with bovine κ-casein-(103–108)-hexapeptide analogues. *Biochim. Biophys. Acta*, **481**, 171–76.
Visser, S., van Rooijen, P.J. & Slangen, C. (1980). Peptide substrates for chymosin (rennin). Isolation and substrate behaviour of two tryptic fragments of bovine κ-casein. *Eur. J. Biochem.*, **108**, 415–21.
Voutsinas, L.P. & Nakai, S. (1983). Coagulation of skim milk with proteases immobilized on hydrophobic carriers. *J. Dairy Sci.*, **66**, 694–703.
Voynick, I.M. & Fruton, J.S. (1971). The comparative specificity of acid proteinases. *Proc. Nat. Acad. Sci. (U.S.)*, **68**, 257–9.
Vreeman, H.J., Both, P., Brinkhuis, J.A. & van der Spek, C. (1977). Purification and some physicochemical properties of bovine κ-casein. *Biochim. Biophys. Acta*, **491**, 93–103.
Vreeman, H.J., Visser, S., Slangen, C.J. & van Riel, J.A.M. (1986). Characterization of bovine κ-casein fractions and kinetics of chymosin-induced macropeptide release from carbohydrate-free and carbohydrate-containing fraction determined by high-performance gel-permeation chromatography. *Biochem. J.*, **240**, 87–97.
Wake, R.G. (1957). The action of rennin on casein. *Aust. J. Sci.*, **20**, 147–8.
Wallace, D.L. (1981). Marzyme II—an effective coagulant for US cheesemakers and whey procesors. In *Proceedings from the Second Biennial Marschall International Cheese Conference*. Marschall Products, pp. 289–96.
Walstra, P. (1979). The voluminosity of bovine casein micelles and some of its implications. *J. Dairy Res.*, **46**, 317–23.
Walstra, P., Bloomfield, V.A., Wei, G.J. & Jenness, R. (1981). Effect of chymosin action on the hydrodynamic diameter of casein micelles. *Biochim. Biophys. Acta*, **669**, 258–9.

Wheelock, J.V. & Kirk, A. (1974). The role of β-lactoglobulin in the primary phase of rennin action on heated casein micelles and heated milk. *J. Dairy Res.*, **41**, 367–72.

Wilson, G.A. & Wheelock, J.V. (1972). Factors affecting the action of rennin in heated milk. *J. Dairy Res.*, **39**, 413–19.

15

HEAT STABILITY OF MILK

H. SINGH

Department of Food Technology, Massey University, Private Bag, Palmerston North, New Zealand

&

L.K. CREAMER

New Zealand Dairy Research Institute, Private Bag, Palmerston North, New Zealand

1. INTRODUCTION

In the past, milk was consumed fresh or preserved by fermentation to cheese or by conversion to butter. With the advent of newer preservation techniques, it was found that milk could be sterilized and preserved in liquid form. Sterilized liquid milk products have been produced in North America and Europe since the middle of the 19th century. These include condensed milk and evaporated milk. More recently, this range has extended to ultra-high-temperature (UHT)-sterilized milks and creams and a number of milk products made by recombining or reconstituting dried milk solids with water close to the site of consumption. The keeping qualities of these products have made long distance transport easier, particularly to areas where the supply of fresh milk is inadequate or uneconomic.

The major problems encountered in the production of heat-sterilized milk products are coagulation during the sterilization process and gelation and/or sedimentation during storage. At its normal concentration, milk is usually capable of withstanding both conventional and UHT sterilization processes. However, sterilization of concentrated milks or reconstituted/recombined concentrated milks often causes partial or complete coagulation during processing. Wide variations occur in the stability of milk to heat sterilization processes, and these are affected by the composition of the milk and the processing treatments to which it has been subjected.

Investigation of these problems has led to an extensive literature on the effects of heat on milk and this has been summarized in a number of reviews (Rose, 1963; Fox & Morrissey, 1977; Fox, 1981*a*, 1982; Singh, 1988; Singh & Fox, 1989; Singh *et al.*, 1989).

This chapter is intended to provide a general survey of the heat stability of normal and concentrated milks, with emphasis on developments that have occurred during the last two decades. It also includes the physical and chemical reactions that occur in milk at sterilization temperatures. In the final section, possible mechanisms of heat coagulation of both normal and concentrated milks are discussed.

2. Determination of Heat Stability

The term 'heat stability' refers to the relative resistance of milk to coagulation when it is heated at sterilization temperatures. It can be defined as the length of time that elapses between placing a container of milk in an oil bath at a definite temperature and the onset of coagulation as indicated by flocculation, gelation or changes in protein sedimentability (Fox & Morrissey, 1977). The temperature required to induce coagulation in a fixed time can also be taken as a measure of heat stability but, for convenience, a fixed temperature (120, 130 or 140°C) is usually chosen. The most widely used method to assess heat stability, at least for research purposes, involves sealing a milk sample (1–2 ml) in a glass tube which is clipped onto a platform and placed in a temperature-controlled oil bath, usually at 140°C for milks of standard concentration or at 120°C for concentrated milks. The platform is rocked at a definite rate until particles of coagulated protein can be seen in the flowing milk (Miller & Sommer, 1940; Cole & Tarassuk, 1946; Davies & White, 1966). This method is commonly referred to as a 'subjective method' and is affected by assay temperature, agitation rate, degree of tube fill and angle of tilt. Consequently, comparison of the results from different laboratories or investigations cannot be made in detail.

A method based on protein sedimentation has been reported (Whitney *et al.*, 1952; White & Davies, 1966); the percentage of total protein sedimentable at low centrifugal force after heating under defined conditions is measured and a sharp increase in sedimentation is obtained at the point of coagulation. This method is referred to as an 'objective method' and the endpoint is more precise than in the subjective method. However, it is exceedingly time consuming and is unsuitable for routine use.

Several methods have been suggested for estimating the heat stability of milk. These include the ethanol test (Sommer & Binney, 1923), the colorimetric test (Leighton & Mudge, 1923), a 'whitening' test (Burton, 1956) and viscosity determinations (Ball, 1955). However, the correlations between data obtained by these tests and by the 'oil bath' method have not been satisfactory.

Foissy and Kneifel (1984) have developed an automatic method in which the heat stability is objectively determined by an electromagnetic measuring system. The method has been shown to correlate well with the oil-bath method. A new computer-linked device, called a Klaro-graph, for measuring heat stability has been developed at the NIZO (de Wit *et al.*, 1986). It is essentially a falling-ball viscometer and the falling time of the glass ball is indicative of changes in viscosity during heating; a sharp increase in viscosity occurs at the point of coagulation.

The heat coagulation times obtained by any of the above methods are often poorly correlated with the stability of milk on commercial sterilization. From the commercial point of view, more reliable results can be obtained by using a pilot-scale sterilizer, but such a method is laborious and requires large samples, and hence it is unsuitable for scientific investigations or quality control laboratories.

3. Factors That Influence Heat Stability

A number of factors influence the heat stability of milk (both normal and concentrated). These include compositional factors, such as pH, concentrations of salts and concentrations of proteins, and milk processing parameters, such as preheating and concentration.

3.1. Compositional Factors

3.1.1. Effect of pH

Some of the early work on the influence of pH on heat stability showed that most milks are destabilized by acidification (Sommer & Hart, 1922; Webb, 1928). However, several investigators noted that the heat stability of some milks was increased by the addition of small amounts of acid (Frazier, 1925; Webb, 1928). Rose (1961*a*,*b*) clarified the relationship between heat stability and pH by showing that most individual-cow milks possess a sharp heat stability maximum at a pH of about 6·7 and a minimum at a pH of about 6·9; stability increases again at higher pH values (Fig. 1). The original pH of the milk could be above, below or at the pH of maximum stability, so that some milks are stabilized by small additions of acid while others are stabilized by small additions of base. Tessier and Rose (1964) found that the milks from individual cows could be classified as one of two types, as shown in Fig. 1. Type A milk shows a pronounced maximum and minimum in the heat stability-pH profile, whereas the heat stability of Type B milks increases gradually throughout the pH range. Approximately two-thirds of milks of individual Canadian cows were of Type A, while almost all of the milks tested in Australia, Ireland, Scotland, the Netherlands and New Zealand were of Type A.

The heat stability of concentrated milk is much lower than that of unconcentrated milk but the effect of pH is very similar to that for unconcentrated milk, with a heat stability maximum usually occurring in the pH range 6·4–6·6. However, concentrated milk remains unstable at all pH values greater than 6·8 (Figs. 1 and 2) (Rose, 1961*c*; Schmidt & Koops, 1965; Muir & Sweetsur, 1978).

3.1.2. Effect of Season/Lactation

A number of studies have shown that there is a marked seasonal/lactational variation in the heat stability of milk. In dairying countries with seasonal production patterns, e.g. New Zealand, the problem of low stability occurs primarily at the beginning and end of the dairying season. During the season, the natural pH of milk may or may not coincide with the pH of the maximum in the heat stability-pH

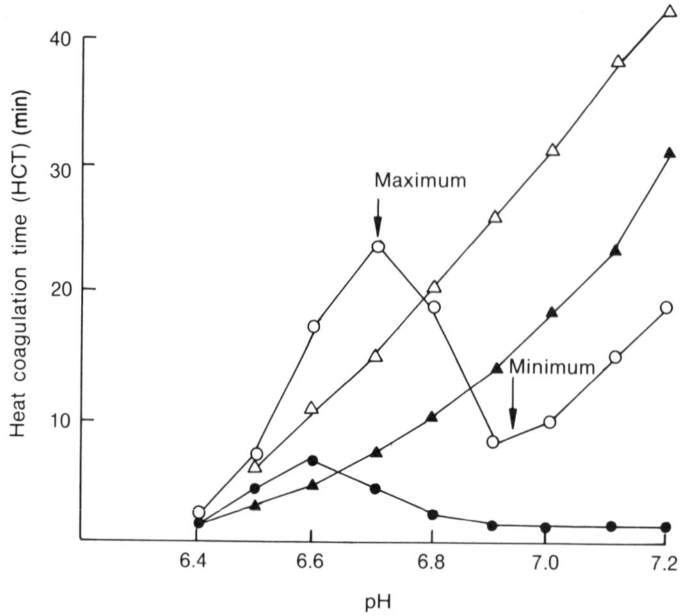

Fig. 1. Heat stability (HCT) versus pH curves for normal skim milk heated at 140°C. ○–○, Type A milk showing a maximum and a minimum in the stability–pH curve; △–△, Type B milk showing gradually increasing stability with pH; ▲–▲, milk from which whey proteins were removed prior to heat treatment or serum-protein-free casein micelle (SPFCM) dispersions; ●–●, concentrated skim milk (20% total solids). Note that HCT does not increase at pH values > 7.

profile (see Fig. 1). For example, in Scottish milks, the natural pH of milk coincides with the pH of maximum stability for the greater part of the dairying season, but occurs in the minimum region at the beginning of the season (Holt *et al.*, 1978*a*).

Seasonal variations in the heat stabilities of either concentrated or normal milks have been extensively reported for bulk Dutch milk (de Koning *et al.*, 1974), creamery milk supplies in south-west Scotland (Holt *et al.*, 1978*a*; Sweetsur & Muir, 1982), bulked Irish creamery milks (Kelly *et al.*, 1982), bulked Australian milks (Feagan *et al.*, 1972) and bulked New Zealand milks (Newstead *et al.*, 1975).

3.1.3. Effect of Milk Salts

For many years, variations in the heat stability of milk were attributed to differences in the salt composition of milk, largely as a result of the theory of Sommer and Hart (1919). According to this theory, there is an optimum value for the ratio of total calcium plus magnesium to total phosphate plus citrate in milk and any deviation from this value, occurring naturally or due to the addition of salts, results in a decrease in heat stability. In most cases, low stability was attributed to an excess of calcium and magnesium.

The usefulness of the salt balance theory was criticized on the grounds that there was no direct relationship between the heat stability of milk and the concentrations

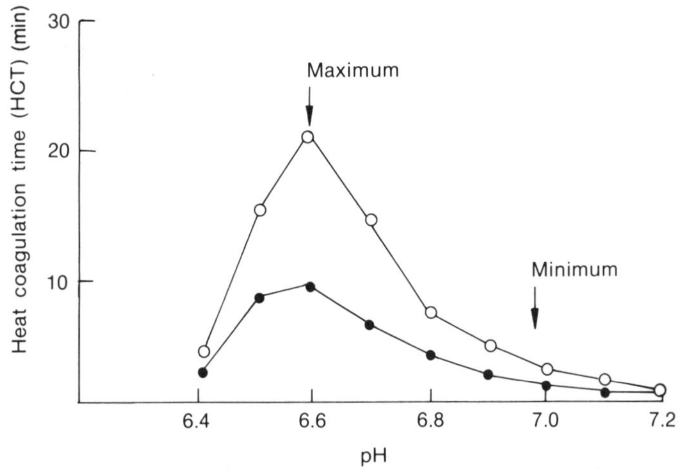

Fig. 2. Heat stability (HCT) versus pH curves for concentrated milks (20% total solids) heated at 120°C. ○–○, Concentrated skim milk, showing a maximum: the skim milk was preheated (forewarmed) prior to concentration; ●–●, concentrated skim milk: no preheating before concentration.

of milk salts (Rogers *et al.*, 1921; Holm *et al.*, 1923). These authors pointed out that for each milk there is a combination of pH and salt balance that gives maximum stability. Most of these studies were carried out using concentrated milk. Processors modified the heat stability of concentrated milk by the use of calcium salts, phosphates and citrates without a full understanding of the relationship between the heat stability and pH.

3.1.3.1. Colloidal salts. Colloidal calcium phosphate (CCP) plays an integrating role in the structure of casein micelles in milk. Several workers showed that removal of CCP from milk, using an acidification and dialysis technique (Pyne & McGann, 1960), markedly increased heat stability (Pyne, 1958; Pyne & McGann, 1960; Rose 1962). Fox and Hoynes (1975) showed that removal of 40% of the CCP increased the heat stability in the region of minimum stability, but removal of greater than 60% of the CCP had a destabilizing influence at pH values above 7·1. Doubling the level of CCP had a slight destabilizing effect over the entire pH range.

3.1.3.2. Soluble salts. Soluble salts, especially calcium and phosphate, are thought to play an important role in the heat stability of unconcentrated and concentrated milks (Pyne & McHenry, 1955; Pyne, 1958; Fox & Morrissey, 1977). Generally, addition of calcium salts to milk decreases heat stability, increases the calcium ion concentration, increases CCP and decreases pH, whereas reduction of calcium concentration by the addition of citrate or phosphate, which also reduces CCP, or by dialysis generally has a stabilizing effect.

Morrissey (1969) showed that if the concentration of soluble calcium plus magnesium in normal milk was reduced from 13 to 11 mM, the stability in the region of the minimum was increased markedly whereas stability in the region of the maximum decreased, thus converting Type A milk to Type B. Conversely, increasing the calcium concentration decreased stability in the region of the minimum. Sweetsur and White (1974) confirmed that increasing the concentration of serum salts accentuated the minimum but, rather surprisingly, increased the heat stability at pH values below 6·7 and introduced a second minimum at pH values above 7·3. Fox and Hearn (1978*a*) dialysed milk against distilled water for 3 h and found that normal Type A milk was converted to Type B milk by reducing stability in the region of the maximum while increasing stability in the region of the minimum. Dilution of milk with water (40 ml water/100 ml milk) increased stability in the region of the minimum without significantly affecting stability at the maximum. The stabilizing effect of brief dialysis on the heat stability–pH profile is probably due to a decrease in the concentrations of soluble calcium and phosphate (Fox & Hearn, 1978*a*) whereas the decrease in stability in the region of the maximum may be due to slower diffusion of calcium compared with phosphate and citrate during dialysis (Fox & Nash, 1979; Fox, 1982).

Soluble salts also influence the heat stability of concentrated milks. Newstead (1977) showed that reducing the concentration of soluble salts in milk increased the heat stability of the resulting concentrate. Partial removal of soluble salts by dialysis of milk against water for short periods increased heat stability when milk was subsequently concentrated (Muir & Sweetsur, 1978). Sweetsur and Muir (1980) also showed that concentrated milks prepared by ultrafiltration were more heat stable than those prepared by conventional evaporation, probably because of their lower soluble salt content. More recently, Hardy *et al.* (1984) confirmed that differences in the concentration of soluble calcium and phosphate had a pronounced effect on the heat stability of concentrated milks.

3.1.4. Effect of Milk Proteins

3.1.4.1. Whey proteins. Several studies have shown that whey proteins have a major influence on the pH dependence of heat stability (Rose, 1961*b*; Tessier & Rose, 1964; Fox & Hoynes, 1975). The heat stability of serum-protein-free casein micelles (SPFCM) prepared by ultracentrifugation and resuspended in milk diffusate increased continuously with increasing pH (Fig. 1) (Rose, 1961*b*). The effect of adding β-lactoglobulin to SPFCM dispersions on the pH sensitivity of the heat stability depended upon the amount of β-lactoglobulin added, especially in the range from 0·06% to 0·36% (w/v). Addition of less than 0·06% (w/v) β-lactoglobulin did not significantly affect the heat stability of the dispersion. Addition of larger amounts increased heat stability in the pH range 6·5–6·7 but decreased it at slightly more alkaline values. The heat stability–pH curve of the original milk was mimicked by the addition of 0·25% (w/v) β-lactoglobulin.

Addition of sulphydryl blocking agents, such as *N*-ethylmaleimide (NEM) or iodoacetamide, markedly reduced the maximum in the heat stability–pH curve at pH 6·7 (Rose, 1962) and it seemed clear that interaction between heat-denatured β-lactoglobulin and κ-casein via disulphide bonding was somehow responsible for the pH sensitivity of the heat stability of milk.

Fox and Hearn (1978*b*) found that the effect of pH on the heat stability of SPFCM dispersions could be modified by a number of other whey proteins: α-lactalbumin had essentially the same effect on heat stability as β-lactoglobulin. However, Schmidt and Poll (1986) were unable to confirm this finding. Bovine serum albumin, ovalbumin and lysozyme destabilized milk and SPFCM dispersions throughout the pH range 6·4–7·5. The destabilizing effect of these proteins on heat stability was also inhibited by NEM, suggesting the involvement of intermolecular disulphide bonds (Fox & Hearn, 1978*b*).

The effect of whey proteins on the heat stability of concentrated milk appears to be different from the effect of whey proteins on the heat stability of unconcentrated milk. Newstead *et al.* (1977) showed that the heat stability of concentrated milk decreased with increasing whey protein concentration (between 10 and 150% of the original) but the shape of the heat stability–pH profile of concentrated milk remained unaffected. Muir and Sweetsur (1978) found that concentrated milks prepared with a high level of β-lactoglobulin were significantly less stable than those with less β-lactoglobulin. Addition of β-lactoglobulin to concentrated milk also destabilized it over the entire pH range studied (Muir & Sweetsur, 1978; Pearce, 1979).

3.1.4.2. Caseins. Rose (1962) demonstrated that reaction between heat-denatured β-lactoglobulin and κ-casein was a major factor responsible for the complex heat stability–pH patterns of most milks. The minimum in the heat stability–pH curve of a type A milk was reduced or completely eliminated on enrichment of the milk with κ-casein (Tessier & Rose, 1964). Tessier and Rose (1964) suggested that the amount of κ-casein on the surface of casein micelles affects the heat stability–pH pattern, and that Type B milks, i.e. those with no minimum, contain an excess of κ-casein on the micellar surfaces. In support of this suggestion, Type B milk was converted to Type A by removing some micellar κ-casein by treatment with NaCl or by increasing the level of soluble β-lactoglobulin, suggesting that the heat stability patterns of milk are controlled by the proportions of surface κ-casein and soluble β-lactoglobulin.

Fox and Hearn (1978*c*) treated milk with rennet to hydrolyse κ-casein before determination of heat stability; maximum heat stability was not significantly reduced until after prolonged pre-renneting, but limited proteolysis of κ-casein caused marked destabilization at pH values alkaline to the normal minimum. These authors suggested that κ-casein is especially important for the heat stability of normal milk at pH values above the minimum (i.e. greater than 6·9).

The addition of κ-casein to milk before concentration increases the heat stability of the resulting concentrate over the entire pH range (Muir & Sweetsur, 1978). Snoeren *et al.* (1978) found that addition of α_{s2}-casein (0·3%, w/v) to skim milk did not affect the maximum stability but reduced the heat stability at pH values above 6·9, thus extending the minimum to cover a wider pH range. Addition of α_{s2}-casein (0·3%) to SPFCM dispersions, with or without β-lactoglobulin (0·3%, w/v), reduced heat stability over the pH range 6·5–7·0 (Snoeren & van Riel, 1979). The heat stabilities of artificially-formed casein micelles and of α_{s2}-casein/β-lactoglobulin mixtures were markedly reduced by incorporation of increasing amounts of α_{s2}-casein into the system (Kudo, 1980*a*).

The effect of α_{s2}-casein on the heat stabilities of concentrated milk has not been studied.

α_{s1}-Caseins and β-caseins appear to increase the heat stability of synthetic micelle systems. Schmidt and Koops (1977) and Schmidt (1979) reported that a high ratio of α_{s1}- to β-casein in synthetic micelles led to higher heat stabilities. When the ratio of α_{s1}-casein to β-casein was low, heat stability increased with increasing pH (Type B) even when β-lactoglobulin at a concentration of 0·34% (w/v) was added. It was concluded that Type B behaviour was restricted not only to a definite κ-casein/β-lactoglobulin ratio but was dependent also on the proportions of α_{s1}- and β-caseins. However, if β-casein was absent from the micelle system, normal Type A behaviour was observed.

3.1.5. Effect of Urea

The ability of urea to increase the heat stability of normal milk is well established (Pyne, 1958; Robertson & Dixon, 1969; Muir & Sweetsur, 1976, 1977; Holt *et al.*, 1978*a*). Addition of urea to milk at low concentrations (less than 7 mM) does not significantly affect heat stability in the region of maximum stability, but addition at higher concentrations increases the heat stability of milk throughout the pH range 6·4–7·3 (Muir & Sweetsur, 1977).

Fox *et al.* (1980) suggested that the most likely mechanism by which urea increases heat stability is via heat-produced ammonia from urea which neutralizes the acidity produced during heating. Sweetsur and Muir (1981) proposed that stabilization is due to the formation of a cyanate–casein complex, possibly via ε-amino groups of lysine residues. Reducing sugars, such as lactose, glucose, galactose and fructose, act synergistically with urea to increase the heat stability of milk and artificial casein micelle systems (Kudo, 1980*b*; Shalabi & Fox, 1982*a*).

In contrast to normal milks, urea does not increase the heat stabilty of concentrated milks (Muir & Sweetsur, 1977). However, the addition of urea to concentrated milk prepared by reconstitution of 'high heat' skim milk powder at 20% (w/v) was shown to increase its heat stability (Kelly, 1982).

3.1.6. Effect of Lactose

Generally, lactose is considered to be a destabilizing factor in the heat stability of milk. Increasing the lactose concentration to 150% (about 65 g/litre) of the normal concentration in milk reduced heat stability over the whole pH range and shifted the pH of the minimum to slightly more alkaline values (Sweetsur & White, 1975). However, Kudo (1980*b*) observed that lactose had a stabilizing effect in artificial casein micelle systems, the maximum effect being observed at 20 g/litre. Removal of lactose from milk by dialysis against a synthetic salts buffer reduced the heat stability (Shalabi & Fox, 1982*a*) but Sweetsur and White (1975) found that removal of lactose stabilized Type A milk. Urea stabilized milk only in the presence of lactose (Kudo, 1980*b*; Shalabi & Fox, 1982*a*).

Further work is needed to clarify the role of lactose in heat stability. The effects of lactose on the heat stability of concentrated milk have not been reported.

3.2. Relationship Between Heat Stability and Overall Milk Composition

There is considerable variation between cows and with season in the heat stability of milk at its natural pH (Holm *et al.*, 1932; White & Davies, 1958) and in its heat stability–pH profile (Rose, 1961*a*). Various early investigators showed that, although the heat stability of milk responded to trace additions of various salts, it did not correlate with any known constituent of milk (White & Davies, 1958; Pyne, 1959). Since the demonstration by Rose (1961*a*) that pH has an overhwelming influence on heat stability, few attempts have been made to correlate heat stability with milk composition. Although Rose (1962) found that the heat coagulation time at the pH of maximum stability (HCT_{max}) was positively correlated with the β-lactoglobulin concentration ($r = +0{\cdot}50$) and that the heat coagulation time at the pH of minimum stability (HCT_{min}) was negatively correlated ($r = -0{\cdot}52$), this relationship has not been found in any other study. For example, no statistically significant correlation between heat coagulation time and any compositional factor, other than urea, was found in Scottish creamery bulk milk samples at their natural, unadjusted pH (Holt *et al.*, 1978*a*). It was suggested that about 40% of the seasonal variation in heat stability could result from changes in the urea level (Muir & Sweetsur, 1976).

Significant correlations were found between 'soluble' casein and heat coagulation times at natural pH (HCT_{nat}) (Morrissey *et al.*, 1981), between casein number and HCT_{nat} (Holt *et al.*, 1978*a*; Kelly *et al.*, 1982) and between casein number and HCT_{max} (Kelly *et al.*, 1982). However, the relationship between casein number and HCT_{nat} was not considered to be causal because there was an equally simple relationship between casein number and urea content. Morrissey *et al.* (1981) showed that in mid-lactation milks HCT_{nat} and HCT_{max} were correlated with soluble phosphate ($r = +0{\cdot}37$) and with soluble calcium ($r = -0{\cdot}38$). This observation supported the hypothesis put forward by Rose (1961*a*) who showed that a significant correlation existed between the ratio of calcium ions to soluble inorganic phosphate and HCT_{max}. Kelly *et al.* (1982) found a significant negative correlation ($r = -0{\cdot}74$) between soluble calcium and HCT_{max}.

Feagan *et al.* (1972) were unable to show any differences between HCT_{nat} and milk protein genetic variations, but found that HCT_{max} was affected by the genetic variants of κ-casein and β-lactoglobulin present. Recently, McLean *et al.* (1987) showed a significant relationship between HCT_{nat} and κ-casein genetic variants, milks containing κ-casein AB being more stable than those containing κ-casein AA. However, there was no significant correlation between HCT_{nat} or HCT_{max} and the genetic variants of β-lactoglobulin.

Thompson *et al.* (1969) found a positive correlation between the degree of hydration of casein micelles and heat stability but Morrissey *et al.* (1981) could not confirm this.

Few significant correlations have been demonstrated between the heat stability of concentrated bulk milk and the composition of raw milk. Morrissey *et al.* (1981) were unable to show any relationship between the heat stability of concentrated milk and any milk component other than citrate. Sweetsur and Muir (1982) confirmed the above findings but also showed a weak correlation between the concentration of soluble calcium and the heat stability of concentrated milk. In contrast to unconcentrated milk, no correlation was found between the heat stability of concentrated milk and its urea concentration (Muir & Sweetsur, 1978; McLean *et al.*, 1987). Newstead *et al.* (1974) found a negative correlation ($r = -0{\cdot}59$) between HCT_{nat} and the content of whey proteins in bulk milk collected over two lactation seasons. However, McLean *et al.* (1987) were unable to confirm this but did show a significant negative correlation between whey protein concentration and HCT_{max}. de Koning *et al.* (1974) were unable to establish any correlation between the concentrations of individual milk proteins or the proportions of SH/SS-containing proteins and the heat stability of concentrated milk. However, more recently, McLean *et al.* (1987) demonstrated that HCT_{max} and HCT_{nat} were positively correlated with β-casein and κ-casein concentrations and were negatively correlated with α_{s1}-casein and β-lactoglobulin concentrations. A negative correlation between the amount of non-sedimentable (soluble) protein in milk and the heat stability of concentrated milk (HCT_{nat}) was demonstrated by Singh and Tokley (1990).

There appears to be no relationship between the heat stability of raw milk and the concentrate made from it (Holm *et al.*, 1932; Webb & Holm, 1932; McLean *et al.*, 1987).

3.3. Effects of Additives

The heat stability of milk can be modified by several chemical reagents, although there are only a few legally permitted food grade additives that can be used commercially to increase stability. The main aim of studying the effects of chemical reagents has been to understand more fully the mechanism of heat coagulation.

3.3.1. Detergents

The heat stability–pH curve of milk can be modified by various surface active agents. Anionic detergents, such as sodium dodecyl sulphate (SDS) increased stability in the region of the maximum and shifted the heat stability–pH curve to more acidic values, whereas cationic detergents, such as cetyltrimethylammonium bromide, progressively shifted the heat stability–pH curve to alkaline pH values while causing a moderate increase in the stability in the region of the maximum (Fox & Hearn, 1978*b*; Shalabi & Fox, 1982*b*). The non-ionic detergents Triton X and Tween 60, added to milk at a level of up to 3% (w/v), had no significant effect on heat stability throughout the pH range 6·4–7·3 (Shalabi & Fox, 1982*b*). Addition of SDS at 0·05% and 0·1% (w/v) to SPFCM dispersions shifted the heat stability–pH curve to more acidic values but at higher concentrations of SDS (0·15%–0·25%, w/v), marked destabilization occurred in the pH range 6·4–6·9, producing a curve similar to that for SDS-treated milk (Shalabi & Fox, 1982*b*).

3.3.2. Aldehydes, Ketones and Sugars

Addition of formaldehyde and other low molecular weight aldehydes increases the heat stability of unconcentrated and concentrated milks over a wide pH range (Nelson, 1954; Holt *et al.*, 1978*b*; Singh & Fox, 1985*a*). The minimum in the heat stability–pH curve of normal milk can be completely eliminated by formaldehyde treatment (250 mg/litre). Urea acts synergistically with aldehydes in stabilizing such systems (Muir *et al.*, 1979; Shalabi & Fox, 1982*a*). The increase in heat stability on addition of aldeydes is due to modification of the ε-amino groups of lysine (Holt *et al.*, 1978*b*) and to cross-linking of polypeptide chains (Singh & Fox, 1985*a*).

Low molecular weight ketones, such as dihydroxyacetone, stabilized lactose-free milk and their stabilizing effect was greatly increased in the presence of urea (Shalabi & Fox, 1982*c*).

Diketones, such as diactyl and 1,2-cyclohexanedione, stabilized milk only in the region of maximum stability, whereas dialdehydes, such as glyoxal and substituted glyoxals, stabilized milk throughout the pH range 6·4–7·3 (Shalabi & Fox, 1982*c*). Diacetyl increased the heat stability of concentrated milks (20% total solids) reconstituted from skim milk powder but had no effect on the stability of concentrated milks prepared by evaporation (Shalabi & Fox, 1982*d*). As these reagents modify arginine residues, more or less specifically, Shalabi and Fox suggested that arginine residues are important in heat stability and that the synergistic action of urea and dicarbonyls may be due to the conversion of lysine by urea to homocitrulline, a structural analogue of arginine, with which a dicarbonyl may react by a mechanism similar to that with arginine (Shalabi & Fox, 1982*c*).

Addition of simple sugars, e.g. glyceraldehyde and erythrose, increased the heat stability of both normal and concentrated milks throughout the pH range; however, ribose was less effective although its 2-deoxy derivative was very effective (Holt *et al.*, 1978*b*). Hexoses and their deoxy derivatives did not affect heat stability, possibly because they are not sufficiently reactive as they occur in a hemi-acetal ring structure.

3.3.3. Other Additives

κ-Carrageenan at concentrations of 0·05% stabilized milk in the region of minimum stability without a significant effect on the stability at the pH of maximum stability (Shalabi & Fox, 1982*e*). Higher concentrations (especially above 0·4%) caused marked destabilization throughout the pH range 6·4–7·3. Other hydrocolloids, e.g. carrageenan K-100, carboxymethyl cellulose and Manucol ester E/RK, had little effect on heat stability.

κ-Carrageenan also stabilized concentrated milk. Soy bean lecithin markedly increased the heat stability of full cream concentrated milk (Hardy *et al.*, 1985) and recombined evaporated milk (Singh *et al.*, 1992). Addition of buttermilk increased the heat stability of recombined evaporated milk (Singh & Tokley, 1990). Small amounts of H_2O_2 (0·05%) improved the heat stability of concentrated skim milk and shifted the pH of maximum stability to more alkaline values, whereas higher concentrations of H_2O_2 (above 1·0%) had a destabilizing effect at all pH values (Koops & Westerbeek, 1970).

3.4. Processing Factors

3.4.1. Preheating

Early workers observed that preheat treatment of milk has a considerable destabilizing effect on the heat stability of normal milk (Leighton & Deysher, 1923; Webb & Holm, 1932; Davies, 1959). However, Rose (1962) showed that the true effect of preheating on the heat stability of unconcentrated milks is to shift the heat stability–pH curve slightly to more acid pH values, so that the observed change in heat stability is a function of the pH decline during preheating. If the pH of milk is increased prior to preheating, the pH of maximum stability is progressively shifted to more acidic values (Morrissey & O'Mahony, 1976; Singh & Fox, 1985*b*). Preheating milk at ultra-high temperatures, e.g. 120°C for more than 20 min (Fox, 1981*a*) or 150°C for 1·5 min (Sweetsur & Muir, 1981), increased heat stability in the minimum region and converted Type A milks to Type B milks.

It has been known since the 1930s that preheating milk, usually at 100°C for 5 min, before concentration, increases the heat stability of the resulting concentrate (Webb & Holm, 1932; Webb & Bell, 1942) and is used commercially in the manufacture of evaporated milk and heat-stable powders. Webb and Holm (1932) showed that skim milk that had been preheated and concentrated to less than 13% total solids had a lower heat stability than non-preheated milk; however, if milk was further concentrated to 18% total solids, the heat stability was improved by suitable preheating.

Although preheating at ultra-high temperatures for short times produces a concentrate that is very stable to subsequent heat processing (Webb & Bell, 1942), UHT preheating has not been used commercially. In Australia, preheating temperatures of 80–90°C for 30 min are commonly used (Kieseker, 1977), pre-heating conditions used in Ireland are usually 95°C for 10 min (Fox, 1982), and 105–120°C for 30–180 s is favoured in New Zealand (Newstead *et al.*, 1975).

3.4.2. Homogenization

Homogenization of skim milk has little effect on its heat stability but the stability of whole milk is reduced by homogenization, the magnitude of the effect increasing with increasing homogenization pressure in the range 3·5–43·5 MPa (Sweetsur & Muir, 1983*a*). Homogenization temperature does not greatly affect the maximum heat stability. The general destabilizing effect of homogenization is lessened by reducing the homogenization pressure (Henry *et al.*, 1944), using two-stage homogenization (Sweetsur & Muir, 1983*a*) or by addition of phosphates (Sweetsur & Muir, 1983*a*). Additions of β-lactoglobulin, calcium phosphate, sulphydryl blocking agents or oxidizing agents affected the stability of homogenized milk in a similar way to unhomogenized milk (Sweetsur & Muir, 1983*a*). The ability of urea to increase the heat stability of homogenized milk decreased with increasing severity of homogenization, whereas formaldehyde stabilized homogenized milk only above pH 7·0 (Sweetsur & Muir, 1983*a*).

As with normal milk, homogenization reduced the heat stability of full-cream concentrated milk (Sweetsur & Muir, 1982) and recombined evaporated milk (Newstead *et al.*, 1979). Addition of β-lactoglobulin to milk before preheating reduced heat stability of concentrates prepared from it, and homogenization of the concentrate accentuated the destabilizing influence of β-lactoglobulin (Sweetsur & Muir, 1983*b*). Addition of sulphydryl blocking agents (e.g. NEM) or oxidizing agents (e.g. potassium iodate) caused a large increase in the heat stability of homogenized concentrated whole milk, suggesting the important role of sulphydryl group interactions (Sweetsur & Muir, 1983*b*). The adverse effects of homogenization on the heat stability of concentrates were reduced by traditional preheating (90°C for 10 min), preheating at high temperatures (120°C for 120 s–145°C for 5 s), two-stage hoogenization, multiple pass homogenization and addition of phosphates (Newstead *et al.*, 1979; Sweetsur & Muir, 1982). The destabilization caused by homogenization is a major limitation in the development of new type of sterile milk products such as full cream evaporated milk. Further studies are needed in this area.

3.4.3. Evaporation and Drying

Removal of water from milk by evaporation (concentration) causes close packing of the casein micelles, which are themselves about two-thirds water, and higher concentrations of whey proteins, lactose and soluble and colloidal salts. The pH of milk decreases almost linearly with increasing concentration of total milk solids (TS) in the range from 9 to 40%. The pH of reconstituted skim milk containing 9% TS is about 6·6, while that of concentrates containing 23 and 40% TS is approximately 6·2 and 6·0, respectively (Howat & Wright, 1934). There is generally an inverse linear relationship between heat stability and the concentration of solids-not-fat (Holm *et al.*, 1923; Webb & Holm, 1932). The destabilizing effects of concentration are essentially reversed by dilution to normal solids concentration. However, if milk is heated at high temperatures prior to concentration, irreversible destabilization occurs in the region of minimum stability (Muir *et al.*, 1978; Singh & Creamer, 1991*a*).

There appears to be very little information on the effect of commercial drying on the heat stability of milk. Muir *et al.* (1978) reported that spray dried powders, after reconstitution to skim milk at 10% TS, were more heat stable than the concentrates from which they were prepared, after dilution to the same level of TS. Singh and Creamer (1991*a*) confirmed the above findings and also showed that the heat stabilities of concentrates (45% TS) diluted to 20% TS were 2–3 times lower than those of their corresponding reconstituted skim milk powders. This suggests that spray drying induces changes in the milk protein system that reverse the destabilizing effect of concentration. It is interesting to note that the addition of diacetyl to 20% TS milk, reconstituted from powder, caused a very large increase in heat stability but had no effect on the heat stability of concentrated milk prepared by diluting the 45% TS concentrate to 20% TS (Shalabi & Fox, 1982*d*). Addition of

TABLE 1
Physico-chemical Changes in Milk During Heating

Changes to casein micelles
– Aggregation and disaggregation of casein micelles
– Changes in zeta potential and hydration
– Dephosphorylation and proteolysis of caseins
Interactions of casein micelles with whey proteins
– Formation of a κ-casein/β-lactoglobulin complex
– Dissociation of the complex from the casein micelles
Transfer of soluble calcium and phosphate to colloidal calcium phosphate
Reaction between lactose and protein (Maillard reactions)
Formation of covalent bonds
Decrease in pH

urea to milk before spray drying had a stabilizing effect on 20% TS milk reconstituted from the powder (Kelly, 1982) whereas addition of urea to concentrated milk made directly by evaporation had no effect (Muir *et al.*, 1978). These results indicate that milk reconstituted from powder behaves differently from 20% TS milk concentrate. Clearly, further work is required to investigate the effects of spray drying on the milk protein system.

4. PHYSICO-CHEMICAL CHANGES

Heating milk at ultra-high temperatures causes a number of physical and chemical changes in milk as a consequence of the modification of its components. Some of these changes are listed in Table 1.

4.1. Changes in Casein Micelles

4.1.1. Structural Changes

Heating milk at temperatures up to 90°C causes only minor changes in the size of casein micelles but severe heat treatments, such as UHT sterilization, increase the micelle size (Wilson & Herreid, 1961; Hostettler *et al.*, 1965; Josephson *et al.*, 1967; Carroll *et al.*, 1971). Recently, Mohammad and Fox (1987*a*), using electron microscopy, showed that heating milk at 140°C for 10 min caused an increase in the diameter of the casein micelles and also reduced the range of micellar sizes. Dalgleish *et al.* (1987*a*) measured the diameter of casein micelles in heated milk using photon correlation spectroscopy; average diameters increased slowly during the initial stages of heating at 130°C until finally a rapid increase in diameter occurred just prior to visible coagulation.

The pH at heating also affects casein micelle size. Dalgleish *et al.* (1987*a*) showed that the average diameter of casein micelles increased as the pH at heating was increased from 6·7 to 6·9 but further increases in pH to 7·2 caused a decrease in the micellar diameter, confirming data obtained by electron microscopy (Creamer & Matheson, 1980; Mohammad & Fox, 1987*a*).

It has also been observed that the increase in micelle size during heating is accompanied by a large increase in the number of small particles, possibly formed by disaggregation of casein micelles (Hostettler *et al.*, 1965; Morr, 1969, 1973; Creamer *et al.*, 1978; Creamer & Matheson, 1980; Mohammad & Fox, 1987*a*). Aoki *et al.* (1974, 1975) found that heating whey-protein-free milk at temperatures greater than 110°C caused a substantial increase in the level of 'soluble' (non-sedimentable at 109 000 g for 60 min) casein, of which about 40% was κ-casein. The extent of release was related to the time and temperature of heating, and to the concentration of casein micelles and soluble salts. Kudo (1980*c*) showed that when milk was heated at 140°C at a pH less than 6·5, the amount of 'soluble' (non-sedimentable at 100 000 g for 60 min) proteins in a heated sample was less than that in a corresponding unheated sample, whereas at higher pH values, the amount of 'soluble' proteins was greater than that in the unheated control. Singh and Fox (1985*b*, 1986) found that in milk the dissociation of κ-casein-rich protein from the micelles was dependent on the pH at heating; below pH 6·7 no dissociation of micellar κ-casein occurred on heating at 140°C for 1 min, but at higher pH values (above 6·9) dissociation of micellar κ-casein increased with increasing pH. Dissociation of micellar κ-casein on heating SPFCM dispersions increased with increasing pH at heating throughout the pH range 6·4–7·3; the addition of β-lactoglobulin to SPFCM dispersions before heating reduced the amount of dissociated κ-casein at a pH less than 6·7 but increased it at higher pH values (Singh & Fox, 1987*a*). Reducing the CCP concentration in skim milk to 40% or increasing it to 120% of the original level did not significantly affect the heat-induced dissociation of micellar κ-casein at pH 7·3 (Singh & Fox, 1987*b*). However, increasing soluble calcium or decreasing soluble phosphate concentrations (increasing the Ca : P ratio) reduced the dissociation of κ-casein from the micelles (Singh & Fox, 1987*b*).

The reason why κ-casein dissociates from the micelles on heating at slightly alkaline pH values is not known. Singh and Fox (1986) suggested that when the negative charge on the micellar surface increases to a certain value, hydrophobic bonds are insufficient to hold κ-casein on the micelles and dissociation of κ-casein then occurs due to electrostatic repulsion. This view was based on the effects of ionic detergents and protein modification on the dissociation of micellar κ-casein; addition of anionic detergents (SDS) or succinylation of milk protein intensified the heat-induced dissociation of micellar κ-casein at any particular pH, whereas cationic detergents or amidation of milk protein had the opposite effect (Singh & Fox, 1986; Singh & Fox, 1987*a*; for a discussion on the forces binding κ-casein to casein micelles, see Creamer, 1991).

Studies involving ultracentrifugation, viscosity measurement and gel permeation chromatography showed that the micelles aggregated initially on heating but then dissociated before the final aggregation at the onset of coagulation (Fox, 1981*a*; Mohammad & Fox, 1987*b*).

The aggregation and dissociation of casein micelles in concentrated milk appears to be different from that in unconcentrated milk. Singh and Creamer (1991*b*) showed that increasing the concentration of milk solids gradually lowered the pH at which micellar κ-casein dissociated on heating. Heating 20% TS milk at 120°C at its normal pH (6·6) caused dissociation of κ-casein-rich protein from the casein micelles; the extent of dissociation increased with increasing pH and heating time. The proportions of the components in the dissociated protein were: κ-caseins > β-caseins > α_s-caseins (Singh & Creamer, 1991*c*). The dissociated caseins existed as aggregates of varying size, including some monomers. Most of the κ-casein was linked to whey proteins. The dissociated κ-casein appeared to have a charge distribution 'different' from that of native κ-casein, but α_s- and β-caseins were essentially the same as in their native state as indicated by their separation on urea-containing polyacrylamide gels. Singh and Creamer (1991*c*) speculated that the mechanism of dissociation of κ-casein in concentrated milk probably involves modification of its charge by some heat-induced effects, e.g. reaction between lysine and lactose.

In contrast to unconcentrated milk, the aggregation of casein micelles in concentrated milk appears to increase gradually with increasing intensity of heating, as indicated by turbidity, ultracentrifugation and gel filtration chromatography (Singh & Creamer, 1991*c*,*d*).

4.1.2. Physico-chemical Changes

4.1.2.1. Zeta potential. Many changes that might be expected to reduce the micellar zeta potential occur in milk on heating at high temperatures: hydrolysis of κ-casein, dephosphorylation of α_{s1}-, α_{s2}- and β-caseins, reduction of pH and precipitation of calcium phosphate (Fox, 1981*a*, 1982). However, Darling and Dickson (1979) showed that no significant changes occurred in the zeta potential (measured at room temperature) on heating milk at 135°C up to the coagulation point. Schmidt and Poll (1986) showed that heating SPFCM dispersions at pH 6·7 for 10 min at 120°C had only a slight effect on their zeta potential at room temperature; addition of whey proteins to the SPFCM dispersions before heating increased their zeta potential in the pH range between 6·6 and 6·9 but reduced it at higher pH values (7·1–7·2), and this effect became more pronounced with increasing whey protein concentration. As it is not possible to measure the zeta potential at high temperatures (120–130°C), and some of the changes that affect it may be reversed on cooling; the potential of several heat-induced changes to alter the micellar zeta potential is unknown.

4.1.2.2. Hydration. The hydration of casein micelles prepared by ultracentrifugation (90 000 g for 2 h) was substantially reduced by heating at 140°C for periods up to 15 min at a starting pH of 6·8 (Fox, 1981*a*). Ruegg *et al.* (1979) found from water sorption isotherms that the equilibrium water content of casein micelles prepared from heated milk was markedly lower than that of casein micelles prepared from unheated milk. However, Creamer and Matheson (1980) showed that micelle hydration decreased with a decrease in pH and, if the reduction in pH that occurs during heating was taken into account, heat treatment did not change micelle hydration substantially.

4.1.2.3. Dephosphorylation and proteolysis. The hydrolytic cleavage of peptide and phosphate bonds in casein is a consequence of either enzymatic action or thermolysis at the elevated temperatures. Early studies showed that sodium caseinate dephosphorylated faster than casein in casein micelles in milk at high temperatures, e.g. milk heated at 120°C for 20 min lost 10% of its serine phosphate whereas sodium caseinate lost 15% (Howat & Wright, 1934; Pyne & McHenry, 1955; Davies & White, 1959; Belec & Jenness, 1962*a*,*b*). Both reactions followed first-order kinetics with an activation energy of 25–29 kcal/mol. Nevertheless, milk concentrated two-fold dephosphorylated faster (Belec & Jenness, 1962*b*); preheating at 90°C for 10 min prior to concentration slightly decreased the rate of dephosphorylation of casein in the concentrate (Belec & Jenness, 1962*b*).

Dalgleish *et al.* (1987*b*) reported that approximately two-thirds of the casein phosphates were hydrolysed after heating for 60 min at 130°C. These authors noted that if the native micelle structure is dependent on links between the CCP and the casein through the serine phosphates, then thermal dephosphorylation of the casein would be expected to lead to micelle dissociation.

The mechanism of high temperature dephosphorylation may be similar to that reported by Manson and Carolan (1980) for dephosphorylation at higher pH and moderate temperatures. In this, the phosphoserine loses phosphate to generate a reactive intermediate, dehydroalanine, which can react with lysine or cysteine to give lysinoalanine or lanthionine, respectively. If the same mechanism operates when milk is heated, then it would be expected that some of these cross-linked amino acids would be present.

Little change occurs in non-protein nitrogen (NPN) during pasteurization or UHT sterilization of unconcentrated milk but more severe heat treatments increase the NPN almost linearly with time (Howat & Wright, 1934); for example, heating milk at 135°C for 60 min converts 10–20% of the total nitrogen to NPN. This could be due to hydrolysis of peptide bonds or release of ammonia from the amide groups.

4.2. Interactions of Casein Micelles with Whey Proteins

It is well established that heating milk at temperatures of about 90°C denatures the whey proteins and they then associate with the casein micelles. The extent and type of association between whey proteins and casein micelles depends on the severity of the heat treatment (Hostettler *et al.*, 1965; Josephson *et al.*, 1967). κ-Casein is involved in the interaction between casein micelles and whey proteins and disulphide interchange reactions are involved in this interaction (Sawyer *et al.*, 1963; Smits & van Brouwershaven, 1980; Singh & Fox, 1987*a*).

In the absence of β-lactoglobulin, α-lactalbumin hardly associates with casein micelles on heating (Smits & van Brouwershaven, 1980), although Baer *et al.* (1976) reported that it associated with β-lactoglobulin and with the casein micelles via

β-lactoglobulin/α-lactalbumin complex formation (Elfagm & Wheelock, 1977). However, Noh *et al.* (1989*a*), using radio-labelled proteins, found no evidence for any specific β-lactoglobulin/α-lactalbumin complexes in heated model systems based on casein micelle dispersions. Creamer *et al.* (1978) isolated a protein complex from heated milk that had been dispersed in citrate buffer, and showed that its amino acid composition was consistent with it being a mixture of all the whey proteins and κ-casein. Creamer (1984) later showed that the size, measured by chromatography on Sephacryl S-1000, of these protein complexes in a SDS buffer system decreased with increasing pH of heating. Snoeren and van der Spek (1977) found that α_{s2}-casein also formed complexes with β-lactoglobulin after UHT treatment of milk, but Noh *et al.* (1989*b*) reported that α_{s2}-casein was relatively unreactive in model systems heated at 90°C for 20 min.

Denatured whey proteins complexed onto the micellar surfaces, involving κ-casein, on heating milk at pH values below 6·7 in the temperature range 90–140°C, but at higher pH values denatured whey proteins remained in the intermicellar fluid as fibrous strands (Creamer & Matheson, 1980; Singh & Fox, 1985*b*, 1986; Mohammad & Fox, 1987*a*). Electron microscopic examination of milk that had been heated at pH 6·7 showed many ‘appendages’, probably whey proteins, on the micellar surfaces and small micelles bridged together by these ‘appendages’. On heating at a higher pH value (7·3), whey proteins did not attach to the micelles and many of the micelles were dissociated and had eroded surfaces (Singh, 1985; Mohammad & Fox 1987*a*). However, Singh and Fox (1985*b*) showed that the interaction between whey protein and κ-casein in the casein micelles occurred at higher pH values and that this complex tended to dissociate from the casein micelles, although cooling of the milk and adjustment of the pH caused reassociation of the complex with the casein micelles (Singh & Fox, 1986). Van Boekel *et al.* (1989*a*) suggested that the reassociation of a whey protein–κ-casein complex with the casein micelles can also occur during heating; the amounts of dissociated κ-casein and whey proteins reached a maximum within the first 2 min of heating milk at 140°C at pH 7·05. Thereafter, a gradual decline in the concentration of these proteins in the serum phase occurred, especially when the pH had dropped below 6·6.

In concentrated milk systems, the degree of association of whey proteins with the casein micelles was less than that in unconcentrated milks; approximately 50% of the original whey proteins remained in the serum phase after heating concentrated milk at 120°C at pH 6·55 and this increased to 75% at pH 6·85 (Singh & Creamer, 1991*c*). Whey proteins that remained in the serum phase were complexed with κ-casein via disulphide bonding (Singh & Creamer, 1991*c*).

4.3. Covalent Modification of Proteins

During heat treatment of proteins, side chains of some amino acids, especially that of lysine, become very reactive and combine with carbohydrates or other residues of the same or other protein molecules, forming intra- and inter-molecular bonds. The browning that occurs when milk is heated at temperatures above 100°C is a consequence of a series of reactions, known as the Maillard reaction, involving the aldehyde group of lactose and the ε-amino group of protein-bound lysine. Fox (1981*a*) reported that the dye-binding capacity of milk heated at 140°C for 20 min at pH 6·7 or pH 7·2 was reduced by 15% and 25%, respectively, reflecting the lower availability of lysine as a result of interaction with lactose. Lysine can also become unavailable through the formation of lysinoalanine when protein solutions are heated at moderately high pH values, as described earlier (Manson & Carolan, 1980). Severe heat treatments (greater than 120°C) of milk at neutral pH have also been shown to produce lysinoalanine (Lorient, 1979).

Covalent bond formation, as measured by the turbidity of heated milks dispersed in a dissociating buffer (urea, mercaptoethanol, EDTA/oxalate), increased gradually (after a lag period of 10–20 min) with heating time at 130°C (Dalgleish *et al.*, 1987*a*; van Boekel *et al.*, 1989*b*). Dalgleish *et al.* (1987*a*) reported that the increase in turbidity was independent of the initial pH of the milk, but van Boekel *et al.* (1989*b*) found that covalent bonds were formed more quickly at lower pH values. However, the nature of the polymerizing reactions involving lysinoalanine in heated milks remains unexplored to date.

4.4. Changes in Salt Equilibria

Heat treatment of milk reduces the concentrations of soluble phosphate and of both soluble and ionic calcium, the extent being dependent on the severity of the heat treatment (Davies & White, 1959; Rose & Tessier, 1959). Soluble calcium and phosphate precipitated as CCP and was protected against sedimentation through association with casein micelles (Evenhuis & de Vries, 1956). On subsequent cooling, some or all of the precipitated material redissolved (Kannan & Jenness, 1961). However, when the temperature of heating was greater than 110°C, the precipitated material did not redissolve.

Contrary to the conclusions in the above reports, Dalgleish *et al.* (1987*b*) reported that heating did not cause deposition of calcium phosphate from the serum onto the micelles, but a considerable quantity deposited on the walls of a stainless steel holding tank. Later, Dalgleish (1989) suggested that the discrepancies between the two types of experiment (heating in a stainless steel vessel versus heating in a glass container) may be due to fewer deposition sites in glass containers than on the walls of a stainless steel heating system.

Heating has little effect on monovalent ions, sodium, potassium and chloride, but its effect on citrate is not so clear. Most investigators found no evidence of deposition of citrate onto casein micelles (Evenhuis & de Vries, 1956; Davies & White, 1959; Boulet & Marier, 1960) or on the walls of the container (Dalgleish *et al.*, 1987*b*), although a recent study (Pouliot *et al.*, 1989) does not support this view.

Very few studies have been performed on the heat-induced changes in milk salts in concentrated milk. Recently, Hardy *et al.* (1984, 1985) and Nieuwenhuijse *et al.* (1988) studied the changes in soluble calcium and phosphate during the manufacture of sterile concentrated milk. There were gradual decreases in the concentrations of both soluble calcium and soluble phosphate during concentration and heat sterilization; more phosphate than calcium became colloidal. The concentration of ionic calcium also decreased during the manufacturing process.

The structure and composition of heat-precipitated calcium phosphate and the nature of its association with casein micelles are not clear. It has been postulated that calcium phosphate precipitates as amorphous calcium phosphate (Evenhuis & de Vries, 1956; Visser *et al.*, 1986) or tricalcium phosphate (Lyster, 1965), although other studies (Nieuwenhuijse *et al.*, 1988; Pouliot *et al.*, 1989) suggest that heat-precipitated calcium phosphate resembles dicalcium phosphate.

The native calcium phosphate of casein micelles dissolves when the pH is lowered to about 4·9 (Pyne & McGann, 1960). However, when concentrated milk was heated at 120°C, a proportion of micellar calcium phosphate did not dissolve on acidification (Singh & Creamer, 1991*d*). The greater the heating time, the greater was the amount of 'insoluble' calcium phosphate in the micelles. This may support the view that after heat treatment at least a part of the deposited material or the indigenous micellar calcium phosphate may become structurally 'different' from the native calcium phosphate in unheated milk.

4.5. Development of Acidity

The pH of milk decreases from about 6·7 during severe heat treatments and is commonly between 5·5 and 6·0 at the point of coagulation (Pyne & McHenry, 1955; Pyne, 1958). The following reactions are considered to contribute about 30%, 20% and 50% to acidity development, respectively.

(i) Precipitation of primary and secondary calcium phosphate as tertiary phosphate with the release of H^+.
(ii) Hydrolysis of organic (casein) phosphate ester bonds and the subsequent precipitation of calcium phosphate.
(iii) Formation of organic acids, mainly formic, from thermal breakdown of lactose (Gould, 1945).

The decrease in pH is twice as large in milk as in lactose-free milk, and has a Q_{10} of about 2. Acidity development is also affected by the amount of oxygen available (Sweetsur & White, 1975).

Heat-induced changes in pH due to shifts in the equilibrium of the milk salts are, to some extent, reversible, especially if the heat treatment is not very severe (Rose & Tessier, 1959). Early workers (Pyne & McHenry, 1955; Pyne, 1958; Sweetsur & White, 1975) measured the pH after cooling the heated milk samples to 20°C, but Fox (1981*b*) recorded the pH of samples after heating at 140°C and then cooled to temperatures ranging from 20 to 90°C. The pH of the milk changed in an approximately linear manner as the heated milk was cooled from 90 to 60°C but further cooling caused only a slight additional increase in pH. Extrapolation of the data to 140°C showed that the pH at 140°C at the point of coagulation (20 min) would have been about 4·9. Fox (1981*a,b*) confirmed the earlier findings of Pyne (1958) that if the pH of heated milk is readjusted occasionally to its original value, it will withstand heating at 140°C for at least 3 h, suggesting that the decrease in pH on heating is the single most important factor leading to the coagulation of milk during heating.

Darling (1980) demonstrated that the rate of acid production increased as a function of the initial pH to a maximum in the region of the maximum in the heat stability–pH curve, whereas in the region of the minimum there was a sharp increase in the rate of acid production.

A more recent study (van Boekel *et al.*, 1989*a*) showed, however, that the decrease in pH during heating is not linear; a rapid decrease in pH occurred during 2 min of heating at 140°C, after which the decrease was slower. In contrast to the findings of Darling (1980), van Boekel *et al.* (1989*a*) found no evidence of anomalous behaviour of pH decrease in the regions of maximum and minimum stability. These authors suggested that the initial decrease was probably due to a shift in the calcium and phosphate equilibria and that the further gradual decrease was largely due to the formation of organic acids.

5. Possible Mechanism(s) of Coagulation

When milk is heated, the stable colloidal particles (micelles) are destabilized to form a coagulum. The structure and stability of casein micelles has been widely investigated and the current views are discussed by Creamer (1991) and in Chapter 3. The stability of casein micelles is maintained by electrostatic repulsions due to electrical double layers (the zeta potential at 20°C is − 13 mV), and steric effects due to the C-terminal segment of κ-casein, which also carries a relatively high negative charge (−9). Other factors important in maintaining casein micelle integrity are CCP, which is bound to the phosphoserine residues of the caseins, and hydrophobic intereactions.

Hydrophobic interactions are involved in stabilizing the casein micelles, since dissociating agents, such as SDS and urea, which act to reduce hydrophobic interactions, tend to disrupt micellar structure. These interactions are temperature dependent; they are minimal below 5°C and increase with increasing temperature up to 60–70°C but decrease at higher temperatures (Scheraga *et al.*, 1962; Brandts, 1964; Mozhaev & Martinek, 1984). The stabilization caused by the hydrophobic effect will decrease at higher temperatures and may be non-existent at the temperatures used in heat coagulation studies (greater than 110°C). Therefore, it would be expected that casein micelles would dissociate extensively on heating. However, most of the data on size distribution of casein micelles, after heating and cooling, indicate that micelles do not lose their structural integrity on heating at temperatures greater than 100°C. Further investigation of the role of hydrophobic interactions in casein micelles heated at high temperatures appears to be warranted.

A number of changes that may affect the electrostatic and steric interactions involved in maintaining the integrity of casein micelles occur in milk on heating (Fox, 1982):

(i) decrease of pH,

(ii) deposition of calcium phosphate onto or into casein micelles,
(iii) dephosphorylation of casein,
(iv) dissociation and hydrolysis of κ-casein, and
(v) several protein polymerization reactions.

The decrease in pH during heating is likely to reduce electrostatic repulsions, thus gradually destabilizing the casein micelle system. If the pH of heated milk is readjusted periodically to its original value, coagulation is delayed more or less indefinitely, suggesting that acidity may be the single most important factor leading to coagulation (Fox, 1982). However, a heat-induced coagulum is not simply an acid coagulum, because:

(i) the coagulum does not dissolve in solvents such as 6 M urea, guanidine HCl, or at extremes of pH, suggesting the involvement of polymerization reactions in the coagulation process (Fox, 1982);
(ii) the apparent activation energies for acid production and heat coagulation are different; a Q_{10} of about 2 for the pH decrease compared with a Q_{10} of 3 for heat coagulation (van Boekel *et al.*, 1989*a*);
(iii) the pH at coagulation is not constant but depends on the initial pH (van Boekel *et al.*, 1989*a*).

The transfer of calcium phosphate from the serum to the casein micelles is expected to shield the negative charges on the micelles, reducing the zeta potential and thus causing a decrease in electrostatic repulsions. As a consequence, the casein micelle may be rendered unstable but there is no substantial evidence to support this hypothesis. The transfer of calcium phosphate from the soluble phase to the casein micelles occurs at an early stage of heat treatment and then remains more or less constant up to the coagulation point (Kudo, 1980*c*), suggesting that this is unlikely to be the sole causative factor in heat coagulation.

Dephosphorylation of casein influences the micelle structure because the phosphoserine residues are involved in binding CCP in the micellar structure. Loss of phosphate groups from the caseins reduces the binding of CCP to the caseins, resulting in the dissociation of caseins from the micelles (Dalgleish *et al.*, 1987*b*). Recently, Aoki *et al.* (1990) showed that the proportions of casein aggregates cross-linked by CCP did not change significantly after heating whey protein-free milk at 90°C for up to 10 min, although the CCP content increased. However, the rate of dephosphorylation of casein in milk does not appear to correlate with the rate of heat coagulation.

When casein micelles are heated in the absence of whey proteins, soluble casein containing about 40% κ-casein is formed (Aoki *et al.*, 1975) but, when milk is heated the heat-induced dissociaton of κ-casein-rich protein from the micelles is pH dependent; at pH values below 6·7, little casein dissociates but at higher pH dissociation increases with an increase in pH (Singh & Fox, 1985*b*). Dissociation of κ-casein-rich protein from the micelles would be expected to reduce the overall negative charge and decrease the steric effects, thus reducing the stability of the micelles. As the disssociated caseins contain relatively less phosphate, Dalgleish *et al.* (1987*b*) suggested that dephoshorylation of caseins during heating causes the dissociation of caseins from the micelles. However, the phosphoserine residue of κ-casein is located in the negatively charged region of the molecule and, unlike the situation with the other caseins, is probably not involved in binding to the CCP. Moreover, the rate of dissociation of κ-casein from the micelles is much more rapid than the rate of dephosphorylation.

Association of denatured whey proteins with casein micelles appears to be dependent on the pH at heating; at pH values below 6·7 whey proteins complexed with κ-casein are on the casein micelle surface, whereas at pH values above 6·8, they remain in the serum phase as complexes with κ-casein. It is not clear whether complexes of κ-casein and whey proteins are formed on the micelle surface and then dissociate into the serum phase or whether complexes are actually formed in the serum phase. The fate of these whey protein–κ-casein complexes during prolonged heating is not clear. There is some evidence that the complexes dissociated above pH 6·8 can reassociate with the κ-casein-depleted micelles after cooling the milk and readjusting the pH to less than 6·7 (Singh & Fox, 1986). Van Boekel *et al.* (1989*a*) reported that the reassociation can occur at high temperatures as the pH is decreased during heating as a result of heat-induced acidity. The sensitivity of casein micelles to calcium is altered by heat treatment (Singh & Fox, 1986). Calcium sensitivity of micelles isolated from milk heated at 90°C for 10 min at pH 6·6 (i.e. whey-protein-coated micelles) was essentially similar to that of unheated casein micelles, but casein micelles isolated from milk heated at pH 7·3 (κ-casein-depleted micelles) were more sensitive to calcium than unheated micelles (Singh & Fox, 1986). The increased sensitivity is presumably due to the loss of κ-casein, which exposes calcium-sensitive caseins.

Polymerization reactions involving covalent bond formation between protein chains appear to increase gradually with heat treatment but, because the extent of covalent bond formation at the point of coagulation is not constant (Dalgleish *et al.*, 1987*a*), it is unlikely that the covalent polymerization is directly correlated with the coagulation time. However, the nature of these polymerization reactions, e.g. formation of lysinoalanine, lanthionine and isopeptides, in relation to heat coagulation has not been explored as yet.

5.1. pH Dependence of the Heat Stability of Milk

Based on the above observations, the pH dependence of the heat stability of normal milk can be explained. The occurrence of a maximum in the heat stability–pH profile of milk systems at pH 6·7 is essentially caused by whey proteins, especially β-lactoglobulin, through the formation of a complex between β-lactoglobulin and κ-casein on the surface of the casein micelles; as a result, the surface properties of the micelles are altered considerably. The increased stability of the micelles in the region of the maximum can be attributed to the following factors:

(i) Complexation of whey proteins with κ-casein on the micellar surface may increase surface charge and hydration (Fox & Hearn, 1978*b*; Singh & Fox, 1985*b*). This view is supported by the following observations: (a) the addition of SDS, which adsorbs hydrophobically to caseins and increases their surface charge, increases the heat stability of casein micelles (Shalabi & Fox, 1982*b*); (b) positively charged derivatives of β-lactoglobulin, prepared by amidation or esterification of the carboxyl groups, reduce the heat stability of SPFCM dispersions in the pH range 6·4–7·4 (no maximum) whereas negatively charged derivatives (acetylated or succinylated) increase the heat stability in the region of the maximum (Singh & Fox, 1987*a*). Moreover, Schmidt and Poll (1986) reported that whey proteins increase the zeta potential of the micelles in the region of maximum heat stability.

(ii) The association of whey proteins with the casein micelles prevents the dissociation of micellar κ-casein during heating, thus retaining the protective effect of this protein (Singh & Fox, 1987*b*).

The actual mechanism of coagulation of casein micelles in the region of the maximum (whey-protein-coated micelles) is not fully understood. The size of the casein micelles appears to increase only slowly during heat treatment until the coagulation time, when a sudden, rapid increase occurs (Dalgleish *et al.*, 1987*a*). Mohammad and Fox (1987*b*) presented evidence that casein micelles in milk initially increase in size, but then decrease slightly before the final aggregation that leads to the onset of visible coagulation. More work is needed to determine how and why shifts in the size distribution of the casein micelles occur during heating at the pH of maximum stability.

Electron microscopic examination of milks coagulated in the region of maximum stability showed that the micelles were joined together through attached material and that the coagulum was relatively open (Mohammad & Fox, 1987*a*). The surfaces of the micelles appeared to be intact and the particles joining the micelles together appeared to be whey protein aggregates (Mohammad & Fox, 1987*a*). The coagulum formed in the region of the maximum is quite insoluble in 6 M urea, suggesting a high degree of covalent polymerization at the point of coagulation.

It can be proposed that with continued heating at the maximum, the protective effect of the whey protein–κ-casein complex gradually diminishes, probably as a result of increased acidity and modification by dephosphorylation and other reactions. For these reasons, the whey-protein-coated micelles can approach more closely or make more frequent contact; whey protein aggregates attached to the surface of the micelles can form covalent cross-links, thus promoting micellar aggregation. It is also possible that casein micelles disintegrate during extended heating and that some of the dissociated protein is involved in forming covalent cross-links, as suggested by Dalgleish *et al.* (1987*a*).

The occurrence of a minimum in the heat stability–pH profile of milk has received considerable research attention. Based on the early work, Rose (1963) postulated that the minimum in the heat stability–pH curve was caused by the failure of heat-denatured β-lactoglobulin to form a complex with κ-casein. More recent findings have shown that β-lactoglobulin and casein micelles interact via sulphydryl–disulphide interchange at any pH between 6·5 and 7·5 (Singh & Fox, 1987*a*). Also, milks that have been preheated above pH 6·9 show maxima–minima on subsequent reheating at 140°C after pH adjustment (Morrissey, 1969; Singh & Fox, 1985*b*). Since soluble calcium and soluble phosphate are also required for the occurrence of a minimum in the heat stability–pH curve, Morrissey (1969) suggested that the minimum is due to heat-induced precipitation of calcium phosphate on the casein micelles/β-lactoglobulin complex. More recently, Singh and Fox (1985*a*,*b*, 1986) proposed that dissociation of micellar κ-casein complexed with whey proteins at pH 6·9 alters the surface of the micelles; presumably, the micellar surface charge and hydration decrease due to loss of hydrophilic κ-casein, and the calcium sensitivity of the residual micelles increases due to loss of the calcium-insensitive fraction, κ-casein. Steric repulsion also decreases as a result of loss of the C-terminal region of casein. Thus, the minimum in the heat stability–pH curve is a result of coagulation of κ-casein micelles with modified surface properties caused by the depletion of the κ-casein. This is shown diagrammatically in Fig. 3.

Electron microscopic examination of casein micelles in milks heated in the region of the heat stability minimum showed that the micelle surfaces were not intact, that most of the micelles appeared to be dissociated or shattered, and that the coagulum consisted of large chains and clusters of fused micelles built into a close gel network (Mohammad & Fox, 1987*a*). It can be concluded that micelles in the region of minimum stability have less κ-casein and that their hydration, zeta potential and steric repulsions are sufficiently reduced to allow coagulation by calcium-mediated aggregation reactions or probably calcium phosphate precipitation (Singh & Fox, 1987*b*). As coagulation is essentially caused by calcium, it is relatively easy to redisperse the coagulum in dissociating buffers (e.g. sodium citrate, 6 M urea). Reducing the soluble calcium or phosphate levels in milk eliminates the minimum because there is not enough calcium to precipitate κ-casein-depleted micelles. Ionic species that increase the net negative charge on the micelle surface would shift the heat stability–pH curve to acidic pH values and the dissociation of micellar κ-casein would occur at lower pH values. Ionic species that decrease the net negative charge would have the opposite effect. The dissociation of micellar κ-casein appears to be reversible, at least to some extent (Singh & Fox, 1986). The reversibility may explain why milks preheated in the region of the minimum show maxima/minima on subsequent reheating at 140°C.

Covalent attachment of κ-casein to other caseins by cross-linking agents, such as formaldehyde, prevents the dissociation of micellar κ-casein above pH 6·9 and thus eliminates the minimum (Singh & Fox, 1985*a*). Reducing agents increase the dissociation of micellar κ-casein by reducing the disulphide-linked polymers of κ-casein (Singh & Fox, 1987*a*). Increased amounts of β-lactoglobulin enhance the dissociation of micellar κ-casein, probably by complexing with greater amounts of micellar κ-casein (Singh & Fox, 1987*a*), which may explain why the addition of increasing amount of β-lactoglobulin to milk accentuate the minimum (Rose, 1961*b*). Addition of κ-casein to milk before heating increases the serum κ-casein level, nullifies the effect of β-lactoglobulin on micellar κ-casein at pH 6·9 and eliminates the minimum, possibly by reacting with a proportion of the β-lactoglobulin before it can react with micellar κ-casein. The elimination of the maximum/minimum from the heat stability–pH profile of milk by very severe pre-heat

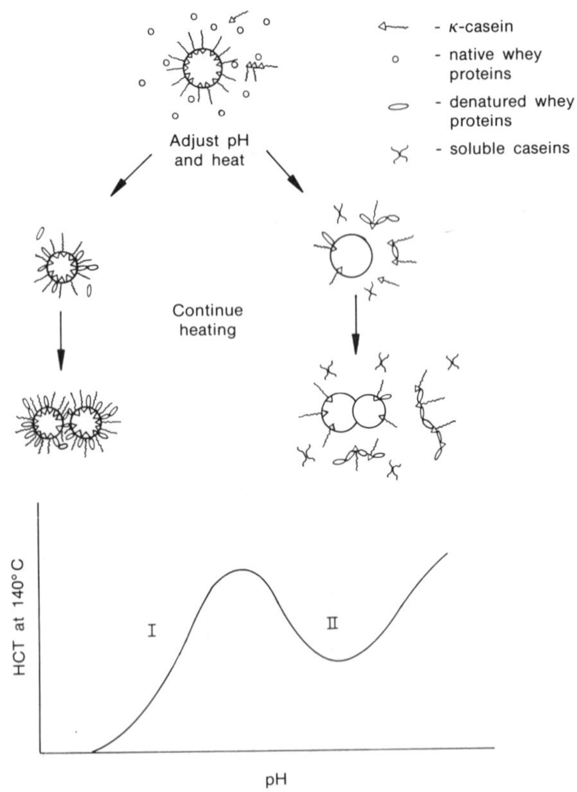

Fig. 3. Diagrammatic representation of the effect of heat on the proteins of normal skim milk at different pH values. (I) At lower pH values, the whey proteins denature and form a complex with κ-casein, which adheres to the micelle surface (whey protein coated micelles). Further heating causes lowering of the pH, dephosphorylation of caseins and other reactions. The coagulation is caused by the formation of covalent cross-links between the whey protein aggregates that adhere to the micelle surface. (II). At higher pH values, the whey protein–κ-casein complex is largely free in the serum. Note that the micelles have swollen a little and that soluble casein is present. The micelles have less κ-casein and are sensitive to calcium ion concentrations. Coagulation is caused by the association of κ-casein depleted micelles by calcium ions.

treatments (140°C for 10–60 min) at a pH near the maximum (Fox, 1981*a*) cannot be readily explained. Perhaps severe preheat treatment precipitates sufficient calcium phosphate to cause a significant reduction in the ionic calcium, or κ-casein becomes covalently cross-linked to the micelles so that it does not dissociate on subsequent heating.

Thus, two types of reaction that cause coagulation of casein micelles appear to occur in the region of the minimum:

(i) a sulphydryl–disulphide interchange reaction between whey protein and micellar κ-casein, followed by dissociation of these complexes from the micelles;

(ii) precipitation of κ-casein-depleted micelles by calcium ions.

The sequence of events involved in the pH dependence of the heat stability of normal milk is summarized in Fig. 3.

On heating milk, the formation of a complex between whey proteins and micellar κ-casein protects the casein micelles against coagulation at pH values less than 6·8 (region I, Fig. 3), essentially by preventing the dissociation of micellar κ-casein and by increasing the net negative charge (perhaps there is an increasd steric repulsion as well) on the micellar surface. Below pH 6·5, the protective effect of this complex is insufficient to overcome the effect of increasing heat-induced acidity and high calcium ion concentrations.

At pH values above 6·9 (region II, Fig. 3), stability decreases due to the dissociation of the κ-casein–whey protein complex, thus reducing the stabilizing effect (electrostatic and steric) of κ-casein sufficiently to induce coagulation in the presence of calcium. Presumably, calcium ion activity is still high enough in this region. At higher pH values (above 7·1), although dissociation of micellar κ-casein increases, stability increases, probably due to increased protein charge and decreased calcium ion activity. In this region, the dissociated protein material may reassociate with the casein micelles as the pH becomes lower, thus resulting in more stable casein micelles (van Boekel *et al.*, 1989*a*).

6. Coagulation of Concentrated Milk

In contrast to normal milk, the heat-induced changes in concentrated milk have not been studied extensively. As the coagulation times of concentrated milks are much shorter, the degree of dephosphorylation and the decrease in pH at the coagulation point are far smaller than in unconcentrated milk (Singh, H., unpublished), although the starting pH is much lower (~6·5 for a 20% TS concentrate). The coagulum formed in concentrated milk is relatively soluble in dissociating buffers, suggesting that little or no covalent polymerization occurs in the coagulation process. The size of the casein micelles appears to increase almost linearly with the severity of the heat treatment (Singh & Creamer, 1991*a*). Dissociation of κ-casein-rich protein from the micelles occurs on heating at lower pH values (6·6) compared with those in normal milk (pH above 6·8) and the extent of dissociation increases at higher pH (Singh & Creamer, 1991*a*) (Fig. 4). Approximately 50–60% of the micellar κ-casein appears to dissociate at the point of coagulation. It can be speculated that the dissociation of κ-casein from the micelles causes coagulation by altering the surface of the micelles, a situation somewhat comparable with that existing for the coagulation of normal milk within the region of minimum stability. There is some evidence that the coagulation process for concentrated milk resembles that of unconcentrated milk within the region of the minimum in some important aspects. Concentrated milks coagulate via a two-stage process and are not stabilized by urea (Muir & Sweetsur, 1978). Addition of β-lactoglobulin reduces the stability of concentrated milk throughout the pH range, whereas it reduces the stability of unconcentrated milk only in the region of the minimum. It

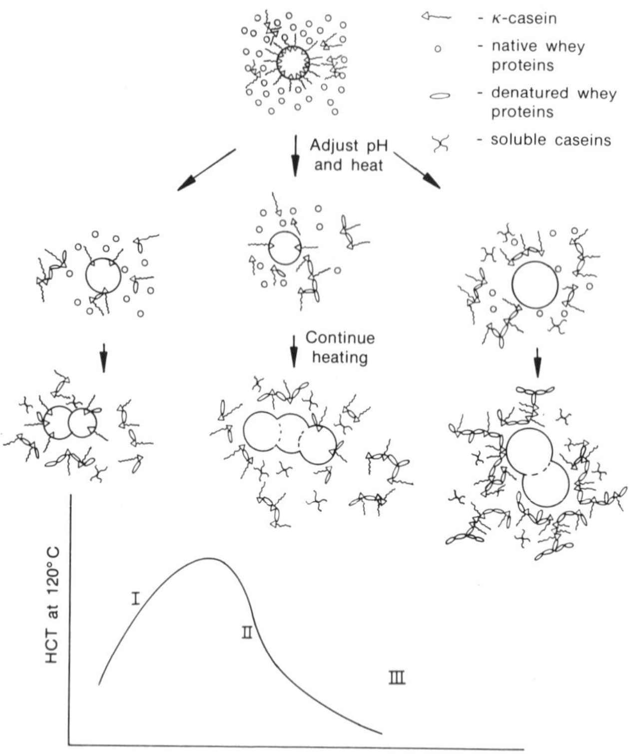

Fig. 4. Diagrammatic representation of the effect of heat on the milk proteins in concentrated skim milk at different pH values. (I) Dissociation of κ-casein from the micelles occurs at lower pH values; aggregation by association of κ-casein depleted micelles. (II) Aggregation similar to I with greater loss of κ-casein from the micelles. (III). Aggregation by gelation of material lost from the micelles. There is some micelle aggregation and much swelling of the micelles. In both II and III, there are few or no small casein micelles.

seems likely that the infuence of β-lactoglobulin on micellar κ-casein dissociation is extended to lower pH values in concentrated milk. The pH-dependent dissociation of κ-casein gradually moves to lower pH values with increasing concentration of milk solids (Singh & Creamer, 1991*b*).

The picture that emerges from the above considerations is that when concentrated milk is heated extensively at its normal pH (Fig. 4, region II), there is an initial association of whey proteins with the κ-casein in the serum and onto the micellar surface. This is followed by dissociation of κ-casein-rich protein from the micelles. Simultaneously, there is an increase in the size of the protein particles. This suggests that these phenomena may be related and that the dissociation of κ-casein-rich protein may create sites on the casein micelles favourable for micelle–micelle intereactions. Calcium phosphate or ionic calcium is likely to be involved in the enlargement of the micelles, attaching the κ-casein-depleted micelles to one another. The aggregation and dissociation reactions continue until at a critical level of κ-casein, visible coagulation occurs. Preheat treatment of milk prior to concentration does not greatly affect the dissociation of micellar κ-casein but reduces the soluble and ionic calcium concentrations and therefore reduces the rate of aggregation of κ-casein-depleted micelles (Singh & Creamer, 1991*c*).

At pH values above the normal (Fig. 4, region III), the casein micelles are highly dissociated and coagulation is more like gelation. Probably, whey proteins and dissociated protein material act as bridging agents. It has been shown that if the concentration of β-lactoglobulin is increased three-fold, casein micelle dispersions can form a gel after heating at 90°C for 10 min at pH 7·3 (Singh & Fox, 1987*b*).

7. Concluding Remarks

The heat stabilty of milk has been subjected to extensive investigations for over 60 years. Most of this research can be broadly separated into two categories:

(i) applied research concerning the factors (including processing parameters) that influence the heat stability of milk, and
(ii) basic research concerning the physico-chemical reactions that occur when milk is heated.

Practical knowledge of heat stability has expanded considerably and most of the problems related to products can now be solved by manipulation of processing and compositional variables or by using permitted additives. However, certain problems regarding the heat stability of concentrated milk, especially full-fat homogenized concentrated milk, remain unsolved. Factors that influence the heat stability of these systems need to be studied in greater depth.

With respect to heat-induced changes in milk, considerable progress has been made during the last 25 years. More important findings include association of whey proteins with the casein micelles, shifts in the calcium phosphate equilibrium, and aggregation and dissociation of casein micelles. The role of these changes in the heat coagulation process has not been clearly defined. The precise nature of heat-precipitated calcium phosphate and dissociated caseins has not been defined with certainty. Assessment of the actual state of proteins and salts in milk at high temperatures has been impossible, mainly because of the experimental problems. In order to study these changes more effectively, further research should involve development of new experimental techniques. A study of the kinetics of heat-induced changes should be particularly useful in understanding the mechanisms of heat coagulation.

The mechanisms of heat coagulation of full-cream homogenized milks are virtually unknown. The structural changes in milk brought about by homogenization and subsequent effect of heat treatments on homogenized milks need further work.

References

Aoki, T., Suzuki, H. & Imamura, T. (1974). Formation of soluble casein in whey protein-free milk heated at high temperature. *Milchwissenchaft*, **29**, 589–94.

Aoki, T., Suzuki, H. & Imamura, T. (1975). Some properties of soluble casein in heated concentrated whey protein-free milk. *Milchwissenchaft*, **30**, 30–5.

Aoki, T., Umeda, T. & Kako, Y. (1990). Cleavage of the linkage between colloidal calcium phosphate and casein on heating milk at high temperature. *J. Dairy Res.*, **57**, 349–54.

Baer, A., Oroz, M. & Blanc, B. (1976). Serological studies on heat-induced interactions of α-lactalbumin and milk proteins. *J. Dairy Res.*, **43**, 419–32.

Ball, C.O. (1955). Studies on high temperature sterilization of evaporated milk with emphasis on stability aspects. *Food Res.*, **20**, 351–70.

Belec, J. & Jenness, R. (1962*a*). Dephosphorization of casein by heat treatment. I. In caseinate solutions. *J. Dairy Sci.*, **45**, 12–19.

Belec, J. & Jenness, R. (1962*b*). Dephosphorization of casein by heat treatment. II. In skimmilks. *J. Dairy Sci.*, **45**, 20–6.

Boulet, M. & Marier, J.R. (1960). Solubility of tricalcium citrate in solutions of variable ionic strength and in milk ultrafiltrates. *J. Dairy Sci.*, **43**, 155–64.

Brandts, J.F. (1964). The thermodynamics of protein denaturation. II. A model of reversible denaturation and interpretations regarding the stability of chymotrypsinogen. *J. Amer. Chem. Soc.*, **86**, 4302–13.

Burton, H. (1956). A new approach to the problem of heat stability in evaporated milk. *Proc. 14th Intern. Dairy Congr. (Rome)*, **1**(2), 342–9.

Carroll, R.J., Thompson, M.P. & Melnychyn, P. (1971). Gelation of concentrated skim milk: electron microscopic study. *J. Dairy Sci.*, **54**, 1245–52.

Cole, W.C. & Tarassuk, N.P. (1946). Heat coagulation of milk. *J. Dairy Sci.*, **29**, 421–9.

Creamer, L.K. (1984). Fractionation of casein micelles and whey protein aggregates on Sephacryl S-1000. *J. Chromat.*, **291**, 460–3.

Creamer, L.K. (1991). Aspects of casein micelle structure. In *Macromolecular Interactions and Food Colloid Stability*, ed. N. Parris & R.A. Bardford. American Chemical Society, Washington, DC, pp. 148–63.

Creamer, L.K. & Matheson, A.R. (1980). Effect of heat treatment on the proteins of pasteurized skim milk. *N.Z. J. Dairy Sci. Technol.*, **15**, 37–49.

Creamer, L.K., Berry, G.P. & Matheson, A.R. (1978). The effect of pH on protein aggregation in heated skim milk. *N.Z. J. Dairy Sci. Technol.*, **13**, 9–15.

Dalgleish, D.G. (1989). The behaviour of minerals in heated milks. *Int. Dairy Fed. Bulletin*, **238**, 31–4.

Dalgleish, D.G., Pouliot, Y. & Paquin, P. (1987*a*). Studies on the heat stability of milk. II. Association and dissociation of particles and the effect of added urea. *J. Dairy Res.*, **54**, 39–49.

Dalgleish, D.G., Pouliot, Y. & Paquin, P. (1987*b*). Studies on the heat stability of milk. I. Behaviour of divalent cations and phosphate in milks heated in a stainless steel system. *J. Dairy Res.*, **54**, 29–37.

Darling, D.F. (1980). Heat stability of milk. *J. Dairy Res.*, **47**, 199–210.

Darling, D.F. & Dickson, J. (1979). The determination of the zeta potential of casein micelles. *J. Dairy Res.*, **46**, 329–32.

Davies, D.T. & White, J.C.D. (1959). Determination of heat-induced changes in the protein stability and chemical composition of milk. *Proc. 15th Intern. Dairy Cong. (London)*, **3**, 1677–85.

Davies, D.T. & White, J.C.D. (1966). The stability of milk protein to heat. I. Subjective measurement of heat stability of milk. *J. Dairy Res.*, **33**, 67–81.

Davies, J.M. (1959). Decrease in the heat stability of milk on forewarming. *Proc. 15th Intern. Dairy Congr. (London)*, **3**, 1859–64.

de Koning, P.J., Koops, J. & van Rooijen, P.J. (1974). Some features of the heat stability of concentrated milk. III. Seasonal effects on the amounts of casein, individual whey proteins and NPN and their relation to variations in heat stability. *Neth. Milk Dairy J.*, **28**, 186–202.

de Wit, J.N., Klarenbeek, G. & de Graaf, C. (1986). Klaro-graph: a new approach for the measurement of the viscosity, density and heat stability in milk and milk concentrates at temperatures up to 140°C. *Voedingsmiddelentechnologie*, **6**(3), 25–7.

Elfagm, A.A. & Wheelock, J.V. (1977). Effect of heat on α-lactalbumin and β-lactoglobulin in bovine milk. *J. Dairy Res.*, **44**, 367–71.

Evenhuis, N. & de Vries, Th.R. (1956). The condition of calcium phosphate in milk. III. *Neth. Milk Dairy J.*, **10**, 101–13.

Feagan, J.T., Bailey, L.F., Hehir, A.F., McLean, D.M. & Ellis, N.J.S. (1972). Coagulation of milk proteins. I. Effect of genetic variants of milk proteins on rennet coagulation and heat stability of normal milk. *Aust. J. Dairy Technol.*, **27**, 129–34.

Foissy, H. & Kneifel, W. (1984). An automatic method for measuring the heat coagulation time of milk powder solutions. *J. Dairy Res.*, **51**, 325–9.

Fox, K.K., Harper, M.K., Holsinger, V.H. & Pallansch, M.J. (1967). Effect of high-heat treatment on stability of calcium caseinate aggregates in milk. *J. Dairy Sci.*, **50**, 443–50.

Fox, P.F. (1981*a*). Heat-induced changes in milk preceding coagulation. *J. Dairy Sci.*, **64**, 2127–37.

Fox, P.F. (1981*b*). Heat stability of milk: significance of heat-induced acid formation in coagulation. *Irish J. Food Sci. Technol.*, **5**, 1–11.

Fox, P.F. (1982). Heat-induced coagulation of milk. In *Developments in Dairy Chemistry —I. Proteins*, ed. P.F. Fox. Applied Science Publishers, London, pp. 189–228.

Fox, P.F. & Hearn, C.M. (1978*a*). Heat stability of milk: influence of dilution and dialysis against water. *J. Dairy Res.*, **45**, 149–57.

Fox, P.F. & Hearn, C.M. (1978*b*). Heat stability of milk: influence of denaturable proteins and detergents on pH sensitivity. *J. Dairy Res.*, **45**, 159–72.

Fox, P.F. & Hearn, C.M. (1978*c*). Heat stability of milk: influence of κ-casein hydrolysis. *J. Dairy Res.*, **45**, 173–81.

Fox, P.F. & Hoynes, M.C.T. (1975). Heat stability of milk: influence of colloidal calcium phosphate and β-lactoglobulin. *J. Dairy Res.*, **42**, 427–35.

Fox, P.F. & Morrissey, P.A. (1977). Reviews of the progress of dairy science: the heat stability of milk. *J. Dairy Res.*, **44**, 627–46.

Fox, P.F. & Nash, B.M. (1979). Physico-chemical characteristics of casein micelles in dilute aqueous media. *J. Dairy Res.*, **46**, 357–63.

Fox, P.F., Nash, B.M., Horan, T.J., O'Brien, J. & Morrissey, P.A. (1980). Effect of selected amides on heat-induced changes in milk. *J. Dairy Res.*, **47**, 211–19.

Frazier, W.C. (1925). The influence of some bacterial enzymes on the heat coagulation of milk. *J. Dairy Sci.*, **8**, 370–89.

Gould, I.A. (1945). Lactic acid in dairy products. III. The effect of heat on total acid and lactic acid production and on lactose destruction. *J. Dairy Sci.*, **28**, 367–77.

Hardy, E.E., Muir, D.D., Sweetsur, A.W.M. & West, I.G. (1984). Changes of calcium phosphate partition and heat stability during manufacture of sterilized concentrated milk. *J. Dairy Sci.*, **67**, 1666–73.

Hardy, E.E., Sweetsur, A.W.M., West, I.G. & Muir, D.D. (1985). Heat stability of concentrated milk: enhancement of initial heat stability by incorporation of food grade lecithin. *J. Food Technol.*, **20**, 97–105.

Henry, K.M., Houston, J., Kon, S.K. & Thompson, S.Y. (1944). The effects of commercial processing and of storage on some nutritive properties of milk: comparison of full-cream sweetened condensed milk and of evaporated milk with the original raw milk. *J. Dairy Res.*, **13**, 329–39.

Holm, G.E., Deysher, E.F. & Evans, F.R. (1923). The relationships of concentration and time to the temperature of coagulation of evaporated skim and whole milk. *J. Dairy Sci.*, **6**, 556–8.

Holm, G.E., Webb, B.H. & Deysher, E.F. (1932). The heat coagulation of milk. 1. Variations in the compositions, heat stability, and other tests of milks from four cows during the course of a lactation period. *J. Dairy Sci.*, **15**, 331–43.

Holt, C., Muir, D.D. & Sweetsur, A.W.M. (1978*a*). Seasonal changes in the heat stability of milk from creamery silos in South-West Scotland. *J. Dairy Res.*, **45**, 183–90.

Holt, C., Muir, D.D. & Sweetsur, A.W.M. (1978*b*). The heat stability of milk and concentrated milk containing added aldehydes and sugars. *J. Dairy Res.*, **45**, 47–52.

Hostettler, H., Imhof, K. & Stein, J. (1965). Studies on the effect of heat-treatment and lyophilisation on the state of distribution and physiological properties of milk proteins with special consideration of heat-treatment conditions applied in uperization. I. Effect on the state of distribution of protein. *Milchwissenchaft*, **20**, 189–98.

Howat, G.R. & Wright, N.C. (1934). The heat coagulation of caseinogen I. The role of phosphorus cleavage. *Biochem. J.*, **28**, 1336–45.

Josephson, R.V., Thomas, E.L., Morr, C.V. & Coulter, S.T. (1967). Relation of heat-induced changes in protein-salt consitituents to astringency in milk systems. *J. Dairy Sci.*, **50**, 1376–83.

Kannan, A. & Jenness, R. (1961). Relation of milk serum proteins and milk salts to the effects of heat treatment on rennet clotting. *J. Dairy Sci.*, **44**, 808–22.

Kelly, P.M. (1982). The effect of preheat temperature and urea addition on the seasonal variation in the heat stability of skim milk powder. *J. Dairy Res.*, **49**, 187–96.

Kelly, P.M., O'Keeffe, A.M., Keogh, M.K. & Phelan, J.A. (1982). Studies of milk composition and its relationship to some processing criteria. III: Seasonal variation in heat stability of milk. *Irish J. Food Sci. Technol.*, **6**, 29–38.

Kieseker, F.G. (1977). The manufacture of heat stable powders. In *Proc. Jubilee Conf. Dairy Sci. Technol.*, New Zealand Dairy Research Institute, Palmerston North, pp. 85–8.

Koops, J. & Westerbeek, D. (1970). Some features of the heat stability of concentrated milk. II. Effect of hydrogen peroxide. *Neth. Milk Dairy J.*, **24**, 52–60.

Kudo, S. (1980*a*). The influence of α_{s2}-casein on the heat stability of artificial milks. *N.Z. J. Dairy Sci. Technol.*, **15**, 245–54.

Kudo, S. (1980*b*). Influence of lactose and urea on the heat stability of artificial milk systems. *N.Z. J. Dairy Sci. Technol.*, **15**, 197–200.

Kudo, S. (1980*c*). The heat stability of milk: formation of soluble proteins and protein-depleted micelles at elevated temperatures. *N.Z. J. Dairy Sci. Technol.*, **15**, 255–63.

Leighton, A. & Deysher, E.F. (1923). Factors influencing the heat coagulation of milk and the thickening of condensed milk. *Proc. World's Dairy Congr.*, **2**, 1276–84.

Leighton, A. & Mudge, C.S. (1923). On the endothermic reaction which accompanies the appearance of a visible clot in milk coagulated by heat: a contribution to the theory of heat coagulation of milk. *J. Biol. Chem.*, **56**, 53–75.

Lorient, D. (1979). Covalent bonds formed in proteins during milk sterilization: studies on caseins and casein peptides. *J. Dairy Res.*, **46**, 393–6.

Lyster, R.L.J. (1965). The composition of milk deposits in an ultra-high-temperature plant. *J. Dairy Res.*, **32**, 203–8.

Manson, W. & Carolan, T. (1980). Formation of lysinoalanine from individual bovine caseins. *J. Dairy Res.*, **47**, 193–8.

McLean, D.M., Graham, E.R.B., Ponzoni, R.W. & McKenzie, H.A. (1987). Effects of milk protein genetic variants and composition on heat stability of milk. *J. Dairy Res.*, **54**, 219–35.

Miller, P.G. & Sommer, H.H. (1940). The coagulation temperature of milk as affected by pH, salts, evaporation and previous heat treatment. *J. Dairy Sci.*, **23**, 405–21.

Mohammad, K.S. & Fox, P.F. (1987*a*). Heat-induced microstructural changes in casein micelles before and after heat coagulation. *N.Z. J. Dairy Sci. Technol.*, **22**, 191–203.

Mohammad, K.S. & Fox, P.F. (1987*b*). Heat-induced association–dissociation of casein micelles preceding coagulation. *J. Dairy Res.*, **54**, 377–87.
Morr, C.V. (1969). Protein aggregation in conventional and ultra high-temperature heated skim milk. *J. Dairy Sci.*, **52**, 1174–80.
Morr, C.V. (1973). Milk ultracentrifugal opalescent layer. 2. Physico-chemical properties. *J. Dairy Sci.*, **56**, 1258–66.
Morrissey, P.A. (1969). The heat stability of milk as affected by variations in pH and milk salts. *J. Dairy Res.*, **36**, 343–51.
Morrissey, P.A. & O'Mahony, F. (1976). Heat stability of forewarmed milks: influence of κ-casein, serum proteins and divalent cations. *J. Dairy Res.*, **43**, 267–74.
Morrissey, P.A., Murphy, M.F., Hearn, C.M. & Fox, P.F. (1981). Composition and stability of mid-lactation milks. *Irish J. Food Sci. Technol.*, **5**, 117–27.
Mozhaev, V.V. & Martinek, K. (1984). Review: structure-stability relationships in proteins: new approaches to stabilising enzymes. *Enz. Microb. Technol.*, **6**, 50–9.
Muir, D.D. & Sweetsur, A.W.M. (1976). The influence of naturally occurring levels of urea on the heat stability of bulk milk. *J. Dairy Res.*, **43**, 495–9.
Muir, D.D. & Sweetsur, A.W.M. (1977). Effect of urea on the heat coagulation of the caseinate complex in skim milk. *J. Dairy Res.*, **44**, 249–57.
Muir, D.D. & Sweetsur, A.W.M. (1978). The effect of concentration on the heat stability of skim milk. *J. Dairy Res.*, **45**, 37–45.
Muir, D.D., Abbot, J. & Sweetsur, A.W.M. (1978). Changes in the heat stability of milk protein during the manufacture of dried skim milk. *J. Food Technol.*, **13**, 45–53.
Muir, D.D., Sweetsur, A.W.M. & Holt, C. (1979). The synergic effect of urea and aldehydes on the heat stability of concentrated skim milk. *J. Dairy Res.*, **46**, 381–4.
Nelson, V. (1954). Effects of formaldehyde and copper salts on the heat stability of evaporated milk. *J. Dairy Sci.*, **37**, 825–9.
Newstead, D.F. (1977). Effect of protein and salt concentration on the heat stability of evaporated milk. *N.Z. J. Dairy Sci. Technol.*, **12**, 171–5.
Newstead, D.F., Sanderson, W.B. & Woodhams, D.J. (1974). Effects of composition on the heat stability of forewarmed concentrated skim milk. *Proc. 19th Int. Dairy Congr.*, New Delhi, Vol. 1E, 182.
Newstead, D.F., Sanderson, W.B. & Baucke, A.G. (1975). The effect of heat treatment and pH on the heat stability of recombined evaporated milk. *N.Z. J. Dairy Sci. Technol.*, **10**, 113–18.
Newstead, D.F., Sanderson, W.B. & Conaghan, E.F. (1977). Effects of whey protein concentrations and heat treatment on the heat stability of concentrated and unconcentrated milk. *N.Z. J. Dairy Sci. Technol.*, **12**, 29–36.
Newstead, D.F., Conaghan, E.F. & Baldwin, A.J. (1979). Studies on the induction of heat stability in evaporated milk by preheating: effects of milk concentration, homogenization and whey proteins. *J. Dairy Res.*, **46**, 387–91.
Nieuwenhuijse, J.A., Timmermans, W. & Walstra, P. (1988). Calcium and phosphate partitions during the manufacture of sterilized concentrated milk and their relations to the heat stability. *Neth. Milk Dairy J.*, **42**, 387–421.
Noh, B., Richardson, T. & Creamer, L.K. (1989*a*). Radio labelling study of the heat-induced interactions between α-lactalbumin, β-lactoglobulin and κ-casein in milk and in buffer solutions. *J. Food Sci.*, **54**, 889–93.
Noh, B., Creamer, L.K. & Richardson, T. (1989*b*). Thermally induced complex formation in an artificial milk system. *Agric. Food Chem.*, **37**, 1395–400.
Pearce, R.J. (1979). Heat stability in concentrated and non-concentrated milks—the effect of urea and β-lactoglobulin levels and the influence of preheating. *J. Dairy Res.*, **46**, 385–6.
Pouliot, Y., Boulet, M. & Paquin, P. (1989). Observations on the heat-induced salt balance changes in milk. I. Effect of heating time between 4 and 90°C. *J. Dairy Res.*, **56**, 185–92.
Pyne, G.T. (1958). The heat coagulation of milk. II. Variations in sensitivity of casein to calcium ions. *J. Dairy Res.*, **25**, 467–74.

Pyne, G.T. (1959). Calcium-ion concentration in relation to the heat coagulation of milk. *Proc. 15th Intern. Dairy Congr.*, London, Vol. 3, 1673–8.
Pyne, G.T. & McGann, T.C.A. (1960). The colloidal phosphate of milk. II. Influence of citrate. *J. Dairy Res.*, **27**, 9–17.
Pyne, G.T. & McHenry, K.A. (1955). The heat coagulation of milk. *J. Dairy Res.*, **22**, 60–8.
Robertson, N.H. & Dixon, A. (1969). The nitrogen fractions and heat stability of bovine milk. *Agroanimalia*, **1**, 141–4.
Rogers, L.A., Deysher, E.F. & Evans, F.R. (1921). The relation of acidity to the coagulation temperature of evaporated milk. *J. Dairy Sci.*, **4**, 294–309.
Rose, D. (1961*a*). Variations in the heat stability and composition of milk from individual cows during lactation. *J. Dairy Sci.*, **44**, 430–41.
Rose, D. (1961*b*). Factors affecting the pH-sensitivity of the heat stability of milk from individual cows. *J. Dairy Sci.*, **44**, 1405–13.
Rose, D. (1961*c*). β-Lactoglobulin and pH-sensitivity of the heat stability of evaporated milk. *J. Dairy Sci.*, **44**, 1763.
Rose, D. (1962). Factors affecting the heat stability of milk. *J. Dairy Sci.*, **45**, 1305–11.
Rose, D. (1963). Heat stability of bovine milk: a review. *Dairy Sci. Abstr.*, **25**, 45–52.
Rose, D. & Tessier, H. (1959). Composition of ultrafiltrates from milk heated at 80 to 230°F in relation to heat stability. *J. Dairy Sci.*, **42**, 969–80.
Ruegg, M., Blanc, B. & Luscher, M. (1979). Hydration of casein micelles: kinetics of water sorption of micellar casein isolated from fresh and heat-treated milk. *J. Dairy Res.*, **46**, 325–8.
Sawyer, W.H., Coulter, S.T. & Jenness, R. (1963). Role of sulfydryl groups in the interaction of κ-casein and β-lactoglobulin. *J. Dairy Sci.*, **46**, 564–5.
Scheraga, H.A., Nemethy, G. & Steinberg, I.Z. (1962). The contribution of hydrophobic bonds to the thermal stability of protein conformations. *J. Biol. Chem.*, **237**, 2506–8.
Schmidt, D.G. (1979). Properties of artificial casein micelles. *J. Dairy Res.*, **46**, 351–5.
Schmidt, D.G. & Koops, J. (1965). Some features of the heat stability of concentrated milk. I. Effect of genetic variations of κ-casein. *Neth. Milk Dairy J.*, **19**, 63–8.
Schmidt, D.G. & Koops, J. (1977). Properties of artificial casein micelles. 2. Stability towards ethanol, dialysis, pressure and heat in relation to casein composition. *Neth. Milk Dairy J.*, **31**, 342–51.
Schmidt, D.G. & Poll, J.K. (1986). Electrokinetic measurements on unheated and heated casein micelle systems. *Neth. Milk Dairy J.*, **40**, 269–80.
Shalabi, S.I. & Fox, P.F. (1982*a*). Heat stability of milk: synergic action of urea and carbonyl compounds. *J. Dairy Res.*, **49**, 197–207.
Shalabi, S.I. & Fox, P.F. (1982*b*). Heat stability of milk: influence of cationic detergents on pH sensitivity. *J. Dairy Res.*, **49**, 597–605.
Shalabi, S.I. & Fox, P.F. (1982*c*). Heat stability of milk: influence of modification of lysine and arginine on the heat stability-pH profile. *J. Dairy Res.*, **49**, 607–17.
Shalabi, S.I. & Fox, P.F. (1982*d*). Effect of diacetyl on the heat stability of concentrated milks. *J. Food Technol.*, **17**, 753–60.
Shalabi, S.I. & Fox, P.F. (1982*e*). Effect of κ-carrageenan on the heat stability and rennet coagulation of milk. *Irish J. Food Sci. Technol.*, **6**, 183–8.
Singh, H. (1985). Studies on the heat-induced coagulation of milk. PhD Thesis, National University of Ireland, Dublin, Republic of Ireland.
Singh, H. (1988). Effects of high temperatures on casein micelles. *N.Z. J. Dairy Sci. Technol.*, **23**, 257–73.
Singh, H. & Creamer, L.K. (1991*a*). Denaturation, aggregation and heat stability of milk protein during the manufacture of skim milk powder. *J. Dairy Res.*, **58**, 269–83.
Singh, H. & Creamer, L.K. (1991*b*). Influence of concentration of milk solids on the dissociation of micellar κ-casein on heating reconstituted milk at 120°C. *J. Dairy Res.*, **58**, 99–105.

Singh, H. & Creamer, L.K. (1991*c*). Aggregation and dissociation of milk protein complexes in heated reconstituted concentrated skim milks. *J. Food Sci.*, **56**, 238–46.
Singh, H. & Creamer, L.K. (1991*d*). Changes in size and composition of protein aggregates on heating reconstituted concentrated skim milk at 120°C. *J. Food Sci.*, **56**, 671–7.
Singh, H. & Fox, P.F. (1985*a*). Heat stability of milk: the mechanism of stabilization by formaldehyde. *J. Dairy Res.*, **52**, 65–76.
Singh, H. & Fox, P.F. (1985*b*). Heat stability of milk: pH-dependent dissociation of micellar κ-casein on heating milk at ultra high temperatures. *J. Dairy Res.*, **52**, 529–38.
Singh, H. & Fox, P.F. (1986). Heat stability of milk: further studies on the pH-dependent dissociation of micellar κ-casein. *J. Dairy Res.*, **53**, 237–48.
Singh, H. & Fox, P.F. (1987*a*). Heat stability of milk: role of β-lactoglobulin in the pH-dependent dissociation of micellar κ-casein. *J. Dairy Res.*, **54**, 509–21.
Singh, H. & Fox, P.F. (1987*b*). Heat stability of milk: influence of colloidal and soluble salts and protein modification on the pH-dependent dissociation of micellar κ-casein. *J. Dairy Res.*, **54**, 523–34.
Singh, H. & Fox, P.F. (1989). Heat-induced changes in casein. *Int. Dairy Fed. Bulletin*, **238**, 24–30.
Singh, H. & Tokley, R.P. (1990). Effects of preheat treatments and buttermilk addition on the seasonal variations in the heat stability of recombined evaporated milk and reconstituted concentrated milk. *Aust. J. Dairy Technol.*, **45**, 10–16.
Singh, H., Creamer, L.K. & Newstead, D.F. (1989). Effects of heat on the proteins of concentrated milk systems. *Int. Dairy Fed. Bulletin*, **238**, 94–104.
Singh, H., Sharma, R. & Tokley, R.P. (1992). Influence of incorporation of soya lecithin into skim milk powder on the heat stability of recombined evaporated milk. *Aust. J. Dairy Technol.*, **47**, 33–37.
Smits, P. & van Brouwershaven, J.H. (1980). Heat-induced association of β-lactoglobulin and casein micelles. *J. Dairy Res.*, **47**, 313–25.
Snoeren, T.H.M. & van der Spek, C.A. (1977). The isolation of a heat-induced complex from UHTST milk. *Neth. Milk Dairy J.*, **31**, 352–5.
Snoeren, T.H.M. & van Riel, J.A.M. (1979). The properties of α_{s2}-group casein. *Zuivelziht*, **71**, 766–8.
Snoeren, T.H.M., Koops, J. & Westerbeek, D. (1978). Some features of the heat stability of (concentrated) milk. IV. The effect of α_{s2}-caseins on the heat stability. *Neth. Milk Dairy J.*, **32**, 255–7.
Sommer, H.H. & Binney, T.H. (1923). A study of the factors that influence the coagulation of milk in the alcohol test. *J. Dairy Sci.*, **6**, 176–97.
Sommer, H.H. & Hart, E.B. (1919). The heat coagulation of milk. *J. Biol. Chem.*, **40**, 137–51.
Sommer, H.H. & Hart, E.B. (1922). The heat coagulation of milk. *J. Dairy Sci.*, **5**, 525–43.
Sweetsur, A.W.M. & Muir, D.D. (1980). The use of permitted additives and heat treatment to optimize the heat-stability of skim milk and concentrated skim milk. *J. Soc. Dairy Technol.*, **33**, 101–5.
Sweetsur, A.W.M. & Muir, D.D. (1981). Role of cyanate ions in the urea-induced stabilization of the caseinate complex in skim milk. *J. Dairy Res.*, **48**, 163–6.
Sweetsur, A.W.M. & Muir, D.D. (1982). Natural variation in heat stability of concentrated milk before and after homogenization. *J. Soc. Dairy Technol.*, **35**, 120–6.
Sweetsur, A.W.M. & Muir, D.D. (1983*a*). Effect of homogenization on the heat stability of milk. *J. Dairy Res.*, **50**, 291–300.
Sweetsur, A.W.M. & Muir, D.D. (1983*b*). Influence of sulphydryl group interactions on the heat stability of homogenized concentrated milk. *J. Dairy Res.*, **50**, 301–8.
Sweetsur, A.W.M. & White, J.C.D. (1974). Studies on the heat stability of milk protein. I. Interconversion of type A and type B milk heat stability curves. *J. Dairy Res.*, **41**, 349–58.
Sweetsur, A.W.M. & White, J.C.D. (1975). Studies on the heat stability of milk protein. III. Effect of heat-induced acidity in milk. *J. Dairy Res.*, **42**, 73–88.

Tessier, H. & Rose, D. (1964). Influence of κ-casein and β-lactoglobulin on the heat stability of skim milk. *J. Dairy Sci.*, **47**, 1047–51.
Thompson, M.P., Boswell, R.T., Martin, V., Jenness, R. & Kiddy, C.A. (1969). Casein-pellet-solvation and heat stability of individual cow's milk. *J. Dairy Sci.*, **52**, 796–8.
van Boekel, M.A.J.S., Nieuwenhuijse, J.A. & Walstra, P. (1989*a*). The heat coagulation of milk. 1. Mechanisms. *Neth. Milk Dairy J.*, **43**, 97–127.
van Boekel, M.A.J.S., Nieuwenhuijse, J.A. & Walstra, P. (1989*b*). The heat coagulation of milk. 2. Kinetic studies. *Neth. Milk Dairy J.*, **43**, 129–46.
Visser, J., Minihan, A., Smits, P. Tjan, S.B. & Heertje, I. (1986). Effects of pH and temperature on the milk salt system. *Neth. Milk Dairy J.*, **40**, 351–68.
Webb, B.H. (1928). Heat coagulation of evaporated milk as affected by mixing different grades of raw milk. *J. Dairy Sci.*, **11**, 471–8.
Webb, B.H. & Bell, R.W. (1942). The effect of high-temperature short-time forewarming of milk upon the heat stability of its evaporated product. *J. Dairy Sci.*, **25**, 301–11.
Webb, B.H. & Holm, G.E. (1932). The heat coagulation of milk. II. The influence of various added salts upon the heat stability of milks of different concentrations. *J. Dairy Sci.*, **15**, 345–66.
White, J.C.D. & Davies, D.T. (1958). The relation between the chemical composition of milk and the stability of the caseinate complex. IV. Coagulation by heat. *J. Dairy Res.*, **25**, 281–96.
White, J.C.D. & Davies, D.T. (1966). The stability of milk protein to heat. III. Objective measurement of heat stability of milk. *J. Dairy Res.*, **33**, 93–102.
Whitney, R.McL., Paulson, K. & Murphy, G.K. (1952). A test of the heat stability of milk proteins for use in research. *J. Dairy Sci.*, **35**, 937–48.
Wilson, H.K. & Herreid, E.O. (1961). Photomicrographs of milk protein particles. *J. Dairy Sci.*, **44**, 552–3.

16

ETHANOL STABILITY

D.S. HORNE

The Hannah Research Institute, Ayr KA6 5HL, UK

1. INTRODUCTION

The start of the 1980s saw an upsurge of interest in the alcohol stability of milk with newcomers to the field of dairy research taking advantage of modern physical techniques to probe the influence of ethanol on casein micelle interactions. Research is like a roller-coaster though, and that initial momentum could not be maintained. As the interest presently wanes, it seems an excellent time to review the progress made and, perhaps through greater awareness of the benefits, to stimulate new interest in ethanol stability studies.

The fruits of the current crop can be seen in three main areas. First, there is now a deeper and probably almost complete understanding of the alcohol stability test as applied to milk. Secondly, there came an almost immediate practical application of this new-found knowledge in extending the shelf-life of the decade's novel dairy product, the cream liqueur, converting what was virtually a 'cocktail' drink for immediate consumption into a commercially viable 'long-life' product. The third area saw the effects of ethanol on the casein micelle being probed by a wide panoply of modern physical instrumentation, testing the newly promulgated theories of steric stabilization incorporated in the 'hairy micelle' approach of Holt (1975) and Walstra (1979).

The earlier parts of the recent studies on the alcohol stability test have been reviewed by Horne (1985), while the developments in the area of cream liqueurs were covered by Banks and Muir (1988). The role of ethanol in the context of the 'hairy micelle' has been briefly mentioned in the wider field of micelle stability recently reviewed by Walstra (1990). The discussion that follows will consider some of these areas in detail before drawing them together in a critique of the current models of casein micelle structure.

2. ALCOHOL STABILITY TEST

2.1. Historical Introduction

The earliest accounts of the alcohol stability test for milk appeared in the scientific literature almost a century ago. Particularly in central Europe, the test was often enshrined in city statutes or public health regulations and provided an objective assessment of the keeping quality of milk for retail sale. A review of this early work and the value of the test in this context is contained in a 1915 United States Department of Agriculture Bulletin (Ayers & Johnson, 1915).

Most people employed the simple unsophisticated single alcohol test. If a precipitate formed when an equal volume of alcohol solution, usually 70%, v/v, was added to a milk sample, the batch of milk was rejected. As late as the 1930s, such alcohol tests were accepted as indicators of whether the milk was going sour, or was mixed with colostrum or was contaminated with milk from a cow suffering from mastitis (Padmos, 1930).

In the early part of this century, the growing milk processing industry, particularly in the USA, was also looking towards such assays as the alcohol test or titratable acidity as indicators of the suitability of milk for the production of sterile concentrates. Initial results were favourable, leading Dahlberg and Garner (1921) to conclude that 'the alcohol test shows good possibilities as a practical and reliable test for determining the quality of milk for condenseries' making evaporated milk. This expectation was not fulfilled, however, and within 10 years Ramsdell *et al.* (1931) were able to state that no available test, including the alcohol test, was 'sufficiently definite for advantageous use in grading milk for sterilization purposes'. Many years later this result was substantiated by White and Davies (1956) who found no correlation between the susceptibility of milk to alcohol and its coagulation time in their subjective heat stability test, although it was admitted that this apparent lack of correlation could in part be due to ignorance of the mechanisms of both coagulation phenomena. More recently, Samuelsson *et al.* (1962) found that the results of the alcohol stability test correlated with the extent of sediment formation in milks subjected to UHT processing. Further instances of similarity between alcohol-induced and UHT-induced instability have been noted, particularly with regard to the effects of pH and Ca^{2+} content (Zadow *et al.*, 1983).

2.2. Ethanol Stability/pH Profile

When the alcohol test first came into use, it was regarded largely as a measure of acidity, particularly acidity produced by bacterial fermentation and proteolysis. However, it was soon observed that fresh milk from individual cows frequently coagulated and that the test was dependent on factors other than acidity. The early major studies relating to the alcohol test concentrated on the role of the inorganic components of the milk serum. Sommer and Binney (1923) concluded that alcohol coagulation occurred as a result of an excess of calcium and magnesium over citrate and phosphate in the serum, in keeping with the salt-balance theory for heat stability proposed by Sommer and Hart (1919). Mitamura (1937), in his mammoth study, confirmed the importance of bivalent cations. Davies and White (1958) took this a stage further when they observed that the strength of the ethanol solution required to induce coagulation in an equal volume of milk was inversely related to the concentration of ionic calcium.

With everyone devoting their attention to milk salts, no one considered establishing a direct relation between alcohol stability and milk acidity or pH. Indeed, Dahlberg and Garner (1921) concluded that there was no direct relation between susceptibility to alcohol coagulation and titratable acidity, though, confusingly, they noted that 'milks high in titratable acidity as a result of fermentation will in the large majority of cases show coagulation with alcohol.' It was not until 1980 that Horne and Parker (1980) demonstrated that artificial adjustment of the pH of a milk provided an ethanol stability/pH profile characteristic of that milk; a plot analogous, but in no way identical, to the coagulation time/pH profiles first observed by Rose (1961*a*,*b*) in heat stability studies.

These ethanol stability/pH profiles (typical examples are shown in Fig. 1) were sigmoidal in shape, with a minimum stability at low pH values and a maximum stability at high pH values. Because increasing pH increased the stability of milk to such an extent that pure ethanol often failed to induce precipitation in the single alcohol test with one-to-one volume addition, Horne and Parker (1980) modified the test, adding two volumes of aqueous ethanol solution to one volume of milk. The ethanol stability of the milk at the specified pH was then reported as the concentration of ethanol solution inducing instantaneous precipitation of the milk protein.

Unlike the heat stability coagulation time/pH profile, the ethanol stability/pH profile can be assigned a functional form. Though purely empirical, this mathematical description facilitates comparison between different milk samples and their composition. Thus, the profile data have been routinely fitted to an equation of the form given below relating ethanol stability (S) to pH.

$$\frac{S_{max} - S}{S - S_{min}} = 10^{(pH - pK)/b} \tag{1}$$

or, more usually, by taking logarithms, to the linear equation:

$$pH = pK + b \log_{10}[S - S_{min})/(S_{max} - S)] \tag{2}$$

Thus, any ethanol stability/pH profile can be characterized by four parameters: the asymptotic values, S_{min} and S_{max}, of the two arms of the sigmoid curve at low and high pH, respectively; an inflexion point or effective pK value denoting the position of the profile along the pH-axis and b, the gradient of the linear transform. Fitting is usually carried out by varying S_{min} and S_{max} until a maximum correlation coefficient is achieved for the linear equation formulation, pK being the intercept of this straight line fit on the pH-axis and b its gradient. The full curves in Fig. 1 are the least-squares fits by this procedure to the experimental points.

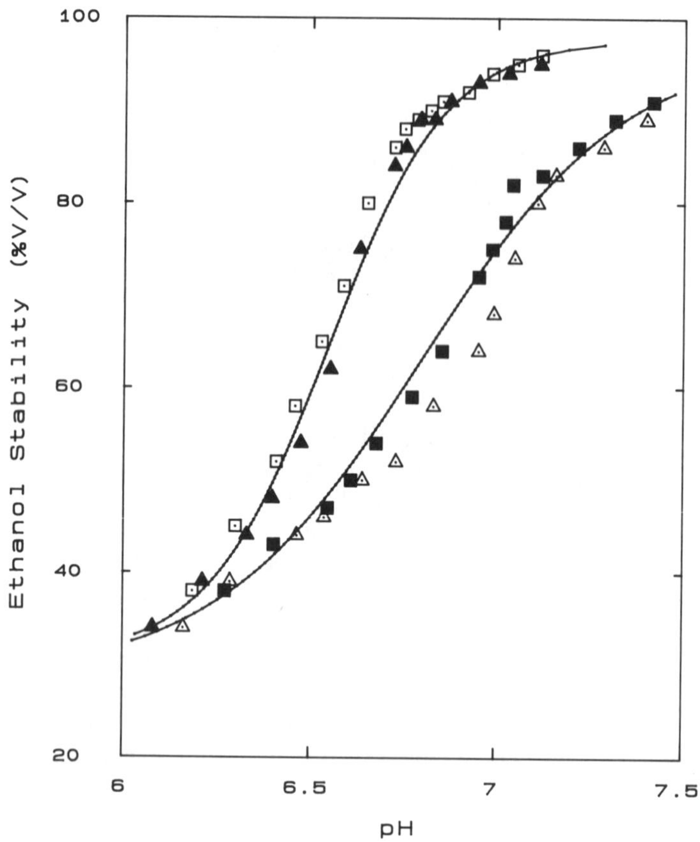

Fig. 1. Typical ethanol stability/pH profiles, illustrating the effect on the profiles of interchanging milk sera by dialysis: ▲, milk A; △, milk A dialysed against milk B; ■, milk B; □, milk B dialysed against milk A. The solid lines are the calculated best fit profiles to the experimental data for milks A and B, computed using the functional form of eqn (1).

2.3. Factors Affecting Profile Parameter Values

Unlike heat-induced coagulation, where two forms of the coagulation time/pH curve are encountered (Rose, 1961*a*,*b*), the sigmoidal shape of the ethanol stability/pH profile has been invariant. Though the majority of determinations have been made on bovine milks from various breeds (generally unknown) and locations, the profiles obtained for goat (Horne & Parker, 1982*a*) and buffalo milks (Unnikrishnan *et al*., 1988) show the same characteristic shape and are located in similar positions along the pH axis.

At least in bovine milk, that location along the pH axis is not fixed but varies from milk to milk from individual cows and systematically during a single cow's lactation (Horne *et al*., 1986). By using a cross-dialysis technique where a small amount of milk A was dialysed against a large volume of milk B and vice versa,

Horne and Parker (1981*a*) were able to demonstrate that the profile position was governed by diffusible components in the milk serum phase. The results of this interchange are shown in Fig. 1 for the two milks designated A and B; virtual superposition of the profiles being obtained. Since the original publication of this finding, there has been a tendency to highlight this observation as simply a lateral shift of the profile, as an interchange of pK values only. The exact superposition meant, however, that the other parameters, S_{min}, S_{max} and b were also either unchanged or switched by the dialysis procedure. Following the examination of many hundreds of profiles over a period of years, only slight variations in the aymptotic parameters, S_{max} and S_{min}, have been noted. However, when the profiles were redrawn in the linear logarithmic format of eqn. (2), a switch in the slope parameter, b, also became apparent (Horne & Parker, 1981*a*). This produces the equivalent of a rotation or slight flattening in the normal format for the profile (Fig. 1) and is not immediately obvious. This observation was not followed up in subsequent studies and the factors governing the value of the slope parameter are still unknown, though an obvious lactational trend has been noted (Horne *et al.*, 1986).

Although not among the diffusible constituents, whey proteins were quickly eliminated as possible regulators of profile parameter values (Horne & Parker, 1981*a*). A previously suggested mechanism for alcohol-induced destabilization envisaged alcohol denaturation of the whey proteins, which then precipitated onto the casein micelles, precipitating these in turn (Leighton *et al.*, 1928). This was ruled out when it was observed that centrifuged micelles, resuspended in skim milk dialysate, gave an ethanol stability/pH profile identical to the original milk (Horne & Parker, 1981*a*). Total removal of the whey proteins had no effect on profile parameter values.

Profile parameter values were also unchanged when lactose and urea contents were increased to three times their naturally occurring average levels (Horne & Parker, 1981*a*). The addition to the milk of Ca^{2+} as calcium chloride displaced the profile to higher pH values, but did not alter its sigmoid character, nor significantly change the values of S_{max} and S_{min}. Reduction of ionic calcium level by treatment with ethylene diamine tetraacetic acid (EDTA) had the reverse effect, shifting the entire profile to lower pH values. Addition of phosphate or citrate also shifted the stability profile to a lower pH, as a result of the chelation of Ca^{2+}, the extent of the shift being in the order of their Ca^{2+}-chelating power when added at the same concentrations.

Such movements of the profile, to higher pH values when Ca^{2+} level was increased and to lower pH values when it was reduced by EDTA addition, were consistent with the earlier findings of Davies and White (1958) that the alcohol stability as measured at the natural milk pH was significantly correlated with the concentration of Ca^{2+} in the milk. These results were also in agreement with the early studies of Sommer and Binney (1923) who found that addition of magnesium or calcium to a milk could render it unstable in the alcohol test and that stability could be restored by the addition of citrate or phosphate.

2.4. Lactational and Compositional Effects

The studies described above suggested that the concentration of ionic calcium in the milk is the dominant factor fixing the position of the ethanol stability/pH profile along the pH axis. Further evidence for this came from a survey of the ethanol stability of Irish creamery milks (Donnelly & Horne, 1986) where concurrent analyses for mineral constituents were made in an attempt to correlate profile parameters with natural variations in milk composition. From linear regression analysis, the strongest relationships ($P < 0{\cdot}001$) were between pK($r \sim 0{\cdot}82$) or S_{max}($r \sim -0{\cdot}72$) and soluble salt balance whether this was expressed as the ratio of calcium plus magnesium to phosphate plus citrate concentration or as the difference between the concentrations of these cations and anions. Other equally strong corrrelations were found between pK($r \sim -0{\cdot}84$) or S_{max}($r \sim 0.61$) and the concentration of soluble inorganic phosphate. Since it can be readily demonstrated algebraically that the concentration of soluble free calcium is governed by the soluble salt balance ratio (Donnelly & Horne, 1986), these relationships provide confirmation from natural compositional variations of previous observations that an increase in calcium shifts the profile pK to more alkaline values, whereas a decrease by sequestrant inorganic phosphate produces a shift to more acid values and a resultant increase in ethanol stability.

The relationships observed for the variation of S_{max}, the asymptotic maximum at high pH, with the salt balance ratio (S_{max} decreases with an increase in the free calcium content) and with inorganic phosphate (S_{max} increases with an increase in the sequestrant) are consistent with a precipitation model where the negative charge of the micelle is under the neutralizing influence of calcium binding. Similar effects were observed with early lactation milks from cows in the Hannah Research Institute herd (Horne *et al.*, 1986). The depressed value of S_{max} found in colostral milks rose rapidly in the first days of lactation, matching the decline normally encountered in the salt balance ratio (White & Davies, 1958). Thereafter, however, the Hannah group detected no further apparent lactational trends in the value of S_{max} (Horne *et al.*, 1986).

Of the other ethanol stability/pH profile parameters, the asymptotic minimum stability (S_{min}) for samples from autumn- and winter-calving cows showed a stepped decrease which could be associated with the transition to summer grazing but manifested no obvious lactational trends (Horne *et al.*, 1986). For these same samples the slope parameter increased slowly during lactation for all eight cows monitored. The remaining profile parameter, the profile pK value, decreased in early lactation but thereafter increased throughout lactation giving the most obvious effect observed in direct measurement, namely an alkaline shift in the profile as lactation progressed (Horne *et al.*, 1986). This movement of the profile pK value mirrored the lactational behaviour of the soluble salt balance ratio as recalculated from the original data presented by White and Davies (1958).

Lactational effects on alcohol stability at the natural pH of milk had been noted previously by Mitamura (1937) and Davies and White (1958). Mitamura

(1937) found milk to be very unstable to alcohol in early lactation, but thereafter to gain stability to a level that remained fairly constant over the major portion of lactation and that was specific for each cow. Towards the end of lactation he found that the milk from some cows became still more stable but a decrease in stability as more common. Davies and White (1958) confirmed the observation that early lactation milks were very unstable to EtOH. They also found that mid-lactation milks showed much higher stability compared with those from early lacation, whereas late lactation milks showed a wide range of stability. Such variations in behaviour during lactation were confirmed in the study by Horne *et al.* (1986) where complete ethanol stability/pH profiles were determined. In early lactation, the most important factor appears to be a low natural milk pH, so that the natural stability is well down the acid arm of the profile. In late lactation, the most important contributory factor is a high soluble salt balance ratio. A shift in the profile to more alkaline pH values is associated with this and, if the natural milk pH does not follow, a decrease in alcohol stability as measured by the single point test must result. If, however, the natural milk pH increases sufficiently to overcompensate for the profile shift, then an increase in alcohol stability at the natural pH will be noted. Hence, the reason for the inconsistent behaviour in the alcohol stability of late lactation milk. On the other hand, if milk pH is adjusted to a chosen fixed value, decreased stability in late lactation milk should be clearly and consistently observed. Thus, Horne *et al.* (1986) found that the ethanol stability calculated at a fixed pH of 6·6 from measured profile parameters passed in early lactation through a maximum value, characteristic for each cow, but declined steadily thereafter. When the ethanol stability behaviour for each cow was normalized to this maximum value, a normalized decay rate of 1×10^{-3} per day, constant to within 10%, fitted the data from all eight cows in the study (Horne *et al.*, 1986). This implies that from a few measurements during the first weeks of lactation it should be possible to predict throughout the remainder of lactation, the ethanol stability of the milk from any individual cow or synchronously calving herd, within a few per cent ethanol.

2.5. Factors Relevant to the Mechanism of Alcohol Destabilization

The discussion so far has demonstrated that the composition of the soluble phase is of paramount importance in determining the ethanol stability of a milk. The manner in which this composition exerts its control, the role of the alcohol in the coagulation reaction, and the part played by the micelle have all been neglected. A full discussion of the mechanism of the precipitation reaction is postponed to the end of the chapter, but further related experimental data are introduced here.

2.5.1. Origin of the pH Transition

The effects of manipulating the ionic calcium level on the profile position have already been documented. When this manipulation was carried out at a fixed pH value, a sigmoidal curve, in essence identical to the ethanol stability/pH profile was obtained (Horne & Parker, 1981*b*). Addition of calcium, as calcium chloride, decreased the ethanol stability but asymptotically to a minimum value, presumably as calcium binding was saturated. On decreasing the effective calcium concentration by inclusion of EDTA, an apparent maximum stability was achieved. These saturation values were close to the asymptotic stabilities in the corresponding full profile for that milk.

In a parallel experiment, removal of the milk's soluble phosphate by dialysis against phosphate-free sera induced an ethanol stability/pH profile where stability monotonically increased with pH but which lacked the typical signmoidal step transition over a narrow pH range (Horne & Parker, 1981*b*). Conversely, stability behaviour identical to that of milk could be induced in caseinate solutions. Sodium caseinate alone showed the same monotonically increasing stability with pH as milk devoid of soluble phosphate, but at a much higher stability level. Addition of calcium to that solution decreased stability across the whole pH range, giving a profile parallel to that of sodium caseinate with the sensitivity to alcohol decreasing with increasing calcium concentration. The inclusion of phosphate in these calcium caseinate mixtures prior to determination of their alcohol stability induced the typical sigmoid transition into the stability/pH profile (Horne & Parker, 1981*b*).

The pH transition in the ethanol stability profile, therefore, owes its existence to the sequestering effects of the doubly-ionized form of phosphate in the milk. As pH is increased and the phosphate anion passes through this ionization to the much more effective calcium sequestrant form, the free calcium level of the milk is reduced and ethanol stability increased. In contrast, decreasing the pH frees ionic calcium from calcium phosphate and leads to a decrease in ethanol stability.

2.5.2. Role of the Alcohol

Of the forces that stabilize the unique folded structure of proteins in solution—hydrophobic interactions, hydrogen bonds and electrostatic forces—each is affected in some way by the addition of an organic solvent to the solution. Such additions can lead to denaturation, inhibition of enzyme reactions or in the case of milk, loss of the stability of the micellar system and subsequent coagulation. The mechanism proposed by Leighton *et al.* (1928) wherein the coagulating effect of the alcohol was attributed to its denaturing properties has already been mentioned. Herskovits *et al.* (1970) demonstrated that the effectiveness of alcohols as protein denaturants increased with increasing chain length or hydrocarbon content. As a rule, branching of the hydrocarbon portion tended to reduce this effectiveness, which was also reduced by increasing hydroxyl content. Broadly speaking, Horne and Parker (1981*c*) found that the effectiveness of alcohols as milk precipitants paralleled the results of Herskovits *et al.* (1970), the effectiveness clearly increasing with increasing hydrocarbon content. When the alcohol stability values were expressed as a function of the dielectric constant of the alcohol/water mixture, rather than as alcohol content, a universal precipitation curve was obtained, either as a function of pH or as a function of added calcium at a fixed pH (Horne & Parker 1981*c*). The stability of the casein micelle system is therefore governed by the dielectric constant of the medium. In contrast, if the results of Herskovits *et al.* (1970) are taken, the conversion of the denaturing concentrations of alcohol to mixture dielectric constants gives a range of values, implying that the extent of protein denaturation is not simply a function of dielectric constant but that more specific interactions with the alcohols are involved.

2.5.3. Chemical Modification of Micellar Charge

The demonstration that the role of the alcohol is apparently to reduce the dielectric constant of the mixture to a critical level to induce precipitation (Horne & Parker, 1981*c*) indicates strongly the importance of a repulsive electrostatic component in the intermicellar potential. Following their earliest studies, Horne and Parker (1981*b*) suggested that the effects of varying soluble calcium concentration were exerted through their influence on the micellar charge. Increased levels of ionic calcium gave higher levels of calcium binding to micellar casein, lowered micellar charge and induced instability. Conversely, lower calcium levels decreased calcium binding and increased stability. Following up these ideas on the role of micellar charge, they investigated the changes in ethanol stability of skim milk consequent upon the chemical modification of the milk protein (Horne & Parker, 1982*b*). Conversion of positively-charged lysine residues to neutral N-acetyl moieties by reaction with acetic anhydride, effectively increasing the net negative micellar charge, increased the ethanol stability. Amidation of protein carboxyl groups, thereby reducing the micellar negative charge, destabilized the milk. Further, the lysine modifications, i.e. increasing negative charge, counteracted destabilization induced by ionic calcium addition. The latter balancing effects were reconciled in a simple (possibly simplistic) model in terms of the changes wrought in micellar charge.

Non-specific chemical modifications with aldehydes (glyceraldehyde, formaldehyde and glyoxal) all increased ethanol stability, though their effectiveness depended on the duration and temperature of heat treatment. No effects were observed when milks were heated with the various sugars which Holt *et al.* (1978) found effective in increasing the heat stability of unconcentrated skim milk, though the heat treatments employed were not as intensive as in the heat stability tests. Holt *et al.* (1978) considered that the increase in heat stability which they observed was probably a result of reaction between the aldehydes and protein amino groups. The observations of Horne and Parker (1982*b*) regarding their effect on ethanol stability are consistent with this reaction pathway.

The ethanol stability test is a catastrophic one; that is to say, the strength of the alcohol solution is increased until immediate, instantaneous precipitation is induced. Under these critical conditions these chemical modification studies imply an important role for micellar charge and residual electrostatic repulsion.

2.5.4. Species of Mammal

As already noted, the ethanol stability/pH profiles for goat (Horne & Parker, 1982*a*) and buffalo (Unnikrishnan *et al.*, 1988) milks have the same characteristic sigmoidal shape as that obtained with bovine milk. Both buffalo and goat milk are, however, much less stable than bovine milk. Cross-dialysis experiments of the type described earlier, indicated for both species that this lower stability was not due simply to different concentrations of serum ionic species. Though Horne and Parker (1982*a*) observed a movement of the caprine profile towards that of the bovine, Unnikrishnan *et al.* (1988) found no such movement with buffalo milk. Indeed their results showed that replacement of bovine micelles by buffalo casein micelles in bovine serum gave a stability profile almost identical to that of the original buffalo milk. The corresponding interchange of cow for buffalo micelles in buffalo serum gave the bovine profile also. The interspecies differences in stability between buffalo and bovine and between caprine and bovine milks are therefore due primarily to intrinsic differences in the casein micelles. As far as goats' milk was concerned, Horne and Parker (1982*a*) attributed the low ethanol stability to the presence of different proportions of the individual caseins, particularly the higher levels of the β-casein homologues and the lower levels of α_{s1} homologue in caprine milk, again pointing to the importance of micellar charge.

2.6. Effects of Processing Milk

The manipulation of milk composition and the chemical modification of the protein, both described above, were deliberate operations designed to provide further understanding of the ethanol stability test. Probably all milk reaching the consumer is processed in some manner. Some operations, such as pasteurization and cooling, are relatively simple; others, such as concentration and conversion to milk powder, are relatively complex in their effects on the properties of the finished product. The effects of two such processes, forewarming and concentration, on the ethanol stability behaviour of milk have been studied in depth (Horne & Parker, 1981*d*, 1983).

2.6.1. Effects of Forewarming Milk

Forewarming is a technique frequently used by milk processors to improve the heat stability of milk subsequently concentrated. In this operation, the milk is heated to 90 or 95°C and held there for 10 to 30 min. The increase in the heat stability of concentrated milk has been attributed mainly to serum protein denaturation (Ramsdell & Hufnagel, 1953) but significant changes in mineral equilibria also occur (Rose & Tessier, 1959). Horne and Parker (1981*d*) studied the effects of forewarming at 90°C for 30 min on the ethanol stability of skim milks of normal concentration. They found that (i) the sigmoidal shape of the ethanol stability/pH profile was retained, (ii) the asymptotic maximum and minimum stabilities were largely unchanged from the values of the original milk, and (iii) the entire profile was shifted laterally to more acid pH values by approximately 0·15 pH units, giving an apparently increased ethanol stability at the natural milk pH. This alteration in the position of the pH profile was found to be insensitive to the presence of sulphydryl-blocking agents, thus apparently ruling out the involvement of denatured serum protein in the effect (Horne & Parker, 1981*d*). Among other reactions occurring in heated milk is the precipitation of insoluble calcium phosphate (Hilgemann & Jenness, 1951). That such reactions occurred under their conditions of forewarming was confirmed by Horne and Parker (1981*d*) by direct measurement of soluble calcium concentration as a function of pH. The observed effects on the ethanol stability of the forewarmed milk were wholly consistent with such a reduction by precipitation in the level of Ca^{2+}, as viewed from the model suggested by Horne and Parker (1981*b*).

Buffalo milk showed a similar response to forewarming but also demonstrated an increase in the value of S_{max} (Unnikrishnan *et al.*, 1988). On storage, however, these effects were partially reversed, particularly the behaviour of the alkaline arm of the profile. Moreover, the extent of the stability improvements and of their rate of reversion varied widely between milk samples. Such behaviour was considered consistent with slow resolubilization of the calcium phosphate precipitated on heating, as reported by Jenness and Patton (1959).

2.6.2. Effects of Concentrating Milk

Concentration also produced marked changes in the ethanol stability/pH profile of a milk. The main effect was seen in the stability at high pH values, which decreased progressively as the milk was concentrated by evaporation (Horne & Parker, 1983). At low pH values, on the acid side of the natural pH, the ethanol stability was largely unaffected by concentration. At pH values of 7·0 and above, Horne and Parker (1983) found that the ethanol stability decreased linearly with the chloride ion content of the concentrate. By adding NaCl to the orignal milk they reproduced this linear decay behaviour. In concentrated samples whose chloride ion content had been subsequently reduced by dialysis, ethanol stability values were found to be as predicted by the remaining chloride level. By adding other monovalent salts, the effect was shown to be non-specific (Horne and Parker, unpublished). Empirically, therefore, the concentration-dependent behaviour of the alkaline arm of the ethanol stability/pH profile was found to be governed largely by the ionic strength of the system. Because of the multiplicity of ionic strength-dependent reactions possible in milk, it is difficult to reconcile these observations within any simple theoretical framework. Nevertheless they do appear to be useful from two contrasting viewpoints: first, predictive, since by adding NaCl to milk, the ethanol stability of its evaporative concentrate can be predicted; secondly, conservative, since if that ethanol stability is found to be unacceptably low, it can be increased by incorporating a short dialysis stage to reduce the chloride concentration, prior to concentration.

2.7. Ethanol Stability and Heat Stability

A detailed comparison of the ethanol and heat stability tests formed the basis of a recent review by Horne and Muir (1990). The main thrust of their argument may be summarized as follows. The ethanol stability test requires addition of increasing concentrations of ethanol solution until instantaneous precipitation is induced. In the conventional heat stability test, heating of the sample at a preselected temperature is prolonged until coagulation occurs, the actual duration of heating, or clotting time, being defined as the measure of heat stability. This extended heating period allows a multiplicity of concurrent and sequential reactions to take place, as summarized by Fox (1981). Some of these reactions are deleterious, some advantageous from the heat stability viewpoint, but each has its own concentration and temperature dependent characteristics. It is little wonder that no correlation has been observed between the results of the alcohol stability test and the measurement of heat coagulation time.

When the basis of the heat stability test is altered, however, to one of measuring the temperature at which instantaneous coagulation occurs, a different picture emerges. Miller and Sommer (1940) measured the heat stability of milk as a function of the degree of acidification using just such a procedure. As a function of pH, their coagulation temperatures showed signmoidal behaviour (Miller & Sommer, 1940), remarkably similar to the ethanol stability/pH profiles. Indeed, Horne and Muir (1990) were able to demonstrate that an equation of the form used to fit ethanol stability profiles could also be applied to the data of Miller and Sommer (1940). Moreover, sufficient results were contained in the paper of Miller and Sommer (1940) to demonstrate that these heat stability profiles responded to the addition of calcium and phosphate in the manner expected of ethanol stability/pH profiles, i.e. an addition of calcium pushed the profile to more alkaline pH values whereas addition of a calcium sequestrant, phosphate or citrate, shifted the profile to more acidic values. The lack of correlation between ethanol and heat stability behaviour seems to be rooted in test protocols rather than in a fundamental difference in the destabilization reactions. The exact nature of those reactions whilst now reasonably certain in the ethanol stability case is still speculative from the heat stability standpoint and must await further research, though Horne and Muir (1990) propose a not unreasonable solution.

3. Ethanol Stability and Casein Micelle Structure

3.1. Casein Micelle Stability

The technological properties of milk are very strongly influenced by the stability of the casein micelle system or rather by the fact that this stability can be lost as a result of any of several treatments, including acidification, heating, addition of ethanol, limited protolysis and addition of calcium. Each of these processes modifies the structure of the micelle in such a fashion as to render it unstable. Understanding how this is brought about provides a valuable contributon to our fundamental knowledge of casein micelle structure and stability which ultimately must reap dividends in optimizing process control and manufacturing efficiency.

In some respects, scientific research can be regarded as a jigsaw puzzle with each contribution adding to and dovetailing with that which has gone before. There is a tendency to regard such a picture as unchanging and immutable. Each piece adds a brick to a solid, firm edifice until a complete well-founded theory is constructed. There comes a time, however, when the strength of that structure is tested by a more than usual stress. Our wall or edifice may then take on all the characteristics of a house of cards and simply collapse. Such an event gives us the opportunity to question and justify all assumptions made in establishing foundations, to reconsider, to reassess and to realign each piece or stone of evidence and data. The author thinks we have reached such a juncture in our consideration of casein micelle stability and structure. The house has not collapsed yet but it is becoming decidedly shaky. In summarizing the contributions that research on ethanol stability has made to our knowledge of the casein micelle, the author would like to take the opportunity to test basic assumptions and justify them where possible. It would be useful to review the problem of micellar interactions and address the current micelle 'model' situation, discarding those features which unnecessarily complicate that model and make it untenable in practice.

3.1.1. Micellar Interactions

All current discussions of casein micelle stability begin with the tacit assumption that the micelles can be treated as hard-sphere, colloidal particles. Justificaton for this is usually given as a statement of particle size, this falling within the size range attributable to colloids. Thus, casein micelles are large aggregates, ranging in diameter from 20 to 600 nm, averaging approximately 200 nm (McGann *et al.*, 1980). That these particles can be treated as 'hard' or even 'soft' spheres has yet to be fully tested. Caseinate solutions at the concentrations normally extant in milk are fairly viscous. Packaged into the casein micelle system, the same casein protein concentration gives a milk whose viscosity shows no appreciable difference from Newtonian behaviour (Scott-Blair, 1953) and a value not much higher than water. It is this behaviour and the stability of the system that justify its characterization as a hard sphere-like dispersion in which only relatively short-range interactions occur. More rigorous testing of this assumption requires measurement of viscosity as a function of micellar volume fraction and applied stress over a much greater range of both parameters than hitherto covered.

The stability of colloidal dispersions is determined by the forces acting between the particles. Schmidt (1982) suggested the involvement of three types of interactions between casein micelles, namely van der Waals' attraction, electrostatic repulsion and forces due to macromolecular adsorption, the latter now commonly grouped under the heading 'steric stabilization' (Napper, 1983). More recently it has become common convention to include a fourth type of force due to solvation and arising from the structuring or ordering of liquid molecules when two surfaces move close together (Israelachvili & McGuiggan, 1988). These forces can be attractive, repulsive or oscillatory. The repulsive and oscillatory forces, also commonly referred to as hydration or structural forces, have been observed in 'force-balance' experiments where two molecularly smooth surfaces are brought together (Israelachvili & Pashley, 1983; Pashley & Israelachvili, 1984) at separations of a few solvent molecule diameters. Such forces are obviously of very short range, far shorter in fact than the anticipated roughness of the casein micelle surface. Moreover, between rough surfaces a much reduced force is determined even when the roughness is only of the order of a few angstroms (Christenson, 1986). Such short range structural forces need not therefore be considered further in the context of micellar stability. More relevant to the present discussion, however, is the detection in aqueous solutions by these 'force-balance' experiments of a smoothly varying component of this hydration force which can be either repulsive or attractive, depending on the nature of the interacting surfaces. Between two hydrophobic surfaces, the force is attractive, its strength depending on the hydrophobicity of the surfaces or surface groups (Israelachvili & Pashley, 1982, 1984; Pashley *et al.*, 1985; Claesson *et al.*, 1986). This hydrophobic interaction is a long range force and appears to have an exponential distance dependence with a characteristic decay length of about 1 nm (Israelachvili & Pashley, 1982, 1984). Its strength and range mean that it could be far stronger than van der Waals' attraction between surfaces out to distances of at least 10 nm (Israelachvili, 1987). Unfortunately, no calculations have yet been made of the likely contribution such interactions would make to the stability of a colloidal dispersion—a gap which is sure to be filled in the near future.

Of these forces, which are operative in the casein micelle system? Van der Waals' forces are monotonically attractive and occur between all molecules. Their existence has been known and accepted for many years but it was not until the 1950s and 1960s that a general macroscopic theory of these dispersion forces between surfaces or colloidal particles was published (Lifshitz, 1956). Empirically, these forces and their dependence on the interaction distance can be treated in terms of the equation derived earlier by Hamaker (1937), incorporating only a single interaction parameter, the so-called Hamaker constant, whose value is determinable in principle by fitting suitable experimental data. Such an equation is incorporated into the DLVO theory (Verwey & Overbeek, 1948) which until recently formed the sole theoretical framework for analysing the properties of colloidal systems. This theory considers stability to be the result of repulsive interactions of electrostatic origin dominating over attractive van der Waals' forces. Theories of electrostatic interactions in ionic solutions are based on the Poisson-Boltzmann equation (Verwey & Overbeek, 1948). These reveal that the electrostatic repulsion between two charged surfaces or particles is screened by the ions in solution forming the so-called 'double-layer'. The strength of this interaction decays exponentially with distance from the surface; the value of its characteristic decay length or 'double-layer' thickness being the reciprocal of the Debye-Huckel parameter, κ, and therefore ionic strength dependent.

Attempts have been made to describe the stability of the casein micelle system in terms of this classical theory (Green & Crutchfield, 1971; Payens, 1977, 1979). Using their measured zeta potential values, Green and Crutchfield (1971) demonstrated that the DLVO theory predicted stability, i.e. a sufficiently high energy barrier, for native casein micelles but a much lower barrier, and therefore probable instability, for rennet-treated micelles. Unfortunately, as Payens (1977) pointed out after performing similar calculations, this energy barrier is located at such a small inter-surface distance ($\sim 0{\cdot}1$ nm) as to be physically meaningless, lying well within the orbit of surface roughness and short-range forces. Another major failure

(Payens, 1979) of the DLVO approach is the inability of the casein micelle system to respond to electrolyte addition in the manner predicted by the Schulze-Hardy Rule, a corollary of the DLVO theory. Such observations were taken to mean that electrostatic repulsion alone was insufficient to account for the stability of the casein micelle system and that other repulsive forces should be invoked (Holt, 1975; Payens, 1979).

In retrospect, however, these failures tend to confirm the inappropriateness of the DLVO theory as a descriptive tool in the context of casein micelles, for casein micelles simply do not conform to the basic requirements of that theory. It is not so much that the theory has failed but rather that the theory does not apply. The casein micelles are neither smooth nor uniformly conductive. They owe their charge, in part, to specific binding of divalent ions, a charge-controlling mechanism whose secondary response to ionic strength changes is outwith the scope of DLVO theory. It is little wonder then that the micellar system can tolerate divalent ion concentrations greater than the critical level of 1 mM predicted by the Schulze-Hardy Rule (Payens, 1979). Similarly, at high ionic strengths the micelles tend to dissociate (Saito, 1973); a pathway not allowed in a theory which considers the particle as immutable. Charge effects remain as an important contribution to casein micelle stability. There are too many indicators of this for it to be otherwise. The previous discussion in this chapter on the role of ionic calcium in controlling ethanol stability is only one of many pieces of overwhelming evidence. Equally certain is the conclusion that this electrostatic contribution cannot be quantified within the scope of the DLVO theory and that some other theoretical framework must be sought.

The failure of a DLVO description can be considered as evidence, if only negative, for the existence of another stabilizing force in the micelle system. This is the third force cited by Schmidt (1982); the steric stabilizing component. This force arises from the interaction of loops or tails on the particle surface. As the micelles approach closely, the protruding chains, often called hairs, interpenetrate or become compressed. The loss of conformational entropy contributes to interparticle repulsion and gives rise to so-called steric stabilization. Such a hairy coat can be invoked to explain the inherent stability of the micelle despite a relatively low zeta potential (from the DLVO viewpoint) (Holt, 1975). Walstra (1979) suggested that such a coat was necessary to explain the high values measured for casein micelle voluminosity and the discrepancies between these values when different measuring techniques were employed. He further proposed that the hairs comprised the C-terminal macropeptide of κ-casein.

Like all good theories, this one is readily put to the test. The surface location of the κ-casein suggests that cleavage by chymosin and loss of the hair should lead to micellar coagulation. That coagulation follows enzyme action is not of itself evidence of hairs and steric stabilization, since previous measurements indicated that enzyme action leads to a decrease in micellar zeta potential (Green & Crutchfield, 1971; Dalgleish, 1984). Convincing evidence came when Walstra *et al.* (1981) directly measured changes in the hydrodynamic size of casein micelles as a result of chymosin treatment. A reduction in diameter of about 10 nm was observed, consistent with the removal of a macropeptide hairy layer. At longer times they observed an increase in apparent size as coagulation ensued. Independent confirmation of the depilatory action of chymosin has since been provided by several laboratories (Horne, 1984; Holt & Dalgleish, 1986; Griffin, 1987). Other evidence in support of this steric stabilization model comes from proton NMR studies which reveal considerable flexibiliy and high levels of mobility in the C-terminal of the κ-casein polypeptide chain (Griffin & Roberts, 1985; Rollema *et al.*, 1988). Yet more evidence in favour of this model is presented later when the response of casein micelles to sub-critical levels of ethanol is considered.

Theories for the calculation of the steric component are now available (Scheutjens & Fleer, 1985) but they are insufficiently refined as yet to permit their application to micellar interactions. For example, the effect of charge along the polypeptide chain has yet to be satisfactorily realized for the theory to have predictive power.

Little can be said about the fourth force, the long-range solvation or hydrophobic interaction, other than that it is almost certainly involved in micellar interactions. Hydrophobic interactions are frequently invoked as the sources of intramicellar bonding, of bonds between the protein molecules making up the micellar aggregate. If they are present there, have they no role to play in the interactions of protein molecules located in different micelles when coagulation is induced? It seems unlikely that they should be absent but once again calculations of their contribution is beyond presently available theory.

3.2. Ethanol Addition and Steric Stabilization of Casein Micelles

It is well known that, in an unfavourable solvent environment, proteins and other polymers undergo a conformational transition to a collapsed form, accompanied by a dramatic reduction in the effective volume of the molecule. Solvent conditions that lead to collapse are often similar to those leading to aggregation (Post & Zimm, 1982). It was therefore postulated that the action of ethanol in micellar aggregation might be to collapse the 'hairs' and remove the steric stabilizing component from the system. To test this hypothesis, Horne (1984) used photon correlation spectroscopy to measure the micellar hydrodynamic radius in the presence of ethanol up to those ethanol levels that induce coagulation. At these sub-critical ethanol levels, the radius of micelles, diluted from a micellar fraction of narrow size distribution, exhibited a maximum decrease of approximately 10 nm from the value measured in the absence of ethanol. The hydrodynamic radius passed through this minimum and increased dramatically as the ethanol concentration was increased through the critical level, the time-dependent increase in radius indicating micellar aggregation.

It was recognized that the decrease in radius could be the result either of a uniform global shrinkage of the micelle or of a conformational change in the 'hairy' outer layer. Further support for the occurrence of a purely surface phenomenon came from an experiment similar to that of Walstra *et al.* (1981), where aggregation was induced by chymosin addition. In the absence of ethanol, a decrease in radius of approximately 5 nm was observed, confirming the previous observations of Walstra *et al.* (1981). In the presence of 16% ethanol (v/v), with no rennet added, the micellar radius was reduced by approximately 8 nm. On adding rennet to this solution, no further change in radius was recorded before aggregation induced by enzyme action ensued. The reaction still exhibited a delay period or clotting time dependent on enzyme concentration, indicating that although the 'hairs' had been collapsed by the ethanol and the steric component thus removed from the interaction equation, reduction of some other repulsive component (presumably electrostatic) through enzyme action was still required to allow aggregation to proceed.

Following these initial experiments which demonstrated so directly the contribution of a steric stabilizing component to micellar interaction, an in-depth study of the role of ethanol was undertaken by the Hannah group. The explanation proposed above for the change of micellar radius on exposure to ethanol was a collapse of the protein hairs onto the micellar surface. An equally applicable explanation would be desorption of those proteins into the serum phase. However, no increases were observed in serum protein levels, even at micellar concentration levels 20 times those used in the light-scattering measurements (Horne, 1986). It was recognized, however, that this was not totally convincing evidence of the absence of desorption, since the possession of only a few hairs would be sufficient to produce significant increases in hydrodynamic radius whilst their putative loss in ethanol solution could easily lie within detection limits on serum protein levels.

Further arguments in favour of hair collapse rather than desorption were advanced as follows. The observed change in micellar radius occurred very rapidly, and showed no time dependence. The value of the radius measured immediately after mixing and correlation function accumulation remained constant over a substantial period of time. Indeed, rennet action was required to induce further change in micellar radius (Horne, 1984). Equally rapid was the reversibility of the decrease in radius, achieved simply by reducing the ethanol concentration (Horne, 1986). No hysteresis was detected and the radius measured at any given sub-critical ethanol level was the same, no matter which path was followed to that concentration. Equally important, the radius value measured at any particular ethanol level remained constant with time. The immediacy of response and its lack of time dependence, the absence of hysteresis on cycling the ethanol levels are evidence against desorption, since the rate of re-adsorption ought to be demonstrably a function of the micellar concentration and hence the extent of dilution employed.

The constancy of micellar radius with time applied only to measurements in what were termed sub-critical levels of ethanol. Beyond a critical ethanol concentration, which was dependent on buffer pH and calcium ion concentration, a time-dependent increase in radius was observed. Since this increase in radius was, at least in its initial stages, accompanied by an increase in light scattering intensity, and since the rate of increase was also dependent on both the micellar and ethanol

TABLE 1
Effect of Casein Micelle Concentration on the Variation of Hydrodynamic Radius (nm) with Added Ethanol

Ethanol concentration (% v/v)	*Volume of milk (μl) added to 2 ml 5 mM Ca^{2+} buffer solution*			
	1	*2*	*4*	*6*
Zero	97·3	95·3	96·4	96·1
14	78·7	80·3	82·0	82·6
16	81·2	80·4	80·9	80·0
18	78·6	78·1	76·5	79·9
20	77·8	77·6	79·8	78·4
22	78·9	79·9	83·9	117
24	92·8	94·5	106	156
26	133	169	Aggregated	Aggregated

concentrations (the latter being above the critical level), this behaviour could only be the result of an aggregation process.

3.3. Micellar Core Radius and its Consequences

The above studies were carried out as a function of micellar concentration not to measure the rate of aggregation but to ascertain whether, by reducing the concentration-dependent rate of the aggregation to a negligible level, the micellar radius collapsed to a definite minimum or whether further collapse could be induced at higher ethanol levels. The results presented in Table 1 indicate that decreasing the micellar concentration, thereby eliminating the subsequent aggregation stage, did not bring about a continued decrease in micellar size at higher ethanol levels. Instead, it produced a range of ethanol concentrations over which this collapsed micelle was stable. This suggests that with total collapse of the 'hairs', the micelle radius can only decrease to a certain minimum value. This was defined as a 'core' radius to distinguish it from a simple minimum, and its measurement allowed an estimate of barrier layer hydrodynamic thickness to be deduced.

The casein micelle system is very polydisperse, with particle sizes ranging from 30 to 600 nm (McGann *et al.*, 1980). Since only an average hydrodynamic radius can be measured using photon correlation spectroscopy, in order to eliminate effects due to possible changes in the micellar size distribution, the majority of barrier thickness measurements were carried out on 'monodisperse' micellar fractions, prepared by resuspension of pellets produced in a series of successive centrifugation steps, always using the supernatant of the previous stage (Horne & Dalgleish, 1985). In this manner, a series of 'monodisperse' slices of the micellar size distribution can be obtained. The barrier layer thickness measured for such fractions increased monotonically with micellar size, i.e. small micelles had thin barriers whereas larger micelles had thicker barriers. Such an observation does not at first sight seem consistent with the simple picture of an outer layer of κ-casein providing a stabilizing sheath. In that model, the same barrier thickness no matter the size of the particle might be anticipated. Faced with similar behaviour from the polymer PVA when adsorbed onto polystyrene latex particles, Garvey *et al.* (1976) developed a model which took the curvature of the particle surface into account. When more extensive studies were made on the response of barrier thickness to change in buffer conditions using the same micellar fraction, it was soon recognized that such a model was not applicable to the casein micelle situation.

All theories of steric stabilization highlight the essential role of the stabilizing polymer and the thickness of the stabilizing layer. Solvent quality, which defines that conformation and thickness, is a function also of the ionic constituents of the buffer system. This is of particular importance where the polymeric stabilizer is a polyelectrolyte species such as a protein. Horne and Davidson (1986) found that the correlation between layer thickness and the level of ethanol required to achieve total collapse to the core, previously observed for varying micelle size (Horne, 1986), was maintained when changes in barrier thickness were induced by varying the pH, ionic strength or calcium ion concentration of the ethanolic buffer system into which the casein micelles were diluted.

Qualitatively, the observed changes in critical ethanol level could be explained on the basis of a mechanism involving electrostatic repulsion between charged polyelectrolyte hairs (Horne & Davidson, 1986). This is a major increase in complexity from the simple model for micellar interaction where the various contributions are treated as additive. Rather, now we envisage the introduction of cross-terms into the interaction equation. Charge and its magnitude influence the strength of the steric stabilization contribution though present theories are as yet incapable of quantitatively estimating such effects. The observed variations with calcium level, pH and ionic strength are, however, consistent with the qualitative predictions of a charged hair model. Thus, the effect of calcium ion concentration, where increasing the calcium level decreases the critical ethanol concentration for total collapse, could be envisaged as resulting from specific calcium ion binding reducing the net negative charge of the protein, decreasing inter-hair repulsion and making them more susceptible to collapse (Horne & Davidson, 1986), similar effects to those observed previously in calcium-induced aggregation of individual caseins (Horne & Dalgleish, 1980; Dalgleish *et al.*, 1981). Similarly, the increase in barrier strength with pH, its increasing resistance to collapse, was viewed as being consistent with the increase in charge brought about by greater ionization of the casein phosphate groups which have a pK value of around 6·5 (Horne & Davidson, 1986). The increase in stability recorded with increasing ionic strength required a more convoluted argument to fit this model, since it was to be expected that increased ionic strength would reduce the range of the putative inter-hair repulsion, thus weakening the steric stabilization barrier—the exact opposite of the behaviour observed. Recourse was made to the observations of Dalgleish and Parker (1980) that calcium binding to individual caseins decreases as ionic strength increases. Decreases in calcium binding would mean an increase in protein charge and result in an increase in inter-hair repulsion and in barrier strength. Domination of this binding effect over the contraction of the Debye-Huckel range of electrostatic repulsion between the hairs would therefore lead to an increase in the strength of the barrier with ionic strength. The reversal in the ethanol stability behaviour at the highest ionic strength employed could then be interpreted as the contraction process eventually becoming the dominant partner in the balance (Horne & Davidson, 1986).

Whilst such a 'charged hair' model accommodates the behaviour observed when ionic species within the buffer solution are varied, it does not address the changes in barrier thickness that also result from these variations. The mechanism envisaged has nothing to contribute on this subject of barrier thickness and would be equally valid in a situation where barrier thickness remained constant. Barrier thickness did not remain constant, however, and consequently a more refined model was required. Moreover, when detailed plots of core radius were included as a function of the ionic variable, it became apparent that the decrease in barrier thickness arose not only from the decrease in hydrodynamic radius from the value observed in alcohol-free solutions, but also from an increase in the actual core radius (Horne & Davidson, 1986). Indeed, increases in core radius were the main source of decreases in barrier thickness when calcium level was increased (Horne & Davidson, 1986). In such situations the micelle is apparently being hardened and its steric stabilizing coat is being thinned from within.

To accommodate these observations it was proposed that the steric stabilizing layer possessed some of the properties of a weak gel. In reality with casein micelles, there is no actual adsorbed layer to be treated as a gel and this refinement is tantamount to describing the casein micelle as a microgel particle formed by extensive cross-linking of the constituent protein molecules. Other implications of this model will be considered later in greater detail. For the moment we consider only how it allows us to explain the variations in micelle size and core radius when the casein micelle is exposed to ethanol.

This microgel model is attractive because it allows a more physically meaningful picture in which the protein hairs are linked not only to the core surface but also to each other. Moreover, the effects of decreased charge, described previously as permitting hair collapse, can be incorporated into this model as promoting increased cross-linking and incorporation of the surface layers into the more rigid structure that is the micelle core. In other words, this model allows for increases in core radii through the formation of further bonds in the region of the outer layers; bonds which are unlikely to differ in chemical or physical nature from those already within the micelle interior. Similarly, the effects of increased charge, with pH for example, would be explicable as the promotion of greater repulsion leading to a reduction in cross-linking, the protein hairs now being free to extend and give rise to the increased hydrodynamic radii observed in alcohol-free buffers (Horne & Davidson, 1986).

A further test of this microgel model is provided by its prediction that in situations where high levels of cross-linking are operative, the stabilizing sheath should be less responsive to the action of ethanol and vice versa. When the stabilizing layer is thin and the system is consequently less stable, the presence of the cross-links renders the hairs less readily collapsible and less susceptible to the effects of added ethanol, compared to the converse situation where the system is sterically stable but has few inter-hair cross-links so that paradoxically the hairs collapse more readily in the presence of the ethanol.

Such behaviour was in fact observed (Horne & Davidson, 1986), assuming the initial gradient of the plot of hydrodynamic radius against ethanol concentration to indicate the susceptibility of the steric-stabilizing sheath to alcohol-induced collapse. The magnitude of this gradient increased linearly with buffer pH. As qualitatively predicted by the microgel model, the steric stabilizing barrier was becoming softer due to the decreasing number of cross-links weakening the gel sheath as pH was increased. The increased softness was not accompanied by decreased stability, however, because as the barrier became softer it also became thicker and that total thickness was the controlling factor which regulated the overall stability of the micellar system. In the final analysis, greater concentrations of ethanol were required to totally collapse the thicker barrier and allow aggregation to proceed. This microgel model was also used to explain the observations noted when ionic calcium concentration was varied. The ethanol response gradient or collapse susceptibility decreased as the calcium concentration was increased (Horne & Davidson, 1986). The gel sheath therefore became more rigid as the increased calcium level neutralized greater amounts of charge, reduced localized inter-hair repulsion and promoted cross-links between hairs. The barrier became thinner as the inner layers became more rigid and were incorporated into the micelle core. Though it became progressively more difficult to induce conformational collapse in the thinning-gel sheath as calcium level was increased, the overall stability of the micellar system declined, because again the overriding parameter controlling that stability was the barrier thickness alone.

3.4. Implications for Structural Models of the Casein Micelle

The modern concept of the internal structure of the casein micelle is based on a model composed of submicelles. This is almost, but not quite, universally accepted. Though the actual model has evolved over many years with various authors contributing refinements to the original proposal, the current view has changed little since the review of casein micelle structure published in 1982 (Schmidt, 1982).

Evidence for a submicellar structure came originally from electron microscopy pictures (Shimmin & Hill, 1964; Schmidt & Buchheim, 1970; Buchheim & Welsch, 1973; Schmidt *et al.*, 1973) but neutron and X-ray scattering data have also been interpreted in subunit terms (Stothart & Cebula, 1982; Pessen *et al.*, 1989). In reality, what all of these techniques are saying is that neither the surface nor the interior of the casein micelle is uniform and homogeneous but that there are regions of particularly high density and scattering power interspersed with areas that have an especially low density. These clusters or clumps of material are best fitted by treating them as spherical subunits with an average diameter in the range 10–15 nm. Complementary to these observations are NMR relaxation measurements on casein micelles, which suggest that the water within them is not in a homogeneous continuum which would permit rapid molecular motion but is trapped in discrete volumes of diameter approximately 7 nm, a value close to, but somewhat smaller than, other estimates of subunit size (Kumosinski *et al.*, 1987; Farrell *et al.*, 1989). What these techniques are unable to demonstrate is that these submicelles are capable of a discrete, separate existence or that their formation is a necessary step in casein micelle production in the mammary gland.

Some evidence from this viewpoint is derived from the dissociation behaviour of micelles where, following the application of a variety of disrupting agents, aggregated particles of diameter ~10 nm are observed (Schmidt & Buchheim, 1970; Carroll *et al.*, 1971; Buchheim & Welsch, 1973; Pepper & Farrell, 1982). However, the self-association tendencies of the casein are well-documented (Schmidt, 1982) and such polymer aggregates are readily produced by mixing appropriate proportions of solutions of the individual caseins under the same ionic strength and pH regime (Slattery & Evard, 1973). Such a reassociation of the individual molecules could form a segment of the overall micellar dissociation pathway and the presence of aggregates cannot be taken as evidence of their identity as subunits from the intact micelle. Rollema and Brinkhuis (1989) have compared the NMR spectra of association products and the caseinate polymers produced by calcium depletion and disintegration of micelles with the signals generated by micellar casein. Both aggregates gave similar spectra but were quite different from the spectrum of casein micelles. Within the aggregates, the flexible regions of the caseins possessed much greater mobility, whereas in the micelle a much more rigid and immobilized structure was apparent (Rollema & Brinkhuis, 1989), arguing against acceptance of the caseinate aggregate as a subunit within the casein micelle.

Criticism can also be levelled at the model for micellar assembly suggested by Slattery (1976, 1977) to generate a subunit structure. In this model, the casein monomers polymerize to form the submicelles which then interact via calcium phosphate linkages—a later refinement proposed by Schmidt (1980), to create casein micelles. To push the equilibrium towards generating aggregates of subunit size from the individual caseins requires relatively high concentrations of the proteins. This presumably must occur in the absence of phosphate to avoid the aggregation reaction that forms the second stage of micellar assembly. However, inorganic phosphate cannot be excluded from these initial stages since synthesis of lactose, a reaction catalysed by divalent Ca^{2+} and producing inorganic phosphate as a by-product at concentrations extant in milk (Kuhn & White, 1977), occurs in the Golgi apparatus concurrently with protein emergence and micellar assembly. The phosphorylated individual caseins emerge into a bath of calcium and inorganic phosphate. Conditions are therefore favourable for the formation of a phosphate/casein/calcium coprecipitate. Horne (1982) demonstrated that such coprecipitation occurred with α_{s1}-casein at protein concentrations below those at which precipitation of α_{s1}/Ca^{2+} complex took place in the absence of phosphate and, more importantly, at concentrations well below those required to produce aggregates of submicellar dimensions from individual proteins. The Slattery/Schmidt step-wise model for micellar assembly does not therefore seem to be a viable option in these circumstances.

Recently, a computer simulation of micellar assembly from individual casein molecules was presented (Horne *et al.*, 1989). This simulation reproduced many features and properties of the casein micelle. Based on a polyfunctional condensation polymerization approach, it assigned α_s-casein a trifunctional chain-branching role, treated β-casein as a bifunctional chain propagator and introduced κ-casein as a monofunctional chain terminator. Thus it overcame a major failing of earlier chain polymer models (Garnier & Ribadeau-Dumas, 1970)—that they had no mechanism for stopping aggregate or micellar growth.

The structures grown by the simulation procedure were very open, highly ramified and porous (Horne *et al.*, 1989). The deep fjords developed would allow penetration of small reagents to the very heart of the micelle, as required to satisfy the experimental observations of Garnier and Ribadeau-Dumas (1970). The interstitial water trapped in these internal open spaces would be hydrodynamically entrained, contributing to a high micellar voluminosity. Equally, areas of high local density were observed in these clusters, around perhaps a particularly reactive accumulation of particles. This argues against the necessity of postulating the existence of discrete submicellar entities when the requisite heterogeneous structure develops as a consequence of the random nature of the polycondensation process.

In the simulation, the monofunctional units (κ-casein) terminated chain growth, but the growing cluster was permitted to circumvent the blocked site. Hence, although the majority of the κ-casein was found on the cluster perimeter, thus fulfilling another model requirement imposed by electron microscopy of micelles with labelled κ-casein (Schmidt & Both, 1982; Carroll & Farrell, 1983), a significant fraction was buried within the micelle. The effects of varying the proportions of the individual caseins within the pool of monomers were also investigated. Only small clusters could be grown with high proportions of κ-casein; all growth sites being rapidly closed off. Reducing the number of κ-casein units allowed greater growth, matching the observed proportionality between κ-casein content and surface-to-volume ratio of casein micelles (Donnelly *et al.*, 1984; Dalgleish *et al.*, 1989). Moreover, because saturation of functionalities on a given monomer also stopped chain extension from that monomer, β- and α_s-caseins also contributed to the outer 'closed' surface of the computer-simulated micelle cluster. All these components were therefore allowed some surface locations on the micelle, consistent with their accesssibility to insoluble cross-linked papain as observed by Ashoor *et al.* (1971) and with the surface composition data of Dalgleish *et al.* (1989).

Qualitatively, therefore, this computer simulation of casein micelle assembly from monomeric units reproduced many of the salient features of the casein micelle. More relevant to the present discussion on ethanol stability and its contribution to our understanding of micelle structure and stability, is the fact that in this polyfunctional model the micelle was assembled by cross-linking these units. The size and varying reactivity of the monomers not only gave rise to internal heterogeneity but also created the dangling surface loops and tails necessary for steric stabilization. In the simulated structures, these are rigid and immobile, but in the micelle itself their mobility, coupled with the emergence of reactive sites along their extent as electrostatic barriers are neutralized, would permit the further cross-linking and growth of the micellar core envisaged in the previous section by the assignation of gel-like properties to surface layers. Quantification of all these predictions is urgently required, but it is another favourable attribute of this simulation that such calculations should be possible, albeit with extensive repetition to achieve statistical significance.

4. Mechanism for Alcohol Destabilization of Milk Protein

The experiments described in the last section confirm the contribution of steric stabilization to micellar stability but relate to the behaviour of the casein micelle in an artificial environment. This was dictated by the requirements of the light scattering technique used to determine micellar hydrodynamic radius. The data from these experiments are most readily interpreted when only single scattering prevails and consequently a high degree of dilution of the scattering particles is required. Whilst the discussion of these results concentrated on the thickness of the stabilizing sheath and the response of the micelle to changing environmental conditions, the experiments also provided measurements of the critical ethanol levels that induce micellar aggregation under these buffer conditions. As a measure of the ethanol stability of the micelle in these environments, these results showed disturbing differences from the behaviour recorded in ethanol stability tests on undiluted skim milk. Overall stability was much reduced and the typical sigmoidal response to pH adjustment was lost when the micelles were diluted into ethanolic calcium buffers for light scattering measurements. Moreover, increasing the ionic strength of these buffers up to 200 mM NaCl at pH 7·0 produced an increase in apparent stability, whereas adding the same levels of salt to milk decreased its stability at that pH (Horne, 1987).

The main difference between the buffer and milk is the absence of phosphate from the buffer. As reported by Pierre (1985), the solubility of the mineral components of milk is also diminished by the addition of ethanol. This led Pierre to suggest that the precipitation of calcium phosphate is the root of the destabilization of casein micelles in milk; the precipitation occurring onto the micelle and somehow neutralizing the micellar charge. If this mechanism were in operation, removal of the phosphate from the suspension medium would eliminate the mineral precipitation stage and should increase micellar stability markedly. This is effectively the procedure followed when the micelles are diluted into non-phosphate buffers for light-scattering measurements but, instead of increasing stability as predicted by the Pierre mechanism, aggregation is induced in these buffers at ethanol levels far below the ethanol stability of the original skim milk (Horne, 1986; Horne & Davidson, 1986).

Far from destabilizing the micellar system, it appears that the precipitation of calcium phosphate in the milk situation exerts a stabilizing influence. Horne (1987) has proposed a mechanism where this role can be fulfilled, incorporating all the known features of micellar destabilization and alcohol stability. In studies of separated micelles in buffer systems, Horne and Davidson (1986) demonstrated that lowering the buffer calcium level increased the thickness of the stabilizing hairy layer and made the micelle more stable. In milk the same mechanism could be operating with the alcohol-induced precipitation of calcium phosphate decreasing the free calcium level, drawing away caseinate-bound calcium, increasing the thickness of the steric layer and also increasing micellar charge, both effects contributing to increased system stability. The higher the alcohol level, the faster the precipitation of calcium phosphate and the greater its extent. The ensuing adjustments in protein charge and conformation, though relatively rapid, as observed in 'hairy layer' studies (Horne, 1986), still require a finite time to respond. Countering them, however, are the destabilizing effects on the micelles themselves, promoting cross-linking and causing the collapse of the steric stabilizing layers. This removal of a major repulsive contribution to the energy barrier might speed up the micellar aggregation rate by orders of magnitude. When the coagulation reaction is occurring faster than the response time for the adjustment of charge and conformation resulting from calcium phosphate precipitation, the aggregation reaction wins and micellar precipitation follows.

The origin of the sigmoidal pH profile was also explained through the effect of pH on calcium phosphate solubility (Horne, 1987). The argument was stated as follows. It is known that on acidification of milk, micellar calcium phosphate is solubilized. It is also known that this calcium phosphate is mainly composed of small crystallites (Lyster *et al.*, 1984). The large exposed surface area of these small crystals would account for the ease of dissolution and, perhaps of greater importance here, of the reprecipitation of calcium phosphate on exposure to ethanol. Increasing pH promotes the calcium phosphate precipitation reaction, enhancing its stabilizing effect by requiring more ethanol to overcome the increasing energy barrier being erected.

Similarly, it was suggested that the observed decrease in the ethanol stability of milk as ionic strength is increased could be accounted for by the inhibition of calcium phosphate precipitation at higher ionic strengths (Horne, 1987), leading to a reduced demand for calcium from caseinate and a consequent decrease in the level of ethanol required to destabilize the system. These proposals are based qualitatively on the effects of ionic strength on mineral solubilities and phosphate pK values in the absence of ethanol. As was pointed out by Pierre (1985), the solubility of salts is also dependent on the dielectric constant of the mixture. The domination of mineral precipitation behaviour would then satisfactorily explain the relation observed by Horne and Parker (1981*c*) between ethanol stability and the dielectric constant of the precipitating mixture.

The proposed role of calcium phosphate precipitation as a stabilizing influence would provide a route by which the composition of the serum phase could exert control over micellar stability, consistent with the relationships found by Donnelly and Horne (1986) and earlier by Davies and White (1958). Similarly, when milk is forewarmed, the resulting precipitation of calcium phosphate would enhance the effectiveness of ethanol-induced mineral precipitation by shifting the stability profile in the direction of decreasing availability of calcium, thereby producing the observed increases in micellar stability (Horne & Parker, 1981*d*).

Finally, what is effectively a competition between mineral precipitation and micellar aggregation reactions for calcium in the system could also explain the anomalous destabilization behaviour of trifluoroethanol (Horne & Davidson, 1987). With this alcohol it was found that after passing through a critical range of trifluoroethanol concentrations producing protein coagulation, higher levels gave rise to micellar dissociation. In the mechanism proposed, this would be the result of the mineral precipitation behaviour proceeding so fast that the micelle disintegrated before the aggregation reaction occurred, so much calcium was being drawn off the casein.

5. Practical Applications

Aside from the suggestions as to how ethanol stability of concentrates can be improved and apart from the increased comprehension of micellar stability and structure derived from observations of ethanol effects on hydrodynamic size, some eminently practical benefits have flowed from ethanol stability studies.

Though not adopted commercially to date, it has been suggested that ethanol destabilization of milk at a slightly lower pH than the natural value would provide a relatively cheap method for milk protein isolation and recovery (Hewedi *et al.*, 1985). Good yields of protein were obtained and the product retained some micelle-like characteristics. In particular, its lower viscosity, in comparison to sodium caseinate, would be advantageous where high protein concentrations were required. It was also apparent that the conformations of the proteins were influenced by the isolation procedures, so that other variations in untested functional properties may be anticipated for these products.

5.1. Cream Liqueurs

The main product to benefit from our increased knowledge of ethanol stability has been the cream liqueur. Introduced in the mid-1970s, this beverage gained immediate consumer acceptance. The growing literature on the product has been reviewed by Banks and Muir (1988). What follows is a précis of how the burgeoning knowledge of ethanol stability of milk was translated into improving what was essentially a cocktail drink into a stable long-life product.

The typical cream liqueur is an oil-in-water emulsion, consisting of casein-stabilized fat droplets in an aqueous dispersion and containing 10–20% (w/w) ethanol, usually with up to 20% added sugar. The early varieties suffered from defects which led to a limited shelf-life. Two major technical problems contributed. In the first, a fat plug formed in the neck of bottles of cream liqueur during storage. Muir and Banks (1986) recognized that this was caused by poor homogenization of the emulsion which could be solved by greater control of conditions and an increase in the severity of homogenization. More recently, Dickinson *et al.* (1989) have suggested that other factors, notably low pH, high calcium content and low concentration of caseinate, all contributing to low emulsion stability, are likely to enhance and aggravate the problems of fat plug formation.

The second major defect in these products manifests itself as complete destabilization of the emulsion. The droplets are stabilized by a coating of casein. Casein micelles could be regarded as protein droplets similarly stabilized by a protein coat, again casein. From studies on the alcohol stability of milk, it was known that ionic calcium exerted a controlling influence on that stability. The cream used to manufacture the liqueur was prepared from milk and contained significant levels of residual soluble milk components, including calcium. Although analysis showed the calcium to be at levels only 1/6th those found in milk, this was found to be sufficiently high to affect the long-term stability of the product liqueur. The solution devised to overcome this calcium-associated instability was to incorporate a calcium sequestrant, such as citrate, into the formulation (Banks *et al.*, 1981). Additions of citrate in the 1–5 mM range were found to ensure a shelf-life of at least 2 years at ambient temperature. Similar increases were obtained by using washed cream or anhydrous milk fat as the fat component, each of these options cleansing the system of residual calcium (Banks & Muir, 1988).

Another aspect of cream liqueurs that bears a strong resemblance to behaviour encountered in ethanol stability studies of milk, particularly with milk concentrates, is the sensitivity of high-proof liqueurs to ionic strength effects. Not only were such high-proof liqueurs sensitive to the presence of calcium, but all ions were found to have a detrimental effect on emulsion stability and product quality (Horne & Muir, 1990). This paralleled effects seen in milks where ionic strength played a dominant role in stable regions in which high concentrations of ethanol were required to induce precipitation (Horne & Parker, 1983). To overcome this instability problem in cream liqueur manufacture, two solutions devised to deal with soluble calcium were found equally applicable, either to remove the ionic constituents of the cream by washing with water or to use anhydrous milk fat.

6. Concluding Remarks

Unravelling the mechanism of the coagulation of casein micelles, by whatever route, is a complex topic. Progress is being made in better defining the casein micelle system and this chapter has summarized the contribution of ethanol stability studies to the understanding of micellar interactions and structure. The subject is not without its controversies, however, and the opinions expressed herein may merely add fuel to the flames.

It is now generally accepted that classical DLVO colloidal stability theory is inadequate to explain casein micelle stability. The simple modification of introducing an additive steric stabilization component has yet to be fully quantitatively tested but qualitative studies already suggest that this might not completely describe the situation. As we have seen in studying the role of pH, calcium and ionic strength in determining the thickness of the stabilizing sheath, electrostatic and steric effects are never wholly independent of one another. The hairs giving rise to the steric effects are charged polyelectrolytes whose electrostatic interaction will influence the steric stabilization contribution to the inter-micellar potential. Conversely, the presence of the steric stabilizing barrier will push the electrostatic double layer further out from the micellar core surface, so that cross-influences will also be observable in the electrostatic contribution. There is, of course, also the possiblity that the attractive side of the stability equation will be influenced by variations in the other components in the equation. Even if one is only estimating a Hamaker constant based on an onion skin model, the removal or collapse of the protein hairs should alter the contribution of the outer layer to the overall attractive potential.

Our understanding of the stability of the casein micelle system is, therefore, by no means complete. We have progressed beyond the simple models and could be described as having arrived at the end of the beginning. To proceed further, having opened up a whole class of problems, previously undefined, will require as much, if not more, input from theoreticians as from experimentalists. This will undoubtedly entail the experimentalist Mahomet scaling the theoretical mountain rather than the theoretician Moses descending with the relevant stability equations engraved on tablets of stone.

In all of the studies of heat, rennet or ethanol stability, the micelle has to be modified in some way before aggregation can proceed. It is not the stability of the original micelle and its response to the reaction conditions that is being measured but rather an extensively modified micelle, stripped of its steric stabilizing sheath in the cases of rennet and ethanol-induced coagulation. The absolute magnitude of the interaction potential energy and how it varies in response to these perturbations is therefore inaccessible to us in the type of experiment carried out to date. Such information may, however, be available from studies of the extent of protein precipitation as a function of ethanol concentration, a form of nitrogen-depletion curve (Horne & Davidson, 1987).

The ultimate goal of all of this work is not, of course, simply understanding why ethanol precipitates casein proteins. Rather it is that, through understanding the casein micelle system, the manufacturer of dairy products will have greater control of the efficiency of his processes and the quality of his output, possibly extending his product range and the uses of milk components.

References

Ashoor, S.H., Sair, R.A., Olson, N.F. & Richardson, T. (1971). Use of a papain superpolymer to elucidate the structure of bovine casein micelles. *Biochim. Biophys. Acta*, **229**, 423–30.

Ayers, S.H. & Johnson, W.T. Jr. (1915). The alcohol test in relation to milk. US Dept. Agric. Bulletin No. 202. Government Printing Office, Washington.

Banks, W. & Muir, D.D. (1988). Stability of alcohol-containing emulsions. In *Advances in Food Emulsions and Foams*, ed. E. Dickinson & G. Stainsby. Elsevier Applied Science, London, pp. 257–83.

Banks, W., Muir, D.D. & Wilson, A.G. (1981). Extension of the shelf-life of cream-based liqueurs at high ambient temperatures. *J. Food Technol.*, **16**, 587–95.

Buchheim, W. & Welsch, U. (1973). Evidence for the submicellar composition of casein micelles on the basis of electron microscopical studies. *Neth. Milk Dairy J.*, **27**, 163–80.

Carroll, R.J. & Farrell, H.M. Jr. (1983). Immunological approach to location of κ-casein in the casein micelle by electron microscopy. *J. Dairy Sci.*, **66**, 679–86.
Carroll, R.J., Farrell, H.M. Jr. & Thompson, M.P. (1971). Electron microscopy of the casein micelle—forces contributing to its integrity. *J. Dairy Sci.*, **54**, 752 (Abst.).
Christenson, H.K. (1986). Interactions between hydrocarbon surfaces in a nonpolar liquid: effect of surface properties on solvation forces. *J. Phys. Chem.*, **90**, 4–6.
Claesson, P.M., Blom, C.E., Herder, P.C. & Ninham, B.W. (1986). Interactions between water-stable hydrophobic Langmuir-Blodgett monolayers on mica. *J. Colloid Interface Sci.*, **114**, 234–42.
Dahlberg, A.O. & Garner, H.S. (1921). The alcohol test as a means of determining quality of milk for condenseries. US Dept. Agric. Bulletin No. 944.
Dalgleish, D.G. (1984). Measurement of electrophoretic mobilities and zeta potentials of particles from milk using laser Doppler electrophoresis. *J. Dairy Res.*, **51**, 425–38.
Dalgleish, D.G. & Parker, T.G. (1980). Binding of calcium ions to bovine α_{s1}-casein and precipitability of the protein-calcium ion complex. *J. Dairy Res.*, **47**, 113–22.
Dalgleish, D.G., Paterson, E. & Horne, D.S. (1981). Kinetics of aggregation of α_{s1}-casein/Ca^{2+} mixtures: charge and temperature effects. *Biophys. Chem.*, **13**, 307–14.
Dalgleish, D.G., Horne, D.S. & Law, A.J.R. (1989). Size-related differences in bovine casein micelles. *Biochim. Biophys. Acta*, **991**, 383–7.
Davies, D.T. & White, J.C.D. (1958). The relation between the chemical composition of milk and the stability of the casein complex. II. Coagulation by ethanol. *J. Dairy Res.*, **25**, 256–66.
Dickinson, E., Narham, S.K. & Stainsby, G. (1989). Factors affecting the properties of cohesive creams formed from cream liqueurs. *J. Sci. Food Agric.*, **48**, 225–34.
Donnelly, W.J. & Horne, D.S. (1986). Relationship between ethanol stability of bovine milk and natural variations in milk composition. *J. Dairy Res.*, **53**, 23–33.
Donnelly, W.J., McNeill, G.P., Buchheim, W. & McGann, T.C.A. (1984). A comprehensive study of the relationship between size and protein composition in natural bovine casein micelles. *Biochim. Biophys. Acta*, **789**, 136–43.
Farrell, H.M. Jr., Pessen, H. & Kumosinski, T.F. (1989). Water interactions with bovine caseins by Hydrogen-2 nuclear magnetic resonance relaxation studies: structural implications. *J. Dairy Sci.*, **72**, 562–74.
Fox, P.F. (1981). Heat-induced changes in milk preceding coagulation. *J. Dairy Sci.*, **64**, 2127–37.
Garnier, J. & Ribadeau-Dumas, B. (1970). Structure of the casein micelle. A proposed model. *J. Dairy Res.*, **37**, 493–504.
Garvey, M.J., Tadros, Th.F. & Vincent, B. (1976). A comparison of the adsorbed layer thickness obtained by several techniques of various molecular weight fractions of poly (vinyl alcohol) on aqueous polystyrene latex particles. *J. Colloid Interface Sci.*, **55**, 440–53.
Green, M.L. & Crutchfield, G. (1971). Density-gradient electrophoresis of native and rennet-treated casein micelles. *J. Dairy Res.*, **38**, 151–64.
Griffin, M.C.A. (1987). Structural studies of casein micelles using photon correlation spectroscopy. *J. Colloid Interface Sci.*, **115**, 499–506.
Griffin, M.C.A. & Roberts, G.C.K. (1985). A ^{1}H-NMR study of casein micelles. *Biochem. J.*, **228**, 273–6.
Hamaker, H.C. (1937). The London-van der Waals attraction between spherical particles. *Physica*, **4**, 1058–72.
Herskovits, T.T., Gadegbeku, B. & Jaillet, H. (1970). On the structural stability and solvent denaturation of proteins, I. Denaturation by the alcohols and glycols. *J. Biol. Chem.*, **245**, 2588–98.
Hewedi, M.M., Mulvihill, D.M. & Fox, P.F. (1985). Recovery of milk protein by ethanol precipitation. *Ir. J. Fd. Sci. Technol.*, **9**, 11–23.

Hilgemann, M. & Jenness, R. (1951). Observations on the effect of heat treatment upon the dissolved calcium and phosphorus in skim milk. *J. Dairy Sci.*, **34**, 483–4.

Holt, C. (1975). The stability of casein micelles. In *Proc. Int. Conf. Colloid Surf. Sci., Budapest*, vol. 1, ed. E. Wolfram. Akademia Kiado, Budapest, pp. 641–4.

Holt, C. & Dalgleish, D.G. (1986). Electrophoresis and hydrodynamic properties of bovine casein micelles interpreted in terms of particles with an outer hairy layer. *J. Colloid Interface Sci.*, **114**, 513–24.

Holt, C., Muir, D.D. & Sweetsur, A.W.M. (1978). The heat stability of milk and concentrated milk containing added aldehydes and sugars. *J. Dairy Res.*, **45**, 47–52.

Horne, D.S. (1982). Calcium-induced precipitation of α_{s1}-casein: effect of inclusion of citrate or phosphate. *J. Dairy Res.*, **49**, 107–18.

Horne, D.S. (1984). Steric effects in the coagulation of casein micelles by ethanol. *Biopolymers*, **23**, 989–93.

Horne, D.S. (1985). The ethanol stability of milk. In *Hannah Research 1984*. Hannah Research Institute, Ayr, pp. 89–100.

Horne, D.S. (1986). Steric stabilization and casein micelle stability. *J. Colloid Interface Sci.*, **111**, 250–60.

Horne, D.S. (1987). Ethanol stability of casein micelles—a hypothesis concerning the role of calcium phosphate. *J. Dairy Res.*, **54**, 389–95.

Horne, D.S. & Dalgleish, D.G. (1980). Electrostatic interaction and the kinetics of protein aggregation: α_{s1}-casein. *Int. J. Biol. Macromol.*, **2**, 154–60.

Horne, D.S. & Dalgleish, D.G. (1985). A photon correlation spectroscopy study of size distributions of casein micelle suspensions. *Eur. Biophys. J.*, **11**, 249–58.

Horne, D.S. & Davidson, C.M. (1986). The effect of environmental conditions on the steric stabilization of casein micelles. *Colloid & Polymer Sci.*, **264**, 727–34.

Horne, D.S. & Davidson, C.M. (1987). Alcohol stability of bovine skim milk. Anomalous effects with trifluoroethanol. *Milchwissenschaft*, **42**, 509–12.

Horne, D.S. & Muir, D.D. (1990). Alcohol and heat stability of milk protein. *J. Dairy Sci.*, **73**, 3613–26.

Horne, D.S. & Parker, T.G. (1980). The pH sensitivity of the ethanol stability of individual cow milks. *Neth. Milk Dairy J.*, **34**, 126–30.

Horne, D.S. & Parker, T.G. (1981*a*). Factors affecting the ethanol stability of bovine milk. I. Effect of serum phase components. *J. Dairy Res.*, **48**, 273–84.

Horne, D.S. & Parker, T.G. (1981*b*). Factors affecting the ethanol stability of bovine milk. II. The origin of the pH transition. *J. Dairy Res.*, **48**, 285–91.

Horne, D.S. & Parker, T.G. (1981*c*). Factors affecting the ethanol stability of bovine casein micelles: 3. Substitution of ethanol by other organic solvents. *Int. J. Biol. Macromol.*, **3**, 399–402.

Horne, D.S. & Parker, T.G. (1981*d*). Factors affecting the ethanol stability of bovine milk. IV. Effect of forewarming. *J. Dairy Res.*, **48**, 405–15.

Horne, D.S. & Parker, T.G. (1982*a*). Some aspects of the ethanol stability of caprine milk. *J. Dairy Res.*, **49**, 459–68.

Horne, D.S. & Parker, T.G. (1982*b*). Factors affecting the ethanol stability of bovine milk. V. Effects of chemical modification of milk protein. *J. Dairy Res.*, **49**, 449–57.

Horne, D.S. & Parker, T.G. (1983). Factors affecting the ethanol stability of bovine milk. VI. Effect of concentration. *J. Dairy Res.*, **50**, 425–32.

Horne, D.S., Parker, T.G., Donnelly, W.J. & Davies, D.T. (1986). Factors affecting the ethanol stability of bovine skim milk. VII. Lactational and compositional effects. *J. Dairy Res.*, **53**, 407–17.

Horne, D.S., Parker, T.G. & Dalgleish, D.G. (1989). Casein micelles, polycondensation and fractals. In *Food Colloids*, ed. R.D. Bee, P. Richmond & J. Mingins. Special Publication No. 75, Royal Society of Chemistry, London, pp. 400–5.

Israelachvili, J.N. (1987). Solvation forces and liquid structure, as probed by direct force measurements. *Acc. Chem. Res.*, **20**, 415–21.

Israelachvili, J.N. & McGuiggan, P.M. (1988). Forces between surfaces in liquids. *Science*, **241**, 795–800.

Israelachvili, J.N. & Pashley, R.M. (1982). The hydrophobic interaction is long-range decaying exponentially with distance. *Nature*, **300**, 341–2.

Israelachvili, J.N. & Pashley, R.M. (1983). Molecular layering of water at surfaces and origin of repulsive hydration forces. *Nature*, **306**, 249–50.

Israelachvili, J.N. & Pashley, R.M. (1984). Measurement of the hydrophobic interaction between two hydrophobic surfaces in aqueous electrolyte solutions. *J. Colloid Interface Sci.*, **98**, 500–14.

Jenness, R. & Patton, S. (1959). *Principles of Dairy Chemistry*, Wiley, New York.

Kuhn, N.J. & White, A. (1977). The role of nucleoside diphosphatase in a uridine nucleotide cycle associated with lactose synthesis in rat mammary-gland Golgi apparatus. *Biochem. J.*, **168**, 423–33.

Kumosinski, T.F., Pessen, H., Prestrelski, J.T. & Farrell, H.M. Jr. (1987). Water interactions with varying molecular states of bovine casein: ^{2}H-NMR relaxation studies. *Arch. Biochem. Biophys.*, **257**, 259–68.

Leighton, A., Benton, A.G., Palmer, L.S. & Matheson, K. (1928). Coagulation of milk. In *Fundamentals of Dairy Science*, ed. Associates of L.A. Rogers. Chemical Catalog Co., New York, pp. 188–222.

Lifshitz, E.M. (1956). The theory of molecular attraction forces between solid bodies. *Soviet Phys., JETP (Eng. Trans.)*, **2**, 73–83.

Lyster, R.L.J., Mann, S., Parker, S.B. & Williams, R.J.P. (1984). Nature of micellar calcium phosphate in cows' milk as studied by high-resolution electron microscopy. *Biochim. Biophys. Acta*, **801**, 315–17.

McGann, T.C.A., Donnelly, W.J., Kearney, R.D. & Buchheim, W. (1980). Composition and size distribution of bovine casein micelles. *Biochim. Biophys. Acta*, **630**, 261–70.

Miller, P.G. & Sommer, H.H. (1940). The coagulation temperature of milk as affected by pH, salts, evaporation and previous heat treatment. *J. Dairy Sci.*, **23**, 405–21.

Mitamura, K. (1937). Studies on the alcohol coagulation of fresh cow milk. *J. Fac. Agric. Hokkaido Univ.*, **41**, 97–362.

Muir, D.D. & Banks, W. (1986). Multiple homogenization of cream liqueurs. *J. Food Technol.*, **21**, 229–32.

Napper, D.H. (1983). *Polymeric Stabilization of Colloidal Dispersions*. Academic Press, New York.

Padmos, W.H. (1930). On the value of the alcohol test for the appraisal of milk. *J. Dairy Res.*, **1**, 201 (Abst.).

Pashley, R.M. & Israelachvili (1984). Molecular layering of water in thin films between mica surfaces and its relation to hydration forces. *J. Colloid Interface Sci.*, **101**, 511–23.

Pashley, R.M., McGuiggan, P.M., Ninham, B.W. & Evans, D.F. (1985). Attractive forces between uncharged hydrophobic surfaces: direct measurements in aqueous solutions. *Science*, **229**, 1088–9.

Payens, T.A.J. (1977). On enzymic clotting processes. II. The colloidal instability of chymosin-treated casein micelles. *Biophys. Chem.*, **6**, 263–70.

Payens, T.A.J. (1979). Casein micelles: the colloid-chemical approach. *J. Dairy Res.*, **46**, 291–306.

Pepper, L. & Farrell, H.M. Jr. (1982). Interactions leading to formation of casein submicelles. *J. Dairy Sci.*, **65**, 2259–66.

Pessen, H., Kumosinski, T.F. & Farrell, H.M. Jr. (1989). Small-angle X-ray scattering investigation of the micellar and submicellar forms of bovine casein. *J. Dairy Res.*, **56**, 443–51.

Pierre, A. (1985). Milk coagulation by alcohol. Studies on the solubility of the milk calcium and phosphate in alcoholic solutions. *Le Lait*, **65**, 201–12.

Post, C.B. & Zimm, B.H. (1982). Theory of DNA condensation: Collapse versus aggregation. *Biopolymers*, **21**, 2123–37.

Ramsdell, G.A. & Hufnagel, C.F. (1953). Effect of heat on the physical and chemical composition of skim milk. *Proc. XIII Int. Dairy Congr. (The Hague)*, **3**, 1025–31.
Ramsdell, G.A., Johnson, Jr., W.T. & Evans, F.R. (1931). A test for the detection of milk unstable to heat. *J. Dairy Sci.*, **14**, 93–106.
Rollema, H.S. & Brinkhuis, J.A. (1989). A ^{1}H-NMR study of bovine casein micelles; influence of pH, temperature and calcium ions on micellar structure. *J. Dairy Res.*, **56**, 417–25.
Rollema, H.S., Brinkhuis, J.A. & Vreeman, H.J. (1988). ^{1}H-NMR studies of bovine κ-casein and casein micelles. *Neth. Milk Dairy J.*, **42**, 233–48.
Rose, D. (1961*a*). Variations in the heat stability and composition of milk from individual cows during lactation. *J. Dairy Sci.*, **44**, 430–41.
Rose, D. (1961*b*). Factors affecting the pH-sensitivity of the heat stability of milk from individual cows. *J. Dairy Sci.*, **44**, 1405–13.
Rose, D. & Tessier, H. (1959). Composition of ultrafiltrates from milk heated at 80 to 230°F in relation to heat stability. *J. Dairy Sci.*, **42**, 969–80.
Saito, Z. (1973). Electron microscopic and compositional studies of casein micelles. *Neth. Milk Dairy J.*, **27**, 143–62.
Samuelsson, E.-G., Gynning, K. & Olsson, H. (1962). Sediment in sterile milk manufactured by the steam injection system. *Proc. XVI Int. Dairy Congr. (Copenhagen)*, **A**, 685–96.
Scheutjens, J.M.H.M. & Fleer, G.J. (1985). Interactions between two adsorbed polymer layers. *Macromolecules*, **18**, 1882–900.
Schmidt, D.G. (1980). Colloidal aspects of casein. *Neth. Milk Dairy J.*, **34**, 42–64.
Schmidt, D.G. (1982). Association of caseins and casein micelle structure. In *Developments in Dairy Chemistry. 1. Proteins*, ed. P.F. Fox. Elsevier Applied Science, London, pp. 61–86.
Schmidt, D.G. & Both, P. (1982). Location of α_{s1}-, β- and κ-casein in artificial casein micelles. *Milchwissenschaft*, **37**, 336–7.
Schmidt, D.G. & Buchheim, W. (1970). An electron-microscopical investigation of the sub-structure of the casein micelles in cow's milk. *Milchwissenschaft*, **25**, 596–600.
Schmidt, D.G., Walstra, P. & Buchheim, W. (1973). The size distribution of casein micelles in cow's milk. *Neth. Milk Dairy J.*, **27**, 128–42.
Scott-Blair, G.W. (1953). *Foodstuffs: Their Plasticity, Fluidity and Consistency*. North Holland Publishing Co., Amsterdam.
Shimmin, P.D. & Hill, R.D. (1964). An electron microscope study of the internal structure of casein micelles. *J. Dairy Res.*, **31**, 121–3.
Slattery, C.W. (1976). Casein micelle structure; An examination of models. *J. Dairy Sci*, **59**, 1547–56.
Slattery, C.W. (1977). Model calculations of casein micelle size distributions. *Biophys. Chem.*, **6**, 59–64.
Slattery, C.W. & Evard, R. (1973). A model for the formation and structure of casein micelles from subunits of variable composition. *Biochim. Biophys. Acta*, **317**, 529–38.
Sommer, H.H. & Binney, T.H. (1923). A study of the factors that influence the coagulation of milk in the alcohol test. *J. Dairy Sci.*, **6**, 176–97.
Sommer, H.H. & Hart, E.B. (1919). The heat coagulation of milk. *J. Biol. Chem.*, **40**, 137–51.
Stothart, P.H. & Cebula, D.J. (1982). Small-angle neutron scattering study of bovine casein micelles and submicelles. *J. Mol. Biol.*, **160**, 391–5.
Unnikrishnan, V., Bhavadasan, M.K. & Rama Murthy, M.K. (1988). Alcohol stability of buffalo milk. *Ind. J. Dairy Sci.*, **41**, 421–6.
Verwey, E.J.W. & Overbeek, J.Th.G. (1948). *Theory of Stability of Lyophobic Colloids*. Elsevier, Amsterdam.
Walstra, P. (1979). The voluminosity of bovine casein micelles and some of its implications. *J. Dairy Res.*, **46**, 317–23.
Walstra, P. (1990). On the stability of casein micelles. *J. Dairy Sci.*, **73**, 1965–79.

Walstra, P., Bloomfield, V.A., Wei, G.J. & Jenness, R. (1981). Effects of chymosin action on the hydrodynamic diameter of casein micelles. *Biochim. Biophys. Acta*, **669**, 258–9.

White, J.C.D. & Davies, D.T. (1956). The relation between the rate of heat coagulation of milk and its protein and mineral compositon. *Proc. XIV Int. Dairy Congr. (Rome)*, **1**(II), 288–301.

White, J.C.D. & Davies, D.T. (1958). The relation between the chemical composition of milk and the stability of the caseinate complex. I. General introduction, description of samples, methods and chemical composition of samples. *J. Dairy Res.*, **25**, 236–55.

Zadow, J.G., Hardham, J.F., Kocak, H.R. & Mayes, J.J. (1983). The stability of goat's milk to UHT processing. *Aust. J. Dairy Technol.*, **38**, 20–3.

17

AGE GELATION OF STERILIZED MILKS

V.R. HARWALKAR
Centre for Food and Animal Research, Agriculture Canada, Ottawa, Canada K1A 0C6

1. INTRODUCTION

The importance of milk in human nutrition has been recognized for a long time. Raw milk is a highly perishable product and could be potentially unsafe for consumption without further treatment. In addition to being a potential source of pathogenic bacteria, unprocessed raw milk is also susceptible to rapid spoilage by the action of naturally occurring enzymes and contaminating microorganisms. These undesirable attributes can be controlled by the age-old practice of boiling milk shortly after milking and before consumption. This practice is still prevalent in many parts of the world where regional handling and distribution of milk are done on a small scale and where refrigeration is not commonly available. Pasteurization, which entails milder heat treatment than boiling, ensures destruction of pathogenic bacteria and inactivation of some undesirable enzymes. This treatment is more suitable for handling and distribution of larger quantities of milk over wider areas but refrigeration and a good transportation system are essential if the product is to be stored for significant length of time (up to two weeks) before distribution. Such facilities are not readily available in countries that are technologically less developed.

In industrialized countries, especially those producing milk far in excess of their requirements, there is an increasing trend to process milk in a limited number of centralized dairy plants. The distribution of milk to wider and remote areas, particularly where refrigeration is not available, necessitates the prolongation of the shelf life of the milk beyond the limits afforded by pasteurization.

Several methods are effective in preserving milk for long periods, for example, freezing milk (or concentrates), addition of sucrose to concentrated milk (condensed milk) and sterilization.

(i) Milk preserved by freezing retains good flavour but exhibits gradual loss of stability during storage (Doan & Warren, 1947; Leviton *et al.*, 1962; Muir, 1984*a*). The feasible application of freezing as a method of preservation depends upon the availability of refrigeration facilities and economic factors.

(ii) Sweetened condensed milk does not require refrigeration and therefore is still a favoured product in many tropical countries (Hunziker, 1944; Webb, 1970). Its composition restricts its use mainly to that of an ingredient in other foods, but rapid changes in colour and flavour result from browning reactions because the product is stored at room temperature. This product also suffers from another serious defect, namely, gelation during storage (Leviton *et al.*, 1962; Parry, 1974).

(iii) Unlike the above preservation methods, in which microbial growth is suppressed, sterilizing treatments destroy microorganisms that can proliferate. Although microbiological sterility is the primary objective of the sterilizing treatment, the accompanying physical and chemical changes are also of great importance because they affect quality (flavour and nutritional) and shelf life of the product (Burton, 1984, 1988). The shelf life is considered in terms of retention of acceptable quality and stability of the product during storage.

Product stability is a critical aspect of quality during post-processing storage and was discussed by Harwalkar (1982). Specific aspects of the stability of sterilized milk products were emphasized. Unsterilized long-life products were discussed briefly for the purpose of comparing changes and mechanisms involved in storage stability. The present chapter has been revised to include major studies on the subject, particularly relating to the mechanisms of age gelation, since the last review (Harwalkar, 1982).

2. Method of Sterilization

Milk can be sterilized by various techniques such as ionizing radiation, chemical treatment or heating. Sterilization by methods other than heating is only of theoretical interest and needs further research to be of any practical use.

2.1. Ionizing Radiation

Ionizing radiation destroys bacteria without generating heat but causes several undesirable changes that result in off-flavours and physical instability of the product during storage (Hoff *et al.*, 1958, 1960*a*,*b*; Edmondson, 1970). Radiation is less effective than heat treatment in inactivating enzymes (Goldblith, 1967) and additional treatment is required to inactivate enzymes. Effects of radiation on the properties of milk are not yet fully understood. A better understanding of the effects of radiation on the physicochemical changes in milk is necessary before its practical application materializes.

2.2. Chemical Treatment

Hydrogen peroxide and antibiotics have been explored as sterilizing agents. By themselves, these agents are ineffective in sterilizing milk but can reduce the requirements of heat treatment in achieving commercial sterility (Edmondson, 1970).

2.3. Sterilization by Heat

Effective sterilization of fluid milk products can be achieved by suitable combinations of time and temperature of heating. The main considerations in the selection of heating conditions are optimal destruction of microorganisms, particularly resistant spores, inactivation of deleterious enzymes and retention of desirable attributes of quality in terms of colour, flavour, stability during storage and nutrition. Destruction of spores is exponentially related to the temperature of heating; sterilization efficiency (SE) is expressed by the formula (Galesloot, 1956; Burton, 1969).

$$\mathrm{SE} = \log \frac{\text{Number of spores before sterilization}}{\text{Number of spores after sterilization}}$$

One needs only to consider 'commercial sterility' in which the proportion of microorganisms either surviving the sterilization treatment or entering as post-sterilization contaminants is so negligibly small that it poses no public health hazard. Complete sterilization is theoretically possible at higher temperatures but the adverse flavour changes in the product outweigh the beneficial sporocidal efficiency and make such treatments impractical.

In practice, heat sterilization of fluid milk falls mainly into two categories.

(i) Retort sterilization, in which the product in sealed containers is heated at 110–120°C for 5–20 min. Commercial production and marketing of evaporated milk processed by this method is a long-established practice all over the world (Goldblith, 1967; Edmondson, 1970; Graf & Bauer, 1976; Ashton, 1979).

(ii) Continuous flow sterilization is achieved in heat exchangers followed by cooling and aseptic packaging. Sterilization temperatures range from 130 to 150°C with a holding period of at least 1 s. Continuous flow sterilization is generally referred to as 'ultra-high temperature' (UHT) or 'high-temperature short-term' (HTST) sterilization. In earlier work in the USA, temperatures close to or slightly below 130°C were used to produce HTST sterilized evaporated milk (Ellertson & Pearce, 1964; Burton, 1969), but the trend in the USA and Canada is to accept and adopt the European view of UHT treatment using temperatures ranging from 135 to 150°C. UHT sterilization of milk is achieved either by direct contact with steam or by indirect heating in heat exchangers. Most of the developmental work on UHT sterilized milk was carried out in Europe, whereas most of the work on HTST sterilization of concentrated milk was done in the USA.

The essential steps in UHT processing of milk can be described broadly as follows. Raw milk is first preheated to 70–80°C for a short period, sterilized, subsequently cooled, homogenized and aseptically packaged.

UHT or HTST sterilized concentrated milks are prepared by first forewarming the milk in a vat (90–100°C for up to 10 min) or by UHT methods (130–150°C for 1–5 s), then sterilized by direct or indirect methods (130–150°C for 1–5s), cooled, homogenized and packaged aseptically. Homogenization may precede sterilization. The details of UHT processing conditions, types of equipment used in preheating and sterilization, cooling, homogenization and aseptic packaging are adequately described elsewhere (Ashton, 1965; Burton, 1965, 1969; Edmondson, 1970; Anon., 1972; Seehafer, 1972; Ashton, 1979; Mehta, 1980; Muir, 1984*b*) and are beyond the scope of this chapter. The production and marketing of UHT sterilized milk or partially skimmed milk products is now practised in Europe, USA and Canada on an increasing scale (Ashton, 1979; Mehta, 1980). The commercial production of concentrated milk sterilized by the UHT or HTST methods is still not widespread (Ellertson & Pearce, 1964; Edmondson, 1970).

Both the retort and UHT methods of sterilization are applicable to unconcentrated, as well as concentrated, milk and skim milk. Technological, manufacturing, marketing, economical and nutritional aspects of these products have been extensively reviewed (Goldblith, 1967; Edmondson, 1970; Anon., 1972).

3. Characteristics of Retort and UHT Sterilized Milk

Fluid milk products sterilized by the retort and UHT methods vary considerably in their initial quality and storage behaviour.

Retort sterilized milk or concentrated milk, generally, suffers from browning discoloration, cooked or caramelized flavour and loss of some nutrients (Ashton, 1979). The product is not suitable as beverage milk but may be used as coffee milk or as an ingredient in food formulations (Edmondson, 1970; Graf & Bauer, 1976). Provided that they are processed properly, retort sterilized products remain stable for long periods of storage without refrigeration; only slight thinning has been observed (Beeby *et al.*, 1971; Parry, 1974) (Table 1).

In contrast to retort sterilized products, UHT sterilized products prepared by heating in the range 135–150°C for a few seconds have the advantage of achieving an equal or better sporicidal effect without adversely affecting the colour, flavour or nutritional qualities. This is possible because at higher temperatures the rate of spore destruction is considerably faster than the rate of most chemical changes (Burton, 1965, 1969; Edmondson, 1970; Mehta, 1980). For example, the ratio of the bacteriocidal action to the browning reaction increases exponentially above 135°C and varies from 1000 to 5000 in the range of 135–150°C (Burton, 1965; Edmondson, 1970). However, with UHT sterilized products, loss of stability (in terms of sedimentation and gelation) during storage is a serious problem. Undesirable flavour changes during storage are no less important, but these can be modified by selecting proper processing conditions. In UHT processed milk,

TABLE 1
Effect of Processing and Storage Conditions upon the Shelf Life (Gelation-Free)[a]

Process	*Temperature or energy input*	*Time*	*Shelf life (incidence of gelation)*	
			At 4°C	*At 20°C*
Milk				
Pasteurization	62°C 72°C	30 min 16 s	5–15 days	1 day
Sterilization				
Direct	130–150°C	1–5 s	4–12 months	3–6 months
Indirect			> 12 months	9–12 months
Autoclaved	117–120°C	10–20 min	> 12 months	> 12 months
Ionizing radiation	$1{\cdot}8 \times 10^{-6}$ rad	—	> 2 months	< 1 month
Concentrated milk (26% TS)				
Sterilization				
HTST or UHT	130–150°C	1–60 s	> 6 months	2–6 months
Autoclave	117–120°C	10–20 min	> 12 months	< 12 months
Ionizing radiation	$1{\cdot}8 \times 10^{-6}$ rad	—	2–5 days	10–15 days

[a]Taken from Harwalkar (1982).

gelation is the most important single problem because it signifies the final limit of shelf life (Burton, 1969; Edmondson, 1970). The problem of gelation has been the major hindrance to the widespread commercial exploitation of the UHT or HTST methods of sterilizing concentrated milk (Edmondson, 1970; Seehafer, 1972).

Milks sterilized by different methods vary considerably in stability during storage. Approximate shelf life, as indicated by the gelation-free periods during storage, is shown in Table 1 for some heated and sterilized milk products. Although not produced commercially, radiation-sterilized samples are also included because, similarly to the UHT treated samples, they may also gel during storage. Under normal conditions, gelation is not a problem with retort sterilized samples, but under some conditions, retort sterilized, evaporated milk does undergo gelation during storage (Heintzberger *et al.*, 1972; Harwalkar *et al.*, 1983).

4. STORAGE STABILITY: GELATION

Loss of stability in sterilized milk during storage is manifested by the separation of fat, sedimentation of protein and gelation (Ellertson & Pearce, 1964; Edmondson, 1970; Hostettler, 1972; Seehafer, 1972).

Fat separation in sterilized milk (or concentrated milk) is not a major problem since it can be controlled by proper heat treatment, homogenization and control of the initial viscosity (Swanson *et al.*, 1965; Hostettler, 1972; Seehafer, 1972). Likewise, sedimentation, which results in separation and settling of solids to the bottom of the container, is also controlled by adjusting processing conditions

(Ellertson & Pearce, 1964; Zadow, 1971; Hostettler, 1972; Seehafer, 1972). Homogenization following sterilization, for example, minimizes the sedimentation problem (Leviton *et al.*, 1963; Hostettler, 1972). Except in extreme instances, where the solids sediment and leave a clear liquid at the top, sedimentation is not as serious a problem as gelation. Unlike sedimentation, which indicates settling or precipitation of colloidal particles, gelation is characterized by loss of fluidity of the product as a result of changes during storage. Gelation has been described by terms such as coagulation (Samuelsson & Holm, 1966; Snoeren *et al.*, 1979*b*), sweet curd formation (van Kreveld, 1950), thixotropic gel formation (Swanson *et al.*, 1965), age thickening (Morr, 1969; Hostettler, 1972; Seehafer, 1972; Schmidt, 1980), partial gelation (Samuelsson & Holm, 1966; Hostettner, 1972) or lumpiness (Tarassuk & Tamasma, 1956).

Gelation is strictly a storage defect and should not be confused with thickening during sterilization nor with coagulation observed during incubation of certain evaporated milks at 55°C for 8–10 days. During storage, UHT sterilized milk (or concentrated milk) undergoes changes in viscosity; in the initial stages, thinning of the product occurs, followed by a fairly long period during which very little change in viscosity is observed, terminated by a period during which there is a sudden rise in viscosity, culminating in gel formation within a short period (1–3 weeks). Samples that are viscous or are at an early stage of gelation can be easily dispersed to a thin consistency by agitation. However, these samples revert to the viscous state within 24–48 h (Ellertson & Pearce, 1964; Harwalkar & Vreeman, 1978*a*). When gelation occurs, the product exhibits a custard-like consistency and the process is irreversible at this stage. The gels are characterized, in most instances, by the absence of syneresis (Tarassuk & Tamasma, 1956) but on further storage syneresis does occur in gelled products (Harwalkar & Vreeman, 1978*a*,*b*).

Changes in viscosity and gelation during storage were first recognized by Deysher *et al.* (1944) and Bell *et al.* (1944) during their early work on sterile concentrated milk prepared by HTST processes. These workers observed that evaporated milk (26% TS) sterilized by the HTST method (135°C for 30 s) gelled within 6–16 weeks of storage at 30°C. Subsequently, these findings were confirmed by numerous workers who observed similar viscosity changes during storage of UHT sterilized milk (Samuelsson & Holm, 1966; Aoki & Imamura, 1974b; Blanc *et al.*, 1980), concentrated milk (Tarassuk & Tamasma, 1956; Leviton & Pallansch, 1961*a*,*b*; Ellertson & Pearce, 1964; Snoeren *et al.*, 1976) and concentrated skim milk (Harwalkar & Vreeman, 1978*a*).

A typical viscosity changes during storage of HTST sterilized evaporated milk is illustrated in Fig. 1. The sample sterilized at 117°C (243°F) for 13 min showed only the initial thinning but no increase in viscosity for a long period during storage. Many studies carried out on gelation, particularly in unconcentrated milks, do not report viscosity data, in the absence of which it is difficult to compare results from different laboratories since it is not certain whether the gelation processes decribed by various authors are similar. Description of storage destabilization by such terms as 'wheying off' or 'clearance' is indicative of sedimentation rather than gelation (Aoki & Imamura, 1974*a*; Snoeren *et al.*, 1979*b*). A method suggested by

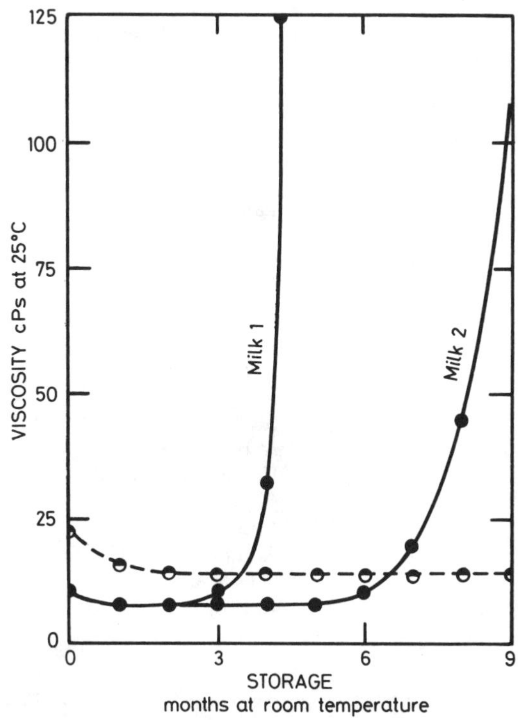

Fig. 1. Viscosity of sterilized evaporated milk (26% TS). ●, Milk samples 1 and Milk 2 sterilized by UHT at 130°C for 1 min; ⊖, milk sample sterilized at 117° for 13 min. Data from Tarrasuk & Tamasma (1956).

Kohlmann *et al.* (1988) measures gelation by pouring off the fluid portion and weighing the non-fluid gel. This gives a quantitative measure of protein aggregation but qualitatively may not be indicative of the processes in which gelation is manifested by total loss of fluidity.

5. Factors Affecting Gelation

The problem of gelation in UHT sterilized milk is less critical than with UHT sterilized concentrated milk; in fact, the gelation problem in the former was recognised much later than in the latter. For this reason, storage gelation of UHT sterilized concentrated milk has been studied more extensively than of sterilized milk (Burton, 1969; Graf & Bauer, 1976), although in recent years an increased concern regarding the gelation of sterilized milk is evident in the literature (Mehta, 1980; Schmidt, 1980; de Koning *et al.*, 1985; Manji *et al.*, 1986). The time at which gelation occurs during the storage of UHT sterilized concentrated and unconcentrated milks is influenced by many variables (Table 2), some of which have been studied extensively. Although most of the factors indicated are applicable to both

TABLE 2
Factors Affecting Gelation of UHT Sterilized Milk or Concentrated Milk

	References
Processing conditions	
Forewarming	Bell *et al.*, 1944; Deysher *et al.*, 1944; Tarassuk & Tamasma, 1956; Leviton & Pallansch, 1961*b*, 1962; Seehafer *et al.*, 1962, Leviton *et al.*, 1963; Ellertson & Pearce, 1964; United States Steel Corporation, 1965; Wheelock & Hindle, 1971; Morr, 1975.
Time and temperature of sterilization	Ellerton & Pearce, 1964; United States Steel Corporation, 1965; Samuelsson & Holm, 1966; Alfa Laval, 1967; Hostettler, *et al.*, 1968; Burton, 1969; Corradini, 1971; Seehafer, 1972; Zadow & Chituta, 1975; Graf & Bauer, 1976.
Direct or indirect UHT treatment	Corradini, 1967; Burton, 1969; Lyster *et al.*, 1971; Perkin *et al.*, 1973; Blanc *et al.*, 1980; Manji *et al.*, 1986.
Homogenization	Edmondson, 1959; Leviton *et al.*, 1963; Ellertson & Pearce, 1964; Snoeren *et al.*, 1976.
Sequence of operations	Leviton *et al.*, 1963; Hostettler, 1972.
Composition of milk	
Total solids	Tarassuk & Tamasma, 1956; Calbert & Swanson, 1959; Leviton *et al.*, 1963; Ellertson & Pearce, 1964; Parry, 1974; Pyne, 1962; Ellertson & Pearce, 1964; Aoki & Imamura, 1974*a,b*; Zadow & Chituta, 1975.
Protein composition	Beeby & Loftus Hills, 1962; Ellertson & Pearce, 1964.
Seasonal variation	Ellertson & Pearce, 1964; Zadow & Chituta, 1975; Graf & Bauer, 1976.
Quality of milk	Swartling, 1968; Zadow & Chituta, 1975; Graf & Bauer, 1976.
Additives	
Mono- and di-sodium phosphate	Edmondson, 1959; Leviton & Pallansch, 1961*a*, Leviton *et al.*, 1962, 1963; Swanson *et al.*, 1965; Seehafer 1972, Harwalkar & Vreeman, 1978*a*; Snoeren *et al.*, 1979*b*.
Citrate	Samuelsson & Holm, 1966.
Carbonate	Samuelsson & Holm, 1966; Seehafer, 1972.
Polyphosphate	Wilson *et al.*, 1961; Leviton *et al.*, 1962; Schmidt, 1968; Seehafer, 1972; Harwalkar & Vreeman, 1978*a*; Snoeren *et al.*, 1979*b*.
Polyhydric alcohols	Leviton *et al.*, 1962.
Minerals (Ca, Mn, etc.)	Deysher *et al.*, 1944; Leviton & Pallansch, 1961*a*; Leviton *et al.*, 1962; Ellertson & Pearce, 1964; Samuelsson & Holm, 1966.

TABLE 2
Continued

	References
Oxidizing or reducing agents H_2O_2, antioxidants	Nakai *et al.*, 1965*a*; Harwalkar & Vreeman, 1978*b*; Harwalkar *et al.*, 1983.
Phosphatides	Leviton *et al.*, 1962.
Sulphydryl reagents	Nakai *et al.*, 1965*a*,*b*.
Proteolytic enzymes and inhibitors	de Koning *et al.*, 1985; Kohlmann *et al.*, 1988; Manji & Kakuda, 1988.
Temperature of storage	Hostettler *et al.*, 1957; Leviton & Pallansch, 1961*b*; Leviton *et al.*, 1962, 1963; Ellertson & Pearce, 1964; Carroll *et al.*, 1971; Corradini, 1971; Samel *et al.*, 1971; Aoki & Imamura, 1974*a*; Adams *et al.*, 1975; Zadow & Chituta, 1975; Andrews *et al.*, 1977; Farah, 1979*a*,*b*; Blanc *et al.*, 1980; Manji *et al.*, 1986.

unconcentrated and concentrated sterilized products, the extent and specific application vary somewhat.

5.1. Heat Treatment

The major factor contributing to storage gelation in unconcentrated milk is the severity of the heat treatment to which the milk is subjected. It is now well recognised that retort sterilized milk (115–120°C for 15–20 min) remains resistant to gelation for long periods of storage, whereas UHT sterilized milk gels during storage. The period of gelation was found to depend upon the time and temperature of the heat treatment either before or during sterilization. Samuelsson and Holm (1966) observed that sterilization by heating at 152°C gave a milk with a longer gelation-free shelf life than sterilization by heating at 142°C; an increase in the time of heating from 6 s to 12 s increased resistance to gelation. Zadow and Chituta (1975) observed a moderate delay of the gelation time by pre-heating at 72°C for 30 s or at 80°C for 30 min and by increasing the sterilization temperature and the holding time, as illustrated in Table 3. Similar increased protection against gelation by increasing the severity of sterilization was observed by others as referenced in Table 2. The protection against gelation has been correlated to the extent of whey protein denaturation (Manji & Kakuda *et al.*, 1988).

Gelation is also influenced by the UHT sterilizing method (Corradini, 1967; Corradini, 1971; Blanc *et al.*, 1980). Sterilization by direct heating methods has been shown to offer less protection against gelation during storage than sterilization by indirect methods. This is believed to result from the inherent difficulty in controlling the severity of heat treatment in the indirect method. The wide variation in the length of the gelation time, ranging from less than 3 months to over 12 months, as observed by many investigators (Alfa Laval, 1967; Hostettler, 1972; Andrews, 1975; Corradini, 1975; Zadow & Chituta, 1975; Harwalkar & Vreeman, 1978*a*;

TABLE 3
Effect of Preheating, Sterilization and Storage Temperatures on the Gelation Time of UHT Milk[a]

Pretreatment	*UHT processing conditions*		*Storage temperature (°C)*	*Time of gelation (days)*
	Temperature (°C)	*Time (s)*		
Nil	140	3	30	96–99
72°C/30 s	140	3	30	110–113
80°C/30 min	140	3	30	117–120
	135	3	30	96–99
	140	2	30	96–99
	140	3	30	96–99
	140	5	30	110–113
	140	3	2	> 208
	140	3	15	130–133
	140	3	20	130–133
	140	3	25	96–99
	140	3	30	96–99
	140	3	35	110–113
	140	3	40	> 208

[a]Data from Zadow & Chituta (1975).

Farah, 1979*a*; Snoeren *et al.*, 1979*b*), could be attributed, at least in part, to the different methods of sterilization.

In concentrated milk, heat treatment, generally, affects gelation in a manner similar to that in milk. However, forewarming, a beneficial step in improving stability of concentrates during sterilization in containers by retort methods, plays a very important role in the gelation phenomenon of HTST sterilized, concentrated milk. HTST methods do not have the stability problem during sterilization because of the short holding periods at high temperatures. Forewarming treatments at a higher temperature and for longer periods delay gelation but sedimentation defects and flavour deterioration resulting from excessive heat treatment limit its use (Swanson *et al.*, 1965; Burton, 1969; Mehta, 1980). A typical beneficial effect of forewarming is illustrated in Table 4.

Sterilization (HTST) of concentrated milk at higher temperatures and longer holding times also retards gelation. However, at equivalent sterilizing effectiveness, higher sterilization temperatures with shorter exposure times result in reduced resistance to gelation (Ellertson & Pearce, 1964; Swanson *et al.*, 1965; Alfa Laval, 1967). Heat treatments in excess of those required for sterilization are needed to achieve greater stability against gelation. For example, Bell *et al.* (1944) observed that a 2:1 milk concentrate (26% TS), processed at 135°C for 30 s and subsequently heated at 115°C for 2, 6, 12 and 18 min, started to thicken after 32, 36, 41 and 51 weeks of storage, respectively.

TABLE 4
Effect of Forewarming Treatment of Milk on the Time Required to Thicken Concentrated Milk Sterilized at 135°C for 30 s[a]

Total solids (%)	*Forewarming*		*Time required to thicken during storage at 30°C (weeks)*
	Temperature (°C)	*Time (min)*	
26	65	10	16
26	120	4	24
32·5	65	10	10
32·5	120	4	13

[a]Data derived from Bell *et al.* (1944).

5.2. Homogenization and Sequence of Operation

Homogenization, used to control fat separation and sedimentation in unconcentrated milks processed by UHT or HTST methods, also influences their gelation behaviour. The placement of homogenization, as well as other processing steps, in the sequence of operations is an important factor in the storage stability of UHT treated products. There are fewer processing steps in UHT treatment of unconcentrated than of concentrated milk; consequently, the sequence of carrying out the different steps has considerable influence on the storage stability of concentrated milk. Placing the homogenization step before concentration and sterilization gives a product with reduced stability against gelation (Leviton *et al.*, 1963). Concentration before sterilization gives a product more stable to gelation (Leviton *et al.*, 1963). Typical effects of changing the sequences of the processing steps are illustrated in Table 5. Factors that prevent gelation often promote sedimentation during storage; an inverse relationship exists between susceptibility to sedimentation and susceptibility to gelation (Calbert & Swanson, 1959; Seehafer, *et al.*, 1962; Burton, 1969; Mehta, 1980). Application of high pressure homogenization prior to in-can sterilization promoted age-gelation in an infant formula (Polliot *et al.*, 1990). The high pressure homogenization is believed to cause increased adsorption of protein on fat surfaces. The modified surface, in combination with changes in mineral balance, is believed to make it more susceptible to aggregation and gelation.

Another processing step that significantly influences the gelation behaviour is holding the sterile concentrate at 94°C to allow development of a critical viscosity. The incipient coagulum is then destroyed by homogenization under aseptic conditions (Giroux *et al.*, 1958; Calbert & Swanson, 1959). This treatment has been shown to retard gelation during storage. However, the problem is one of determining the optimum viscosity at which homogenization is carried out. Homogenization of an excessively thick product results in graininess and sedimentation whereas homogenization below the optimum level of thickening results in gelation.

TABLE 5
Effect of Sequence of Operations and Additives on the Shelf Life of HTST Sterilized Milk at Different Storage[a] Temperatures

Storage temperature (°C)	*Additive[b]*	*Storage life (days)*	
		FCASH[c]	*FHASC[d]*
21	Control	45	7
21	Monophosphate	16	3
21	Polyphosphate	441	240
30	Control	30	5
30	Monophosphate	13	3
30	Polyphosphate	220	76
37·2	Control	26	6
37·2	Monophosphate	10	3
37·2	Polyphosphate	84	36

[a]Data taken from Leviton *et al.* (1963). All samples contained 36% total solids and were forewarmed and sterilized at 138°C for 15 s.
[b]In the proportion 0–4 lb additive/100 lb milk solids (1 lb = 0·454 kg).
[c]Forewarming–concentration–additive–sterilization–homogenization.
[d]Forewarming–homogenization–additive–sterilization–concentration.

5.3. Total Solids

Gelation of concentrated milk, sterilized by HTST or UHT methods, is hastened by increasing total solids (Bell *et al.*, 1944; Tarassuk & Tamasma, 1956; Calbert & Swanson, 1959; Leviton *et al.*, 1963; Ellertson & Pearce, 1964; Parry, 1974). Increased severity of heat treatment has less influence on retarding the gelation of concentrates containing higher total solids than of less concentrated milks. The data in Table 6 illustrate the effect of total solids and severity of sterilizing treatment on the time of gelation. It is also interesting to note that the heat treatment given to over-concentrated milk (32–45% TS), followed by dilution to 26% TS, imparts to the diluted milk a greater stability against gelation (Stewart *et al.*, 1959).

TABLE 6
Effect of Total Solids and Sterilization Conditions upon Gelation Time during Storage at Room Temperature[a]

Total solids (%)	*Sterilizing time at 132°C (s)*	*Time to gel (weeks)*
26	35	33
26	70	42
36	35	12
36	70	16–20

[a]Data derived from Figs 5 and 6 of Ellertson & Pearce (1964).

5.4. Composition of Milk

Since sterilized skim milk concentrates are highly susceptible to gelation, hastening of gelation by increasing the total solids content inUHT sterilized milk is generally assumed to be due to the solids-not-fat portion, whereas fat is not directly involved in gelation (Tarassuk & Tamasma, 1956; Graf & Bauer, 1976).

Different milk proteins added to milk affect the gelation behaviour of UHT sterilized products. Addition of dialysed acid whey to milk to increase whey protein content by 10% and subsequently processing the milk into evaporated milk (26% TS) sterilized by the HTST method, markedly hastens the onset of gelation. Doubling the sterilizing time is relatively ineffective in overcoming the increased rate of gelation (Ellertson & Pearce, 1964). Additions of α_s-, β- and κ-caseins to milk influence the characteristics of the casein micelles but do not affect the gelation period of UHT sterilized concentrated milk (Schmidt, 1969).

5.5. Quality of Milk

Some factors that influence the composition of milk may indirectly affect the gelation behaviour of sterilized milk. For example, seasonal variation, known to affect composition, influences gelation (Graf & Bauer, 1976). Summer milk gives more stable products than winter milk (Zadow & Chituta, 1975) and UHT-treated mastitic milk is more susceptible to gelation than normal milk (Swartling, 1968). Evaporated milk from Jersey cows, with a higher protein/water ratio, is more susceptible to gelation than evaporated milk from other breeds (Beeby & Loftus Hills, 1962). Early lactation milk is more susceptible to storage gelation of UHT products (Zadow & Chituta, 1975; Graf & Bauer, 1976).

The bacteriological quality of raw milk is also important in gelation; sterilized milk from poor quality milk is very susceptible to gelation during storage (Swartling, 1968; Zadow & Chituta, 1975; Speck & Adams, 1976; Law *et al.*, 1977; Snoeren *et al.*, 1979*b*). The types of microorganisms or spores present in poor milk are important (Zadow & Chituta, 1975; Snoeren *et al.*, 1979*b*); organisms that produce heat-stable enzymes, e.g. proteinases, cause most serious gelation problems.

The widespread practice of holding cooled raw milk in bulk tanks for extended periods increases the possibility of growth of psychrotrophs that produce heat-stable proteinases, which could affect storage stability. A thermization treatment, which consists of subjecting raw milk to a quick heat shock (63–65°C for 12–15 s) before storage in bulk tanks, has been recommended to minimize the growth of phychrotrophs (Ashton, 1979). This treatment is less severe than pasteurization and is effective against phycrotrophs but not against their proteolytic enzymes.

5.6. Enzyme Treatment

Modification of milk by pretreatment with rennet or neuraminidase before sterilization has been reported to increase stability against gelation (Morr, 1975; Graf & Bauer, 1976).

Addition of heat-resistant proteinases or inoculation with psychrotrophs that produce heat-resistant proteinases hastens gelation; this aspect will be discussed later. Manji and Kakuda (1988) added plasmin to achieve different degrees of proteolysis and observed gelation to be independent of the degree of proteolysis although a certain minimum of proteolysis was a prerequisite to gelation. They controlled further proteolysis by adding proteinase inhibitors. Adding plasminogen (0·3 mg/litre) to UHT-sterilized milk promoted age-gelation whereas adding plasmin at 0·3 or 0·15 mg/litre did not result in age-gelation, presumably because too extensive and rapid proteolysis occurred to allow gelation (Kohlmann *et al.*, 1988).

However, West *et al.* (1978) have suggested a low-temperature treatment to inactivate some types of heat resistant proteinases; about 90% of the proteolytic activity is destroyed when UHT treated milk is cooled to about 55°C and held for 60 min, apparently by mechanisms different from the inactivation by heat treatment. Kocak and Zadow (1985*a*) also observed a high level of inhibition of proteolysis by low temperature inactivation treatment. The level of protection against gelation was not as high as that observed by West *et al.* (1978) and was therefore considered to be of limited value for the control and prevention of age-gelation.

5.7. Additives

Extensive studies have been made to develop additives that would be effective in controlling the gelation of sterilized milk (or concentrate) during storage. Additions of sodium phosphate and citrate hasten, while polyphosphates delay, the gelation process in both UHT sterilized milk (Samuelsson & Holm, 1966; Snoeren *et al.*, 1979*b*) and concentrated milk (Leviton & Pollansch, 1962; Leviton *et al.*, 1962, 1963; Wilson *et al.*, 1963; Schmidt, 1968; Schmidt & Buchheim, 1968; Harwalkar & Vreeman, 1978*a*). Sodium phosphate and citrate are known to improve the heat stability of concentrated milk during retort sterilization (Parry, 1974). Polyphospates, on the other hand, significantly delay gelation; the extent of protection against gelation increases with the chain length and concentation of the polyphosphate; polyphosphates with an average of 4·8 phosphorus atoms per chain were effective against gelation (Leviton & Pallansch, 1962; Leviton *et al.*, 1962, 1963). Typical effects of orthophosphate that hasten gelation and of polyphosphate that delay gelation are evident from date shown in Table 5. The phosphates also show similar effects on the stability of concentrates preserved by freezing or by high concentrations of sugar (Doan & Warren, 1947; Leviton *et al.*, 1962; Webb, 1970). Polyphoshates also delay rennet coagulation (Odagiry & Nickerson, 1964). Cyclic condensed phosphates are more effective against gelation than linear polyphosphate because the former are less susceptible to hydrolysis than the latter (Leviton *et al.*, 1962; Leviton & Pallansch, 1962; Glandorf & Thomasow, 1977). A mixture of monophosphate and polyphosphate augments gelation (Leviton *et al.*, 1962; Leviton & Pallansch, 1962; Graf & Bauer, 1976).

Addition of manganous sulphate is also effective in delaying gelation (Leviton & Pallansch, 1962). Interestingly, gelation of milk concentrates sterilized by ionizing radiation is also inhibited by polyphosphate and manganese salts (Hoff *et al.*, 1960*a*,*b*), suggesting a common mechanism of gelation of both heat and radiation sterilized concentrates.

Other additives that delayed the gelation of sterile concentrates are polyhydric compounds like lactose, sucrose, dextrose and sorbitol (Leviton & Pallansch, 1962). Added phosphatides also increase storage life (Leviton & Pallansch, 1962). Sulphydryl blocking agents, such as PCMB, *N*-ethylmaleimide and iodacetamide, retard gelation, while disulphide reducing agents, such as mecaptoethanol, thioglycolate and gluthathione, promote the gelation of sterile concentrates (Nakai *et al.*, 1965*a*). Addition of hydrogen peroxide to UHT sterilized skim milk concentrate hastens gelation (Harwalkar & Vreeman, 1978*b*).

Addition of sodium hydroxide (Board & Mullet, 1970) or sodium carbonate (Seehafer, 1972) to adjust the pH of a 3 : 1 sterile concentrate to near neutrality delays gelation but adjusting the pH of unconcentrated milk before UHT sterilization has no effect on gelation (Zadow & Chituta, 1975).

5.8. Temperature of Storage

Time of gelation of UHT sterilized milk is markedly influenced by the temperature of storage. However, there is not unanimous agreement between different investigators regarding the effect of storage temperature on gelation. Samel *et al.* (1971) observed the gelation time (13 months) to be independent of storage temperature between 4 and 30°C, and at 37°C there was no gelation for two years. Improved resistance to gelation during storage at 35°C and above was also observed by others (Hostettler *et al.*, 1957; Andrews, 1975; Zadow & Chituta, 1975; Andrews *et al.*, 1977; Farah, 1979*a*). Zadow & Chituta (1975) observed minimum gelation-free storage life at 25–35°C but the storage life was extended considerably at 2°C or at 40°C, as shown in Table 3. Andrews *et al.* (1977) noted that samples of UHT sterilized milk stored at 4°C gelled in 19 months, but samples stored at 30–35°C had not gelled in 28 months. Hostettler *et al.* (1957) observed that UHT sterilized milk, stored in glass containers, gelled in 10 months at room temperature but remained liquid at 35°C. The sample in glass at 35°C turned brown but remained white when stored in tinned containers at 35°C and gelled at about the same time as the sample stored at room temperature. Similar observations were made by Manji *et al.* (1986), with direct UHT-processed milk.

There is better agreement regarding the effect of storage temperature on gelation time of concentrated milk sterilized by UHT HTST methods. Generally, the higher the temperature of storage, the earlier the gelation of samples (Leviton *et al.*, 1963; Ellertson & Pearce, 1964) but the extent of temperature dependence of gelation time varies. The gelation rate of HTST sterilized evaporated milk (25% TS) was about three to four times higher on storage at 38°C (100°F) than at 21°C (70°F) (Ellertson & Pearce, 1964). The data of Leviton *et al.* (1963), shown in Table 5, indicate in control samples a lesser effect of storage temperature on the gelation time of samples containing higher total solids (36%).

The effect of storage of HTST concentrate at 35°C and above is different from that of unconcentrated milk sterilized by UHT treatment. In contrast to the sterilized unconcentrated milk, the sterile concentrates were less stable at 38°C than at lower temperatures (compare data in Tables 3 and 5).

6. Physical and Chemical Changes Resulting From UHT Sterilization Treatment and Storage

Although various processing and storage conditions are known to affect the gelation of UHT sterilized milk, as shown in the preceding section, the precise mechanism of gelation is still not fully understood. It is now generally accepted that storage gelation is not caused by microbial activity of either surviving heat-resistant spores or post-sterilization contaminants (van Kreveld, 1950; Hostettler, 1972). In an effort to gain a better understanding of the gelation phenomenon, numerous investigations have been made on the physical and chemical changes that occur immediately following sterilization and during storage.

6.1. Changes in Milk Immediately After Sterilization

Several changes may result from heat treatment of milk during sterilization. The chemical and biochemical reactions that occur when milk is heated have been extensively reviewed (Burton, 1984). The extent of these changes is generally less with UHT sterilization treatments than with retort sterilization. The changes that could be of possible significance for gelation are, particularly, those that affect proteins, enzymes and the mineral balance.

6.1.1. Proteins

As a result of heat treatment, i.e. both the forewarming treatment and sterilization, an association takes place between whey proteins (β-lactoglobulin) and casein (McGugan *et al.*, 1954; Trautman, 1958: Sawyer *et al.*, 1963; Sawyer, 1969). Whey proteins either become denatured or interact with components of casein micelles (κ-casein) and become sedimentable with casein or co-precipitable at the isoelectric point of casein (Hostettler, 1972; Parry, 1974). The extent of association between whey proteins and casein varies with the severity of the heat treatment (Melechouris & Tuckey, 1966; Burton, 1969; O'Sullivan, 1970; Lyster *et al.*, 1971; Perkin *et al.*, 1973; Morr, 1975, Manji & Kakuda, 1988). For example, heating at 146°C for 16 s causes less denaturation than heating at 90°C for 10–30 min (Morr, 1975). UHT sterilization by direct heating methods denatures less whey protein (60–70%) than indirect heating (70–80%) of milk (Burton, 1969, 1984). The heat-induced changes in whey proteins cause shifts in nitrogen distribution as measured by the different methods of partioning proteins in milk (Rowland, 1938; Aschaffenburg & Drewry, 1959; Harwalkar & Vreeman, 1978*a*). The extent of complex formation between denatured β-lactoglobulin and κ-casein, predominantly through disulphide bridges (Trautman, 1958; Sawyer *et al.*, 1963; Sawyer, 1969), significantly modifies the properties of casein micelles. Heat-altered casein micelles are more resistant to coagulation by heat (Rose & Tessier, 1959; Beeby *et al.*, 1971) and enzymes (Kannan & Jenness, 1961; Burton, 1969; Hindle & Wheelock, 1969, 1970; Samel *et al.*, 1971; Hostettler, 1972; Humbert & Alais, 1976; Möller *et al.*, 1977), and to storage gelation (Table 2) and less sensitive to treatment with oxalate that removes calcium ions (Morr, 1975). A higher proportion of whey protein complexed at the surface of casein micelles offers better protection against gelation (Hostettler *et al.*, 1968; Burton, 1969; Corradini, 1971; Manji & Kakuda, 1988). β-Lactoglobulin is also reported to accelerate heat-inactivation of plasmin and plasminogen by specific interaction between denatured β-lactoglobulin and plasmin/plaminogen via SH–SS interaction (Rollema & Poll, 1986).

UHT treatment of milk is also known to cause some rearrangements of casein components in the micellar structure through a simultaneous process of disaggregation and aggregation reactions (Wilson *et al.*, 1961, 1963; Hostettler *et al.*, 1965, 1968; Fox *et al.*, 1967*a*; Morr, 1969; Aoki *et al.*, 1977; Rüegg & Blanc, 1978). Thus, in UHT sterilized milk or concentrated milk, while there is an increased amount of casein (submicelles) not sedimentable on high speed centrifugation ($> 100\,000$ g), there is also an increase in the size of the casein micelles. Harwalkar *et al.* (1989) demonstated by electron microscopy that high temperature (150–200°C) treatment did not cause extensive disintegration of casein micelles. They used a sealed y-shaped glass vial to fix the casein micelles at the high temperatures. The non-sedimentable casein particles (submicelles) have diameters of less than 20 nm. The increase in the size of sedimentable casein micelles is slight in UHT sterilized milk (Hostettler *et al.*, 1965; Rüegg & Blanc, 1978) but a two- or three-fold increase is observed in sterilized, concentrated milk (Hostettler *et al.*, 1968; Schmidt, 1968; Burton, 1969; Harwalkar & Vreeman, 1978*b*). The size enlargement results from a coalescence process (Schmidt, 1968; Carroll *et al.*, 1971; Harwalkar & Vreeman, 1978*b*; Rüegg & Blanc, 1978; de Koning *et al.*, 1985). The importance of increased dimensions of casein micelles for the gelation process is not known.

6.1.2. Mineral Balance

Changes in the inorganic constituents of milk occur as a result of heat treatment. Decreases in ionic calcium and magnesium result from their precipitation as phosphate and citrate salts (van Kreveld & van Minnen, 1955; Pyne, 1962; Fox *et al.*, 1967*b*; Carrol *et al.*, 1971; Aoki & Imamura, 1974*a*). Reduction in the level of soluble calcium by UHT treatment may be insufficient to be of any significance to gelation. Furthermore, the role of calcium in the gelation is not clear. Additives such as phosphates or citrates, that reduce the availability of calcium, promote gelation. Moreover, increasing the level of calcium by 6% did not affect gelation

(Ellertson & Pearce, 1964); small amounts of calcium delay gelation (Graf & Bauer, 1976) but larger increases in calcium hasten it (Leviton & Pallansch, 1961*a*; Samuelsson & Holm, 1966). Another indication that calcium plays an uncertain role in gelation is the protection against gelation afforded by polyphosphates. Cyclic polyphosphates complex calcium to a lesser extent than linear polyphosphates, yet the former retard gelation more effectively (Leviton & Pallansch, 1962; Leviton *et al.*, 1962). The effectiveness of polyphosphates is believed to result from their ability to complex with casein micelles (Odagiri & Nickerson, 1965; Melnychyn & Wolcott, 1967; Beeby *et al.*, 1971), thus preventing their interaction by charge repulsion.

6.1.3. Enzyme Activity

Proteinases that are naturally present in milk or those of bacterial origin are not always completely inactivated by the heat treatment in the UHT process. Proteolytic enzymes that survive UHT sterilization have been isolated from UHT treated milk (Corradini & Pecis, 1979; Snoeren *et al.*, 1979*a*). On the other hand, no proteinase was detected in HTST concentrated milk (Nakai *et al.*, 1964*a*) or in 10% UHT cream (Cheng & Gelda, 1974), even though protein breakdown was evident during storage. Manji *et al.* (1986) observed the conversion of plasminogen to plasmin during storage. Proteolytic enzymes of psychrotrophic bacteria are heat-resistant. Indigenous milk proteinase, plasmin (Snoeren & Van Riel, 1979; Schmidt, 1980) is less heat-stable (Driessen & van der Waals, 1978). Manji *et al.* (1986) measured plasmin and plasminogen activities and found higher levels in direct as compared with indirect UHT-processed milks. Although Manji *et al.* (1986) observed age-gelation only in direct UHT-processed milk, they found no relation between extent of proteolysis and gelation. They observed plasminogen conversion to plasmin during storage. The action of these enzymes during storage is of considerable significance to the problem of gelation and will be discussed in greater detail in following sections.

6.2. Changes During Storage

Sterilized milk undergoes several physical and chemical changes during storage. The changes that affect organoleptic and nutritional quality will not be discussed here since they are presumed not to be directly involved in gelation. Some of the changes that are implicated in the storage gelation of sterilized milks are listed in Table 7. Though the changes occur qualitatively in both non-concentrated and concentrated milks which had been sterilized by UHT treatments, the extent of these changes differs in quantitative terms.

Changes during the storage of sterilized milk affect the stability of the colloidal calcium–caseinate–phosphate complex of milk. The sensitivity of this complex to coagulation by such agents as ethanol, rennet and calcium are indicative of general stability. Data presented in Table 8 show typical behaviour of sterilized milk during storage at 20°C.

TABLE 7
Physical and Chemical Changes during Storage of UHT Sterilized Milk

Property	*Unconcentrated milk*	*Concentrated milk*
Physical		
Stability to enthanol	↓	↓
Stability to calcium ions	↓	↓
Rennet coagulation time	↑	↑
Aggregation of proteins	↑	↑
Non-sedimentable casein	↑	↑
Ca/N ratio of the sediment	↑	↑
P/N ratio of the sediment	↑	↑
Viscosity (relative)	↑	↑
Chemical		
pH	unchanged	unchanged
Maillard-type reaction (browning)	↑	↑
Sulphydryl groups	↓	↓
Protein breakdown	↓	↓
Casein nitrogen		
Non-casein nitrogen	↑	↑
Non-protein nitrogen	↑	↑
N-Acetylneuraminic acid	↑	↑
Dephosphorylation (P)	unchanged	unchanged

Note: ↑, Increase; ↓, decrease.

6.2.1. Stability to Ethanol

Alcohol stability is determined as the minimum percentage of ethanol which when added to an equal volume of milk causes its coagulation; milk coagulating at a lower concentration of ethanol is less stable. During storage, UHT sterilized milk

TABLE 8
Stability of Proteins in Milks Sterilized by UHT or Autoclaving and Stored at 20°C

Stability of protein	*Storage (months)*				
	0	*2*	*5*	*9*	*13*
Stability to ethanol (%)					
UHT milk	96	96	83	62	gelled
Autoclaved	>96	>96	>96	>96	>96
Stability to calcium (ml 0·1 M CaCl)					
UHT	2·8	1·4	0·8	0·2	gelled
Autoclaved	2·1	1·8	2·0	1·8	2·0
Rennet coagulation time (min)					
UHT	50	50	12	4	gelled
Autoclaved	> 120			120	120
Raw milk	7				

Data from Samel *et al.* (1971).

shows a gradual loss of stability to alcohol (Table 8). This loss of stability appears to correlate with gelation during storage, since retort sterilized (autoclaved) milk which did not gel during storage, remains unchanged in its stability to ethanol. At a lower storage temperature the stability to ethanol decreases more rapidly in UHT sterilized milk than in autoclaved milk (Samel *et al.*, 1971). A decrease in alcohol stability was also observed during storage of UHT sterilized concentrated milk (Harwalkar & Vreeman, 1978*a*). UHT sterilized concentrates are less stable than diluted UHT sterilized milk. Samples containing polyphosphates, which delay gelation, are more stable to ethanol than samples in which the onset of gelation is hastened.

6.2.2. Stability to Calcium

Rennet-treated milk shows a decreased stability to added calcium ions (Hostettler *et al.*, 1957; Pyne, 1962; Beeby *et al.*, 1971; Samel *et al.*, 1971), i.e. the amount of calcium required to cause precipitation of casein is reduced. The decreased stability to calcium has been attributed to a loss of the protective action of κ-casein due to its proteolytic breakdown. Similar loss of stability to calcium has been observed with aged sterile milk (Hostettler *et al.*, 1957; Nakai *et al.*, 1964*b*; Samel *et al.*, 1971) although to a lesser extent than in rennet-treated milk (Hostettler *et al.*, 1957; Nakai *et al.*, 1964*b*). Samel *et al.* (1971) observed a gradual decrease in stability to calcium during the storage of UHT sterilized milk but not in retort sterilized milk (data in Table 8 are for samples stored at 20°C). At a higher storage temperature the loss of stability to calcium was faster than at lower temperatures (Andrews & Cheeseman, 1971), which was attributed to the increased chemical modification at higher temperatures by Maillard-type reactions between ε-NH_2 of protein and reducing groups of lactose or its degradation products (Andrews & Cheeseman, 1971). The calcium sensitivity of autoclaved milk stored at 4°C decreased but at a a slower rate than that of UHT sterilized milk (Samel *et al.*, 1971). The UHT treated sample gelled in 13 months but autoclaved milk did not. Thus, there is no consistent relationship between increased calcium sensitivity and gelation.

6.2.3. Rennet Coagulation Time (RCT)

Heat treatment of milk increases the RCT to varying extents depending upon the severity of heat treatment. The RCT is considerably longer for UHT sterilized milk than for raw milk but markedly lower than for autoclaved milk (Table 8). Casein micelles in skim milk sterilized by indirect UHT treatment, when compared with direct UHT treatment, were more stable during storage as well as to the action of rennet (Guthy & Horak, 1980). The RCT is affected by the formation of a heat-induced complex between β-lactoglobulin and κ-casein which may hinder the availability of the sensitive Phe–Met bond in κ-casein to attach by rennet (Kannan & Jenness, 1961; Zittle *et al.*, 1962; Parry, 1974; Morr, 1975). Also, shifts in ionized and colloidal milk salts may alter the charge and surface properties of casein micelles (van Kreveld & van Minnen, 1955; Hostettler *et al.*, 1957; Beeby *et al.*, 1971; Schmidt, 1980).

Samel *et al.* (1971) observed that the RCT progessively decreased in UHT sterilized milk during storage; after six months of storage, the RCT of UHT milk was the same as for unheated milk. The RCT of autoclaved milk was unchanged (Table 8).

The reduction of RCT during storage of UHT sterilized milk is indicative of some subtle, reversible and as-yet poorly understood changes in the surface of the casein micelles; the unchanged RCT of autoclaved milk during storage is indicative of irreversible changes. These considerations regarding the rennet coagulation behaviour of stored sterilized milk are consistent with the hypothesis of Hostettler (Hostettler *et al.*, 1968; Hostettler, 1972) in which storage gelation is linked to incomplete or reversible interaction between whey protein and casein. However, it is not certain if the change in RCT and gelation during storage are governed by a common physical condition of the casein micelles.

6.2.4. Shifts in Mineral Balance

During storage of UHT sterilized milk many changes are observed in the distribution of mineral components. Initially, some of the calcium phosphate, precipitated by the UHT treatment, dissociates (Rose & Tessier, 1959; Hostettler *et al.*, 1968; Parry, 1974) and binding of calcium and phosphate by the casein is readily reversed. Prolonged storage, on the other hand, causes precipitation of some forms of calcium phosphate (Fox *et al.*, 1967*b*; Aoki & Imamura, 1974*b*). Since colloidal calcium phosphate is necessary for the stability of casein micelles, changes in the mineral balance affect stability.

The manner in which these shifts in mineral balance affect stability is not clear. Increased dissociation of casein micelles is caused by removal of calcium (Fox *et al.*, 1967*a*; Morr, 1967; Parry, 1974) and therefore the increased non-sedimentable casein observed during storage could arise from such a removal. However, Aoki and Imamura (1974*a*,*b*) observed that Ca/N and P/N ratios of the casein fraction sedimentable by ultracentrifugation (> 100 000 g) increased during the storage of UHT skim milk. This increase was of smaller magnitude than in UHT concentrated skim milk. Aoki and Imamura (1974*a*,*b*) attributed the destabilization of the casein complex during storage to the incorporation of calcium and phosphate. However, an increase in Ca/N ratio in the ultracentrifugal casein complex was also observed in stored autoclaved milk and this makes the explanation less tenable. There is no discernible pattern of mineral distribution that satisfactorily explains the mechanism of gelation during storage.

6.2.5. Disaggregation, Aggregation and Microstructure of Casein Micelles

During storage of UHT milk, caseinate particles undergo a number of changes as measured by ultracentrifugal sedimentation, gel permeation chromatography (Andrews & Cheeseman, 1972; Andrews, 1975) and light (Wilson *et al.*, 1961, 1963) and electron microscopy (Hostettler *et al.*, 1988; Schmidt, 1968; Schmidt & Buchheim, 1968; Hostettler, 1972; Andrews *et al.*, 1977; Aoki *et al.*, 1977; Harwalkar & Vreeman, 1978*b*; Farah, 1979*a*; Blanc *et al.*, 1980). First, an increase in the amount of non-sedimentable casein (at 100 000 g) is observed. The increase in non-sedimentable casein is assumed to result from a partial disaggregation of the casein micelles (Wilson *et al.*, 1963; Morr, 1969) and is seen by electron microscopy as fine particles or subunits of casein (Wilson *et al.*, 1963; Odagiri & Nickerson, 1964; Nakai *et al.*, 1965*b*; Schmidt, 1968; Glandorf & Thomasow, 1977; Harwalkar & Vreeman, 1978*b*). The dissociation of large casein particles is more rapid at 4°C than at 20 or 37°C. The forces that lead to such disaggregation are not, as yet, well understood but shifts in the mineral balance seem to be involved, as indicated in the preceding section. Secondly, increased formation of protein polymers is observed (Aoki & Imamura, 1974*a*,*b*; Andrews, 1975) during storage. These polymers result from covalent binding or cross-linking of casein polypeptide chains through carbonyl intermediates resulting from Maillard-type reactions (Andrews & Cheeseman, 1971, 1972; Andrews, 1975). The polypeptide chains may also be linked through disulphide bonds. The extent of polymer formation is dependent upon time and temperature of storage. After six months of storage, the proportion of milk proteins existing as covalently bound polymers is 50% at 37°C, 40% at 30°C, 26% at 20°C and 21% at 4°C. The extent of polymer formation does not appear to be related to gelation time since UHT milk stored at 4°C gelled sooner than milk stored at 30 or 37°C, although the milk stored at 4°C contained the lowest amount of polymers (Andrews, 1975).

In addition to the above two changes, which progress independently of viscosity changes, there are changes in the structure of casein micelles that coincided with the observed changes in viscosity or gelation. Examination of microstructure by electron microscopy shows that, during storage, the casein micelles associate increasingly during the period in which a rapid increase in viscosity and gelation occur (Harwalkar & Vreeman, 1978*b*; Farah, 1979*a*). The changes in microstructure observed during storage of UHT treated milk are gradual; those occurring in UHT sterilized concentrated skim milk are illustrated in Fig. 2. During the initial period, when there is no change in viscosity, the micelles remain spherical and well separated from each other but the surface appearance changes and the filamentous appendages, attributed to denatured β-lactoglobulin (Davis *et al.*, 1978), or 'hairiness' (Harwalkar & Vreeman, 1978*b*) on the micelle surface become less prominent. After 10 weeks of storage, when viscosity starts to rise, the casein micelles show a slight distortion and thread-like tails appear on the micelle surfaces (Fig. 2(b)). Similar deformation of the micelle surface and tendrils prodruding from it were also observed before gelation in UHT milk (Andrews *et al.*, 1977; Blanc *et al.*, 1980). After 13 weeks storage of concentrated skim milk, when the milk had become viscous, pairs or triplets of casein micelles were observed (Fig.2(c)). The micelles were attached either by fusion between micelles or by thin, fibre-like material, analogous to the bridging material described by Carroll *et al.* (1971). At this stage, the micelles still retained their identity and several unattached individual micelles were present. The pairs and triplets of micelles observed in this sample were perhaps equivalent to the increase in micelle size observed by Andrews *et al.* (1977) and to the increased amount of casein particles sedimented by low-speed centrifugation (3000 g) as observed by Aoki and Imamura (1974*b*) in stored UHT milk. After 17 weeks of storage, when the sample had gelled, there was an indication of severe distortion and aggregation of casein micelles into chains which

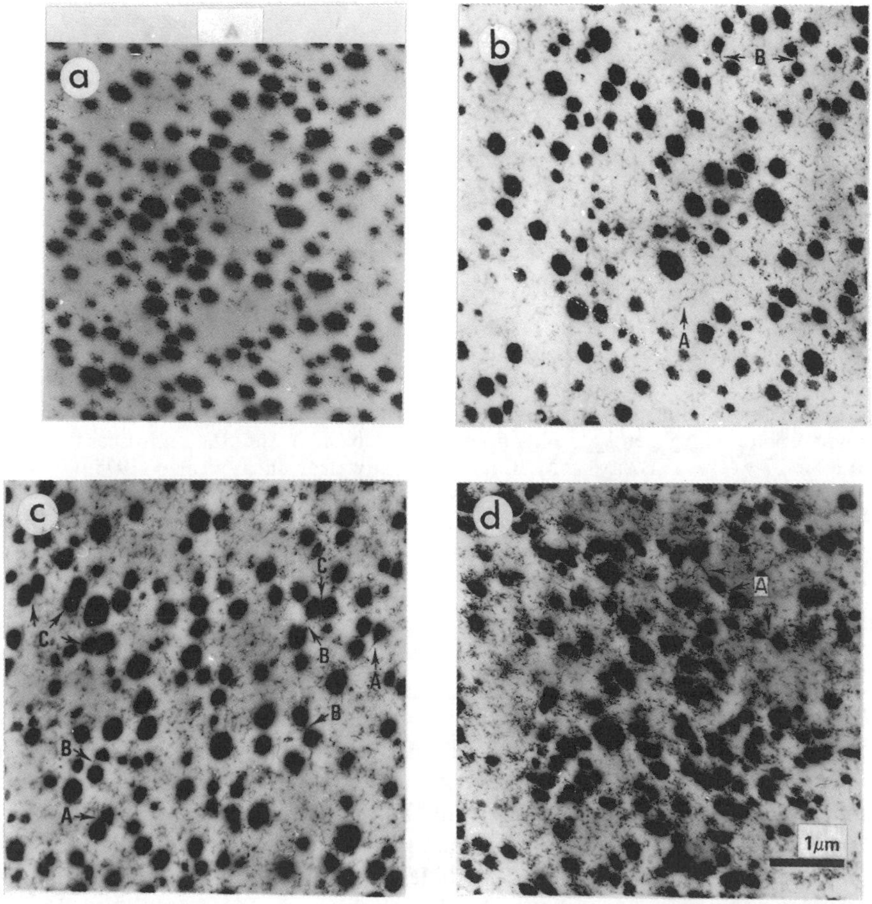

Fig. 2. Changes in the microstructure of UHT sterilized concentrated skim milk stored at 28°C. (a) Immediately after sterilization; viscosity = 21°M. (b) After 10 weeks: viscosity = 26°M; A, non-micellar particles; B, distorted casein micelles with thread-like tails. (c) After 13 weeks: viscosity = 120°M; A, distorted casein micelles with thread-like tails; B, fibre-like material joining casein micelles; C, clusters of fused casein micelles. (d) After 17 weeks: solid gel; A, fibre-like material joining clusters of casein micelles. Data from Harwalkar and Vreeman (1978*b*). (°M = relative viscosity using a Mojonnier torsion viscometer.).

were connected through fibre-like material to form a continuous network (Fig. 2(d)). Network formation by linkages between casein micelles has been observed by others in gelled concentrated (Wilson *et al.*, 1963; Hostettler *et al.*, 1968; Schmidt, 1968; Schmidt & Buchheim, 1968; Carroll *et al.*, 1971; de Koning *et al.*, 1985) and unconcentrated milks (Andrews *et al.*, 1977; Farah, 1979*a*; Blanc *et al.*, 1980; de Koning *et al.*, 1985). Gelation results from gradual changes in the casein micelles rather than from an abrupt coalescence of micelles shortly before gelation (Schmidt, 1968; Schmidt & Buchheim, 1968). The abundance of non-micellar particles increases with storage but their involvement in gelation is uncertain. In

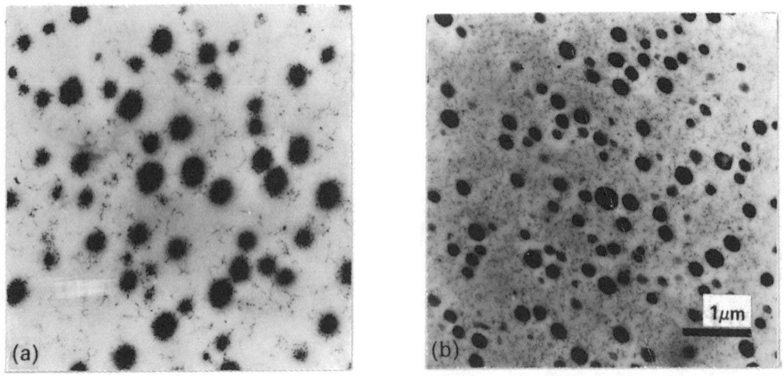

Fig. 3. Changes in the microstructure of UHT sterilized concentrated skim milk containing polyphosphate (1·5 g per kg milk). (a) Immediately after sterilization, viscosity 17°M; (b) after 17 weeks, viscosity 15°M. Data from Harwalkar and Vreeman (1978*b*). (°M = relative viscosity using a Mojonnier torsion viscometer).

contrast to the changes in microstructure of UHT-sterilized concentrated milk during storage, the gelation of UHT-sterilized unconcentrated skim milk was accompanied by thread-like structures while the casein micelles were almost fully disintegrated (de Koning *et al.*, 1985).

Samples of UHT sterilized concentrated skim milk, containing orthophosphate, also showed changes in microstructure at various stages of gelation (Harwalkar & Vreeman, 1978*b*). These samples became viscous or gelled sooner than the samples without orthophosphate and changes in their microstructure also were seen earlier. Samples containing hexametaphosphate showed little change in microstructure during storage and the micelles remained mostly unchanged and well separated. Compared to a sample examined immediately after sterilization, samples stored for 17 weeks contained micelles that had a smoother surface and were of a smaller size (compare Fig. 3(a) and Fig. 3(b)). Samples of retort sterilized milk and samples of UHT milk stored at 45°C, neither of which gelled during storage, also contained micelles that remained well separated from each other (Farah, 1979*a*).

The characteristics of the microstructure of gels from stored UHT milk or concentrated skim milk are quite different from the microstructure of gels produced by the action of chymosin. The latter gels show only clumping of casein micelles and do not show the surface deformation and thread-like bridging between micelles that is typical of storage-gelled products (Andrews *et al.*, 1977; Green *et al.*, 1978; Harwalkar & Vreeman, 1978*b*; de Koning *et al.*, 1985).

6.2.6. Protein Breakdown

Destabilization of casein micelles during storage of sterilized milk could result from either dephosphorylation or cleavage of peptide bonds of proteins. Dephosporylation, which is observed in milk given excessive heat treatment (Belec & Jenness, 1962; Parry, 1974), has not been demonstrated to occur during storage (Harwalkar & Vreeman, 1978*a*). Evidence of proteolysis is voluminous, and therefore consider-

TABLE 9
Changes in Nitrogen Distribution in UHT Sterilized Milk during Storage at Room Temperature[a]

	Nitrogen distribution[b] *(%)*				
	Before sterilization	*After sterilization*	*Storage (months)*		
			2	*4*	*5*
Casein N	80·3	84·4	80·8	75·5	72·1
Non-protein N	6·2	6·2	6·3	8·1	8·8
Proteose-peptone	1·1	1·8	2·7	4·7	8·9
Globulin N	0·8	0·7	1·2	1·3	1·4
β-Lactoglobulin N	7·0	3·1	4·3	3·1	0·9
Albumin N	4·6	3·5	4·7	7·3	7·9

[a]Data taken from Corradini (1975).
[b]Determined as described by Aschaffenburg and Drewry (1959).

able attention has been directed to find a relationship between proteolysis of sterilized milk and its gelation during storage. During their early investigations, Hostettler *et al.* (1957) observed decomposition of casein in UHT sterilized milk that had gelled during storage and Samuelson and Holm (1966) observed an inverse relationship between the increased level of non-protein nitrogen and time of onset of gelation. Both groups of investigators attributed the decomposition of proteins to reactivated proteolytic enzymes. Reactivation of phosphatases in heated milk is known (Write & Trammer, 1953; Edmondson *et al.*, 1966). Subsequently, numerous investigators have observed protein breakdown during storage of both unconcentrated (Murthy *et al.*, 1958; Samel *et al.*, 1971; Björck, 1973; Andrews, 1975; Corradini, 1975) and concentrated milk (Wilson *et al.*, 1963; Hostettler *et al.*, 1968; Hostettler, 1972; Harwalkar & Vreeman, 1978*b*; Snoeren *et al.*, 1979*b*) sterilized by UHT treatment. During storage, a progressive decrease in the amount of casein nitrogen and a corresponding increase in non-casein and non-protein nitrogen was observed in these samples. Typical changes during storage of UHT sterilized milk in the fractions partitioned by the scheme of Aschaffenberg and Drewry (Aschaffenberg & Drewry, 1959; Corradini, 1959) are illustrated in Table 9. Although others have observed similar changes, there is considerable variation in the extent and rate of proteolytic change. The differences are assumed to result from variations in the conditions of processing treatment and the quality of milk. UHT treatment by the 'direct' method results in a higher degree of proteolysis during storage than in milk sterilized by the 'indirect' method or by autoclaving (Samuelson & Holm, 1966; Samel *et al.*, 1971; Farah, 1979*b*; Blanc *et al.*, 1980). The survival of heat-stable enzymes is determined not only by the severity of heat treatment but also by the nature and quantity of the enzymes present. Poor quality milk, which contains increased levels of proteinases of bacterial origin, shows extensive degradation of protein and a short gelation time during storage (Swartling, 1968; Zadow & Chituta, 1975; Driessen & van der Waals, 1978; Snoeren *et al.*, 1979*b*). There is a wide variation in the heat stability

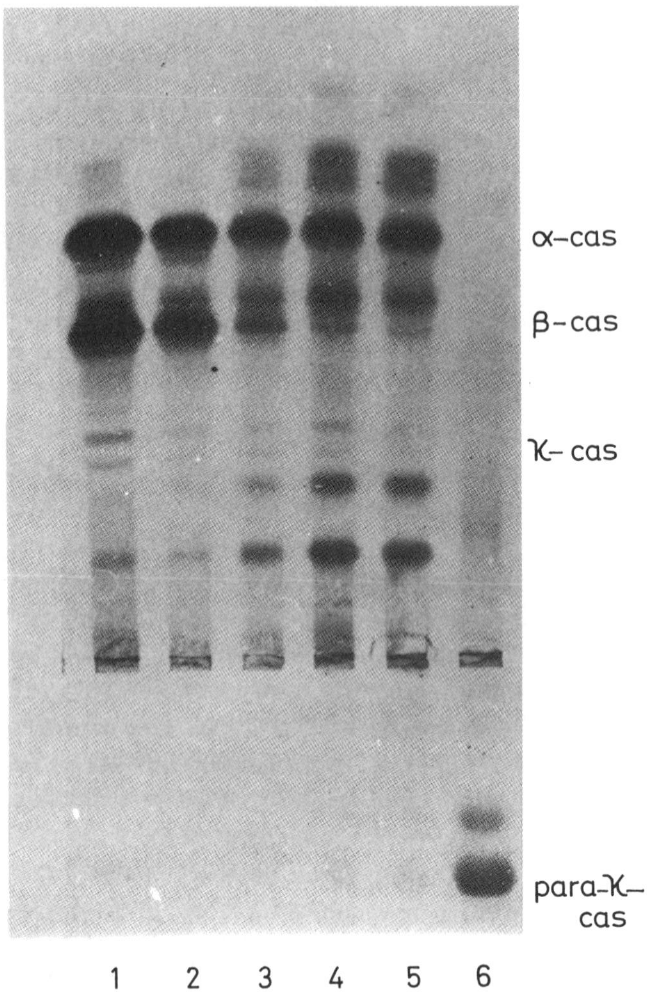

Fig. 4. Starch–urea gel electrophoretograms of casein from UHT sterlized milk stored at room temperature: 1, before sterilization; 2, after sterilization; 3, stored 3 months; 4, stored 5 months; 5, stored 6 months; 6, rennet-treated milk. Data from Corradini (1975).

of proteolytic enzymes of bacterial origin. Thermal inactivation data for some proteinases are known (Speck & Adams, 1976; Ashton, 1979; Driessen, 1989) but such data for all the proteinases is, practically, not possible.

Those proteinases that are natural to milk or are of bacterial origin vary considerably in their specific action on individual proteins that are discernible by electrophoretic methods. Many of the earlier storage studies on UHT sterilized milk indicated major changes only in the β-casein fraction along with a concomitant increase in the γ-casein fractions that are now known (Whitney *et al.*, 1976) to result from proteolysis of β-casein. Electrophoretic patterns presented in Fig. 4

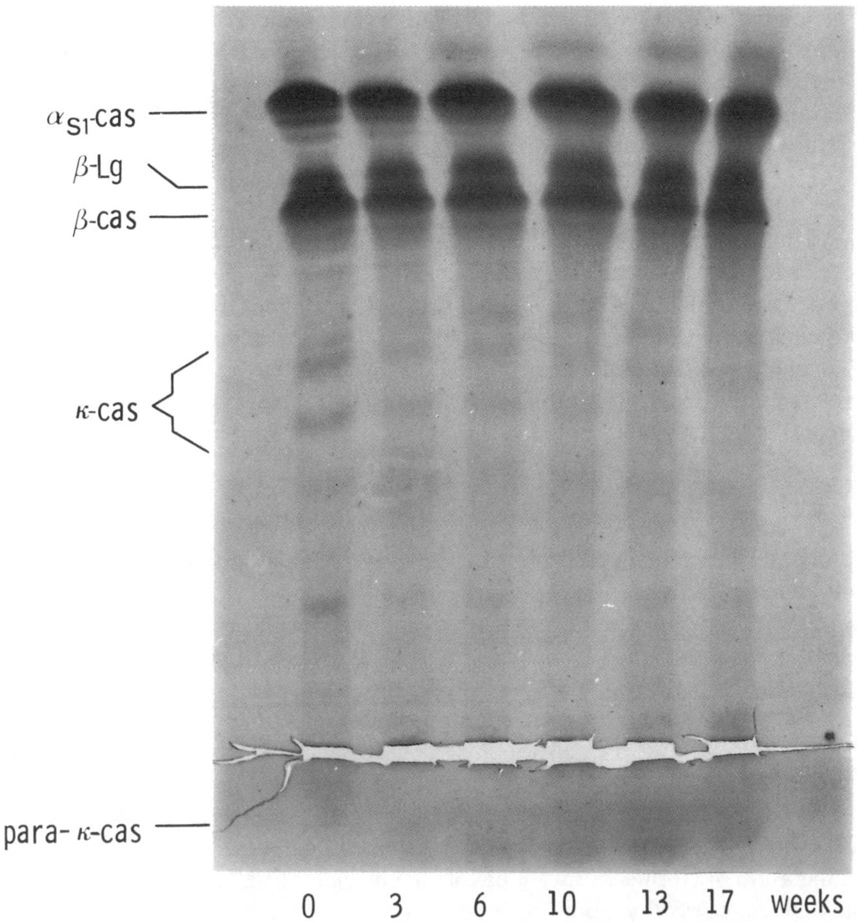

Fig. 5. Starch–urea gel electrophoretograms of casein from UHT sterilized concentrated skim milk containing polyphospate during storage for up to 17 weeks. Data from Harwalkar & Vreeman (1978*a*).

illustrate the typical changes in casein fractions during storage of UHT sterilized milk. These patterns indicate a decrease in the relative concentrations of α_{s1} –, α_{s2} –, β – and κ-caseins and a corresponding increase in γ-casein fractions and para-κ-casein. These changes are consistent with the observations of Snoeren and Van Riel (1979) that the principal milk proteinase, plasmin, is equally specific for β- and α_{s2} –caseins although their action on other caseins is slower.

The mode of action of these proteinases is influenced by some other factors. For example, Harwalkar and Vreeman (1978*a*) found that the increase in γ-casein was less during storage of UHT concentrated skim milk than in UHT milk (Fig. 5) although there was an increase in proteose-peptone and, as pointed out by Snoeren and Van Riel (1979), a decrease in α_{s2}-casein. An increase in the proteose-peptone fraction is also observed in stored UHT milk (Fig. 4 and Table 9) and is attributed to the formation of proteose-peptones 8 and 5 which are N-terminal serments of β-casein (Andrews, 1978*a*,*b*). The basic fractions observed in Fig. 5 appear to be cleavage products of α_{s2}-casein (Snoeren & Van Riel, 1979).

Numerous studies have indicated that proteinases of microbial origin survive UHT treatment (Harper *et al.*, 1970; Choudhery & Mikolajcik, 1971; Bengtsson *et al.*, 1973; Meyerhofer *et al.*, 1973; Adams *et al.*, 1975; Barach *et al.*, 1976; Speck & Adams, 1976; Law *et al.*, 1977; Driessen & van der Waals, 1978; West *et al.*, 1978; Richardson & Newstead, 1979; Snoeren *et al.*, 1979*a*; Driessen, 1989). The action of bacterial proteinases on individual proteins is different from that of indigenous proteinases. In contrast to indigenous milk proteinase, bacterial proteinases appear to degrade κ-casein (Adams *et al.*, 1975; Law *et al.*, 1977; Snoeren *et al.*, 1979*b*) preferentially. However, information regarding the specificity of bacterial proteinases is limited. Undoubtedly, the bacterial and the milk proteinases influence the storage behaviour of UHT products to different extents because they differ in their mode of attack on individual proteins. Snoeren *et al.* (1979*b*) observed that a sample of milk that was deliberately contaminated with psychrotropic bacteria before UHT treatment, gelled on storage sooner than aseptically drawn milk. The contaminated milk had the appearance of rennet curd whereas the uncontaminated milk, containing only milk proteinases, coagulated without typical gel formation. Gelation of both samples was accompanied by extensive proteolysis, the rate of which was higher in contaminated milk. In contrast to this, Koops, quoted by Payens (1978), observed gelation of products manufactured from aseptically drawn milk without an accompanying increase in the amount of non-protein nitrogen.

Storage temperature is another factor that influences the rate of proteolysis in UHT milk. Samel *et al.* (1971) observed increased breakdown of protein at a higher temperature of storage but were unable to correlate the extent of proteolysis with the time of onset of gelation. Samples stored at 4, 20 or 30°C gelled at about the same time (13 months) in spite of varying levels of protein breakdown. Furthermore, samples stored at 37°C showed the most extensive proteolysis but did not gel.

During the storage of UHT-sterilized concentrated milk, considerably slower proteolysis is observed than in UHT treated milk (Hostettler *et al.*, 1957; Harwalkar & Vreeman, 1987*b*). Hostettler *et al.* (1957) observed low levels of protein breakdown and no enzyme reactivation. Harwalkar and Vreeman (1978*a*) also observed a low rate of protein breakdown as indicated by the decrease in casein nitrogen and increase in non-casein and non-protein nitrogen and by gel electrophoresis. They further observed that the rates of protein breakdown were unaffected by the addition of orthophosphate or hexametaphosphate; orthophosphate hastened gelation while hexametaphosphate delayed it. Similar observations regarding the effect of added phosphates upon changes during shortage of unconcentrated milk were made by Snoeren *et al.* (1979*b*). Interestingly, the gelation time of poor-quality milk, sterilized by UHT treatment, was unaffected by the addition of phosphates. The reason for this behaviour is not known.

Wide variations have been observed by different investigators in the rate and extent of proteolysis during the period in which onset of gelation takes place. Data in Table 10 are drawn from different studies and show the variations in the time and gelation and the decrease of casein content (as a percentage of total protein) and the increase in non-protein nitrogen (mg/100 ml milk) at gelation time. The relative rates, shown as an approximate decrease of percentage of casein per month or increase of non-protein nitrogen per month, correlate well with the gelation time of either unconcentrated or concentrated product; however, there seems to be no critical level of protein breakdown at which gelation could be observed. The UHT treated, concentrated milk gelled sooner than unconcentrated milk even though the rate and extent of protein breakdown were considerably lower than in milk. Also, in UHT milk from different laboratories, which showed approximately the same degree of protein breakdown (Hostettler *et al.*, 1957; Corradini, 1975), the time of gelation was markedly different.

7. Gelation of Retort Sterilized Evaporated Milk

Gelation is not usually a problem during storage of evaporated milk, retort sterilized in containers. However, Heintzberger *et al.* (1972) occasionally observed gelation in these products. They investigated this problem and demonstrated conclusively that gelation was caused by prolonged (2–3 days) cold storage (4°C) of the concentrate before canning and sterilization. The products sterilized after cold storage invariably gelled after about 7 months of storage at 28°C, but those sterilized without cold storage remained liquid for more than 12 months. Heintzberger *et al.* (1972) tested evaporated milk of different composition and observed gelation only in concentrates with Fat : SNF ratio of 9/22 or 10/23 but not in 8/18 standard samples. Typical results obtained with 9/22 samples are shown in Table 11 (samples 1 and 2). Additional heat treatment (115°C for 0·5–1 min) of cold-stored concentrate before sterilization provided a slight protection gelation but did not prevent it (sample 5). Orthophosphate was not considered to be involved in gelation (samples 3 and 4).

Harwalkar *et al.* (1983), working with 8/18 standard evaporated milk and evaporated skim milk (19% TS), have confirmed the usual observation by Heinzberger *et al.* (1972) of gelation in retort sterilized evaporated milk. The 8/18 standard concentrates held at 4°C for 3 days before sterilization significantly increased in viscosity or gelled within 6–9 months but remained liquid for more than 12 months when sterilized without cold storage. An explanation for the unusual gelation problem was sought by examination of possible involvement of (i) alteration of casein micelles and (ii) oxidation/reduction reactions:

(i) Some of the changes in the casein micelles resulting from cold storage are known to be reversed by heating at 55–60°C for 30 min (the so-called 'anticool-ageing' treatment). Cold storage of milk is known to delay its coagulation by rennet (Fox, 1969). The delayed coagulation time, presumed to be the result of changes in casein micelles during cold storage, is reversed

TABLE 10
Relationship between Gelation and Changes in Casein and Non-casein Nitrogen in UHT Sterilized Milk and Concentrated Milk

Casein (%)			*Non-protein nitrogen (mg/100 ml)*				*References*
Initial	*At gelation*	*Change per month*	*Initial*	*At gelation*	*Change per month*	*Time of gelation (months)*	
UHT sterilized milk							
85·2	62·0	7·8	37	80	14·4	3	Snoeren *et al.* (1979*b*)
85·0	70·0	1·5	32	52	3·0	10	Hostettler *et al.* (1957)
84·7	72·1	2·5	35	50	3·0	5	Corradini (1975)
84·3	76·0	0·6	26	60	2·6	13	Samel *et al.* (1971)
—	—	—	28	120(78)[a]	10·0	5	Samuelsson & Holm (1966)
—	—	—	28	78(67)[a]	5·6	> 7	Samuelsson & Holm (1966)
—	—	—	28	54(58)[a]	2·9	> 10	Samuelsson & Holm (1966)
Concentrated milk							
85·0	80·0	1·4	24	32	0·2	4	Harwalkar & Vreeman (1978*a*)
89·2	82·0[b]	1·0	21	28	0·1	7	Hostettler *et al.* (1968)

[a]The non-protein nitrogen values were obtained after nine months of storage. The values in parentheses are estimates for the time of gelation, assuming a linear increase of this fraction with time of storage.
[b]The percentage of casein remained unchanged during the first $2\frac{1}{2}$ months, then started to decline.

TABLE 11
Age-Thickening and Gelation of Sterilized Milk Stored at 20°C

No.	*Concentrate fat/SNF (%/%)*	*Holding at 4°C (days)*	*Treatment*	*Relative viscosity after storage (months)*		
				0	*6*	*9*
1	9/22	0	None	−−	−−	−−
2	9/22	3	None	−−	−−	++
3	9/22	0	Na_2HPO_4, 75 g/100 kg	−−	−−	−−
4	9/22	3	Na_2HO_4, 75 g/100 kg	−−	++	+++
5	9/22	3	Additional heating 115°C/1 min	−−	−−	+
6	8/18	0	None	6·7	11·4	13·6
7	8/18	3	None	12·6	75·3	89·1
8	8/18	3	Anticool-aging	13·8	81·7	> 100
9	8/18	0	Aeration	11·4	15·2	33·0
10	8/18	3	Aeration	9·8	51·5	69·2
11	8/18	0	0·1% H_2O_2	5·4	21·0	48·0
12	8/18	3	0·1% H_2O_2	6·1	> 100[b]	> 100
13	8/18	3	Deaeration	11·4	59·7	74·8
14	8/18	3	0·2% BHT	7·8	53·1	69·0
15	0/19	0	None	2·5	3·9	> 100
16	0/19	3	None	3·8	31·2	> 100
17	0/19	0	0·1% H_2O_2	2·8	> 100[c]	> 100
18	0/19	3	0·1% H_2O_2	3·1	> 100	> 100

[a]Data for samples 1–5 taken from Heintzberger *et al.* (1972), data for samples 6–18 taken from Harwalkar *et al.* (1983).
The viscosity ratings for samples 1–5 were as follows: −−, unchanged; +, viscous; ++, strongly viscous; +++, solid gel. These correspond to < 20; 20–50; 50–100 and > 100 relative viscosity units respectively in samples 6–18.
[b]The sample had a relative viscosity of 48·6 after 3 months of storage.
[c]The sample had a relative viscosity of 4·5 after 3 months of storage.

by the 'anticool-ageing' treatment. Such an 'anticool-ageing' treatment was ineffective in preventing gelation and possibly hastened it slightly (sample 8). Therefore, the change in casein micelles leading to storage gelation must be of a different nature from those affecting rennet coagulation time.

(ii) Aeration or incorporation of H_2O_2 in concentrated milk, sterilized without cold storage, increased its viscosity during storage (samples 9 and 11 in Table 11). Treatment of cold-stored samples with H_2O_2 accelerated gelation (sample 12), whereas aeration or deaeration of the cold-stored sample did not affect gelation significantly (samples 10 and 13). Likewise, treatment with antioxidant was ineffective in preventing the age-thickening process (sample 14).

Retort sterilized evaporated skim milk samples were more susceptible to gelation than evaporated milk but the cold-storage treatment and incorporation of H_2O_2 had similar effects on both samples (compare samples 15–18 with samples 11 and 12).

It was observed in these experiments that samples that had become viscous or gelled did not contain microorganisms capable of growing and showed no significant change in pH or proteolysis.

The above observations strongly suggest a physicochemical rather than an enzymatic basis for the gelation of retort sterilized evaporated milk. The results demonstrate the significance of oxidizing/reducing type reactions in the gelation phenomenon but the mechanism by which these reactions influence gelation is not clear.

8. Mechanisms of Gelation

Since the last review on this subject, there still remain gaps in the knowledge of the mechanism of age-gelation. Although involvement of proteolysis in gelation, in at least some cases, has been demonstrated, the information is limited and definitive views on the mechanism of gelation have not emerged yet, particularly regarding the extent and nature of the physicochemical forces involved in the gelation. The following discussion derived mostly from earlier comments (Harwalkar, 1982) on the mechanisms of age-gelation of sterilized milk products is still pertinent and includes some of the recent findings that are significant to the understanding of the mechanism of gelation.

It is now generally accepted that gelation of stored, sterilized milk results from direct interaction between casein micelles and their linkage into a three-dimensional network. Although some subunits of casein micelles dissociate from the micelles during storage, there is no evidence to support the concept that gelation results from complete disintegration of the micelles followed by restructuring of the subunits to form a gel (Harwalkar & Vreeman, 1978*b*; Payens, 1978). The interaction between micelles is preceded by changes at the surface of the micelles, as a result of which they become more reactive, i.e. become susceptible to interaction and destabilization. The forces that lead to the weakening of the micelle surface are not fully understood as yet. Several models have been hypothesized to explain the mechanism of gelation and, in general, the changes that trigger the loss of stability of the casein micelles fall into two categories:

(i) changes that arise from proteinase activity (i.e. with an enzymatic basis);
(ii) changes that arise from non-enzymatic reactions (i.e. with a physicochemical basis).

8.1. Proteinase Hypothesis

Storage gelation of UHT sterilized milk is accompanied by an increased level of proteolytic breakdown products and by decreased stability to calcium ions (Hostettler *et al.*, 1957; Samel *et al.*, 1971). These changes, also characteristic of rennet coagulation of milk, have led many investigators (Bell *et al.*, 1944; Hostettler *et al.*, 1957; Samuelson & Holm, 1966; Adams *et al.*, 1975; Corradini, 1975; Law *et al.*, 1977; Payens, 1978; Snoeren *et al.*, 1979*b*) to suggest that the gelation of aged sterile milk (or concentrates) takes place by a mechanism analogous to that of rennet coagulation in which an enzyme-triggered primary protein degradation step is followed by a secondary aggregation reaction, resulting in gelation.

The enzyme hypothesis is further supported by the striking resemblance of the pattern of viscosity changes before gelation and the coagulation kinetics of rennet-treated and aged sterilized milks (Payens, 1978; Schmidt, 1980). On the basis of extensive theoretical treatment, Payens (1978) considers that the phenomenon of the initial age-thinning, followed by an explosive growth of the average particle weight, results from the action of a clotting proteinase. He rules out non-enzymatic coagulation on the ground that such coagulation would be accompanied by a linear increase in weight-average molecular weight (M_w) with time, t, according to the equation of Payens (1978) and as simplified by Schmidt (1980):

$$\frac{M_w}{M_{t=0}} = 1 + At \tag{1}$$

where $M_{t=0}$ is the molecular weight at zero time and A is a constant. The enzyme-mediated clotting, which is preceded by an explosive growth of average particle weight, is explained satisfactorily by the following equation (Payens, 1978; Schmidt, 1980):

$$\frac{M_w}{M_{t=0}} = 1 - A\left(\frac{t}{\tau}\right) + B\left(\frac{t}{\tau}\right)^3 \tag{2}$$

where τ, the so-called enzyme clotting time, is equal to $(K_s V/2)^{\frac{1}{2}}$; V being the rate at which coagulation particles are formed and K_s the coagulation rate constant (Payens, 1976; Payens, 1977).

Equation (2) predicts that when t is very small compared with τ, M_w will decrease but when $t \geqslant \tau$, the molecular weight will increase explosively. This predicted behaviour was observed by turbidity and viscosity measurements also in model experiments on clotting of casein by chymosin or trypsin but the time-scale was considerably smaller (Payens, 1978).

A study of the coagulation kinetics can explain satisfactorily certain forms of gelation in sterilized milk but is not applicable to some other types of gelation, for example in samples of sterilized milk, where gelation was observed without accompanying proteolysis (Payens, 1978, de Koning & Kaper, 1981; Harwalkar *et al.*, 1983; de Koning *et al.*, 1985) or in samples where gelation could not be correlated with the rate and extent of proteolysis, as influenced by storage temperature (Wilson *et al.*, 1961; Samel *et al.*, 1971; Manji & Kakuda, 1988) or the presence of additives such as phosphates, citrates and hydrogen peroxide, etc. (see Section 5). Also, the observed gelation of evaporated milk that was cold-stored before retort sterilization (Leviton *et al.*, 1963; Heintzberger *et al.*, 1972) and the failure to prevent the gelation of HTST sterilized concentrates by an additional retort sterilization treatment (Heintzberger *et al.*, 1972) are incompatible with the proteinase hypothesis. Furthermore, in a recent study of physicochemical changes during storage of UHT-sterilized concentrated casein micelles from aseptically drawn milk, de Koning and Kaper (1981, 1985) reported typical age-thinning followed by gelation without accompanying proteolysis. They controlled proteolysis by addition of proteinase inhibitors. Age-gelation was also observed in UHT sterilized evaporated skim milk in which proteolysis was prevented by the addition of protease inhibitors (de Koning *et al.*, 1985). The initial decrease in viscosity during storage of milk has been linked to heat treatment rather than proteolysis (de Koning *et al.*, 1985). They observed no age-thinning in raw preserved milk with or without added proteinase inhibitor.

De Koning *et al.* (1985) have suggested two separate mechanisms for the age-gelation of unconcentrated and concentrated UHT-sterilized skim milk. Age-gelation of unconcentrated skim milk is preceded by proteolysis whereas gelation of concentrated skim milk occurs even when the proteolysis is prevented. Manji and Kakuda (1988) made similar observations regarding the different age-gelation behaviour of unconcentrated and concentrated sterilized milks. They achieved different levels of proteolysis by adding plasmin and adding proteinase inhibitors to stop further proteolysis. They observed that a certain minimum level of proteolysis was required for the age-gelation of unconcentrated milk although higher levels of proteolysis did not accelerate gelation further. Sterilized concentrated milk gelled, even without induced proteolysis, although low levels of proteolysis accelerated gelation. Kohlmann *et al.* (1988), however, were unable to observe age-gelation of unconcentrated milk treated with higher levels of plasmin ($>0{\cdot}3$ mg/litre) but observed gelation with very low levels of plasmin (0·15 mg/litre) and plasminogen. The visual characteristics and microstructure by electron microscopy of the gelled UHT-sterilized unconcentrated milk were different from those for concentrated milk. Perhaps this is due to loss of micellar integrity in the gels from unconcentrated milk as seen in micrographs (de Koning *et al.*, 1985). It is not, however, clear from de Koning's data at what state of storage the micellar integrity is lost, although micellar integrity is retained during storage when proteolysis was prevented.

Although proteinases have been isolated from some UHT-sterilized milks (Corradini & Pecis, 1979; Snoeren *et al.*, 1979*a*), it has not been possible to demonstrate their presence in other sterilized products, for example in UHT-sterilized concentrated milk (Nakai *et al.*, 1964*b*) or 10% cream (Cheng & Gelda, 1974), possibly due to lack of sensitivity of the assay methods. Furthermore, there is a wide variation in the nature and extent of proteolysis and the occurrence of gelation could not be related to any specific degree of protein breakdown (Table 10). This apparent lack of relationship between gelation and degree of proteolysis could be the result of wide variations in the specificity of the proteinases that are indigenous to milk or that are of bacterial origin. Some of these proteinases lack the highly-specific proteolytic action of rennet, which primarily attacks the Phe–Met bond between residues 105 and 106 of κ-casein to yield a (glyco)macropeptide and para-κ-casein. In rennet-treated milk, 80–90% of κ-casein is degraded before an increase in viscosity is registered (Green *et al.*, 1978; Green & Morant, 1981) whereas in sterilized milk (or milk concentrates), only 5–10% of κ-casein is degraded, as evidenced by the observation of low levels of macropeptide at the time of gelation (Samel *et al.*, 1971; Harwalkar & Vreeman, 1978*a*). Presumably, these enzymes affect the surface of the casein micelles differently from rennet to create sites at which interaction and aggregation reactions occur. However, experimental evidence to support such an hypothesis is lacking.

8.2. Non-enzymatic Basis for Gelation

The failure to correlate the rate and extent of proteolysis to gelation in concentrated sterilized milk products has led to the suggestion that gelation results from a modification of the surface properties of casein micelles by non-enzymatic, physicochemical processes (Hostettler *et al.*, 1968; Samel *et al.*, 1971; Hostettler, 1972; de Koning *et al.*, 1985). The exact nature of the physicochemical processes are poorly understood. Hostettler *et al.* (1957) and Hostettler (1972) observed that extensive proteolytic breakdown accompanies the gelation of unconcentrated milk but were unable to observe such proteolysis during the gelation of UHT-sterilized concentrated milk (Hostettler *et al.*, 1968). These observations were similar to those of de Koning *et al.* (1985). Hostettler *et al.* (1968) suggested that gelation of concentrated products was affected by the nature and extent of heat induced complex formation between the whey proteins and caseins. In milk subjected to a severe heat treatment, e.g. autoclaved milk, a complete and irreversible interaction occurs between whey proteins and casein and the resultant complex protects the micelles from further changes and interaction with each other during storage. A milder heat treatment, as in UHT treatment by direct steam injection, leads to the formation of an incomplete and reversible complex that does not protect the micelles from further changes and interaction during storage. A similar suggestion was also made by Graf and Bauer (1976) who consider the UHT heat treatment to be insufficient to 'complete' the heat-induced micellar transformation. The subsequent rapid cooling 'freezes' an unfinished micellar state which is 'unstable'. The incomplete nature of the complex is evident from the experimental observation of residual undenatured whey proteins in UHT-sterilized milk, but the reversible dissociation of the complex, as suggested by Hostettler *et al.* (1968) and Hostettler (1972), has not been substantiated. The hypothesis suggesting protective action of the complex against gelation is less tenable because gelation was observed in some evaporated milk samples that were cold-stored before sterilization by the retort method (Heintzberger *et al.*, 1972; Harwalkar *et al.*, 1983). In these samples, the sterilization temperature is sufficient to form an irreversible complex. Furthermore, increasing the level of whey proteins (Ellertson & Pearce, 1964), which presumably increases the degree of complex formation, hastens gelation of HTST-sterilized concentrated milk.

A chemical modification of casein micelles by Maillard-type (Andrews & Cheeseman, 1971, 1972; Andrews, 1975; Möller *et al.*, 1977) reactions or by disulphide bonding (Nakai *et al.*, 1965*b*; Patrick & Swaisgood, 1976) has been implicated in the mechanism of gelation. Andrews and Cheeseman (1971, 1972) and Andrews (1975) suggested that the gelation of UHT-sterilized milk is linked with the polymerization of casein and whey proteins through Maillard-type reactions which are promoted as the temperature of storage is increased. However, the failure to observe gelation during storage of sterilized milk at temperatures above 35°C is not consistent with their suggestion. Substituting sorbitol for lactose did not prevent the gelation of UHT-sterilized concentrated suspensions of casein micelles (de Koning & Kaper, 1981, 1985). Others (Hostettler *et al.*, 1957; Samel *et al.*, 1971) have suggested that the blockage of ε-NH_2 groups of lysine residues by interaction with lactose or its degradation products prevents casein micelles from interacting and gelling. These reactions presumably modify the charge on the casein micelles but their significance in the mechanism of gelation is not clear as yet.

Modification of the surface properties of casein micelles may also be caused by partial dissociation of the micelles during storage (see Section 6). This process of dissociation or 'shedding' could expose regions on the surface of casein micelles that promote interaction between casein micelles. However, the forces that lead to the 'shedding' of subunits are not known. The reaction of the exposed regions could result from slow conformational changes at the surface of the casein micelles (Fox, 1969; Samel *et al.*, 1971; Harwalkar & Vreeman, 1978*b*). The deformation at the surface of the casein micelles observed during storage by electron microscopy (Andrews *et al.*, 1977; Harwalkar & Vreeman, 1978*b*; Blanc *et al.*, 1980; de Koning *et al.*, 1985) would tend to support this notion but more definitive experimental substantiation of such conformation changes at the surface of casein micelles is needed.

Another hypothesis of gelation, based on non-enzymatic modification of casein micelles, is considered in terms of changes in free energy (Graf & Bauer, 1976). The micelles in sterilized milk are perceived as being in a metastable state with a high surface potential. During storage, a progressive, spontaneous transformation of the micelles from a 'high' potential to more stable micelles with lower surface potentials occurs by random potential jumps. The decrease in potential energy of some micelles creates a potential difference which promotes aggregation of the micelles, depending upon the probability of contacts and the number of low-potential micelles; both of these increase with time. Initially, the aggregation is insignificant but when the average surface potential of the micelles attains a critical level, aggregation becomes evident as a pronounced increase in viscosity (Graf & Bauer, 1976). The stabilizing influence of an intense heat treatment and of additives is believed to result from a condition of the micelles in which changes in the surface potential and racemization of the micelles are reduced. The factors that cause stabilization against gelation are believed to enhance such changes. Such a hypothesis based on potential energy changes is highly speculative but merits further consideration through experimental approach.

9. Conclusion

Although the fundamental understanding of the mechanism of gelation in sterilized milk (or concentrates) is not adequate at present, there are several empirical approaches that are useful in extending gelation-free shelf life of sterilized products in practical situations:

(i) *Assurance of good-quality milk for the manufacture of sterilized products.* Microbial contamination should be kept to a minimum by maintaining good hygienic conditions throughout production and processing of milk by minimizing the cold storage of raw milk and by thermization treatment of milk before cold storage. The use of milk that is likely to contain high levels of proteolytic enzymes (e.g. late lactation, mastitic infection) for the manufacture of UHT-sterilized products should be minimized.

(ii) *Adequacy of heat treatment during forewarming and sterilization.* What consitutes adequate heat treatment is difficult to determine because of wide variations in the quality of raw milk processed in different geographic regions and at different times. Generally, a slightly more severe heat treatment than is necessary to achieve commercial sterility delays gelation. However, such beneficial effects must be weighed against undesirable changes in flavour.

(iii) *Maintenance of lower storage temperatures ensures longer gelation-free shelf life.* The sterilized products should be stored, where possible, at cooler temperatures ($< 20°C$). Storage at refrigeration temperatures is beneficial from the point of view of flavour changes and gelation but it nullifies the primary advantage of UHT-sterilized products, namely, the possibility of storage at ambient temperatures.

(iv) *Incorporation of additives.* Polyphosphates, thus far, appear to offer best protection against gelation.

Acknowledgements

This is contribution No. 875, Food Research Centre, Agriculture Canada, Ottawa, K1A OC6 Canada.

The authors sincerely thank Dr M.R. Sahasrabudhe, Miloslav Kalab and J.D. Jones for making useful comments and reviewing the chapter, and D. Raymond, J. Obas and C. Lafontaine for their assistance in the preparation of the manuscript.

References

Adams, D.M., Barach, J.T. & Speck, M.L. (1975). Heat resistant proteases produced in milk by psychrotrophic bacteria of dairy origin. *J. Dairy Sci.*, **58**, 828–34.

Alfa Laval (1967). A method of sterilizing milk. British Patent No. 1 058 030.

Andrews, A.T. (1975). Properties of aseptically packed ultra-high-temperature milk. III. Formation of polymerized protein during storage at various temperatures. *J. Dairy Res.*, **42**, 89–99.

Andrews, A.T. (1978*a*). The composition, structure and origin of proteose-peptone component 5 of bovine milk. *Eur. J. Biochem.*, **90**, 59–65.

Andrews, A.T. (1978*b*). The composition, structure and origin of proteose-peptone component 8f of bovine milk. *Eur. J. Biochem.*, **90**, 66–71.

Andrews, A.T. & Cheeseman, G.C. (1971). Properties of aseptically packed UHT milk: Casein modification during storage and studies with model systems. *J. Dairy Res.*, **38**, 193–207.
Andrews, A.T. & Cheeseman, G.C. (1972). Properties of aseptically packed ultra-high-temperature milk. II. Molecular weight changes of the casein components during storage. *J. Dairy Res.*, **39**, 395–408.
Andrews, A.T., Brooker, B.E. & Hobbs, D.G. (1977). Properties of aseptically packed ultra-heat-treated milk. Electron microscopic examination of changes occurring during storage. *J. Dairy Res.*, **44**, 283–92.
Anon. (1972). Monograph on UHT milk. FIL-IDF Ann. Bull. Part V. International Dairy Federation, Brussels.
Aoki, T. & Imamura, T. (1974*a*). Changes of the casein complex in sterilized concentrated skim milk during storage. *Agric. Biol. Chem.*, **38**, 309–14.
Aoki, T. & Imamura, T. (1974*b*). Changes of casein complex during storage of sterilized skim milk. *Agric. Biol. Chem.*, **38**, 1929–34.
Aoki, T., Hatanaka, C. & Imamura, T. (1977). Large casein micelles and soluble casein in heated concentrated whey protein-free milk. *Agric. Biol. Chem.*, **41**, 2349–55.
Aschaffenburg, R. & Drewry, J. (1959). New procedure for the routine determination of the various non-casein proteins of milk. *Proc. IX Intern. Dairy Congr.*, London, Vol. 3, Section 5, pp. 1631–7.
Ashton, T.R. (1965). Practical experience: The processing and aseptic packaging of sterile milk in the United Kingdom. *J. Soc. Dairy Technol.*, **18**, 65–85.
Ashton, T.R. (1979). Factors affecting the keeping quality of heat-treated milk. International Dairy Federation, FIL-IDF B-DOC. 76.
Barach, T.J., Adams, D.M. & Speck, M.L. (1976). Low temperature inactivation in milk of heat resistant proteases from psychrotrophic bacteria. *J. Dairy Sci.*, **59**, 391–5.
Belec, J. & Jenness, R. (1962). Dephosphorylation of casein by heat treatment. 1. In casein solutions. *J. Dairy Sci.*, **45**, 12–19.
Bell, R.W., Curan, H.R. & Evans, F.R. (1944). Effects of temperature and time of sterilization upon properties of evaporated milk. *J. Dairy Sci.*, **27**, 913–19.
Beeby, R. & Loftus Hills, G. (1962). Gelation in evaporated milk—The effect of protein content. *Proc. XVI Intern. Dairy Congr.*, Copenhagen, B, pp. 1019–23.
Beeby, R., Hill, R.D. & Snow, N.S. (1971). Recent advances in milk protein chemistry are assessed in relation to problems of milk technology. In *Milk Proteins: Chemistry and Molecular Biology*, vol. II, ed. H.A. McKenzie. Academic Press, New York, pp. 421–65.
Bengtsson, K., Gardhage, L. & Isaksson, B. (1973). Gelation in UHT treated milk, whey and casein solution. The effect of heat resistant proteases. *Milchwissenschaft*, **28**, 495–9.
Björck, L. (1973). Modifications of casein in ultra-high-temperature treated milk during storage. *Milchwissenschaft*, **28**, 291–3.
Blanc, B., Fluckiger, R., Ruegg, M. & Steiger, C. (1980). Changes in UHT milk during storage. *Alimenta-Sonderausgabe*, 27–47.
Board, P.W. & Mullett, J.T. (1970). An improved method for canning 3-fold concentrated whole milk. *J. Dairy Res.*, **37**, 505–11.
Burton, H. (1965). Ultra-high temperature processing and aseptic packaging in the dairy industry. *J. Soc. Dairy Technol.*, **18**, 58–65.
Burton, H. (1969). Ultra-high temperature processed milk. *Dairy Sci. Abstr.*, **31**, 287–97.
Burton, H. (1984). Reviews of the progess of dairy science. The bacteriological, chemical and biochemical and physical changes that occur in milk at temperatures of 100–150°C. *J. Dairy Res.*, **51**, 341–63.
Burton, H. (1988). *Ultra-High Temperature Processing of Milk and Milk Products.* Elsevier Applied Science Publishers, London.
Calbert, H.E. & Swanson, A.M. (1959). Techniques for processing sterile concentrated whole milk. *Proc. XV Intern. Dairy Congr.* London, I, pp. 442–7.

Carroll, R.J., Thompson, M.P. & Melnychyn, P. (1971). Gelation of concentrated skimmilk: electron microscopic study. *J. Dairy Sci.*, **54**, 1245–52.

Cheng, W.S. & Gelda, C.S. (1974). Proteolysis in ultra-high temperature treated and canned ten percent cream. *J. Dairy Sci.*, **57**, 1502–4.

Choudhery, A.K. & Mikolajcik, E.M. (1971). Bacilli in milk. 2. Sweet curd formation. *J. Dairy Sci.*, **54**, 321–5.

Corradini, C. (1967). Researches on sterilized milk. III. Gel formation during conservation. *Scienza Tec-Latt-Casear*, **18**, 277–86.

Corradini, C. (1971). Factors influencing gel formation in UHT-sterilized milk. *Scienza Tec-Latt-Casear*, **22**, 243–8.

Corradini, C. (1975). Gel formation behaviour in UHT-sterilized milk. *Milchwissenschaft*, **30**, 413–16.

Corradini, C. & Pecis, P.P. (1979). Proteolytic activity in UHT-sterilized milk. *J. Dairy Res.*, **46**, 227–9.

Davis, F.L., Shankar, P.A., Brooker, B.E. & Hobbs, D.G. (1978). A heat-induced change in the ultrastructure of milk and its effect on gel formation in yoghurt. *J. Dairy Res.*, **45**, 53–8.

de Koning, J. & Kaper, J. (1981). The effects of protease inhibitor and the Maillard reaction on the gelation of UHTST-concentrated casein micelle solution. *Neth. Milk Dairy J.*, **35**, 323–7.

de Koning, P.J. & Kaper, J. (1985). Effect of some proteinase inhibitors and the Maillard reaction on the process of age-thinning and gelation of UHTST-sterilized concentrated casein micelle dispersions. *Neth. Milk Dairy J.*, **39**, 37–47.

de Koning, P.J., Kaper, J., Rollema, H.S. & Driessen, F.M. (1985). Age-thinning and gelation in unconcentrated and concentrated UHT-sterilized skim milk. Effect of native milk proteinase. *Neth. Milk Dairy J.*, **39**, 71–87.

Deysher, E.F., Webb, H.B. & Holm, G.E. (1944). The viscosity of evaporated milks of different solids concentration. *J. Dairy Sci.*, **27**, 345–55.

Doan, F.J. & Warren, F.G. (1947). Observations on the instability of the protein phase of frozen concentrated milk. *J. Dairy Sci.*, **30**, 837–48.

Driessen, F.M. (1989). Inactivation of lipases and proteinases (indigenous and bacterial). International Dairy Federation, Brussels, Bulletin no. 238, pp. 71–93.

Driessen, F.M. & van de Waals, C.B. (1978). Inactivation of native milk proteinase by heat treatment. *Neth. Milk Dairy J.*, **32**, 245–54.

Edmondson, L.F. (1959). A study of the effect of variations in the initial viscosity of HTST sterilized evaporated milk at the time of aseptic homogenization and subsequent gelling in storage. *J. Dairy Sci.*, **42**, 910 (abstr.).

Edmondson, L.F. (1970). Sterilized products. In *By-products From Milk*, 2nd edn, ed. B.H. Webb & E.V. Whittier. The AVI Publishing Company, Inc., Westport, CT, pp. 226–73.

Edmondson, L.F., Avants, J.K., Douglas, F.W. & Easterley, D.G. (1966). Phosphatase and lipase reactivation in high-temperature short-time sterilized milk. *J. Dairy Sci.*, **49**, 708 (abstr.).

Ellertson, M.E. & Pearce, S.J. (1964). Some observations on the physical–chemical stability of sterile concentrated milks. *J. Dairy Sci.*, **47**, 564–72.

Farah, Z. (1979*a*). Electron microscope studies on gelation by UHT-treated milk during storage. *Lebensm-Wiss, U. Technol.*, **12**, 169–71.

Farah, Z. (1979*b*). Changes in proteins in UHT-heated milk. *Milchwissenschaft*, **34**, 484–7.

Fox, K.K., Harper, M.K., Holsinger, V.H. & Pallansch, M.J. (1967*a*). Effects of high-heat treatment on stability of calcium caseinate aggregates in milk. *J. Dairy Sci.*, **50**, 443–50.

Fox, K.K., Holsinger, V.H., Posati, L.P. & Pallansch, M.J. (1967*b*). Composition of granules in evaporated milks stored at low temperatures. *J. Dairy Sci.*, **50**, 1032–7.

Fox, P.F. (1969). Effect of cold-aging on the rennet-coagulation time of milk. *Irish J. Agr. Res.*, **8**, 175–82.

Galesloot, T.E. (1956). A simple method for estimating the bacteriological effect of sterilizing process for milk, applied to the sterilization of milk in ultrahigh temperature-short time (UHTST) and 'in bottle' sterilization. *Neth. Milk Dairy J.*, **10**, 79–100.

Gebre-Egziabher, A., Humbert, E.S. & Blankenagel, G. (1980). Hydrolysis of milk proteins by microbial enzymes. *J. Food Protection*, **43**, 709–12.

Giroux, R.N., Calbert, H.E. & Swanson, A.M. (1958). A study of the processes of gelation, flocculation, and caramelization as related to heat stability in concentrated milk. *J. Dairy Sci.*, **41**, 710 (abstr.).

Glandorf, K. & Thomasow, J. (1977). The behaviour of polyphosphates during production and storage of long keeping milk products. *Z. Lebensmittel-Unters. Forshung*, **163**, 178–82.

Goldblith, S.A. (1967). Radiation processing of foods and drugs. In *Fundamentals of Food Processing Operations*, ed. J.L. Heid & M.A. Joyslyn. AVI Publishing Company, Westport, CT, pp. 328–73.

Graf, E. & Bauer, H. (1976). Milk and milk products. In *Food Emulsions*, ed. S. Friberg. Marcel Dekker Inc., New York, pp. 295–383.

Green, M.L. & Morant, S.V. (1981). Mechanism of aggregation of casein micelles in rennet-treated milk. *J. Dairy Res.*, **48**, 57–63.

Green, M.L., Hobbs, D.G., Morant, S.V. & Hill, V.C. (1978). Intermicellar relationships in rennet-treated separated milk. II. Process of gel assembly. *J. Dairy Res.*, **45**, 413–22.

Guthy, K. & Horak, P. (1980). On the stability of casein micelles during storage of directly or indirectly heated UHT skim-milk. *Milchwissenschaft*, **35**, 600–3.

Harper, W.J., Hildago, J.E. & Mikolajcik, E.M. (1970). *Bacillus cereus* protease as a factor in gelation of sterilized dairy products. *Proc. Intern. Dairy Congr.*, Sydney, IE., 192.

Harwalkar, V.R. (1982). Age gelation of sterilized milks in developments. In *Dairy Chemistry 1—Proteins*, ed. P.F. Fox. Applied Science Publishers, London, pp. 229–69.

Harwalkar, V.R. & Vreeman, H.J. (1978*a*). Effect of added phosphates and storage on changes in ultra-high temperature short-time sterilized concentrated skim-milk. 1. Viscosity, gelation, alcohol stability, chemical and electrophoretic analysis of proteins. *Neth. Milk Dairy J.*, **32**, 94–111.

Harwalkar, V.R. & Vreeman, H.J. (1978*b*). Effect of added phosphates and storage on changes in ultra-high temperature short-term sterilized concentrated skim-milk. 2. Micelle structure. *Neth. Milk and Dairy J.*, **32**, 204–16.

Harwalkar, V.R., Beckett, D.C., McKellar, R.C., Emmons, D.B. & Doyle, G.E. (1983). Age thickening and gelation of sterilized evaporated milk. *J. Dairy Sci.*, **66**, 735–42.

Harwalkar, V.R., Allan-Wojtas, P. & Kalab, M. (1989). Effect of heating to 200°C on casein micelles in milk: A metal shadowing and negative staining electron microscopy study. *Food Microstructure*, **8**, 217–24.

Heintzberger, H., Koops, J. & Westerbeek, D. (1972). Gelation of sterilized canned evaporated milk. *Neth. Milk and Dairy J.*, **26**, 31–40.

Hindle, E.J. & Wheelock, J.V. (1969). Action of rennin on sterilized milk. *Biochem. J.*, **115**, 19P (abstr.).

Hindle, E.J. & Wheelock, J.V. (1970). The primary phase of rennin action in heat-sterilized milk. *J. Dairy Res.*, **37**, 389–96.

Hoff, J.E., Wertheim, J.H., Roychodhury, R.N., Deolalikar, S.T., Proctor, B.E. & Goldblith, S.A. (1958). Radiation preservation of milk and milk products. VII. Storage studies on milk subjected to cobalt 60 radiation distillation. *Food Tech.*, **12**, (12), 648–54.

Hoff, J.E., Sunyack, J., Proctor, B.E. & Goldblith, S.A. (1960*a*). Radiation preservation of milk. X. Studies on the radiation induced gelation of concentrated milk. Methods and effect of some additives. *Food Tech.*, **14**, (1) 24–6.

Hoff, J.E., Sunyach, J., Proctor, B.E. & Goldblith, S.A. (1960*b*). Radiation preservation of milk. XI. Studies on the radiation-induced gelation of concentrated milk. The effect of polyphosphates. *Food Tech.*, **14** (1) 27–9.

Hostettler, H. (1972). Appearance, flavour and texture aspects. History of the development of UHT processes. IDF monograph on UHT milk. FIL-IDF Annual Bull., Part V, pp. 169–74.

Hostettler, H., Stein, J. & Bruderer, G. (1957). The age thickening of sterilized milk. *Landw. Jahrb. Schweiz*, **71**, 143–63.

Hostettler, H., Imhof, K. & Stein, J. (1965). Studies on the effect of heat treatment and lysophilisation on the state of distribution and physiological properties of milk proteins with special consideration of heat-treatment conditions applied in uperization. I. Effect on the state of distribution of milk proteins. *Milchwissenschaft*, **20**, 189–98.

Hostettler, H., Stein, J. & Imhof, K. (1968). Research is undertaken on the processes in relation to gelation and age thickening of sterilized unsweetened milk concentrates. *Mitt. Geb. Lebensm., Unters., U. Hug.*, **59**, 60–5.

Humbert, G. & Alais, C. (1976). Enzymatic coagulability of milks after different types of heat treatment. *Industria Lechera*, **9**, (647) 9–12.

Hunziker, D.F. (1944). *Condensed Milk and Milk Powder*, 7th edn. D.F. Hunziker, LaGrange, IL.

Kannan, A. & Jenness, R. (1961). Relation of milk serum proteins and milk salts to the effects of heat treatment of rennet clotting. *J. Dairy Sci.*, **44**, 808–22.

Kocak, H.R. & Zadow, J.G. (1985*a*). The effect of low temperature inactivation treatment on age gelation of whole milk. *Aust. J. Dairy Technol.*, **40**, 53–8.

Kocak, H.R. & Zadow, J.G. (1985*b*). Age gelation of UHT whole milk as influenced by storage temperature. *Aust. J. Dairy Technol.*, **40**, 14–21.

Kohlmann, K.L., Nielsen, S.S. & Ladisch, M.R. (1988). Effect of serine proteolytic enzymes (trypsin and plasmin), trypsin inhibitor and plasminogen activator addition to ultra-high temperature processed milk. *J. Dairy Sci.*, **71**, 1728–39.

Law, B.A., Andrews, A.T. & Sharpe, M.E. (1977). Gelation of ultra-high temperature-sterilized milk by proteases from a strain of *Pseudomonas flourescens* isolated from raw milk. *J. Dairy Res.*, **44**, 145–8.

Leviton, A. & Pallansch, M.J. (1961*a*). High temperature short-time sterilized evaporated milk. II. Laboratory techniques for the preparation and study of sterile evaporated milk. *J. Dairy Sci.*, **44**, 442–50.

Leviton, A. & Pallansch, M.J. (1961*b*). High temperature short-time sterilized evaporated milk. III. The influence of the lipid phase on heat and storage stability. *J. Dairy Sci.*, **44**, 633–42.

Leviton, A. & Pallansch, M.J. (1962). High temperature short-time sterilized evaporated milk. IV. The retardation of gelation with condensed phosphates, manganous ions, polyhydric compounds and phosphatides. *J. Dairy Sci.*, **45**, 1045–56.

Leviton, A., Pallansch, M.J. & Webb, B.H. (1962). Effect of phosphate salts on the thickening and gelation of some concentrated milks. *Proc. XVI Intern. Dairy Congr.*, Copenhagen, B64, pp. 1009–16.

Leviton, A., Anderson, H.A., Vettel, H.E. & Vestel, J.H. (1963). Retardation of gelation in high-temperature short-time sterile milk concentrates with polyphosphates. *J. Dairy Sci.*, **46**, 310–19.

Lyster, R.L.J., Wyeth, T.C., Perkins, A.G. & Burton, H. (1971). Comparison of milks processed by the direct and indirect methods of ultra-high temperature sterilization. *J. Dairy Res.*, **38**, 403–8.

Manji, B. & Kakuda, Y. (1988). The role of protein denaturation, extent of proteolysis and storage temperature on the mechanism of age-gelation in a model system. *J. Dairy Sci.*, **71**, 1455–63.

Manji, B., Kakuda, Y. & Arnott, D.R. (1986). Effect of storage temperature on age-gelation of ultra-high temperature milk processed by direct and indirect heating systems. *J. Dairy Sci.*, **69**, 2944–3001.

McGugan, W.A., Zehren, V.P., Zehren, V.L. & Swanson, A.M. (1954). Interaction between casein and β-lactoglobulin on heating. *Science*, **120**, 435 (abstr.).

Mehta, R.S. (1980). Milk processed at ultra-high temperature—a review. *J. Food Protection*, **43**, 212–25.
Melechouris, N.P. & Tuckey, S.L. (1966). Denaturation of whey proteins in milk heated at high temperatures for short times. *J. Dairy Sci.*, **49**, 1154–6.
Melnychyn, P. & Wolcott, J.M. (1967). Simple procedure for isolation of α_s-caseins. *J. Dairy Sci.*, **48**, 780 (abstr.).
Meyerhofer, H.J., Marshall, R.T., White, C.H. & Lu, M. (1973). Characterization of a heat stable protease of *Pseudomonas flourescens*. *Appl. Microbiol.*, **25**, 44–8.
Möller, A.B., Andrews, A.T. & Cheeseman, G.C. (1977). Chemical changes in ultra-heat-treated milk during storage. 1. Hydrolysis of casein by incubation with pronase and peptidase mixture. *J. Dairy Res.*, **44**, 259–66.
Morr, C.V. (1967). Some effects of pyrophosphate and citrate ions upon the colloidal caseinate-phosphate micelles and ultrafiltrate of raw and heated skimmilk. *J. Dairy Sci.*, **50**, 1038–44.
Morr, C.V. (1969). Protein aggregation in conventional and ultra-high temperature heated skimmilk. *J. Dairy Sci.*, **52**, 1174–80.
Morr, C.V. (1975). Chemistry of milk proteins in food processing. *J. Dairy Sci.*, **58**, 977–84.
Muir, D.D. (1984*a*). Reviews of the progress of dairy science: Frozen concentrated milk. *J. Dairy Res.*, **51**, 649–64.
Muir, D.D. (1984*b*). UHT-sterilized milk concentrate: a review of practical method of production. *J. Soc. Dairy Tech.*, **37**, 135–41.
Murthy, L., Herreid, E.O. & Whitney, R.M. (1958). Electrophoretic properties of casein from sterilized milk stored at different temperatures. *J. Dairy Sci.*, **41**, 1324–41.
Nakai, S., Wilson, H.K. & Herreid, E.O. (1964*a*). Assaying sterile concentrated milk for native proteolytic enzymes. *J. Dairy Sci.*, **47**, 754–7.
Nakai, S., Wilson, H.K. & Herreid, E.O. (1964*b*). Measurement of protein stability in sterile concentrated milk. *J. Dairy Sci.*, **47**, 1056–61.
Nakai, S., Wilson, H.K. & Herreid, E.O. (1965*a*). Effect of changes in sulfur compounds on stability and gelation of caseins and of sterile concentrated milk. *J. Dairy Sci.*, **48**, 431–7.
Nakai, S. Wilson, H.K. & Herreid, E.O. (1965*b*). Effect of denaturing agents of stability of casein, milk, and sterile concentrated milk. *J. Dairy Sci.*, **48**, 186–90.
Odagiri, S. & Nickerson, T.A. (1964). Hexametaphosphate, oxalate, citrate, and EDTA in milk. 1. Effects of complexing agents on turbidity and rennet coagulation. *J. Dairy Sci.*, **47**, 1306–9.
Odagiri, S. & Nickerson, T.A. (1965). Complexing of calcium by hexametaphosphate, oxalate, citrate and EDTA. 2. Dialysis of milk containing complexing agents. *J. Dairy Sci.*, **48**, 19–22.
O'Sullivan, A.C. (1970). Whey protein denaturation in conventional versus UHT sterilized milks. *Proc. XVIII Intern. Dairy Cong.*, Sidney, 1E, p. 187.
Parry, R.M. (1974). Milk coagulation and protein denaturation. In *Fundamentals of Dairy Chemistry*, 2nd edn, ed. B.H. Webb, A.H. Johnson & J.A. Alford. AVI Publishing Co., Westport, CT, pp. 603–61.
Patrick, P.S. & Swaisgood, H.E. (1976). Sulfhydryl and disulfide groups in skim milk as affected by direct ultra-high temperature heating and subsequent storage. *J. Dairy Sci.*, **59**, 594–600.
Payens, T.A.J. (1976). On the enzyme-triggered clotting of casein; a preliminary account. *Neth. Milk Dairy J.*, **30**, 55–9.
Payens, T.A.J. (1977). On enzymatic clotting processes. 1. Kinetics of enzyme-triggered coagulation reactions. *Biophys. Chem.*, **6**, 253–61.
Payens, T.A.J. (1978). On different modes of casein clotting; the kinetics of enzymatic and non-enzymatic coagulation compared. *Neth. Milk Dairy J.*, **32**, 178–83.
Perkin, A.G., Henschel, M.J. & Burton, H. (1973). Comparison of milks processed by the direct and indirect methods of ultra-high temperature sterilization. *J. Dairy Res.*, **40**, 215-20.

Polliott, Y., Britten, M. & Latreille, B. (1990). Effect of high pressure homogenization on a sterilized infant formula: Microstructure and age gelation. *Food Microstructure*, **9**, 1–8.

Pyne, G.T. (1962). Reviews of the progress of dairy science. Section C. Dairy chemistry. Some aspects of the physical chemistry of the salts of milk. *J. Dairy Res.*, **29**, 101–30.

Richardson, B.C. & Newstead, D.F. (1979). Effect of heat stable proteases on the storage life of UHT milk. *New Zealand J. Dairy Sci. Technol.*, **14**, 273–9.

Rollema, H.S. & Poll, J.K. (1986). The alkaline milk proteinase system: kinetics and mechanism of heat-inactivation. *Milchwissenschaft*, **41**, 536–40.

Rose, D. & Tessier, H. (1959). Composition of ultrafiltrates from milk heated at 80 to 230°F in relation to heat stability. *J. Dairy Sci.*, **42**, 969–80.

Rowland, S.J. (1938). The precipitation of the proteins in milk. I. Casein II. Total proteins III. Globulin IV. Albumin and proteose peptone. *J. Dairy Res.*, **9**, 30–41.

Rüegg, M. & Blanc, B. (1978). Influence of pasteurization and UHT processing upon the size distribution of casein micelles in milk. *Milchwissenschaft*, **33**, 364–6.

Samel, R., Weaver, R.W.V. & Gammack, D.B. (1971). Changes on storage in milk processed by ultra-high temperature sterilization. *J. Dairy Res.*, **38**, 323–32.

Samuelson, E.G. & Holm, S. (1966). Technological principles for ultra-high heat-treatment of milk. *Proc. XVIII Int. Dairy Congr.*, Munich, B. pp. 57–65.

Sawyer, W.H. (1969). Complex between β-lactoglobulin and κ-casein. A review. *J. Dairy Sci.*, **52**, 1347–55.

Sawyer, W.H., Coulter, S.T. & Jenness, R. (1963). Role of sulfhydryl groups in the interaction of κ-casein and β-lactoglobulin. *J. Dairy Sci.*, **46**, 564–5.

Schmidt, D.G. (1968). Electron-microsopic studies on the gelation of UHTST sterilized concentrated skim milk. *Neth. Milk Dairy J.*, **22**, 40–9.

Schmidt, D.G. (1969). Effect of the addition of α_s-, β- and κ-casein on the size of the casein micelles in UHTST-sterilized concentrated skim milk. *Neth. Milk Dairy J.*, **23**, 128–35.

Schmidt, D.G. (1980). Colloidal aspects of casein. *Neth. Milk Dairy J.*, **34**, 42–64.

Schmidt, D.G. & Buchheim, W. (1968). Electron-microscopic studies on the casein particles in sterile concentrated milk. *Milchwissenschaft*, **23**, 505–9.

Seehafer, M.E. (1972). The development and manufacture of sterilized milk concentrate. *FAO Agriculture Stud.*, **72**, 1–52.

Seehafer, M.E., Swanson, A.M. & Calbert, H.E. (1962). Effect of forwarming temperature on some properties of sterilized concentrated milks. *J. Dairy Sci.*, **45**, 656 (abstr.).

Snoeren, T.H.M. and Van Riel, J.A.M. (1979). Milk proteinase, its isolation and action on αs_2- and β-casein. *Milchwissenschaft*, **34**, 528–30.

Snoeren, T.H.M., Koops, J. & Westerbeek, D. (1976). Effect of the molecular weight of κ-carrageenan on the retardation of fat separation in evaporated milk. *Neth. Milk Dairy J.*, **30**, 29–40.

Snoeren, J.H.M., Van Riel, J.A.M. & Both, P. (1979*a*). Some properties of a milk proteinase isolated from UHT treated milk. *Zuive/Zicht*, **72**, 43–3.

Snoeren, J.H.M., van der Spek, C.A., Dekker, R. & Both, P. (1979*b*). Proteolysis during the storage of UHT-sterized whole milk. 1. Experiments with milk heated by the direct system for 4 seconds at 142°C. *Neth. Milk Dairy J.*, **33**, 31–9.

Speck, M.L. & Adams, D.M. (1976). Impact of heat stable microbial enzymes in food processing. *J. Dairy Sci.*, **59**, 786–9.

Stewart, A.P., Jr., Johnson, R.A. & Wilcox, D.F. (1959). Evaporated milk of improved taste. US Patent No. 2 886 450.

Swanson, A.M., Seehaffer, M.E., Calbert, H.E. & Geissler, J.J. (1965). Sterilized milk concentrate. 1. Efforts of processing treatments on physical properties and flavour. *Res. Bull. Agric. Exp. Stn, University of Wisconsin*, **256**, 1–60.

Swartling, P. (1968). UHT sterilization and aseptic packaging of milk. Technical Publication, Austalian Soc. Dairy Technol., **19**, 18–25.

Tarassuk, N.P. & Tamasma, A.F. (1956). Milk sterilization. Control of gelation in evaporated milk. *J. Agric. Food Chem.*, **4**, 1033–5.

Trautman, J.C. (1958). Heat induced complex formation in milk. PhD Thesis, University of Wisconsin.
United States Steel Corporation (1965). Processing sterilized milk concentrates of beverage quality. Pittsburgh, PA (Monograph).
Van Kreveld, A. (1950). Correlation between bacterial infection and sweet curd in evaporated milk. *Neth. Milk Dairy J.*, **4**, 115–20.
Van Kreveld, A. & van Minnen, G. (1955). Calcium and magnesium ion activity in raw milk and processed milk. *Neth. Milk Dairy J.*, **9**, 1–29.
Webb, B.H. (1970). Condensed products. *By-Products From Milk*, ed. B.H. Webb & E.V. Whittier. AVI Publishing Co., Westport, CT, pp. 83–123.
West, F.B., Adams, D.M. & Speck, M.L. (1978). Inactivation of heat resistant proteases in normal ultra-high temperature sterilized skim milk by a low temperature treatment. *J. Dairy Sci.*, **61**, 1078–84.
Wheelock, J.W. & Hindle, E.J. (1971). The effect of ageing on heat-sterilized milk. *J. Dairy Res.*, **38**, 145–9.
Whitney, R.McL., Brunner, J.R., Ebner, K.E., Farrell, H.M. Jr., Josephson, R.V., Morr, C.V. & Swaisgood, H.E. (1976). Nomenclature of the proteins of cow's milk: Fourth revision. *J. Dairy Sci.*, **59**, 795–815.
Wilson, H.K., Yoshino, U. & Herried, E.O. (1961). Size of protein particles in ultra-high temperature sterilized milk as related to concentration. *J. Dairy Sci.*, **44**, 1836–42.
Wilson, H.K., Vetter, J.L., Sasago, K. & Herried, E.O. (1963). Effects of phosphates added to concentrated milks before sterilization at ultra high temperatures. *J. Dairy Sci.*, **46**, 1038–43.
Wright, R.C. & Trammer, J. (1953). Reactivating of milk phosphatase following heat treatment. 1. *J. Dairy Res.*, **20**, 177–88.
Zadow, J.G. (1971). The influence of pH and heat treatment on the colour and stability of ultra-high-temperature sterilized milk. *J. Dairy Res.*, **38**, 393–401.
Zadow, J.G. & Chituta, F. (1975). Age gelation of ultra-high-temperature milk. *Aust. J. Dairy Technol.*, **30**, 104–6.
Zittle, C.A., Thompson, M.P., Custer, J.H. & Cerbulis, J. (1962). κ-Casein-β-lactoglobulin interaction in solution when heated. *J. Dairy Sci.*, **45**, 807–10.

18

ASPECTS OF PROTEINS IN MILK POWDER MANUFACTURE

H. Singh

Department of Food Technology, Massey University, Palmerson North, New Zealand

&

D.F. Newstead

New Zealand Dairy Research Institute, Palmerson North, New Zealand

1. Introduction

Milk powder has been known in some parts of the world for many centuries. References to milk powder date back to the 13th century when Marco Polo reported that the soldiers of Kubla Khan carried sun-dried milk on excursions. During the latter half of the 19th century, 10 patents (5 British and 5 American) were granted for various processes to produce milk powder. Developments in milk drying reached commercial reality early in the 20th century, when the drum drier with two rollers was invented (Hunziker, 1949). Spray-drying equipment was invented a few years later and was successfully applied to skim milk. Several improvements and modifications to the spray-drying process were developed during the following years. Until about 1947, roller-drying and spray-drying processes were equally used. In recent years, as the production units have become larger, there has been a complete shift to spray drying. Other drying techniques, such as foam drying and freeze drying, although feasible, have not become commercially viable. 'Instant' skim milk powder with improved solubility properties was first produced in the 1950s in the United States, as a result of the pioneering work of Peebles (1956). The basis of this new product was the formation of clusters of dried milk particles by agglomeration following rewetting of normal powder with steam. Widespread distribution of instant skim milk powder started in the USA during the 1960s, and in the course of a few years instant skim milk powder replaced ordinary spray-dried powder for home use. The manufacture of 'instant' whole milk powder was not successful until the early 1970s; the 'instant' properties were achieved by treating the agglomerated powder with lecithin. Processing equipment and procedures used in the manufacture of both normal and 'instant'

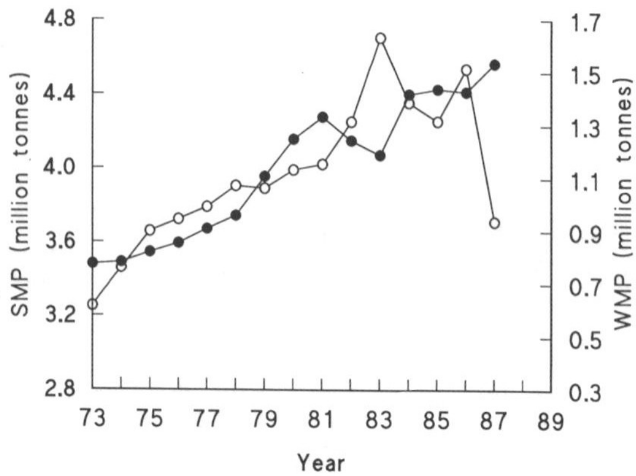

Fig. 1. Production trends of whole milk powder (WMP) and skim milk powder (SMP). —○—, Skim milk powder (data for 24 countries including the EEC, USA, Japan, Eastern Europe, Australia and Japan); —●—, whole milk powder (data for 20 countries). Source: International Dairy Federation Bulletins. *Meat and Dairy Products*, published by Commonwealth Secretriat, Marlborough House, London and *World Dairy Situation, Dairying in the Nineties*, International Dairy Federation Bulletin No. 243 (1989), pp. 12–13.

milk powders have been described comprehensively (Hall & Hendrick, 1971; Masters, 1972; Jensen, 1975).

Until the 1960s, evaporated and sweetened condensed milks were the most important export items of the US and European dairy industries, especially to countries with warm climates. It was found that costs could be reduced, especially for transport, by exporting skim milk powder and butterfat, recombining them into evaporated milk or sweetened condensed milk and filling them into cans close to the site of consumption. In recent years, several recombining plants have been built in the Far East, Middle East and Africa.

World production of milk powder totalled around 6 million tonnes in 1988, of which around 4 million tonnes was skim milk powder. Production trends for whole and skim milk powders are shown in Fig. 1. Available data indicate that the production of skim milk powder in the 'developed' countries increased steadily up to 1983 but has declined in recent years, mainly as a result of measures to control milk production in the EEC and the USA. The production of whole milk powder, however, has increased considerably during the last 15 years. This increase has been due mainly to the development of more sophisticated processing and customer-oriented marketing to produce powders with desired physico-chemical and functional characteristics. The growing importance of recombining operations in 'developing' countries has greatly increased the demand for milk powders.

Proteins are the main functional components in both whole and skim milk powders. For the most part, the caseins retain their integrity and thus their unique functional properties, i.e. curd formation, heat stability and water holding capacity. The whey protein components may be more or less denatured, depending on the heat treatment of the milk during manufacture, and thus affect the disper-

TABLE 1
Specific Grading Requirements for Milk Powders (Standard Grade, ADMI, 1971)

	Whole milk powder	*Skim milk powder*
Milkfat (%)	min. 26·00	max. 1·50
Moisture (%)	max. 3·00	max. 5·00
Titratable acidity (%)	max. 0·17	max. 0·17
Solubility index (ml)	max. 1·00	max. 2·00
Bacterial estimate (per g)	max. 100 000	max. 100 000
Scorched particles	Disc C (22·5 mg)	Disc C (22·5 mg)
Copper (ppm)	max. 3·00	–
Iron (ppm)	max. 15·00	–

sibility and solubility of these powders. Although an increasing amount of research into milk powders has been directed towards factors that control their physical properties, e.g. solubility and dispersibility (Kieseker, 1976; Sanderson, 1978), much less information is available on the physico-chemical reactions that occur during processing and the effects these reactions have on the functionality of the final powder.

2. TYPES AND USES OF MILK POWDERS

Skim and whole milk powders are bought and sold on the basis of grade and are available in 'instant' and 'normal' forms. The grades may follow the standards of the American Dry Milk Institute (ADMI, 1971) or the British Standards Institution. Basically, the grades specify the quality of the milk powder and are determined on the basis of flavour, odour, physical appearance, bacterial load, scorched particle content, fat content, dispersibility, solubility, acidity and moisture content (Table 1).

Skim milk powder is also classified according to the content of undenatured whey proteins, which depends on the heat treatment given to the milk before making the powder. Thus, high-, medium- and low-heat skim milk powders contain less than 1·5, 1·5–5·9, and more than 6 mg, respectively, of undenatured whey protein nitrogen per gram of skim milk powder. Many buyers impose additional specifications, depending on the end-use of the product. These may include water absorption characteristics or the viscosity of the reconstituted powder, a limit on the number of thermoduric or themophilic microorganisms, the quantity of free fat in whole milk powder or the density and flow characteristics of the powder. Several other types of milk powders with specialized properties and applications are available: heat-stable powders; filled milk powders containing various levels of fat; bakery powders with added shortenings, emulsifiers and potassium bromate; powders enriched with casein or whey proteins; powders with increased or reduced levels of lactose or milk salts.

Milk powder is the most widely used dairy-based ingredient for the food industry; its properties make it suitable for use in a wide range of foods (Table 2). Various components of milk powders impart some characteristic sensory, func-

TABLE 2
General Uses of Milk Powders

Reconstituted and recombined dairy products
Baked products, including cakes, cake mixes, bread, biscuits
Confectionery, including chocolate
Soups
Meat products
Cultured milk products, including yoghurt
Desserts, including mousse and ice cream
Diet beverages
Processed cheeses and cheese products
Baby foods
Animal feed

tional and nutritional properties to the food. A large quantity of high-heat skim milk powder is used in the bakery industry. The high water absorption ability of the high-heat powder is also useful in confectionery products and comminuted meat products. Medium-heat powders are used in products, such as ice cream, where both flavour and water binding are important. Low-heat powder is preferred for liquid products, e.g. beverages, because of its good flavour and solubility properties. Instant breakfast beverages, diet supplements and a wide range of similar products, including low calorie products, are essentially low-heat skim milk powder blended with sweeteners, thickeners, flavours, vitamins and minerals.

The main application of whole milk powder in non-dairy foods is in confectionery products, such as candy and chocolate, in cakes, cookies and biscuits, and in convenience foods, e.g. soups. Both whole and skim milk powders are used as ingredients in the formulation of various filled and imitation products, e.g. frozen desserts, sour creams, various cheeses, (cream cheese, ricotta cheese), soups, etc. Viscosity, gelation, emulsification and fat binding properties of the milk proteins are important in these products.

In countries where the fresh milk supply is inadequate, the principal use of skim milk powder is in reconstituted/recombined milk products. About 70% of the skim milk powder produced in New Zealand is exported for recombining.

3. THE PROCESSING SEQUENCE FOR MILK POWDER MANUFACTURE

Figure 2 shows the basic plant layout for the manufacture of milk powders. The properties of the incoming raw milk are influenced by the breed of cow, by the stage of lactation and by the time in the season, as well as by other aspects of farm management, including nutrition and mastitis control. The milk is then affected by ageing, through the growth and action of microorganisms, and through lipolysis and proteolysis due to the action of indigenous milk lipase and proteases and enzymes secreted by the microbial flora. There are also other effects of ageing, e.g. β-casein dissociates from the casein micelles at low temperatures and redoposits on or reassociates with the micelles only slowly on rewarming.

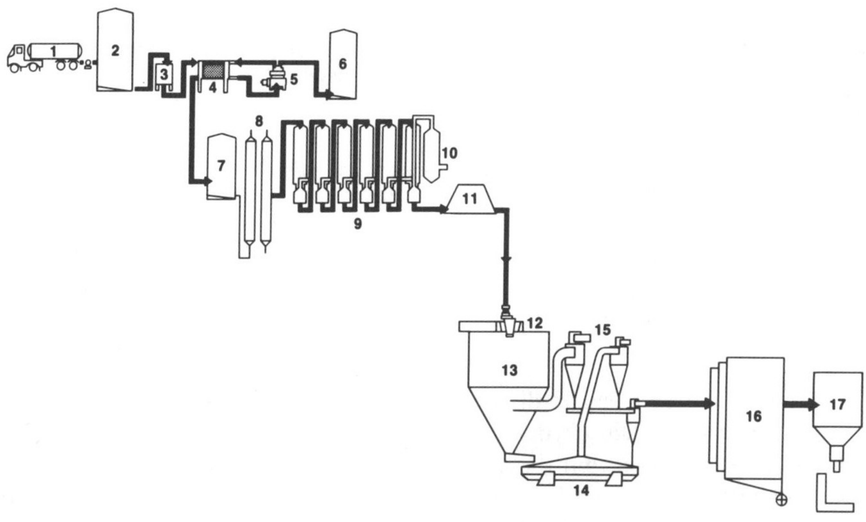

Fig. 2. Flow diagram for manufacture of milk powder: 1, incoming raw milk; 2, reception silo; 3, balance tank; 4, plate heat exchanger; 5, separator; 6, cream silo; 7, standardizing silo; 8, preheater; 9, evaporator; 10, condenser; 11, homogenizer; 12, atomizer; 13, drier; 14, fluid bed; 15, cyclones; 16, powder silo; 17, packer.

The stored milk is separated into skim milk and cream. The skim milk may then be dried, or the cream and skim milk (or whole milk and skim milk) streams may be blended to give whole milk with a standardized fat content ready for drying. Supplements, such as vitamins or minerals, may be added at this stage.

The skim milk or standardized whole milk is subjected to a preheat treatment and then passed to the evaporator (preheat treatment merely implies heating before the evaporator). The variables in the preheating process are temperature and holding time, and may be influenced by the concentration of the milk. The reasons for preheat treatment are to meet the microbiological requirements (e.g. low spore counts), to give functional properties for particular end-uses, and, the most basic, to ensure that the milk will boil when it is exposed to low pressure in the first stage of the evaporator (otherwise the evaporator will not work).

From the preheaters, the milk passes directly to a multi-stage evaporator, where it is subjected to a further heat treatment during the process itself and is concentrated to high total solids. A decrease in pH results from both the heating and the concentration.

The concentrate emerges from the evaporator at about 50°C and 46–50% total solids. It must be passed to the drier with minimal holding time. Excessive holding may result in an undesirable increase in viscosity and in the growth of thermophilic microorganisms. Whole milk concentrate must be homogenized to avoid high levels of free fat in the final product.

The concentrate is then pumped through the spray nozzle to form fine droplets in the hot air stream in the drying chamber. Atomization conditions are manipulat-

TABLE 3
Effects of Seasonal Changes in Milk Composition on the Composition of Skim Milk Powder Produced in New Zealand

Composition (g/100 g)	*November*		*April*	
	Skim milk	*Skim milk powder*	*Skim milk*	*Skim milk powder*
Water	91·80	3·7	81·50	3·7
Fat	0·08	0·8	0·08	0·8
Protein	3·70	39·0	4·50	45·0
Lactose	4·70	49·0	4·20	43·0
Ash	0·74	7·7	0·76	7·7

ed to control the particle size distribution, which is also affected by the viscosity of the concentrate. The drying air flow and temperature are matched to the concentrate throughout, but these also affect the moisture content of the powder and can affect its solubility. Following the drying of the spray in the chamber, the powder particles fall into the vibrating fluid beds where the powder is dried further to the required moisture content. The two-stage drying method gives more efficient energy usage and may improve powder quality.

After drying, the powder must be cooled before storage in bins or packing, otherwise, it may cake or its solubility may be impaired.

4. SIGNIFICANCE OF MILK COMPOSITION AND SEASONAL/LACTATIONAL CHANGES

In dairying countries with seasonal milk production, there are marked changes in the composition of the milk with season/lactation, which are reflected in changes in the gross composition of milk powder. As the milk production falls at the end of the season, the absolute concentration of lactose in the milk decreases whereas those of protein and fat increase. The relative effects, expressed on a dry matter basis, on the composition of powders are shown for November (mid-season) and April (late-season; New Zealand) in Table 3.

The protein content can reach high levels (sometimes over 50% in a skim milk powder basis) and this can cause difficulties with the drying of a high quality product, particularly whole milk powder. Problems include 'burn-out' in the evaporators and reduced solubility of some types of milk powders. These difficulties can be overcome by standardizing the protein content of the milk by adding lactose at the end of the season but the consequent effects on the functional properties, such as heat stability, of powder tend to be unpredictable.

Other minor components also vary with season/lactation. It is probable that some of the difficulties experienced with early lactation milk result from the presence of high levels of immunoglobulins.

The effects of other seasonal/lactational changes in the composition of the milk are less well understood. However, the increase in sodium and the decrease in

TABLE 4
Effect of Various Operations in the Manufacture of Milk Powder on the Milk Components

Processing operation	*Effect on milk components*
Fat standardization	Negligible
Preheat treatment	Destruction of bacteria; inactivation of enzymes; denaturation of whey proteins; formation of complex between κ-casein and β-lactoglobulin; transfer of soluble salts to the colloidal phase; modification of casein micelle structure; Maillard browning reaction between protein and lactose; decrease in pH.
Evaporation	Concentration of solids; further denaturation of whey proteins; increase in colloidal salts; increase in casein micelle size; further decrease in pH.
Homogenization	Increase in number of fat globules; adsorption of casein onto fat surface; decrease in protein stability.
Spray drying	Further removal of water; additional heat treatment; relatively minor further changes in proteins and salts.

potassium at the end of the season may be important when manufacturing infant foods, which have a maximum specification for sodium content.

5. Changes in Milk Components During the Manufacture of Milk Powder

The manufacture of milk powder involves heating at several stages (e.g. preheating, evaporation, drying) and it is often difficult to separate the individual effects of the processing sequence on the milk constituents (Table 4). The heat received at all stages must be considered in the overall process, although in some cases one stage overshadows the others.

5.1. The Preheating Process

Preheat treatment, prior to concentration, is the most important tool in the manufacture of specialized milk powders. Many of the physical, chemical and functional properties of milk powders for particular end-uses are determined mainly by the preheating conditions used. Preheating temperatures vary widely, ranging from pasteurization (72°C for 15 s), through low-temperature, long-time heating (e.g. 85°C for up to 30 min) to high-temperature, short-time heating (120°C for 2 min) using direct (steam injection) or indirect (plate heat exchanger) heating. Many physical and chemical reactions occur when milk is preheated and many reactions involve more than one of the milk constituents. The effects of heat treatment on milk have been reviewed (Fox, 1981, 1982; Singh, 1988; Singh & Fox, 1989; Singh *et al.*, 1989) and are discussed in Chapter 15.

5.1.1. Protein Denaturation

The proteins in milk vary widely in their susceptibility to denaturation by heat. Caseins are particularly insensitve, whereas the various whey proteins are more or less heat labile. Denaturation of whey proteins commences at temperatures greater than 65°C and is followed by aggregation and precipitation. In the milk system these denatured whey proteins remain in suspension, becoming attached to the casein micelles. The thermal denaturation of whey proteins has been the subject of considerable research and has been reviewed recently by Mulvihill and Donovan (1987). Since β-lactoglobulin is the principal whey protein, it tends to dominate the overall behaviour of the total whey protein system. The order of sensitivity of the various whey proteins to heat has been reported to be immunoglobulins > blood serum albumin > β-lactoglobulin (variant A > B) > α-lactalbumin, as determined using protein precipitation methods (Larson & Rolleri, 1955; Dannenberg & Kessler, 1988). However, α-lactalbumin has been shown to be the least stable protein using differential scanning calorimetry (Ruegg *et al.*, 1977; Bernal & Jelen, 1985). It is thought that, to a large extent, the heat sensitivities of these proteins are responsible for the different kinds of properties of milk powders when different preheat treatments are applied.

The extent of denaturation of the whey proteins is affected markedly by pH, ionic strength and the concentrations of protein and calcium (Hillier *et al.*, 1979; Park & Lund, 1984; Dannenburg & Kessler, 1988). The rates of denaturation of both β-lactoglobulin and α-lactalbumin are higher at pH 6·5 than at pH 4·5; however, denaturation of both these proteins is much slower at pH 2·5 (Harwalkar, 1986). Increasing the calcium concentration up to 4 mg/ml tends to retard the denaturation of both proteins, but further increase has little effect. Increasing the protein concentration decreases the rate of denaturation of β-lactoglobulin but increases the rate of denaturation of α-lactalbumin (Hillier *et al.*, 1979). Heat denaturation of whey proteins is also influenced by lactose and other sugars, polyhydric alcohols and protein modifying agents (Hillier *et al.*, 1979; Bernal & Jelen, 1985; Donovan & Mulvihill, 1987).

The kinetics of the denaturation of whey proteins appear to be complex, with the reaction characteristics showing marked changes above 90–100°C, although the denaturation of α-lactobumin appears to be first order while that of β-lactalglobulin appears to be second order (Lyster, 1970: Hillier & Lyster, 1979; Park & Lund, 1984). Recently, Dannenburg and Kessler (1988) confirmed that the denaturation of α-lactalbumin can be described by first order kinetics but found that denaturation of β-lactalglobulin depends on the reaction medium: the reaction order was 1·5 in milk and 2 in whey. The apparent activation energies reported by various workers (Dupont & Yon, 1961; Gough & Jenness, 1962; Lyster, 1970; Hillier & Lyster, 1979; Dannenburg & Kessler, 1988) are in the ranges 260–280 kJ/mol for β-lactoglobulin and 270–280 kJ/mol for α-lactalbumin at temperatures below 90°C. At higher temperatures, the activation energy is lower, ranging from 54 to 60 kJ/mol for β-lactoglobulin and from 55 to 70 kJ/mol for α-lactalbumin. The temperature of denaturation for α-lactalbumin is lower (63°C) than for β-lactoglobulin (74°C) but extensive renaturation (80–90%) of α-lactalbumin can occur after cooling (Ruegg *et al.*, 1977; de Wit *et al.*, 1985). This renaturation may account for the observation that α-lactalbumin is the most resistant of the whey proteins to denaturation by heat. The reversible denaturation of α-lactalbumin appears to be due to calcium ion dissociation and reassociation from the protein. α-Lactalbumin is a metallo-protein and contains two calcium binding sites (Hiraoka *et al.*, 1980). Removal of calcium from α-lactalbumin by EDTA causes a decrease of denaturation temperature and loss of ability of protein to renature (Hiraoka *et al.*, 1980; Bernal & Jelen, 1984).

Methods for assessing the denaturation of whey proteins include measurement of loss of solubility, reactivity of thiol groups, electrophoretic analysis, loss of antigenic activity and differential scanning calorimetry. As these methods are based on different physical or chemical properties of the protein, it is difficult to decide whether these measurements of denaturation are comparable under different experimental conditions. The current methods for determining denaturation in commercial applications are based on quantitation of the protein precipitated by acid or salt. The method of Harland and Ashworth (1947) is used to quantify undenatured whey protein in milk powders and is the official method used by the ADMI (1971) to classify milk powders on the basis of heat treatment. Basically, the powder is dissolved in water and then saturated with NaCl solution, followed by filtration which retains denatured whey proteins and casein. The filtrate contains undenatured whey proteins. On the basis of this test, skim milk powders can be classified into low-, medium- and high-heat types (whey protein nitrogen index, WPNI).

The method, although widely used, has limitations as a guide to heat treatment owing to natural, lactational or seasonal variations in the level of whey proteins in the raw milk and changes in the extent of denaturation under varying conditions of heating. Furthermore, the susceptibility to denaturation varies with the composition of the milk, e.g. pH, calcium, lactose. A wide range of heat treatments could produce the same WPNI, i.e. increasing the temperature and decreasing the holding time, but give milk powders with different functional properties. It should be emphasized at this point that the WPNI is not the most useful indicator of the suitability of a milk powder for a specific application; knowledge of physical and functional properties is of greater value.

There appears to be a need for a rapid quantitative method for assessing the extent of heat denaturation of whey proteins in milk powders. Measurement of casein number, i.e. the percentage of total nitrogen in milk that is precipitated by the adjustment of the pH to 4·6, gives a more reliable estimate of the true value of whey protein denaturation than the WPNI. Since denatured whey protein coprecipitates with casein at pH 4·6, the presence of denatured whey proteins in milk results in an increase in casein number over that of milk that has received no heat treatment. Casein number, however, gives no better indication of functional properties than does WPNI, since both methods measure the same indirect parameter in different ways. The undenatured whey proteins remaining in the filtrate, after precipitation of heated milk with acid or NaCl, can be quantified or fractionated by gel filtration, if separation and identification are required.

5.1.2. Casein/Whey Protein Interactions

The formation of a complex between β-lactoglobulin, the principal whey protein, and κ-casein, a cysteine-containing casein, has long been recognised (Zittle *et al.*, 1962; Long *et al.*, 1963; Sawyer *et al.*, 1963). This interaction has been shown to involve sulphydryl–disulphide interchange and hydrophobic interactions (Sawyer *et al.*, 1963; Hartman & Swanson, 1965; Doi *et al.*, 1983; Haque & Kinsella, 1988). Direct interaction between α-lactalbumin and κ-casein when heated is limited, if it occurs at all, but the complex formed between α-lactalbumin and β-lactoglobulin is able to interact with κ-casein (Elfagm & Wheelock, 1977).

Similar reactions occur when milk is heated, and heat-induced complexes of κ-casein/β-lactoglobulin/α-lactalbumin have been isolated from heated milks (Morr, 1973; Snoeren & van der Spek, 1977; Creamer *et al.*, 1978). When milk is heated at above 90°C, whey proteins denature and interact with κ-casein on the surface of the casein micelles, giving the appearance, under an electron microscope, of thread-like appendages, protruding from the micelles (Creamer & Matheson, 1980; Mohammad & Fox, 1987). However, not all the denatured whey proteins complex with the casein micelles and some remain in the serum. Farah (1979) found that 68% of the total whey proteins were attached to the casein micelles after heating for 17 min at 120°C. Singh and Creamer (1991) reported that 40–45% of β-lactoglobulin and 25% of α-lactalbumin were associated with the casein micelles after heating milk in a commercial plant at 120°C for 2 min. The extent of association of denatured whey protein with casein micelles is markedly dependent on the pH prior to heating. Heating at pH values less than 6·7 results in a greater quantity of denatured whey proteins associating with the casein micelles, whereas at higher pH values whey protein/κ-casein complexes dissociate from the micelle surface, apparently due to dissociation of κ-casein (Singh & Fox, 1985, 1986). This aspect has been discussed in Chapter 15. It would be interesting to know whether altering the pH of milk prior to preheating affects the functionality of milk powders. Newstead and Conaghan (1978) have shown that the heat stability of recombined evaporated milk increases as the pH of skim milk at preheating is increased from 6·5 to 6·6 but that a further increase in pH decreases the heat stability. The interactions between denatured whey proteins and casein micelles markedly affect the heat stability and the rennet clotting behaviour of the reconstituted powders.

5.1.3. Development of Antioxidant Activity

The development of antioxidant or reducing substances is another important function of preheating. Whole milk powder with low to medium preheat treatments (equivalent to a WPNI of about 3–5) has a greater tendency to undergo lipid oxidation, with distinctive tallowy, musty flavours, than powders made with higher heat treatments (equivalent to WPNI values of 2·5 or less).

Early studies (Gould & Sommer, 1939; Thomas, 1954) showed that the formation of reactive sulphydryl (—SH) groups, resulting from the heat treatment of milk, are responsible for preventing oxidized flavour development. Free —SH groups act as free radical scavengers, and thus as antioxidants, in biological and other systems. Treatment of raw or heated milk with sulphydryl blocking agents, e.g. iodoacetamide, reduces both the —SH content and the antioxidant activity (Thomas, 1954). In heated liquid products, the active —SH groups gradually disappear, largely by oxidation, and the product becomes susceptible to oxidation. Whole milk powder made using a high preheat treatment and thus containing a high level of —SH groups, is resistant to oxidation, but in this case the —SH groups are not oxidised during storage (Harland *et al.*, 1952). Taylor and Richardson (1980) demonstrated that reactive —SH groups are not the sole antioxidant in heated milk; other factors are also involved. For instance, β-lactoglobulin, which accounts for most of the sulphydryl groups in milk, is a weak antioxidant, whereas the caseins, which have few or no sulphydryl groups, have significant antioxidant activity. Recently it has been shown that antioxidants involving Maillard browning reactions can be formed during heating (Hansen & Hempill, 1984; Lingerst & Hall, 1986). It is likely that both kinds of reaction contribute to the overall antioxidant effect in milk powders.

5.1.4. Changes in Antigenicity

One aspect of the heat denaturation of milk proteins, which is of great interest for milk powders used in infant formulae, is the destruction of antigenic properties. The antigenic properties of the immunoglobulins and serum albumin are destroyed by heating at 70–80°C for 15 min but β-lactoglobulin and α-lactalbumin remain active after heating up to 100°C for 15 min. Casein retains its antigenicity on heating up to 120°C for 15 min (Hanson & Mansson, 1961). Heat-treated milk (127°C for 8·5 min) induces a diminished immunogenic response in comparison with pasteurized milk (May *et al.*, 1982). Thus, it would appear that the antigenic properties of the major milk proteins are not destroyed by the commonly-used preheat treatments.

5.1.5. Inactivation of Enzymes

Inactivation of enzymes by heat treatment usually depends on conditions such as pH, ionic strength and water activity as well as the time and temperature of heat treatment. Milk contains naturally occurring proteases and lipases, which degrade protein and fat, and may contain similar enzymes derived from contaminating microorganisms. These require more or less heat for their inactivation, depending on their type.

Bacterial proteases and lipases produced by some *Pseudomonas* species are extremely heat resistant; their occurrence and characteristics have been reviewed by Cogan (1977), Cerf (1981) and McKellar (1989). These enzymes are capable of surviving UHT processing and cause quality problems during subsequent storage; however, they are inactivated to a much greater extent by extended heating at temperatures close to 55°C. Residual protease activity causes an increase in non-protein nitrogen, gelation and bitterness, whereas lipase activity can result in free fatty acids, acidity and rancidity, and other off-flavours. Most of the indigenous milk lipase is inactivated by conventional pasteurization (complete inactivation; 73°C for 11·8 s) (Driessen, 1989) and hence no problem is to be expected from that source once the milk has been pasteurized. However, some residual activity has been observed after pasteurization if the fat globule membrane has been damaged (Linden, 1986).

The principal milk protease, plasmin, is fairly heat stable; it is capable of surviving UHT processing but is extensively inactivated by in-container sterilization (Driessen & van der Waals, 1978; Snoeren *et al.*, 1979; Grufferty & Fox, 1988; Driessen, 1989). Many other enzymes, such as catalase, xanthine oxidase, lactoperoxidase and alkaline and acid phosphatase, exist naturally in milk; the heat inactivation of these enzymes has been summarized by Walstra and Jenness (1984) and Linden (1986).

The difference between the two phosphatases is noteworthy; acid phosphatase is very heat resistant, 100°C for at least 2 min being required for complete inactivation, whereas alkaline phosphatase is completely inactivated by the conventional pasteurization process (72°C for 15 s) (Andrews, 1974). Inactivation of alkaline phospatase has long been used as an index of adequate pasteurization because it is inactivated by treatments similar to those required to destroy tuberculosis bacteria. Certain other enzymes have been suggested as indicators of heat treatment, but no such test has gained wide acceptance. In any test involving an enzyme, there is a possibility of some reactivation after heat treatment, especially after high-temperature short-time treatments. Inactivation of most enzymes is irreversible but certain enzymes, e.g. alkaline phosphatase and plasmin, show some reactivation (Linden, 1986). The extent of reactivation, although small in comparison with the activity in raw milk, is nevertheless sufficient to seriously negate the use of the tests based on residual enzyme activity.

5.1.6. Distribution of Salts

The distribution of milk salts between the colloidal and soluble phases is affected by preheat treatments. Heating milk reduces the ionic calcium, soluble calcium and phosphate concentrations by converting them to the colloidal state. As a result of their association with casein micelles, heat-precipitated calcium phosphate does not sediment (Evenhuis & de Vries, 1956). On subsequent cooling, some of the indigenous or heat-precipitated calcium phosphate may redissolve, especially if the heating temperatures are less than 85°C (Kannan & Jenness, 1961). However, after higher heating temperatures, the precipitated material does not resolubilize. Heat treatments have little effect on the monovalent ions, Na^+, K^+ and Cl^-. It is not known what subsequent effects changes in the equilibria of salts and ions have on the processing properties of milk or the functional properties of the final powder.

5.1.7. Interactions of Lactose

Many components of milk can interact with lactose during preheat treatments, although most of the changes associated with lactose involve the Maillard reaction. This reaction involves condensation of lactose with the free amino groups of protein, with subsequent rearrangements and degradations leading to a variety of brown-coloured compounds. Manifestations of the Maillard reactions, which are important in milk powder, include brown colour, off-flavour, loss of available lysine, lowered nutritional value, reduced digestibility and reduced solubility. The rate of reaction is strongly dependent on pH, time and temperature of heating, water activity and temperature during storage (Burton, 1983). If the temperature of heating is greater than 100°C, lactose can be converted to acids, especially formic acid (Gould, 1945), thus decreasing the pH of the milk, and to lactulose (Adachi, 1958). The lactulose content of milk has been suggested as an indicator of the heat treatment of milk (Martinez-Castro & Olano, 1980). Since the formation of lactulose or acids from lactose is fairly slow below 100°C, the influence of preheating in the usual range, except for heat stable powders, should not be significant.

5.2. The Evaporation Process

The removal of water from milk occurs freely during the initial stages of evaporation; as the concentration of milk solids increases, the water is held by the various constituents and the rate of evaporation decreases. If the temperature rises or heating is prolonged, the proteins may be heat-damaged to varying extents, particularly as the degree of concentration rises. For this reason, evaporation is performed under the lowest practicable pressure (and hence lowest temperature).

Falling film evaporators, which are predominant in modern industry, are designed to remove as much water as possible from the milk at a low energy cost and with minimal heat-damage to the milk constituents. In the traditional design, the milk is boiled under vacuum in the first effect through the application of externally supplied heat to the first stage. Subsequent stages (or effects) are heated by the vapour generated from evaporation in the preceding effects (see Fig. 2).

In normal milk, gross alterations to the more susceptible proteins, the whey proteins, do not usually occur below about 65°C. The effects of preheating have already been described but the heat load on the product during its passage through the evaporator must also be considered. The residence time in each stage is typically 1 min, so that in the first effect the milk is held for about 1 min at 70–72°C, which is a rather more severe heat treatment than pasteurization (typically 72°C for 15 s) (Table 5). Subsequent stages are at lower temperatures, and would be expected to have less effect. Unfortunately, few data on the effects of heat on milk

TABLE 5
Heat Treatment Received by Milk Passing through a Four-Effect Falling Film Evaporator and a Multiple-Pass Mechanical Vapour Recompression Evaporator

Effect	*Temperature (°C)*	*Residence time (s)*	*Concentration (% total solids)*
1st	72	60	9·5–12·5
2nd	64	60	12·5–16·7
3rd	56	60	16·7–25·0
4th	50	60	25·0–48·0
Mechanical vapour recompression	60	600	9·5–48·0

of intermediate concentration over this temperature range are available. It can only be inferred from experience that the effects are usually less than those incurred during forewarming.

It must also be remembered that, in order for the evaporator to function properly, the milk must flash boil as it enters the first effect and must therefore be fed in at temperatures above the temperature of the first effect. So, when producing a very-low-heat milk powder, the holding time at 72°C, between the preheater and the first effect, must not be ignored; it may be a significant part of the total heat treatment. In cases where there are special requirements for a low degree of heat treatment, evaporators have been built to run at lower first-effect temperatures.

The situation for evaporators of the mechanical vapour recompression type is rather different; essentially, there is one effect and one boiling temperature, usually about 60°C, with multiple passes of the product (which becomes more concentrated with each pass) determining the holding time.

5.2.1. Changes to Milk Components During Evaporation

There appear to be no systematic studies on the effects of evaporation on milk constituents. In addition to concentration, evaporation causes numerous other changes in the milk system. The changes depend to some extent on other conditions, such as preheat treatment, temperature during concentration and time elapsed after concentration. The pH of milk decreases during concentration from an average initial value of 6·7 to approximately 6·3 at 45% total solids (Howat & Wright, 1934). This is partly due to changes in salt equilibria as more calcium phosphate is transferred from the soluble to the colloidal state, with a concomitant release of hydrogen ions. Le Graet and Brule (1982) have shown that when the milk is concentrated about five-fold by evaporation, soluble calcium and soluble phosphate increase by a factor of about two; the remainder of the soluble calcium and phosphate being transformed into the colloidal state. The activity of calcium ions increases only slightly but the ratio of monovalent to divalent cations increases markedly (Nieuwenhuijse *et al.*, 1988). Casein micelles may increase in size due to the increase in colloidal calcium phosphate or due to coalescence of micelles (Walstra & Jenness, 1984). Heat denaturation of β-lactoglobulin decreases but the denaturation of α-lactalbumin increases with increasing concentration of milk solids (Hillier *et al.*, 1979). The sensitivity of casein micelles to heat increases almost linearly with increasing concentration (Webb & Holm, 1932). Maillard browning and the insolubilization of protein begin to occur if the concentration factor is too high, particularly at high temperatures.

The viscosity of the milk increases with increasing concentration and the effect is most marked when the concentration of milk solids exceeds 45%. The viscosity increase reflects the increase in the extent of inter- and intramolecular contact in the milk brought about by concentration. At 45% total solids, the casein micelles alone occupy about 30% of the total volume of milk. Concentrate viscosity can have a major influence on the size of the droplets formed during atomization and hence on the rate of drying and final powder particle size distribution. These factors, in turn, affect the physical properties of the dried product (de Wilder & Marten, 1974; King *et al.*, 1974; Baldwin *et al.*, 1980).

Several factors influence concentrate viscosity; the more important of these include the composition of the milk, the preheat treatment given before evaporation and the temperature and holding time of the concentrate. The marked increase in concentrate viscosity towards the end of the dairying season in New Zealand has been attributed largely to the increased protein content (Bloore & Boag, 1981). None of the other compositional factors investigated have a significant effect on the viscosity of the concentrate (Bloore & Boag, 1981). Generally, raising the preheat temperature increases concentrate viscosity and the effect is greater for longer preheat holding times.

Heating the concentrate at temperatures up to 60°C causes a gradual decrease in concentrate viscosity, but if the concentrate is held for long periods, the viscosity increases (Muir, 1980; Bloore & Boag, 1981). It is good practice to keep the balance tank between the evaporator and the drier as empty as possible, because this keeps the residence time to a minimum. If the residence time is unduly prolonged, particularly if the holding temperature exceeds 50°C, aggregation of protein may occur. Aggregation is accompanied by an increase in viscosity and extensive aggregation can lead to gelation (Muir, 1980). The practical result of aggregation/gelation of this type is that the powder prepared from the concentrate has poor solubility characteristics. The aggregation phenomenon occurs quite rapidly at temperatures above 60°C and when the total solids concentration is over 46%. It is likely that this type of aggregation and gelation is caused by the precipitation of calcium phosphate on the surface of the casein micelles. Whey proteins may also be involved in gel formation, especially at higher temperatures.

Aggregation/gelation of this type is rare in whole milk concentrate, partly because the concentration of proteins and salts in the aqueous phase is much less than that in skim milk concentrate at a similar nominal concentration and partly because the structure is disrupted by the presence of fat globules.

As a final quality assurance measure, the concentrate may be heated to pasteurizing temperatures on its way to the drier to reduce bacterial numbers. The danger of thickening and gelation is less when the concentrate holding time is comparatively short, as the concentrate is subjected to flow and shear.

5.2.2. Concentrate Homogenization

Previously, only concentrate to be dried in driers with centrifugal atomizers was homogenized. It was thought that the shear in high pressure nozzles was sufficient to achieve the desired homogenizing effect. Also, whole milk powders had a tendency to cake and lump, particularly if the cooling of powder was inadequate. It was often found that caking and lumping were associated with high levels of free fat. The free fat on the surface of the powder particles, not covered by a protective layer of protein, tended to agglomerate the particles, especially at temperatures near the melting point of the fat. Nowadays, there are tight specifications for free fat in all whole milk powders (less than 2% of the powder) and the problems have been reduced. This has been achieved by homogenization of the whole milk concentrate before drying so that the level of free fat in the powder is greatly reduced.

The effects of homogenization on milk have been extensively reviewed by Mulder and Walstra (1974). The surface area of the fat is greatly increased and is covered by plasma components, the major material adsorbed being protein. The quantity of protein adsorbed varies with the homogenizing conditions such as homogenization pressure and temperature. The major protein adsorbed is casein but whey proteins, as well as the natural globule membrane material, are also present. Mol (1975) showed that the adsorption of casein on to the fat globule during homogenization of the concentrate adversely influences the solubility of the final powder, because the casein micelles adsorbed on to the fat globules are less stable during drying than those normally present in milk plasma.

5.3. The Drying Process

Spray drying is now used almost exclusively in milk powder manufacture. Roller drying (also known as drum drying) is not now used because of its marked destabilizing effect on milk protein. The heat treatment during roller drying is severe because drying takes a longer time and occurs at a higher temperature than in spray drying.

The principle of spray drying is to remove the water from the milk concentrate as fast and at as low a temperature as possible so as to minimise heat damage to the milk solids. This is achieved by spraying the milk as very fine droplets into a hot (above 200°C) dry air stream, thus exposing a large evaporative surface.

Most driers used for milk and milk products are of the parallel flow type, where the drying air and product travel in the same direction. In counterflow driers, the dry particle meets the hot air so that considerable heat damage may occur. Atomization of the concentrate should be such that uniform particle size distribution is obtained; small particles may overheat, whereas large particles must be heated excessively to attain the desired moisture level.

As the moisture evaporates, removing latent heat of evaporation, the concentrate droplets remain relatively cool and the temperature starts to rise only after most of the moisture has evaporated. However, at that stage the particle should have reached a cooler region of the drier, where the air temperature is near 100°C. In addition, the milk solids seem to be less susceptible to heat damage at these very high concentrations (i.e. 90% total solids, or 10% moisture, and higher). The milk solids are most susceptible to heat at a droplet moisture content of 13–16% (Webb, 1950). If particles in this state of dryness are swirled back into the hot air stream, at 200°C, then poor solubility may result.

5.3.1. Changes to Milk Components During Drying

The changes induced in milk systems by drying are not well understood. The heat exposure of milk during spray drying may vary considerably, depending on the design of the drier, the operating conditions and the length of time the powder is held before cooling. The native properties of the milk components are essentially unmodified by moderate drying conditions. The normal size distribution of the casein micelles and their heat stabilities and renneting characteristics are substantially recovered on reconstitution of spray-dried milk (King, 1965). Using electron microscopy, Graf and Bauer (1976) observed a more dense arrangement of the subunits of the micelles in reconstituted dried milk than in fresh milk.

During spray drying, the increases in protein denaturation and aggregation are likely to depend on the temperature of the air into which the milk is sprayed (inlet air temperature), the degree of concentration and the temperature of the concentrate prior to drying, the size of the drying droplets and the temperature of the air/power mixture exiting from the drier (outlet air temperature). Drying is usually very rapid and the temperature of the milk droplets does not exceed 70°C until they have lost almost all their water. The temperature of the droplets approaches that of the outlet air as the drying process nears completion (Parry, 1974). For this reason, the outlet air temperature is a critical parameter controlling heat damage to dry milk products. At relatively high outlet temperatures, the protein may become insoluble at a rate that is approximately logarithmic with the temperature (Parry, 1974). The insoluble components of dry milk are mainly casein, denatured whey proteins and calcium phosphate. Under normal spray drying conditions, whey protein denaturation is negligible, and most enzymes remain active (Walstra & Jenness, 1984). However, some large droplets may be exposed to more severe heating conditions during drying and therefore some powder particles may contain considerable amounts of denatured whey proteins (Walstra & Jenness, 1984). In contrast to spray drying, roller drying can cause extensive denaturation of whey proteins and considerable browning. The amino acid composition of the milk proteins is not greatly affected by the drying process; normal drying conditions cause only small losses of lysine (up to 5% in spray drying and 3–15% in roller drying) (Renner, 1983).

Changes may occur in the salts equilibria during spray drying. The process of drying would be expected to produce the same types of changes in salt equilibria as does evaporation, i.e. an increase in colloidal calcium phosphate and a decrease in pH. It has been shown that the concentrations of soluble calcium and soluble phosphate in reconstituted skim milk are about 20% lower than those in the original milk (Le Graet & Brule, 1982).

6. Changes in Milk Powder During Storage

The changes that occur in milk powder during storage are, to a large extent, dependent on the processing conditions, the final composition of the powder and the storage conditions. A gradual decrease in solubility occurs during the manufacturing process; low and medium preheating treatments generally have little effect but high preheat treatments may result in lower solubility and the solubility continues to decrease during storage (Litman & Ashworth, 1957; Kieseker, 1970). The moisture content of the powder and the storage temperature are critical; high moisture levels, as a result of inadequate dehydration or reabsorption of atmospheric moisture, promote protein insolubilization at relatively low storage temperatures. Milk powders with low moisture contents (below 5% w/w) or water activities of less than 0·6 show little change in solubility even at temperatures approaching 40°C, whereas the temperature coefficient of the reactions leading to insolubility in high moisture powders exceeds a value of 5 at temperatures above 20°C (Henry *et al.*, 1948; Kieseker & Clarke, 1984). Consequently, the solubility of moist powders may remain unchanged for long periods of storage at 20°C, but falls rapidly at 37°C.

Concurrent with the loss of solubility, a number of significant changes occur in high moisture powders (moisture content above 5%). The lactose is gradually bound by the protein, via Maillard reactions, which leads to the development of brown-coloured compounds. Various intermediate compounds in the reaction can be measured and their concentrations can be used as an indication of heat treatment and storage. Some of the principal Maillard reaction products are hydroxymethylfurfural (HMF), furosine and lysinoalanine (Kenney & Bassette, 1959; Annan & Manson, 1981). HMF has been the most widely measured of these compounds and its concentration has been shown to increase with time and storage temperature of the milk powders. However, Burton (1983) concluded that total HMF, as conventionally determined, cannot be used satisfactorily as a measure of heat treatment, measurement of furosine seems to be a more reliable predictor.

Another major change during the storage of whole milk powder is oxidation of fat, which significantly affects the sensory quality of milk. This aspect will not be considered here.

7. Physico-chemical and Functional Properties of Milk Powders

The physico-chemical and functional properties of milk powders are important when the powders are used for recombining or in the manufacture of various food products. For use in most foods, the milk powders should be light in colour, devoid of off-flavours and easy to hydrate and disperse in solution. The physico-chemical properties are critical for classifying milk powders for reconstitution.

7.1. Particle Size

The particle size of powders affects appearance, reconstitution and flow characteristics. The common method of measurement involves sieving a sample of the powder through standard screens and weighing the retained particles. The particle size of a powder depends on the atomization conditions and the viscosity of the concentrate; a high atomizing pressure and a low concentrate viscosity reduce particle size (Woodhams & Murray, 1974). The best particle size for rapid dispersion is 150–200 μm diameter (Varshriey & Ojaha, 1975).

TABLE 6
Factors Affecting Ease of Reconstitution of Milk Powders

(1) Structure and physical properties of the powder, e.g. particle size, shape, density, porosity, specific volume.
(2) Chemical properties of the powder, e.g. free fat, extent of protein denaturation.
(3) Conditions of reconstitution, e.g. temperature and quality of water, time and speed of agitation.
(4) Age of powder and conditions of storage.

7.2. Reconstitution Properties

An important function of many milk powders is reconstitution to a form of milk suitable for further processing. Reconstitutability refers to the ease with which a powder can be dissolved in water. It may take up to 24 h before some reversible changes that occur during evaporation and drying are reversed; this mainly concerns the distribution of minerals between the colloidal and soluble phases, and the size and structure of the casein micelles (Walstra & Jenness, 1984). The irreversible changes (off-flavours, loss of nutrients) that have occurred during processing or storage remain in the reconstituted product. Factors that affect the reconstitution properties of powders are summarized in Table 6.

The reconstitution properties are also affected by the processing conditions and the type of equipment: drier and system of atomization, preheat treatment of the milk and concentrate, total solids of the concentrate, outlet air temperature, and storage time and temperature (Woodhams & Murray, 1974; Bloore & Boag, 1981).

The reconstitution tion process involves a number of steps requiring the following measurable milk powder properties (Crosby & Marshall, 1958; King, 1966; Pisecky, 1980): wettability; sinkability; dispersability; solubility. There is no sharp borderline between these individual reconstitution steps, i.e. they proceed both consecutively and simultaneously and the individual properties influence each other. Therefore, it is difficult to determine the individual properties independently of each other. Numerous methods for measuring the various reconstitution properties of milk powders have been proposed (Neff & Morris, 1968; Prentice, 1972; Mol, 1974; International Dairy Federation, 1970). Ideally, high rates of wettability and rapid solubilization with a minimum of scum remaining are required.

7.2.1. Wettability

Wettability is a measure of the ability of a powder to absorb water on the surface to be wetted. The ease and rapidity of hydration of the powder particles determines whether an intractable, pasty, clumpy mass or a smooth, uniform dispersion is obtained. The wettability of dry particles is determined by the contact angle (θ) between the powder particles and water. If θ is less than 90°, the particles are wettable. For dried milk, it is difficult to measure contact angles accurately because the surface of the powder particles is rough and uneven. van Kreveld (1974) reported θ values of 20° for skim milk powder and 50° for whole milk powder, showing the effect of fat on the interfacial equilibria. However, the effective θ

value for the wetting of a mass of powder could be much higher than that for the wetting of a single powder particle. Capillary attraction between powder particles and water either causes the particles to move into the water or the water to move into the pores of, and between, the particles. The rate of movement of water into the powder depends on the width of the pores between the particles; small particle size and symmetrical shape enhance close packing of the particles, thus slowing down water penetration and wetting. In normal skim milk powder, the pores are too narrow to allow rapid penetration of water. As soon as water enters the pores, the rate of movement is hindered because the capillary action of the penetrating water pulls the wetted particles closer to each other, thus narrowing the pores. The increase in viscosity on dissolution of lactose and salts, and the swelling of protein molecules, may cause further narrowing of the pores. The wettability of whole milk powder is poor at water temperatures below the melting point of fat because the surface of the particles is always covered with fat, forming a water-repellent layer around the dry powder particles. It is possible to overcome this water-repelling effect by treating the surface with a surface active agent. Lecithin is the most commonly used surface active agent, mainly because it is an acceptable food additive and has high functional performance. Generally, the wettability of dry powder particles depends on the surface activity of the particles, surface area, surface charge, particle size, density, porosity and the presence of amphipathic substances (Lascelles & Baldwin, 1976; Sanderson, 1978; Jensen & Nielsen, 1982).

7.2.2. Sinkability

When the powder particles have been wetted, the individual components of milk powder start dissolving and dispersing and at the same time particles start sinking to the bottom. Sinkability is the ability of powder particles to overcome the surface tension of water and sink into the water after passing through the surface. This is expressed as milligrammes of powder that sink per minute per square centimetre of surface area. Sinkability reflects the forces tending to submerge a particle on the surface and depends mainly on the density of the particles, which depends on the mass of the particles and the quantity of occluded air. During the initial stages of reconstitution, the density of the particles decreases, mainly because the heavier components, lactose and minerals, dissolve faster than the other components. At the same time, the density of the solution being formed increases. Consequently, the density difference between the particles and the medium decreases, so that particles start to rise again after the initial sinking. There is an increasing tendency to sink as the particle density increases and the quantity of occluded air decreases.

7.2.3. Dispersibility

Dispersibility reflects the ability of the wetted aggregates of powder particles to become uniformly dispersed when in contact with water. This property has been chosen to measure whether or not the product is 'instant'. Various factors affect dispersibility; increasing the particle size gradually improves the dispersibility of whole milk powder, whereas there is a progressive decrease in dispersibility as the percentage of fine particles (below 90 μm) increases. If there is an excessive number of fine particles, there is a tendency for lumps of milk powder to be cemented together by thick layers of wet particles. Dispersibility shows a marked optimum at 60°C and decreases at lower temperatures; most powders show low dispersibility below 40°C (Lascelles & Baldwin, 1976). The dispersibility of milk powder can be improved by: (1) keeping the heat treatment at preheating to a minimum; (2) minimizing the holding time and heat treatment of the concentrate; (3) increasing the particle size of the powder by decreasing the atomization pressure in pressure atomizing spray driers or by alteration to the disc (speed, design) in disc atomizing driers (Baldwin & Woodhams, 1974).

7.2.4. Solubility

Solubility is a measure of the final condition to which the constituents of the powder can be brought in solution or stable suspension. The 'solubility index' is used as a measure of the undissolved residue of milk powder. This index is determined by dispersing a known quantity of milk powder in water under defined conditions and separating the insoluble material by centrifugation. The solubility index is the volume (ml) of sediment remaining at the bottom of the centrifugation tube. The insoluble material consists mainly of denatured whey proteins, casein and minerals. In the drying of milk, low solubility occurs when the concentration of total solids exceeds 50% because of the ease of protein interparticle cross-linking. The solubility index is most directly affected by heat treatments during powder manufacture and the temperature of the water used for reconstitution.

Agglomeration of milk powder and the reduction of the proportion of fine particles are extremely important for enhancing the reconstitution properties. Agglomeration alters the physical state of the powder to such an extent that the rates of wetting, sinking and dispersion are so increased that reconstitution occurs very rapidly. During the agglomeration (instantizing) process, the original small particles are transformed into porous agglomerates. When these porous agglomerates come in contact with water, water quickly enters the pores to wet the particles. Subsequently, the particles sink into the water, disperse and finally dissolve. Basically, two principal processes are used to produce agglomerated powders.

(1) the 'straight-through' process;
(2) the 're-wet' process.

In the 'straight-through' instantization process, the powder leaves the spray-drying chamber at a moisture content 2–4% higher than that desired in the final powder. Drying is then completed in a vibro-fluidizer at temperatures suitable for the required moisture content. The powder is cooled in the last section of the fluid bed and the smallest and lightest particles, the so-called 'fines', are separated and recirculated to form agglomerates in the primary drier.

In the 're-wet' instantization process, the powder is moistened with water, either in the liquid or in the vapour state. The particles are then brought into intimate contact with each other. Optimum rewetting or hydration is obtained at 12% moisture. Lactose and some of the proteins dissolve in the added water, with the result that the particles become sticky and bridges are formed between them. The clusters are then dried to the desired moisture content and then cooled in a vibro-fluidizer.

7.3. Hydration Properties

The absorption of water by milk powders is of practical importance in selecting drying procedures, suitable packaging materials and storage conditions. Functional properties, such as swelling, viscosity, gelation and foaming and emulsifying capacity are affected by protein–water interactions in the food systems themselves (Kinsella, 1984). These interactions are usually expressed as water-sorption–desorption isotherms, which measure the degree of hydration of a protein as a function of the water activity (A_w). When a protein is exposed to increasing equilibrium relative humidity, or A_w , a sigmoidal sorption isotherm is usually obtained, reflecting the progressive increase in the amount of water associated with the protein. At low values of A_w (less than 0·3), water is selectively hydrogen bonded to charged and polar residues to form a monolayer of bound water. At A_w values of between 0·3 and 0·9, multilayers of water build up beyond this vicinal layer. At higher values of A_w, bulk water may associate loosely with the protein and may be trapped within the matrix of the polypeptides or in the capilliaries between protein particles. Water of these three types represents so-called 'bound' water. Water binding is usually affected by protein composition, pH, salt, ionic strength and temperature.

Only a limited number of sorption isotherms have been published for different milk powder preparations, Heldman *et al.* (1965) determined the sorption isotherms for skim milk powders, prepared using low, medium and high preheat treatments, at storage temperatures of 2, 16 and 30°C. Preheat treatment affected moisture sorption at all A_w values and at all storage temperatures. The equilibrium moisture content increased progressively from the low-heat to the high-heat preheat treatments. The moisture contents corresponding to the monomolecular layer state, calculated using the BET equation, for high-, medium- and low-heat skim milk powders equilibrated at 16°C were 11·34, 10·0 and 9·3 respectively. Generally, at A_w values between 0·2 and 0·8, there was a significant decrease in the equilibrium moisture content with increasing storage temperature, especially at an A_w value of around 0·5. The monolayer water content decreased with increasing storage temperature.

The presence of fat in milk powder apparently reduces the rate and magnitude of water sorption. Generally, spray-dried milk powders have higher water contents than freeze-dried milk powders in the A_w range from 0·5 to 0·6 (Berlin *et al.*, 1968). The presence of lactose caused a break in the isotherms of milk powders, usually at A_w values of between 0·5 and 0·6; removal of lactose from skim milk powder results in a typical continuous isotherm (Berlin *et al.*, 1968). At an A_w value below 0·2, the rate of moisture uptake by milk powders in which the lactose had been hydrolyzed was lower than that of normal milk powder, whereas at an A_w above 0·4, hydrolyzed lactose milk powder absorbed twice as much water as normal milk powder. There is a progressive order of moisture sorption by the components of milk powders with increasing A_w; proteins absorb preferentially at low A_w and the usual order for milk powders is casein (A_w = 0·01–0·3) > whey proteins (A_w = 0·05–0·35) > lactose (A_w = 0·2–0·6) and milk salts (A_w = 0·6–1·0) (Berlin *et al.*, 1968; Harmansson, 1977).

Thus, the method of preparation of milk powders, i.e. the drying method, and the composition of the powders influence moisture sorption.

7.4. Heat Stability

The heat stability characteristics of reconstituted milk powders are important when the powders are to be used in hot beverages, custard, white sauce mixes, bakery products and, most importantly, the manufacture of recombined sterilized evaporated milk. In machines for vending hot beverages, milk powder (or coffee whiteners) and coffee powder are dispensed into a disposable cup and very hot water is then added. The resulting combination of acidity (from the dispersed coffee solids) and high temperature severely test the stability of milk protein. Skim milk powder, being the principal ingredient for recombined evaporated milk, must be sufficiently heat stable to withstand the sterilization temperatures applied to the reconstituted canned product.

Various aspects of the heat stability of milk have been covered in Chapter 15. In addition to factors such as lactation, breed of cow and weather conditions, there is a marked seasonal variation in the heat stability of milk which extends through to the reconstituted powder. Under New Zealand and Australian conditions, heat-stable powders are readily produced from milk produced from October to March. However, this period may be extended by paying greater attention to processing conditions, followed by close screening of samples.

Apart from the inherent variability in the heat stability of milk itself, the main factor influencing the heat stability of reconstituted powders is the level of heat treatment applied during the preheating stage. This alone, however, is not a guarantee of sufficient heat stability. Preheat treatments of 80–90°C with holding periods of 10–20 min are commonly used. However, it has been shown that preheating temperatures in excess of 100°C with shorter holding times produce more stable powders. Heat treatments in the region of 110–125°C for 30–240 s are favoured in New Zealand for the manufacture of heat-stable powders.

The pH of milk also exerts an important influence on its heat stability (see Chapter 15). It has been shown that the pH of maximum heat stability does not always coincide with the natural pH of milk. In fact, adjustment of the pH, either before or after preheat treatment, has been used with variable success to improve the heat stability of powders for recombination (Newstead *et al.*, 1975; Newstead, 1977*a*).

Other factors that may influence the heat stability of reconstituted powder include adjustment of the mineral content (Newstead, 1977*b*), addition of aldehydes and sugars (Holt *et al.*, 1978), addition of urea (Kelly, 1982), hydrolyzing the lactose (Kelly, 1981) and addition of lecithin or buttermilk (Singh & Tokley, 1990).

Of all the known variables that influence the heat stability of reconstituted milk powders, preheating continues to be the most reliable method for controlling of the heat stability of skim milk powder within the processing plant.

7.5. Emulsifying Properties

The emulsifying properties of milk proteins are important in several food products such as soups, coffee whiteners, comminuted meat products and recombined milk products. Skim milk powders have been the major source of milk protein ingredience to perform the above functions in food products in the past, but in recent years they have been replaced by a number of relatively new and more versatile milk protein products, e.g. caseinates and whey protein concentrates.

No systematic studies on the emulsifying properties of skim milk powder have been reported. Pearson *et al.* (1965) found that skim milk powder showed good emulsifying properties over a range of pH values and ionic concentrations. The emulsifying capacity of skim milk powder was greater than that of caseinates at a pH of 5·4, regardless of the ionic strength, but the caseinates were superior in the alkaline pH range. This may be due to the formation of a more compact and mechanically strong interfacial film by skim milk powder protein particles at lower pH than by caseinates.

8. Recombined and Reconstituted Milk Products

In the 7th edition of the Code of Principles concerning milk and milk products (FAO/WHO, 1973), definitions of recombined and reconstituted products are as follows:

> Reconstituted product is the milk product resulting from the addition of water to the dried or condensed form of product in the amount necessary to re-establish the specified water–solids ratio, whereas a recombined product is the milk product resulting from the combining of milk fat and milk solids-not-fat in one or more of the various forms with or without water.

This combination must be made so as to re-establish the product's specified fat to solids-not-fat ratio and solids-to-water ratio. The processes used to produce reconstituted milk and recombined milk are shown diagramatically in Fig. 3. The manufacturing processes and properties of various recombined milk products have been reviewed extensively (Jensen & Nielsen, 1982; Sanderson, 1982; Kieseker, 1983; Al-Tahiri, 1987).

Appropriately prepared milk powders provide the basis for most recombined milk products. In addition to their nutritional value, they are, in the main, responsible for the functional properties of the recombined products (Table 7). Until recently, whole milk powders were not used extensively in the manufacture of recombined milk products, mainly because of cost and oxidation problems during the storage of these powders. Changes in the relative value of milkfat and more

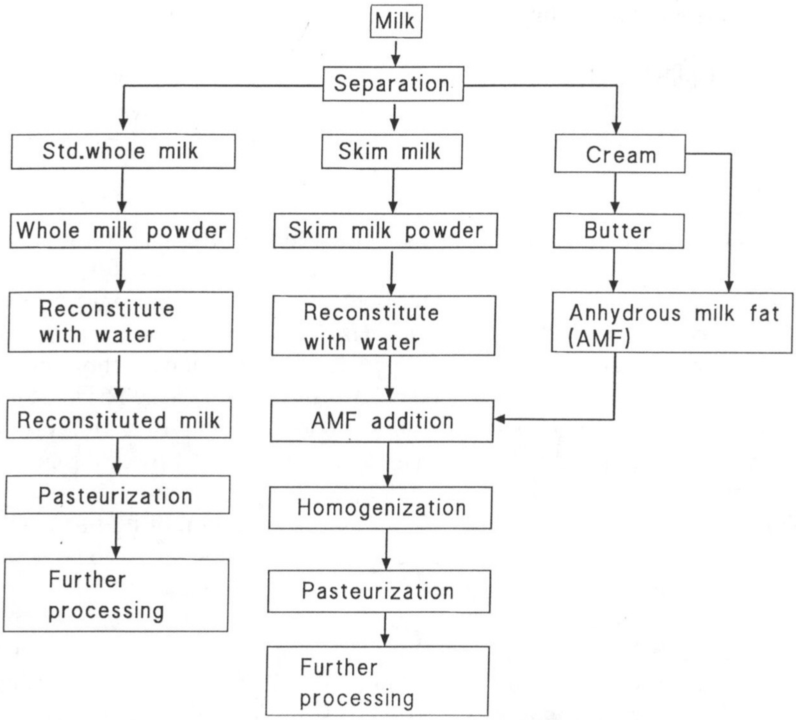

Fig. 3. Diagrammatic representation of the processes used in the manufacture of recombined/reconstituted milks.

TABLE 7
Functional Requirements of Milk Powders for Recombined Milk Products

Product	*Type of powder*	*Functional requirements*
Milk, cream	Extra-low-, low- and medium-heat	Reconstitutability, emulsifying capacity, fat emulsion stability, flavour stability.
UHT milk, UHT cream	Low- and medium-heat	Reconstitutability, heat stability, fat emulsion stability.
Buttermilk, yoghurt, lebnah	Medium- and high-heat	Water binding capacity, viscosity properties.
Cheese	Extra-low, and low-heat	Rennetability, gelation.
Evaporated milk	High-heat	Heat stability, viscosity properties.
Sweetened condensed milk	Medium-heat	Reconstitutability, viscosity properties.
Ice cream	Low- and medium-heat	Whipping properties, emulsifying capacity.

efficient transport and technical control have now made whole milk powder an attractive raw material for use in reconstituted products. With few exceptions, a full range of milk products can be prepared using recombining or reconstitution technology. The composition of various products is controlled by local legislation or international standards and by consumer demands. These products include pasteurized milk, sterilized milk, UHT-sterilized milk, evaporated sterilized milk, sweetened condensed milk, cream, cheese and fermented milk products. Low- to medium-heat skim milk powders are used in the manufacture of milks of normal concentration, whereas high-heat powders are more suitable for evaporated sterilized milks. Low-heat powders are used for cheesemaking, since heat treatments increase rennet clotting times, decrease curd strength and diminish whey drainage. The heat stability of skim milk powder is of the utmost importance in the manufacture of recombined evaporated milk, as discussed earlier. The most important quality aspect of recombined sweetened condensed milk is its viscosity which is determined by the quality of the skim milk powder and the processing conditions after recombining, such as pasteurization and homogenization.

Milk proteins play an important role in the emulsification of fat in recombined milks. This is mainly because the material of the natural fat globule is almost entirely absent; essentially all of the fat globule surface is covered with casein and some whey proteins. Fats other than milkfat are also used to make recombined milks. If these fats contain significant quantities of phospholipids or other polar lipids, casein can be displaced from the interface. The structure and composition of fat globule membranes in recombined milks were studied by Oortwijn *et al.* (1977), Oortwijn and Walstra (1979) and Walstra and Oortwijn (1982). The fat droplets in recombined milks are covered predominantly with protein, i.e. casein micelles, casein submicelles and serum proteins. The protein load is of the order of 10 mg/m^2 fat surface area and is dependent on the concentration of the dispersed phase and the conditions of homogenization. Preheating of milk before homogenization leads to higher protein loads, which in turn affect the properties of the product. The properties of adsorbed protein layers are expected to play an important role in the heat stability and storage stability of recombined milks, but very little work has been done in this area.

References

Adachi, S. (1958). Formation of lactulose and tagatose from lactose in strongly heated milk. *Nature*, **181**, 840–1.

Al-Tahiri, R. (1987). Recombined and reconstituted milk products. *N.Z. J. Dairy Sci. Technol.*, **22**, 1–23.

ADMI (American Dry Milk Institute) (1971). Standards for grades of dry milk including methods of analysis. *Bull. Amer. Dry Milk Inst.*, **916**, 7–12.

Andrews, A.T. (1974). Bovine milk acid phospatase. II. Binding to casein substrates and heat inactivation studies. *J. Dairy Res.*, **41**, 229–37.

Annan, W.D. & Manson, W. (1981). The production of lysinoalanine and related substances during processing of proteins. *Food Chem.*, **6**, 255–61.

Baldwin, A.J. & Woodhams, D.J. (1974). The dispersibility of skim milk powder at high total solids. *N.Z. J. Dairy Sci. Technol.*, **9**, 140–51.

Baldwin, A.J., Baucke, A.G. & Sanderson, W.B. (1980). The effect of concentrate viscosity on the properties of spray dried skim milk powder. *N.Z. J. Dairy Sci. Technol.*, **15**, 289–97.

Berlin, E., Anderson, B.A. & Pallansch, M.J. (1968). Water vapor sorption properties of various dried milks and wheys. *J. Dairy Sci.*, **51**, 1339–44.

Bernal, V. & Jelen, P. (1984). Effect of calcium binding on thermal denaturation of bovine α-lactalbumin. *J. Dairy Sci.*, **67**, 2452–4.

Bernal, V. & Jelen, P. (1985). Thermal stability of whey proteins—a calorimetric study. *J. Dairy Sci.*, **68**, 2847–52.

Bloore, C.G. & Boag, I.F. (1981). Some factors affecting the viscosity of concentrated skim milk. *N.Z. J. Dairy Sci. Technol.*, **16**, 143–54.

Burton, H. (1983). Bacteriological, chemical, biochemical and physical changes that occur in milk at temperatures of 100–150°C. *Int. Dairy Fed. Bull.*, **157**, 3–16.

Cerf, O. (1981). Factors affecting the keeping quality of heat treated milk: UHT treatment. *Int. Dairy Fed. Bulletin*, **130**, 42–57.

Cogan, T.M. (1977). A review of heat resistant lipases and proteinases and the quality of dairy products. *Irish J. Food Sci. Technol.*, **1**, 95–105.

Creamer, L.K. & Matheson, A.R. (1980). Effect of heat treatment on the proteins of pasteurized skim milk. *N.Z. J. Dairy Sci. Technol.*, **15**, 37–49.

Creamer, L.K., Berry, G.P. & Matheson, A.R. (1978). The effect of pH on protein aggregation in heated skim milk. *N.Z. J. Dairy Sci. Technol.*, **13**, 9–15.

Crosby, E.J. & Marshall, W.R. (1958). Effects of drying conditions on the properties of spray dried particles. *Chem. Eng. Prog.*, **54** (7), 56–63.

Dannenberg, F. & Kessler, H.G. (1988). Reaction kinetics of the denaturation of whey proteins in milk. *J. Food Sci.*, **53**, 258–63.

de Wilder, J. & Marten, R. (1974). Influence of homogenization of the concentrated milk on some whole milk powder characteristics. *Proc. 19th Intern. Dairy Cong. (New Delhi)*, **1E**, 609–10.

de Wit, J.N., Klarenbeek, G. & Hontelez-Backx, E. (1985). Evaluation of functional properties of whey protein concentrates and whey protein isolates. 1. Isolation and characterization. *Neth. Milk Dairy J.*, **37**, 37–48.

Doi, H., Ideno, S., Ibuki, F. & Kanamori, M. (1983). Participation of hydrophobic bond in complex formation between κ-casein and β-lactoglobulin. *Agric. Biol. Chem.*, **47**, 407–9.

Donovan, M. & Mulvihill, D.M. (1987). Effects of chemical modification and sodium dodecyl sulphate binding on the thermostability of whey proteins. *Irish J. Food Sci. Technol.*, **11**, 77–85.

Driessen, F.M. (1989). Inactivation of lipases and proteinases (indigenous and bacterial). *Int. Dairy Fed. Bulletin*, **238**, 71–93.

Driessen, F.M. & van der Waals, C.B. (1978). Inactivation of native milk proteinase by heat treatment. *Neth. Milk Dairy J.*, **32**, 245–54.

Dupont, M. & Yon, J. (1961). Effect of NaCl on the rate of heat denaturation of bovine β-lactoglobulin at pH 6·9. *J. Chem. Phys.*, **58**, 682–9.

Elfagm, A.A. & Wheelock, J.V. (1977). Effect of heat on α-lactobumin and β-lactoglobulin in bovine milk. *J. Dairy Res.*, **44**, 367–71.

Evenhuis, N. & de Vries, Th.R. (1956). The condition of calcium phosphate in milk. III. *Neth. Milk Dairy J.*, **10**, 101–13.

FAO/WHO (1973). Code of Principles, concerning milk and milk products, International standards and standard methods of sampling and analysis of milk products. 7th ed., pp. 127.

Farah, Z. (1979). Changes in proteins in UHT-heated milk. *Milchwissenschaft*, **34**, 484–7.

Fox, P.F. (1981). Heat-induced changes in milk preceding coagulation. *J. Dairy Sci.*, **64**, 2127–37.

Fox, P.F. (1982). Heat-induced coagulation of milk. In *Developments in Dairy Chemistry —1. Proteins*, ed. P.F. Fox. Applied Science Publishers, London, pp. 189–228.

Gough, P. & Jenness, R. (1962). Heat denaturation of β-lactoglobulin A and B. *J. Dairy Sci.*, **45**, 1033–9.

Gould, I.A. (1945). The formation of volatile acids in milk by high temperature heat treatment. *J. Dairy Sci.*, **28**, 379–86.

Gould, I.A. & Sommer, H.H. (1939). Effect of heat on milk with special references to cooked flavour. *Michigan State College Agric. Exp. Stn. Tech. Bull.*, No. 164, East Lansing, Michigan.

Graf, E. & Bauer, H. (1976). Milk and milk products. In *Food Emulsions*, ed. S. Friberg. Marcel Dekker, New York, pp. 295–377.

Grufferty, M.B. & Fox, P.F. (1988). Milk alkaline proteinase. *J. Dairy Res.*, **55**, 609–30.

Hall, C.W. & Hendrick, T.I. (1971). *Drying Milk and Milk Products*, 2nd edn. The AVI Publishing Co., Westport, CT.

Hansen, A.P. & Hempill, P.L. (1984). Utilization of heat to increase shelf life of blended acid whey and butter milk powder for frozen desserts. *J. Dairy Sci.*, **67**, 54.

Hanson, L.A. & Mansson, I. (1961). Immune electrophoretic studies of bovine milk and milk products. *Acta Pediatr.*, **50**, 484–90.

Haque, Z. & Kinsella, J.E. (1988). Interaction between κ-casein and β-lactoglobulin: predominance of hydrophobic interactions in the initial stages of complex formation. *J. Dairy Res.*, **55**, 67–80.

Harland, H.A. & Ashworth, U.S. (1947). A rapid method for estimation of whey proteins as an indication of baking quality of nonfat dry milk solids. *Food Res.*, **12**, 247–51.

Harland, H.A., Coulter, S.T. & Jenness, R. (1952). The inter-relationship of processing treatment and oxidation-reduction systems as factors affecting the keeping quality of dry whole milk. *J. Dairy Sci.*, **35**, 643–54.

Harmansson, A.M. (1977). Functional properties of proteins for food—water vapour sorption. *J. Food Technol.*, **12**, 177–87.

Hartman, G.H. & Swanson, A.M. (1965). Changes in mixtures of whey protein and κ-casein due to heat treatments. *J. Dairy Sci.*, **48**, 1161–7.

Harwalkar, V.R. (1986). Kinetic study of thermal denaturation of proteins in whey. *Milchwissenschaft*, **41**, 206–9.

Heldman, D.R., Hall, C. & Hedrick, T.J. (1965). Vapor equilibrium relationship of dry milk. *J. Dairy Sci.*, 845–52.

Henry, K.M., Kon, S., Lea, C.H. & White, J.C.D. (1948). Deterioration on storage of dried skim milk. *J. Dairy Res.*, **15**, 293–363.

Hillier, R.M. & Lyster, R.L.J. (1979). Whey protein denaturation in heated milk and cheese whey. *J. Dairy Res..*, **46**, 95–102.

Hillier, R.M., Lyster, R.L.J. & Cheeseman, G.C. (1979). Thermal denaturation of α-lactobumin and β-lactoglobulin in cheese whey: effect of total solids concentration and pH. *J. Dairy Res.*, **46**, 103–11.

Hiraoka, Y., Segawa, T., Kuwajima, K., Sugai, S. & Murai, N. (1980). α-Lactalbumin: a calcium metalloprotein. *Biochem. Biophys. Res. Commun.*, **95**, 1098–103.

Holt, C., Muir, D.D. & Sweetsur, A.W.M. (1978). Seasonal changes in the heat stability of milk from creamery silos in south-west Scotland. *J. Dairy Res.*, **45**, 183–90.

Howat, G.R. & Wright, N.C. (1934). Factors affecting solubility of milk powders. III. Some physico-chemical properties of concentrated solutions of milk solids. *J. Dairy Res.*, **5**, 236–44.

Hunziker, O.F. (1949). Milk drying systems. In *Condensed Milk and Milk Powder*. La Grange, Illinois, pp. 383–98.

International Dairy Federation (1970). Provisional screening test for non-instant dried skimmed milk. In *Recent IDF Work in Relation to Dried Milk*. International Dairy Federation, Brussels, Belgium, Document No. **59**, pp. 14–17.

Jensen, G.K. & Nielsen, P. (1982). Milk powder and recombination of milk and milk products. *J. Dairy Res.*, **49**, 515–44.

Jensen, J.D. (1975). Some recent advances in agglomerating, instantizing and spray drying. *Food Technol.*, **29**, 60–70.

Kannan, A. & Jenness, R. (1961). Relation of milk serum proteins and milk salts to the effects of heat treatment on rennet clotting. *J. Dairy Sci.*, **44**, 808–22.

Kelly, P.M. (1981). Heat-stable milk powders. *J. Soc. Dairy Technol.*, **34**, 157–62.

Kelly, P.M. (1982). The effect of preheat temperature and urea addition on the seasonal variation in the heat stability of skim milk powder. *J. Dairy Res.*, **49**, 187–96.

Kenney, M. & Bassette, R. (1959). Detection of intermediate compounds in the early stages of Maillard browning reaction in milk products. *J. Dairy Sci.*, **42**, 945–60.

Kieseker, F.G. (1970). The manufacture of high solids recombined evaporated milk. *Proc. 18th Intern. Dairy Cong.*, **1E**, 265.

Kieseker, F.G. (1976). Milk proteins in manufacture of milk powders. Milk Protein Workshop, Tanunda, South Australia, H1–H14.

Kieseker, F.G. (1983). Recombined dairy products. *CSIRO Food Res. Q.*, **43**, 25–37.

Kieseker, F.G. & Clarke, P.T. (1984). The effect of storage on the properties of nonfat milk powders. *Aust. J. Dairy Technol.*, **39**, 74–7.

King, D.W., Sanderson, W.B. & Woodhams, D.J. (1974). Spray drying of dairy products. *Dairy Industries*, **39**, 13–16.

King, N. (1965). The physical structure of dried milk. *Dairy Sci. Abstr.*, **27**, 91–104.

King, N. (1966). Dispersibility and reconstitutability of dried milk. *Dairy Sci. Abstr.*, **28**, 105–18.

Kinsella, J.E. (1984). Milk proteins: physicochemical and functional properties. *CRC Crit. Rev. Food Sci. Nutr.*, **21**, 197–226.

Larson, B.L. & Rolleri, C.D. (1955). Heat denaturation of the specific serum proteins in milk. *J. Dairy Sci.*, **38**, 351–60.

Lascelles, D.R. & Baldwin, A.J. (1976). Dispersibility of whole milk powder in warm water. *N.Z. J. Dairy Sci. Technol.*, **11**, 283–4.

Le Graet, Y. & Brule, G. (1982). Effect of concentration and drying on mineral equilibria in skim milk and retentates. *Lait*, **62**, 113–25.

Linden, G. (1986). Monograph on pasteurized milk: biochemical aspects. *Int. Dairy Fed. Bull.*, **200**, 17–21.

Lingnerst, H. & Hall, G. (1986). Formation of antioxidative Maillard reaction products during food processing. In *Developments in Food Science. Vol. 13. Amino-Carbonyl Reaction in Food and Biological Systems*, ed. M. Fujimaki, M. Namiki & H. Kato. Elsevier, New York, pp. 273–9.

Litman, I.I. & Ashworth, U.S. (1957). Insoluble scum-like materials in reconstituted whole milk powders. *J. Dairy Sci.*, **40**, 403–9.

Long, J.E., van Winkle, Q. & Gould, I.A. (1963). Heat induced interaction between crude κ-casein and β-lactoglobulin. *J. Dairy Sci.*, **46**, 1329–39.

Lyster, R.L.J. (1970). The denaturation of α-lactalbumin and β-lactoglobulin in heated milk. *J. Dairy Res.*, **37**, 233–43.

Martinez-Castro, I. & Olano, A. (1980). Influence of thermal processing on carbohydrate component of milk: formation of epilactose. *Michwissenschaft*, **35**, 5–8.

Masters, K. (1972). *Spray Drying*. Leonard Hill Books, London.

May, C.D., Fomon, S.J. & Remigio, L. (1982). Immunological consequences of feeding infants with cow milk and soy product. *Acta Paed. Scand.*, **71**, 43–51.

McKellar, R.C. (ed.) (1989). *Enzymes of Psychrotophs in Raw Food.* CRC Press, Boca Raton, FL.

Mohammad, K.S. & Fox, P.F. (1987). Heat-induced microstructural changes in casein micelles before and after heat coagulation. *N.Z. J. Dairy Sci. Technol.*, **22**, 191–293.

Mol, J.J. (1974). Effect of two-stage drying process on the properties of whole milk powder. *Proc. 19th Intern. Dairy Cong. (New Delhi)*, **1E**, 605–6.

Mol, J.J. (1975). The milk fat globule membrane and the solubility of whole milk powder. *Neth. Milk Dairy J.*, **29**, 212–24.

Morr, C.V. (1973). Milk ultracentrifugal opalescent layer. I. Composition as influenced by heat, temperature and pH. *J. Dairy Sci.*, **56**, 544–52.

Muir, D.D. (1980). Concentration and milk powder quality. In *Milk and Whey Powders*. Soc. Dairy Technol., Middlesex, UK, pp. 73–84.

Mulder, H. & Walstra, P. (1974). *The Milk Fat Globule: Emulsion Science as Applied to Milk Products and Comparable Foods*. Commonwealth Agricultural Bureaux, Farnham Royal, UK.

Mulvihill, D.M. & Donovan, M. (1987). Whey proteins and their thermal denaturation —a review. *Irish J. Food Sci. Technol.*, **11**, 43–75.

Neff, E. & Morris, A.L. (1968). Agglomeration of milk powder and its influence on reconstitution properties. *J. Dairy Sci.*, **51**, 330–8.

Newstead, D.F. (1977*a*). Heat stability of evaporated milk. In *Proceedings of the Jubilee Conference on Dairy Science Technology*. New Zealand Dairy Research Institute, Palmerston North, New Zealand, pp. 89–93.

Newstead, D.F. (1977*b*). Effect of protein and salts concentrations on the heat stability of evaporated milk. *N.Z. J. Dairy Sci. Technol.*, **12**, 171–5.

Newstead, D.F. & Conaghan, E.F. (1978). On the heat stability of recombined evaporated milk—influence of pH alteration of the skim milk prior to processing. *Deutsche Molkerei-Zeitung*, **99**, 1688–90.

Newstead, D.F., Sanderson, W.B. & Baucke, A.G. (1975). The effects of heat treatment and pH on the heat stability of recombined evaporated milk. *N.Z. J. Dairy Sci. Technol.*, **10**, 113–18.

Nieuwenhuijse, J.A., Timmermans, W. & Walstra, P. (1988). Calcium and phosphate partitions during the manufacture of sterilized concentrated milk and their relations to the heat stability. *Neth. Milk Dairy J.*, **42**, 387–421.

Oortwijn, H. & Walstra, P. (1979). The membranes of recombined fat globules. 2. Composition. *Neth. Milk Dairy J.*, **33**, 134–54.

Oortwijn, H., Walstra, P & Mulder, H. (1977). The membranes of recombined fat globules. 1. Electron microscopy. *Neth. Milk Dairy J.*, **31**, 134–47.

Park, K.H. & Lund, D.B. (1984). Calorimetric study of thermal denaturing of β-lactoglobulin. *J. Dairy Sci.*, **67**, 1699–706.

Parry, R.M. (1974). Milk coagulation and protein denaturation. In *Fundamentals of Dairy Chemistry*, 2nd edn, ed. B.H. Webb, A.H. Johnson & J.A. Alford. AVI Publishing Co., CT, pp. 603–61.

Pearson, A.M., Spooner, M., Hegarty, G. & Bratzler, L. (1965). The emulsifying capacity and stability of soy sodium proteinate, potassium caseinate and non-fat dry milk. *Food Technol.*, **19**, 103–7.

Peebles, D.D. (1956). The development of instant milk. *Food Technol.*, **10**, 64–5.

Pisecky, J. (1980). Instant whole milk powder. *Aust. J. Dairy Technol.*, **35**, 95–9.

Prentice, M.J. (1972). Measuring the rate of solution of dried milk. *Dairy Industries*, **37** (1), 18–19.

Renner, E. (1983). *Milk and Dairy Products in Human Nutrition*. Volkswirtschaftlicher Verlag, Munich.

Ruegg, M., Moor, U. & Blanc, B. (1977). A calorimetric study of the thermal denaturation of whey proteins in simulated milk ultrafiltrates. *J. Dairy Res.*, **44**, 509–20.

Sanderson, W.B. (1978). Instant milk powders: manufacture and keeping quality. *N.Z. J. Dairy Sci. Technol.*, **13**, 137–43.

Sanderson, W.B. (1982). Recombination of milk and milk products—principal processes and equipment. *Int. Dairy Fed. Bulletin*, **141**, 18–21.

Sawyer, W.H., Coulter, S.T. & Jenness, R. (1963). Role of sulfydryl groups in the interaction of κ-casein and β-lactoglobulin. *J. Dairy Sci.*, **46**, 564–5.
Singh, H. (1988). Effects of high temperatures on casein micelles. *N.Z. J. Dairy Sci. Technol.*, **23**, 257–73.
Singh, H. & Creamer, L.K. (1991). Denaturation, aggregation and heat stability of milk protein during the manufacture of skim milk powder. *J. Dairy Res.*, **58**, 269–83.
Singh, H. & Fox, P.F. (1985). Heat stability of milk: pH-dependent dissociation of micellar κ-casein on heating milk at ultra high temperatures. *J. Dairy Res.*, **52**, 529–38.
Singh, H. & Fox, P.F. (1986). Heat stability of milk: further studies on the pH-dependent dissociation of micellar κ-casein. *J. Dairy Res.*, **53**, 237–48.
Singh, H. & Fox, P.F. (1989). Heat-induced changes in casein. *Int. Dairy Fed. Bulletin*, **238**, 24–30.
Singh, H. & Tokley, R.P. (1990). Effects of preheat treatments and buttermilk addition on the seasonal variations in the heat stability of recombined evaporated milk and reconstituted concentrated milk. *Aust. J. Dairy Technol.*, **45**, 10–16.
Singh, H., Creamer, L.K. & Newstead, D.F. (1989). Effects of heat on the proteins of concentrated milk systems. *Int. Dairy Fed. Bulletin*, **238**, 94–104.
Snoeren, T.H.M. & van der Spek, C.A. (1977). The isolation of a heat-induced complex from UHTST milk. *Neth. Milk Dairy J.*, **31**, 352–5.
Snoeren, T.H.M., van der Spek, C.A., Dekker, R. & Both, P. (1979). Proteolysis during the storage of UHT-sterilized whole milk. I. Experiments with milk heated by direct steam for 45 s at 142°C. *Neth. Milk Dairy J.*, **33**, 31–9.
Taylor, M.J. & Richardson, T. (1980). Antioxidant activity of skim milk: effect of heat and resultant sulphydryl groups. *J. Dairy Sci.*, **63**, 1783–95.
Thomas, A. (1954). A study of the mechanism of sulphydryl action in milk. PhD Thesis, University of Minnesota, USA.
van Kreveld, A. (1974). Studies on the wetting of milk powders. *Neth. Milk Dairy J.*, **28**, 23–45.
Varshriey, N.M. & Ojaha, T.P. (1975). Particle size and size distribution of dried baby food. *Aust. J. Dairy Technol.*, **30**, 22–6.
Walstra, P. & Jenness, R. (1984). *Dairy Chemistry and Physics*. John Wiley & Sons, New York, pp. 317–20.
Walstra, P. & Oortwijn, H. (1982). The membranes of recombined fat globules. 3. Mode of formation. *Neth. Milk Dairy J.*, **36**, 103–13.
Webb, B.H. (1950). Dried products. In *By Products From Milk*, ed. E.O. Whittlier & B.H. Webb. Reinhold Publishing Co., New York, pp. 93–129.
Webb, B.H. & Holm, G.E. (1932). The heat coagulation of milk. II. The influence of various added salts upon the stability of milk of different concentrations. *J. Dairy Sci.*, **15**, 345–66.
Woodhams, D.J. & Murray, M.J. (1974). Preparation of spray dried milk powders. *N.Z. J. Dairy Sci. Technol.*, **9**, 172–8.
Zittle, C.A., Thompson, M.P., Custer, J.H. & Cerbulis, J. (1962). κ-Casein/β-lactoglobulin interaction in solution when heated. *J. Dairy Sci.*, **45**, 807–10.

INDEX